Horticultural Research International

Horticultural Research International

Directory of horticultural research institutes and their activities in 63 countries

Published by the International Society for Horticultural Science

Wageningen 1986

Text typed by:

Lia Verwaal
Manon Benda
Els van der Borg
Vera de la Ferté
Carol 't Mannetje

Maps by:

Th.W. van Betuw

Organizing editors:

Ir. H.H. van der Borg
M. Koning-van der Veen

First edition 1966
Second edition 1972
Third edition 1981
Fourth edition 1986

ISBN 90 6605 332 1

Price for non-members of ISHS: Dfl. 300,—

Printed in the Netherlands
Drukkerij Avanti
Schimmink 11
5301 KR Zaltbommel

CONTENTS

		Page
Preface		7
Directions for Use		8
The ISHS		9
Argentina	AR	11
Australia	AU	27
Austria	AT	62
Belgium	BE	71
Brazil	BR	86
Bulgaria	BG	115
Canada	CA	122
Colombia	CO	148
Costa Rica	CR	152
Cyprus	CY	157
Czechoslovakia	CS	159
Denmark	DK	172
Egypt	EG	181
Ethiopia	ET	184
Finland	FI	188
France	FR	193
Germany, Democratic Republic	DD	223
Germany, Federal Republic	DE	225
Ghana	GA	242
Greece	GR	247
Hungary	HU	258
Iceland	IS	280
India	IN	281
Indonesia	ID	316
Ireland	IE	319
Israel	IL	324
Italy	IT	335
Japan	JP	355
Kenya	KE	423
Korea	KR	426
Libya	LY	439
Malawi	MW	441
Mexico	MX	442
Morocco	MA	454
Netherlands	NL	457
New Zealand	NZ	475
Nigeria	NG	489
Norway	NO	494
Peru	PE	502
Philippines	PH	505
Poland	PL	512
Portugal	PT	535
Romania	RO	539
Senegal	SN	557
Seychelles	SC	559
South Africa	ZA	561
Spain	ES	565
Sri Lanka	LK	572
Sudan	SD	578
Surinam	SR	580
Sweden	SE	581
Switzerland	CH	589

CONTENTS

		Page
Taiwan (China)	TW	593
Tanzania	TZ	599
Thailand	TH	602
Tunisia	TN	608
Turkey	TR	610
Union of Soviet Socialist Rep.	SU	619
United Kingdom	GB	624
United States of America	US	651
- New England States	US-A	652
- North Atlantic States	US-B	659
- Middle Atlantic States	US-C	673
- South Atlantic States	US-D	680
- South Central States	US-E	701
- North Atlantic States	US-F	707
- Great Plains States	US-G	719
- Southwestern States	US-H	732
- Mountain States States	US-I	742
- Pacific Northwest States	US-J	752
- California	US-K	759
- Alaska	US-L	773
- Hawaii	US-M	776
- Puerto Rico	US-N	780
Venezuela	VE	782
Yugoslavia	YU	784
Zimbabwe	ZI	801
Index of names of places		805
Index of names of research workers		813

PREFACE

As intended we can present you with the fourth edition of Horticultural Research International in a shorter period than we have had between the second and third edition. We started with the preparation of this fourth directory in April, 1984, and it is published in July 1986. This edition is prepared by the secretariat of ISHS by using personal computers with specific software programmes. The printing is realized by electronic systems starting from our definite text on the computer equipment.
We would like to thank the coordinators, and administrators of the Universities, Institutes and Departments, also Government Officials and some other helpful volunteers, who were kind enough to do the collecting of data or just to pass on letters with requests.

Although we decided to delete three countries, we were able to get material from five new countries. To obtain this new material we have approached 21 new countries. With 12 of those we had intensive contacts but unfortunately only five of them succeeded to deliver us the data of their country in time for this edition. The present edition covers some 16650 scientists at 1250 institutes in 63 countries; in the IIIrd edition this was 14000, 1400 and 61 respectively.
Many of the maps of the countries are reviewed and elaborated with wavy lines as an indication for sea or lakes along its borders to facilitate reading the maps, and we have standardized the scale as much as possible. As new information you can find for many places elevation in meters (m), rainfall in millimeters per year (mm), hours of sunshine per year (h) and for tropical regions also percentage of relative humidity (%). When available we have listed phone and telex numbers of the institutions.

As the ISHS is now the publisher of the book we present you with, you will find a short piece of information on our Society in the first pages. To inform you about our members there is an asterisk (*) after the names in the descriptive part of the text and also in the index of names.

We are convinced that the fourth edition will be useful to many people interested in horticultural research as scienctist or user. As we want to continue improving the contents of this directory, do not hesitate to contact us with remarks or suggestions.

We hope that this book will again contribute to better contacts between research workers all over the world and the bringing together of scientists, technicians and production people.

Wageningen, May 1986

Ir. H.H. van der Borg
Mevr. M. Koning-van der Veen

Organizing editors

DIRECTIONS FOR USE

Contents:
All countries are mentioned alphabetically with standard abbreviation in codeletters followed by the page number. The used abbreviation is from ISO (International Organization for Standardization).
The United States of America are divided into 14 groupes of States.

Index of names of places:
Each name is followed by the relevant country codeletters and its number in the country. The names are not always arranged alphabetically descriptive in the definitive text.

Index of names of research workers:
An alphabetical list beginning with the surname, followed by the initials, the country codeletters and the numbers of places and institutes.
Surnames starting with "de" are to be found under D, "van", "von" under V and "la", "le" under L.

We have tried to get full addresses, phone and telex numbers of each institute or station. Unfortunately the information is not complete.

Also data like elevation in meters (m), rainfall in millimeters (mm), hours of sunshine (h) and relative humidity (rh in %) have been introduced but we could not complete them fully.

Elevation is always above sea level, unless otherwise indicated. Rainfall and sunshine hours are usually averages per year; if otherwise it is indicated. Relative humidity is only mentioned for tropical countries.

The maps are taken from two recently published internationally well-known, atlases, i.e. "Le Monde" and "Time/Life". The wavy lines indicate bordering sea or lake; land frontiers are also marked.

The editors hope these additional data will be helpful to the users of this book.

It would be much appreciated if the readers will send us suggestions for further improvement of this directory.

THE INTERNATIONAL SOCIETY FOR HORTICULTURAL SCIENCE - ISHS

Short History

Long ago in 1864 horticultural scientists realized that an exchange of views and experiences should be promoted on a world wide scale rather than merely on a local, regional or national scale.
In the beginning this led to the organization of four in-official international horticultural congresses followed by the first official one which was held in 1869.
Once in every five years an international horticultural congress was organized.
In 1923 during the international horticultural congress which took place in Amsterdam, an international committee for horticultural congresses was founded.
This committee realized that international co-operation also had to be encouraged in the period between the congresses. In 1959 it was decided to replace the international committee with the International Society for Horticultural Science abbreviated ISHS. In 1984 ISHS celebrated it's 25th anniversary and in 1986 the XXIInd Congress was held in Davis, California, USA.

General Objective and Means

The main objective of this scientific society is the advancement of horticulture through international co-operation in science and technology, mainly achieved by a good familiar atmosphere which is the basis for international co-operation.
The means available to ISHS for the achievement of this objective are:

. the holding of an	. international horticultural congress once in every five years
. the setting up of	. sections dealing with groups of horticultural plants
	. commissions engaged in various scientific and technical aspects of horticulture
	. working groups of the sections and the commissions for scientists to co-operate in specific fields of research
. the organizing of	. symposia, workshops and seminars on specific topics for scientists and other specialists

Another way ISHS employs to achieve its objectives is the publishing of the Society's publications which are:

o Chronica Horticulturae, the bulletin of the Society which keeps the members of ISHS informed of it's activities and about other matters related to the objectives of the Society.

o Acta Horticulturae, a world-wide known series of scientific and technical publications mainly devoted to proceedings of ISHS Symposia.

o Horticultural Research International (HRI).

o Scientia Horticulturae, a joint venture of ISHS and Elseviers Science Publishers. The ISHS is responsible for the scientific contents and the appointment of the editorial board, while Elsevier is responsible for commercial matters.

o Proceedings of International Horticultural Congresses, published by ISHS or the national organizing committees are obtainable from the ISHS Secretariat.

Structure of the Society

The organization of ISHS is in the hands of the council and the board. The council, formed by the representatives of the member countries (50), meets every two years to assure the realization of the objectives of the Society. The members of the board of the Society existing of the president, vice-president, past-president and the secretary-general are also members of the council.
The council is assisted by an executive committee which is composed of the chairmen of the four sections and nine commissions and the board.
The executive committee, which meets annually, shall direct the Society and shall carry out the permanent activities of the Society.

These activities are carried out by the sections and commissions and their specific working groups. At present the sections and commissions are as follows:

Sections

Fruits
Vegetables
Ornamentals
Medicinal & Aromatic Plants

Commissions

Engineering
Plant Protection
Plant Substrates
Urban Horticulture
Protected Cultivation
Education & Training
Economics & Management
Nomenclature & Registration
Tropical & Subtropical Horticulture

The secretariat of the Society is working according to the directives of the council and under the direction of the executive committee. It registers the Society's members and assists in organizing symposia and congresses and publishing the Society's publications.

Membership

Besides membership of countries, ISHS distinguishes other kinds of membership:

- Individual membership, applicable for any person engaged or interested in scientific, technological, economic, educational, recreational or amateurish aspects of horticulture.
- Affiliated membership of organizations, applicable for any institute, college, university, ministry, society, association, institution, organization, agency, firm or section of any of these, active in some aspect of horticulture or concerned with the advancement of horticulture.

To become a member of a section or commission one has to be appointed by the council and to become a member of a working group one has to be appointed by the section or commission concerned.

The advantages connected with an individual membership are:
o reduction in the fee of each symposium organized by ISHS
o free volume of Chronica Horticulturae per annum which contains four issues
o discount of 20% on the price of all publications of Acta Horticulturae
o discount of 10% on the price of Horticultual Research International
o discount of 10% on the subscription price of Scientia Horticulturae

The advantages connected with an affiliated membership are:
o reduction in the fee for one representative for each symposium and congress organized by the ISHS
o free volume of Chronica Horticulturae per annum which contains four issues
o discount of 35% on the price of all publications of Acta Horticulturae
o discount of 10% on the price of Horticulturae Research International
o discount of 10% on the subscription price of Scientia Horticulturae

When you are convinced of the attractiveness of belonging to our Society and you want to adhere, application forms and other information can be obtained from the ISHS Secretariat.

ISHS Secretariat, De Dreijen 6, 6703 BC WAGENINGEN, Netherlands.

ARGENTINA

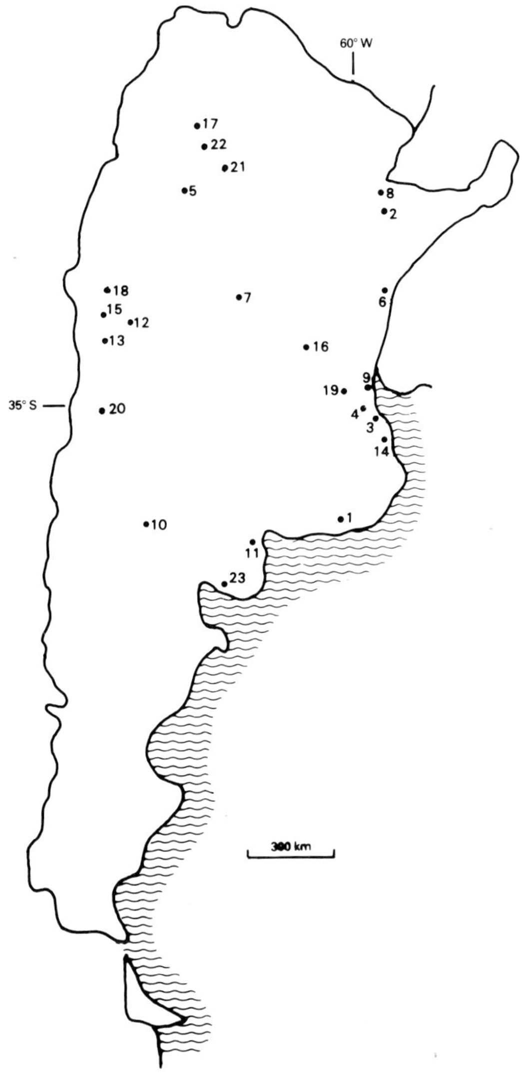
60° W
35° S
300 km
17
22
21
5
8
2
18
15
12
13
7
6
16
9
19
4
3
14
20
1
10
11
23

1 Balcarce
2 Bella Vista
3 Buenos Aires
4 Castelar
5 Catamarca
6 Concordia
7 Córdoba
8 Corrientes
9 Delta del Paraná
10 General Roca
11 Hilario Ascasubi
12 Junín
13 La Consulta
14 La Plata
15 Mendoza
16 Rosario
17 Salta
18 San Juan
19 San Pedro
20 San Rafael
21 Santiago del Estero
22 Tucumán
23 Viedma

Survey

All institutions dealing with Horticultural Research in Argentina can be divided in three groups:
- National Institute for Agricultural Technology (Instituto Nacional de Tecnologia Agropecuaria) (INTA)
- National Universities
- Provincial Institutions

The first one (INTA) has one National Center for Agricultural Research at Castelar, near Buenos Aires, where the basic research is done, and 35 Agricultural Experiment Stations distributed all over the country, with very various orientation of work, mostly looking for practical solutions. Only 17 Experiment Stations are dealing with Horticultural Research. Among them every one has its proper main orientation of work, for example vegetable growing, pome fruit growing, citrus fruit growing, stone fruit growing, viticulture, sometimes with a further subdivision like breeding of new varieties, or cultural practices and management of a certain crop, etc. In some cases this main line of work is quite remote from horticulture, horticultural research being limited to some crops of rather local interest.

INTA, although a governmental institution, is supported by direct contributions of the producers on the basis of exports of agricultural and live stock. The producers, on the other hand, take part in the government of the Institute through their representatives on

Prepared by Ing. Agr. José Crnko*, Consultor Director, INTA, Mendoza.

the Directory Council and through members of local councils at each Experiment Station.
Publication of INTA containing articles to horticultural research are:
- Informativo de Investigaciones Agropecuarias (IDIA). Editor: Ing. Agr. Leopoldo F. Brugnoni, Chile 460, Buenos Aires.
- Revista de Investigaciones Agropecuarias (RIA). Serie 2. Biologiá y Producción Vegetal. Serie 5. Patologiá Vegetal. Editor: Ing. Agr. Leopoldo F. Brugnoni, Chile 460, 1098 Buenos Aires.
- Moreover: Memorias Anuales: Boletines técnicos; Divulgaciones Agropecuarias, etc.
In the National Universities, besides its educational work, some Horticultural Research is done through its specialized Institutes or Departments. Here the work is mostly on basic research.
Some few Provinces perform Horticultural Research, looking for solutions of rather local problems, through its proper experimental stations.

1 Balcarce

1.1 Estación Experimental Regional Agropecuaria, INTA (Regional Experiment Station for Agriculture and Cattle)

Casilla de Correo 276
7620 - Balcarce, Buenos Aires
Phone: 2/2040, 2041, 2042
Dir.: Ing.Agr.M.Sc.O. Costamagna

Facultad de Ciencias Agrarias de la Universidad Nacional Madel Plata (University of Mar del Plata, Faculty of Agricultural Sciences, Department of Vegetable Crops)

Dean: Ing.Agr. J.M. Lahitte

elevation: 130 m
rainfall : 858 mm
sunshine : 2445 h

Main orientation of work: live stock breeding, forage crops and pastures, plant production and breeding with emphasis on potatoes and wheat; soil conservation and management; agricultural economics. Area 1200 ha.

Vegetables
Crops adaptation, varietal studies, improvement of cultural practices and seed production

Ing.Agr. J.L. Marrapodi
Ing.Agr. D. Tosoni
Ing.Agr. M. Espinillo
Ing.Agr. D. Tosoni
Ing.Agr. E. Marrapodi
M.Sc. M. Verellen

2 Bella Vista

2.1 Estación Experimental Agropecuaria, Bella Vista, INTA (Agricultural Experiment Station)
elevation: 70 m
rainfall : 1180 mm
sunshine : 2641 h

Casilla de Correo 5
3432 Bella Vista, Corrientes
Phone: 42
Dir.: Ing.Agr. D.S. Rodriguez

This station is the centre of citriculture and vegetable crops research in the Province of Corrientes. Other fields of research work are tobacco crop and forestry.

Citrus crops
Breeding — Ing.Agr.M.Sc. H.M. Zubrzycki
Entomology — Ing.Agr. S. Cáseres
Plant Pathology — Ing.Agr. B.J. Canteros
Plant Virology — Ing.Agr.M.Sc. A.D. Zubrzycki
Vegetable Crops
Breeding and Management — Ing.Agr. H. Ishikawa

Entomology
Plant Pathology

Ing.Agr. S. Cáseres
Ing.Agr. M. Del H.C. de Ramirez

3 Buenos Aires

3.1 Universidad Nacional de Buenos Aires, Facultad de Agronomia, Departamento de Producción Vegetal (University of Buenos Aires, Faculty of Agricultural Sciences, Agricultural Department)
elevation: 25 m
rainfall : 1076 mm
sunshine : 2400 h

Av. San Martin 4453
1417 Capital Federal
Phone: 51 - 0084
Dean: Ing.Agr. J.H. Lemcoff

3.1.1 Chair of Vegetable crops

Crop management and varietal studies. Champignon, plastics in vegetable crops. — Ing.Agr. J.G. Ringeisen
Onion, garlic — Ing.Agr. J. De Sancho
Asparagus — Ing.Agr. C.G. Barón
Lettuce, sweet corn — Ing.Agr. H. Vallejo
Potato — Ing.Agr. J. Fernàndez Lozano
Tomato, pepper, soil fertility, pulse, collection of species — Ing.Agr. E. Kramarovsky; Ing.Agr.C.A.Mundt/Ing.Agr.S.Souto; Ing.Agr.D.Diaz/Ing.Agr.C.Felpeto
Cole crops — Ing.Agr. J.C. Limongelli
Marketing, nutrition value — Ing.Agr. A. Chiesa; Ing.Agr. M.E. Daorden
Storage, field and postharvest diseases — Ing.Agr. M.J. Vigliola; Ing.Agr. L. Calot

3.1.2 Chair of Fruticulture

Propagation of peach and citrus. Thinning of fruits in plum — Ing.Agr. J.F. Ferreira; Ing.Agr. H. Polero; Ing.Agr. C.T. Barilari

Modification of germination of citrus seeds through G.A. treatment. Floral microfenology of peaches and nectarines, varietal studies of pomological collection — Ing.Agr.S.Ochatt/Ing.Agr.E. Faita; Ing.Agr.J.Calvar; Ing.Agr.L.Berasategui
Relation between rootstock and graft of different citrus species and cultivars, propagation of different plum cultivars — Ing.Agr. F. Covatta
Crossings between different plum cultivars and compatibility studies — Ing.Agr. A.O. Lorusso; Ing.Agr. M. Peralta

4 Castelar

4.1 Centro de Investigaciones en Ciencias Agronómicas, INTA (Agronomic Sciences Research Center)

Casilla de Correo 25
1712 Castelar, Buenos Aires
Phone: 621 - 1876

4.1.1 Department of Genetics — Head: Ing.Agr. E.A. Favret
Phone: 621 - 1876, - 0805, - 0772

Vegetable improvement — Ing.Agr. J. Devcic
Methods in fruit breeding — Ing.Agr. L.C. de Terraciano

4.1.2 Agricultural Machinery Department — Head: Ing.Agr.M.Sc.J.M.Casares

Vegetable sowing machinery
Phone: 665 - 0495, - 0450, - 0541
Ing.Agr. R. De La Fosse

4.1.3 Microbiology Department
Head: Dra. N.S. de Nuñez
Phone: 621 - 0670, - 1701

Rhizobiology
Ing.Agr. M.Sc. R. Dieguez

4.1.4 Department of Plant Pathology
Head: Ing.Agr. R Rizzo
Phone: 621 - 1683, - 1534

Biology and control of diseases — Ing.Agr.M.Sc. C. Fortugno
Bacteriology — Ing.Agr. L.A. Rossi
Control of vegetable diseases — Ing. Agr. M.Sc. G. Baigorria
Biological pest control — Dra. I.S. Crouzel
Methods of integral control of fruit flies in citrus regions — Dr. A. Turica
Plant therapeutics — Ing.Agr. E. Touron

5 Catamarca

5.1 Estación Experimental Agropecuaria, INTA (Agricultural Experiment Station Catamarca)
elevation: 526 m
rainfall : 403 mm
sunshine : 2828 h

Casilla de Correo 25
4700 San Fernando del Valle de Catamarca
Phone: 28192
Dir.: Ing.Agr. N. Ansótegui

Research work is concerned with fruit crops (table grapes, walnut, olive, citrus, peach, plum, almond, fig) and vegetable crops (tomato, pepper, garlic, onion, lentil, pea, bean, vegetable cowpea, lettuce).

Fruticulture
Fruit crops — Ing.Agr.M.Sc. A.G. Prataviera

Vegetables
Variety studies and seed production — Ing.Agr. H.A. Ratti

Plant protection — Ing.Agr. J.E. Pinilla

Viticulture
Training systems for table grapes, variety studies — Ing.Agr. J.M. Denett, Ing.Agr. V.N. Páez

Soils and irrigation
Soil fertility, management and conservation — Ing.Agr. M.A. Correa de Sal

6 Concordia

6.1 Estación Experimental Agropecuaria, INTA (Agricultural Experiment Station)
elevation: 47 m
rainfall : 1100 mm
sunshine : 2664 h

Casilla de Correo 34
3200 Concordia, Entre Ríos
Phone: 21 - 4027
Dir.: Ing.Agr. D.R. Hogg

This station is the main centre of citrus research in Argentina. Other fields of research include forestry and honey production.

Citriculture
Citrus crops — Ing.Agr. H.N. Beñatena, Ing.Agr. C.M.A. de Marcó

Citrus by-products (Citrus juice and oils) — Ing. Quim M.Sc. R.W. Drescher
Citrus management — Ing.Agr. M.L. Ragone

Plant Protection	
Entomology	Ing.Agr. N.C. Vaccaro
	Ing.Agr. M.M.P. de Beñatena
Epidemiology	Ing.Agr. E. Danós
Plant Pathology	Ing.Agr. G.M. Marcó
	Ing.Agr. C.M. Casafús
	Ing.Agr. N.B. Costa
	Ing.Agr.S.M.Garran/Dr.M.A.Messina
	Ing.Agr.D.Vazquez/Ing.Agr.A.Robles
Pesticides	Ing.Agr. F.J. Valsangiácomo
Soils	Ing.Agr. A.S. Schatz
Soil fertility	Ing.Agr. G.N. Banfi
Forestry	Ing.Agr. M.A. Marcó
Eucaliptus and Pines	Ing.Agr. M. Sanchez Acosta
Agricultural Economics and Statistics	
Marketing, production economics, statistics	Dr. L.H. Larocca

7 Cordoba

7.1 Universidad Nacional de Cordoba, Facultad de Ciencias Agropecuarias (University of Cordoba, Faculty of Agricultural Sciences)
elevation: 425 m
rainfall : 677 mm
sunshine : 2659 h

Casilla de Correo 509
5000 Cordoba
Phone: 051 - 65130
Telex: 51822 Bucor
Dean: Ing. H. Contin

General: Obtainment and propagation of virus free material of potato, garlic, sweet potato, strawberry and stone fruit trees. Genetic and cultural improvement of vegetables and pulse.

Crop management	
Vegetables	Ing.Agr. M.Sc. J.L. Burba
	Ing.Agr. H.M. Fontan
	Ing.Agr. M.J. Buteler
	Ing.Agr. M.P. Blanco
Fruit crops	Ing.Agr.M.Sc. R.E. Bengoa
	Ing.Agr.L.E.Olocco/Ing.Agr.M.Flores
Plant Breeding	
Vegetables	Ing.Agr.M.Sc. H. Ceballos
	Ing.Agr.M.Errasti/Ing.Agr.C.Nazar
Legumes	Ing.Agr.M.Sc. E. Biderbost
	Ing.Agr.M.Sc. D. Peiretti
	Ing.Agr. J. Carreras
Plant Physiology	
Vegetables	Ing.Agr.M.Sc. J. Arguello
	Ing.Agr.D.Moriconi/Ing.Agr.R.Luna
	Biol.A.Ledesma/Biol.M.Aiazzi
Legumes	Ing.Agr.M.Sc. R. Racca
	Ing.Agr.D.Collino/Biol.T.Gonzalez
Fruit trees	Ing. Agr. R. Taborda
Plant protection	
Entomology	Ing.Agr.M.Sc. H. Sosa
	Ing.Agr. D. Igarzabal
Virology	Ing.Agr.M.Sc. F. Nome

Ing.Agr.M.Sc. D. Docampo
Ing.Agr.M.Sc. J. Muñoz
Ing.Agr. D. Ducasse
Ing.Agr.G.Zumelzú/Biol.G.Laguna
Biol.V.Conei/Ing.Agr.G.Truol

Bacteriology — Ing.Agr. V. Yossen
Pesticides — Ing.Agr.M.Sc. R. Novo
Ing.Agr.A. Cavallo
Ing.Agr. C. Cragneolini
Herbicides — Ing.Agr. R. Nobile/Ing.Agr. G. Luque

8 Corrientes

8.1 Universidad Nacional del Nordeste, Facultad de Ciencias Agrarias (National University of N.E. Argentina, Faculty of Agronomy Science)

Sargento Cabral 2131
3400 - Corrientes

Vegetables
Fertilization in tomatoes and peas — Prof.Ing.Agr. DM. Crocci
Action of chemical substances in sex expression and yield of cucurbits — Ing.Agr. G.E. Vallebella
Fruticulture
Management in pineapple, banana tree and genip tree crops — Prof.Ing.Agr. F.M. Pacayut, Ing.Agr. J.R. Niveyro

9 Delta del Paraná

9.1 Estación Experimental Agropecuaria Delta del Paraná INTA (Agricultural Experiment Station Delta del Paraná)
rainfall : 950 mm
sunshine : 2550 h

Casilla de Correo 14
2804 Campaña, Pcia. de Bs. Aires
Phone: 749 - 9162
Dir.:Ing.Agr.M.Sc.A.F. Garciá

This Station is the main centre of poplar research in Argentina. Other research work is done with fruit crops (pecan, plums, lemons), aquatic weeds and agricultural economics.

Fruit crops — Prof.M.Bakarcic/Ing.Agr.R.Reichart
Plant protection — Ing.Agr. H.A. Toscani
Agricultural economics — Ing.Agr. F. Mujica
Climatology — Lic. G. de Sosa Gonzalez

10 General Roca

10.1 Estación Experimental Regional Agropecuaria Alto Valle de Rió Negro, INTA (Regional Agricultural Experiment Station)
elevation: 242 m
rainfall : 188 mm
sunshine : 2810 h
Main orientation of work: Pomology

Casilla de Correo 52
8332 General Roca Rió Negro
Phone: 250017, 250027, 250037
Dir.: Ing.Agr. C. Casamiquela

Agricultural econimics
Marketing, production economics — Ing.Agr.M.Sc. A. Bongiorno
Fruticulture
Pomology — Ing.Agr.M.Sc. H. Castro

Post - harvest physiology	Ing.Agr. R. Rodriguez Ing.Agr. C. Benitez Ing.Agr. S. Francile
Plant protection	
Plant Pathology	Ing.Agr.D.Bergna/Ing.Agr.M.Rosini
Entomology	Ing.Agr. J. Vermeulen Ing.Agr. M.A. Tassara
Soils and irrigation	
Irrigation practices	Ing.Agr. J. Nolting Ing.Agr. J.M. Requena
Nutrition	Ing.Agr. C.R. Bestvater Ing.Agr. F. Sanchez
Viticulture	
Vinifera grapes	Ing. Agr. A. Cassino Ing.Agr. A. Llorente
Vegetable crops	
Breeding (tomatoes)	Ing.Agr. D.J. Calvar Ing.Agr. A. Sansinanea
Crop management	Ing.Agr. C. Argerich

11 Hilario Ascasubi

11.1 Estación Experimental Agropecuaria INTA (Agricultural Experiment Station)
elevation: 22 m
rainfall : 492 mm

8142 Hilario Ascasubi, Pcia. de Buenos Aires
Phone: 0928 - 91011
Dir.: Ing.Agr. C.J. Moschetti

Research work is concerned with: forage crops for seed production, cereals under irrigation, cattle production, vegetables (tomato, garlic, onion, potatoes), fruit crops (apples, pears, peaches).

Vegetable production	Ing.Agr.M.Sc. A. Lopez Camelo Bach.Agr. O. Caracotche
Fruticulture	Ing.Agr. C.D. Garcia

12 Junin

12.1 Subestación Experimental Agropecuaria, INTA (Agricultural Experimental Substation)
elevation: 653 m
rainfall : 216 mm
sunshine : 3102 h

Casilla de Correo 78
5570 General San Martín, Mendoza
Phone: 0623 - 22203
Dir.: Ing.Agr. O.D. Wouters

This Station is Argentina's main research center of canning clingstone cultivars of peaches. Also research is done in other stone fruit crops: nectarines, almond, plum, apricot and olives. Other fields of research are related with table grape cultivars, management and their storage.

Fruticulture	
Stone fruit production	Ing.Agr. O.D. Wouters Ing.Agr. S.C. de Toloza Ing.Agr. A.J. Reta
Viticulture	
Table grape variety trials, training and management of vineyards	Ing.Agr. M.L. Nazrala Ing.Agr. H. Martinez Pelaez
Industrial processing	
Stone fruits and olives	Ing.Agr. E.M. Salvarredi

13 La Consulta

13.1 Estación Experimental Agropecuaria, INTA (Agricultural Experiment Station)
elevation: 940 m
rainfall : 296 mm
sunshine : 2928 h

Casilla de Correo 8
5567 La Consulta, Mendoza
Phone: 130
Dir.: Ing.Agr.Dr. M.A. Jauregui

This Station is Argentina's main centre research station in vegetable crops under irrigation (tomato, onion, pepper, lettuce, carrot, squash, pulse, muskmelon) and vegetable seed production (elite and original seeds).

Agrometeorology
Bioclimatological requirements for vineyards, vegetables and fruit crops — Ing. Agr. H.L. Pasqualotto
Crop management
Chemical weed control in irrigated vegetable and fruit crops — Ing.Agr.M.Sc. O.G. Campeglia; Ing.Agr. R. Borgo
Plant breeding
Breeding of vegetable varieties for resistance to diseases, variety trials and introduction of new varieties:
lettuce, pulse — Ing.Agr. N.J.G. de Millán
onion, squash, carrot, muskmelon — Ing.Agr. R.N. Oliva
tomato, pepper — Ing.Agr.M.Sc. A.C. Senetiner; Agr. G.S. Gallardo

Plant Pathology
Vegetable crops — Ing.Agr. R.J. Piccolo
Soils
Soil fertility — Ing.Agr.Dr. M.A. Jauregui
Seed Production and Technology
Vegetable seed production (carrot, lettuce, onion, pea, pepper, squash, tomato, pulse and muskmelon) — Ing.Agr. J.C. Gaviola
Seed Pathology
Testing and quality control of vegetable seeds — Ing.Agr. M.A. Makuch

14 La Plata

14.1 Estación Experimental Gorina, Ministerio de Asuntos Agrarios, Provincia de Buenos Aires (Experiment Station Gorina. Ministry of Agricultural Affairs, Province of Buenos Aires)
elevation: 15 m
rainfall : 1100 mm
sunshine : 2600 h

Calle 501 y 149
1896 Joaquín Gorina
Phone: 84 - 0443
Dir.: Ing.Agr. O. Martinez Quintana

The Experiment Station is situated in the heart of the vegetable growing belt of Buenos Aires area. The research work is carried out with the main vegetable crops of the area: tomato, celery, artichoke, lettuce, also with crops such as cucurbits peas, pepper eggplant, and coles.

Plant Breeding
Vegetable crops — Ing.Agr. J. Castro Gamarra; Ing.Agr. J. Bulnes Mendoza

Plant Protection
General Plant Protection — Ing.Agr. O. Larroque

Plant Pathology	Ing.Agr. B.L. Ronco
	Ing.Agr. B.S. Gamboa
Herbicides	Ing.Agr. S.A. Passalacqua
Entomology	Ing.Agr. J.E. Roán
Crop management	
Vegetable crops	Ing.Agr. O. Martinez Quintana

15 Mendoza

15.1 Estación Experimental Regional Agropecuaria, INTA (Regional Agricultural Experiment Station)
elevation: 935 m
rainfall : 220 mm
sunshine : 2800 h

Casilla de Correo 3
5507 Luján de Cuyo - Mendoza
Phone: 960300, 960546
Telex: 55472 INTAL AR
Dir.:Ing.Agr.M.Sc.H.R. Galmarini

Consultor Dir.: Ing.Agr. J. Crnko*

The Station is the main centre of viticulture and enology research in Argentina. Other fields of research are concerned with fruit crops (cherry, almond, nectarine, apricot, apple, quince) and vegetable seed production (tomato, onion, pepper, lettuce) and machinery equipment.

Agricultural economics and statistics	
Marketing, production economics	Lic. M.D. Rodriguez
	Ing.Agr. G.M. Martín
Biostatistics	Ing.Agr.M.Sc. P. Gomez Riera
Crop management and field machinery	
Vegetable, vineyard and fruit crops	Ing.Agr. R.F. Del Monte
Fruticulture	
Pome fruit production	Ing.Agr.M.Sc. J.J. Camba
Stone fruit production	Ing.Agr. S.N. Gonzalez Caldwell
Plant Protection	
Entomology	Ing.Agr. J.C. Espul
Biological control	Ing.Agr.Dr. M.F. Garciá
Plant virology	Ing.Agr.M.Sc. O. Gracia
Plant pathology	Ing.Agr. E.J.A. Oriolani
	Ing.Agr. M.E. Gatica de Mathey
Pest ecology	Prof. A.H. Riquelme
Pesticides	Ing.Agr. N.J.A. Cucchi
Pesticides residues	Ing.Agr. A.E. Puiatti
Soils and irrigation	
Irrigation and drainage	Ing.Agr. M.J.C. Oriolani
Soil fertility	Ing.Agr. J.L. Luque
	Ing.Agr. M.L. Gonzalez
	Ing.Agr. M.E. Quiroga de Oriolani
Soil taxonomy, salt affected soils	Ing.Agr. R.R. Hudson
Viticulture and enology	
Viticulture	Ing.Agr. L.J. Laborde
	Ing.Agr. C.R. Tizio Mayer
	Ing.Agr. C.D. Catania
	Enól. M.S. Avagnina de Del Monte

15.2 Universidad Nacional de Cuyo, Facultad de Ciencias Agrarias (National University of Cuyo, Faculty of Agricultural Sciences)
elevation: 940 m

Alte. Brown 500
5505 Chacras de Coria, Mendoza
Phone: 960406

rainfall : 220 mm
sunshine : 2800 h

15.2.1 Chair of Fruticulture	Head: Prof.Ing.Agr. E.M.L. Welkerling de Tacchini

This Chair is concerned with all problems of temperate climate fruits production, their breeding up to their harvest.

Fruit production	
Pome fruits	Ing.Agr. E.M.L. Welkerling de Tacchini/Ing.Agr. R. Pozzoli Ing.Agr. E. Belleli
Stone fruits (other than olive)	Ing.Agr. E.M. Zaina Ing.Agr. A. Bustamante
Olive	Ing.Agr. R. Pelliciari
Micropropagation	Ing.Agr. C. Arjona
Plant protection	Ing.Agr. L. Sarcinella

15.2.2 Instituto de Suelos y riego (Institute of Soils and Irrigation)	Casilla de Correo 7 Chacras de Coria, Mendoza Phone: 061 - 960004, - 960431 Dir.:Ing.Agr.L.Nijensohn,Prof.Emer

This Institute is devoted to teaching and research activities concerning soil, water problems and management in irrigated agriculture: Viticulture, fruticulture, and vegetable crops.

Section of soils	Head: Ing.Agr. F.S. Olmos
Section of Agricultural Chemistry	Head: Ing.Agr. M.O. Avellaneda
Section of Irrigation	Head: Ing.Agr. N. Ciancaglini

15.2.3 Instituto de Viticultura (Institute of Viticulture)	Almirante Brown 500 Casilla de Correo 7 5505 Chacras de Coria, Mendoza Phone: 960004, 960396 Dir.: Ing.Agr. E.M. Zuluaga

The Insitute is the important academic center of viticulture and enology in Argentina. Besides teaching, major emphasize is given to research for improving grape quality (flower biology and use of plant growth regulators in *Vitis vinifera L.c.v.)* clone selection of *Vitis vinifera L.* varieties. Studies of vine growing as well as vine ecology to determine the Argentina's grape growing regions and viticultural agroecosystems are being made, to get the national viticultural type.

Staff	
Professor	Ing.Agr. E.M. Zuluaga
Associated Professor	Ing.Agr. F.J de la Iglesia
Assistant Professor	Ing.Agr. I.M. Galarraga
Laboratory Assistant	Ing.Agr. M.M. Ocvirk de Simón
Laboratory Assistant	Ing.Agr. L.G. Borsani
Research Assistant	Ing.Agr. M.S. Matus

15.2.4 Instituto de Horticultura (Institute of Vegetable crops)	Almirante Brown 500 5505 Chacras de Coria, Mendoza Dir.: Ing.Agr. L.O. Melis

Breeding work in cucurbits, beans and tomatoes — Prof. Ing.Agr. L.O. Melis; Ing.Agr. C.V. Bartuchotto

15.2.5 Instituto de Ciencia y Tecnología de Alimentos (Institute for Food Science)

Casilla de Correo 7
5505 Chacras de Coria, Mendoza
Phone: 960433
Dir.: Ing.Agr. Máximo F. Bocklet

Juices of different fruits	Ing.Agr. A.O Mei
Dehydration of fruits and vegetables	Ing.Agr. E.B. Revillard
Analysis of foods	Lic.R.M.de Dias/Ing.Agr.H.Estrella
Keeping of fruits and vegetables through refrigeration and freezing	Ing.Agr. H. Roby
Machinery and Equipment	Ing.Chem. J. Dávila
Milk and deriveds	Ing.Agr. A.M.D. de Baldini
Experimental factory	Ing.Agr. E. Cerchiai

15.3 Centro de Investigación Tecnologíca de frutas y hortalizes (CITEF - INTI) (Technological Research Center for fruits and vegetables CITEF - INTI)
elevation: 935 m
rainfall : 220 mm
sunshine : 2800 h

Casilla de Correo 15
5505 Chacras de Coria, Mendoza
Phone: 960702
Telex: 21859 INTI AR

Equipment: - 4 Laboratories for research.
- Experimental factory for fruit and vegeable canning, pulps of tomato, peaches, apples, pears, quince and olives, obtainment of concentrated products. Research on varietal performance and adaptability for above mentioned purposes. Teaching to technical personnel of canning industry.

Staff: Ing.Agr. A. Rearte
Ing.Agr.J.Berman/Ing.Agr.A.Mei
Ing.Agr. R.F. Bonino
Ing.Agr. D.E. Aranda
Ing.Agr. E. Revillard
Lic.Qca. M.P. de Silvestri
Lic.Qca. M.E.B. de Manzino
Enol.C.Martinez/Eno.L.Giarrizzo

16 Rosario

16.1 Universidad Nacional de Rosario. Facultad de Ciencias Agrarias (National University of Rosario. Faculty of Agricultural Sciences)

Santa Fé 2051
2000 Rosario, Prov. de Santa Fé
Phone: 210107, 42477, 67893

Chair of Vegetable crops	
Crop management, artichoke	Ing.Agr. J.C. Zembo
Crop management under plastics	Ing.Agr.J.C.Zembo/Ing.Agr.J.Ferrato
Cucurbits, Swiss chard, cruciferous	Ing.Agr. J.T. Firpo
Artichoke, cruciferous	Ing.Agr. S.M. García
Spinach, cruciferous	Ing.Agr. L.M. Torres
Plant growth regulators in tomato	Ing.Agr. M. Panelo
Micropropagation of artichoke	Ing.Agr. R. Murray
Chair of Genetics: Breeding work in tomato	Ing.Agr. E. Prado
Chair of Plant Physiology	
Micropropagation of artichoke	Ing.Agr. Ç. Severin
Plant growth regulators in beans	Ing.Agr. A. Salinas
Chair of Phytopathology	
Virology (tomato and artichoke)	Ing.Agr. M. Gonzalez

17 Salta

17.1 Estación Agropecuaria Regional INTA (Regional Agricultural Experiment Station)

Casilla de Correo 228
4400 Salta (Provincia de Salta)

elevation: 1250 m
rainfall : 692 mm
sunshine : 2287 h

Phone: 087 - 221585
Dir.: Ing.Agr.M.Sc. F.B. Bravo

At this Station the horticultural research is concerned with dry beans, winter legumes (lentils, broad beans, chickpea) and vegetables (tomato, sweet pepper, hot pepper).

Agricultural economics and Statistics

Production economics
Ing.Agr.M.Sc. R.A.E. Diedrich
Cont.Públ.M.Sc. M.A. Elena

Marketing
Ing.Agr. L.R. Fellín

Biostatistics
Lic. V.P. Brescia

Legumes (dry beans, lentils, chickpeas, broad beans)

Plant Breeding
Ing.Agr.M.Sc. R.D. Sosa Quiroga

Crop management
Ing.Agr. C.P. Pastrana

Vegetables (tomato, peppers)

Plant Breeding
Ing.Agr.M.Sc. E.M.G. de Saravia
Ing.Agr. N. Colombo

Plant Protection

Plant Virology
Ing.Agr. R. Giroto

Weed Control
Ing.Agr. J.A. Arias

Entomology
Ing.Agr. J.M. Benavent

Soils and irrigation

Irrigation and drainage
Ing.Agr. C.E. Yañez
Ing.Agr. L.E. Blanco

Soil fertility
Ing.Agr. P.N. Figueroa
Ing.Agr. J.L. Tau

Soil taxonomy
Dr. J.R. Vargas Gil

Soil management
Ing.Agr. R.F. Roman

Agrometeorology
Prof.A.Bianchi/Tech.Agr.J.J.Nievas

18 San Juan

18.1 Estación Experimental Agropecuaria San Juan, INTA (Agricultural Experiment Station San Juan)
elevation: 618 m
rainfall : 92 mm
Equipment: - soil and irrigation Laboratory, Agrometeorological Station, Library

Apartado Interno
5427 Villa Aberastain - San Juan
Phone: 921079
Dir.: Ing.Agr. E.C. Fontemachi

Due to the dry climate conditions and irrigation in the zone, the viticulture is its principal activity, growing varieties of excellent quality with different purposes: table grapes, raisins and wine production. Other research work is concerning with stone fruit crops (almond, olive, peach, apricot and plums), quince and fig. Besides, vegetable crops like tomato, onion, eggplant, pepper, watermelon, muskmelon and lettuce, specially on the seed production of them, also produce alfalfa seed. A new crop was introduced a few years ago, with very good performance, cotton (long fiber: 29 mm).

Agrometeorology

Applied agrometeorology
Enól. R. Cornejo

Agricultural economics

Marketing, production economics
Per.Merc. M.R. Benito

Fruticulture

Stone fruit production
Enól. H.N. Andrada

Plant Protection

Weed, Pest and Disease control
Ing.Agr. J.E. Bustos

	Ing.Agr. E.C. Fontemachi
Soils and irrigation	
Irrigation and drainage	Ing.Agr. T.S. Castro Tecn.Hidrául. M.A. Liotta
Soil and water analysis	Tecn.Quim. P. Gil
Recognition and soil classification	Enól. J.O. Bocelli
Vegetable crops	
Breeding new varieties. Crop management, mainly in onions, lettuce, tomatoes, artichoke, peppers and lentil	Ing.Agr. A.R. Acosta Enól. R.O. Moreno
Viticulture	
Crop management. Introduction of new varieties and its adaptation	Ing.Agr. E.M. Cáceres
Vegetable and Forrage Seed Production	
Forrage seed production	Ing.Agr. E.M. Echeverria
Vegetable seed production	Enól. C.E. Cerezo

19 San Pedro

19.1 Estación Experimental Agropecuaria San Pedro, INTA (Agricultural Experiment Station San Pedro)
elevation: 29 m
rainfall : 1055 mm

Casilla de Correo 43
2930 San Pedro, Buenos Aires
Dir.: Ing.Agr.M.Sc. A. Boy

This Experimental Station is located in the most important area of vegetable production for fresh marketing. The main orientation of work concerns with vegetable crops and fruit growing (stone and citrus).

Vegetable crops	
Breeding and crop management:	
Peas and Sweet corn	Ing.Agr.E.Riva/Agr.P.Bianchini
Lentils	Ing.Agr. E. Riva
Potatoes	Agr. P. Bianchini Ing.Agr. M.J. Stoppani
Sweet potato	Ing.Agr.M.Sc. A. Boy Agrónomo P. Bianchini
Table beets, swiss chard, spinach and summer squash	Ing.Agr. J.P. Rodriguez Ing.Agr. M. Nakama
Strawberry, tomatoes and asparagus	Ing.Agr. J.P. Rodriguez Ing.Agr. M.J. Stoppani
Tissue culture in strawberry and sweet potato	Ing.Agr. N. Hompanera
Fruit crops	
Breeding and crop management in peaches	Ing.Agr.C.Torroba/Ing.Agr.H.Frangi Ing.Agr.M.Zapletal/Agr.J.Biglia
Rootstocks and dormancy of peaches	Ing.Agr.H.Frangi/Ing.Agr.C.Torroba Ing.Agr. M. Zapletal/Agr.J.Biglia
Breeding and cultural practices in plum	Agr.R.Gamietea/Ing.Agr.H.Frangi
Soil management and fertility	
Tillage and organic wastes	Ing.Agr.J.Gonzalez/Ing.Agr.A.Amma
Soil Fertility	Ing.Agr.A.Amma/Ing.Agr.J.Gonzalez
Plant protection	
Vegetable diseases and control	Ing.Agr.J.Martinengo de Mitidieri
Weed biology and control	Ing.Agr. A. Mitidieri
Pest biology studies and its control in vegetables, citrus and peach orchards	Ing.Agr. H. Bimboni

20 San Rafael

20.1 Estación Experimental Agropecuaria Rama Caída, INTA (Agricultural Experiment Station Rama Caída)
elevation: 692 m
rainfall : 300 mm
sunshine : 2583 h

Casilla de Correo 79
5600 San Rafael, Mendoza
Phone: 0627 - 21559
Dir.: Ing.Agr.M.Sc. W. Cinta

The principal fields of research are: viticulture (breeding of new grape varieties, *vitis vinifera L.)*; fruit growing (physiology, growth regulators, propagation, nursery); nematology in vegetable crops, orchards and vineyards; bioclimatology.

Viticulture and enology	
Viticulture (grape breeding)	Ing.Agr. A.A. Gargiulo
Enology	Enól. J.C. Fuentes
Fruticulture	
Pomefruit and stone fruit production	Ing.Agr.Dr. L.N. Di Césare
Plant protection	
Nematode's identification and their control. Virology	Ing.Agr. E. Vega
Agricultural meteorology	
Microclimate studies	Bach.Sc. P.A. Worlock

21 Santiago del Estero

21.1 Estación Experimental Agropecuaria Santiago del Estero, INTA (Agricultural Experiment Station Santiago del Estero)

Independencia 341
Casilla de Correo 268
4200 Santiago del Estero
Dir.: Ing.Agr. F. Cantos

The main fields of research work: alfalfa, cotton, natural resources, live stock production, cultivated forrage crops, vegetables.

Vegetable breeding:	
Cucurbits and onion	Ing.Agr.F.Cantos/Ing.Agr.M.J.Baez
Sweet potato	Ing.Agr. F.M. Fernandez Ing.Agr. F. Cantos
Variety studies:	
Sweet potato, tomato for canning, onions and lettuce	Ing.Agr. F.M. Fernández Ing.Agr. M.J. Baez
Crop management and weed control in vegetables	Ing.Agr. F.M. Fernández
Irrigation and fertilization	
In vegetable crops, citrus orchards and vineyard	Ing.Agr. A.C. Lozano Cruzado Ing.Agr. L.H. Ochoa
Vegetable seed production	Ing.Agr. F.M. Fernández Ing.Agr.F.Cantos/Ing.Agr.M.J.Baez

22 Tucumán

22.1 Estación Experimental Regional Agropecuaria Famaillá, INTA (Regional Agricultural Experiment Station Famaillá)
elevation: 363 m
rainfall : 1269 mm

Casilla de Correo 9
4000 S.M. de Tucumán
Phone: 081 - 221510
0863 - 61048/61049
Dir.: Ing.Agr. J.A. Dominguez

Citriculture and tropical fruits
Breeding, colection of species and varieties, root-

stocks for commercial production, virology	Ing.Agr.M.Sc. M.A. Garciá
Cultural management	Ing.Agr. G. Martínez Ing.Agr. F. Fernández
Harvesting and postharvest processing. Physiology	Ing.Agr. M. Frutal Ing.Agr. M.Sc. F.J. Tan Jun
Plant protection:	
Entomology	Ing.Agr. M.E.B. de Pacheco
Plant Pathology	Ing.Agr. M.A. González Agr. A. Chavarria
Soils and nutrition	Ing.Agr. M.L. Casen Ing.Agr.Quím. H.G. Ayala Pto.Sac. C.A. Torne

22.2 Universidad Nacional de Tucumán Facultad de Agronomía y Zootecnia (National University of Tucumán, Faculty of Agronomy and Zootechnics)
elevation: 400 m
rainfall : 900 mm

Av. Roca 1900
4000 S.M. de Tucumán
Phone: 081 - 242155
Dean: Ing.Agr. R. Zuccardi

Chair of vegetable crops:	
Agrotechnics and breeding:	
Sweet potatoes, potatoes and tomatoes	Ing.Agr. J. Ploper Ing.Agr. R. Fernández Ing.Agr. E. Brandan Ing.Agr. M.T.D. de Ricci
Onion	Ing.Agr.J.Ploper/Ing.Agr.E.Brandan
Beans	Ing.Agr. M.T.D. de Ricci Ing.Agr. J. Ploper
Agrotechnics of vegetable crops under plastics	Ing.Agr. R. Fernández Ing.Agr.E.Brandan/Ing.Agr.J.Ploper
Chair of Fruticulture:	
Citrus crop management and variety trials	Ing.Agr. J. Palacios Ing.Agr. E. Padilla
Stone fruit management	Ing.Agr. J. Palacios Ing.Agr. S. Campos
Chair of plant protection	
Biological control of fruit flies	Ing.Agr.A.J.Nazca/Ing.Agr.A.Terán Ing.Agr. R.V. Fernández
Virus and fungus diseases of solanaceae and leguminous crops	Ing.Agr.A.Mena/Ing.Agr.C.Ramallo Ing.Agr. S. Whett Ing.Agr. B. de Diaz Botto

22.3 Estación Experimental Agro-Industrial "Obispo Colombres" (Agro-Industrial Experiment Station "Obispo Colombres")
elevation: 400 m
rainfall : 900 mm

4000 S.M. de Tucumán
Casilla de Correo 71
Phone: 081 - 216561, - 225475
Dir.: Ing.Agr. V. Hemsy

Citrus and Avocado:	
Breeding	Agr. J.L. Foguet Ing.Agr. J.L. González Ing.Agr. B.E.S. De Lizarraga
Cultural management	Ing.Agr. H. Vinciguerra Pto.Agr. S. Alvarez
Soils and nutrition	Ing.Agr. P.J. Aso Ing.Agr. M.R. Casanova

Plant Protection

Entomology	Ing.Agr. M. Costilla
	Ing.Agr.H.J.Basco/Lic.E.Willink
	Lic. N. de Martínez
Pathology	Ing.Agr. N.V. de Ramallo
	Ing.Agr. D.L. Ploper
	Ing.Agr. G.M. Salas
	Ing.Agr. E. Würschmidt
Weed control	Ing.Agr. J.M. Alonso
	Ing.Agr. R.S. Loria
Economics and Statistics	Ing.Agr. C.A. Gargiulo
	Lic. C. Gonzalez Terán
Potatoes	Ing.Agr. E. Rojas (Coordinator)
	Ing.Agr.N.Zamudio/Ing.Agr.R.Orell
Dry beans	Ing.Agr. J.R. Ricci (Coordinator)
	Ing.Agr. O.N. Vizgarra

23 Viedma

23.1 Instituto de Desarrollo del Valle Inferior del Río Negro, IDEVI (Institute for development of Eastern Valley of Río Negro)
elevation: 4 m
rainfall : 399 mm

Belgrano 536
Casilla de Correo 353
8500 Viedma, Río Negro
Phone: 0920 - 22233, - 22365
Dir.: Ing.Agr. E. Alvarez Costa

Research and experimental work is related with fruit trees (peaches, cherries, almonds, plums, walnuts, filberts, apples, pears and quince), vegetables (tomatoes, onions, peppers, potatoes), cereals and oil crops (wheat, corn, soybean). There is also research area devoted to animal production under irrigation (cattle, sheep, swine). Beekeeping.

Fruit trees and crops	Ing.Agr. J.P. Rolca
	Ing.Agr. L. Jannamico
Vegetables	Ing.Agr. M. Gorrochetegui
	Ing.Agr. R. Rossini
Plant Protection	Ing.Agr. A. dall'Armellina
	Ing.Agr. M.C. Pozzo Ardizzi

Ecology Department:

Meteorology	Técn.Agr. N. Callupil
Soils	Ing.Agr. M.E. Ozcariz
	Ing.Agr. S. Plunkett
Soils Laboratory	Ing.Agr. R. Martínez
	Ing.Agr. L. Casamiquela

Irrigation and Drainage Department

Irrigation and Drainage	Ing.Agr. N. P. Funes

AUSTRALIA

130° E
20° S
500 km
62
55
64
53
58
54
61
60
59
57
51
52
56
50
21
17
20
13
14
24
16
22
23
19
15
18
25
11
2
1
41
40
38
4
6
10
7
3
5
8
9
42
36
37
33
12
34
29
32
65
39
26,27
30
28
44
49
42
47
48
46
45

New South Wales: 1 Alstonville, 2 Armidale, 3 Bathurst, 4 Dareton, 5 Gosford, 6 Griffith, 7 Orange, 8 Richmond, 9 Sydney, 10 Wagga Wagga, 11 Wollongbar, 12 Yanco, **Queensland**: 13 Ayr, 14 Bowen, 15 Brisbane, 16 Bundaberg, 17 Cairns, 18 Cleveland, 19 Gatton, 20 Innisfail, 21 Mareeba, 22 Maryborough, 23 Nambour, 24 Rockhampton, 25 Stanthorpe, **Victoria**: 26 East Melbourne, 27 Burnley, 28 Frankston, 29 Shepparton, 30 Knoxfield, 31 Irymple, 32 Tatura, 33 Mildura, 34 Swan Hill, 35 Healesville, **South Australia**: 36 Adelaide, 37 Glen Osmond, 38 Loxton, 39 Struan, 40 Nuriootpa, 41 Roseworthy, 42 Waite, **Tasmania**: 43 Devonport, 44 Forth, 45 Grove, 46 Hobart, 47 Launceton, 48 Scottsdale, 49 Burnie, **Western Australia**: 50 Albany, 51 Bunbury, 52 Busselton, 53 Carnarvon, 54 Geraldton, 55 Kununurra, 56 Manjimup, 57 Medina, 58 Midland, 59 Perth, 60 Stoneville, 61 Swan, **Northern Territory**: 62 Darwin, 63 Katherine, 64 Alice Springs, **Plant Quarantine Branch**: 65 Canberra

Survey

Organisations involved in horticultural research in New South Wales and their activities are as follows:

1. State Department of Agriculture - The first horticultural plantings at the Tropical Fruit Research Station, Alstonville, were made in 1960. Research is concerned with bananas, the most important tropical fruit grown in NSW, and a range of other tropical and sub-tropical fruits and vegetables. Banana investigations involve physiology and growth/climate studies, nutrition, soil/plant/water relations, and pest/disease management. Research with other fruit crops includes avocado rootstock propagation, evaluation of avocado varieties and the effect of rootstocks on varietal performance, root rot management and control in avocadoes; clonal selection of papaws; disease control and its influence on fruit set of mangoes; evaluation of varieties of lychee and custard apples and selection of guava clones; evaluation of macadamia varieties, irrigation and mechanical harvesting of macadamias. Vegetable research activities involve

Prepared by N.D. Honan, Dept. of Primary Industrie, Canberra New South Wales evaluation of sweet potato varieties and nutrition research, especially in relation to tomato production on krasnozems.

In 1976, a major trial to evaluate five different training systems for intensively planted dessert peaches was established at the Bathurst Agricultural Research Station. Other studies include evaluation of stone fruit varieties, investigation of integrated pest control programmes for apples, spray technology, evaluation of herbicides in orchard soil management systems, and research into post-harvest handling and rot control of dessert stone fruits and pome fruits. A collection of miscellaneous berry fruits has been established, including red and black currants, cape gooseberries and various bramble berries.

Situated in the arid south western corner of NSW on the Murray River, the Agricultural Research and Advisory Station, Dareton was established to investigate problems of citrus growing and grape vine culture. The initial trial plantings were made in 1955. Current citrus projects are concerned with variety and rootstock assessment, container propagation, planting density and tree size control, nutrition, irrigation and salinity management. Viticultural projects include studies on irrigation requirements and salt tolerance of sultanas, and the influence of trellising systems on summer pruning and trellis drying. Trials with avocadoes are being carried out on variety and rootstock evaluation, and management of mature plantings. Mangoes, lychees, macadamias, pistachios and pomegranates are being evaluated as possible alternative crops for the region.

The Gosford Horticultural Post Harvest Laboratory was established on the central coast of NSW in 1948, in response to urgent industry pressures for research on citrus handling and decay control. It is operated in co-operation with the CSIRO and staffed by this Department. Evaluation of fungicides for decay control in fruit and vegetables remains an important feature of the research and since 1976 research on postharvest handling and treatment of vegetables has been undertaken. Another aspect of the Laboratory's research programme has been the examination of disinfestation treatments for fruit and vegetables against Queensland fruit fly *(Dacus tryoni)* to meet quarantine requirements for export and interstate trade. Fumigation treatments have been developed for a range of commodities, including citrus, bananas, apples, pears, cherries, tomatoes and capsicums. The use of irradiation is under study.

The Horticultural Research Station, Gosford, originally a viticultural nursery, has since developed a major citrus research programme. More recently programmes have been developed with strawberries, vegetables, miscellaneous fruit crops and ornamentals. Field trials with the above crops are mostly carried out by a sub-station located nearby at Somersby. Citrus investigations include rootstock scion relationships as it relates to tree size, planting densities, virus status, susceptibility to root rot diseases, fruit production and quality; nutrition. Activities with strawberries, miscellaneous fruit crops such as blueberries, kiwifruit, pepino, pecan nuts and vegetables are mostly centred on varietal evaluation. A major ornamental horticulture programme has developed including investigations into potting mixtures, nutrition, shelf life of cut flowers and flowering potted plants, herbicide usage, determining and developing commercial potential of Australian native flora. Entomological investigations developing biological pest management programmes for strawberries and ornamental crops are being conducted.

The Viticultural Research Station, Griffith is the main centre for viticultural research in N.S.W., and is expanding into citrus research. Irrigated land area is 28 ha with 12 ha planted to grape vines, wine and table grape types. The viticultural research programme, supported by a small scale wintery facility, deals with all aspects of viticulture, including irrigation, vine improvement and mechanical pruning. The Station supplies foundation planting material of improved clones and varieties to the N.S.W. wine and table grape industries.

The Agricultural Research Centre, Orange, was originally established in 1963 for potato research, but subsequently was also developed for deciduous fruit tree plantings. More recently, it has been selected as a base for a research team investigating ecology and control of weeds and it is the headquarters of the Central West, South East and Illawarra Regions. Major fruit trials being conducted include evaluation of tree training systems for three major varieties of apples on four rootstocks and a comparison of tree spacings within apple hedgerow plantings. As well, there are several small demonstrations of rootstocks and three training systems for apples and pear including systems for mechanization of apple growing. A small trial to compare four rootstocks for cherries has been planted. A collection of pome fruit varieties free of all known viruses provides a source of virus-tested propagation material for nurserymen. Variety evaluation of apples, pears, cherries, walnuts, chestnuts and filberts are also established. Plant protection investigations include apple scab prediction, control of postharvest rots in stone and pome fruit and integrated pest control in apples utilizing predatory mites and codling moth sex pheromone traps.

The Horticultural Research Unit, Richmond, has been the Department's main centre of vegetable breeding for many years. Breeding programmes are currently being undertaken to develop indeterminate hybrids and fixed lines of fresh market tomatoes resistant to Fusarium, Verticillium and nematodes as

well as cold tolerant varieties of fresh and processing tomatoes.
The Biological and Chemical Research Institute, Rydalmere (Sydney) is divided into three sections, viz. Biology Branch (Plant Pathology, Microbiology, Virology, Nematology), Chemistry Branch and Entomology Branch. Most officers of the Institute travel throughout the State to participate in research projects whilst some are stationed at Regional Research Centres.
The Division of Plant Industries, Sydney, is jointly responsible with Regional Directors for the technical direction and administration of the Department's research programmes in fruit, vegetables and ornamental horticulture at all Stations and Laboratories.
The Agricultural Research Centre, Wollongbar, near Alstonville, is the headquarters of the North Coast Regional Research Directorate. Research is being undertaken on macadamia nutrition and nut quality; diseases of sub-tropical and tropical fruit, including banana bunchy top virus, avocado sun blotch virus, avocado root rot and mango and avocado anthracnose; and pest management in fruit and vegetables.
Horticultural Research at the Agricultural Institute, Yanco, headquarters of the Murray and Riverina research region, covers citrus fruit and canning stone fruit culture, and vegetable agronomy research. A major aspect of citrus research is the evaluation of the use of dwarf trees for intensive planting - commercial aspects of dwarf tree plantings, evaluation of selected dwarf bud lines and transmission of the dwarf character by bud inoculation. Stone fruit research involves variety testing and investigations of tree training methods for high density plantings of peaches. Vegetable research includes processing tomatoes, potatoes, onions, garlic and asparagus. Disease and insect pests of fruit and vegetables are also studied.

2. The University of Sydney, Sydney - Undergraduate courses in horticulture are taken by candidates for the degree of Bachelor of Science in Agriculture. Postgraduate courses include the Diploma in Horticultural Science, Diploma in Agricultural Science, Master of Agriculture, Master of Science in Agriculture and Doctor of Philosophy. Research in production horticulture is concerned with a variety of fruit, vegetable and ornamental species. In landscape horticulture special attention is being given to the use of native Australian plants. Physiology of flowering, plant regeneration in vitro and studies on the plant symplast are subjects of major research programmes. An agronomy field unit (14 ha) and a horticulture field unit (14 ha) are situated at the University Farms, Camden, about 60 kilometres from Sydney.

3. Macquarie University, North Ryde (Sydney) - A number of studies of fruit and vegetable crops are being made within the School of Biological Sciences, often in conjunction with officers of the N.S.W.. Department of Agriculture and with the CSIRO Division of Food Research, Plant Physiology Group, which is located within the School. These include physiology of pests, notably aphids, and basic aspects of physiology and biochemistry such as photosynthesis and amino acids metabolism, responses to water stress of peaches, growth and development of tomatoes, nutrition of onion, etc.

4. The University of New South Wales, Kensington (Sydney) - Studies in the field of postharvest physiology and biochemistry of fruit and vegetables cover possible mechanisms of fruit ripening, the use of calcium slats to extend storage life and the mode of action of calcium in fruit and vegetable tissue. These studies are currently concentrated on tropical fruit, however postharvest storage and pathology of root crops is also being investigated.

5. University of New England, Armidale - A final year elective Horticultural Science course was established in the Agronomy Department in 1973 for undergraduates and for the following nine years one semester contained course work and the second was devoted to a research project. The course emphasizes propagation and husbandry of flower crops and fruit species. Twelve students have had horticultural topics for the honours or postgraduate diplomas. Currently two people are on horticultural Ph.D. programmes, while four higher degrees on tissue culture emanated from the Botany and Agronomy Departments. There are two research programmes involving a study of endogenous cytokinins and abscisic acid in relation to the dormancy of bulbs and corms. Other research topics are rootstock-blossom relationships in pears, anti-senescence factors in cut flowers and the physiology and tissue culture of papaws by three external higher degree students. The Agronomy Department possesses glasshouses, a fruit nursery and an orchard of temperate fruits and berries.

6. Hawkesbury Agricultural College, Richmond - The College was established in 1891 as a centre of agricultural learning and research. Previously attached to the Department of Agriculture it now operates as an autonomous College with a major interest in horticultural education. Staff associated with the horticultural teaching programme carry out research activities in areas of plant improvement, plant physiology, entomology and plant pathology. Evaluation of new varieties of vegetables and fruits, pesticide and herbicide testing, breeding of native plants, hydroponics, weed control and nutrition of container grown plants are major research interests. Of recent development are the areas of mushrooms, permaculture, tissue culture and organic farming.

7. Riverina College of Advanced Education (Wagga Agricultural College), Wagga Wagga - The College, through the School of Agriculture, carries

out teaching and research into Horticulture (Viticulture and Amenity Horticulture) and Oenology. It has a commercial winery, 13.5 ha of vineyards under irrigation, a registered nursery with the capacity of 50.000 plants and will soon establish an irrigated source area for ornamental trees and shrubs.

1 Alstonville

1.1 Tropical Fruit Research Station
elevation: 110 m to 170 m
rainfall : 1695 mm
sunshine : 2750 h

Alstonville, 2477
New South Wales
Phone: 066-280604
A/Manager: G.N. Hill, HDA

Banana variety comparison and yield forecast modelling, development of optimal nutrition strategies for vegetable crops, evaluation of varieties/seedling selections of macadamia, avocado, lychee, mango, guava, longan, carambola, wax jambu, stone fruits; maturity standards for lychee and custard apple; avocado root management and control.

Culture and management of tropical fruit lychees, guavas, custard apples, mangoes, new fruits arboretum, low chill peaches.	D.J Batten, B.Sc.Agr., Ph.D.
Vegetables - varietal evaluation,nutrition	D.O. Huett, M.Sc.Agr., Ph.D.
Physiology and nutrition of bananas	G.G. Johns, B.Rur.Sc., Ph.D.
Irrigation, pest and disease management of subtropical fruits, macadamia varieties and culture.	T. Trochoulias*, B.Sc.Hort.

2 Armidale

2.1 University of New England
Department of Agronomy and Soil Science
elevation: 980 m
rainfall : 1508 mm
sunshine : 2700 h

Armidale, 2351
New South Wales
Phone: 067-732829
Telex: 66050
Head: Dr. R.S. Jessop

Undergraduate course in propagation and fruit culture. Research specializing in pear and papaw physiology and hormones in relation to the dormancy of bulbs and corms.

Snr. Lecturer: Dr. N.G. Smith
Demonstrator: Mrs. Margaret Smith B.Sc. (Hort.), N.D.H.

3 Bathurst

3.1 Agricultural Research Station
elevation: 713 m
rainfall : 626 mm

Bathurst, 2795
New South Wales
Phone: 063 311988
Manager: R. Wickson

Total area 66 ha with 14 ha of apples and 15 ha of stone fruit (peaches, nectarines, plums and apricots). Investigations of spray technology, post harvest disease control, stone fruit tree training, growth regulators and variety evaluation.

Entomology: Integrated orchard mite control	Dr. C.C. Bower
Pathology: Post harvest diseases	W. Koffmann
Culture: Peach, nectarine and apricot variety and peach rootstock evaluation	W. Koffmann
Plum variety and rootstock	R. Wickson
Tree training, growth regulators and blueberry variety and culture evaluation	A.R. Menzies*
General: Spray technology	W.G. Thwaite

4 Dareton

4.1 Agricultural Research and Advisory Station
elevation: 50 m
rainfall : 277 mm
sunshine : 2925 h

Dareton, 2717
New South Wales
Phone: 050 274409
Manager: W.J. Esler

Research on citrus and avocado culture; irrigation and salinity management for citrus and dried vine fruits; centre for citrus budwood and seed distribution.

Citrus improvement, crop productivity	Dr. K.B. Bevington
Irrigation requirements, salt tolerance of citrus and grapevines	Dr. A.M. Grieve
Irrigation requirements, salt tolerance of citrus and grapevines	Ms. L.D. Prior

5 Gosford

5.1 Horticultural Research Station
Research on citrus culture, strawberry selection; development of Australian native ornamental plants; ornamental nursery propagation and cultural research, vegetable varietal and cultural investigations.

Gosford, 2250
New South Wales
Phone: 043 283177
Manager: W.A. Trimmer

Citrus improvement and physiology, planting density, nutrition and water relations, strawberries	R. Sarooshi*, M.Sc.Agr.
Pests of strawberries and ornamentals	S. Goodwin, M.Agr.Sc., Ph.D.
Nursery propagation and floriculture	R.J. Worrall*, B.Sc.Agr.
	G.P. Lamont*, B.Sc.Agr.

6 Griffith

6.1 Viticultural Research Station
elevation: 120 m
rainfall : 408 mm
sunshine : 2686 h

Griffith, 2680
Phone: 069 621855
Manager: D.R. Blundell

Vine improvement and citrus research	C.R. Turkington, B.Sc.Agr.
	R.D. Oen
Vine physiology	Dr. B.M. Freeman, Ph.D.

7 Orange

7.1 Agricultural Research and Veterinary Centre*
elevation: 800 m
rainfall : 788 mm
sunshine : 2729 h

Orange, 2800
New South Wales
Phone: 063 636700
Telex: NSWAGO AA63730
Manager: R. Murray-Prior

Propagation, stock establishment and management of deciduous tree fruits.

Pome fruits - varieties, rootstocks, pruning, training and spacing.	Mrs. J.E. Campbell*, M.Sc.Hort.

Plant protection - deciduous fruits | C.C. Bower, BSc.(Hons.), Ph.D.
W.G. Thwaite, BSc.Agr., M.Sc.
L.J. Penrose, M.Sc.Agr.

8 Richmond

8.1 Hawkesbury Agricultural College
elevation: 22 m
rainfall : 801 mm
sunshine : 2600 h

Bourke St., Richmond, 2753
New South Wales
Phone: 045 701333
Telex: National
INSTY 020 71104101 + HAWKSCOL
International
INSTY AA10101 HAWKSCOL
Principal: Dr. F.G. Swain

The College was established in 1891 as a centre of agricultural learning and research. Previously attached to the N.S.W. Department of Agriculture, it now operates as an autonomous College with a major interest in horticultural education. Offers courses which lead to specialist horticultural degrees and diplomas. Staff have research interests in the areas of plant improvement, plant physiology, entomology, plant pathology, integrated pest management, post-harvest studies, soil-less culture.

Horticulture section

Administration, plant improvement, soil-less culture, vegetable production	Head: A.G. Biggs
Entomology, integrated pest management	R.N. Spooner-Hart
Fruit crops, plant improvement	G.D. Richards
Post-harvest studies, plant physiology	G.N Richards
Nursery and ornamental crops, tissue culture	M.R. McConchie
Australian native plants, antique plants, roses, herbs, permaculture, garden history	Dr. J.A. McLeod
Landscape construction, ornamental plant use	M. Wade
New crop protection chemicals	P. Loneragan

8.2 Horticultural Research Unit
Vegetable research by the N.S.W. Department of Agriculture, including general culture, breeding of tomatoes, weed control in vegetables.

Based at Hawkesbury Agricultural College (as above)

Vegetable research	L. Mutton/ V. Nguyen*
Weed research	J.T. Toth

9 Sydney

9.1 Biological and Chemical Research Institute
elevation: 60 m
rainfall : 1081 mm
sunshine : 2240 h

Rydalmere, 2116
New South Wales
Phone: 02 6300251
Telex: NSWRYD AA71399
Dir.: Dr. A.M. Smith

The Institute is the principal centre of the NSW Department of Agriculture for research into the nutrition, pest and disease control and pesticide residue problems of fruit, vegetables and other crops. The Institute also provides diagnostic services and a Plant, Soil and Water Samples Testing Service.

9.1.1 Biology Branch
Administration | A/Dir.: Dr. R.J. Roughley

Administration	Asst. Dir.: R.J. Conroy
Citrus and nursery diseases	Mrs. P. Barkley
Ornamental and flower diseases	A.L Bertus
Bacterial diseases	Dr. P.C. Fahy
Vegetable diseases	D.B. Letham
Field crop and pasture diseases	G.E. Stovold
Crop disease management	Dr. P.F. Kable
Nematology	R.W. McLeod
Virology	Dr. R.D. Pares
9.1.2 Chemistry Branch	
Administration	Dir.: Dr. F.R. Higginson
Administration	A/Asst. Dir.: Dr. H. Baker
Pesticides	W.S. Gilbert
Soils	Dr. T.S. Abbott
Plant nutrition and fertilisers	R.G. Weir
9.1.3 Entomology Branch	
Administration	Dir.: M. Casimir
Administration	Asst. Dir.: Dr. V.E. Edge
Citrus pests	Dr. G.A.G. Beattie
Citrus and vegetable pests	J.G. Gellatley
Vegetable pests	J.T. Hamilton
9.2 Division of Plant Industries	N.S.W. Dept. of Agriculture P.O. Box K220, Haymarket, 2000 N.S.W. Phone: 02 2176666 Telex: 24991 Chief: R.A. Claxton, BSc. (Agr.)
Administration	Deputy Chief: J.K. Long*, B.Sc.Agr.
Administration and Research Direction	Principal Officer: J.A. Seberry*, B.Sc.Agr.
9.3 University of Sydney Faculty of Agriculture Department of Agronomy and Horticultural Science	Sydney, 2006 New South Wales Phone: 02 6923367 Telex: AA20056 Head: Prof. M.G. Mullins*
Genetic improvement of fruit crops	Prof. M.G. Mullins*
Ornamentals, plant physiology	Dr. P.B. Goodwin
Fruit quality, pollution studies	Dr. W.J Greenhalgh*
Landscape horticulture	Dr. J. Clemens
9.4 Macquarie University School of Biological Sciences	North Ryde, 2113 New South Wales Phone: 02 888-8000 Telex: MACQUINI AA22377 Head: Prof. D.W. Cooper
Fruit and vegetable physiology and	Dr. R.G. Hiller, B.A. Ph.D.
Biochemistry	Dr. R.W. Hinde, B.Sc. Ph.D.
Environmental physiology of fruit and vegetable crops	Dr. E.W.R. Barlow, M.Rur.Sc. Ph.D.
Aphid physiology	Dr. D.R. Hales, B.Sc. Ph.D.
9.5 University of New South Wales	Kensington, 2033

School of Food Technology

New South Wales
Phone: 02 6630351
Telex: AAZ6054

Postharvest handling, physiology and pathology of fruit and vegetables such as the use of calcium salts to delay ripening and senescence; enzyme involvement in fruit ripening; storage characteristics and pathology of root crops. Offers M.Sc. and Ph.D. research programs and M.Applied Sc. course work.

R.B.H. Wills, B.Sc. Ph.D.
F.M Scriven, B.Sc. Ph.D.

10 Wagga Wagga

10.1 Riverina College of Advanced Education
elevation: 350 m
rainfall : 586 mm
sunshine : 2359 h

P.O. Box 588, Wagga Wagga
New South Wales
Phone: 069 232222
Telex: RIVCOL 69050

The College, through the School of Agriculture, carries out teaching and research into Horticulture (Viticulture and Amenity Horticulture) and Oenology. It has a commercial winery, 13.5 ha of vineyards under irrigation, a registered nursery with the capacity of 50,000 plants, and will soon establish an irrigated source area for ornamental trees and shrubs.

Physiology of vine propagation methods	Dr. L.F. De Filippis
Micropropagation of horticultural plants	
Heavy metals in horticulture	
Testing of wine grape varieties	M.A. Loder
Survey and incidence of vine mites	
Clonal selection of vine varieties	
Grape fungal diseases	Dr. A.J. Markides
Structural performance of trellises	M.D. Rebbechi
Grafting and propagation of Pistachio	R.K. Withey

11 Wollongbar

11.1 Agricultural Research Centre
elevation: 150 m
rainfall : 1650 mm
sunshine : 2750 h

Wollongbar, 2480
New South Wales
Phone: 066 297511
Mgr.: M.R. Bellert, QDA, QDH

Administration and research direction	Reg. Dir.: D. Leggo, B.Sc.Agr.
Root diseases horticultural crops, avocados	R.N. Allen, M.Sc.Agr. Ph.D.
Plant nutrition	D.R. Baigent, B.Sc. Ph.D.
Tropical plant pathology	R.D. Fitzell, B.Sc.Agr. MSc.Agr.
Insect pests	N.L. Treverrow, M.Rur.Sc.

12 Yanco

12.1 Agricultural Institute
elevation: 146 m
rainfall : 429 mm
sunshine : 2644 h

Yanco, 2703
New South Wales
Phone: 069 530211
Mgr.: W.J. Ryan

Administration and research direction	Reg. Dir.: E.J. Corbin
Citrus - High density planting, graft, transmissible dwarfing and mechanical harvesting	R.J. Hutton*
Stonefruit - Tree training of canning peaches;	J. Slack

evaluation of prune varieties and clones	
Vegetables - Evaluation of planting density, time for onion varieties, garlic culture, asparagus varieties	P.T. Sinclair
Irrigation of processing tomatoes	R. Hermus
Irrigation and evaluation of new potato varieties	B. Logan
Entomology - fruit and vegetables	E. Jones
Pathology - vegetables	E. Cother
Prune rust	P. Ellison

Queensland

Survey

The major responsibility for horticultural research in Queensland, comprising pre-harvest, post-harvest and processing investigations, rests with the Horticulture Branch of the Department of Primary Industries in conjunction with its horticulture advisory service. Entomological, pathological and certain soils research is carried out by the Entomology, Plant Pathology and Agricultural Chemistry Branches of that Department. Occasionally, specific horticultural research projects are undertaken by post-graduate students in the Faculty of Agriculture of the University of Queensland.

Horticultural research carried out by the Department of Primary Industries is primarily problem-oriented, directed towards investigation of existing problems and the development of improved cultural methods in the fairly wide range of tropical, sub-tropical and temperate fruit, vegetable and ornamental crops grown in Queensland. Investigations are also directed to improvement in quality, packaging, transport, marketing and processing outlets.

Initially, attention was largely confined to the introduction and regional testing of a range of potential horticultural crops conducted by a limited staff. More intensive cultural and nutritional research commenced in the 1930's with increased staff and the establishment of research centres in the major horticultural districts. Since 1945, a number of Horticultural Research Stations, with associated laboratory and glasshouse facilities, have been established. Their work is supplemented and further extended by regionals trials in appropriate districts. Horticultural research and extension is directed and administered by a Director of Horticulture and two Assistant Directors located and the Head Office of the Department in Brisbane. The State is sub-divided into 7 Horticultural Regions under the control of a senior research officer, who is also responsible for the operation of the Horticultural Research Station in the Region. The Sandy Trout Food Preservation Research Laboratory is concerned with post-harvest physiology, food technology and processing. It is located in Brisbane.

1. Granite Belt - situated on the southern highlands of Queensland, produces pome and stone fruits, together with grapes and a limited range of summer-grown vegetables. The extension staff are located at Stanthorpe and the research staff are 3 miles from Stanthorpe. The Granite Belt Horticultural Research Station - was established in 1935 as a Field Station of the Commonwealth Scientific and Industrial Research Organisation to undertake apple rootstock investigations. In 1962 it was transferred to the Queensland Department of Primary Industries and was re-named. The scope of its activities was expanded to include breeding, nutritional, water relations, physiological, entomological and plant pathological research in pome, stone and vine fruits. Facilities include appropriate laboratories, a glasshouse, an insectary and cool storage chambers. The staff comprises of horticulturists, plant breeders, an entomologist, and a pathologist, together with non-graduate technicians, and assistants.

2. South Moreton - comprising the main vegetable and small crop growing areas of south-eastern Queensland supplying metropolitan and southern markets, has its headquarters at the Redlands Horticultural Research Station near Cleveland where both extension and research staff are located. Advisory staff are also centred at Gatton and Coolangatta. The Redlands Horticultural Research Station was established in 1947 in the heart of the Brisbane metropolitan vegetable growing area as the main centre for research in vegetables. Facilities have been gradually improved since its inception, culminating in the completion in 1969 of a well-equipped laboratory and administrative building to accommodate the combined research and extension staff. Research facilities include horticulture, plant physiology, and entomology laboratories, together with controlled environment chambers and glasshouse. In addition, it is provided with modern packaging research equipment, including a compression tester and shock testers. Research embraces nutrition, physiological disorders, weedicides and breeding of the major vegetable crops, such as tomatoes, beans, cucurbits and crucifers, as well as strawberries and sundry other crops, including ornamentals and flowers. The packaging research unit is concerned with the whole range of fruits and vegetables. The staff comprises of horticulturists, plant breeders, a chemist and an entomologist, together with non-graduate technicians and assistants. Gatton Office - situated in the grain and field crop belt of

southern Queensland is now the centre for an expanding vegetable processing industry. The prominent vegetable crops are peas, beans, potatoes, onions, beetroot and carrots. A horticulturist and one technical assistant are involved in research in the district, and an extension service operates from Gatton. This office is administered by the Redlands Horticultural Research Station.

3. North Moreton - extending from Brisbane north to Gympie is a major producer of pine-apples, bananas, macadamias, avocados and some citrus, vegetables and low chill stone fruit. The headquarters is at Nambour, with advisory staff located at Caboolture, Gympie, Nambour and Kingaroy. It is served by research staff from the Maroochy Horticultural Research Station. The Maroochy Horticultural Research Station - situated near Nambour, was established in 1945 as an adjunct to the long established Nambour research centre to provide improved facilities for research in plantation crops, namely: pineapple, banana and papaw. New facilities opened in 1983 include well-equipped office, laboratory and glasshouse complex with controlled environment facilities. Research, covering the fields of nutrition, cropping control, herbicides, plant breeding and pest and disease control has been expanded and now includes ginger, citrus, macadamia and a range of sub-tropical tree crops. The staff working on the Station, comprise of horticulturists, entomologists, a plant pathologist, and non-graduate technicians and assistants.

4. Burnett - comprising the main citrus growing areas of the State at Maryborough and Gayndah, is centred at Maryborough, with advisory staff at Gayndah and Bundaberg. Of recent years, there has been considerable development of vegetable growing at Bundaberg for the fresh market. Facilities include a plant physiology and horticulture laboratory at Maryborough. Research staff comprises of 2 horticulturists at Maryborough and one at Bundaberg.

5. Central - through which passes the Tropic of Capricorn, is a major producer of pineapples and papaws, together with a range of vegetables. The headquarters is at Rockhampton from where an Extension service operates. A horticulturist is engaged in applied research in the district in conjunction with, and supported by, research at the Maroochy Horticultural Research Station.

6. Dry Tropics - extending from Mackay to north of Townsville, includes the irrigated vegetables and small crops areas of Bowen and the Burdekin and the citrus-growing area at Charters Towers. Amongst other things the Burdekin is the main producer of bean seed for Australia. The headquarters is the Bowen Horticultural Research Station and advisory staff are located at Ayr and Townsville. The Bowen Horticultural Research Station - was established in 1965 for research primarily in vegetables under tropical conditions and functions in conjunction with the Redlands H.R.S. It has limited laboratory facilities, with a staff comprising horticulturists together with non-graduate technical assistants.

7. Wet Tropics - comprising the banana growing area of Mareeba and the Atherton Tablelands, is centred at Cairns. The research centre is at the Kamerunga Horticultural Research Station, near Cairns, and advisory staff are located also at Innisfail and Mareeba. A horticulturist and technical assistant are stationed at Innisfail.

The Kamerunga Horticultural Research Station - was originally established in 1889 as the Kamerunga State Nursery to serve as a plant introduction centre. It became a Horticultural Research Station after World War II but for many years its activities were hampered by lack of adequate research facilities. These have been gradually improved and now include horticulture laboratories together with a research glasshouse and a plant quarantine glasshouse. It is the centre for research in all tropical horticultural crops, of which banana is the most important at present. The staff working at the Station comprises of 2 horticulturists and together with 2 non-graduate technicians and assistants.

8. Sandy Trout Food Preservation Research Laboratory, Brisbane - was established and erected in 1961 to undertake research into the post-harvest physiology and handling and processing of fruits and vegetables. It functioned as a separate Branch of the Department of Primary Industries until it was re-incorporated in the Horticulture Branch in 1969. The laboratory is well-equipped with low temperature storage facilities and sophisticated analytical equipment. It has a staff of physiologists, food technologists, and non-graduate laboratory technicians and assistants. The Market Extension Group services the packaging, pre-cooling and storage of fruit and vegetables. Wine technology is handled by the food technologist.

13 Ayr

13.1 Department of Primary Industries
elevation: 10 m
rainfall : 1023 mm
r.h. : 72 %

P.O. Box 591, Ayr, 4807
Phone: (077) 832 355

Insect Pests of tomato and mango	I. Kay, M.Sc.

14 Bowen

14.1 Department of Primary Industries - Bowen Horticultural Research Station elevation: 10 m rainfall : 1036 mm r.h. : 72 %	P.O. Box 538, Bowen, 4805 Phone: (077) 852 255
Tomato breeding	D.J. McGrath, B.Agr.Sc.
Vegetable agronomy	R.M Wright, B.Agr.Sc.

15 Brisbane

15.1 Department of Primary Industries - Horticulture Branch, Head Office elevation: 38 m rainfall : 1147 mm	P.O. Box 46, Brisbane, 4000 Phone: (07) 224 0414 Telex: 44969
Organisation and administration of horticulture research and extension in Queensland, through district offices and horticultural research stations.	Dir. of Hort.: N.S. Kruger, M.Sc. Res. Asst. Dir.: R.E. Barke, MSc. E.T. Carroll, B.Agr.Sc., B.A. P.E. Page, B.Sc.
Extension and field services	Asst. Dir.: K.R. Jorgensen, B.Sc. Grad.Dip.Bus.Admin.: J.R. Blake*, B.Sc. M.Pub.Admin.: G.I. Jamieson, B.Sc.Agr., M.Agr.Sc.
15.2 Department of Primary Industries - Sandy Trout Food Preservation Research Laboratory	19 Hercules Street, Hamilton, 4007 Phone: (07) 268 2421
Fruit and Vegetable Processing:	
All aspects	R.P Bowden, B.Sc.
Techniques and quality	B.F. Bradley, Dip.Ind.Chem.
Quality, vegetables, bacteriology	A.R. Isaacs, B.App.Sc.
Quality, new products, can technology, wine and grape quality	P.D. Scudamore-Smith, B.App.Sc.
Post-harvest physiology:	
Tropical fruit, fruit maturation	B.C. Peacock*, B.Sc.
Fruit Quality Maintenance:	
Tropical fruit	B.I. Brown, M.Sc.
Transport and temperature management	R.A. Jordan, B.Sc. R.L. McLauchlan, B.App.Sc.
Pineapples, apples	L.G. Smith, B.App.Sc.
Cooling and temperature management, packaging	J.B. Watkins, B.Sc.
Marketing Factors:	
Price/quality relationships, packaging	S.N. Ledger, B.Agr.Sc.
15.3 Department of Primary Industries - Entomology Branch	Meiers Road, Indooroopilly, 4068 Phone: (07) 377 9311
Insect pest of ornamentals	N. Gough, Ph.D.

15.4 Department of Primary Industries - Plant Pathology Branch	Meiers Road, Indooroopilly, 4068 Phone: (07) 377 9311
Fungal diseases of sub-tropical fruits	K.G. Pegg, MSc.
Vegetables	R.G. O'Brien, M.Agr.Sc.
Bacterial diseases of fruit and vegetables	M.L. Moffett, M.Sc. Ph.D.
Post-harvest diseases of fruits and vegetables	I.F. Muirhead, M.Agr.Sc., Ph.D.
Virus diseases:	R.S. Greber, M.Sc. J.L. Dale, B.Sc.Agr., Ph.D. J. Thomas, Ph.D.
Nematode diseases of fruits and vegetables	R.C. Colbran, M.Agr.Sc., Ph.D. G.R. Stirling, Ph.D.
Banana tissue culture	Wee Chong Wong, Ph.D.
Diseases of ornamentals	I.K. Hughes, M.Sc.
15.5 University of Queensland*	Faculty of Agricultural Science St. Lucia, Brisbane, 4067 Phone: (07) 377 3621 Telex: 40315
Supervision of Post-Graduate horticultural research projects	C.J Asher, Ph.D./ W.G. Slater, Ph.D./ D.G. Edwards, Ph.D.

16 Bundaberg

16.1 Department of Primary Industries elevation: 14 m rainfall : 1146 m	P.O. Box 1143, Bundaberg, 4670 Phone: (071) 712 761
Vegetable agronomy	J.A. Barnes, B.Sc.

17 Cairns

17.1 Department of Primary Industries elevation: 3 m rainfall : 2036 mm r.h. : 72 %	Kamerunga Hort. Res. Stn., Harley Street, Redlynch, 4872 Phone: (070) 551 023
Tropical fruit and vegetables	B.J. Watson, B.Agr.Hort.
Mango, coffee	E.C. Winston, M.Sc.

18 Cleveland

18.1 Department of Primary Industries Redlands Horticultural Research Station elevation: 11 m rainfall : 1329 mm	Delancey Street, Ormiston, 4163 Phone: (07) 286 1488
Vegetable breeding	P.R. Beal, M. Agr. Sc.
Container design, packaging and transport	D. Schoorl, M.Agr.Sc.
Tissue culture, herbicides	R.A. Drew, BAgr.Sc.
Plant breeding	P.J. Farlow, BAgr.Sc.
Vegetable crop improvement	A.M Hibberd, B.Agr.Sc. M.E. Herrington, B.Agr.Sc.
Ornamental crops	S.A. Lacey, BAgr.Sc.
Vegetable agronomy, herbicides	J.F. Gage*, B.Sc.Hort.,M.Agr.Sc.

Pests of vegetables and passionfruit	J.R. Hargreaves, MSc.
Plant chemistry	M.L. Carseldine, Dip.Ind.Chem.

19 Gatton

19.1 Department of Primary Industries elevation: 95 m rainfall : 856 mm	P.O. Box 245,Gatton,443 Phone: (075) 621 377
Vegetable Agronomy	CR McMahon, B.App.Sc.

20 Innisfail

20.1 Department of Primary Industries elevation: 7 m rainfall : 3757 mm r.h. : 82 %	Phone: (070) 642400
South Johnstone Research Station	P.O. Box 20, South Johnstone, 4859 Phone: (070) 64 2400
Banana production and cucurbits	J.W. Daniells*, B.Agr.Sc.

21 Mareeba

21.1 Department of Primary Industries elevation: 335 m rainfall : 962 mm r.h. : 72 %	P.O. Box 149, Mareeba, 4880 Phone: (070) 921 555
Mango and lyche pests	I.C. Cunningham, M.Agr.Sc., Q.D.A. Ph.D.
Pests of bananas and cucurbits	B. Pinese, BAgr.Sc.
Diseases of sub-tropical fruit and vegetables	R.A. Peterson, MAgr.Sc. M. Ramsay, BAgr.Sc.

22 Maryborough

22.1 Department of Primary Industries elevation: 11 m rainfall : 1187 mm	P.O. Box 432, Maryborough, 4650 Phone: (071) 212 343
Citrus culture and management	J.C. Chapman, BApp.Sc.
Sub-tropical fruits and vegetables	L.S. Lee, B.App.Sc.

23 Nambour

23.1 Department of Primary Industries Maroochy Horticultural Research Station elevation: 30 m rainfall : 1776 mm	P.O. Box 83, Nambour, 4560 Phone: (071) 412 211
Plantation crops	B.W. Cull, M.Agr.Sc.
Sub-tropical tree crops; physiology	K.R. Chapman, MAgr.Sc.
Sub-tropical crops; production	A.P. George, BAgr.Sc.
Pineapples and bananas	J.D. Glennie, B.Agr.Sc.
Plant improvement, sub-tropical tree fruits	C.W. Winks, B.Sc.

Macadamia production and physiology	R.A. Stephenson, B.Agr.Sc. Ph.D.
Avocado and ginger	A.W. Whiley, B.Hort.
Sub-tropical fruit and vegetables - diseases	P. Mayers, B.Agr.Sc.
Citrus and French bean pests	D. Smith, B.Sc.
Pineapple and passionfruit and strawberry pests	G. Waite, M.Agr.Sc.
Macadamia pests	D.A. Ironside, QDA

24 Rockhampton

24.1 Department of Primary Industries
elevation: 10 m
rainfall : 856 mm
r.h. : 69 %

P.O. Box 689, Rockhampton 4700
Phone: (079) 361 011

Pineapples and papaws	F.A. Aquilizan, M.Sc.

25 Stanthorpe

25.1 Department of Primary Industries
Granite Belt Horticultural Research Station
elevation: 870 m
rainfall : 784 mm

P.O. Box 10, Applethorpe, 4378
Phone: (076) 81 1255

Deciduous fruits, nutrition and management	B.C. Dodd, B.Agr.Sc.
Deciduous fruit, plantation management	S.G. Middleton, B.Agr.Sc.
Deciduous fruit, plant breeding	L.B. Baxter, B.Appl.Sc.
Deciduous fruits, irrigation	P.S. Crew, B.Agr.Sc.
Deciduous fruits, plant breeding	B.L. Topp, B.Agr.Sc.
Diseases of deciduous fruit and vegetables	J.B. Heaton, B.Sc.
Deciduous fruit pests	B.F. Ingram, B.Sc.

Victoria

Survey

Research in fruit and vegetable production and handling by the Department of Agriculture is now conducted almost entirely by the Division of Plant Research and Development, however the Division of Crop Industry Services and the Plant Standards Branch have roles related to research.

The Division of Plant Research and Development consists of seven research institutes, four research stations and an agricultural engineering centre. Four of the institutes and two of the research stations have some research in horticulture. These units and their major responsibilities are as follows:

Plant Research Institute*, Burnley - Plant Protection. Diagnosis and control of pests and diseases of all agricultural and horticultural crops. Insect biology, pest management, pesticides, mycology, bacteriology, nematology, virology.

Horticultural Research Institute, Knoxfield - Horticulture. Fruit Research for southern Victoria, post-harvest fruit handling and storage. Propagation and growth of ornamental and nursery crops.

Potato Research Station, Healesville - Potato agronomy. Breeding of improved potato varieties and coordination of district research. Multiplication of pathogen-tested potato and cool temperate fruit crops.

Vegetable Research Station, Frankston - Vegetable agronomy. Selection of improved varieties and development and extension of management methods in vegetable production. Utilization of waste water.

Sunraysia Horticultural Research Institute*, Irymple - Horticulture. Research on production and plant protection of grape, citrus and other fruit crops for the lower Murray Valley.

Irrigation Research Institute, Tatura - Irrigation. Research in soils (including salinity control) agronomy, crop physiology for improved production of horticultural and irrigated field crops.

The Crop Industry Services Division includes units having interests in Fruit Crops (pome and stone fruits, grapes, citrus, berries, nuts), Row Crops (vegetables including potatoes) and Ornamentals. In relation to research, the roles of these units are directed towards: 1. research needs, resources and planning and 2. research results and their use by the extension service and growers.

The Plant Standards Branch is largely concerned with regulatory work and horticultural inspection, but it does stimulate research on fruit and vegetable disinfestation and through its Plant Improvement Group is responsible for development work in pathogen-tested schemes and seed testing.
Publications - Research results are published in national and international scientific journals and in annual or biennial reports from research establishments. Arising from a new publications system introduced by the Department in 1979, research results are also made available in Agnotes (fact sheets), bulletins and booklets in the Research Project and Technical Report Series.

26 East Melbourne

26.1 Division of Plant Research and Development (Department of Agriculture)	166 Wellington Place East Melbourne, Vic. 3001 G.P.O. Box 4041, Melbourne, Victoria 3001 Phone: (03) 651 7011 Chief: J.W. Meagher
26.2 Division of Crop Industry Services	Chief: J.V. Mullaly
Fruit Crop Industry Services	J.D.F. Black
Row Crop Industry Services	P.F. Shea
Ornamentals and Recreational Turf Industry Services	R.D. Price*
26.3 Plant Standards Branch	Dir.: M.N. Kinsella
Plant Standards Control	R.R. Hedding
Plant Improvement	J.M. Blackstock

27 Burnley

27.1 Plant Research Institute* (Department of Agriculture) elevation: <200 m rainfall : 630 mm	Burnley Gardens, Swan St. Burnley, Vic. 3121 Phone: (03) 810 1511 Dir.: P.T. Jenkins
Pathology	P.R. Smith
Mycology	P. Merriman/ F. Greenhalgh W. Washington/ I. Pascoe I. Porter/ R. Clarke D. Harrison
Pathogen-tested Schemes	R. Osborn
Bacteriology	S. Wimalajeewa
Virology	R. Sward/R. Garrett/ G. Guy E. Bjarnason
Nematology	R. Brown/ R. Winoto-Suatmadji
Entomology	T. Amos
Horticulture	J. MacFarlane
Pesticides	M. Campbell/ J. Whan
Taxonomy	L. Crawford
Biological Control	C. Reinganum/ S. Gagen
Quarantine	M. Mekhamer
27.2 Seed Testing Station	Officer in Charge: E.M. Felfoldi

28 Frankston

28.1 Vegetable Research Station	Ballarto Rd., P.O. Box 381

(Department of Agriculture) elevation: <200 m rainfall : 700 mm	Frankston, Vic. 3199 Phone: (03) 786 3311 Officer in Charge: A.S. Morgans
Agronomy; herbicides	A.S. Morgans
Soils; irrigation	F.G. Kaddous
Seeds; crop establishment	W.C. Morgan
Plant spacing; crop establishment	D.J. McGeary

29 Shepparton

29.1 District Centre (Department of Agriculture) elevation: <200 m rainfall : 500 mm	Public Offices, Welsford St., P.O. Box 862, Shepparton, Vic. 3630 Phone: (058) 21 4788 Extension Dir.: D. Wauchope
Fruit Crops	L. van Heek/ H.G. Schneider J.C. Read

30 Knoxfield

30.1 Horticultural Research Institute (Department of Agriculture) elevation: <200 m rainfall : 720 mm	Burwood Highway, P.O. Box 174 Ferntree Gully, Vic. 3156 Phone: (03) 221 2233 Dir.: G. Frith*
Deciduous tree fruits; management, training	W. Thompson/ K. Clayton-Greene* J. Raff
Deciduous tree fruits, variety improvement, quality of propagation material	B. Morrison/ L. Jager H. Craig-Brown
Deciduous fruits; growth regulators	P. Miller
Pollination	J. Raff/ K. Clayton-Greene*
Berry Fruits	B. El-Zeftawi*/ B. Morrison K. Clayton-Greene*/ F. Goubran
Minor fruits	F. Goubran/ B. El-Zeftawi*
Propagation, ornamentals and fruit	G. Thomson
Cool storage; maturity, post harvest physiology of fruit and vegetables	L. Dooley*/ C. Little/ L. Brohier
Packaging, handling and transport of fruit and vegetables	I. Peggie*/ A. Wollin B. Cumming
Post harvest; cut flowers	J. Faragher
Tissue culture	B. Hanger*/J. Hutchinson
Ornamentals; hormones, roots	D. Richards
Ornamentals; selection, breeding	D. Beardsell
Ornamentals; nutrition	L. James
Ornamentals; growth regulators	I. Wilkinson
30.2 Fruit and Vegetables Extension Group	Senior District Extension Officer:
Pome fruit	R.O. Soderlund
Stone fruit, kiwifruit, Irrigation	H.B. Scott
Berries	R. Cahill
Grapevines	K.H. Kroon
Post harvest	G.D. Godden*
Other tree crops/ economics	R.J. Holmes K.B. Higgins

31 Irymple

31.1 Sunraysia Horticultural Research Institute	P.O. Box 460, Irymple, Vic. 3498

(Department of Agriculture) rainfall : 300 mm	Phone: (050) 24 5603 Dir.: B.K. Taylor
Viticulture, plant protection	G.A. Buchanan
Dried vine fruit, wine grapes	J.R. Whiting
Dried vine fruit quality	M.A. Brockhus
Table grapes	T.W. Deer
Vine improvement	G.C. Fletcher
Wine grapes	J.B. Flehr
Plant pathology	R.W. Emmett
Nematology	M.E. Edwards
Citrus, alternative crops, irrigation, mechanical harvesting	M.O. Dale
Irrigation; salinity; crop physiology	S. Nagarajah
Citrus productivity and storage	R.T. Dimsey
Avocados and other alternative crops	D.G. Madge
Citrus improvement	I.R. Thornton
Vegetables, nut crops	R.M. Blennerhassett

32 Tatura

32.1 Irrigation Research Institute (Department of Agriculture) rainfall : 490 mm	Tatura, Vic. 3616 Phone: (058) 24 1344 Dir.: D.J. Chalmers
Plant pathology	S. Flett
Entomology	I. Barrass
Plant physiology	P.H. Jerie/ I. Dann
Deciduous tree fruit mechanisation	G. Young
Deciduous tree fruit close planting	B. van den Ende
Irrigation strategies	P. Mitchell
Peach breeding	L. Issell
Tomato agronomy	W. Ashcroft

33 Mildura

33.1 Sunraysia Department of Agriculture elevation: <200 m rainfall : 300 mm	Public Offices, 252 Eleventh St. Mildura, Vic. 3500 Phone: (050) 23 0256 Ext. Dir.: B.A. Harford
Economics	R. Witcombe
Vines	K. Leamon
Citrus and alternative crops	R. Cadman
33.2 Robinvale	Perrin St., Robinvale, Vic. 3549 Phone: (050) 26 3905
Fruit crops	R. Johns

34 Swan Hill

34.1 District Centre (Department of Agriculture) elevation: <200 m rainfall : 320 mm	McCallum St. Swan Hill P.O. Box 501, Swan Hill, 3585 Phone: (050) 32 4461
Fruit crops	A. Heslop/ D.A. Kneen A.R. Polglase

35 Healesville

35.1 Potato Research Station
(Department of Agriculture)
elevation: 580 m
rainfall : 1400 mm

Healesville, Vic. 3777
Phone: (059) 62 9218
Officer in charge: A.W. Kellock

Potato breeding — R.P. Kirkham
Potato processing — R.W. de Jong
Berry fruits — G.R. McGregor

South Australia

Survey

Several organisations are involved in South Australia in carrying out research into problems of the horticultural industries. There is no formal grouping of these organisations into one association or council but close liaison does occur between them and in various specialised fields co-ordinating committees provide a formal grouping of research personnel for discussion and liaison. The several organisations and their particular fields of concern are as follows:

1. Department of Agriculture of the South Australian Government - Established in the 1890's this organisation is the principal body concerned with horticulture in South Australia. It carries out mainly applied research and extension and has regulatory functions in all fields of agriculture-animal production, field crops, horticulture, soils and economics. The Department has general laboratory, post harvest, plant improvement/plant quarantine facilities located at Northfield (Adelaide), but its main horticultural research facilities are on research centres located at Lenswood in the Adelaide Hills, at Nuriootpa in the Barossa Valley and at Loxton in the Murray Lands irrigation areas. These three sites are representative of the three major regions of horticultural production in the State. Lenswood, the high rainfall area where apples, cherries and pears predominate; Nuriootpa the centre of the non-irrigated viticultural region; and Loxton representing the irrigated horticultural areas along the River Murray, with citrus, stonefruit and vines the principal crops. The Lenswood Research Centre was commenced in 1963. The area being nearly 80 hectares is situated in the wetter hills (mean rainfall 875 mm) in which the apple and cherry industries have concentrated. While the earlier stations specialised in varieties, the Lenswood Centre is premarily concerned with management aspects and the provision of alternative crops. Loxton Research Centre started in 1960, is largely spray irrigated (overhead and undertree sprinklers). Research facilities have expanded and the centre is now the Horticultural Research facility for the Murray Lands region i.e. an arid zone irrigated horticultural area. The Nuriootpa Research Centre is solely concerned with viticultural research and since its establishment in 1938 has contributed much to the wine grape industry in providing improved clonal planting material and vine management methods.

2. Waite Agricultural Research Institute - within the University of Adelaide and primarily concerned with teaching and fundamental research at undergraduate and post graduate levels in the Faculty of Agricultural Science. Departments of Entomology, Plant Pathology and Plant Physiology are involved with Horticultural projects.

3. Australian Wine Research Institute - Principally concerned with research into aspects of the wine making process and characters of the grape affecting wine quality. The Institute was created under a special Act of Federal Parliament in 1955 and is financed by the Industry and Commonwealth.

4. Roseworthy Agricultural College - An autonomous College of Advanced Education established in 1883, primarily concerned with teaching but also research and extension in the technology of crops and live-stock production industries of southern Australia. Has the only Faculty of Oenology and Viticulture in Australia. Provides tertiary level courses leading to awards in agricultural production, farm management and wine marketing at Associate Diploma level, in applied science in agriculture, including production, horticulture; in natural resources management including amenity horticulture and in oenology and viticulture at Degree level; and in horticulture or viticulture, farm management and agricultural extension at Graduate Diploma level.
Research projects relevant to horticulture are conducted in the three faculties in Departments of Plant Production (Production Horticulture), Applied Sciences (Crop Protection) Natural Resources (Energy Crops, Land rehabilitation) Oenology and Viticulture (Grape quality and wine quality) and Management and Education (Rural Sociology).
Rainfall and climatic data were obtained from the Bureau of Meteorology, S.A.

36 Adelaide

36.1 Central Region Office elevation: 47 m rainfall : 557 mm sunshine : 2796 h	Department of Agriculture, Box 1671 G.P.O., Adelaide, S.A. 5001 Phone: (08) 227 9911 Telex: AA88442
Temperate tree crops	R.T.J. Sloane, Dip.Hort., B.Sc. (Hort).
Glasshouse crops and vegetables	B. Philp, R.D.A. R.D.A.T.
Vegetables	I.S. Rogers, B.Agr.Sc.
Horticultural farm management	G. Ronan, B.Ag.Ec.
Wine grape economics	G.D. McClean, B.Ag.Ec.
36.2 Northfield Laboratories elevation: 77.0 m rainfall : 518 mm sunshine : 2796 h	Department of Agriculture, Box 1671 G.P.O., Adelaide, S.A. 5001 Phone: (08) 266 0911 Telex: AA88442
Plant nutrition	J.B. Robinson*, B.Agr.Sc. (Hons.) Ph.D.
Pests of apples and vegetables	P.E. Madge, B.Agr.Sc., Ph.D.
Pests of vines and brassica vegetables	G.J. Baker, B.Agr.Sc.
Post-harvest horticulture	B.L. Tugwell, B.Agr.Sc. A.P. Dahlenburg*, B.Agr.Sc.
Vegetables	N.A. Maier, B.Sc., B.Agr.Sc.
Quarantine plant pathology	D. Cartwright, B.Ag.Sc.(Hons.).
Plant pathology	T.J. Wicks, B.Agr.Sc.
Ornamental horticulture	G. Barth, M.Sc. (Hort.).
36.3 Soils Branch	Department of Agriculture, Box 1671 G.P.O. Adelaide, S.A. 5001 Phone: (08) 266 0911, Telex: AA88442
Irrigation methods	M.R. Till, B.Agr.Sc.
Irrigation requirements of horticultural crops	G. Schrale, Ph.D.
36.4 Economics Division	Department of Agriculture, Box 1671 G.P.O. Adelaide, S.A. 5001 Phone: (08) 227 9911, Telex: AA88442
Horticulture modelling	B.R. Hanson, B.Ag.Ec., M.Ec.
36.5 Lenswood Research Centre* elevation: 550 m rainfall : 1015 mm sunshine : 2609 h	Department of Agriculture, Lenswood, South Australia 5240. Phone: (08) 389 8302 Telex: AA88442 A/g Manager: M. Whitehead, R.D.A.T.
Pome fruits and cherries	L. Nitschke, B.Ag.Sc.
Vegetable crops	G.J Lomman, R.D.A.
Potatoes	C.M.J. Williams, B.Sc.Agr. (Hons.), Ph.D.
Temperate fruit and berry crops	L.C. McMaster, B.Ag.Sc., B.A.

37 Glen Osmond

37.1 Australian Wine Research Institute	Private Bag - Glen Osmond,

elevation: 122 m rainfall : 627 mm sunshine : 2290 h	South Australia, 5064. Phone: (08) 796817 Telex: CCISA AA88370 (sub 5) Dir. : T.H. Lee, B.Sc. Ph.D.
Energetics of fermentation, malo-lactic fermentations, wine yeasts	P.R. Monk, Ph.D.
Pigments and tannins of red wines, wine maturation	T.C. Somers, D.Sc.
Aroma volatiles in white wines	R.F. Simpson, B.Sc. Ph.D.
Grape flavour compounds	P.J. Williams, B.Sc. Ph.D.

38 Loxton

38.1 Murray Lands Region	Department of Agriculture, Loxton, South Australia, 5333
Advisory officers at Loxton, Renmark, Waikerie, Berri and Murray Bridge	Phone: (085) 847241 Telex: AA88442
Extension	R.L. Wishart*, RDA
Horticulture farm management	A.V. Cook, M.Ec. M. Krause, B.Ag.Ec., M.Ec.
Vegetables	D. Henderson, B.Sc.
38.2 Loxton Research Centre elevation: 66 m rainfall : 281 mm sunshine : 2924 h	Department of Agriculture, Loxton, South Australia, 5333. Phone: (085) 847315 Telex: AA88442.
Soils and irrigation	P. Cole, B.Agr.Sc.
Biological control of citrus pests and technology of pesticide application to horticultural crops	G.O. Furness, B.Agr.Sc. M.Agr.Sc.
Citrus	P.T. Gallasch, RDA, B.Agr.Sc.
Stonefruit production	F.J. Gathercole, B.Agr.Sc.
Plant pathology (vines)	P.A. Magarey, B.Agr.Sc.(Hons.)
Vine improvement	P. Nicholas, B.Agr.Sc.
Nematology/soil borne diseases	G. Walker, B.Sc. (Hons.)
Irrigation and drainage	K.A. Watson, B.Agr.Sc.
Nutrition (analysis)	T. Glenn, Cert. Anal.
Vegetables	S. Warriner, M.Agr.Sc.

39 Struan

39.1 South Eastern Region elevation: 37 m rainfall : 554 mm sunshine : 2345 h	Department of Agriculture, Struan, South Australia, 5271. Phone: (087) 647419 Telex: AA88442
Extension	(Mt. Gambier): D.E. Moss, R.D.A.

40 Nuriootpa

40.1 Nuriootpa Research and Advisory Centre elevation: 274 m rainfall : 508 mm	Department of Agriculture, Nuriootpa, South Australia 5355. Phone: (085) 621355 Telex: AA88442

sunshine : 2628 h — Extension D. Hodge, Dip.Hort.

Viticulture — R.M. Cirami, B.Sc., M.Sc.
R.W. Radford, R.D.A.
M.G. McCarthy, B.Agr.Sc.

Horticultural crop improvement — P.A. James, R.D.A.

41 Roseworthy

41.1 Roseworthy Agricultural College
elevation: 115 m
rainfall : 440 mm
sunshine : 2813 h

Roseworthy Agricultural College,
Roseworthy, South Australia 5371.
Dir.: Dr. B. Thistlethwayte
Assoc. Dir.: M.B. Spurling

Academic Secretariat:
Educational technology in horticulture — M.B. Spurling, M.Ag.Sc. Grad. Dip.Ed.Tech. FAIAS
Department of Agricultural Engineering:
Mechanical harvesting — P.W. Rowland, B.Sc.(Eng.) M.I. Mech. E.

Department of Applied Sciences:
Biological control of pests — N.L. Richardson, B.Ag.Sc. M.Sc. Ph.D. Dip.Ed.
Chemistry of pheromones — B.D. Williams, Dip.App.Chem. B.Sc. (Hons.) Ph.D.

Department of Field Services:
Orchard management — A.C. Jenkins, RDA

Department of Management and Education:
Horticultural management — I.M. Cooper, B.Ag.Sc. M.Sc.
Agricultural education and extension — B.T. Sheahan, B.Ag.Sc. Adv.Dip.Ed. Dip.Ag.Ext.

Department of Natural Resources:
Energy crops, windbreaks, reafforestation — F.J. van der Sommen, B.Sc.(For.) M.Env.Sc.
Plant propagation — K.R. Cowley, D.App.Sc. Cert.Orn.Hort.

Department of Oenology and Viticulture:
Viticulture and wine quality — A.J.W. Ewart, B.Hort.Sc.(Linc.)M.S.
Viticulture - grafting, pruning — P.R. Dry, B.Ag.Sc.
Oenology — P.G. Iland, B.App.Sc.

Department of Plant Production:
Almond research - clonal selection and pollination — A.J.W. Short, B.Ag.Sc.

42 Waite

42.1 Waite Agricultural Research Institute
elevation: 122 m
rainfall : 627 mm
sunshine : 2290 h

University of Adelaide, Private Bag
Glen Osmond, South Australia.
Phone: (08) 797901
Telex: UNIVAD AA89141
Dir.: Prof. J.P. Quirk, B.Sc.Agr. (synd.) Ph.D. D.Sc.(Lond.) FAIAS FRACT FAA

Studies involving gummosis of apricots, ecology of insect pests and growth studies of plants and fruits with emphasis on formation and distribution of assimilates in perennial plants.

Entomology Department — D.A. Maelzer, M.Sc. Ph.D.

Plant pathology Department	Head: Prof. H.R. Wallace, B.Sc. Ph.D. D.Sc. FAA
Pathology	M.V. Carter, Ph.D.
	A. Kerr, B.Sc. Ph.D.
Virology	J.W. Randles, M.Agr.Sc.
	R.I.B. Francki, Ph.D.
Nematology	J.M. Fisher, B.Sc.Agr.Ph.D.
Plant physiology Department	Head: Prof. L.G. Paleg, B.A.Ph.D.
Viticulture	B.G. Coombe*, Ph.D. M.Agr.
Apples, stone fruits	G.R. Edwards, M.Sc.,B.Agr.Sc. Ph.D.
Stress physiology	P.E. Kriedemann*, B.Agr.Sc.(Hons.) Ph.D.

Tasmania

Survey

Tasmania is the Island State of Australia, situated at the South East corner of the Continent. In land area it approximates double that of the Netherlands but much is non arable. Tasmania is very mountainous, and most horticulture is limited to coastal strips and lower reaches of river valleys in the North West, North East and South East. The West, South West, and Central Highland regions are unsuited to horticulture.
Horticultural areas of the State have a coastal maritime temperate climate. Production is limited to temperate crops. Inland frosts limit expansion of many crops away from the sea coast.
Rainfall varies 3,000 mm in the West to 500 mm in rain shadow areas in the South East and East. Many of the latter areas have good climatic attributes other than rainfall for temperate horticulture. However, soils are very variable in type and topography and broad acre enterprises cannot be established easily.
The horticultural areas in Tasmania on red kraznozen soils along a narrow strip of the North West Coast and a similar but smaller segment of the North East Coast are perhaps the best vegetable production areas of Australia. Rainfall is higher and more reliable than most other arable areas of the State. However, it still has to be supplemented with summer irrigation to produce the quality produce demanded by the market. These regions have been developed into broad acre producers of peas, beans, potatoes, brassicas, alkaloid poppies, and other minor crops for a number of processing and freezing enterprises.
The lower reaches of the Tamar, Derwent and Huon Rivers and surrounding coastal sites suited to fruit growing contain the other major horticultural enterprises. The main crops are apples, pears, hops and berry fruits. There is interest in reviving a once significant stone fruit industry. The old industry was based on processing (mostly jams) and the proposal is to develop a fresh fruit trade based on high quality produce. This policy is already paying dividends for the berry fruits industry.
New and potential industries based on this concept of high quality are wines, flowers, and a range of essential oils. The unusual combination of latitude and climate in the horticultural areas of Tasmania are conductive to qualities, flavours and aromas suited to servicing expanding sophisticated markets for temperate fruit, vegetable, and related crops.
The State Department of Agriculture has the prime role of applied research for horticulture in Tasmania. Administration of inputs is divided between Hobart (New Town) for Fruit and Ornamentals, Devonport (Vegetables and Allied Crops) and Launceston (Mt Pleasant) for essential oils. More basic research is however conducted at the University of Tasmania in Hobart, particularly on the potential for essential oils and related products.
Lavender is the one fully developed industry of this type, located at Lilydale in the North of the State.
Field research is conducted on Research Stations and growers' properties. Vegetables and Allied Crops are primarily investigated at the Forthside Research Station near Devonport, and Fruit at the Huon Research Station near Hobart.
The University of Tasmania has recently established a field station to the east of Hobart.

43 Devonport

43.1 Department of Agriculture - Vegetables and Allied Crops Branch

Stoney Rise Centre, P.O. Box 303,
Devonport, Tasmania 7310
Phone: (004) 240201

Branch direction and management, research supervision	B.D. Frappell*, B.Agr.Sc.,M.Sc.

ıranch management, industry liaison	M.R. Walker, Dip.Ag.
'lant improvement, crop establishment	B.M. Beattie*, B.Agr.Sc.
'lant nutrition, plant spacing	K.S.R. Chapman, B.Agr.Sc. (Hons.) Ph.D.
Vater requirements	B. Chung*, B.Agr.Sc. (Hons.)M.Phil.
'lant nutrition	J.C. Laughlin*, B.Agr.Sc.,M.Agr.Sc. B.Ec. Dip.Pub.Admin.
Advisory	J.I. Cox, H.D.A., G.D.E.
Development Research and Advisory	J.R. Maynard, B.Agr.Sc.

43.2 Department of Agriculture
Plant Health Branch

Diseases of vegetables	Vacant
Insect pests of vegetables	Vacant

43.3 Department of Agriculture - Weeds Section — P.O. Box 303, Devonport Tasmania 7310.

Weed control in vegetables	Vacant

44 Forth

44.1 Forthside Vegetable Research Station — Forth - Tasmania 7310, Phone: (004)282217

Management of research station	J.J. Forbes, H.D.A.

45 Grove

45.1 Huon Horticultural Research Station — Grove - Tasmania 7106, Phone: (002) 66 4305

Management of research station

45.2 Department of Agriculture - Fruit and Ornamentals Branch — Huonville - Tasmania 7109, Phone: (002) 64 1066

Stone fruits, Orchard systems	W. Boucher, B.Agr.Sc. Ph.D.

46 Hobart

46.1 Department of Agriculture - Plant Services — G.P.O. Box 192B - Hobart Tasmania 7001. Phone: (002) 30 3050, Dir.: E.J. Martyn, M.Agr.Sc.

46.2 Department of Agriculture - Fruit and Ornamental Branch — New Town Research Laboratories, St. Johns Avenue, New Town, Tasmania 7008. Phone: (002) 28 4851

Administration	Chief of Branch: R.J. Hardy H.D.A. B.Agr.Sc.
Harvesting, post-harvest handling, physiology and storage, packaging and transport, disinfestation studies, plant nutrition, crop regulation, pruning, propagation, rootstocks variety of selection berry fruits, hazel nuts - varieties, physiology and nutrition. Stone fruits, varietal assessment.	J.B. O'Loughlin, B.Agr.Sc.B.Ec. K.M. Jones, B.Sc. P. Jotic*, B.Sc.(Hort.) S.J. Wilson*, B.Agr.Sc. (Hons.) C. Young*, B.Agr.Sc. R. Richards, B.Agr.Sc.

46.3 Department of Agriculture - Plant Health Branch — New Town Research Laboratories, St. Johns Avenue, New Town,

(Incorporating Entomology and Plant Pathology Sections)	Tasmania 7008.
Administration - Chief of Branch	I.D. Geard, B.Agr.Sc. M.Ag.Sc.
46.4 Department of Agriculture - Entomology Section	New Town Research Lboratories, St. Johns Avenue, New Town, Tasmania 7008.
Integrated control of fruit pests, disinfestation studies	A. Terauds, B.Sc.
Predatory mites (apples), Eriophyid mites (blackcurrants)	Mrs. M. Williams M.Sc.B.Sc.
46.5 Department of Agriculture - Plant Pathology	New Town Research Laboratories, St. Johns Avenue, New Town, Tasmania 7008.
Diagnosis, laboratory and glasshouse studies of fruit and vegetable diseases; electron microscopy; serological and glasshouse virus testing; biological control of soil borne Sclerontinia diseases	P.J. Sampson, B.Agr.Sc. G.R. Johnstone, B.Ag.Sc. Ph.D. J. Wong, B.Sc. (Hons.) Ph.D. D. Munro, B.Ag.Sc.
46.6 C.S.I.R.O. Division of Entomology	Stowell Avenue, Hobart Phone: (002) 23 5555
Entomophagus Nematodes	Dr. R. Bedding
46.7 Department of Agriculture - Weed Section	New Town Research Laboratories St. Johns Avenue, New Town Tasmania, 7008
Weed control in berry fruit	Dr. A.R. Harradine B.Sc.Agr. Ph.D."
46.8 University of Tasmania Department of Agricultural Science	University of Tasmania Sandy Bay, Tasmania 7005 Phone: (002) 20 2101
Physiology of fruit, vegetable and Essential oil crops	R.C. Menary, B.Sc.,M.Sc.,Ph.D.

47 Launceston

47.1 Department of Agriculture - Chemistry and soils section	Mt. Pleasant Laboratories Launceston South, Tasmania 7250 Phone: (003) 32 2101
Soil water studies	D.W. Armstrong, BAgr.Sc.
Soil chemistry	M.G. Temple-Smith*, B.Agr.Sc.,Ph.D.
Soil physics	D.N. Wright, B.Agr.Sc.
47.2 Department of Agriculture - Fruit and ornamentals branch	P.O. Box 407, Launceston 7250
Grape industry research	F. Peacock, B.Ag.Sc.

47.3 Department of Agriculture, Pastures and field crops branch — Mt. Pleasant Laboratories, Launceston South, Tasmania 7250

Administration - Chief of branch — J.G. Stephens, B.Sc. Hons.
Essential oils — M. Hart, B.Ag.Sc. Hons.

47.4 Department of Agriculture - Plant Health Branch: Diagnosis research, development and extention with vegetable and fruit diseases — Mt. Pleasant Laboratories, Launceston South, Tasmania 7250

Plant pathology section — Miss M.S. Grice, B.Agr.Sc.
Entomology section — Vacant

48 Scottsdale

48.1 Department of Agriculture - Vegetables and Allied Crops Branch — Ellenor Street, Scottsdale, Tasmania 7254, Phone: (003) 32 2588

Advisory — A.R. Walker, D.D.A.

49 Burnie

49.1 Department of Agriculture - Vegetables and Allied Crops Branch — Reece House, Cnr. Mount and Cattley Streets, Burnie 7320, Phone: (004) 30 2202

Advisory — P.M. Hood, D.D.A.
Development research and advisory — P.A. Regel, B.Agr.Sc.

Western Australia

Survey

The W.A. Department of Agriculture was established in 1898 when it took over the activities of the old Bureau of Agriculture. Horticultural research commenced in 1912 when a commissioner was appointed to serve the fruit industries. An Officer in Charge, Fruit Industries, was appointed in 1918 and was followed by the first Chief of the Division of Horticulture in the early 1950's.

The combined value of horticultural production was $ 129 million in the 1983/84 season. Exports amount to $ 24 million. The area under production is approximately 15 500 hectares.

1. Temperate fruit - Apples are the main fruit crop accounting for about 60% of the gross value of fruits which includes pears, stone fruits and citrus. Minor fruit crops include passion fruit, strawberries, avocados, Chinese gooseberries and nuts. Fruit production is concentrated in the Darling Range immediately east of Perth and through the south west areas of Dwellingup, Donnybrook and Manjimup.

Fruit research is carried out at Stoneville, Manjimup and Gascoyne (Carnarvon) research stations. New varieties of fruit and rootstocks are evaluated in relation to tree performance and the effects of density of planting on yields and fruit quality. Investigations are being conducted into soil management methods, nutrient requirements, soil moisture relations, pest and disease control and training methods of fruit trees. Special attention is given to the improvement of handling, packaging and storage techniques to enable local growers to compete effectively with overseas and interstate exports.

2. Tropical fruit - Bananas are the main tropical fruit crop and supply about 60% of the State's needs. The majority are grown at Carnarvon, which is situated at the mouth of the Gascoyne River, some 900 km north of Perth. Small areas of mangoes and papayas are grown also. Research is carried out at the Gascoyne Research Station and in the Kununurra area.

3. Vegetables - The main vegetable producing areas are the Perth Metropolitan region, the South West near the towns of Harvey, Busselton, Donnybrook, Manjimup and Pemberton and the South Coast near Albany and Denmark. Carnarvon is an important

area for winter vegetable production. Potatoes are grown in the Perth Metropolitan area and in the Busselton, Donnybrook, Manjimup, Pemberton, Denmark and Albany areas. The total production of potatoes was 67 000 tonnes worth $ 17 million in 1983/84. Peas and beans are grown for processing in the Manjimup and Albany districts. Production of cauliflowers and onions for the export trade is increasing in the Manjimup District. Vegetable research is conducted at Medina, Manjimup and Gascoyne research stations. The programme includes the testing and selection of vegetable varieties, the study of management techniques, and the assessment of the effects of spacing, fertilizer and irrigation treatments on the yield and quality of vegetables. Foundation seed of Westralia runner beans and of Spearwood brown globe onions is maintained in certification programmes. New chemicals are evaluated for pest, disease and weed control and for residual effects. Seedling potatoes and named varieties from local, interstate and overseas breeding programmes are selected at the Manjimup research station. Clones with acceptable agronomic features are compared with standard commercial varieties in yield trials on the Station and on grower's farms. Culinary, processing and storage trials are carried out at Medina Research Station and in conjunction with processing companies and the Potato Marketing Board. The increased production of vegetables for export has led to an expansion of research and extension of improved-packaging and handling techniques for perishable produce.

4. Viticulture - Production of grapes in Western Australia is centred in the Swan Valley region near Perth, in the Busselton-Margaret River and Mount Barker-Frankland River regions in the South West of the State. Grapes are grown for the fresh market, dried fruits and wines in the Swan Valley and for wine production in the South West. The value of production in terms of grape production is estimated to be $ 5 million. The value of wine produced in WA. is estimated to be in excess of $ 20 million. Viticultural research is carried out on the Swan, Mount Barker and Manjimup research stations and on grower's properties. The programmes include the evaluation and distribution of new varieties and rootstocks under different environmental conditions, the study of new production techniques for quality in table and wine grapes and dried vine fruits and the control of pests and diseases. Studies into regional problems are initiated as required.

5. Floriculture - Floriculture is centred in the Perth Metropolitan region. Here many nurseries produce potted plants, native and exotic, for both indoor and outdoor gardens and landscaped areas. Cut flower producers provide supplies of blooms for the florist trade from both greenhouse and field plantings. The production of indigenous flowers from commercial plantings of wildflowers is increasing and an expanding export trade has developed for both fresh and dried produce. Interest in the production of all types of floricultural produce is also developing in the Manjimup and Albany areas. Floriculture research is located at South Perth and the Medina Research Station and includes developmental work on potting media for nursery plants, exotic cut flower production methods and the selection of wildflowers for increased productivity and disease resistance.

6. Horticultural Training - The Department of Horticulture at the Bentley Technical College is the main centre for urban Horticultural Training in W.A. It offers a range of part and full-time courses for the Turf, Nursery, Landscape and Gardening trades. Research projects are undertaken in a number of fields, predominantly involving indigenous flora. The university of Western Australia's Faculty of Agriculture supervises post-graduate and undergraduate research projects in the pathology of horticulturally important crops.

50 Albany

50.1 Regional Office
elevation: 13 m
rainfall : 815 mm
sunshine : 2190 h

Department of Agriculture,
Albany Highway, Albany 6330
Phone: (098) 41 2166

Adviser - Vegetables — Vacant

51 Bunbury

51.1 Regional Office
elevation: 4 m
rainfall : 881 mm
sunshine : 2555 h

Department of Agriculture,
80 Spencer Street,
Bunbury 6230
Phone: (097) 21 4166

Adviser - Fruit — G.L. Godley, B.Sc.Agric.,Dip.Hort.
Research officer - Fruit and vegetables — N.D. Delroy, B.Agr.Sc. (Hons.)
Adviser - Vegetables — H.V. Gratte, B.Sc.Agric.

52 Busselton

52.1 District Office
elevation: 4 m
rainfall : 838 mm
sunshine : 2555 h

Department of Agriculture,
Queen Street, Busselton 6280
Phone: (098) 52 1688

Adviser - Viticulture — P.B. Gherardi, M. Sc. Agric., B.App.Sc.

53 Carnarvon

53.1 Regional Office
elevation: 4 m

Department of Agriculture,
Carnarvon 6701
Phone: (099) 41 8103

Adviser - Tropical fruit and vegetables — J.R. Burt*, B.Sc. Agric. (Hons.)

53.2 Gascoyne Research Station
elevation: 4 m
rainfall : 233 mm
sunshine : 3287 h

Department of Agriculture,
Carnarvon 6701
Phone: (099) 41 8103

Vegetables and semi tropical fruit — Manager: A.T. Muller

54 Geraldton

54.1 Regional Office
elevation: 37 m
rainfall : 477 mm
sunshine : 2920 h

Department of Agriculture,
Marine Terrace, Geraldton 6530
Phone: (099) 21 2500

Adviser — I.B. Jenkins, B.Sc. Agric.

55 Kununurra

55.1 Regional Office
elevation: 45 m
rainfall : 813 mm
sunshine : 3285 h

Department of Agriculture,
Coolibah Street,
Kununurra 6743
Phone: (091) 68 1166
Telex: AA99414

Research Officer - Tropical fruit and vegetables — B.L. Toohill, B.Sc. Agric.
Research Officer - Tropical fruit and vegetables — Vacant

56 Manjimup

56.1 District Office
elevation: 280 m
rainfall : 1055 mm
sunshine : 2190 h

Department of Agriculture,
Rose Street, Manjimup 6258
Phone: (097) 71 1299

Adviser - vegetables	M.G. Webb, B.Sc. Agric. T.R. Hill, B.Sc. Agric.
Adviser - fruit	R. Paulin, B.Sc. Hort.
Research officer	A. McKay
56.2 Research Station elevation: 250 m rainfall : 838 mm sunshine : 2224 h	Department of Agriculture, Smith Brook Road, Manjimup 6258 Phone: (097) 72 3544
Fruit and vegetables and viticulture	Manager: R.H.T. Pearce, M.D.A.

57 Medina

57.1 Vegetable Research Station elevation: 20 m rainfall : 773 mm sunshine : 2900 h	Department of Agriculture, Abercrombie Road, Medina 6167
Vegetables and floriculture	Manager: V.A. Swain, Dip. Agric.

58 Midland

58.1 District Office elevation: 15 m rainfall : 837 mm sunshine : 2920 h	Department of Agriculture, Railway Parade, Midland 6056 Phone: (09) 274 5355
Adviser - Fruit	N.H. Shorter, B.Sc. Agric.
Senior technical officer - Viticulture	I.J. Cameron, Dip. Agric. Tech.

59 Perth

59.1 Department of Agriculture Head office elevation: 20 m rainfall : 883 mm sunshine : 2920 h	Jarrah Road, South Perth, Western Australia 6151 Phone: (09) 367 0111 Telex: AA93304 Chief of Div.: B.A. Stynes Ph.D. Adelaide; M.Sc. Agric. Sydney
Horticulture Division Vegetables:	
rincipal officer	D.C. Hosking, B.Sc. Agric.
Senior adviser	M.G. Hawson, B.Sc. Agric., Dip. Trop. Agric. (Trin.)
Advisers	R.M. Floyd, B.Sc. Agric., M. Phil. D.R. Phillips, B.Sc. Agric. (Hons.)
Viticulture: Principal viticulturist	A.C. Devitt,B.Sc.Aric.(Hons)., R.D. Oen. (Hons.)
Adviser	J.F. Elliott, B.Sc. Agric., B.App. Sc. (Oen).
Floriculture: Adviser	A. Reid
Fruit: Principal officer	K.T. Whitely, B.Sc.Agric.

Post Harvest:	
Advisers	R.T. Gwynne, B.Sc. Agric. G.A. Ward, B.Sc. Agric. (Hons)., Assoc. Mech. Eng.
Horticulture Division	
Senior reseach officer	J.E.L. Cripps, B.A., B.Sc. Hort., M.Sc. Agric.
Plant Pathology Branch	
Senior plant pathologist - Fruit, ornamentals	R.F. Doepel, M.Sc. Agric.
Plant pathologist - Vegetables	E.M. Carter, M.Sc. Agric.
Weed Agronomy Branch	
Research officer - Horticultural crops	D.J. Gilbey, B.Sc. Agric., Dip. Agric. Sc.
Entomology Branch	
Senior entomologist	A.N. Sproul, B.Sc. Agric.
Resource Management Division	
Irrigation and Water Resources Branch	
Senior adviser	K.S. Cole, B.Sc. Agric.
Senior research officer	.A.F. Laing, B.Sc.Agric.
59.2 Bentley Technical College	Jarrah Road, East Victoria Park 6101 Phone: (09) 362 1088
Department of Horticulture	Head: C. Oliver, Dip. Hort., T.T.T.C.
Urban Horticultural Training	C. Crocker, dip. Hort., Dip.Ag. Tech., T.T./ A. Lobo, B.Ag.Sc., M.Bus., B.Ed., Dip.Ed.Admin. A. Blake, Dip.Hort., Dip.Hort.Mgmt. L. Dungate, Cert.Hort., T.T. E. Pervan, Cert. Hort. & Park Admin., T.T./ P. Pleskus, Cert. Hort (U.K.)/ S. Ganeson, B.Sc. (Hons)., Dip.Hort.Sc., M.Hort.Sc., T.T./ S. Snook, B.A., T.T. R. Cary, Cert.Hort., T. & G. N. Parker, T.T, T.G./ L. Plowman, Cert. Hort./ P. Dean, R.H.S. ULCI City and Guilds NCH R. Lullfitz*, B.Sc., T.T. B. Bowen, B.Ed. (Sydney) P. Graham, Tr.D. Cert. (N.Z.) A. Sims, B.App.Sc. (WAIT), Cert. Hort. Dip. Hort. J. Viska, B.Ed. (W.A.)
59.3 University of Western Australia	Nedlands 6009 Phone: (09) 390 3838, Telex: AA92992
Faculty of Agriculture Soil Science and Plant Nutrition Group - pathology of horticultural crops	
Lecturer	K. Sivasithamparam, MSc., Ph.D.

Honorary Fellow — K.W. Dixon, Ph.D.
Research Officer — Helen Savage, BSc.

60 Stoneville

60.1 Research Station
elevation: 274 m
rainfall : 940 mm
sunshine : 2881 h

Department of Agriculture,
Anketell Road, Stoneville 6554
Phone: (09) 295 1137

Fruit — Manager: D. Johnston

61 Swan

61.1 Research Station
elevation: 46 m
rainfall : 775 mm
sunshine : 2900 h

Department of Agriculture
Railway Parade, Upper Swan 6056
Phone: (09) 296 4269

Viticulture — Manager: V. Rakich

Northern Territory

Survey

The major responsibility for fruit and vegetable research and extension in the Northern Territory rests with the Department of Primary Production of the Northern Territory Government. The Department provides plant pathology, entomology, soil chemistry and advisory services to the industry and conducts research into new varieties, crop agronomy, pest and disease control, post-harvest requirements and production economics. It operates research and development sites on the Berrimah and Coastal Plains Research Stations near Darwin, Katherine Rural Education Centre and in the Arid Zone Research Institute near Alice Springs.
Climatic Data - The Northern Territory extends over 1600 kilometers from N-S and 930 kilometers E-W, with 80 percent of the area lying within the tropics. All of the Territory is characterised by a pattern of wet summers and dry winters although this is less pronounced in the southern arid zone. Rainfall varies from a November-March average of 1500 mm in Darwin to about 250 mm in Alice Springs.
Average maximum temperature range from 34°C to 28°C and the minimum average temperatures range from 23°C at Darwin to 13°C at Alice Springs.
The MacDonnell Ranges have the highest elevation of 520 m, and generally areas north of these Ranges may be considered frost free. Evaporation rates throughout the Territory are very high resulting in depleted soil moisture reserves for most of the year in the north and all year round in the south - 2000-3000 mm.
In the north, humidities range from 30-80 percent, in the south humidity is low throughout the year at about 30 percent.

62 Darwin

62.1 Department of Primary Production

P.O. Box 4160, Darwin, NT 5794
Phone: (089) 22 1211

Tropical fruits and nuts — I. Baker, M.Ag.Sc. B.App.Sc.(Hort).
Tropical vegetables — K. Blackburn, B.Sc.Agr.
Post-harvest research
Market research
Administration — Senior Horticulturist: T. Piggott, B.Ag.Sc.

63 Katherine

63.1 Department of Primary Production

Katherine NT 5780
Phone: (089) 72 1722

Tropical fruits
Citrus
Vegetables
Market research

B. Merrett, B.Ag.Sc. (OIC)

64 Alice Springs

64.1 Department of Primary Production

P.O. Box 2134, Alice Springs,
NT 5750
Phone: (089) 52 2344
C. McColl, B.Ag.Sc. (OIC)
F. McEllister, L.D.A.

Table grapes
Dates
Citrus
Nut crops
Vegetables
Market research

CSIRO

Survey

Commonwealth Scientific and Industrial Research Organization (CSIRO) - CSIRO is an Australian government statutory organization concerned with research for Australian industry, agriculture and the community. Divisions and laboratories are located in all States of Australia, but the head office of CSIRO is located in Canberra. The Divisions concerned with horticulture are Horticultural Research, Food Research, Entomology, Water and Land Resources and the Centre for Irrigation Research.

1. Division of Horticultural Research - The Division has laboratories in Adelaide, South Australia (headquarters), at Merbein, Victoria, and staff are located at the CSIRO laboratory in Darwin (Northern Territory). Research aims at the improvement of woody perennial crops in Australia; specifically subtropical and tropical fruit and nut species and grapevines. The Division's major field laboratory and field plantings are located at Merbein, near Mildura in the irrigated region of north-western Victoria. Three farms, totally 84 ha, are operated by CSIRO in this region, together with field plantings at a number of sites in northern and central Australia. Research is concerned principally with the improvement, by breeding and selection, of grapevines, citrus, pistachio nuts, avocados and a range of subtropical and tropical species with potential for cultivation in the inland irrigated regions or in northern Australia. Emphasis is placed on the selection of salt tolerant and disease resistant plants and on understanding plant response to environmental stresses.

2. Division of Food Research - A major part of the work of the Division is the study of postharvest handling, transport and storage of fruits and vegetables and their preservation by canning and drying and processing into juices. The physiological and biochemical principles underlying the behaviour of fruits and vegetables postharvest is an integral part of this research which ranges from practical methods applicable to industry (especially in cooperation with the NS W. Department of Agriculture) to molecular and cellular biochemistry of plants and their organs. Of particular interest are studies on the physiology, atmosphere control, disorder control and quality attributes of tropical and sub-tropical fruits such as mango, banana, avocado and tomato as well as a number of minor, but developing crops such as kiwifruit, custard apple, lychee, guava and sweet potato. Some of the research on tropical fruits is being carried out in collaboration with scientists from several SE Asian countries supported in part by grants from the Australian Centre for International Agricultural Research. Fundamental studies concentrate on the nature of chilling injury and the ripening and senescencing processes in plants and their organs at the molecular level with the aim of providing the necessary biochemical information to enable plant breeding or genetic engineering methods to be applied to control or modify these processes.

The chemistry, composition and flavour of fruit products, and methods for improvement in product quality and for optimum extraction of the soluble fruit constituents is being studied. The transport of horticultural products, such as peas, onions, coffee and cocoa in containers, and procedures for the disposal and utilization of fruit processing wastes are being investigated. Related topics included studies on the allergens of peanuts, and the deterioration of products by fungal attack and factors affecting the production of mycotoxins.

Research carried out in collaboration with other organizations includes: Macquarie University (North

Ryde): Plant biochemistry.
N.S.W. Department of Agriculture: Nature and control of post harvest wastage of citrus fruits and vegetables; tomato breeding.
Commonwealth Department of Primary Industry and State Departments of Agriculture: Postharvest disinfestation of fresh fruit against fruit flies, and other pests of quarantine importance.

3. Division of Entomology - This Division of CSIRO has its headquarters in Canberra and laboratories in various other centres around the country. Groups of staff in Canberra and Sydney are concentrated with aspects of horticultural entomology. Work on insect sex pheromones centres around evaluation of the prospects of using these secretions in the control of a number of lepidopterous pests of field and orchard crops. Studies on integrated control of pests of glasshouse and other horticultural crops aim to minimize the use of pesticides in protected cropping systems and other intensively sprayed horticultural crops by either introducing, or selectively protecting, various biological agencies, especially predators, parasites, and pathogens, for the control of major pests such as mites, aphids, mealy-bugs and whiteflies. Many of these pests are resistant to the majority of modern pesticides. 'Current target species are two-spotted mite, European red mite and grape phylloxera. Fruit fly investigations, based in Sydney, are working towards the development of improved methods of detection and suppression of Queensland fruit fly. The objective of current work on biological control of scale insects is to assess the effectiveness of parasites introduced to control white wax scale, and to monitor the abundance and spread of other scale insects infesting citrus in eastern Australia.

The CSIRO Division of Entomology in Hobart Tasmania collaborates with Departmental Officers in the study of entomophagus nematodes as control agents of pests of horticultural crops.
4. Centre for Irrigation Research - The Centre is located at Griffith in the Murrumbidgee Irrigation Area of south central New South Wales. Research is concerned with improving the productivity of irrigated crops, and also with minimising the deterioration in water quality which occurs during re-use of water on a local or regional scale. The main research programs are concentrated on the management of water resources to maintain the long-term productivity of irrigated crops, and to minimise damage to the aquatic environment including rivers and wetlands. These studies include investigations into the physical, chemical and biological processes in the soil-water-root zone system in irrigated fine-textured soils. Other research is concerned with energy-conserving methods for greenhouse crop production and with oilseed breeding.
5. Division of Water and Land Resources - This Division's main functions are to investigate the physical, biological and other processes important in water and land management and to develop methods for acquiring, processing and applying natural resources data for planning and other decision-making purpose at national, regional and local levels. Research on horticulture involves the development of computer-based techniques of evaluating land for various horticultural uses. New methods using climatic and other geographic data enable estimates to be made of key variables, such as temperature or amount of sunlight received, for any point in Australia. Co-ordinated sets of site-weather-crop management data are collected for use in modelling whole cropping systems.

65 Canberra

65.1 CSIRO Division of Horticultural Research

GPO Box 350 Adelaide,
South Australia 5001.
Phone: (08) 274 9244
Telex: 88000
Chief: J.V. Possingham*, DPhil., DSc.,FTS, FAIAS

Interaction between the chloroplast and nucleus; chloroplast growth and replication	J.V. Possingham*, DPhil., DSc., FTS, FAIAS N.S. Scott, B.Agr.Sc., Ph.D. M.E. Lawrence, B.Sc.(Hons), Ph.D. D.L. Whisson, B.Sc.(Hons), Ph.D. P.A. Cain, B.Sc.
Photosynthesis, plant response to salinity and water stress	W.J.S. Downton, B.Sc., Ph.D.
Plant hormone metabolism, regulation of photosynthesis and water use	B.R. Loveys, B.Sc., Ph.D.
Photosynthetic gas analysis	W.J.R. Grant, B.Sc. (Hons).
The biochemistry of photosynthesis	S.P Robinson, BSc. (Hons),Ph.D.

Flower physiology, pollination	M. Sedgley*, B.Sc. (Hons), Ph.D.
Flower physiology, microscopy	M.A. Blesing, B.Sc.
In vitro culture, tissue differentiation	K.G.M. Skene, B.Agr.Sc., Ph.D.
65.2 CSIRO Division of Horticultural Research*	Merbein Laboratory, Private Bag, Merbein, Victoria 3505. Phone: (050) 256201 Telex: 55581 Chief: J.V. Possingham*, DPhil,DSc, FTS, FAIAS (located in Adelaide)
Nematode taxonomy	Off. in Ch.: M.R. Sauer, B.Agr.Sc.
	C.M. Sykes, BA (Hons)
Subtropical and tropical tree fruits	D.McE. Alexander, B.Sc., MS
Floral biology, pollination of subtropical fruit trees	H.I.M.V. Vithanage, B.Sc. (Hons), Ph.D.
In vitro culture of woody fruit species; virus elimination	M. Barlass*, B.Sc.(Hons), M.Sc., Ph.D.
Screening for disease resistance in grapevines, in vitro	R.M. Miller, B.Sc.Agr. (Hons)
Vine improvement, vine management	P.R. Clingeleffer, B.Agr.Sc.(Hons)
Vine improvement	H.P. Newman, B.Sc.Ag.
Small-scale winemaking	G.H. Kerridge
Physiology of salt tolerance	R.R. Walker, B.Ag.Sc.(Hons), Ph.D.
Properties of root membranes with respect to salt tolerance	T.J. Douglas, B.Sc.(Hons), Ph.D.
Root anatomy and the cellular location of salts	R. Storey, B.Sc., Ph.D.
Genetics of salt tolerance	S.R. Sykes, B.Sc.(Hons),MSc,Ph.D.
Physiology of salt tolerance	M. Schache, B.Sc. (Hons)
65.3 CSIRO Darwin Laboratories	Private Mail Bag No. 44 Winnellie, N.T. 5789 Phone: (089) 84 3611 Telex: 85294 Chief: J.V. Possingham*, DPhil,DSc, FTS, FAIAS (located in Adelaide)
Field assessment, vegetative and reproductive physiology of tropical fruits	P.B. Scholefield, B.Ag.Sc., Ph.D.
Field assessment, vegetative and reproductive physiology of tropical fruits	D.R. Oag, B.Sc.
65.4 CSIRO Division of Food Research	P.O. Box 52, North Ryde, New South Wales 2113 Phone: (02) 887 8333 Telex: 23407 Chief: J.H.B. Christian, Ph.D., FAIFST, FTS
Chemistry of fruit products	B.V. Chandler, B.Sc., M.Sc., Ph.D.
	R.L. Johnston, M.Sc.,Dip.Ed.,Ph.D.
Flavour chemistry	F.B. Whitfield, M.Sc., ASTC, Ph.D.
	C.R. Tindale, B.Ap.Sc.
	G. Stanley, B.Sc., ASTC
Container transport	A.K. Sharp, BE, MEngSci., Ph.D.
	A.R. Irving, B.Sc.

Fruit waste disposal and utilization	A.G. Lane, M.Sc., Ph.D. P. Gwatkin, B.Sc.
Peanut allergens	D. Barnett, BA
Fruit juice extraction and processing studies	D.J. Casimir, M.Sc.,Dip.Ed.,Ph.D. P.W. Board, B.Sc.
Mycotoxins and fungal studies	J.I. Pitt, M.Sc., ASTC, Ph.D. A.D. Hocking, B.Sc. K. Wheeler, B.Sc.
Temperature stress in plants	J.K. Raison, B.Sc., Ph.D. (located at Macquarie University) D.G. Bishop, M.Sc., Ph.D. D. Graham, B.Sc., Ph.D. D.G. Hockley, B.Sc. B.D. Patterson, B.Sc., Ph.D. R.M. Smillie, M.Sc., Ph.D., D.Sc.
Fruit ripening and senescence	C.J Brady, M.Sc.Agr., Ph.D., (located at Macquarie University) E. Lee, B.A. (located at Macquarie University) J.A. Pearson, M.Sc. J. Spiers, M.Sc., Ph.D. W.B. McGlasson*, BAgSc., Ph.D. S.K Meldrum, B.Ap.Sc.
Postharvest fruit and vegetable storage	J.E. Algie, BE, ASTC, M.Sc. D. de l'U (Paul Sabatier) A.J. Shorter, B.Sc. G.R. Chaplin, BScAgr., M.Sc. S.H. Satyan, M.Sc., Ph.D. E. Kavanagh, B.Sc. S. Cole, B.Sc.
65.5 CSIRO Division of Entomology	Box 1700 GPO, Canberra, Australian Capital Territory 2601 Phone: (062) 46 4911, Telex: 62309 Chief: M.J. Whitten, B.Sc.(Hons), BA, Ph.D.
Integrated control of pests	J.L. Readshaw, B.Sc.(Hons), Ph.D.
Sex pheromone studies on fruit moths	R.J. Bartell, B.Sc. (Hons), Ph.D. T.E. Bellas, B.Sc. (Hons), Ph.D. W.V. Brown, B.Sc. (Hons), Ph.D. M.J. Lacey, M.Sc., Ph.D. G.H.L. Rothschild,B.Sc.(Hons) Ph.D. E.R. Rumbo, BA (Hons), Ph.D. R.A. Vickers, B.Sc. (Hons) C.P. Whittle, B.Sc. (Hons), Ph.D.
65.5.1 Entomology Research Station	'Kooyong' 55 Hastings Road Warrawee New South Wales 2074 Telephone: (02) 487 1756
Biological control of scale insects	R.G. Lukins, B.Sc.
Fruit fly ecology	M.A. Bateman, B.Sc.(Hons), Ph.D. B.S. Fletcher, B.Sc.(Hons),Ph.D.
65.6 CSIRO Centre for Irrigation Research	Griffith, New South Wales 2680 Phone: (069) 621 700

	Telex: 6990 Officer in charge: D.S. Mitchell, B.Sc., Ph.D.
Aquatic plant physiology	P.J.M. Sale*, B.Sc., Ph.D.
Environmental plant physiology	G.I. Moss, B.Sc., Ph.D.
Aquatic plant growth	P.R. Cary*, B.Sc., Ph.D.
Environmental engineering	K.V. Garzoli, BMechEng., MEng.
Sunflower nitrogen and root physiology	B.T. Steer, B.Sc., Ph.D.
65.7 CSIRO Division of Water and Land Resources	GPO Box 1666, Canberra, Australian Capital Territory 2601 Phone: (062) 46 4911, Telex: 62337 Chief: R.J. Millington, M.Sc., Ph.D., FTS
Resource management in horticulture	C. Hackett*, B.Sc., Ph.D. K. Rattigan*, B.App.Sc. R.M. Scott, B.Sc. J.R. Ive, B.Agr.Sc., LDA (Hons), BEc.

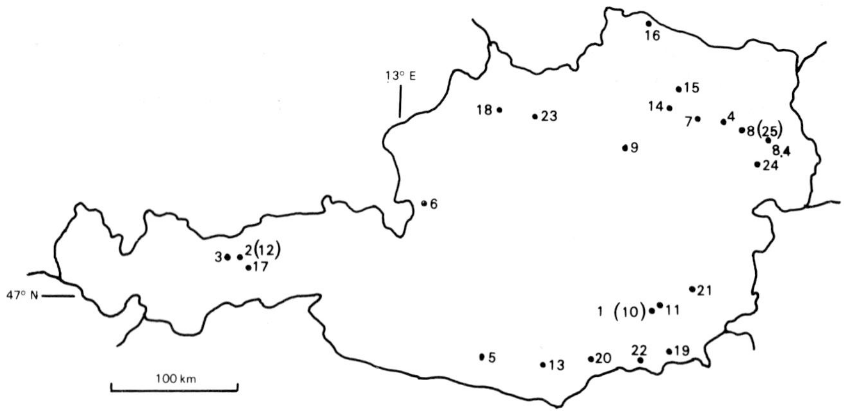
13° E
47° N
16
15
18
23
14
7
4
8(25)
8.4
9
24
6
3
2(12)
17
21
1 (10)
11
5
13
20
22
19
100 km

1 (10) Graz, 2 (12) Innsbruck, 3 Kematen, 4 Klosterneuburg, 5 Pitzelstätten, 6 Salzburg, 7 Sitzenberg, 8 (25) Wien, 8.4 Gross Enzersdorf, 9 Wieselburg, 11 Haidegg, 13 Klagenfurt, 14 Krems, 15 Langenlois, 16 Retz, 17 Rinn, 18 Ritzlhof, 19 Silberberg, 20 St. Andrä, 21 Wetzawinkel, 22 Wies, 23 Linz, 24 Seibersdorf

Survey

The research and experimental activities in the field of horticulture, fruit growing and viticulture in Austria is primarily performed by the research institutes of the confederacy and the Federal States. The research carried out by the Chambers of Agriculture of the States, the Universities and some private enterprises are of less importance. The main research institutes concerned are the Federal Highschool and Research Institute for Horticulture in Vienna - Schönbrunn, the Federal Highschool and Research Institute for Viticulture and Fruit Growing in Klosterneuburg and the State Experimental Station for Spezialized Cultures in Wies-Burgstall (Styria).

Federal Institutes

1 Graz

1.1 University of Graz, Naturwissenschaftliche Fakultät
Institut für Botanik
elevation: 377 m
rainfall : 890 mm

Holteigasse 6, 8020 Graz
Phone: 0316/3880, 5646

Botanical Garden of University of Graz

Phone: 0316/31189
Ing. T. Stehr

2 Innsbruck

2.1 University of Innsbruck, Naturwissenschaftliche Fakultät, Botanisches Institut
elevation: 580 m
rainfall : 900 mm

Innrain 52, 6020 Innsbruck
Phone: 05222/26741

Plant physiology of alpine plants, temperature requirements of horticultural plants

Botanical Garden of University of Innsbruck

Botanikerstr. 10
Phone: 05222/82715
Dr. G. Gärtner

Prepared by Dr. H. Halbmayr, Federal Highschool and Research Institute for Horticulture, Wien

3 Kematen

3.1 Hohere Bundeslehranstalt für landwirtschaftliche Frauenberufe (Federal Highschool for Agricultural Female Professions)
elevation: 590 m
rainfall : 860 mm

Birkenweg 6, 6175 Kematen
Phone: 05232/2319
Dir.: Mag. H. Bachmann

Education of female students in farming business, home economics, fruit growing, vegetables and general horticulture

Departments of Horticulture — Ing. L. Bazand

4 Klosterneuburg

4.1 Höhere Bundeslehr- und Versuchsanstalt für Wein- und Obstbau mit Institut für Bienenkunde (Federal Highschool and Research Institute for Viticulture and Fruit Growing
elevation: 180 m
rainfall : 600 mm

Wienerstr. 74,
3400 Klosterneuburg
Phone: 02243/2159
Dir.: Dipl.Ing. J. Haushofer

Training facilities for experts in wine and fruit growing. Apiculture research. Control of wines, origin classification in connection with the export control. Fruit growing, varieties and rootstock, experiments with intensive planting systems. Viticulture, wine making, selection of clone, mutation breeding, diseases. Winechemistry, economy. Publication.

5 Pitzelstätten-Wölfnitz

5.1 Höhere Bundeslehranstalt für landwirtschaftliche Frauenberufe (Federal Highschool for Agricultural Female Professions
elevation: 380 m
rainfall : 850 mm

9061 Pitzelstätten-Wölfnitz
Phone: 04222/49391
Dir.: Dipl.Ing. H. Koll

Education of female students in farming business, home economics and general horticulture.

Department of Horticulture — Ing. R. Nagel

6 Salzburg

6.1 University of Salzburg, Naturwissenschaftliche Fakultät, Institut für Botanik (Botanical Institute)
elevation: 435 m
rainfall : 1330 mm

Freisaalweg 16, 5020 Salzburg
Phone: 996/445110/239
Prof.Dr. Kiemayer

7 Sitzenberg

7.1 Höhere Bundeslehranstalt für landwirtschaftliche Frauenberufe (Federal Highschool for Agricultural Female Professions)
elevation: 300 m
rainfall : 690 mm

3454 Sitzenberg-Reidling
Phone: 02276/335
Dir.: Dipl.Ing. G. Fruhlich

Education of female students in farming business, home economics, fruit growing and horticulture

Department of Horticulture — Ing. E. Gundacker

8 Wien

8.1 Höhere Bundeslehr und Versuchsanstalt fur Gartenbau (Federal Highschool and Research Institute for Horticulture
elevation: 200 m
rainfall : 690 mm
sunshine : 1600 h

Grunbergstr. 24
1131 Wien-Schönbrunn
Phone: 0222/833535
Dir.: Prof.Dr. L. Urban

General Horticulture under glass and in the open, floriculture, vegetable variety testing, rootstocks, seed testing, plant protection. Research, teaching and advising. Publication of the annual research report.

Department of Tree Growing	
Experimental Tree Growing Hetzendorf	Dr. H. Pirc
Department of Botany	Mag. H. Reif
Department of Chemical and Soil Culture	Dr. F. Klinger
Department of Vegetable Growing	Dr. K. Danek-Jezik
Department of Vegetable Growing and Varieties	Dipl.Ing. T. Reeh
External Department Neusiedl am See	Phone: 02167/648
elevation: 135 m	
rainfall : 619 mm	
External Department Zinsenhof	Phone: 02756/2814
elevation: 250 m	
rainfall : 868 mm	
Department of Ornamentals under glass	Dr. H. Halbmayr
Experimental Greenhouses Schönbrunn	

8.2 Verwaltung der Bundesgarten (Management of the Federal Gardens)

Schloss Schönbrunn, 1130 Wien
Phone: 0222/833646
Dir.: Dipl.Ing. E. Kaven

Management of all Austrian Federal Gardens and parcs. Big collections of orchids, cacti, ericaceous plants and alpin plants. In vitro cultures of orchids and carnivorous plants.

Reservegarten Schönbrunn	Ing. O. Rinnerbauer
Burggarten	Ing. E. Klenkhart
Belevedere	Ing. W. Ludwig
Alpengarten	R. Klaus
Augarten	Ing. D. Kainrath
Hofgarten Innsbruck	Ing. O. Koppensteiner

8.3 Universität für Bodenkultur (University for Soil Culture)

Peter Jordanstr. 82, 1190 Wien
Phone: 0222/314511/362

Research greenhouses FGH
Applied research for all institutes

Institute for Landscape Architecture and Horticulture — Gregor Mendel Strasse 33, 1190 Wien

	Phone: 0222/314541 Prof.Dr. F. Woess
Experimental Garden Essling	Schlachthammerstrasse, 1220 Wien Phone: 0222/2241
Institute for Fruit Growing	Feistmantelstrasse 4, 1190 Wien Phone: 0222/7342500 Prof.Dr. K. Pieber*
Experimental Garden Gerasdorf	Gerasdorferstr. 131, 1210 Wien Phone: 0222/392555

The Experimental Garden carries out: Selection of stone fruit trees, testing of national and foreign fruit tree cultivars, development of treeforms suitable for mechanical harvesting, improvement of walnut cultivars, walnut grafting, apricot production, vegetative propagation of tree fruit rootstocks.

Institute for Plant Growing and Breeding Cultivation of bush tomatoes, multiplication of vegetable varieties, chicory breeding, sprinkling in vegetable cultures, cultivation and breeding of vegetables.	Schlosshoferstr. 31, 2301 Gross Enzersdorf Phone: 02249/2302, 2593 Prof.Dr. G. Storchschnabel
8.4 Versuchswirtschaft der Universität für Bodenkultur (Experiment Station of the University for Soil Culture)	Schlosshoferstr. 31, 2301 Gross Enzersdorf Phone: 02249/2593 Prof.Dr.Dr. O. Steineck

Realisation of particular operations in connection with the research of institutes and acceptance of research commissions and agricultural fields. Demonstrations for students and participants of congresses and special courses.

8.5 Technische Universität Wien (Technical University of Vienna, Institute for Landscape and Garden Planning)	Karlsplatz 13, 1040 Wien Phone: 0222/56010 Prof.Dr. R. Galzer

Landscape ecology, garden construction and technology, history of landscape and garden architecture, methodes for landscape analysis and evaluation.

8.6 Universität Wien (University of Vienna) Formal- und Naturwissenschaftliche Fakultät-Biologiezentrum	Althanstr. 4, 1090 Wien Phone: 0222/314510
Institute for Plant Physiology Horticulture under glass, plant physiology of horticultural plants growing chambers, photosynthesis	Prof.Dr. H. Kienzel Prof.Dr. H. Bolhar-Nordenkampf*
Experimental Garden Augarten	Obere Augartenstr. 1-4, 1020 Wien Phone: 0222/335584
Institute for Plant Sociology	Prof.Dr. G. Wendlberger
Botanical Institute and Botanic Garden of the University	Renngweg 14, 1030 Wien Phone: 0222/787102 Prof.Dr. F. Ehrendorfer Ing. K. Liebeswar
8.7 Bundesanstalt für Pflanzenbau (Federal Institute for Plant Growing)	Alliiertenstr. 1-3, 1020 Wien Phone: 0222/241511 Dir.: Dr. R. Meinx

Control of the germination of horticultural and agricultural seeds. Variety testing and registration. Breeding of cereals.

Department of Seed Testing and Certification	Dr. F. Fiala
Registration research	Dr. J. Steinberger
Department of Growth Trials	Dipl.Ing. K. Nagl
Department of Chemical Technology	Dr. K. Waltl
Cereal cultivation and breeding, Varietylist Committee	Dipl.Ing. R. Hron
Maize cultivation	Dipl.Ing. J. Hinterholzer
Potato	Dipl.Ing. G. Kweta
Sugarbeet	Prof.Dr. A. Graft
Forage crops, oilcrops and leguminous crops	Dr. D. Wolffhardt

8.8 Stadtgartenamt Wien, MA 42 (Municipal Garden Administration of Vienna)

Heumarkt 2b, 1030 Wien
Phone: 0222/722171
Dir.: Ing. P. Schiller

Management of all municipal gardens and parcs. Glasshouses for flowers and tree nurseries. Education of the junior gardeners of Vienna and Burgenland in cooperation with the Ministry of Education.

Reservegarden Hirschstetten	Ing. Podsednik
School for gardeners and florists, schoolgarden	Ing. Membier

8.9 Naturhistorisches Museum (Museum of Natural History)

Burgring 7, 1014 Wien
Phone: 0222/9345410
Dir.: Dr. R. Paket

Herbarium and plant collections. Research about chromosome number and plant systematics.

Botany Department	Doz.Dr. Riedl/Dr. Pollatschek

8.10 Landwirtschaftlich chemische Bundesanstalt (Federal Agricultural Chemical Research Institute)

Trunnerstr. 1-3, 1020 Wien
Phone: 0222/241511
Dir.: Dr. W. Beck

Official analyses of foodstuffs, wines and spirits, fertilizers, milk. Testing and certification of seed, sanitary control of seed potatoes, pollution control of sewage sludges. Plant nutrition, agronomy and ecology of the soil plant-system. Libraries.

Institute for Plant Nutrition and Soil Culture	Dr. H.E. Oberländer
Institute for Viticulture	Dr. J. Boiler
Institute for Analysis and Biochemie	Dr. R. Bankl
Institute for Agricultural Biology	Winnengerstr. 8, 4020 Linz Phone: 997/81261 Dr. J. Gusenleitner

8.11 Bundesanstalt für Pflanzenschutz (Federal Institute for Plant Protection)

Trunnerstr. 5, 1020 Wien
Phone: 0222/241511
Dir.: Dr. K. Russ

Central expert institution of the Austrian plant protection research and administration of the plant protection service. Integrated prevention and control of plant diseases. Inspection of infected plants and public information. Testing of agents and equipment for pest control. Cooperation in plant protection legislation. Random supervision of tree nurseries. Two periodicals: Pflanzenschutzberichte (Plant Protection Reports) and Pflanzenschutz.

Department of Zoology	Dr. H. Berger/Dr.P. Fischer-Kolbri Dr. H. Schönbeck
Department of Herbology	Dr. H. Neururer
Department of Botany	Dr. G. Vukowitz/Dr. R. Krexner
Chemical Department	Dr. E. Glofke/Dr. W. Zislowsky Mag. H. Kohlmann/Dr. P. Fida
Austrian Plant Protection Service	Dr. R. Maser

9 Wieselburg

9.1 Bundesanstalt für Landtechnik (Federal Institute for Research and Testing of Agricultual Machinery
elevation: 260 m
rainfall : 805 mm

Rottenhauserstr. 1
3250 Wieselburg/Erlauf
Phone: 0716/2175
Dir.: Dr. E. Reichmann

Work study and technical research in collaboration with the management of the Federal Gardens. Automation in greenhouses, internal means of transport and labour chain. Testing of horticultural and agricultural machinery.

Department of Machinery Testing	Dr. J. Zehetner
Department of Engineering Research	Dipl.Ing. Pernkopf
Department of Measure- and Engineering Development	Dr. J. Schrothmayer
Department of Agricultural Labour Management	Ing. A. Wernisch
Department of Mountain- and Soil Culture Techniques	Ing. R. Sieg

Institutes of the State

10 Graz

10.1 Landwirtschaftliche Berufsschule für Gartenbau Lehr und Versuchsgärtnerei Raiffeisenhof (Junior Gardener School and Experimental Garden)
elevation: 365 m
rainfall : 890 mm

Krottendorferstr. 79-81
8052 Graz-Wetzelsdorf
Phone: 993/292366
Dir.: Ing. J. Kriegl

Education of the junior gardeners of Styria. Tests with various plants for window boxes. Floriculture trials.

10.2 Landwirtschaftlich-chemische Versuchs- und Untersuchungsanstalt des Landes Steiermark (Agricultural - Chemical Research and Experimental Institute of Styria)

Burggasse 2, 8010 Graz
Phone: 993/70312433
Dir.: Dr. H. Kesselring

Department of Viticulture — Dipl.Ing. W. Puchwein
Systematic research of Styrian wines, wine analysis, wine making

11 Haidegg - Graz

11.1 Obstbauversuchsstation Haidegg (Research Station for Fruit Growing)
elevation: 370 m
rainfall : 910 mm

Ragnitzstr. 193, 8010 Graz
Phone: 993/301507
Dir.: Ing. F. Strempfl

Testing of new varieties of fruits. Collecting varieties

12 Innsbruck

12.1 Berufsschule für Gartenbau, Tirol (Junior Gardener School of the State Tyrol)
elevation: 580 m
rainfall : 855 mm

Trientlgasse 2, 6020 Innsbruck
Phone: 05222/45411
Dir.: Dr.Ing. A. Pühringer

12.2 Stadtgartenamt Innsbruck (Municipal Garden Office)

Trientlgasse 13, 6020 Innsbruck
Phone: 05222/45432
Dir.: Ing. E. Falch

13 Klagenfurt

13.1 Landwirtschaftliche Landesberufsschule für Gartenbau (Junior Gardener School of the State Carinthia)
elevation: 450 m
rainfall : 950 mm

Haupstr. 119, 9020 Klagenfurt
Phone: 04222/43296
Dir.: Dipl.Ing. F. Ehrenthaler

Education of the Carinthian junior gardeners, experiments with flowers and vegetables

14 Krems

14.1 Landwirtschaftliche Fachschule Krems (Agricultural Vocational School)
elevation: 210 m
rainfall : 540 mm

t07Wienerstr. 101, 3500 Krems
Phone: 901/7516
Dir.: Dipl.Ing. F. Epp

Fruit growing, rootstock of morellos and apples, mulch in orchards, thinning with chemicals on apples, wine making

15 Langenlois

15.1 Landeskursstätte für Obst-Wein und Gartenbau (School for Fruit, Wine Growing and Horticulture of the State Lower Austria)
elevation: 220 m
rainfall : 540 mm

Am Rosenhügel 15
3550 Langenlois
Phone: 02734/2106
Dir.: Dipl.Ing. E. Ettenauer

Simplification of manuring methods in fruit growing, rootstock and variety trials on apples and pears, pruning of stone and pome fruits, trials with wine varieties and rootstock, hail control with silverjodine rockets.

15.2 Landwirtschaftliche Fachschule, Fachrichtung Gartenbau (Horticultural Vocational School)

Phone: 02734/2206
Dir.: Dipl.Ing. E. Ettenauer

Experiments with flowers, rootstocks, hydroculture

15.3 Landwirtschaftliche Berufsschule für Gartenbau (Junior Gardener School for Horticulture)

Phone: 02734/2206
Dr.Ing. F. Weigl

16 Retz

16.1 Landwirtschaftliche Weinbauschule Retz (Viticultural Vocational School)

Seeweg 2, 2070 Retz
Phone: 02942/2202

elevation: 260 m
rainfall : 500 mm

Dir.: Ing. J. Weiser

Education of registered masters of viticulture. Alternative ways of agricultural production, wine growing, cellerage fruit cultures, viticulture trials.

17 Rinn

17.1 Landesanstalt für Pflanzenzucht und Samenprüfung (Institute for Plant Growing and Seed Testing of the State Tyrol)
elevation: 590 m
rainfall : 855 mm

Versuchsfeld Nr. 2, 6074 Rinn in Tirol
Phone: 05223/8117
Dir.: Dr. L. Köck

Testing of agricultural and horticultural seeds. Variety testing, breeding.

18 Ritzlhof (Haid)

18.1 Landwirtschaftliche Fachschule Ritzlhof, Fachrichtung Gartenbau (Horticultural Vocational School)
elevation: 297 m
rainfall : 865 mm

Kremstalstr. 125, 4053 Haid-Ansfelden
Phone: 07229/88312
Dir.: Dr. L. Klug

19 Silberberg

19.1 Landwirtschaftliche Fachschule für Obst- und Weinbau (Agricultural Vocational School for Fruit Growing and Viticulture)
elevation: 450 m
rainfall : 830 mm

Kogelberg 16, 8340 Leibnitz-Siberberg
Phone: 03452/2339
Dir.: Dipl.Ing. R. Eder

Plant protection of grapes, pruning systems, variety trials

20 St. Andrä

20.1 Obstbauversuchsanlage der Landwirtschaftskammer für Kärnten (Fruit Experiment Station of the Chamber of Carinthia)
elevation: 430 m
rainfall : 690 mm

Langen 6, 9433 St. Andrä/Lavanttal
Phone: 04358/2296
Dir.: Ing. H. Gartner

Fruit growing, tests of fruit varieties for the conditions in Carinthia and the Alps region. Test of tree forms and pruning methods. Fruit sorting, storage and processing. Multiplication of blueberries and cranberries.

21 Wetzawinkel - Gleisdorf

21.1 Landwirtschaftliche Fachschule Wetzawinkel (Agricultural Vocational School)
elevation: 380 m
rainfall : 890 mm

8200 Wetzawinkel - Gleisdorf
Dir.: Dipl.Ing. M. Steurer

Technics in fruitculture and in plant protection

22 Wies

22.1 Landesversuchsanlage für Spezialkulturen (Experimental Garden for Spezialized Cultures of Styria)
elevation: 371 m
rainfall : 830 mm

8551 Wies-Burgstall
Phone: 03465/2423
Dir.: Dr. E. Müller
Ing. H. Pelzmann*

Variety trials of vegetables and ornamentals, cultivation trials of medicinal plants and spices, processing of new vegetables, breeding of pumpkin and lettuce

Private Firms and Institutions

23 Linz

23.1 Biologische Forschung der Chemie Linz AG (Biological Research Section of the Austrial Nitrogen Works)
elevation: 280 m
rainfall : 860 mm

St. Peter Strasse 25, 4020 Linz
Phone: 997/5910
Dir.: Dr. W. Wendtland

Nitrogen fertilizers on fruits, vines and ornamentals, growth regulators plant protection, new pesticides.

24 Seibersdorf

24.1 Osterreichisches Forschungszentrum Seibersdorf G.m.b.H. (Austrian Research Center)
elevation: 210 m
rainfall : 615 mm

Lenaugasse 10, 1082 Wien
Phone: 0222/427511
Prof.Dr. E. Haunold

Agricultural Institute

2244 Siebersdorf
Phone: 02254/800

Research and development with environment-related aspects, of plant production. Soil-plant analysis, toxic elements in plants, soil, water and air. Water stress in relation to plant production, humus analysis, production and propagation of diseasefree plants in tissue cultures. Development of environmental and agrometerological monitoring systems for special applications.

25 Wien

25.1 Österreichische Düngeberatungsstelle (Austrian Advisory Service for Fertilizers)
elevation: 200 m
rainfall : 670 mm

Auenbruggergasse 2, 1030 Wien
Phone: 0222/752135
Prof.Dr. W. Rückenbauer

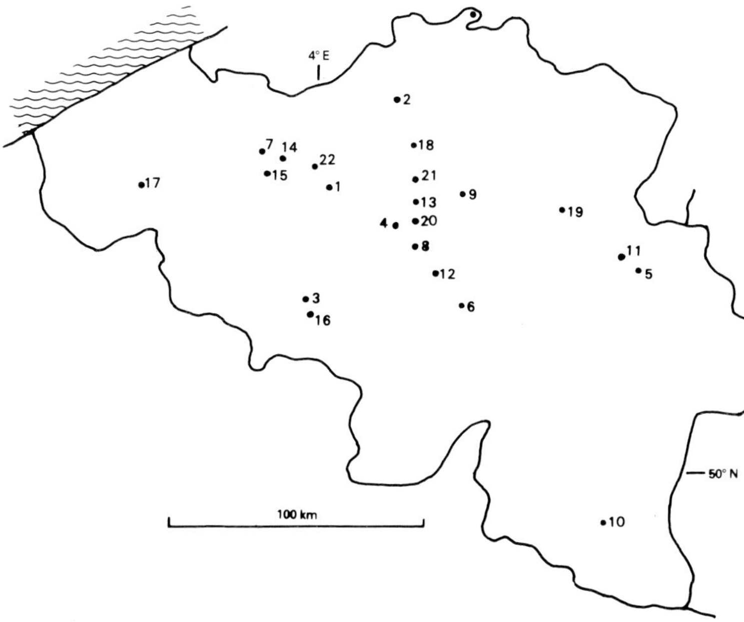
4° E
2
7
14
22
18
17
15
1
21
9
13
4
20
19
8
11
12
5
3
16
6
50° N
100 km
10

1 Aalst, 2 Antwerpen, 3 Ath, 4 Brussel-Bruxelles, 5 Cerexhe-Heuseux, 6 Gembloux, 7 Gent, 8 La Hulpe, 9 Leuven, 10 Libramont, 11 Liège, 12 Louvain-la-Neuve, 13 Meise, 14 Melle, 15 Merelbeke, 16 Ormeignies, 17 Rumbeke, 18 St. Katelijne-Waver, 19 St. Truiden, 20 Tervuren, 21 Vilvoorde, 22 Wetteren

Survey

Despite the fact that Belgian horticulture has been in an enviable position since the beginning of the 19th century because of its various specialities, the research in this field - undertaken by both official and semi-official institutions - is of more recent date. As a matter of fact, it is only since World War II that the competent authorities have taken an active part in the development of scientific horticultural research. Nearly a century ago however, a few leading nurseries had already made a start with research work and practical improvement. Owing to the limited knowledge of fundamental sciences at that time, such work was of a rather elementary nature. Nevertheless, the leading nurseries of the period can be considered as the pioneers of the present-day horticultural research work in Belgium.
Just before World War II Belgium already had a series of institutions actively engaged in agricultural research. Most of them were located in or around both the State Agricultural Colleges (University) of Ghent (founded 1920) and of Gembloux (founded 1860), as well as in the 'Institut agronomique de l'Université libre de Louvain' (1878). Horticultural research goes back to the pre- and post-World War II years. From 1946 onwards Research Stations and Laboratories - entirely or partially devoted to horticultural research - came into being. In 1957 the Department of Agricultural Research was founded at the Ministry of Agriculture which led to co-ordination of the work of State Research Stations.
Agricultural research is carried out by different bodies, i.e.: Government institutions, by far the most important of all, Institutions supported by the government, Official, or private institutions, supported by the IRSIA (Institut pour l'encouragement de la Recherche Scientifique dans l'Industrie et l'Agriculture) or by the FNRS (Fonds National de la Recherche Scientifique), Institutions supported either by the provincial or city authorities in the case of special research and Private Institutions.
Government Institutions - Research institutions of the Ministry of Agriculture now number 22 and are spread over two important agricultural research centres: Ghent, with 10 stations, and Gembloux with 12. Each centre has a certain autonomy, even for budgetary matters. The research programmes are drawn up in conformity with science, practice and extension services. The basic principle is that, for the actualization of the programmes, the personality of the research worker, is respected as far as possible. The

Prepared by Prof. J.G. van Onsem, National Institute of Ornamental Plant Growing, Melle research stations of the Ministry of Agriculture have to work for the general benefit in the widest sense of the word. Their activity is aimed mainly at those problems from which solutions are directly profitable to the agricultural economy.The stations that work entirely or partially in the field of horticultural research at the Ghent Centre are the Government Research Station for Ornamental Plants, for Nematology and Entomology, for Phytopathology and for Agricultural Engineering. At the Gembloux Centre there are the Government Research Station for Fruits and Vegetables, for Applied Zoology, for Phytopathology, the Phytopharmacy Station and the Government Research Station for Agricultural Engineering. The Department of Agricultural Research in Brussels directs the work of the Ghent and Gembloux research stations, as well as that of the State Botanical Gardens at Meise and some other agricultural research centres. The Chairs and Laboratories of the Agricultural Colleges belonging to the Ministry of Education are engaged in horticultural researchwork. Their main aim is fundamental research, very often forming part of higher education. The research programmes are completely optional. Fundamental horticultural research is carried out at the State Agricultural University of Ghent and Gembloux, the Liège State University and the Universities of Brussels and Louvain.

State-supported Institutions - Certain private or independent institutions obtain financial aid from the Ministry of Agriculture, intended for agricultural scientific research. The most important institutions belong to the Chatolic University of Louvain, where horticultural research is mainly carried out by the laboratory for phytopathology and plant protection, the laboratory for applied genetics, the laboratory for pedology and the centre for soil research in horticulture.

Official, independent or private institutions, supported either by the IRSIA or the FNRS. IRSIA (Institute for the encouragement of industrial and agricultural scientific research) is a semi-official institution created by an Act of Parliament of the 27th December 1944, in order to promote and encourage, by means of grants, scientific and technical research leading to advances in both industry and agriculture. The funds are provided for in the budget of the Ministry of Agriculture for the agricultural and horticultural research projects and in the budget of the Ministry of Economic Affairs for the industrial projects. The institute supports research a.o. on the production of fruits, vegetables and ornamentals; storage of fruits, substrates, hydroponics, growth regulation, in vitro techniques, diseases in horticultural crops, application of electricity in horticulture, biometrics and biochemical identification of cultivars.

FNRS (Fonds national de la recherche scientific) - Established by private initiative (University Foundation). The National Fund has been approved by Royal Decree, dated June 2, 1928. Its funds are used to grant subsidies to research workers or to provide laboratory equipment, for development of scientific research and fundamental scientific research in particularly.

Special Research supported by Provincial or Municipal Authorities, consists mainly of trial gardens set up by the provincial authorities. There is one specialised Provincial Institute i.e. in Western-Flanders dealing with applied research and also advisory work especially on vegetables (Rumbeke).

Independent Research Institutions are being engaged completely or partially in agricultural research: Institutions are supported principally by private enterprises, professional associations and similar bodies.

Grouped Institutions - Since 1959 a new organization came into being which gives a new shape to the whole scientific policy of Belgium i.e.: The ministrial Committee of Scientific Policy, which is presided over by the Prime Minister and co-ordinates the activities of the interested ministries; An interministrial Commission, of high ranking officials, entrusted with the adjustment of the measures required to be taken by different ministries; The National Council of Scientific Policy outlines the scientific policy and checks its development. In addition, in 1968, the Ministry of Scientific Policy came into being. The National Council of Scientific Policy is subordinated to it. As far as agricultural research is concerned, the activities of the Ministry of Agriculture are co-ordinated by the "Consultative Council for Agricultural Research". The latter advises the Ministry of Agriculture on all problems connected with agricultural research. The Council includes delegates from the administrations, scientific institutions, advisory services and professional associations. The Council is presided over by the Secretary General of the Ministry of Agriculture.

1 Aalst

1.1 Tuinbouwstichting Aalst en Omgeving.
(Horticultural Foundation Aalst and Surroundings)

Albrechtlaan 117, 9300-Aalst
Phone: 053-701247
Dir.: K. De Bondt

Private organisation

Private organisation with financial aid from the Ministry of Agriculture, the Ministry of Regional Planning, the Farmers Union (Belgische Boerenbond), the Province, the City of Aalst and the Auction Markets. Trial garden. Rose variety testing, research on rootstock and fertilisation: Technique of cultivation and research of fertilisation on some bulbous plants and some cutflowers.

2 Antwerpen

2.1 University of Antwerp, Department of Biology,
Laboratory of Plant Ecology
elevation: 8 m
rainfall : 725 mm
sunshine : 1432 h

Universiteitsplein 1
B-2610 Wilrijk-Antwerp
Phone: 03-8282528
Telex: 33646 UIA
Head: Prof. Dr. Ir. I. Impens
Dr. R. Ceulemans*

Ecophysiological gas exchange measurements (CO_2 gas, water vapour) of ornamental and horticultural plants in relation to abiotic factors (atmospheric conditions, substrate nutrient status), cultural applications, growth and productivity models.

3 Ath

3.1 Centre Agronomique de Recherche Appliquée du Hainaut (CARAH) (Hainaut Agronomic Centre of Applied Research)

Rue de l'Agriculture, 301
B-7800 Ath
Phone: 32-68221281
Dir.: M. van Koninckxloo

Vegetable cultivation - New varieties trials, weed control, fungicides (onion, carrot, potato, Brussels cichory). Medicinal plants - Mechanisation of yield (angelia, tobacco, valerian). Fruit cultivation - Old cherries variety.

4 Brussel - Bruxelles

4.1 Institute of Agricultural Economics

21, Avenue du Boulevard
1000 Brussels
Phone: 02-2117622
Dir.: Ir. A. Villers

The Institute is attached to the Administration of Agricultural Research, Ministry of Agriculture.
The activities are focused mainly on statistics, market analyses and rural sociological investigations (Department of Macro-Economics), on farm costs and returns, farm management (Department of Micro-Economics).
The research includes market studies for the horticultural sectors and financial results of horticultural holdings.

Department of Macro-Economics
Head: Ir. G. Pevenage
Dr. L. Muermans/Lic. L. Nicolaus

Department of Micro-Economics
Head: Ir. R. Goffinet
Ir. A. Kempenaers/Ir. D. van Lierde

4.2 Université Libre de Bruxelles
(Free Brussels University)

Av. Paul Héger, 28
1050 Bruxelles
Phone: 32-26490030

Plant Physiology Dept.
Diversification of Agricultural production by introducing vegetables as primary and secundary cultivation (physiological and phytotechnical aspects)
Conception and building of a pilot-unit for industrial production of Pleurotus
Conception of pilot-units for experimentation in hydroponics

Prof. R. Lannoye
J. Crabbé*/Ir. J.P. Delhaye
L. Perez-Aranda

5 Cerexhe-Heuseux

5.1 "Profruit"

Rue des Pépinières, 41
4632 Cerexhe-Heuseux
Phone: 32-44771270
Dir.: Ir. A. Hallet

Research on industrial fruits; experimentation of new clones of "table" fruits; new variety trials; experimentation of cherry dwarf rootstocks

6 Gembloux

6.1 Station des Cultures Fruitières et Maraîchères
(Research Station for Fruits and Vegetables)

Chaussée de Charleroi, 234
5800 Gembloux
Phone: 32-81612935
Dir.: Ir. A. Monin

Fruit Culture Department — Ir. A. Monin
Dwarfing cherry trees — Ir. R. Trefois
Training of sour cherry trees
Pruning of sour cherry trees
Rootstocks and intermediate graft of appletrees — Ir. A. Monin
Clones of industrial plums
Apple tree varieties
Peartree varieties
Training of apple and pear trees planted in high densities
Mechanical harvesting of industrial stone fruits
Vegetable Culture Department
Breeding of new strawberry varieties — Ir. R. Linden
Breeding of greenhouse tomato — Ir. W. Plumier
Containers cultures and artificial substrates
Breeding of peas and beans
Breeding of muskmelon "Epritel"
Improvement methods applied to Brussels Chicory — Ir. W. Plumier/R. Valette
(in cooperation with UCL Ir. B.P. Louant)
Tissue culture — Ir. P. Boxus*
Behaviour of micropropagation strawberry plants
Early production of strawberry runners

6.2 Station de Phytopathologie
(Phytopathology Research Station)

Av. Maréchal Juin, 13
5800 Gembloux
Phone: 32-81612099
Dir.: Ir. G. Parmentier

Mycology
Fruit trees old varieties — Ir. C. Populer/Ir. C. Delmotte
Toxine resistant tissue culture — Dr.Ir. A. Dutrecq
Virology
Hop — Ir. G. Krug
Fruit trees, cereals — Ir. C. Maroquin

6.3 Station de Phytopharmacie
(Research Station for Phytopharmacy)

Rue du Bordia, 11, 5800 Gembloux
Phone: 32-81612971
Dir.: Ir. L. Detroux

Analytical methods (Laboratory of analytic chemistry) — Ir. J-C. van Damme
Laboratory of physico-chemistry — Ir. J. Henriet

Laboratory of insecticides	Ir. E. Seutin
Laboratory of fungicides	Ir. P. Meeus
Laboratory of herbicides	Ir. J-F. Salembier
6.4 Station de Zoologie Appliquée (Research Station for applied Zoology)	Chemin de Liroux, 8, 5800 Gembloux Phone: 32-81611104 Dir.: Ir. J. Bernard
Bioecology of Starling	Ir. S. Tahon
Protection of cultivation (Starling)	Ir. J-M. Hoyoux
Arthropodes of apple orchards	Ir. C. Fassotte
6.5 Comité pour l'Etude des Cultures Maraîchères (Committee for Study of Vegetable Culture) IRSIA, Res. Stat. Fruit and Vegetables and Dpt Horticult. Fac. of Agronomy)	Phone: 32-81612958 Dir.: Dr. A. Nisen/Ir. A. Monin
Brussels Chicory	Ir. R. Valette
Energy saving in cultivation of strawberry, tomato (Research conducted in Phytotron)	Ir. M-Th. Ferauge/Ing. J-P. Smal Av. Faculté d'Agron., 2 5800 Gembloux
6.6 Groupe de Travail pour l'Etude du Cerisier (Working group for Cherry study)	Chaussée de Charleroi, 234 5800 Gembloux Dir. A. Monin
Dwarfing cherry trees Tissue culture of woody plants	Ir. P. Druart*
6.7 Centre d'Etude des Pesticides Agricoles (Centre of Pesticide Research in Agriculture)	Rue du Bordia, 11, 5800 Gembloux Dir.: Ir. L. Detroux
Experimental methods in insecticides, fungicides, herbicides and other phytopharmaceutical products	Ir. W. Haquenne
Analytical methods of residues (pesticides)	Ir. M. Galoux
6.8 Faculté des Sciences Agronomiques (Faculty of (Faculty of Agronomy)	Phone: 32-81612958
6.8.1 Chaire des Cûltures Fruitières et Maraîchères*, (Fruit and Vegetable Department)	Head: Prof.Dr.Ir. A. Nisen
Growth of vegetables Greenhouse construction Natural and artificial illumination of plants Plastics in protected cultivation Greenhouse covering and shading materials	Dr. A. Nisen
Growth study in apple trees	Ir. H. Magein
6.8.2 Chaire des Cultures Ornamentales*, (Ornamental Plants Department)	Head: Prof.Dr.Ir. G. Neuray
Landscape study Salinity in ornamental cultivation Physiology of growth of Cyclamen	Ir. G. Neuray
Purification and use of residual waters	Ir. P. Vandamme
Viral purification in tissue culture	Dr.Ir. A. Toussaint
6.8.3 Chaire de Phytopathologie, (Phytopathology Dept.)	Prof.Dr.Ir. J. Semal
Selection of resistant genotypes	Dr.Ir. P. Lepoivre
Peas protection	

Virus diseases of horticultural plants	Dr.Ir. J. Vanderveke
6.8.4 Chaire de Phytopharmacie, (Phytopharmacy Department)	Prof.Dr.Ir. J. Fraselle
Coating of seeds	Ir. B. Schiffers
6.8.5 Comité pour l'Etude des Economies d'Energie en	Dir.: Prof.Dr. J. Deltour
Culture Protégée (IRSIA), (Committee for study	Prof.Dr.Ir. A. Nisen
of energy saving in protected cultivation)	
Horticultural and Physics Department	
Luminous and thermical balances of greenhouses	Ir. J. Nijskens
Radiometric properties of materials for greenhouse covering, shading	Ir. S. Coutisse
6.8.6 Comité d'Etudes de la Réproduction Végétale (Committee for study of plant reproduction) (IRSIA)	Dir.: Prof.Dr.Ir. J. Crabbé*
Intensification of fruit cultivation	Ir. C. Papeians
Behaviour of fruit cultivation	
Multiplication of woody ornamental plants	
6.8.7 Centre de Lutte Intégrée en Phytopathologie (Centre for integrated control in plant pathology)	Dir.: Prof.Dr.Ir. J. Semal
Section 1. (CLIP IRSIA)	Ir. M. Meulemans
Study of host-parasite relationships	Ir. J. Viseur
Somaclonal variations in plant regenerated in vitro from calluses, protoplasts and meritems.	
Selection of potato, wheat and pear variants for resistance to various pathogens	
6.8.8 Centre de Phytovirologie (Centre for Plant virology)	Dir.: Prof.Dr.Ir. J. Semal
Section 3 (IRSIA)	Dr.Ir. J. Kummert
Diagnosis of plant viruses and their control: use of antiviral substances.	
Viruses of fruit trees	
Resistance of potato somaclonal variants to viruses	
Study and control of fungal transmitted viruses.	
6.8.9 Comité de Recherche pour l'Utilisation des Pesticides en Agriculture (CRUPA-IRSIA) (Committee for pesticides utilization in Agriculture)	Dir.: Prof.Dr.Ir. P. Nangniot
Residues control and analyse for pesticides in vegetable cultivation	
6.8.10 Comité de Recherche pour l'Amélioration des	Dir.: Prof.Dr.Ir. J. Semal
Techniques de Traitements Phytosanitaires (IRSIA)	Prof. J. Fraselle
(Committee for Research in the improvement of the techniques of crop protection treatments)	Prof.Dr.Ir. S. Dautrebande
Caracterization and degradation of pesticides deposits	Ir. R. Caussin
Improvement of crop protection techniques and machinery in agriculture and horticulture	
6.8.11 Comité d'Application des Méthodes Isotopiques (CAMIRA-IRSIA) (Committee for application of isotopic methods to agricultural Researches)	Dir.: Prof.Dr. J. Deltour
Translocation of nutritive elements in vegetable cult.	Ir. P. Dreze/Ir. M.C. Gasia
Migration of labelled pesticides and metabolics in soil	
Injection of nutritive solutions into the trunk of trees	
Improvement of methods of pulverisation	
6.8.12 Bureau de Biométrie (IRSIA) (Centre of Biometry)	Dir.: Prof.Dr.Ir. P. Dagnelie
Statistical and informatical treatment of datas	Ir. P. Ramelot/Ir. K. In
from researches in agriculture and horticulture	Ir. G. Carletti
6.9 Industrial Institute	Rue Verlaine, 7, 5800 Gembloux Phone: 32-81611931 Dir.: Ir. J. Gaspard

Horticulture Unit

In vitro multiplication of different ornamental species — Ing. J. Roggemans

Pedagogical researches on in vitro technics

Demonstration Centre for fruit, vegetable and ornamental cultivation — Ing. J. Roggemans/J. Gillet

Study Centre for recuperation of residual energy — Dir.: Ir. A. Sansdrap

Vegetables, strawberry, eggplant, tomato — Ir. A. Sansdrap

Mushrooms, optimal growth conditions — Ir. M. Clignez

6.10 Association pour la Promotion de l'Horticulture en Wallonie (APHW - TCT)

Av. de la Faculté, 2
5800 Gembloux
Phone: 32-81615474

Draw-up of an inventory of horticulturists and research stations — Dir.: Remy-Paquay

Prof. G. Neuray

New markets for horticultural products — Ir. V. Simon

7 Gent

7.1 Laboratorium voor Tuinbouwplantenteelt* (Laboratory for Horticultural Sciences)
Faculty of Agricultural Sciences of the State University of Ghent)
elevation: 11 m
rainfall : 725 mm
sunshine : 1480 h

Coupure Links 653, B-9000 Gent
Phone: 091-236961
Telex: 12.754 RUGENT
Dir.: Prof.Dr.Ir. G. Boesman*

Growth and flowering of ornamentals and vegetables; propagation; tissue culture, growth regulators, nutrition; domestication of edible mushrooms

7.2 Centrum voor de Studie van de Sierplantenteelt - Sectie V (Centre for Study of Ornamental Plants - section V) I.W.O.N.L. - I.R.S.I.A.

Dir.: Prof.Dr.Ir. G. Boesman*

Growth and flowering and nutrition of selected potplants; outdoor plants for cutflowers

7.3 Centrum voor de Studie van de Voortplanting bij Tuinbouwgewassen (Centre for Study of Propagation of Horticultural Plants) I.W.O.N.L. - I.R.S.I.A.

Dir.: Prof.Dr.Ir. G. Boesman*
Dir.: Dr.Ir. P. Debergh

Tissue culture of horticultural crops: micropropagation, industrialization, fundamental aspects

7.4 Sierteeltonderzoek, Leerstoel voor Bodemfysica (I.W.O.N.L.), Horticultural Research Unit (I.R.S.I.A.), Dependence of the I.R.S.I.A.
This department, on the campus of the University of Ghent, faculty of agriculture, is the main centre of soil physics research in Belgium. This means also activities in soil conditioning and horticultural soil science.

c/o State University of Ghent
Faculty of Agricultural Sciences
Department of Soil physics, soil conditioning and Horticultural Soil Science
Head: Prof.Dr.Ir. M. de Boodt

Valorization of solid waste materials in horticulture and agriculture as soil conditioners — Ir. R. Penninck

7.5 Centrum voor de Studie van de Sierplantenteelt (Centre for Study of Ornamental Plant Breeding - IRSIA - IWONL)	Dir.: Prof.Dr.Ir. A. Gillard Dir.: Prof.Dr.Ir. C. Pelerents
Entomology, acarology, nematology and applied soil zoology in ornamental plants culture	Dr.Ir. A. Heungens
7.6 Leerstoel voor Dierkunde (Chair of Zoology), Faculty of Agricultural Sciences of the State University of Ghent	head: Prof.Dr.Ir. A. Gillard Assoc.: Prof.Dr.Ir. C. Pelerents
Applied zoology, entomology and nematology	Dr.Ir. A. De Grisse
7.7 Laboratoria of Phytopathology and Phytovirology elevation: 6,5 m rainfall : 820 mm sunshine : 1588 h	Gent-University Faculty of Agronomy Dir.: Prof.Dr.Ir. W. Welvaert

These laboratoria are incorporated into our Chair of Plant Diseases. The chair itself has didactic specialization on all the disciplines in plant pathology. The laboratoria are divided into Phytovirology and general Phytopathology with classic lab-equipment and special experimental greenhouses for virus and phytomycoses, with research work on plant sanitation and biological control.

Section Phytovirology	Head: Prof.Dr.Ir. W. Welvaert
Virus detection and plant sanitation	Ir. G. Samyn
Viroid characterisation	Ir. M. van Labeke
Obtaining virus-free nuclear stock	Ing. E. van Wymersch
Soil and nematode transmitted viruses	Ing. A. de Vleeschauwer
Section Phytomycology	Head: Prof.Dr.Ir. W. Welvaert
Artificial fructification of Hymenomycetes	Dr.Ir. J. Poppe
Mycorrhiza for growth optimalisation	Ir. G. De Prest
Fungal diseases on ornamental plants	Ir. G. De Prest

7.8 Research Centre for Ecological Chemistry, Laboratory for Analytical and Agrochemistry, State University of Ghent	Dir.: Prof.Dr.Ir. A. Cottenie
Soil fertility, soil-plant inter-relationships, crop quality and pollution problems	Prof.Dr.Ir. A. Cottenie Dr.Ir. M. Verloo/Ir. G. Willaert Ir. F. De Spiegeleer
soil fertility, soil-plant inter-relationships, crop quality and pollution problems	Prof.Dr.Ir. A. Cottenie Dr.Ir. M. Verloo/ Ir. G. Willaert Ir. F. De Spiegeleer
7.9 Centrum voor Onkruidonderzoek (Centre for Weed Research)	Dir.: Prof.Dr.Ir. J.M.T. Stryckers
Funds made available by the IRSIA. Weed research in horticultural crops: vegetables, fruits, ornamental crops, medicinal and aromatic crops, forest nurseries, lawns; in agronomic crops and grassland. Weed problems in the maintenance of waterways. Weed control on non-cropped land.	Ir. M. van Himme/ Dr.Ir. R. Bulcke
7.10 Seminarie voor Marktkunde van Land- en Tuinbouwprodukten (Department Marketing of Agricultural and Horticultural Products)	Dr.Ir. J. Viaene

7.11 Seminarie voor Landbouweconomie (Department of Agricultural Economics)	Dir.: Prof.Dr.Ir. L. Martens Ir. L. Van Huylebroek
7.12 Laboratorium voor Plantenfysiologie (Laboratory for Plant Physiology) Rijksuniversiteit Gent, Faculty of Sciences	Ledeganckstraat 35, B-9000 Gent Phone: 22.78.21 Dir.:-
Photocontrol of light-requiring seed germination in relation to phytochrome and interaction with plant growth substances; energy metabolism in imbibing and germination seeds (In collaboration with Prof.Dr. J. De Greef, U.I.A.)	Prof.Dr. H. Fredericq Dr. R. Rethy/ Dr. A. Dedonder E. de Petter Ind.Ir. L. van Wiemeersch
Physiological aspects of phyto-bacterioses	Dr. M. de Cleene, in collab. with Lab. Microbiol. and Microbio. Gen. Dir.: Prof.Dr. J. de Ley

8 La Hulpe

8.1 Station Provinciale de Recherches et d'Analyses appliquées en Agriculture et Horticulture (Agricultural and Horticultural Applied Research Station)	Rue St Nicolas, 17, 1310 La Hulpe Phone: 32-26534097 Dir.: Ir. LeComte
Chemical soil analysis Water for plants Physical soil analysis Agrotechnic for ornamentals (open air and protected cultivation)	Ir. R. Delmotte
Researches on pests (nematodes)	Ir. A. Descamps

9 Leuven

9.1 Katholieke Universiteit Leuven (Catholic University Leuven)	Kardinaal Mercierlaan 92 3030 Leuven (Heverlee) Phone: 00-16220931
Laboratory of Phytopathology and Plant Protection	Head: Prof. C. van Assche
Phytopathological and phytiatrical aspects of biotic and abiotic entities on vegetables	Prof. C. van Assche
Epidemiology and control of nematodes on vegetables, fruit and cutflowers	Dr. J. Coosemans
Chemical and physical alternatives for crop growing substrate disinfestation	Ir. E. van Wambeke
Research on chemical soil disinfestation	Ing. A. Vanachter*
Laboratory of Soil Fertility and Soil Biology	Head: Prof.Dr. K. Vlassak
Nitrates in vegetables	Ir. J. Delvaux
Laboratory of Soil and Water Engineering	Head: Prof.Dr. J. Feyen
Development of a microcomputer controlled unit for the automation of the irrigation timing of greenhouse crops	Prof.Dr. J. Feyen/ Ing. D. Crabbé
Laboratory of Food Technology	Head: Prof.Dr.Ir. P. Tobback
Technology of fruit and vegetable preservation	Lic. C. Cardinaels
New product development (fruit, vegetables)	Ir. M. Hendrickx
Cold- and frost resistance of vegetables	Mrs. J. van Cutsem
9.2 Bodemkundige Dienst van België (Belgian Soil Service)	W. de Croylaan 48, 3030 Heverlee Dir.: Dr.Ir. M. Geypens
Chemical soil analysis	Ir. J. de Venter

Plant and feed analysis	
Advisory service on soil-fertility	Ir. J. Wauters
Fertilisation advices for horticultural crops	Ir. E. Pasture
Studies on soil fertility, fertilisation and plant responses	Ir. R. Boon
Study of soil use	Dr. W. Boon
Feasibility studies on soil improvement techniques	
Plant protection. Implementation of EPIPRE in winterwheat	Dr.Ir. M. Geypens

10 Libramont

10.1 Station de Haute Belgique	Rue de Serpont, 48, 6600 Libramont Phone: 061-223721 Dir.: Ir. L. Nys
Chemistry - Technology: Storage conditions, nutrition, food quality, forage and industrial quality	Ir. R. Biston
Rural engineering information	Ir. P. Dardenne
Developed cereal sector	
Quantitative analysis of proteins by infra-red	
Phytotechny	Ir. L. Couvreur
Potato - General problems, production of elite plants	Ir. G. Fouarge
Virology and pathology	
Micropropagation	

11 Liège

11.1 Centre de Physiologie Végétale Appliquée (Centre of Applied Plant Physiology), IRSIA	Université de Liège Dept. Botanique B. 22 Sart Tilman 4000 Liège Dir.: Prof. G. Bernier
Flowering of leek, tomato and strawberry	Dr. J-M. Kinet*
Curd initiation and maturation of cauliflower	Ir. A. Parmentier
Fruit set in tomato	
Flowering of Azalea, Weigelia and other ornamental woody plants	Dr. M. Bodson*

12 Louvain-la-Neuve

12.1 Faculté des Sciences Agronomiques Université Catholique de Louvain (Catholic University of Louvain, Faculty of Agricultural Sciences).	Place Croix du Sud, 3 1348 Louvain-la-Neuve Phone: 32-10433746
Horticultural Department	Prof. M. Verhoyen
Unite de Phytopathologie (Phytopathology Unit)	
Effects on pesticides on horticultural species	Prof. J. Meyer/Ir. J. Rouchaud
Viruses and bacteria of horticultural species	Prof. M. Verhoyen/Ir. G. Legrand
(Vegetables and ornamentals)	Ir. S. Meunier/ Ir. M. Laroche
Unité de Phytotechnie Tropicale (Unit for tropical cultures)	
Selection and breeding *Cichorium intybus*	Prof. B. Louant/Ir. B. Longly
Unité d'Ecologie des Prairies (Unit for ecology of meadows)	
Soil analysis in Horticulture	Prof. J. Lambert
Ecology in Horticulture	Prof. J. Ledent

Unité des Eaux et Forêts
(Forestry Unit) — Prof. P. André
Micropropagation of conifers — Ir. De Cannière
Unité de Physiologie Biochimique — Prof. M. Briquet
Study of cytoplasmic heredity — Ir. J.P. Goblet

12.2 Botany (Faculté des Sciences)
Universite Catholique de Louvain

Place de la Croix du Sud, 5
1348 Louvain la Neuve
Phone: 32-10433935
Prof. J. Bouharmont

Unité de Génétique
(Unit of Genetics)
Tissue culture and micropropagation of
Azalea, Ananas, Fuchsia — Ir. P. Dabin

13 Meise

13.1 Jardin Botanique National de Belgique/
Nationale Plantentuin van België
(National Botanical Garden of Belgium)
(Ministry of Agriculture, Department of
Agricultural Research)
elevation: 40 m
rainfall : 780 mm
sunshine : 1555 h

Domein vanBouchout
B-1860 Meise
Phone: 02-2693905
Dir.: Dr. E. Petit

Living Collections Section — Curator: Ir. E. Lammers

This Section is engaged in the conservation, systematics, phenology and ecology of 12.000 species and varieties of tropical and subtropical plants (greenhouses: 1.35 ha) and 8.500 species and varieties of herbacious and woody plants in the open (93 ha).

Determination and systematics — Mrs. F. Billiet
Micropropagation and plant patholgy — Dr. Ir. J. van Waes
Gene bank activities — Ir. L. van der Veken
Systematics, phenology and ecology of woody plants — Ir. de Meyere
Management of the greenhouse collections — Ing. J. Heylen/Miss Ing. V. Leyman
Management of the collections in the open — Ing. V. de Wandeleer
Technical assistance and automatisation — Ing. G. de Bont

14 Melle

14.1 Rijksstation voor Sierplantenteelt
National Institute of Ornamental Plant
Growing
elevation: 25 m
rainfall : 723,7 mm
sunshine : 1470,8 h

Caritasstraat 21
9230 Melle
Phone: 091-521052
Dir.: Prof.Ir. J.G. van Onsem

00 32 9 2722900

Government institution depending on the Department of Agricultural Research of the Ministry of Agriculure. Breeding work and improvement of growing techniques of economic important ornamental crops."In vitro"-culture of foliage plants, bromelias, ornamental trees and shrubs. Importation of botanical species and varieties to be evaluated on their ornamental value. Glasshouses: 10.000 m^2, experimental plots: 12 ha A gene pool of *Begonia, Azalea (Rhododendron simsii Planch.), Bromeliaceae* and hothouse plants is at the disposal of research workers. Equipment: 2 atomic absorption spectrophotometers, 3 growth chambers, chromatographic facilities (HPLC), a CO^{60} source and X-ray installation, computers.

Bulbous and tuberous plants: tuberous begonia hybrids — Dr. Ir. J. Haegeman

Ericaceae: *Azalea indica (Rhododendron simsii Planch.), Rhododendron* — Dr. Ir. J. Heursel

Hothouse and foiliage plants — Ir. O. Mekers

Bromeliaceae, "in vitro" culture — Ing. F. Thomas

Roses, ornamental trees and shrubs, and cut flowers — Ir. I. Meneve

Plant nutrition, irrigation water quality, substrates, ecophysiology — Ing. W. Istas

Mutation breeding, irradiation, biochemical characterization of cultivars (IRSIA) — Dr.Ir. R. Gabriels/ Ing. H. Engels

Working group on ornamental plant growing (subgroup-pedology) — Ing. W. van Keirsbulck

Soil suitability for ornamental plants — Dr. Ir. R. de Loose

Horticultural substrates

Composting of organic waste materials

Dr. Ir. O. Verdonck*
Coupure Links 653
B-9000-Gent
Phone: 091-236961-ext. 501

15 Merelbeke

15.1 Rijksstation voor Landbouwtechniek*
(National Institute of Agricultural Engineering)
rainfall : 723 mm
sunshine : 1470 h

B. van Gansberghelaan 115
B-9220 Merelbeke, Belgium
Phone: 091-521821
Dir.: Dr.h.c. Dr. Ir. A. Maton

The unit belongs to the government (Ministry of Agriculture).

Dept. of Horticultural Engineering:

Section for Horticultural Technics — Head: Ir. W. Taveirne

Construction

Heating and automation in glasshouses

Section for Work Organization in Horticulture — Head: Ir. J. Lips

Mainly 'witloof'

Study on horticultural crops (a.o. Chrysanthemums, tuberous Begonias)

15.2 Rijksstation voor Nematologie en Entomologie
(Government Research Station for Nematology and Entomology)

B. Van Gansberghelaan 96
9220 Merelbeke
Dir.: Dr. Ir. O. van Laere

Government institution depending on the Department of Agricultural Research of the Ministry of Agriculture

Research on plantparasitic nematodes and insects in agriculture and horticulture

Research on problems in apiculture and bee-plant relationship

Three sections: phytonematology, entomology, and apicultural research

Dr. Ir. R. de Clercq
Dr. W.A. Coolen/Ir. L. de Wael

15.3 Rijksstation voor Plantenziekten (Government Research Station for Phytopathology)

B. Van Gansberghelaan 96
9220 Merelbeke
Phone: 091-522083
Dir.: Ir. R. Veldeman

Goverment institution depending on the Department of Agricultural Research of the Ministry of Agriculture. The study and research of the most important damages on agricultural and horticultural plants on poplar and ornamental trees.

Mycology — Dr.Ir. R. Veldeman
Physiology of fungi-biological control — Dr.Ir. O. Kamoen
Bacterial diseases — Ir. P. Bosman/ Dr.Ir. J. Geenen
Ir. J. van Vaerenbergh

16 Ormeignies

16.1 Centre d'Essai Horticole du Hainaut
(Horticultural Research Centre)

Chemin des Serres, 14
8605 Ormeignies
Phone: 32-68221160

Ornamental Nursery - specific ornamental culture: chrysanthemum, pelargonium

Ing. A. Jaivenois

17 Rumbeke

17.1 Provinciaal Onderzoek- en Voorlichtingscentrum voor Land- en Tuinbouw (Provincial Research and Advisory Centre for Agriculture and Horticulture)
elevation: 10 m
rainfall : 700-720 mm

Ieperweg 87, Beitem
B-8810 Roeselare (Rumbeke)
Phone: 051-203218/ 203219
209831/ 209832
Dir.: Ir. L. Bockstaele

Provincial institute
Research on cultural techniques on commercial crops, vegetables (outdoor and protected crops), mushrooms, witloof (chicory), strawberries

Ir. K. Maddens
Ir. G. Vulsteke*/ Ir. J. Derolez
Ir. G. Ampe/ Ir. R. Sarrazyn
Ir. L. Vanparys/ Ir. P. Bleyaert
techn.ing.: A. Overstyns

Laboratories for quality control on vegetables. Research Station for mushrooms. Trial garden for fruit and alternative vegetables.

18 Sint-Katelijne-Waver (Mechelen)

18.1 Vegetable Research Station
elevation: 9 m
rainfall : 723 mm
sunshine : 1417 h

Binnenweg 6, B-2580 Sint-Katelijne-Waver
Phone: 015-290592
Head: F. Benoit* (Senior Researcher Horticulturist, Biologist, Research Coordinator)

Private institute supported by I.R.S.I.A. Experimental fields: 6 ha; glasshouses: 7.500 m^2; plastic green-houses: 2.500 m^2. The research is coordinated of 20 institutes belonging to universities and industries.

Research on physiological, thermoperiodical and phenological problems of the most cultivated vegetables in open air, under plastic protection and in glasshouses (soilless culture: NFT, rockwool, etc.)
Advisory service: assistance to practical problems of farmers; field experiments in various ecological sites; energy saving problems

Research Horticulturist:
N. Ceustermans/W. Adriaensens
K. Glorie/A. van den Eyden
F. van Linden

19 Sint-Truiden

19.1 Opzoekingsstation van Gorsem VZW

Brede Akker 3, B-3700 Sint-Truiden

(Research Station of Gorsem)
elevation: 60 m
rainfall : 720 mm
sunshine : 1500 h

Phone: 011-682019
Telex: Opzogo 39719
Dir.: Ir. C. Verheyden

The Research Station is a non-profit association including representatives from industry, horticultural associations and public authorities. It is in fact an official institution under the aegis of the IRSIA (Institute for the Encouragement of Scientific Research in Agriculture and in Industry). Spezialized in Phytopathology, Physiology, Entomology, Plant-Management, Virology in hard and soft fruit.

Entomology
Integrated pest control, resistance to insecticides, biology of pests and predators, warnings phenology, orchard mites — G. Vanwetswinkel/E. Paternotte

Phytopathology
Mycorrhizae, modellisation in parasitology, resistance problems, mycology, inoculation of fungus, fruitrots, curative control of scab, side-effects of fungicides — Ir. P. Creemers, J. Bollen-Vandergeten
Diseases and pests in strawberry culture, soil-fungi — Ir. G. Gilles*/Ing. E. Bal

Plant management
Pruning, pollinisation, behaviour of new varieties and rootstocks, nucellus research — Ir. W. Porreye*/Ir. T. Deckers
Fire Blight, storage advices on behalf of leaf and fruit analysis

Physiology
Fruit ripening, post-harvest physiology, physiology of photosynthesis, growth regulation, water relations, influence of pruning and bending on physiology, bureau for fruit and leaf analysis — Dr. R. Marcelle*/Dr. P. Simon, Ing. W. Brugmans

Virology
Inventory of virus in fruit growing, sanitation and thermotherapy, control of sanitated material and selection of pomologic value — Ir. G. Gilles*/Ing. H. Bormans

20 Tervuren

20.1 Institute for Chemical Research

Museumlaan 5, 1980 Tervuren
Phone: 02-7675301
Dir.: J. Istas

Study and investigations related to damages on vegetables, fruit, ornamental plants, crops and vegetation in general, resulting from atmospheric pollutions. Heavy metals in vegetables and soils. Desinfection of soil (methylbromide); influence on the environment. Use of bio-accumulators and bio-indicators to control the pollution around industries and in the rural environment.

21 Vilvoorde

21.1 Research Committee for Storage of Horticultural Products (VCTV-IWONL), Hoger Rijksinstituut voor Tuinbouw (Institute of Horticulture)

De Bavaylei 116, 1800 Vilvoorde

Post-harvest problems of horticultural products; storage of fruits and vegetables — Ir. M. Herregods/R. Vanderwaeren, Ir. G. Goffings
Handling and packaging of fruits and vegetables
Fruit ripening and fruit quality

22 Wetteren

22.1 Proefstation voor de Tuinbouw*
(Horticultural Experiment Station)
elevation: 17 m
rainfall : 760 mm
sunshine : 1580 h

Stookte 1, 9200 Wetteren
Phone: 091-690427
Dir.: E. Stautemas

Private Institution. Subsidized by the IRSIA, Brussels, the Provincial Chamber of Agriculture, the Belgische Boerenbond (Belgian Farmers Union), the Ministry of Agriculture, various municipalities and societies. Research on cultural techniques of ornamental plants.

Physiology: Nurseries, seed research, cutflowers, perennials	Ir. R. Blomme
Potplants and *Ericaceae:* Physiology	Ir. E. Beel
Physiology of *Azalea, Rhododendron sp., Rhod. japonicum* Vegetative reproduction, cutting, grafting, pinching Substratum research, fertilisers, flowering	Ing. G. Piens
Physiology of potplants: reproduction substrata fertilizers, flowering, energy saving	A. Schelstraete
Physiology: Nurseries. Vegetative reproduction, cutting, grafting, fertilizers. Use of glasshouses in nurseries, Container culture	J. van Wezer
Physiology: seed research: forestry, ornamentals	Ing. L. Degeyter
Physiology: Cutflowers, perennials: reproduction, substrata, fertilizers	Ing. P. Dambre

BRAZIL

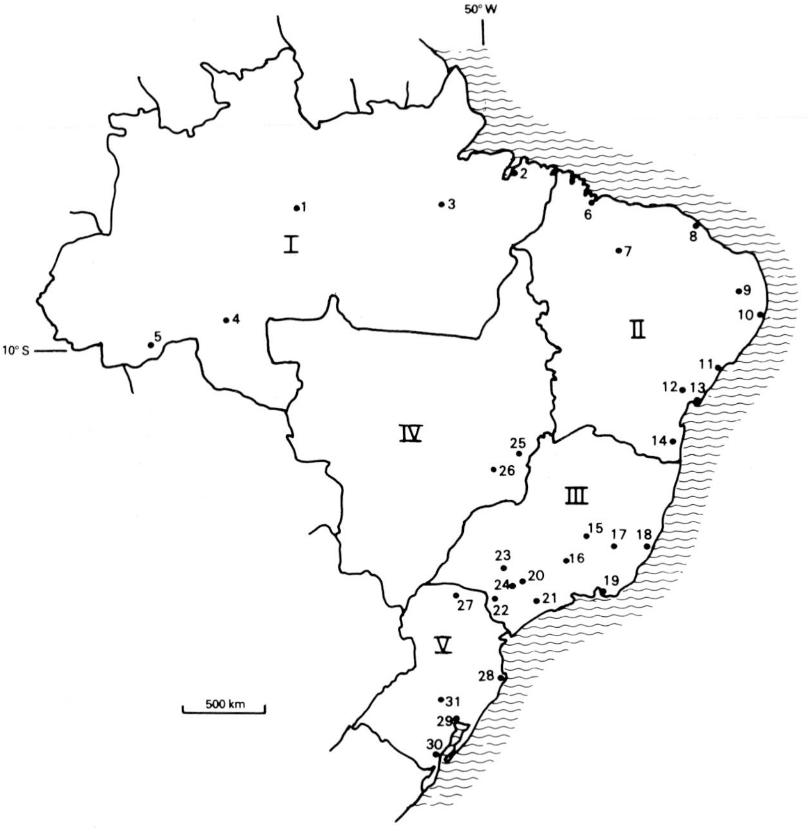
50° W
10° S
500 km
I
II
III
IV
V
1
2
3
4
5
6
7
8
9
10
11
12
13
14
15
16
17
18
19
20
21
22
23
24
25
26
27
28
29
30
31

I North Region, **Amazon**: 1 Manaus, **Pará**: 2 Belém, 3 Altamira, **Rondônia**: 4 Porto Velho, **Acre**: 5 Rio Branco

II Northeast Region, **Maranhão**: 6 São Luis, **Piauí**: 7 Teresina, **Ceará**: 8 Fortaleza, **Paraíba**: 9 Alagoinha, **Pernambuco**: 10 Recife, **Sergipe**: 11 Aracaju, **Bahia**: 12 Cruz das Almas, 13 Salvador, 14 Itabuna

III Southeast Region, **Minas Gerais**: 15 Belo Horizonte, 16 Lavras, 17 Viçosa, **Espírito Santo**: 18 Vitória, **Rio de Janeiro**: 19 Niterói, **São Paulo**: 20 Campinas, 21 São Paulo City, 22 Botucatu, 23 Jaboticabal, 24 Piracicaba

IV Centrewest Region, **Federal District**: 25 Brasilia, **Goiás**: 26 Goiânia

V South Region, **Parana**: 27 Londrina, **Santa Catarina**: 28 Florianópolis, **Rio Grande do Sul**: 29 Porto Alegre, 30 Pelotas, 31 Bento Gonçalves

Survey

Horticultural research in Brazil is sponsored and carried out by Federal Government, and State Research Institutions including Universities. Federal Institutions belong mostly to the Brazilian Enterprise for Agricultural Research (EMBRAPA) which is under the Ministry of Agriculture; all EMBRAPA's unities will be referred here as (E). Other Federal Institutions doing research in Horticulture are: Cocoa Research Centre (CEPEC), Ministry of Agriculture, Brazilian Coffee Institute (IBC) which is a part of the Ministry of Industry and Commerce because of export implications. The State' Institutions are mainly Agronomic Research Institutes, State Enterprises for Agricultural Research and Universities. The establishment of EMBRAPA in 1973 has changed completely the organization of Agricultural research in Brazil; besides a national network of 14 research centers specialized on commodities, there are also five natural resources centers, a Soil Service, a Basic Seed Production Service, an Agricultural Pesticide Center, a Genetic Resources Center and a Food Technology Center. EMBRAPA's programme is well coordinated with State's research institutions and Universities funding a considerable number of projects. Reference

Prepared by Dr. F.R. Ferreira, CENARGEN, Brasilia.

should be made also to the National Council for the Development of Science and Technology (CNPq) which supports financially several special projects on Horticultural research.
The information on Horticultural research in Brazil will be presented in a Region basis with respective States where horticultural programmes or projects are operational:
I. North Region: States of Amazon (AM), Acre (AC), Pará (PA), Rondônia (RO) and Amapá (AP) Federal Territory Roraima (RR).
II. Northeast Region: States of Maranhão (Ma), Piaui (PI), Ceará (CE), Rio Grande do Norte (RN), Paraíba (PB), Pernambuco (PE), Alagoas (AL), Sergipe (SE) and Bahia (BA).
III. Southeast Region: Minas Gerais (MG), Espírito Santo (ES), Rio de Janeiro (RJ) and São Paulo (SP).
IV. Centrewest Region: Distrito Federal (DF), Goiás (GO), Mato Grosso (MT) and Mato Grosso do Sul (MS).
V. South Region: Paraná (PR), Santa Catarina (SC) and Rio Grande do Sul (RS).

I. North Region - The North Region is well known as the Amazonian: total area is 3.578.500 km^2 which corresponds to 42% of Brazil's territory; population, seven million. Most of the region is still covered by tropical rain forest. Soils are predominantly tropical red and yellow latosols with alluvial valley along major rivers. There are several native species of horticultural importance in the Amazonian Region; a good number of them are under evaluation; special reference should be made to species of *Hevea, Theobroma, Bertholetia, Manihot, Annona, Ananas, Bromelia, Platonia, Rheedia, Lecythis, Eugenia, Elaeis, Euterpe, Bactrys, Mauritia, Oenocarpus, Astrocaryum, Passiflora, Genipa, Manilkara, Pouteria,* etc. The Ornamental plants should also be mentioned, mainly flowering trees, shrubs and vines, orchids, Bromeliads, *Helioconia, Marantha, Calathea,* inumerous species of *Gesneriaceae, Acanthaceae, Araceae, Piperaceae, Gengiberaceae* and ferns.
Horticultural research started in 1939 at the former Instituto Agronômico do Norte, now the Agricultural Research Center for the Humid Tropics (CPATU), (E). Present horticultural programme includes research on black pepper, Brazil nut, African palm oil, rubber tree, native fruits and palms, cocoa, cassava, guarana and cowpea. Predominant horticultural crops in the region are: African palm-oil, black pepper, cocoa and rubber tree; there are about 110 native fruit species some of them with commercial value mainly açai (*Euterpe oleraceae*), bacuri (*Platonia insignis*), Brazil nut (*Bertholetia excelsa*), mangaba (*Hancornia speciosa*), murici (*Byrsonima crassifolia*) and Taperaba (*Spondias lutea*); the main cultivated fruits are: avocado, banana, bread fruit, cashew, citrus (lime, mandarin and sweet orange), coconut, capuaçu (*Theobrama grandiflora*), guava, jack fruit, soursop, pineapple, mango, papaya, sapodilla (*Manilkara achras*); special reference should be made to the guarana (*Paullinia cupana* var. *sorbilis*) which produces a berry from which a powder is produced for a soft -drink; it is becoming a major crop because of increasing demand in Brazil and international market. Vegetables: aroids, cabbage, cassava (leaves and roots), cowpea, chayote, cucumber, egg plant, green pepper, hearts of palm, herbs, hot pepper, kale, lettuce, melon, tomato, squash, sweet potato, watermelon, yam.

Amazon

1 Manaus

1.1 Centro nacional de Pesquisa de Seringueira e Dendê (CNPSD) (E) (National Rubber and Oil Palm Research Center)
elevation : 50 m
rainfall : 2200 mm
temperature: 27.5°C

C. Postal 319, 69000 - Manaus - AM
Phone: (092) 233-5298
Telex: 043-2208
Chief: L.A. Silva Melo

Research on problems of rubber cultivation includes breeding for resistance to South American leaf blight, physiology, biochemistry, propagation, crop management, diseases control and latex processing. A gene bank of clones and wild species of *Hevea* is operational.

Hevea breeding	J.R. de Paiva
Physiology	V.H. de Moraes/H.E. Oliveira da Conceição
Soil fertility	A.V. Pereira
Phytopathology	H. Martins e Silva/L. Gasparotto
Entomology	P. Celestino
Economy	F. Mendes Rodrigues

Oil palm breeder	P. Braz Tinôco
Genebanks of rubber tree and oil palm	M. de Miranda Santos
Statistics	L.O.A. Teixeira/M.M. Santos
	A.G. Rossetti/Maria E. da C. Vasoncellos

1.2 Instituto Nacional de Pesquisa da Amazônia (INPA) (National Research Center for the Amazon)
elevation : 40 m
rainfall : 2100 mm
temperature: 28°C

Estrada do Aleixo 1756,
69000 - Manaus - AM
Phone: (092) 236-5700
Telex: 092226
Dir.: R. dos Santos Vieira

Under the CNPq. Research on natural resources of the Amazonian Region and some crops. Plant taxonomy.

Evaluation of native tropical fruits	Ch.R. Clement/G.A. Venturieri
Vegetable breeding (cucurbits, Solanaceae)	H. Noda
Leafy vegetables	W.S. de Oliveira

1.3 UEPAE Manaus (E)
Unidade de Execução de Pesquisa de Âmbito Estadual (Unity of Research Execution at State Level)

C. Postal 455, 69000 - Manaus - AM
Phone: (092) 236-2044
Telex: 0922440
Chief: E. de Moraes

Research on African oil-palm, beans, guarana, cassava, soybean, vegetable and other non-horticultural crops.

Vegetable crops (cowpea, tomato, cabbage, green pepper, squash and sweet potato)	Ana L. Carvalho Guedes
Phytopathology	E. Rose Carneiro
Entomology	J. da Silva
Guarana	C. da Silva Martins
Soil fertility	J. Braga Bastos

Para

2 Belém

2.1 Centro de Pesquisa Agropecuária do Trópico Ûmido (CPATU) (E) (Agricultural Research Center for the Humid Tropics)
elevation : 14 m
rainfall : 2930 mm
temperature: 26°C

C. Postal 49, 66000 - Belém - PA
Phone: (091) 226-1941
Telex: 0911210
Chief: E. Botelho de Andrade

Horticultural perennial crops constitute a part of the research programme; main crops are: African oil-palm, black pepper, Brazil nut, cocoa, guarana and rubber tree, gene banks of black pepper and guarana.

Gene banks of black pepper and guarana	F.C. de Albuquerque
	A.A. Müller
Brazil nut and tropical fruits	C.H. Müller
Cowpea and bean	A.F.F. de Oliveira
Guarana and gene bank	A. Kouzo Kato
Natural resources of the Amazon	I.C. Falesi/Kin-Ichi Hashimoto
Vegetable crops	S. Cheng

2.2 Museu Paraense Emílio Goeldi (Emilio Goeldi Museum of Pará State) Under INPA Programme (CNPq).	C. Postal 399, 66000 - Belém - PA Dir.: Dr. L.M. Scaff
Taxonomy of native tropical fruits from the Amazon	P.B. Cavalcante
2.3 Faculdade de Ciências Agrárias do Pará	C. Postal 917, 66000 - Belém - PA Phone: (091) 226-3493 Telex: 091-935 Dir.: A.C. Albério
Tropical fruits: taxonomy of Amazonian species, selection, propagation and growing methods	B.B.G. Calzavara
Tomato breeding for tropical conditions	T.S. Oliveira Lima
Cowpea, cucumber, okra, cabbage	A.A. Moussallem Pantoja Pimentel

3 Altamira

3.1 UEPAE/Altamira (E) elevation : 80 m rainfall : 1680 mm temperature: 26°C	Rua 1° de Janeiro 1586, C. Postal 0061, 68370 - Altamira - PA Phone: (091) 515-1085 Telex: 091-2549 Chief: R.R. Lopes Vilar

The UEPAE of Altamira has 3 experimental stations, with 100 ha each. The research includes the following products: rice, banana, coffee, citrus, cowpea, bean, guarana, maize, cassava, vegetable crops, black pepper, rubber and soyabean.

Rural economy	A.C.P.N. da Rocha/R.A. Carvalho
Cultural practices	E.J. Maklouf Carvalho J.R. Viana Corrêa Maria do Socorro Andrade Kato O. Ryohei Kato
Plant breeding	F.R.S. de Souza M. Costa Poltronieri
Phytopathology	L.S. Poltronieri

Rondônia

4 Porto Velho

4.1 UEPAE/Porto Velho (E) elevation : 98 m rainfall : 2230 mm temperature: 24°C	Av. Pinheiro Machado 2129, 78900 - Porto Velho - RO Phone: (069) 22-2751 Telex: 069-2258 Chief: Dr. J.F. Bezerra Mendonça

This experiment station works with beans, coffee, rubber tree, soybeans and fruit crops; mainly variety trials and development of production systems.

Phytopathology	C.A. Monteiro Sobral
Crop management	J.E. Rodrigues/S. de Melo Lisboa W. Veneziano

Entomology — Maria A. Santos Oliveira
Plant breeding — S. Maeda
Soil fertility — S.I. Ribeiro

Acre

5 Rio Branco

5.1 UEPAE/Rio Branco (E)
elevation : 160 m
rainfall : 2000 mm
temperature: 26°C

Rua Sergipe 216, C. Postal 392,
69900 - Rio Branco - AC
Phone: (068) 224-4035
Telex: 069-2589
Chief: V.H. de Oliveira

This experiment station works with cassava, beans, coffee and rubber; mainly variety trials and development of production systems.

Rubber (phytopathology) — A.C. Rebouças Lins
Crop management — F. das Chagas A. Paz
G. de Melo Moura
Maria U. Correia Nunes
Crop management (rice) — I. Soares Campos
Phytopathology (bean) — J.E. Cardoso
Plant breeding — J.E. de Lima Mesquita
T.S. de Oliveira Lima
Crop management (coffee) — V.H. de Oliveira

II. Northeast Region - This region has an area of 1.548.672 km^2 which is 18,20% of the country's territory and about 38 million inhabitants. Climate is mostly tropical, being humid in the coast and semi-arid in the interior. Where irrigation is practiced, conditions are adequate for the production of high quality tropical and sub-tropical fruits and vegetables. This region is a major producer of cocoa, papaya, pineapple, passion-fruit, onion, tomato, mango and cashew-nut. In the San Francisco Valley, irrigated table grape reaches high yield of good quality with two crops a year; this Valley is also a great onion and melon producer, with a recognized potential for out-of-season vegetable production and seed industry. Tomato production for processing has been a traditional horticultural crop for the last twenty years. Because of great possibility for horticultural production, plus the availability of labor, the Northeast is developing a wide research programme on fruits and vegetables.

Maranhão

6 Sao Luis

6.1 Empresa Maranhense de Pesquisa Agropecuária (EMAPA) (Maranhão Agricultural Research Enterprise)
elevation : 4 m
rainfall : 1715 mm
temperature: 26°C

Rua Henrique Leal 149, C. Postal
176, 65000 - Sao Luis - MA
Phone: (098) 221-2833
Telex: 098-2283
Pres.: C.A. dos Santos

This agricultural research enterprise, works in close ties, as all other state research enterprises, with EMBRAPA. Main horticultural crops are beans, cowpea, banana, black pepper, cassava and soybeans.

Black pepper, variety trials and production system — T. Tokokura
Entomology — E.F. das Chagas
Phytopathology — Eliane A. Silva Prazeres
Eliane Alvares P. Souza
G. Soares da Silva

Vegetable: variety trials for the tropics — G. Correa
Soil fertility — C.A. Costa Veloso/J.F. Ribeiro
J.M. Japhar Berniz/M.J.F. Porto
R. Nunes Fernandes
Crop management — F. Tien Liao
Soil science — H.F. Leite

6.2 Universidade Federal do Maranhão
Largo dos Amores 21
65000 - Sao Luis - MA
Phone: (098) 221-5433

Tropical fruits and vegetables — Dr. W.E. Kerr

Piauí

7 Teresina

7.1 UEPAE/Teresina (E)
elevation : 72 m
rainfall : 1297 mm
temperature: 27°C

Av. Duque de Caxias 5650
C. Postal 01, 64000 - Teresina - PI
Phone: (086) 225-1611
Telex: 086-2337
Chief: H. Ferrer de Almeida

Develops studies on sweet potato, tomato cultivars, tomato pests, carrot, collection of Cucurbitaceae germplasm and adaptation of vegetables under babassu palm shade.

Vegetable improvement — Sieglinde Brune
Soybean variety trials — G.J. de Azevedo Campelo
Cassava — J.N. de Azevedo
Hydrology — J.R. Cortez Bezerra
Entomology — P.H. Soares da Silva
Babassu palm gene bank — J.M. Ferro Frazão

Ceará

8 Fortaleza

8.1 Empresa de Pesquisa Agropecuária do Ceará (EPACE) (Ceara Agricultural Research Enterprise)
elevation : 20 m
rainfall : 1248 mm
temperature: 26°C

Av. Rui Barbosa 1246
60000 - Fortaleza - CE
Phone: (085) 244-4166
Telex: 085-1195
Pres.: Dr. E. Matos Cavalcante

Wide agricultural research programme mainly on irrigated crops; major horticultural crops: cowpea, beans, cassava, coffee, tropical fruits: avocado, banana, cashew nut, guava, mango, papaya, passion-fruit, soursop, citrus, peanuts and vegetable crops.

Coffee — J. Torres Filho
Banana — J. Gonçalves Barreira
Cassava — G.M de Queiroz/L. Nunes de Pinho
Cashew nut — L. de Moura Barros
J.I.L. de Almeida
Vegetable crops — J.O. de L. Muniz
E. Farias Bezerril
L.A. da Silva

Tropical fruits	G.C. de Araujo Filho J.G. Vasconcelos Lopes
Cowpea breeding and germplasm	P.D. Barreto/Maria de Fatima P. Sá
Corn breeding and germplasm	J.F. Antero Neto/A.A.T. Monteiro
Phytopathology	Lianna M. Saraiva Teixeira
Entomology	R.D. Cavalcante Quélzia M. s. Melo F.E. de Araújo Mary-Ann W. Quinderé

8.2 Centro de Ciências Agrárias, Universidade Federal do Ceará (CCA/UFC) (Science Agricultural Center, Federal University of Ceara)
elevation : 20 m
rainfall : 1248 mm
temperature: 26°C

Av. da Universidade 2853,
C. Postal 3038, 60000 - Forteleza- CE
Phone: (085) 243-3011
Telex: 085-1077
Head: Dr. J. Anchieta E. Barreto

Research and teaching on tropical fruits and vegetable crops. Lettuce, onion, cowpea, cauliflower, string-beans, melon, cucumber and tomato.

Vegetable crops	Erimá Cabral do Vale H.G. de Oliveira/J.T. Alves Costa R.F. Pinheiro Maciel
Vegetable diseases	J.J. da Ponte Filho
Cowpea breeding and germplasm	J. Braga Paiva
Tropical fruits	Erimá Cabral do Vale H.G. de Oliveira/J.T. Alves Costa R.F. Pinheiro Maciel
Pest control of vegetable crops	J.H. Ribeiro dos Santos
Tropical fruit processing	L. Holanda

Paraíba

9 Alagoinha

9.1 UEPAE/Alagoinha (E)

Rodovia PB-75, km 12
58390 - Alagoinha - PB
Chief: D.J. Vieira

Research programme on banana, beans, cassava, citrus, pineapple, potato.

Pineapple fertilizing and variety trials	S. Abreu Choairy
Pineapple Fusarium control	E.B. Lopes
Crop management	A. Soares de Melo/C.F.O. Franco L.C. Silva/R. Torres Soares

Pernambuco

10 Recife

10.1 Empresa Pernambuca de Pesquisa Agropecuária (IPA) (Pernambuco Agricultural Research Enterprise)

C. Postal 1022,
50000 - Recife - PE
Phone: (081) 227-1903

elevation : 6 m
rainfall : 2153 mm
temperature: 25°C

Telex: 081-2283
Pres.: P.E.S. Araujo

The former Instituto de Pesquisa Agronômica, now integrating the state agricultural research enterprise, is a prosperous institution with a dynamic research programme in horticultural crops mainly on banana, beans, cashew nut, cassava, coffee, peanuts, pineapple, minor tropical fruits: guava, papaya, soursop, sapodilla, mango.

Onion breeding	D. Menezes
Tomato breeding for processing	E. Ferraz
Tomato, onion and carrot, growing practices	L.J. da Gama Wanderley
Tomato and banana diseases	U. Cabus Maaze
Bean breeding	G.R. do A. Lima
Papaya breeding	I.E. Lederman
Citrus	A. Coelho Pedrosa
Tropical fruits: banana, mango and sapodilla	R.J. de Melo de Moura
	A. Tiburcio
Guava, passion fruit	L. Gonçalves Neto
Coffee	J.L. Almeida Jr.
Hydrology	C. de Araújo Torres
Climatology	A.C. Souza Reis
Tropical fruits	A. Pinheiro Dantas
Entomology	A.F. Souza L. Veiga
	Elizabeth Pinto de Araujo
	J. Vargas Oliveira/G.P. Arruda
Soil science	A.V. de M. Netto
Crop management (coffee)	C.C.A. Lima
Plant breeding	D. Nascimento
Phytopathology	E.H. de A. Maranhão
Crop management (vegetable)	E.B. Correo
Production systems	H. Almeida Burity
Crop management (tropical fruits)	J.E.F. Bezerra/L. Abramof
Soil fertility	J. Barbosa Cabral
Crop management (cassava)	J.M. Garcia Bessa
Crop management (rubber)	J.R.B. Marques/M.A. de C. Fonseca
Crop management (cocoa)	L.J. Oliveira Accioly
Crop management (bean)	Maria C.L. da Silva
Plant breeding (cowpea)	P. Miranda

Sergipe

11 Aracaju

11.1 Centro Nacional de Pesquisa do Côco (CNPCo) (E) (National Coconut Research Center)
elevation : 3 m
rainfall : 1069 mm
temperature: 25°C

Av. Beira Mar s/n
C. Postal 44, 49000 - Aracaju - SE
Phone: (079) 222-8977
Telex: 079-2318
Chief: Dr. J. do Prado Sobral

Research on coconut which is an important tree crop in the Northeast Region. It deals with breeding, tissue culture, soil fertility, pests and diseases control and cultural practices.

Coconut:	
Phytopathology	C. Ran
Plant breeding and tissue culture	E. Ramos de Siqueira
	M.S. França-Dantas
Plant physiology	E.E. Melo Passos
Rural extention	E.R.C. Donald
Cultural practices	H. Rollemberg Fontes
Entomology	Joana M. S. Ferreira
	M. Ferreira de Lima
Soil fertility	L.F. Sobral
	Zorilda Gomes dos Santos
Water and soil management	R.G. Costa
IRHO consultant	J.P. Morin/R.R. Manciot

Bahia

12 Cruz das Almas

12.1 Centro Nacional de Pesquisa de Mandioca e Fruticultura (CNPMF) (E) (National Research Center for Cassava and Fruit Crops)
elevation : 220 m
rainfall : 1197 mm
temperature: 24°C

C. Postal 007,
44380 - Cruz das Almas - BA
Phone: (075) 721-1210
Telex: 071-2201
Chief: L. da Silva Souza

This research center specializes on cassava, tropical and sub-tropical fruits: research on cassava has received considerable impetus because the possibility of alcohol production from cassava. Priority fruits are: banana, citrus, mango and pineapple; minor fruits: guava, papaya, sugar apple and soursop mainly for variety improvement. Five gene banks are installed at CNPMF, such as banana, citrus, mango, cassava and pineapple. Cassava collection has 640 accessions including several wild relatives of *Manihot*. Data on characterization and evaluation has already been processed and stored at the Germplasm Data Bank of EMBRAPA in Brasilia.

Cassava research programme genebank:	S. de Oliveira/S. Sampaio
Diseases	C. Fukuda
Physiology	M. c. Marques Porto
Pests	A.R. Nunes Faria
Plant nutrition	F.N.E. Sulyras/M.C. Motta Macedo
Soil fertility	J. Cerqueira Gomes
Crop management	J.L. Loyola Dantas
	J.E.B. de Carvalho
	P.C.I.L. de Carvalho
	P.A. de Almeida/P.L.P. de Matos
Breeding	Wania M.G. Fukuda/A. Bueno
Irrigation	S.L. de Oliveira
Citrus research programme:	M. Almeida Oliveira
Physiology	Y. da Silva Coelho
Citrus breeding and genebank:	A.P. da C. Sobrinho
Soil fertility	A.F. de J. Magalhães
Pests	A. Souza do Nascimento
Diseases	H. Peixoto S. Filho
Breeding	W. dos Santos s. Filho
Germplasm	V. Medina
Banana research programme:	F.L.O. Cintra
Banana	E.J. Alves
Irrigation	S.L. de Oliveira
Diseases control	Z.J.M. Cordeiro
Pests control	A. Lindenberg M. Mesquita
Breeding and genebank	K. Shepherd

Soil fertility	Ana Lucia
Pineapple research programme:	G. Pinto da Cunha
Breeding and genebank	G.F. Souto/J.R. Santos Cabral
Diseases control	A. Pires de Matos
Pests control	N.F. Sanchez
Soil fertility	L.F. Souza
Mango research programme:	J.A. Rodrigues
Varieties and genebank	J.M.M. Sampaio

12.2 Escola de Agronomia, Universidade Federal da Bahia (Agronomy School, Federal University of Bahia)

44380 - Cruz das Almas - BA
Dean: Dr. B.L. Sampaio Seixas

Research on cassava, tropical fruits, cowpea and vegetables.

Cassava variety improvement	A.J da Conceição
Tropical fruits	C. Vaz Sampaio
Vegetable crops: carrot, peppers, cowpea, okra, tomato	L. Costa Pinto de Araujo
Seed technology	L. Soares de V. Sampaio

13 Salvador

13.1 Empresa de Pesquisa Agropecuária da Bahia (EPABA) (State of Bahia's Agricultural Research Enterprise)
elevation : 35 m
rainfall : 1923 mm
temperature: 24°C

C. Postal 1222,
40000 - Salvador - BA
Phone: (071) 247-9067
Telex: 071-7140
Pres.: Dr. R. de Pinto Pereira

This state's enterprise executes an agricultural research programme in 15 experiment stations, four of which work with tropical fruits. Main horticultural crops under programme are: cassava, coconut, cowpea, beans, papaya and vegetable crops.

Crop management (pineapple and vegetables)	N.M. da Silva
Papaya breeding and tropical fruits	J.V. Uzêda Luna
Crop management (garlic and potato)	R. Goto
Tomato and green pepper variety trials	N. Duarte Costa
Crop management (carrot and sweet potato)	Irisdalva Ferreira Mota

13.2 Centro de Pesquisa e Desenvolvimento (CEPED) (Research and Improvement Center)

C. Postal 1606,
40000 - Salvador - BA
Dir.: J.L. Perez Garrido

This center carries research on technology and is under the Secretariat of Planning, Science and Technology of Bahia State. It is a high level new institution with well-trained staff and excellent laboratory facilities. Research on tropical fruit processing.

Horticultural research leader	C.R.N. de Aquino
Post harvest physiology	J.C.P. Santos Neta
Fruit processing	J.F.L. Rosa
Fruit marketing (internal and export)	L.C.S. Araujo
Quality control	J.S. de Carvalho Neto

14 Itabuna

14.1 Centro de Pesquisa do Cacao (CEPEC)

Km 22 Rodovia Ilhéus-Itabuna

(Cocoa Research Center)

apto. 7, 45600 - Itabuna - BA
Dir.: J.M. de Abreu
Technical Coordinator:
Dr. P de T. Alvim

The Cocoa Research Center (CEPEC) is one of the most modern and well-equiped research institutions in the world, devoted to tropical agriculture, its major interest being cocoa. CEPEC is one of the five departments of CEPLAC, a semi-private institution linked to the Federal Government. A network of 13 Experimental Stations of 5,057 ha has 995 ha as experimental areas in the region, and additional sites in Amazonia. It has been responsible for the modern technology applied to Brazilian cocoa cultivation. Its goal is to generate or create, through experimentation and research, cocoa production systems for increasing yield and consequently Brazilian cocoa production. Research to improve production systems of rubber, oil palm, coconut, clove, black-pepper and guarana are carried out additionaly by CEPEC through technical cooperation with other agricultural institutions. The facilities of CEPEC is 15 thousand m^2 of construction, housing 11 technical divisions besides the laboratories, data processing center, auditorium, herbarium and a technical library containing more than 110 thousand publications of books, periodicals, pamphlets and maps. It maintains both national and international exchange programs. There are 112 researchers from the scientific staff of CEPEC, the majority with post graduate training. They are involved in studies of plant physiology, plant pathology, genetics, soils, botany, entomology, rural engineering, food technology, agricultural economics, animal husbandry, biochemistry, crop diversification and rural sociology including socio-economic surveys. CEPEC also uses the specialized services of the plant quarantine, climatology (with 16 meteorological observation posts and a satellite tracking station), photointerpretation and bibliographic documentation, soil analysis and improved seed branch.

Agronomy	P.R. Siqueira
Irrigation and drainage	P.R. Siqueira
Weedkillers and crop protection	A.F. de S. Pinho
Plant breeding and genetics	J.R. Garcia
Crop management	L.A. dos S. Dias/L.B. Freire
Plant physiology and weed control	M.W. Muller/R.M de O. Leite
Plant pathology	J.L.M. Pereira
Pesticides application technology	J.L.M. Pereira/J. de O. Cezar
Epidemiology of *Phytophthora spp.*	A.M.F.L. Campelo
Diseases of diversified crops	A. Ram
Epidemiology, control of coconut "lixa" (scab)	D.P. de Oliveira
Cocoa resistance to *Phytophthora sp.,*	E.D.M.N. Luz
Epidemiology and control of *Corticium salmonicolor*	
Biological and chemical control of *Phytophthora sp.*	J.M. de Figueirebo
Fungal diseases of oil palm and coconut	J.L. Bezerra
Epidemiology of *Microcyclus ulei*	L.C.C. de Almeida
Epidemiology and control of *Verticillium wilt*	M.L. de Oliveira
Other diseases of cocoa	W.T. Lellis
Soil science	F.P.C. Rosand
Mineral nutrition of perennial crops and	F.P.C. Rosand
phosphorus chemistry	
Soil classification and survey	A.A. de O. Melo/A.C. Leão
Soil hydric and physical relations, management	A.C. Zevallos
Soil acidity	C.J.L. de Santana
Mineral nutrition and fertilizing of rubber	E.L. Reis
and sugar cane	
Soil conservation	J. de O. Leite
Mineral nutritional and fertilizing of cocoa	M.B.M. Santana
Chemistry and mineralogy of soils	M.F. Soares
Mineral nutrition and fertilizing of spices	R.E.C. Silva
and stimulants	
Mineral nutrition and fertilizing of pastures	R.B. Cantarutti
Genetics	A.H. Mariano
Cocoa breeding	A.H. Mariano/G.A. Carletto

Entomology	E.C. de A. Ferraz
Insect control in cocoa	E.C. de A. Ferraz/P.F.N. da Cruz
Biology and ecology of *Percolaspis ornata* (Coleoptera)	E.M. de O. Ferronatto
Pollinating insects and insect control with synthetic sexual pheromones	F.P. Benton
Quality control and residue of agricultural pesticides	P.R.F. Berbert
Biology of *Scolytidae* (Coleoptera)	P. dos S. Terra
Agro-forest systems	R. Alvim
Oil palm	A. de S. Maia/C.K. do Sacramento
Rubber tree	A. de C.V. Filho/A.R. Sena Gomes
Coconut	A.R. de Carvalho
Black-pepper	J.V. Ramos

III. Southeast Region - Total area of Southeast is 924.935 km², corresponding to 10,86% of Brazil; this is the most populated region of Brazil with 59 million inhabitants. Climate is mostly subtropical because of altitute, but humid tropical in the coast. The Southeast is the most important horticultural region of Brazil, highest producer of citrus fruits, banana, potato, onion, coffee, tomato and other vegetables. The deciduous fruit production is increasing steadily supported by research, mainly breeding for low chilling requirement. Ornamentals and cut flower industry are expanding in this region because of the increasing demand at Rio de Janeiro and Sao Paulo market as well as export to Europc, mainly roses.

Minas Gerais

15 Belo Horizonte

15.1 Empresa de Pesquisa Agropecuária de Minas Gerais (EPAMIG) (State of Minas Gerais' Agricultural Research Enterprise)

C. Postal 515,
30000 - Belo Horizonte - MG
Phone: (031) 226-4740
Telex: 031-1366
Pres.: Dr. M.J.A. Neto

The main purpose of EPAMIG is the developing of new agricultural technologies mainly for the State of Minas Gerais. There are 6 Experiment Stations over the State and 13 Experimental Farms. The horticultural research programmes of EPAMIG include: fruit research on pineapple, peach, grape, plum, papaya, mango, banana, fig and citrus: pest control, bromatology, plant nutrition, food processing; vegetable research on garlic, onion, green pepper, black pepper, squash, potato, beet, tomato, carrot, watermelon, chayote, yam and cabbage: variety trials, germplasm, breeding, growing problems, weed control, plant nutrition, diseases, genetics, bromatology, crop management.

Crop management	D.J. Pardal Nogueira
Plant nutrition and soil fertility	F. Dias Nogueira
Plant breeding	J.G. de Pádua
Plant pathology	F. Assis Paiva/Sara M. Chalfoun de Souza/Maria A. de Souza Tanaka
	O. Almeida Drummond/Maria Araujo
Soil fertility	Miralda Bueno de Paula
Plant breeding and plant management	V.H Vargas Ramos
Entomology	P.R. Reis/J.C. Souza/W. Botelho
Deciduous fruits	E. Abrahão
Sub-tropical fruits	L. Rios de Alvarenga
	M. Leite Meirelles
	M. Albuquerque Regina
Bromatology	Vânia D. de Carvalho
Genetics and plant breeding	F.A. D'Araujo Couto
Crop management	F. de Paula Godinho/R.A. Nazar
	C.M.F. Pinto

Irrigation and drainage	T.J. Caixeta/L. Silva
Crop production	R. Coelho/J.B. da Silva
	J.M. Bendezu
Vegetables	F.A. Ferreira
	G. Milanez de Rezende
Plant nutrition and soil fertility	P.C. Rezende Fontes
Weedkeeler and weed control	Maria H. Takim Mascarenhas
Tissue culture	Ilza M. Sittolin
Plant pathology	Heloisa M. Saturnino

16 Lavras

16.1 Escola Superior de Agricutura de Lavras (ESAL) (Agricultural College of Lavras)
elevation : 801 m
rainfall : 996 mm
temperature: 19°C

C. Postal 37, 37200 - Lavras - MG
Phone: (035) 821-3700
Telex: 031-935
Dir.: LA. de Paula Lima

Research and teaching on temperate, tropical and subtropical fruits, potato, tomato, onion and garlic.

Tropical and subtropical fruits	M. de Souza/T. de Pádua
Tissue culture	M. Paiva
Seed pathology	J. da Cruz Machado
Virology: citrus and potatoes	Antonia dos R. Figueira
Mycorrhiza	P. de Souza/J.O. de Siqueira
Pest management	L.O. Salgado
Biological control	A.I. Ciociola
Deciduous fruits	N.N.J. Chalfun

17 Viçosa

17.1 Universidade Federal de Viçosa (UFV) (Federal University of Viçosa)
elevation : 658 m
rainfall : 1315 mm
temperature: 19°C

36570 - Viçosa - MG
Phone: (031) 891-1790
Telex: 031-1587
Pres.: G. Martins Chaves

Research and teaching on bean, soybean, green pepper, garlic, tomato, brassicas, lettuce, coffee, tropical, sub-tropical and deciduous fruits and ornamentals.

Agricultural Sciences Center	Head: J. Campos
Bean breeding and gene bank	C. Vieira
Coffee breeding for resistance to coffee rust	G. Martins Chaves
Citrus	F.C.C. Silva
Sub-tropical fruits	R.V.R. Pinheiro
Tropical fruits	F.C.C. Silva
Deciduous fruits	J.M. Fortes
Ornamentals	L.C. Lopes
Plant physiology	M. Maestri
Soybean breeding	T. Sediyama
Vegetable crops: green pepper, garlic, pumpkin, taro and genebank	V.W.D. Casali/K. Matusoka
Tomato, egg plant breeding and cultural practices	J.P. Camos/K. Matsuoka

Weed control	J.F. da Silva
Breeding - tissue culture	Aquira Mizubuti/S. Lopes Teixeira
Brassica, lettuce and plant nutrition	P.H. Monnerat
Seed: production and technology	R. Ferreira da Silva

Espírito Santo

18 Vitória

18.1 Empresa Capixaba de Pesquisa Agropecuária (EMCAPA) (State of Espirito Santo's Agricultural Research Enterprise)
elevation : 20 m
rainfall : 1238 mm
temperature: 23°C

C. Postal 391,
29000 - Vitória - ES
Phone: (027) 222-3188
Telex: 027-935
Pres.: F.X. Hemerly

The basic aim of EMCAP is to adopt and to develop new agricultural technology, mainly for the State of Espirito Santo. EMCAPA has 11 experimental units over the State: 3 Experimental Stations and 8 Experimental Farms. They were selected as priorities for research on: rice, corn, beans, cassava, black pepper, banana, pineapple, papaya, deciduous fruits, citrus, rubber tree, potato and vegetable crops. Research staff are located in the Experimental Stations but they develop their research program both in the Stations and Farms. The Experimental Station of Linhares develops the research program for the State North Region. This research program includes the rational exploitation of natural resources. Food crop and alternative crops. The research program of Experimental Station of Mendes da Fonseca includes deciduous fruits, vegetable crops, rice, beans, cassava and maize, mainly for the highlands. The Experimental Station of Bananal do Norte carries on researchers for the South, mainly on crops and tropical fruits.

Papaya variety trials	S.L.D. Martin
Papaya insect control	C.J Fantom
Bean variety trials	B.E. Vieira Pacova
Bean basic seed multiplication	E. Dan/Eliana Lopes Ban
Bean disease	A. de O. Athayde
Cassava genebank	A.V. Pereira
Soil microbiology	A.A. Teixeira Vargas
	J.S.M. da Silva
Disease evaluation and control	J.A. Ventura
Insect control	M. Fornavier
Irrigation	L.L. Fores de Castro
Variety trials	A.N. da Costa
Weed control	L.R. Ferreira
Garlic variety trials	C.A.S. do Carmo
Onion variety trials	C.A.S. do Carmo
Green pepper variety trials	J.C.A. Viana
Potato variety trials	A.P.M. de Andrade Neto
	C.A.S. do Carmo
Carrot variety trials	L.R. Ferreira
Tomato variety trials	L.R. Ferreira
Weed control	L.R. Ferreira
Disease control	J.A. Ventura/D. Cassetari
Insect control	M. Fornavier
Bean variety trials	B.E. Vieira Pacova
Pineapple crop management	A. Cassiano da Rocha
Pineapple fusarium rot control	J. Aires Ventura
Banana fertilizing and variety trials	A.C. Nóbrega
Banana crop management	J.A. Gomes
Banana insect control	R.J. Arleu

Citrus variety trials	F. de Lima Alves
Bean intercropping	J.F. Candal Neto
Cassava crop management	M.J. Furtado
Cassava variety trials	M.J. Furtado

Rio de Janeiro

19 Niterói

19.1 Empresa de Pesquisa Agropecuária do Estado do Rio de Janeiro (PESAGRO) (State of Rio de Janeiro's Agricultural Research Enterprise)
elevation : 14 m
rainfall : 1224 mm
temperature: 23°C

Alameda Sao Boaventura 770,
24000 - Niterói - RJ
Phone: (021) 717-1709
Pres.: L.A. Leite

PESAGRO-Rio deals with research on horticultural crops mainly tomato, green pepper, okra, carrot, potato, gilo, string bean, pineapple, banana and citrus.

Fruit crops	A. Vieira
Pineapple breeding germplasm	R.S.S. Gadelha
Banana growing	J.F.M. Maldonado
Citrus breeding and growing	J.C.M. de Barros
	H. de O. Vasconcellos
	J. Graça
Pineapple, citrus and banana pests	Regina C.P. Alves
Pineapple, citrus and banana diseases	A. de Goes
Vegetable crops:	
Tomato and okra breeding and genebank	N.R. Leal
Diseases control	J.T. de Athayde
	Maria do C.F. Esteves
Pest control	Alda M. de Oliveira
	Celma de A. da Cruz
Seed technology	Odette H.T. Liberal
	Rozane da Cunha
Crop management	R.G. Coelho/Maria L. de Araujo
Post-harvest physiology	H.C. Saleck
Eletrophoresis	Márcia Gomes Marques
Plant breeding	N.R. Leal/M. Teixeira Liberal
Nitrogen fertilization	P.A. da Eira

19.2 Universidade Federal Rural do Rio de Janeiro (UFRRJ) (Federal University of Rio de Janeiro)

Km 47 Rod. Rio-Sao Paulo
20000 - Rio de Janeiro
Dean: Dr. A.P. Barcelos

Research and teaching of fruit, vegetable and industrial crops.

Agronomy Institute	A.C.X. Velloso
Department of Agronomy	C.A. Lopes
Fruit crops: Citrus	C.M. Araújo
Guava, papaya, pineapple	C.A.M. Pace
Banana, mango, avocado	R. Nei Briançon Busquet
Vegetable crops:	
Squash, cucumber and okra breeding	A.L. Pereira

Potato breeding	M.S. Parraga
Plant propagation	R. Motta Miranda
Seed physiology	Marlene de Matos Malavasi
Seed production and technology	Elisabet M.N. da Silva
Weed science	R. Tozani
Bean, maize	G.G. Pessanha
Cassava	C.A. Lopes
Coffee, cocoa	R. Tozani
Soybean	T. Hara
Department of Soil Science	D.P. Ramos
Institute of Biology	A. Sales
Department of Plant Biology	E.C. Ribeiro
Plant pathology	Ch.F. Robbs
Vegetable breeding	R. de Lucena Duarte Ribeiro
	F. Akiba/O. Kimura
	P.S. Torres Brioso
	A. Oliveira de Carvalho
Nematology	J.P. Pimentel
Entomology	E.B. Menezes
Biology control and dynamic of populations	E.B. Menezes/P. Cesar
	R. Cassino

São Paulo

20 Campinas

20.1 Instituto Agronômico (IA)*
(Agronomic Institute)
elevation : 674 m
rainfall : 1353 mm
temperature: 20°C

C. Postal 28,
13100 - Campinas - SP
Phone: (019) 231-5422
Telex: 019-1059
Dir.Gen.: Dr. N. Paulieri Sabino

Belongs to the Secretariat of Agriculture, São Paulo State. This State Institute has 17 experimental stations spread all over São Paulo State. Its horticultural programme includes several fruits, vegetables and ornamentals.

Division of Horticulture	Hiroshi Nagai
Citrus branch	J. Pompeu, Jr.
Citrus genebank	Rose Mary Pio
Tropical fruit branch:	
Avocado, papaya and mango	N.B. Soares
Banana	R.S. Moreira
Pineapple	E.J. Giacomelli
Deciduous fruit branch	M. Ojima
Viticulture branch	M.M. Terra
Vegetable branch:	
Tomato, pepper, cucurbits	H. Nagai
Onion, garlic	R.S. Lisbão
Cabbage, carrot	J.B. Fornasier
Strawberry	F.A. Passos
Floriculture and ornamental branch	L.A. Matthes
Division of plant biology	O.P. Filho
Economic botany branch	C. Aranha
Plant pathology branch	J. Soave
Entomology branch	C.J. Rossetto
Genetic branch	L.C. Fazuoli
Virology branch	G.W. Müller
Plant physiology branch	Maria Luiz Carvalho Carelli

Citology branch	Neuza Diniz da Cruz
Seed branch	L.F. Razera
Division of food crops	L. D'Artagnan de Almeida
Division of agricultural basic activities	A.R. Pereira
Division of industrial crops	C.R. Bastos
Division of agricultural publications	C.V. Pommer
Division of soils	J.M.A.A. Valadares

20.2 Instituto de Tecnologia de Alimentos (ITAL) (Food Technology Institute)

Av. Brasil 2880, C. Postal 139
13100 - Campinas - SP
Phone: (019) 241-5222
Telex: 019-1009
Dir.: Dr. R.O. Teixeira Neto

Under the Secretariat of Agriculture, São Paulo State. Horticultural research programme includes fruit and vegetable processing.

Fruit processing and storage	D.C. Quast
By-products and residues	T.J.B. Menezes
Banana processing: juice	R.P. Tocchini
Banana processing: dehydration	D.A. Travaglini
Cassava processing	P. Vitti
Pineapple processing	S.D. da Silva
Apple processing	E.A.G. Salomon
Guava and papaya processing	Z.J. de Martin
Passion-fruit juice processing	M.F. de F. Leitão
Cold storage of fruits and vegetables	E.W. Bleinroth
Avocado oil processing	L.C. dos Santos
Soybean processing	S.I. da Costa
Guarana processing	R.P. Tocchini
Hearts of palm processing	V.L.P. Ferreira/S.D. da Silva
Grape wine	T. Hashizume
Vegetable processing-biochemistry	I. dos S. Dreatta
Asparagus processing and onion	Y. Yokomizo
Garlic processing	J.L.M. Garcia
Coffee processing	N.K. Sabbahg
Vegetable processing	J.E. Paschoalino
Citrus juice processing	J.C.C. Lara
Vegetable processing	L.W. Bernhardt/J.E. Paschoalino
Postharvest and cold storage of vegetables	E.W. Bleinroth/J.M.M. Sigrist Josalba V. de Castro
Tomato processing	L.F.C. Madi

21 São Paulo City

21.1 Instituto de Botânica
(Botanical Institute)
elevation : 674 m
rainfall : 1365 mm
temperature: 20.6°C

C. Postal 4005,
01000 - São Paulo - SP
Dir.: Dra. Vera L. Ramos Bononi

Belongs to the Secretariat of Agriculture of São Paulo State. Research on biology and taxonomy of orchids, other ornamental plants and edible mushrooms.

Ornamental branch	R. Bergmann de A. Silveira Izaura Telles de Menezes Pereira A.L. Gonçalves/F.F. Alves Aguiar
Orchids	Maria Sakane/L.F. de Barros V. Luiz Gil
Edible mushrooms	Vera L. Ramos Bononi

21.2 Instituto Biológico (IB)
(Biological Institute)

Av. Cons. Rodrigues Alves 1252
03014 - São Paulo - SP
Phone: (011) 570-4234
Telex: 01-1935

This Institute is under the State of São Paulo's Secretariat of Agriculture. It has been a leading institution in Brazil on research on plant diseases and their control.

Division of plant pathology	Head: Veridiana V. Rossetti
Diseases of basic food crops and vegetables	E. Issa
Vegetable diseases	B.C. Barros
Diseases of fruit crops branch	J.A. Martinez
Citrus viruses and *Phytophthora spp.*	Veridiana V. Rossetti
Diseases of industrial crops branch	P. Figueiredo
Coffee diseases	P. Figueiredo
Plant bacteriology branch	M. Barreto Figueiredo
Fruit trees	K. Watanabe
Plant pathology biochemistry branch	Walkyria B.C. Moraes
Plant resistance to coffee rust	Walkyria B.C. Moraes
Serology	A.P.C. de Alba
Citrus canker	Elisabete Bath
Mycology of plant pathology branch	Regina E. de Melo Amaral
Seed pathology	Celia C. Lasca
Citrus diseases	E. Feiehtenberger
Ornamentals	Rosa M. Gaioso Cardoso
Cocoa diseases	Maria I. Feitosa
Plant virology and physiology branch	Marly V.L. Barbosa de Oliveira
Potato viruses	P.R. Malazzi
Center for research on plant cancer	A.A. Bitencourt (emer. res.)
Division of plant parasitology	A. Ferreira Cintra
General entomology branch	N.P. Mendonça
Nematology branch	S.M. Curi
Pests of food and vegetable crops branch	S.F. do Amaral
Pests of industrial crops branch	G. Calcagnolo
Pests of fruit tree crops branch	N. Suplicy Filho
Banana pests	A.F. Sampaio
Biological control branch	Zuleide A. Ramiro
Division of pesticides	O. Jannotti
Chemistry branch	J.R. Piedice
Insecticide branch	P.R. de Almeida Conrado
Fungicide branch	A.A. Campacci
Herbicide branch	L. Leidermann
Residue branch	P. Pigati
Pilot center for formulation	H.B.F. Barreto

22 Botucatu

22.1 Faculdade de Ciências Agronômicas

C. Postal 237

(Agronomic Sciences University)
elevation : 830 m

Belongs to the University of São Paulo-UNESP.

18600 - Botucatu - SP
Phone: (014) 922-3883
Telex: 014-2107
Dir.: Dr. R.A. de Arruda Veiga

Department of Horticulture	F.A.D. Conceição
Fruit crops:	
Citrus, avocado macadamia	A.A. Salibe
Grapes, apple, strawberry	R. Carbonari
Citrus, banana, passion-fruit	E. Cereda
Papaya, pineapple, strawberry	R.J.P. Cunha
Vegetable crops:	
Production - tomato, onion, garlic, cabbage, cucumber	T. Kimoto
Processing - tomato, onion, carrot, cucumber, strawberry	F.A.D. Conceição
Seed production - tomato, pepper, pumpkins, egg-plant, okra	A.C.W. Zanin
Floriculture and landscaping:	
Carnation, ornamentals	Maria A. de L.B. Sousa

23 Jaboticabal

23.1 Faculdade de Ciências Agrárias e Veterinárias-UNESP (Agricultural and Veterinary Sciences University)
elevation : 610 m
rainfall : 1285 mm
temperature: 22°C

C. Postal 145,
14870 - Jaboticabal - SP
Phone: (016) 322-0814
Telex: 016-13935
Dir.: V.J. de Melo

Research and teaching on fruit crops (citrus, banana, pineapple, mango etc.) and vegetable crops (tomato, onion, carrot and lettuce etc.).

Fruit crops:	
Citrus, avocado, mango	L.C. Donadio
Banana, pineapple, papaya	C. Ruggiero
Grapes, guava, persimon, macadamia nut, figs	F.M. Pereira
Passion-fruit	J.C. Oliveira
Vegetable crops:	
Tomato, onion, carrot	M.G.C.C. Masca
Cabbage, cucumber, lettuce	P.D. Castellane
Floriculture and landscaping	Maria E. Soares P. Dematté
	T. Testes Graziano

24 Piracicaba

24.1 Escola Superior de Agricultura "Luiz de Queiroz" (ESCALQ) ("Luiz de Queiroz" Agricultural College)

C. Postal 09,
13400 - Piracicaba - SP
Phone: (019) 422-5926
Telex: 019-1141
Dir.: Dr. J.C. Engler

A leading college in horticultural research and teaching in Brazil, with post-graduate training at M.S. and Ph.D. levels, under the University of São Paulo State.

Department of Horticulture:	
Avocado and banana	V.R. Sampaio
Citrus	C.S. Moreira

Mango — S. Simão
Vegetable crops:
Cultural practices — E.F.C. Vasconcellos
Onion and tomato breeding — C.P. da Costa
Plant nutrition: lettuce, tomato and strawberry — H.P. Haag
Carrot, cauliflower, strawberry, variety trials — K. Minami
Weed control — R.V. Filho

24.2 Centro de Energia Nuclear na Agricultura (CENA) (Nuclear Energy in Agriculture Center)

C. Postal 96,
13400 - Piracicaba - SP
Dir.: Dr. E. Salatti

This is a research center in Brazil, with post-graduate training at M.S. level, under the University of São Paulo State (USP), that works in close cooperation with the International Atomic Energy Agency (IAEA).

Induced mutation on beans — A. Tulmann Neto
Tissue and cell culture — D. Martins Silva
Microbiology — A.P. Ruschel

IV. Centrewest Region - The Centrewest Region has an area of 1.879.456 km² which represents 22% of the country's total area and a population of 9 million inhabitants. This region is booming since 1961 when the new capital, Brasilia was inaugurated. A large part of this region enjoys a tropical and sub-tropical climate with a dry and mild winter because of altitude from 700 to 1.200 meters. Soils are mostly red latosols, with a savana-like vegetation known as "cerrado"; these soils have a low pH, poor fertility and an aluminium phytotoxicity problem, but they are well suitable for reclamation because of good physical conditions, becoming then adequate for growing fruit trees and vegetables. At present, this is mainly a grazing region but annual crops, such as rice, beans, soybean and maize are produced significantly.

Federal District

25 Brasília

25.1 Empresa Brasileira de Pesquisa Agropecuária (EMBRAPA) (Brazilian Agricultural Research Enterprise)
elevation : 1030 m
rainfall : 1574 mm
temperature: 20°C

C. Postal 101316,
70000 - Brasília - DF
Phone: (061) 225-3870
Telex: 061-1620
Pres.: Dr. Rivaldo

The headquarters of EMBRAPA are located at Brasília where administration and technical supervision are centralized.

Departments:
Technology Difusion — J. Batista da Silva
Studies and Research
Quantitatives Methods
Research on Programming — Sulz Gonçalves
Human Resources — J.E. Schneider

25.2 Centro de Pesquisa Agropecuária dos Cerrados (CPAC) (E) (Cerrado Agricultural Research Center)
elevation : 1077 m
rainfall : 1475 mm
temperature: 22°C

Rod. BR 020 km 18
73300 - Planaltina - DF
Phone: (061) 596-1171
Telex: 006-1621
Chief: Dr. R. Pontes Nunes

This research center deals with research on natural resources of the "cerrado". Research programme includes evaluation and use of natural resources, farming systems, energetic forests, irrigation, crop pests and diseases, soil microbiology.

Soil fertility	L. Vilela
Soil biology	J.R.R. Peres
Soil conservation and management	J. Pereira
Water deficiency	E. Medrado da Silva
Annual crops	G. Urben Filho
Cassava - variety trials	S. Perim
Perennial crops	J.B.R. Sampaio
Citrus and avocado	P. de Carvalho Genu
Mango and annonas	A.C. de Queiroz Pinto

25.3 Centro Nacional de Recursos Genéticos (CENARGEN) (E) (National Center for Genetic Resources)
elevation : 1030 m
rainfall : 1574 mm
temperature: 20°C

C. Postal 102372
70000 - Brasilia - DF
Phone: (061) 273-0100
Telex: 061-1662
Chief: Dr. J. Silva

This center coordinates, at national level, the activities related to plant introduction, exchange, germplasm characterization, evaluation and conservation. A network of 70 genebanks located mostly at EMBRAPA's research units spread all over the country have been organized. A data bank on germplasm is also included for processing, storage and retrieval of information on germplasm. Research programmes on tissue culture and genetic engineering are full operational.

Introduction and post entry quarantine	Coordinator: A.C. Rebouças Lins
Phytosanitary laws	J.M. Lemos Fonseca
Micology	A. Fontes Urben
Nematology	Renata Cezar V. Tenente
Virology	Maria de Fátima Batista
Bacteriology	A. Soares dos A. Santos
Tissue culture	E.L.R. Filho
Plant exploration	Coordinator: L. Coradin
Plant exploration and taxonomy	J.F.M. Valls/A. Costa Allem
	E. Lleras Perez
Germplasm conservation	Coordinator: C. Terra Wetzel
Seed storage, conservation and quality control	C. Bragantini
	Clara Oliveira Goedert
	Maria M.V.S. Wetzel
Criopreservation	A.R. de Miranda
Tropical fruits germplasm	F.R. Ferreira
Germplasm data bank	Coordinator: E.A.V. Morales
Cassava germplasm	Curator: R.A. Mendes
Forestry germplasm	Coordinator: J. Alves da Silva
	A. Gripp
Medicinal plants	R. Fontes Vieira
Genetic engineering	Coordinator: L.A.B. de Castro
	Maria I.C. Santos Gama
	P.A. Pinheiro/Maria J.A.M. Sampaio
Tissue culture laboratory	J.B. Teixeira/K. Matsumoto

25.4 Universidade de Brasília (UnB) (Brasilia University)

Campus Universitário Asa Norte
70000 - Brasilia - DF
Pres.: vacancy

Research and teaching on fruits crops

(papaya, citrus, pineapple etc.) and vegetable crops.

Department of Agronomy	H. de Miranda Flor
Vegetable crops	J. Kleber de Abreu Mattos
Fruit crops	Tereza Vaz Parente
Department of Plant Biology	Terezinha Paviani
Tissue culture	Linda S. Caldas
Cassava diseases	A. Takatsu
Virus diseases of vegetable crops	F.P. Cupertino
Nematode on vegetable crops	C.S. Huang
Department of Cell Biology	E.W. Kitajima
Virology	E.W. Kitajima
Micology	J. Carmine Dianese

25.5 Centro Nacional de Pesquisa de Hortaliças (CNPH) (E) (National Vegetable Research Center)	C. Postal 07018
elevation : 1013 m	70358 - Brasilia - DF
rainfall : 1483 mm	Phone: (061) 556-5011
temperature: 21°C	Telex: 061-2445
	Chief: R.L. Vaz

The general objectives of CNPH research activities are to solve basic problems of the vegetable grower as: improve productivity, quality, producer's income and national nutrition; reduce imports of vegetables and create new options for export. The strongest concentration of research is on breeding, diseases control and plant nutrition, in view of the general need of the producers.

Physiology - post harvest	A. Gimenes Calbo
Cropping systems - potato	A.A. Fedalto/O. Furumoto
Phytopathology - virology	A.C. de Avila
Vegetable seed technology	A.C. Guedes/Cl. Andreoli
Physiology	A.C. Torres/Ecilda L.S. Souza
Plant nutrition	A.F. Souza/J. de Almeida Lima
	J.R. de Magalhães
	M.V. de Mesquita Filho
	R. Rezende Fontes
Phytopathology	C.A. Lopes/C. Bittencourt da Silva
	F.J.B. Reifschneider
	J.A. Espinal Aguilar/M.A. Beek
Irrigation	C.A.S. Oliveira
Entomology, pest control	F.H. França
Breeding potato	F. de La Puente Ciudad
Economics	G.C. Gomes
Breeding carrot	J.V. Vieira
Cropping systems, garlic	J. Alves de Menezes Sobrinho
Breeding tomato, sweet potato	J.E. Cabral de Miranda
Phytopathology, nematology	J.M. Charchar
Breeding cucumber	J.F. Lopes
Breeding cauliflower	J.A. Buso
Breeding cauliflower, cabbage, peas, mustard and brocoli	L. de Brito Giordano
	N. da Silva
Breeding onion	M. de Targa Araújo
Agroclimatology	N. Vianna Barbosa dos Reis
Technology diffusion	N. Makishima
Breeding cucumber, melon, carrot, tomato	H. Bittencourt Salazar V. Pessoa
Difusion of technology	R.V. Cobbe
Phytopathology, nematology	S. Pin Huang

Soil microbiology, string beans and peas	W. Coelho e Silva
Physiology-weed control	W. Pereira
Cropping systems, tomato, pepper	Y. Horino

Goiás

26 Goiânia

26.1 Centro Nacional de Pesquisa de Arroz e Feijão (CNPAF) (E) (National Center for Rice and Bean)
elevation : 800 m
rainfall : 1647 mm
temperature: 22°C

C. Postal 179,
74000 - Goiânia - GO
Phone: (062) 261-3383
Telex: 062-2241
Chief: vacancy

The research programme of this center includes bean, cowpea and rice.

Phytopathology (bean)	A. Sartorato
Plant breeding	A. de Matos Lopes/A.B. Freire
Crop management	A. Silveira Filho
Entomology (cowpea)	B. Pereira das Neves/E. Ferreira
Physiology (bean)	C. Moraes Guimarães
	T. de Aquino P. Castro
Phytopathology (cowpea)	G. Pereira Rios
Crop management (bean)	H. Aidar/R. Faria Vieira
Plant breeding (bean)	Iraja Ferreira Antunes
	M. Grande Teixeira
	Maria Jose de O. Zimmermann
Soil Science	I. Pereira de Oliveira
Crop management (cowpea)	J.P.P. de Araujo
Plant breeding	J.G.C. da Costa
Phytopathology	J. Correia de Faria
Irrigation and drainage	P. Marques da Silveira
Bean genebank	Marlene Silva Freire

26.2 Empresa Goiana de Pesquisa Agropecuária (EMGOPA) (State of Goiaz's Agricultural Research Enterprise)
elevation : 730 m
rainfall : 1358 mm
temperature: 23°C

C. Postal 49,
74000 - Goiânia - GO
Phone: (062) 225-4755
Telex: 062-2925
Pres.: C. da Silva Moreira

This State's research institution deals with several commodities including fruits and vegetables.

Team leader for fruit research	T. Ogata
Tropical fruits	W. Bacelar
Temperature fruits	R. Vaz
Team leader for research on vegetable crops	F.A.R. Filgueira
Bean research	L.G. Dutra
Soybean research, variety trials	A.V. Costa
Plant breeding (soybean)	A. Vasconcelos Costa

V. South Region - The South Region has an area of 577.723 km² which corresponds to 7% of the total area of Brazil; this region is densily populated and has the largest number of immigrants, mainly Italians, Germans and

Japanese; population: 21 million inhabitants. The region is a major producer of coffee and soybean, as well as wine, canned peaches and vegetables. Apple and pear production is booming in the high altitude where enough chilling occurs. Climate is mostly subtropical in land, temperate in altitude above 1000 meters and tropical in the coast where banana, pineapple, citrus and other tropical fruits are grown successfully.

Paraná

27 Londrina

27.1 Centro Nacional de Pesquisa de Soja (CNPS) (E) (National Center for Soybean Research)
elevation : 585 m
rainfall : 1388 mm
temperature: 21°C

C. Postal 1061
86100 - Londrina - PR
Phone: (043) 223-1830
Telex: 043-2208
Chief: vacancy

This center coordinates the soybean and sunflower research national programs in Brazil.

Soybean breeding	R.A.S. Kiihl/A. Dall'Agnol
Soybean genebank	Mercedes C.C. Panizzi
Crop management	N. Neumaier/A. Garcia
Physiology	Gamin Ma Wang/Shin R. Wang
Weed control	D.L.P. Gaaziero/E. Voll
Soil fertility	G.J. Sfredo/A.F. Lantmann
Soil microbiology	R.J. Campo
Seed technology	J.B. França Neto N. Pereira da Costa
Plant pathology	J.T. Yorinori/A. Henning A.M.R. Almeida
Nematology	Helenita Antônio
Entomology	Beatriz S.C. Ferreira/A.R. Panizzi
Economics	A.C. Roessing
Technology diffusion	P.R. Galerani/J.G.M. Andrade
Information/documentation	Leocádia M.R. Mecenas/A.B.A. Lima

27.2 Instituto Agronômico do Paraná (IAPAR) (Parana's Institute of Agronomy)

C. Postal 1331,
86100 - Londrina - PR
Phone: (043) 222-5388
Telex: 043-2122
Dir.: F. de Assis Lemos de Souza

This institution, inaugurated in 1975, belongs to the Secretariat of Agriculture of Paraná State; it is considered among the most modern agricultural research institutions in Brazil.

Fruit nutrition	M.A. Pavan
Fruit crops: table grapes	A.Y. Kishino
Variety evaluation: apple, peach, plum	M. Tsuneta
Tropical fruits	S.L.C. Carvalho
Vegetable crops diseases	Lucila Maschio/N.R.X. Nazareno
Entomology	N.L. Domiciano/W.J. dos Santos
Agronomy	M.A. Hoepfner/N.L. Brenner
Seed technology	R.L. Schinzel
Agrometerology	L. Grodzki
Breeding	C.A. Scotti

Santa Catarina

28 Florianópolis

28.1 Empresa Catarinense de Pesquisa Agropecuária S.A. (EMPASC) (State of Santa Catarina's Agricultural Research Enterprise)
elevation : sea level
rainfall : 1584 mm
temperature: 24°C

C. Postal D-20,
88000 - Florianópolis - SC
Phone: (048) 233-1344
Telex: 048-2242
Pres.: J.O. Kurtz

EMPASC is a governmental institution linked to the State Department of Agriculture. It is responsible for developing research on agricultural products of major socio-economic importance to the State. The horticultural research programmes developed by EMPASC include: deciduous fruits: apple, pear, grape, plum and peach; tropical and subtropical fruits: citrus and banana; vegetables: garlic, onion, cucumber, lettuce, tomato, potato and cabbage. Research is carried out in four Experiments Stations located in the following towns: Caçador, Itajai, Videira and São Joaquim.

Apple research programme
Breeding and genebank
Propagation and rootstocks
Plant physiology and nutrition
Diseases control
Pest control

Cultural practices and budwood
Post-harvest physiology programme
Pear breeding and genebank
Grape: variety trials and rootstocks
Citrus research programme:
Variety trials and rootstocks
Diseases control
Banana research programme:
Variety trials
Soil fertility and fertilizing
Diseases control
Vegetable research programme:
Garlic: variety trials
Diseases control
Pest control
Weed control
Seed multiplication
Onion and lettuce

Potato and Capsicum
Carrot and tomato
Tomato disease control
Statistical analysis and data processing
Seed technology
Agricultural ecology
Agricultural economics

Diffusion of technology

F. Denardi/A.P. Camilo/M. Pasqual
A.C. Driessen
A. Ebert/C. Basso/A. Suzuki
J.I.S. Bonetti/Y. Katsurayama
P.A. Ribeiro/A.I. Orth
L.A. Palladini
M. Losso
R.J Bandu/Z.S. Raasch
E. Brighenti
C.S. Matos

O.L. Koller*
J.F. Frosi

L.A. Lichtemberg
J.L. Malburg
J.F. Rosi

I.D. Faoro
W.F. Becker
M.A. Mandelli
S. Mueller
L.F. Thomazelli
A.C.F. da Silva/S. Nokoyama
J.J. Muller
Z.S. Souza
V. Bonin
J.I.S. Bonetti
R.C. Diltrich/D.L Segalin
J.A. Lanini Neto
V.M. Thomé
M.C. Silva/E.O. Bulblitz
W.J. Sorrenson
A. Buss/V.T.M. Cardoso
H.H. Tassinari

Rio Grande do Sul

29 Porto Alegre

29.1 Instituto de Pesquisa Agronômicas do Rio Grande do Sul (IPAGRO) (State of Rio Grande do Sul's Agricultural Research Institute)

Rua Gonçalves Dias 579
90000 - Porto Alegre - RS
Phone: (051) 233-5411
Telex: 051-1211
Dir.: J.L.D.H. Aldegretti

This Institute coordinates all the horticultural research programmes under the Secretariat of Agriculture of Rio Grande do Sul State.

Potato, variety trials	G.M. da Cunha
Potato, seed production	A.L.P. Winandy
Cassava, variety trials	E.L. Machado
Soybean breeding	E.R. Hilggarg
Soybean variety evaluation	B.H. de Souza
Weed control on soybean field	L.R.C. Venturela
Soybean cultural practices	H. Bergamaschi
Soybean seed quality	W. da Silva Fulko
Soybean breeding for resistance to *Meloidogyne javanica*	J.E. da Silva Gomes
Soybean diseases control	W. Bangard
Soybean insect control	S. Moruzini
Soybean seed *Rhizobium* inoculation	J. Kolling
Onion breeding	B. Bendjouya
Garlic: variety trials	R.M. Ramos
Citrus varieties and rootstocks	O. de Menezes Porto
Citrus genebank	S.R. Reck

29.2 Instituto de Pesquisa de Recursos Naturais Renováveis (Renewable Natural Resources Rcsearch Institute)

Rua Gonçalves Dias 570
90000 - Porto Alegre - RS
Phone: (051) 233-5411
Telex: 051-2935
Dir.: E. Correia

This research institute is under the Secretariat of Agriculture of Rio Grande do Sul State; its main objective is to conduct research on the area of renewable natural resources.

Identification and evaluation of native forestry trees	J.M.O. Vasconcellos
Erosion control	E.A. Cassol
Research on "erva-mate" tea	M.T.T. Ferreira
Research on forestry seeds	V.R.B. Galardo
Research on "palmito" *Euterpe edulis*	D.I. Amaral
Flora of Rio Grande do Sul	J.R. Mattos

30 Pelotas

30.1 Centro Nacional de Pesquisa de Fruteiras de Clima Temperado (CNPFT) (E) (National Research Center for Temperate Climate Fruit)
elevation : 36 m
rainfall : 1528 mm
temperature: 17°C

C. Postal 403, 96100 - Pelotas - RS
Phone: (053) 221-2121
Telex: 053-2301
Chief: L. Nunes e Nunes

This Center, founded in 1984, carries out research programmes on peach breeding for varieties suitable for processing and fresh consumption; strawberries, asparagus, apple, onion, tomato, fig, garlic, quince, peas, plum, potato and peach research programmes.

Breeding variety trials and genebank	B. Hideyuki Nakasu
Orchard management	N.L. Finardi/A.R.M. de Medeiros
	J.F. Martins Pereira
Tree nutrition	C.J. da Silva Freire/M. Magnani
	Eva Choer Moraes
Diseases control	C. Castro
Insect control	L.A. Benincá de Salles
	R.P.L. Carvalho
Apple: breeding and variety trials	B. Hideyuki Nakasu
Orchard management	A. Kovaleski/J.R. Tiscornia
	A.R.M. de Medeiros/N.L. Finardi
	J.F. Martins Pereira/T. Iuchi
Plant nutrition	C.J. da Silva Freire/M. Magnani
	Eva Choer Moraes
Disease control	J. Figueiredo Fortes
Insect control	L.A. Benincá de Salles
	R.P.L. Carvalho
Strawberry: breeding, variety trials and genebank	A. Machado dos Santos
Plant nutrition	J.F. Dynia
Diseases control, virology	C. Castro/J. Daniels
Quince: orchard management	J.F. Martins Pereira
Diseases control	J. Figueiredo Pereira
Pear: orchard management	N.L. Finardi
Plum: orchard management	N.L. Finardi
Diseases control, virology	J. Daniels
Pecan nut: breeding, variety trials and genebank	B. Hideyuki Nakasu
Potato: breeding, variety trials and genebank	D. Mota da Costa
Plant nutrition	Eva Choer Moraes
Diseases control: virology	C. Castro/J. Daniels
Post harvest physiology	R.F. Flores Cantillano
Onion: breeding, variety trials and genebank	A. Garcia
Plant nutrition	C.J. da Silva Freire
	Eva Choer Moraes/J.F. Dynia
Disease control	C. Castro
Garlic: variety trial and genebank	A. Garcia
Plant nutrition	Eva Choer Moraes
Insect control	R.P.L. Carvalho
Diseases control	J. Figueiredo Fortes
Asparagus: breeding, variety trials and genebank	Eliane Augustin Oliveira
Plant nutrition	Eva Choer Moraes
Tomato: crop management	A. Lima Pereira
Plant nutrition	J.F. Dynia
Diseases control	C. Castro
Fig: orchard management	A.R.M. de Medeiros
Plant nutrition	C.J da Silva Freire/M. Magnani
Diseases control	J. Figueiredo Fortes
In vitro propagation	M. Assis
Food technology	J.L. Silva Vendruscok

30.2 Centro de Pesquisa Agropecuária de Terras Baixas de Clima Temperado (E) (Agricultural Research Center for Low Lands)

C. Postal 553,
96100 - Pelotas - RS
Phone: (053) 221-1248
Telex: 053-2627

This research unit is under EMBRAPA; it works with several crops, including soybean.

Chief: A. da Silva Gomes

Soybean	F. de Jesus Vernetti
Entomology	A.B. Menschoy
Soybean	F. de Jesus Vernetti
Entomology	A.B. Menschoy
Breeding varieties for low lands	M.F. da Cunha Gastal
Phytopathology	C.R. Casela
Crop management	A.A.A. Raup
	Veronia Peixoto Vernette
Soil fertility	D. da Silva Cordeiro
Weed control	G.L. Broener
	V. Anunciação de Andrade

30.3 Faculdade de Agronomia Eliseu Maciel ("Eliseu Maciel" Agronomy Faculty)

C. Postal 767, 96100 - Pelotas - RS
Phone: (053) 221-1496
Telex: 053-2935
Dir.: Prof. G. Azambuja Centeno

This college of agriculture is a part of the Federal University of Pelotas. Vegetable and fruit crops.

Fruit processing	A. Barreira Bilhalva
Fruit nutrition and fertility	A. Duarte Cruz
Fruit production and physiology	A.A.F. Ferreira
Fruit physiology	B. Gomes dos Santos Filho
Vegetable physiology	Dora Suely Santos
Deciduous fruit production	E. Kersten/J.C. Fachinello
Pest control of fruits and vegetables	E. Salazar Cavero
Diseases control of vegetables	G.C. Luzzardi
Vegetables production	Gilda Pinheiro Nunes
	J. Silva Filho
Vegetable breeding and production	Heloisa Santos Fernandes
Fruit and vegetable physiology	J.A. Peters
Disease control of fruits	N.L. Garibaldi
Vegetable nutrition	Vera L. Gonçalves de Assis

31 Bento Gonçalves

31.1 Centro Nacional de Pesquisa de Uva e Vinho (CNPUV) (E) (National Grape and Wine Research Center)

C. Postal 130,
95700 - Bento Gonçalves - RS
Phone: (054) 252-2144
Telex: 054-935
Chief: J.G. Filho

This new Center deals mainly with research on grapes and Oenology. The Center has 100 ha, 15 are experimental vineyards and 7 ha with foundation mother plants. There are 5 greenhouses for studies on virus diseases of grapes.

Variety trials and genebank	U. Almeida Camargo
Fungal diseases control	A. Grigoletti Jr.
Virus diseases	G. Barcelos Kuhn
Plant nutrition	J.C. Fráguas
Plant physiology	A. Miele/J.C. Haas
Grape production systems	S. Manfredini
Production costs	Loiva M. de Mello Freire

Rural economy	J. de Melo Freire
Oenology	J.G. Filho/L.A. Rizzon
Agricultural microbiology	G. Almeida da Silva
	Maria A. Amazonas A. da Silva
Cultural practices	J. Tonietto/L. Paixão Passos
Insect control	S. de Jesus Soria Vasco

In Brazil there are scientific and technical societies which promote horticulture in the country. A National Society of Olericulture, founded in 1960 has 815 members and biannual National Congresses. The Brazilian Society of Fruitculture, founded in 1970, has 800 members and biannual National Congresses of Fruitculture. The Brazilian Society of Floriculture and Ornamental Plants is the newest, was founded in 1978 and has 237 members, also biannual Congresses. All three Societies publish proceedings in Portugese with a summary in English.

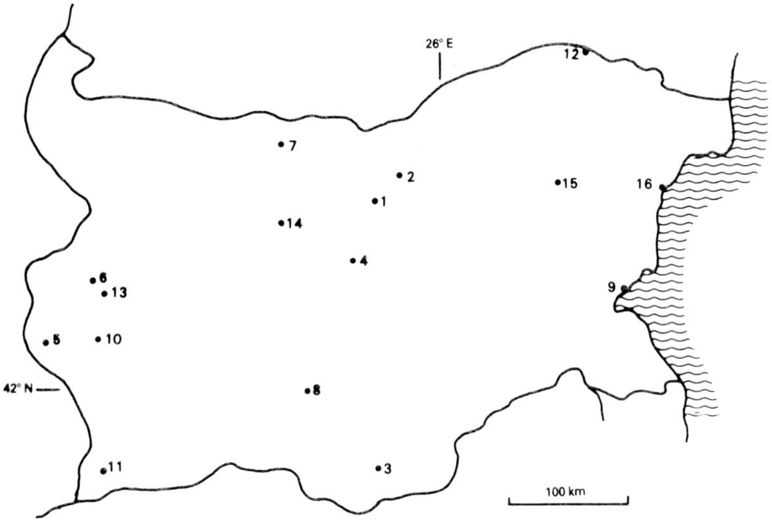
26° E
42° N
100 km
1
2
3
4
5
6
7
8
9
10
11
12
13
14
15
16

1 Dryanovo, 2 Gorna Oryahovitsa, 3 K"rdzhali, 4 Kazanl"k, 5 Kyustendil, 6 Kostinbrod, 7 Pleven, 8 Plovdiv, 9 Pomorie, 10 Samokov, 11 Sandanski, 12 Silistra, 13 Sofia, 14 Troyan, 15 Tsar Krum, 16 Varna

Syrvey

Bulgaria is a country with climatic and soil conditions favourable for growing different horticultural plants. Grapes, fruits, and lately also flowers, are widely cultivated. The Bulgarians are known as master-gardeners and horticultural research has a long tradition. For viticulture the first Experiment Station was founded in 1902, for fruit in 1929 and for vegetables in 1930. After the 9th September 1944 horticultural research workers were challenged to solve the new problems arising from the collectivization of the land and the formation of large orchards, vineyards, vegetable and flower plantations. At present research institutes and experiment stations engaged in agriculture are directed by the National Agro-Industrial Union (NAIU). There are three leading institutes in the field of horticulture: the 'Maritsa' Vegetable Crops Research Institute and the Fruit Growing Institute in Plovdiv and the Institute of Viticulture and Enology in Pleven. Each one of these institutes has at its disposal several experiment stations situated in the varying soil and climatic zones of the country. The leading institute directs, coordinates and controls research work in this branch. The Agricultural Academy directs also horticultural research which is done in the 'V. Kolarov' Higher Institute of Agriculture in Plovdiv.

The institutes and experiment stations are in different regions of the country, so that horticultural research can include the varying climatic conditions of the plants, high mountains, sea coast and the hot southern zones. Research projects are planned by the groups of associates in the institutes and experiment stations in accordance with the problems arising in horticultural practice. The President of the Agricultural Academy confirms or rejects the projects.
Publications: Genetika i selektsiya, Lozarstvo i vinarstvo, Mekhanizatsiya na selskoto stopanstvo, Ovoshtarsto, gradinarsvo i konservna promishlenost, Rastenievadni nauki.

1 Dryanovo

1.1 Experiment Station of Pomology
elevation: 275 m
rainfall : 745 mm

5370 Dryanovo
Phone: 23-41
Dir.: Dr. M. Yoncheva

The station is the main centre of research with the plum-variety trials, breeding, cultural practices.

Prepared by Dr. N. Yordanov, 'Maritsa' Vegetable Crops Research Institute, Plovdiv.

Plum breeding — Dr. M. Yoncheva/ P. Iliev
Methods of growing — Dr. M. Yoncheva
Leaf analysis and nutrition — Dr. I. Vitanova
Biochemistry — Dr. Y. St"rkova

2 Gorna Oryachovitsa

2.1 Experiment Station for Vegetable Crops
elevation: 50 m
rainfall : 605 mm

5100 Gorna Oryachovitsa
Phone: 5-62-39
Dir.: Dr. St. Bachvarov

The station is the main centre of research with the onion, melons, squash-variety trials, breeding, growing methods and certificated seed production. In addition, it serves the central areas of North Bulgaria in all problems concerning the vegetable production.

Biology and breeding of onion and garlic — Dr. St. Bachvarov
Breeding of cucumbers — Dr. T. Petkova
Breeding of melons and squash — Dr. P. Lozanov
Growing methods in vegetables — Dr. M. Petkov/ B. Christov

3 K"rdzahli

3.1 Agricultural Experiment Station
elevation: 230 m
rainfall : 663 mm

6600 K"rdzahli
Phone: 20-06
Dir.: Dr. G. Anadoliev

The station is a centre of research with nuts and almond in the region of the Rhodopen

Variety trials and breeding — Dr. G. Anadoliev
Propagation and growing practices — Dr. G. Anadoliev/ Dr. Z. Anadoliev

4 Kazanl"k

4.1 Institute of oil bearing rose, ehtereal oil and medicinal plants
elevation: 350 m
rainfall : 580 mm

6100 Kazanl"k
Phone: 2-14-91
Dir.: V. Topalov

The institute is the only research centre for ethereal oil and medicinal plants in Bulgaria. There is a rich museum for the history of the rose and rose oil production, visited by a lot of foreign tourists. There are over 2000 varieties of ornamental roses in the collection garden.

Rose breeding — Dr. R. Tsvetkov
Lavender breeding — Dr. B. Boyadzhiev
Mint breeding — Dr. D. Stanev
Biochemistry and physiology — Dr. D. Kosseva/ Dr. R. Detcheva
Industrial methods of growing — Dr. V. Topalov
Plant protection — Dr. I. Tanev
Technologies of processing ethereal and medicinal plants — Dr. A. Balinova

5 Kyustendil

5.1 Institute of Pomology
elevation: 518 m
rainfall : 608 mm

2500 Kyustendil
Phone: 2-26-12
Dir.: Dr. K. Doytchev

The Institute serves S. West Bulgaria on problems concerning fruit production. Breeding and research with apples, pears, cherries.

Breeding and variety trials	Dr. V. Georgiev
Fruit growing practices	Dr. B. Milanov
Irrigation	Dr. K. Doytchev

6 Kostinbrod

6.1 Institute of Plant Protection
elevation: 535 m
rainfall : 600 mm

2230 Kostinbrod
Phone: 87-72-25
Dir.: Dr. G. V"lkov

The Institute is the main center of the plant protection research on all crops, including horticultural ones

Phytopathology	Dr. M. Mladenov
Plant immunity	Dr. Chr. K"rjin
Entomology	Dr. Fr. Straka
Virology	Dr. M. Yankulova
Weed control	Dr. I. Lubenov
Toxicology	Dr. I. Balinov
Biomethods	Dr. E. Videnov

6.2 Small Fruit Experiment Station
The Station is a center of research with small fruit crops (strawberry, raspberry, dewberry), variety trials, breeding, cultural practices. Production of virus free plant material by tissue culture methods.

2230 Konstinbrod
Phone: 20-70
Telex: 23751
Dr. V. Ivanov

Strawberry breeding	Dr. L. Popova/ Dr. D. Djelianov
Raspberry and dewberry breeding	Dr. L. Christov/ Dr. N. Borcheva
Cultural practices	Dr. V. Ivanov
Tissue culture	Dr. S. Stoyanov/ V. Velchev

7 Pleven

7.1 Institute of Viticulture and Enology
elevation: 163 m
rainfall : 600 mm

5800 Pleven
Phone: 2-24-68
Dir.: Dr. I. Malenin

The Institute coordinates and directs research work in the field of viticulture in Bulgaria. There are 75 ha of experimental plantations and an experimental winery with the annual production of 1 mln litres top-quality selected wines. The experiment base of the Institute produces annually 2 mln grafted grapevines.

Genetics and breeding	Dr. V. V"lchev
Physiology and ampelography	Dr. A. Donchev
Biology	Dr. K. Katerov
Growing practices	Dr. J. Magrisso
Mechanization	Dr. A. Petrakiev/ A. Georgiev
Plant protection	Dr. I. Malenin
Economics and management	Dr. T. Cholakov
Enology	Dr. G. Petkov
Planting material	Dr. P. Mamarov

8 Plovdiv

8.1 "Maritsa" Vegetable Crops Research Institute
elevation: 160 m
rainfall : 520 mm

4003 Plovdiv
Phone: 5-22-96
Telex: 44525 IZK
Dir.: Dr. V. Karaivanov

The Institute coordinates and directs research work on vegetables and potatoes in Bulgaria. There are 300 ha irrigated area, 4 ha of glasshouses, a modern laboratory for the production of virus-free planting material of potatoes and flowers through the tissue-culture method, an experiment laboratory for processing and technological evaluation of vegetable varieties. Super elite and elite vegetable seeds are produced here.

Genetics and breeding of tomatoes - male sterility, diseases resistance, heterosis	Dr. M. Yordanov/ Dr. L. Stamova
Pepper - breeding for heterosis and resistance	Dr. Y. Tododrov/ Dr. D. Kostova
Peas, beans - breeding for resistance and mechanized harvesting	Dr. I. Poryazov/ E. Uzunova
Cucumber - breeding for female type varieties and heterosis, disease resistance	Dr. M. Alexandrova
Potatoes - breeding for resistance and earliness	Dr. St. Kalfov
Phytopathology	Dr. D. Bachariev
Entomology	Dr. S. Stefanov/ E. Loginova
Growing practices	Dr. G. Dimitrov
Irrigation	Dr. I. Dimov
Mechanization of the processes in vegetable growing	Dr. B. Kumanov/ Dr. D. Popov
	Dr. St. V"lchev
Crop rotation	Dr. P. Sourlekov
Glasshouse cultivation	Dr. S. Spassov/ Dr. Ch. Simitchiev
Plastics	Dr. G. Tsekleev
Laboratory of biochemistry and quality	Dr. Kh. Manuelyan
Laboratory of physiology	Dr. S. Genchev
Laboratory of agrochemistry	Dr. V. Rankov/ Dr. V. Kanasirska
Tissue culture laboratory	Dr. V. Rodeva
Economics and organization	Dr. V. Karaivanov/ Dr. E. Tongova
	Dr. D. Kostov

8.2 Fruit Growing Institute

4000 Plovdiv
Phone: 7-13-49
Telex: 48440
Dir.: Prof.Dr. V. Belyakov

The Institute coordinates and directs research work on fruits in Bulgaria. There are 400 ha of irrigated land, a laboratory for the production of virus-free planting stock of fruits and strawberries, and a nursery for elite planting stock.

Peach breeding	Dr. Y. Grigorov/ Dr. St. Dabov
Apple breeding	Dr. V. Djuvinov
Walnut breeding	Dr. N. Nedev
Grapewine breeding	Prof.Dr. M. Kondarev
Interspecies hybridization	Dr. Ch. Baev
Mechanization	Prof.Dr. V. Belyakov/ Dr. K. Gogova
Plant protection	Dr. Sp. Ivanov/ Dr. R. Radev
Virology	Dr. M. Topchiiska/ Dr. R. Gbova
Physiology and biochemistry	Dr. A. Petrov
Agrochemistry	Dr. G. Stoilov/ Dr. I. Yovchev
Herbicides	Dr. I. Stamatov
Irrigation	Dr. D. Dochev/ Dr. M. Gospodinova

Economics and organization — Dr. A. Grozdin
Planting stock — Dr. G. Trachev
Cutting and shaping — Dr. I. Jelev/ Dr. N. Maximov
Microreproduction and tissue culture — Dr. K. Kornova

8.3 "Vassil Kolarov" Higher Institute of Agriculture, Faculty of Viticulture and Horticulture

4000 Plovdiv, ul. Mendeleev 12
Phone: 2-41-00
Rector: Prof. Dr. Ts. Petkov

The Institute is the only educational institution in Bulgaria where engineer agronomists with a higher education are trained. It has well-equipped classrooms, an experiment base and a hostel. There are 4 departments - viticulture and horticulture, field crop, plant protection and subtropical agriculture. Research work on horticulture is done in close contact with other research institutes.

Department of vegetable and flower - breeding and growing methods — Prof. Dr. P. Kartalov, Prof. V. Angeliev, Prof. Dr. Ch. Petrov/Dr. N. Nikolova'

Department of pomology - breeding and growing methods — Prof. S. Popov/Prof. Dr. T. Angelov, Prof.Dr. P. Mitov

Subtropical crop — Prof. Dr. Tz. Tzolov

Department of viticulture - breeding and growing methods, ampelographic studies — Prof. Dr. D. Babrikov, Prof. Dr. B. Tzankov, Prof. Dr. L. Radulov, Prof. Dr. Z. Zankov

9 Pomorie

9.1 Experiment Station for Pomology and Viticulture
elevation: 20 m
rainfall : 470 mm

8200 Pomorie
Phone: 22-91
Dir.: Dr. A. Boychev

The station elaborates specific problems of viticulture and fruit growing in the southern part of the Black Sea shore.

Fruit tree biology with special interest to breeding and variety studies of almond — Dr. N. Lalev
Growing peaches and almonds — Dr. Y. Gürcheva/A. Djeneva
Growing methods and variety trials with winegrape — Dr. G. Draganov/Dr. B. Rangelov
Herbicides — Dr. A. Boychev
Plant protection — Dr. I. Iliev

10 Samokov

10.1 Potato Experiment Station
elevation: 1030 m
rainfall : 653 mm

2000 Samokov
Phone: 20-25
Dir.: Dr. P. V"alchev

Breeding of potatoes — Dr. P. V"alchev
Agrotechniques of potatoes — Dr. St. Dimitrov

11 Sandanski

11.1 Agricultural Experiment Station
elevation: 191 m
rainfall : 534 mm

2800 Sandanski
Phone: 21-91
Dir.: Dr. S. Bakhchevanova

The station works on specific problems of viticulture

and horticulture in the south-western part of Bulgaria, characterized by a warm and mild climate.

Viticulture	Dr. R. Stoeva
Grape growing methods	Dr. Y. Atanassov/S. Tchingov
Growing methods and variety trials with pears, peaches, strawberry	Dr. S. Sapundjieva
Variety trials with vegetables	Dr. S. Bakhchevanova
Early vegetable production - growing methods	Dr. I. Belichki

12 Silistra

12.1 Agricultural Experiment Station
elevation: 20 m
rainfall : 546 mm

7500 Silistra
Phone: 2-59-61
Dir.: Dr. P. Marinov

The station is associated with the Agro-Industrial Complexe - Silistra.

Breeding and growing of apricot, pest and disease control	Dr. P. Marinov/N. Nikolov

13 Sofia

13.1 Institute of Genetics
elevation: 550 m
rainfall : 570 mm

1113 Sofia
Phone: 76-80-75
Dir.: Prof. Dr. M. Stoilov

The Institute is under the direction of the Bulgarian Academy of Sciences. It carries out fundamental research on genetics, breeding and heterosis of cultivated plants. Great attention is being paid to the problems of tomatoes and pepper.

Genetics and breeding of tomato	Dr. Ch. Georgiev/Dr. B. Vladimirov
Genetics and breeding of pepper	Dr. St. Daskalov/Dr. L. Milkova
Heterosis in tomato and pepper	Dr. Ch. Georgiev/Dr. M. Popova
Interspecific hybridization in tomato	Dr. Z. Achkova
Cytological and cytoembryological studies of tomato and pepper	Dr. E. Molhova/M. Abadjieva
Tissue culture	Dr. A. Atanassov

13.2 Experiment Station of Floriculture and Ornamentals

1758 Negovan, Sofia
Phone: 39-46-39
Dir.: Dr. M. Dinova

The station directs research work on floriculture in Bulgaria. It has a well-equipped tissue-culture laboratory.

Breeding of carnation	Dr. A. Boykov
Variety trials and growing methods	Dr. M. Dinova/Dr. D. Kirilov
Pot flowers	Dr. Al. Gürov
Agrotechniques of roses	Dr. I. Groshkov
Tissue culture	Dr. I. Chavdarov

14 Troyan

14.1 Experiment Station of High-Mountain Agriculture

5600 Troyan

elevation: 422 m
rainfall : 746 mm

Phone: 42-71
Dir.: Dr. P. Donchev

Plum breeding — Dr. P. Mondeshka

15 Tsar Krum

15.1 Agricultural Experiment Station
elevation: 120 m
rainfall : 670 mm

9863 Tsar Krum, District of Shumen
Phone: 5-71-09
Dir.: Dr. K. Kunev

Fruit tree growing methods — Dr. K. Kolev
Viticulture - growing methods — Dr. H. Alishev
Plant protection — Dr. K. Kolev

16 Varna

16.1 Experiment Station of Viticulture
elevation: 20 m
rainfall : 498 mm

9000 Varna
Phone: 3-00-78
Dir.: Dr. B. Danailov

Viticulture: growing methods and variety trials — Dr. B. Dimchev
Enology — Dr. N. Spirov
Plant protection — Dr. P. V"alcheva
Plant nutrition — Dr. B. Danailov

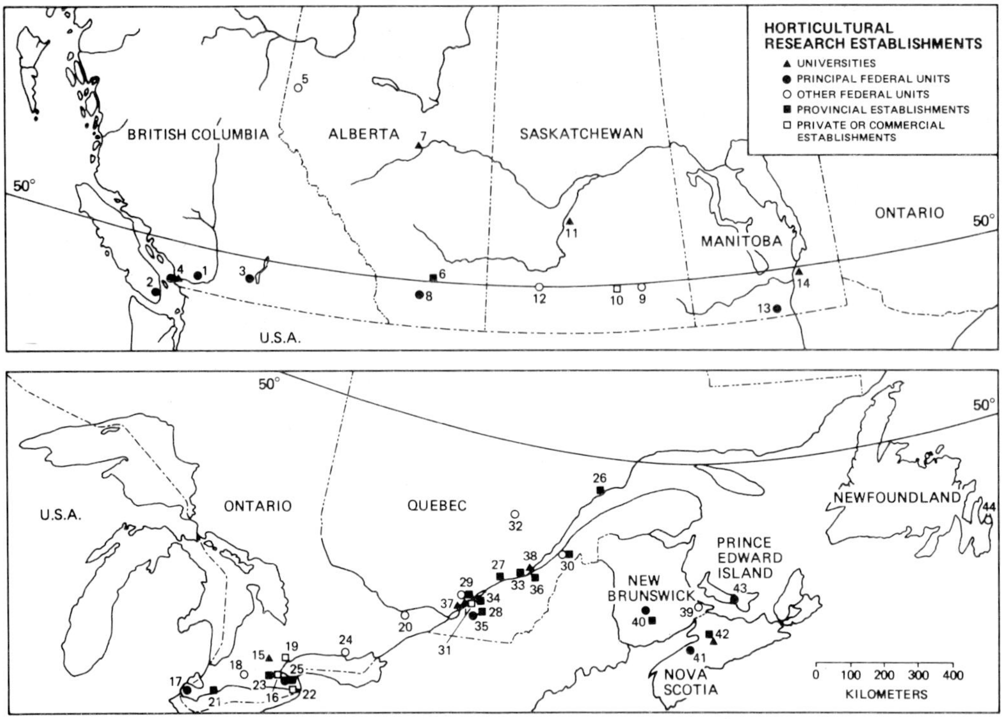
HORTICULTURAL RESEARCH ESTABLISHMENTS
▲ UNIVERSITIES
● PRINCIPAL FEDERAL UNITS
○ OTHER FEDERAL UNITS
■ PROVINCIAL ESTABLISHMENTS
□ PRIVATE OR COMMERCIAL ESTABLISHMENTS
BRITISH COLUMBIA
ALBERTA
SASKATCHEWAN
MANITOBA
ONTARIO
U.S.A.
50°
QUEBEC
NEW BRUNSWICK
PRINCE EDWARD ISLAND
NOVA SCOTIA
NEWFOUNDLAND
0 100 200 300 400
KILOMETERS

British Columbia: 1 Agassiz, 2 Sidney, 3 Summerland, 4 Vancouver, **Alberta**: 5 Beaverlodge, 6 Brooks, 7 Edmonton, 8 Lethbridge, **Saskatchewan**: 9 Indian Head, 10 Regina, 11 Saskatoon, 12 Swift Current, **Manitoba**: 13 Morden, 14 Winnipeg, **Ontario**: 15 Guelph, 16 Hamilton, 17 Harrow, 18 London, 19 Mississauga, 20 Ottawa, 21 Ridgetown, 22 St. Catharines, 23 Simcoe, 24 Trenton, 25 Vineland Station, **Quebec**: 26 Chûtes-aux-Outardes, 27 Deschambault, 28 Farnham, 29 L'Assomption, 30 La Pocatière, 31 Montreal, 32 Normandin, 33 St-Augustin, 34 St-Hyacinthe, 35 St-Jean-sur-Richelieu, 36 St-Lambert-de-Levis, 37 St-Anne-de-Bellevue, 38 Ste-Foy, **New Brunswick**: 39 Bouctouche, 40 Fredericton, **Nova Scotia**: 41 Kentville, 42 Truro, **Prince Edward Island**: 43 Charlottetown, **Newfoundland**: 44 St. John's

Survey

the horticultural industry of Canada is restricted in the crop species represented due to its location in the northern temperate climate zone. This restriction is off-set by the diversity of production areas in the different regions of the country. The adaptation of crop species to these varied conditions frequently requires extensive plant breeding programs or new methods of culture, because available cultivars are generally from countries with quite contrasting climates. Emphasis must be placed on horticultural crop cultivars that are resistant to disease, drought, low winter temperature, and are adapted to a relatively cool, short season for growth and maturity. There are five reasonably distinct horticultural crop production regions in Canada. Listed from west to east these are British Columbia, the Prairie Provinces, Ontario, Quebec and the Atlantic Provinces.

British Columbia - The coastal region of British Columbia is the mildest production area in Canada and special winter-tender ornamentals such as holly and flowering dogwood grow well there. Berry crops are the principal fruits and the mid-Fraser Valley is one of the world's most favorable areas for raspberry production. Vegetables are also important in the Fraser Valley and on Vancouver Island. Blueberries (highbush) and cranberries are grown on organic soils in the delta region, as well as some of the vegetable crop production. Most of British Columbia is very mountainous and the province has less than five

Prepared by Dr. C.J. Bishop, Research Branch Headquarters, Agriculture Canada, Ottawa percent arable land. In the interior, the southern mountain valleys running north-south are suited to fruit production, especially the Okanagan Valley which grows all types of pome and stone fruits, though with emphasis on apples and peaches.

Prairie Provinces - The climate is severe in the winter with temperatures dropping to −40°C or less and relatively light snowcover. Perennial plants must therefore be extremely winter hardy to survive. The principal fruit is the crab apple and hardy cultivars are grown quite generally as backyard garden trees. Amelanchier is a popular wild shrub that is now starting to be grown domestically for its berry-like fruits, and some wild Prunus species are also of limited interest. Strawberries and raspberries do well where there is adequate snowcover. Vegetable crops must be suited to the relatively short growing season, 110 to 125 frost-free days, although normal summer temperatures prevail. Some regions further north may have frost any month during the summer, but production in these areas is also favored by the long, almost continuous days. Nights tend to be cool in all of the Prairies because of the drier, cloud-free weather, and sweet corn and tomato production, for example, require the earliest cultivars. Potato production is well adapted to the area and is fairly general. Most horticultural crops require supplementary irrigation, especially in the south-western Prairies.

Ontario - The southern part of Ontario is the most favored horticultural area of Canada, and the Niagara Peninsula is one of the best micro-climates on the continent for the production of peaches, cherries, grapes and other tender fruits. Unfortunately, the land is restricted to a lakeshore zone about 80 km long and 8 km wide. Even within this area urban development is a continual threat to its existence. The other particularly favorable area in Ontario is the extreme southwestern part of the province. It is especially suited for vegetable crops, e.g. processing tomatoes, but some of the lighter soils are used for apples, peaches, other tree fruits and grapes. Most of the remainder of the southern part of the province is suitable for horticulture, and potatoes, processing vegetables, apples and pears, etc., are grown in many areas. Limited deposits of organic soils are used for fresh vegetable production, especially onions, carrots, lettuce, celery, radishes, etc. Eastern Ontario just north of the St. Lawrence River was the site of origin of the famous Canadian apple, the McIntosh.

Quebec - Southwestern Quebec is also a suitable horticulture area, though somewhat less favorable than southern Ontario, primarily because of its heavier soils and more extreme winter temperatures. The tree fruit most grown commercially is the apple. Hardy cultivars do well on the gravely slopes of hills in the southern part of the province. Processing vegetables are important in Quebec, especially peas, green and wax beans, sweet corn and some tomatoes. The province is the major Canadian producer of canned green and wax beans. This area is the site of major organic soil deposits, and such vegetables as carrots, onions, lettuce and cole crops are of major importance. Carrot preservation in jacketed storages has made it possible to market this vegetable through much of the winter period.

Atlantic Provinces - A relatively temperate summer and winter characterizes the climate, although not as mild as the west coast. There is a relatively high snowfall and adequate summer rainfall, although conditions are changeable. Tree fruits, principally apples, are grown in the Annapolis Valley of Nova Scotia and to a limited degree in the Saint John River Valley of New Brunswick. Small fruits are important, especially strawberries and blueberries. The latter are principally in "managed" plantations of the wild lowbush blueberry and there is extensive production of this crop. The species is mainly *Vaccinium angustifolium*. Potato production is of major importance in New Brunswick and Prince Edward Island. Both provinces have extensive processing industries and produce frozen "French fries "for sale throughout Canada and for export. These two provinces are also the major source of seed potatoes. Internationally, Canada is the second largest shipper of seed potatoes with markets in many parts of the world. Other vegetables grown include processing peas and beans, and cole crops, almost entirely on mineral soils.

Horticultural Research in Canada - The centers of horticultural research in Canada may be grouped in four categories: (1) university departments, (2) the federal establishments of Agriculture Canada, (3) provincial government establishments, (4) private and commercial institutions.

Universities - Nine universities have departments of horticulture or departments of plant or biological sciences in which horticulture forms a part. These are the Universities of British Columbia, Alberta, Saskatchewan, Manitoba, Guelph (Ontario Agricultural College), McGill University (MacDonald College), Montreal, Laval University and Nova Scotia Agricultural College. All support some horticultural research as well as teaching. These universities, except Montreal, give agricultural degrees in which, as one possible option course, emphasis may be in horticulture. Post-graduate training to the Master's and Doctorate level is also given in all cases, except at the Nova Scotia Agricultural College. At Laval and Montreal Universities lectures are given in French; the other institutions are English speaking. Money for university research is received principally from federal and provincial government sources.

Agriculture Canada - Responsibility for federal research in horticulture is under the Research Branch of Agriculture Canada. Its Research Stations and

Experimental Farms, 48 in number, are situated in all major agricultural areas of Canada. The major federal Research Stations involved with horticulture include Agassiz, Sidney, Summerland and Vancouver, BC; Lethbridge, Alta.; Morden, Man.; Harrow, and Vineland, Ont.; St. Jean-sur-Richelieu, Que.; Fredericton, N.B.; Charlottetown, P.E.I.; and Kentville, N.S.

Provincial Government Establishments - The Alberta Horticultural Research Center, Brooks, Alberta directs its research toward the growing vegetable industry of southern Alberta, as well as the improvement of farm home gardens and grounds. Handling and storage methods for fresh vegetables are of particular interest. Propagation of trees for shelterbelts and roadsides as well as ornamental plantings are also given consideration. Cultivar testing involves *Malus, Prunus, Rubus, Fragaria,* and *Ribes,* and many species of vegetables and ornamentals. The Horticultural Research Institute of Ontario, Ontario Ministry of Agriculture and Food, Vineland Station, Ontario, is the major provincial research centre. The Institute continues to introduce commercially important varieties of fruits and vegetables. The Institute also coordinates province-wide research on total production systems for pome fruits, peaches and grapes, dealing with such aspects as land use and mechanization as well as variety testing and cultural systems. Work is also carried out on ornamentals, small fruits, mushrooms, and greenhouse vegetables, floriculture and energy conservation. Horticultural research in Ontario is also conducted at Colleges of Agricultural Technology located at Ridgetown, in the southern most part of the province, at Kemptville in eastern Ontario and at New Liskeard in northern Ontario. Vegetable crops are the major area of research at Ridgetown. Potato research is the new thrust at New Liskeard with some work on cole crops and small fruits, while Kemptville has a modest research program in fresh market vegetables, fruit and maple syrup technology.

The Quebec Department of Agriculture supports research at several stations in the province. A Quebec Crop Protection Service conducts research at several sub-stations in connection with the prevention, detection and control of plant diseases, destruction of weeds, and the biology and control of insects and other pests harmful to horticultural crops. A notice network advises the growers at appropriate times during the season about pest control measures and covers the following crops: small fruits, vegetables, potatoes, apples and sweet corn. The Quebec Soil Research Service conducts fertility and cultural trials on vegetable crops.

Private and Commercial Institutions - A few private and commercial institutions, and botanical gardens, conduct independent horticultural research for specific purposes in such fields as ornamental displays and nursery stocks, fruit and vegetable crops for processing, and chemicals for pest control. Others cooperate with or participate in work conducted at Canadian universities and government institutions or at centres outside Canada. In addition, a small number of amateur and professional horticulturists conduct specialized breeding work with certain ornamental and fruit species.

Scientific Societies - The Agricultural Institute of Canada is a professional organization, with approximately 5500 members engaged in all aspects of agriculture in Canada. Affiliated with it (among others) are the following specialized scientific societies whose members may be concerned with horticultural crops: Canadian Society for Horticultural Science, Canadian Society for Agronomy, Canadian Society for Soil Science, Canadian Society for Agricultural Engineering, Canadian Pesticide Management Society and Canadian Society of Extension.

The Institute has a national General Manager and service office in Ottawa and arranges facilities for technical sessions at the time of the Institute's annual meeting. At the provincial level, legally chartered Institutes of Agrologists maintain professional standards and provide a medium of contact through local Branches.

The Canadian Society for Horticultural Science is a society of professional horticulturists which exists to promote and foster the science of horticulture in Canada. The present membership is about 235. Annual and regional meetings are held each year. The Western Canadian Society for Horticulture maintains committees on fruits, vegetables, ornamentals and for extension work. Both the Canadian Society for Horticultural Science and the Western Canadian Society for Horticulture issue proceedings of annual meetings; these are distributed to members giving preliminary, up-to-date findings on much of the current horticultural research.

International Society for Horticultural Science - Many professional horticulturists in the country also hold individual membership in the ISHS and thereby maintain contact with fellow horticultural scientists throughout the world. At present these number about 100. Some 50 Canadians are also involved in the various ISHS Sections, Commissions, and Working Groups. Canada is a country member of the ISHS, and ten Canadian organizations are also members.

The Canadian Horticultural Council is an industry-based organization supported by associations of growers, shippers, wholesalers and processors, by marketing boards and by provincial government departments of agriculture. The Council comprises approximately 75 representatives and includes almost that number of member organizations. Its Committee on Horticultural Research compiles an annual report on Canadian horticultural research for the information of Council members and collaborating research laboratories (federal, provincial and university). This informal report provides, to those engaged in horticultural research in Canada, a convenient guide to most of the work in progress across the country each year.

Publications - Two Journals of the Agricultural Institute of Canada publish, in English or French, the results of original scientific research including those related to horticulture: the Canadian Journal of Plant Science and the Canadian Journal of Soil Science are published quarterly. Other journals of original research papers, such as the Canadian Journal of Botany and the Canadian Journal of Genetics and Cytology, published monthly and bi-monthly by the National Research Council of Canada, also frequently include research with horticultural plants. Periodicals of an extension or news type which might be of interest to horticulturists include: Agriculture (Ordre des Agronomes du Quebec); Agrologist (Agricultural Institute of Canada); Alberta Horticulturist (Alberta Horticultural Association); Annual Proceedings of the Nova Scotia Fruit Growers Association; British Columbia Orchardist; Bulletin des Agriculteurs; Canadian Florist; Gardenland; Greenhouse and Nursery; Canadian Food Industries; Canadian Fruitgrower; Canadian Plant Disease Survey (Agriculture Canada); The Grower (Ontario Fruit and Vegetable Grower's Association); Horticulture Review; Landscape Ontario; Landscape Trades; Lawn and Garden Trade; Nursery Trade; Prairie Landscape Magazine; Quebec Vert; Western Producer.

British Columbia

1 Agassiz

1.1 Research Station (Agriculture Canada)
elevation: 15 m
rainfall : 1638 mm
sunshine : 1447 h

Agassiz, British Columbia V0M 1A0
Phone: (604) 796-2221
Dir.: Dr. J.M. Molnar*

Plant physiology, crop management, soil and water management, and weed control for small fruits (raspberries, strawberries), vegetables and turf. Operates a substation at Abbotsford.

Small-fruit physiology (raspberries and strawberries), weed control (small fruits, vegetables)	Dr. J.A. Freeman*
Management systems (turf)	Dr. S.G. Fushtey
Soil and water management	Dr. J.C.W. Keng
Crops fertility requirements (small fruits, vegetables, filberts)	Dr. C.G. Kowalenko
Management systems, physiology and quality (vegetables)	A.R. Maurer
Post-harvest physiology (small fruits, vegetables)	Dr. P.W. Perrin

2 Sidney

2.1 Saanichton Research and Plant Quarantine Station (Agriculture Canada)
elevation: 30 m
rainfall : 640 mm
sunshine : 2200 h

8801 East Saanich Road, Sidney
British Columbia V8L 1H3
Phone: (604) 656-1173
Dir.: vacant

Plant science as related to physiology and nutrition of greenhouse vegetables and ornamentals, propagation of woody ornamentals, tissue culture of small fruits, as well as woody and herbaceous ornamentals, biological control of diseases and insect pests, hydroponic crop production, solar greenhouse technology and computer control of greenhouses.
Since 1966, the station has cooperated with the Plant Health Division, Food Production and Inspection Branch, Agriculture Canada in the operation of their Post Entry Quarantine Station for tree fruits and grapes at the station.

Ornamentals Section	
Woody ornamentals	Head: Dr. C.J. French
Floriculture	Dr. W.C. Lin
Tissue culture	Dr. P.J. Monette
Virology	Dr. A.W. Chiko
Vegetable Section	
Physiology	Head: Dr. E.M. Van Zinderen Bakker
Pathology	Dr. H. Hartmann
Entomology	Dr. D. Gillespie
Nutrition	C. Kempler
Quarantine Section	A/Head: Dr. W. Lanterman
Grape viruses	R.C. Johnson
Tree fruit viruses	D. Thompson

3 Summerland

3.1 Research Station (Agriculture Canada)	Summerland British Columbia V0H 1Z0
elevation: 454 m	Phone: (604) 494-7711
rainfall : 283 mm	Telex: 048-88174
sunshine : 2023 h	Dir.: Dr. D.M. Bowden

Production and processing of tree fruits and grapes. Of 70 ha under sprinkler irrigation, 42 are planted with tree fruits (apples, apricots, cherries, peaches, pears and prunes), two ha with grapes. Equipment: a three ha isolation orchard with special greenhouse for studies of virus diseases of fruit trees. Gammacell 220' with a 1200 Curie source of $Co.^{60}$. Freezing unit for hardiness testing of fruit tree stock, well-equipped fruit processing laboratory. Gas-liquid chromatographic equipment and atomic absorption apparatus for analysis. Facilities for studying light and other environmental factors, also growth regulating substances. Operates a sub-station at Kelowna for cultural, nutritional and entomological studies on tree fruits.

Entomology-Plant Pathology	
Bionomics of pear psylla	Head: Dr. R.D. McMullen
Orchard mite control, San Jose scale	Dr. N.P.D. Angerilli
Stone fruit insects	F.L. Banham
Apple pest management	Dr. J. Cossentine
Management of codling moth	Dr. V.A. Dyck
Tree fruit virus diseases	Dr. A.J. Hansen
Post harvest diseases	Dr. P. Sholberg
Tree fruit fungus diseases	Dr. R.S. Utkhede
Food Processing	
Food technology	Head: Dr. D.B. Cumming
Food chemistry and analytical methods	Dr. H.J.T. Beveridge
New product development	Dr. A. Paulson
Enology and food technology	G.E. Strachan
Pomology and Viticulture	
Herbicides, vegetation management, nutrition	Head: Dr. E.J. Hogue
Viticulture, grape breeding, tree varieties and training	L.G. Denby
Apple and cherry breeding	Dr. W.D. Lane*
Pomology, plant physiology and growth regulators	Dr. N.E. Looney*
Fruit storage and biochemistry	Dr. M. Meheriuk*
Orchard management, hardiness	Dr. H.A. Quamme*
Grape management	Dr. A.G. Reynolds
Soil Science and Agricultural Engineering	
Soil moisture	Head: Dr. D.S. Stevenson
Pesticide and environmental chemistry	Dr. A.P. Gaunce
Soil chemistry and management	Dr. P.B. Hoyt
Agricultural equipment - development and assessment	Dr. A.L. Moyls
Soil fertility and plant nutrition	Dr. G.H. Neilsen

Agricultural equipment - development and assessment	P. Parchomchuck

4 Vancouver

4.1 Research Station (Agriculture Canada)
elevation: 87 m
rainfall : 1200 mm
sunshine : 1925 h

6660 N.W. Marine Drive, Vancouver
British Columbia V6T 1X2
Phone: (604) 224-4355
Telex: 04-55210
Dir.: Dr. M. Weintraub

The Research Station at Vancouver conducts research on berries (strawberries, raspberries, cranberries and blueberries) and vegetables (potatoes, cole crops, onions, lettuce and asparagus). The Station is the main centre for plant virus research in Canada, with strong programs in both basic and applied areas. A major program for breeding strawberries and raspberries is located there, with cooperating researchers in other disciplines. Extensive research is also being carried out on integrated pest management programs, pesticide chemistry and nematology. Located on the campus of the University of British Columbia, it serves the coastal area on all problems concerning insects and plant disease of horticultural crops, including purification and characterization of viruses and the development of sensitive tools for diagnosis.

Entomology section	
Virus vectors	Head: Dr. A.R. Forbes
Aphid ecology	Dr. D.B. Frazer
Berry insects, leaf hopper vectors	J. Raine
Soil insects, biological control of weeds	A.T.S. Wilkinson
Biological control	Dr. D.A. Raworth
Pesticide chemistry	S.Y.S. Szeto
Insect virology	D.A. Theilmann
Vegetable insects	Dr. R.S. Vernon
Plant pathology section	
Raspberry viruses, virus characterization	Head: Dr. R. Stace-Smith*
Plant breeding, small fruits	Dr. H.A. Daubeny*
Nematology	Dr. T.C. Vrain
Strawberry viruses	Dr. R.R. Martin
Root rots	Dr. H.S. Pepin
Potato viruses	P.J. Ellis
Potato diseases, serology	Dr. N.S. Wright
Bacterial diseases	Dr. S.H. De Boer
Virus chemistry and physiology section	
Chemistry and ultrastructural cytopathology of viruses	Head: Dr. H.W.J. Ragetli
Virus interactions, seed transmission	Dr. R.I. Hamilton
Biochemical virology	Dr. G.G. Jacoli
Biophysical virology	Dr. J.H. Tremaine
Cytopathology, electron microscopy	Dr. M. Weintraub
Plant viruses, molecular biology	D'A. Rochon

4.2 University of British Columbia, Faculty of Agricultural Sciences, Department of Plant Science
elevation: 93 m
rainfall : 1250 mm
sunshine : 1870 h

Vancouver, British Columbia V6T 1W5
Phone: (604) 228-4384
Head: Dr. V.C. Runeckles

Research in pathology, physiology, entomology, genetics and weed sciences of tree fruits, small fruits, vegetables and ornamentals.

Plant pathology	Dr. R.J. Copeman
Tree fruits, small fruits, reproductive physiology	Dr. G.W. Eaton*

Entomology, integrated pest control	Dr. M.B. Isman
Vegetable genetics	Dr. N.R. Knowles*
Environmental physiology, growth analysis	Dr. P.A. Jolliffe*
Landscape architecture and horticulture weed science	Dr. M.K. Upadhyaya
Growth regulation and air pollution	Dr. V.C. Runeckles
Insect pests, slugs	Dr. W.G. Wellington
Tissue culture, seed physiology	Dr. C.R. Norton
Landscape architecture	L. Diamond
Landscape architecture	Dr. P.A. Miller
Landscape architecture	D.D. Paterson
Landscape architecture	M. Quayle
Taxonomy	Dr. R.L. Taylor

Alberta

5 Beaverlodge

5.1 Research Station (Agriculture Canada)
elevation: 745 m
rainfall : 467 mm
sunshine : 2126 h

Beaverlodge, Alberta T0H 0C0
Phone: (403) 354-2212
Dir.: Dr. J.D. McElgunn

The Beaverlodge Research Station is the most northerly one in Canada. Fruit research is conducted on strawberry, *Amelanchier* (saskatoon) and apple to provide new northern cultivars and/or new management practices for commercial growers and home gardeners in northwestern Canada. Operates the Fort Vermillion Experimental Farm.

Fruit management and breeding	Dr. J.G.N. Davidson

6 Brooks

6.1 Alberta Horticultural Research Centre (AHRC)
elevation: 756 m
rainfall : 335 mm
sunshine : 2334 h

Bag Service: 200, Brooks,
Alberta T0J 0J0
Phone: (403) 362-3391
Dir.: Thomas R. Krahn

Provincial institution (405 ha) administered as a unit of the Plant Industry Division of Alberta Agriculture. Responsible for commercial horticulture research and extension in Alberta with production field plots at Brooks, Bow Island, Strathmore, Edmonton and Peace River.
Provides research and diagnostic facilities for irrigated crop production in southern Alberta. Emphasis of horticulture research is on the production, protection, storage and utilization of existing and potential crops including potatoes, processing and fresh market vegetables, small fruits, greenhouse crops, native fruits, shelterbelt trees and ornamental shrubs and flowers. There is also a unit working on annual forages and other special crops. The growing season is 135 days.

Tree and bush fruit: selection and production	Dr. S. Mahadeva
Ornamentals: hardiness and propagation	E.B. Casement
Greenhouse crops: management, floriculture and vegetables	G. Grant
Potato: commercial, processing and variety development	C.A. Schaupmeyer
Potato: seed production and improvement	J. Letal (located at Olds, Alberta)
Commercial vegetables: production of processing and fresh vegetables	P. Ragan
Market gardening: strawberries, raspberries and vegetable production	L.G. Hausher
Special crops: dry beans, corn, pulses, cereals	Dr. R. Gaudiel
Irrigation water use: assigned staff of Irrigation	Dr. R.C. McKenzie

Division	
Post harvest physiology: filacell storage	Dr. K. Mallett
Food Processing and product development: cultivar evaluation and native fruit processing	Dr. T. Smyrl
Plant pathology: greenhouse and crop diseases	Dr. R.J. Howard
Entomology: insecticide efficacy and diagnosis	Dr. U. Soehngen
Weed control: herbicide efficacy and application techniques	R. Essau
Plant diagnostics	M. Dykstra

7 Edmonton

7.1 University of Alberta, Faculty of Agriculture and Forestry, Department of Plant Science
elevation: 665 m
rainfall : 447 mm
sunshine : 2315 h

Edmonton, Alberta T6G 2E3
Phone: (403) 432-3239
Head: Dr. W. Van den Born

Controlled environment production of food crops; physiology of earliness; cold hardiness; influence of environmental factors on growth and development, with special emphasis on long days, limited heat units and relatively short season; pathological problems with special emphasis on virus and mycoplasma diseases.

Vegetable and potato physiology	Dr. W.T. Andrew*
Virus and mycoplasma diseases of vegetables and ornamental crops	Dr. C. Hiruki
Landscape architecture, turfs, woody ornamentals and fruits	Vacant
Floriculture, physiology of protected ornamentals	Dr. E.W. Toop*

7.2 University of Alberta, Faculty of Agriculture and Forestry, Department of Food Science

Edmonton, Alberta T6G 2E3
Phone: (403) 432-3236
Head: Dr. F. Wolfe

Post harvest handling and physiology; storage and processing of fruits and vegetables. Chemistry of raw and processed potatoes.

Starch characteristics; natural phenolics of potatoes; use of surfactants in potato processing	Dr. D. Hadziyev
Aqueous freezants; water and solute movements in fruits and vegetables	Dr. M. LeMaguer
Development of "instant potato" products	Dr. B. Ooraikul
Glycoalcaloids in potatoes; flavor enhancement in fruits and vegetables	P. Sporns

8 Lethbridge

8.1 Research Station (Agriculture Canada)
elevation: 923 m
rainfall : 405 mm
sunshine : 2370 h

Lethbridge, Alberta T1J 4B1
Phone: (403) 327-4561
Telex: 03-849343
Dir.: Dr. D.G. Dorrell

Breeding of potatoes for the prairie region and disease control in potatoes and processing peas. Potato breeding program acquires new crosses from Fredericton Research Station (see 40.3) and several northern U.S. States and coordinates testing and selection in Alberta, Saskatchewan and Manitoba. Breeding for yield, quality and disease resistance and epidemiology study of bacterial ring rot. Study of seedling diseases and root rot in processing peas. Also research on soil, water and fertility relationship, weed, disease and insect control.

Plant Science Section — Head: Dr. D.B. Wilson
Potato breeding for the prairies — Dr. D.R. Lynch
Plant Pathology Section
Control of root rot and seedling diseases in peas — Head: Dr. F.R. Harper
Control of bacterial ring rot in potatoes — Dr. G.A. Nelson

Saskatchewan

9 Indian Head

9.1 Tree Nursery Division
elevation: 613 m
rainfall : 391 mm
sunshine : 2333 h

Indian Head, Saskatchewan S0G 2K0
Phone: (306) 695-2284
Sup't.: Dr. J.A.G. Howe

Administered by the Prairie Farm Rehabilitation Administration of the Department of Agriculture (Federal). Nursery production and distribution of tree and shrub seedlings for farm, field and roadside shelterbelts. Reclamation and conservation plantings in the Prairie region, with applied studies in propagation and storage, entomology, herbicides, soil nutrition and shelterbelt studies.

Nursery production section — Foreman: D.J. Gruber
Tree distribution section — Head: H. Fox
Ass.Head: R. White
Investigation section — Head: Dr. B. Neill
Ass.Head: W. Schroeder
Shelterbelt Studies: J. Kort

10 Regina

10.1 Hoechst Canada Inc. (Agriculture Division)

295 Henderson Drive, Regina
Saskatchewan S4N 6C2
Phone: (306) 924-1500
Telex: 071-3114

Hoechst Canada Inc. maintains a network of research farms across Canada for the testing of experimental crop protection agents. Horticultural research is conducted in southwestern Ontario with beans, corn and potatoes being the major crops. Research on grapes and tree fruits is carried out in the Niagara Peninsula and a number of vegetable crops (e.g., cole crops, carrots, onions) are tested at the Bradford Marsh. In western Canada, horticultural research is conducted at Portage La Prairie, Manitoba (beans, peas, potatoes) and Lumsden, Saskatchewan (cole crops). In addition, Hoechst Canada Inc. cooperates with and/or participates in pest control research at universities and government institutions.

11 Saskatoon

11.1 University of Saskatchewan
Department of Horticulture Science
elevation: 497 m
rainfall : 370 mm
sunshine : 2318 h

Saskatoon, Saskatchewan S7N 0W0
Phone: (306) 966-5855
Telex: 07742659

Laboratories equipped for cryopreservation, — Head: Prof.Dr. C. Stushnoff
Cold hardiness, tissue culture and adenine nucleotide research — Prof.Dr. S.H. Nelson*
Prof.Dr. E.A. Maginnes*
Breeding *(Amelanchier, Fragaria, Malus, Rubus)* — Prof. D.H. Dabbs*/A.T. Ward

Vegetable culture and weed control, potato starch. Energy conservation greenhouses. Woody ornamental cultivar testing, hardiness breeding.

12 Swift Current

12.1 Research Station (Agriculture Canada)
elevation: 754 m
rainfall : 360 mm
sunshine : 2200 h

P.O. Box 1030, Swift Current
Saskatchewan S9H 3X2
Phone: (306) 773-4621
Dir.: vacant

Variety evaluation and cultural studies on a small scale include fruits, vegetables and flowering and woody ornamentals.

Manitoba

13 Morden

13.1 Research Station (Agriculture Canada)
elevation: 299 m
rainfall : 528 mm
sunshine : 2197 h

Morden, Manitoba R0G 1J0
Phone: (204) 822-4471
Dir.: Dr. D.K. McBeath

Varietal improvement of hardy ornamentals for the Prairie Provinces (roses, chrysanthemums, lilies, Heuchera, Monarda, Penstemon, street trees, willow, poplar); production practices for potatoes, cole and root vegetables; weed control; processing of food crops. Equipment: A well equipped food processing laboratory conducts quality studies of potatoes, vegetables, new crops and essential oils; greenhouses; analytical laboratories; a 30-ha arboretum of hardy trees and shrubs includes 5000 taxa.

Horticultural Crops Section

Vegetable crops - management	Head: Dr. B.B. Chubey
Woody ornamental - breeding	C.G. Davidson
Roses, herbaceous ornamentals - breeding	L.M. Collicutt
Potatoes - management	B.L. Rex
Weed control	D.A. Wall
Food processing	Dr. G. Mazza

14 Winnipeg

14.1 University of Manitoba, Department of Plant Science (Horticulture Section)
elevation: 215 m
rainfall : 500 mm
sunshine : 2200 h

Winnipeg, Manitoba R3T 2N2
Phone: (204) 474-8221
Telex: 07-587-721
Head: Dr. M.K. Pritchard*

A university department with training and research in the plant sciences including horticulture. There are two ha for testing cultivars of annual and perennial flowers and vegetables. A new test arboretum for woody species incorporated into two km of native stand along a riverbank is being established. Training of graduate students in areas of ornamental horticulture, potatoes and vegetables.

Vegetable and potato trials and storage	Dr. M.K. Pritchard*
Physiology and greenhouse research	Dr. L.J. LaCroix
Ornamental horticulture	L.M. Lenz
Fruit crops	J.A. Menzies
Woody ornamentals	Dr. W.R. Remphrey

Ontario

15 Guelph

15.1 Department of Horticultural Science

Guelph, Ontario N1G 2W1

University of Guelph*
elevation: 78 m
rainfall : 833 mm
sunshine : 2035 h

Phone: (519) 824-4120
Head: Dr. I.L. Nonnecke

Conducts teaching of horticultural science at the diploma, degree and graduate levels. Research includes environmental stress, photo-respiration physiology, plant breeding, genetics, plant nutrition, post-harvest physiology relating to CA, low oxygen, herbicides and growth regulators. All production areas are examined on a commodity basis including apples, grapes, small fruits, asparagus, potatoes, rutabagas, Brassica sp., muck crops, minor vegetables, nursery and turf, floriculture, roses and geraniums. Departmental laboratories relating to physiology, biochemistry, plant nutrition, tissue culture, plant breeding, cytogenetics, environmental stress. Forty growth rooms, 2500 m² of greenhouses, 500 m² storage rooms, 49 ha of field research at Cambridge, Ontario.

Perennial crop herbicides	N. Cain, Coord. ext.
Continuing education - teaching	H.R. Crawford
Turf research	Dr. J.L. Eggens
Small fruit breeding - cytogenetics	Vacant
Photo-respiration	Dr. B. Grodzinski
Plant genetics and breeding	Dr. P.M. Harney
Post-harvest physiology	Dr. E.C. Lougheed*
Nursery physiology	Dr. G.P. Lumis
Continuing education - teaching	Vacant
Post-harvest physiology/hormones	Dr. D.P. Murr
Seed physiology-production	Dr. I.L. Nonnecke, Chairman
Environmental stress physiology	Dr. D.P. Ormrod*
Pomology	Dr. J.T.A. Proctor*, Coord. Fruit Research
Plant nutrition	Dr. J.W. Riekels
Vegetable physiology	Vacant
Molecular genetics - vegetable breeding	Dr. V. Shattuck
Nitrogen metabolism	Dr. B. Shelp
Genetics and physiology - vegetable herbicides	Dr. V. Souza Machado
Vegetable production	Dr. H. Tiessen*, Coord. Veg. Res.
Floriculture physiology	Dr. M.J. Tsujita, Coord. Orn. Res.
Biochemistry	Dr. A. Zitnak
Potato breeding (Agriculture Canada)	Dr. R. Coffin
Floriculture (Ontario Ministry of Agriculture and Food)	J. Hughes, Ext. Spec.
Bio-physic-water relationship	Dr. M. Dixon

16 Hamilton

16.1 Royal Botanical Gardens
elevation: 50 m
rainfall : 839 mm
sunshine : 2050 h

Hamilton, Box 399, Ontario L8N 3H8
Phone: (416) 527-1158
Dir.: A.P. Paterson

Scientific, educational and cultural institution funded by tax-based grants from local jurisdictions, individual and corporate contributions and the Province of Ontario. Site has 800 ha including arboretum and horticultural gardens displaying living collections of ornamentals and 480 ha of natural areas preserving and interpreting various local environments. Research and teaching in horticulture, arboriculture, plant taxonomy, ornamental plant breeding and natural history. Centre for Canadian Historical Horticultural Studies (CCHHS). Province-wide extension program provided by R.B.G. Outreach. Supporting resources including library, herbarium and greenhouses. A cool greenhouse complex (opened 1986) displays species from around the world with a Mediterranean climate.
Collections: Wide range of genera including *Clematis, Iris, Magnolia, Malus, Prunus, Rhododendron, Rosa* and *Syringa* (International Registration Authority for *Syringa).* Topical collections include arboretum, climbers, hedges, herbaceous plants, medicinal plants, native trees and shrubs, scented garden.

Publications: Pappus (includes The Gardens' Bulletin); Technical Bulletin; Special Bulletin (annual report); Canadian Horticultural History, an interdisciplinary journal (CCHHS).

Collections	F. Vrugtman*
Conservation and plant protection	Dr. P.F. Rice
Education	J. Lord
Environmental biology	W.L. Simser
Horticulture	C. Graham
Library and CCHHS	Ina Vrugtman
Plant breeding	H. Pearson
Plant taxonomy	Dr. J.S. Pringle

17 Harrow

17.1 Research Station (Agriculture Canada)
elevation: 191 m
rainfall : 837 mm
sunshine : 1991 h

Harrow, Ontario N0R 1G0
Phone: (519) 738-2251
Dir.: Dr. C.F. Marks

This station serves a major vegetable and fruit production area in southwestern Ontario. Interdisciplinary research is conducted on tree fruits: peach, nectarine, apricot, pear, apple; grapes; field vegetables: tomatoes, vine crops, cole crops; and greenhouse vegetables: tomatoes, cucumbers. Studies are conducted on the breeding and evaluation of tree fruits, emphasizing increased yields, cold hardiness and disease resistance; and in the breeding of processing and fresh fruit market tomatoes. Studies on the biology and control of diseases, insects and weeds of vegetable and fruit crops are also emphasized. Crop management research is undertaken including irrigation, soil fertility and integration of various crop management practices for optimum production of horticultural crops. A modern research laboratory and greenhouse complex, growth chambers and extensive field plots are available for research on horticultural crops.

Horticultural Science Section	
Tree fruit breeding	Head: Dr. R.E.C. Layne*
Vegetable evaluation and management	R.W. Garton
Vineyard management	D.M. Hunter
Orchard management	Dr. F. Kappel
Vegetable physiology	Dr. A. Liptay
Grape cultivar and rootstock evaluation	R. Michelutti
Greenhouse crop management	Dr. A.P. Papadopoulos*
Vegetable breeding	Dr. V. Poysa
Soil Science Section	
Soil fertility	Head: Dr. W.I. Findlay
Greenhouse energy engineering	T.J. Jewett
Soil moisture and agrometeorology	Dr. C.S. Tan*
Weed Science and Chemistry Section	
Weed science	Head: Dr. A.S. Hamill
Herbicide chemistry	Dr. J.D. Gaynor
Weed physiology	Dr. P.B. Marriage
Weed ecology	Dr. S.E. Weaver
Plant Pathology Section	
Vegetable diseases - biocontrol	Head: Dr. W.R. Jarvis
Bacterial diseases - fruit	Dr. W.G. Bonn
Bacterial diseases - vegetables	Dr. B.N. Dhanvantari
Vegetable diseases	Dr. L.F. Gates
Tree fruit diseases	Dr. J.A. Traquair
Pea root rot	Dr. J.C. Tu
Entomology Section	
Insect pathology	Head: Dr. R.P. Jaques
Vegetable insects	Dr. G.J.R. Judd

Greenhouse and field vegetable insects — Dr. R.J. McClanahan

18 London

18.1 Research Centre (Agriculture Canada)
elevation: 278 m
rainfall : 909 mm
sunshine : 1894 h

London, Ontario N6A 5B7
Phone: (519) 679-4452
Dir.: Dr. H.V. Morley

This centre addresses current health and environmental concerns regarding the use of pesticides for crop protection. Multidisciplinary teams carry out research in integrated pest management and environmental toxicology.

Crop Protection - Entomology

Stored products - fumigation	Dr. E.J. Bond
Insect toxicology	Dr. C.H Harris
Physiology	Dr. D.G.R. McLeod
Insect attractants and repellants	Dr. A.M. Starratt
Applied entomology	Dr. J.H. Tolman
Pesticide ecology	Dr. A.D. Tomlin
Insect pathogens	Dr. C.M. Tu
Insect toxicology	S.A. Turnbull

Crop Protection - Pathology

Phytobacteriology - tomato disease	Dr. D.A. Cuppels
Fungicides - potato diseases	Dr. G. Lazarovits
Biochemistry - growth regulators	Dr. T.T. Lee
Biochemistry - tomato diseases	Dr. C. Madhosingh
Fungicide transport	Dr. D.M. Miller
Herbicides - microflora interaction	Dr. E.B. Roslycky
Phytoalexins - resistance	Dr. E.W.B. Ward
Fungicide mode of action	Dr. G.A. White

19 Mississauga

19.1 Ciba-Geigy Canada Ltd. and Green Cross, Division of Ciba-Geigy Canada Ltd.

6860 Century Avenue, Mississauga,
Ontario L5N 2W5
Phone: (416) 821-4420
Telex: 06-217752
Dir. R&D: Dr. D.R. Ridley

Ciba-Geigy handles agricultural products which also include horticultural products. Green Cross handles the home and garden (including horticultural) products as well as turf products. Research and development of fungicides (potatoes, tree fruits and vegetables), insecticides and herbicides (potatoes, peas, beans, corn, strawberries, turf, etc.). Entomology, chemistry and plant pathology.

20 Ottawa

20.1 Research Branch Headquarters (Agriculture Canada)

Central Experimental Farm, Ottawa
Ontario K1A 0C5
Phone: (613) 995-7084
Telex: 053-3283
Ass.Dep.Min.: Dr. E.J. LeRoux

The headquarters of the Research Branch of the federal Department of Agriculture is located in the Sir John Carling Building at the Central Experimental Farm. The broad administration and coordination of the national research program is directed from this centre through the Branch Management Committee (including the regional Directors General) and the Coordination Directorate.

Research Coordinator (horticultural crops) — Dr. C.J. Bishop*

20.2 Research Station (Agriculture Canada)
elevation: 80 m
rainfall : 846 mm
sunshine : 1989 h

Central Experimental Farm, Ottawa
Ontario K1A 0C6
Phone: (613) 995-5287
Telex: 053-3283
Dir.: Dr. A.I. de la Roche

Although the main objective of the Ottawa Research Station is the breeding of superior feedgrain and forage cultivars involving genetics, cytology, physiology, pathology and entomology, considerable research is directed to increasing efficiency of production in ornamentals and the development of new cultivars of winterhardy roses and flowering shrubs. Floricultural physiology, effect of mycorrhiza on nursery crops, and diseases of greenhouse and nursery plants constitutes a major part of the ornamentals program. The Station also is responsible for the Plant Gene Resources of Canada, and for development and maintenance of the Dominion Arboretum and ornamental display and test gardens.
Equipment: Extensive experimental field, greenhouse and growth facilities, long-term seed storage units, well-equipped laboratories and libraries.

Ornamentals Section

Pathology	Head: Dr. A.T. Bolton
Physiology and floriculture	Dr. J.A. Simmonds
Plant breeding	Dr. F. Svejda*
Nursery research	Dr. S. Nelson
Arboretum and test gardens	Curator: T.J. Cole
Plant gene resources	Head: Dr. R. Loiselle

21 Ridgetown

21.1 College of Agricultural Technology, Ontario
Ministry of Agriculture and Food
elevation: 206 m
rainfall : 820 mm

Ridgetown, Ontario N0P 2C0
Phone: (519) 674-5456
Principal: D.W. Taylor

Provincial institution, administered by the Ontario Ministry of Agriculture and Food: research, teaching and extension. Processing crops: variety trial studies, crop protection (insect, disease and weed control).

Horticulture and Biology Section	Head: R.H. Brown
Cultivar trials: peppers, tomatoes, cole crops, cucumbers, grapes, tree fruits and sweet corn	J.K. Muehmer*
Weed control: asparagus, lawns, tomatoes, beans, potatoes, cole crops and cucurbit crops	C.J. Swanton/R.H. Brown
Disease control: tomatoes and potatoes	Dr. R.E. Pitblado
Insect control: potatoes, tomatoes, peppers, sweet corn, cucumbers and snap beans	J.G. Morton
Nutrition research: tomatoes, cauliflower and peppers	J.G. Morton

22 St. Catherines

22.1 Bright's Wine Limited
elevation: 98 m
rainfall : 819 mm
sunshine : 2048 h

R.R. Apt. 4, Niagara-on-the-Lake
Ontario L0S 1J0
Phone: (416) 358-7141
Res.Dir.: G.W.B Hostetter

Vineyard Research Division with established experimental plots of *Vitis vinifera* and *Vinifera hybrids*. Research on all aspects of grape vine culture. Grape breeding, nutrition and rootstocks. Viticulture: grape pruning, rooting of cuttings, insect and disease control, herbicides, trellising.

Viticulturist — J. MacFarlane

22.2 Jordan and Ste-Michelle Cellars Ltd.
elevation: 105 m
rainfall : 825 mm
sunshine : 2048 h

Bag Service: 3022, 120 Ridley Road
St. Catherines, Ontario L2R 7E3
Phone: (416) 688-2140
Res.Dir.: J. Zimmermann

Evaluation of grape cultivars using various cultural techniques in the Niagara Peninsula (St. Catherines and Vineland, Ontario) and the Okanagan Valley (Surrey, British Columbia). Shoot tip multiplication, of material from around the world, to distribute to grower field trials. Studying the feasibility and effectiveness of various field grafting techniques as an alternative to replanting vineyards of low quality cultivars.

23 Simcoe

23.1 Horticultural Experiment Station
elevation: 241 m
rainfall : 907 mm
sunshine : 1915 h

Box 587, Simcoe, Ontario N3Y 4N5
Phone: (519) 426-7120
Dir.: A. Loughton*

Section of the Horticultural Research Institute of Ontario, Vineland Station (see 25.1). Research with vegetable and fruit crops (culture and breeding) for production in the central countries along the northern shore of Lake Erie with province wide relevance for apples, tomato breeding and herbicides.

Apples	Dr. D.C. Elfving*
Small fruits	Dr. A. Dale
Potatoes	Dr. A.W. McKeown
Tomatoes	Dr. W.H. Courtney
Asparagus, baby carrots, cole crops and alternative crops for tobacco farms	A. Loughton*
Peppers and pickling cucumbers management, herbicides	Dr. J. O'Sullivan

24 Trenton

24.1 Smithfield Experimental Farm (Agriculture Canada)
Operated under the Vineland Research Station
(see 25.2)
elevation: 125 m
rainfall : 940 mm
sunshine : 1960 h

P.O. Box 340, Trenton
Ontario K8V 5R5
Phone: (613) 392-3527
Sup't.: Dr. S.R. Miller

Management, nutrition, physiology, breeding and processing of fruit (apples, pears) and vegetable crops (peas, tomatoes, cucumbers and other processing crops).

Pomology	Dr. S.R. Miller
Food processing	Dr. W.P. Mohr
Tomato breeding	J.G. Metcalf
Vegetable management	N.J. Smits
Pomology	J. Warner

25 Vineland Station

25.1 Horticultural Research Institute of Ontario
Ontario Ministry of Agriculture and Food
elevation: 79 m
rainfall : 797 mm
sunshine : 2030 h

Vineland Station, Ontario L0R 2E0
Phone: (416) 562-4141
Telex: 06 22546
Dir.: Dr. F.C. Eady*

Responsible for integration of all horticultural research in Ontario supported by the Ontario Ministry of Agriculture and Food and for research programs at the Grape Research Station (14 ha), Vineland Horticultural

Experiment Station (87 ha), Simcoe Horticultural Experiment Station (80 ha, see 23.1), the Muck Research Station, Bradford Marsh (4 ha) and the Horticultural Products Laboratory. Research projects include development of new cultivars of fruits, vegetables and ornamentals better suited to Ontario climatic and market requirements; cultivar testing, production research, plant nutrition, plant population, soil management, hardiness, pruning, propagation, weed control, mechanization and plant analysis. Food processing, post-harvest physiology and storage, and wine production are studied by the Horticultural Products Laboratory. The Muck Research Station conducts studies with vegetable crops (carrots, onions, potatoes, lettuce, celery, cauliflower and parsnips) on organic soils. Special facilities include 3700 m^2 of research greenhouses, a mushroom research facility and a protected environment (.2 ha) for fruit production.

Horticultural Experiment Station	
Production research, vegetables	Head: Dr. F.J. Ingratta*
Ornamentals	Dr. C. Chong
Peaches	Dr. N.W. Miles
Grapes, apricots	K.H. Fisher*
Pears, cherries, plums	Dr. G. Tehrani*
Mushrooms	Dr. D.L. Rinker
Floriculture	Dr. T.J. Blom
Plant nutrition	Dr. R.A. Cline/H.J. Reissmann
Plant pathology	Dr. Z.A. Patrick (Univ. of Toronto)
Horticultural Products Laboratory	
Microbiology	Head: Dr. R.V. Chudyk
Biochemistry	Dr. T. Fuleki
Food science	Dr. S. Wang
Post-harvest physiology	Dr. R.B. Smith*/Dr. C.L. Chu
Oenology	Dr. W. Edinger
Engineering	E.M. Lauro
Sensory evaluation	V.P. Gray
Muck Research Station	
Vegetable crops	Head: M. Valk

25.2 Research Station (Agriculture Canada)
elevation: 77 m
rainfall : 817 mm
sunshine : 2030 h

Vineland Station, Ontario L0R 2E0
Phone: (416) 562-4113
Dir.: Dr. D.R. Menzies

Protection of tree fruits, grapes, berries, vegetables and ornamental crops from insects, mites, diseases (including viruses and nematodes) with emphasis on the development of pest management programs. Operates the Smithfield Experimental Farm (see 24.1).

Entomology	
Toxicology	Head: Dr. D.J. Pree
Vegetable pest management	Dr. A.B. Stevenson
Ornamental entomology	Dr. A.B. Broadbent
Fruit pest management	Dr. E.A.C. Hagley
Acarology	H.M.A. Thistlewood
Bioclimatology	Dr. R.J.M. Trimble
Nematology, Chemistry and Computer Science	
Nematode ecology and chemical control	Head: Dr. J.W. Potter
Residue chemistry	Dr. M. Chiba
Chemistry	B.D. McGarvey
Host-parasite relations	Dr. T.H.A. Olthof
Nematode ecology	J.L. Townshend
Mathematics and computing	Dr. J. Yee
Plant Pathology	
Fruit and soil-borne viruses	Head: Dr. W.R. Allen
Tree fruit diseases	Dr. A.R. Biggs

Vegetable diseases — Dr. R.F. Cerkauskas
Ornamental pathology — Dr. J.A Matteoni
Fruit mycology — Dr. J. Northover
Vegetable mycology — Dr. A.A. Reyes
Grapevine viruses — Dr. L.W. Stobbs

Quebec

26 Chûtes-aux-Outardes

26.1 Les Buissons Potato Research Station
elevation: 15 m
rainfall : 416 mm/season
sunshine : 1043 h/season

P.O. Box 55, R.R.
Chûtes-aux-Outardes (Saguenay)
Quebec G0H 1C0
Phone: (418) 567-2235
Dir.: Dr. G. Banville

Originally opened to explore the possibility of seed potato production on the North shore of the Lower St. Lawrence River, research is now oriented toward the control of the potato disease *Rhizoctonia solani,* cultivar and management trials of potatoes.

26.2 Manicouagan Elite Potato Seed Farm
elevation: 15 m
rainfall : 416 mm/season
sunshine : 1043 h/season

P.O. Box 257, R.R. Apartment 1
Chûtes-aux-Outardes (Saguenay)
Quebec G0H 1C0
Dir.: M. Tennier

Production of Elite seed potatoes for Quebec producers since 1961. Maintainance of disease free nuclear stock in vitro and in growth chambers. Use of serological tests and growing tests in Florida, U.S.A., during the winter.

27 Deschambault

27.1 Agricultural Research Station, Quebec Ministry of Agriculture, Horticulture Section
elevation: 15 m
rainfall : 521 mm/season
sunshine : 1137 h/season

P.O. Box 123, Deschambault
(Portneuf) Quebec G0A 1S0
Phone: (418) 286-3351
Dir.: J. Genest

Variety trials and cultural practices including weed and insect control (potatoes, tomatoes, cole crops, lettuce, leeks, asparagus, sweet corn, raspberries and strawberries). Propagation of virus-free raspberry and strawberry plants.

Vegetables and small fruits — R. Rushdy/S. Bégin

28 Farnham

28.1 Orchard Protection Section
elevation: 69 m
rainfall : 536 mm/season

1400 St-Paul Street North
Farnham (Missisquoi) Quebec J2N 2R5
Phone: (514) 293-6072
Dir.: M. Mailloux

Operated under the Quebec Crop Protection Service, conducts research on the protection of apples.
Insect control in apples. Apple protection warning system — M. Mailloux

29 L'Assomption

29.1 Crop Protection Research Station, Quebec Ministry

867 L'Ange-Gardien Blvd.

of Agriculture
elevation: 21 m
rainfall : 447 mm/season
sunshine : 1156 h/season

L'Assomption, Quebec J0K 1G0
Phone: (514) 589-4780
(418) 643-2380
Telex: 052-4863
Dir.: Dr. G. Emond

Insect control: asparagus, cabbages, cucumbers, strawberries and raspberries: Dr. G. Mailloux/L.M. Tartier
Disease control: potatoes: D. Vezina/M. Giroux/Dr. C. Ritchot

29.2 Experimental Farm (Agriculture Canada)
Operated under the Saint-Jean-sur-Richelieu Research Station (see 35.1)
elevation: 21 m
rainfall : 965 mm
sunshine : 1992 h

L'Assomption, Quebec J0K 1G0
Phone: (514) 589-4775
Sup't.: F. Darisse

Management, production and protection of ornamental plants.

Plant pathology, ornamentals M. Caron
Management: ornamentals C. Richer-Leclerc

30 La Pocatière

30.1 Agricultural Research Station, Quebec Ministry of Agriculture, Horticulture Section
elevation: 30 m
rainfall : 435 mm/season
sunshine : 1117 h/season

Route 230, La Pocatière
(Kamouraska), Quebec G0R 1Z0
Phone: (418) 856-1110
Telex: 051-3445
Dir.: J. Archambault

Cultivar trials and cultural practices: sweet corn, broad beans and potatoes. J. Archambault/L. Lord

30.2 Experimental Farm (Agriculture Canada)
elevation: 30 m
rainfall : 967 mm
sunshine : 1952 h

P.O. Box 400, La Pocatière
Quebec G0R 1Z0
Phone: (418) 856-3141
Sup't.: J.E. Comeau

Operated as a sub-station of Ste-Foy Research Station, the Experimental Farm is the main centre for potato breeding in Quebec. In collaboration with the Food Production and Inspection Branch of Agriculture Canada, the Experimental Farm is also responsible for disease-freeing of potatoes to produce nuclear stocks. It also collaborates in cultivar testing of potatoes from the Fredericton Research Station (see 40.3) and of fruits (apples, pears and plums) with the St-Jean-sur-Richelieau Research Station (see 35.1).

Potato breeding A. Frève
Disease-freeing of potatoes P. Sylvestre

31 Montreal

31.1 Botanical Garden
elevation: 50 m
rainfall : 1071 mm
sunshine : 1997 h

4101 Sherbrooke Est, Montreal
Quebec H1X 2B2
Phone: (514) 252-8861
Dir.: P. Bourque

This city-operated botanical garden collaborates closely with the Botanical Institute (see 31.2). Some 25000 species or varieties are maintained in various collections including several collections of tropical plants of African,

American or Asian origins. Some 49 greenhouses are in use, of which nine are devoted to exhibitions. Research involves personnel from both the Botanical Garden and the Botanical Institute.

Orchids, library	C. Arsenault
Taxonomy, fruit development	D. Barabé
Conservation, ecology	Dr. A. Bouchard
Taxonomy, cytogenetics of vascular plants	Dr. L. Brouillette
Genetics, tissue culture	Dr. M. Cappadocia
Taxonomy, tropical plants, weeds	N. Cornellier
Morphology, natural habitats	Dr. M. Famelart
Taxonomy, Marie-Victorin herbarium	S. Hay
Production and research	E. Jacqomain
Nursery production	M. Labrecque
Plant pathology	M. St-Arnaud
Tissue culture, anatomy and morphology	Dr. J. Vieth
Weeds, taxonomy, aquatic plants	G. Vincent

31.2 Botanical Institute, Department of Biological Sciences
elevation: 50 m
rainfall : 1071 mm
sunshine : 1997 h

University of Montreal
4101 Sherbrooke Est, Montreal
Quebec H1X 2B2
Phone: (514) 256-7511
Ass.Dir.: Dr. J. Vieth

Located on the site of the Montreal Botanical Garden (see 31.1), this institution emphasizes research on tissue culture, especially of *Gerbera jamesonii, Pelargonium peltatum* and *Cynara scolymus*.

In vitro culture	L. Chrétien
Histology	Dr. M.A. Dubuc-Lebreux
Pathology	Dr. P. Newmann
Anatomy, morphology	Dr. J. Vieth

31.3 Department of Biological Sciences, University of Montreal
elevation: 135 m
rainfall : 1020 mm
sunshine : 2003 h

P.O. Box 6128, Succ. A Montreal
Quebec H3C 3J7
Phone: (514) 343-6875
Head: Dr. R. Carbonneau

Teaching conducted in French. Graduate training also offered. Research relating to horticultural plants include post-harvest physiology, tissue culture, anther culture, somatic hybridization, pathology and mycology.

Genetics	Dr. M. Cappadocia
Plant pathology, mycology	Dr. P. Newmann
Physiology, metabolism	Dr. C.T. Phan

32 Normandin

32.1 Experimental Farm (Agriculture Canada)
elevation: 137 m
rainfall : 866 mm
sunshine : 1800 h

1472 St-Cyrille Street, Normandin
Quebec G0W 2E0
Phone: (418) 274-3378
Sup't.: J.M. Wauthy

Work at Normandin Experimental Farm is centered on cultivar trials (potatoes, strawberries, lowbush blueberries and broad beans), cultural practices specially on broad beans and evaluation of ornamental trees and shrubs in collaboration with l'Assomption Experimental Farm (see 29.2). Operated as a sub-station of Ste-Foy Research Station.

Cultivar trials and Cultural practices	R. Drapeau

33 St-Augustin

33.1 Crop Protection Research Station
elevation: 58 m
rainfall : 578 mm/season
sunshine : 1137 h/season

567, RR 138, St-Augustin
(Portneuf), Quebec G0A 3E0
Phone: (418) 878-2753/ 643-2348
Dir.: M.A. Richard

Operated under the Quebec Crop Protection Service, conducts research on protection of vegetables and small fruits.

Insect Control, potatoes	M.A. Richard
Virus diseases, potatoes, tomatoes, onions, leeks and shallots	Dr. J.D. Brisson
Disease Control, broccoli, cauliflower and strawberries	P.O. Thibodeau
Weed control, cabbages, turnips, onions, strawberries, blueberries and asparagus	B. Maltais
Weed ecology, sweet corn	C.J. Bouchard
Weed toxonomy and survey, vegetables	Dr. D. Doyon
Bird control, corn	C. Bouchard

34 St-Hyacinthe

34.1 Agricultural Research Station, Quebec Ministry of Agriculture, Horticulture Section
elevation: 31 m
rainfall : 527 mm/season
sunshine : 1080 h/season

3300 Sicotte Street, St-Hyacinthe
Quebec J2S 7B8
Phone: (414) 774-0660
Telex: 05-830561
Dir.: P. Lavigne

Breeding cultivars of tomatoes and swede turnips. Testing and evaluation of insecticides, herbicides, and disease control methods related to various vegetable cultivars (tomatoes, red beets, cabbages, sweet corn, turnips) and small fruits (raspberries and strawberries).

Improvement and cultivar trials, tomatoes, swede turnips and red beet	R. Doucet
Insect control, sweet corn	Dr. C. Ritchot
Herbicides, strawberries, cabbages and tomatoes	L. Tartier
Disease control, potatoes	L. Vezina

35 St-Jean-sur-Richelieu

35.1 Research Station* (Agriculture Canada)
elevation: 38 m
rainfall : 391 mm/season

Saint-Jean-sur-Richelieu, Quebec
J3B 6Z8
Phone: (514) 346-4494
Dir.: Dr. C.B. Aube
Ass.Dir.: R. Crête

The station carries out research on problems affecting the production and protection of fruits (apples, strawberries, raspberries, high and lowbush blueberries), vegetables (cole crops, onions, carrots, potatoes, asparagus, garlic), herbs and ornamental plants. Also studies on storage of vegetables, crop nutrition, pesticide retention in organic soils, pesticide residues in crops, and energy conservation. Operates an Experimental Farm at L'Assomption (see 29.2) and three sub-stations for research on vegetables and herbs on mineral soils (at L'Acadie) and on organic soils at (Ste-Clotilde) and for research on tree fruits and small fruits (at Frelighsburg).

Fruit Section

Small fruits, blueberries	Head: M.J. Lareau
Breeding, strawberries, raspberries	Dr. D. Bagnara

Mites, apples	Dr. N.J. Bostanian
Plant pathology, apples	Dr. L.J. Coulombe
Plant physiology, apples	Dr. R.L. Granger
Plant pathology, apples	J.R. Pelletier
Breeding, apples	Dr. G.L. Rousselle
Entomology, apples	Dr. C. Vincent
Biotechnology	J.C. Côté
Vegetable Section	
Entomology	Head: Dr. J. Belcourt
Nematology	G. Bélair
Weed science	D. Benoit
Post-harvest physiology	Dr. L. Bérard*
Entomology	Dr. G. Boivin
Breeding, cole crops	Dr. M.S. Chiang
Entomology	M. Hudon
Biotechnology	B. Landry
Entomology, toxicology	Dr. P. Martel
Management	B. Vigier
Herbs	A. Conti
Engineering and Soils Section	
Pesticide chemistry	Head: Dr. A. Bélanger
Engineering, energy	R. Chagnon
Engineering, aerial sprayings	B. Panneton
Engineering, mechanization	Dr. R. Thériault
Soils, hydrology	M. Trudelle
Soils, hydrology	Dr. J. Millette

36 St-Lambert-de-Lévis

36.1 Soil Research Station, Horticulture Section
elevation: 122 m
rainfall : 298 mm/season
sunshine : 1064 h/season

Rang Saint-Patrice, Saint-Lambert
(Lévis) Quebec G0S 2W0
Phone: (418) 889-9950
Dir.: A. Dubé

Operated under the Quebec Soil Research Service, this station conducts fertility and cultural practices trials on vegetable crops.

Fertility trials: onions, cabbage, lettuce, pimento and potatoes	B.T. Cheng/Dr. M. Tabi/M. Giroux

37 Ste-Anne-de-Bellevue

37.1 Macdonald College, McGill University
elevation: 40 m
rainfall : 920 mm
sunshine : 1952 h

Macdonald College, 21111 Lakeshore Road, Ste-Anne-de-Bellevue
Quebec H9X 1C0
Phone: (514) 457-2000
Telex: 05 821788
Dean: Dr. R. Buckland

Agricultural Faculty of McGill University. Research on field production of fruits and vegetables, including apples, raspberries, strawberries, tomatoes, peppers, leeks, melons. Protected cropping and hydroponic culture of salad crops. Studies on winter hardiness of fruit crops.

Department of Entomology	
Population dynamics and control of field pests	Head: Dr. R.K. Stewart
Soil micro-arthopods	Dr. S.B. Hill
Bionomics of orchard pests	Dr. S.B. Hill

Insect morphology	Dr. D.K.E. McKevan
Toxicology of pesticides	Dr. W.N. Yule
Department of Plant Science	Head: Dr. H.R. Klinck
Fruit crops physiology and management	Dr. D.J. Buszard*
Fruit and vegetable storage and processing	Dr. J. David
Plant viruses	Dr. J.F. Peterson
Diseases of horticultural crops	Dr. R.D. Reeleder
Host pathogen interactions	Dr. S.A. Sparace
Vegetable crops physiology and management	Dr. K.A. Stewart*
Weed biology, herbicides	Dr. A.K. Watson
Department of Renewable Resources	Head: Dr. R.D. Titman
Vegetable crops fertilization	Dr. A.F. MacKenzie
Maples and sap yields	A.R.C. Jones

38 Ste-Foy

38.1 Plant Science Department
elevation: 75 m
rainfall : 1174 mm
sunshine : 1852 h

Faculty of Food and Agriculture Science, Laval University, Ste-Foy
Quebec G1K 7P4
Phone: (418) 656-2165
Head: J.-M. Girard

The horticulture section is concerned with teaching and research. It cooperates with Food Science and Technology and Human Nutrition Departments. Equipment: research laboratories, several growth chambers, cold storages, 1200 m^2 of glasshouses, 2300 m^2 of plastic greenhouses, 20 ha of on-campus fields. Botanical garden and arboretum.

Greenhouse vegetable production: lighting, mineral nutrition, management	Dr. M.J. Trudel/Dr. A. Gosselin*
Garden and greenhouse flower production	Dr. B. Dansereau
Nursery management and weed control	Dr. J.A. Rioux
Small fruit and vegetable production	Dr. R. Bédard
In vitro propagation	Dr. A. Gosselin*/Y. Desjardins
Post-harvest physiology, fruit and vegetable processing	Dr. F. Castaigne/Dr. R.E. Simard Dr. M. Boulet
Fruit and vegetable nutritive and organoleptic value	Dr. J. Zee/M. Raymond
Horticultural crop diseases	Dr. A. Asselin/Dr. J.-G. Parent

New Brunswick

39 Bouctouche

39.1 Hervé J. Michaud Experimental Farm
Operated under the Fredericton Research Station (see 40.3)
elevation: 30 m
rainfall : 666 mm/season
sunshine : 2000 h

P.O. Box 667, Bouctouche, New Brunswick E0A 1G0
Phone: (506) 743-2464
Sup't.: Dr. G.L. Rousselle*

Cultural practices studies and cultivar trials of tree fruits, berry and vegetable crops.

Tree fruits and small fruits	M. Luffman
Vegetables	P.V. Leblanc

40 Fredericton

40.1 New Brunswick Department of Agriculture Plant

P.O. Box 6000, Fredericton, New

Industry Branch

Brunswick E3B 5H1
Phone: (506) 453-2108
Dir.: E.T. Pratt

Involved in a range of adaptive research projects directed towards improved crop management and the adaptation of horticultural crops in New Brunswick. The New Brunswick Horticulture Centre in Hoyt is a substation used for plant propagation, hardiness testing, and crop management studies.

Horticultural Section	
New crops and ornamentals	Head: B.W. Dykeman*
Vegetable crops	Dr. T.J. Johnson
Fruit crops	R. Tremblay
Blueberries	S.G. Michaud
Greenhouse crops	A.W. Currie
Weed control	D.J. Doohan
Plant pathology	K.V. Lynch
Insect control	D.B. Finnamore
Land Resource Section	
Soil fertility and crop nutrition	Head: Dr. I.B. Ghanem
Soil management	K.T. Michalica
Climatology and winter hardiness	G.S. Read

40.2 New Brunswick Department of Agriculture Potato Industry Division

P.O. Box 6000, Fredericton, New Brunswick E3B 5H1
Phone: (506) 453-2108
Dir.: Dr. C.E. Smith
Ass.Dir.: R.E. Routledge

Involved in a range of applied research activities directed toward quality seed production and improved crop and storage management. Staff are located in Fredericton, Wicklow, and Bonaccord. Facilities include the Bonaccord Seed Farm substation, the Wicklow Potato Storage Lab, and the plant propagation (tissue culture) centre, Fredericton.

Plant pathology	A.F. Perley
Pest management	Dr. R.H. Parry
Soil fertility	R.P. Hinds
Storage management	J.R. Walsh
Seed potato production	W.W.M. Schrage
Potato production	G.E. Bernard
Potato production	G. Pelletier
Seed potato production	G.M. Barclay
In-vitro propagation	S.E. Coleman

40.3 Research Station (Agriculture Canada)
elevation: 25 m
rainfall : 436 mm/season
sunshine : 1878 h

P.O. Box 20280, Fredericton, New Brunswick E3B 4Z7
Phone: (506) 452-3260
Dir.: Dr. Y. Martel

The Research Station at Fredericton is the major centre in Canada for potato breeding research, and also carries a multi-disciplinary program on table stock and processing potatoes. It is the site for a cooperative program with Plant Health and Plant Products of Agriculture Canada on the production of seed tubers in a special Elite seed program. Particular attention is given in the Station's potato research program to the biology of all types of tuber borne diseases and the production of virus-free seed. The Station also conducts limited research on fruit crops such as the apples, blueberries, strawberries and raspberries. These and other tree fruit, berry and vegetable crops are the focus of experiments and cultivar trials at the Hervé J. Michaud Experimental Farm, in Bouctouche (see 39.1).

Potato Breeding	
Potato breeding and genetics	Head: Dr. D. Young
Potato breeding	R. Coffin
Diploid breeding and genetics	H. DeJong
Pathology	A.M. Murphy
Propagation methods	J.E.A. Seabrook
Quantative genetics	G.C.C. Tai
Cytogenetics	T.R. Tarn
Potato Pest Management	
Virus diseases	Head: Dr. R. Singh
Virus epidemiology and resistance	R.H. Bagnall
Potato insect ecology	G. Boiteau
Biochemistry of disease resistance	M.C. Clark
Physiology	W.K. Coleman
Late blight forecast. Fungicides	J. Holley
Residue chemistry	R.R. King
Tuber-borne pathogens	A.R. McKenzie
Chemical ecology and behaviour of potato insects. Aphid taxonomy	Y. Pelletier
Engineering, Horticulture and Soils	
Harvesting and storage engineering	Head: Dr. G.C. Misener
Soils hydrology	T.L. Chow
Fruit crop management	E.N. Estabrooks*
Agricultural mechanization engineering	C.D. McLeod
Soil engineering	P.H. Milburn

Nova Scotia

41 Kentville

41.1 Research Station (Agriculture Canada)
elevation: 30 m
rainfall : 1073 mm
sunshine : 1793 h

Kentville, Nova Scotia B4N 1J5
Phone: (902) 678-2171
Dir.: Dr. G.M. Weaver

Major location for fruit and vegetable research in the Atlantic Provinces. One of the pioneer centres for the development of ecological studies on orchard insects and the biological factors influencing their control.
Principal Station function is to conduct fundamental and applied studies on problems of growers and processors of tree fruits, berry crops and vegetables. Emphasis is placed on the development of an integrated pest control program, pesticide residues, fruit and vegetable storage, processing, crop diseases, plant nutrition, plant breeding, management and cultivar evaluation. Equipment: Laboratories for food technology, plant physiology, plant nutrition, pesticide-residue determinations in plants and soils, plant pathology, entomology and food technology. Facilities for controlled atmosphere storage, cold storage, plant growth chambers and greenhouses.

Crop Protection Section	
Weed physiology	Head: Dr. K.I.N. Jensen
Plant pathology	Dr. M.G. Anderson
Toxicology, residue chemistry	S.O. Gaul
Insect ecology	Dr. J.M. Hardman
Plant pathology	Dr. P.D. Hildebrand
Residue chemistry	E.R. Kimball
Plant pathology	C.L. Lockhart
Fruit maggot control	W.T.A. Neilson
Plant pathology	Dr. N.L. Nickerson*
Residue chemistry	Dr. M.T.H. Ragab
Orchard insects control	K.H. Sanford

Fruit and vegetable crops control	Dr. H.B. Specht
Crop Protection Section	
Nutrition and physiology	Head: Dr. C.R. Blatt
Tree fruit breeding and physiology	Dr. A.D. Crowe*
Tree fruit management	C.G. Embree*
Blueberry physiology	Dr. I.V. Hall*
Plant breeding, berries	A.R. Jamieson
Physiology, ornamentals	Dr. P.R. Hicklenton
Crop management, vegetables	Dr. C.L. Ricketson*
Nutrition of tree fruits	Dr. D.H. Webster
Crop Storage Section	
Storage physiology of tree fruits	Head: Dr. P.D. Lidster
Storage physiology, vegetables	P.A. Poapst*
Food Processing Section	
Food technology	Head: Dr. R. Stark
Food engineering	J.M. Burrows
Food microbiology	Dr. E.D. Jackson
Food engineering	R.A. Lawrence

42 Truro

42.1 Horticulture and Biology Services, Nova Scotia Department of Agriculture
elevation: 40 m
rainfall : 1139 mm
sunshine : 1759 h

P.O. Box 550, Truro, Nova Scotia
Phone: (902) 895-1571
Telex: 019-34532
Dir.: D.M. Sangster*

The Horticulture and Biology Branch provides an advisory service to the Nova Scotia Horticulture industry. Commodity specialists, in cooperation with Branch entomologists, a plant pathologist and an apiarist provide information and advice on all aspects of Horticulture.

Tree fruits	W.E. Craig
Berry crops	R.A. Murray
Blueberries	J.D. Sibley
Vegetables	C. Thompson
Ornamentals	R.S. Morton
Greenhouses	B. Toms
Greenhouses	B. MacPahil
Apiarist	L.M. Crozier
Maple syrup	D.W. McIsaac
Tree fruit entomology	Dr. R.J. Whitman
Home Gardens	M.J. Blenkhorn
Other entomology	J.M. Hollett
Plant pathology	R.W. Delbridge

42.2 Nova Scotia Agricultural College
elevation: 40 m
rainfall : 1139 mm
sunshine : 1759 h

Truro, Nova Scotia B2N 5E3
Phone: (902) 895-1571
Telex: 019-345-32
Dir.: Dr. H.F. MacRae

Campus and college farm consists of 223 ha. College offers degree, technology, technician and vocational courses in agriculture. Horticultural plots consist of three ha of vegetables, two ha of tree and small fruits and one ha of turfgrass and ornamentals. Research greenhouses, labs and controlled environment units are on campus. Provincial horticultural extension specialists are located on campus.

Plant Science Department — Head: Dr. R.K. Prange
Potato physiology and management
Fruit specialist — Dr. H.-Y. Ju
Ornamentals — W.J. Higgins
Tissue culture — Dr. J. Nowak
Turf specialist — R.W. Daniels
Vegetable specialist — T.H. Haliburton
Crop management (herbicides) — L. Mapplebeck

Prince Edward Island

43 Charlottetown

43.1 Research Station (Agriculture Canada)
elevation: 15 m
rainfall : 1100 mm
sunshine : 1835 h

P.O. Box 1210, Charlottetown
Prince Edward Island C1A 7M8
Phone: (902) 892-5461
Dir.: Dr. L.B. MacLeod

This Station, bordering on the city of Charlottetown, serves the province of Prince Edward Island where potatoes are the main horticultural crop. This is a major potato station for research conducted on the control of viral and fungal diseases, fertilization, weed control, haulm dessication and crop production techniques. Beside potatoes, other crops include peas, rutabagas and cole crops, as well as strawberries and blueberries.

Nutrition potatoes — Dr. R.P. White
Management and nutrition, potatoes — J.B. Sanderson
Diseases, potatoes — Dr. H.W. Platt
Virus diseases, potatoes — Dr. J.G. McDonald
Insect control, potatoes — Dr. L.S. Thompson
Weed control — Dr. J.A. Ivany
Micronutrients — Dr. U.C. Gupta
Nematodes — Dr. J. Kimpinski
Nutrition and management, vegetables — J.A. Cutcliffe*
Bioactivity and degredation of pesticides — Dr. D.C. Read

New Foundland

44 St. John's

44.1 Research Station (Agriculture Canada)
elevation: 14 m
rainfall : 1602 mm
sunshine : 1532 h

P.O. Box 7098, St. John's
Newfoundland A1E 3Y3
Phone: (709) 772-4619
Dir.: Dr. H.R. Davidson

This Station conducts research into production of horticultural and forage crops, with special emphasis on utilization of peat soils and management of native blueberry stands. Plant pathology and plant breeding studies are mainly directed towards investigation of potato wart disease and the development of potato cultivars resistant to this disease, and to potato cyst nematode. Breeding of rutabagas for resistance to clubroot disease is also undertaken. Two sub-stations are operated at Avondale and Colinet.

Agronomy — Head: A.F Rayment
Entomology — Head: Vacant
Horticulture — Head: B.G. Penney*
Plant breeding — Head: K.G. Proudfoot
Plant pathology — Head: M.C. Hampson

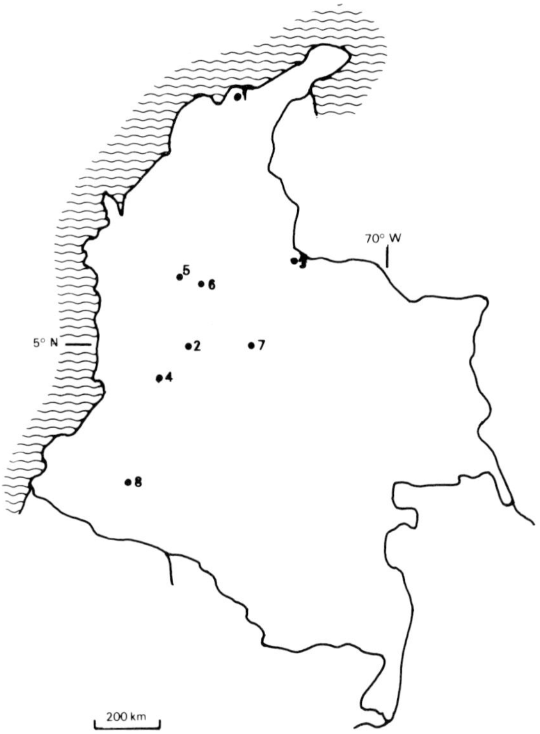
70° W
5° N
1
2
3
4
5
6
7
8
200 km

1 Cienaga
2 Espinal
3 Bucaramanga
4 Palmira
5 Bello
6 Rionegro
7 Bogotá
8 Pasto

Survey

Colombian agricultural research is undertaken by the Instituto Colombiano Agropecuario, ICA.The Institute is a specialized agricultural agency of the government, established in 1963 and with functions in research, extension and education. It has its Executive Office in Bogotá and nine Regional Offices throughout the country. For the success in agricultural research ICA operates special programmes according to the more important crops cultivated in Colombia, and covering such fields as Genetics, Plant Physiology, Entomology, Soils and Crop Pathology.
National Program of Fruit - The Fruit Program was started in 1929 at Palmira in order to improve fruits adapted to its climatic conditions. However, because of the numerous fruit crops adapted to the different Colombian soils and climates, it was necessary, in 1963, to operate in other places with different ecological conditions.
The research results obtained are published in annual progress reports and in technical publications as "Revista ICA", one official organ of the ICA itself. In 1983 it was calculated that in Colombia 35.000 ha was planted with fruits. Important fruits are: Citrus, pineapple, guava, papaya, avocado, mango, yellow passion-fruit and grapes. The location and the climatic characteristics of ICA experimental stations in which fruit research was conducted in 1983 are presented.
Research on vegetables - Research on vegetables, carried out by the material program of vegetables which belongs to the Instituto Colombiano Agropecuario, ICA. The programme of the institute in charge of agricultural research in the country is represented in research stations in four mayor horticulture areas. It tries to cope with the main problems of the more important vegetables in Colombia. Last year were planted some 50.000 ha with different species. The research results are published every year in the Annual Report and articles of interest.
There are several types of publications.

1. Cienaga

1.1 Centro Regional de Investigación Caribia
elevation: 18 m

Santa Marta - Cienaga
Apartado Aéreo 654

Prepared by Messrs. R.T. Monedero and J. Jaramillo V., I.C.A., Palmira

rainfall : 1393 mm
r.h.: 84%

Phone: 5162
Head: I.A. G.Calderón

Variety trials and cultural practices with citrus, mango, papaya and guava. Production of basic propagation material.

2. Espinal

2.1 Centro Regional de Investigación Nataima
elevation: 400 m
rainfall : 1282 mm
sunshine : 2298,5 h

Ibagué - Espinal
Apartado Aéreo 527
Phone: 34304
Head: I.A. D. Vega
I.A. J.R.Cartagena

Physiological studies. Variety trials and cultural practices. Production of basic propagation material. Mango, citrus and avocado.

3. Bucaramanga

3.1 Bucaramanga
elevation: 800 m
rainfall : 1200 mm

Bucaramanga
Apartado Aéreo 1017
Phone: 55185
Head: I.A. O.R. Martínez

Cultural practices, physiological and nutrition studies on pineapple and papaya.

4. Palmira

4.1 Centro Nacional de Investigación Palmira
elevation: 1006 m
rainfall : 1022 mm
sunshine : 2153,5 h
r.h.: 72%

Palmira
Apartado Aéreo 233
Phone: 28163
Fruit Dir.: I.A.M.Sc. R.E.Torres
I.A.M.Sc. R. Salazar
I.A. L.A.Sánchez

This Station is the main centre of fruit research in Colombia. Investigations are concerning with introduction of new varieties; breeding and release of new varieties; production of basic propagation material; physiological studies; pests and diseases control; virus diseases studies on citrus and papaya. General research with the following fruits: citrus, mango, pineapple, yellow passion-fruit, papaya, grape and guava.

Fruit pathology section — I.A. P.J.Tamayo/Dr.G.A. Granada
I.A.M.Sc. F.de Agudelo

Entomology section — I.A.M.Sc. B.de Gutierrez
Irrigation section — I.A.M.Sc. M.Molano

4.2 Research Station (ICA)
(Vegetables)

Palmira - Valle
Phone: 28164
P.O. Box 233
Dir.: Ir.H. Chamorro

Breeding of which 450 ha is covered with a wide range of matters on agricultural and livestock; cacao, corn, rice, beans, platano and vegetables. Cattle, dairy and poultry management.

Vegetable Section
Tomato breeding; yield, TMV resistance

National Coordinator:
Ir.M.Sc. J.Jaramillo

Vegetable variety trials
Squash breeding and seed production of biannuals. — Ir.Y. Palacios
Germplasm Bank of *Capsicum, Lycopersicon* and *Cucúrbita*.

5. Bello

5.1 Estacion Experimental Tuloi Ospina
elevation: 1470 m
rainfall : 1447 mm
sunshine : 1971 h

Medellin
Apartado Aéreo 51764
Phone: 75100
Head: I.A. A.Mafla

Introduction and selection of avocado varieties. Production of basic propagation material of citrus, avocado, tree tomato *(Cyphomandra betacea)*, lulo *(Solanum quitoense)*.

6. Rionegro

6.1 Centro Regional de Investigación La Selva
elevation: 2200 m
rainfall : 1831 mm
sunshine : 1861,5 hrs
r.h.: 81%

Medellin - Rionegro
Apartado Aéreo 51764
Phone: 751200
Head: I.A. A.Mafla

Problems concerning with planting fruits of cold climate areas: tree tomato, lulo and fig. Selection of avocado varieties for cold climate.

6.2 La Selva. Research Station (ICA)
(Vegetables)
This station with 40 ha is dedicated to breeding of corn, beans and potatoes.

Rionegro - Antioquia
Phone: 720311
P.O. Box 51764 - Medellin
Dir.: Ir. M. Sc. J. Llano

Vegetable Section — Ir.M.Sc. E. Girard
Variety trials in cabbage, carrot, lettuce
Cultural practices
Seed production of biannuals
Physiological studies
Germplasm Bank of *Allium fistolosum*

7. Bogotá

7.1 Centro Nacional de Investigación Tibaitatá
elevation: 2550 m
rainfall : 662 mm
sunshine : 2170 h
r.h.: 73%

Bogotá
Apartado Aéreo 151123
Phone: 813277

Virus diseases studies on citrus — Biologa M.Sc. C. de Luque
Tissue culture — Quim. V. de Kekan

7.2 Research Station (ICA)
(Vegetables)

Bogotá - Cundinamarca
P.O. Box : 151123 Bogotá
Phone: 2813088
Dir.: Vet. U. Ariza

Work is done mainly on breeding of potatoes, corn, wheat, barley, legumes, vegetables and dairy breeding. Here is centralized the statistical analysis system for the whole Institute. It has an extension of 420 ha.

Vegetable Section
Breeding on faba beans, bush onion, cucurbit variety trials, fertilizer and cultural practices on different vegetables.
Germplasm Bank on Tibaitatá

Ir. M.Sc. F.Higuita
Ir. M.Sc. J. Osorio

8. Pasto-Obonuco

8.1 Research Station (ICA)
elevation: 2720 m
rainfall : 890 mm
sunshine : 2170 h

Pasto - Nariño
P.O. Box 339 Pasto
Phone: 2318
Dir.: Ir.D.H.Zambrano

Research on potatoes, wheat, barley, legumes, corn for yield, resistance to diseases and management of dairy and sheep. It has 320 ha of extension.

Vegetable Section
Faba breeding for yield, diseases resistance
Variety trials
Germplasm Bank of garlic and *Vicia*

Vacant

COSTA RICA

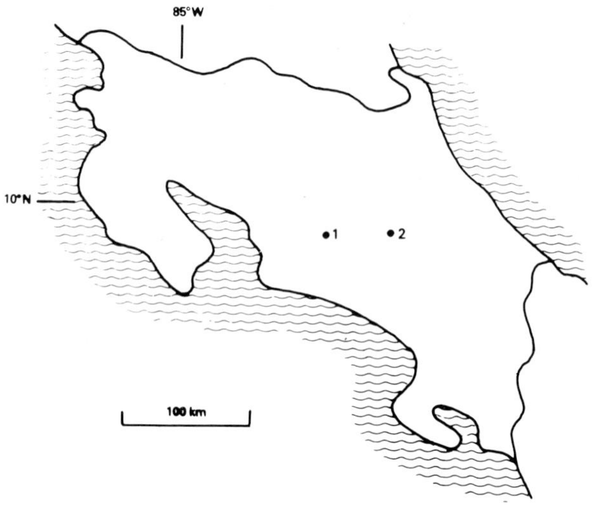
85°W
10°N
1
2
100 km

1 San José
2 Turrialba

Survey

The Inter-American Institute for Cooperation on Agriculture, IICA, is an international intergovernmental organization specialized in agriculture. It is governed by its own Convention and has been recognized as a specialised Inter-American agency, under the Charter of the Organization of American States.

The Institute was founded in 1942 as the Inter-American Institute for Agricultural Sciences, and its Convention was opened to signature in 1944. It was ratified that same year. Thirty-five years later, on March 6, 1979, a new Covention was opened to the signature of the American States. It was ratified on December 8, 1980. Under this new Convention, the Institute changed its name to the Inter-American Institute for Cooperation on Agriculture, expanded its purposes, and altered its institutional structure.

According to its new Convention, the purposes of ICCA are to "encourage, promote, and support the efforts of the Member States to achieve their agricultural development and rural well-being". In order to reach these goals, IICA seeks agreement with the countries to carry out technical co-operative actions designed to enhance the operating capacity in the organizations responsible for agricultural development and rural well-being. It operates as a multinational vehicle so that the member countries may avail themselves to its services when they require joint actions. It is a forum and an instrument for the exchange of ideas, experiences, and cooperation among the countries and the organization or agencies active in the agricultural field.

Documents. The Convention on the Inter-American Institute for Cooperation on Agriculture, and the Rules of Procedure of the Inter-American Board of Agriculture, the Executive Committee and the General Directorate together appear in Volume 22 of IICA's Official Documents Series. Two additional documents that describe the Institute's present action are: General Policies of IICA and The Medium-Term Plan for 1983-1987. Copies of these documents may be found in IICA's Offices in the Member Countries.

General Policies of IICA. The purpose of General Policies of IICA is to trace a long term, general policy for the Institute and to provide a framework within which the General Directorate can regularly issue medium term plans, programs and budgets. IICA's history is summarized and gives information on economic, social, and political problems in the region. It discusses existing constraints on the agricultural sector in Latin America and the Caribbean, and examines the outlook for agriculture in the region for the 1980's.

The document also specifies the nature, purposes, and functions of IICA. It outlines objectives and strategies and describes the Institute's work and the participation of the member countries. Before taking concrete action, IICA concurs with its Member States on possible technical cooperation services. The resulting cooperation takes place by means of concentrated actions performed in a decentralized framework, for the purpose of obtaining meaningful results.

Medium-Term Plan. The Medium-Term Plan provides a framework to guide Institute actions taking place from 1983 to 1987, on the basis of guidelines included in the General Policies of IICA. It stipulates that the basic tools for Institute action are: technical support, studies, direct action, technical and scientific brokerage, and reciprocal technical cooperation.

Prepared by J. André Ouelette, IICA, San José.

Programs. IICA'S ten hemispheric programs are: 1. Formal Agricultural Education. 2. Support of National Institutions for the Generation and Transfer of Agricultural Technology. 3. Conservation and Management of Renewable Natural Resources. 4. Animal Health. 5. Plant Protection. 6. Stimulus for Agricultural and Forest Production. 7. Agricultural Marketing and Agroindustry. 8. Integrated Rural Development. 9. Planning and Management for Agricultural Development and Rural Well-Being. 10. Information for Agricultural Development and Rural Well-Being.
Adresses of the National Offices:

Argentina
Ing. M. Paulette, Sarmiento 760, 9 Piso,
1041 Cap. Fed. Buenos Aires, Phone: 394-1319,
1324, Telex: 121197 OEAARG

Barbados
M. Moran, PO Box 705-C, Bridgetown,
Phone: 51432, 51433, 51434, Telex: 2446 IICAWB

Bolivia
Econ. M. França, Casilla 6057, La Paz,
Phone: 37-4988, 35-2086, 37-1892,
Telex: 5741 IICA BV

Brazil
A. Centrángolo, Enc., Caixa Postal. 09-1070,
71600 Brasília, Phone:248-6799,
Telex:611959INAGBR

Canada
Dr. L. McLaren, Carling Executive Park,
5th Floor, 1565 Carling Avenue,
Ottawa, Ontario, Canada K1Z 8R1,
Telex: 197649IICAUT

Colombia
Dr. M. Blasco, Apartado Aéreo,
14592 Bogota, Phone:244-28-68, 244-90-05 al 09,
Telex: OEACOL 44669

Chile
Ing. E.M. da Costa Fiori, Casilla 3631,
Santiago, Phone: 2238255, 495613, 2231400,
Telex: 240644IICA CL

Dominica
Dr. R. Pierre, C/- Division of Agriculture,
Botanical Gardens Roseau, Phone: 449-4502,
Telex: 6302 OAS 8634 OAS DOM DO

Ecuador
F.R. Cantoral, Apartado 201-A, Quito
Phone: 524-238, Telex: 2837 IICA ED

El Salvador
Dr. R. Soikes, Apartado (01) 78, San Salvador
Phone: 23-25-61, 23-37-74, 23-54-46,
Telex: 20246 OEAES

U.S.A.
Dr. G. Páez, 1889 F. Street, N.W., Suite 820,
Washington DC 20006-4499, Phone: (202)789-3767,
789-3768, 789-3769, Telex: 197649 IICAUT

Grenada
Dr. R. Pierre, PO Box 228, St. George's,
Phone: 4547, Telex: 3427 OASGA

Guatemala
Dr. R. Clifford, 1815, Guatemala
Phone: 64304, 62306, 62795
Telex: 5977 OEAGUA

Guyana
Dr. F. Alexander, PO Box 10-1089
Georgetown, Phone: 68347, 68835
Telex: 2279 IICAGY

Haïti
Dr. P. Aitken-Soux, BP 2020, Port-au-Prince
Phone: 5-3616, 5-1965, Telex: IICA 2030511
(Horario: 8 a 16)

Honduras
Ing. A. Franco, Apartado 1410, Tegucigalpa
Phone: 22-58-00, 22-58-02, Telex: 1131 OEAHON

Jamaica
PO Box 349, Kingston 6, Phone: 9276462,
9274837, 9270007, Telex: 2270 OASJAM

Mexico
Dr. E. Salvadó, Apartado 4830, México
Phone: 598-55-89, conmutador, 598-60-00 ext. 165
Telex: 17-72-485 DGA México

Nicaragua
Dr. J.M. Montoya, Apartado 4830, Managua
Phone: 51-4-43, 51-7-57, Telex: 2274 IICANN NK

Panama
Ing. G. Guerra, Apartado 10731, Panamá
Phone: 69-5308, 59-5779, Telex: 2713 OEAPAN
**NO ETAT PRIORITY

Paraguay
Ing. S. González, Cassilla de Correos: 287,
Asunción, Phone: 41-650, Telex: 165 OEAASN

Peru
Dr. H. Chaverra, Apartado 11185, Lima 14

Phone: 22-28-33, 40.57-04, Telex: 25281 OEAPE

Republica Dominicana
Ing. H. Morales, Apartado 711, Santo Domingo
Phone: 533-7522, 533-2797, Telex: 3460350 IUI-CARD

Saint Lucia
Dr. R. Pierre, PO Box 972, Castries
Phone: 25482, 21688, Telex: 6302 OASLU

Suriname
Ing. G. Villanueva, PO Box 1895, Paramaribo
Phone: 72710, Telex: 167 SURTOR

Trinidad and Tobago
Dr. C. Brathwaite, PO Box 1318, Tacarigua
Telex: OEA 22244

Uruguay
Ing. E. Montero, Casilla de Correos 1217, Montevideo, Phone: 95-93-26, 95-92-80, 95-93-80
Telex: 6457 OEA

Venezuela
Dr. M. Segura, Apartado 5345, Caracas
Phone: 571-8211, 571-8844, 571-8055
Telex: 21533 OEAVEN

1 San José

1.1 Inter-American Institute of Agricultural Sciences

Apartado 55, 2200 Coronado,
San José, Costa Rica
Phone: 29-02-22, 29-18-59, 29-12-46
Telex: 2144 IICA
Dir.Gen.: Dr. M.E. Piñeiro

The central office of IICA is located in San Isidro de Coronado, in the Province of San José. It is the headquarters of the Institute's managerial and administrative operations carried out by the Director General, the Deputy Director General, the Assistant Depty Directors General for Operations, Program Development and External Affairs, and the Directore of technical and support services, all with the necessary support personnel. Approximately three hundred people work in the Central Office.

Director General	M.E. Piñeiro
Deputy Director General	L.H. Davis
Assistant Deputy Director General for Program Development	J. Soria
Assistant Deputy Director General for Operations	J.A. Torres
Director of External Financing	J. Werthein
Director of Communications	J.A. Ouellette
Director of the Internal Auditing	J. Acosta
Director, Center for Investment Project (CEPI)	J.A. Aguiree

The National Offices are incorporated into Areas and are answerable to their particular Area Directors. Area I, the Central Area, is headquartered in Costa Rica. It is made up of El Salvador, Guatemala, Honduras, Mexico, Nicaragua, Panama, Dominican Republic and Costa Rica. Area II, the Caribbean, is headquartered in Jamaica. It is made up of Barbados, Dominica, Grenada, Guyana, Haiti, Saint Lucia, Suriname, Trinidad and Tobago and Jamaica. Area III, Andean, is headquartered in Peru. It is made up of Bolivia, Colombia, Ecuador, Venezuela and Peru. Area IV, Souhern, is headquartered in Uruguay. It is made up of Argentina, Brazil, Chile, Paraguay and Uruguay.
There are also three Specialized Centers. Two are direct branches of IICA and are located in the Central Office: the Inter-American Agricultural Documentation and Information Center, CIDIA, and the Investment Projects Center, CEPI.

Director, Centro Interamericano de Documentacion e Informacion Agricola, (CIDIA)	Dr. M. Kaminsky Apartado 55, 2200 Coronado, San José
Director, Oficina del IICA en Costa Rica, (Area I)	Dr. C.E. Fernández Carretera a Coronado, San José Phone: 29-02-22

2 Turrialba

2.1 Centro Agronomica Tropical de Investigacion y Ensenanza (CATIE)
(Tropical Agricultural Research and Training Center)
elevation : 625 m
rainfall : 2670 mm
temperature : 22°C

Turrialba, Costa Rica
Phone: 56-64-31, 56-01-69
Telex: 8005 CATIE CR
Dir.: Dr. R.T. Ponce
Dep.Dir.: Dr. C.S. Pacheco

CATIE is located at Turrialba, Costa Rica, a distance of 70 km from San José and has an area of 950 ha. CATIE is an autonomous non-profit scientific and educational regional center, established in 1973 as a civil association and constituted between IICA and the Costa Rican Government. CATIE employed in 1985 the following personnel: professional staff 189, auxiliary (support) staff 351, permanent field laborers 341, graduate students 68, and professional staff residing in member countries 38. The Governments of Panama, Nicaragua, Guatemala, Honduras and the Dominican Republic joined CATIE within a few years. CATIE hopes to form a cooperative network with national universities in the area for its graduate studies program. CATIE's research work, is oriented toward the development of technologies applicable to the socio-economic environment of the small and medium-sized farmer of the Central American area. It trains professionals responsible for the technological development of that sector, both at the graduate and non-graduate level. In Turrialba, CATIE has ample laboratory facilities, a library and experimental fields. It also has programs and supporting offices in El Salvador, Guatemala, Honduras, Nicaragua, Panama and the Dominican Republic. CATIE also provides technical assistance to these countries.

Head of Technical and Financial External Cooperation	A. Chibbaro
Head of Department of Graduate Studies and Training	Dr. J.L. Parisí
Head of Department of Plant Production	Dr. R. Martínez
Head of Department of Renewable Natural Resources	Dr. G. Budowski
Head of Department of Animal Production	Dr. J.A. Zaglul
Head of Department of Administration and Finances	Lic. O. Campos

2.2 Department of Plant Production

The Department works with perennial plants such as cacao, coffee, "pejibaye", palm and the macadamia nut. In annual food crops the Department works with root crops and tubers which includes cassava, sweet potato, taro and yam. In its research on cropping systems, the Department works with such food crops as maize, beans, upland rice, corn, cucurbits. The work covers cultivar evaluation and horticultural practices as they affect polycultural systems.

Agricultural economics	Ec.Ag.G.Calvo/Ec.Ag.W.González, Dr. J. French, Ing. Margarita Meseguer (M.S.), Ing. Anabelly Rodríguez
Cacao	Dr. G. Enríquez, Ing. L.A. Paredes/Ing. A. Mora, Ing. W. Phillips (M.S.), Ing. V. Porras (M.S.)
Coffee	Ing. L. Avendano, Ing. J. Echeverri (M.S.)
Communications	Lic. H. Chavarria, Biol. E. Rodríguez
Cropping systems	J. Arze (M.S.)/W.Bejarano(M.S.), Dr. S. Bradfield/Dr.D.Kass
Entomology	Ing. T. Coto/Dr. J.R. Quesada, Dr. J. Saunders/P.Shannon(M.S.)
Genotype evaluation	Dr. F. Rosales
Horticulture	Dr. R. Martínez
Integrated pest control	Dr. J. Lastra/Dr.M.Pareja,

	Dr. J. Pinochet/R.Meneses(M.S.), Dr. F. Alonzo/Dr. D. Monterroso, G. Von Lindeman (M.S.), Ing. M. Carballo (M.S.)
Meteorology	Ing. F. Jiménez
Plantains	Ing. J. Jiménez/Ing. L. Morales
Plant breeding	Ing. F. Herrera
Plant pathology	Dr. E. Bustamante, Dr. J.J. Galindo (cacao), Dr. O. Trocmé (cacao)
Plant physiology	Dr. J. Fargas
Root crops	Ing. W. Rodríguez (M.S.)
Soils analysis	Ing. Robert
Soils management	Dr. C. Burgos
Tissue culture	Dr. L. Muller/Ing. N. Guzmán (coffee)/Ing. J.Sandoval, Ing. Inés Mora
Training	Ing. E. Ledezma
Weed control	Dr. A. Beale/Ing.R.De La Cruz Ing. A. Merayo

2.3 Germplasm Project — Head:Dr. H. Fromberg

As part of the Department of Plant Production, it is a regional center in exploration; collection, maintenance and characterization of plants from Mexico and Central America, which is known as one of the most important centers of plant genetic diversity in the world. Financed by CATIE and GTZ (Federal Republic of Germany). Plants covered belong to the root crops (for example: *Colocasia, Dioscorea, Manihot, Impomoea,* etc); vegetable crops *(Cucurbita, Capsicum, Sechium, Phaseolus,* etc.); fruit crops *(Anona, Eugenia, Bactrys,* etc.) and plantation crops (Coffee and *Theobroma).*

Documentalist	Ing. C. Astorga
Plant breeder	Ing. J. Salazar
Plant genetic resources	Ing. Ana Guadalupe/Dr.A.Stolberg, Ing. Marlen Vargas
Seed technology	Ing. J. Arce
Tissue culture	Bach. S. Salazar

Publications of CATIE

"Silvoenergia" (monthly, in Spanish), Informe Anual (Annual Report in Spanish), "Actividades en Turrialba" (Quarterly information bulletin in Spanish). Special documents prepared by the Programs (usually mimeographed publications-bulletins); a list is available from the Natural Renewable Resources and Plant Production Departments.

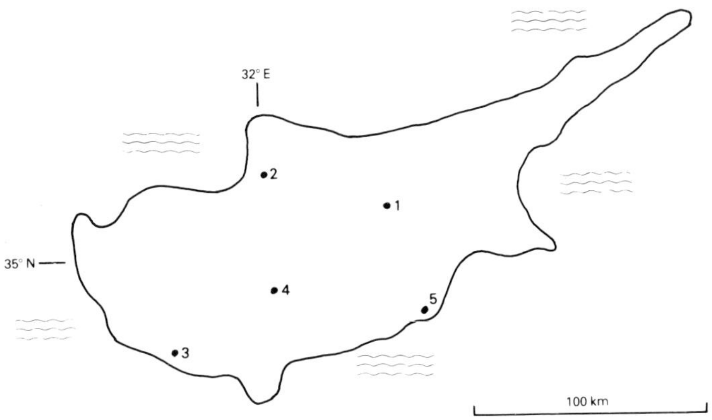
32° E
35° N
1
2
3
4
5
100 km

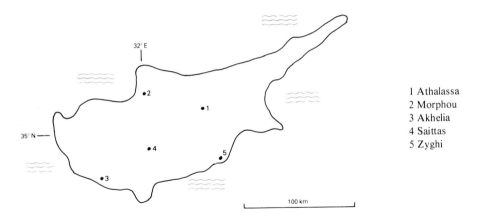

1 Athalassa
2 Morphou
3 Akhelia
4 Saittas
5 Zyghi

Survey

In Cyprus horticultural research is carried out by the Agricultural Research Institute of the Ministry of Agriculture and Natural Resources. The Agricultural Research Institute, situated at Athalassa, near Nicosia, is the only institution in Cyprus carrying out agricultural research. It carries out mainly applied research in all fields of agriculture - plant production, animal production and agricultural economics. Horticultural research comprises an important part of the Institutes programme due to the significance of horticultural crops, mainly citrus, vegetables, potatoes, carrots, grapes and vine products to the overall economy of Cyprus. The research programme was intensified after 1974 to meet with the requirements of the irrigation development schemes of the government of Cyprus and the needs which resulted due to the Turkish invasion and the occupation of almost 40% of the Cyprus territory.

Horticultural production in the occupied area represented 80% of citrus, 70% of carrotes, 25% of potatoes, and 30% of all other vegetables of Cyprus. Horticultural research covers a wide range of activities and crops such as citrus (oranges, grapefruit, lemons, soft citrus), deciduous crops (apples, pears, cherries, peaches, apricots, almonds), tropical and subtropical crops (avocadoes, olive trees, bananas, mangoes), vegetables (open air and protected crops), potatoes, table grapes, oenology and postharvest handling of fruits and vegetables.

Publications: The results of horticultural research like all other experimental results of the Institute are published in the Annual Report, Technical Paper and Technical Bulletin series of the Institute and in international scientific journals and magazines. All the Institutes' publications are circulated free to institutions, research stations and Journals abroad.

1 Athalassa

1.1 Headquarters of the Agricultural Research Institute
elevation: 150 m
rainfall : 314 mm
sunshine : 5334 h

Ministry of Agriculture,
Athalassa, Nicosia, Cyprus
Phone: (02) 403431

Horticultural Science — Dir.: C.V. Economides
Citrus
Deciduous and sub-tropical fruit trees — C.V. Economides
Vegetables, potatoes and post-harvest handling of fruits and vegetables — Dr. N. Vakis
Vegetables (protected cultivation)
Deciduous, post-harvest handling of fruits — T. Kokkalos*
Viticulture and oenology — T. Myrianthousis
Viticulture — Dr. I. Aziz
Citrus and sub-tropical fruits — C. Gregoriou

Prepared by Dr. N. Vakis, Agricultural Research Institute, Athalassa

Floriculture

G. Philippou
Eleven agricultural research technicians are also employed by the Section

1.2 Athalassa Experimental Farm

Athalassa, Nicosia

Vegetables, sub-tropical fruits

2 Morphou (under Turkish occupation)

2.1 Citrus Experimental Station
elevation: 30 m
rainfall : 315 mm

Morphou, Nicosia

3 Akhelia

3.1 Vegetable Experimental Station
elevation: 50 m
rainfall : 468 mm
sunshine : 3268 h

Akhelia, Paphos
Phone: (061) 34821

Vegetables and cut-flowers

3.2 Horticultural Research Station

Citrus, tropical, sub-tropical and deciduous fruit trees and vines

During the last 5 years the banana production expanded rapidly in this area.

4 Saittas

4.1 Horticultural Research Station
elevation: 730 m
rainfall : 776 mm
sunshine : 3106 h

Saittas, Limassol

Deciduous crops

5 Zyghi

Zyghi, Larnaca

5.1 Experiment Station
elevation: sea level
rainfall : 390 mm
sunshine : 3300 h

Phone: (0433) 2440

Vegetable experiments in the open air and protected cultivation
Avocadoes, olives, strawberries, citrus

In addition to these stations experiments on potatoes are carried out in grower's fields in the main potato growing area.

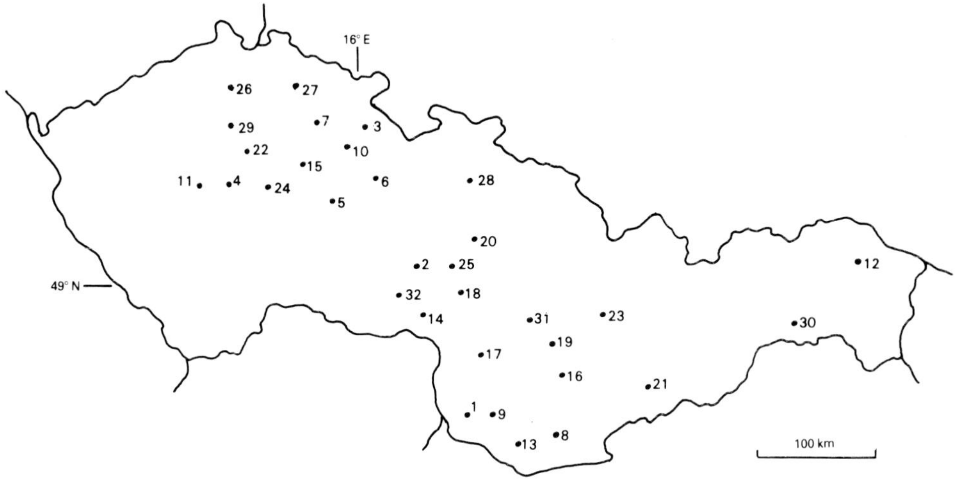
16° E
49° N
•26
•27
•29
•7
•3
•22
•10
•15
11
•4
•24
•6
•28
•5
•20
•2
•25
•12
•32
•18
•14
•31
•23
•30
•19
•17
•16
•21
•1
•9
•13
•8
100 km

1 Bratislava, 2 Brno, 3 Ceská Skalice, 4 Dobřichovice, 5 Heřmanův Městec, 6 Holice, 7 Holovousy, 8 Hurbanovo Sesileś, 9 Ivanka, 10 Jaroměř, 11 Karlštejn, 12 Klčov, 13 Kvetoslavov, 14 Lednice, 15 Lysá nad Labem, 16 Mlyňany, 17 Modra, 18 Mutěnice, 19 Nitra, 20 Olomouc, 21 Orechová, 22 Praha, 23 Prievidza, 24 Průhonice, 25 Smržice, 26 Těchobuzice, 27 Turnov, 28 Velké Losiny, 29 Veltrusy, 30 Velký Krtíš, 31 Veselé, 32 Želešice

Survey

Science and research are generally directed by the Czechoslovak Academy of Sciences and by Universities. The agriculture and foodstuff research is organized and guided by the respective Ministries of Agriculture and Foodstuff Industry; scientific establishments in Czech Socialistic Republic belong to the Ministry of CSR, the establishments in Slovak Socialistic Republic to the Ministry of SSR and some of the Institutes are controlled by the Federal Ministry of Agriculture and Foodstuff Industry. The research centres in horticulture are: Research and Breeding Institute for Ornamental Horticulture in Průhonice (26), Research and Breeding Institute for Vegetable Crops in Olomouc (19), Research and Breeding Institute of Pomology in Holovousy (7) and Research and Breeding Institute of Wine and Viticulture in Bratislava (1).

Research and Breeding Institute for Ornamental Horticulture in Průhonice is engaged with the problems connected with landscape architecture and flower growing. There is an ample dendrology collection in the park founded by Sylva Tarouca. At present the park is the botanical garden. A numerous assortment of ornamentals is concentrated in this institute; breeding of flowers and trees (Rhododendron, Rosa), physiology and growing bulbs and flower nutrition are being studied there.

Research and Breeding Institute for Vegetable Crops in Olomouc is concentrated mainly on breeding, genetics, physiology, biochemistry, cultural practice, nutrition and mechanization. Considerably advanced is the research of vegetable forcing in greenhouses and plastic greenhouses.

Research and Breeding Institute of Pomology in Holovousy is mainly centred on breeding, physiology, cultural practice, protection of fruit trees, especially apple trees and on intensive methods in pomology. The institute possesses large research plantations for investigation of up-to-date methods of fruit growing.

Research and Breeding Institute of Wine and Viticulture in Bratislava is aimed at the grape-vine growing and wine technology. The problems under investigation are genetics, breeding, physiology, cultural practice, ecology, protection of grape-vine and biochemistry, microbiology and wine technology. The institute disposes of grape-vine assortment and research plantations on its stations. Following institutes are important from the point of view of horticulture: Research Institute of Plant Production in Prague-Ruzyně,directed on research in genetics, physiology, agrochemistry, microbiology, pedology and plant protection. The institute is the International centre for seed and plant material exchange. The Institute for Scientific and Technical Informations is a significant establishment

Prepared by Dr. F. Mareček, Research Institute of Plant Production, Praha

for documentation in the field of agriculture; it communicates the exchange of scientific literature and includes an agricultural library with more than 1 million of scientific and professional volumes.
The Universities of Agriculture in Prague, Brno and Nitra established their Departments of Horticulture dealing with general horticulture and viticulture. The Faculty of Horticulture, University of Agriculture Brno is placed in Lednice. About 40 students graduate from this Faculty every year after 5 year's study. Research activity is directed at the physiology and cultural practice of fruit and vegetable growing and viticulture. Extensive horticultural plantings and a collection of ornamental trees and flowers belong to the Faculty. Horticultural teaching is given at Horticultural high schools (6 schools with 4 year courses). About 1000 students pass horticultural schools every year. Varietal testing is carried out by organizations of the Ministry of Agriculture and Foodstuff Industry, i.e. Central Agricultural Control and Testing Institute in Prague and Bratislava. For varietal testing of garden crops there are specialized stations. The Czechoslovak Academy of Agriculture, which includes prominent research workers, holds scientific meetings and symposia in agriculture and horticulture.
Publications: Results of horticultural research work are published in scientific journals: Rostlinná výroba, Gentika a šlechtění, Ochrana rostlin. Special publications with the title Vědecké práce (Research work) are published by respective horticultural institutes.
For the benefit of practice the following journals are issued: "Záhradnictvo" and "Vinařství".

1 Bratislava

1.1 Research Institute of Wine and Viticulture
elevation: 150 m
rainfall : 650 mm
sunshine : 2.200 h

800 00 Bratislava, Mátuškova 21
Phone: 462 24
Dir.: Prof.Ing. A. Veres DrSc.

Belongs to the Ministry of Agriculture and Foodstuff Industry of SSR. Breeding of grape vines, cultural, practice, grape vine protection and nutrition, biochemistry and wine fermentation, vineyard technology and economy.

Agroecology and ecophysiology of grape vine	Prof.Ing. A. Vereš DrSc.
Research of grape nutrition on vine quality	Prof.Ing. M. Fic CSc.
Influence of micro-nutrients on wine quality	Ing. J. Tesař CSc.
Influence of irrigation on grape yields and wine quality	Ing. D. Kubečka CSc.
Grape vine nursery technologies	Doc.Ing. K. Dobrovoda CSc.
Grape-vine variety collection, grape-vine genetics, breeding of wine varieties	Dr. D. Pospíšilová CSc.
Grape-vine physiology, pruning, growth dynamics, inflorescence differentation	Ing. A. Żembery CSc.
Research of Oidium, pesticides residue in wine and grape-vine; research of virus diseases and thermotherapy	Ing. P. Ragala CSc.
Economy of grape harvest and wine production, stabilization of low alcohol wines, yeast ecology in wine production, influence of ecological conditions on wine quality, yeast collection	Ing. V. Minarik CSc. Ing. J. Navara CSc.
Economy of large scale technologies	Ing. D. Kubečka CSc.

1.2 Department of Varietal Testing of the Central Agricultural Control and Testing Institute

800 00 Bratislava, Matúškova 21
Phone: 434 41
Telex: 93246
Dir.: Ing. A. Hlavička CSc.

Varietal experiments with peppers, tomatoes, melons, cucumbers, irrigation experiments	Ing. A. Pisklová
Variety and rootstock experiments, wine technology	Ing. J. Záruba

2 Brno

2.1 University of Agriculture, Department of Horticulture, Faculty of Agronomy
elevation: 240 m
rainfall : 550 mm
sunshine : 1.800 h

600 00 Brno-Černá Pole,
Zemědělska 1
Phone: 604
Telex: 62489
Head: Doc.Ing. K. Křikava CSc.

Physiology and vegetable nutrition. Plastics in vegetable growing. Variety collection and growing of roses — Doc.Ing. B. Jaša CSc.

2.2 Department of Variety Testing of the Central Agricultural Control and Testing Institute
elevation: 196 m
rainfall : 550 mm
sunshine : 1.800 h

600 00 Brno-Pisárky
Hroznová 2
Phone: 33 48 11
Telex: 63062
Dir.: Ing. J. Trnka CSc.

Variety experiments with beans, peas, onions, tomatoes cucumbers — Ing. E. Stuchlíková CSc.

3 Česká Skalice

3.1 Research and Breeding Station for Flowers of the Research and Breeding Institute of Ornamental Horticulture in Průhonice
elevation: 284 m
rainfall : 800 mm
sunshine : 1.600 h

552 03 Česká Skalice
Phone: 0441/527 55
Dir.: R. Adam

Breeding of Gerbera, Primula, Cyclamen, Salvia, seed production

4 Dobřichovice

4.1 Department of Varietal Testing of the Central Agricultural Control and Testing Institute in Prague
elevation: 205 m
rainfall : 550 mm
sunshine : 1.800 h

252 29 Dobřichovice
Phone: 21 80
Dir.: Ing. M. Melichar

Varietal experiments with vegetables, small fruits and flowers
Varietal testing of cabbage crops, leaf and root vegetables — Ing. M. Hnízdil CSc.
Varietal testing of currant, gooseberries, strawberries — Ing. J. Dlouhá CSc.
Varietal testing of bulbs and perennial plants — Ing. M. Klapzubová

5 Heřmanův Městec

5.1 Research and Breeding Station for Flowers of the Research and Breeding Institute of Ornamental Horticulture in Průhonice
elevation : 280 m
rainfall : 650 mm
sunshine : 1.800 h

538 03 Heřmanův Městec
Phone: 0455/952 13
Dir.: Ing. J. Novotný

Breeding of Dahlia, Gladiolus, Tulipa — Ing. M. Václavík
Production of Lilia — Ing. J. Jóst

Breeding of cabbage and carot hybrids — Ing. K. Křivský CSc.

6 Holice

6.1 Research and Breeding Station for Flowers of the Research and Breeding Institute of Ornamental Horticulture in Průhonice
Elevation: 244 m
rainfall : 700 mm
sunshine : 1.800 h

534 01 Holice v Čechách
Phone: 26 58
Dir.: Ing. J. Mottl

Breeding of cyclamen, varietal collection, seedling production — Ing. J. Mottl

7 Holovousy

7.1 Research and Breeding Institute of Pomology
elevation: 306 m
rainfall : 720 mm
sunshine : 1.600 h

507 51 Holovousy v Podkrkonoší
Phone: 921 21
Telex: 194679
Dir.: Ing. J. Drobný CSc.
Dep.Dir.: Ing. J. Blažek CSc.

Belongs to the Ministry of Agriculture and Foodstuff Industry of CSR. Breeding of fruit-trees, methods of intensive fruitgrowing, varietal collection, nursery

Varietal collection of apple, pear, cherrytrees, goose-berries, currants, apple and pear breeding rootstock of Prunus	Ing. J. Blažek CSc. Ong. V. Kosina CSc.
Physiology of growth and fertilization	Fr. P. Vit
Tissue cultures, growth regulators	Ing. K. Ludvík
Methods of intensive fruit-growing	
Technology in nurseries	
Nutrition of fruit-trees	Ing. G. Hudská CSc.
Virus diseases. bacterial and fungal diseases of fruit trees, herbicides	Ing. M. Erbenová CSc. Ing. J. Staněk CSc.
Mechanization of harvest, small-fruits, stone fruits	Ing. J. Klimpel CSc.
Storage of fruit	Ing. V. Zika
Specialization and concentration of fruit production, statistics	Ing. K. Kricnar CSc.

8 Hurbanovo-Sesíleš

8.1 Research and Breeding Institute for Vegetables
elevation : 115 m
rainfall : 550 mm
sunshine : 2.200 h

947 01 Hurbanovo
Phone: 23 35
Telex: 98359
Dir.: Ing. V. Streclec CSc.
Dep.Dir.: Ing. T. Tóth CSc.

Belongs to the Ministry of Agriculture and Foodstuff Industry of SSR. Breeding and cultural practice of thermophilic vegetables

Nutrition of tomatoes, red pepper, cucumbers	Ing. K. Černá
Growing of thermophilic vegetables. Application of plastics for vegetable growing, irrigation	Ing. H. Prussová CSc.
Mechanization of tomato, cucumber, and red pepper harvest	

Vegetable storage, transportation. Physiology of tomatoes and red pepper after harvest — Doc.Ing. K. Kopec CSc.
Breeding of red pepper, tomatoes, cucumber, onion, garlic and cabbage — Ing. K. Michálek CSc.

9 Ivanka pri Dunaji

9.1 Institute of Experimental Plant Pathology and Entomology of the Slovac Academy of Sciences in Bratislava
elevation: 135 m
rainfall : 600 mm
sunshine : 2.200 h

900 28 Ivanka pri Dunaji
Phone: 94 32 51
Dir.: Doc.Dr. A. Huba CSc.

Research of interactions between plant pathogens, host plants and environment

Department of Entomology
Effect of abiotic factors on harmful insects — Ing. J. Jasič CSc.
Pathophysiological changes of plants caused by infection — Ing. J. Hässler CSc.

10 Jaroměř

10.1 Research and Breeding Station for Flowers of the Research and Breeding Institute of Ornamental Horticulture in Průhonice
elevation: 270 m
rainfall : 650 mm
sunshine : 1.800 h

551 00 Jaroměř
Phone: 0441/23 12
Dir.: Ing. J. Přikazský

Breeding of Petunia, Begonia, heterosis — Ing. J. Černý/ Dr. J. Černý

11 Karlštejn

11.1 Research Station of Viticulture of the Research Institute of Plant Production in Prague-Ruzyně
elevation: 345 m
rainfall : 500 mm
sunshine : 1.700 h

267 18 Karlštejn
Phone: 941 31
Dir.: Ing. V. Hubáček

Problems of viticulture of northern areas — Ing. J. Krpeš
Breeding of grape vine, varietal and rootstock experiments, pruning of grape vine, nutrition, protection, varietal collection. Technology of wine production — Ing. V. Hubáček

12 Klčov

12.1 Research and Breeding Station of the Research Institute of Fruit and Ornamental Trees in Prievidza
elevation: 493 m
rainfall : 800 mm
sunshine : 2.000 h

053 02 Klčov, p. Spišský Hrhov
Phone: 0966/622 31
Dir.: Ing. M. Župník CSc.

Breeding of small-fruits, apples, apple-rootstock, collection of East Carpathian types of *Malus silvestris*

13 Kvetoslavov

13.1 Research and Breeding Station of Vegetables of the Research and Breeding Institute of Vegetables in Hurbanovo
elevation: 127 m
rainfall : 600 mm
sunshine : 2.200 h

930 41 Kvetoslavov
Phone: 24 43
Dir.: Ing. K. Michálek CSc.

Breeding of tomatoes, red pepper — Ing. K. Michálek CSc.
Breeding of cabbage, cauliflower, cucumber — Ing. J. Machurek
Breeding of onion, carrot — Ing. V. Rušnák CSc.

14 Lednice

14.1 Faculty of Horticulture, University of Agriculveture Brno
elevation: 173 m
rainfall : 550 mm
sunshine : 1.800 h

691 44 Lednice na Moravě
Phone: 0627/982 11
Telex: 62838

General horticulture, 5-years study, teaching facilities

14.1.1 Department of Vegatables
Cultural practice of tomatoes, pepper, cucumbers — Prof.Ing. J. Štambera CSc.
Vegetable forcing, plastics in horticulture, seed production of vegetables

14.1.2 Department of Fruit Growing
Cultural practices of peach- and apricot-tree growing — Doc.Ing. F. Pospišil CSc.
Fruit-tree physiology, irrigation, water management — Doc.Ing. Z. Vachún CSc.
Varietal collection

14.1.3 Department of Ornamental Gardening
Garden architecture, landscape architecture, dendrology, ornamental nurseries park, ornamental plant collection, greenhouse flower growing — Prof.Ing. B. Wágner CSc; Doc.Ing. J. Machovec CSc.

14.1.4 Department of Garden Plant Breeding
Breeding of resistant vegetables, heterosis of onion, cucumbers, seed production of annual flowers — Doc.Ing. J. Lužny, CSc.
Aromatic and medicinal plants — Doc.Ing. M. Křikava CSc.

14.1.5 Department of Viticulture
Wine technology, microbiology — Ing. V. Švejcar CSc.
Genetics, physiology and breeding of grape vine, ecology, varietal collection — Doc.Ing. V. Kraus CSc.

14.1.6 Department of Fruit and Vegetable Technology
Physiology of harvested fruits and vegetables — Doc.Ing. J. Goliáš CSc.
Biochemistry of stored fruits and vegetables. Cold storage

14.1.7 Department of Mechanization and Technics
Harvest mechanization of tomatoes, red-pepper, cucumbers, plastic materials — Prof.Ing. V. Fic CSc.
Harvest mechanization of small-fruits and stone fruits

14.2 Research Station "Mendeleum" of University of Agriculture in Brno

691 44 Lednice na Moravě
Phone: 0627/982 80
Dir.: Doc.Dr. J. Vožda CSc.

Genetics of garden crops, cytology, cucumber and watermelon heterosis — Doc.Dr. J. Vožda CSc.; Ing. M. Průdek CSc.

Resistance breeding of tomatoes — Dr. G. Voždová CSc.

15 Lysá

15.1 Research and Breeding Station of Vegetables
elevation: 178 m
rainfall : 620 mm
sunshine : 1.800 h

289 22 Lysá nad Labem
Phone: 972 51
Dir.: Ing. J. Pícha

Belongs to the Research Institute of Vegetable crop in Olomouc

Breeding of vegetable crop and production of fruit rootstocks — Ing. V. Smékal

Breeding of onion, cucumber hybrids, cauliflower, cabbage, production of apple rootstocks — Ing. P. Votruba

16 Mlyňany

16.1 Arboretum Mlyňany. Institute of Dendrobiology, Slovac Academy of Sciences
elevation: 165 m
rainfall : 600 mm
sunshine : 2.200 h

951 52 Mlyňany-Vieska, p. Slepčany
Phone: 814/942 11
Dir.: Doc.Ing. F. Benčať CSc.

Systematics, ecology, physiology, and genetics of ornamental trees
Research of planting of evergreen trees

16.1.1 Department of Systematics and Introduction
Adaptation and acclimatization of East Asia trees. Dendrology — Doc.Ing. F. Benčať CSc.

16.1.2 Department of Tree Psysiology and Genetics
Translocation of assimilates in evergreens. Influence of anthropomorphic factors on the metabolism of evergreens — Dr. G. Steinhübel CSc.

17 Modra

17.1 Research Station of Viticulture
elevation: 270 m
rainfall : 650 mm
sunshine : 2.200 h

900 01 Modra
Phone: 26 85
Dir.: Ing. L. Lošonský CSc.

Belongs to the Research Institute of Wine and Viticulture in Bratislava. Cultural practice of grape vine. Basic cultural practice of grape vine. Basic cultural practice, terracing, physiology of seedlings pruning and grape vine nutrition

18 Mutěnice

18.1 Research Station of Viticulture
elevation: 183 m
rainfall : 600 mm
sunshine : 1.800 h

696 11 Mutěnice
Phone: 972 32
Dir.: Prof.Ing. V. Fic CSc.

Belongs to the Research Institute of Wine and Viticulture in Bratislava. Mechanization in viticulture, nutrition and protection of vine, economy in viticulture.

19 Nitra

19.1 University of Agriculture, Department of Horticulture
elevation: 143 m
rainfall : 600 mm
sunshine : 2.200 h

949 01 Nitra
Phone: 235 01
Head: Doc.Ing. J. Hajnal CSc.

Genetics and physiology of red-pepper, agroecology of vegetables — Doc.Ing. E. Pevná CSc.
Breeding and flower seed production (annual flowers) — Ing. K. Oberthová CSc.
Cultural practice in fruit growing: apple-, apricot- and peach-trees — Doc.Ing. J. Hričovský CSc.

20 Olomouc

20.1 Research and Breeding Institute of Vegetable Crops
elevation: 211 m
rainfall : 650 mm
sunshine : 1.800 h

770 00 Olomouc-Holice
Phone: 280 51
Telex: 66282
Dir.: Doc.Ing. F. Vlček CSc.
Dep.Dir.: Doc.Ing. L. Toul CS

Belongs to the Ministry of Agriculture and Foodstuff Industry of CSR.General problems of vegetable growing, research of breeding methods, biochemistry and nutrition of vegetables, agricultural practice.

Breeding of onion and cabbage hybrids — Ing. J. Jiřík CSc.
Genetics of *Capsicum anuum* and *Solanum lycopersicum* — Ing. J. Betlach CSc.
World variety collection of 10 000 species of vegetables — Ing. J. Moravec CSc.
Breeding of carrots and garden beans — Ing. V. Molkup CSc.
Seed production of vegetables — Ing. V. Novák CSc.
Breeding of aromatic crops — Ing. J. Dušek
Vegetable forcing, hydropinics — Doc.Ing. F. Vlček CSc.
Seedling production, plastic greenhouses — Ing. C. Chmela CSc.
Nutrition and fertilization of vegetables — Ing. J. Polách CSc.
Mechanization and technics
Storage of vegetables — Doc.Dr. A. Hovadík CSc.
Influence of nutrition and chemical treatment on vegetable quality — Doc.Ing. L. Toul CSc.
Residue of pesticides in vegetables — Dr. J. Pospíšilova CSc.
Photosynthesis in brassica, cucumber, tomato
Influence of nutrition on physiological functions in vegetables — Dr. J. Fridrch CSc.
Growth regulators in vegetable growing — Ing. H. Kratochvílová CSc.
Herbicides, diseases in bulb vegetables, tissue cultures of bulb vegetables (garlic) — Ing. A. Janýška CSc.; P. Láska CSc./ Dr. J. Rod CSc.
Economy of large scale production technology — Ing. M. Jarošová CSc.

21 Orechová

21.1 Research Station of Viticulture
elevation: 119 m
rainfall : 600 mm
sunshine : 2.000 h

930 02 Orechová
Phone: 35 97
Dir.: Ing. A. Semjon CSc.

Belongs to the Research Institute of Wine and Viticulture in Bratislava.

Problems of viticulture of Tokay area	
Nutrition of vine, pruning and training of vine.	Ing. M. Jenčo CSc.
Varietal and rootstock experiments.	
Technology of Tokay wine production	

22 Praha

22.1 University of Agriculture, Department of Horticulture, Faculty of Agronomy	165 00 Praha 6 - Suchdol
elevation: 265 m	Phone: 32 36 41
rainfall : 500 mm	Telex: 122323
sunshine : 1.800 h	Head: Prof.Ing. J. Holub CSc.
Physiology of fruit trees, systematics of fruit trees, varietal collection	
Physiology of vegetable nutrition, water management in vegetables	Doc.Ing. J. Duffek CSc.
Biochemistry, growth regulators in horticulture	Ing. M. Povolný CSc.
Landscape architecture, dendrology	Ing. J. Tobiášek CSc.
22.2 Institute of Experimental Botany, Czechoslovak Academy of Sciences	160 00 Praha 6, Na Karlovce 1
elevation: 221 m	Phone: 34 18 51
rainfall : 500 mm	Dir.: Dr. F. Pospíšil DrSc.
sunshine : 1.800 h	
Genetics and breeding of apple trees	Ing. J. Tupý CSc.
Growth regulators in fruit growing	Ing. L. Chvojka CSc.
Regulation of blossoming	Dr. J. Krekule CSc.
Tissue cultures of vegetables, fruit trees	Dr. L. Novák CSc.
	Dr. K. Opatrný CSc.
Dormancy of fruit trees	Dr. L. Černý CSc.
Natural sources of viruses, virus and mycoplasma diseases in garden plants.	Dr. J. Brčák DrSc.
Biological therapy of plant viruses	
22.3 Research Institute of Fruit and Vegetables Preservation	140 00 Praha 4, Bránická 114
elevation: 199 m	Phone: 46 14 51
rainfall : 500 mm	Dir.: Ing. B. Špaček CSc.
sunshine : 1.800 h	
Belongs to the Distilling and Preserving Trust. Preserving of fruit and vegetables, preserving technology, packing technology. Variety suitable for preserving technology, research of preserving and fermenting technologies	
Rheological characters of fruit and vegetable	Ing. J. Nosek CSc.
Research of fermentation	Ing. O. Žváček CSc.
Department of Package Technique	Ing. J. Stárek CSc.
Department of Construction and Machinery	Ing. J. Šilhan CSc.
Department of Technique and Economy	Ing. O. Hrdlička CSc.
Department of Standardization	Ing. F. Šimek CSc.
22.4 Research Institute of Plant Production	161 06 Praha 6 - Ruzyně
elevation: 326 m	Phone: 36 08 51

rainfall : 500 mm
sunshine : 1.800 h

Telex: 123093
Dir.: Prof.Ing. A. Kováčik DrSc.

Belongs to the Ministry of Agriculture and Foodstuff Industry of CSR. Genetics and breeding, physiology, plant protection and plant nutrition

Research of heterosis, breeding of tomato hybrids, cucumber hybrids — Doc. E. Troníčková CSc.
Cold-resistance of onions — Ing. A. Procházková CSc.
Breeding of kohlrabi hybrids — Ing. V. Kučera CSc.
Breeding of cauliflower — Ing. F. Mareček CSc.
Breeding of apple and pear trees, varietal and root-stock experiments with apple and pear trees
Research of dormancy in grape vine — Ing. M. Hubáčková CSc.
World collection of vatieties (10 000 vegetable varieties, 2 000 fruit trees, 2 000 grape vines, 7 000 ornamental plants). Department is the Czechoslovak center for international variety exchange — Ing. I. Bareš CSc.

23 Prievidza

23.1 Research Institute for Fruit and Ornamental Trees
elevation: 298 m
rainfall : 800 mm
sunshine : 2.000 h

971 01 Prievidza
Phone: 708/24 43
Telex: 72522
Dir.: Doc.Ing. I. Hričovský CSc.

Belongs to the Ministry of Agriculture and Foodstuff Industry of SSR.Breeding of fruit trees, mainly of small-fruits, stone fruits, apple trees. Production of virus free seedlings.

Breeding of currant, gooseberry, strawberry, hip *(Rosa oinnufera L.), Sambucus nigra L.,* pears, cherries — Ing. F. Hnízdil CSc.
Projection and management of orchards — Ing. J. Smetana CSc.
Research of fruit nursery methods — Ing. P. Hudec CSc.

24 Průhonice

24.1 Research and Breeding Institute of Ornamental Horticulture
elevation: 306 m
rainfall : 550 mm
sunshine : 1.700 h

252 43 Průhonice
Phone: 75 95 07
Dir.: Doc.Ing. J. Mareček CSc.

Belongs to the Ministry of Agriculture and Foodstuff Industry of CSR. Landscape architecture, ornamental plant nurseries, greenhouse flower growing, breeding and seed production, nutrition and protection of ornamental plants.

Department of Landscape Architecture and Dendrology — Ing. Z. Bouček CSc.
Research of general landscape architecture — Ing. D. Šonský CSc.
Landscape architecture in towns, parks — Ing. J. Ondřej CSc.
Landscape architecture in country and agricultural landscape, protection of environment. Parks and their maintainance — Ing. P. Bulíř CSc.

Bioclimatic function of green plants	Dr. Z. Suchara CSc.
Department of Technology	Ing. M. Wolf CSc.
Forcing of roses, carnations, Gerbera, Chrysanthemum	Ing. R. Votruba CSc.
Physiology of bulb flowers	Ing. E. Petrová
Nutrition of ornamental plants, substrates	Ing. J. Soukup CSc.
Multiplation of ornamental plants	Ing. J. Obdržálek CSc.
Protection, virus and bacterial diseases	Dr. V. Mokrá CSc.
Department of Breeding	Ing. J. Matouš CSc.
Genetics of ornamental plants - (Viola)	Dr. M. Novotná CSc.
Breeding of Azalea, Rhododendron, Gerbera, Tulipa	Ing. J. Matouš CSc.
Breeding of perennial flowers	Ing. J. Opatrná CSc.
Breeding of ornamental trees	Ing. K. Hieke CSc.
Seed production of ornamental plants	
Department of Economics and Statistics	Ing. J. Pavlík CSc.
Economics of flower farms and nurseries	Ing. M. Pinc CSc.
Economics and management of parks	Ing. V. Ondřejová
Production costs and prices of ornamental plants	

24.2 Institute of Botany, Czechoslovak Academy of Sciences	252 43 Průhonice Phone: 75 05 22 Dir.: Dr. S. Hejný Dr.Sc.
Basic research of plant associations, systematics, flora of ČSSR, dendrology	
Department of Geobotany	Dr. J. Moravec CSc.
Department of Taxonomy	Dr. B. Slavík CSc.
Department of Anthropophyta	Dr. Z. Kropáč CSc.
Botanical garden of 300 ha. Collection of garden plants, dendrological collection, breeding of fruit trees	Dr. S. Hejný CSc.
Department of Forest Biology	Dr. V. Rypáček DrSc.
Department of Hydrobotany	Ing. S. Přibyl CSc.
Department of Ecology	Dr. M. Rychnovská CSc.

24.3 Institute of Landscape Ecology, Czechoslovak Academy of Sciences	252 43 Průhonice Phone: 75 96 21 Dir.: Dr. J. Pospíšil CSc.
Center of research work in landscape ecology and protection of life environment	
Department of Geoecology	Ing. J. Vaněk CSc.
Department of Anthropoecology	Ing. J. Stoklasa CSc.
Department of Country Landscape	Dr. E. Nováková CSc.

25 Smržice

25.1 Research and Breeding Station for Vegetables elevation: 216 m rainfall : 600 mm sunshine : 1.800 h	798 16 Smržice, p. Čelechovice na Hané Phone: 40 17 Dir.: Ing. V. Slavík
Belongs to the Research and Breeding Institute for Vegetable Crops in Olomouc	
Breeding of pea, cucumber hybrids, tomatoes, carrots onion	Ing. M.Holmann CSc/ Dr.A.Lebeda CSc Ing. J. Zavadil CSc.

26 Těchobuzice

26.1 Research and Breeding Station for Fruit Trees
elevation: 243 m
rainfall : 550 mm
sunshine : 1.600 h

411 42 Těchobuzice, p. Ploskovice
Phone: 86 87
Dir.: Ing. A. Pavlůsek

Belongs to the Research and Breeding Institute of Pomology
Breeding of rootstocks and virus free material

Breeding of apple and pear rootstock — Ing. A. Dvořák CSc.
Breeding of apple trees
Breeding of pear trees, varietal collection of pear trees — Ing. J. Vondráček CSc.
Laboratory of tissue culture, thermotheraphy — Ing. F. Zimandl CSc.
Isolates of virus free varieties of apple and pear trees — Doc.Ing. C. Blattný CSc.

27 Turnov

27.1 Research and Breeding Station for Vegetables
elevation: 260 m
rainfall : 800 mm
sunshine : 1.600 h

511 01 Turnov
Phone: 213 43
Dir.: Ing. J. Košek

Belongs to the Research and Breeding Institute for Vegetable Crops in Olomouc
Breeding of vegetables (lettuce, carrots, Brussels sprouts, kohlrabi) — Ing. Z. Staněk CSc.
Breeding of fruit trees (cherry, morello, plum trees), strawberry — Ing. J. Konečný CSc. / Ing. S. Pavlišta CSc.
Breeding of flowers (Cyclamen, Dianthus, Dahlia) — Ing. V.Bernard/ Ing. P.Stiborek CSc
Biochemistry, quality of new vegetable varieties — Ing. M. Hadincová CSc.

28 Velké Losiny

28.1 Research and Breeding Station for Small-Fruits
elevation: 406 m
rainfall : 850 mm
sunshine : 1.600 h

788 15 Velké Losiny
Phone: 9649/94 93 64
Dir.: Ing. L. Šenk CSc.

Belongs to the Research and Breeding Institute of Pomology

Breeding of small-fruits, currants, gooseberry, strawberry, *Sorbus moravica* — Ing. M. Holubová CSc.

29 Veltrusy

29.1 Research and Breeding Station for Flowers
elevation: 172 m
rainfall : 550 mm
sunshine : 1.800 h

277 46 Veltrusy
Phone: 812 23
Dir.: Ing. J. Váňa

Belongs to the Research and Breeding Institute of Ornamental Horticulture in Průhonice

Breeding anual, bianual and perennial ornamental plants — Ing. J. Porš/ Ing. M. Doubková

30 Velký Krtíš

30.1 Research Station of Viticulture
elevation: 193 m
rainfall : 600 mm
sunshine : 2.000 h

990 01 Velký Krtíš
Phone: 225 16
Dir.: Ing. J. Liska CSc.

Belongs to the Research Institute of Wine and Viticulture in Bratislava. Breeding of vine and vine nursery. Varietal and rootstock experiments. Technology of vine multiplication. Nutrition and pruning of vine.

31 Veselé

31.1 Research and Breeding Station for Fruit Trees
elevation: 160 m
rainfall : 650 mm
sunshine : 2.200 h

922 08 Veselé pri Piešťanoch
Phone: 838/961 15
Dir.: Ing. L. Keblovský CSc.

Founded by the Research Institute of Fruit and Ornamental Trees in Bojnice, 1978
Research of cultural practice, breeding of apricots, peaches cultural practice in nurseries and fruit tree rootstock

Ing. M. Nitranský CSc.
Ing. J. Cifranič CSc.
Ing. H. Lúčanská CSc.

32 Żelešice

32.1 Department of Varietal Testing of Fruit Trees
elevation: 210 m
rainfall : 550 mm
sunshine : 1.800 h

664 43 Żelešice
Phone: 33 99 49
Dir.: Ing. J. Kalášak CSc.

Belongs to the Central Agricultural Control and Testing Institute in Prague

Variety experiments on peaches
Apricot, morello, cherry
Apple and pear trees

Ing. J. Kalášek CSc.
Ing. J. Richter CSc.
Ing. M. Bárta

DENMARK DK

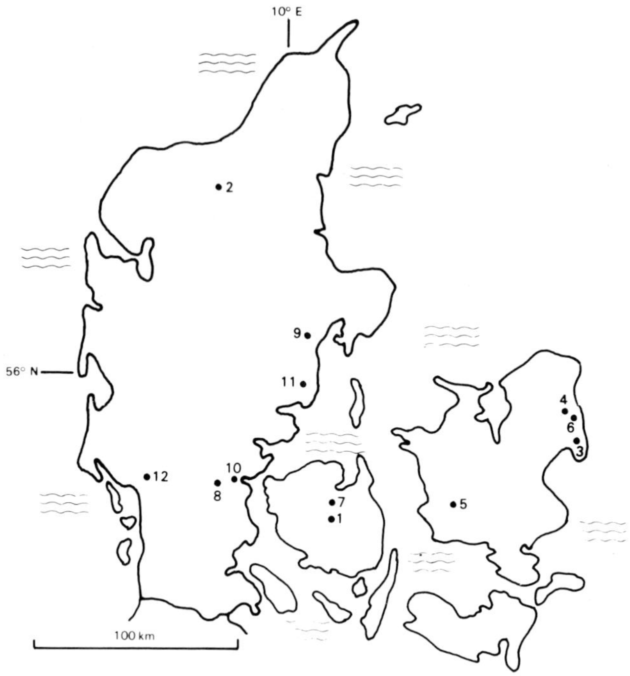
10° E
56° N
100 km

1 Aarslev
2 Hornum
3 København
4 Lyngby
5 Flakkebjerg
6 Hellerup
7 Odense
8 Lunderskov
9 Arhus
10 Kolding
11 Odder
12 Bramminge

Survey

Education - The centre of higher horticultural education is the Royal Veterinary and Agricultural University of Copenhagen, founded in 1858. Here students may follow courses in horticulture at university level and graduate with the degree of Cand.hort., "hortonom", after 4 years. This first degree may be followed by a degree called "Licentiat" (2-2.5 years), comparable with Ph.D., and after submitting a thesis the degree of "Dr.agro." may be conferred.

Research - Horticultural research is mainly conducted by two groups of institutions:

a. Departments of the Royal Veterinary and Agricultural University of Copenhagen, or, at a smaller scale, other universities, all belonging under the Ministry of Education.

b. Research Institutes under the Danish Research Service for Plant and Soil Science, which belongs under the Ministry of Agriculture.

Research work at the Royal Veterinary and Agricultural University of Copenhagen is carried out at the different institutes of the University by staff members and graduate students. The research institutions of the Department of Horticulture are placed at the Research Farm "Højbakkegaard", Taastrup (20 km from Copenhagen). Research work is carried out in co-operation with other departments of the University e.g. plant physiology, plant pathology, hydrology, taxonomy, genetics and atomic energy. At the departments of biological science of the universities of Copenhagen and Aarhus, some biological research connected with horticulture is also carried out.

Danish Research Service for Plant and Soil Science (Statens Planteavlsforsøg) concerning agriculture was founded in 1886 and since 1915 also horticultural experiments have been carried out. The Ministry of Agriculture appoints the main body called "Statens Planteavlsudvalg" (The State Board Committee), which organizes and finances the institutes, each with a special field of research activity. Recently the institutes and activities have been organized in 4 centres:

1. The Administrative Centre connected with the Secretariat for Variety Testing, Laboratory for Data Analysis, Agricultural Meteorology Service and the State Bee Disease Committee.

2. The Research Centre for Agriculture with 10 institutes and 2 laboratories for soil and crop research.

3. The Research Centre for Plant Protection with 3 institutes, 1 branch institute (weed) and 1 laboratory for pesticide analysis.

4. The Research Centre for Horticulture with 4 institutes concerning vegetables (Aarslev), glasshouse crops (Aarslev), pomology (Aarslev) and landscaping plants (Hornum) respectively.

Connected to the Ministry of Agriculture are also the

Prepared by Mrs. Grethe Bjerregaard, Gartnerinfo, København

State Seed Control Laboratory ("Statsfrøkontrollen"), the Institute for Agricultural Engineering ("Statens Jordbrugstekniske Forsøg") and The State Plant Protection Service ("Statens Plantetilsyn"). At the Research Centre for Horticulture the crops research work is connected with quality determinations in laboratories, with mechanical harvesting, storage capacity, preservation and energy savings in the glasshouse sector. A number of experiments are carried out as co-institutional projects as well as in cooperation with institutes outside the centre. The research activities are planned in close contact with the growers associations, the advisory service and the industry. An increasing amount of projects are directly planned and financed on a 50 percent basis in cooperation with growers associations, industry a.o.

Joint Committees - In order to supplement activities such as listed above, a number of joint committees have been formed. The first one, founded in 1919, is the "Joint Committee for Test Growings of Vegetables" ("Faellesudvalget for Prøvedyrkning at Køkkenurter"), where official and private enterprises co-operate. After Denmark's accession to the EEC the Joint Committee was reconstructed and is now responsible for the variety tests of vegetables in the open and under glass. The variety tests consists of a test for distinctness, homogenity and stability and a test for value or for Plant Breeders' Rights, all according to the legislation and the directives of the EEC and UPOV. The test growings are carried out in collaboration with the State Research Stations.

For those varieties that fulfill the requirements of the Joint Committee and the Secretariat for Test Growing a recognition is granted. In case of testing for value, the recognition is indicated by adding the owner's trade mark followed by an S and the two last figures of the year of recognition to the variety name. The post control of recognized varieties is taken over by the State Seed Control Laboratory ("Statsfrøkontrollen").

Since 1980 a plant health control has been carried out in all plant producing firms by the Danish Plant Protection Service ("Statens Plantetilsyn"). The Danish Research Service for Soil and Plant Sciences produces plants with defined genetic properties as well as freedom for defined pathogens by request of the Horticultural Plant Control Commission ("Gartnerikontrolkommissionen"). These plants are propagated at the Danish Growers Elite Plant Station ("Planteopformeringsstationen").

Extension - Advisory work and extension service is carried out by associations such as the Danish Market Grower's Association ("Dansk Erhvervsgartnerforening"), Denmarks Association of Producers of Nursery Stocks ("Dansk Planteskoleejerforening"), the Association of Commercial Fruit Growers of Denmark ("Danmarks Erhvervsfrugtavlerforening") and Denmark's Flower Bulb Association ("Danmarks Blomsterløgavlerforening"). Although advisory service is the main purpose of these associations, some experiments are carried out on specialized subjects.

Publications - Reports and results of research and experiments are published in "Tidsskrift for Planteavl" (Danish Journal of Plant and Soil Science), "Ugeskrift for Agronomer og Hortonomer" (weekly Journal of Agronomy, Horticulture), "Gartner Tidende" (weekly magazine for growers), "Frugtavleren" (monthly magazine for fruit growers) or as leaflets from the Danish Research Service for Plant and Soil Science.

Climatic Conditions - Denmark is, with the exception of Greenland and the Faroe islands, where no horticultural research until now has been performed, situated a few meters above sea level. At Bramminge 5 meters and at Arhus 63 meters. The annual mean precipitation varies between 726 mm at Bramminge and 602 mm at Copenhagen and is spread equally over the year. Regarding sunshine there is a variation between the parts of the country amounting to more than 350 hours. At Copenhagen the mean registration of sunshine is 1301 hours and at Aarslev, the research center for horticulture, a mean of 1750 hours of sunshine is registered.

1 Aarslev

Havebrugscentret (Research Centre for Horticulture)
Comprising four institutes, three of which situated at Aarslev. For the fourth institute, see 2. Hornum

Kirstinebjergvej 12,
DK-5792 Aarslev, Phone: 459-991766
Dir.: E. Poulsen, M.Sc.agr.

1.1 Institut for grønsager (Institute of Vegetables)

Kirstinebjergvej 6,
DK-5792 Aarslev, Phone: 459-991766
Dir.: M. Blangstrup Jørgensen*
M.Sc.hort.

Established in 1915 at Spangsbjerg. Moved to Aarslev in 1972. Acerage 45 ha. 1740 m² greenhouse area. Research in field vegetables, strawberries and flower bulbs. Breeding of certain vegetables and berries. Mechanical harvesting.

Vegetables, production systems, plastic covering, plant establishment	K. Henriksen, M.Sc.agr.
Vegetables, DVS-testtiming programs, once-over harvest, industrial relations	J. Jensen, Degr.cand.hort.
Vegetables, test for value	Jette Larsen, cand.sci.
Vegetables, growth physiology	N. Poulsen, cand.sci.
Vegetables, storage	P. Molls Rasmussen, Degr.cand.hort.
Vegetables, nutrition, plant analysis	J. Nygaard Sørensen, Degr.cand. hort.
Flower bulbs, field and forcing experience	E. Rasmussen, Degr.cand.hort.
Strawberries, breeding, cultivation and mechanical harvesting	A. Thuesen, M.Sc.hort.

1.2 Institut for vaeksthuskulturer* (Institute of Glasshouse Crops)

Kirstinebjergvej 10,
DK-5792 Aarslev, Phone: 459-991766
Dir.: H.E. Kresten Jensen, Ph.D.

Established in 1929 in Virum. Moved to Aarslev in 1976. Glasshouses covering a total area of 4800 m^2. Growth chambers, phytotron and propagation unit. Research on pot plants, cut flowers and vegetables under glass.

Growth regulators for ornamental plants	E. Adriansen, Degr.cand.hort.
Standard growing programmes for pot plants	Henny Andersen, Degr.cand.hort.
Glasshouse environment, energy saving	M.G. Amsen*, Ph.D.
Use of artificial light	N.E. Andersson, Degr.cand.hort.
Selection of clones	C.O. Ottosen, cand.sci.
Cut flowers: blueprint cropping, storage and product development	N. Bredmose*, Degr.cand.hort.
Perennials and shrubs suitable as pot plants and cut flowers, grown under outdoor and greenhouse conditions	Linda Noack Kristensen*, M.Sc.hort.
Plant plants: production planning, growth and flowering, selection and nuclear stock	O. Voigt Christensen*, M.Sc.hort.
Development of new pot plants	Kirsten Friis, Degr.cand.hort.
Development of new cut flowers	Vibeke Geertsen, Degr.cand.hort.
Vegetables under glass: cucumber diseases and watering systems for vegetables	Grethe Haupt, Degr.cand.hort.
Propagation, hormones, dormancy, hypobaric storage of cuttings	J. Hansen, Ph.D.hort.
Keeping quality of pot plants	L. Høyer, Degr.cand.hort.
Electronics, process control	P. Konradsen, Electr.eng.
Energy saving in glasshouses, technical	O. Frøsig Nielsen, Electr.eng.
Protection of new plant varieties	K. Kristiansen, Degr.cand.hort.
Development of new vegetables, variety trials	Kirsten Rasmussen, Degr.cand.hort.
Thermal environment and energy saving	J.S. Strøm*, Electr.eng.
Substrates, water, nutrition, hydroponics	J. Willumsen*, M.Sc.hort.

1.3 Institut for fruit og baer* (Institute of Pomology)

Kirstinebjergvej 12,
DK-5792 Aarslev, Phone: 459-991766
Dir.: J. Vittrup Christensen*, Dr.Sc.hort.

Established in 1915 at Blangstedgaard. Moved to Aarslev in 1983. Acreage 28 ha. Research in tree fruit and small fruit, cultivation, storage and preservation.

Cherry cultivars and clones, filberts and walnut	J. Vittrup Christensen*, Dr.Sc.hort.
Rootstocks, planting systems, mechanical harvesting, Rubus varieties	O. Callesen*, M.Sc.hort
Food technology, texture, colour	P.E. Christensen, Civ.eng.
Apple cultivars, growth regulators, frost protection,	J. Grauslund*, M.Sc.agr.

fruit thinning and set
Food technology, flavour, taste and Elderberry — K. Kaack, M.Sc.agr.
Nutrition, small fruits, pesticide application — O. Vang-Petersen, Degr.cand.agro.
Storage of fruit and vegetables, pear varieties — P. Molls Rasmussen, Degr.cand.hort.

2 Hornum

2.1 Institut for Landskabsplanter (Institute of Landscape Plants)

Hornum. DK-9600 Aars
Phone: 458-661333
Dir.: I. Groven*, M.Sc.hort.

Established in 1916. Acreage 43 ha. 2000 m² of glasshouses. Main functions: Selection of optimal cultivars and seed sources of landscape plants. Methods of propagation, cultivation, production and establishment.

Clonal selection and breeding. Testing and trials of cultivars and seed sources — P.E. Brander, M.Sc.hort.
Growing medium, container systems, root development — O. Bøvre, Degr.cand.hort.
Experiments on shelterbelts — I. Groven*, M.Sc.hort.
Nutrition, irrigation, root vitality, plant and seed physiclogy — F. Knoblauch*, M.Sc.hort.

Propagation techniques storage of plant material, cultivar trials of low garden roses — O. Nymark Larsen, Degr.cand.hort.
Propagation and cultivation of *Miscanthus sinensis* for production of plant fibres — Anne Sloth, Degr.cand.hort.
Incompatibility studies of selected clones and seed sources, pollination, breeding — Birgit Hansen, Degr.cand.hort.

3 København

3.1 Den kgl. Veterinaer- og Landbohøjskoles Havebrugsinstitut (Horticultural Institute of the Royal Veterinary and Agricultural University)

Rolighedsvej 23
DK-1958 Copenhagen V
Phone: 451-351788

Floriculture — Prof. A. Skytt Andersen*, Ph.D.
Greenhouse climate — Aa. Andersen*, Lic.agro.
Glasshouse vegetables, nutrition — P. Karlsen, Degr.cand.hort.
Plant breeding — B. Farestveit, Lic.agro.
Glasshouse construction and heating — O. Skov*, Tech.eng.
Mushroom research — Kaj Bech, Degr.cand.hort.
Botanical garden — J. Mosegaard, Degr.cand.hort.
Pomology, fruit growing and small fruit — Prof. P. Hansen*, Dr.agro.
Nursery crops, propagation — E.N. Eriksen, Lic.agro.
Tropical and subtropical horticulture — Kirsten Lundsten, Degr.cand.hort.
Vegetable crops — F. Larsen, Lic.agro.
Ornamental plant collection — M. Fonnesbech*, Ph.D.

3.2 Den kgl. Veterinaer- og Landbohøjskoles Institut for Have og Landskab (Department of Landscape Architecture)

Rolighedsvej 23, I.
DK-1958 Copenhagen V
Phone: 451-351788

Landscape construction, plant material and use — Prof. J.P. Schmidt
Plant material and use — Annemarie Lund, Degr.cand.hort.
Contemporary and historical landscape architecture — Jette Abel, Degr.cand.hort.
Landscape planning and design — Malene Hauxner, Degr.cand.hort.
Karen Altwell, M.Sc.(Landsc.)
P. Stahlschmidt, M.Sc.(Landsc.)

Landscape preservation	N. Elers Koch, Degr.cand.hort.
Administration and process engineering	Susanne Guldager, Degr.cand.hort.
3.3 Arboretum	Arboretet, Kirkegårdsvej 3A DK-2970 Hørsholm Phone: 452-860641 Dir.: Dr.agro. B. Søegaard
Resistance breeding in forest trees, Thuja	
Dendrology, Registration, Japanese trees	F.G. Christensen, cand.(silv.)
Breeding of pines, extraction and storing of pollen, Korean trees	L. Feilberg, M.Sc.(silv.)
Breeding of larches, forest tree breeding in the tropics	H. Keiding, M.Sc.(silv.)
Forestry history, dendrology, phenology	P.C. Nielsen, M.Sc.(silv.)
Grafting and applied techniques in forestry	K. Naess-Schmidt, M.Sc.(silv.)
Breeding of douglas fir and firs, vegetative propagation	H. Roulund, dr.agro.
Breeding of spruces, induction of flowering, isoenzymes	H. Wellendorf, M.Sc.(silv.)
Dendrology, general botany, plant geography, the subarctic	S. Ødum, cand.mag.
Taxonomy, forestal history	P.Chr. Nielsen
3.4 Den kgl. Veterinaer- og Lanbohøjskoles Afdeling for Fysiologisk Botanik (Department of Plant Physiology and Anatomy)	Thorvaldsensvej 40 DK-1871 Copenhagen V Prof.: Dr.phil. R. Rajagopal
Plant anatomy	Dr.phil. S. Allerup/I. Móller, M.Sc.(pharm.)
General plant physiology, growth regulators	B. Veürshov, Ph.D
Nutrient uptake	Lic.agro. G. Jensen
Photosynthesis	A.C. Madsen
Photosynthesis	Dr.phil. B. Lindberg Møller
3.5 Den kgl. Veterinaer- og Landbohøjskoles Plante-patologiske Afdeling (Department of Plant Pathology of the Royal Veterinary and Agricultural University)	Thorvaldsenvej 40, III DK-1871 Frederiksberg C Phone: 451-351788
Pathogenesis, diseases of agricultural plants	Prof. V. Smedegaard-Petersen, Dr.Agro.
Epidemiology, diseases of agricultural plants	Lisa Munk, Ph.D.
Soil-born diseases, biological control, diseases of horticultural plants	J. Hockenhull*, Ph.D.
Mycology, diseases of forest trees	J. Koch, M.Sc.(silv.)
Plant diseases in relation to bromatology	G. Kovács, Ph.D.
General plant virology, virus diseases	Lic.agro. T. Lundsgaard, Ph.D.
3.6 Den kgl. Veterinaer- og Landbohøjskole, Botanisk Institut (Botanical Institute of the Royal Veterinary and Agricultural University)	Rolighedsvej 23 DK-1958 Copenhagen V Phone: 451-351788
Silvicultural botany, dendrology	Prof. H. Vedel
Taxonomy	Prof. N. Jacobsen/C. Baden
Plant ecology	Prof.Dr. K. Hansen/J. Jensen
Botany of food plants	O. Høst
Silvicultural botany	L. Rastad
3.7 Dansk Erhvervsgartnerforenings Konsulentvirksomhed (Advisory Service of the Danish Market Growers	Anker Heegaards Gade 2 DK-1572 Copenhagen V

Association)
District office Zealand
Chief advisor
Technical advisor

Phone: 451-158530
Telex: 19230 DEG DL
F. Busk Madsen, Tech.Eng.,
B.D. Olesen, Tech.Eng.

Glasshouse crops: cucumber, Chrysanthemum
pot plants

N.P. Holmenlund, Degr.cand.hort.
G. Priisholm Andersen, Degr.cand.hort.,
Margit Andersen, Degr.cand.hort.

General crops: lettuce, tomatoes
Production planning
Economy

E. Hvalsøe, Degr.cand.hort.
T. Lippert, Degr.cand.hort.
M. Blåbjerg, Degr.cand.hort.

3.8 Dansk Planteskoleejerforenings Sekretariat (Danish Nurserymen's Association)

Anker Heegaards Gade 2, III
DK-1572 Copenhagen V
Phone: 451-145658

3.9 GartnerINFO (HortINFO)

Phone: 451-158530

Informationservice covering glasshouse crops, vegetables in the open and nursery crops. The purpose is to increase the knowledge of foreign and domestic research and experimental results within the horticultural area.

Secr.: Grethe Bjerregaard, Degr.cand.hort.
Ulla Wicksell, Degr.cand.hort.

4 Lyngby

4.1 Institute of Plant Pathology

Lottenborgvej 2, DK-2800 Lyngby
Phone: 452-872510
Dir.: H. Rønde Kristensen*

Botany Department
Aim of work: Procuring knowledge about diseases in agricultural and horticultural plants caused by fungi, bacteria and physiogenic disorders.

Head: A. Jensen
Scientific staff: L. Buchwaldt
Ib.G.Dinesen/K.Bolding Jørgensen
H.A. Jørgensen/B. Løschenkohl
S. Stetter/B. Welling

Virology Department
Aim of work: Development of reliable, quick and cheap diagnostic methods. Increased knowledge of transmission and spread of virus diseases. Elucidation of the distribution and damage of virus diseases. Procurance of knowledge about therapeutic treatments. Production of healthy nuclear stocks.

Head: H. Rønde Kristensen*
Scientific staff: J. Begtrup
B. Engsbro/M. Herde/N. Paludan
A. Thomsen

Zoological Department
Aim of work: To produce knowledge about distribution, biology and the economic severity of pests. To work with development of integrated control measures against pests. To examine soil samples for potato cyst nematodes. To test new potato crosses for resistance against potato cyst nematodes.

Head: J. Jacobsen
Scientific staff: P. Esbjerg
L. Monrad Hansen
Lise Stengaard Hansen
F. Lind/J. Reitzel
Lise Samsøe-Pedersen

Advisory Service
Aim of work: To give information and answer enquiries about pests and diseases of agricultural and horticultural plants.

Scientific staff: G.C. Nielsen
L.A. Hobolth

4.2 Pesticide Research Institute

Lottenborgvej 2, DK-2800 Lyngby
Phone: 452-872510

Aim of work: To ascertain
- that trials and investigations are carried out to obtain a sufficient basis for approval of pesticides, plant growth regulators and spray additives.
- that qualitative and quantitative efficacy data are supplied to and evaluated by the Institute in connection with the registration of pesticides and plant growth regulators by the National Agency of Environmental Protection.

Dir.: E. Nøddegaard
Scientific staff: B. Bromand
Kirsten Junker/L.N. Jørgensen
B.J. Nielsen/S. Lykke Nielsen
A. Nørh Rasmussen/E. Schadegg

5 Flakkebjerg

5.1 Institut for ukrudtsbekaempelse (The National Weed Research Institute)

Flakkebjerg, DK-2400 Slagelse
Phone: 453-586300
Dir.: K.E. Thonke

This station has its own fields and laboratories for trials and investigations, but co-operates with other State Experiment Stations concerning agricultural and horticultural experiments.

G. Noyé, Degr.cand.hort.

6 Hellerup

6.1 Statens Plantetilsyn (Danish Plant Protection Service)

Gersonsvej 13, DK-2900 Hellerup
Phone: 451-620787
Dir.: Degr.cand.silv. P. Regenberg

The Danish Plant Protection Service carries out plant health control in all Danish plant producing firms, inspects all potato seed crops, is responsible for the domestic control of a number of plant pests, inspects fruits, vegetables and potatoes for export and homemarket and checks imported plant material for quality and health.

7 Odense

7.1 Dansk Erhvervsgartnerforenings konsulentvirksomhed Advisory Service of the Danish Market Grower's Association) District office Fünen

Middelfartvej 9-11,
DK-5000 Odense C, Phone: 459-126939
Chief adv.: I. Stenberg Pedersen*, Degr.cand.hort.

Glasshouse crops: vegetables — E. Jensen*, Degr.cand.hort.
pot plants — J. Solvang, Degr.cand.hort., J. Rystedt, Degr.cand.hort.
cut-flowers and flower bulbs — A. Pilgaard*, Degr.cand.hort.
General crops: vegetables — H.P. Mathiassen*, Degr.cand.agro.
nutrition, analyses — M. Hansen*, M.Sc.
Production planning — E. Moes, Degr.cand.hort., A. Beck Larsen, Degr.cand.hort.
Economy — H. Søndergaard Nielsen, Degr.cand.hort., O. Baerenholdt-Jensen, Degr.cand.hort., O. Christiansen, Degr.cand.hort., J. Grønborg Eskesen, Degr.cand.agro.
Socio economy — Sv. Lundgaard, Com.Degr.
Technical advisors — J. Hjelmdal Jensen, Tech.Eng., J. Kronmann, Tech.Eng.

7.2 Dansk Erhversfrugtavl (Association of Commercial Fruit Growers of Denmark)
Vindegade , DK-5000 Odense C
Phone: 459-126362
Secr.: E. Burgaard, Degr.cand.hort.

Advisors:
S. Thorup Degr.cand.hort.,
S. Oluf Ramborg, Degr.cand.hort.,
E. Larsen, Degr.cand.hort.,
Vibeke Langer, Degr.cand.hort.,
P.E. Jørgensen, Degr.cand.hort.,
Christa Heegaard, Degr.cand.hort.,
G. Borg, Degr.cand.hort.

7.3 Havebrugets Plantebeskyttelsesudvalg (Committee for Trial of Horticultural Chemicals)
Vindegade 72, DK-5000 Odense C
Phone: 459-126362

8 Lunderskov

8.1 Planteopformeringsstationen (The Danish Growers Elite Plant Station)
Nagbøl Kirkevej 12
DK-6640 Lunderskov
Phone: 455-586026
Dir.: P. Mortensen, Degr.cand.hort.

Private foundation established in 1980 by the Association for Danish Horticultural Producers and the Danish Fruit and Vegetables Producers Association. Area: 12 ha, 1200 m^2 glasshouses. Aim: to propagate and distribute plant material, which has been approved by the Horticultural Plant Control Commission. Furthermore to help purchasing and distributing of plantmaterial from foreign countries. The Elite Plant Station holds motherplants of fruittrees, fruit bushes, stocks, strawberries, ornamentals, perennials and glasshouseplants. Furthermore several seed sources with known characteristics are at disposal.

9 Århus

9.1 Dansk Erhvervsgartnerforenings konsulentvirksomhed (Advisory Service of the Danish Market Growers Association) District Office Jutland
Finlandsgade 15, DK-8200 Aarhus N
Phone: 456-164166
Chief adv.: A. de Lasson*, Dr.agr.

Glasshouse crops: pot plants — E. Riis Lavsen, Degr.cand.hort., O.H. Mathiassen, Degr.cand.hort.
vegetables — L. Bay, Degr.cand.hort.
General crops: vegetables — B. Olsen, Degr.cand.hort.
Production planning — N. Bloch, Degr.cand.hort.
Economy — A. Svendsen, Degr.cand.hort., K. Steffensen, Degr.cand.hort.
Technical advisor — K. Madsen, Tech.Eng.
Laboratory of soil and plantanalysis — K. Lystlund, Degr.cand.hort.

9.2 Det Faglige Landscenter (The Danish Agricultural Advisory Centre)
Belongs to Danish Farmers' Unions and Danish Smallholders' Unions.
Kongsgårdsvej 28, DK-8260 Viby J
Phone: 456-110888
Telex: 68691 LAFALA DK

Landskontoret for Planteavl (Crop Husbandry Department)
Field Vegetables
Senior advisor: T. Huus-Bruun*, Degr.cand.hort.

10 Kolding

10.1 Bioteknisk Institut* (Biotechnical Institute)
Private foundation, affiliated to the Danish Academy of Technical Sciences
Holbergsvej 10, DK-6000 Kolding
Phone: 455-520433
Telex: 16600 TLGR BIOTECH
Manager of Veg.Div.:

Afd. for vegetabilske råvarers Teknologi (Division of Vegetable Technology) conducts experimental and extension work on fruit and vegetables.
Main topics: storage, handling and packing of fresh fruit and vegetables.
Facilities: storage bins, testing facilities, analytical laboratory

P. Vendelbo, M.Sc., Chem.
G. Nissen, M.Sc., Chem., Dipl. in Comm.
T. Pedersen, Degr.cand.hort.
H. Aabye Jensen, M.Sc., Chem.

11 Odder

11.1 Dansk Planteskoleejerforening (Danish Nurserymen's Association)

Hølkenhavevej 21, DK-8300 Odder
Phone: 456-556714
Gen.adv.: A. Bertram, Degr.cand.hort.

12 Bramminge

12.1 Dansk Planteskolejerforening (Danish Nurserymen's Association)

Nr. Ilstedmarkvej 4, DK-6740 Bramminge, Phone: 455-172023

PR and marketing

Lis Langschwager, Degr.cand.hort.

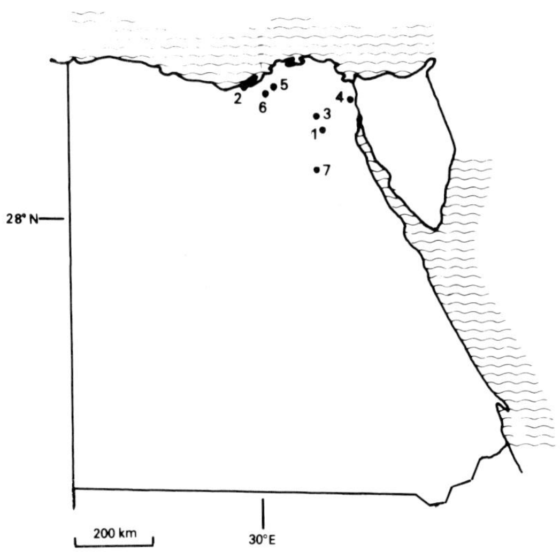
28° N
200 km
30°E
1
2
3
4
5
6
7

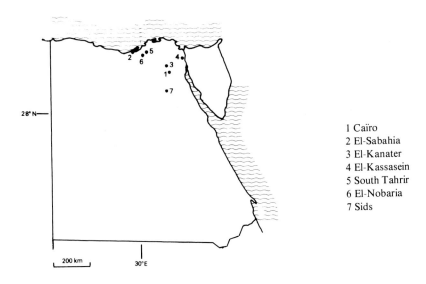

1 Caïro
2 El-Sabahia
3 El-Kanater
4 El-Kassasein
5 South Tahrir
6 El-Nobaria
7 Sids

Survey

the main responsibility of the Horticulture Research Institute of the Ministry of Agriculture and Food Security is to work on improving production of horticultural crops, developing their techniques, and solving various problems of production.
The Institute cooperates in research work with 13 colleges of agriculture located in different provinces of Egypt, and also the National Research Center located in Caïro.

The Headquarters of the institute is in Cairo. However, research work is carried out in 6 Experiment Stations in upper and lower Egypt by a technical staff belonging to the Horticulture Institute. These stations are the El-Sabahia, El-Kânater, Sids, South Tahrir, El-Kassasein (Ismailia) and El-Nobaria Research Stations.

The Horticulture Research Institute is one of 14 Institutes belonging to the Agricultural Research Center of the Ministry of Agriculture and Food Security.
The Horticulture Research Institute consists of four groups of research sections: Fruit Research Sections, Vegetable Production, Medicinal and Aromatic Plants Research Sections, Ornamental and Timber Trees Research Sections and Processing Research Sections.
The Stations conduct experiments in cooperation with the different sections of the Horticultural Research Institute, to solve local problems. Also to have close contacts with the extension service staff and the growers regarding agricultural problems and their solutions; recommendations and their application for the purpose of developing horticultural production in different locations.

1 Caïro

1.1 Horticultural Research Institute
Agricultural Research Center (ARC)
(Ministry of Agriculture and Food Security)

Giza, Orman, Caïro
Phone: 720617
Dir.: Dr. Salah Baha El-Din*
Dep.Dir.: Dr. Hosni Ali Kamel, Samy Abd El-Moneim

The ARC is considered one of the important scientific centers in Egypt, working on developing plant and animal production. The ARC participates in planning the National Agriculture policy and is responsible for its implementation within the economic policy of the government, for the purpose of securing local market requirement and to fulfil the projected export plan of Agricultural products. The ARC consists of thirteen Research Institutes and two central laboratories, i.e.: Horticulture Research Institute; Cotton Research Institute; Field crops Research Institute; Soil and Water Research Institute; Plant protection Research Institute; Plant

Prepared by Dr. Salah Baḥa El-Din, Horticultural Research Institute Giza, Caïro

Pathology Research Institute; Animal Research Institute; Animal Health Research Institute; Agriculture Economic Research Institute; Agriculture Mechanization Research Institute; Sugar Crops Research Institute; Animal reproduction Research Institute; Animal Vaccine Research Institute and two Central Laboratories, one for statistical analyses and one for pesticides.
The first start of the ARC was in 1897 under the name of the Royal Agriculture Egyptian Society, including technical sections for plants and animal production. Horticulture Section was established in 1911, developed later to the Horticulture Research Institute (HRI) as one of the ARC important institutions and is mainly concerned with research work and studies to improve horticultural crops and developing production in addition to indicating the most apropriate technology for the processing of horticultural products.
Horticultural crops, which are the main concern of the Institute are identified as follows: Fruit crops, vegetable crops, ornamentals, forestry, medicinal, and aromatic plants. The relative importance of horticultural crops in Egyptian agriculture might be explained. The total acreage of horticultural crops is about 1.600 million feddan distributed as follows: 1100000 vegetables, 450000 fruits and 50000 medicinal and aromatic plants. The total cropping area in Egypt is about 11 million feddan. Field crops represent most of the agriculture land estimated at about 90% of the total cropping area.
The importance of the role of the ARC and its activities can be recognized if we know that about 55% of the population work in agriculture. Also it represents about 30% of the national income, and about 80% of return from export products. Technical staff amounts to 301 and supporting staff personnel to 2755.

Research Sections:

1.2 Fruit Research Sections:

Citrus research	Dr. M.A. El-Nokrashi
Viticulture research	Dr. Foad Fauzy
Deciduous fruit research	Dr. Mohamed Zaki
Tropical and minor fruits research	Dr. Mamdouh Riyad
Olives and dry farming research	Seif El-Deen, Abou- Baker Sari El-Deen
Fruit post-harvest research	Dr. Bahia Abd El-Aziz
Fruit nutrition research	Dr. Hoda Habib Hanin

1.3 Vegetable Research Sections:

Tomato and selfpollinated vegetable research	Dr. Ahmed Zein El-Abdein
Cucurbits and cross pollinated vegetable research	Samy Abd El-Moneim
Potatoes and vegetative propagated research	Dr. Sherifa Foda
Seed production technology research	Dr. Moktar El-Sherbini
Vegetables post harvest research	Dr. A. Fayza
Protected cultivation research	Dr. Arafa Imam Arafa
Medicinal and aromatic plants research	Dr. Kamal El-Din Awad

1.4 Food Processing Research Sections:

Fruit processing research	Dr. Foad El-Ahwah
Vegetable processing research	Dr. M. El-Shiaty
Plant oil research	Dr. J. Samuel
Meat and fish research	Dr. Samir El-Daslerity

1.5 Ornamental and Forestry Research Sections:

Botanical gardens research	Dr. Hosni A. Kamel
Ornamental plants research	Dr. M. Meleegy
Landscape gardening research	Dr. Hosni A. Kamel
Wood trees research	Ibrahim Heikal
Ornamental and plant taxonomy research	Dr. Hosni A. Kamel

2 El-Sabahia

2.1 El-Sabahia Experiment Station
elevation: 17.8 m
rainfall : 1.325 mm
sunshine : 12.16 h

Dir.: Dr. Magdi Nagib

3 El-Kanater

3.1 El-Kanater Experiment Station
elevation: 64.12 m
rainfall : 0.04 mm
sunshine : 12.12 h

Barrage
Dir.: Dr. Ahmed Khalifa

4 El-Kassasein

4.1 El-Kassasein Research Station
elevation: 11.15 m
rainfall : 0.03 mm
sunshine : 12.16 h

Ismailia
Dir.: Dr. Ibrahim Moussa

5 South Tahrir

5.1 South Tahrir Experiment Station
elevation: 42.5 m
rainfall : 0.05 mm
sunshine : 12.12 h

Dir.: M.H. Said-Allah

6 El-Nobaria

6.1 El-Nobaria Experiment Station
elevation: 17.8 m
rainfall : 1.325 mm
sunshine : 12.16 h

Dir.: Dr. A. Moustafa

7 Sids

7.1 Sids Experiment Station
elevation: 21.0 m
rainfall : 0.02 mm
sunshine : 12.17 h

Dir.: Dr. M.H. Abo-Zeid

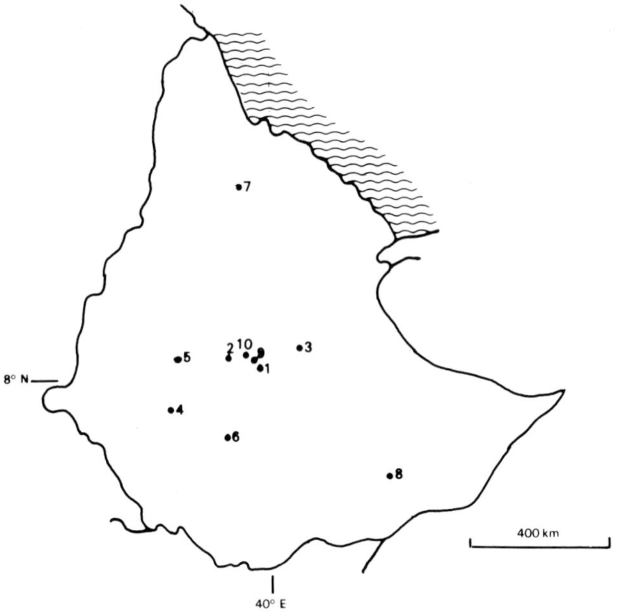

•7
2
10
•3
•5
•1
8° N
•4
•6
•8
400 km
40° E

1 Nazareth
2 Holetta
3 Melka Werer
4 Jimma
5 Bako
6 Awassa
7 Mekele
8 Gode
9 Debre Zeit
10 Addis Ababa

Survey

The Institute of Agricultural Research, (IAR), was established in 1966. It is supported by the Ethiopian Government and UNDP. It has eight Research Departments of which Horticulture is one. The research is carried out in eight main and several sub-stations. The research sites are representative of the different ecological areas of the country. Recent programmes of the Horticulture Department (IAR) - Emphasis of research is given to the locally important and export potential fruits and vegetables. Varieties of some crops have been recommended and released for production. Previous trails on artichoke, asparagus, strawberry and egg plant are terminated. Planting material increment and techniques of seed production of most vegetables are in progress in IAR and other organization sites. Collections of grape vine varieties from Yugoslavia and California are on evaluation for disease resistance, high quality and high yield for fresh consumption. Winery and raisin trials are in progress at Abaddir and Dukem.

1 Nazareth

1.1 Research Station (Coordinating Station)

Nazareth Research Station,
PO Box 103
Horticultural Dept. Co-ord.:
Seifu G/Mariam
Station Repr.: Terefe Belehu
O.I.C.: Imru Assefa

Situated about 95 km SE of Addis Ababa: established 20 years ago. The research sites Koka and Melkassa are located at 1500 and 1550 m altitude respectively. Most of the collection and evaluation as well as major trials are extensively facilitated with both locally important and minor fruit and vegetable crops, rainfed and irrigated. Most of the pest, disease and processing trials (dehydration, preservation etc) are conducted here.
Observation and evaluation of different rootstock/scion combination of citrus trees (orange, mandarin, grape fruit, lemon and lime). Propagation of budded trees for trial and growers. Adaptation trials of tropical and subtropical fruits and nuts, including mango, avocado, guava, Jack fruit, loquat, passion fruit, Annona sp. and cashewnuts. Nursery for the propagation and distribution of temperate fruit trees to research sites and growers. Variety screening, sowing date, pruning effect, hardening before transplanting, seed production techniques etc. on onion and tomato. Crops have been on trial for a long time and varieties of tomato for consumption and processing are

Prepared by T.H. Jackson, Horticultural Development Department, Addis Ababa recommended. Sweet potato is the most accepted crop in low rainfall and marginal soil areas. Variety screening, methods of land preparation for moisture conservation, intercropping with sorghum are in progress. Collection and selection for disease and other desirable characteristics, cultural practices and minor improvement trials on hot pepper *(Capsicum spp.)* Collection and screening of cultural practice of white potato. Processing trials for storing perishable horticultural products, dehydration (onion, tomato, white and sweet potatoes, cassava, chilies, etc) and preservation.

2 Holetta

2.1 Research Station

Holetta Research Station
c/o Institute of Agricultural
Research, PO Box 2003,
Addis Ababa
Station Repr.: Yilma Abebe
Staff: Tameru Mehrete
O.I.C.: Dr. Taye Teferedeg

Located about 45 km W of Addis Ababa; altitude 2390 m. Variety observation of different rootstock/scion combination on peach, apple, plum, quince, pear, apricot, almond, fig, nectarine and walnut. Research on cool season vegetable crops. Collection and screening, cultural practice of white potatoes.

3 Melka Werer

3.1 Research Station

Melka Werer Research Station
c/o Institute of Agricultural
Research, PO Box 2003
Addis Ababa
Station Repr.: Belachew Haile
O.I.C.: Ahmed Sherif

Located in the Awash Valley (Great Rift Valley, SE). Altitude 750 m. All trials are conducted under irrigation. Major activities are variety evaluation and propagation on citrus, banana, papaya and various high temperature requiring crops. Propagation of budded trees for trial and growers and papaya seed production. Collection and screening, cultural practice of cassava, taro, yam, etc.

4 Jimma

4.1 Research Station

Jimma Research Station, PO Box 192,
Jimma, Keffa
Station Repr.: W/t. Hanna Assefa
Staff: Dr. L. Louis/Teklu Negash
O.I.C.: Gebre Mariam Shikur

Located SW, about 350 km from Addis Ababa. Altitude 1700 m, adequate rainfall most part of the year. Collection and evaluation of herbs and spices, method of propagation and yield determination of ginger, corriander, fenugreek, basil, black cumin, tumeric, thyme and some minor ones. Extensive pineapple variety trials. Evaluation and adaptation trials of tropical and subtropical fruit trees (loquat, mango, passion fruit, avocado, guava, macademia nut, tree tomato, key apple, casimiroa, etc). Papaya seed production and banana variety trials. Collection and screening, cultural practice of root and tubers (cassava, taro, yam etc).

5 Bako

5.1 Research Station

Bako Research Station, PO Box 3,
Bako, Wollega
Station Repr.: W/o. Debritu
Staff: Getachew Mengestu
O.I.C.: Dr. Beru Abebe

Located in SW of Ethiopia, about 250 km from Addis Ababa; altitude 1680 m. Observation and evaluation of different rootstock/scion combination of citrus trees since 1969. Pineapple variety and cultural practice studies including peach, apple, plum, quince, pear, apricot, fig, nectarine and walnut. Hot peper *(Capsicum Spp.)* collection and selection for disease resistance and other desirable characteristics. This crop is in high demand in local markets and by the Ethiopian Spices Extraction Co. for export of Eleorasin. Minor trials of citrus and vegetables (tomato, sweet and white potato, snap beans, onion, cassava, banana and beet root) are in progress. Most of the trials are conducted under rainfall.

6 Awassa

6.1 Research Station

Located SW of the Capital; altitude 1700 m. This station came under IAR in 1978. Collection, variety trial and cultural practices on hot pepper, tomato, onion, white potato and observation on other vegetables and fruit.

Awassa Research Station, PO Box 6
Awassa, Sidamo
Station Repr.: Gezaheg Taddesse
Staff: W/o Kimya Mohammed
O.I.C.: Abduraheman Ali

7 Mekele

7.1 Research Station

Located NE of Addis Ababa. This is mainly a cereals and pulse station. The horticultural research started recently with observation on some vegetable crops.

Mekele, Tigre

8 Gode

8.1 Research Station

Located in the Eastern part of the country. Evaluation and cultural practice as well as variety screening trials of banana and other vegetables.

Gode, Bale

9 Debre Zeit

9.1 Agricultural Research Centre
elevation : 1850 m
rainfall : 800-900 mm
sunshine : 65 h
r.h. : 66 %

PO Box 32
Phone: 33.80.26
Dir.: Dr. Asfaw Zelleke

The Debre Zeit Agricultural Research Centre is located about 45 km south of Addis Ababa. The area is well suited for the production of cereal grains, pulse crops and some horticultural crops. The rainfall is distributed in two periods, i.e., short rainy season (March-April) and main rainy season (June-September). The main soil type of the area is vertisol and sandy loams are found to a lesser extent. The Debre Zeit Agricultural Research Centre is a branch of the Alemaya Agricultural College (the new Agricultural University). The centre has Crop Science and Animal Science Departments.

Horticulture Section
Agrobotanical studies or *Phytolacca dodecandra* (endod) and grapes.
Additional research work is also being conducted on Ensete, potato, onion, garlic and other vegetable crops

Head: Tigst Demeke
A. Zimmerman (Viticulturist)
Mrs Shewaye Haile Michael - Senior Technical Assistant

10 Addis Ababa

The Horticultural Development Department (HDD), formerly Horticultural Development Agency (HDA), was set up by the Ethiopian Government in 1976 with the objective of developing all aspects of the horticultural industry,

which is recognised to have a very great potential for improving the economy of the country. HDD is presently supported by the Government of the Federal Republic of Germany. Implementation of the 10 Year National Horticultural Development Plan is largely by the Horticultural Development Corporation (HDC) with other Agricultural Development Corporation's participating. The large scale farms are now being utilised as nucleus estates for development of horticulture in the peasant sector.
HDD has planned an adaptive research programme and is advising HDC on its execution on several large scale irrigated farms in different climatic zones and appropriate results are also being applied to the peasant sector. The principal objectives of the programme include the following:-
Development of large scale methods of production adapted to local conditions; introduction of high quality planting material of perennial crops, particularly citrus, mango, avocado, guava, papaya, grapes and temperate fruits, and the establishment of a permanent national collection; establishment of large scale nurseries and study of propagation techniques; introduction and trial of improved vegetable cultivars; development of seed production; integrated development of raw material production for processing and the processing industry; introduction and development of new crops, particularly for fresh export, such as, strawberry, asparagus, cut flowers, green beans, solo papaya; co-operation is maintained with other national and international research organisations.

10.1 Horticultural Development Department (HDD)	PO Box 62320, Addis Ababa Phone: 153611/153458/153560 Telex: 21106 ETFRUIT Dir.: Samu-Negus Haile-Mariam*
Ag. Project Leader, Adviser on Fruit Production	R. Schall
Senior Technical Adviser	T.H. Jackson*
Fruit Expert	Ato Zewdie Ayele
Vegetable Expert	Ato Ayele Onke
Adviser on Agro-Industry	V. Hache
Food Technologist	R. Herbst
Economist	Ato Abraham Demere
Adviser on Horticulture in the Peasant Sector	Dr. I.U. Lewis*
10.2 Horticultural Development Corporation (HDC)	PO Box 60061, Addis Ababa Phone: 159550 Telex: 21106 ETFRUIT Dir.: Ato Fana Wolde-Giorgis Dep.Gen.Mng.: Ato Demissie Retta
Head of Agricultural Department	Ato Fekadu Tadesse
Manager of Nura Era Enterprise	Ato Agegnehu Sissay

FINLAND

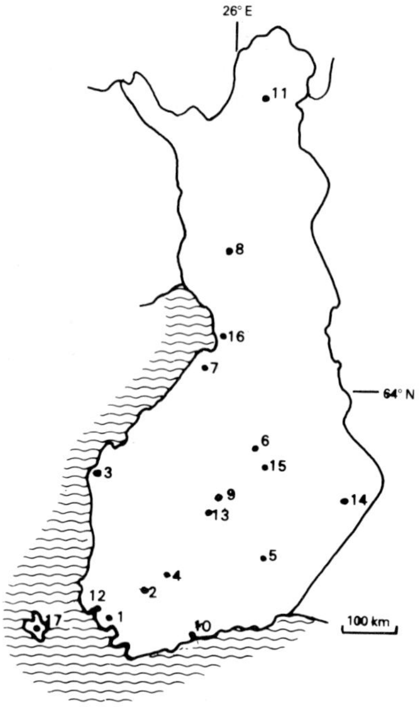
26° E
64° N
100 km
1
2
3
4
5
6
7
8
9
10
11
12
13
14
15
16
17

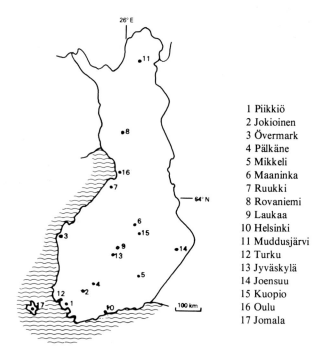

1 Piikkiö
2 Jokioinen
3 Övermark
4 Pälkäne
5 Mikkeli
6 Maaninka
7 Ruukki
8 Rovaniemi
9 Laukaa
10 Helsinki
11 Muddusjärvi
12 Turku
13 Jyväskylä
14 Joensuu
15 Kuopio
16 Oulu
17 Jomala

Survey

The horticultural research activity in Finland is mainly carried out by the Agricultural Research Centre (under the Ministry of Agriculture and Forestry) and also by the University of Helsinki. The universities of Turku, Jyvaskyla, Joensuu, Kuopio and Oulu have included some horticultural projects in their programs. Some economical and greenhouse-technical investigations are conducted by the Finnish Commercial Growers Association. In addition, a remarkable contribution to horticultural research is still made by the industry, different advisory and commercial communities, and by private enterprise. The Department of Horticulture at the Agricultural Research Centre at Piikkio in south-west Finland handles the main research activity in variety testing and in the study of the growing techniques of horticultural crops. The following fields of activity are included in the program: Tree fruit, small fruit, vegetables in the open, floriculture and dendrology, and in addition the breeding of apples, apple rootstocks and small fruit, such as *Fragaria, Ribes, Rubus, Vaccinium* and *Hippophae.* Owing to the hard climate conditions the work is directed towards breeding and selecting plant material, which is winter hardy and adaptable to a short and cool growing season. In the study of growing techniques the aim is also to develop methods, which will promote the winter hardiness of the plants. During the last few years the breeding activity has produced new varieties of apples, apple rootstocks, the nectar raspberry, highblush blueberry and the strawberry. The vegetable research is concentrated on testing varieties for keeping a register of varieties recommended for cultivation in Finland. Besides this the raising of plant material for further planting in the field and fertilizing is studied. Furthermore, an important research aim is to find methods to help vegetable cropping in a short and cool season. In floriculture the main object is the timing of the flowering of different cut flowers, e.g. roses, carnations, chrysanthemums and freesias. The raising technique has to be regulated and adapted to the light conditions, which differ from those prevailing in other countries. The program also includes searching for and domestication of new flower species for ornamental purposes. The ornamental trees and shrubs adaptable to parks, landscapes and home surrounding are tested and their propagation is studied. The research station for greenhouse vegetables at Narpio in western Finland went into operation in 1984. Its task is to test varieties, develop growing techniques, and study the use of peat as growth substrate. The Propagation Unit for Healthy Plants at Laukaa, in central Finland raises elite material of fruit, small fruit and some ornamentals for further propagation and delivery. The research stations at Pälkäne, Mikkeli, Maaninka, Ruukki and Rovaniemi assist in the horti-

Prepared by dr. H. Hiirsalmi, Agric. Research Centre, Piikkiö cultural program. Pesticides for horticulture are investigated and tested at the Departments of Plant Pathology and Pest Investigation at the Agricultural Research Centre at Jokioinen. Special attention is given to the biological control of pests. Since 1956 the Institute of Horticulture at the University of Helsinki has, besides its educational activity, carried out research into longterm projects with special plants and plant groups. The work is thus concentrated only on a limited number of field experiments and test plots making up the projects. The main topics of research are:

- improved methods in the production of apples, berries, vegetables, ornamentals, and spice crops with the emphasis on a more profitable use of energy,
- the influence of ecological factors on the quality of edible crops,
- the culture and endurance of landscape plants under different conditions.

The investigations are partly carried out in the environment of southern Finland at Viikki, in Helsinki, partly in the subarctic zone at Muddusjarvi Experiment Station.

The Department of Plant Breeding at the University of Helsinki is concerned with breeding work in agriculture, horticulture and forestry. The breeding in horticulture is concentrated on some berry species and on woody ornamentals, particularly Rhododendron species and varieties. Special emphasis is given to modern biotechnological methods.

Publications: The most important periodicals containing publications from horticultural research in Finland are Annales Agriculturae Fenniae and Journal of Agricultural Science in Finland.

Rainfall: 30 years averages of rainfall per growing season (from the 1st of May to the 30th of September) in millimeters

Hours of sunshine: 10 years averages of hours of sunshine per growing season (from the 1st of May to the 30th of September)

1. Piikkiö

1.1 Agricultural Research Centre
Department of Horticulture
elevation: 6 m
rainfall : 275 mm
sunshine : 1157 h

SF-21500 Piikkiö
Phone: 921-727806
Dir.: Prof.Dr.J. Säkö

Tree fruits and small fruits	Prof.Dr.J. Säkö*
	Ms. E. Laurinen, M.Sc.*
Small fruit breeding	Dr.H. Hiirsalmi*
	Ms.S. Junnila, M.Sc.
Vegetables in the open	Ms. R. Pessala, M.Sc.*
	Ms. P. Huuhtanen, M.Sc.
Floriculture	T. Pessala, M.Sc.*
	Ms. S. Juhanoja, M.Sc.
Dendrology	A. Lehmushovi, M.Sc.*

2. Jokioinen

2.1 Agricultural Research Centre
elevation: 104 m
rainfall : 286 mm
sunshine : 1126 h

SF-31600 Jokioinen
Phone: 916-84411
Telex : 6741 mttk sf

2.1.1 Department of Plant Pathology — Dir.: Prof.Dr.E. Seppänen
2.1.2 Department of Pest Investigation — Dir.: Prof.Dr.M. Markkula

3. Övermark

3.1 Martens Vegetable Research Station
(Agricultural Research Centre)

SF-64610 Övermark
Phone: 962-53338
Dir.: Ms. L. Kurki, M.Sc.*

Greenhouse vegetable culture and varieties for Finnish conditions — Ms. L. Kurki, M.Sc.*
Growing substrates for greenhouses — Ms. M-L. Bartosik, M.Sc.*
Horticultural economics — Ms. M-L. Bartosik, M.Sc.*

4. Pälkäne

4.1 Häme Research Station
(Agricultural Research Centre)
elevation: 103 m
rainfall : 279 mm

SF-36600 Pälkäne
Phone: 936-2214
Dir.: M. Takala, M.Sc.

5. Mikkeli

5.1 South Savo Research Station
(Agricultural Research Centre)
elevation: 138 m
rainfall : 300 mm

SF-50600 Mikkeli
Phone: 955-30028
Dir.: Prof.Dr.E. Huokuna

Small fruits — Ms. P. Dalman, M.Sc.*

6. Maaninka

6.1 North Savo Research Station
(Agricultural Research Centre)
elevation: 88 m
rainfall : 281 mm

SF-71750 Maaninka
Phone: 971-511162
Dir.: K. Rinne, M.Sc.

7. Ruukki

7.1 North Pohjanmaa Research Station
(Agricultural Research Centre)
elevation: 48 m
rainfall : 288 mm

SF-92400 Ruukki
Phone: 982-71371
Dir.: H. Hakkola, M.Sc.

8. Rovaniemi

8.1 Lapland Research Station
(Agricultural Research Centre)
elevation: 103 m
rainfall : 283 mm
sunshine : 1173 h

Apukka, SF-97999 Rovaniemi
Phone: 960-83217
Dir.: O. Nissinen, Agr.Lic.

9. Laukaa

9.1 Propagation Unit for Healthy Plants
(Agricultural Research Centre)

SF-41340 Laukaa
Phone: 941-633740
Dir.: Ms. M. Uosukainen, M.Sc.

10. Helsinki

10.1 University of Helsinki
elevation: 5 m
rainfall : 278 mm
sunshine : 1247 h

SF-00710 Helsinki
Phone: 90-378011
Telex : 124690 unih sf

10.1.1 Department of Horticulture, Viikki — Dir.: Prof.Dr.E. Kaukovirta*
Small fruits — Ms. I. Nyman, M.Sc.*

Vegetables	Ms. S. Ahonen, M.Sc. Ms. K. Osara, M.Sc.* Ms. S. Hälvä, M.Sc.
Ornamental horticulture	A. Pajunen, M.Sc.
Flower crops	Prof.Dr.E. Kaukovirta*
Plant propagation	T. Haapala, M.Sc.
10.1.2 Department of Plant Breeding	Dir.: Prof.Dr.P.M.A. Tigerstedt
10.1.3 Botanical Garden	Unionink. 44, SF-00170 Helsinki Phone: 90-650188 Dir.: Prof.Dr.L. Hämet-Ahti

11. Muddusjärvi

11.1 The Subarctic Experiment Station (University of Helsinki) elevation: 149 m rainfall : 231 mm sunshine : 1019 h	SF-99910 Kaamanen Phone: 9697-52751 Dir.: Prof.Dr.V. Ryynänen

12. Turku

12.1 University of Turku Botanical Garden elevation: 6 m rainfall : 279 mm sunshine : 1199 h	Ruissalo, SF-20100 Turku Phone: 921-645111 Telex : 62123 tyk sf Dir.: Aman. M. Yli-Rekola, Lic.

13. Jyväskylä

13.1 University of Jyväskylä Botanical Garden elevation: 115m rainfall : 316 mm	Seminaarink. 15, SF-40100 Jyväskylä Phone: 941-291211 Telex : 28219 jyk sf Dir.: Ms. T. Raatikainen, M.Sc.

14. Joensuu

14.1 University of Joensuu Botanical Garden elevation: 116 m rainfall : 316 mm sunshine : 1167 h	P.O.B. 111, SF-801/01 Joensuu Phone: 973-26211 Telex : 46224 univj sf Dir.: position vacant P. Piironen

15. Kuopio

15.1 University of Kuopio Botanical Garden elevation: 94 m rainfall : 291 mm sunshine : 1188 h	P.O.B. 6, SF-70211 Kuopio Phone: 971-162211 Telex : 42218 kukko sf Dir.: Prof.Dr.L. Kärelampi

16. Oulu

16.1 University of Oulu Botanical Garden elevation: 12 m	Linnanmaa, SF-90570 Oulu Phone: 981-353611 Telex : 32375 oylin sf

rainfall : 273 mm
sunshine : 1199 h

Dir.: Prof.Dr.S. Eurola

17. Jomala

17.1 Aland Experiment Station
elevation: 19 m
rainfall : 244 mm
sunshine : 1289 h

SF-22150 Jomala
Phone: 928-31010
Dir.: Ms. U. Boman, M.Sc.

FRANCE

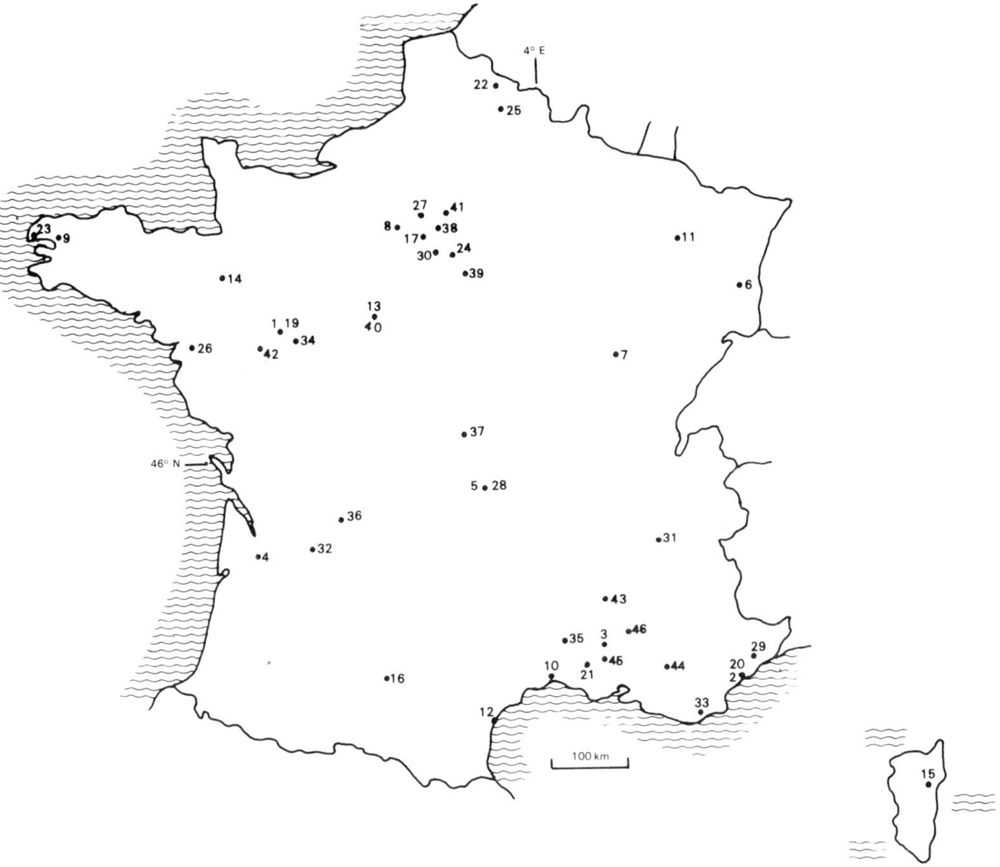
4° E
46° N
100 km
22
25
27
41
8
17
38
30
24
39
11
6
23
9
14
13
40
1 19
34
26
42
7
37
5 28
36
32
4
31
43
35
3
46
10
21
45
44
20
2
29
16
12
33
15

INRA: 1 Angers (Maine-et-Loire), 2 Antibes (Alpes-Maritimes), 3 Avignon (Vaucluse), 4 Bordeaux (Gironde), 5 Clermont-Ferrand (Puy-de-Dôme), 6 Colmar (Haut-Rhin), 7 Dijon (Côte d'Or), 8 Grignon (Yvelines), 9 Landernau (Finistère), 10 Montpellier (Herault), 11 Nancy (Meurthe-et-Moselle), 12 Narbonne (Aude), 13 Orléans (Loiret), 14 Rennes (Ille-et-Vilaine), 15 San-Giuliano (Corse), 16 Toulouse (Haute-Garonne), 17 Versailles (Yvelines), 18 Petit-Bourg (Guadeloupe).

AUTRES ORGANISMES: 19 Angers (Maine-et-Loire), 20 Antibes (Alpes-Maritimes), 21 Bellegarde (Gard), 22 Bersée (Nord), 23 Brest (Finistère), 24 Brétigny-sur-Orge (Essonne), 25 Cambrai (Nord), 26 Carquefou (Loire-Atlantique), 27 Chambourcy (Yvelines), 28 Clermont-Ferrand (Puy-de-Dôme), 29 La Gaude (Alpes-Maritimes), 30 Gif-sur-Yvette (Essonne), 31 Grenoble (Isère), 32 La Force (Dordogne), 33 La Londe (Var), 34 La Ménitré (Maine-et-Loire), 35 Lédenon (Gard), 36 Malemort-sur-Corrèze (Corrèze), 37 Malicorne (Allier), 38 Meudon-Bellevue (Haut-de-Seine), 39 Milly-la-Forêt (Essonne), 40 Orléans (Loiret), 41 Paris (Seine), 42 Les Ponts de Cé (Maine-et-Loire), 43 Saint-Marcelin (Isère), 44 Saint-Paul-lez-Durance (Bouches-du-Rhône), 45 Saint-Rémy-de-Provence (Bouche-du-Rhône), 46 Sarrians (Vaucluse)

Introduction

L'Institut National de la Recherche Agronomique (INRA) est l'organisme de recherches français qui consacre la totalité de son activité à l'agriculture et aux industries qui lui sont liées. Etablissement privé à caractère scientifique et technologique, il est placé sous la double tutelle du Ministère de la Recherche et de la Technologie et du Ministère de l'Agriculture. Ses missions sont multiples: amélioration des produc-

Preparé par Mme. D. Blanc, INRA, Antibes

tions animales et végétales; conservation et transformation des produits agricoles en produits alimentaires; biotechnologie en rapport avec l'agriculture; protection, sauvegarde et gestion des ressources naturelles; problèmes socio-économiques du monde agricole et rural. Ceux de ses services qui consacrent leur activité, en totalité ou pour partie, à l'horticulture sont recencés dans la première partie de cet inventaire (N° 1 à 18).
Des équipes de recherches dépendant du Centre National de la Recherche Scientifique (CNRS), et de l'enseignement supérieur agronomique ou universitaire s'intéressent également aux productions horticoles. Par ailleurs, des firmes privées disposent de leurs propres laboratoires de recherche. Enfin les Instituts Techniques de la profession agricole, tels le Centre Technique Interprofessionnel des Fruits et Légumes (CTIFL) et le Comité National Interprofessionnel de l'Horticulture (CNIH) qui constituent un maillon intermediaire entre la Recherche Agronomique et le Developpement agricole, contribuent à l'application de la Recherche et poursuivent leurs propres actions dans le domaine de la Recherche appliquée.
L'ensemble de ces unités de recherche sont répertoriées dans la deuxième partie de cet inventaire (N° 19 à 46).

INRA

1 Angers (Maine et Loire)

1.1 Centre de Recherches I.N.R.A. Angers — Route de St. Clément, Beaucouzé
altitude : 45 m — 49000 Angers
pluviométrie : 591 mm — Tél.: 41 48 51 23
ensoleillement: 1931 h — Télex: 720565
Dir.: P. Bondoux

1.1.1 Station de Recherches d'Arboriculture Fruitière* — Dir.: L. Decourtye*

Etude de collections: pommier (pommes de table - pommes d'industrie) — M. Le Lezec
Poirier, cassis, groseillier, framboisier — B. Thibault*/ B. Lantin
Sélection sanitaire: obtention de clones indemnes de maladies à virus — J. Lemoine
Amélioration par hybridation: poirier: obtention de variétés à floraison tardive, à maturité de consommation tardive et très précoce, de variétés résistantes au feu bactérien. — B. Thibault*
Pommier: obtention de variétés résistantes à la tavelure, peu sensibles à l'oidium et au chancre à nectria. — Y. Lespinasse
Cassis: obtention de variétés résistantes à la rouille, à l'oidium, peu exigentes en froid hivernal et de bonne qualité aromatique. — B. Lantin
Induction de mutations à l'aide de rayons gamma chez le poirier, le pommier et le cherisier — L. Decourtye*, Mlle. E. Chevreau, Mme. I. Pochon
Amélioration des porte-greffes: cognassier (porte-greffes clonaux du poirier), *Pyrus communis*, pommier (expérimentation des porte-greffes clonaux).
Recherches de physiologie: facteurs de l'induction florale, effets des régulateurs de croissance, chez le poirier. — L. Hermann
Techniques culturales pour le cassis: taille, dates de cueillette, récolte mécanique. — J.C. Michelési

1.1.2 Laboratoire d'Amélioration des Arbustes Ornementaux — Dir.: L. Decourtye*

Amelioration des plantes de haies — L. Decourtye*
Obtention de nouvelles variétés de *Berberis* et de *Pyracantha*. — A. Cadic
Mutagenèse sur *Forsythia* et *Weigela*
Cultures de plantes in-vitro — M. Duron

1.1.3 Station de Pathologie Végétale et de Phyto-bactériologie — Dir.: M. Ridé

Bactériologie — Mlle. M.R. Barzic/ B. Digat
L. Gardan/ J. Luisetti/ C. Manceau
X. Nesm/ J.P. Paulin*/ B Rat
Mme. R. Samson

Etude biosystématique et collection systématique (2 500 souches) des bactéries phytopathogènes. Méthodologie du diagnostic. Epidémiologie: feu bactérien des pomoidées, dépérissement du pêcher, bactériose du noisetier, nécrose bactérienne de la vigne. Structure et dynamique des populations phytopathogènes. Etiologie des maladies infectieuses en cultures maraîchères, florales, fruitières, forestières et vigne. Bactéries transmises par les semences. Recherches de critères de résistance *(Xanthomonas populi, X. ampelina, Erwinia amylovora)*. Lutte biologique (Crown Gall, taches bactériennes du champignon de couche)

Mycologie — P. Bondoux/ M. Bourgeois
Maladies de conservation des poires et des pommes: étude physiologique
Variété des populations de champignons pathogènes du pommier *(Venturio inaequalis, Podosphaera leucotricha)*
Dynamique en fonction des systèmes de lutte génétique et chimique. Maladies de dépérissement des arbres fruitiers

Virologie — J.C. Morand
Viroses des arbres ornementaux et forestiers.
Sélection sanitaire

1.1.4 Station d'Agronomie — Dir.: J. Salette

Horticulture ornementale: recherches sur les substrats, caractéristiques physiques et évolution, nutrition minérale et fertilisation (en particulier important du rapport NO_3/NH_4 de la solution nutritive matière organique — F. Lemaire*/A. Dartigues

Vigne: essai terroir: caractérisation des sols viticoles; étude de l'enracinement; étude de la chlorose — R. Morlat/J. Salette

1.1.5 Station de recherches oenologiques — Dir.: C. Asselin

Recherche sur les relations milieu physique - vigne - type de vin. Elaboration des vins effervescents — C. Asselin/ Mme. H. Léon

Le Centre dispose de 3 domaines expérimentaux: Bois l'Abbé, La Rétuzière et Montreuil-Bellay.

2 Antibes (Alpes Maritimes)

2.1 Centre de Recherches INRA Antibes
altitude : 70 m
pluviométrie : 0.770 mm
ensoleillement: 2810 h

41, Boulevard du Cap, BP 78
06602 - Antibes Cedex
Tél.: 93 61 55 60
Télex: INRAANT 461434F

2.1.1 Station d'Agronomie et de Physiologie Végétale

45, Boulevard du Cap, B.P. 78
06602 Antibes Cédex
Dir.: A. Morisot

Nutrition minérale de la plante entière: métabolisme des nitrates: absorption, accumulation, réduction, nitrate réductase. — S. Adamowicz

Valeur nutritionnelle des productions maraîchères (tomate) — D. Blanc*/ S. Mars/ A. Morisot

Influence de l'état physiologique ou nutritionnel de la plante sur sa réaction à des agents pathogènes, recherche de méchanismes d'action (calcium, tabac, mosaïque) — S. Ferrario

Système de culture: conduite de culture, approche physiologique de la plante entière — A. Champéroux

Qualité de la production	S. Mars
Cultures hors sol maraîchères et florales: méthodologie des analyses	C. Otto
Caractérisation physique des substrats de culture	R. Gras
Mise au point de systèmes de culture nouveaux	R. Brun
Cultures d'appoint méditerranéennes, plantes aromatiques et médicinales	G. Gilly
2.1.2 Laboratoire de Science du Sol	45 Boulevard du Cap, BP 78 06602 - Antibes Dir.: J.C. Arvieu
Proprietes d'echange cationique de la tourbe	J.P. André
Réactivité de la màtiere organique au cours de l'humification	
Interactions pesticides-sol	J.C. Arvieu
Désinfection chimique des sols et substrates - Résidues bromés.	R. Durand
2.1.3 Station de Botanique et de Pathologie végétale Villa Thuret	62 Boulevard du Cap, B.P. 78 06602 Antibes Dir.: J. Ponchet*
Mycologie: Maladies des cultures florales et ornementales (rosier, oeillet, gerbéra, anémone, renoncule, cyprès). Diagnostic, biologie, lutte.	R. Tramier S. Mercier
Maladies transmises par les sols et les substrats de culture. Lutte biologique.	J.C. Pionnat C. Andréoli
Virologie: Diagnostic sérologique, sélection sanitaire, certification (maladies des cultures florales). C.M.V., T.M.V. (immunologie).	J.C. Devergne L. Cardin
Physiopathologie: Mécanismes de défense des plantes à l'infection (virus, cryptogames).	A. Poupet/ M. Ponchet
Phytoalexines, phénolamides, protéines "b." Génétique du pouvoir pathogene chez *Phytophthora*.	N. Maïa/ Ph. Bonnet
Micropropagation *in vitro*	
Botanique: Acclimation et expérimentation de végétaux exotiques ligneux en région méditerranéenne. Cyprès, Eucalyptus, etc. Arboretums d'essai.	P. Allemand/ P. Augé C. Billette
2.1.4 Station de Zoologie et de Lutte Biologique	37, Boulevard du Cap 06602 Antibes Dir.: J.C. Onillon
Lutte biologique (utilisation d'entomophages) et intégrée contre les ravageurs animaux des cultures protégées	P. Jourdheuil
Lutte contre les acariens *(Tetranychus)* et les lépidoptères tortricidae	M. Pralavorio/ J.C. Malausa
Lutte contre les aleurodes	A. Panis
Lutte contre les aochenilles	J.M. Rabasse
Lutte contre les aucerons	
Lutte contre les aiptères et notamment contre *Liriomyza trifolii*	J.P. Lyon
Expérimentation de la lutte biologique et intégrée	P. Millot
La Station dispose d'une annexe insectarium	Route de Biot, 06560 Valbonne Tél.: 93 42 02 35
2.1.5 Station de Recherches sur les Nématodes et de génétique moléculaire de invertébrés	123, Boulevard Francis Meilland BP. 78 - 06602 Antibes Cédex

	Dir.: A. Dalmasso
Nématodes des plantes pérennes (arbres fruitiers, vigne, forêt)	C. Scotto la Massesse G. de Guiran
Nématodes des champignons de couchets et champignons nématophages	J.C. Cayrol
Nématodes entomophages	C. Laumond
Nématodes des cultures maraîchères et florales	A. Dalmasso
Méchanismes génétique et moléculaire du pouvoir pathogène des nématodes phytophages	M.C. Cardin
Génétique moléculaire de la résistance aux pesticides	J.B. Bergé/ A. Cuany C. Mouches/ D. Fournier

2.1.6 Station d'Amélioration des Plantes Florales
Etudes de génétique et de sélection. Création variétale pour fleurs à couper

La Gaudine, 83600 Frejus
Tél.: 94 51 43 40
Dir.: E. Berninger*

J. Meynet/ J. Marty/ C. Rameau

Anémone: création de variétés tétraploïdes
Chrysanthèmes: obtention de clones pour chauffage réduit
Gerbera: clones; tolérance à *Phytophthora cryptogea;*
Lignées homozygotes par haplométhode.
Glaïeul: hybrides pour floraison d'hiver à partir d'espèces sud-africaines.
Renoncule: hybrides de clones; physiologie.
Rosier: hérédité et emploi de critères physiologiques; techniques de sélection.
Strelitzia: hybrides de clones.
Gazons: collections et essais en conditions d'arrosage réduit
Serres: potentialités microclimatiques et comportement de quelques cultures.

2.1.7 Unité Experimentale du G.E.V.E.S.

Groupe d'Etudes des Variétés et des Semences)
altitude : 30 m
pluviométrie :0.800 mm
ensoleillement: 2810 h

La Baronne, RD 2209,
BP 55 Cédex
06702 - St. Laurent du Var
Tél: 93 31 15 12
Télex: INRAANT 46 1434
Dir.: F. Ferrèro

Etude et conservatoire des variétés nouvelles des espèces ornementales pour lesquelles peut être délivré un Certificat d'Obtention Végétale (C.O.V.). Travaux menés actuellement, plus particulièrement, sur rosiers en plein champ et oeillets sous abri: taxonomie, nomenclature, descriptions variétales et étude approfondie des variétés mutantes par une approche biochimique.

3 Avignon (Vaucluse)

3.1 Centre de Recherches Agronomiques d'Avignon
altitude : 24 m
pluviométrie : 654 mm
ensoleillement: 2757 h

B.P. 91 Domaine Saint Paul
84140 Montfavet
Tél.: 90 88 91 45
Télex: INRAAVI 432870F

3.1.1 Station d'Agronomie*
Effets des techniques culturales sur les rendements

Dir.: J-G. Huguet

Objectifs: détermination des techniques culturales qui optimisent la production en quantité, qualité et prix de revient, dans le respect de l'environnement.	Ph. Bussières P. Cornillon Y. Dumas
Arboriculture fruitière Tomate de plein champ; maraîchage sous abri	R. Habib/ Cl. Huguet

3.1.2 Station d'Amélioration des Plantes Maraîchères*
Création de nouvelles variétés de légumes, mieux

Dir.: R. Dumas de Vaulx

adaptées aux conditions de production et d'utilisation.
Objectifs: amélioration de la qualité des récoltes, résistance génétique aux maladies, adaptation des plantes au milieu (en particulier aux serres peu chauffées), diminution du prix de revient

A. Bonnet/ D. Chambonnet
P. de Cockborne/ M-C. Daunay
R Dumas de Vaulx/ H. Laterrot
A. Palloix/ P. Pécaut
J. Philouze/ M. Pitrat
E. Pochard/ G. Risser
Chef d'exploitation: C. Vizier

Artichaut, carotte, radis, tomate, piment, aubergine, melon, courgette, fraise.

3.1.3 Station de Recherches Fruitières Méditerranéennes

Dir.: P. Crossa-Raynaud

Création et sélection de variétés et de porte-greffes adaptés au Midi Méditerranéen.
Objectifs: amélioration de la qualité et de la quantité des récoltes, résistance génétique aux maladies et aux insectes, diminution des coûts de production
Abricotier, amandier, pecher, poirier, pommier.

J-M Audergon
Ch. Grassely/J.L.Poessel,
J.C. Nicolas

3.1.4 Station de Bioclimatologie et S.T.E.F.C.E.
Service Technique d'Etudes des Facteurs Climatiques de l'Environnement

Dir.: P-G. Schoch

S.T.E.F.C.E.: service technique qui assure la bonne marche des 120 stations climatiques de tout l'I.N.R.A., gere la banque de toutes ces données (journalières) et donne son appui technique à toute unité de l'I.N.R.A. ayant besoin de mesures climatiques particulières.

Responsable: A. Pinguet

Influence du climat sur la production agricole et sur l'environnement.
Objectifs: caractérisation des microclimats créés par le relief, les aménagements du paysage (brisevents, implantations industrielles, etc...) ou les abris (tunnels plastiques, serres). Détermination pour les cultures sous serres des besoins en eau, en énergie (chauffage et ventilation) en vue de réduire les coûts de production. Etude des réactions des plantes (vitesse de croissance, de floraison et de production) aux divers microclimats. Utilisation des satellites pour la surveillance des grandes cultures et forêts, pour la détermination de leur déficit en eau et pour la prévision de leur rendement. Cartographie des potentialités de production d'une région.
Tomate et plantes ornementales (sous serre), blé, mais, olivier.

R. Antonioletti/ A. Baille*,
C. Baldy/ Th. Boulard/ Y. Brunet,
R. Delecolle/ C. Gary,
M. Guerif/ J-P. Guinot/ G. Guyot,
J-P. Lagouarde/ J-J. Longuenesse*,
Ph. Malet/ M. Mermier
P. Reich/ B. Seguin/ M. Verbrugghe
O. de Villèle

3.1.5 Station de Pathologie Végétale

Dir.: J-M. Lemaire

Lutte biologique, chimique et culturale (lutte intégrée) contre les champignons, bactéries et virus des cultures.
Objectifs: identification et caractérisation des parasites et de leur cycle biologique; définition de leur potentiel de variabilité: maïtrise des facteurs agissant sur les épidémies, stratégies de combinaison des résistances naturelles contre les parasites ou leur propagateur; utilisation des mecanismes d'interaction entre souches ou micro-organismes pour la protection biologique.

Tomate, piment, poivron, melon, courgette, concombre, laitue, chicorée, artichaut, asperge, fraise, abricotier, cerisier, pêcher, poirier, platane, blé

D. Blancard/ C. Castelain,
B. Delecolle/ A. Glandard,
C. Grosclaude/ M. Jacquemond,
H. Lecoq/ H. Lot/ G. Marchoux,
P. Mas/ P-M. Molot/ G. Morvan*,
K. Gebre Selassie/ J-P. Prunier

3.1.6 Station de Phytopharmacie

Dir.: L. de Cormis
Dir. adjoint: M. Bounias

Propriétés des produits chimiques utilisés en agriculture. Conséquences de leur utilisation sur l'environnement. Etude des pollutions atmosphériques.
Objectifs: détermination des normes d'utilisation des produits phytosanitaires, à la fois efficaces et sans nocivité pour le consommateur et l'environnement.
Productions fruitières, vigne, tomate, salade et autres légumes.

L. Belzunces/ M. Luttringer

3.1.7 Station de Science du Sol

Dir.: P. Stengel

Physique et mécanique des sols appliquées au choix des techniques culturales et des systèmes de culture, à la conservation des sols, à la gestion de l'eau et des engrais.
Objectifs: optimisation des techniques de travail et d'entretien des sols, des apports d'eau et d'engrais (en particulier dans les systèmes d'irrigation localisée fertilisante); conservation de la fertilité physique des sols; évaluation des contraintes physique d'exploitation des terrains cultivés (calcul du nombre de jours disponibles pour les travaux).
Systèmes céréaliers, vegers, légumes de plein champ, systèmes fourragers intensifs.

P. Bertuzzi/ L. Bruckler,
B. Cabibel/ A. Chanzy,
A-M. de Cockborne/ A. Faure,
J-C. Fies/ R. Guennelon,
V. Hallaire/ F. Lafolie,
G. Monnier/ P. Renault,
N. Souty

3.1.8 Station de Technologie des Produits Vègètaux

Dir.: P. Ferry

Qualité des fruits et légumes à l'etat frais ou après conditionnement. Techniques de conservation et utilisations industrielles.
Objectifs: mise au point des méthodes d'appréciation de la qualité et des nouveaux procédés industriels de conservation ou de transformation des fruits et légumes.
Artichaut, asperge, carotte, aubergine, tomate, melon, salade, haricot vert, champignon, amande, abricot, pêche, olive, cerise, limes, fraise, raisin, fruits tropicaux.

M-J. Amiot/ S. Aubert/ L. Breuils
M. Buret/ M. Cabibel/ Y.Chambroy
F. Duprat/ B. Fils/ C. Nguyen The
J.Nicolas/ M.G. Nicolas/ C.Pelisse
Ch. Rothan/ M. Souty/ P. Varoquaux

3.1.9 Station de Zoologie et Apiculture

Dir.: S. Poitout

Lutte intégrée (biologique, chimique et par les techniques culturales) des insectes ravageurs des cultures à partir de la connaissance précise de leur cycle de développement.
Apiculture: amélioration génétique de l'abeille: mesure de son rôle dans la pollinisation des vergers. Caractérisation morphologique (aspects extérieurs) biochimique des races d'abeilles. Insémination artificielle. Analyse technologique des miels.
Objectifs: détermination des niveaux démographiques critiques des ravageurs qui nécessitent des moyens de lutte. Mise au point de nouvelles méthodes de lutte, utilisant le plus possible la lutte biologique et les techniques culturales.
Lépidoptères (papillons) nuisibles aux arbres fruitiers *(Tortricidae)*, aux cultures maraîchères *(Noctuidae)*, aux graminées *(chenilles endophytes)*.
Psylles, etc... pour leur importance dans les stratégies de protection intégrée.

H. Audemard/ R. Bues
A. Burgerjon/ R. Causse/ R. Cayrol
J-M. Cornuet/ J. Fresnaye
P. Galichet/ M. Gonnet/ Y. Loublier
C. Pelissier/ R. Rieux
J-P. Torregrossa

3.1.10 Unité du Service de Recherches Integrees sur la Production Végétale

Dir.: P. Dauplé
Dir. adjoint: P. Jullian
P. Vergniaud/ G. Ginoux

Innovations techniques en production de légumes et possibilites de leur introduction dans les divers types d'exploitations.

Objectifs: adaptation des techniques culturales (alimentation minérale, besoins en eau, conduite des plantes, etc...) aux nouveaux types de serres, d'abris permanents et de couvertures temporaires. Lutte contre les parasites du sol (désinfection, techniques du greffage). Adaptation de la conduite des légumes de plein champ à la mécanisation (choix des variétés, désherbage, techniques de semis). Tomate de marché et d'industrie, aubergine, oignon, ail, asperge.

3.1.11 Groupe d'Etude et de Contrôle des Variétés et de G.E.V.E.S.	Domaine d'Olonne, Les Vignères 84300 Cavaillon Tél.: (90) 71 26 85 Dir.: R. Brand G. Breuils
Contrôle l'originalite et l'homogénéité de toute variété nouvelle d'espèce légumière en demande d'inscription au Catalogue Officiel des Variétés. Propose la délivrance d'un titre de protection de la nouvelle obtention.	

Objectifs: garantie de l'originalité de toute variété et protection des obtenteurs contre les contrefaçons. Recherche de tous nouveaux caractères fiables permettant de mieux différencier les variétés d'une même espèce entre elles. 32 espèces légumières, lavande, lavandin, thym, soja, tournesol.

3.1.12 Domaine Expérimental de Gotheron (Unité de Recherches Intégrées en Production Végétale)	26320 Saint Marcel les Valence (Drôme), Tél.: 75 58 70 25 Dir.: P. Atger
Arboriculture fruitière: pêcher; plantes protéagineuses: pois, lupin, soja; plantes fourragères: contrôle pour l'inscription au Catalogue Officiel des Variétés, adaptation aux conditions régionales	Dir. adjoint: C. Billot
Gazon: expérimentation de diverses espèces et des techniques culturales à leur appliquer.	M. Arnoux/ C. Bussi
3.1.13 Domaines Expérimentaux d'Arboriculture Fruitière	Les Garrigues - 30129 Manduel et Les Pins de l'Amarine - 30127 Bellegarde, Tél.: 66 74 50 62
Pêcher, pommier, poirier	
3.1.14 Station Expérimentale Horticole du Mas Blanc	Alenya - 66200 Elne Tél: 68 22 27 98 Dir.: R. Brun
Légumes sous serres, sous abris, en plein champ, Comparaison et amélioration des différents types d'abri	B. Jeannequin
Expérimentation de divers substrats pouvant remplacer le sol (sous serre) et de l'alimentation hydroponique (eau enrichie en substances nutritives)	J. Peyrière/ G. Ridray
3.2 Recherches Forestières	Avenue A. Vivaldi, 84000 Avignon Tél.: 90 89 33 25
3.2.1 Station de Sylviculture Mediterraneenne Evaluation et contrôle des risques d'incendies de forêts, croissance des arbres et des peuplements forestiers.	Dir.: Y. Birot

Objectifs: évaluation des risques d'incendies de forêts et des effets des débroussaillements de protection; établissement des lois de croissance et de développement des arbres et des peuplements forestiers en fonction des contraintes du milieu méditerranéen, des especes et de leur écophysiologie, des traitements; précision des conditions de regénération naturelle des forêts.

Chêne pubescent et vert, cheñe kermès, Cèdre et pin noir, sapin, maquis et garrigues	M. Ducrey/ J-P. Guyon, J. Toth/ J-Ch. Valette
3.2.2 Station de Zoologie Forestière Protection des forêts contre les insectes ravageurs.	Dir.: D. Schvester
Objectifs: mise au point des méthodes de lutte, possibles techniquement et économiquement, dans les forêts. Développement de la lutte intégrées.	
Processionnaire du pin, dépérissement du pin maritime	J-F. Cornic/ G. Demolin

(en Provence), scolytide (creusant des galeries) des conifères, insectes du cèdre, insectes des peupliers, tordeuse du sapin. — J-P. Fabre/ P. du Merle

3.2.3 Domaine Expérimental du Ruscas — Forêt du Dom, 83230 Bormes les Mimosas, Tél.: 94 71 15 69, Dir.: P. Ferrandes

Sélection des espèces forestières pour le reboisement:
Pin, sapin, douglas, cèdre, cyprès, eucalyptus.

4 Bordeaux

4.1 Centre de Recherches INRA de Bordeaux — La Grande Ferrade - Pont de la Maye, 33140 Villenave d'Ornon, Tél.: 56 37 44 44, Télex: INRABDX 541 521 F

altitude : 25 m
pluviometrie : 841,2 mm
ensoleillement: 1 872 h

4.1.1 Station de Recherches d'Arboriculture Fruitière*. — Dir.: R. Bernhard

Génétique et sélection: cytogénétique du genre Prunus; étude de collections (pêcher, abricotier, prunier, cerisier, amandier, noisetier); sélection clonale et sanitaire (pêcher, abricotier, prunier, noyer); création de variétés nouvelles après hybridation (pêcher, cerisier, châtaignier, prunier, noyer, noisetier). Sélection de porte-greffes (pêcher x amandier, Damas, hybrides de St Julien, francs, Reine-Claude, Myrobolan, prunier domestique, pêcher, prunier Mahaleb, noyers hybrides).
Recherches sur les maladies: détermination des viroses et dépérissements, méthodes d'indexage, thermothérapie. Qualité des fruits en fonction des variétés, dates de récolte.
Physiologie: Multiplication végétative, compatibilité au greffage, problèmes de mise à fruit et de fructification (chute de bourgeons floraux, alternance chez le prunier), influence du porte-greffe, biologie florale, incompatibilité d'auto-pollinisation ou de fécondation.
La Station dispose de domaines expérimentaux: Grande Ferrade, 33140 Pont de la Maye - Ile d'Arcins, 33360 Latresne - La Tour de Rance, Bourran 47320 Clairac, Les Jarres, Toulenne 33210 Langon.

Noyer et noisetier	E. Germain
Physiologie du prunier	J. Couranjou
Méthodologie de la sélection	P-L. Lefort
Pêcher: génétique et sélection	R. Monet/ R. Saunier
Cytogénétique et sélection des porte-greffes	G. Salesses/ R. Bernhard
Cerisier	R. Saunier
Etude des maladies de dépérissement et sélection sanitaire	Mme. F. Dosba
Châtaignier	G. Salesses/ J. Chapa
Prunier	R. Renaud
Pommier: sélection et modes de conduite	J-M. Lespinasse
4.1.2 Station d'Agronomie	Dir.: C. Juste
Nutrition minérale et fertilisation de la vigne	J. Delas/ J.-P. Soyer
Ecophysiologie de la vigne - Techniques culturales	C. Molot
Nutrition minérale et fertilisation petits fruits rouges (framboisier)	E. Lubet/ C. Juste
4.1.3 Station de Recherche sur les Champignons	Dir.: J.M. Olivier
Cultures des espèces saprophytes	J. Laborde
Microbiologie - pathologie	J.M. Olivier
Microbiologie - biochimie	J.P. Goulas
Ecologie et domestication des especes mycorhiziennes	N. Poitou
Laboratoire de Génétique Moléculaire et Amélioration des Champignons Cultivés (laboratoire associé avec l'Université de Bordeaux II)	J. Labarère (Université)
4.1.4 Station de Physiologie Végétale	Dir.: A. Pradet
Mécanismes d'alimentation en oxygène des racines et des semences. Asphyxie racinaire.	P. Saglio/ F. Bruzau

Métabolisme énergétique au cours de la germination des semences	A. Pradet/ P. Raymond
Mobilisation des réservés carbonées et qualité des semences	P. Saglio/ P. Raymond
Biologie moléculaire de l'adaptation aux stress	B. Mocquot/ X. Gidrol/ B. Pacard
Formation des semences. Culture d'embryons	M. Rancillac
4.1.5 Laboratoire de Biologie Cellulaire et Moléculaire	Dir.: J.M. Bové
Recherches sur les mycoplasmes. Replication des virus de végétaux. Maladies infectieuses des agrumes	
4.1.6 Station de Pathologie Végétale	Dir.: J. Dunez
Maladies cryptogamiques de la vigne (lutte chimique, biologique et génétique; lutte intégrée)	J. Bulit/ M. Clerjeau/ B. Dubos F. Jailloux/ R. Lafon
Maladies à virus des arbres fruitiers (etiologie, détection)	J. Dunez/ G. Tavert
Etude et contrôle du pouvoir pathogène des virus et des viroïdes	Th. Candresse/ G. Macquaire, M. Monsion
4.1.7 Station de Zoologie	Dir.: P. Anglade
Recherches sur la résistance d'arbres fruitiers à des pucerons.	G. Massonié
4.1.8 Station de Recherches de Viticulture	Dir.: J. Pouget
Raisins de table: recherche de variétés de raisin sans pépin pour la zône de Chasselas et de la prune de sécherie	J.P. Doazan/ M. Ottenwaelter

5 Clermont Ferrand (Puy de Dôme)

5.1 Centre de Recherches Agronomique du Massif Central	Domaine de Crouelle
altitude : 330 m	63039 Clermont-Ferrand Cédex
pluviométrie : 592 mm	Tél.: 73 92 81 48
ensoleillement: 1.900 h	Dir.: J.C. Mauget

Laboratoire de Bioclimatologie

Biologie des arbres fruitiers dans ses relations avec les facteurs du climat. Espèces étudiées: pêcher, noyer, pommier.	
Phénomènes de développement et morphogénèse: dormance des bourgeons, ramification, dynamique de la population des organes reproducteurs	M. Bonhomme/ J.C. Mauget R. Rageau
Photosynthèse et transpiration de l'arbre	F.A. Daudet
Fonctionnement hydrique à l'échelle de la plante	P. Cruiziat
Dynamique des réserves carbonées: élaboration, mobilisation	A. Lacointe
Fonctionnement racinaire dans l'ensemble plante entière	J.S. Frossard

6 Colmar (Haut Rhin)

6.1 Centre de Recherches INRA de Colmar	28 Rue de Herrlisheim
altitude : 200 m	68021 - Colmar
pluviométrie : 0,53 mm	Tél.: 89 41 11 68
ensoleillement: 1.606 h	Télex: 880.657

6.1.1 Station d'Agronomie	Dir.: P. Girardin
Nutrition minérale et culture houblon	R. Marocke
Valorisation petits fruits spontanés	
Physio-écologie et phytotechnie espèces médicinales	

6.1.2 Station de Pathologie Végétale — Dir.: C. Putz
Viroses de la vigne — B. Walter

6.1.3 Station de Zoologie — Dir.: M. Stengel
Lutte intégrée en verger. Effets secondaires — P. Blaisinger
Relations plantes-insectes. Semio et allélochimique — P. Robert
Cultures protégées froides — M. Stengel

6.1.4 Station de Viticulture — Dir.: C. Schneider
Création variétale — A. Bronner
Sélection clonale — J. Balthazard
Phytotechnie et ecologie de la vigne — C. Schneider

6.1.5 G.R.I.S.P. Alsace — Dir.: G. Cloquemain
Viroses vigne - houblon

7 Dijon (Côte d'Or)

7.1 Centre de Recherches de Dijon
altitude : 430 m
pluviométrie : 719 mm
ensoleillement: 1902 h

Adresse postale BP 1540
21034 - Dijon Cédex

7.1.1 Dijon ville

17, Rue Sully, 21034 - Dijon
Tél.: 80 65 30 12
Télex: INRADIJ: 350507

7.1.2 Laboratoire d'Agronomie — Dir.: J. Concaret
Cultures sous serres et en plein champ: fertilisation, irrigation, entretien des sols, cultures hors sol. Concombre, tomate, aubergine, framboisier, étude de la qualité. — J.M. Lefèbvre

7.1.3 Station de Recherches sur la flore pathogène dans le sol — Dir.: J. Louvet
Maladies cryptogamiques dues au sol (fontes de semis, nécroses du collet, fusarioses vasculaires) en pépinières forestiéres et en cultures maraîchères et florales. — C. Alabouvette
Analyse sanitaire des sols. Epidémiologie — D. Bouhot
Résistance des sols aux maladies. Lutte biologique — P. Camporota/ Y. Couteaudier
Fatigues des sols — R. Perrin

7.1.4 Laboratoire de Recherches sur les Arômes — Dir.: J. Adda
Arômes de cassis, framboise, fraise. Sélection et transformation — A. Latrasse/ E. Guichard

7.1.5 Domaine d'Epoisses
altitude: 250 m

Bretenieres, 21110 - Genlis
Tél.: 80 39 80 22
Télex: INRADIJ: 350448

Station d'Amélioration des Plantes — Dir.: J. Picard

7.1.6. Labo Génétique et Mutagénèse — A. Cornu/ H. Dulieu/ A. Berville
Etude du mode d'action des agents mutagènes physiques ou chimiques.
Etablissement de la carte génique et chromosomique du Petunia.
Analyse génétique et biochimique la biosynthèse des anthocyanes.
Etude du contrôle génétique de la recombinaison méiotique (Petunia).
Etude de la variabilité génétique consécutive à la culture de protoplastes *in vitro*.
Analyse du contrôle génétique de l'aptitude à la

régénération à partir de cals issus de protoplastes (Petunia)
Recherches sur le déterminisme génétique de différents types de stérilité mâle cryptoplasmique: relations avec le génome mitochondrial (différentes espèces horticoles ou de grande culture).

7.1.7 Labo Phytoparasitologie

Biologie des mycorhizes (rôle dans la nutrition minérale et la résistance aux pathogènes du sol). — S. Gianinazzi
Méthodes de mycorhization *in vitro* et post *in vitro* (application à la micropropagation) — V. Gianinazzi-Pearson
Analyse génétique et moléculaire de la résistance aux virus. — S. Gianinazzi

7.1.8 Laboratoire de Malherbologie — Dir.: G. Barralis

Taxonomie, écologie, cycle de développement des mauvaises herbes; évolution des communautés adventices; interaction entre espèce d'une communauté.

7.1.9 Laboratoire des Herbicides — Dir.: R. Scalla

Mode d'action des herbicides

7.1.10 Station de Physiopathologie Végétale — Dir.: Cl. Martin

Virus. Répression de la synthèse, hypersensibilité, métabolisme des végétaux virosés, physiopathologie des végétaux parasités par des champignons.
Guérison par culture de méristèmes. — J. Martin-Tanguy
Mécanismes fondamentaux responsables de l'immunité des méristèmes et des gamètes — J. Negrel/ M. Paynot
Multiplication *in vitro* de plantes ornementales (rosier...) fruitières (framboisier, pêcher, pommier..) et forestières (merisier.) — P. Mussillon
Etude du développement sur leurs propres racines des plantes classiquement greffées (morphologie, mise à fruits, physiologie générale).

8 Grignon (Yvelines)

8.1 Laboratoire de Recherches de la Chaire de Chimie Biologique et l'Institut National Agronomique
altitude : 80 m
pluviométrie : 650-700 mm
ensoleillement: 2000 h

78850 - Thiverval-Grignon
Tél: 30 56 45 10 Poste 253
Dir.: C. Costes

Photophysiologie: application aux cultures sous serres.

9 Landernau (Finistère)

9.1 Station d'Amélioration de la Pomme de terre et des Plantes à bulbes
altitude : 85 m
pluviométrie : 1,130 mm
ensoleillement: 1.760 h

29207 - Landerneau
Tél.: 98 83 61 76
Dir.: P. Pérennec

Biologie de la plante entière: croissance, tubérisation et floraison chez les plantes à tubercules (pomme de terre) et à bulbes (glaïeul, iris, tulipe, narcisse, echalote, etc.), problème de conservation des organes de multiplication (tubercules, bulbes) et du forçage des espèces florales.
Amélioration génétique de la pomme de terre: utilisation de l'haploïdie dans la sélection, problèmes de résistance aux maladies (virus, nématodes, mildiou, etc.), obtention de variétés adaptées à divers modes d'utilisation (primeurs, transformation, conservation), expérimentation variétale.

Amélioration génétique des espèces florales à bulbes: selection pour l'adaptation à une production de fleurs à contresaison et à faibles exigences thermiques problèmes de résistance aux maladies (virus, botrytis, fusariose, etc.).
Mise au point de techniques de culture *in vitro:* multiplication rapide (pomme de terre, tulipe, glaïeul), obtention et utilisation de clones haploïdes dans la se-lection.
Etude et maintien de collections varie-tales: pomme de terre, tulipe, glaïeul.

Pomme de terre — P. Rousselle/ F. Rousselle, D. Ellissèche
Tulipe, iris bulbeux — M. Le Nard
Glaïeul, échalote — J. Cohat

10 Montpellier (Hérault)

10.1 Centre de Recherches Agronomiques de Montpellier (C.R.A.M.)
altitude : 45 m
pluviométrie : 766,4 mm
ensoleillement: 2.613 h

Ecole Nationale Supérieure Agronomique de Montpellier
9, Place Viala
34060 Montpellier Cédex
Tél.: 67 63 03 15
Télex: INRAMON 490818 F

10.1.1 Laboratoire d'Arboriculture fruitière
Tél.: 67 63 38 35
Dir.: J. Hugard*
P. Villemur/J.M. Legave

Recherches sur la mutagenèse induite sur abricotier et amandier. Etude des phénomènes de croissance et d'initiation florale, de leurs anomalies et pertubations chez l'abricotier et l'olivier. Recherche sur la conduite et la ramification du pommier en pépinière grâce à l'utilisation de régulateurs de croissance. Recherche sur certaines techniques culturales (éclaircissage chimique) - utilisation de ralentisseurs de croissance dans les verger de pêcher et de pommier. Recherche sur les techniques de multiplication végétative de l'olivier. Amélioration variétale par hybridation intraspécifique sur abricotier et pêcher.

10.1.2 Laboratoire de Recherches de la Chaire d'Ecologie animale et de Zoologie agricole
Tél.: 67 61 24 89
Dir.: F. Leclant

Etude bioécologique et biosystématique d'Arthropodes piqueurs (acariens phytophages, pucerons et cicadelles) — G. Fauvel
Recherches sur le fonctionnement des acarocénoses en verger et en vignoble. — G. Labonne/ S. Kreiter
Epidémiologie des maladies à virus et à mycoplasmes sur arbres fruitiers et en cultures légumières. — J. Gutierrez
Mise en oeuvre de méthodes de lutte dans le respect du concept de lutte intégrée.

10.1.3 Station de Recherches Viticoles
altitude : 5 m
pluviométrie : 0,608 mm
ensoleillement: 2671 h

Domaine du Chapitre, B.P. 13
34750 Villeneuve-lès-Maguelonne
Tél.: 67 69 48 04
Dir.: R. Wagner

Amélioration du raisin de table: recherches de variétés précoces à grosses baies, de muscats à grosses baies, de variétés tardives de bonne conservation; création de variétés résistantes aux parasites cryptogamiques par hybridation *Vitis vinifera x Muscadinia rotundifolia.* — A. Bouquet/ Ing. P. Truel, C. Samson/ A. Vergnes

11 Nancy (Meurthe et Moselle)

11.1 Centre de Recherches Forestières
altitude : 240 m
pluviométrie : 780 mm

Champenoux, 54280 Seichamps
Tél.: 83 31 70 80
Télex: INRANCY 960458 F
Dir.: M. Bonneau

11.1.1 Station Sols Forestiers et Microbiologie — Dir.: F. Le Tacon
Fertilisation des pépinières — G. Lévy
Mycorhization des plants forestiers — F. Le Tacon/ J. Garbaye/ F. Martin
Elevage sur tourbe — F. Le Tacon/ J. Garbaye
11.1.2 Station de Sylviculture et de Production — Dir.: G. Aussenac
Désherbage chimique des pépinières — H. Frochot
11.1.3 Unité d'Amélioration des Arbres forestiers — Dir.: M. Vernier
Conservation et germination des graines — Mme. C. Muller
11.1.4 Laboratoire de Pathologie Forestiere — Dir.: C. Delatour
Identification des maladies cryptogamiques des plants forestiers — M. Morelet

12 Narbonne (Aude)

12.1 Institut des Produits de la Vigne
Station Expérimentale de Pech-Rouge
altitude : 2 a 50 m
pluviométrie : 587,61 mm
ensoleillement: 2400 h

11430 Gruissan
Tél.: 68 49 81 44/ 68 49 85 05/ 68 49 80 08
Télex: INRAGRU 505102 F
Dir.: P. Bénard

Etudes sur la maturation du raisin. Expérimentation sur les techniques de vinification et le vieillissement. Etude sur les jus de raisin, concentrés et produits nouveaux. Les fermentations. Les techniques à membranes

13 Orléans (Loiret)

13.1 Centre de Recherches
altitude : 95 m
pluviométrie : 845,9 mm
ensoleillement: 1872,7 h

Ardon et Saint Cyr en Val
45160 - Olivet
Tél.: 38 63 02 06
Télex: 760359

13.1.1 Departement des Recherches Forestieres — Chef du Département: J.F. Lacaze
13.1.2 Station d'Amélioration des Arbres Forestiers — M. Lemoine
Choix des espèces de reboisement amélioration des peupliers; physiologie du développement; technologie des graines; conservation levée de dormance.
13.1.3 Station de Sylviculture — D. Auclair
Production de biomasse à des fins énergétiques
Mise en place et suivi des programmes portant sur la production de biomasse — L. Bouvarel
13.1.4 Station de Zoologie Forestière — J. Levieux
Etude détaillée de la biologie des ravageurs des forêts (scolytes, lophyres) en vue de l'élaboration de techniques de lutte et de prévention écologique adéquates des pullulations.
13.1.5 Laboratoire d'Economie Rurale — A. Brun
Atlas de la France rurale agricole et forestière. Pluriactivité des familles agricoles.
Service d'Etude des Sols et de la carte pédologique de France — M. Jamagne
Inventaire et caractérisation des sols. Etude de la dynamique de l'eau dans les sols. Etablissement de la carte pédologique de France au 1.100 000

14 Rennes (Ille et Vilaine)

14.1 Centre de Recherches INRA de Rennes
Ecole Nationale Supérieure Agronomique (ENSA)

65, Route de St Brieuc
35042 Rennes Cédex

altitude : 37 m
pluviométrie : 633 mm
ensoleillement: 1630 h

Tél.: 99 59 04 68
Télex: 730866 F

14.1.1 Station d'Amélioration des Plantes
1.a: Rennes - Amélioration du chou-fleur d'automne

Dir.: J. Morice
Y. Hervé

1.b: Laboratoire d'Amélioration des Plantes maraîchères

Plougoulm - 29150 St Pol de Léon
Tél.: 98 29 98 87/ 98 29 96 43
Dir.: Y. Hervé

Amélioration variétale du chou fleur d'hiver et de l'artichaut - Sélection conservatrice de l'échalote

14.1.2 Station de Pathologie Végétale

Dir.: B. Jouan
F. Rouxel

Maladies des crucifères maraîchères de la carotte
Maladies de la pomme de terre — B. Jouan/ B. Tivoli
Maladies et viroses des *Allium* — C. Kerlan/ A. Migliori
Bactérioses des cultures maraîchères (haricot, pomme de terre) — M. Lenormand/ L. Hingand

14.1.3 Laboratoire de Recherches de Zoologie, rattaché à la chaire de Zoologie de l'ENSA

Dir.: J. Missonnier

Diptères des cultures maraîchères — E. Brunel/ J. Missionnier
Nématodes des cultures maraîchères — M. Bossis/ G. Caubel/ D. Mugniery
Lépidoptères (noctuelles, teignes) — R. Rahn
Pucerons de la pomme de terre et de l'artichaut — Y. Robert

14.1.4 Station de Recherches Cidricoles

Dir.: J.F. Drilleau

Technologie du cidre et des divers dérivés de la pomme, aptitudes des variétés aux différents usages, conservation des fruits après la récolte.

15 San Giuliano (Corse)

15.1 Station de Recherches Agronomiques I.N.R.A.-I.R.F.A.
altitude : 10 m
pluviométrie : 880 mm
ensoleillement: 2570 h

San Giuliano - 20230 San Nicolao
Tél.: 95 38 02 14
Télex: 460 683 F
Dir.: F. Lelièvre

15.1.1 Laboratoire d'Amélioration des Plantes

D. Tisne
C. Jacquemond

Collection de 1 200 variétés d'agrumes
Sélection génétique de variétés et porte-greffe du clémentinier
Sélection d'agrumes divers
Sélection de l'avocatier, de l'olivier et du feijoa
Multiplication de matériel végétal de base (graines, greffons)

15.1.2 Laboratoire de Pathologie Végétale

R. Vogel

Indexage des agrumes par plantes indicatrices
Sélection sanitaire des agrumes par micrograffage in-vitro
Etude des maladies à virus et à viroides des agrumes, et de leur transmission; section par insectes

15.1.3 Laboratoire de Zoologie

P. Brun

Etude de la biologie des ravageurs des agrumes et arbres méditerranéens
Etude de la vection des maladies par les insectes
Méthodes de lutte biologique et intégrée contre les ravageurs des agrumes

15.1.4 Station d'Agronomie

Biologie florale des agrumes; taille et densité de plantation	F. Lelièvre
Substances chimiques sur agrumes: hormones, éclaircissage chimique	H. Vannière/ M.P. Davous
Entretien et irrigation des agrumes	J. Cassin
Techniques d'entretien du sol en vergers; effets à long terme sur la fertilité du sol et la productivité du verger	G. Vullin
Systèmes de culture maraîchères sous serre et en plein champ (tomate, melon, aubergine, poivron, salade, artichaud)	Ph. Morel
Cultures fourragères et systèmes fourragers méditerranéens	

16 Toulouse

16.1 Centre de Recherches de Toulouse
altitude : 150 m
pluviométrie : 684 mm
ensoleillement: 2040 h

Chemin de Borde-Rouge
Auzeville, BP 12
31320 - Castanet-Tolosan
Tél.: 61 73 81 75
Télex: INRATSE 52009 F

Laboratoire de Technologie des Produits Végétaux — Dir.: M. Jouret
J.P. Roson

Activités développées en complémentarité avec les travaux de l'Institut des produits de la vigne (Centre INRA de Montpellier) et en concertation avec la profession agricole.
Eaux de vie de vins et de fruits: fermentation et mécanisme du vieillissement en fûts. Valorisation des produits de cépages locaux.
Etude sur les composés polyphénoliques et les composés aromatiques dans la prune d'Ente.

17 Versailles (Yvelines)

17.1 Centre National de Recherches Agronomiques
altitude : 90 m
pluviométrie : 600 mm
ensoleillement: 1750 h

Route de Saint-Cyr
78000 Versailles
Tél.: (1) 30 21 74 22
Télex: INRAVER 695269F

17.1.1 Station d'Amélioration des Plantes* — Dir.: R. Cousin

Augmentation de la productivité par l'utilisation de la vigueur hybride grâce à la stérilité ou l'auto-incompatibilité chez la laitue, l'endive, l'oignon, le poireau, l'asperge; régulation des rendements et augmentation de la qualité des produits grâce à la création de variétés résistantes ou tolérantes aux parasites: haricot (anthracnose, graisse, virus 1 et 2), pois (mildiou, oïdium, anthracnose, fusariose, mosaïque commune, jaunisse apicale, streak, virus du concombre), laitue (mosaïque, Bremia); aux accidents de climat: froid de l'hiver (pois), froid du printemps (haricot).
Sélection pour la qualité, soit pour la consommation en frais, soit pour la conservation (appertisation, surgélation): endive, haricot, pois.
Adaptation aux techniques culturales et de récolte (pois, haricot), à la culture sous-serre (haricot, laitue). Transfert de cytoplasme mâle stérile (radis à choux). Utilisation de divers niveaux de ploïdie (asperge).

Haricot serre, endives	H. Bannerot
Asperge	Mme. L. Corriols-Thevenin
Pois	R. Cousin
Oignon, poireau, ail	B. Schweisguth
Laitue	B. Maisonneuve
Haricot	G. Fouilloux
Choux	L. Boulidard

17.1.2 Station de Pathologie Végétale | Dir.: Mme. M. Lemattre

Maladies des plantes florales (Chrysanthème, Orchidee, Pelargonium, Saintpaulia, Begonia, Dieffenbachia, Hortensia, Cyclamen). Bulbs (Dahlia, Tulipe) et arbustes d'ornement (Conifères, Erica, Rhododendron). Diagnostic et sélection sanitaire (méthodes appliquées à l'indexage) vis-a-vis des champignons, virus bactéries mycoplasmes. Certification. Réceptivité des substrats vis-à-vis des bactéries et des Fusarium vasculaires. Thermothérapie dans les cas des mycoses: Chrysanthème, Glaïeul, Pelargonium.

Lutte biologique vis-à-vis d'*Agrobacterium tumefasciens*	A. Faivre-Amiot
Maladies à mycoplasmes	Mme. M-T. Cousin
Maladies virales	Mme. J. Albouy
Maladies fongiques des arbustes d'ornement	I. Vegh
Maladies fongiques des plantes en pots et des bulbes	Mme. D. Grouet
Maladies bactériennes des arbustes plantes en pots, plantes vertes et bulbes	Mme. M. Lemattre/ A. Faivre
Maladies cryptogamiques du pois et du haricot	C. Allard
Viroses du champignon de couche	H. Lapierre/ G. Molin
Viroses du cresson	J. Bertrandy
Bactéries de l'endive	Mme. M. Lemattre
17.1.3 Station de Zoologie	Dir.: Mlle. N. Hawlitzky
Systématique des insectes	A. Bessard
Identification des espèces nuisibles aux cultures	J.-P. Chambon
Lutte intégrée en vergers	H. Milaire
Insectes vecteurs de viroses et mycoplasmoses	J.-P. Moreau
Tomate: lutte biologique contre l'aleurode des serres	J.-C. Onillon
Hortensia: vecteurs de maladies	J-P. Moreau
17.1.4 Laboratoire de Biologie Cellulaire	Dir.: J.-P. Bourgin
Génétique cellulaire végétale (tomate)	J.-P. Bourgin/ Y. Chupeau Mme. C. Missonnier
Agrobactériologie	
Rhizosphère (tomate)	D. Tepfer
17.1.5 Laboratoire du Métabolisme et de la Nutrition des Plantes	Dir.: E Jolivet J.-F. Morot-Gaudry
Alimentation en eau et en ions minéraux en culture hydroponique	
Plantes ornementales et légumes	Mme. C. Lessaint/ L. Roux C. Tendille
Etude de la composition biochimique des organes de réserves (endive, bulbes)	E. Jolivet/ V. Fiala
17.1.6 Station de Recherche de Lutte Biologique	Domaine de la Minière 78280 Guyancourt Tél.: (1) 30 43 81 13 Télex: INRAMIN 698450 F Dir.: P. Ferron
Utilisation des micro-organismes en lutte biologique contre les insectes	
Utilisation des champignons	J. Fargues/ G. Riba
Othiorhynque du fraisier	P. Ferron
Utilisation des bactéries	D. Martouret
Utilisation des virus	Mme. J. Chaufaux
Noctuelles des crucifères	G. Biache
17.1.7 Station de Bioclimatologie	78850 Thiverval-Grignon Tel.: (1) 30 56 44 44 Dir.: B. Itier
Facteurs du climat et du sol affectant l'efficience de l'eau, irrigation, stress hydrique	N. Katerji
Gelées de printemps, prévision, lutte préventive et active, seuils de résistance	B. Itier

Modélisation
Inventaire potentiel et bases physiques pour la production de biomasse utilisable à des fins énergétiques — G. Gosse

17.2 Ecole Nationale Supérieure d'Horticulture*

4 Rue Hardy
78009 Versailles Cédex
Tél.: (1) 39 50 60 87

Laboratoire de la Chaire d'Agronomie des Plantes Horticoles et Paysagères — Dir.: A. Anstett
Fertilisation des plantes horticoles: plantes à massif
Laboratoire de Recherche de la Chaire de Génétique — Dir.: M. Mitteau
Recherche de stérilité mâle pour la production d'hybrides F1: Saintpaulia, Streptocarpus
Recherche de types à moindre exigence en chaleur: Matthiola, Muflier, Tagete.
Laboratoire de la Chaire de Cultures Légumières et Grainières — Dir.: C. Foury
Laboratoire de la Chaire de Botanique et Malherbologie — Dir.: Janzein
Taxinomie: détermination
Ecologie: relation plante-milieu
Phytogéographie: répartition
Laboratoire de Recherches d'Economie et de Sociologie Rurale — Dir.: Ph. Mainié
Aménagement paysagers
Economie de la production horticole
Commercialisation des produits horticoles
Laboratoire de la Chaire de Cultures Ornementales — Dir.: P. Lemattre*
Etude de déterminisme de la floraison: Bougainvillier, Clerodendron, Nerium
Régulation de la ramification: Cordyline, Dieffenbachia, Fatshedera
Nomenclature des végétaux d'ornement

18 Petit Bourg (Guadeloupe)

18.1 Centre de Recherches INRA des Antilles-Guyane

Domaine Duclos - Petit Bourg
97170 (Guadeloupe)
Tél.: 590 85 20 40
Télex: INRA AG 919 867 GL

Le Centre étudie les problèmes particuliers des Antilles françaises et poursuit des recherches sur l'amélioration, les maladies et les ravageurs de certaines cultures légumières de la zone tropicale.
Les Stations de Recherches concernëes sont:
Station d'Agronomie — Dir.: Y.M. Cabidoche
Station de Bioclimatologie — Dir.: Ch. Valancogne
Station d'Amëlioration des Plantes — Dir.: G. Anais
Station de Pathologie Végétale — Dir.: J. Fournet
Station de Zoologie — Dir.: A. Kermarrec

Autres Organismes

19 Angers (Maine et Loire)

19.1 Comité National Interprofessionnel de l'Horticulture Florale et Ornementale et des Pépinières (CNIH)

20, Boulevard Lavoisier
Belle Beille, 49000 Angers
Tél.: 41 48 19 32

Station Val-de-Loire altitude : 49 m pluviométrie : 0,683 mm ensoleillement: 1928 h	Chef de Station: Cl. Lajoux
Spécialités: plantes en pot, bulbes. Expérimentation et application des résultats de la recherche:	
Régénération de plantes horticoles par culture de méristèmes, thermothérapie et indexage (Chrysanthème, Pelargonium, Hortensia, Cymbidium)	J. Granger
Techniques nouvelles permettant d'améliorer les productions horticoles	J.P. Maillet
Serre, matériel, mécanisation, économie d'énergie	J.C. Laury
Qualités des plantes en pot, certification	J. Daguenet
Bulbiculture (production, conservation, forçage)	A. Mével
19.2 Institut de Recherche de Biologie et Physiologie Végétales des Pays de Loire Association de 1901 - Conseil d'Administration Assemblée Générale, Bureau, Comité Technique	16, Boulevard Lavoisier 49000 Anges - Belle-Beille Tél.: 41 48 21 10/ 41 48 67 52 Dir.: Prof. Y. Demarly Chargé de Direction: R. Letouzé
Culture in-vitro	
Régénération par multiplication végétative	R. Letouzé
Sylviculture, horticulture florale et maraïchère	M. Duron
Arbustes ornementaux, arbres fruitiers	C. Poulain
Plantes médicinales, plantes des zones arides	C. Bureau/ M. Montfort
Action des lumières monochromatiques	R. Letouzé
Phytométabolisme et vitrification	F. Daguin
Culture de cellules	
Culture de protoplastes	F. Lavergne
Symbiose et micropropagation	D.G. Strullu/ B. Grellier/ C. Roman
19.3 Pépinières MINIER S.A. altitude : 34 m pluviométrie : 600 mm ensoleillement: très variable	Les Fontaines de l'Aunay, BP 35 49250 Beaufort en Vallee Tél.: 41 47 41 65 Télex: 720311 F Dir. Tech.: Cl. Bellion
Mutagénèse expérimentale, arbustes d'ornement. Multiplication de souches saines. Etude de la nutrition des plantes en cultures hors sol. En relation avec INRA et CNRS	
19.4 Université d'Angers. Institut de Recherches Scientifiques et Techniques	Boulevard Lavoisier 49045 Angers Tél.: 41 48 32 34 Dir.: M. Astié
Laboratoire de Biologie Végétale Multiplication végétative et contrôle de la floraison sur plantes entières et organes isolés en culture in-vitro.	
Recherches sur la multiplication végétative et l'obtention de la floraison en culture in-vitro à partir de microspores, d'organes isolés (pièces florales de	M. Astié/ C. Coquen J.D. Viémont

Lilium candidum, fragments de feuilles, tiges, hampes inflorescentielles de *Digitale, Cleome, Streptocarpus*, etc.)

Contrôle de la floraison de plantes entières *(Cappa-ridacées, Primulacées, Gesnériacées, Liliacées*, etc.). Recherches sur le photopériodisme et thermopériodisme.

Etude de l'action de substances de synthèse (pesticides) sur la croissance et la floraison — M. Astié/ C. Coquen, J.D. Viémont

19.5 Ecole Nationale d'Ingenieurs des Techniques Horticoles* — 2 Rue Le Notre, 49045 Angers Cédex, Tél.: 41 48 36 24, Dir.: J.P. Bigre

Laboratoire de Chimie et Sciences du Sol — Professeur: L.M. Rivière*

Connaissance du comportement de l'eau dans les matières organiques (tourbes, composts, etc.) — L.M. Rivière*

Mise au point de supports de cultures hors sol et maitrise de l'irrigation — L.M. Rivière*

Potentiel osmotique de la phase liquide: influence sur le comportement hydrique du couple substrat/plante — S. Charpentier

Drainage des sols de vergers — J.P. Rossignol

Laboratoire d'Horticulture Ornementale — Professeur: H. Vidalie

Hortensia: influence de différentes durées de traitement au froid avant forçage — H. Vidalie

Chrysanthème: floraison programmée sous abri-tunnel

Gaz carbonique: influence en production de plantes en pots

Gerbera: cultures hors sol sur différents substrats chauffés ou non

Protection des obtentions végétales: collection de référence — H. Bertrand

Laboratoire d'Arboriculture Fruitière — Professeur: Y. Guiheneuf

Etude de l'évolution de critères d'appréciation de la qualité gustative et du calibre des pommes aux différents stades de croissance du fruit — Y. Guiheneuf

Dynamique de l'enracinement du pommier 'Lysgolden' greffé sur Malling 26 avec et sans paillage plastique de la plantation à l'âge adulte — P. Raimbault

Laboratoire de Physique et Génie Rural — Professeur: G. Chasseriaux

Création d'un logiciel pour l'estimation des besoins en chauffage, des serres de production

Etude du comportement thermique en culture "hors-sol"

Etude de l'évapotranspiration en conteneurs

Microclimat en culture *in vitro*

Laboratoire de Productions Légumiere et Grainière — Professeur: J.Y. Péron*

Diversification des productions légumières, y compris celles destinées aux salles de forçage d'endive — J.Y. Péron*

Création d'un légume nouveau: le Crambé Maritime

Contribution à la réémergence d'un légume oublié: le cerfeuil tubéreux. Recherche d'embryons somatiques

Etude de la floraison du melon in vitro, production d'embryons immatures après fécondation in vitro

Production du melon hors sol en serre, en ambiance tempérée — J.Y. Péron*/ P. Lemanceau

Etude du rôle éventuel de la compétition vis-à-vis

du fer sur la résistance du sol de Chateaurenard au Fusarium Oxysporum — P. Lemanceau
Essais de lutte biologique contre le Fusarium oxysporum en culture hors sol — P. Lemanceau/ L.M. Rivière*

Laboratoire de Sciences Biologiques — Professeur: R. Augé

Micropropagation *in vitro: Iris germanica, Actinidia chinensis, Hippeastrum* — R. Augé
Mise au point d'hybrides somatiques par fusion de protoplastes chez la pomme de terre — J. Boccon-Gibod
Effet de l'acide gibbérellique sur la floraison des plantes ornementales — R. Augé
Etude des adventices des cultures de vigne dans la vallée de la Loire — J. Grelon

20 Antibes (Alpes Maritimes)

20.1 Universal Rose Selection Meilland
altitude : 70 m
pluviométrie : 0.770 mm
ensoleillement: 2810 h

134, Boulevard du Cap
06600 Cap d'Antibes, BP 225
06601 Antibes Cédex
Tél.: 93 61 30 30
Dir.: A. Meilland

Recherche et expérimentation de roses nouvelles — J.L. Gorda

21 Bellegarde (Gard)

21.1 Centre C.T.I.F.L. de Balandran
altitude : 50 m
pluviométrie : 694 mm
ensoleillement: 2617 h

30127 Bellegarde
Tél.: 66 01 10 54
Télex: 480935
Resp.: G. Planton/ A. Osaer

Etudes du matériel végétal, des techniques de production, de la défense des cultures sur arbres fruitiers, légumes de serre et de plein champ et notamment:

Climat, culture sans sol et lutte intégrée en serre — M. Musard/ C. Wacquant, J.P. Thicoïpé/Y. Trottin-Caudal
Variétés, porte-greffes, modes de conduite, irrigation fertilisante, substances de croissance sur arbres fruitiers — P. Soing/ J. Vidaud
Variétés, desherbage, bâches au sol et techniques diverses sur légumes de plein champ — G. Joubert/ H. Zuang
Micropropagation des arbres fruitiers et du fraisier — J.C. Navatel
Production de matériel de base d'abres fruitiers, de fraisier et d'allium en vue de la certification
Principales espèces étudiées: pêche, cerise, pomme, poire, tomate, melon, poivron, aubergine, fraise

22 Bersée (Nord)

22.1 Maison André Blondeau, Semences Sélectionnées
altitude : 50 m
pluviométrie : 0,7 mm

B.P. N° 1 - 59235 - Bersée
Tél.: 20 59 20 33
Télex: 110331
Dir.: G. Leclercq

Génétique: amélioration du pois — A. Verhaegen
Amélioration du haricot, type flageolet, chevrier -type mangetout — F. Bassème

23 Brest (Finistère)

23.1 Laboratoire de Microbiologie Appliquée,
Université de Bretagne Occidentale
altitude : 98 m
pluviométrie : 1,1 mm
ensoleillement: 1548 h

Faculté des Sciences et Techniques
6, Avenue Le Gorgeu
29287 - Brest Cédex
Tél.: 98 03 16 94
Dir.: Mme. M. Moreau

Maladies des sols et pathologie vasculaire: *Fusarium, Verticillium, Phialophora*

Mme. M. Moreau/ M. El Mahjoub
D. Le Picard/ R. Lefèvre
J.-L. Tanguy

Quelques aspects de lutte biologie

B. Trique/ Y. Tirilly
Mlle. B. Picard

24 Brétigny sur Orge (Essonne)

24.1 L. Clause - Société Anonyme pour la Culture des Graînes d'Elite

1, Avenue Lucien Clause
91221 - Brétigny-sur-Orge Cédex
Tél.: (1) 60 84 95 84
Telex: 692 159 F
Dir.Scien.: L. Merchat*

Recherches, création, sélection et production de variétés potagères, graminées à gazons, pomme de terre, pour le professionnel et l'industrie de la conserve, et florales pour l'amateur et le professionnel. Ces travaux sont exécutés en général sur l'amélioration des rendements, en pleine terre ou en serre, la résistance aux maladies, la recherche du cycle végétatif plus court par la création de variétés fixées ou d'hybrides F1.

25 Cambrai (Nord)

25.1 Seminor - Société Anonyme
altitude : 77 m
pluviométrie : 677 mm
ensoleillement: 1500 h

50 bis, Allée Saint-Roch,
59402 - Cambrai Cédex
Tél.: 27 83 18 00
Prés. Dir.: G. Euverte

Recherche dirigées sur tous les types de pois et haricot présentant un intérêt pour l'appertisation, la surgélation et lyophylisation.
Etude des meilleurs types d'épinard et création de variétés répondant aux besoins de l'industrie.

Ing. G. Meslin/ Ing. J.M. Pilas

26 Carquefou (Loire Atlantique)

26.1 Station d'Expérimentation sur les cultures légumières (C.T.I.F.L.)
altitude : 39 m
pluviométrie : 800 mm
ensoleillement: 1900 h

Allée des Sapins, 44470 Carquefou
Tél.: 40 50 81 65
Resp.: J. Pelletier

26.1.1 Serres et grand tunnels

M. Létard

Cultures hors sol: amélioration des techniques (substrats, solution fertilisante, sur tomate, melon, concombre).
Economie d'énergie: utilisation de la pompe à chaleur, air/eau; chauffage basse température.
Interaction température air/température sol.
Diversification des cultures: aubergine, poivron, céleri, fenouil.

26.1.2 Plein champ et petits tunnels — J. Le Bohec
Amélioration des techniques culturales: carotte primeur, poireau, mâche, navet, céleri branche.

26.1.3 Lutte contre les mauvaises herbes — J. Pelletier
Désherbage chimique des principales espèces légumières.
Etude de la rémanence des herbicides.

27 Chambourcy (Yvelines)

27.1 Comité National Interprofessionnel de l'Horticulture, Station Ile de France
altitude : 90 m
pluviométrie : 639 mm
ensoleillement: 1674 h

Domaine de la Jonction
78240 Chambourcy
Tél.: (1) 39 73 40 77
Chef de Station: J.C. Foucard

Techniques culturales - innovations - diversification — Ing. P. Locher
Cultures hors-sol en pépinières. Recherche de substrats. Désherbage chimique. Irrigation.
Etude de l'influence de la nutrition minérale des pieds-mères sur l'aptitude au bouturage des ligneux d'ornement *(X Cupressocyparis)*. Utilisation des régulateurs de croissance sur plantes vivaces ornementales. Mise au point de méthodes de lutte contre les dépérisments du *Prunus Laurocerasus*. Collection de végétaux de substitution aux espèces sensibles au feu bactérien. Economie d'énergie - valorisation des temps de travaux. Biotechnologie et sélection sanitaire. Qualité - contrôle et maintenance du produit. Certification du matériel fruitier de reproduction.

28 Clermont Ferrand (Puy de Dôme)

28.1 Laboratoire de Physiologie Végétale et de Phytomorphogenèse
altitude : 330 m
pluviométrie : 592 mm
ensoleillement: 1900 h

4, Rue Ledru - 63000 Clermont-Ferrand
Tél.: 73 93 35 71
Dir.: P. Champagnat,
Prof. J.C. Courduroux

UER Sciences - Université de Clermont-Ferrand et Laboratoire associé au CNRS.
Corrélations entre organes (corrélations entre feuilles et bourgeons axillaires: dominance apicale et phénomènes analogues).
Dormance des bourgeons: mècanisme, liens avec la morphogenèse; acrotonie et basitonie des végétaux ligneux; tubérisation; biologie des bulbes; acides nucléiques, dormance et morphogenèse.
Totipotence cellulaire: régéneration; production d'embryoïdes, multiplication végétative des orchidées.

Corrélations feuille/bourgeons	N. Boyer/ M.O. Desbiez/ M. Rémy P. Champagnat
Dominance apicale	M. Rémy/ E. Hugon/ P. Champagnat
Mécanisme de l'acrotonie et la basitonie. Dormance des bourgeons	S. Lavarenne/ P. Barnola P. Champagnat
Dormance et tubérisation (rôle des acides nucléiques)	J.C.Courduroux/ C. Teppaz-Misson
Mécanisme de l'action des hormones	M. Gendraud/ D. Charnay M.F. Trouslot/ M. Bongen
Croissance rythmique des ligneux en milieu uniforme	S. Lavarenne/A. Lavarenne P. Barnola/ P. Champagnat/ D.Lamond
Multiplication végétative des orchidées; totipotence cellulaire; embryons adventifs des crucifères	M. Champagnat/ M. Delaunay P. Ballade
Morphogenèse et fonctionnement des systèmes racinaires	M. Lamond/ D. Lamond/ M. Sévert

29 La Gaude (Alpes Maritimes)

29.1 Comité National Interprofessionnel de l'Horticulture Station Midi
altitude : 30 m

La Baronne, 7630, CD 2209
06610 - La Gaude
Corr.: BP 31 - 06021 Nice Cédex

pluviométrie : 800 mm
ensoleillement: 2810 h

Tél.: 93 31 22 05
Chef de Station: R. Jacquemont

Localisation du chauffage et économies d'énergie pour les rosiers et les fleurs coupées ainsi que pour les potées fleuries. — D. Delmon

Multiplication *in vitro* et usage des produits issus de la technique: fleurs coupées, plantes en pots pépinière — J.P. Onesto

Physiologie et conservation des fleurs coupées
Régénération et sélection sanitaires ainsi que contrôle et certification sanitaire sur oeillet — R. Poupet

Valleur sanitaire et productivité des variétés: oeillet
Comportement des végétaux ligneux méditerraneens (laurier rose - cyprès) — J. Bontemps

Analyses nématoligique sur toutes les cultures agricoles et horticoles — D. Esmenjaud

30 Gif sur Yvette (Essonne)

30.1 Institut de Physiologie végétale (Phytotron)
altitude : 90 m
pluviométrie : 600 mm
ensoleillement: 1750 h

Centre National de la Recherche
Scientifique, Avenue de la Terrasse
91190 - Gif sur Yvette
Tél.: (1) 39 07 78 28
Télex: CNRSGIF 691137F
Dir.: R. Jacques

Le laboratoire traite différents aspects de la biologie et de la physiologie végétales en rapport avec les facteurs de l'environnement (lumière, température, eau) pour une meilleure connaissance des bases physiologiques de la production végétale. Le laboratoire est structure en 2 départements.

30.1.1 Département a: — Responsable: R. Jacques

Objectif général: etude *in vivo* et *in vitro* des processus de la croissance et du développement
Secteurs principaux:

Déterminisme hormonal de la croissance et du développement cette recherche est basée sur l'utilisation d'anticorps dirigés contre les principales hormones endogènes connues en vue de leur dosage et de leur localisation chex les végétaux herbacés et ligneux. — R. Maldiney/ B. Sotta, L. Sossountzov

Photorégulation de la croissance en particulier étude du mode d'action du phytochrome. — A. Lecharny

Etude du contrôle et de la régulation cellulaire de la différenciation *in vivo* et *in vitro* grâce à la mise au point de systèmes expérimentaux simplifiés chez *N. tabacum* et *N. plumba* ginifolia. — K. Tran Thanh Va

Etude du développement de deux végétaux ligneux: le Pseu Tsuga Menziesii et le Terminalia superba — M. Jacques/ E. Miginiac

Action du froid sur le développement des plantes (Blé d'hiver, maïs, perilla). — F. Blondon

30.1.2 Département b: — Responsable: M.O. Queiroz

Objectif général: étude des mécanismes moléculaires et cellulaires de l'adaptation à l'environnement (en particulier à la sècheresse et à la pollution par le SO_2)
Secteurs principaux:

Détection et transduction des signaux externes: photopériodisme et adaptation à la sècheresse. — J. Brulfert
L'existence d'un marqueur enzymatique, la pepcarboxylase constitue dans le cas des plantes à métabolisme de type CAM (Crassulacean Acid Metabolism) (sur *Kalanchoe Blossfeldiana)* un avantage et permet une approche des mécanismes d'action de la photopériode au plan moléculaire.
Coordination du fonctionnement métabolique: par étude de systèmes intégrés favorables et la mise au point de systèmes simplifiés (choroplastes mitochondries, thylacoides isolés, protoplastes...) — J.N. Pierre/ C. Queiroz
Analyse des réponses des plantes aux conditions extrêmes de milieu (sècheresse, salinité, pollution par SO_2) — C. Hubac
Haricot, soja, opuntia sp. cotonnier, colza. — A. Tremolieres

31 Grenoble (Isère)

31.1 Centre d'Etudes Nucléaires de Grenoble Laboratoire de Biologie Végétale
altitude : 223 m
pluviométrie : 985 mm
ensoleillement: 2076 h

38, Avenue des Martyrs
Corr.: CEN 6 85 X
38041 - Grenoble Cédex
Tél.: 76 88 44 00
Télex: Energat Greno N° 320.323
Chef de Dép.: M.J. Winter

Biostatistiques — B. Lachet
Radiobiologie des cellules végétales, chlorelles — R. Gilet/ J.C. Roux/ J. Drimon
Physiologie des racines, pénétration par voie foliaire — J. Gagnaire/ A. Bonicel
Microlocalisation des éléments minéraux. — A. Chamel*
Utilisation horticole des calories perdues — A. Fourcy/ J.P. Garrec

32 La Force

32.1 Centre C.T.I.F.L. de Lanxade*
altitude : 30 m
pluviométrie : 760 mm
ensoleillement: 1861 h

Prigonrieux - BP 21
24130 La Force
Tél.: 53 58 00 05
Responsable: M. Savio

Experimentation Fruitière
Etude du matériel végétal, au niveau national: pommier, poirier, cerisier, prunier, noisettier, noyer — A. Chartier
Etudes technique ou expérimentation régionale (relations avec les centres régionaux: CIREF-CIREA)
Production de Matériel Certifié INFEL
Sélection conservatrice: contrôle de l'état sanitaire, et de l'authenticité des cultivars — J.C. Desvignes
Prémultiplication du matériel certifié: distribution de matériel, contrôle chez les multiplicateurs, délivrance d'une étiquette certifiée INFEL — M. Labergère

33 La Londe (Var)

33.1 Laboratoire de Physiologie Végétale Barberet & Ducloux S.C.S.
altitude : 25 m

83250 - La Londe
Tél.: 94 66 81 12
Dir.: M. Rudelle

pluviométrie : 800 mm
ensoleillement: 2885 h

Oeillet:

Régénération des variétés commerciales par culture de méristèmes in-vitro	Florence Ducloux
Sélection clonale des variétés régénérées, création de nouveautés	Y. Mitteau
Multiplication *in-vitro* des nouveautés, recherches biotechnologiques	Françoise Jouannic
Gerbéra:	
Création de variétés nouvelles	M. Rudelle
Multiplication *in-vitro* commerciale et amélioration des techniques de micropropagation	Françoise Jouannic

34 La Ménitré (Maine et Loire)

34.1 Institut de Recherche Vilmorin
altitude : 25 m
pluviométrie : 58 mm

La Ménitré
49250 Beaufort-en-Vallée
Tél.: 41 47 52 21
Télex: 720 850
Dir.: J.N. Plages

Résistance génétique aux viroses et mycoses. Inoculation artificielle, culture in vitro	A. Feutry
Amélioration du haricot *(Phaseolus vulgaris)* et du pois *(Pisum sativum)*	B. Bosc/ J.P. Gé
Amélioration concombre *(cucumis sativus)*, épinard *(spinacia oleracea)*	J.N. Plages/ B. Lefèvre
Amélioration laitue *(Lactuca sativae)* et radis *(Raphanus sativus)*	H. Michel/ B. Ricard
Amélioration chou *(Brassica oleracea)*, oignons *(Allium cepa)*, graminées à gazon *(Lolium, Festuca)*, betteraves potagères	A. Chesnel/ R. Croisé A. Greffet
Amélioration des plantes florales *(Impatiens, Primula, Viola, Salvia)*	A. Boggio/ M.F. Bourrigault
Pathologie et analyse des semences	A. Bidaud/ C. Candelá/ R. Germain

35 Lédenon (Gard)

35.1 Station d'Amélioration des Plantes de la Société Anonyme Vilmorin-Andrieux, Producteurs de Semences Sélectionnées
altitude : 70 m
pluviométrie : 0,72 mm
ensoleillement: 2650 h

Vilmorin-Andrieux, Centre de Sélection de la Costière
Lédenon, 30210 Remoulins
Tél.: 66 57 57 56
Télex: VILRECH 480870 F
Dir.: Ing. G. Berric

Résistance génétique aux viroses et mycoses. Amélioration des espèces légumières méditérranéennes: aubergine *(Solanum melongena)*, courge *(Cucurbita pepo)*, *melon (Cucumis melo)*, piment *(Capsicum annuum)*, tomate *(Lycopersicum esculentum)*, carotte *(Daucus carotta)*, asperge *(Asparagus officinalis)*
Création et expérimentations de variétés pour cultures de légumes frais en plein champ et sous abris.

J. Goulpaud/ J. Sjerps
G. Simon/ A. Borgel
D. Gabillard/ JW. Hennart
M. Jacquet/ A. Amiraux
C. Robledo

36 Malemort sur Corrèze

36.1 Station C.T.I.F.L. de Novert
altitude : 175 m
pluviométrie : 1012,6 mm
ensoleillement: 1943,50 h

BP 7, 19360 Malemort sur Corrèze
Tél.: 55 23 76 21
Responsable: Y. Jestin

Noix et Châtaignes: variétés, portes-greffes, techniques culturales, récoltes et séchage, conservation.
Noisettes: variétés, densités, techniques culturales.
Truffes: techniques culturales: mise au point, comparaison des types de plants truffiers.
Unité d'expérimentation technologique:
Aptitudes des variétés des marrons, à l'épluchage mécanique. Mise au point de chaînes de fabrication. Mise au point de produits nouveaux à base de marron. La technologie de la transformation du marron. Orientation vers la 4ème gamme de légumes.

37 Malicorne (Allier)

37.1 Pépinières et Roseraies Georges Delbard
altitude : 400 m
pluviométrie : 758 mm
ensoleillement: 1850 h

Malicorne, 03600 Commentry
Tél.: 70 64 33 34
Télex: 391 632
Dir.: J.-P. Barbe

Recherches et sélections variétales rosiers et fruitiers — Mireille Feliziani, Frédérique Souq
Laboratoire de production *in-vitro* — Y. Mazière/ B. Détienne

38 Meudon Bellevue (Hauts de Seine)

38.1 Laboratoire de Physiologie des Organes Végétaux après Récolte, Centre National de la Recherche Scientifique (CNRS)
altitude : 50 m
pluviométrie : 600 mm
ensoleillement: 1850 h

4 ter, route des Gardes
92190 Meudon
Tél.: (1) 15 34 75 50
Télex: 204135 LABOBEL
Dir.: Prof. D. Côme

Physiologie de la maturation et de la sénescence des fruits. Application à la conservation par réfrigération et en atmosphère controlée. Rôle et métabolisme de l'éthylène. — P. Marcellin (Dir. de Recherche CNRS), C. Leblond (Ing. CNRS)

Mécanismes de la sénescence des fleurs coupées. Mise au point de solutions nutritives — A. Paulin (Maître de recherche CNRS)

Congélation des fruits et légumes — J. Philippon (Ing. CNRS), Mme. M.A. Rouet (Ing. CNRS)

Physiologie des organes végétatifs et cryoconservation — P. Soudain (Ing. CNRS), J. Dereuddre (Maître-Assistant, Université Paris 6)

Protéines végétales — J. Daussant (Maître de recherche CNRS)/ Mme. C. Laurière (Chargé de recherche CNRS)

Physiologie des semences: germination et dormances — D. Côme (Prof. Univ. Paris 6), Mme. C. Thèvenot (Ass. Univ. Paris 6)/ Mlle. F. Corbineau (Chargé de recherche CNRS)

39 Milly La Forêt

39.1 Société Civile Darbonne

6, Boulevard du Maréchal Joffre,

altitude : 50 m
pluviométrie : 600 mm
ensoleillement: 1850 h

BP 8, 91490 - Milly La Forêt
Tél.: (1) 64 98 95 95
Télex: 690 373 Darbonne F
Dir.: H. Darbonne

Recherches sur asperges et fraisiers. Etude technique cultures nouvelles (production et transformation). Etude herbicide, fongicide, nématicide en collaboration avec des Centres d'Etudes Officiels et Firmes de Produits Chimique.

39.2 Institut Technique des Plantes Médicinales Aromatiques et Industrielles

Route de Nemours, BP 38
91490 - Milly La Forêt
Tél.: (1) 64 98 83 77
Dir.: P. Maghami*

Recherche: modernisation des méthodes culturales et de conditionnement des plantes médicinales et aromatiques cultivées. Acclimatation des espèces nouvelles. Entretien de collection source de semences. Vulgarisation: assistance technique, publications fourniture de semences. Relations nationales et internationales avec les Syndicats de Recherches.

40 Orléans (Loiret)

40.1 Syndicat pour l'Amélioration des Sols et des Cultures S.A.S.
altitude : 125 m
pluviométrie : 600 mm
ensoleillement: 1854 h

Ardon, 45160 Olivet
Tél.: 38 69 26 31/ 38 69 20 10
Dir.: M. Lépine

Laboratoire agronomique national, avec un service d'analyses spécifiques pour l'horticulture, assurant des délais très rapides aux horticulteurs (Terres, substrats, solutions nutritives).
Bureau d'ingénieurs-conseil en matière de conduite de sols et des substrats, de nutrition de la plante et de l'animal; fertilisation, travail du sol, drainage, irrigation. Départment horticole et maraîcher: toutes cultures, cultures sous serres, cultures hors sol, étude de substrats.

41 Paris (Seine)

41.1 Museum National d'Histoire Naturelle Service des Cultures
altitude : 50 m
pluviométrie : 600 mm
ensoleillement: 1850 h

43, Rue Buffon, 75005 Paris
Tél.: (1) 43 36 12 33
Dir.: J. Mongour

Grand etablissement scientifique dépendant du Ministère de l'Education Nationale (Direction des Enseignements Supérieurs)
Conservation, enrichissement, présentation des collections végétales vivantes du Museum National d'Histoire Naturelle
Ressources génétiques: banque de gènes végétaux en association avec le Laboratoire de Palynologie du Museum

41.2 Centre Technique Interprofessionnel des Fruits et Légumes (C.T.I.F.L.)

22, Rue Bergère, 75009 Paris
Tél.: (1) 47 70 16 93
Dir.: Françoise Rastoin

Etudes Economiques: Etudes économiques sur l'appareil de production et sur les marchés, études sur l'evolution des modes et des circuits de distribution, analyses qualitatives et quantitatives sur la consommation et son évolution.
Analyses Statistiques et Documentation:
Analyses statistiques - banque de donnees chiffrées, service documentaire et diffusion des publications

Responsable: M. Michel

Formation et Animation: Actions de formation pour les metiers du commerce, actions de formation pour les techniciens et les producteurs agricoles du Secteur Fruits et Légumes, appui méthodologique pour les actions générales de diffusion, animation de la distribution pour la relance de la consommation

Responsable: M. Laborde

42 Les Ponts de Cé (Maine et Loire)

42.1 Les Graines Caillard S.A.

Chemin de Pouillé, P.O. Box 30
F.49130, Les Ponts de Cé
Dir.: J.L. Desclaux

Département de Sélection et de Recherche
Sélection d'haricot vert — E. Marx
Sélection de chou-fleur et carotte — B. Smets

43 Saint Marcellin (Isere)

43.1 Station C.T.I.F.L. de Chatte
altitude: 320 m

"Les Colombières"
38160 Saint Marcellin
Tél.: 76 38 30 63
Responsable: P.M. Grospierre

Noyer: variétés et portes-greffes, techniques culturales: mode de plantation - taille - fertilisation désherbage chimique - irrigation, protection phytosanitaire, matériel de récolte
Châtaignier: variétés, reconversion des tailles en vergers
Framboisier: variétés: 60 variétés et hybrides, comparaison de souches saines obtenues par des filières différentes
Matériel de récolte, matériel de préconditionnement et de séchage

44 Saint Paul lez Durance (Bouches du Rhône)

44.1 Commissariat a l'Energie Atomique, Centre d'Etudes Nucléaires de Cadarache
altitude : 235 m
pluviométrie : 646 mm
ensoleillement: 2700 h

B.P. n° 1
13 115 Saint-Paul-lez-Durance
Tél.: 42 25 70 88
Télex: 440678 F
Dir.: L. Saint-Lèbe
Resp. Scien.: A. Silvy

Service de Radioagronomie, ame0lioration des plantes a multiplication végétative par radiomutagenèse et culture in vitro

45 Saint Rèmy de Provence (Bouches du Rhône)

45.1 Station de Conservation C.T.I.F.L.
altitude : 24 m
pluviométrie : 654 mm
ensoleillement: 2757 h

Route de Mollégès
13210 Saint Rémy de Provence
Tél.: 90 92 05 82
Responsable: P. Moras

Conservation en frais des fruits et légumes. — P. Moras
Qualité.
Traitement phytosanitaire de post récolte. — J.F. Chapon
Perfectionnement des systèmes de production du froid.
Maturation. Economies d'énergies.

46 Sarrians (Vaucluse)

46.1 Les Graines Caillard S.A.*
altitude : 36 m
pluviométrie : 635 mm
ensoleillement: 2767 h

Domaine du Moulin, F. 84260
84260 - Sarrians
Tél.: 90 65 41 67
Télex: 431889
Dir.: J.L. Desclaux

Département de Sélection et de Recherche	
Sélection de tomate et melon	L. Hedde
Sélection de laitue, chicorée et oignon	B. Moreau
Sélection de piment, aubergine et courgette	J.L. Nicolet
Laboratoire de culture in vitro	M. Cabannes

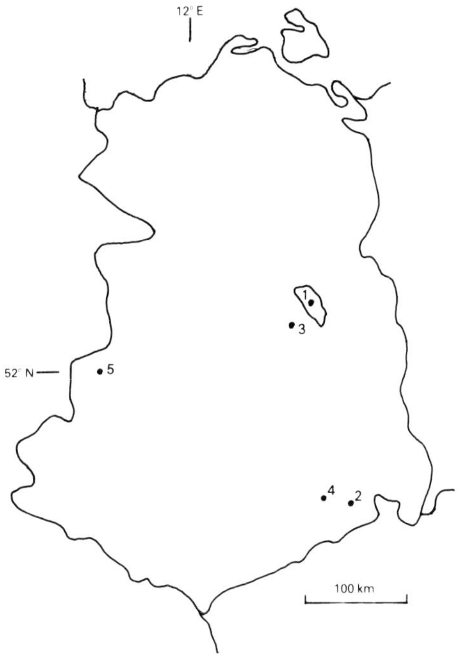
12° E
52° N
1
2
3
4
5
100 km

1 Berlin
2 Dresden
3 Grossbeeren
4 Nossen
5 Quedlinburg

Survey

Akademie der Landwirtschaftswissenschaften der Deutschen Demokratischen Republik (Academy of Agricultural Sciences of the German Democratic Republic) is the central institution of agricultural sciences in this country. It plans the research and development activities of its various institutions. At the same time, it coordinates the research work of its institutions with that of the Universities, Colleges and other Academies and of institutions coming under other branches of the national economy. Vegetable research is the responsibility of Institut für Gemüseproduktion Grossbeeren der Akademie der Landwirtschaftswissenschaften der Deutschen Demokratischen Republik (Institute of Vegetable Production Grossbeeren of the Academy of Agricultural Sciences of the German Democratic Republic) engaged in research on field and greenhouse vegetable production, of Institut für Züchtungsforschung Quedlinburg der Akademie der Landwirtschaftswissenschaften der Deutschen Demokratischen Republik (Institute of Plant Breeding Research Quedlinburg of the Academy of Agricultural Sciences of the German Democratic Republic) concerned with vegetable breeding, and of Sektion Gartenbau der Humboldt-Universität zu Berlin (Horticultural Section of Humboldt University of Berlin). Fruit research is concentrated in Institut für Obstforschung Dresden-Pillnitz der Akademie der Landwirtschafswissenschaften der Deutschen Demokratischen Republik (Institute of Fruit Research Dresden-Pillnitz of the Academy of Agricultural Sciences of the German Democratic Republic) and in Sektion Gartenbau der Humboldt-Universität zu Berlin.

Research on ornamental plants is carried out mainly by Sektion Gartenbau der Humboldt-Universität zu Berlin.

Students for certificated horticultural engineer (Diplomgartenbauingenieur) are trained at Sektion Gartenbau der Humboldt-Universität zu Berlin. Studies there extend over four years and a half for full-time students, and five years for students participating in correspondence courses.

Students for horticultural engineer (Gartenbauingenieur) are trained at Ingenieurschule für Gartenbau Erfurt (School of Horticultural Engineering at Erfurt, Ingenieurschule für Gartenbau Werder (School of Horticultural Engineering at Werder, near Potsdam), and Fachschule für Zierpflanzenwirtschaft Bannewitz (Technical School of Ornamentals Growing at Bannewitz, near Dresden). Results of research work are published mainly in the journals "Archiv für Gartenbau" (incl. summaries in Russian and English) and "Gartenbau" (incl. contents in Russian and English).

Prepared by Prof.Dr. G. Vogel, Institute of Vegetable Production, Grossbeeren.

1 Berlin

1.1 Sektion Gartenbau der Humboldt-Universität zu Berlin* (Horticultural Section of Humboldt University of Berlin)
rainfall : 560 mm
sunshine: 1710 h

Invalidenstrasse 42
DDR - 1040 Berlin
Phone: 28970
Telex: 112823
Dir.: Prof.Dr.habil. E. Rempel

Vegetable production, Fruit production, Ornamentals production

2 Dresden

2.1 Institut für Obstforschung Dresden-Pillnitz der Akademie der Landwirtschaftswissenschaften der DDR (Institute of Fruit Research Dresden-Pillnitz of the Academy of Agricultural Sciences of the GDR)
rainfall : 720 mm
sunshine: 1570 h

Pillnitzer Platz 2
DDR - 8057 Dresden
Phone: 39346
Telex: 2478
Dir.: Prof.Dr.sc. W. Fehrmann

Fruit production, Fruit breeding

3 Grossbeeren

3.1 Institut für Gemüseproduktion Grossbeeren der Akademie der Landwirtschaftswissenschaften der DDR (Institute of Vegetable Production Grossbeeren of the Academy of Agricultural Sciences of the GDR)
rainfall : 540 mm
sunshine: 1710 mm

Theodor-Echtermeyer-Weg
DDR - 1722 Grossbeeren
Phone: 80
Telex: 15493
Dir.: Prof.Dr.sc. G. Vogel

Field vegetable production
Greenhouse vegetable production

4 Nossen

4.1 Zentralstelle für Sortenwesen der Deutschen Demokratischen Republik (National Variety Testing and Registration Centre of the German Democratic Republic)
rainfall : 630 mm
sunshine: 1550 h

Waldheimer Strasse 219
DDR - 8255 Nossen
Phone: 7721
Telex: 277257
Dir.: Dipl.-Landwirt H. Witt

Variety testing

5 Quedlinburg

5.1 Institut für Züchtungsforschung Quedlinburg der Akademie der Landwirtschaftswissenschaften der DDR (Institut of Plant Breeding Research Quedlinburg of the Academy of Agricultural Sciences of the GDR)
rainfall : 530 mm
sunshine: 1560 h

Ethel-und-Julius-Rosenberg-Strasse 22-23
DDR - 4300 Quedlinburg
Phone: 470
Telex: 48546
Dir.: Prof.Dr. J. Dehne

Vegetable breeding, Plant breeding research

GERMANY, FEDERAL REPUBLIC OF DE

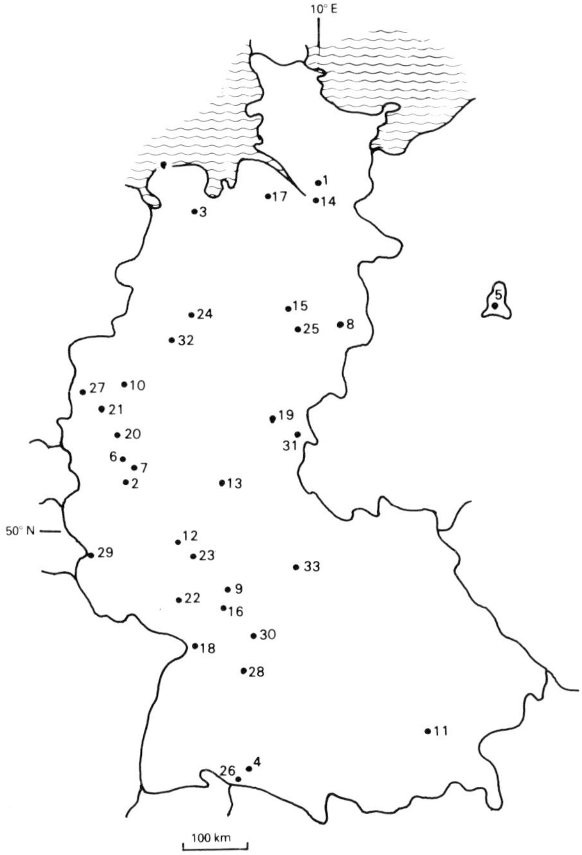

10° E
50° N
100 km

1 Ahrensburg
2 Bad Neuenahr-Ahrweiler
3 Bad Zwischenahn
4 Bavendorf
5 Berlin
6 Bonn
7 Bonn-Bad Godesberg
8 Braunschweig
9 Dossenheim
10 Essen
11 Freising-Weihenstephan
12 Geisenheim
13 Giessen
14 Hamburg
15 Hannover
16 Heidelberg
17 Jork
18 Karlsruhe
19 Kassel
20 Köln
21 Krefeld
22 Neustadt
23 Oppenheim
24 Osnabrück
25 Sarstedt
26 Schlachters
27 Straelen
28 Stuttgart
29 Trier
30 Weinsberg
31 Wittenhausen
32 Wolbeck
33 Würzburg-Veitshöchheim

Survey

In the Federal Republic of Germany (Bundes-republik Deutschland) research in general is decentralized. Nevertheless, research work is largely coordinated in different ways: cooperation is achieved through the individual sections of the German Society for Horticultural Sciences, through different working groups, and by the individual institutes being specialized in different fields. An annually published documentation of the current research activities concerned with agricultural and horticultural sciences, human nutrition, forestry, as well as veterinary medicine helps to avoid duplications. This catalogue is revised by the Zentralstelle fur Agrardokumentation und -information (Central Agency of Agricultural Documentation and Information) at Bonn-Bad Godesberg, which is financed by the Federal Ministry of Food, Agriculture, and Forestry.

The Federal Research Centres. Federal Research Centres such as the ones for horticultural plant breeding, for agricultural sciences, for human nutrition, and for plant protection problems (Biologische Bundesanstalt - Federal Biological Institute) are maintained by funds provided by the Federal Ministry of Food, Agriculture, and Forestry. These Research Centres mainly study actual problems for the benefit of agriculture and horticulture. The Max-Planck Society, independent of any ministry, is engaged in basic research and in different projects on horticultural plant breeding. Both the Federal Research Centres and the Max Planck Institutes are central institutes, which are purely research stations without any teaching activities.

The Universities - Combined with their teaching activities the universities engage in research, too. The main part of horticultural research in the Federal Republic of Germany is covered there. The degree given

Prepared by Prof.Dr. W. Rothenburger, Deutsche Gartenbauwissenschaftliche Gesellschaft, Freising-Weihenstephan.

to their graduates is "Dipl.Ing.Agr." which is the prerequisite for graduation as "Dr.Agr." and "Dr.rer.hort.." although the universities draw their grants from the respective Ministry of Education and Culture of the Federal States, they are mainly concerned with research of a wider scope. This is particularly true of the University of Hannover (Faculties of Horticulture and Landscape Planning), of the Technical University of München (Faculty of Agriculture and Horticulture), and of two faculties of the Technical University of Berlin. Several institutions of the agricultural faculties at the Universities of Bonn, Stuttgart-Hohenheim, and Giessen provide facilities for horticultural research and contribute to the training of agricultural students.
The Technical Colleges (Fachhochschulen, FHS), too, are sponsored by the State Ministries of Education and Culture. The degree achieved there is that of a "Diplo.-Ing.(FH)". The Technical Colleges of Osnabruck, Geisenheim, Kassel and Weihenstephan also maintain research institutes. The other FH's with departments of Horticulture, Landscape architecture and maintainance are in Berlin, Essen, Nürtingen-Stuttgart.
The State Institutes are financed by Agricultural Boards or by the Ministry of Agriculture of the respective Federal State and are engaged in regional and advisory work. Some of them are purely experimental and research stations connected with advisory boards; horticultural schools are in some cases attached to them.
Private research centres conducted by industrial firms carry out valuable horticultural research in their laboratories, experimental and research stations.
Scientific Organisations - The German Society for Horticultural Science, Herrenhäuser str. 2, D-3000-Hannover 21, under the guidance of her president, Prof.Dr. W. Rothenbürger, represents the majority of the horticultural scientists in the Federal Republic of Germany. There are other scientific societies, too, in which horticultural scientists participate. All of these societies are united in the Dachverband Wissenschafl. Gesellschaften der Agrar-, Forst-, Ernährungs-, Veterinär- und Umweltforschung e.v. (Federation of scientific societies of agricultural, forestry, food, veterinary, and environmental research), at München.
Bibliography - Internationales Verzeichnis wissenschaftlicher Institutionen und Forschungsinstitute, 33. Ausg., Minerva, Walter de Gruyter, Berlin - New York 1972.
Vademecum Deutscher Lehr- und Forschungsstätten - Vdlf, 6. Aufl. 1973, Bertelsmann Universitätsverlag, Düsseldorf. Kürschners Deutscher Gelehrten-Kalender, 12. Auf., Walter de Gruyter, Berlin 1976.
Results of horticultural research and investigations are mainly published in scientific journals:
Gartenbauwissenschaft. Verlag Eugen Ulmer, Stuttgart. Angewandte Botanik. Verlag Paul Parey, Berlin und Hamburg. Zeitschrift für Pflanzenernährung und Bodenkunde. Verlag Chemie GmbH., Weinheim/Bergstr. Zeitschrift für Pflanzenkrankheiten und Pflanzenschutz. Verlag Eugen Ulmer, Stuttgart. Phytopatologische Zeitschrift. Verlag Paul Parey, Berlin und Hamburg. Zeitschrift für Pflanzenzüchtung. Verlag Paul Parey, Berlin und Hamburg. Garten und Landschaft. Georg Callwey Verlag, München. Landschaft und Stadt. Verlag Eugen Ulmer, Stuttgart. Landwirtschaftliche Forschung. Sauerländer's Verlag, Frankfurt.
Several technical journals - Der Erwerbsobstbau. Verlag Paul Parey, Berlin und Hamburg. Gemüse. BLV München und Verlag Eugen Ulmer, Stuttgart. Gb + Gw, Gärtnerbörse und Gartenwelt. Verlag Paul Parey, Berlin-Hamburg und Georgie, Aachen (with emphasis on ornamental plants). Deutsche Baumschule. Verlag Deutsche Gärtnerbörse, Aachen. Landtechnik. Verlag Beckmann, Lehrte/Hannover. Obstbau. Verlag Bundesausschuss Obst und Gemüse, Bonn-Bad Godesberg. Mitteilungen des Obstbauversuchsringes des Alten Landes. Jork Kr. Stade. Deutscher Gartenbau. Verlag Eugen Ulmer, Stuttgart.

1 Ahrensburg

1.1 Bundesforschungsanstalt für gartenbauliche Pflanzenzüchtung (Federal Research Center for Horticultural Plant Breeding)

Bornkampsweg, 2070 Ahrensburg
Dir.: Prof.Dr. R. Reimann-Philipp*

2 Bad Neuenahr-Ahrweiler

2.1 Landes- Lehr- und Versuchsanstalt für Weinbau, Gartenbau und Landwirtschaft (Teaching and Experimental Station for Viticulture, Horticulture and Agriculture)

Walporzheimerstr. 48, 5483 Bad Neuenahr-Ahrweiler
Dir.: Dr. G. Stumm

3 Bad Zwischenahn

3.1 Lehr- und Versuchsanstalt für Gartenbau der Land-

2903 Bad Zwischenahn-Rostrup

wirtschaftskammer Weser-Ems (Teaching and Experimental Station for Horticulture of the Chamber of Agriculture Weser-Ems)
elevation: 9 m
rainfall : 800 mm
sunshine : 1500 h

Phone: (04403) 7221- 71200
Dir.: Dr. H.H. Witt

Plant nutrition and water management — Dr. H.H. Witt
Plant protection and growth regulating methods — Dr. R. Härig
Culture of pot azaleas — G. Struppek
Culture and breeding of Rhododendron — W. Schmalscheidt

4 Bavendorf

4.1 Schumacherhof (Experimental Station of the University of Hohenheim-Stuttgart, see 28.1.3)

Schumacherhof - 7980 Ravensburg-Bavendorf

5 Berlin

5.1 Fachbereich Internationale Agrarentwicklung der Technischen Universität Berlin* (Department of International Agricultural Development, Technical University Berlin)
elevation: 42 m
rainfall : 596 mm
sunshine : 1672 h

5.1.1 Institut für Nutzpflanzenforschung (Institute of Crop Science)
Fachgebiet Gemüsebau (Division of Vegetables)

Königin-Luise-Strasse 22
1000 Berlin 33 (Dahlem)
Phone: (030) 314 712 61

CO_2 exchange of plants, CO_2 fertilization, ecophysiology in greenhouses — Prof.Dr. H.J. Daunicht*
Flower biology and fruit load capacity of tropical vegetables — M. Hermann
Growth rhythm of sweet plants — G. Awad*
Response of aroids to antitranspirants — R. Gawish

Fachgebiet Obstbau (Division of Pomology)

Albrecht-Thaer-Weg 3
1000 Berlin 33 (Dahlem)
Phone: (030) 314 711 56

Breeding research in woody perennials — Prof.Dr. A. Karnatz*
Root temperature effects on vegetative and reproductive growth in deciduous and tropical plants — D. Besold
Effects of growing conditions and pre-harvest treatJ.Döring/M. Soylu
ments on quality and post-harvest life of fruits — Prof.Dr. P. Lüdders*
Salt resistance of fruit trees
Effect of fertilization on growth and quality of fruit plants — F. Schembecker

5.2 Fachbereich Landschaftsentwicklung der Technischen Universität Berlin (Department of Landscape Development, Technical University Berlin)

5.2.1 Institut für Ökologie (Institute of Ecology)
Fachgebiet Zierpflanzenbau (Division of Floriculture)

Königin-Luise-Strasse 22
1000 Berlin 33 (Dahlem)

Effect of growth regulators and environment on flower formation and development — E. Preller

Effects of temperature on growth — H. Patzer

Fachgebiet Freilandpflanzenkunde (Division of Plant Use in Landscape Architectute)

Winterhardiness of trees and shrubs — Prof.Dr. W. Heinze

Plant use on rooftops — K. Ludwig/ P. Schönfeld

Dry hardiness of trees and shrubs

Institut für Landschafts- und Freiraumplanung (Institute of Landscape and Open Space Planning)

Franklinstr. 28/29, 1000 Berlin 10
Dir.: Prof. J. Wenzel
Prof. H. Loidl

5.3 Fachbereich Lebensmitteltechnologie und Biotechnologie (Department of Food Processing and Biotechnology)

5.3.1 Institut für Lebensmitteltechnologie. Fachgebiet Frucht- und Gemüsetechnologie (Institute for Food Processing, Division of Fruit and Vegatable Processing)

Königin-Luise-Strasse 22
1000 Berlin 33 (Dahlem)
Phone: (030) 314 712 50
Prof.Dr. H.-J. Bielig
Prof.Dr. D. List

5.4 Biologische Bundesanstalt für Land- und Forstwirtschaft, Institut für Pflanzenschutz im Zierpflanzenbau (Federal Biological Research Center for Agriculture and Forestry, Institute for Plant Protection in Flower Production and Ornamental Plant Industry)

Königin-Luise-Strasse 19
1000 Berlin 33 (Dahlem)
Phone: (030) 8304-1
Head: Prof.Dr. W. Sauthoff
Dr. U. Brielmaier/ Dr. V. Köllner

Parasitic and non-parasitic diseases of ornamental plants. Ecology and epidemiology of fungal and bacterial diseases, animal pests. Improvement of disease and pest control.

6 Bonn

6.1 Institut für Obstbau und Gemüsebau der Universität Bonn (Institut for Fruit and Vegetable Crops of the University of Bonn)

Auf dem Hügel 6, 5300 Bonn
Dir.: Prof.Dr. F. Lenz*
Prof.Dr. J. Henze/ Dr. H. Baumann
Dr. G. Naumann/ Dr. G. Noga
Dr. H. Werner

Improvement of yield and quality, in particular keeping quality of fruit and vegetables. Relationships between vegetative and reproductive growth. Post harvest physiology, cold storage, controlled atmosphere, storage, equipment. Fruit virus, methods to obtain virus free plant material. Soil management, nutrition, fruit quality.

6.1.1 Obst-Versuchsanlage Klein-Altendorf (Fruit Experiment Farm Klein-Altendorf)

5308 Rheinbach
Dr. G. Engel

Varieties, rootstocks, soil management, nutrition, pruning, harvesting methods for apples, pears and cherries.

6.1.2 Versuchswirtschaft für Obst- und Gemüse, Marhof (Experimental Farm for Fruit and Vegetables Marhof)

Sechtemer Strasse 29
5047 Wesseling
Dr. H.J. Darfeld/ Dr. Th. Wendt

Cultivation methods, irrigation, plant water relationships in vegetables. Soil and air temperature effects on vegetables.

7 Bonn-Bad Godesberg

7.1 Lehr- und Versuchsanstalt Friesdorf für Zierpflanzenbau, Baumschulen und Floristik der Landwirtschaftskammer Rheinland (Teaching and Experiment Station Friesdorf for Ornamental Plants, Arboriculture and Flower-art of the Chamber of Agriculture Rheinland)
elevation: 60 m
rainfall : 650 mm
sunshine : 1300 h

Langer Grabenweg 68, 5300 Bonn 2
Phone: (0228) 376802
Dir.: G. Bouillon

from 1.7.86 new address:
Gartenstrasse, 5000 Köln 71

Experimental Station — Dr. A. Papenhagen

8 Braunschweig

8.1 Biologische Bundesanstalt für Land- und Forstwirtschaft, Institut für Pflanzenschutz im Gartenbau (Federal Biological Research Center for Agriculture and Forestry, Institute for Plant Protection in Horticultural Crops)
elevation: 81 m
rainfall : 622 mm
sunshine : 1517 h

Messenweg 11-12, 3300 Braunschweig
Head: Dr. G. Crüger*

9 Dossenheim

9.1 Biologische Bundesanstalt für Land- und Forstwirtschaft, Institut für Pflanzenschutz im Obstbau (Federal Biological Research Center for Agriculture and Forestry, Institute for Plant Protection in Fruit Crops)

Schwabenheimerstr. 101
6901 Dossenheim bei Heidelberg
Phone: (06221) 85238
Head: Prof.Dr. A. Schmidle

The institute investigates diseases and pests of pome, stone and small fruit and develops methods to control them.

Virus diseases and MLO:
Pome and stone fruit — Dr. L. Kunze
Small fruit, histology — Dr. H. Krczal
Bacterial and fungus diseases:
Pome and stone fruit — Dr. A. Schmidle
Small fruit, histology — Dr. E. Seemüller
Pests — Dr. E. Dickler

10 Essen

10.1 Lehr- und Versuchsanstalt für Garten- und Landschaftsbau und Friedhofsgärtnerei der Landwirtschaftskammer Rheinland (Teaching and Experimental Station for Horticulture and Garden Art and Cemetery Horticulture of the Chamber of Agriculture Rheinland)

Kühlshammerweg 22, 4300 Essen
Dir.: Dr. J. Rehbogen

11 Freising-Weihenstephan (near München)

11.1 Fakültat für Landwirtschaft und Gartenbau der

8050 Freising-Weihenstephan

Technischen Universität München (Faculty of Agriculture and Horticulture of the Technical University of Munich)
elevation: 460 m
rainfall : 814 mm
sunshine : 1786 h

Phone: (08161) 71-1

11.1.1 Institut für Landwirtschaftlichen und Gärtenerischen Pflanzenbau (Institute for Agricultural and Horticultural Plant Cultivation)

Lehrstuhl für Obstbau (Chair of Fruit Growing)
Breeding of stone fruit rootstocks, physiology and biochemistry of incompatibility, propagation of rootstocks, grafting techniques, tissues culture.

Phone: (08161) 71-234
Dir.: Prof.Dr. W. Feucht*
Dr. P. Schmid/ Dr. Gebhardt
H. Schimmelpfeng/ E. Christ

Lehrstuhl für Gemüsebau (Chair of Vegetable Growing)
Field production and protected cultivation of vegetables: quality and nutritional value, physiology, agrotechnique, storage (c.a.) and post-harvest physiology, fertilization and soil fertility, spice plants and medicinal plants.

Phone: (08161) 71-427
Dir.: Prof.Dr. P.D. Fritz*
Prof.Dr. F. Venter*,
Dr. Ch. Franz*,
Dr. J. Weichmann*

Lehrstuhl für Zierpflanzenbau (Chair of Floriculture Crops)
Production of pot plants, cut flowers and bedding plants. Soilless culture, plant tissue culture (propagation and mutagenesis), selection of low temperature tolerant types and related physiological research. Research on horticultural plant breeding. Statistical methods and experimental design in horticulture.

Phone: (08161) 71-416
Dir.: Prof.Dr. W. Horn*
Dr. P. Lange*,
Dr. G. Schlegel*,
G. Sackmann

11.1.2 Institut für Wirtschafts- und Sozialwissenschaften (Institute of Economic and Social Sciences)

Lehrstuhl für Wirtschaftslehre des Gartenbaues (Chair of Horticultural Economics)

Phone: (08161) 71-481
Dir.: Prof.Dr. W. Rothenburger*

Section Production Economics and Council Methods

Prof.Dr. W. Rothenburger*
B. Felbinger

Accountancy

Dr. Ch. Herrmann

Section Economic of Landscape Management

Ch. Goppel

Section Market Policy and Marketing

Prof. Dr. H.D. Ostendorf*

11.1.3 Institut für Bodenkunde, Pflanzenernährung und Phytopathologie (Institute of Soil Science, Plant Nutrition and Phytopathology)

Lehrstuhl für Bodenkunde (Chair of Soil Science)

Phone: (08161) 71-476
Dir.: Prof.Dr. U. Schwertmann

Lehrstuhl für Pflanzenernährung (Chair of Plant Nutrition)
Horticultural plant nutrition: nitrogen fertilizer, nitrification inhibitors, oganic fertilizers, influence on nitrate leaching, quality and concentration in plants. Trace elements, fertilization, physiological effects. Temperature nutrition, relationships. Bark potting media.

Phone: (08161) 71-390
Dir.: Prof.Dr. A. Amberger
Dr. Gutser,
Dr. A. Wünsch,
Dr. H. Goldbach

Lehrstuhl für Phytopathologie (Chair of Phytopathology)

Phone: (08161) 71-681
Dir.: Prof.Dr. G.M. Hoffmann*
Dr. V. Zinkernagel*

11.1.4 Lehrstuhl für Allgemeine Chemie und Biochemie (Chair of Chemistry and Biochemistry)

Phone: (08161) 71-253
Dir.: Prof.Dr. H.L. Schmidt

11.1.5 Institut für Landespflege und Botanik (Institute for Landscaping and Botany)

Lehrstuhl für Botanik (Chair of Botany)
Phone: (08161) 71-395
Dir.: Prof.Dr. B. Hock
Prof.D. W. Huber

Lehrstuhl für Landschaftsarchitektur (Chair of Landscape Architecture)
Phone: (08161) 71-248
Dir.: Prof. P. Latz
Prof. Ch. Valentin

Lehrstuhl für Landschaftsökologie (Chair of Landscape Ecology)
Phone: (08161) 71-495
Dir.: Prof.Dr. W. Haber
Prof.Dr. J. Pfadenhauer

Institut für Landtechnik (Institute for Agricultural Engineering)
Phone: (08161) 71-440
Dir.: Prof.Dr. L. Wenner

11.2 Fachhochschule Weihenstephan and associated Experimental Establishment of Horticulture
8050 Freising-Weihenstephan
Phone: (08161) 71-360
Prasident: Prof.Dr. K. Seidel

11.2.1 Institut für Bodenkunde und Pflanzenernährung (Institute of Soil Science and Plant Nutrition)
Plant nutrition and manuring of pot plants, cut flowers, vegetables, fruit, woody plants, media for ornamental horticulture, hydroponics.
Phone: (08161) 71-658
Head: Prof.Dr. P. Fischer
Dr. A. Schumm*,
L. Forchthammer

11.2.2 Institut für Botanik und Pflanzenschutz (Institute of Botany and Plant Protection)
Problems of plant protection, chemical weed control, virus, fertilization, ecology in fruit growing
Phone: (08161) 71-360
Head: Prof.Dr. J. Völk
E. Lohweg

11.2.3 Institut für Garten- und Landschaftsgestaltung (Institute of Garden and Landscape Design)
Testing of ornamental plantations under different conditions
Phone: (08161) 71-351
Head: Prof.Dr. G. Richter

11.2.4 Institut für Gärtnerische Betriebslehre (Institute for Horticultural Management)
Evaluation of enterprises, investigation into the economics of enterprise, marketing problems, mapping of horticultural production in European countries.
Phone: (08161) 71-350
Head: Prof.Dr. E. Schürmer
Prof. G. Kirchgatter

11.2.5 Institut für Gemüsebau (Institute of Vegetable Crops)
General problems of plant cultivation in the open and under glass
Phone: (08161) 71-366
Head: Prof.Dr. F.-W. Frenz*

11.2.6 Institut für Obstbau und Baumschulwesen (Institute of Fruit Growing and Arboriculture)
Planting systems for pome, stone and berry fruit, rootstock and variety testing
Phone: (08161) 71-355
Head: Prof.Dr. H. Kettner

11.2.7 Institut für Obst- und Gemüseverarbeitung (Institute of Fruit and Vegetable Processing
Fruit juices and ciders, fruit wines, liquid and frozen preserves, jams
Phone: (08161) 71-358
Head: Prof.Dr. K. Schmidt
Dr. A. Kraus

11.2.8 Institut für Stauden, Gehölze und Angewandte Pflanzensoziologie (Institute for Herbaceous Perennials and Woody Plants and Applied Plant Sociology)
Survey and testing of varieties of herbaceous and woody plants and turfs for garden and landscape plant communities and landscape protection
Phone: (08161) 71-371
Head: Prof.Dr. F. Kiermeier
Prof.Dr. J. Sierber/H. Müssel

11.2.9 Institut für Zierpflanzen unter Glas (Institute of Ornamental Plant Growing under Glass)
Cultivation techniques, decisions on cultural and varie-
Phone: (08161) 71-363
Head: Prof.Dr. R. Röber*

tal details, testing of provenance of pot plants and cut flowers

11.3 Bayerische Hauptversuchsanstalt für Landwirtschaft (Bavarian Central Experimental Station for Agriculture
Analysis of feed stuffs, soils, agricultural products and pesticide residues

8050 Freising-Weihenstephan
Phone: (08161) 71-381
Prof.Dr. K. Ranfft

11.4 Agrarmeteorologische Forschungsstelle des Deutschen Wetterdienstes (Agrarmeteorologic Research Institute of the German Weather Bureau)

8050 Freising-Weihenstephan
Phone: (08161) 71-246
Dir.: Dr. H. Häckel

12 Geisenheim

12.1 Forschungsanstalt für Weinbau, Gartenbau, Getränketechnologie und Landespflege (Research Station of Viticulture, Horticulture, Beverage Technology and Landscape Development
elevation: 109 m
rainfall : 535 mm
sunshine : 1648 h

Von-Lade-Strasse 1, Postfach 1154
6222 Geisenheim
Phone: (06722) 502-1
Dir.: Prof.Dr. H.H. Dittrich

12.1.1 Institut für Weinbau (Institute of Viticulture)

Phone: (06722) 502-250
Head: Prof.Dr. W. Kiefer
Dr. B. Steinberg/Dr. W. Bettner

Relations between quantity and quality. Effect of spacing and pruning. Influence of soil water content, of organic fertilizer, of cultivation methods on growing and fruiting. Ecological management

12.1.2 Institut für Rebenzüchtung und Rebenveredlung (Institute of Grape Vine Breeding and Grafting)
Improvement by clone selection and cross breeding of new varieties of grapevines and rootstocks. Research work and development of grafting and propagation of vines

Phone: (6722) 502-251
Head: Prof.Dr. H. Becker
Dr. W. Schenk/Dr. W. Ries

12.1.3 Institut für Kellerwirtschaft (Institute of Cellar Techniques)
Muration, stabilisation, bottling and storage of wines; examination of containers, materials, implements, bottle-closures and winetreating-agents; actual processing techniques

Phone: (06722) 502-253
Head: Prof. H. Haubs
Dr. M. Perscheid/Dr. F. Zürn

12.1.4 Institut für Weinchemie und Getränkeforschung (Institute of Wine Chemistry and Beverage Research)
Technology, stabilisation and pasteurisation of beverages, chemical-physical analysis and development of analytical methods

Phone: (06722) 502-254
Head: Prof.Dr. K. Wucherpfennig

Dr. H. Dietrich/ Dr. H. Kern,
Dr. K. Otto/ K.D. Millies

12.1.5 Institut für Mikrobiologie und Biochemie (Institute of Microbiology and Biochemistry)
Microbiology of wine and beverages, physiology of fermentation

Phone: (06722) 502-255
Head: Prof.Dr. H.H. Dittrich
Dr. W. Sponholz/ Dr. K. Wenzel

12.1.6 Institut für Gemüsebau (Institute of Vegetable Crops)
Physiological problems on protected vegetables (water, temperature) and outdoor vegetables, seed production, asparagus

Phone: (06722) 502-258
Head: Prof.Dr. H.D. Hartmann*
Dr. M. Kretschmer*

12.1.7 Institut für Obstbau (Institute of Fruit Growing)

Phone: (06722) 502-256
Head: Prof.Dr. H. Jacob

Rootstocks for fruit, variety testing, pruning of fruit trees, diseases, parasites of fruit crops	Dr. M. Bauckmann
12.1.8 Institut für Zierpflanzenbau (Institute of Ornamental Plant Growing	Phone: (06722) 502-260 Head: Prof.Dr. W.-U. v. Hentig*
Vegetative propagation and cultivation problems under glass (day length, temperature, nutrition, growth regulation). Introduction of new ornamental plants. Perennials and woody ornamentals	Dr. H.D. Molitor, M. Fischer, I. Hass, Dr. J. Rohde
12.1.9 Institut für Landschaftsbau (Institute of Landscape Architecture)	Phone: (06722) 502-262 Head: ....
Landscape plants (native and exotic)	E. Stähr
12.1.10 Institut für Botanik (Institute of Botany)	Phone: (06722) 502-264 Head: Prof.Dr. G. Reuther
Ecophysiology and growth regulation in horticulture and viticulture, germination physiology, tissue culture, Iris-breeding, histology	Dr. T. Geier, Dr. H. Wienhaus
12.1.11 Institut für Bodenkunde und Pflanzenernährung (Institute of Soil Science and Plant Nutrition)	Phone: (06722) 502-263 Head: Prof.Dr. K. Schaller
Bark as a substrate in horticulture. Water and nutrient requirements. Nitrate in soils, vegetables and wine	Dr. U. Brückner, Dr. E. Leidenfrost
12.1.12 Institut für Phytomedizin und Pflanzenschutz (Institute of Phytomedicine and Plant Protection)	Phone: (06722) 502-266 Head: Prof.Dr. H. Holst
Plant pathology, bacteria and fungus diseases in viticulture and fruit-growing; phenology, pheno-pathology, viruses, pests and integrated plant protection in viticulture and fruit-growing	Dr. G. Brendel/Dr. E. Hofman, Dr. W. Wohanka
12.1.13 Institut für Betriebswirtschaft und Marktforschung (Institute of Management, Economics and Market Research)	Phone: (06722) 502-268 Head: Prof.Dr. H. Kalinke*
Economic problems of production of viticulture, marketing; economic problems of production of horticulture	Dr.H.-H. Bock/Dr.D. Hoffmann, Dr. G. Timm
12.1.14 Institut für Technik (Institute of Engineering)	Phone: (06722) 502-267 Head: Prof.Dr. W. Rühling
Viticultural machinery and equipment	Dr. G. Bäcker
Horticultural machinery and equipment; greenhouse design, climatisation, watering	Dr. K. Mackroth*
12.2 Agrarmeteorologische Beratungs- und Forschungsstelle des Deutschen Wetterdienstes (Information Center and Research Station for Agricultural Meteorology of the German Weather Service)	Kreuzweg 25, 6222 Geisenheim Phone: (06722) 8372 D. Hoppmann
12.3 Bundesanstalt für Ernährung, Aussenstelle Geisenheim (Federal Research Center for Nutrition, Field Station)	Rüdesheimer Str. 12-14 6222 Geisenheim Phone: (06722) 8001 Dir.: ...

13 Giessen

13.1 Institut für Pflanzenbau und Pflanzenzüchtung II* - Obstbau und Obstzüchtung der Universität Giessen (Institute for Plant Culture and Plant Breeding - Fruit Growing and Fruit Breeding of the University of Giessen)	Ludwigstrasse 27, 6300 Giessen Phone: (0641) 7026010 Dir.: Prof. D.W. Gruppe
Breeding of rootstocks (stone fruit, apomictic apples)	

and small fruits (Ribes), rootstock-scion relations, growth physiology

13.2 Institut für Pflanzenbau und Pflanzenzüchtung I der Universität Giessen, Abt. für Arznei- und Gewürzpflanzen (Institute of Plant Production and Breeding, Dept. Medicinal and Aromatic Plants). Versuchsstation Rauischholzhausen (Experimental Station Rauischholzhausen)

D-3557 Ebsdorfergrund 4
Teleph.: (06424) 301352
Dir.: Prof.Dr. A. Vömel

Production and breeding of medicinal plants

14 Hamburg

14.1 Hamburgische Gartenbauversuchsanstalt Fünfhausen (Hamburg's Experimental Horticultural Station Fünfhausen)

Ochsenwerder Landscheideweg 277
2050 Hamburg 80
Phone: (040) 7372310
Dir.: U. Schmoldt/ T. Miske

15 Hannover

15.1 Fachbereich Gartenbau* der Universität Hannover (Department of Horticulture of the University of Hannover)

Herrenhäuser Str. 2
3000 Hannover 21
Phone: 0511 7621

15.1.1 Institut für Angewandte Genetik (Institute of Applied Genetics)
Techniques in breeding hybrid varieties in vegetables and ornamentals; Analysis of mitochondrial DNA; Incompability; Genetic analysis of isozymes; Development and application of in vitro-techniques in breeding research; Breeding for disease resistance; Selection theory on quantitative traits (Vegetables, ornamentals, crop plants)

Head: Prof.Dr. G. Wricke
Phone: (0511) 762-2670
Prof.Dr. J. Grunewaldt,
Dr. W.E. Weber,
Dr. T. Tatlioglu

15.1.2 Institut für Gartenbauökonomie (Institute of Horticultural Economics)

Phone: (0511) 762-4185
Head: Prof.Dr. R. von Alvensleben

Production economics
Production planning
Farm management, Accounting

Prof.Dr. H. Storck*,
Prof.Dr. E.-W. Schenk

Labour management:
Market economics:
Market research, Market Policy
Marketing in developing countries

Prof.Dr. G. Stoffert*
Prof.Dr. R. von Alvensleben,
Dr. D.M Hörmann

15.1.3 Institut für Gemüsebau (Institute of Vegetable Crops)

Phone: (0511) 762-2634

Parameters of actions interactions of climatic growth factors, photoperiodism and vernalization, growth models, improvement of cultivation procedures

Prof.Dr. H. Krug*/ Dr. E. Fölster,
Dr. H.-P. Liebig*/ R. Habegger,
E. Lederle/ W. Mann/ B. Pfeufer,
H. Rippen/ T. Schaer

15.1.4 Institut für Obstbau und Baumschule (Institute of Fruit Science and Arboriculture)

Haus Steinberg, 3203 Sarstedt
Phone: (05066) 82-6122
Head: Prof.Dr. W.-D. Naumann

Cultural systems and rootstocks, growth and physiology of pome and stone fruits, post-harvest physiology, nursery management. Experimental fields (20 ha) at

Prof.Dr. G. Bünemann*,
Dr. D. MacCarthaigh/ Dr. K. Roemer,
Dr. V. Behrens/ D. Bläsing,

Ruthe near Sarstedt
H. Freund

15.1.5 Institut für Pflanzenernährung (Institute of Plant Nutrition)
Phone: (0511) 762-2626
Head: Prof.Dr. J. Wehrmann*

Evaluation of the nutritional status of plants. Determination of N, P and K fertilizer requirements. NO_3 content of vegetable as affected by NO_3, NH_4 and Cl nutrition. Nitrogen uptake and transport in soil. Relationship between plant diseases and nutritional status of plants

Prof.Dr.M. Schenk*/ G. Baumgärtel, H. Coldewey-zum Eschenhoff, A. Dobritz/ B. Heins/ H. Kuhlmann

15.1.6 Institut für Pflanzenkrankheiten und Pflanzenschutz (Institute of Plant Diseases and Plant Protection)
Phone: (0511) 762-2641
Head: Prof.Dr. F. Schönbeck

Fungal diseases, plant protection, virus diseases

Prof.Dr. H. Buchenauer,

15.1.7 Institut für Technik in Gartenbau und Landwirtschaft (Institute of Horticultural and Agricultural Engineering)
Phone: (0511) 762-2646
Head: Prof.Dr. H.-J. Tantau

Glasshouse construction, installation and machinery, heat requirement and energy saving, glasshouse climate control, investigations on growing rooms and artificial light, light transmission of glasshouses and utilization of solar energy for greenhouse heating, heat recovery, utilization of waste heat

Prof.Dr. C. von Zabeltitz, Dr. B. von Elsner/ Dr. J. Meyer, A. Baytorun/ G. Müller/ M. Parlitz C. Reuter/ M. Rüther, A. Steinrücken/ G. Weimann,

15.1.8 Institut für Zierpflanzenbau (Institute of Ornamental Crops)
Phone: (0511) 762-2661
Head: Prof.Dr. K. Zimmer*

Physiology of germination of ornamental plants. Growth and development (e.g. vernalization, dormancy, growth, flowering). Tissue culture methods. Testing and classification of cultivars of annuals. International Registration Authority for Petunia, Callistephus, Tagetes

Dr. E. Bachthaler/ J. Hölters, H. Töpperwein/ A. Zens

15.1.9 Lehrgebiet Berufsdidaktik des Gartenbaus (Didaktics in Horticulture)
Appelstrasse 23, 3000 Hannover 1
Phone: (0511) 762-3345
Head: Prof.Dr. C. Jürgensen*

Development of curricula for horticulture vocational schools; analysis of professional occupations; scientific experimentation on horticulture education

15.1.10 Dokumentationsstelle Gartenbau (Documentation Center for Horticulture)
Nienburger Strasse 7
3000 Hannover 1
Phone: (0511) 762-2686
Head: Dr. W. Hildebrandt*

Documentation of literature on horticulture. Contributions to AGRIS, ELFIS

15.2 Fachbereich Landespflege der Universität Hannover (Department of Landscape Architecture of the University of Hannover)

15.2.1 Institut für Freiraumentwicklung und Planungsbezogene Soziologie (Institute for Open Space Development and Planning Related Sociology)
Appelstrasse 23, 3000 Hannover 1
Phone: (0511) 762-5528
Head: Prof.Dr. U. Herlyn

Residential quarters in the 1920ies and 1960ies; history of open space planning; housing and open space evaluation; open space use in highrises; allotment gardening; static camping/caravanning; open space in urban renewal areas; open space administration

Prof.W.Y. Wolff, Ph.D.

15.2.2 Institut für Grünplanung und Gartenarchitektur (Institute of Architecture and Landscape Design)
Herrenhäuser Strasse 2
3000 Hannover 21
Phone: (0511) 762-2694
Head: Prof.Dr. D. Hennebo
Prof.Dr. H.J. Liesecke,
Prof. G. Nagel

15.2.3 Institut für Landesplanung und Raumforschung (Institute of Land Planning and Region Research)

Phone: (0511) 762-2660
Head: Prof.Dr. H.-G. Barth
Prof.Dr. D. Fürst

15.2.4 Institut für Landschaftspflege und Naturschutz (Institute for Landscape Management and Nature Protection)
Landscape-planning. -ecology, -design, nature protection

Phone: (0511) 762-2651
Head: Prof.Dr. H. Kiemstedt

Prof.Dr. H.-R. Höster,
Prof.Dr. H. Langer,
Prof.Dr. U. Schlüter,
Prof.Dr. H.-H. Wöbse,

15.3 Fachbreich Biologie der Universität Hannover (Department of Biology of the University of Hannover)

15.3.1 Institut für Biophysik (Institute of Biophysucs)

Herrenhäuser Strasse 2
3000 Hannover 21
Phone: (0511) 762-2603
Head: Prof.Dr. E.-G. Niemann

Ecological biophysics, chlorophyll fluorescence, indicator activation analysis, quantitative microscopy, pattern kinetics of biological molecules, radioecology

Prof.Dr. H. Glubrecht,
Prof.Dr. D. Ernst/ Dr. A. Anders,
Dr. E. Burger/ Dr. D. Christoffers,
Dr. L. Kiselnic

15.3.2 Isotopenlaboratorium Gemeinsame wissenschaftliche Einrichtung der Fachbereichs Biologie und Gartenbau (Isotope Laboratory, Common Scientific Unit of the Departments of Biology and of Horticulture)
Application of ionizing radiation and tracer techniques in biological and horticultural research, indicator activation analysis, radioecology, biological nitrogen fixation of non-legumes

Phone: (0511) 762-2604
Head: Prof.Dr. E.-G. Niemann
Prof.Dr. J. Grunewaldt,
Prof.Dr. M. Klenert,
Prof.Dr. M. Schenk*/ Dr. I. Fendrik,
Dr. M. Knälmann/ Dr. H.W. Seibold

15.3.3 Institut für Botanik (Institute of Botany)

Herrenhäuser Strasse 2
3000 Hannover 21
Phone: (0511) 762-2631
Head: Prof.Dr. G. Richter
Prof.Dr. F. Herzfeld,
Prof.Dr. K. Kloppstech/ Dr. H. Jansen,
Dr. R. Niemeyer

15.3.4 Institut für Mikrobiologie (Institute of Microbiology)

Schneiderberg 50, 3000 Hannover 1
Phone: (0511) 762-4359
Head: Prof.Dr. G. Auling
Prof.Dr. H. Diekmann,
Dr. H.J. Plattner

Structure elucidation and biosynthesis of low-molecular weight metabolites from microorganisms, special problems of biotechnology, microorganisms from sewage plants, transport and mechanism of action of iron ionophores from microorganisms, DNA precursor biosynthesis in bacteria, transport of manganese in microorganisms, taxonomy of Pseudomonas, thermophilic anaerobic bacteria

15.3.5 Institut für Geobotanik (Section of Plant Ecology)

Nienburger Strasse 17
3000 Hannover 1
Phone: (0511) 762-3632
Head: ...

Vegetation in different regions of FRG, ecological research in woods, ecology and plant production in lakes, pollen analysis, syntaxonomy of weed and nitrophile brink associations in Central Europe

Dr. H. Leippert/ Dr. S. Leippert,
Dr. H. Möller

15.3.6 Fachgebiet Zoologie-Entomologie (Section of Zoology-Entomology)

Herrenhäuser Strasse 2
3000 Hannover 21

Phone: (0511) 762-5549
Head: Prof.Dr. G.H. Schmidt
Dr. A. Melbert

15.4 Institut für Bodenkunde der Universität Hannover (Institute for Soil Science of the University Hannover)

Phone: (0511) 762-2622
Head: Prof.Dr. J. Richter
Prof.Dr. K.-H. Hartge,
Prof.Dr. H. von Reichenbach,
Prof.em.Dr. P. Schachtschabel

15.5 Institut für Meteorologie und Klimatologie der Universität Hannover (Institute of Meteorology and Klimatology of the University of Hannover)

Phone: (0511) 762-2677
Head: Prof.Dr. R. Roth

vaporation, boundary layer meteorology, agrometeorology, topoclimatology — Prof.Dr. D. Etling/ Dr. F. Wilmers, Dr. M. Elmdust/ Dr. G. Tetzlaff, A. Frenzel

15.6 Lehr- und Versuchsanstalt für Gartenbau der Landwirtschaftskammer Hannover (Teaching and Experimental Station for Horticulture of the Chamber of Agriculture Hannover)

Harenberger Strasse 130
3000 Hannover 91 - Ahlem
Phone: (0511) 762-40050
Head: Dr. H.C. Scharpf

Influence of climatic growth factors on flower plants; nitrogen nutrition of crops; nutrition and watering of pot plants; culture substrates — Dr. L. Hendriks

16 Heidelberg

16.1 Staatliche Lehr- und Versuchsanstalt für Gartenbau (State Teaching and Experiment Station of Horticulture)
elevation : 111 m
rainfall : 760 mm
sunshine : 1665 h

Diebsweg 2, 6900 Heidelberg
Phone: (06221) 73011
Dir.: Dr. H. Bahnmüller

Production methods, soils and plant nutrition, greenhouse technics, quality improvement, testing of species and varieties in glasshouses and open ground

Vegetables — Dr. E. Wilhelm
Ornamental crops — H. Loeser
Garden and landscape — J. von Malek
Soil and plant nutrition — E. Schwemmer

17 Jork

17.1 Obstbauversuchsanstalt Jork der Landwirtschaftskammer Hannover (Experimental Fruit Growing Station Jork of the Chamber of Agriculture Hannover)

Westerminnerweg 22, 2155 Jork
Phone: (04162) 7004
Head: Dr. K.-H. Tiemann*

Influence of soil and fertilizing on growth and yield of fruit trees — Dr. P. Quast
Testing of pesticides — G. Palm, H. Hauschildt
Use of growth-regulators — Dr. H. Graf
Technic in fruit-production — W. Bockstedte
Fruit-storage — H.-G. Blank

Selection of varieties and rootstocks of apples, pears, cherries, plums, strawberries and raspberries	F.-G. Zahn, W. Blank

18 Karlsruhe

18.1 Bundesforschungsanstalt für Ernährung (Federal Research Center for Nutrition)	Engesserstrasse 20 7500 Karlsruhe 1 Phone: (0721) 60114 Dir.: Prof.Dr. J.F. Diehl
18.1.1 Institut für Biologie (Institute of Biology)	Dir.: Prof.Dr. H.K. Frank
Laboratorium für Obst und Gemüse (Laboratory of Fruit and Vegetables)	Head: H. Hansen
Laboratorium für Pflanzenphysiologie (Laboratory of Plant Physiology)	Head: Dr. H. Bohling*
18.1.2 Aussenstelle Geisenheim (Field Station at Geisenheim) see 12.3	Rüdesheimer Strasse 12-14 6222 Geisenheim

19 Kassel

19.1 Lehr- und Versuchsanstalt für Gartenbau (Teaching and Experimental Station for Horticulture) elevation: 160 m rainfall : 647 mm sunshine : 1596 h	Oberzwehrener Strasse 103 3500 Kassel Phone: (0561) 402034 Head: W. Papke
Floriculture and pot plants	Dr. W. Hurka
Vegetable growing	R. Gerlach
Tree nursery and fruit growing	W. Meiss
Garden-and landscape construction	R. Nowitzki
Virus testing	W. Papke

20 Köln

20.1 Versuchsanstalt für Obst- und Gemüsebau der Landwirtschaftskammer Rheinland (Experimental Station for fruit and Vegetable Growing, Chamber of Agriculture Rheinland)	Gartenstrasse, 5000 Köln 71 - Auweiler Phone: (0221) 5901041 Head: Dr. G. Hohmann Dr. H. Rüger

21 Krefeld

21.1 Versuchanstalt für Pilzenbau der Landwirtschafts-kammer Rheinland (Experimental Station for Mush-room Growing, Chamber of Agriculture Rheinland)	Hüttenallee 235, 4150 Krefeld Phone: (02151) 58005 Head: Dr. J. Lelley

22 Neustadt

22.1 Landes- Lehr- und Forschungsanstalt für Landwirt-schaft, Weinbau und Gartenbau (Teaching and Research Center for Viticulture and Horticulture)	Breitenweg 71, 6730 Neustadt, Weinstrasse 16 Phone: (06321) 671-1 Dir.: Dr. K. Adams
Viticulture	Dr. W. Fader
Horticulture	Dr. K.-G. Schwarz/ Dr. H.-P. Lorenz
Agriculture	Dr. W. Wagner
Botany	Dr. N. Beran

Chemistry on vine	Dr. L. Jakob
Phytomedicine	Dr. K.-W. Eichhorn

23 Oppenheim

23.1 Landes- Lehr- und Versuchsanstalt für Landwirtschaft, Weinbau und Gartenbau (Teaching and Experimental Station for Agriculture, Viticulture and Horticulture)	Zuckerberg 19, 6504 Oppenheim Phone: (06133) 2098 Dir.: Dr. H. Finger
Section Viticulture	Head: Dr. H. Lott Ch. Krienke/ H. Schlamp, K.H. Laubenheimer/ F. Pfaff, H. Friess
Section Horticulture	Head: Dr. K. Hein E. Boy/ J. Thal/ S. Nitschke, H.R. Rohlfing
School, training facilities	Head: H. Schmidt

24 Osnabrück

24.1 Fachhochschule Osnabrück*

Fachbereich Gartenbau (Department of Horticulture)	Oldenburger Landstrasse 24 4500 Osnabrück Phone: (0541) 608-5111
School course including ornamental horticulture, fruit and vegetable growing, arboriculture, seed growing. Research in tree and berry fruit growing, in vegetable and ornamental plants.	
Fachbereich Landespflege (Department of Landscape Design) School course covering landscape architecture, landscape design, and landscape construction	Am Krümpel 33, 4500 Osnabrück Phone: (0541) 608-5160

25 Sarstedt

25.1 Institut für Obstbau und Baumschule der Universität Hannover (Institute of Fruit Science and Arboriculture of the University of Hannover) see 15.1.4	Haus Steinberg, 3203 Sarstedt

26 Schlachters

26.1 Lehr- und Versuchswirtschaft für Obst- und Gartenbau (Teaching and Experimental Plant for Fruit Growing and Horticulture)	Schlachters 65, 8995 Sigmarszell

27 Straelen

27.1 Lehr- und Versuchsanstalt fur Gemüse- und Gartenbau der Landwirtschaftskammer Rheinland (Teaching and Experiment Station for Vegetables and Horticulture of the Chamber of Agriculture Rheinland)	Hans-Tenhaeff-Strasse 40/42 4172 Straelen Phone: (02839) 6091 Dir.: H. Müller

28 Stuttgart

28.1 Universität Hohenheim (University of Hohenheim)

28.1.1 Institut für Obst-, Gemüse- und Weinbau	Schloss Westhof, Postfach 700562

(Institute for Fruit, Vegetable and Grape Growing)

7000 Stuttgart 70
Phone: (0711) 4501-2350
Dir.: Prof.Dr. R. Stösser*
Prof.Dr. G. Buchloh,
Prof.Dr. F. Bangerth*,
Dr.W. Hartmann*/ Dr.J Neubeller,
Dr. S.F. Anvari/ Dr. G. Buflerof

Pomology, selection of fruiting plants; physiology and nutrition of fruiting plants; fruit physiology; anatomy and morphology of fruiting plants and fruits; yield capacity and evaluation of fruit varieties; physiology vegetable plants

28.1.2 Institut für Agrartechnik, Abteilung Technik im Gartenbau (Institute for Engineering, Section of Horticultural Engineering)

Garbenstrasse 9, 7000 Stuttgart 70
Phone: (0711) 4501-2530
Head: Prof.Dr. E. Moser

28.1.3 Schumacherhof (Experimental Station of the University of Hohenheim)

Schumacherhof, 7980 Ravensburg - Bavendorf
Phone: (0751) 91291
Head: Prof.Dr. F. Winter*
Dr. H. Janssen*/ Dr. W. Kennel,
Dr. H. Link*/ Dr. J. Streif,
Dr. R. Silbereisen

29 Trier

29.1 Landes- Lehr- und Versuchsanstalt für Landwirtschaft, Weinbau und Gartenbau (Teaching and Experiment Station for Agriculture, Viticulture and Horticulture)

Egbertstrasse 18-19, 5500 Trier
Phone: (0651) 49061
Head: Dr. K. Faas

30 Weinsberg

30.1 Staatliche Lehr- und Versuchsanstalt für Wein- und Obstbau (State Teaching and Experimental Station for Viticulture and Fruit Growing)

Hallerstrasse 6, 7102 Weinsberg
Phone: (07134) 6121
Dir.: Dr. G. Gotz

31 Witzenhausen

31.1 Gesamthochschule Kassel, Fachbereich Internationale Agrarwirtschaft, Abteilung für Tropische und Subtropische Landwirtschaft (University of Kassel, Department of International Agriculture, Section of Tropical and Subtropical Plants)

Steinstrasse 19, 3430 Witzenhausen
Phone: (05542) 503-0
Prof.Dr. M. Rommel
Dr. C. Hoeppe

Methods of cultivation and plant protection under glasshouse conditions. Generative and vegetative propagation of crop plants and fruit trees. Greenhouse with collection of subtropical and tropical cultivated plants.

32 Wolbeck

32.1 Lehr- und Versuchsanstalt für Gartenbau der Landwirtschaftskammer Westfalen-Lippe (Teaching and Experimental Station of Horticulture of the Chamber of Agriculture Westfalen-Lippe)

Münsterstrasse 24
4400 Münster-Wolbeck
Phone: (02506) 1041
Dir.: Dr. H. Peper

Research station of vegetables, ornamental plants and flower production; special: climate controlling including CO_2-fertilization, hydroculture of pot plants, culture techniques of Phalaenopsis, variety testing.

33 Würzburg-Veitshöchheim

33.1 Bayerische Landesanstalt für Weinbau und Gartenbau

Residenzplatz 3, 8700 Würzburg

(Bavarian State Institute of Viticulture and Horticulture

Phone: (0931) 9002-1
Pres.: Dr. G. Scheuerpflug

Research station for viticulture, fruit and vegetable growing, pot plants and cut flowers production, native outdoor plants

W. Müller-Haslach/
Dr. P. Reimher,
Dr. W. Kolb

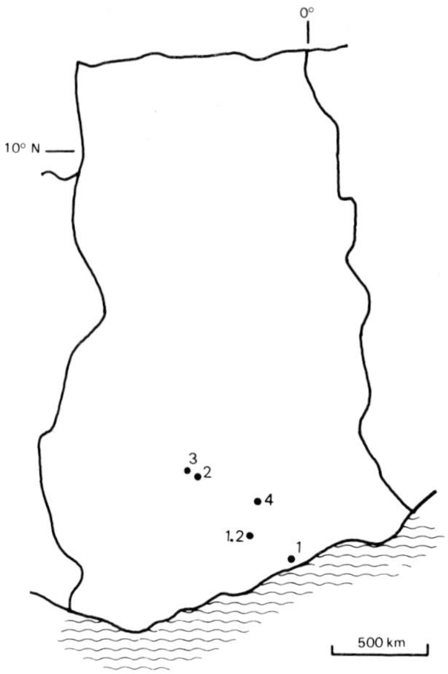
0°
10° N
3
2
4
1.2
1
500 km

1 Legon
1.2 Kade
3 Kumasi
4 Tafo

Survey

The institutions of Ghana which are concerned with horticultural research activities are:
- the Department of Crop Science and the Agricultural Research Station, Kade of the University of Ghana.
- the Department of Horticulture and the Department of Crop Production of the University of Science and Technology.
- the Horticulture, Plant Physiology, Plant pathology and Entomology Sections and the Oil Palm Research Centre, Kusi, all of the Crops Research Institute, Kwadaso, and
- the Cocoa Research Institute, Tafo.

Publications: Anual Reports published by the Research Stations

1 Legon

1.1 University of Ghana, Department of Crop Science

Legon - Accra
Head: Dr (Mrs) E. Blay Ph.D.

A new vegetable farm was opened in 1966 and in addition two growth rooms were constructed. The vegetable farm has helped to do intensive research with much more detailed and acurate results. Crops are grown in the growth rooms for teaching and research purposes. Research activities include flower induction in onions and shallots. Photoperiod and temperature effect on vegetables. Plant-growth regulators on vegetables, legumes and some cereals and field crops including cassava, yams, sweet potato and cocoyam.

Breeding and genetics, root crops, maize and grain legumes — Prof. E.V. Doku
Breeding and genetics, soybean, millet — R.B. Dadson, Ph.D.

Prepared by Dr. F. Agble, Crops Research Institute, Kumasi.

Breeding and genetics, solanaceous vegetables	E. Blay (Mrs) Ph.D.
Post-harvest diseases of fruit crops	Dr K.A. Oduro
Horticultural crops: vegetable crops, plant growth regulators floriculture	
Plant pathology including virology and nematology	P. Lamptey, Ph.D.
1.2 Agricultural Research Station	P.O. Box 43 - Kade Head: S.K. Karikari M.Sc., FGhIH
Nematode control in plantains. Screening winged beans Effect of seed size and harvesting periods on yield of cocoyams	S.K. Karikari, M.Sc.
Rootstock trials on citrus. Selection for fruit quality in avocado pears	A.K. Abaka-Gyenin, M.Sc.

2 Kumasi

2.1 University of Science and Technology, Department of Horticulture	Kumasi Head: O.K. Atubra, B.Sc., M.Sc.

The Department offers a 2-year Diploma Course in Tropical Horticulture. It also teaches horticulture to students reading for a 4-year BSc. (Agric) Hons. Degree Course of which horticulture is one of the fields of specialization in the fourth year. Facilities exist for post-graduate studies leading to MSc. and PhD degrees. Facilities for research include a rapidly developing horticultural science laboratory, a 20-hectare orchard, a 4-hectare budwood garden, a 2-hectare ornamental nursery, a 6-hectare site for vegetable crops, two plant houses and a lath house. The beautifully and well-landscaped University campus serves as a good ground for the training of landscape horticulturists and designers.

Post-harvest physiology (vegetables and fruits); Seed production and ornamental horticulture. Soil mixes	Dr (Mrs) N.S. Olympio, M.Sc. Ph.D.
Fruit crops, plant propagation, citrus virology, citrus indexing, avocado pear (*Persea americana*) screening for commercial production	O.K. Atubra, M.Sc.
Vegetable agronomy; vegetable selection for adaptation	P.Y. Boateng, B.Sc.
2.2 Department of Crop Production	
2.2.1 Crop Protection Section	Head: J.S. Kankam, Ph.D.
Foliage diseases of tomato	J.S. Kankam, Ph.D.
Breeding of egg-plants and soya beans	G.F. Nsowah, Ph.D.
Nematology of vegetable crops	B. Hemeng (Mrs.), M.Sc., MGhIH

3 Kwadaso

3.1 Crops Research Institute, Horticulture Section	Head: W.S.Y. Abutiate, M.Sc., FGhIH
Fruit crops - citrus, mango, avocado pear and pine- - their culture and propagation	W.S.Y. Abutiate, M.Sc.,FGhIH
Tomato breeding - cultivars suitable for cultivation in the humid tropical conditions of Ghana. Long shelf-life and heat tolerant tomato. Improvement of other vegetables - pepper, okra, eggplant and onion	F. Agble, Ph.D.*
Vegetative propagation of fruit tree crops eg. citrus, mango and avocado pear	K.A. Addae-Kagyah, M.Sc.,Ph.D.

3.2 Plant Pathology Section — Head: E.A. Addison, M.Sc.

Control of *Sclerotium rolfsii* in okra, tomato and egg-plant — E.A. Addison, M.Sc.

3.3 Entomology Section

Pest problem on egg-plant, tomato and cabbage — Y.A. Duodo, Ph.D.

3.4 Oil Palm Research Centre, Kusi — Head: J.B. Wonkyi-Appiah, Ph.D.

Master register for crops grown in Ghana — M.A. Adansi
Citrus budwood certification
Oil palm physiology — J.B. Wonkyi-Appiah, Ph.D.

4 Tafo

Cocoa Research Insititute of Ghana

P.O. Box 8, Tafo (Akim)
Exec.Dir.: Veronica A. Martinson

The Cocoa Research Institute of Ghana, popularly known as CRIG, began as the Central Cocoa Research Station of the then Gold Coast Department of Agriculture. The Research Station had then been created by the Gold Coast Government for the specific purpose of cocoa research, mainly to combat the swollen-shoot disease which was first recognised in 1936 in the cocoa-growing area of the Eastern Province around Tafo and thereby to help maintain the levels of cocoa production. At that time, production in the Eastern Province was on the decline; areas of hitherto normal productivity were being abandoned and exports were being maintained only by the yields from farms in Ashanti and the Western Province. Thus without facilities for research the Department of Agriculture was ill-fitted to give advice on the reestablishement of cocoa in the derelict areas or even determine whether rehabilitation was possible. Later in 1944, increase in the incidence of cocoa pests and diseases in the Gold Coast and Nigeria led to the creation of the West African Cocoa Research Institute (WACRI) to serve all the cocoa-growing areas in the former British West Africa. This new research organization formed a solid basis on which to build and expand further research into coca diseases and pests. The Central Cocoa Station at Tafo thus became the Head-quarters of WACRI on account of existing facilities for research and its proximity to the areas being devasted by the swollen-shoot disease. WACRI Sub-Station was later set up in Ibadan, Nigeria. The attainment of Independence and National Sovereignty by Ghana, Nigeria and Sierra Leone, three of the four countries which with Gambia participated in the inter-territorial organization, led to the break-up of WACRI as an inter-territorial organization and from October 1962 each country assumed full responsibility for any WACRI facilities within her boundaries. In Ghana, WACRI at Tafo was reconstituted as the Coca Research Institute of Ghana and became part of the then National Research Council which later changed to the Ghana Academy of Sciences, now Council for Scientific and Industrial Research. In July 1972, the Cocoa Research Institute became part of the Ghana Cocoa Marketing Board which also became part of the Ministry of Cocoa Affairs in 1975.
There are six research Divisions: Agronomy, Plant Physiology and Biochemistry, Soil Science, Entomology, Plant Breeding and Plant Pathology, supported by Administration, Plantation Management and Works Divisions.

4.1 Agronomy Division

Cocoa agronomy, plant density, shade tree studies
Establishment and maintenance, cocoa rehabilitation
Weed biology and control — F.K. Oppong, B.Sc.
Cocoa microclimate
Periodicity of flushing and flowering
Intraspecific and interspecific competition
Cola agronomy, cola storage — K. Opoku Ameyaw, B.Sc.
Establishment and maintenance, coffee agronomy
Nursery studies, phenology, statistics, experimental design, statistical analysis, computer programming

4.2 Physiology and Biochemistry Division — Ag. Head: D. Adomako, Ph.D.

The biochemstry of cocoa fermentation, optimum conditions for the curing of cocoa beans and research on cocoa swollen shoot virus disease (in collaboration with Pathology Division) formed a major part of the research programme of the Biochemistry Section in the past. The programme has been expanded in recent times to include research on cocoa by-products, pesticides, bio-chemistry of cocoa insect pests and the black pod disease caused by Phytophthora palmivora, analysis of shea butter and Pentadesma butyraceae fats.
The mineral nutrition and effects of shade regimes on cocoa establishment, growth and reproduction. Water relations and photosynthesis of cocoa in relation to various environmental factors.

Seasonal variation in the carbohydrage content of mature cocoa trees is being studied to discover factors affecting flowering, pod growth and cherelle wilt — D. Adomako, Ph.D.

4.3 Soil Science Division — Head: Y. Ahenkorah, Ph.D.

The determination of the effects of fertilizer x shade interaction on the yield of cocoa; nutrient cycle in a cocoa ecosystem and foliar analyses, supplemented by soil analyses, as diagnostic technique in cocoa nutrition form the major research programme of the Soil Science Division of this Institute. Significant progress on the type of fertilizers suitable for cocoa production and the economic limit of fertilizer utilization have been made. Work on "soil test" is in progress and trials are located in all the cocoa production areas of the country. — B.J. Halm, Ph.D./M.R. Appiah,M.Sc.

4.4 Entomology Division — Ag. Head: E. Owusu-Manu, Ph.D.

There are two main aspects of research carried out by the Division. Capsid control and ecological studies on mealy-bugs. On capsid control, researach is carried out on screening insecticides for the control of lindane resistant capsids, the long term effect of insecticides on the cocoa insect fauna and capsids movement in relation to their control. Natural enemies and population fluctuation in capsids are also being studied. Research work on mealy-bugs include the studies on the fecundity of various mealy-bugs, abundance of cocoa verietal resistance to mealy-bugs and ants/mealy-bug relationship and stability of ant mosaic. The Division also carries out research work on the population studies and natural enemies of *Bathycoelia* and cocoa leaf-eating caterpillars and coffee entomology.

Insecticide screening	E. Owusu-Manu, M.Sc. PhD.
Capsid population studies	R.K. Mensah, B.Sc.
Bathycoelia studies, caterpillar studies	
Long-term effect of insecticides	B. Padi (Mrs.), M.Sc.
Capsid movement studies, coffee entomology	
Natural enemies	A.H. Brew, M.Sc./S.K. Asante B.Sc.
Mealy-bug/virus outbreak, ants/mealy bug relationship	
Varietal resistance, mealy-bug fecundity	

4.5 Plant Breeding Division — Head: V.A. Martinson, Ph.D.

Cytogenetics of cultivated species of *Theobroma* and *Coffea*	V.A. Martinson, Ph.D.
Breeding and selection for black pod resistance	J.D. Amponsah, M.Sc.
Screening tests and trials	Y. Adu-Ampomah, B.Sc.Ph.D.
Drought resistance breeding	A.Abdul-Karim, B.Sc.
Block planting trials CRIG/Cocoa Division project	
Studies on natural pollination	
Cocoa swollen shoot virus resistance; trials and progeny tests	B. Adomako,B.Sc.,M.Sc.,DIC
Introduction of germplasm, compatibility studies	
Series II hybrids	
Determination of pod and bean characters	
Development of clonal plots of selected parents for future seed gardens, cola breeding	

4.6 Plant Pathology Division — Head: J.T. Dakwa, Ph.D., MIBiol.

Study of the epidemiology of swollen shoot disease; in particular the distribution of latent infection associated with natural outbreaks and the reinfection of young cocoa planted in epidemic areas. The screening and selection of cocoa progenies resistant to swollen shoot disease. The study of cocoa viruses.

Fungal studies. Sources of primary field inoculum of *Phytophthora palmivora*. Factors affecting stem and node infections by *P. palmivora*. Factors affecting infection of pods by inoculum in the stem. Level of stem and node infections in cocoa farms. Longevity of cushion infections and cankers. Disease and cropping patterns and factors affecting them. Disease incidence in pods of different maturity. Differences in pod susceptibility among Series II varieties. Chemical control of *P. palmivora*. Screening of candidate fungicides. Breeding for resistance to *P. palmivora* (with Plant Breeding Division). Nematode studies. Transmission of cocoa. Necrosis virus with nematodes. Population dynamics of nematodes in cocoa farms. — A.Asare-Nyako, Ph.D.

Meteoropathology of cocoa black pod disease with a view to developing a forecasting system. — J.T. Dakwa, Ph.D.,MIBiol.

Survey of black pod disease in Ghana. The occurrence of *P. palmivora* in soil and its effect on the establishment and development of transplanted cocoa seedlings. Establishment, development and eradication of cankers caused by *P. palmivora*.

The epidemiology and control of cocoa leaf blight caused by *Colletotrichum gloesporoides* through screening of candidate fungicides. The epidemiology and control of blight in gauze-house and nurseries caused by Cladosporium sp. Resistance trials of cocoa swollen shoot. Virus disease. Tissue culture of cocoa. — J.T. Dakwa, Ph.D.,MIBiol.

Improvement of the extraction, purification and mechanical transmission of CSSV. Characterisation of CSSV isolates. Virus-vector relationships. Purification and properties of cocoa nectosis virus (CNV). Occurrence of the disease. Search for natural vector and sources of infection — G.K. Owusu, Ph.D.

Epidemiology of cocoa viruses, particularly the distribution of latent infections. Adaptation of hybrid seedlings to epidemic areas. Study of cocoa viruses. — L.A. Ollenu,l B.Sc., M.Sc., Ph.D.

24° E
40° N
100 km
1
2
3
4
5
6
7
8
9
10
11
12
13
14
15
16
17
18
19
20
21
21
22
23
24
25
26

1 Ag. Anargyroi, 2 Ag. Deca, 3 Ag. Paraskevi, 4 Aliartos, 5 Arta, 6 Athens, 7 Chaidari, 8 Chanea, 9 Corfou, 10 Heraklion, 11 Gastouni, 12 Ierapetra, 13 Kalamata, 14 Kifissia, 15 Lamia, 16 Larissa, 17 Lesbos, 18 Lycovryssi, 19 Naoussa, 20 N. Moudania, 21 N. Filothei, 22 Patra, 23 Pyrgos, 24 Rhodes, 25 Thessaloniki, 26 Volos

Survey

Most of the research institutes belong to the Ministry of Agriculture. These were reorganized in 1961, but 5 Agricultural Research Centers (ARC) were established in 1977, each of them including the following institutes:

ARC - Athens: Institute of Viticulture, Wine Institute, Institute of Technology of Agricultural products, Institute of Soil Science, Institute of Agricultural Machinery and Structures (Ag. Anargyroi), Section of Floriculture (Nea Philothei) and Station of Aliartos.

ARC - Thessaloniki: Institute of Plant Protection, Institute of Soil Science, Section of Vegetables, Institute of Deciduous Fruit trees (Naoussa) and Station of N. Moudania.

ARC - Larissa: Institute of Plant Protection (Volos), and Station of Vardates.

ARC - Patra: Plant Protection Institute, Vineyard and Vegetable Institute (Gastuni), Olive-tree Institute (Corfu), Olive, Fruit-tree and Vegetable Institute (Ka-

Prepared by Dr. Ch.C. Kitsos and G. Andritsos, Ministry of Agriculture, Athens

lamata) and Station of Arta.
ARC - Chania: Institute for Subtropical Plants and Olives, Institute for Viticulture, Vegetable Crops and Floriculture (Heraclion), Institute of Plant Protection (Heraclion), Station of Messara, Station of Ierapetra, Station of Rhodes and Station of Lesbos.

1 Ag. Anargyroi

1.1 Agricultural Machinery and Structures - Ministry of Agriculture

GR 135 10 Ag. Anargyroi
Phone: 2611011
Dir.: G. Souvatzis

Testing of horticultural machinery. Research on harvesting machinery for fruit trees — N. Kafetzakis M.Sc.
I. Kyriakopoulos
Research in mushroom growing — N. Vaindirlis

2 Ag. Deca

2.1 Agricultural Research Station
Ministry of Agriculture
elevation: 130 m
rainfall : 220 mm
sunshine : 3.000 h

GR 711 10 Heraklion-Crete
Phone: 0892-31288
Dir.: J. Tzompanakis

Orchards of olive, citrus, avocado and some other subtropical fruit trees (120 ha), plus 6.000 m^2 of plastic greenhouses.
Experiments on olive, citrus and avocado trees. — J. Tzompanakis
Vegetables under plastic — J. Tzompanakis/ G. Pediaditakis

3 Ag. Paraskevi

3.1 National Center for Research in Natural Sciences "Demokritus"
elevation: 300 m
rainfall : 400 mm
sunshine : 2730 h

GR 153 Ag. Paraskevi
Phone: 6513111
Telex: 21 61 99

3.1.1 Laboratory of Entomology — Coord.: Dr. J. Tsitsipis
Biotechnical Genetics, Pest control of fruit-trees *(Dacus, Ceratitis, Rhagoletis, Zeuzera, Cossus, Margaronia)* — Dr. A. Economopoulos/ Dr. Haniotakis
Dr. G. Tsiropoulos/ Dr. G. Zervas
Dr. Manoukas/ Dr. B. Mazomenos

3.1.2 Laboratory of Soil Science — Dir.: Dr. E. Papanicolaou
Fertilizer efficiency studies in fruit trees (citrus, olives) and vegetables — C. Apostolakis/ C. Nobeli
V. Skarlou

3.1.3 Laboratory of plant Pathology — Dir.: Dr. Ch. Christias
Physiology of the development of the resting phase (sclerotia) of soil phytopathogenic fungi

4 Aliartos

4.1 Agricultural Research Station
Ministry of Agriculture
elevation: 220 m
rainfall : 430 mm
sunshine : 2.650 h

GR 320 01 Aliartos
Phone: 0268-22537
Dir.: Dr. C. Christoulas

Working mainly on field crops (120 ha). In addition it serves some problems concerning vegetable crops, especially onions and vegetables for processing.

Vegetable crops — C. Georgiadis M.Sc.

5 Arta

5.1 Agricultural Research Station
Ministry of Agriculture
elevation: 40 m
rainfall : 1.080 mm
sunshine : 2.646 h

GR 471 00 Arta
Phone: 0681-41395
Dir.: G. Kalyvas

The station has 3 farms with 67 ha of fields. Vineyard with stock plants for propagation material production. Citrus, walnut and other deciduous fruit-trees. Soil laboratory.

Cultural problems of citrus trees — Chr. Papageorgiou
Cultural problems of other fruit trees — G. Kalyvas, DEA/ E. Tassiopoulos
Irrigation of citrus-trees

6 Athens

6.1 Agricultural College of Athens
State Institution

Iera Odos 75, GR 118 55 Athens

Dir.: Prof.Dr. H. Dekazos

6.1.1 Laboratory of PomologyPhone: 3467350

Dir.: Prof.Dr. H. Dekazos

Post harvest physiology of fruits, especially lemon and apricot. Nutrition of fruit trees (Pistacio). Micropropagation of walnut, R/S studies of
H. Tsantili
clementine, dormancy of peach seeds. Study of meadow orchard system with peach. Developmental studies of pistacio inflorescence.

Dr.H. Dekazos/Th. Conteas
Soula Tzoutzoukou/E. Efthimiadou

6.1.2 Laboratory of Ampelology — Phone: 3464994
Dir.: Prof. P. Lelakis

Grape plant physiology, breeding, improvement, analysis and propagation — Dr. P. Lelakis/ E. Michaelidis
Control of diseases and airpollution — Dr. P. Lelakis
Vineyard protection techniques — Dr. E. Stayrakakis/K. Zacharakis
Ampelography of grape CV.S and R/S,s — Dr. E. Stayrakakis/E. Michaelidis

6.1.3 Laboratory of Vegetable Production — Phone: 3466890
Dir.: Prof.Dr. Ch.M. Olympios*

Research on vegetables in the open and under protected cultivation — C. Akoumianakis/ G. Kapotis

6.1.4 Laboratory of Floriculture and Landscape Architecture — Phone: 3468235
Dir.: N. Kantartzis

Propagation of ornamental shrubs, Landscape Architecture, postharvest physiology of cut flowers — Dr. N. Kantartzis
Dr. I.Chronopoulos/ N.Akoumianakis

6.1.5 Laboratory of Plant Breeding and Biometrics — Dir.: Prof.Dr. P. Kaltsikis

Breeding of tomato — E. Raftopoulou/ P. Bebeli

6.1.6 Laboratory of Entomology and Agricultural zoology — Dir.: Prof.Dr. K. Pelekassis

Study of the occurence of bud mite *Calomerus vitis,* cause of spot on the berries of Rozaki grapes — Dr. K. Pelekassis/Dr. N. Emanuel
Biolog. studies on the phenology of Dacus for estimating the timing of sprays — Dr. K. Pelekassis/N. Roditakis

6.1.7 Laboratory of Plant Pathology — Phone: 3468437
Prof. S.G. Georgopoulos

Fungal Diseases of Fruit, Vegetable, and

Ornamental Crops, particularly as regards to chemical control — Dr. P. Stathis/ Dr. B.N. Ziogas, Dr. A. Grigoriu

Fungal wilt diseases — Prof. S.G. Georgopoulos, Dr. V.Emmanuil/ Miss. K. Kouvaraki

Bacterial diseases of fruit and ornamental crops — Prof. C.G. Panagopoulos

Ecology and biological control of crown gall — D. Panoutsos

6.1.8 Laboratory of Farm Structures — Phone: 3460025; Dir: Prof.Dr. S. Kyritsis

Greenhouse construction, environment regulation, energy conservation systems: use of renewable energies, waste heat use, energy saving, geothermal energy — Dr. G. Marogiannopoulos

7 Chaidari

7.1 Botanical Garden of Diomedes
Non-profit organization under the supervision of the University of Athens. Its construction is in evolution in an area of 150 ha.

Iera Odos, GR 124 61 Chaidari
Phone: 581 557
Dir.: Prof.Dr. C. Mitrakos

Ornamental plant section. Historical plant section. Systematic Botany section and Medicinal Plant section under construction and plant Geography Section to be constructed. — G. Priebe/ A. Vasilatou

8 Chanea

8.1 Agricultural Research Center
Ministry of Agriculture

GR 731 00 Chanea - Crete
Phone: 57157
Dir.: N. Psiyllakis

8.1.1 Subtropical Plants and Olive Tree Institute
elevation: 20 m
rainfall : 700 mm
sunshine : 2.943 h

Phone: 0821-57157
Dir.: N. Michelakis

The Institute has 5 farms totalling 90 ha and 5.000 m^2 of greenhouses. Olive-tree, citrus, avocado and actinidia orchards. These orchards and vineyards are used for experiments, cultivar collection and as a source of propagation material. Well-equipped laboratories for leaf and soil analysis, olive-oil processing, irrigation techniques, entomology, biological control of insects with insectary.

Olive-tree cultural techniques — J. Metzidakis

Citrus tree cultural problems — Dr. S. Lionakis/Dr. E. Protopapadakis

Avocado and actinidia culture problems — S. Lionakis/ E. Protopapadakis

Irrigation of orchards — N. Michelakis

Nutritional problems — Dr. J. Androulakis/ M. Loupassaki, C. Economakis M.Sc.

Olive-processing — A. Koutsaftakis

Table-olive processing — D. Vamvoukas/ S. Bekiari

Insect pest control, biological control — Dr. S.Michelakis/ Dr.B.Alexandrakis, Dr. N. Paraskakis

Disease control — Dr. E.Bourbos/ M.Scountridakis M.Sc.

9 Corfu

9.1 Olive Tree Institute
Ministry of Agriculture
elevation: 25 m
rainfall : 1.170 mm
sunshine : 2.300 h

GR 491 00 Kerkyra
Phone: 0661-30462
Dir.: G.C. Karvounis

Three farms totalling 30 ha of olive-tree orchards. Experimental olive-oil extraction laboratory, well equipped for pest and disease control.

Olive tree cultural problems
Olive-oil technology
Olive tree insect control

Olive tree disease control

G. Karvounis/ D. Kassimis
A. Kantas
Dr. E. Kapatos/ E. Stavropoulou
M. Makropodi
E. Trimeri-Makri

10 Heraclion

10.1 Institute of Viticulture, Vegetable Crops and Floriculture - Ministry of Agriculture
elevation: 15 m
rainfall : 510 mm
sunshine : 2.850 h

GR 713 06 Heraclion - Crete
Phone: 081-286891
Dir.: M. Petrakis

Forty ha of land including 1.000 m^2 of glasshouses and 4.000 m^2 of plastic houses. Vineyards. Well-equipped laboratory for soil and leaf analysis.

Breeding and variety selection of vegetable crops
Forcing vegetables under cover

Soil and leaf analysis for vegetable crops
Composting of agricultural wastes
Grape cultural practices

Dr. N. Fanourakis
M. Petrakis/ C. Smardas
Dr. B. Manios
Dr. P. Tsikalas/ D. Linardakis
B. Manios
E. Vardakis/ E. Argyrakis
M. Nikolantonakis

10.2 Plant Protection Institute

GR 713 06 Heraclion
Phone: 282 716
Dir.: N. Baltzakis

Research programmes for plant protection
Diseases of vegetables and vine

Viruses of vegetables and vine
Pests of vegetables and vine

Dr. N. Malathrakis
Dr. B. Bakalounakis
Dr. A. Avgelis
Dr. N. Roditakis

11 Gastouni

11.1 Vine and Vegetable Institute
Ministry of Agriculture
elevation: 60 m
rainfall : 750 mm
sunshine : 2850 h

GR 273 00 Gastouni
Phone: 0623-23058
Dir. C. Bacalacos

The Institute serves problems of vegetable crops, especially breeding and cultivar selection. It contains fifty five ha of land and 5.000 m^2 of plastic greenhouses

Cultural problems of vegetable crops
Breeding and genetics
Irrigation and soil fertility problems
Technological problems on proc. vegetables

Z. Vasiliou
P. Christakis M.Sc.
C. Bacalacos
Dr. Th. Kondilis

12 Ierapetra

12.1 Agricultural Research Station

GR 722 00 Ierapetra - Crete

Ministry of Agriculture
elevation: 15 m
rainfall : 207 mm
sunshine : 3050 h

Phone: 0842-22788
Dir.: J. Tsampanakis*

Five ha of land including 10.000 m² of plastic houses.

Vegetables and flower crops under plastics — J. Tsampanakis*/N. Drosos
Use of solar energy for greenhouse heating — J. Tsampanakis*

13 Kalamata

13.1 Olive, Fruit tree and Vegetable Institute
Ministry of Agriculture
elevation: 25 m
rainfall : 840 mm
sunshine : 2900 h

GR 241 00 Kalamata
Phone: 0721-29812
Dir.: V. Pontikis

Institute of Research in olive, fig and other tree crops. With approximately 0.5 ha of plastic-covered greenhouses for vegetable research. Equipment includes a mist propagation unit for tree propagation. Laboratories for soil analysis and oil determination. Total area of farm 35 ha.

Olive tree cultural problems — B. Pontikis/ J. Karydes
Cultivar selection of figs — J. Karabetsos M.Sc.
Vegetable crops under cover — A. Papandreou/ P. Karavitis
Soil analysis — S. Petsas

14 Kifissia

14.1 Benaki Phytopathological Institute*
elevation: 291 m
rainfall : 420 mm
sunshine : 2730 h

Delta 8, GR 145 61
Phone: (01) 8012-376
Dir.: Dr. P.A. Mourikis
Dep.Dir.: Dr. C.D. Cholevas

Non profit organization sponsored by the Ministry of Agriculture. Among other subjects, pests diseases and weeds on fruit trees, vegetables and ornamentals, as well as their chemical control are included.

14.1.1 Plant pathology

Fungus diseases: citrus, olive, stone and seed fruit trees, vegetable and ornamental and other horticultural plants — Dr. A. Chitzanidis/Dr. E. Tzamos; Dr. A. Pappas
Bacterial diseases: grape, potato, almond and other horticultural plants — Dr. P.G. Psallidas
Mycoplasma diseases — Dr. A.S. Alivizatos
Virus diseases: citrus, vegetables and other hort.plants — Dr. P.E. Kyriakopoulou/ Dr. F.P.Bem
Meristem culture of citrus — Dr. V.A. Plastira
Non parasitic diseases: olive, seed fruits, pepper and other horticultural plants — Dr. C.D. Cholevas

14.1.2 Entomology, Agriculture and Zoology

Entomology: *Capnodis, Zeuzera, Sesia, Prays* and other pests of horticultural crops — P. Alexopoulou; Dr. A.S. Drosopoulos
Nematology: potato, grape, citrus and other horticultural crops — C. Gazelas/Dr. M. Antoniou
Acarology: various horticultural crops — P. Souliotis
Biological control: insect pests of citrus, olive — Dr. L.C. Argyriou/Dr. H.G. Stavraki

and other horticultural crops	
Microbiological control of crop pests: bacteria, viruses, fungus for control of pests of olive, apple, symbiotic bacterium of *Dacus*	Dr. Ch.N. Yamvrias/M. Anagnou
Insect physiology: biology and control of Leptinotarsa (potato)	Dr. E. Fytizas/Aik. Mantaka
14.1.3 Weed science (Weed control)	Dr. C.N. Giannopolitis Dr. E.A. Paspatis
14.1.4 Pesticide control and Phytopharmacy	
Insecticides	Dr. P.G. Patsakos/ P.E. Kalmoukos Dr. E.G. Kapetanakis
Fungicides	Dr. N. Petsikou/ Dr. M. Chryssagi
Chemical analysis	G.S. Spryropoulos/ Dr. A Chourdaki
Residues of pesticides	C.Z. Zafiriou
Toxicology of pesticides	Dr. R. Fytizas/ Dr. G.V. Vassiliou

15 Lamia

15.1 Agricultural Research Station Ministry of Agriculture elevation: 30 m rainfall : 550 mm sunshine : 2.600 h	GR 351 00 Lamia Phone: 0231-41246 Dir.: N. Katranis
A farm of 24 ha, devoted mainly to field crop research including 1 ha of olive tree orchard and 1 ha of fruit tree collection.	
Olive tree problems	N. Katranis
Fruit tree problems	D. Rouskas M.Sc.

16 Larissa

16.1 Agricultural Research Center Ministry of Agriculture	GR 413 35 Larissa Phone: 239711 Dir.: E. Stylopoulos

17 Lesbos

17.1 Agricultural Research Station Ministry of Agriculture elevation: 85 m rainfall : 650 mm sunshine : 3053 h	GR 811 00 Mytilene - Lesbos Phone: 0251-93463 Dir.: E. Tsirtsis
A farm of 20 ha, planted with olive trees.	E. Tsirtsis
Olive tree cultural problems, olive technology	S. Thalassinos

18 Lycovrissi

18.1 Agricultural Research Center Ministry of Agriculture elevation: 160 m rainfall : 400 mm sunshine : 2730 h	Soph. Venizelou 1 GR 141 10 New Heraclion Phone: 2819019 Dir.: N. Karantonis
18.1.1 Institute of Viticulture	Phone: 01-2816978 Dir.: N. Karantonis

Six ha of vine yards with cultivar collection. Well-equipped laboratory for leafanalysis.	
Grape vine botany and ampelography	N. Karantonis/ B. Michos
Cultural practices, growth regulators	D. Pimpli/ B. Daris/ Dr. B. Michos
Propagation of grape vine	N. Karantonis/ C. Panagiotou
Plant analysis, minor elements	B. Pippous
18.1.2 Institute of Technology of Agricultural Products	Phone: 01-2816985 Dir.: E. Emanouilidis
Applied research on preservation and processing of agricultural products in general.	
Preservation of fruits and vegetables (cold storage-controlled atmosphere)	Dr. E. Manolopoulou
Freezing and frozen storage	K. Katsaboxakis M.Sc.
Canning of fruits and vegetables	Dr. K. Mallidis
Microbiological problems	Dr. F. Samaras
Biotechnology	Dr. K. Israilides
Byproducts, better utilization and processing	J. Arkoudilos M.Sc.
Olives and olive-oil technology	D. Kontonikolaou
Food quality control	R. Papadopoulou M.Sc.
18.1.3 Soil Science Institute	Phone: 2816974 Dir.: E. Gougas
Responsible for nutrition-fertilization of all crops; special programmes for pistachio, olive, tomato, potato and vegetables under cover: tomato, cucumber, pepper	E. Gougas/G. Chardas Dr. I. Mitsou/M. Christou M.Sc./ Dr. A. Kollias
18.1.4 Wine Institute	Phone: 2819094 Dir.: N. Danilatos
Research in wine and wine distillates	S. Sotiropoulos
Wine making, distilling, processing and analysis	M.Vassiliou/ I.Nolas/ E.Tsoutsouras P.Lanaridis/ M.Tsoukalas/ M.Salacha E. Tzourou

19 Naoussa

19.1 Institute of Deciduous Tree Fruits Ministry of Agriculture elevation: 320 m rainfall : 510 mm sunshine : 2450 h	GR 592 00 Naoussa Phone: 0332-41548 Dir.: G. Sirianidis
Three farms with 100 ha orchards of deciduous trees. Research on cultural problems, cultivar selection and improvement, insect pests, etc.	
Stone fruit trees	D. Stylianides/ E. Mouchtouri E. Karagianni
Cherry trees	J. Hatziharissis
Pome fruit trees	A. Dinopoulos/ A. Maganaris
Nut fruit trees	A. Mainou
Propagation of deciduous trees	C. Tsipoulidis

20 Nea Moudania

20.1 Agricultural Research Station Ministry of Agriculture elevation: 10 m rainfall : 460 mm sunshine : 2550 h	GR 632 00 Nea Moudania Phone: 0373-21422 Dir.: N. G. Moustakas

Working mainly on field crops. Olive tree orchard.
Olive tree cultural problems — N. Stavropoulos

21 Nea Philothei

21.1 Agricultural Research Center of Athens
Section of Floriculture. Ministry of Agriculture
elevation: 136 m
rainfall : 400 mm
sunshine : 2730 h

GR 151 23 Maroussi
Phone: 01-6812851
Head: Dr. NG Koutepas*

Research on cultural problems of flower crops, 4 ha of fields and 2.000 m^2 of plastic greenhouses. Equipment: Basic laboratory equipment for plant growth regulator research.

Cut flower crops and pot plants	N.G. Koutepas*
Mites of flower and other crops	E. Hatzinikolis

22 Patra

22.1 Agricultural Research Center
Ministry of Agriculture

GR 261 10 Patra
Phone: 061-421264
Dir.: (Dep.) G. Kitsos

22.1.1 Plant Protection Institute

Phone: 061-422142
Dir.: G. Kitsos

Research programmes for plant protection, epidemiological studies of plant diseases, bioecology of pests and weeds of crops and research for more efficient methods for controlling diseases and pests of crops.

Diseases of fruit-trees, vegetables and vine	G. Kitsos/ Dr. I. Theocharis
Viruses of fruit-trees, vegetables and vine	Dr. P. Panayiotou
Physiological, non-parasitic diseases of vegetables and vine	D. Bonatsos M.Sc.
	L. Panagiotopoulos M.Sc.
Pests of fruit-trees	Dr. E. Konstantinidou

23 Pyrgos

23.1. Institute of Black Corinth (raisin)

GR 271 00 Pyrgos
Phone: 0621-22551
Dir.: G. Papafitsoros

Belongs to the Central Currant office, a non-profit organization

Productivity of Black Corinth vine yards. Improvement of produce	G. Mpithas/ G. Giallelis
	Th. Athanasopoulos
Substation of Filiatra-Messinia	P. Kalofonos
Substation of Kokonion-Corinthia	G. Anastasiou
Substation of Messini-Messinia	G. Giokas

24 Rhodes

24.1 Agricultural Research Station
Ministry of Agriculture
elevation: 10 m
rainfall : 800 mm
sunshine : 3000 h

GR 851 00 Rhodes
Phone: 0241-91230
Dir.: A. Konstantakis

Two farms with 40 ha of land, of which approximately 8 ha are citrus and olive tree orchards.

Citrus cultivar selection and cultural practices.	X. Papanicolaou
Fruit tree cultivar collection	A. Haritou

25 Thessaloniki

25.1 Aristotle University - Thessaloniki School of Agriculture elevation: 70 m rainfall : 486 mm sunshine : 2555 h	GR 540 06 - Thessaloniki Phone: 992581

The University Farm, including 16 ha of orchards, 1 ha of vegetables, 4 ha of Landscape plants and 0,2 ha of greenhouses, is next to the farm of ARC-Thessaliniki

25.1.1 Laboratory of Pomology	Dir.: Dr. E. Sfakiotakis
Post Harvest Physiology	
Mineral nutrition of fruit trees / propagation	Dr. J. Therios
Flower physiology of fruit trees	Dr. M. Vasilakakis
Flower incompatibility	V. Tsiracoglou M.Sc.
Tissue culture of fruit trees	K. Dimasi-Theriou
25.1.2 Laboratory of Vegetable Crops	
Vegetable nutrition, storage and quality aspects	Dr. C. Dogras/ E. Koufakis
25.1.3 Laboratory of Ornamentals	
Floricultural crops. Tissue culture	Dr. A. Economou*
Woody ornamentals	Chr. Georgakopoulou-Voyiatzi
Landscape architecture	Dr. I. Tsalikidis
25.1.4 Laboratory of Biology of Horticulture	Dir.: Prof.Dr. I.C. Porlingis*
Physiology of growth substances. Physiology and	Dr. D. Voyiatzis*
propagation of the olive tree	M. Koukourikou-Petridou
25.1.5 Laboratory of Viticulture	Dir.: Dr. M. Vlachos
Breeding new grape varieties. Growth regulators. Ampelographical studies. Bud differentation.	
Propagation. Cultural practices	Dr. D. Stavrakas
Propagation using rootstocks	N. Nicolaou
Tissue culture	E. Zioziou
25.1.6 Laboratory of Applied Zoology and Parasitology	Dir.: Dr. M. Tzanakakis
Insects of fruit trees. *Dacus oleae*	Dr. D. Stamopoulos
	Dr. V.Katsoyiannos/ Dr. D.Profitou
	M. Savopoulou
25.1.7 Laboratory of Plant Pathology	Dir.: Dr. D. Zachos
Diseases of fruit trees, vegetable and floricultural crops	Dr. C. Tzavella-Klonari

25.2 Agricultural Research Center Ministry of Agriculture elevation: 70 m rainfall : 486 mm sunshine : 2.555 h	GR 54510 Thessaloniki Phone: 471206 Dir.: Z. Bagtzoglou
25.2.1 Section of Vegetable Crops	Phone: 031-471439 Head: N. Relias

This Section is attached to the Cereal Institute,

the farm of which it uses for the experiments and 6.000 m^2 of greenhouses.

Cultural problems of vegetable crops	S. Georgiadis/ E. Mikros
Vegetables for processing	N. Relias
Greenhouses, uses of solar and geothermical energy	Dr. M. Grafiadellis*

25.2.2 Plant Protection Institute

Phone: 471577
Dir.: A. Atzemis

Research programmes for plant protection, epidemiological studies of vegetable diseases, bioecology of pests and weeds, vegetable crops and research for more efficient methods for controlling diseases and pests of crops.

Pests of olive	A. Atzemis
Soil pests of vegetables and fruit trees	D. Stathopoulos
Weed of crops weed killers	E. Kotoula-Syka
Viruses of fruit trees	D. Tsialis
Pests of fruit trees and vegetables	I. Evagelopoulos/ Dr. S. Paloukis

25.2.3 Soil Science Institute

Phone: 411429
Dir.: N. Koroxenidis

Plant nutrition and fertilization of seed, stone fruits and vegetables	Dr. L. Simonis/ P. Koukoylakis M.Sc Dr. G. Schoinas

26 Volos

26.1 Plant Protection Institute

GR 381 10 Volos
Phone: 61087
Dir.: I. Ioannidis

Research programmes for plant protection, epidemiological studies of plant diseases, bioecology of pest and weeds of crops and research for more efficient methods for controlling diseases and pests of crops.

Diseases fruit-trees and vegetables	D. Biris/ G. Gouramanis
Diseases and viruses of fruit-trees, vegetables and grape	Dr. I. Roubos
Bacterial diseases of vegetables	Dr. I. Tsiantos
Pests of fruit trees	A. Bacogiannis
Wood-boring insects of fruit trees	Dr. A. Koutroubas

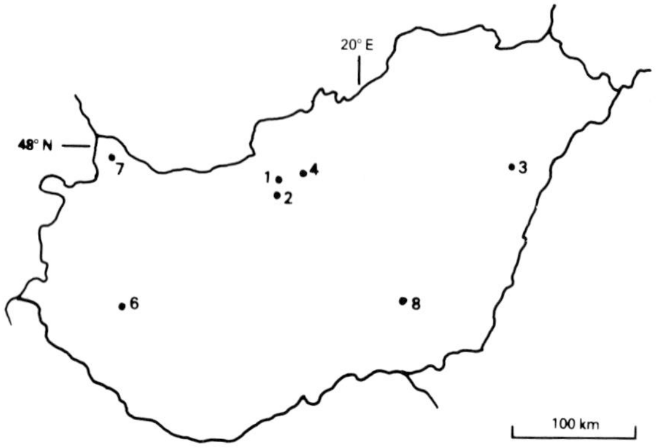
20° E
48° N
7
1
4
2
3
6
8
100 km

1 Budakalász, 2 Budapest, 3 Debrecen, 4 Gödöllő, 5 Kecskemét, 6 Keszthely, 7 Mosonmagyaróvár, 8 Szentes

Survey

Horticultural research work is carried out by the departments of the University of Horticulture, by the horticultural departments of the Agricultural Universities, by the horticultural research institutes and by some departments of other research institutes. These institutions are supervised by the Ministry of Agriculture and Food. The Department for Agricultural Sciences of the Hungarian Academy of Sciences has no horticultural research institutes of its own but gives special grants to some research projects and has a great influence on scientific programs and education. The research projects are determined by the Ministry of Agriculture and Food on the basis of proposals of project teams in which representatives of workers engaged in the principal research projects and representatives of growers cooperate. The institutes have special farms for research and education.

1 Budakalász

1.1 Gyógynövény Kutato Intézet
(Research Institute for Medicinal Plants)
elevation: 103 m
rainfall : 700 mm
sunshine :2000 h

2011 Budakalász, P.O.B. 11
Phone: (26) 20203
Telex: 224829 medpl h
Dir.: Prof.Dr. P. Tétényi

More than nine tenths of the activity in this Institute serves the pharmaceutical industry, approximately 30% basic research, 35% applied research and 30% development work. The research work done with medicinal plants, spices and plants furnishing essential oils covers the propagation, utilization, biology, inheritance, breeding and the determination of the best cultural regions of species cultivated or proposed for cultivation. About half of the research efforts are aimed at the study of three most important Hungarian medicinal plants: poppy, ergot and foxglove.
As the competent authority, the Institute has been charged with the quality control of medicinal plants by the Ministry of Agriculture and Food and by the Ministry of Public Health.

Prepared by Prof.Dr. A. Somos and Prof.Dr. S. Balázs, University of Horticulture, Budapest

Research work is conducted in two sections: in the Section of Biology and in the Section of Chemistry. In this survey only the first one is dealt with.

Agrotechnical Department — Head: Dr. D. Földesi
This department is engaged in the elaboration of up to date agrotechnical methods and in their application at the state farms and cooperatives.

Plant Breeding Department — Head: Dr. J. Bernáth
Breeding of medicinal plants and spices important for the Hungarian pharmaceutical industry with over 10 species and more than 40 state certified varieties selected by the Institute.
Elite and super-elite seed production.

2 Budapest

2.1 Gyümölcs és Disznövénytermesztési Fejlesztö Vállalat
(Enterprise for Development in Fruitgrowing and Ornamentals)
elevation: 100 m
rainfall : 550 mm
sunshine :2100 h

1223 Budapest, Park utca 2
Phone: 264-068
Telex: 22-4790
Dir. gen.:Dr. B. Molnár

Main activities are the breeding, selection and the propagation of temperate top and soft fruits; cultural techniques; plant protection; storage of fruit; economics of fruit growing.
Breeding of roses, annual and perennial flowers. Four research stations located in major fruit growing regions of the country belong to the Enterprise.

2.1.1 Research Station Cegléd
elevation: 100 m
rainfall : 520 mm
sunshine :2110 h

2700 Cegléd, Szolnoki ut 52
Phone: (20) 10-597
Telex: 22-4840
Dir.: F. Nyujtó

Scientific work is done mainly on apricot and plum: breeding and selection of scion and rootstock cultivars. Experimental orchards: 76 ha.

Breeding and introduction of apricot cultivars	Head: F. Nyujtó/ Miss M.M. Kerek
Plum variety trials	Dr. E. Tóth/Dr. Z. Erdős
Crop regulation in apricot, plum and peach	Dr. D. Surányi
Selection of rootstocks	F. Nyujtó

2.1.2 Research Station Érd
elevation: 150 m
rainfall : 625 mm
sunshine :1860 h

1223 Budapest, Park utca 2
Dir.: Dr. I. Gergely

The Station's main objectives are breeding and/or variety research in pears, stone fruit and nuts; elaboration of technologies and economy of fruit production; virus elimination and propagation of top fruit scions and rootstocks. Equipment: Experimental orchards at Érd over 200 ha and at Kecskemét-Szarkás 66 ha. Laboratory and greenhouses for virus elimination work and micropropagation.

Department of Variety Research

Selection of clonal rootstocks for stone fruit and pear	Head: Dr. P. Nagy
Walnut breeding, variety research in chestnut and hazel	Dr. P. Szentiványi
Variety research in peach	Dr. I. Gergely

Sour cherry breeding	Dr. J. Apostol
Sweet cherry and almond breeding and variety research	Mrs. E. Apostol
Variety research in pear and plum	L. Varga
Propagation by cuttings of stone fruit and pear rootstocks; hand grafting	Dr. G. Mezei
Propagation in vitro	Miss Dr. J. Vértesy
Embryo culture	Mrs. I. Balla
Plant Protection Department	
Infectious and physiological diseases of stored fruit	Head: Mrs. Dr. M. Kemenes
Diseases caused by fungi, root-rot of fruit trees	Miss Dr. K. Véghelyi
Orchard pests, integrated control	Dr. Gy. Sziráki
Serology of fruit tree viruses	Miss Dr. V. Schuster
Diagnosis and biological test of viruses and MLO-s	Dr. L. Megyeri
Physiology and technology of storage of pome fruit	Dr. T. Kállay
Specific replant disease	Dr. K. Magyar
Department for Development in Fruit Growing	
Economics and ecology of fruit growing	Head: Mrs. Dr. E. Kállay
Mechanization in fruit growing and nursery	Dr. D. Andor
Physiology of intensive orchards	Dr. T. Brunner
Soil management and nutrition	Dr. E. Szücs
Training and pruning	Dr. J. Mihályffy
Organization and mechanization of harvest	Dr. G. Kollár
Rentability of nursery production. Technology of walnut propagation	Miss Dr. M. Szabó
Computer use in fruit production and in the organization of operations	Gy. Szenci

2.1.3 Research Station Fertőd
elevation: 116 m
rainfall : 620 mm
sunshine :1860 h

9431 Fertőd, Madách sétány 4
Phone: (99) 45-921
Telex: 24-9187
Dir.: Dr. K. Szilágyi

The objectives of the Research Station are breeding, virus eradication and solving of cultural problems of soft fruit. Equipment: Experimental plantations 21 ha. Tissue culture laboratory, 400 m^2 greenhouse and 3500 m^2 isolating net serve the virus eliminating work.

Strawberry breeding	Dr. K. Szilágyi
Raspberry and blackberry breeding	Dr. L. Kollányi
Breeding of currants	Dr. A. Porpáczy
Nutrition, irrigation	Dr. A. Kiss
Virus elimination	Mrs. S. Porpáczy
Plant protection	F. Bakcsa
Micropropagation	J. Zatykó
Cytology	I. Simon
Plant physiology	Dr. M. László

2.1.4 Research Station Ujfehértó
elevation: 120 m
rainfall : 580 mm
sunshine :2120 h

4244 Ujfehértó
Phone: 38
Telex: 73-457
Dir.: Dr. L. Harmat

The station is located within the major apple growing region in Hungary. Main activities are testing of cultivars and developing cultural techniques applied to the demands of the different apple cultivars; studying problems of gooseberry production. Orchard surface over 200 ha, planted mainly with apples.

Introduction of apple cultivars	Head: Dr. L. Harmat
Selection of apple cultivars	Dr. T. Szabó

Plant protection specific to cultivars	J. Bartha
Crop regulation	Dr. I. Zatykó
Growth regulation	Dr. T. Bubán
Pruning	Dr. I. Gonda

2.1.5 Ornamentals Department — Head: Dr. Z. Kováts

Breeding, maintenance-breeding and variety trials with roses, annuals and perennials; improvement of cultural techniques. 5 ha of annuals and perennials plus 6 ha rosarium with 2400 cultivars in variety trials and for the preservation of genetic resources.

Breeding of annual flowers	Dr. Z. Kováts/ Mrs. B. Nagy Mrs. L. Ganczaugh
Breeding and testing of perennials	Miss K. Zatykó/ Miss N. Koch
Roses: variety trials and genetic resources	Mrs. M. Lehmann/ Miss E. Ács

2.2 Kertészeti Egyetem
(University of Horticulture)
elevation: 120 m
rainfall : 550 mm
sunshine : 1950 h

1118 Budapest
Villányi ut 35-43
Phone: 850-666
Telex: 226011 uhort h
Rector: Prof.Dr. I. Dimény

The responsibility of the University of Horticulture is the education of experts on the highest level. It has three faculties: the Faculty of Horticultural Production, Faculty of Food Preservation and the College of Horticulture (see Kecskemét 5.1). Graduates obtain university degrees as certificated horticultural engineers, certificated canning industry engineers, horticultural production engineers or canning industry production engineers, after 5 or 3 years' courses, respectively. Post-graduate courses are organized for the education of specialized engineers in several fields of horticulture or food preservation. Higher scientific degrees can be obtained by individual post-graduate training, supervised by the University of Horticulture and by the Hungarian Academy of Sciences. All departments of the University are engaged in basic and/or applied research and development work. The University has model farms and experimental farms in different settlements near Budapest. Data on the experimental fields and equipment are listed under Departments.

2.2.1 Institute of Food Technology

1118 Budapest, Ménesi ut 43-45
Phone: 664-635
Telex: 226011 uhort h
Dir.: vacant

Education: food, refrigerating and canning technology. Research and development in the same fields. The Institute has regular contacts with the factories and cold storage plants. The departments have modern apparatuses, e.g. gas cromatographer, photometers, colorimeters, rheological instruments, laboratory freeze dryer, microwave oven and microwave thawing apparatus etc.; a CA storage system with small boxes and an experimental and demonstration plant.

Department of Food and Refrigerating Technology

New frozen foods. Processing of potato.	Mrs. Dr. L. Erdélyi
Fruit juices	
Storage of fruit and vegetables, CA storage	Dr. T. Sáray/ Dr. Cs. Balla
Testing chemicals and packaging materials for storage. Cryogene freezing	Dr. T. Sáray
Measuring heat production of fruit and vegetables in storage	Dr. Cs. Balla
Testing new fruit and vegetable varieties for freezing	Mrs. Dr. V. Bognár Mrs.Dr. L. Polyák/ Miss E. Zackel
Dietetic frozen foods: natural agents for substituting sugar in frozen foods	Mrs. I. Faludi
Chemical analysis of frozen foods. Gas cromatography of frozen foods	Mrs. K. Koncz

Packaging of frozen foods	Miss. E. Kriszta
Natural food colours for frozen products	Mrs. Dr. L. Polyák
Chemical analysis of foods and raw materials	Mrs. Dr. E. Rácz
Freeze-drying. Testing physical properties of food and raw materials	Dr. Gy. Urbányi
Histological changes during freezing and freeze-drying	Miss E. Zackel
Department of Canning Technology	Phone: 667-435
Biologically active components in canned foods	Head: Prof.Dr. Sz. Török
Quality regulation and quality control	E. Hergár
Machinery and equipment	Dr. J. Barta
Extraction by diffusion	Miss Dr. Gy. Pátkai
Organisational problems in tomato processing	Mrs. Dr. J. Monszpart-Sényi
Dietetic foods. Foods of special composition	Mrs. Dr. E. Horvath-Kerkai
Manufacturing fruit juices and nectares	Dr. I. Rák

2.2.2 Department of Fruit Growing
elevation: 103 m
rainfall : 520 mm
sunshine :1900 h

1118 Budapest, Villányi ut 35-43
Phone: 664-683
Telex: 226011 uhort h
Head: Prof.Dr. F. Gyuró

(Szigetcsép Model Farm)
Education: bases of fruitgrowing, technology of fruit growing, propagation of top and soft fruit crops, nursery growing. Research: New, intensive production systems. Training and pruning methods. Resistance breeding. Variety research. Biology of fruit setting. Nutrition problems in apple, raspberry, strawberry. Methods of fruit storage. Harvesting date optima (apple, pear, peach and plum). Experimental fields at the Szigetcsép Model Farm of the University of Horticulture.

New intensive planting systems. Training and pruning methods. Differentiated agrotechnics in order to meet the demands of the cultivars. Regulation of growth, yield and maturity by chemicals. One-variety intensive commercial plantations of apple and pear.	Prof. Dr. F. Gyuró
Mildew resistance breeding in apple. Monilia resistance breeding in sour cherry	Dr. S. Kovács
Adaptation of apple, pear, cherry, plum, apricot and pear varieties to Hungarian natural conditions Biology of flowering and fruit setting. Combination of varieties in the orchard	Dr. J. Nyéki
Apple and cherry breeding. Adaptation of foreign varieties	Mrs. Dr. M.G. Tóth
Nutrition problems in fruit crops of apple, peach and small fruit	Dr. J. Papp
Fruit storage: biological and technological aspects of apple, pear and stone fruit. Storage under normal/controlled conditions	Dr. P. Sass

2.2.3 Department of Chemistry

1118 Budapest
Villányi ut 35-43
Phone: 664-272
Telex: 226011 uhort h
Head: Prof.Dr. I. Pais
Prof.Dr. G. Gombkötó
Mrs. Dr. M. Fehér-Ravasz
Mrs. Dr. E. Tóth-Farkas

The role of titanium-chelate in the living systems.
The role of micronutrients in the biochemical processes of horticultural plants
Biochemistry of anthocyanine compounds
The role of titanium-chelate in the carbohydrate synthesis

Problems of the very accurate ICP-analysis of horticultural products — Dr. P. Fodor
The analysis of aromatic compounds in horticultural products — A. Tóth
Chemical analysis of pollution of environment — A. Tóth/ Dr. P. Fodor

2.2.4 Institute of Landscape Planning
elevation: 110 m
rainfall : 550 mm
sunshine :1950 h

1118 Budapest, Villányi ut 35-43
Phone: 664-333
Telex: 226011 uhort h
Dir.: Prof.Dr. B. Nagy

(Soroksár Experimental Field)
Intensive research work is done in the selection of native trees, shrubs and perennial flowering plants of ornamental value; in the elaboration of their production technologies and application. The Institute maintains and develops a national dendrological collection. New introductions from other countries are tested, evaluated and added to the collection. Arboretum: 7 ha at the centre of the Institute. Experimental field 50 ha (plantations completed on 12 ha). Glasshouses 500 m^2, plastic covered structures 1000 m^2, shelters 500 m^2. Because of the rather limited protected area, experiments with ornamental plants grown under glass are made also in production enterprises. Tissue culture laboratory.

Department of Ornamental Plant Growing and Dendrology — Head: Prof.Dr. B. Nagy
Technology of large-scale ornamental plant growing under glass and plastics: testing native substrates for flower production — Prof.Dr. B. Nagy/ Mrs. K. Márta
Selection and controlled growing of Kalanchoe — Mrs. Dr. M. Szántó
Production technology of succulents suitable for commercial production — F. Incze
Tissue culture methods for ornamental plants — Mrs. Dr. E. Jámbor-Benczur
Evaluation, breeding and production of native and exotic hardy plants and perennials tolerating adverse climatic conditions: propagation of Tilia argentea by cuttings — Dr. G. Schmidt
Large-scale production methods of Paeonia varieties — Dr. Gy. Lászay
Evaluation of the ecological conditions in different parts of the country — Dr. J. Galambos
Use of herbicides in the nursery for hardy ornamental plants. Energy-saving methods of forcing — L. Komiszár

Department of Landscape Architecture

1118 Budapest, Villányi ut 35-43
Phone: 665-283
Telex: 226011 uhort h
Head: Dr. I. Jámbor

The Department's major tasks are education, research and development work concerning the green areas of settlements and the theoretical and practical aspects of landscape architecture. The disciplines taught are: methodology of garden design, history of landscape architecture, geodesy, planning the green area system. Research work is connected with national problems in environmental protection, urban development and landscape architecture.

Development of the green area system in settlements — Dr. I. Jámbor/ Dr. A. Perjés, Mrs. I. Szász-Kozma
Investigations in the history of landscape architecture — Mrs. Dr. I. Balogh-Ormos
Realization in landscape architecture — Dr. A. Perjés/ I. Demjén
Instruments for remote sensing — Dr. L. Babos

Department of Landscape Planning	1118 Budapest, Villányi ut 35-43 Phone: 850-666/228 Telex: 226011 uhort h Head: Dr. A. Csemez

Education: Landscape planning. Nature conservation planning. Land levelling. Horticultural geodesy. Visualization of the landscape architectural aspects of garden, landscape and landscape-types.
Research: This is the only place for research in landscape planning in Hungary. The Department participates in landscape planning (regional, general and detailed plans) and environmental protection linked with the new developments of national importance; in environmental impact studies. Methods of landscape evaluation and for determining the carrying capacity of the landscape are also dealt with.

Landscape planning. Landscape evaluation. Multipurpose use of the landscape. Rehabilitation of disturbed land. Urban fringe	Dr. A. Csemez
Landscape planning. Carrying capacity of the landscape. Planning of recreation areas	Dr. P. Csima
Land levelling. Geodesy	J. Virág/ Cs. Tóth
Visual aspects of setting developments into landscape	Mrs. I. Kecskés-Szabó
Landscape planning: multipurpose use of backwater	Mrs. M. Csillik

2.2.5 Department of Horticultural Engineering	1114 Budapest, Villányi ut 29 Phone: 666-749 Telex: 226011 uhort h Head: Prof.Dr. J. Karai

Developing energy saving technologies in horticultural production and processing. Automatization of the operation of greenhouses. Utilization of different energies and waste energies in horticultural production.

Automatization in greenhouses: Irrigation and climatization systems, installations for irrigation with fertilizer solutions and for plant protection	Prof.Dr. J. Karai Dr. J. Sz. Lukács
Automation of the mechanized grape and fruit production without soil damage	Prof.Dr. J. Karai
Cucumber harvest mechanization. Automation of the construction and maintenance of parks and green areas	A.Z. Horváth
Complex mechanization of material transport in greenhouse units and in small fruit production	Mrs. J. Magyar-Bándi
Measuring and utilization of agrophysical properties of horticultural products.	Dr. Z. Láng Mrs. S. Deák-Karácsonyi
Designing energy saving harvesters suitable for harvesting vegetables for the fresh market	
Energy saving mechanized technology of small fruit production	Dr. E. Horváth
Application of solar and geothermic energy in horticultural production. Increasing the heat content of low potential waste energy by heat pumps	Dr. S. Nagy/ S. Kurtán
Automation of energy saving technologies in horticultural production and processing with the aid of microprocessors and data collecting software. Development of methods and instruments for measuring agrophysical properties of the crop	Dr. B. Gyulai

2.2.6 Department of Genetics and Plant Breeding	1118 Budapest, Menési ut 44 Phone: 664-120
Research work: genetics of horticultural plants.	Telex: 226011 uhort h

Development of breeding methods. Resistance breeding. Crossing of species and genera. Experimental Station	Head: Prof.Dr. I. Tamássy
Szigetcsép: 41 ha experimental fields. Co-60 irradiating garden.	
Mutation breeding: supermutagenes and X-rays	Dr. I. Tamássy Mrs. Dr. I. Tamássy
Vine breeding: breeding for resistance to frost, *Peronospora, Botrytis, Uncinula necator, Agrobacterium tumefaciens and Viteus vitifolii*	Dr. I. Koleda/ J. Korbuly
Fruit breeding: apple, peach, apricot, almond, cherry and hazel	Dr. I. Tamássy/ A. Pedryc
Breeding watermelons and other vegetables for resistance to Fusarium	Dr. K. Mozsár/ P. Muzik
Plant physiology: ecophysiology of horticultural crops. Tissue culture, micropropagation	Dr. A. Máthé/ Dr. J. Szalai Mrs. M. Hemle-Jánosi
2.2.7 Department of Botany and Botanical Garden Soroksár	1118 Budapest, Ménesi ut 44 Phone: 665-494 Telex: 226011 uhort h Head: Prof.Dr. A. Terpó Dir. of the Botanical Garden Soroksár

Cytological, organological, taxonomical, arealgeographical, ecological, phytocoenological and archeobotanical investigations in cultivated plants and in wild growing plants of economic value. Studies on the origin of cultivated plants and on the immigration of weeds.
Botanical Garden Soroksár (Hortus Botanicus "Soroksarensis" Budapest): Living plant collections of *Vitis, Pyrus, Hedera, Rosa, Prunus, Cerasus, Pinus nigra, Arum, Paeonia* etc. Area of the Botanical Garden: 60 ha. A part of it has been declared Nature Conservation Area because it is covered by original steppwood plants like *Festuca vaginata, Secale silvestre, Acer tataricum, Fraxinus pannonica* etc. and by sandy soil vegetation.

Taxonomical investigations in *Vitis, Pyrus, Prunus, Arum, Ornithogalum, Panicum, Avena spp.*	Prof.Dr. A. Terpó
Investigations in the flora and vegetation of industrial areas and settlements	Prof.Dr. A. Terpó Mrs. Dr. K.E. Bálint
Reclamation of industrial areas (red mud impoundments). Flora and vegetation research in abandoned vineyards and stone quarries.	Mrs. Dr. K.E. Bálint
Study of the biology of flowering	
Genus *Rosa*: taxonomy, arealgeography, coenology, living collections. Archeobotany. Carpology	Dr. G. Facsar
Histology of vegetables	Mrs. Dr. E. Felhős-Váczi
Selection, propagation and arealgeography of the genus *Hedera*	Mrs. Dr. M. Bényei-Himmer
Biology of mushrooms, arealgeography and coenology	Dr. I. Rimóczi
Cytology of cultivated plants	Miss dr. M. Tömösközi
Fenology of evergreens: *Ilex aquifolium, Laurocerasus officinalis, Viburnum rhitidophyllum, Berberis spp.*	Prof.Dr. A. Terpó Mrs. Dr. M. Benyei-Himmer Miss Dr. M. Tömösközi Dr. K.E. Bálint
2.2.8 Department of Plant Protection	1118 Budapest, Ménesi ut 44 Phone: 850-666/321 Telex: 226011 uhort h Head: Prof.Dr. M. Glits

Nematodes in horticulture	Dr. K. Farkas
Insect pests in horticultural crops	Dr. B. Pénzes
Phytophagous mites in horticulture	Mrs. Dr. K. Kerényi-Nemestóthy
Scale insects. Woolly aphids	Miss Dr. G Ördögh
Weed control	Dr. A. Meszleny
Diseases in vegetable and fruit crops	Prof.Dr. M. Glits
Botrytis spp.	
Diseases in ornamental plants	Dr. Gy. Folk
Smut fungi	Mrs. Dr. K. Imre
Virus diseases in ornamental plants	Miss Dr. Zs. Némethy
2.2.9 Institute of Economy Education and research on management, labour organization, business administration, marketing in horticultural farms and in the processing industry.	1114 Budapest, Villányi ut 29 Phone: 851-919 Telex: 226011 uhort h Dir.: Prof.Dr. I. Dimény
Department of Agricultural Production Agricultural and horticultural production relationship in farm management	Phone: 664-355 Head: V. Balla
Department of Farm Management Economic problems of horticultural engineering. Economy of production systems in horticulture	Phone: 851-919 Head: Prof.Dr. I. Dimény
Economy of vegetable growing	Prof.Dr. I. Rédai
Economy of fruit production: methodology of economic studies in vertical integration and investments	Dr. L.Z. Kiss
Farm organization. Labour study. Personal computers	Dr. J. Bálint
Field production of vegetables. Processing, marketing. Vegetable seed growing	Mrs. Dr. P. Vig/ Dr. B. Kolozsvári
Utilization systems of glasshouse and plastic tunnel units	Mrs. Dr. K. Csóke
Economy of viticulture and enology: technology, mechanization and technical development	Mrs. Dr. A. Klenczner
Economy in ornamental plant growing. Production, marketing. Economy of landscape architecture, planning, building and maintenance of parks	I. Boross/ Mrs. Dr. Gy. Jáni
Economy in canning industry: technical development, management. Methodology of economic studies in production and processing	Dr. A. Szabadkai
Drip irrigation systems	Dr. Zs. Horánszky
Department of Agricultural Economics Marketing and consumption of vegetables and fruit	Phone: 664-958 Head: Dr. O. Pozsgay Prof.Dr. Tomcsányi Dr. G. Szabó/ Dr. G. Totth Mrs. Dr. K. Szép
Applied statistics: statistical methods in agricultural production	Dr. O. Pozsgay
Department of Computer Techniques Applied mathematics, computer sciences, operation research	Phone: 669-273 Head: Dr. F. Szidarovszky Dr. K. Szenteleki/ Dr. A. Ferenczy
2.2.10 Department of Viticulture	1118 Budapest, Villányi ut 35-43 Phone: 664-650 Telex: 226011 uhort h Head: Prof.Dr. P. Kozma

Research work in progress includes: Studies on the physiology of grapevine. Phytotechnical operations, nutrient supply and their effects on wine quality (organoleptic properties).
Breeding, cultivar testing. Equipment: Experimental Station Szigetcsép: 40 ha of grafted and own rooted vine with 82 cultivars, 200 hybrids, 6670 seedlings. 650 m^2 glasshouses used for model experiments with different soils. 1000 m^2 plastic covered greenhouses of which 450 m^2 are used as nursery for the new hybrids.

Biological bases of viticulture. Examination of the variability of grape varieties. Breeding of grape varieties by selection and crossing. Nutrition and fertilization of vine in relation to wine quality	Prof.Dr. P. Kozma
Variety tests. Phytotechniques of vine. Training and pruning methods. Methodology of testing vine varieties and phytotechnical operations	Dr. P. Csepregi
Physiology of vine: Effects of fertilization on the nutrient uptake and chemical composition of the plant. Modelling of the uptake and translocation of mineral elements in vine	Dr. D. Polyák
Summer pruning. Wood to herbaceous shoot grafting of vine. Breeding by selection and crossing	Dr. L.Sz. Nagy
Effects of training and pruning methods on the biology of vine. Study of the regeneration of vine after frost	Dr. I. Balogh
Economy of viticulture. Climatic conditions and vine growing relationships. Prognostication of yield by bud testing	Dr. F. Bényei
Physiology of vine: Mineral element content of the plant as affected by foliar fertilization	A. Lőrincz
Pollen studies. Examination of micromorphological characteristics of pollen in grapevine species and varieties by scanning and transmitting electron microscopy	Mrs. Dr. B. Tompa
Breeding of grape varieties by selection and crossing: testing resistance	Mrs. Dr. L. Seszták
Biochemistry of vine: Examination of biogene amines, aminoacids and protein in vine by chromatographical and electrophoretical methods. Chemical composition of the plant: its influence on the organoleptic properties of must and wine	Mrs. Dr. O. Juhász

2.2.11 Department of Soil Science

1118 Budapest, Villányi ut 35-43
Telex: 226011 uhort h
Head: Prof.Dr. L. Hargitai

Soil chemistry and biochemistry. Availability of nutrient elements and nutrient supply of horticultural plants. Nitrogen forms, availability and distribution of N in soils and movement of N forms in horticultural plants. Man-made soils, humus research, peat in horticulture. Nutrient requirement of ornamental plants. N supply in orchards. Soils and environmental protection	Prof.Dr. L. Hargitai
Fertilization problems in vineyards. P and K dynamics in sandy soils. P fertilization and microelement supply	Mrs. Dr. K. Tóth-Surányi
N requirement of ornamental plants. Soil and	Mrs. Dr. E. Vass

plant N forms and their distribution in ornamental plants and vegetables	
Mobility and adsorption of microelements in soils. Adsorption of microelements and toxic heavy metals on soil organic matters: its importance in the nutrient supply of horticultural plants and in environmental protection	Mrs. Dr. M. Takács
N dynamics: soil-plant relationships in vegetable crops. Nutrient movement and availability in man made soils	Mrs. E. Forró
Soil-plant relationships in horticultural crops: soil sulphur content and sulphur fractions movement and availability. Sulphur contamination of soils as a problem in environmental protection.	Dr. Bialy A. Ragab
Soil-plant relationships: N dynamics and their effect on the quality of fruits. Transformation of nitrogen forms and the biological activity of soils	Mrs. Dr. M. Garami

2.2.12 Institute of Vegetable Growing

1118 Budapest, Ménesi ut 44
Phone: 667-282
Telex: 226011 uhort h
Dir.: Prof.Dr. S. Balázs

Department of Vegetable Growing

Head: Prof.Dr. S. Balázs

Education covers all aspects of vegetable growing. The main fields of research and development are: Physiology of vegetable crops. Vegetable growing in the field and under protection. Breeding of several vegetable species.
Equipment at the centre of the Institute: phytotron 4 chambers of 5m^2, temperature range 5 to 45°C, illumination up to 40000 lux, humidity up to 100%. Plant tissue culture laboratory.
Laboratory for phytochemistry. At the Experimental Field Budapest-Soroksár: experimental fields 4 ha; glasshouses 1 ha; heated plastic covered structures 1 ha; plastic covered mushroom growing units 1000 m^2; rooms for mushroom growing substrate preparation (dry substrate, 100°C) 500 m^2; soilless culture units 120 m^2; soil analytical laboratory for advisory service to farms producing vegetables under protection.

Soil and nutrient requirements and fertilization of vegetables grown under protection. Symptoms of nutrient deficiency and overdosage. Physiological disorders. Factors affecting nitrate content of vegetables	Dr. F. Tarjányi/ Dr. I. Terbe
Heat and light: causes of bolting in cole crops. Flower fertilization and fruit setting in paprika	Mrs. Dr. E. Kristóf-Kégl
In vitro cell-, tissue- and organ culture of vegetables: micropropagation of onions, leek and paprika	Dr. J.I. Nagy
Production technology of field onions	Dr. F. Tarjányi Mrs. J. Támas-Nyiri
Peas	Dr. J. Nagy
Large-scale production systems. Relations between vegetable production and processing industries. Economy of green bean production	Mrs. Dr. J. Dimény
Vegetable growing under protection: profitable cropping systems for the production units	Prof.Dr. S. Balázs
Development of plastic covered structures. Energy saving heating methods	Dr. I. Turi/ J. Gyurós
CO_2 application. Economy of vegetable growing under plastics	Dr. F. Zatykó
Cultivation in containers. Melons, zucchini	Dr. J. Nagy

Plant growing substrates. Advice on soil fertilization. Fruit setting agents in tomato culture — Dr. I. Terbe

Paprika growing under plastics. Breeding of paprika varieties for protected growing — Dr. I. Turi

Research on the introduction of mushroom species not grown so far — Prof.Dr. S. Balázs, Dr. I. Szabó

Production technology of *Pleurotus spp.* — Dr. I. Szabó/ J. Gyurós

Breeding of *Pleurotus spp.* Production technology of *Agaricus bisporus* — Dr. I. Szabó

Department of Medicinal Plant Growing
Research work is done with medicinal and aromatic plants and spices. Equipment:
10 ha experimental station Budapest-Soroksár

1114 Budapest, Villányi ut 29
Phone: 664-998
Telex: 226011 uhort h
Head: Dr. L. Hornok

Ecology of aromatic plants and spices. *Umbelliferae, Labiatae.* Introduction and production technology of medicinal plants — Dr. L. Hornok

Medicinal plants of tonic effect — Mrs. Gy. Csáki

Nutrient uptake of spices — Mrs. M. Heltmann-Tulok, Mrs. K. Halász-Zelnik

2.3 Konzerv- és Paprikaipari Kutatóintézet
(Research Institute for Canning and Paprika Manufacture)

1097 Budapest, Földvári u. 4
Phone: 335-750
Telex: 224408
Dir.: Mrs. Dr. M. Szenes

Investigations on raw material suitability. Research and development in technology and technics. Development of products and packaging. Technological model trials. Advisory service.

Complex study and modernization of paprika manufacture — I. Szilágyi

Department for Research and Development (pilot plant)

Testing and evaluation of cultivars of crops grown for the canning industry from the aspects of processing technologies. Elaboration of processing technologies for new raw materials of vegetable origin. — Head: Mrs. E. Sárosi-Tánczos

2.4 Központi Élelmiszeripari Kutató Intézet
(Central Food Research Institute)

1022 Budapest
Herman Ottó ut 15
Phone: 354-569
Telex: 22-4709 KÉKI
Dir.: Dr. P. Biacs

Organization and coordination of the food scientific and food technological researches in Hungary. Food scientific and food technological and also research-econometric and micro-economic tasks. Comparative analysis of results in national and international research and in management. Information, data-provision.

Food Chemical Division
Elaboration of instrumental and enzymic analytical methods. Development of new technological processes based on biochemical principle — Head: Miss Dr. A. Halász, Mrs. Dr. M. Petró-Turza

Food Physical Division
Determination of the physical properties of — Head: Dr. K. Kaffka/ Dr. J. Gönczy

foods following their changes during processing and storage. Computer control of the food industrial production

Food Biology Division
Elaboration of biotechnological and food preservational procedures. Radiation energy and its combined treatments
Head: Dr. I. Kiss
Miss Dr. É. Gelencsér

Food Technological Division
Economic production of foods of low energy and high dietary fibre content and completed in proteins
Head: Dr. B. Czukor
Mrs. Dr. K. Zetelaki-Horváth
Miss Dr. J. Petres
Mrs. Dr. Zs. Sallay-Horváth

Section of Food Enzymology
Elaboration of enzymic processes. Development of modern instrumental analytical methods
Head: Dr. Á. Hoschke

2.5 Magyar Tudományos Akadémia Növényvédelmi Kutatóintézete
(Plant Protection Institute, Hungarian Academy of Sciences)
1525 Budapest
Herman Ottó u. 15. Pf. 102
Phone: 358-137
Telex: 224709 KÉKI
Dir.: Prof.Dr. Z. Király

This is the national institute for research work in the field of plant protection, dealing with research on plant pathology, disease resistance, entomology, weeds and pesticide chemistry.

Department of Zoology
Investigations in the apple orchards' and maize stands' ecosystems
Head: Dr. F. Kozár
Mrs. Dr. K. Balázs/ Dr. G. Jenser
Dr. F. Kozár/Dr. G. Lóvei
Dr. Z. Mészáros/ Dr. B. Nagy
Mrs. Dr. V. Rácz/ F. Szentkirályi
Mrs. Dr. É. Visnyovszky

Department of Plant Pathology
Head: Prof.Dr. J. Vörös

Dieback and canker (apoplexy) diseases on stone fruit *(Pseudomonas syringae, Cytospora cincta, Eutypa armeniacae)*
Prof.Dr. Z. Klement
Mrs. Dr. Zs. D. Rozsnyay

Fungi causing bark decay on fruit trees
Dr. L. Vajna

Powdery mildew diseases of vegetable crops and ornamental plants
Mrs. Dr. Gy. Zs. Nagy

Viruses and virus diseases of leguminous crops, vegetables and grapevine
Dr. L. Beczner/ Dr. J. Burgyán
Dr. J. Lehoczky/ Dr. S.P. Salamon

Department of Pathophysiology
Head: Prof.Dr. Z. Klement

Genetic background of the pathogenicity of *Pseudomonas phaseolicola* (halo-blight disease) on bean
G. Somlyay/ Mrs. M. Hevesi
Prof.Dr. Z. Klement

2.6 Mezógépfejlesztö Intézet
Development Institute for Agricultural Engineering)
1016 Budapest
Krisztina krt. 55
Phone: 556-122
Telex: MEFI 22-4532
Dir.: A. Szalka

Design and development of agricultural and horticultural machines and machine systems. Machine testing. Instrument development. Manufacturing experimental model machines.

Department of Horticultural Machinery
Design of electrohydraulic equipment for selective pruning in viticulture and fruit growing
Head: F. Kocsis
A. Kovács

Design of systems for processing pome and stone fruit	J. Balsay
Harvesting and utilization of fruit-tree trimmings	Gy. Szabó/ P. Bede
Department of Harvesters	Head: Dr. J. Janzsó
Tomato harvesters, combing-picking type	
Harvesting and preparing horticultural by-products for thermal utilization and other purposes	L. Lakos
Harvesting vine-shoots	K. Gergely

2.7 Növénytermesztési és Minösitö Intezet (Institute for Plant Production and Qualification)

1024 Budapest
Kisrókus u. 15/a
Phone: 153-840
Telex: 224860 noemi h
Dir.: Dr. B. Szalóczy

The Institute's major tasks are: The state supervision of variety testing; making proposals for licensing the introduction of varieties; maintaining national collections of genetic resources of field crops and vegetables; conducting tests related to the protection of the new varieties; the supervision of seed growing and vegetative propagation material production and qualification. The Institute has 15 research stations of 1500 ha altogether, for variety testing and 12 regional centres in the country.

Division for Horticulture and Forestry	Dir.: Prof.Dr. P. Tomcsányi
Department of Variety Testing	Head: Dr. Z. Faluba
Stone fruits	J. Harsányi
Pome fruits and nuts	Mrs. Dr. I. Bödecs
Table grapes, small fruit	L. Majoros
Wine-grapes	Gy. Kotmajer
Major Department of Vegetables and Medicinal Plants	Head: S. Tuza
Tomatoes, cucumbers	Mrs. E. Györi
Onion, green pepper, asparagus	A. Fehér
Cole crops	Mrs. Cs. Sánta
Root crops, leaf vegetables	Mrs. K. Bogdán
Cucurbits, sweet corn, red pepper, mushroom	B. Mártonffy
Green peas, beans	Mrs. Á. Schmelcz
Seed supervision. Variety tests with medicinal plants	O. Köck
Variety Testing Group for Ornamental Plants and Forest Trees	Head: I. Bach
Ornamental trees and shrubs	Miss M. Telbisz
Annuals, perennials, pot and greenhouse plants and roses	L. Kocsis
Forest trees	I. Bach
Testing seed vigour	Mrs. K. Ertsey
Testing flower seeds	Mrs. G. Horváth
Testing seeds of forest trees	Mrs. M. Csapai
Laboratory for Cytology	2200 Monor, Madách u. 6 Phone: 262 Head: Mrs. Zs. Nagyjános
Agrobotanical Centre	2766 Tápiószele, Phone: 41 Telex: 22-69-81
National collection of genetic resources for vegetable crops	Dir.: Dr. J. Unk
Central Fruit and Vine Research Station	6034 Helvécia Phone: (76) 22-682

Testing grapevine and fruit varieties — Head: Cs. Bartha
Quality of fruit varieties for processing — I. Fazekas
Research Station — 8929 Pölöske, Phone: 2
Testing grapevine, fruit and vegetable varieties — Head: Dr. D. Pálfi

3 Debrecen

3.1 Agrártudományi Egyetem Debrecen
(University of Agriculture Debrecen)
elevation: 122 m
rainfall : 580 mm
sunshine :2020 h

4015 Debrecen
Böszörményi ut 138
Phone: (52) 17-888
Telex: 72211
Rector: Prof.Dr. G. Halász

3.1.1 Department of Horticulture — Head: Prof.Dr. F. Pethő

Education of students of agriculture in all branches of horticulture. Research work on fruit crops and vine with particular attention to the special problems of the district. Vegetable breeding, cultivar testing. Experimental station 69 ha.

Evaluation of apple rootstocks and varieties — Prof.Dr. F. Pethö/ Miss M. Ress
Testing the productivity and yield stability of new grapevine varieties selected for the South-Nyirség district — Dr. L. Géczi
Technology of horse-radish growing — J. Haraszthy
Pea breeding for canning — Prof.Dr. A. Ács/ Miss Gy. Csontos
Green pepper, radish, celery, garden sorrel breeding — Mrs. Dr. I. Harmat
Mrs. M. Samek
Cabbage, squash, *Canna generalis breeding* — I. Nagy

4 Gödöllő

4.1 Agrártudományi Egyetem Gödöllö
(University of Agriculture Godollö)
elevation: 207 m
rainfall : 560 mm
sunshine :1960 h

2103 Gödöllö
Phone: (28) 10-200/634
Telex: 224892
Rector: Prof.Dr. F. Biró

4.1.1 Department of Horticulture — Head: Prof.Dr. L. Cselőtei

Education of students of agriculture in vegetable, fruit and vine growing. Fundamental research work in plant/soil/water/meteorological factors relations. Applied research work in the development of irrigation systems for horticultural crops (methods, doses, frequency of irrigation). Equipment: 15 ha experimental field. 1500 m^2 glasshouses, plastic covered structures and laboratories for culture pot experiments.

Bases of the irrigation of horticultural crops. Effects of environmental factors on the water uptake and transpiration of vegetables. Irrigation of vegetable crops — Prof.Dr. L. Cselőtei
Interactions of water and temperature in relation to the development and yield of vegetable crops — Dr. Gy. Varga
Water relations of orchards as affected by environmental factors — Dr. S. Nyujtó
Water relations of vineyards as affected by environmental factors — A. Csáky

4.2 MÉM Müszaki Intézet
(National Institute of Agricultural Engineering)

2101 Gödöllő, Tessedik S.u 4
Phone: (28) 20-644

Formerly: Mezőgazdasági Gépkisérleti Intézet

Telex: 022-5816
Dir.: Dr. Gy. Bánházi

Tasks of the Institute: Applied research and development work. Development of agricultural machine systems. Testing new Hungarian made and imported agricultural machines. Elaborating development programs for agricultural engineering. Introduction of new machines, machine systems and technologies to the farms. The Major Department for Horticultural Engineering is engaged in solving engineering problems and in elaborating mechanized technologies for all branches of horticultural production.

Major Department of Horticultural Engineering	Head: I. Bakos
Department for the Mechanization in Vegetable Production	Head: L. Szepes
Mechanization of onion and root crop growing	F. Jakovác
Mechanization of the production of leguminous crops	P. Nádas
Mechanization of vegetable growing under protection	S. Tatár
Department for the Mechanization of Fruit Growing	Head: S. Velich
Department for the Mechanization of Vine Growing	Head: Dr. G. Göblös
Mechanization in vineyards on slopes	Dr. L. Némethy
Department for the Mechanization of Plant Protection	Head: Dr. Gy. Dimitrievits
Development of plant protection technologies	Mrs. Dr. J. Varga
Department of Horticultural Establishments	Head: F. Bakos
Mechanization in small farms	Mrs. G. Kasza-Kerék
Department of Economy	Head: L. Berczi
Bureau for the Co-operation in Horticultural Engineering	Head: L. Csukás

5 Kecskemét

5.1 Kertészeti Egyetem Kertészeti Föiskolai Kara
(University of Horticulture, College of Horticulture)
elevation: 120 m
rainfall : 520 mm
sunshine :2050 h

6000 Kecskemét
Erdei Ferenc tér 1-3
Phone: (76) 20-655
Telex: 26-239 uhort h
Dir.: Prof.Dr. I. Filius

The College is engaged in the education of horticultural production engineers specialized for fruit and vine growing or vegetable and ornamental plant growing. Parallel to education, research work is done in the above mentioned branches of horticulture, aimed at production technology development. Experimental fields have been established on a part of the 300 ha model farm of the College. Research work is in progress in glasshouses, phytotrons and laboratories, too.

5.1.1 Institute for Social Sciences and Economy — Dir.: Dr. L. Király

Department of Agricultural Economy	Head: Dr. L. Király
Investigations on energy utilization in different horticultural technologies. Economy of the utilization of horticultural by-products. Development of management science	Mrs. Dr. S. Tajthy Dr. I. Hévizi Mrs. Dr. I. Nagy Mrs. Dr. N. Hamar T. Ferenczy/ Dr. I. Verók

5.1.2 Institute for Fruit Growing and Viticulture — Dir.: Dr. M. Soltész

Department of Fruit Growing	Head: Dr. M. Soltész
Production techniques adapted to cultivars in intensive orchards. Composition of varieties in	I. Szita/ Mrs. Á. Megyeri Mrs. I. Horváth/ E. Misurák

fruit plantations. Higher yield stability in fruit plantations in the open and under plastics. Fruit growing in containers — F. Csorbai

5.1.3 Institute for Vegetable and Ornamental Plant Growing — Dir.: Prof.Dr. I. Filius

Department of Vegetable Growing — Head: Prof.Dr. I. Filius
Physiological demands of vegetables. Plastics in vegetable growing. Energy utilization in vegetable production — Dr. L. Dobos/ Dr. A. Kovács, Dr. E. Gólya/ Mrs. D. Somogyi, J. Fejes
Department of Ornamental Plant Growing — Head: Mrs. B. Lőrincz
Flower arrangement. Production of dry flowers. Ornamental plant growing under plastics. Turfing — Dr. P. Lévai/ Mrs. F. Csorbai

5.1.4 Institute for Natural Sciences and Engineering — Dir.: Dr. I. Sohajda

Department of Engineering — Head: Dr. S. Szabó
Evaluation of model technologies in mechanized horticultural production with particular attention to rentability — Dr. Gy. Kovács/ Dr. I. Pekáry, Miss A. Tóth/ M. Viola
Department of Agriculture and Plant Protection — Head: Dr. I. Petrányi
Planting density and energy saving cultivation in asparagus production. Electrostatic treatment of potato tubers. Population dynamics and signalization of moth species with the aid of light traps and feromon traps. Exact methods for measuring pest populations. Collection of data on the etology of pests with the aim of increasing the effectiveness of defense. Energy-saving methods of turfing, with particular respect to soil and nature conservation — Mrs. B. Fehér/ Dr. J. Jarfas, Dr. E. Tomcsányi/ O. Szabó
Department of Natural Sciences — Head: Dr. I. Sohajda
Tissue culture propagation methods. Systematization of pear species. Modern methods of solving organizational problems in horticultural production — Dr. T. Fehér/ Dr. F. Frigyesy, Mrs. Dr. M. Soltész, Mrs. J. Baranyi/ J. Nagy, J. Szalai

5.2 Kertészeti Egyetem Szőlészeti és Borászati Kutató Intézete (Research Institute for Viticulture and Enology of the University of Horticulture)

6001 Kecskemét-Kisfái 182
Phone: (76) 22-066
Telex: 26518
Dir.: Dr. J. Zilai

This is a national institute for grapevine breeding and for research in viticulture and enology. Equipment: 30 ha vineyards for breeding and research work and 150 ha for large-scale experiments. Phytotron contruction in progress.

Major Department of Viticulture — Head: Dr. L. Szőke
Department of Breeding and Genetics — Head: Dr. P. Kozma jr.
Breeding of grapevine varieties by crossing. Breeding for resistance to frost and foliar diseases caused by fungi
Clonal selection, acclimatization — Dr. O. Luntz
Breeding of winegrapes by crossing. — Miss Dr. E. Hajdu

Frost resistance breeding	
Department for Variety Maintenance and Evaluation	
Setting up and maintenance of virus-free stock	Head: Dr. O. Luntz
Variety evaluation	Dr. J. Zilai
Setting up and maintenance of stock showing no symptoms of virus infection	T. Simon
Department of Plant Protection	Head: Dr. J. Mikulás
Plant protection in viticulture	
Viruses	J. Lázár/ Dr. J. Lehoczky
Agrobacterium tumefaciens	Dr. E. Szegedi
Pests	G. Szabó
Weeds in vineyards	Dr. E. Pölös
Department of Physiology and Ecology	Head: S. Misik
Frost tolerance of vine	Mrs. É. Nagy-Radnai
Growth regulators	Mrs. I. Szabó-Murányi
Photosynthesis	Miss B. Bálo
In vitro culture	Miss Dr. N. Jákó
International ecological test	Dr. L. Diófási (Station Pécs)
Soil Conservation Department	Head: Dr. L. Szöke
Manuring and fertilization: Nutrient accumulation and the maintaining of the nutrient level in the soil	
EUF soil testing method	Mrs. Dr. Zs. Várnay
Foliar analysis	Mrs. B. Környei (Station Pécs)
Department of Production Technique	Head: Dr. J. Füri
Water balance and irrigation	
Intensive methods of pruning and training	Dr. L. Diófási (Station Pécs)
Soil management. Green manuring	B. Tantó
Re-cultivation on hillsides	A. Bányai
Mechanization in vine plantations	Dr. S. Varga

5.3 Zöldségtermesztési Kutató Intézet
(Vegetable Crops Research Institute)

6001 Kecskemét
Mészöly Gy. ut 6
Phone: (76) 27-633
Telex: 26330
Dir. Gen.: Prof.Dr. S. Balázs

The Institute is the national centre of scientific investigations related to vegetable growing. In its capacity of research program leader the Institute coordinates the research work of the cooperating institutions. Its main responsibilities are vegetable breeding, maintenance breeding and production development, including seed production. Three stations belong to the central institute. The stations and station-departments are located in different vegetable production districts of the country. Their activity is related to the special problems in the districts.

5.3.1 Institute centre — Kecskemét

Research work is conducted in the following fields: Breeding of tomato, cucumber, root vegetables, sweet corn, popcorn, mushrooms and several minor vegetable crops. Resistance breeding. Virus research. Production technology. Advisory service. Large-scale seed production. Equipment: 500 ha arable land, 1.2 ha glasshouses, 3 ha plastic covered structures. Phytotron. Laboratory for canning technology. Chemical laboratory. Soil laboratory. Tissue culture laboratory. Laboratory for the production of mushroom spawn.

Research Division	Head: Dr. N. Hamar
Breeding and Stock Maintenance Department	
Tomato breeding	Head: Dr. J. Farkas
	Miss Cs. Komlósi
Cucumber breeding	Dr. P. Milotay/ Mrs. L. Körös

Sweet corn and popcorn breeding	Dr. L. Daniel/ Miss Dr. I. Bajtay
Watermelon breeding	T. Féher
Minor vegetable crops	Dr. I. Cserni/ Dr. S. Hodossi
Breeding of root vegetables	Miss Dr.G. Bujdosó/ Mrs. I. Hraskó
Mushroom breeding	Mrs. Dr. M. Gyenes
Plant Protection and Resistance Research Department	
Resistance breeding	Head: Dr. S.A. Hódosy
	Mrs. Dr. F.H. Kiss
Virus research	Mrs. Dr. Zs. Basky
Chemical weed control	Mrs. I. Szundy
Agrotechnical Department	Head: Dr. I. Cserni
Nutrient and water regime of vegetables	Dr. N. Hamar
Production technologies, seed treatment,	Dr. N. Hamar
sowing, planting methods	
Soil and plant analysis, nutrition advice	Dr. K. Prohászka
Mushroom production	Prof.Dr.S. Balázs/Mrs.Dr.M. Gyenes
Computer methods	Dr. L. Kecskeméti
Laboratory for canning technology	Head: L. Bontovits/K. Gyuris
Development of processing methods.	
Product qualification	
Biochemical research	Dr. L. Vidéki
Production Division	Head: Dr. I. Ackerl
Large-scale seed production	
Production Development Division	Head: J. Bittsánszky
Advisory service: Cucumber (gherkin) growing for pickles	
Vegetable growing under protection	Gy. Malatinszki
Vegetable production for processing, canning	P. Biró
Vegetable growing under protection	A. Szijjártó
Field vegetables. Asparagus	F. Bauer
Plant protection in field vegetable growing	P. Klimaj
Onion production	Dr. S. Györffy
Seed production	G. Vurai
Tomato and pepper growing in the field	Mrs. G. Zala

5.3.2 Research Station Ujmajor	3024 Ujmajor - Selyp
elevation: 127 m	Phone: Lórinci 263
rainfall : 570 mm	Telex: 25-311
	Stat. Dir.: Dr. Gy. Botos

Breeding and production technics of green peas, pepper, cucumber, onion, melon, green beans. Fundamental research in virology, entomology, plant pathology, tissue culture. Area: 143 ha.

Growing and breeding of green peas	Head: Dr. A. Lászlóffi
	Dr. L.Csizmadia/Mrs. H. Dienes

Department Budatétény	1775 Budapest P.O.B. 95
elevation: 100 m	Phone: 264-060
rainfall : 550 mm	Telex: 22-6088
sunshine :2100 h	Head: Dr. Gy. Botos

Vegetable pepper breeding	Dr. L. Zatykó/ Mrs. Dr. A. Moór
Genetics of pepper	G. Csilléry
Nutrient relations in pepper	Miss Dr. I. Fischer
Cucumber breeding	Dr. P. Pados/ P. Balogh
Technology of cucumber growing	Dr. Gy. Botos
Biochemistry	Mrs. Dr. É. Baboth

Onion breeding	Mrs. I. Gubicza/Miss Zs. Füstös
Genetics of onion	Dr. A. Andrásfalvy
Melon breeding	Mrs. Dr. E. Zatykó
Green bean breeding	Dr. I. Velich
Entomology	Dr. V. Martinovich
Virology	Dr. I. Tóbiás
Bacterial diseases	Dr. I. Szarka
Fungal diseases	Dr. P. Lukács
Plant tissue culture	Dr. M. Fári

5.3.3 Red Pepper Research Station
elevation: 93 m
rainfall : 577 mm
sunshine :2090 h

6300 Kalocsa, Obermayer tér 9
Phone: Kalocsa 50
Telex: 26360
Dir.: Dr. K. Kapeller

The station and its department Szeged are located in the Hungarian red pepper production district. They are concerned in red pepper breeding and production development. Area: 100 ha.

Breeding: quality improvement, resistance. Production technology	Dr. K. Kapeller
Breeding, genetics	Dr. F. Márkus
Irrigation, production technology	Dr. M. Bérenyi
Direct seeding	Dr. J. Kapitány

Red Pepper Department Szeged
elevation: 87 m
rainfall : 573 mm
sunshine :2100 h

6728 Szeged, Külterület 7
Phone: (62) 61-433
Head: Dr. Gy. Somogyi

Irrigation and water regime of red pepper varieties	Dr. Gy. Somogyi
Environmental, chemical and physiological factors influencing colour in red pepper	Dr. J. Mécs
Red pepper breeding	K. Szepesy

5.3.4 Research Station Makó
elevation: 82 m
rainfall : 526 mm
sunshine :1815 h

6901 Makó, Széchenyi u. 89
P.O.B. 66
Phone: (65) 12-455
Telex: 82-477
Stat. Dir.: Dr. F. Szalay

The station, situated in the traditional district of the two-year production method of onions, is engaged in the development of this traditional system and in developing the one-season production method, too. For each system, breeding of varieties and maintenance breeding are in progress. Area: 25 ha.

Onion breeding	Dr. F. Szalay
Genetics, variety maintenance	Dr. A. Barnóczki
Plant protection. Resistance breeding	Dr. E. Stoylova

6 Keszthely

6.1 Agrártudományi Egyetem Keszthely
Készthelyi Mezógazdaságtudományi Kar
(University of Agriculture Keszthely)
(Faculty of Agriculture Keszthely)
elevation: 116 m

8361 Keszthely, Deák F. u. 16
Phone: 12-330, 11-290
Telex: 35242
Dean: Prof.Dr. Z. Kardos

rainfall : 700 mm
sunshine :1985 h

Department of Horticulture — Head: Prof.Dr. Mrs. K. Kótun

Horticultural education in vegetable, fruit and vine growing of students of agriculture. Research work in the same branches of horticulture. Equipment: 16 ha orchard, 12 ha vineyard, 12 ha field vegetables, 3000 m^2 plastic covered structures, 500 m^2 glasshouses.

Collections of apple and pear cultivars as genetic resources	Prof. Mrs. Dr. K. Kótun
Regulation of growth and fruit setting in pome fruits	Mrs. Dr. Gy. Borka
Grapevine breeding	Dr. K. Bakonyi
Breeding of grapevine rootstock varieties and clones	Mrs. D.A. Körmendi
Vegetable breeding: virus resistance breeding in green pepper and beans	Dr. J. Kovács
Lettuce breeding. Storage of vegetables	Dr. J. Farkas

7 Mosonmagyaróvár

7.1 Agrártudományi Egyetem Keszthely, Mezögazdaságtudományi Kar Mosonmagyarkóvár
(University of Agriculture Keszthely, Faculty of Agriculture Mosonmagyaróvár)
elevation: 118 m
rainfall : 594 m
sunshine : 1916 h

9200 Mosonmagyaróvár
Vár 2
Phone: (98) 15-911
Telex: 24211
Dean: Prof.Dr. J. Schmidt

Kertészeti Tanszék
(Department of Horticulture)
The Department is engaged in the horticultural education of students of agriculture and in some special problems of horticultural research.

9200 Mosonmagyaróvár
Lenin ut 80
Phone: (98) 11-067
Telex: 24211
Head: Prof.Dr. Gy. Nagy

Acclimatization, breeding and maintenance of squash and pumpkin varieties. Vegetative propagation of rootstock for chestnut of poor growth type. Maintaining a collection of genetic resources of cultivated blueberry.	Prof.Dr. Gy. Nagy

8 Szentes

8.1 Vetómag Vállalat Kutató Állomása
(Seed Producing and Trading Co. Research Station)
elevation: 98 m
rainfall : 550-650 mm
sunshine :2050-2200 h

6601 Szentes, P.O.B. 41
Phone: 445
Telex: 82392
Dir.: Dr. L. Faragó

Breeding, maintenance and seed production of pepper, cole crops, radish, melon, medicinal plants, spices and several agricultural crops. Equipment: 342 ha arable land, 2 ha plastic covered area, 0.14 ha thermal water heated glasshouses.

Vegetable Crops Department — Head: Dr. M. Pesti

Breeding for resistance to fungal and bacterial diseases of vegetables	Dr. M. Pesti
Sweet and hot pepper breeding	Dr. A. Barta

Breeding of cole crops	Dr. B. Maczák
Breeding of minor vegetables, medicinal plants and spices	Dr. Gy. Horváth/ L. Fodor
Gel-electrophoretic seed quality determination	Dr. H. Niemi
Plant tissue culture	Dr. M. Viola

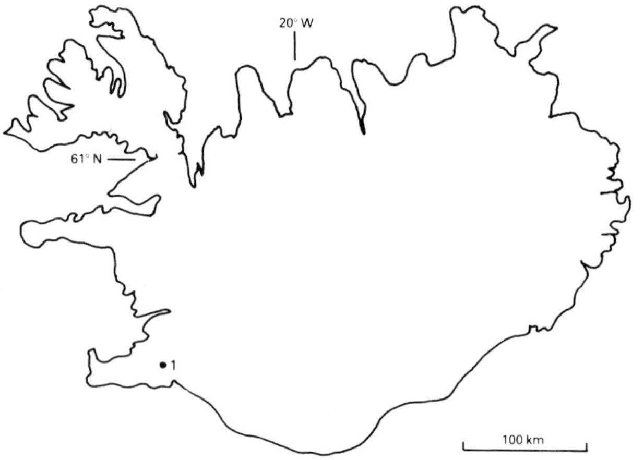
20° W
61° N
1
100 km

1 Hveragerdi

Survey

The Research Institute Nedri-As is an incorporated non-profit organization, located in the lowland of south Iceland, where geothermal activities are common.
Geographic location is 64°00' North and 21°12' West.
Main activities are research in the field of geomorphology, soil sciences, botany, geothermal organisms, algology, plant breeding and genetics of vegetable crops, resistance to viruses and parasitic nematodes.
Annually, several foreign scientists and post-graduates visit the Institute and work on independent research problems.
Publication: The Bulletin of Research Institute Nedri-As.

1 Hveragerdi

1.1 Research Institute Nedri-As
elevation: 15 m
rainfall : May - Sept. 453 mm

810 Hveragerdi
Phone: (99)-4185
Dir.: G. Sigurbjornsson

Geormorphology	Prof. Dr. E. Schunke Dr. K. Richter
Soil sciences	Dr. J.-F. Venzke Mrs. K. Venzke dipl.-geogr.
Agricultural Meteorology	Dr. H. Liebricht Dr. T. Jakobsson
Botany	Prof. H. Böttcher Prof. Dr. E. Stahl Prof. Dr. Lj. Kraus Prof. Dr. L. Steubing
Geothermal organismes	Dr. A. Binder Dr. R.A.D. Williams Dr. D.M. Ward Dr. P. Füglistaller
Algology	Dr. I.M. Munda
Plant breeding and genetics	Dr. E.I. Siggeirsson
Greenhouse production	O. Hjartarson O. Asgrimsson

Prepared by Agr. Dr. E.I. Siggeirsson, Research Institute Nedri-As

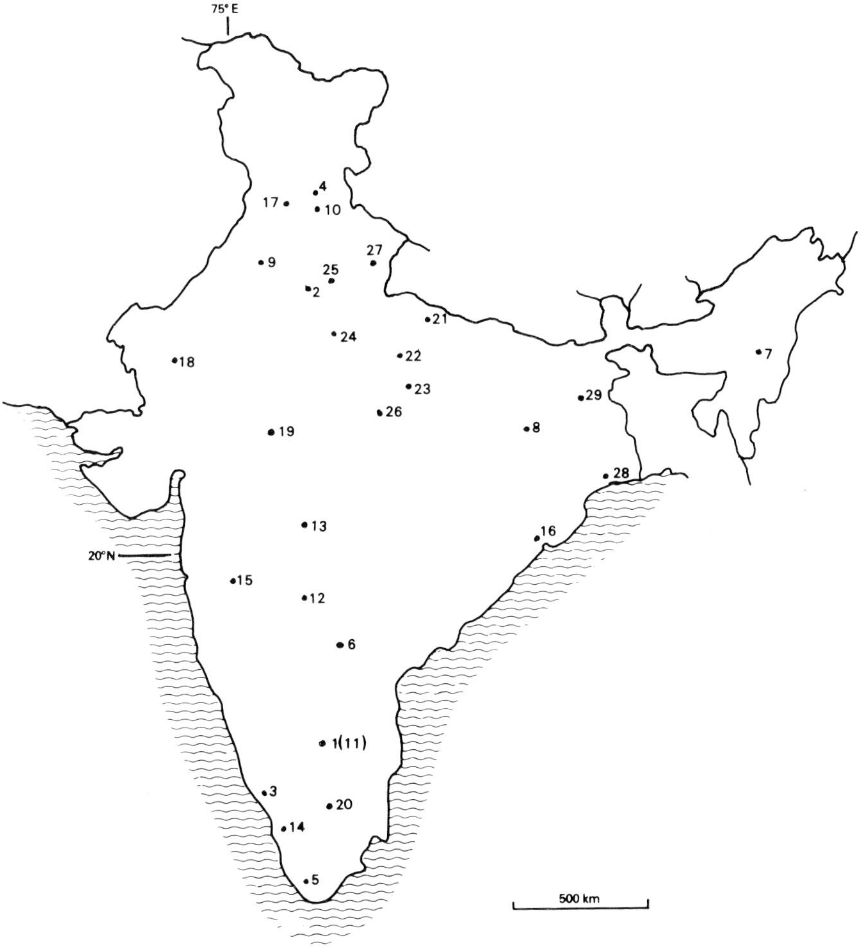
75° E
20°N
17
4
10
9
25
2
27
21
24
22
18
23
26
19
29
8
7
28
13
16
15
12
6
1(11)
3
20
14
5
500 km

Karnataka: 1 (11) Bangalore, **Haryana**: 2 New Delhi, 9 Hissar, **Kerala**: 3 Kasaragod, 5 Trivandrum, **Himachal Pradesh**: 4 Simla, 10 Solan, **Andhra Pradesh**: 6 Hyderabad, **Assam**: 7 Jorhat, **Bihar**: 8 Sabour, **Maharashtra**: 12 Parbhani, 13 Akola, 14 Rahuri, 15 Dapoli, **Orissa**: 16 Bhubaneswar, **Punjab**: 17 Ludhiana, **Rajasthan**: 18 Jodhpur, 19 Udaipur, **Uttar Pradesh**: 20 Pantnagar, 21 Faizabad, 22 Kanpur, 23 Allahabad, 24 Chaubattia, 25 Saharanpur, 26 Lucknow, 27 Varanasi, **West Bengal**: 28 Kalyani, 29 Malda

Survey

India is fortunately endowed with a diversity of soils and climates, which permit cultivation of temperate, tropical and sub-tropical fruit in various regions of the country. Some of these fruits like mango and some species of citrus and banana are indigenous to the country, whereas many fruits have been introduced from abroad. In recent years, there has been great enthusiasm for the extension of areas under apples in the Western Himalayas, under grapes both in the peninsular and northern India and pineapples in the southern states of India. Potato and vegetables have also been receiving more importance because of the urgency of improving nutritional content of the commonly eaten food. As a result during 1983-84, the area under fruit increased up to ± 2.25 million ha and that of vegetables and potato to ± 4.89 million ha. As the areas have increased progressively, the various problems relating to different crobs have also increased and require urgent attention. To cater to these requirements, horticultural research in the country is now being carried out by a number of organisations, namely, Indian Council of Agricultural Research (ICAR), State Agricultural Universities (SAU) and private and semi-government organisations like the Indian Coffee Board, the Indian and United Tea Planters Association, Tea Research Association etc.

Prepared by Dr. G.L. Kaul, Ass. Director-General of Horticulture, ICAR, New Delhi, India

No commercial or private organisation undertakes any research on horticultural crops. In 1935 four regional stations were established in different parts of the country, i.e. Sabour (Bihar), Kodur (Madras), Krishnagar (Bengal) for research on sub-tropical and tropical fruit and Chaubattia (UP) for temperate fruit. It was, however, only after independence (1947) that horticultural research received some importance. A first step in this direction was establishment of the Central Potato Research Institute in 1949 at Patna, which was subsequently shifted to Simla in 1956. In 1956, an independent Division of Horticulture was created in the Indian Agricultural Research Institute, New Delhi. This Division has now been further split into three Divisions viz. horticulture and fruit technology, olericulture, floriculture. Later, the Central Tuber Crop Research Institute was established in 1963 at Sreekaryam, near Trivandrum for research on tuber crops (other than potato). During the 3rd Five-Year-Plan-period a significant development took place with the establishment of the Institute of Horticultural Research (IIHR) at Hessaraghatta near Bangalore (Karanataka) by the ICAR in 1967 to work on major problems of horticultural industry. This is the first Institute of its kind in the country which has now developed into an internationally known institute devoted to a large variety of horticultural crops covering all major disciplines. The Institute has 7 regional stations in different parts of the country. One of its Stations at Rehmankhera, Lucknow (UP) has now been upgraded recently (1984) into a fullfledged institute named as Central Institute of Horticulture for the Northern Plains with mandate for working on mango as lead crop and other horticultural crops important for the region. The Central Plantation Crops Research Institute (CPCRI), Kasaragod came into existance in 1970 with the merging of two Coconut Research Stations, one at Kasaragod and the other at Kayangulam, as also the Central Arecanut Research Station, Vittal and its Regional Stations. The Institute is working on coconut, arecanut, cashew, cacao, oil palm, pepper, cardamom, ginger, turmeric and tree spices.

During the third plan-period, major development also took place in many States where agricultural universities came into existance on the pattern of Land Grant Institution of U.S.A. Such universities, now exist at least one in each of the major states, except Maharashtra, Uttar Pradesh and Bihar, which have 4, 3 and 2 universities respectively. Full-fledged horticulture departments now exist in these universities and have given a big boost to horticultural research.

Another significant development in horticultural research has been the organisation of research on a co-ordinated basis. At present All India Coordinated Research Projects exist on fruits, vegetables, tuber crops, potato, coconut and arecanut, spices and cashew, medicinal and aromatic plants, floriculture, mushrooms and post-harvest technology. For this purpose full-time Project Co-ordinators who are senior research scientists, have been made responsible for co-ordination of the project at the centres all over the country. All these projects are proposed to be continued during the Seventh Plan (1985-90), besides having new projects on tissue culture and temperate fruits.

In addition to these, ICAR has been sanctioning ad-hoc schemes in horticultural crops to work on specific topics in a time-bound period for which assistance is given in full by the Council.

The research on improvement of vegetable crops was initiated in 1947-48 at the Indian Agricultural Research Institute, New Delhi, and in different states like Punjab, Uttar Pradesh, West Bengal, Maharashtra, Himachal Pradesh, Jammu and Kashmir and Tamil Nadu under ad-hoc schemes sponsored by the ICAR. Vegetable Breeding Station at Katrain in Kulu Valley (Himachal Pradesh) was established in 1949, subsequently transferred to the Indian Agricultural Research Institute in 1955. This centre is carrying out research on temperate vegetables, besides growing these crops for seed production. Work on the vegetable crops was also taken up in the Agricultural Universities and also at the Indian Institute of Horticultural Research, Hessarghatta (Bangalore), after these were established. All India Coordinated Vegetable Improvement Project was started by the ICAR in 1970-71 with 7 main centres and 10 sub-centres located in various agro-climatic regions of the country. Presently, the research on these crops is being carried out at the ICAR Institutes like IARI, New Delhi, IARI Regional Station, Katrain (H.P.), I.I.H.R., Bangalore, Vivekanand Parvatiya Krishi Anusandhan Shala, Almora, and ICAR Research Complex for the North Eastern Hill Region as well as in the Agricultural Universities. The All India Coordinated Vegetable Improvement Project is now operating at 22 cooperating centres and 24 voluntary centres covering about 20 vegetable crops. In addition a number of ad-hoc research schemes on the selected vegetable crops are also supported by ICAR operating in the Institute/Universities.

The ICAR has launched systematic research work on mushrooms as a new addition to its horticultural research activity during the Sixth Plan (1980-85). In this one the National Research Centre at Solan (Himachal Pradesh) for doing basic work, and a separate All India Coordinated Project for locating specific work at 6 centres, have been established.

During the Seventh Plan (1985-90) period, it is proposed to add a few more centres under the Project to cover major vegetable producing areas like Gujarat and Andhra Pradesh.

1 Bangalore

1.1 Indian Institute of Horticultural Research*
elevation: 863 m
rainfall : 900 mm

Bangalore 560 089, Karnataka State
Phone: 384356/32486
Telex: BG-534
Dir.: Dr. K.L. Chadha, M.Sc. (AgO PhD.)*

Breeding, production technology and processing of fruits and vegetables, breeding of ornamentals, production of mushrooms, fruits, including mango, guava, papaya, citrus, banana, pineapple, grape, pomegranate, zyziphus and custard apple. Vegetables including, tomato, gourds, egg plant, okra, onion, green and bell pepper and leafy vegetables. Experimental area 220 ha. Specialised laboratory for tissue culture and residues of insecticides. Equipment: Gamma chamber with Co_{60} source, mass spectroscopy, electron microscope, auto analyser growth chamber, glasshouse facilities available.

1.1.1 Fruit Crops

Breeding of mango, guava, and papaya	C.P.A. Iyer/M.D. Subramanyam
Citrus scion breeding	V.V. Sulladmuth/Narayanappa
Citrus rootstock breeding	P.K. Agarwal
Banana varietal evaluation	P.K. Agarwal
Grape breeding	Rajendra Singh
Improvement of semi arid fruits	S.H. Jalikop
Rootstocks for mango	R.R. Kohli/Y.T.N. Reddy
Mango nutrition	K.L. Chadha*/R.R. Kohli Y.T.N. Reddy
Guava agrotechniques	R. Chittirai Chelvan S.D. Shikhamany/K.C. Dubey*
Citrus rootstocks	S.A. Haleem/V.V. Sulladmath M.M. Ganapathy
Citrus nutrition	S.A. Haleem/K. Anajaneyulu K.A. Nanaya/M.M. Ganapathy
Cultural practices in citrus	M.M. Mustaffa
Banana agrotechniques	B.M.C. Reddy/M.M. Mustaffa M.M. Ganapathy
Spacing and fertilizer trials in pineapple	B.M. Reddy/K.L. Chadha* M.M. Mustaffa
Rootstocks, training van pruning in grape	B.M.C. Reddy
Grape nutrition	S.D. Shikhamany/R. Chittirai Chelvan/K.L. Chadha*
Propagation of fruit crops	Y.N. Reddy
Horti-Silvi Pastoral Systems	G.B. Raturi/S.S. Hiwale

1.1.2 Vegetable Crops

Tomato breeding	S.K. Tikoo/N. Anand
Improvement of *Capsicum* and chillies	A.A. Deshpande/C.S. Pathak D.P. Singh
Brinjal and okra breeding	O.P. Dutta
Muskmelon breeding	K.R.M. Swamy/O.P. Dutta
Improvement of melons and gourds	O.P. Dutta/K.R.M. Swamy V.S.R. Krishna Prasad
Peas breeding for yield, quality and disease resistance	A.B. Pal/Akella Vani M.B.N.V. Prasad/S. Kumar
Breeding of French bean, dolichos, cowpea and minor beans	Akella Vani
Onion breeding	C.S. Pathak/A.A. Deshpande D.P. Singh
Improvement of leafy and cruciferous vegetables	S.C. Pandey/M.B.N.V. Prasad

Agro-techniques and nutrition of vegetables	V. Shukla/B.S. Prabhakar L.B. Naik
Multiple cropping in vegetables	B.S. Prabhakar/V. Shukla
1.1.3 Ornamental Crops	
Improvement of chrysanthemum, China aster and gladiolus	S.S. Negi/S.P.S. Raghava
Improvement of roses and bougainvillea	R.N. Bhat/Meenakshi Srinivas
Improvement of bulbous ornamentals	J.L. Karihaloo
Orchid breeding	Foja Singh
Agro-techniques for rose, gladiolus and tuberose	A. Mukhopadhyay
Agro-techniques for jasmine	G.K. Sharma/R. Jayaseelan
1.1.4 Medicinal and Aromatic Crops	
Dioscorea improvement	V. Rama Rao
Improvement of *Catharanthus* and steroid bearing *solanum*	R. Krishnan/Subhas Chander
Improvement of jasmine	H.C. Srivastava/P.G. Karmakar
Mutation breeding in scented geranium and breeding of Patchouli	V.R. Naragund/T.V. Kumar
1.1.5 Post Harvest Technology	
Storage of fruits	Shantha Krishnamurthy
Storage of vegetables	K.P. Gopalakrishna Rao
Product improvement and new product development in fruits	Amba Dan
Packaging and storage of fruit products	Amba Dan
Preparation and quality improvement of wines	E.R. Suresh/S. Ethiraj
Microbial spoilage of fruit and vegetable products	S. Ethiraj/E.R. Suresh
1.1.6 Plant Pathology	
Fungal diseases of fruit crops	R.D. Rawal/B.A. Ullasa M. Narayanappa
Guava wilt	S. Kumar
Diseases of fruits in transit and storage	B.A. Ullasa/R.D. Rawal
Fungal diseases of vegetables	T.S. Sridhar/ S. Kumar
Fungal diseases of leguminaceous crops and seed pathology	M.N. Maholay
Studies on fungal diseases of ornamental crops	C.I. Chacko/N.N. Raghavendra Rao
Studies on fungal diseases of medicinal and aromatic crops	N.N. Raghavendra Rao/C.I. Chacko
Virus diseases of fruit crops	S.R. Sharma/K.L. Manjunath
Virus and mycoplasma diseases of vegetable crops	S.J. Singh
Bacterial diseases of fruit and vegetable crops	Ram Kishun
Cultivation of European and oyster mushroom	R.P. Tiwari/ V. Girija
1.1.7 Entomology and Nematology	
Pest management in mango and citrus	P.L. Tandon/B.A. Bhumannavar
Pest management in solanaceous and leguminous vegetables	G.C. Tewari/P.N. Krishna Murthy
Pest management in okra and cruciferous vegetables	K. Srinivasan/N.K. Krishnakumar
Pest management in ornamental, medicinal and aromatic crops	Sujaya Udayagiri/N. Jaganmohan
Biological control of fruit and vegetables pests and weeds	K.P. Jayanth/S.P. Singh A. Krishnamurthy/B.S. Bhumannavar K. Narayanan

Parasitic nematodes of fruit and vegetable crops	P.P. Reddy
Fishery	D.S. Krishna Rao

1.1.8 Plant Physiology and Biochemistry

Physiological growth analysis of vegetable crops	N.K. Srinivasa Rao
Biological nitrogen fixation in vegetables	Sukhada Mohandas
Endogenous growth substances in citrus fruit	G.S.R. Murti
Uneven ripening of grape	C.K. Mathai
Tissue culture	R. Doreswamy
Chemical weed control	Prabha Challa/D. Leela
Growth and development and aroma constituents	D.K. Pal/Y. Selvaraj
Triterpeniods of cucurbitaceous crops	M.V. Chandravadana
	M. Subhas Chander
Biochemical basis of disease resistance in vegetable crops	M. Subhas Chander
	M.V. Chandravadana
Pesticide residues	M.D. Awasthi
Fruit lipids	A.K. Ahuja

1.1.9 Soil Science

Available NPK in soils for fruit and vegetable crops	T.R. Subramanian/M.H.P. Rao
Salt tolerance in fruit and vegetable crops	R. Palanaippan
Micronutrient studies in fruit and vegetable crops	M. Edward Raja/K.A. Nanaya
	K. Anjaneyulu
Leaf nutrient guides in fruit crops	B.S. Bhargava/K. Anjaneyulu
	K.A. Nanaya
Fertilizer use effeciency	B.R.V. Iyengar/S.V. Kesava Murthy
Water management	D.M. Hegde
Mycorrhizal fungi	H. Onkarayya
Storage requirements of fruit and vegetable seeds	S.D. Doijode
Pollen preservation in fruit and vegetable crops	S. Ganeshan/M.P. Alexander

1.1.10 Economics and Statistics

Economics of production and marketing of vegetable crops	K.V. Subrahmanyam
Optimum size and shape of plots	V.R. Srinivasan/S.R. Biswas
	R.P. Singh
Yield forecasting in fruit trees	K.S. Shamasundaran
Meteorological data of horticultural regions	P.R. Ramachander/V.R. Srinivasan

2 New Delhi

2.1 Indian Agricultural Research Institute	New Delhi-110012, Haryana State
elevation: 2286 m	Phone: 585214
rainfall : 7086 mm	Dir.: Dr. A.M. Michael
Division of Fruits and Horticultural Technology	Head: Dr. R.M. Pandey

This Division deals with fruit crops and horticultural technology. It has evolved two mango varieties 'Mallika' and 'Amrapali'. The latter variety appears dwarf and shows a possibility of being used in high density planting. Recently, a landmark has been achieved in developing true-to-the-type plants of papaya through tissue culture. The International Check-list of Mango Cultivars comprising chief characteristics of 793 cvs. from India, Florida, Philippines, Puerto Rico, Sri Lanka, South Africa has been published. It has three substations. The substation at Simla (Himachal Pradesh) is engaged in the assessment and utilization of wild species of pome and stone fruits. At Pusa (Bihar), a breeding programme in papaya has been set up and five outstanding selections from sib mating have been made out which are in the process of release. These selections are named as Pusa Giant, Pusa Majesty, Puse Delicious, Pusa Dwarf and Pusa Nanha. At sub-station Karnal (Haryana) plant multiplication programme is in progress. Six major research programmes are in progress in the main Division at Delhi.

2.1.1 Improvement of plant types	
Improvement of mango by breeding	Dr. D.K. Sharma/Dr. P.K. Majumder
Improvement of grapes by breeding	Dr. P.C. Jindal/Dr. S.N. Pandey*
	Kashmir Singh
Improvement of minor fruits (ber, phalsa)	Dr. V.P. Sharma
2.1.2 Propagation and rootstock studies	
Tissue culture in improvement of horticultural crops (papaya and mango)	Dr. R.M. Pandey/Dr. D.K. Kishore*
	Dr. H.C. Sharma
Propagation techniques in mango and guava	Dr. P.K. Majumder/Dr. P.C. Bose
	Shri Chokha Singh
Standarization of citrus rootstock	Dr. S.K. Saxena/Dr. A.M. Goswami
Standarization of guava rootstock	Dr. P.K. Majumder/Dr. V.K. Sharma
2.1.3 Physiological studies on flowering and fruit growth in some major fruits	
Mango	Dr. R.M. Pandey/Dr. Y.K. Sharma
Grape	Dr. R.M. Pandey/Dr. S.N. Pandey
Studies on salt tolerance in mango	R.C. Khanna/Chokha Singh
2.1.4 Orchard Managament	
Improved orchard management of citrus	Dr. B.B. Sharma/R.C. Khanna
	Chokha Singh/Dr. A.K. Sinha (W.T.C.)
Establishment of high density orchard in citrus	Dr. A.M. Goswami/Dr. S.K. Saxena
Improved guava orcharding	Dr. P.C. Bose/Dr. L.D. Tewari
	Dr. Vishwanath
2.1.5 Taxonomical studies	
Establishment of National Registration Authority on some important fruits in India (including Internation Registration of Mango Cultivars)	Dr. S.N. Pandey
2.1.6 Fruit and Vegetable Preservation	Dr. S.K. Roy*
All India Coordinated Research programme on Post-Harvest Technology of horticultural crops	Dr. S.E. Maini*
Preservation of fruits and vegetables with chemical additives	Mrs. Dr. Vijay Sethi*
Studies on rural oriented preservation of fruits and vegetables	Dr. S.K. Roy*/Dr. D.S. Khurdiya
Regional Stations	
Utilization of native wild species as rootstocks, breeding for disease resistance and developing techniques for their propagation	Dr.S.S. Randhawa
Improvement of papaya by breeding	Dr. Mansha Ram*
2.2 ICAR Headquarters	Krishi Bhawan Dr. Rajendra Prasad Road New Delhi-110001 Ass. Dir.Gen.(Hort): Dr. G.L. Kaul

The Indian Council of Agricultural Research is responsible to undertake, aid, promote and coordinate agricultural and animal husbandry research, education and its application throughout the country since 1929. Its headquarters is located in New Delhi. The Assistant Director General (Horticulture) is responsible for guiding, supervising and coordinating research projects in the field of horticulture including plantation crops, medicinal and aromatic plants and spices. He is also responsible for monitoring all the research schemes and 12 co-ordinated projects on different horticultural crops, operated through Council's assitance at Agricultural Universities and ICAR Institutes. He is also looking after the technical programmes of the work being done in the five ICAR Research Institutes viz.

Indian Institute of Horticultural Research, Bangalore, Central Institute of Horticulture for Northern Plains, Central Potato Research Institute, Simla, Central Tuber Crops Research Institute, Trivandrum and Central Plantation Crops Research Institute, Kasaragod (Kerala). At present he is assisted by the following five scientists at the headquarters of the Council looking after the subjects shown against each.

Fruits, hill and tribal area projects in horticulture, Indian Institute of Horticultural Research, Bangalore and Central Institute of Horticulture for Northern Plains, Lucknow	C.P. Dutta*
Vegetables, potato, floriculture, Central Potato Research Institute, Simla	Ram Phal*
Plantation crops and spices, commodity boards on coffee, tea, rubber and cardanom, Central Plantation Crops Research Institute, Kasaragod	T.A. Sriram
Tuber Crops (except potato), Medicinal and Aromatic Plants, Post-Harvest Technology, Central Tuber Crop Research Institute, Trivandrum	R.S. Arya
Mushroom	V.C.Gupta

3 Kasaragod

3.1 Central Plantation Crops Research Institute (ICAR) elevation: 10 m rainfall : 3492 mm sunshine : 2524 h	Kasaragod 670 124, Kerala State Phone: 94 Telex: 0842-248 CASC IN

This Institute came into existence in 1970 by merging the Central Arecanut Research Station, Vittal (Karnataka) and its five Regional Stations with the two Central Coconut Research Stations at Kasaragod and Kayangulam (both in Kerala). It was then charged with the responsibility of conducting and coordinating research on cashew, cacao, oil palm, spices (black pepper, cardanom, ginger, turmerie, cinnamon, nutmeg, clove and all spice) in addition to those on coconut and arecanut. It is a constituent unit of ICAR, Ministry of Agriculture, Government of India. It has a completent of 225 scientists and 1115 supporting, administrative, technical and auxillary staff. The work of the Institute on various crops is conducted under 14 disciplines in 15 regional stations, research centres, seed farms, and research complexes.

Kasaragod (Head Quarters)	Director: Dr. K.V.A. Bavappa
Agronomy Division	Head: Dr. R.C. Mandal
Genetics Division	Head: Dr. R.D. Iyer
Soil Science Division	Head: Dr. C.C. Biddappa
Plant Pathology Division	Head: Dr. K.K.N. Nambiar
Agril. Economics Division	Head: Dr. P.K. Das
Agril. Extension Division	Head: M.K. Muliyar
Agril. Statistics Division	Head: Jacob Mathew
CPCRI Regional Station	Calicut 673 012, Kerala Jt. Dir.: Dr. M.K. Nair
CPCRI Regional Station	Kayangulam, Krishnapuram 690 533, Kerala Jt.Dir.: Dr. N.P. Jayasankar
Agril. Entomology Division	Head: Dr. Chandy Kurian
Nematology Division	Head: Dr. P.K. Koshy
Plant Physiology Division	Head: Dr. K.D. Patil
Microbiology Division	Head: Dr. Suresh Kumar Ghai
CPCRI Regional Station	Vittal 574 243, Karnataka Jt.Dir.: Dr. K. Shama Bhat
CPCRI, Research Centre	Palode, Pacha 695 562 Trivandrum District, Kerala Sc.-in-ch.: Dr. K.U.K. Nampoothiri

CPCRI, Research Centre	Peechi, Kannara 680 652 Trichur District, Kerala Sc.-in-ch.: A.A. Mohammed Sayed
CPCRI, Research Centre	Hirehalli 572 131 Tumkur District, Karnataka Sc.-in-ch.: Dr. N. Sannamarappa
CPCRI, Research Centre	Appangala, Heravanad 571 201 Coorg District, Karnataka Sc.-in-ch.: Dr. R. Naidu
CPCRI, Research Centre	Kahikuchi, Gauhati 781 017 Assam., Sc.-in-ch.: A.K. Ray
CPCRI, Research Centre	Mohitnagar 715 101 Jalpaiguri District, West Bengal Sc..in.ch.: R.K. Singh
World Coconut Germplasm Centre (CPCRI)	Port Blair 744 101, Andamans Sc.-in-ch.: Dr. Josy Joseph
CPCRI, Field Station	Punkunnam, Trichur 680 002 Trichur District, Kerala Sr.Sc.-in-ch-: Mrs. Dr. K. Radha
ICAR Research Complex for Goa (CPCRI)	Ela, Old Goa 403 402 Sc.-in-ch.: Dr. A.R. Bhattacharyya
ICAR Research Complex for Lakshadweep	Minicoy 673 559 Sc.in-ch.: Chacko Mathew
CPCRI, Seed Farm	Kidu, Nettana 574 230 South Kanara Dist. Karnataka Sc.-in-ch.: N. Yadukumar
CPCRI, Cashew Seed Farm	Narimogru, Shantigodu 574 204 South Kanara District, Karnataka Farm Superintendent: E. Mohan

In addition, this Institute is also the headquarters of three All India Coordinated Projects on Coconut and Arecanut; Cashew and Spices, and a centre for All India Coordinated Projects on Biological control, Rodent Biology and Post Harvest Technology. While research on coconut in this Institute is over 67 years old, and that on arecanut 28 years old, work on other crops has been initiated only since 1970. The foremost achievement of this Institute is considered to be the observation made in early 1930's that the hybrids between darf and tall forms of coconut exhibit hybrid vigour. This finding has since been confirmed in all major coconut growing countries and hybrid coconut seedlings are now being used all over the world for replanting purposes. This Institute has established Elite Seed Farm at Kidu in Karnataka and also helped the State Govts. in Kerala, Karnataka, Tamilnadu, Orissa and West Bengal to set up similar gardens for production of coconut hybrids by supplying the parental materials. The Institute has also developed improved cultivars in several crops as arecanut, pepper, ginger and turmerie.

As recent as 1982, the constant association of Mycoplasma like organisms (MLOs) with root (wilt) diseases has been proved through EM studies of diseased palm tissues and of the brain and salivary gland tissues of the vector, *Stephanitis typicus*. The loss caused by this disease assessed in a recent survey (1984) is to the tune of 901 million nuts valued at 3000 millions rupees annually.

In 1984 the Institute has achieved a significant breakthrough in coconut tissue culture with the production of clonal plantlets through direct somatic embryogenesis from tender leaf segments in a defined medium.

The Institute conducts comprehensive research on developing integrated control measures for the important pests and diseases of all agricultural plantation crops. Another area of thrust is tree crop based farming systems for making small and mariginal holdings as viable units of production.

4 Simla

4.1 Central Potato Research Institute	Simla-171 001, Himachal Pradesh Dir.: Dr. N.M. Nayar*

One of the 50 research institutions of the ICAR, which functions under the Ministry of Agriculture and Rural

Development, Government of India, New Delhi. It is responsible for conducting and coordinating research on all aspects of the potato, producing breeders' seed of cultivars included in the National Potato Seed Certification Programme and acts as the source of information on potato. The Institute has its headquarters at Simla and the following 11 research stations in different agro-climatic regions of the country: Kufri-Fagu (HP), Jalandhar (Punjab), Patma (Bihar), Modipuram-Macchari and Mukteswar (UP), Rajgurunagar-Pune (Maharashtra), Gwalior (MP), Ootacamund (TN), Darjeeling (West Bengal), Shillong (Meghalaya) and Bangalore (Karnataka). The staff in position consists of 135 scientific, 108 technical, 114 administrative, 227 supporting and 51 auxiliary personnel. The sanctioned strength consists of 210 scientific, 200 technical, 55 auxiliary, 144 administrative and 278 supporting staff. The work in the Institute is conducted in the following disciplines. In addition, the All India Coordinated Project begun in 1971 has its headquarters in the Institute. It conducts coordinated research programmes in association with the State Agricultural Universities (SAUs). It has five centres in CPRI and 13 centres in SAUs.

Division of Genetics	Head: P.C. Misra
Division of Physiology and Biochemistry	Head: Dr. N.P. Sukumaran
Division of Crop and Soil Science	Head: Dr. J.S. Grewal
Division of Plant Pathology	Head: Dr. S.K. Bhattacharayya
Division of Entomology and Nematology	Head: Dr. K.S. Krishna Prasad
Division of Seed Production and Certification	Head: H.S. Chauham
Division of Post Harvest Technology	Head: Vacant
Division of Social Sciences	Head (in-ch.): V.P. Malhotra
All India Coordinated Potato Improvement Project	Project Coordinator: Dr. K.P. Sharma

Research Stations:	Officers-in-Charge:
Central Potato Research Station, Kufri-171 009, Simla Hills, HP	Jagpal Singh
CPRS, Shillong-793 009, Meghalaya	Vacant
CPRS, Mukteswar-263 138, Nainital, UP	G.C. Upreti
CPRS, Muthorai-643 053, Nilgiris, TN	C.P. Gajaraja
CPRS, JC Bose Road, Manor Lodge, Darjeeling-743 101, West Bengal	Vacant
CPRS, Model Town PO, Jalandhar-144 003 Punjab	Dr. M.S. Rana
CPRS, Gwalior-474 006, MP	Dr. Chokhey Singh
CPRS, Sahaynagar-801 506, Patma, Bihar	B.B. Das
CPRS, Modipuram-250 110, Meerut, UP	S.M. Verma
CPRS, Rajgurumagar-410 505, Pune, Maharashtra	M.N. Akhade
CPRI Karnataka Centre, c/o Indian Institute of Horticultural Research, 255 Upper Palace Orchards, Bangalore-560 080, Karnataka	Dr. A.B. Gadewar

The Institute has evolved 23 high yielding and improved varieties of potato suitable for cultivation in different agro-climatic zones of the country. The Institute also produces about 2000 tonnes breeder seed of 8 cultivars. The Institute organises annually an International Training on Current Trends in Potato Production (four weeks) in collaboration of International Potato Centre, Lima (Peru), and three courses (ten days each) on Seed Production and Certification.

National Centre for Mushroom Research and Training, Chambaghat, Solan District, Himachal Pradesh	Officer on Special Duty: Dr. H.S. Sohi

This National Research Centre was set up by the ICAR in 1983 to conduct research on all aspects of mushroom production, preservation and utilization and impart training on various aspects of mushroom culture. An All India Coordinated Mushroom Improvement Project (AICMIP) was set up in 1982 for conducting multilocational research on mushroom. It has six centres located in State Agricultural Universities. The staff in position consists of 7 scientific, 1 technical, 4 administrative and 3 supporting staff.

KERALA STATE

5 Trivandrum

5.1 Central Tuber Crops Research Institute
elevation: 30 m
rainfall : 1600 mm
sunshine : 1620 h

Sreekariyam, Trivandrum 695 017
Kerala State
Phone: 8551-8554
Telex: 0884-247 ROOT IN
Dir.: S.P. Ghosh

The Institute was established in 1963 and is located in an area of 42 ha on the outskirts of the city of Trivandrum, the capital of Kerala State. The objectives of the Institute are to conduct and coordinate research on all aspects of all the tropical tuber crops, other than potato, like cassava, sweet potato, aroids and yams. It also acts as a clearing house for all the informations on these crops.

Division of Genetics and Plant Breeding — Dr. G.G. Nayar

Genetic improvement of tuber crops involving introduction of genetic variability by enriching the germplasm collection, breeding high yielding, better quality, disease and pest resistant and drought and shade tolerant varieties.

Division of Agronomy and Soils — Dr. N.G. Pillai

Developing suitable cropping systems and working out the best standards of culture for all tuber crops, water management in tuber crops cultivation and study on the nutritional requirements and fertility management of tuber crops.

Physiology and Biochemistry — Dr. T. Ramanujam

Investigation of the mechanism of tuberisation in all the important tropical tuber crops and to work out the suitable environment conclusive for tuber initiation and development. Identification of suitable yield determining physiological parameters for large scale screening of genetic materials for yield potential and also to develop crop model ideal to fit in different cropping systems. Collection of basic information on photosynthesis, light utilisation, canopy development, partitioning in dry matter etc. with a view to understand the physiological basis of yield variation among genotypes. Identification of suitable physiological characters contributing for tolerance towards moisture stress and adoptation to shaded conditions.
Studies on biosynthesis and catabolism of secondary compounds in tuber crops. Investigations on the mechanism of spoilage in tuber crops. Studies on the biochemical basis for resistance to pests and diseases.

Pathology — Dr. M. Thankappan

Survey and identification of diseases of tuber crops screening of germplasms against diseases and formulation of control measures for diseases.

Entomology — Dr. K.S. Pillai

Identification of pests problems of tuber crops. Survey on the intensity and distribution of different pests. Formulation of effective control measures against the major pests of tuber crops.

Technology — Dr. C. Balagopalan

Post harvest processing of tuber crops. Development of technology for the utilitsation of tuber crops in food, feed and industry.

Agricultural Economis, Statistics and Extension — Dr. T.K. Pal

Evaluation and monitoring of adoption and socio-economic changes introduced by tuber crops technology.

Market information and assessment of investment potential of cassava based industries. Training and dissemination of information and transfer of technology in tuber crops. Designing, experiments, statistical analysis and interpretation of data.

5.2 Kerala Agricultural University
College of Agriculture (Vellayani)
elevation: 29 m
rainfall : 2001.4 mm

Vellayani - 695 522
Trivandrum (Dist) Kerala State
Phone: 73021
Dean: Dr. N. Sadanandan

Imparting scientific knowledge in agriculture at the undergraduate and post graduate level and research in agricultural subjects. There are fourteen departments. The campus has an area of 243 ha. The total academic staff strength of the college is about 127.

5.3 Kerala Agricultural University
elevation: 2 m
rainfall : 3177 mm

Vellanikkara, Trichur - 680 654
Phone: 22497
Dir. of Research - Dr. P.C.S. Nair
Assoc.Dir.: Dr. C.C. Abraham,
Prof. P.N. Pisharody,
Dr. (Mrs) M.K. George
Dr. M. Subramaniam

The directorate is responsible for formulation, implementation and monitoring and evaluation of the research programmes in the College as well as in the research stations under the Kerala Agricultural University. In addition the KAU is also undertaking coordinated research projects (27 nos) financed by the ICAR and the State Govt. on the basis of 75:25. Ad hoc research projects (12 nos) fully financed by the ICAR are also being implemented under this University. National Agricultural Research Project funded by the ICAR with the help of the World Bank is also being implemented by this University.

5.3.1 College of Horticulture

Vellanikkara, Trichur
Phone: 21822
Dean: Dr. P.K. Gopalakrishnan

Imparting scientific knowledge in agriculture, horticulture, at the undergraduate and post-graduate level and do research in Agricultural Subjects. There are fifteen departments. The college of co-operation and Banking at Mannuthy also functioned as part of the academic programme of the institution. Total area of 129 ha comprising of two instructional farms are at Mannuthy (33.7) and the other at Vellanikkara (95.3) is under the institution. The total staff strength of the college is 139.

5.3.2 Agricultural Research Station

Vellanikkara, Trichur
Phone: 20726
Head: Prof. T.F. Kuriakose

Research on agro-techniques of rice cultivation. Out of the total area of 95.35 ha under the station, 5 ha of land is under experiments.

5.4 Regional Agricultural Research Station
elevation: 15 m
rainfall : 3570 mm

Pilicode, Cannanore Dist.
Kerala State
Phone: - 232 - Cheruvathur
Head: Dr. R.R. Nair

The major work in the station is for the improvement of coconut and for development of agro-techniques for coconut and coconut based farming systems. Also identification of locating specific problems and to find solutions. A total area of 56.90 ha is spread at two campuses viz. Pilicode and Nileshwar which comprise of 4.0 ha of wet land and 52.9 ha of garden land. 31.43 ha area is under experiments. A total of 64 field experiments are under operation.

5.5 Regional Agricultural Research Station
elevation: 0.6 m
rainfall : 2794 mm

Kumarakom, Kottayam Dist. - 686 566
Kerala State, Phone: 21
Head: Prof. U. Muhammed Kunju

Research on coconut and coconut based farming system with special reference to integrated farming of coconut, livestock and fish culture, underlying the principle of organic recycling. Also problem oriented location specific research on all crops in the problem region of Kerala. Total area is 44.76 ha of which 27.12 ha area is put under experiments. Total of 42 experiments are in progress.

5.6 Regional Agricultural Research Station
elevation: 974 m
rainfall : 2400 mm

Sultan Battery, Ambalavayal
Wynad District, Kerala State
Phone: 21
Head: Prof. K. Kannan

Research on different aspects of tropical and subtropical fruits and vegetables, spices, paddy, and coconut are being undertaken in addition to studies on coffee based on cropping system. Total area of 87.03 ha is under the station out of which 25.0 ha are spread under experiments. Total of 36 experiments are in progress.

ANDHRA PRADESH

6 Hyderabad

6.1 Agricultural Research Institute
Andhra Pradesh Agricultural University

Rajendranagar
Hyderabad - 500030, Andhra Pradesh

Andhra Pradesh Agricultural University (APAU) was established in 1963 and has three campuses at Rajendranagar (Hyderabad), Bapatla and Tirupathi and a number of Research Stations in various agro-climatic regions. It is the sole agency for the entire research work in the field of Horticulture in the State. Work is in progress on fruit, vegetable, flowers, spices and condiments, and plantation crops.

Agricultural College - Department of Horticulture	Head: B. Ranga Reddy
Studies on hybrid seed production in round fruit types of brinjal	Dr. B. Ranga Reddy
Mycorrhizal studies in Citrus	Dr. Hari Babu/B.V. Rama Rao
Horticultural Research and Development	Head: Dr. G. Satyanarayana
Vegetable seed production	Dr. K.V. Subba Reddy
Research on tuber crops	Smt. Afzalunisa
Grape Research Station	Head: V. Padmanabham
Pathology	K. Chandrasekhar Rao
Entomology	Hari Rao
Research on varietal performance and effect of plant growth regulators on fertility and yield of different varieties.	
Agricultural College	Tirupathi, Chittoor Dist., A.P.
Department of Horticulture	Head: M. Rama Rao
Pathology	Dr. Siddalinga Reddy
Research on vegetable crops and citrus bud certification scheme.	
Agricultural College	Bapatla, Guntur Dist., A.P.
Department of Horticulture	Head: Dr. V. Suryanarayana
Vegetable Research	
Cashew Research Station	Bapatla, Guntur Dist., A.P.
Fertilizer trials and commercial nursery production	Head: Dr. Rao Rama Rao
Fruit Research Station	Sanga Reddy, Medak, A.P.
Research on varietal performance, hybridization, commercial nursery production in major tropical fruit crops	Head: Dr. Vinod Kulkarni
Coconut Research Station	Ambajipet, East Godhavari Dist., A.P.
Commercial Nursery and coconut parasite breeding	Head: T. Rama Rao
Banana Research Station	Kovvur, West Godhavari, A.P.
Varietal performance and fertilizer trials on banana	Head: Prabhakar Rao
Betelvine Research Station	Utukur, Cuddapah Dist., A.P.

Millet Research Station Spices and condiments Chillies breeding	Lam, Guntur Dist., A.P. M. Venkat Rao N. Ramachandra Murthy
Horticultural Research Station Fruits and vegetables research	Vijayarai, West Godhavari, A.P. Dr. P. Venkat Rao
Agricultural Research Station Breeding in chillies	Malyal, Warangal Dist., A.P. Dr. K. Malla Reddy
Fruit Research Station Breeding in mango, guava and nursery production in citrus	Anantharajpet, Cuddapah Dist., A.P. S. Madhavachari
Agricultural Research Station Semi arid fruit research	Ananthapur Dist., A.P. D.V. Satyanarayana Rao
Agricultural Research Station	Rajole, East Godahavari Rajamannar, A.P.
Coconut Disease Investigation Centre Agricultural Research Station	Chintapalli, Visakhapatham Dist., A.P.
Pepper research	Dr.P. Radhakrishna Murthy
Agricultural Research Station Vegetable seed production	Nandyal, Kurnool Dist., A.P. V.K. Murthy

ASSAM STATE

7 Jorhat

7.1 Assam Agricultural University
Department of Horticulture
elevation: 90 m
rainfall : 1600 mm
sunshine : 1800 h

Jorhat, Assam State
Head: Prof. Dr. S. Barooah

Horticultural research covering varietal, agronomic, wed control insect pest and disease aspects on pineapple, vegetables (tomato, okra, brinjal, colocrops, cucurbits) and tuber crops (tapioca, sweet potato, cococasia, *pachyrrhizus Dioscorea,* yam); survey, collection and evaluation of jackfruit germplasm is carried out in the Department under the four All India Coordinated project of the ICAR.
The research on the fruit aspects not covered under the projects is handled by the teaching staff and post-graduate students. The other fruits covered are guava, banana, citrus, papaya, litchi etc. and the vegetables are onion, cucumber, peas *Momordica cochinchinensis, Coccinia, Trichosanthus* etc.

Citrus Research Station

Tinsukia
Off.in Ch.: Dr. K.N. Bhagawati

This station is one of the centres of All India Coordinated Fruits Improvement project on Citrus. The research covers varietal, rootstock, manurial, micro-nutrient and weed control affects of mandarins, citrus decline and rejuvenation of old citrus orchards. The research work is being extended on lemons and pomelos.

Horticultural Resarch Station

Kahikuchi, Assam State
Off.in Ch.: Dr. B. Chakarvarty

The station handles varietal and manurial aspects of banana and coconut research, agronomic and hormonal aspects of pineapple, observational trials on citrus, guava, litchi and mango, varietal and agronomic aspects of cauliflower, beans and brinjal, and entomological and pathological experiments on fruits and vegetables. All aspects of coconut and arecanut research is done under All India Coordinated Research project. Recent introduction is Floricultural Research Project on Tropical and Subtropical flower crops.

Coconut Research Sub-station

Kharuwa, Assam State
Off.in Ch.: R. Das

The station tackles various aspects of coconut research. The particular emphasis is on varietal selection, causes and control of crown rot malady.

Regional Agricultural Research Station

Diphu, Assam State
Horticulturist: Dr. G. Medhi

The multidisciplinary station covers research on horticultural crops of tribal hilly belt. The crops covered are pineapple, banana, citrus, cucurbits, colocasia, tapioca, sweet potato, Brinjal, okra, beans etc. The station aims at devising scientific cropping systems alternative to shifting cultivation. Hill banana research is carried out. Research on flower crops introduced recently.

BIHAR STATE

8 Sabour

8.1 Research Station
Department of Horticulture (Fruits)
Bihar Agricultural College
elevation: 46.2 m
rainfall : 902 mm

Sabour, Bihar State
Phone: 11
Head: Dr. Ram Kumar

Teaching undergraduate and post graduate students. Conducting research work on important fruits such as mango, litchi, guava, citrus and few minor fruits as bael, amla, ber, custard apple, jackfruit etc. encluding utilization of these fruits. Production of grafts gooties. Processing of fruits and vegetables.

Fruit Research and Regional Fruit Research Section — R.K. Singh/B. Mandal
Collection and maintenance of various fruits — R. Singh/S. Tahkur
Hybridization of mango and litchi
Nutritional trial on fruit crops
Production on grafts and gooties
Coordinated Fruit Improvement Project (Mango and Guava) — Dr. B.P. Jain*
Survey, collection and maintenance of germplasm — S.C. Mandal
Hybridization study — M.N. Hoda
Propagational trial — Dr. J. Singh
Regulation of bearing
Fruit Preservation Section
Post harvest studies on mango, litchi, — Dr. T. Singh/Dr. M.M. Pujari
Standardization of fruit product utilization of minor fruits. Commercial production on fruit products
Horticulture teaching Section — R.K. Sharma/V.S. Brahmachari*
U.P. Singh

HARYANA STATE

9 Hisar

9.1 Haryana Agricultural University
Department of Horticulture
(Fruit Crops, Floriculture and Fruit Technology)
elevation: 215 m
rainfall : 400 mm
sunshine : 1900 h

Hisar, Haryana State
Head: Prof. Dr. R. Yamdagni

Improvement of fruit crops — Dr. B.S. Daulta
Fruit tree nutrition — Sh. V.P. Ahlawat

Post-harvest handling of fruits	Dr. O.P. Gupta/Dr. Ran Singh
Salinity tolerance of fruit crops	Dr. P.K. Mehta
Fruit technology	Dr. S.S. Dhawan
Floriculture	Dr. S.S. Sharma

Research Station
Dryland horticulture
Research Station
Horticulture

Bawal (Distt. Mahendragarh)
Off.in Ch.: Dr. A.K. Gupta
Gurgaon
Off.in Ch.: Dr. M. Makhija

HIMACHAL PRADESH

This state is primarily a horticultural state, having a variety of agro-climatic conditions, with reputation of producing the best quality of temperate fruits in the country. Research work in this state is being conducted on important fruit crops like apple, pear, peach, plum, apricot, almond, grapes, walnut, pecan nut, olive and persimmon.

10 Solan

10.1 College of Agriculture, Department of Horticulture and Fruit Technology

Krishi Vishva Vidyalaya
Himachal Pradesh
Head: Prof. Dr. T.R. Chadha,
Sub-Proj. Coord., UNDP Project

At this department research work is conducted mainly on pome fruits, stone fruits and nuts. The department has a research station named S.N. Stokes Hort. Complex, Oachghat, Solan. This Horticultural Complex has a sizeable collection of germplasm of apples, pears, plum, peach apricot, almond, pecan, walnut, olive, pomegranate, persimmon, fig and strawberry. This complex has been recognized by Indian Council of Agricultural Research as he national germplasm centre for stone fruits in particular and temperate fruits in general.
Research work is being carried out on nutritional and irrigation requirements of stone fruits, vegetative propagation of walnut, pecannut, and olive. Rootstock studies in stone fruit, fruit breeding and post-harvest physiology. ICAR/UNDP Project on Centre of Advanced Studies in temperate horticulture is in operation to impart post-graduate education and provide advanced research facilities for post-graduate students. Agro-techniques of temperate fruits propagation, rootstocks, training and pruning pollination studies. Apple breeding cytology, cold storage studies are in progress.

Fruit physiology, growth regulators	Dr. K.K. Jindal
Physiology of dwarfing	
Nutrition, irrigation, vegetative propagation and rootstock studies	Dr. J.S. Chauhan*
Nutrition	Dr. J.N. Bhargava
Fruit technology	Dr. B.B. Lal
Post-harvest physiology	Dr. S.K. Chopra
Post-harvest pathology	Dr. J.L. Kaul
Fruit breeding	Dr. S.D. Sharma

10.2 Regional Fruit Research Station
Himachal Pradesh Agricultural University
elevation: 2286 m
rainfall : 1500 mm

Mashobra, H.P.
Simla-171 007
Phone: 8249,8261
Chief Scientist: Dr. R.P. Awasthi

This is the main centre on temperate fruit, particularly apple, pear and cherry with an area of 64 acres. Work on different aspects, such as varietal evaluation, breeding, orchard management, and plant protection measures, is under progress.

Pomology	G.C. Thakur/S.S. Thakur
Fruit breeding	Yoginder Sharma

Fruit nutrition	Dr. Hari Singh Verma/B.K. Karkara C.S. Tomar
Plant physiology	Dr. S.P. Bhartiya/Dr. P.S. Chauhan
Entomology	Dr. Mahabir Singh/Dr. Ramesh Chander Dr K.L. Verma/S.P. Bhardwaj
Fruit pathology	Dr. G.K. Gupta/Dr. K.D. Verma
Agrostology	Janak Raj Sharma

10.3 Palampur, College of Agriculture, Department of Pomology and Fruit Technology	Palampur, H.P.
Research on citrus, pecannuts, chestnuts and tea husbandry	Dr. V.P. Bhutani/K.L. Sharma S.C. Lakhanpal/Dr. J. Badiyal
Dhaulakuan, Regional Fruit Research Station The research station is primarily engaged in research work on sub-tropical fruits namely citrus, guava, litchi, mango, sub-tropical peach etc.	Dhaulakuan, Dist. Sirmaur (H.P.)
Agro-techniques in citrus and guava, post-harvest physiology of citrus and propagation and weed control	Dr. V.K. Sharma/Dr. U.U. Khokhar R.P. Agnihotri
Kotkhai, National Hortorium Collection, introduction, evaluation and systematic studies on various species and varieties of temperate fruits. Propagation and evaluation of clonal rootstocks for apple, pears, peaches and plums.	Kotkhai, Dist. Simla, H.P. M.P. Dwivedi
Sharbo, High Altitude Dry Zone Research Station This station is engaged in research on grapes, almond, walnut, apricot and apple grown under high altitude and dry zone.	Sharbo, Dist. Kinnaur, H.P. Dr. S.A. Ananda/Dr. A.S. Kashyap Dr. H.L. Farmahan
Bajaura, Regional Research Station Research work on stone fruits, almond, apple, cherry and evaluation of germplasm	Bajaura, Dist. Kulu, H.P. Dr. R.L. Sharma/Dr. N.K. Joolka
Kandaghat, Regional Research Station Research work on various aspects of plum, apricot, peach, persimmon, walnut, pecannut etc. is in progress	Kandaghat, Dist. Solan, H.P. Dr. C. Parmar/G. Sud/R.K. Natyal

KARNATAKA STATE

11 Bangalore

11.1 Horticultural Research Station University of Agricultural Sciences elevation: 930 m rainfall : 875 mm	Bangalore-560065, Karnataka State Gandhi Krishi Vignana Kendra Phone: 366753 Head: Dr. U.V. Sulladmath

Horticultural Research in Karnataka State is carried out by the University of Agricultural Sciences, Bangalore at its campuses at G.K.V.K., Bangalore and Dharwad, Regional Research Stations, Mudigere and Raichur, and at other Research Stations at Arasikere, Mangalore (Ullal and Brahmavar), Chintamani and Hassan (Madenur). In addition, the Indian Institute of Horticultural Research, Hessaraghatta, carries out research on horticultural crops at Hessaraghatta, Chethalli and Gonicoppal. Research mainly on plantation crops is being carried out by the Central Plantation Crops Research Institute, Research Station, Vittal at its other stations at Hirehalli and Appangala. Research work on coffee is being carried out at the Central Coffee Research Institute, Balehonnur.

11.2 Division of Horticulture

Improvement of citrus fruits, mango, nutritional studies on sapota, guave and grapes	Dr. U.V. Sulladmath
Moisture regimes for fruits, silvihorticultural intercropping	Dr. K.M. Bojappa

Cultural studies on vegetable crops and varietal evaluation	Dr. P. Muddappa Gowda
Improvement of mango, clonal propagation of semi-wild fruits	Dr. K.R. Thimma Raju
Cultural studies on papaya grapes	Dr. N. Vijaya Kumar
Tissue culture of fruit plants	
Cultural studies in cashew, citrus fruits	Dr. K.R. Melanta
Cultural studies of commercial flower, improvement of *Hibiscus sp.*	J.V. Narayana Gowda
Improvement and cultural studies on Jasmines and landscaping	Dr. B.G. Muthappa Rai
Cultural studies on aromatic and medicinal crops	A.A. Farooqui
Improvement of onion, eggplant, tomato and gourds	Dr. B.B. Madalageri
Post-harvest technology fruits and vegetables	A.G. Huddar
Cultural studies on vegetable crops	R. Samiullah
Post-harvest physiology of cut-flowers and other horticultural crops	Dr. T. Venkatarayappa*
Improvement of potato and cultural studies on potato	K.S. Krishnappa
Cultural studies on plantation crops	T.H. Ashok

11.3 Department of Horticulture
College of Agriculture
elevation: 670 m
rainfall : 1200 mm

Dharwad 580 005, Karnataka State
Phone: 80191
Head: Dr. U.V. Sulladmath

Improvement of ornamental plants and cultural studies on commercial flowers	Dr. U.G. Nalawadi
Cultural studies on mango, citrus and sapota	Dr. M.M. Rao
Cultural studies on vegetable crops	V.M. Bankapur
Cultural studies on improvement of guava and plantation crops (oil palm/betelvine)	Dr. N.C. Hulamani
Improvement of eggplant and onion	A.A. Patil
Cultural studies on fruit crops	S. Jaganath
Improvement of citrus fruits	A.N. Mokashi
Improvement of sapota	A.K. Rokhade

11.4 Regional Research Station
elevation: 1100 m
rainfall : 2500 mm

Mudigere, 577132, K.S.
Phone: 46
Reg. Ass. Dir.: Dr. H.V. Pattanshetti

Improvement of certain plantation crops and cultural studies on fruit and plantation crops	
Improvement of cardanom and cultural studies on cardanom	Dr. G.S. Sulikhere

11.5 Regional Research Station
elevation: 800 m
rainfall : 600 mm

Raichur, 584 101, K.S.
Phone: 8378

Cultural studies on citrus, mango, fig, certain flower and vegetable crops	P. Narayana Reddy

11.6 Agricultural Research Station
elevation: 875 m
rainfall : 900 mm

Arasikere, 573 103, K.S.
Phone: 465
Superintendent: P.B. Shanthappa

Improvement of coconut and cultural studies on coconut	V.V. Sulladmath

11.7 Agricultural Research Station — Ullal, 574 159, K.S.
elevation: 15 m — Phone: 6249
rainfall : 3100 mm — Head: Dr. M.M. Khan

Improvement of cashew, cultural studies on cashew, improved vegetative techniques in cashew — Dr. M.M. Khan
Pests and disease managament of cashew — Dr. I.G. Hiremath
Studies on evaluation of pest damages and their control in cashew — Dr. B. Mallik

11.8 Agricultural Research Station — Brahmavar, 576 213, K.S.
elevation: 20 m — Phone: 11
rainfall : 3000 mm — Head: Dr. Balakrishna Rao, Regional Ass. Director

Introduction and evaluation of major fruits and certain plantation crops in coastal Karnataka — Dr. B.C. Uthaiah
Introduction and evaluation of annual summer vegetables, and cultural studies on fruit and vegetable crops — B. Lingaiah

11.9 Agricultural Research Station — Chintamani, 563125
elevation: 850 m — Phone: 218
rainfall : 600 mm — Head: M.A. Narayana Reddy

Collection and evaluation of cashew and fruit crops — M.A. Narayana Reddy
Cultural studies on fruit crops — N.C. Narase Gowda

11.10 Agricultural Research Station — Madenur, Hassan, 573220, K.S.
elevation: 900 m — Phone: 25
rainfall : 500 mm

Varietal evaluation of potato and cultural studies — S. Vijayakumar
Studies on pests of potato and their control — B.S. Nandihalli

MAHARASHTRA STATE

12 Parbhani

12.1 Marathwada Agricultural University — Parbhani, Maharashtra State
Department of Horticulture — Dean: Dr. V.K. Patil
elevation: 409 m — Head: Dr. P.A. Deshmukh
rainfall : 824 mm
sunshine : 12 h

Nutritional studies in fruit crops — Dr. V.K. Patil
Standardization on agro-techniques in chrysanthemum — Dr. P.A. Deshmukh
Improvement of fruit crops by selection and breeding — Dr. V.R. Chakrawar, Sr. Res. Off.
Post-harvest physiology in fruit and vegetables — Dr. D.M. Khedkar, Assoc. Prof.
Standardization of agro-techniques in *Gladiolus* and *Jasminum sambac* — Dr. K.W. Anserwadekar, Professor*
Improvement of melons by hybridization — Dr. N.N. Shinde, Sr. Res. Off.
Improvement of chillies by breeding and other methods — Dr. M.B. Sontakke
Standardization of agro-techniques in potato — Dr. J.T. Nankar
Development of package of practices for turmeric and ginger — Prof. S.G. Rajput, Ass. Prof.
Salinity and alkalinity research in citrus — Prof. B.A. Kadam
Growth and development studies in mango and banana — Dr P.R. Narwadkar
Standardization of plant propagation techniques in

tropical and subtropical fruit crops	Prof. S.N. Gunjkar
Standardization of package of practices for cabbage and cauliflower	Prof. S.P. Jinturkar
Standardization of package of practices for leafy vegetables	Prof. R.M. Kulkarni, Ass. Prof.
Standardization of package of practices in merigold	Prof. U.G. Deshmukh
Development of nutritional standards in oranges grown on various rootstocks	Dr. G.S. Shinde

12.2 Fruit Research Station — Aurangbad, M.S.
elevation: 581 m
rainfall : 726 mm
sunshine: 12 h

Development of agro-techniques in mango and banana	Prof. A.L. Ballal
Improvement of tomato by hybridization	Prof. N.D. Budrukkar
Research on fungicidal effects on fruits and vegetables	Prof. J.U. Ashtaputre
Improvement of brinjal and tomato by hybridization	Prof. Darbind
Research on minor fruit crops like tamarind, jamun, feronia etc.	Prof. R.G. Nilangakar

12.3 Vegetable Improvement scheme — Ambajogai, M. State

Standardization of agro-techniques in chilli — Prof. A.S. Mandge

13 Akola

13.1 Punjabrao Krishi Vidyapeeth
Department of Horticulture
elevation: 307.4 m
rainfall : 837.4 mm
sunshine: 2380 h

P.K.V. Akola, Maharashtra State
Phone: 4471 or 4472
Telex: 725/215
Head: Dr. P.P. Deshmukh

Since 1969, this department undertakes teaching, research and extension activities on all the tropical and sub-tropical fruit, vegetable and flower crops commercially important in the region. Teaching includes - teaching for horticultural subjects as per curriculum of M.Sc. (Agril.) in horticulture degree course. The major and commercial crops dealt with in this department are mandarins, lime, sweet orange, banana, guava, pomogranate, mango, fig, chiku, and other dryland fruit crops; chilli, tomato, brinjal, peas, cowpea, beans, cluster beans, potato, cabbage, cauliflower, cucurbitaceous crops and leafy vegetables like fenugreek, spinach, indian spinach etc. and roses, tuberose, jasmine, gaillardia, merigold and other annual flower crops. Total farm area of the horticultural farm is 78 ha of which 20 ha is under perennial fruit orchard.

13.2 Citrus Die Back Sub-Centre
(Under All India Coordinated Fruit Improvement Project on Citrus Die Back)

Central Research Farm
P.K.V., Akola, M.S.
Sr. Hortic.: Dr. G.B. Ohekar

This Sub-Centre of ICAR, New Delhi undertakes research activities as approved by the Project Coordinator as well as those pertaining to local Citrus region with reference to citrus die-back. Under this centre, there is a well-established citrus orchard in an area of 22.0 ha.

13.3 Commercial Fruit Nursery Unit

P.K.V., Akola, M.S.
Jr. Hortic.: B.D. Shelke

This Nursery Unit undertakes multiplication of quality plants of various commercial fruit crops of this region. There are well established progeny orchards of these fruit crops available. These activities are carried out in an area of 5 ha.

13.4 Chilli Research Unit

P.K.V., Akola, M.S.
In charge: Prof. M.M. Damke

This Research Unit undertakes researches on agronomical, breeding, plant protection aspects of chilli crop which is one of the major commercial crops of the region. The activities of the unit are concentrated in an area of 1.75 ha.

13.5 Horticultural Section

College of Agriculture, Akola, M.S.
Prof.in Ch.: M.P. Kawathalkar

This Section undertakes teaching activities of horticultural courses as per curriculum of B.Sc. (Agril.) and B.Sc. (hort.) both 4 years degree course.

13.6 Regional Fruit Research Station
elevation: 417 m
rainfall : 1060 mm

Katol, Dist. Nagpur, M.S.
Phone: 91
Dir.of Hortic.: Dr. T.R. Bagade

The activities of this Research Centre are mostly concentrated on citrus fruit crops including mandarins and other citrus. The important research aspects like evaluation of different varieties, suitable rootstocks, agronomy and plant protection etc. are handled by this station. Besides, multiplication of good quality mandarin and lime plants for distribution to orchardists are also taken up.

13.7 Horticulture Section
elevation: 321.3 m
rainfall : 1136.9 mm

College of Agriculture, Nagpur,
M.S., Phone: 23315
Head: Prof. S.B. Umale

This Section has an area of 41 ha of which 16 ha is under perennial fruit crops. This section undertakes teaching activities of horticultural courses as per curriculum of B.Sc. (Agril.) 4 years degree course and M.Sc. (Agril.) in Horticulture. Besides, some research and extension activities are also being undertaken on horticultural crops.

14 Rahuri

14.1 Mahatma Phule Agricultural University
Department of Horticulture
elevation: 500 m
rainfall : 510 mm

Rahuri-413722, Maharashtra State
Phone: Krishivid
Head: Dr. K.G. Choudhari
Vice-Chanc.: Prof. D.K. Sulunkhe

Post-graduate teaching on fruits and vegetables; research on fruit (mango, grape, pomegranate, citrus, ber, anona, guava, chiku, and fig). Vegetables and flower crops: Central Horticultural Farm covers 475 ha and includes full-fledged nursery production.

Pomology	K.G. Choudhari/A.S. Kokate
	K.N. Wahwal/U.T. Desai
	B.G. Keskar/A.R. Karale
Vegetables	P.N. Kale/Z.V. Deshmukh
	K.B. Choudhari/J.S. Desle
	S.B. Raijadhav/E.D. Yadav
	S.D. Sarode
Floriculture	N.R. Bhat
Pathology	P.Y. Patil/S.A. Mamane
	D.N. Padule
Entomology	D.B. Pawar
Farm	S.M. Choudhari

14.2 College of Agriculture, Horticulture Section
elevation: 559 m
rainfall : 800 mm

Pune-411 005, M.S.
Phone: Prinagri
Head: Dr. B.C. Patil

Undergraduate teaching in horticulture; research on fruit, vegetables and flowers; nursery production	Prof.: D.A. Rane
Pomology	M.R. Gaikwad/G.Y. Desle S.B. Gurav/D.B. Ranade
Vegetables	K.B. Patil/M.T. Patil
Extension (Hort.)	R.N. Nagwade
14.2.1 Regional Fruit Research Station	Ganeshkhind, M.S.
Pomology, Garden Superintendent:	N.S. Shirsath/J.J. Patil
Vegetable	K.E. Lawande
Floriculture	B.A. Patil/S.S.D. Patil
14.3 College of Agriculture, Horticulture Section elevation: 550 m rainfall : 930 mm	Kolhapur, M.S. Phone: Prinagri Head: Dr. B.B. More
Undergraduate teaching; research on mango, coconut; technical advise and extension	Prof.: R.B. Jadhav
Pomology	S.A. Dhumal/S.D. Warade S.M. Ghunke/Devgire
Vegetable	P.Y. Kadam/S.D. Warade
Agriculture School	K.K. Mangave
Horticulture Extension	S.B. Mogdam
14.4 College of Agriculture, Horticulture Section elevation: 258 m rainfall : 650 mm	Dhule, M.S. Phone: Prinagri Head: Dr. D.B. Mogal
Undergraduate teaching; research on vegetables, fruits and technical advise (extension)	Prof.: A.S. Naphade
Pomology	K.V. Sanghavi/M.N. Desai
Vegetables	N.D. Jogdande
Extension (Hort)	Y.S. Patil/D.V. Kothwade
14.5 College of Horticulture	Pune, M.S. Phone: Prinagri
Undergraduate teaching in horticulture	Proj. Off.: Dr.D.A. Rane
Horticulture	G.Y. Desle
14.6 Citrus Die-back Research Station elevation: 750 m rainfall : 625 mm	Ahmednagar, M.S. Shrirampur Dist. Off.in Ch.: D.P. Bhore
Research on citrus, (Production physiology; Citrus die-back complex); and technical advice	
Citrus pathologist	D.M Sawant
Entomologist	B.S. Shewale
14.7 Banana Research Station	Yawal, M.S. Off.in Ch.: M.B. Sable
Research on banana, extension	
Physiologist	S.S. Deshmukh

Pathologist, Entomologist — L.K. Patil

14.8 Grape Research Station
elevation: 260 m
rainfall : 750 mm

Pimpalgaon, Dist. Nasik, M.S.
Off.in Ch.: S.N. Kaulgud

Research on grape; technical advice

14.9 Onion Research Station

Pimpalgaon Baswant
Dist. Nasik, M.S.
Off.in Ch.: J.G. Patil

Research on Onion; and technical advice

14.10 Turmeric Research Station
elevation: 580 m
rainfall : 765 mm

Digraj, Dist. Sangli, M.S.
Off.in Ch.: P.D. Pujari

Research on turmeric production and storage

14.11 Betelvine Research Station
elevation: 400 m
rainfall : 650 mm

Vadnar Bhaimo
Dist. Nasik, M.S.
Off.in Ch.: A.B. Pawar

Research on betelvine agronomy, pathology, technical advice.

Agronomy — J.D. Patil
Pathology — N.S. Madne
Entomology — Jagdale

15 Dapoli

15.1 College of Agriculture, Department of Horticulture

Konkan Krishi Vidyapeeth, Dapoli
Dist. Ratnagiri, Maharashtra State
Vice Chanc.: Dr. P.V. Salvi
Dean: Dr. S.B. Kadrekar
Head: Dr. M.J. Salvi
Prof.: M.M. Patil
Assoc. Prof.: Dr. G.D. Joshi

At this department work is conducted mainly on mango, cashewnut, coconut, arecanut, spices and the minor fruits like jackfruit, kokum, sapota, etc. The crop-wise research work is being carried out as follows: Mango - experiment on fertilizer requirement, yield maximisation trial, studies on mango hybrids, and poly embryonic rootstock trials. Casewnut - observational trial on cashew varieties. Coconut - fertilizer trial. Kokum - evaluation of kokum seedlings. Ber - observation trial. Jackfruit - evaluation of seedlings trees of jackfruit. Spices - propagation studies. Propagation - research on propagation of major and minor fruits as stated above. Research work on post-harvest technology of horticultural crops. There are different Research Stations in the jurisdiction of this university, working on various aspects of fruit and vegetable cultivation.

Regional Mango Research Station

Vengurla, M.S.
Assoc. Dir.: Dr. R.T. Gunjate (NARP)
Hortic. (Mango): Dr.M.B. Magdum

Fertilizer requirement of mango, effect poly-embryonic rootstock and breeding work in mango

Regional Cashew Research Station

Vengurla, M.S.
Hortic. (Cashewnut): R.N. Nawale

Breeding work in cashewnut, manurial and irrigation requirement of cashewnut, propagation of cashewnut

Mango Research Sub station

Girye, M.S.
Jr. Hortic.: A.B. Tawade

Cultivation practices in Alphonso in Deogad area, studies on flowering period, growth flushes of Alphonso and growth of different grafts of mango

Regional Coconut Research Station

Bhatye, M.S.
Agronomist: J.L. Patil

There is a good coconut germplasm collection. Breeding and manurial trial on coconut

Central Experiment Station

Wakawali, M.S.
Dir.: S.V. Majgaonkar
Farm Supdt.: K.M. Dangale

Spacing trial in mango, varietal trial in cashewnut, mulching trial in mango and cashewnut and irrigation requirement of sapota

Tuber Crop Improvement Project, CES

Wakawali, M.S.
Jr. Agronomist: A.R. Karnik

There is a good germplasm collection of different tuber crops. Work on evaluation of different tuber crop varieties, irrigation and manurial requirement evaluation of varieties of different tuber crops.

Vegetable Improvement Scheme, CES

Wakawali, M.S.
Vegetable Specialist: M.M. Patil

At this station there is a good germplasm of different vegetables. Work on various cultural aspects of different vegetables, on vegetable breeding and on seed production

Arecannut Research Station

Shriwardhan, M.S.
Jr. Horticulturist: D.D. Nagawekar

Control of band, koleroga and anabe disease, varietal trial and evaluation of arecanut types

Minor Fruit Improvement Scheme

Dapoli, M.S.
Chief Invest.: Prof. M.M. Patil

Evaluation of different seedling type of the minor fruits

Spices Propagation Scheme

Chief Invest.: A.G. Desai

Propagation of spices crop

Strengthening of Nursery for Horticultural Crops

Dapoli, M.S.
Plant Physiologist: B.L. Lad

Propagation of various fruit crops by vegetative means. Standardization of propagation by means of tissue culture methods.

All India Coordinated Research Programme on Post-Harvest Technology of Horticultural Crops Centre

Dapoli, M.S.
Assoc. Prof. of Food Technology: Dr. G.D. Joshi

Maturity indices, pre-and post-harvest treatments, packaging and transports, drying and storage of mango and banana

Strengthening of Horticultural and Plantation Crop

Off.in Ch.: Dr. M.J. Salvi

ORISSA STATE

16 Bhubaneswar

16.1 Regional Centre of the Central Tuber Crops Research Institute (CTCRI)

316-A Kharvel Nagar
Bhubaneswar 751 001, Orissa State
Phone: 50337
Telex: TUBER SEARCH
Sc.in Ch.: Dr. S.P. Varma

This Centre was set up as a Plant Introduction Centre in 1976 and it has been developed subsequently as a Regional Station of the CTCRI to intensify research on tuber crops in South-Eastern and Eastern India with special reference to sweet potato and aroids.

16.2 Department of Horticulture, Orissa University of Agriculture and Technology

Bhubaneswar, Orissa State
Dean: Dr. R.C. Das, Extension Education and Prof. Department of Horticulture
Head of Department: G.C. Das

Fruit crops - varietal trials on mango, citrus, banana and papaya, manurial trials on litchi, orange, sapota and cultural trials on pineapple.
Vegetable crops - varietal trials on okra, tomato, potato, sweet corn, cucumber, gourds, peas, beans, radish, cabbage and cauliflower, nutritional studies on onion, garlic and brinjal, propagational studies on pointed gourd, cultural trials on tomato and growth regulators studies on pea.
Cultural and propagational studies on chrysanthemum and other ornamental crops. Work on storage life of mango, mango, guava, orange, banana and tomato.

T. Konahar/Dr. R.S. Misra
T. Moharana/P. Mohapatra
S.N. Patra/J.M. Panda/C.S. Panda
S.N. Moshra/Dr. P. Das/B.D. Samanta
A.K. Das/Dr. T.K. Das/Dr. D.P. Rao
K.K. Patnaik/J. Das/N.B. Patnaik
B. Mohanty/Dr. S. Rath/A.K. Patnaik
Dr. B.K. Das/C.R. Mohanty
H.N. Mishra/B. Naik/P.C. Lenka
D.K. Das (Fruit preservation and Tech. Section Directorate of Hort.)

The Department of Horticulture is also working in coordination with the ICAR Schemes. Studies on spices like ginger, turmeric, cumin are taken up. Work on medicinal plants have been initiated. Studies on propagation by cutting under mist of jackfruit, guava, mango, Kamarakh, spine gourd (Kankado). Research in various aspects of coffee, pepper and cardamon in the Orissa hills. Nutritional studies on coconut. Studies on cashew cultivation.
The Department of Horticulture is also working in coordination with ICAR on the following projects: Vegetable Improvement Project, Tuber Vegetable Scheme, Potato Research Scheme, Pineapple Scheme, Coconut Scheme, Cashew Nut and Spice Scheme, Beetle vine research project. Farmers' Science Centre (KVKs) Lab-to-Land Project/University Extension Block Programme/SC & OBC Project/Tribal Area Research where horticultural work is carried on including trials on farmers fields. Work on seed physiology of horticultural crops and agro-forestry plants are being initiated by Dr. R.C. Das. Work on tea and rubber is going to be initiated. The research on tuber crops with CTCRI, establishment of horticultural research institute, ICAR at Bhubaneswar is being taken up.

PUNJAB STATE

All the horticultural research in Punjab State is being carried out by the Punjab Agricultural University, Ludhiana. Besides Ludhiana, which is the main campus of the University, research work is being carried out at five Regional Research Stations.

17 Ludhiana

17.1 Punjab Agricultural University
College of Agriculture
Department of Horticulture

Ludhiana, Punjab State
Head: Dr. D.K. Uppal, Sr. Hort.*

Fruit crop improvement	Dr. D.K. Uppal*/Dr. Y.R. Chanana
	Dr. G.S. Dhaliwal/D.S. Dhillon
Rootstocks-propagation and selection	Dr. H.N. Khajuria/Dr. Ajmer Singh
Nursery management	Dr. A.S. Sandhu*/Dr. S.N. Singh
	Dr. I.S. Deol/Dr. M.P. Singh
Micro propagation	Dr. S.S. Gill/Dr. H.R. Chopra
Nutrition	Dr. Raghbir Singh*/Dr. A.S. Dhatt*
	Dr. S.K. Chaudhary/Dr. J.S. Kanwar*
	Sh. A.S. Rehalia/Dr. C.S. Malhi*
	R.P.S. Gill/Dr. S.K. Kalra
Orchard management	R.S. Dhillon/Dr. Gurcharan Singh
Growth regulators and post-harvest physiology	Dr. B.S. Dhillon/Dr. G.S. Bajwa
	Dr. S.S. Sandhu/Dr. J.S. Randhawa
	J.S. Bal
Plant protection	
Virology and plant pathology	Dr. S.P. Kapur/Dr. H.S. Sohi
Entomology	Dr. R.C. Batra/Dr. J.S. Khangura
	Dr. B.S. Sohi
Extension	Dr. S.S. Grewal*
Grapes: fruitfulness, nutrition,	Dr. A.S. Bindra/Dr. Sohan Singh
management practices and quality improvement	Dr. S.S. Brar/Dr. S.S. Cheema

17.2 Abohar Regional Fruit Research Station — Abohar (Ferozepur), Punjab State
Dr. G.S. Chohan

It is situated in the 'Citrus Belt' of the Punjab State. This area is very much suitable for the cultivation of citrus fruits, particularly the sweet orange which is extensively grown. Work on the improvement and introduction of dates was taken up at this station in the year 1955. Grape research was taken up in 1962, as grapes have also been found very successful in this region. Varietal trials and use of growth regulators in grapes.

Citrus decline and dates	G.S. Chohan
Rootstock trials in citrus	V.K. Vij/Dr. J.N. Sharma
Grapes and citrus, nutritional studies in citrus	
Grape pruning, fruitfulness of sweet limes	J.N. Sharma

RAJASTHAN STATE

18 Jodhpur

18.1 Central Arid Zone Research Institute — Jodhpur, Rajasthan State
Dir.: Dr. K.A. Shankarnarayan, M.Sc. Ph.D.

Central Arid Zone Research Institute was established initially as Desert Afforestation and Soil Conservation Station in 1952. Later on Govt. of India established a full-fledged Institute namely Central Arid Zone Research Institute involving development of arid zones in every aspect including afforestation, pasture development, soil conservation and other related aspects of saline water irrigation etc. Out of eight Divisions of the Institute covering different aspects like basic resources survey, plant studies, animal studies, solar and wind power energy, soil-plant-water relationships, human factor studies, economics and statistics and extension activities. Apart from this there are coordinated projects on dry-land agriculture, millet improvement and rodent control.
Division of Plant Studies has in its mandate research on grasses and legumes improvement, range management, forestry genetic improvement of tree species, horticulture, organic chemistry of economic products and plant protection.
The Horticulture Section in this Division was started sometime in 1972 and has been devoting mainly to fruits which can grow in arid zones. As far as facilities are concerned the Division has excellent facilities on analytical aspects. The Section of Horticulture including related subjects like horti-silvi-pastoral system has 12 research projects covering different aspects of research on ber *(Zizyphus nummularia)* datepalm *(Phoenix dactylifera)*, pomegranate *(Punica granatum)*, gonda *(Cordia myxa)*, kair *(Capparis decidua)*, karonda *(Carissa carandas)*.

18.2 Division of Plant Studies	Head: Dr. H.C. Dass
Introduction, evaluation and improvement of fruit	
plant cultivars	H.C. Dass/B.B. Vashishtha
Introduction, evaluation and improvement of ber	B.B. Vashishtha/H.C. Dass
	M.P. Singh/R.R. Bhansali
Introduction, evaluation and improvement of	B.B. Vashishtha/H.C. Dass
datepalm *(Phoenix dactilifera)*	R.R. Bhansali
Introduction, evaluation and improvement of	H.C. Dass/Ashok Gupta
pomegranate *(Punica granatum)*	
Improvement of chillies for increased productivity	B.B. Vashishtha
in arid regions	
Horti-pastoral management in arid regions	S.K. Sharma/B.B. Vashishtha
Studies on rapid regeneration of datepalm	N.L. Kackar/H.C. Dass
through tissue culture	R.R. Bhansali
Standardization of farming system based on ber	H.S. Daulay/H.C. Dass
(A. mauritiana)	
18.3 Regional Research Station	Pali, Rajasthan State
Water requirement and salt tolerant studies	
in some horticultural crops	B.L. Jain/H.C. Dass
Response of grafted ber, *Zizyphus mauritiana*	
to fertilizer and gypsum under saline water	
irrigation	R.S. Goyal
18.4 Regional Research Station	Jaisalmar Chandan, R.S.
Effect of sub-surface barriers on water and	R.S. Mertia/H.C. Dass/H.P. Singh
nutrient retention in datepalm	R.K. Aggarwal/K.A. Shankarnarayan
Water use studies in datepalm	R.S. Mertia/B.B. Vashishtha
Datepalm based farming systems	R.S. Mertia/H.C. Dass
18.5 Regional Research Station	Bikaner, Rajasthan State
Top working on wild ber with cultivated types	Attar Singh

19 Udaipur

19.1 Sukhadia University, Rajasthan College of Agriculture	Udaipur, Rajasthan State
Water relation of horticultural crops	Head: Prof. Dr. M.S. Manohar*
Nutrition of fruit crops	Dr. S.P. Pathak
Nutritional studies in vegetables	Dr. K.C. Pundrik
Breeding of vegetable crops	Dr. R.C. Khandelwal
Nutritional and propagational studies on floricultural	
crop	R.S. Dhaka
Pomology	M.L. Shrimal/Sayeed Ahmed
Floriculture	S.N. Yadav
Growth regulators in floricultural crops	C.L. Nagda
Horticultural crops	K.S. Khangarot
Fruit physiology	M.C. Jain
Vegetable physiology, nutrition and varietal trials	
on potato	Dr. R.P. Singh
Nutrition in floricultural crops	Dr. P.K.S. Kushwaha
Nutritional and cultural practices in vegetable crops	B.V. Dave
19.2 S.K.N. College of Agriculture	Jobner, Rajasthan State
Growth regulators, nutrition of fruit crops	Head: Prof. Dr. H.C. Sharma

Breeding of vegetable crops	Dr. S.S. Verma
Floriculture	V.S. Dixit
Fruits	J.P.S. Pundir/Prabhakar Singh
Horticulture	J.S. Shekhawat
Vegetables	M.M. Acharya

19.3 Datepalm Research Station
College of Veterinary and Animal Science — Bikaner, Rajasthan State

Pre-harvest and post-harvest physiology of fruits	Dr. S.L. Soni
Water relations of horticultural crops	S.P. Purohit/Inder Mohan

19.4 Agricultural Research Station — Durgapura, Rajasthan State

Agronomy of vegetable crops	Dr. Y.S. Babel/T. Pant
Breeding of vegetable crops	Rajesh Charan
Vegetables	L.P. Pareek

19.5 Fruit Rcscarch Station — Banswara, Rajasthan State

Beetlevine cultivation	M.M. Singh
Horticulture	B.S. Verma/T.S. Rawat

19.6 Agricultural Research Sub-station — Chittorgarh, Rajasthan State

Vegetables	S.S. Tyagi

19.7 Krishi Vigyan Kendra — Fatehpur, Rajasthan State

Nutrition in vegetables	S.K. Trivedi

UTTAR PRADESH

20 Pantnagar

20.1 G.B. Pant University of Agriculture and Technology, College of Agriculture
Department of Horticulture

Pantnagar, Dist. Nainital
Uttar Pradesh
Head: Dr. Sant Ram

The department has well established laboratories and a Horticulture Research Centre at Patherchatta situated at about 10 km west of the University. The centre has an all modern laboratory and field facilities to work on cultural, physiological, crop protection, crop improvement aspects including agro-forestry. Horticultural Extension programmes are being operated in 8 hills as well as in 12 plain districts.
Projects on the collection and evaluation of germplasm and cultural studies on following fruit and vegetable crops are being conducted.

Pomology

Mango	Dr. Sant Ram/Dr. L.D. Bist
	S.C. Sirohi/C.P. Singh
Aonla and banana	Dr. Sant Ram/C.P. Singh
Citrus	Dr. Ranvir Singh/K.K. Misra
Guava	Dr. J.P. Tewari/Shant Lal
Litchi	P.P. Rastogi/R.L. Lal
Papaya	Dr. M.L. Lavania/K.C. Upadhyaya
Ber	Dr. Ganesh Kumar
Peach and plum	R.L. Arora

Olericulture	
Potato	Dr. R.P. Singh/Dr. N.P. Singh R.S. Singh
Tomato	Dr. Gulshan Lal/Dr. B.K. Srivastava D.K. Singh/M.P. Singh
Chillies	Dr. Gulshan Lal/Jawahar Lal
Onion	Dr. R.S. Tewari/O.V. Kumar
Brinjal, cauliflower and garden pea	Dr. B.K. Srivastava/M.P. Singh
Okra	Dr. Gulshan Lal/J.L. Yadav
Spices	Dr. R.S. Tewari/O.V. Kumar
Post-harvest physiology, handling and storage of fruits and vegetables	Dr. Ganesh Kumar
Floriculture	Dr. M.L. Lavania
Agro-forestry	Dr. R.P. Singh/Dr. N.P. Singh

21 Faizabad

21.1 Department of Horticulture, N.D. University of Agriculture and Technology — Faizabad, Uttar Pradesh

The department of Horticulture at this University has started during 1977. There are three disciplines i.e. Pomology, Post Harvest Technology and Floriculture. The research programmes pertain to production, post harvest technologies of fruits and ornamental plants. Major emphasis is being given to indigenous fruits and ornamental plants.

Evaluation of promising strains, salt tolerance and standardization of agrotechniques of aonla, bael, ber, guava, mango etc. Propagational and rootstock studies in sub-tropical fruits	Prof.Hort.: Dr. R.K. Pathak*
Post-harvest technology of mango, guava, citrus, grapes, ber, bael, aonla, jamun, papaya and jackfruit. Studies pertaining to ripening, maturity standard, storage, biochemical changes, varietal evaluation for processing and preparation of beverages	Assoc. Prof.: Dr. I.S. Singh
Evaluation of varieties and nutritional studies in banana and papaya, propagational studies, studies in standardization of agro-techniques of opium poppy and other medicinal and aromatic plants.	Hort.: Dr. J. Prasad*
Evaluation of ornamental plants in sodic soils, standardization of propagational studies	Ass. Hort.: Dr. G.N. Tewari
Improvement of opium poppy and other medicinal and aromatic plants	Ass. Plant Breeder: Dr. O.P. Singh
Studies on alkoloid/chemical constituents in opium poppy and other indigenous medicinal and aromatic plants	Jr. Phytoch.: Dr. M.P. Pandey

22 Kanpur

22.1 Chandra Shekhar Azad University of Agriculture and Technology — Kanpur-208002, Uttar Pradesh

Mango	Dr. A. Prasad/Dr. J.B. Singh Dr. S.K.N. Pandey/Dr. S.S. Yadava
Bottle gourd	Dr. Ramjee Prasad/Dr. J.B. Singh Dr. S.R. Verma
Jackfruit	Dr. R.K. Trivedi/Dr. M.M. Singh

Acid lime	Dr. P.R.K. Murty/Dr. P.V. Singh
	Dr. B. Singh
Pineapple	Dr. S.K. Roy
Grape	Dr. L.M. Shukla/Dr. Moti Singh
Luffa species	Dr. C.P. Sharma
Aonla	Dr. P.N. Bajpai/Dr. C.P. Shukla
	Dr. S. Kumar/Dr. B.K. Misra
Carrot	Dr. I.P. Sharma/Dr. V.K. Sachan
Bitter gourd	Dr. R.S. Katiyar/Dr. D.P. Trivedi
Ber	Dr. H. Lal/Dr. R.S. Pandey
	Dr. S.K. Bajpai
Grapefruit	Dr. P.R.K. Murty
Jamun	Dr. R.S. Misra/Dr. J.P. Shukla
Amranthus	Dr. A.K. Misra
Papaya	Dr. B. Swarup/Dr. B.C. Katiyar
Guava	Dr. U.S. Pandey/Dr. A. Kumar
	Dr. P.S. Dubey
Rose	Dr. G. Prasad/Dr. R.M. Dvivedi
	Dr. S.C. Sharma
Tomato	Dr. V.K. Sharma
Tube rose	Dr. N. Roy
Gladiolus	Dr. O.P. Chaturvedi
Bael	Dr. N.C. Pandey
Phalsa	Dr. R.P. Singh
Bhindi	Dr. G.N. Singh
Onion	Dr. V.N. Maurya
Amarylus Lily	Dr. M.C. Nautiyal
Mosambi	Dr. S.G. Gupta
Litchi	Dr. R.K. Shukla/Dr. G.N. Gaur
Banana	Dr. L. Prasad
Loquat	Dr. A. Prasad/Dr. N. Tripathi
	Dr. A.S. Chaudhary

23 Allahabad

23.1 Allahabad Agricultural Institute — Allahabad, Uttar Pradesh

Department of Horticulture

Herbicidal trials for weed control in orchards and nursery, propagation and nursery industry, systematic pomology, fruit phsysiology and nutritional trials in fruit crops — Dr. G. Shankar

Cultural and nutritional requirement of vegetables — J.P. Singh/Dr. D.B. Singh

24 Chaubattia (Almora)

24.1 Horticultural Experiment and Training Centre — Chaubattia, Ranikhet-Uttar Pradesh

The Horticultural Experiment and Training Centre Chaubattia is the oldest research station in temperate fruits and it has done excellent work on improving temperate fruit cultivation in U.P. hills as well as the other temperate regions of the country. The work done at this research station has caused rapid rise in the hectarage of horticultural crops in Kumaon and Uttarakhand divisions. There are five sub-research stations namely Horticultural Sub-Research Station, Dunda (Uttarkashi), Pithoragarh Valley Fruit Sub-Research Station Srinagar (Garhwal), Jeolikote (Nainital) and three field stations at Rudrapur (Nainital), Kotedwar (Pauri) and Kothiyalsain (Chamoli) under this research station.

24.1.1 Horticulture:

Studies on the performance of dwarf rootstock, training pruning in apple, evaluation of pome fruit cultivars and propagational studies in nut fruit, specially in pecannut. Regulation of ripening in persimmon	Dir.: Dr. R.S. Misra, (Hort. cum Off.in Charge)
Propagation and rootstock studies in temperate fruit crops	Dr. N.P. Upadhyaya, Chief Hortic.
Propagational studies in nut fruit: walnut, hazelnut and pecannut, training, pruning of apple. Studies of crop regulation in peaches, evaluation of peach, plum and apricot cultivars in temperate climate.	S.P. Tripathi
Biochemical studies in pome and stone fruits	Dr. J.P. Tewari/R.K. Sexana
Evaluation of peach, plum, apricot cultivars. Propagational and cultural studies in stone fruits. Studies on the spices crops specially turmeric and garlic with special reference to use of plant growth regulators on growth, yield and quality.	K.R. Joshi

24.1.2 Fruit and Vegetable breeding:

Breeding strawberry, fruits, vegetables and ornamental crops	Dr. J.N. Seth/Dr. S.D. Lal
Testing of fruit and vegetable seeds and breeding cucumber	Dr. S.S. Solanki
Breeding apple, apricot, plums and vaselife studies of cut flowers of ornamentals	P.C. Childiyal
Breeding chillies, broad bean and gladiolus	C.C. Pant
Cultural trials on colocasia, crop rotations in different vegetables and breeding summer squash	R.P. Joshi
Breeding radish, gladiolus and carnations	N.S. Danu
Self-incompatibility studies in Brassica and breeding beet root	A.P. Misra
Breeding peas, fertilizer trials on gladiolus and collection of germplasm of gladiolus	Achal Shah

24.1.3 Soil Chemistry:

Nutritional studies in temperate fruits, dwarf apple and soil management	B.L. Divakar
Hill soil management, nutritional studies in apple and maturation standardization of apple	K.S. Adhikari
Biochemical studies on apple and other fruits	R. Kunwar
Nutritional studies in plum and apple through foliar application	N.S. Mehta
Studies on peach nutrition, water holding capacity in relation soil management and nutrient availability	J.C. Tewari
Soil testing and soil survey	K.N. Bhagat
Biochemical studies in hybrid apple and apricot in relation to maturation	S.M. Joshi

24.1.4 Entomology:

Etiology and control of insect pests. Survey and bionomics of insect pests of temperate fruits. Presently detailed investigations on bionomics and control of apple leaf blotch matter, *Ectocdemia atricollis*	Dr. B.P. Gupta
Studies on biology and control of white grubs of beetles and control of defoliating and fruit eating beetles of apple. Studies on the attraction of dif-	G.M. Tripathi

ferent species of beetles to light trap	
Studies on the lepidopteroces pests of temperate fruit trees with particular reference to their biology under laboratory conditions	Km. Rachna Joshi
Apiculture - bee botany including pollination and bee pathology	K.M. Rai
Bee management and bee pathology	M.T. Hamid
24.1.5 Plant Pathology:	
Survey of different diseases of vegetables, olive and strawberry grown in temperate region of Uttar Pradesh, their symptoms, identification and finally the suitable control measures	Dr. Raj Kumar Agarwal
Survey of scab disease of apple control measure of temperate fruit diseases. Resistance of apple root-stocks for collar rot and root rot diseases	A.J. Roy
Chemical control measures for apple, pear, almond and vegetable diseases	K.K. Srivastava
Control, measure of fruit rot of *capsicum,* sooty blotch and fly species diseases of apple, screening of different varieties of winged bean	V.B. Jain
Control measures of powdery mildew of pea, blight disease of tomato and potatoes	J.C. Pandey
24.1.6 Virus, Mushroom and Floriculture Mycology:	
Virus diseases of temperate fruits, survey, identification and control measures. Selection of resistant rootstocks and preparation of virus free mother plants	Dr. R.N. Singh
Systematic mycology	
Survey of diseases of flowering plants, their identification and control measures	J. Upadhyay
Studies on certain cultivation aspects of button mushroom especially composting, spawning and spawn preparation	Dr. I.S. Bisht
Nutritional aspects of button mushroom, preservation and insect control	T.P.S. Bhandari
Survey of virus diseases of temperate fruits and vegetables, collection of indicator plants for indexing, varietal reaction and resistance	G. Singh
Studies on diseases of button mushroom and their control measures	J.P. Kanaujia
24.1.7 Physiology:	
Micro-element nutrition in temperate fruits Micro-element diseases in temperate fruits. Growth studies in apples	Dr. J.D. Tewari
Plant growth regulator studies in apples. Weed control in peach orchard. Bitter pit studies in apples	Mrs. N. Pant
Growth regulator studies in pears. Growth studies in apricot orchard	R.M. Rai
Growth regulator studies in bean. Weed control studies in vegetables	C.P. Pathak
24.1.8 Medicinal plants cultivation:	
Nutritional and growth hormones trials, effect on the active ingredients by chemical analysis	L.K. Gupta
Effect of micro nutrients and fertilizers, survey of medicinal plants	S.C. Sah

Valley Fruit Research Station — Srinagar, Garhwal, Uttar Pradesh

Evaluation of different subtropical fruit crops for the valley areas of hills	Dr. M.M. Sinha
Nutritional and cultural studies in citrus, guava, pomegranate and peaches	
Horticultural Sub Research Station	Dunda, Uttarkashi, Uttar Pradesh
Training, pruning and rootstock studies in apple and almonds. Nutritional studies in citrus. Standardization of propagational technique in nut crops	S.B. Singh
Horticultural Sub Research Station	Pithoragarh, Uttar Pradesh
Training, pruning and rootstock studies in apple. Nutritional studies in citrus fruits. Evaluation of nut fruit crops, specially walnut, hazelnut, pecannut and almonds in the hills of U.P.	Dr. H. Lal/R.M. Rai
Valley Fruit Sub Research Station	Jeolikote, Nainital, Uttar Pradesh
Nutritional and cultural studies in pear, plum and strawberry. Assessment of spice crops, mainly ginger, garlic, turmeric in the hills of U.P.	Dr. D. Pandey

25 Saharanpur

25.1 Horticultural Experiment and Training Centre Research Station elevation: 270.5 m rainfall : 77.8 mm sunshine: 234.9 h	Saharanpur-247001, Uttar Pradesh Phone: 3126 (Office), 3283 (Res.) Dir.: Dr. R.P. Srivastava

The Centre is situated in the heart of the city which is also a public garden covering an area of 64 ha where research mainly on tropical and sub-tropical fruits, floriculture, vegetables and bee-keeping is conducted. The orchardists, farmers/cultivators are being advised technically/imparted proper training in different fields pertaining to horticulture to solve their day-to-day problems.

Vegetable Section:	Head: Dr. I.C. Pandey
Collection of germ-plasm of different varieties of vegetables.	L.R. Patel
Study of manurial, irrigational and cultural aspects of various vegetables	D.S. Yadav
Mycology Section and Entomology Section:	Head: Dr. J.H. Gupta
Control of various diseases, insects and pests of different fruit and vegetable crops	Dr. G.R. Yadav
Mushroom cultivation	R.Y. Singh/Y.P. Singh/B. Prasad
Chemistry and Cold Storage Sections:	Head: Kalu Singh
Nutritional studies of fruit and vegetable crops	R.B. Srivastava
Study of post-harvest physiology of fruit and vegetables	Shailender Singh/Bijender Singh
Biochemistry Section:	In charge: D.R. Singh
Studies on micro-nutrients on fruit and vegetable crops	Dr. P.K. Saxena
Studies on mutation irradiation in vegetables and fruits	Dharampal Singh
Physiology Section:	In charge: M.P. Singh
Physiological studies on various fruit crops	Sukhpal Singh
Statistical Section:	In charge: S.V. Verma
Layout of experiments and analysis of data; help the Ph.D. scholars	
Pomology Section:	Head: Dr. K. Dayal
Collection and selection of fruit crops	Birbal Singh/Dr. R.A. Pathak D.S. Singh/S.K.N. Pandey

Studies on various nutritional and propagational aspects of plant growth regulators in tropical and sub-tropical fruit crops	S.P. Misra/Ilam Chand
Fruit-breeding Section:	Head: S.N. Singh
Breeding of fruit crops	B.N. Singh
Plant propagation studies in fruit crops	R.P. Tomar
Cytogenetic Section:	Head: Dr. S.S. Solanki
Cytogenetical studies of fruit and vegetables	
Bee-keeping Section:	Head: K.P. Mahur
Studies on various aspects of honey bees	

26 Lucknow

26.1 Central Institute of Horticulture for the Nothern Plains — Lucknow, Uttar Pradesh

The Central Institute of Horticulture for the Northern Plains, Lucknow started in 1984, when the Central Mango Research Station which was established in 1972 under Indian Institute of Horticultural Research, Bangalore, was upgraded into a full-fledged Institute under the aegis of ICAR. The overall objective is to conduct researches on various aspects of fruits and vegetables, with mango as the lead crop, to develop horticulture in Northern Plains of India. The laboratories of the Institute are presently housed in Mahanagar, Lucknow while the field experiment station is situated at Rehmankhera on the Lucknow-Hardoi road near Mallihabad, the famous mango belt of India, about 22 km from Lucknow.

Nutritional, propagational, rootstock, intercropping and physiological studies	Dr. K.C. Srivastava*
Improvement of fruit crops by selection and hybridization	Dr. I.S. Yadav*
Nutritional and intercropping studies	M.S. Rajput
Fruit technological, packaging and developmental	Dr. S.K. Kalra
studies in fruit crops	Dr. D.K. Tandon
Storage of mango and guava	Dr. B.P. Singh
Development of fermented beverages of mango	Dr. D.K. Tandon
Physiological studies in mango	Dr. R.N. Pal/Dr. S.A.E. Khader
Nutritional and soil studies	M.G. Boppaiah
Pest management and biological control of mango pests	Dr. S.R. Abbas/A. Verghese Miss M. Fasih
Economical studies in fruit crops	A. Verma
Yield forecasting and biometrical aspects of mango	C.L. Suman
Horticultural Extension	G.K. Gupta

27 Varanasi

27.1 Department of Horticulture
Institute of Agricultural Sciences
Banaras Hindu University

Varanasi, Uttar Pradesh
Head: Prof. A.N. Maurya
Prof. C.B.S. Rajput/Dr. B.P. Singh
Dr. K.P. Singh/Dr. S.P. Singh
Dr. M.M. Syamal/Dr. J.N. Singh

Teaching at B.Sc. (Ag), M.Sc. (Ag) and Ph.D. levels. Research confined to tropical and sub-tropical fruits, vegetables and some ornamental plants.

WEST BENGAL STATE

28 Kalyani

28.1 Bidhan Chandra Krishi Viswavidyalaya

Kalyani - West Bengal State
Main Campus, Cooch Behar

Apart from the Horticultural Research Station at Mondouri, on the main campus, there are also Pineapple Research Station, Mayanaguri, Jalpaiguri Citrus and sub-tropical Research Station, Pedong, Darjeeling; Mango Research Station, Ratua, Manikchak, Malda; Regional Research Station, Jhargram, Midnapur.

Fruits:

Mango: Propagation, rootstock, macro- and micronutrition, rejuvenation, orchard efficiency evaluation, germplasm evaluation, flower induction, control of fruit drop, ripening and storage	T.K. Bose/S.K. Sen B.C. Mazumder/S.C. Maiti S.K. Mitra/P.K. Chattopadhyay S.N. Ghosh/A. Roy D. Ghosh/R.S. Dhua B. Biswas/S. Ghosh R.Y. Ram/T. Hossain B. Banik/M. Hossain S. Mallik/B. Ghosh B. Das/J. Hore
Pineapple: Planting material, plant density, macro- and micronutrition, regulation of flowering, physiology of flowering, fruit growth, chemical control of weeds, ripening and storage	
Guava: Propagation, germplasm evaluation, plant density, macro- and micronutrition, regulation of flowering, control of fruit drop, fruit growth, ripening of storage	
Litchi: Propagation, germplasm evaluation, nutrition, flower induction, control of fruit drop, fruit growth, ripening and storage	
Papaya: Screening of germplasm, nutrition, plant density, sex-expression	
Banana: Planting material, germplasm evaluation, plant density, nutrition flowering, fruit growth, irrigation, chemical control of weeds, intercropping, storage and ripening of banana	
Mandarin orange: Rootstock, orchard efficiency analysis, rejuvenation of declined orchards, nutrition, intercropping, control of fruit drop and storage of mandarin orange	
Pear and Peach: Screening of germplasm, nutrition, plant density and flowering of peach and pear	
Cashew: Screening of germplasm, nutrition, plant density and flowering of cashew	
Coconut: Screening of germplasm, nutrition, plant density of coconut	

Floriculture:

Orchid: Propagation, nutrition, flowering, breeding for development of hybrids	T.K. Bose, R.G. Maity/L.P. Yadav, N. Roychoudhury/P. Pal T.K. Chattopadhyay
Rose: Screening of varieties, rootstock, nutrition, plant density, weed control	
Jasmine: Regulation of flowering, weed control, control of diseases	
Tuberose: Weed control, control of diseases	N. Roychoudhury,
Gladiolus: Evaluation of varieties, breeding, nutrition, weed control, control of diseases	
Chrysanthemum: Screening of varieties, nutrition, control of diseases	
Carnation: Propagation, spacing, pinching, nutrition	
Marigold: Screening of varieties, seed production, weed control	
Aster: Screening of varieties, seed production, weed control	

29 Malda

29.1 Mango Research Station — Malda 732 103, West Bengal State

(State Department of Agriculture)
elevation: 25 m
rainfall : 1146 mm

Phone: Malda Farm 2386
Head: Dr. U. Chakrabarti

Rootstock, fertilizer and propagation trials, rejuvenation work, trial on orchard efficiency of plants propagated by different methods, adaptive and spacing trial, trial on hybrid mangoes and survey of mango varieties.

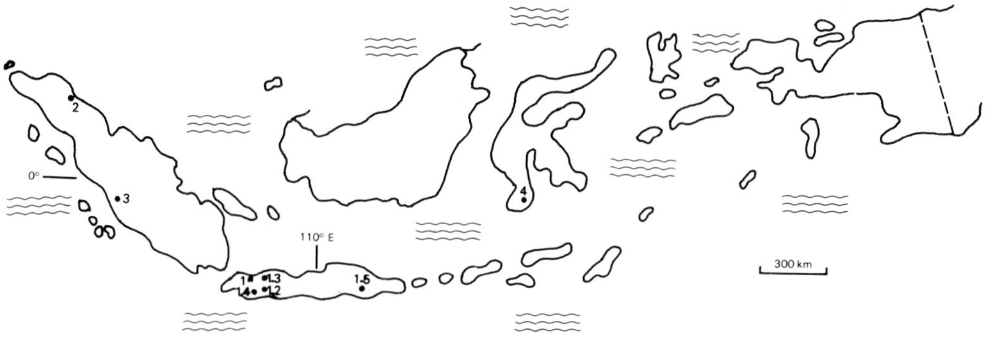

0°
110° E
300 km
1
1.2
1.3
1.4
1.5
2
3
4

1 Jakarta, 1.2 Bandung, 1.3 Segunung, 1.4 Cipanas, 1.5 Malang, 2 Brastagi, 3 Solok, 4 Jeneponto

Survey

General - In Indonesia agriculture provides about 60% of the total employment. There are 17.5 million smallholder families. Agriculture contributes 25% to the GNP, while two thirds of the Indonesian non-oil exports are of agricultural origin. The Indonesian rice yields increased greatly in recent years to such an extend that the country is currently self-supporting.
It is estimated that of the total agricultural area in Indonesia of 18 million ha horticultural crops now cover 10-13%; lowland horticulture occupies at least 80% of that area. Horticulture is important to local small farmers as it improves the financial return and also the nutritional level. With its labour intensive production methods, horticulture fits well into areas with a high population density. Because of a rapidly growing population and an expanding tourist industry the demand for horticultural products is increasing, but vegetable consumption by the indigenous people is still low, particularly in East Java. The total market production of fresh vegetables in Indonesia is estimated at 2 million tons. Post-harvest losses may amount to 20%, thus only 1.6 million tons is consumed. Assuming that about 50 million Indonesians live in (semi)-urban areas and depend on the local markets for their supply of vegetables, then the annual vegetables consumption would be about 33 kg per head or 90g/day. This intake is far below the international standard for a well-balanced diet with rice as the main food source, requiring at least 150 g of vegetables/day, 50 g of which should be leafy vegetables.

Organization - The organization of horticultural research is headed by the Central Research Institute for Horticulture (CRIH), with a national and regional mandate for specific research aspects to be carried out on Java (Pasar Minggu, Segunung, Cipanas, Lembang, Malang and Tlekung), Sumatra (Brastagi, North Sumatra; Solok, West Sumatra) and South Sulawesi (Jeneponto).

The CRIH has two main research branches; the Balai Penelitian Hortikultura Solok (SORIH) for fruit tree research and the Balai Penelitian Hortikultura Lembang (LERIH) for vegetable research.
The CRIH has 4 departments/divisions:
Administration Department (Head: Ir. Daryono), Research Information Dessimination Department (Head: Dr. Syaifullah), Research Planning Department (Head: Ir. Satsijati) and Effectiveness of Research Facilities (Head: Anwar Said BSc.).

Policy - Horticulture recently gained first priority. The policy is aimed at an increase of the production of about 5% per year and a decrease and eventually a termination of horticultural imports. Export potentials for several crops should be raised. Policy goals are to increase farmers' income and to improve the nutrition of low-income groups. Research activities in horticulture are still limited. Only about 4% of the Agency for Agricultural Research and Development (AARD) budget is available for horticultural research. This is less than 0.1% of the current value of horticultural production. Hence, current priorities in research are to be reexamined.

1 Jakarta

1.1 Pusat Penelitian Dan Pengembangan Hortikultura
(Central Research Institute for Horticulture/CRIH)
elevation: 8 m

Head office CRIH, Jl. Ragunan 19,
Pasar Minggu, Jakarta
Phone: 021-781135

Prepared by Dr. Subijanto, CRIH, Jakarta

rainfall : 2173 mm
sunshine : 402 h
rh : 82.4%

Dir.: Dr. Subijanto

Sub-Balai Penelitian Hortikultura Pasar Minggu (Pasar Minggu Sub Research Station for Horticulture - PASRIH)

Jl. Ragunan 29, Pasar Minggu,
Jakarta Selatan, Jakarta
Phone: 021-782570
Head: Laksmi DSP. BSc.

PASRIH carries out research on post-harvest technology of fruits.

Bandung

1.2 Balai Penelitian Hortikultura Lembang (Lembang Research Institute for Horticulture/LERIH)

Jl. Tangkuban Perahu 517, Lembang
6247, Bandung - West Java
Head: Dr. Azis Azirin

LERIH is responsible for the research activities in vegetables and ornamentals.
Research priority includes 5 commodities: potatoes, tomatoes, cabbages, onions and *Capsicum*.
This research program includes aspects of plant breeding, agronomy, crop protection, seed storage, post harvest technology and multiple cropping systems.

Segunung

1.3 Sub-Balai Penelitian Hortikultura Segunung (Segunung Sub Research Station for Horticulture - SESRIH)

Jl. Raya Pacet, Segunung, Cipanas,
West Java
Head: Ir. Suhardi MS.

SESRIH tackles crop protection problems of vegetables and ornamentals.

Cipanas

1.4 Sub-Balai Penelitian Hortikultura Cipanas (Cipanas Sub Research Station for Horticulture/CISRIH)
elevation: 1100 m
rainfall : 3119 mm
sunshine : 278.4 h
rh : 81.6%

Jl.Landbow, Cipanas - West Java
Head: Ir. Holil Sutapraja

CISRIH tackles research on ornamental plants such as orchids, Amaryllis, roses, gladioli, cacti, succulents, and introduction of ornamentals.

Malang

1.5 Sub-Balai Penelitian Hortikultura Malang (Malang Sub Research Station for Horticulture/MASRIH)
elevation: 445 m
rainfall : 1905 mm
sunshine : 390,7 h
rh : 81.9%

Jl. Wilis 10, Malang - East Java
Head: Ir. F. Kasijadi MS

MASRIH carries out research on fruit trees in dry lowland areas including commodities: apples, grapes, pineapples, papayas and water melons.

Tlekung

1.6 Sub-Balai Penelitian Hortikultura Tlekung (Tlekung Sub Research Station for Horticulture/TESRIH)

Tlekung - East Java
Head: Ir. Soenarso

TESRIH carries out fruit research in dry high-land areas including citrus and bananas.

2 Brastagi

2.1 Sub-Balai Penelitian Hortikultura Brastagi (Brastagi Sub Research Station for Horticulture/BASRIH)
elevation: 1300 m
rainfall : 1834 mm
sunshine : 398 h
rh : 83.6%

Brastagi - North Sumatra
Head: Ir. M. Nur HI MS.

BASRIH tackles research on vegetables including potatoes, tomatoes, cabbages, carrots, Chinese cabbage and *Capsicum.*

3 Solok

3.1 Balai Penelitian Hortikultura Solok (Solok Sub Research Institute for Horticulture/SORIH)
elevation: 390 m
rainfall : 2013 mm

Jl. KS Tubun, Kotak Pos 5,
Solok, Sumbar
Head: Dr. M. Winarno

SORIH is only responsible for research activities in fruits. Research priorities cover 6 commodities: bananas, citrus, papayas, mangoes and grapes. This research program also includes: post-harvest technology, agronomy. Vegetables, price development, market information and farm management is furthermore studied.

4 Jeneponto

4.1 Sub-Balai Penelitian Hortikultura Jeneponto (Jeneponto Sub Research Station for Horticulture/JESRIH)
elevation: sea level
rainfall : 3006 mm
sunshine : 354 h
rh : 82.1%

Jeneponto - South Sulawesi
Head: Ir. M. Fadhly MS.

JESRIH carries out only special research on citrus.

IRELAND

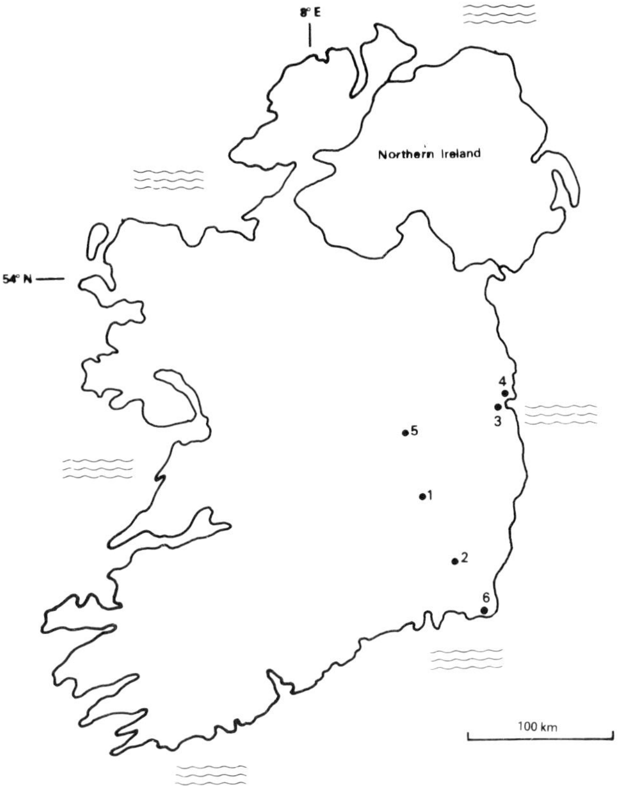
8° E
54° N
Northern Ireland
100 km
1
2
3
4
5
6

1 Carlow
2 Clonroche
3 Dublin
4 Kinsealy
5 Lullymore
6 Wexford

Survey

Horticultural research in the Republic of Ireland began at University College Dublin with the planting of an apple cultivar trial at Glasnevin in 1903. A considerable expansion in the research support available to the horticultural industry took place during the 1960s following the establishment of the Agricultural Institute. In addition a collection of plants suitable for cultivation in Ireland is maintained at the National Botanic Gardens, Dublin.

Climate: Although Ireland lies between latitudes 51° and 56°N on the western edge of a great land mass, the climate is moderated by the warm ocean current from the Gulf of Mexico. The prevailing south-easterly winds are strong at times and wind shelter is required for many horticultural crops. As the winds are conditioned by the Gulf Stream, warm maritime air covers the country for most of the year.

About four-fifths of the country has an annual rainfall of between 750 and 1250 mm. Although this rainfall would be considered only moderate or even deficient in some countries it is ample for many horticultural crops in most seasons because of low evaporation.

Ireland's maritime position is clearly reflected in its temperatures. The average annual range is only 8°C at Valencia in the south-west and 10°C at Dublin on the east coast. Extremely high and low temperatures are unknown. At Dublin the mean temperature is 5.4°C for January and 15.8°C for July. The mild winters and humid atmosphere results in a twelve month growing period for many plants. Weeds continue to grow throughout the year and although the country is further north than Newfoundland, numerous shrubs from Mediterranean countries thrive in coastal areas. Winter light in Ireland is as good as that on the south east of England, the Channel Islands and the southern coast of France and is better than that found in the glasshouse areas of north west Europe.

Location of industry: Commercial horticultural production in Ireland is a relatively recent development. Following the Second World War, a soft fruit industry, based mainly on strawberries became established principally in the south east. A glasshouse industry, with tomatoes as the principal crop developed mainly along the south and east coasts. At present mushrooms are the most important horticultural crops and recent expansion has been based on low cost plastic technology. Vegetables for fresh consumption are

Prepared by Dr.D.W. Robinson, Kinsealy Research Centre

grown principally in areas surrounding main population centres.
As commercial horticulture in the Republic of Ireland is a relatively young industry, the emphasis of the research is on applied problems. This work is undertaken by several sections of the Agricultural Institute (An Foras Taluntais) and University College, Dublin.

1. Carlow

1.1 Oak Park Research Centre
elevation: 61 m
rainfall : 780 mm
sunshine : 1314 hr

Agricultural Institute,
Carlow
Phone: 0505 31425
Dir.: P.J. O'Hare

Dealing mainly with agricultural crops, especially crop husbandry, plant pathology, plant breeding, pesticide residues and short rotation forestry, this Centre also carries out research on mechanisation of horticultural crops.

Agricultural Engineering Department
Mechanisation — B. Rice

2. Clonroche

2.1 Soft Fruit Research Station
elevation: 116 m
rainfall : 1158 mm
sunshine : 1394 hr

Clonroche, Co. Wexford
Phone: 054 44106
Officer in Charge: T.F. O'Callaghan

Involved in applied research on soft fruit production on 33 ha. The work involves strawberry breeding and investigations on cultivars and weed, pest and disease control, nutritional requirements and cultural systems. The strawberry is the main crop under investigation. Black currants, gooseberries, raspberries and blueberries are also included in the research programme. Research in beekeeping includes strain testing, swarm and disease control, comparison of the honey potential in different areas and marketing.

Soft Fruit Department
Breeding and cultivar selection — J.B. MacLachlan
Chemical weed control, physiology, production under plastic — P. MacGiolla Ri

Beekeeping Department
Beekeeping — P. Bennett

3. Dublin

3.1 University College Dublin
elevation: 25 m
rainfall : 760 mm
sunshine : 1385 hr

Belfield, Dublin 4
Phone: 01 693244
Telex : 32693
President: Prof.T. Murphy

Research in University College Dublin is carried out in conjunction with the teaching of horticultural sciences at undergraduate and postgraduate levels. Research and teaching facilities are located on the university campus, Belfield, Dublin, 4 and also at the University's field Station, 11 miles west of the city at Lucan, Co. Dublin. In the commercial sector research is in progress on many aspects of the culture of fruits, vegetables and greenhouse food and ornamental crops; production of virus-free strawberries by meristem culture; factors affecting the development of explants and micropropagules.

In the landscape sector, areas of research and professional interest include the contribution of the agricultural sector to Ireland's landscape; the measurement of environmental impact, degree of visual integration and ecological reinstatement of large-scale developments in rural areas; functional and aesthetic uses of plant materials and establishment; management and maintenance of landscapes.

Department of Horticulture	Head: Prof.J.V. Morgan*
Landscape horticulture	P. Dempsey/ M. Hallinan
Pomology	Dr.M.J. Hennerty*
Physiology of protected cropping	Prof.J.V. Morgan*
Vegetable cropping	Dr.P. Tiernan*
Micropropagation and nursery stock	Dr.A. Hunter*

Department of Agricultural:	
Botany	Head: Dr.P.L. Curran
Chemistry and Soil Science	Head: Prof.D.M. McAleese
Engineering	Head: Prof.P. McNulty
Economics	Head: Prof.S.J. Sheehy
Extension	Head: Dr.J Reidy
Zoology and Genetics	Acting Head: Dr.P. Brennan

Department of Plants Pathology — Head: Prof.J.A. Kavanagh

3.2 Economics and Rural Welfare Research Centre
elevation: 10 m
rainfall : 695 mm
sunshine : 1382 hr

Agricultural Institute,
19 Sandymount Avenue,
Dublin, 4
Phone: 01 684791
Telex : 30459
Dir.: B. Kearney

Situated about 3 km from the centre of Dublin and specialising in agricultural economics, marketing, farm management, rural sociology, farm structures and environment.

Agricultural Marketing Department	
Horticultural economics and marketing	Head: Dr.F. O'Neill
Horticultural marketing	C. Cowan

4. Kinsealy

4.1 Kinsealy Research Centre
elevation: 19 m
rainfall : 719 mm
sunshine : 1431 hr

Agricultural Institute,
Malahide Road, Dublin, 17
Phone: 01 460644
Telex : 31479
Dir.: Dr.D.W. Robinson*

This is the headquarters of the Agricultural Institute's work on horticulture and is the main centre for research on vegetables, greenhouse food and flower crops, mushrooms, nursery stocks, plant bio-technology and turf grass. Much of the work is directed towards obtaining information on suitable cultivars, nutritional requirements, cultural systems, weed, disease and pest control and energy saving.
The Plant Protection Department operates a disease and pest diagnostic service; provides nematological testing for a number of certification schemes of the Department of Agriculture; maintains virus-free nuclear stocks of soft fruit; provides a turfgrass consultancy service.
The food science area includes research on plant products with emphasis on physical and chemical testing of fresh and processed fruit and vegetables, processing, product modification and development, cooling techniques and shelf life studies.

Vegetable Crops Department	
Weed control, crop husbandry	Head: J.C. Cassidy
Strain selection and crop husbandry	R.F. Murphy
Vegetable breeding and strain selection	Ms. M.D. Prendiville

Protected Crops Department	
Mushroom compost, casing materials, environmental control diseases	Head: L. Staunton

Applied physiology of food crops and temperatures, light CO_2 and watering, NFT, nutrition, husbandry and peat culture of food crops	M.J. Maher
Mushroom production systems, harvesting, mechanisation. Marketing of horticultural produce	C. MacCanna
Environmental control and energy conservation	T.T. O'Flaherty
Crop husbandry and nutrition of ornamentals	J.C.R. Seager
Biotechnology, Forestry and Nursery Stocks Department	
Propagation and production nursery stocks and weed control	Acting Head: J.C. Kelly
Forestry and land use research	M. Bulfin
Biotechnology research including micropropagation	Dr.G. Douglas
Diseases of nursery stock and ornamentals and culture of edible fungi, micropropagation	F. O'Riordain
Plant Protection Department	
Diseases of vegetables	Head: Dr. E.W. Ryan
Biological disease control	
Insect pests of horticultural crops	R.M. Dunne
Biological pest control	
Turfgrass	Dr.T. Kavanagh
Plant Pathology Department	
Diseases of soft fruit	
Nematology in relation to all crops	J.F. Moore
Food Science and Technology	
Food Science and technology relating to horticultural crops	Head: Dr.T.R. Gormley
Cereal, milling, baking technology	Ms. B. Dwyer
Cereal, milling, baking technology	G. Downey
Food chemistry aspects of horticultural products	Dr.D. O'Beirne
Soil Physics Department	
Soil physics	Head: W. Burke
Soil structure, irrigation	Dr.R.M. Jelley

5. Lullymore

5.1 Peatland Experimental Station
elevation: 77 m
rainfall : 834 mm
sunshine : 1274 hr

Agricultural Institute
Lullymore, Co. Kildare
Phone: 045 60133
Telex : 31184
Officer in Ch.: A. Cole

This station, the main centre for research on peatland, investigates crop and livestock systems on cut-over peatland and the amelioration of peat for the production of horticultural crops and short rotation forestry.

6. Wexford

6.1 Johnstown Castle Research Station
elevation: 49 m
rainfall : 1022 mm
sunshine : 1504 hr

Agricultural Institute,
Johnstown Castle,
Phone: 053 22888
Telex : 80522
Dir.: Dr.A. Conway

Johnstown Castle is the principal centre for the analysis of soil and plant samples and deals with soil-fertility, soil classification and plant nutrition.

National Soil Survey
Soil survey — Head: J Lee

Soil Fertility, Chemistry and Biology Department
Crop Nutrition — Head: Dr.T. Gately
Dr.P. Blagden

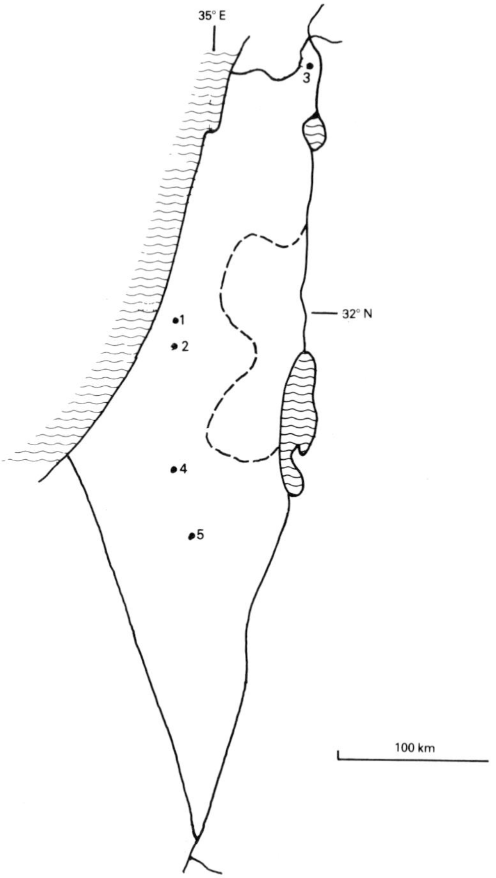
35° E
32° N
1
2
3
4
5
100 km

1 Bet Dagan
2 Rehovot
3 Kiryat Shemona
4 Beer Sheva
5 Sede Boqer

1 Bet-Dagan

1.1 Institute of Field and Garden Crops
elevation: 30 m
rainfall : 538 mm
sunshine : 118 h

Agricultural Research Organization
The Volcani Center, P.O. Box 6
Bet-Dagan 50-250
Phone: (03) - 980483
Telex: 431118BXTV-IL EXT. 5430
Dir.: B. Revzin, Ph.d.
Head: Dr. S. Izhar

1.1.1 Department of Vegetable Crops

Genetics and breeding of petunia	
Cytoplasmic male sterility	
Protoplasts culture and somatic genetics	
Cultural and physiological aspects of strawberry cultivation in winter for export	Mrs. Hadassa Avigdori
	Mrs. Eva Izsak*/ Dr. S. Izhar
Cultural and physiological aspects in host-parasite relationship of Orobanch. Agrotechnical aspects of carrot for export	Dr. R. Jacobsohn
Breeding of new varieties and cultural problem in cucurbits	Dr. Z. Karchi/Dr. H. Paris
	Dr. H. Nerson*

Prepared by Dr. Yoav Sarig, The Volcani Center, Bet Dagan

Cultural physiological aspects of heat stress in potatoes; potato breeding	Dr. D. Levi/Dr. S. Izhar
Testing of new varieties of potatoes; physiological aspects of dormancy	M. Susnovski
Breeding new varieties of onion and garlic	Dr. Z. Mitchnick
Cultural and physiological aspects of eggplant cultivation	Dr. I. Nothmann*
Cultural and physiological aspects in growing celery and cauliflower cultivation, asparagus and leafy vegetables	Dr. E. Pressman
Cultural and physiological aspects in growing protected vegetable crops cultivation	Dr. Irena Rylski
Stress physiology of different crops	Dr. B. Aloni
1.1.2 Department of Ornamental Horticulture	Head: Dr. M. Horowitz
Breeding:	
Gladioli	Dr. A. Cohen*
Carnation	Dr. N. Umiel
Anemones	Dr. N. Umiel/Dr. O. Horowitz
Ranunculus	Dr. N. Umiel
Narcissus	Ing. H. Yahel
Lilies	Ing. H. Yahel
Statice	Dr. A. Cohen*
Propagation, regulation of flowering, horticultural problems:	
Carnation	Dr. N. Umiel
Anemones	Dr. N. Umiel/Dr. Y Ozeri
Ranunculus	Dr. N. Umiel/Dr. Y. Ozeri
	Ing. Y. Elber
Narcissus, Lilies	Ing. H. Yahel
Statice	Dr. A. Cohen*
Roses	Dr. H. Ginsburg/Dr. M. Ravid
Iris	Dr. Y. Ozeri
Hippeastrum	Ing. D. Sandler
Proteacea	Dr. I. Wallerstein
Ornamental foliage plants	Dr. A. Hagiladi/Dr. Y. Ben-Yakov
Physiology:	
Photoperiod of flowering plants	Dr. A. Kadman-Zahavi*
Regulation of dormancy of gladioli	Dr. H. Ginsburg
Regulation of flowering	Dr. I. Wallerstein
Post harvest physiology of cut flowers	Dr. B. Steinitz
Storage and transport of foliage plants	Dr. B. Steinitz/Dr. Y. Ben-Yakov
	Dr. A. Hagiladi
Tissue-culture techniques	Ing.H.Lilien-Kipnis*/Dr.B.Leshem
Energy-saving in glass houses	Dr. H. Ginsburg in cooperation with Ing. N. Zamir* and Ing. N. Levan
Irrigation systems	Dr. H. Sofer
Plant protection:	
Weed control	Dr. M. Horowitz
Phytopathology	Dr. H. Vigodski-Haas (Inst. Plant Protection)
Virus diseases	Dr. A. Riger-Stein/Ing. M. Alper (Inst. Plant Protection)
Introduction:	
Bulb plants	Ing. Y. Elber/Ing. S. Ivker
Foliage and gardening plants	Dr. A. Hagiladi/Ing. Y. Elber
1.1.3 Department of Plant Genetics and Breeding	Head: Prof. R. Frankel
Genetics and breeding of open field fresh market tomatoes	Dr. Dvora Lapushner Prof. R. Frankel

Genetics and breeding of protected fresh market tomatoes	Dr. M. Pilowsky
	Dr. Dvora Lapushner
Genetics and breeding of peppers *(Capsicum sp.)*	Dr. Ch. Shifriss
Genetics and breeding of gerbera	Prof. R. Frankel

1.2 Institute of Horticulture*
elevation: 30 m
rainfall : 538 mm
sunshine : 118 h

Agricultural Research Organization
The Volcani Center, P.O. Box 6
Bet-Dagan 50-250
Phone: (03) - 980483
Telex: 431118BXTV-IL EXT. 5430
Dir.: Y Bet-Tal, Ph.d.

The Institute has five departments covering all commodities of fruit tree growing. Research involves field, laboratory, applied and basic research in breeding, nutrition, developmental physiology, fruit development, orchard management and spacing, irrigation, cultivars and rootstock.

1.2.1 Department of Fruit Tree Breeding	Head: Prof. P. Spiegel-Roy
Mutation breeding, selection cultivars and rootstock, citrus, almonds, grapes	Prof. P. Spiegel-Roy*
Mutation breeding. Radiation embiogenesis in citrus tissue culture	Dr. Gozal Ben-Haim
Breeding of citrus rootstocks and tissue culture	Dr. Aliza Vardi
Breeding of avocado and mango cultivars	Dr. U. Lavi
Grape cultivar breeding	I. Baron
1.2.2 Department of Pomology	Head: Dr. A. Erez*
Deciduous fruit trees and nuts	
Breaking of dormancy, meadow orchard, growth control and physiology	Dr. A. Erez*
Fruit development, thinning fruit drop forming dense plantation	Dr. D. Gaash*
Propagation, tissue culture	Dr. Iona Snir
Stress physiology, zinc metabolism juvenility	Prof. B. Kessler
Cultivars varieties irrigation	Dr. R. Assaf
1.2.3 Department of Citriculture	Head: Prof. A. Bar-Akiva
Nutrition, enzymology in relation to mineral deficiency	Prof. A. Bar-Akiva/Dr. M. Achituv
	R. Lavon M. Sc.
Aging of trees and fruit development. Environment effects	Dr. A. Cohen*
Irrigation under different conditions and indicators for water requirement	Dr. Y. Levy/ A. Goell* M. Sc.
Bearing dwarfing and fruit bud differentiative	Dr. E. Solomon
Morphology and anatomy of citrus fruit	Mrs. H. Safran M. Sc.
Varieties and rootstocks	Dr. A. Shaked
Physiology and development of citruses	Dr. Y. Erner
1.2.4 Department of Olive and Viticulture	Head: Prof. S. Lavee*
Growth physiology, abscission tissue culture, fruit bud differentiation	Prof. S. Lavee
Metabolism of gibberellins and ethylene, fruit abscission and differentiation, flowering	Dr. Y. Bental
Metabolism of ausin and ethylene and vegetative and fruit tissue, propagation	Dr. E. Epstein
Nutrition, nitrogen metabolism in plants and tissue culture	Dr. I. Klein
Fruit development and ripening	Dr. Y. Shulman
Morphogenesis tissue culture	Mrs. N. Adiri M. Sc.
Varieties and tree form	A. Haskel
Oil quality and metabolism	Dr. Maria Wodner

1.2.5 Department of Subtropical Horticulture	Head: Dr. A. Blumenfeld*
Avocado, mango, date, banana etc.	
Litchee and macadamia	Dr. A. Kadman*
Propagation tissue culture, date culture	Dr. O. Reuveni
Cultivars and root stock bearing capacity of avocado	Dr. A. Ben Yaacov*
Introduction and selection	E. Slor/ M. Goren M. Sc.
Nutrition and water requirements, banana	Dr. E. Lahav
Selection and fruit development of mango	Dr. E. Tomer
Anatomy of fruit development	Miss. Y. Roiseman M. Sc.
Fruit development and metabolism	Dr. A. Blumenfeld*
Regional adaptation and growth	Dr. Y. Adatto
Mango, growth and metabolism	Prof. S. Gazitt*
Biochemistry and isosyme identification	Dr. H. Degani

1.3 Institute for Technology and Storage of Agricultural Products
elevation: 30 m
rainfall : 538 mm
sunshine : 118 h

Agricultural Research Organization
The Volcani Center, P.O. Box 6
Bet-Dagan 50-250
Phone: (03) - 980483
Telex: 431118BXTV-IL EXT. 5430
Dir.: Dr. E. Chalutz

The Institute has three divisions: of Fruit and Vegetable Storage, of Food Technology and of Stored Products. Research involves studies on storage problems of fruit and vegetables, food preservation through processing, problems related to storage of grains and dried products.

1.3.1 Division of Fruit and Vegetable Storage	Head: Dr. R. Ben-Arie
Handling and Storage of lemon and easy peel citrus fruit	Dr. E. Cohen
Pathology, disease control, storage and fumigation of citrus fruit	Dr. E. Chalutz
Role of ethylene in disease development	
Prolonging the storage of life of citrus fruit and pepper in sealed plastic wrappings	Dr. S. Ben-Yehoshua
Incorporation of plastic wrapping with fungicides to control decay of sealed fruit	
Storage and keeping quality of tomatoes	Dr. Y. Fuchs*
Production and mode of action of ethylene	
Physiology, technology and storage of avocado and mango	Dr. G. Zauberman
Postharvest handling of lichee fruit	
Pathology and disease control of subtropical and deciduous fruit	Dr. D. Prusky
Strategies for control of decay caused by fungi developing resistance to fungicides	
Handling and storage of persimmon, pome fruit, stone fruit and grape	Dr. R. Ben-Arie
Physiological disorders of stored pome fruit	
Development of plastic packages for the export of fresh fruit and vegetables	
Improving the postharvest quality of sweet corn melon, strawberry, banana and date	Dr. Y. Aharoni
Use of natural volatiles for the control of insects and diseases of fruits and vegetables	
Postharvest treatment of grape and stone fruit with acetaldehyde to improve their quality	Dr. E. Pesis

Storage and export of leafy vegetables	Dr. N. Aharoni
Handling and packaging of cut lettuce	
Control of ethylene biosyntheses and mode of action	
Storage and handling of potato and root vegetables	Dr. A. Apelbaum
Postharvest treatments for preventing spoilage of stored vegetables. Research on pathogens and identification of plant fungi	Dr. R. Barkai-Golan
1.3.2 Division of Food Technology	Head: Dr. I. Rosenthal
Development of processed poultry and fish products	Dr. S. Angel/Dr. Z. Weinberg
Utilization of industrial waste from the food processing industries	A. Levi M. Sc.
Biochemistry of fruit	Dr. V. Kahn/Dr. P. Lindner
Oxidation of Lipids in foods	Dr. J. Kanner
Food microbiology	Dr. B. Juven
Citrus research	Dr. I. Shomer/Dr. N. Ben-Shalom
Packaging and corosion research	Dr. A. Albu-Yaron
Quality evolution	D. Basker B. Sc.
Membrane filtration in food industry	Dr. U. Merin
Food technology, computer optimization	Dr. Saguy
Fruits and vegetables chemistry, dairy research	Dr. I. Rosenthal
1.3.3 Division of Stored Products	Head: Prof. M. Calderon
Biology and physiology of stored products	Dr. S. Navarro
Studies on the effects of controlled atmospheres on insects	E. Donahaye M. Sc.
Stored products microflora and microtoxins	Dr. E. Shaaya
Storage of pecan, groundnut and almond	Dr. N. Paster
Studies for prevention of insect damage in stored dates	
Research on fumigation and the use of insecticides	Y. Carmi M. Sc.
Forage conservation and utilization of agricultural by-products for feedstuff	G. Ashbell/Dr. N. Lisker

1.4 Institute of Agricultural Engineering

Volcani Center, P.O. Box 6
Bet-Dagan
Phone: (03) - 940303
Telex: 341118BXTV-IL EXT. 5430
Dir.: Dr. Y. Alper

The main activities of the institute can be divided into three major areas: Research on basic problems in agricultural engineering; development of new machinery and systems for use in agriculture, or adaptation and improvement of imported machinery to local conditions; and testing of agricultural equipment and machinery.

1.4.1 Department of Field Crops Mechanization — Head: Dr. Y. Alper

Development of a tomato harvester for family farms.
Mechanization of cultivation in low plastic tunnels
Development of a harvester and gleaner for sweet paprika
Development of a harvester for meadow orchard peaches
Mechanization of onion harvesting
Development of vibrating diggers for peanuts
Development of peanut gleaning machine
Mechanization of solar sterilization of soil

Development of planters for seedling through plastic mulch
Development of continuous moving drip irrigation systems
Development of mechanical aids for vegetable picking

1.4.2 Department of Fruit Mechanization — Head: Dr. Y. Sarig

Development of mechanical harvesting equipment for citrus for industry and export
Development of a harvester for jojoba-beans
Mechanical aids for cultivation of dates
Mechanical detachment and processing of pomegranate seeds
Study of mechanical damage incurred in soft fruits during mechanical harvesting

1.4.3 Department of Sizing and Separation — Head: Dr. R. Feller

Separation of clods, by the difference in the coefficient of restitution, from potatoes, flower bulbs and tomatoes for industry
Development of peanut gleaning machine and a mobile unit for clod separation after harvesting, with a special gravity separator developed for field use.
Application of the gravity separator for flower bulbs
Peanut sorting. Development of a screen for accurate sizing
Separation and sorting of herbs and medical plants.
Harvest and separation of asparagus

1.4.4 Department of Handling — Head: Dr. D. Nahir

Mechanization of banana cultivation
Harvesting of herbs in hilly regions
Automatic wrappings of citrus fruits
Packaging and handling of citrus fruits
Utilization of a pulsating air stream for pollination of tomatoes and other agricultural applications
Separation of agricultural products by their surface rigidity
Separation of agricultural products by utilization of fluidized beds
Topping of wide spraying boom balanced in two planes for application of chemicals utilizing controlled droplets in low volumes

1.4.5 Department of Environment and Energy Engineering — Head: G. Felsenstein M. Sc.

Pre-cooling and refrigeration of horticultural products for export
Investigation of environmental conditions during transportation to export of perishable products
Drying of agricultural products
Investigation of heat transfer in fresh products cooled in carton containers
Utilization of solar energy in agriculture
Utilization of warm water from geothermal wells for agricultural applications
Utilization of dry plant-residues for energy production via direct combustion
Harvesting and separation of leaves form medical and herb plants

1.4.6 Department of Pesticide Application — Head: H. Frankel

Development and improvement of pesticide application systems in order to reduce consumption of chemicals and the environment pollution
Development in coorporation with other departments of low-volume sprayers for orchards and field crops
Development of an aerosol chamber for disinfection of bulbs
Development of systems for wax coating of fruits, vegetables and poisonous granules for research purposes
Development of soil sterilization systems by solarization or flaming

1.4.7 Department of Protected Crops Engineering — Head: N. Zamir* M. Sc.

Development of improved systems for environment control in greenhouses, and other agricultural buildings
Development of technologies for utilization of solar energy in greenhouses, hydrosolar greenhouse and a sloped greenhouse
Development of heating systems for fishponds and other agricultural structures
Application of energy storage systems with solar energy utilization in greenhouses
Development of heating systems for fishponds and other agricultural structures
Development of greenhouse structures for severe climate regions with hilly terrain and strong winds

1.4.8 Department of Industrial Engineering and Systems — Head: B.A. Silberstein M. Sc.

Application of system and industrial engineering in agriculture
Feasibility and economic studies of systems, processes, equipment and machinery
Development and application of quality control systems for horticultural products for export
Development of systems for sorting and handling of flowers
Development of devices and systems to facilitate manual operations in agriculture, especially on small farms

1.4.9 Department of Testing and Instrumentation — Head: U.M. Peiper M. Sc.

Testing of equipment and machines in the laboratory and during field operation with respect to performance and reliability
Development of equipment and methods for testing of machinery and equipment during field operation
Ivestigation of new methods for hay and silage harvesting, handling and storage
Investigation of environment control systems in greenhouses
Investigation of energy saving techniques for tractors and agricultural machinery

2. Rehovot

2.1 Department of Horticulture — The Hebrew University of Jerusalem

Research is conducted on citrus, plant propagation, deciduous fruit trees, grapevines, subtropical fruit trees, forestry. Problems of these different branches are studied at physiological and agrotechnical levels.	Faculty of Agriculture P.O. Box 12, Rehovot 76-100 Phone: 951751 Head: Prof. S.P. Monselise*
Citrus (control of growth, development and maturation)	Prof. S.P. Monselise*
Citrus (physiological abscission-enzymatic and hormonal aspects)	Prof. R. Goren*
Citrus (Hormonal aspects of flowering-tissue senescence)	Prof. E.E. Goldschmidt
Plant propagation (bud and tissue cultures, physiological effects of polyamines)	Prof. A. Altman
Plant propagation (liposome incorporation into cells)	Dr. A. Gad
Deciduous and grapevines (effects of environmental stresses on apple roots; herbicides)	Prof. A. Gur*
Deciduous and grapevines (trickle, irrigation, transportation, photosynthesis, photorespiration)	Prof. B. Bravo
Subtropical trees (fruit set and development of avocado, mango, annona, litchi)	Prof. S. Gazit*
Forestry (tree physiology, ethylene and auxin effects)	Prof. J. Riov
2.2 Department of Ornamental Horticulture Control of flowering, postharvest physiology and handling of cut flowers	The Hebrew University of Jerusalem Faculty of Agriculture P.O. Box 12, Rehovot 76-100 Phone: 951751 Head: Prof. S. Mayak
Introduction of new ornamental crops	Prof. A.H Halevy*
Control of growth and flowering of roses Greenhouse management; woody ornamental plants	Prof. N. Zieslin*
Postharvest physiology of cut flowers. Water and low temperature stress; ethylene-membrane interactions	Dr. A. Borochov
Chrysanthemum gypsophylla, foliage and flowering pot plants. Introduction of new ornamental plants	Dr. Ruth Shillo
Ornamental gardening and landscape planning Turfgrass acclimatization	Dr. I. Biran
Postharvest physiology of cut flowers, flowering physiology of roses; introduction of new plants Flowering ornamental bulbs; Physiology of low temperature in flowering plants.	Dr. Y. Mor*
Membrane physical and chemical parameters related to senescence; physiology of bulb and corn dormancy	Dr. U. Borochov
2.3 Department of Field and Vegetable Crops	Head: Prof. A. Dovrat
Genetics and breeding of tomatoes Physiology of potatoes	Dr. N. Kedar
Genetics and physiology of *Allium* Seed production of cucurbits	Dr. H.D. Rabinovitch
Genetics and physiology of tomatoes and cucurbits; Partition of assimilates in potatoes	Prof. J. Rudich*
2.4 Department of Soil Sciences	Head: Prof. N. Lahav
Irrigation of crops; soil fertility	Prof. N. Lahav
Micronutrient supply to crops	Prof. A. Banin

2.5 Department of Entomology	Head: Prof. I. Harpaz
Viruses of vines	Prof. I. Harpaz/Prof. I. Sela
Biological and integrated control of citrus pests; Citrus pests and their natural enemies	Dr. D. Rosen/Dr. H. Podoler
Scale insects and mites infesting fruit trees	Prof. U. Gerson
Deciduous fruit tree pests	Dr. H. Podoler
2.6 Department of Plant Pathology and Microbiology	Head: Prof. A. Dinoor
Taxonomy and mycology	Prof. R. Kenneth
Bacterial and fungal diseases in vegetables and ornamentals	Prof. Y. Henis
Soil-born diseases in vegetables	Prof. Y. Katan
Chemical control of diseases; disease resistance	Prof. A. Dinoor
Diseases of deciduous trees	Prof. A. Sztenjberg
Diseases of ornamental plants	Dr. Nava Eshed
Bacterial diseases of vegetables	Prof. Y. Okon

3 Kiryat Shemona

3.1 Cold Storage Research Laboratory, Israel Fruit Growers Association	Kiryat Shemona 10200 Phone: 067 - 40208 Dir.: A. Sive*

The laboratory was established in 1959 by the Israel Fruit Growers Association to investigate storage problems of apples and pears. The laboratory has since investigated storage questions of a wide range of fruits and vegetables. Its main functions relate to post harvest physiology, controlled atmosphere storage, fruit nutrition and extension.

4 Beer-Sheva

4.1 The Rudolph and Rhoda Bokyko Institute for Agriculture and Applied Biology, The Institute for Applied Research, Ben-Gurion University of the Negev elevation: 282 m rainfall : 200 - 220 mm	P.O. Box 1025 Beer-Sheva 84110 Phone: 057 - 78382 Telex: 341390 RELAY IL EXT. HMP Dir.: Dr. E. Birnbaum
Prolongation of shelf life of tomato and melons	Dr. S. Arad
Research into the firmness of the tomato fruit	
Introduction of horticultural and industrial crops	A. Aronson
Novel Medicinal plants	Dr. A. Benzioni
Irrigation and fertilisation regimes for jojoba	Dr. A. Benzioni/ A. Nerd
Regulation of flowering, fruit set and fruit development	
Mass propagation of plants	Dr. E. Birnbaum
Introduction and production of industrial and fodder plants	M. Forti
Effect of polyploidy on tomato fruits	Dr. V. Kagan-Zur
Effect of air pollution on vegetation	Dr. S. Lerman
Genetics of agricultural crops	Dr. S. Mendlinger
Production of long-shelf-life tomatoes	Dr. Y. Mizrachi
Polyamines in the development of tomato ovaries	
Introduction of wild tropical and subtropical fruit and nut trees to the Negev	
Influence of saline water on the growth and quality of tomatoes	Dr. Y. Mizrachi/Dr. S. Arad
Development of new decorative plants	Dr. D. Pasternak

Agriculture based on saline water irrigation	
Methods for cultivation of medicinal plants	Dr. D. Sitton
Use of vegetable products for production of milk substitutes	Dr. M. Trop
Nutritional potential of desert fodder shrubs	Dr. A. Yaron

5 Sede Boqer

5.1 Jacob Blaustein Institute for Desert Research
Ben-Gurion University of the Negev
elevation: 580 m
rainfall : 90 mm
sunshine June-Sept.: almost 100%
Oct.-May : not available
air humidity, annual mean 51.6%

Beer-Sheva,
Sede Boqer Campus, 84990
Phone: 057 - 88691
Telex: UNASI IL 5253
Dir.: Prof. J. Gale

Harnessing scientific knowledge in the struggle against spreading deserts and forcussing the Nations efforts to develop the Negev.

5.1.1 Controlled Environment Agriculture (C.E.A.)	Head: Prof. J. Gale
Effects of environmental conditions on the whole plant level	Dr. M. Zeroni
Environmental factors on physiological biochemical pathways at cell level	Dr. M. Guy
Technological development of C.E.A.	R. Kopel
5.1.2 Hydrogeology and Water Resources Center	Head: Prof. A. Issar
Salinity and Water Engineering	Head: Prof. J. Ben-Asher
Salinity and mineral nutrition of crops	Prof. J. Ben-Asher
Use of sewage water for irrigation	Dr. G. Oron/Dr. M. Silberbush
Run-off water harvesting by microcatchment	Prof. J. Ben-Asher
Physiological-biochemical studies of salt stress and its interaction with mineral nutrition	Head: Prof. H. Lips
Early marker of salt stress	Prof. H. Lips/Dr. M. Agami
Mechanism of salt tolerance	Dr. Y.M. Heimer/Dr. A. Golan
5.1.3 Runoff Farms (reconstruction of Nabatean Farm)	Head: Prof. M. Evenari
Pistachio growth on runoff	
Pistachio growth on saline water	D. Massig
Botanic garden on runoff irrigation	H. Bruins
Use of runoff farming in Israel (economy)	H. Lovenstein
Use of runoff farming in developing countries (food)	
5.1.4 Applied Geobotanics and Desert Plant Introduction	Head: Prof. Y. Gutterman
Plant survival mechanism	Dr. M. Agami
Plant genetic engineering	
Introduction of desert plants	
Research on botanical gardens under desert conditions	
Establishment of gene bank	
Desert gardening - collection of geophytes and hemicriptophytes	
Research of ecotypes as economic plants	
Study of phenology of ecotypes	
Study of germination under thermo and photoperiodic conditions	
Search for methods of water collection from plants under emergency conditions	

5.1.5 Ecology Center	Head: Prof. U. Safriel
Watershed ecology	Prof. A. Yair
Experiment of tree growth on rocky slopes in the desert	Dr. M. Shachak
5.1.6 Algal Culture in Deserts	Head: Prof. A. Richmond
Algal culture under outdoor conditions	Prof. A. Richmond
Chemicals from algal biomass	DR. A. Vonshak/Dr. Z. Cohen

ITALY

14° E
42°N
100 km

1 Acireale
2 Bari
3 Bologna
4 Cagliari
5 Catania
6 Como
7 Conegliano (Treviso)
8 Cosenza
9 Firenze
10 Milano
11 Ora (Bolzano)
12 Padova
13 Palermo
14 Parma
15 Perugia
16 Piacenza
17 Pisa
18 Portici (Napoli)
19 Roma
20 Salerno
21 S. Michele all'Adige (Trento)
22 Sanremo
23 Sassari
24 Torino
25 Udine
26 Verona
27 Viterbo

Survey

Scientific research in the proper sense of the work, related to fruits, vegetables and ornamentals, has begun only after the first world war. The fact is that before that war research had a practical character only and was carried out by governmental educational institutions and by some private growers. In 1887, the Royal School of Fruits, Vegetables and Gardening in Florence was established. Since 1926, Professor A. Morettini directed this Institute. Later it was transformed into the Agricultural Institute, specializing in horticulture. Many new cultivars emerged, based upon research on genetics and plant breeding.
In 1935 Prof. Morettini proceeded his work at the University of Florence in the Fruit Growing Institute. In 1950, the Centre for Improvement of Fruits and Vegetables was established in Florence. Dependant from the Ministry of Agriculture is the Experimental Institute for Floriculture at San Remo, director Dr. L. Volpi. Work on flowers and ornamental plants under glass and in the open is carried out there.
In 1930, the Station for Electrogenetics and Fruit Growing was established in Rome. Many new cultivars of table grapes, pears and peaches have been bred. Recently, horticultural departments (called institutes) were established in the agricultural faculties of the Universities of Turin, Pisa, Catania and Palermo. The research work in horticulture is gradually intensified and extended by the institutes depending from the Ministry of Education and the Ministry of Agriculture.
The most important scientific journal is 'Rivista dell' Ortoflorofrutticoltura Italiana' founded and edited by Prof. A. Morettini at the Fruit Growing Institute of the University of Florence.

1. Acireale

1.1 Istituto Sperimentale per l'Agrumicoltura — Corso Savoia, 190 Acireale

Prepared by Prof. P.L. Pisani, Institute of Tree Growing Science, Firenze.

(Citrus Experiment Institute)
elevation : 208 m
rainfall : 848 mm

(Catania)
Phone: 095-891555
Dir.: Prof. P. Spina

The Institute depends on the Ministry of Agriculture

Environmental and varieties problems; pruning	Prof. P. Spina
Citrus clonal selection and rootstocks, citrus nucellar selection and breeding	Prof. F. Russo
Biology and plant protection	Prof. E. Di Martino
Propogation, viruses and mycolplasms	Dr. G. Terranova
Cultural technique, pruning and mechanical harvesting	Dr. G. Raciti
Irrigation and fertilizers problems	Dr. A. Scuderi
Nematodes and weed control	Dr. V. Lo Giudice
Tissue culture	Dr. A. Starrantino
Electronic microscopy and isozymes	Dr. G. Reforgiato Recupero
Pathology, mycorrhizae and "malsecco"	Dr. G. Lanza
Viruses and sanitary selection	Dr. A. Caruso
Cultural technique, mineral nutrition	Dr. F. Intrigliolo
Citrus biology and physiology	Dr. E. Di Martino Aleppo

2. Bari

2.1 University of Bari, Faculty of Agricultural Science
Istituto di Coltivazioni Arboree
(Institute of Tree Growing Science)
elevation : 12 m
rainfall : 585 mm
sunshine : 2347 hrs

Via Amendola 165/A, 70126 Bari
Phone: (080) 339967
Telex: 810598 UNIVBA I
Dir.: Prof. A Godini*

Researches on propagation, floral biology, training systems, varietal performances of the main mediterranean tree fruits, such as almond, apricot, cherry, citrus, fig, table- and wine grapes, table- and oil olives, quince. Clonal selection of table- and wine grapes varieties. 25 ha of experimental fields under irrigation planted with the above species.

Arboriculture	Prof. A. Godini*
General pomology	Prof. A. Godini*
Fruit industry	Prof. A. Reina
Oliveculture	Prof. G. Casella
Viticulture	Prof. E.T. Ferrara

2.2 Istituto di Agronomia Generale e Coltivazioni Erbacee (Institute of Agronomy and Field Crops)

Phone: (080) 339551
Dir.: Prof. V. Marzi

Special interest in artichoke (Cynara scolymus L.) production, improvement and biochemistry. The Institute owns a large collection of Italian and foreign populations and cultivars.

Vegetables cultural studies in relation to quality	Prof. V.V. Bianco*
Water relations and irrigation	Prof. A. Caliandro*
Flower production	Prof. M. Cocozza Talia*
Vegetable crops for processing	Prof. G. Damato
Fertilization studies on field crops	Prof. A. De Caro
Greenhouse vegetable production	Prof. V. Dellacecca/Dr. L. Mancini
Atichoke, medicinal and aromatic plants improvement	Prof. V. Marzi
Broad bean and vegetables improvement	Dr. V. Miccolis

Weed control	Prof. P. Montemurro
Histological studies and in vitro culture	Prof. I. Morone Fortunato
Plant ecology	Prof. G. Pacucci

2.3 Centre of Study of Vegetable Crops for Processing
Italian National Researches Council

Phone: (080) 339551
Dir.: Prof. V.V. Bianco*

Special interest on: broccoli for processing variety trials and cultural studies, chemical weed control, vegetal crop rotation, ecophysiological studies on vegetables, studies on evapotranspiration, improvement of vegetable crop quality, biochemical studies on vegetables.

Chemical analysis of vegetable crops	Dr. A. Cascarano
Vegetable crops for processing variety trials and cultural studies	Dr. A. Elia
Soil management	Dr. S. De Franchi
Physiological and biochemical studies on vegetables	Dr. V. Lattanzio
Irrigation methods	Dr. D. Linsalata
Plant nutrition and vegetable fertilization	Dr. V. Magnifico*
Root studies in relation to irrigation	Dr. P. Rubino
Studies on evapotranspiration in relation to irrigation	Dr. E. Tarantino

3. Bologna

3.1 Istituto di Coltivazioni Arboree
(Institute of Pomology and Tree Growing Science)
elevation : 40 m
rainfall : 804 mm
sunshine : 2050 hrs

Via Filippo Re, 6, 40126 Bologna
Phone: (051) 239256, 226478, 229610, 235844
Dir.: Prof. S. Sansavini*

The Institute has laboratories in its main seat at Bologna (via Filippo Re, 6) for biological and anatomical researches and at the Cadriano Experimental Center (via Gandolfi 19) (10 km from Bologna), for genetic and breeding activities, growth regulators assessment, fruit quality evaluation, tissue culture, micropropagation and mechanical work-shop. The research activity of the Institute is divided in four branches: Fruit Science-Pomology, Viticulture, Landscape and Ornamentals, Forestry.
Besides, it has incorporated four separate Research Centers for carrying out important topics of research. They are: Centro Studi Tecnica Frutticola of CNR (which studies growth regulators, mechanization of harvest and pruning of fruit trees, energy consumption and saving, micropropagation). It is directed by Dr. G. Cristoferi; the Centro Miglioramento Varietale in Frutticoltura (directed by Prof. Sansavini), the Centro Ricerche Viticole ed Enologiche, Viticulture Section, directed by Prof. Intrieri and Centro di Studio sulla produzione di biomasse da colture da legno, directed by Prof. U. Bagnaresi. Most of the Institute members are cooperating with these Centers for researches.
The Institute or the connected Centers have three experimental farms: 1) Cadriano (Bologna) for orchards and vineyards; 2) "Bordone" (Cadriano 2) for breeding activity; 3) Tebano (Faenza) and Vignola for vineyards, cherry and peach orchards and table grape; 4) Montecuccolino (Bologna) for ornamental and woody trees. They have also many experimental field plots in public or private farms in Argelato Cesena, Ferrara, Forli, Imola, Modena, Rimini, Castel del Rio, Ravenna, etc., which are the main growing areas.

Forestry management and wood production	Prof. U. Bagnaresi
Fruit tree physiology, fruit trees and vines, mechanical pruning and harvesting, energy problems	Prof. E. Baldini*
Peach, apricot and chestnut fruit production Embrioculture	Prof. D. Bassi
Protected cultures, greenhouses	Prof. R. Bazzocchi
Fruit germoplasm, fruit repository	Dr. F. Camorani
Landscape; garden and park project	Prof. A. Chiusoli*
Fruit thinning; growth regulators; rootstocks; apricot, cherry, plum and plum mechanical harvesting	Prof. G. Costa*

Bioassay and tree physiology	Dr. G. Cristoferi*
Growth regulators application to fruit trees and biosasays	Dr. N. Filiti*
Fruit collections and experimental farms	Dr. M. Grandi
Viticulture: pruning and training systems; grape physiology, rooting, clonal selections; pruning and harvesting mechanization	Prof. C. Intrieri*
Light physiology and respiration in vineyards	Dr. E. Magnanini
Rootstocks for grapes and fruit trees, vinegrapes breeding, pruning and harvesting, carbohydrate physiology	Prof. B. Marangoni*
Micropropagation and tissue culture	Dr. G. Marino
Fruit breeding; strawberry, micropropagation	Dr. P. Rosati*
Fruit industry; pollination, fruit set, pruning systems, growth regulators as pruning aids harvesting, rootstocks; apple and pear production. Peach and nectarine breeding	Prof. S. Sansavini*
Nursery management, methods of propagation; fruit tree mechanical harvesting	Prof. A. Zocca

3.2 Istituto di Agronomia Generale e Coltivazioni Erbacee (Institute of Agronomy and Field Crops, University of Bologna)

Via Filippo Re, 8, 40126 Bologna
Phone: (051) 239001, 233451, 267724, 227539
Dir.: Prof. G. Toderi

The Institute carries out its research and teaching activity in different centres: Bologna (Direction, Seacretariate, Library and Laboratories); Cadriano (Laboratories and experimental farm); Ozzano (Experimental farm). The staff of the Institute consists of 70 persons, among whom 13 are professors and 20 are graduated.

Breeding tomato and eggplant	Prof. S. Conti
Fertilization	Prof. G. Toderi
Weed control	Prof. P. Catizone
Irrigation	Prof. L. Cavazza
Seed production	Prof. A. Lovato
General problems of vegetable crops	Prof. C. Antoniani

3.3 CRIOF - Experimental Center of Pathology for Cold Storage and Processing of Fruits and Vegetables. Offices at the University

Via Gandolfi 19, 40057 Bologna
Phone: (051) 766425
Via Filippo Re, 8, 40126 Bologna
Phone: (051) 221963
Dir.: Prof. G. Goidanich
Vice-Dir.: Prof. G. Pratella

The CRIOF is an experimental Center of the Deparment of Protection and Valorization of Horticultural Food Crops of the University of Bologna. The CRIOF site is in Bologna, at the Agriculture Faculty and at Cadriano (Bologna) where the laboratories and the research plants are situated. The CRIOF activity is carried out mainly, in the following fields: post-harvest physiology, pathology and technology of horticultural and citrus crops. Moreover, the Center is involved in studies on the processing of fruits and vegetables. It carries out an extension and advisory service for the packing houses in the Emilia-Romagna District.

Seed pathology, cold storage of horticultural and citrus crops	Prof. P. Bertolini
Post-harvest diseases, maturity indexes, dehydration methods	Prof. G. Biondi
R.S. and C.A. storage of Chinese gooseberry, ethylene influence on post-harvest diseases and ripening of horticultural crops. Chemical control of Penicillium sp.	Dr. S. Brigati
Temperature and low oxygen influence on the growing of pathogens "in vitro" and "in vivo" and on non-parasitic diseases of fruits and vegetables	Dr. Andrea Cimino

Post-harvest diseases, R.S and C.A. storage of new apple cultivars and persimmons	Dr. A. Folchi
Biological control of diseases, R.S. and C.A. storage of horticultural crops, crioprotection methods of seedlings	Dr. Marta Mari
Post-harvest biochemistry of horticultural produce	Dr. M. Maccaferri
Post-harvest diseases of citrus, nutrient deficiences Pollution diseases	Dr. Anna M. Menniti
Storage and post-harvest diseases of horticultural crops Plant pathology	Prof. G. Pratella
R.S. and C.A. storage of stone fruits, precooling and transport of fruits and vegetables, Botrytis sp. and Monilinia sp. control, quick-freezing of vegetables	Prof. G. Tonini

4. Cagliari

4.1 Centro Regionale Agrario Sperimental
(Regional Agricultural Experiment Center)

Via L.B. Alberti 22, 09100 Cagliari
Dir.: Prof. M. Deidda

As there is no updated material available, the text of HRI III is repeated.

The Center is interested in researches on cultural techniques of vegetables (artichoke and tomato particularly), cut flowers (chrysanthemum, carnation and rose) and fruits (citrus, almond and vine). Specific researches are carried out concerning plant breeding (tomato, vine); high density planting (citrus, protected tomatoes, cauliflowers, carnations); propagation techniques (artichokes, chrysanthemums, roses, almonds and oaks); no tillage in orchards.

Special topics:

Breeding of new tomato cultivars	Prof. M. Deidda
Breeding of self-stopping tomatoes	Prof. M.G. Carletti
Seasonal variations of rooting potential	Dr. G. Serra
Methods of propagation of artichokes	Dr. S. Leoni
High density plantations of citrus	Dr. A. Demuro
Breeding of new vine cultivars	Dr. A. Sechi
Use of solar energy in greenhouse	Dr. G. Serra

5. Catania

5.1 University of Catania, Faculty of Agricultural Sciences, Institute of Tree Growing Science
elevation : 65 m
rainfall : 419 mm
sunshine : 2592 hrs

Via Valdisavoia 5, 95123 Catania
Phone: (095) 354641
Dir.: Prof. P. Damigella

The main research activities of the Institute concern nutrition, irrigation, genetic improvement, tissue culture and rootstocks of Mediterranean fruit trees.

Nutrition, genetics and rootstocks of fruit trees	Prof. P. Damigella
Citrus genetic improvement and rootstocks	Prof. E. Tribulato
Vine, nuts and fleshy fruits	Prof. O. Alberghina
Citrus, olive and subtropical fruits; tissue culture	Prof. G. Continella
Nutrition	Dr. A. Cicala
Irrigation	Dr. G Germanà
Citrus genetic improvement	Dr. G. La Rosa

5.2 Istituto di Orticoltura e Floricoltura
(Institute of Vegetable and Flower Crops)

Phone: (095) 355079
Dir.: Prof. G. La Malfa*

The Institute's experimental activity is carried out in different areas of Eastern Sicily. It concerns didactic and scientific problems of vegetable and flower crops, field and laboratory research on biological and technical topics of vegetable crops (particularly tomato, peper, carrot, eggplant, lettuce, broad and snap bean, strawberry) and flower crops (gerbera, gladiolus, carnation) cultivated in open air and in greenhouses.

General problems of vegetable and flower crops in open air and in protected cultivation, biology of tomato, eggplant, carrot, strawberry, snap bean, gerbera	Prof. G. La Malfa*
Protected cultivation; biology and technical problems of tomato, carrot, strawberry and carnation	Prof. V. Lipari*
Protected cultivation; biology and technical problems of broad bean, pepper, strawberry, gerbera, gladiolus and other bulbs	Prof. G. Noto
Technical problems of cucurbits and allium cultivations	Dr. A. Ruggeri

5.3 Istituto di Agronomia generale e Coltivazioni Erbacee dell'Università degli Studi di Catania (Institute of Agronomy and Field Crops, University of Catania)
Phone: (095) 351361, 351420
Dir.: Prof. S. Foti*

The experimental activity of the Institute concerns: genetic improvement, biology and technical problems of broad bean, processing tomato, Italian broccoli and early potato; use of photoselective plastic materials for protected crops.

Broad bean	Prof. S. Foti*/Prof. V. Abbate
Processing tomato	Prof. G. Restuccia
Italian broccoli	Prof.S.Cassaniti/Prof.P.Signorelli
Early potato	Prof. G. Longo/Prof. P. Signorelli
Photoselective plastic materials for protected crops	Prof. S. Foti*/Prof. V. Abbate

5.4 Centro di Studio sulle Colture Precoci Ortive in Sicilia del Consiglio Nazionale delle Ricerche (Research Centre for Early Vegetable Crops in Sicily of the National Research Council)
Phone: (095) 351361
Dir.: Prof. S. Foti*

The Center is connected to the Institute of Agronomy and Field Crops, University of Catania. The activity concerns environmental factors, genetic improvement, biology and technical problems of the main vegetable crops in Sicily, particularly globe artichoke, early potato, tomato, snap bean, Italian broccoli, eggplant and pepper. Furthermore the Center is carrying out research on the phenologic stages of some vegetable species in relation to microclimatic conditions, in connection with the oriented project "Increment Productivity Agricultural Sources" of the National Council of Research. The research staff, in conformity with the Convention, belongs partly to National Council of Research, partly to the University of Catania.

General problems of early productions	Prof. S. Foti*
Globe artichoke	Prof. S. Foti*/Prof. G. La Malfa*/ Dr. G. Mauromicale
Early potato	Prof. S. Foti*/Prof. P. Signorelli
Tomato	Prof.G.La Malfa*/Dr. G.Mauromicale
Snap bean	Dr. G. Mauromicale/Dr V.Copani
Italian broccoli	Prof. S.Cassaniti/Prof.P.Signorelli
Irrigation in protected crops	Prof.G.Restuccia/Dr. G.Mauromicale
Fenologic stages of tomato and snap bean	Dr. G. Mauromicale/Dr. V. Copani

6. Como

6.1 Fondazione Centro Lombardo per l'Incremento della
Via Raimondi 48, 22070 Vertemate

Floro-Orto-Frutticoltura
(Horticultural Research Institute Minoprio)
elevation : 350 m
rainfall : 1400/1450 mm

con Minoprio (Como)
Phone: (031) 900224
Dir.: Dr. S. Ena

Experimental work in the branch of ornamental species, perennial and annual plant breeding propagation, nutrition, substrates in greenhouses, weed control, plant protection, growth regulators, collection of cultivars. Small fruits collection of cultivars propagation.

Cultural methods	Head: Dr. B. Rusmini
Collection of cultivars	Dr. L. Conti
Greenhouse climate	Dr. L. Oggioni
Plant protection	Dr. F. Righi
Propagation	Dr. D. Beretta
Mineral nutrition and substrates	Dr. G. D'Angelo
Perennial ornamental plants	Dr. C. Vanzulli

7. Conegliano

7.1 Experimental Institute of Viticulture
elevation : 60 m
rainfall : 1260 mm
sunshine : 1812 hrs

Via XXVIII Aprile 21,
31015 Conegliano (Treviso)
Phone: (0438) 61645
Dir.: A. Calò

The Institute's activity consists of seven sections: four at the headquarters in Conegliano and three situated in Asti, Arezzo and Bari. The Institute owns four farms to exploit for experimental purposes, besides co-operating with private and state concerns all over the country. Furthermore, both headquarters and branches are provided with specially equipped laboratories for analyses. The most important Italian and foreign wine grape cultivars and rootstocks belong to these farms' collections. The following tests are carried out over an area of approximately 30 ha of wine grape vineyards: irrigation, fertilization, comparison of planting distances, training systems, grafting combinations. Moreover, other tests regarding disease weed control are also made. Approximately 2000 m^2 of greenhouses are employed to grow quickly the cultivars used for clonal selection, cross-fertilization and study of virus diseases.

Ampelography and genetic improvement:	
Genetic improvement	Dept.Dir.: A. Costacurta
Clonal selection	A. Cersosimo
Ecological tests	S. Cancellier
Ampelography	
Cultivation techniques:	
Mechanization of cultivation	
Growing systems	Dept.Dir.: B. Iannini
Plant fertilization and nutrition	F. Giorgessi
Metabolism of grafting combinations	A. Lavezzi
Vine's physiology	A. Ridomi
Biology and protection:	
Vine diseases	Dept.Dir.: G. Cappelleri
Plant chemicals control of weeds	M. Borgo
Control of nematology virus diseases	
Environmental impact of plant protection products	
Nematology	
Propagation:	
Propagation techniques	Dept.Dir.: T. De Rosa
Nursery cultivation techniques	G. Moretti
Rooting of rootstock cuttings	

7.2 Asti (External Department)

Via Einaudi 60, 14100 Asti
Phone: (0141) 30269

Clonal selection Cultivation techniques Protection	Dept.Dir.: A. Costacurta L. Corino
7.3 Arezzo (External Department)	Via L. Cittadini, 52100 Arezzo Phone: (0575) 20229
Cultivation techniques Protection	Dept.Dir.: L. Egger L. Grasselli
7.4 Bari (External Department)	Via Podgora 31, 70100 Bari Phone: (080) 360055
Irrigation Cultivation techniques Table-grape clonal selection	Dept.Dir.: C. Liuni D. Antonacci M. Collapietra

8. Cosenza

8.1 Istituto Sperimentale per la Olivicoltura (Experimental Institute of Oliveculture)	Via S. Pellico 50, Commenda Rende, 87036 Cosenza Dir.: Dr. Nicola Lombardo

The Institute depends on the Ministry of Agriculture and has been organized in three central operative sections in Cosenza and two peripheric sections in Spoleto and Palermo. The central sections are: Propagation and cultural techniques, cultural olive species description and genetic improvement, biology and pest diseases control.

Peripheric Operative Section of Spoleto, is concerned with cultural techniques of olives in north and center Italy	Head: Prof. G. Petruccioli
Peripheric Operative Section of Palermo deals with problems concerning especially table olives	

9. Firenze

9.1 Istituto di Coltivazioni Arboree (Institute of Tree Growing Science) elevation : 50 m rainfall : 752 mm sunshine : 2193 hrs	Via Donizetti 6, 50144 Firenze Phone:(055) 368166, 361688 Dir.: Prof. P.L. Pisani

Germoplasm collection and breeding for peach, plum, pear, olive, grape, persimmon. Propagation physiology and techniques. Planting density and pruning for grape and peach. Biology of flowering and fruiting in grape and olive. Rootstock for grape. Soil management methods and chemical weed control. Training systems for grapevine.

Pomology and breeding	Prof. E. Bellini*
Olive culture	Prof. G. Bini
Breeding, small fruits	Prof. E. Casini*
Nursery industry	Prof. R. Magherini
Viticulture and planting density	Prof. P.L. Pisani
Propagation and polanting density	Prof. F. Scaramuzzi
Grape pruning and harvesting	Dr. G. di Collalto
Propagation physiology	Dr. A. Fabbri
Grape cultivation and physiology	Dr. E. Rinaldelli

9.2 Istituto per la Propagazione delle Specie Legnose - C.N.R.	Phone: (055) 360048 Dir.: Prof. P. Fiorino

(Institute for the Propagation of Woody Species)
Laboratories: Via Ponte di Formicola 74, 50018 Scandicci (FI) Phone: (055) 754718

All aspects of vegetative propagation, including grafting, micropropagation, cuttings, stock plants, treatments with growth regulators, nutrition, rootstocks, nursery techniques, seed propagation, germplasm collection, replant disease, breeding. The most important species concerned are peach, plum, olive, grapevine, apple, chestnut, quince and ornamentals.

Nursery techniques	Dr. G. Bartolini
Micropropagation	Dr.ssa.S.Castelli/Dr.ssa.A.R.Leva
Olive culture	Dr. A. Cimato
Nutrition	Prof. P. Fiorino/Dr. M. Tattini
Breeding	Dr. G. Roselli

9.3 Istituto Sperimentale per la Zoologia Agraria (Experimental Institute of Agricultural Zoology)
Via Lanciola - Cascine del Riccio
50125 Firenze
Phone: ( ) 209182
Dir.: Prof. R. Zocchi

The Institute depends on the Ministry of Agriculture and has been organized in five central sections and two external sections, in Rome and Padua. The research on fruits, vegetables and ornamentals is concerned with biology and control of insects, mites and nematodes.

Agriculture entomology	
Forest entomology	Head: Dr. M. Covassi
Acarology	Head: Prof. F. Pegazzano
Nematology	Head: Prof. A. Marinari
Plant protection	Head: Dr. R. Ferrari
Apiculture	Head: Dr. M. Tonini
Silkworm breeding	Head: Prof. G. Reali

10. Milano

10.1 Istituto di Agronomia dell' Universitá degli Studi di Milano
elevation : 121.58 mm
rainfall : 1012 mm
sunshine : 2000 hrs

Via Celoria 2, 20133 Milano
Phone: (02) 291264, 230645
Dir.: Prof. P.L. Ghisleni

Experimental work and teaching in the branch of general agronomy, field crops, forage crops, plant breeding, weed control, soil metal pollution, agrometeorology, vegetable crops, ornamental crops.

Iceland poppy in vitro culture	Prof. P.L. Ghisleni
Weed control in vegetables and ornamentals	A.C. Sparacino
Potato breeding	L. Martinetti

11. Ora

11.1 Research Station for Agriculture and Forestry Laimburg
elevation : 220 m
rainfall : 811 mm
sunshine : 1840 hrs

I-49040 Auer-Ora
Prov. Bozen/Bolzano
Phone: (0471) 963609
Dir.: Dr. H. Mantinger

Producing and processing of tree fruits, especially apples and grapes. On the Center of Laimburg there is a farm with 60 ha under sprinkler irrigation: 56 ha apples, 3 ha vinery and 1 ha tree and vine nursery. Laimburg carries out its activity on different farms: Direction, Secretariat and the most divisions or sections are at Laimburg with the main farm of 60 ha - eight other farms in all parts of our region with 100 ha are available for carrying out trials and for producing fruits, grapes and fodder. At the center of Laimburg there are a Chemical Laboratory (soil and leaf analysis, analysis of fruits for quality control, analysis of farm produced forages, research in vine making technology) and a wine cellar. The staff consists of five university graduates, nine specialised technicians, 13 technician aids and 40 workers in the different farms. The research station Laimburg depends on the local government: Assessorato per l'Agricoltura e le Foreste, Via Brennero 6, 39100 Bolzano.

Fruiticulture and plant protection	Head:Dr.H.Mantinger/Ing.R.Stainer Dr. S. Boscheri/Ing. J. Vigl
Storage of fruit and vegetables	Per. agr. C. Nardin
Viticulture	Dr. B. Raifer
Oenology	Ing. K. Platter/Ing. M. Aurich
Vegetables	Ing. F. Schuster
Agriculture, grass and animal husbandry	Ing. H. Bachmann
Forestry	
Chemical laboratory	Dr. W. Huber

12. Padova

12.1 University of Padova, Faculty of Agriculture, Istituto di Coltivazioni Arboree (Institute of Pomology)
elevation : 6 m
rainfall : 1000 mm
sunshine : 2324 hrs

Via Gradenigo 6, 35100 Padova
Phone: (049) 31018
Telex: 430176 UNPADU I
Chairman: Prof. C. Giulivo*

Water relations and irrigation in fruit trees and grapevine	Prof. C. Giulivo*
Effects of rootstocks and water relations in fruit trees	Prof. G. Ponchia
Hormonal regulation and control in fruit trees	Prof. A. Ramina
Growth regulators	Dr. A. Masia

12.2 Institute of Agronomy and Field Crops*, University of Padova
elevation : 6 m
rainfall : 830 mm
sunshine : 2324 hrs

Via Gradenigo 6, 35131 Padova
Phone: (049) 27184
Telex: 930176 UNPADU I
Chairman: Prof. L. Toniolo

Vegetable crops and floriculture: Control of tomato, celery, zucchini, and melon production in open field and protected cultivation. Effects of hormones in severtal ornamental species propagation and cultivation.

13. Palermo

13.1 Istituto di Coltivazioni Arboree (Institute of Tree Growing Science)
elevation : 72 m
rainfall : 721 mm
sunshine : 2607 hrs

Viale delle Scienze, 90128 Palermo
Phone: 423398, 484482
Dir.: Prof. F. Calabrese

Studies on the most important fruit species cultivated in temperate and subtropical regions (almond, apricot, hazell-nut, olive, peach, pear, prune, vine, citrus, avocado, annona, guava)	Staff: Prof. F.G. Crescimanno, Prof. G. Fatta/Prof. I. Sottile, Prof. L. Di Marco/Prof.B. Baratta, Prof. A. De Michele

13.2 Istituto di Orticoltura e Floricoltura (Institute of Vegetable and Flower Crops)	Dir.: Prof. P. Caruso
Breeding, selection and cultural techniques in faba bean (Vicia faba L.)	Prof. P. Caruso
Phenology research on vegetables and ornamental plants; improvement and fertilization in soils of greenhouses; biology and flowering of ornamental bulbs	Prof. A. Sciortino
Vegetative reproduction and agronomy techniques in vegetables cultivated in greenhouses or in the open	Dr. G. Incalcaterra
Fertilization research in natural or artificial soils in greenhouses; breeding and selection of Strelitzia reg.	Dr. U. Amico Roxas
Research on vegetables and ornamental plant forcing	Dr. F. D'Anna
Research on new plastic films; biology of Papaver nudicaule	Dr. G. Curatolo
Research on lily bulbs production	Dr. G. Iapichino

14. Parma

14.1 Stazione Sperimentale per l'Industria delle Conserve Alimentari (Experiment Station for the Food Preserving Industry)	Via le F. Tanara, 31/A Parma Phone: (0521) 72841 Dir.: Prof. A. Porretta

Higher Institute with its own legal status and administrative autonomy, under the supervision of the Ministry of Industry and Commerce. Its purpose is to promote, by studies, surveys, researches and analyses, the technical progress of the food preserving industry and to take care of the specialization of the technical staff. In Salerno there is a separate office with its own laboratories and technicians. It comprises various laboratories or centres with about 60 graduated and diplomated technicians.

Preserved food of vegetable origin	Prof. R. Andreotti/Dr. S. Gherardi, Dr. C. Leoni
Preserved foods of animal origin	Dr. P. Baldini (meat), Dr. G. Baldrati (fish)
Special technologies	Dr. G. Dall'Aglio
Microbiology and sterilization	Prof. A. Casolari/Dr. R. Massini
Containers and packing	Dr. G. Barbieri
Waste waters	Dr. C. Leoni
Publications, library and documentation	Dr. V. Castelli
Section in Salerno	Dr. D. Proto

15. Perugia

15.1 Istituto di Coltivazioni Arboree (Institute of Tree Growing Science) elevation : 462 m rainfall : 920 mm sunshine : 1740 hrs	Borgo XX Giugno, 06100 Perugia Phone: (075) 30711 Dir.: Prof. A. Tombesi
Olive harvesting problems	Prof. N. Jacoboni
Reproduction biology and olive harvesting	Prof. A. Tombesi
Fruit thinning and growth regulators	Prof. A. Antognozzi
Grape clonal selection and propagation	Prof. A. Cartechini.
Filbert and olive clonal selection	Prof. P. Preziosi
Tissue culture of fruit trees	Prof. A. Standardi

The Institute also has a Research Centre of Oleiculture of the National Research Council with the following subjects:

Clonal selection and breeding of olive (table-oil cultivars, rootstocks)	Prof. N. Jacoboni
Vegetable propagation by cuttings	Prof. G. Fontanazza
Tissue culture and growth regulator applications	Dr. E. Rugini
Endogenous growth regulators	Dr. G. Bongi

16. Piacenza

16.1 Institute of Tree Growing Science
elevation : 72 m
rainfall : 794 mm
sunshine : 2041 hrs

Via Emilia Parmense 84,
29100 Piacenza
Phone: (0523) 62600
Telex: 321033 UCATMI I
Dir.: Prof. M. Fregoni

The Institute attends -over-all- to mineral nutrition of vine breeding, pruning and no tillage. It serves the areas of Piacenza, Parma, Pavia, Verona.

Mineral nutrition of vine	Prof. M. Fregoni
Hydric nutrition (vine)	Dr. G. Dorotea
Vine breeding (classification by electrophoresis)	Dr. B. Volpe
Training system, pruning, stalk necrosis (vine)	Dr. M. Boselli
Drought resistance of rootstocks, hormones, no tillage (vine)	Dr. M. Zamboni
Vine breeding	Dr. L. Bavaresco
Fruit trees for processing (peach, cherry, quince, hazel)	Prof. A. Roversi

17. Pisa

17.1 Istituto di Coltivazioni Arboree
(Institute of Tree Growing Science)
elevation : 5 m
rainfall : 1000 mm

Via del Borghetto 80, 56100 Pisa
Phone: (050) 571551
Telex: 590035 UNIVP I
Dir.: Prof. F. Loreti*

Breeding and clonal selection of varieties and rootstocks of the most important fruit species, orchards management systems, vegetative propagation and in vitro culture, growth regulators and water relations in fruit crops, bud dormancy, collection and conservation of genetic resources. The laboratories and the greenhouses of the Institute are in Pisa (anatomical and cytological researches, fruit quality evaluation, vegetative propagation) and in S.Piero a Grado (Pisa) (micropropagation, bud dormancy and physiological researches, water relations in fruit crops studies). The Institute has also three experimental fields: 1) S.Piero a Grado (Pisa) 45 ha of nursery, orchards (pears, apples, cherries, plums), rootstock mother plants and plant germoplasm; 2) Colignola (Pisa) 6.5 ha of orchards (peaches, nectarines, apricots and apples), vineyards and peach and nectarine varieties collections; 3) Venturina (Livorno) 6.5 ha of orchards (apricots, almonds and olives), apricot varieties collections. Besides, the Institute has several experimental fields placed in private farms in Pisa, Livorno and Grosseto provinces.

Growth regulators, fruit tree pruning and training systems, high density planting, vegetative propagation, rootstocks, apple, peach and nectarine production	Prof. F. Loreti*
Breeding and clonal selection of grape varieties, evaluation of winegrape rootstocks	Prof. M. Basso

17.2 Istituto di Orticoltura e Floricoltura
(Institute of Vegetable and Flower Crops)

Viale delle Piagge 23, 56100 Pisa
Phone: 570420
Dir.: Prof. E. Moschini*

Propagation and growing techniques on vegetable and flower crops in the open and greenhouses	Prof. E. Moschini*

Plant propagation, tissue culture and temperature effect in plant growth	Prof. F Tognoni*
Bulb physiology and seed formation; development particularly related to hormonal balance	Prof. A. Alpi
Vegetable crops: plant breeding of tomato and squash; nursery problems	Prof. R. Tesi*
Seed germination, media and fertilization on vegetables	Prof. A. Graifenberg
Phytohormones and seed development	Prof. R. Lorenzi
Phytomorphogenesis and role of phytochrome in photoperiodic control of bulbing in onion plants	Prof. B. Lercari
Phytohormones metabolism in seeds	Prof. N. Ceccarelli
Fruit trees pruning and training systems, stone fruit rootstocks, bud dormancy, breeding of apricot varieties	Prof. R. Guerriero*
Fruit tree water relationships, grape varieties and rootstocks clonal selection, evaluation of winegrape rootstock	Prof. S. Natali
Fruit tree water relationships, irrigation problems	Prof. C. Xiloyannis
Vegetative propagation, micropropagation, high density planting	Prof. S. Morini
Bud dormancy, growth regulators in fruit production	Dr. Scalabrelli
Selection and evaluation of pear rootstocks, evaluation of peach and nectarine varieties, fruit plants fertility	Dr. R. Viti
Breeding, selection and evaluation of resistance to water logging of rootstocks for stone fruits, pruning and training systems for peach	Dr. R. Massai

18. Portici (Napoli)

18.1 Centro Miglioramento Genetico Piante da Orto del CNR (Research Center for Vegetable Breeding)
elevation : 15 m
rainfall : 900 mm
sunshine : 2500 hrs

Via Università 133,
80055 Portici (NA)
Dir.: Prof. L. Monti

The Center was established in 1983 in the frame of a Convention between the National Research Council (CNR) and the University of Naples. The Center is engaged in the genetic improvement of some vegetable species (tomatoes, peas, faba beans, cauliflower and potatoes). The research staff belongs partly to CNR and partly to the University of Naples.

Studies are in course on the role of pectins and other traits in tomatoes in relation with fruit development. Male sterility genes in tomatoes and F_1 hybrids are evaluated. Self-incompatible and autofertile lines of cauliflower are under investigation. Genetic variability induced by tissue culture in peas is studied; a selection programme for stress resistance is carried out on tomatoes and on peas. Aneuploidy and chromosome structural aberrations are identified and utilized for genetic and breeding purposes in peas and in faba beans. Mutants characterized by different leaf structure are investigated in peas and in faba beans to study their metabolic patterns.

Breeding for quality traits and use of male sterility and self-incompatibility	A. Leone
Use of in vitro culture	E. Filippone/C. Conicella
Study and utilization of chromosome	C. Conicella/A. Errico/E. Filippone
Evaluation of new types in grain legumes	L. Frusciante/R. Rao/S. Grillo

18.2 Istituto di Agronomia Generale e Coltivazioni Erbacee (Institute of Agronomy and Field Crops)

Via Università 100,
80055 Portici (NA)
Phone: (081) 274632
Dir.: Prof. L. Postiglione

Special interest in plant breeding, mechanization, nutrition and irrigation of vegetable crops and greenhouse production. Experimental farm: 50 ha partially under sprinkler irrigation and largely covered with horticultural crops. Equipment: 6 plastic and 3 glass greenhouses.

Plant breeding	Prof. L. Monti
Mechanization	Prof. L. Postiglione
Fertilization	Prof. L. Postiglione/A. Duranti, L. Cuocolo
Irrigation	Prof. G. Barbieri
Physiology	Prof. A. Duranti/G. Barbieri
Seed production	Prof. L. Cuocolo
Weed control	Prof. A. Duranti

19 Roma

19.1 Istituto Sperimentale per la Frutticoltura*
elevation : 40 m
rainfall : 750 - 800 mm

Via Fioranello 52, Roma
Phone: (06) 600048, 600251
600252, 600253, 600098
Dir.: Prof. C. Fideghelli*

The Institute belongs to the Ministry of Agriculture and carries out researches on decidous fruits (pome, stone, nut and small fruits). The experimental farms are in Fiorano (2 ha), in Capocotta (30 ha) and in Tormancina (10 ha).

Biology Section	
Climate on phenology of fruit trees	Head: Dr. Paola Cappellini
Source of resistance to fungus, insects and nematodes of stone fruits scions and rootstocks	Dr. Anna M. Simeone*
Fruit Breeding Section	
Stone fruits rootstocks breeding	Head: Dr. A. Nicotra/L. Moser
Peach, apricot, plum breeding	Dr. Roberta Quarta*
Chestnut and fig varieties	Dr. G. Grassi
Propagation Section	
Almond, pistachio, kiwi varieties and propagation	Head: Dr. F. Monastra
Peach, nectarine varieties and propagation	G. Della Strada
Propagation technique and media	Dr. D. Avanzato
In vitro propagation	Dr. M. Antonelli
Orchard Management Section	
Filbert and table grape varieties, cultural practices	Head: Dr. P. Manzo
Irrigation and mechanization	Dr. G. Colorio
Fertilization and irrigation	Dr. G.F. Monastra
Wallnut varieties	G.C. Tamponi
Tree training	Dr. R. De Salvador

19.1.1 Trento Field Station
The species under trial in Trento are: apple, cherry, raspberry, current, blackberry and blueberry The experimental farms are in Pergine (14 ha) and in Borgo Valsugana (10 ha)

Vigalzano di Pergine, Trento
Phone: (0461) 532609
Dir.: Dr. A. Bergamini*

Apple varieties, rootstocks and irrigation	Dr. A. Bergamini*
Cherry varieties and breeding	Dr. A. Albertini
Apple varieties and breeding	Dr. M. Sacco
Raspberry, current, blackberry and blueberry varieties	Dr. Silvia Angelini

19.1.2 Forlì Field Station
The species under trial in Forlì are apple,

P.le Vittoria 15, 47100 Forlì
Phone: (0543) 69256

pear, peach and nectarine, plum, strawberry. The experimental farms are in Diegaro (15 ha), Monticino (2 ha) and Fornace (3 ha)	Dir.: Dr. D. Cobianchi*
Growth regulators, plum varieties	Dr. D. Cobianchi*
Apple and strawberry varieties and breeding	Dr. W. Faedi
Peach and plum varieties and breeding	Dr. A. Liverani
Pear varieties and breeding	L. Rivalta
19.1.3 Caserta Field Station The species under trial in Caserta are peach and nectarine, apricot, filbert, loquat, persimmon. The experimental farms are in Francolise (14 ha) and Caserta (3 ha)	Via Torrino 3, Caserta Phone: (0823) 467203 Dep. Dir.: Dr. C. Damiano*
Strawberry breeding	Dr. C. Damiano*
Strawberry breeding, peach training	Dr. S. Recupero
Filbert orchard management, wallnut varieties	Dr. F. Limongelli
Apricot varieties and breeding	Dr. F. Pennone
Loquat, persimmon, peach varieties	O. Insero
19.2 Istituto Sperimentale per la Patologi Vegetale (Experimental Institute for Plant Pathology)	Via C.G. Bertero 22, 00156 Roma Dir.: Prof. A. Quacquarelli

The Institute is one among 23 research units of the Ministry of Agriculture created in 1967 by the law which re-organised agricultural research. However, the Institute is more deeply rooted into tradition of Plant Pathology, as it originates from the Experimental Station for Plant Pathology founded in 1877. The latter, in its turn, inherited the experience of the Laboratory of Mycology of the Rome University, led by De Notaris. The station, directed by outstanding scientists as Cuboni, Petri and Sibilia, was the melting-pot where generations of plant pathologists met and matured scientifically. Among them we can mention Professors A. Biraghi, R. Ciferri, R. Gigante, V. Grasso, V. Peglion, G. Ruggeri, A. Traverso, A. Trotter, A. Ciccarone, G. Goidanich, and O. Lovisolo.
The Institute is regularly consulted by the Ministry of Agriculture on problems of plant quarantine. The Director and staff are members of national and international Committees (Pesticides registration, Residue analysis methods, Seed analysis methods, Integrated control of crops, etc.).

Biological control of weeds and pathogenic fungi by biological agents, mostly fungi	Dr. M.T. Ialongo
Field control of diseases and pesticide residues in crops	Drs. G. Imbroglini/A. Leandri, Elisa Conte
Diseases of cereals, mainly rusts	Dr. Luciana Corazza
Diseases of grapevines	Dr. Raffaella Nalli
Virus diseases of grapevines and fruit trees and selection of virus free stocks	Drs. Marina Barba/F. De Sanctis
Virus diseases of vegetables and ornamentals	Dr. Maria P. Benetti
Seed and nursery pathology of forest trees	Dr. Emma Motta
Seed pathology of cereals and vegetables crops	Drs. A. Porta Puglia/F. Montorsi
Bacterial diseases and fastidious bacteria	Dr. Francesca J. Casano
Phytotoxins and mycotosins	Dr. Jolanda F. Amici

20 Salerno

20.1 Istituto Sperimentale per l'Orticoltura (Vegetable Crops Research Institute)	Via Cavalleggeri 25, 84098 Pontecagnano (SA) Dir.: Prof. S. Porcelli

The state-supported Institute for experimental research is under the supervision of the Ministry of Agriculture and Forestry, incorporating one main Institute with four Sections in Pontecagnano (SA), and two external branch

Sections located respectively in Ascoli Piceno and Montanaso Lombardo (ML). The Institute also operates a centre for scientific exchanges in Pontecagnano (SA).

Biology, physiology and disease resistance	Head of Section:
Incorporation of genetic resistance to virus and fungus diseases in tomato and egg-plant	Dr. A. Saponaro/ Dr. R.Santoro, Dr. F. Fiume
Biochemistry and technology: biochemical studies in plant and fruits with main interest in tomato and egg-plant	Head of Section: Dr. A. Sozzi
Genetics and breeding: egg-plant, tomato, pepper, lettuce, chicory, marrow and lima bean breeding	Head of Section: Dr. F. Restaino
Agronomics: agronomic technique for main vegetable crops in the open, in plastic houses and soilless	Head of Section:
Studies on alternative energy requirement in some species	Dr. R.D'Amore/Dr. S. Petralia, Dr. C. Perrella
Branch Section (Ascoli Piceno)	Head of Section:
Breeding and agronomic techniques	L. Uncini
Breeding and cultivation techniques of table tomatoes, hot and sweet pepper, melon, water melon, chicory and fennel	Dr. V. Ferrari/Dr. N. Acciarri
Branch Section (Montanaso Lombardo)	Head of Section:
Asparagus, bean, tomato for processing and onion genetic improvement	Dr. A. Falavigna Dr. M.Schiavi/Dr.A.Allavena/ Dr. B. Campion

21 S. Michele all'Adige (Trento)

21.1 Stazione Sperimentale Agraria e Forestale*
(Experimental Station of Agriculture and Forestry)
elevation : 190 m
rainfall : 900 mm
sunshine : 1880 hrs

S. Michele all'Adige (Trento)
Phone: (0461) 650107
Dir.: Dr. G. de Stanchina

Fruit tree pruning and training system	Dr. M. Comai
Ripening and storage of apples	Dr. G. de Stanchina
Grape varieties clonal selection	P.En. I. Roncador
Production of rootstocks and virus free apple varieties	Dr. Elisabetta Vindimian
Agrometeorology	Dr. P. Ferrari.

22 Sanremo

22.1 Istituto Sperimentale per la Floricoltura*
(Agriculture and Forestry
elevation : 50 m
rainfall : 700 mm
sunshine : 2000 hrs

Corso Inglesi 508, 18038 Sanremo
Phone: (0184) 884944
Dir.: Dr. L. Volpi*

At present the Institute carries out studies and researches on the genetic improvement of flowering plants as well as on the techniques of cultivation and protection in the open and in greenhouses, and of "in vitro" or "in vivo" propagation.
Equipment: 1.2 ha of experimental fields with 0.7 ha in greenhouses at Sanremo, 2 ha experimental fields at Prescia and in the neighbourhood of Palermo 2 ha for branch sections. Laboratories for soil and tissue analysis, for fungal pathogens identification and for micropropagation at Sanremo. Growth chambers at Sanremo. Laboratories at Pescia and in Palermo.

Biological Pest and Disease Control Section: — Head: Dr. C. Dalla Guda
Control of Fusarium wilt of carnation

Control of carnation pest Epichoristodes acerbella by biological methods — Dr. C. Pasini
Control of Fusarium wilt of bulbous plants
Study of fungicides resistent strain of pathogenic fungi of ornamental crop
Control of Lyriomiza trifolii

Genetic Improvement Section: — Head: Dr. T. Schiva
Breeding of Gerbera Jamesonii for winter production at low energy requirements
Breeding of Gerbera Jamesonii for dwarf pot plants
Breeding of carnation for genetic resistence against Fusarium oxisporum f. sp. Dianthi
Breeding of Genista monosperma for improved cultivars
Breeding of Chrisanthemum frutescens for new cultivars
Identification of cold shock proteins on Gerbera Jamesonii

Propagation Section: — Head: Dr. C. Damiano*
Propagation of roses, gerbera, Eucalyptus sspp., Genista monosperma, succulent plants by "in vitro" culture
Propagation of several species by conventional methods (grafting, cuttings, layering) — Dr. P. Curir
Comparison between micropropagated and non micropropagated plants

Cultural Techniques Section: — Head: Dr. E. Farina
Timing of production of Euphorbia fulgens, ranunculus, Gypsophila paniculata, carnation, Agapanthus, Alstroemeria
Growth regulators on Gerbera and carnation for winter production
Extension of life of cut flowers (Euphorbia fulgens, Papaver nudicaule)

22.1.1 Branch Section of Pescia: — Head: Dr. M. De Ranieri
Researches on Lilium for cut flower production in summer (timing of flowering, "in vivo" propagation, breeding) — Dr. G. Pergola
Control of Phialophora cinerescens on carnation — Dr. A. Grassotti
Control of Lyriomiza trifolii and Trialeurodes vaporariorum on various ornamental crops — Dr. P. Rumine*
Propagation and timing of production of Gypsophila paniculata
Cultivation of woody species for cut foliage

22.1.2 Branch Section of Palermo: — Head: Dr. M Cirrito
Researches on storage of bulbs, forcing and planting densities of Polianthes tuberosa, and Gladiolus for cut flower production — Dr. M. G. Provenzale, Dr. M. De Vita
Researches on cultivation of Polianthes tuberosa, Gladiolus and Iris for bulb enlargement — Dr. G. Zizzo

23 Sassari

23.1 Istituto di Agronomia Generale e Coltivazioni Erbacee — Via E. de Nicola, 07100 Sardinia Sassari, Phone: (079) 217431, Dir.: Prof. G. Rivoiza
elevation : 225 m
rainfall : 600 m

The activity of the Institute concerns genetic improvement of durum wheat and triticale — Prof. Dr. P. Bullitta

Agronomic trials on soyabean, sufflower and sunflower	Prof.Dr.M.Deidda/Ass.Prof.G.F.Marras
Maize and sorghum irrigation technique	Ass. Prof. A. Caredda
Cultivation technique of rice	Ass. Prof. A. Murtas
Pastureland improvement	Ass. Prof. M. Milia
Grass and legumes evaluation	Ass. Prof. Spanu
Cultivation technique under greenhouse and hydroponic	Dr. G. Attene

24 Torino

24.1 Istituto Coltivazioni Arboree Università di Torino (Fruit Growing Institute, Torino University) elevation : 229 m rainfall : 756 mm sunshine : 2115 hrs	Via P. Giura 1, 10126 Torina Phone: (011) 683571, 655496
Fruit trees Local varieties germplasm preservation, almond, apricot, apple, cherry, citrus, olive	Prof. R. Paglietta Dr. G. Bounous
Small fruits Introduction and evaluation gen. ribes, Rubus, Vaccinium, Eleagnus, Sambucus, Hippophaë, Mespilus, Cornus Obtainment of new cultivars	Prof. R. Paglietta Dr. G. Bounous
Chestnut culture Introduction and elevation of Japanese chestnuts C. crenata Introduction and evaluation of new hybrids C. crenata x C. sativa. Grafting techniques	Prof. R. Paglietta Dr. G. Bounous
Grape vine Seedless table grapes, pruning systems, rootstocks, chemical weed control, ecology, nutrition and VA mycorrhiza, root development, vinery mechanization	Prof. I. Eynard Dr. A. Schubert/Dr. M. Bovio
Kiwi-fruit Introduction and evaluation of new cultivars Obtainment of new cultivars	Prof. I. Eynard
Tissue culture Actinieia, Camellia, Corylus, Ficus, Prunus, Rhodondendron Meristem culture, callus and another culture	Prof. R. Jona
Cell wall histochemistry of fruit pulp	Dr. Silvia Sacerdote
24.2 Centro di Studio per il Miglioramento Genetico della Vite Consiglio Nazionale delle Ricerche (Grapevine Breeding Centre, National Council of Research)	Via P. Giuria 15, 10126 Torino Phone: (011) 6503757 Dir.: Prof. I. Eynard
Dalmasso crosses selection and grapevine germplasm conservation	Dr. M. Bovio
Introduction of grape cultivars from abroad	Prof. I. Eynard
New crosses for wine and table grapes	Dr. V. Novello
Comparison of new cultivars	Prof. R. Carlone
Grapevine cariology and histology	Prof. R. Jona
Mutagenesis	Dr. Rosalina Vallania
Grapevine flower biology	Prof. R. Jona
Growth regulators	Dr. Giuliana Gay
Grapevine propagation	Dr. Anna Schneider

In vitro culture of grapevine — Prof. R. Jona
Clonal selection of wine grape cultivars — Dr. F. Mannini

24.3 Istituto di Scienza delle Coltivazioni, Università di Torino, (Institute of Crop Science, University of Turin)

Via P. Giuria 15, 10126 Torino
Phone: (011) 651669, 6505588
Dir.: Prof. G. Luppi

The experimental work of the Institute is carried out mainly at the Experimental Center of the Faculty of Agriculture located in Carmagnola (Torino) but also in private farms of the Piedmont and Liguria regions and at the laboratories of the Institute itself.

Cut flowers production and keeping quality — E. Accati Garibaldi*
Protected crops-growth regulators — L. Basoccu
Tissue culture — M. de Donato
Hydrology — G. Luppi

24.4 Istituto di Miglioramento Genetico e Produzione delle Sementi, Università di Torino (Institute of Plant Breeding and Seed Production, University of Turin)

Via P. Giuria, 15 - 10126 Torino
Phone: (011) 683969
Dir.: Prof. L. Quagliotti*

Breeding and seed production of vegetables, ornamentals and forest species. Ageing of the seeds. Collection and evaluation of pepper and bean germplasm. Embryoculture of bean. Aploids in eggplant. Natural cross pollination intra and interspecific in bean
Bean, carrot, cyclamen, egg-plant, gerbera okra, pepper, poppy

P. Belletti/S. Lanteri, G. Lepori*/M.O. Nassi, L. Quagliotti*

25 Udine

25.1 Istituto di Produzione Vegetale (Institute of Crop Science), Facoltà di Agraria, Università di Udine
elevation : 110 m
rainfall : 1500 mm

Piazzale Kolbe 4, 1-33100 Udine
Phone: (39) 432/470970
Telex: 450412 UNIVUD I
Dir.: Prof. P. Sequi

The Institute includes the following research fields: pomology, viticulture, agronomy, agricultural chemistry, rural engineering, genetics. Concerning pomology and viticulture sections, there are three laboratories for gaschromatographic and HPLC analysis, biological and anatomical research and micropropagation; research is carried out also in the Experimental Farm of the Institute (30 ha), S. Osvaldo, Udine and in many public or private farms in the Friuli plain and hills

Growth retardants (apple, peach, kiwi) shoot/fruit competition, fruit set and abscission — Prof. G. Costa
Vegetative propagation, high density orchards — Dr. R. Testolin*
Tissue culture, growth regulators physiology — Dr. R. Messina
Physiology and ampelography of wine grape and rootstocks — Prof. B. Marangoni*
Grape physiology and training systems, gas and liquid chromatography, cherry selections — Dr. E. Peterlunger

26 Verona

26.1 Istituto Sperimentale di Frutticoltura (Experimental Institute for Fruit Growing)

Via San Giacomo 25, 37135 Verona
Phone: (045) 32545

elevation : 30 m
rainfall : 900 mm

Dir.: Dr. G. Bargioni

Founded in 1954. Breeding and selection in cherry mechanical harvesting, self-fertility. Breeding and selection in peach, white fleshed cultivars. Training systems and high density planting in peach, nectarines, cherry, and olive. Trials on cultivation and new cultivars evaluation of strawberries. Tissue culture for propagation and breeding of fruit trees.

Cherry, peach and olive	G. Bargioni
Strawberry	T. Tosi
Apple, pear, and tissue culture	F. Cossio*
Peach, kiwi-fruit	G. Baroni
Cherry, apricot and plum	C. Madinelli

27 Viterbo

27.1 Istituto di Ortofloroarboricoltura

Via S. Camillo de Lellis
01100 Viterbo
Dir.: Prof. F. Maggini
Dr. C. Bignami/Dr. A. Jacoboni

Studies on: Biological and agronomical behaviour of most important fruit species, peach rootstock and high density planting, clonal selection and flower biology of chestnuts, flower biology of filbert, germplasm protection, oliviculture

JAPAN JP

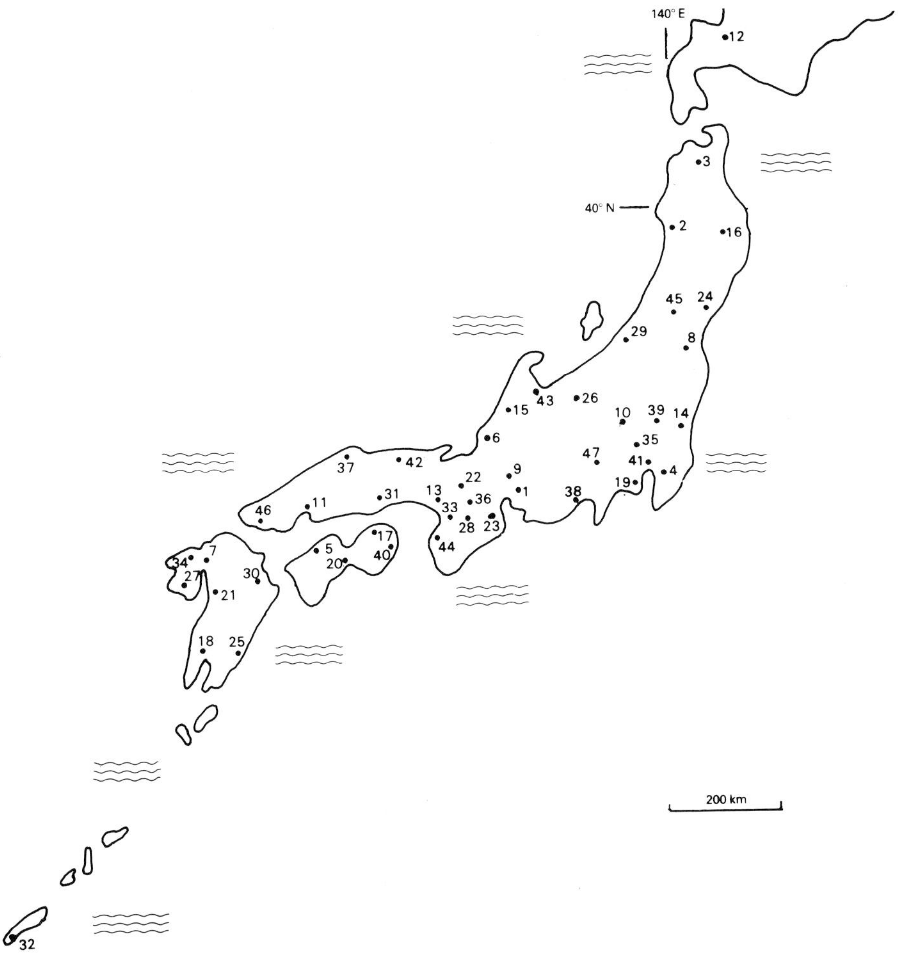
140° E
40° N
12
3
2
16
45
24
29
8
43
26
15
10
39
14
35
6
47
41
4
37
42
9
22
19
1
38
46
11
31
13
33
36
28
23
17
5
40
20
44
7
34
27
30
21
18
25
32
200 km

1 Aichi, 2 Akita, 3 Aomori, 4 Chiba, 5 Ehime, 6 Fukui, 7 Fukuoka, 8 Fukushima, 9 Gifu, 10 Gunma, 11 Hiroshima, 12 Hokkaido, 13 Hyogo, 14 Ibaraki, 15 Ishikawa, 16 Iwate, 17 Kagawa, 18 Kagoshima, 19 Kanagawa, 20 Kochi, 21 Kumamoto, 22 Kyoto, 23 Mie, 24 Miyagi, 25 Miyazaki, 26 Nagano, 27 Nagasaki, 28 Nara, 29 Niigata, 30 Oita, 31 Okayama, 32 Okinawa, 33 Osaka, 34 Saga, 35 Saitama, 36 Shiga, 37 Shimane, 38 Shizuoka, 39 Tochigi, 40 Tokushima, 41 Tokyo, 42 Tottori, 43 Toyama, 44 Wakayama, 45 Yamagata, 46 Yamaguchi, 47 Yamanashi

Survey

Geographical and climatic situation - Japan is narrow, however, extends 3,200 km from north east (46°N, 149°E) to south west (24°N, 124°E), intervened by seas. It has mountainous topographical characters. The ecological condition of the land varies extremely in each region, as do the cultivated crops. The range of yearly average temperatures is 7-17°C, and average precipitation per year ranges from 1,000 to 2,500 mm, with the rainy season between June and July, except in northern Japan. From November to February the regions along the Japan Sea Coast have a great deal of rain and snow, although less rain and more sunshine in the regions along the Pacific Ocean. Sometimes, typhoons pass through from August to

Prepared by Prof. M. Iwata, Tokyo University of Agriculture, Tokyo

October, damaging horticultural productions.

Horticultural industries - Characteristics of horticultural industries in Japan are that many crops from temperate to tropical origins are grown. The planted areas per farmer are small, i.e. less than 1 ha. In addition, diseases, insects, pests and weeds are prevalent owing to relatively high temperatures and a great deal of precipitation in summer. Successful growing is impossible without control by chemicals, plastic coverings, and introduction of resistant cultivars etc. With this type of background, labour growth has been inevitable to increase production per unit area. However, due to recent labour shortage, labour saving mechanization suitable for small scale land has become necessary.

In tree fruit growing, the planted areas and production for 1983 are 396,000 ha and 6.3 million tons, respectively, the main fruits being satsuma mandarin, apple, Japanese pear, grape, persimmon, peach, Japanese apricot and Japanese chestnut. Satsuma mandarin is the most important, occupying 31% of the total planted areas and 45% of the total production of all tree fruits. This is 41% and 57% including other citrus spp. respectively.

Apples follow, occupying 14% an 16% resp. Growing areas of satsuma mandarin are distributed mainly along coastal slope regions of the central and southern part with a yearly average temperature of 15-17°C, main growing prefectures being Shizuoka, Wakayama, Hiroshima, Ehime, Saga, Nagasaki and Kumamoto. Apples are grown in the northern part with a yearly average temperature of 8-11°C, and main producing prefectures are Aomori, Iwate, Akita, Yamagata, Fukushima and Nagano.

The planted areas and production of vegetable crops in 1983 are 644,000 ha and 15.8 million tons, respectively, main vegetables being Japanese radish, carrot, cabbage, Chinese cabbage, Welsh onion, onion, cucumber, watermelon, tomato and eggplant. In olericulture, all-year-round production system of main vegetable crops have been almost established by growing suitable cultivars for respective region and season, in the open or under coverings.

In floriculture, the areas devoted to cut flowers, nursery of ornamental trees and shrubs, and bulb production in 1983 are 12,000 , 16,000 and 1,600 ha, respectively. Cut flowers are mainly chrysanthemum, carnation, lily, rose and stock. Bulb production are for tulip, lily and gladiolus.

As for protected cultivation, the areas of plastic greenhouses in 1983 are 37,000 ha. They are used mainly for vegetable fruits as cucumber, melon, watermelon, tomato, eggplant, pepper, strawberry, and cut flowers as chrysanthemum, rose and also tree fruits as grape and satsuma mandarin. The area of glasshouses is 1,800 ha mainly for netted melon, cucumber, tomato, muscat grape and ornamental potted plants.

Research organization - In Japan, research work in horticulture is mainly done in universities, national research stations and prefectural experiment stations.

Universities - Fundamental researches are performed as well as education of students in universities. About 40 universities have a faculty of agriculture with its university farm, mostly national and one-fourth private or prefectural. However, only one-fourth of these universities have departments of horticulture including 4 laboratories at least, with different specialities such as pomology, olericulture, floriculture and postharvest etc., and the rest are on small scale with one or two laboratories of horticulture not specialized as above. Most universities have master course and about half have doctor course besides undergraduate course. In addition, there are several junior agricultural colleges including horticultural courses.

National research stations (Ministry of Agriculture, Forestry and Fisheries) - In horticulture, there are National Fruit Tree Research Stations, National Vegetable and Ornamental Crops Research Stations and also some National Agricultural Experiment Stations with horticultural divisions as in Hokkaido or Shikoku where horticultural industries are prevailing. In these research stations the problems are studied of country-wide interest or on a large scale.

The main station of National Fruit Tree Research Station is located in Tsukuba, Ibaraki pref., and breeding of Japanese pear, Japanese chestnut and prunus spp. with good quality and resistant to diseases and insect pest are made as well as researches in relation to fruit growing and orchard management. This Station has 4 branch stations; Morioka Branch in Iwate pref. (apple), Okitsu Branch in Shizuoka pref. (citrus and loquat), Akitsu Branch in Hiroshima pref. (citrus, persimmon and grape) and Kuchinotsu Branch (late-maturing citrus).

The main station of the National Vegetable and Ornamental Crops Research Station is located near Tsu in Mie pref., and breeding of vegetable crops resistant to diseases is one of the main subjects of research.

This Station has 3 branch stations; Morioka Branch in Iwate pref. (vegetable and ornamental crops in cool regions), Kurume Branch in Fukuoka pref. (vegetable and ornamental crops in warm regions) and Department of Greenhouse Cultivation in Aichi pref. (horticulture under structure).

Prefectural experiment stations - Fortyseven prefectures have respective prefectural experiment stations. Their organizations are of various forms; horticultural experiment station, fruit tree experiment station, vegetable and ornamental crops experiment station, department of horticulture in agricultural experiment station, or that in sand-dune agricultural

experiment station. In special cases we have Apple Experiment Station in Aomori pref., and Citrus Experiment Station in Shizuoka pref. Practical problems that affect directly production in local areas are mainly subjected to researches in prefectural experiment stations. These stations are also the centers for technical extension services based on their experimental results.
On the problems common to several prefectures, co-operative works of different experiment stations are carried out. Co-operative works are also made between prefectural experiment stations and national research stations, as occasion demands. Combined meeting of national and prefectural experiment stations are held regularly in respective regions, for reviewing of experimental results performed. Sometimes, staffs of universities are joined. In addition, spring and fall meetings of Japanese Society for Horticultural Science and also annual meetings of the chapters of JSHS (5 chapters) provide opportunities for mutual contact and discussion among horticultural researchers.

1 Aichi

1.1 Aichi Agricultural Research Station

elevation: 90 m
rainfall : 1,440 mm
sunshine : 1,890 h

Nagakute, Aichi-gun,
Aichi-pref 480-11
Phone: 05616-2-0085
Dir.: T. Kamura

A vegetable field, 4.0 ha, and a deciduous fruit orchard, 4.3 ha, are being used for the tests of adaptability and the performance of new improved variety (breedings), fertilizing, prevention of various diseases and pests, and labor saving test. Green-houses, 4200 m^2 and plastic greenhouses, 3400 m^2 are being used for the tests of saving petrol-energy and solution culture, and prevention of replant failure.
The tests of nursing non-virus mother plants of strawberry by shoot-apex culture and prevention tests with the inoculation of weak virus of tomato are being performed using facilities for tissue culture of 470 m^2.

1.1.1 Horticultural Institute	Chief: A. Kawabuchi
Vegetable Crop Laboratory	Head: N. Takase
Cultivation and breeding of strawberry	T. Suzuki
Breeding and water culture of tomato	S. Sugahara
Cultivation of cucumber and cabbage	Y. Hotta
Cultivation and breeding of muskmelon	M. Kasuya
Floriculture Laboratory	Head: Dr. K. Yonemura
Orchid culture in greenhouse	K. Nakagami
Rose culture in greenhouse	Y. Katano
Chrysanthemum culture in greenhouse	K. Ohishi
Carnation culture in greenhouse	T. Sakashita
Pomology Laboratory	Head: M. Aoki
Cultivation with dwarf stock of kaki and peach	N. Kimura
Deciduous fruit culture in plastic- greenhouse	N. Manago
Soil water management of grape, kiwi and fig	S. Suzaki
Soil and Plant Nutrition Laboratory	Head: A. Takei
Fertilizer application of vegetable in greenhouse	M. Ogiso
Subsoil improvement of vegetable field	T. Kinoshita
Fertilizer application of flowers in greenhouse	T. Kato
Phytopathology and Entomology Laboratory	Head: T. Miyagawa
Forecasting of occurrence on fruit pests	H. Koide
Ecological control of vegetable and fruit diseases	K. Hirota
Control and ecology of vegetable insect injury	T. Nakagome
Ecological control of vegetable and flower diseases	M. Fukaya
Gamagori Branch Laboratory	Head: M. Katsumine
Citrus growing	K. Tsuda/ M. Kaneko/ M. Kato
Utsumi Branch Field	Head: S. Takase
Citrus and loquat growing	K. Sinkai
1.1.2 Seedling and Sericulture Institute	Chief: A. Hishida
Aseptic Seedling Laboratory	Head: S. Washida
Plant propagation with tissue culture	Y. Sakurai

Plant breeding with tissue culture	K. Yabe
Aseptic seedling production and propagation	T. Iida
1.1.3 Toyahashi Agriculture Technical Center	Chief: K. Ito
Upland Farming Technical Laboratory	Head: T. Kinbara
Deciduous fruit (kaki) growing	M. Tanaka
Vegetable (sweet potato) growing	G. Hayashi/ S. Kato
Protected Cultivation Technical Laboratory	Head: H. Hagiwara
Vegetable growing in greenhouse	M. Murakami/ H. Imagawa
Floriculture in greenhouse	M. Fukuta/ J. Nishio
1.1.4 Yatomi Agriculture Technical Center	Chief: M. Fukunaga
Technical Laboratory	Head: H. Kojima
Floriculture	K. Sakai
Vegetable growing	F. Yamashita
1.1.5 Anjyo Agriculture Technical Center	Chief: Y. Fujikawa
Technical Laboratory	Y. Shimizu
1.1.6 Mountainous Region Experiment Farm	Chief: K. Tanabe
Horticultural Laboratory	Head: M. Nishioka
Flower growing	Dr. M. Morita
Vegetable growing	S. Ema

1.2 Meijo University, College of Agriculture
Laboratory of Horticulture
elevation: 10 m
rainfall : 1,500 mm
sunshine : 2,150 h

Tempaku-ku, Nagoya-shi,
Aichi-pref. 468
Phone: 052-832-1151
Dean: Dr. H. Koyama

Research and education is done in two laboratories of the Department of Agronomy and Horticultural Science in close connection with four sections of horticulture in the Experimental Farm. Of 7.7 ha of experimental field for horticulture, 4 ha are planted with ornamental trees (bamboo, camellia etc.), 2.6 ha with fruit trees (grapes, chestnuts, kaki etc), 1.1 ha with vegetable crops and flower crops. 1,500 m^2 of greenhouses are planted with melons, tomatoes, chrysanthemum, cyclamen and grapes. Doctor's program is arranged in our laboratories.

Olericulture:	
Mineral nutrition, growth analysis, and environmental physiology	Prof.Dr. T. Takano/ Dr. S. Suzuki F. Kawazoe
Pomology:	
Morphogenesis and physiology of fruit	Ass.Prof.Dr. N. Nii/ M. Nakajima
Floriculture:	
Clonal propagation of orchids and bulbs, breeding of chrysanthemum	Ass.Prof. S. Hayashi/ T. Oyamada
Landscape Horticulture:	
Landscape design and architecture	Prof.Dr. S. Nitta
Growing of bamboo, camellia	Y. Tanaka

1.3 Nagoya University, Faculty of Agriculture
Laboratory of Horticulture
elevation: 55 m
rainfall : 1,575 mm
sunshine : 2,143 h

Chikusa-ku, Nagoya-shi
Aichi-pref. 464
Phone: 052-781-5111
Dean: Dr. T. Saito

Principles of cultivation and breeding of horticultural plants are being studied in relation to their practical applications, covering the field of pomology, vegetable crop science, and floricultural science. The research fields in progress are as follows: (a) developmental physiology, (b) flowering physiology, (c) vegetative propagation, (d) self incompatibility, (e) mycorrhizas, nodules and other microbial symbiosis. Research attention has been paid to the plant growth in relation to microbial, light, nutritional and temperature environment.

Physiological studies on plant development, mycorrhizas, nodules and self incompatibility — Prof.Dr. Yukio Yamamoto
Clonal propagation and flowering control of ornamental plants — Ass.Prof.Dr. Shunji Kako
Plant growth in relation to microbial and light environment — r. Takafumi Tezuka
Development physiology of horticultural plants — H. Ohno

1.4 Aichi University of Education — Igaya-cho, Kariya-shi, Aichi-pref. 448

Plant tissue culture — Ass.Prof.Dr. Syoichi Ichihashi

2 Akita

2.1 Akita Prefectural Agricultural Experiment Station
elevation: 9.4 m
rainfall : 1,807 mm
sunshine : 1,751 h

Niida, Akita-shi
Akita-pref. 010-14
Phone: 0188-39-2121
Dir.: Dr. J. Kaneko

Horticultural Division — Chief: J. Hatakeyama
This division is doing the studies on the development of new cropping systems producing virus-free plants, avoiding replant failure; freshness retention of horticultural crops, regulation of flowering in vegetables and ornamental plants.
Vegetable Crops Section — Head: J. Fujimoto
Vegetable growing — Y. Noguchi/ A. Asari
Soil and fertilization — R. Uemura/ H. Kagaya
Flower Crops Section — Head: Y. Mizukoshi
Flower growing — A. Arino/ H. Shibata

2.2 Akita Prefectural Fruit Tree Experiment Station
elevation: 85 m
rainfall : 1,716 mm
sunshine : 1,944 h

Daigo, Hiraka-machi, Hiraka-gun,
Akita-pref. 019-05
Phone: 0182-25-4224
Dir.: H. Suzuki

Apples, the most important, grapes, pears and cherries are grown in 8 ha fields. Main themes of experiments in this Station are the breeding of apples, cultivation of dwarfed apple trees in heavy snowy districts, control of diseases and pests and establishment of soil management. Development of virus-free apple trees by heat treatment is also carried out.

Cultural Division — Chief: M. Kumagai
Breeding — Head: T. Niizuma/ H. Sato
Cultivation — Head: S. Tanno/ Y. Kume
Chemistry — Head: I. Matsui/ Y. Fujii
Plant pathology — Head: S. Takahasi/ M. Asari
Entomology — Head: H. Narita/ S. Osumi
Kazuno Branch — Head: Y. Takahashi
Apple growing — N. Mizuno/ S. Kondo/ S. Kurosawa
Tenno Branch — Head: S. Kato
Fruit growing on sand dunes — T. Taguchi/ M. Fukaya

2.3 Akita Prefectural College of Agriculture
elevation: -2.0 m
rainfall : 1,544 mm
sunshine : 2,018 h

Ohogata-mura, Akita-pref. 010-04
Phone: 018545-2026
Dir.: T. Nakano

The College was founded in 1973 in Ohgata-village which was built up by reclamation of Hachiro-lake in 1966.

Main themes of experiments in the laboratory of horticulture are on cultivation of apples and grapes, cultural systems of vegetables and flowers, vegetable cultivation under structure and water culture of vegetables in snowy and cold districts.

Laboratory of Horticulture	
Vegetables under structure	Prof.Dr. T. Takai
Cultural systemes of vegetables	H. Takahashi
Cultivation of flower and ornamental crops	T. Yamaya
Apple rootstocks, pruning	Prof.Dr. K. Kanbe
Fruit set and development	M. Sato

3 Aomori

3.1 Aomori Agricultural Experiment Station	Sakaimatsu 1-1, Kuroishi-shi
elevation: 40 m	Aomori-pref. 036-03
rainfall : 1,229 mm	Phone: 01725-2-4331
sunshine : 2,068 h	Dir.: S. Chiba

In this Station, improvement of some selected vegetable crops in the Tsugaru-region (western part of the prefecture) is made. In the Sand-dune Branch Station, research activities aim at selection of suitable crops for the low moisture contents of the area; also improved cultivation techniques irrigated by sprinklers and driptubes.

Field-crop, Horticulture and Agricultural Machinery Div.	Chief: T. Sato
Horticulture Section: Highland vegetables	Head: R. Sato
Vegetable fruit and garlic	T. Yamamoto
Edible herbs	T. Ichita
Sand-dune Branch Station	Chief: M. Nakamura
Soil and fertilization	T. Yuza
General vegetable culture	Y. Sakai
Vegetable fruit	H. Hasegawa
Edible roots	K. Kasai

3.2 Aomori Apple Experiment Station	Kuroishi, Aomori-pref. 036-03
elevation: 70 m	Phone: 0172-52-2331
rainfall : 1,225 mm	Dir.: S. Kudō
sunshine : 1,869 h	Vice Dir.: M. Sōma

This Station is responsible for research on apple trees. The aim is improvement of productivity, quality and market and consumer acceptability. The station has 23.2 ha of field and 0.6 ha for laboratories and office. Equipment: greenhouse for studies of virus diseases, phytotron for studies of fruit development, and radioisotope laboratory for studies of plant nutrition.

3.2.1 Cultural Division	Chief: N. Obara
1st Cultivation Section	
Control of growth by pruning	Head: S. Saitō
High density orchard management	C. Kamada
Rootstock-scion interaction	M. Okamoto
Rootstock evaluation and selection	T. Tonosaki
Orchard system, yield and quality	H. Yamaya
Control of variability in yield	H. Ichinohe
Methodology, data analysis	Y. Osanai
2nd Cultivation Section	
Chemical fruit thinning and abscission	Head: N. Kudō
Ripening physiology	T. Kudō
Chemical control of fruit set	K. Imai
Physiology of growth regulators	S. Noro
3.2.2 Breeding and Farm Management Division	Chief: M. Yamada

Breeding Section	
Breeding and genetics	Head: M. Ishiyama
Variety evaluation	C. Suzuki
Yield due to cultivars	H. Kitayama
Breeding of disease resistance	T. Satō
Farm Management Section	
Growing and management	Head: M. Watanabe
3.2.3 Insect Pest and Disease Division	Chief: Y. Tanaka
Insect Pest Section	
Integrated control	Head: Dr. M. Yamada
Use of pheromones in biological control	S. Shirasaki
Assessment of damage	N. Sekita
Orchard mite control	M. Kinoda
Resistance of pesticides	K. Kawashima
Insect ecology and control of pests	H. Aizu
Disease Section	
Biology and control of fungal diseases	Head: C. Fukushima
Control of soil-borne diseases	N. Nakazawa
Control of scab and mildew	N. Suzuki
Virus and mycoplasma diseases	I. Machida
Control of apple canker	K. Yukita
Control of silver leaf	S. Arai
Producting of virus-free clones	A. Saitō
Chemical deposition and persistence	Y. Masuda
Chemical analysis of pesticides	I. Ōkawa
3.2.4 Chemistry Division	Chief: S. Ichiki
Soil Improvement Section	
Specific replant diseases	Head: H. Narita
Irrigation and water use	T. Katō
Control of soil acidity	S. Sakurada
Soil management	T. Kon
Nutrition and Fertilizer Section	
Nutrient uptake and utilization	Head: M. Seitō
Carbohydrate metabolism	J. Kamakura
Physiological disorders	M. Maeda

3.3 Aomori Field Crops and Horticultural Experiment Station
elevation: 135 m
rainfall : 1,080 mm
sunshine : 1,843 h

Gonohe-machi, Sannohe-gun,
Aomori-pref. 039-15
Phone: 0178-62-4111
Dir.: T. Mikami
Vice-Dir.: H. Nasu

The Station consists of Fruit Tree Division and Horticulture Division, located in Sannohe-gun and Kamikita-gun, respectively. Fruit Tree Division of 20 ha are planted with fruits (apples, grapes, cherries, pears, peaches and plums). Facility of plant tissue culture for making virus free and cold storage rooms for fruit preservation are equipped. Horticulture Division of 40 ha are planted with vegetables (Chinese yam, garlic and many vegetables), flowers (chrysanthemum etc.) and field crops (wheat, soybean, rape etc.). Special greenhouses for making virus free plants and pre-cooling facility for vegetables are equipped.

3.3.1 Fruit Tree Division	Chief: K. Segawa/ K. Kuriu
Apple Tree Section	
Dwarfed cultivation of apple	Head: T. Tamada/ T. Yamada
Cold injury and frost protection	T. Imamura
Soil improvement of apple orchard	A. Iwaya
Fruit Tree Section	
Dwarfed cultivation of cherry and grape	Head: I. Nakagawara
Cherry cultivation and soil management	Y. Sato

Grape and peach cultivation and fruit storage	J. Araya
Plant Protection Section	
Pathology of apple and Fusarium disease of vegetables	Head: K. Matsunaka
Insectology of fruit trees	N. Sato
Pathology of apple	K. Fujita
Pests of vegetables and virus vector	M. Ishitani
Pathology of vegetables	S. Sugiyama
Pathology of grape, pear and cherry	S. Noro
Plant tissue culture and meristem culture	H. Niwata
3.3.2 Horticultural Division elevation: 55 m rainfall : 1,115 mm sunshine : 1,909 h	Rokunohe-machi, Kamikita-gun Aomori-pref. 033 Phone: 0176-53-7171 Chief: T. Takemura
Vegetable Section	
Variety of vegetables and growth regulators	Head: S. Ohba
Cultivation of Chinese yam and pre-cooling of vegetables	S. Toyokawa/ M. Yanagida
Field and greenhouse vegetables	T. Iwase
Pre-cooling and preservation	T. Tomizawa
Floriculture Section	
Floriculture suitable for cold region, *Lilium longiflorum* and *Achillea*	Head: S. Hatai
Chrysanthemum for year-round culture and *Limonium sinuatum* and *Gerbera*	H. Hayashi
Seed Production Section	
Multiplication of virus free strains of garlic, Chinese yam and strawberry	Head: M. Matsuda/ E. Chuu
Physiological and performance test of virus free strains of Chinese yam, garlic	T. Hirai/ H. Tsugawa

3.4 Hirosaki University, Faculty of Agriculture elevation: 55 m rainfall : 1,300 mm sunshine : 2,000 h	Bunkyo-cho, Hirosaki-shi, Aomori-pref. 036 Phone: 0172-36-2111 Dean: Prof.Dr. N. Sasaki
Laboratory of Pomology	
Training and pruning of apple tree	Prof.Dr. T. Kikuchi
Measurement of tree vigor	Ass.Prof. T. Asada
Laboratory of Vegetable Crops and Ornamental Plants	
Cold hardiness of ornamental plants	Prof.Dr. M. Okumura
Hormone physiology of vegetable crops	Ass.Prof. I. Okuse
Growth physiology of red pepper plant	Dr. K. Saga
University Farm	
Semi-intensive system of apple growing	Ass.Prof. Y. Shiozaki

4 Chiba

4.1 Chiba Prefectural Agricultural Experiment Station elevation: 49 m rainfall : 1,489 mm sunshine : 1,804 h	Daizenno-cho, Chiba-shi Chiba-pref. 280-02 Phone: 0472-91-0151 Dir.: H. Ishiwata

This Station is the main centre of the general agricultural research in addition to vegetable, deciduous fruit tree and ornamentals in Chiba. It consists of a general affairs division and 20 research sections with 197 persons including 86 researchers, and has a 40.1 ha field for rice, upland field crops and horticultural crops. The main research subjects of the Station are on the breeding and cultivation of rice, peanut and other upland field crops as well as horticultural ones, soil and fertilizer, pests and diseases, air and water pollution, post harvest physiology and farm management.

Deciduous Fruit Tree Section	
Physiology of fruit trees	Head: W. Hitokuwada
Variety, physiological disorder	T. Nagato
Training and pruning	T. Ishida
Breeding and chemical control	M. Kitaguchi
Cultivation in plastic greenhouse	S. Kawase
Ornamental Plant Section	
Composts	Head: M. Endo
Ornamental trees and shrubs	K. Tomioka
Pot and cut flowers	E. Seki/ K. Aoki
Ornamental foliage plants	T. Shibata
Vegetable Crop Section	
Cultivation of vegetable crops	Head: T. Toki
Physiological disorder	K. Kohta
Soilless culture	U. Udagawa
Environment in greenhouse	K. Yuhashi
Energy saving in greenhouse	H. Suzuki
Vegetable Crop Section for 'Nanso Area'	
Cultivation and breeding	Head: Y. Kawasaki/ F. Yamamoto
Cultivation in drained paddy field	K. Arihara
Vegetable Crop Section for 'Sandy Areas'	
Tomato breeding	Head: H. Aoki
Environmental control in greenhouse	M. Inoue
Cultivation of vegetable crops	M. Ishikawa
Vegetable Crop Section for 'Toso Areas'	
Cultivation of vegetable crops	Head: S. Tokoro
Meteorological environment	S. Aoyagi
Physiology of vegetable crops	N. Jinbo

4.2 Chiba Prefectural Foundation Seed and Stock Farm
elevation: 6 m
rainfall : 1,457 mm
sunshine : 2,746 h

Chosei-mura, Chosei-gun,
Chiba-pref. 299-43
Phone: 04753-2-3377
Dir.: Y. Takeichi

Preservation is performed on foundation seeds and stocks of cereal crops, fruit trees, vegetables and ornamental crops. Propagation and distribution of seeds, nursery stocks and tissue cultured virus-free plants are also made.

Vegetable Crop Section	Head: A. Yoshino/ M. Murai
	S. Yoshida
Ornamental Plant Section	Head: Y. Marushima/ K. Kodachi
	H. Katsura
Virus-free Section	Head: S. Jitsukawa/ K. Ohkoshi
Fruit Tree Section	Head: H. Ishibashi/ T. Sase
	Y. Yamamoto/ H. Ito

4.3 Chiba Prefectural Horticultural Experiment Station
elevation: 70 m
rainfall : 2,160 mm
sunshine : 2,415 h

Yamamoto, Tateyama-shi,
Chiba-pref. 294
Phone: 0470-22-2603
Dir.: S. Morioka

The work of the Station is concerned with the cultural techniques and breeding of horticultural crops suitable for production in warmer districts located in the southern area of Chiba prefecture. Total area 17.3 ha with 5.3 ha orchards (citrus, loquat), 2.6 ha vegetable crop fields (strawberry, muskmelon, lettuce, parsley), 1.2 ha floriculture crop fields (rose, carnation, stock, gerbera) containing 0.3 ha greenhouse.

Fruit Tree Section	
Loquat breeding and cultivation	Head: S. Nakai

Citrus cultivation	S. Tachibana
Loquat cultivation and physiological disorder	S. Yahata
Vegetable Crop Section	
Breeding and forcing of fruit vegetables	Head: K. Tanaka
Strawberry forcing and leaf vegetable cultivation	H. Togura
Tissueculture	T. Mihira
Muskmelon Section	
Breeding and cultivation	Head: Y. Urabe
Breeding and greenhouse environment	T. Ōizumi
Floriculture Section	
Bulbous plant cultivation	Head: T. Horikawa
Cut flower breeding and cultivation	A. Tanaka
Cut flower cultivation in greenhouse	M. Hosoya*
Tissue culture	M. Kanda
Plant Protection Section	
Disease control of cut flowers and vegetable crops	Head: S. Onogi
Pest control of fruit trees and vegetable crops	S. Uematsu
Disease control of fruit trees	K. Sekiyama
Soil management of muskmelon and cut flowers	T. Shirasaki
Soil management of cut flowers and fruit trees	T. Matsuo

4.4 Chiba University, Faculty of Horticulture, Department of Horticultural Science
elevation: 25 m
rainfall : 1,500 mm
sunshine : 2,000 h

Matsudo-shi, Chiba-pref. 271
Phone: 0473-63-1221
Dean: Dr. H. Oizumi

In this Faculty, there are seven laboratories in relation with horticulture: Pomology, Vegetable Science, Floriculture, Plant Breeding, Horticultural Engineering, Plant Pathology and Environmental Biology. The faculty posesses 3 farms as follows: Matsudo farm (18 ha), Tone farm (17 ha) and Atagawa farm (5 ha).

Laboratory of Pomology	
Physiology of developing fruit	Dr. N. Hirata*
Hormonal control of fruit drop	Dr. E. Takahashi
Self-incompatibility	S. Hiratsuka
Laboratory of Vegetable Science	
Nitrogen nutrition	Dr. K. Gomi
Photosynthesis and translocation	Dr. T. Ito*
Calcium nutrition	T. Maruo
Laboratory of Floriculture	
Color and pigment distribution	Dr. M. Yokoi
Physiology and classification	Dr. T. Ando*
Classification and breeding	Y. Ueda
Laboratory of Plant Breeding	
Genetics	Dr. M. Iizuka*
Tissue culture	Dr. M. Mii
Laboratory of Horticultural Engineering	
Energy analysis	Dr. I. Watanabe
Control of greenhouse climate	Dr. T. Kozai*
Energy saving method	Dr. M. Hayashi*
Laboratory of Plant Pathology	
Soilborne disease	Dr. W. Iida
Laboratory of Environmental Biology	
Orchard mite control	Dr. N. Shinkaji

5 Ehime

5.1 Ehime Prefectural Agricultural Experiment Station — Dogo, Matsuyama-shi,

elevation: 32.9 m
rainfall : 1,462 mm
sunshine : 2,103 h

Ehime-pref. 790
Phone: 0899-24-8108
Dir.: A. Watanabe

Cultivar selection, breeding and improvement of cultivation methods of vegetables (tomato, strawberry, broad bean, daikon and turnip etc.) and flowers (chrysanthemum and lily etc.) are made in a 1 ha field. Equipment: 8 greenhouses for cultivation experiment, 3 for environmental factor analysis, 1 for photoperiodic control, 1 for mist propagation, 1 for acclimatization and the equipment for analyzing photosynthesis.

Horticulture Group	
Breeding and improvement of cultivation method of vegetables	Head: K. Shinohara/ O. Watanabe Y. Saiki/ H. Kawauchi
Physiological and ecological investigation of vegetables	H. Obayashi/ M. Shimizu
Breeding and improvement of cultivation of flowers	K. Kiyasu
Factor analysis of the environment under structure	T. Kono

5.2 Ehime Prefectural Fruit Tree Experiment Station
elevation: 200 m
rainfall : 1,335 mm
sunshine : 2,140 h

Shimoidai-cho, Matsuyama-shi 791-01
Phone: 0899-77-2100
Dir.: H. Oomori
Manager: S. Mori
A. Oowada

Researches on the improvement of yield, quality of citrus and deciduous fruits, control of diseases and insects of fruit trees. Twenty ha are planted with citrus (mandarin, Iyo tagor, navel-orange, miscellaneous varieties and hybrids), and 5 ha are planted with deciduous fruits (kiwifruit, peach, pear, grape, chestnut and persimmon).

Citrus Growing Group	
Storage, cold hardiness, physiological disorders	Head: E. Beppu
Chemical controls of citrus	H. Nakata
Citrus Breeding Group	
Selection and establishment of new varieties	Head: E. Watanabe
Breeding and weed control	K. Kita
Deciduous Fruit Group	
Selection and establishment of new varieties	Head: T. Yamanaka
Growing and storage of kiwifruit	T. Ninomiya
Chemical controls of deciduous fruits	K. Ishikawa
Soil and Fertilizer Group	
Fertilizer on fruit quality	Head: S. Akamatsu
Diagnosis and nutrition	N. Takagi
Entomology Group	
Ecology and control of citrus pest	Head: H. Ogihara/ Y. Oomasa
Ecology and control of deciduous fruit pest	S. Kubota
Phytopathology Group	
Citrus virology	Head: Y. Tachibana
Citrus canker disease	M. Sagawa
Occurrence forecast of citrus diseases	T. Yano
Management and Mechanization Group	
Orchard machinery and installation	Head: A. Okazoe
Analysis of operating costs	H. Watanabe
Nanyo Branch	
Citrus growing in greenhouse	Head: Y. Ishida
Variety, rootstock and storage of citrus	K. Funagami
Storage and growing of citrus	T. Nishiyama
Iwaki Branch	
Rootstock of lemon and Hassaku	Head: M. Zinno
Lemon growing in field and greenhouse	Y. Waki
Navel orange, Encore and Murcott growing	T. Matsushita

Kikohu Branch	
Selection and growing of chestnut	Head: S. Okuchi
Selection and growing of peach and chestnut	R. Kawakami

5.3 Ehime University, College of Agriculture, Department of Agrobiology and Horticultural Science
elevation: 53 m
rainfall : 1,337 mm
sunshine : 2,134 h

Tarumi-cho, Matsuyama-shi
Ehime-pref. 790
Phone: 0899-41-4171
Dean: Dr. O. Inose

Improvement of fruit growing techniques, especially in citrus, genetic improvement of vegetables and ornamental plants, and environmental control under structure are the principal research subjects. Equipment: high performance liquid chromatograph, atomic absorption spectrometer, mass spectrometer, liquid scintillation spectrometer, air conditioning rooms.

Laboratory of Citriculture	
Fruit quality improvement in citrus	Prof.Dr. K. Matsumoto*
Morphology in citrus	Ass.Prof.Dr. M. Shiraishi
Physiological disorders in citrus	S. Chikaizumi
Laboratory of Horticulture	
Ecological fruit growing	Prof.Dr. K. Kadoya*
Physiology of deciduous fruit trees	Ass.Prof.Dr. F. Mizutani*
Photosynthesis of fruit trees	A. Hino
Laboratory of Plant Breeding	
Cytogenetics of crucifers and Pelargonium	Prof.Dr. S. Tokumasu*
Water relations and stomatal movement	Ass.Prof. S. Jodo
Cytogenetics of crucifers, tissue culture	M. Kato
Physiology of reproduction in Pelargonium	F. Kakihara
Experimental Farm	
Ecology of palms and orchids	Prof.Dr. T. Sento
Fruit setting, yield, and quality of citrus fruit	Ass.Prof. J. Watanabe
Environmental control under structure	T. Fukuyama
Fruit quality and environmental factors	H. Akiyoshi
Plant growth regulators	S. Amano

6 Fukui

6.1 Fukui Prefectural Agricultural Experiment Station
elevation: 10 m
rainfall : 2,475 mm
sunshine : 1,763 h

Ryo-machi henguri, Fukui-shi
Fukui-pref. 910
Phone: 0776-54-5100
Dir.: K. Takahashi

Horticultural Division: Conducting research on vegetables, flower plants, and fruit trees which will help to promote horticulture in this prefecture.

Vegetable and Floriculture Section	Head: T. Yamaguchi
Mass-production technique for grafted seedlings of fruit vegetables	M. Matsuyama
Cultivation techniques for leafy vegetables in greenhouses	Y. Hayakawa
Cultivation techniques for leafy and root vegetables	H. Yamaguchi
Cultivation techniques for cut flowers in greenhouses	T. Kazuma
Breeding of vegetables	T. Tomita
Sakai-hill Upland Field Farming Section	Head: H. Yamazaki
Water relation to vegetables	M. Matsuura

Water relation to vegetables in greenhouses	K. Iino
Fruit Tree Section	Head: K. Tomita
Persimmon cultivation	R. Honda
Chestnut cultivation	Y. Nishiguchi
Pear cultivation	K. Nagasawa

6.2 Fukui Prefectural Horticultural Experiment Station
elevation: 10 m
rainfall : 2,778 mm
sunshine : 1,500 h

Kugushi, Mihama-cho, Mikata-gun,
Fukui-pref. 919-11
Phone: 0770-32-0009
Dir.: K. Miyamatsu

This Station conducts research on fruit trees, vegetable crops, ornamental plants, as well as soil improvement methods especially for Reinan district (southwestern region in Fukui prefecture).

Fruit Tree Section (mainly plum)	
Cultivation techniques for dwarfed plum trees	Head: K. Tanabe
Breeding of plum trees	T. Miyahara
Preventing physiological fruit drop	T. Takano
Vegetable Crops Section	
Cultivation techniques for vegetable crops	Head: S. Morikawa
Breeding of vegetable crops	I. Takezawa
Apical meristem culture for virus-free seedlings	I. Kawahara
Floriculture Section	
Regulation for flowering of *Physanthus narcissus*	Head: S. Morikawa
Cultivation techniques for flowers	N. Nagai
Breeding and performance tests for flowers	T. Tanaka
Environmental Control Section	
Soil improvement of converted paddy fields in Reinan district	Head: S. Sato
Forecasting and monitering of diseases and pests	K. Yamamoto
Diagnosis of nutrient condition in paddy field soil	T. Watanabe
Yield forecasting of rice in Reinan district	S. Nozaki

6.3 Fukui Prefectural College, Department of Agriculture
elevation: 45 m
rainfall : 2,475 mm
sunshine : 1,764 h

Obatake-cho, Fukui-shi,
Fukui-pref. 910
Phone: 0776-54-7611
Dir.: Dr. K. Imabori

This College is the only agricultural education center in Fukui prefecture with 80 students. Tissue culture facilities with a Laminar flow bench and an air conditioned incubating room. Scanning electron microscope. Of two greenhouse of 100 m^2 each in the campus, one is equipped with hydroponic systems. The Experimental Farm, situated about 10 km from the campus possesses 7 ha of field, of which 0.6 ha with vegetables, 0.3 ha with tree fruits (grapes, mumes, pears, persimmons, chestnuts). One greenhouse of 200 m^2 and three plastic greenhouses of 500 m^2 each.

Laboratory of Horticulture	
Tissue culture propagation	Ass.Prof.: Dr. S. Ohki
Fertilization and postharvest relations	D. Miyajima
Experimental Farm	
Plant pathology	Prof.Dr. K. Nasuda
Vegetable breeding	Y. Mori
Machinery in vegetable cultivation	H. Katsuda

7 Fukuoka

7.1 Fukuoka Prefectural Research Center, Institute of

Ashiki, Chikushino-shi

Horticulture elevation: 120 m rainfall : 2,261 mm sunshine : 1,892 h	Fukuoka-pref. 818 Phone: 092-922-4111 Dir.: S. Yoshitake

This Institute carries out practical researches of fruit trees, vegetables and ornamental plants.

Division of Fruit Tree	Chief: M. Shimoohsako
Fruit Variety Section	Head: T.Sumi/ M.Nozuka/ N.Hirakawa
Evergreen Fruit Tree Section	Head: M.Yoshida/ Y.Ōba/ S. Kusano
Deciduous Fruit Tree Section	Head: F.Hamachi/ A.Morita/ S.Himeno
Division of Vegetable and Ornamental Crop	Chief: S. Yoshitake
Vegetable Variety Section	Head: M. Murozono/ H. Fushihara
Vegetable Cultivation Section	Head: Y.Tanaka/ M.Takao/ M.Hayashi
Ornamental Crop Section	Head: Dr.T. Matsukawa/ Y. Kobayashi S. Mamezuka/ H. Kondo
Buzen Branch	Dir.: M. Wada
Deciduous fruit trees	Head: A. Kanafusa/ K. Shoda M. Awamura

7.2 Fukuoka Prefectural Agricultural Research Center, Institute of Farm Management and Environment	Yoshiki, Chikushino-shi, Fukuoka pref. 818 Phone: 092-924-2938 Dir.: M. Miyahara

Division of Agricultural Chemistry	Chief: M. Matsui
Plant Nutrition Section	Head: Y.Nakashima/ Y.Ito/ KKonomi
Division of Phytopathology and Entomology	Chief: H. Sakai
Vegetable Section	Head: S.Tanaka/ H.Ikeda/ T.Nakamura
Fruit Tree Section	Head: M.Noda/ Y.Noguchi/ K.Yamada
Division of Farm Management	Chief: T. Wada
Distribution and Utilization Section	Head: A. Matsumoto/ T. Hirano S. Yamashita

7.3 National and Ornamental Crops Research Station, Kurume Branch elevation: 40 m rainfall : 2,000 mm sunshine : 2,000 h	Mii-machi, Kurume-shi, Fukuoka-pref. 830 Phone: 0942-43-8271 Dir.: F. Honda

This Branch is the southwest center of vegetable and ornamental crops research for the warm region of Japan, and also the national center for the breeding of strawberry, lily and Japanese azaleas.

Laboratory of Breeding for Greenhouse Vegetable	
Melon and tomato breeding	Head: Dr. R. Koyyama
Melon breeding	T. Yoshida
Laboratory of Breeding for Open-air Vegetable	
Strawberry genetics and breeding	Head: S. Yamakawa/ H.Sato/ Y.Noguchi
Laboratory of Vegetable Crop Cultivation	
Vegetable physiology and nutrition	Head:Dr.S.Okitsu/ H.Ikeda/ S.Furuya
Laboratory of Ornamental Crop Breeding	
Lilium breeding	Head: Y. Hirata
Rhododendron breeding and cytology	Dr. S. Yamaguchi
Genetics	Y. Kiyosawa

Laboratory of Plant Pathology
Soil-borne disease — Head: Dr. Y. Sonku
Wilt disease — Y. Nomura
Laboratory of Entomology
Integrated control of vegetable pests — Head: T. Yoshihara
Population biology of insect pests — A. Kawai

7.4 Kyushi University, Faculty of Agriculture
elevation: 2.5 m
rainfall : 1,689 mm
sunshine : 1,973 h

Hakozaki, Higashi-ku, Fukuoka-shi,
Fukuoka-pref. 812
Phone: 092-641-1101
Dean: Dr. H. Miyajima

Laboratory of Horticultural Science
In this laboratory, three projects, namely, plant breeding and genetics, plant physiological analysis and micro-propagation, are practised in fruit tree, vegetables and ornamentals. Genus investigated are the following: Allium, Anthirrhinum, Brassica, Calanthe, Camellia, Chrysanthemun, Citrus, Cucurbits, Cymbidium, Hyacinthus, Lilium, Psophocarpus, Raphanus, Rosa, Tulipa and Vitis.

Physiology of growth cycle in horticultural plants. — Prof.Dr. S. Uemoto*
Micro-propagations
Plant breeding and physiology in fruit trees — Ass.Prof.Dr. S. Shiraishi*
Physiology in bulbous plants — H. Okuba*
Experimental Farm
Genecology in vegetables — Prof.Dr. K. Fujieda
Manurial effects of ammonium ion in vegetables — Ass.Prof.Dr. K. Hanada
Cytogenetics in Citrus — A. Wakana

8 Fukushima

8.1 Fukushima Prefectural Agricultural Experiment Station
elevation: 235 m
rainfall : 1,210 mm
sunshine : 1,957 h

Tomitamachi, Koriyama-shi,
Fukushima-pref. 963
Phone: 0249-32-3020
Dir.: A. Okazaki

Improvement of vegetable crop management practices. Of 1.5 ha, 1.2 ha are open field (cucumbers, tomato, cabbage, lettuce), and 0.3 ha for greenhouses. Equipment: tissue culture laboratory and pre-cooling chambers for postharvest study.

Vegetable Crop Section — Head: E. Kimura
Cherry tomato — S. Sato
Strawberry and weed control — M. Enomoto
Cucumber and postharvest problems — S. Yamauchi
Tomato and physiological disorder — K. Yoshioka
Asparagus and lettuce — M. Sato
Tissue culture and postharvest problems
Iwaki Branch — Taira, Iwaki-shi, Fukushima-pref. 970-01
Head: S. Henmi

Cucumber and tomato — H. Kitahara
Asparagus and strawberry — T. Suda
Lily and gentian — H. Nitta
Azalea and chrysanthemum — S. Numa
Flowering tree and shrub — Y. Toyama

8.2 Fukushima Prefectural Fruit Tree Experiment Station
elevation: 102 m
rainfall : 1,238 mm
sunshine : 1,680 h

Iizaka-machi, Fukushima-shi,
Fukushima-pref. 960-02
Phone: 0245-42-4191
Dir.: M. Kumakura

Producing and utilization of tree fruits and grapes. 7 ha are planted with apples, peaches, Japanese pears and grapes. Equipment: heat therapy room for plant viruses.

Fruit Growing Section	Head: N Hashimoto/ S. Uchiyama S. Inoue/ T. Kunisawa/ K. Abe Y. Matsukawa/ Y. Sawada
Soil and Fertilizer Section	Head: Dr. K. Komamura/ A. Suzuki N. Matsumoto
Postharvest Section	Head: R. Sato/ K. Kato
Phytopathology Section	Head: Dr. M. Ochiai/ S. Hayashi N. Takamura
Plant Virus Section	Head: H. Yamaga/ T. Munakata
Entomology Section	Head: N. Abe/ R. Sato

9 Gifu

9.1 Gifu Prefectural Agricultural Experiment Station
elevation: 15 m
rainfall : 2,011 mm
sunshine : 2,322 h

Matamaru, Gifu-shi,
Gifu-pref. 501-11
Phone: 0582-39-3131
Dir.: M. Nobuta

Breeding of onion, strawberry, Chinese cabbage, radish, Hakuran, chrysanthemum and lily, and improvement in culture of strawberry, eggplant, taro, chrysanthemum and dendrobium are carried out in 0.34 ha of greenhouse and a 0.7 ha field in the first section; breeding of early, maturing sweet persimmon and Japanese pear at a 1.4 ha field in the second section. In the third section the methods for securing production of tomato, cucumber and melon.

Horticulture Division	Chief: T. Futatsudera
Vegetable and Flower Section	
Culture of strawberry, taro	Head: T. Watanabe
Culture of eggplant, radish	Y. Hibino
Breeding of Brassica, strawberry and onion	K. Koshikawa
Breeding and culture of flower crops	K. Tanahashi
Fruit Tree Section	
Culture of persimmon	Head: H. Yai
Breeding of persimmon	T. Go
Culture of Japanese pear	H. Matsumura
Warm Region Agricultural Management Division	Chief: T. Kuwabara
Horticulture Section	
Culture of tomato, cucumber	Head: Y. Maruyama
Culture of melon	K. Tsuda

9.2 Gifu prefectural Highland Agricultural Experimental Station
elevation: 493 m
rainfall : 1,765 mm
sunshine : 1,603 h

Furukawa-machi, Yoshiki-gun,
Gifu-pref. 509-42
Phone: 05777-3-2029
Dir.: H. Yasuda

Experiments are carried out to improve the culture of vegetables in the open, vegetables under cheap and easily built-up greenhouses, fruit trees and flowers on 500 - 1,300 m high areas in Gifu prefecture.

Research Division — Chief: H. Yasuda
Olericultural Section — Head: H.Kato/ T.Nogawa/ T.Fujimoto
Fruit Tree and Floriculture Section — Head: M. Umemaru/ A.Takagi/ T.Kamiya

9.3 Gifu Prefectural Hilly-land Agricultural Experiment Station
elevation: 390 m
rainfall : 1,839 mm
sunshine : 2,056 h

Sendanbayashi, Nakatsugawa-shi,
Gifu-pref. 509-91
Phone: 05736-8-2036
Dir.: F. Fukutomi

This Station is studying on crops of the hilly-land areas (elevation 300 to 500 m). 3.4 ha are planted with chestnuts and tree fruits (apple and peach), 0.4 ha with vegetables (tomato and strawberry) and 0.3 ha of greenhouse with flowers.

Fruit Tree Section
Fruit Growing — Head: M. Goto*/ M. Kawashima
Horticultural Section
Flower breeding and growing — Head: S. Ohno
Vegetable growing — S. Suzuki/ N. Katsuragawa

9.4 Gifu University, Faculty of Agriculture
elevation: 14 m
rainfall : 1,800 mm
sunshine : 2,250 h

Yanagido, Gifu-shi,
Gifu-pref. 501-11
Phone: 0582-30-1111
Dir.: Prof.Dr. I. Isogai

Laboratory of Horticultural Science - Studies are done on growth physiology and flower bud differentiation in vegetables and fruit trees such as tomato, persimmon and apple in relation to endogenous plant growth regulators, methods of cutting propagation of Japanese native Rhododendrons, fruit tissue cultures, ovary cultures and cultural condition of protoplast of orchids.

Fruit development and growth regulators — Prof.Dr. M. Nakamura
Vegetable and orchid growth — Ass.Prof.Dr. S. Matsui
Plant tissue culture — S. Ohta

10 Gunma

10.1 Gunma Prefectural Horticultural Experiment Station
elevation: 83 m
rainfall : 1,016 mm
sunshine : 2,480 h

Azuma-mura, Sawa-gun,
Gunma-pref. 379-22
Phone: 0270-62-1021
Dir.: S. Sato

Main research activities are establishment of cultural techniques and variety tests for improving productivity and quality of horticultural crops.

Vegetable Crop Division — Chief: S. Watanabe
Vegetable Crop Section — Head: K. Kurihara
Vegetable growing — I. Ogiwara/ F. Kabasawa/ U. Kanai
Plant tissue culture — N. Kurihara/ Y. Kimura
Greenhouse Vegetables Section — Head: H. Ota
A. Nishide/ K. Nasu/ U. Yutani
H. Abe/ T. Shiraishi
Fruit Tree and Ornamental Plant Division — Chief: S. Takahashi

Fruit Tree Section — Head: S. Hoshikawa
T. Miyoshi/ K. Muraoka/ T.Matsunami

Ornamental Crop Section — Head: Y. Hanaoka
T. Motegi/ T. Yoshino

11 Hiroshima

11.1 Hiroshima Prefectural Agricultural Experiment Station
elevation: 230 m
rainfall : 1,500 mm
sunshine : 2,750 h

Hachihonmatsu-cho,
Higashihirosima-shi,
Hiroshima-pref. 739-01
Phone: 0824-29-0521
Dir.: N. Takihiro

Studies in this Station are being carried out in order to level-up the productivity of paddy rice, vegetables, flowers, ornamental trees and mat rush grass etc. The Horticultural Section is studying improvement of cultivation methods of vegetables and ornamental crops.

Horticultural Section — Head: H. Korematsu
Vegetable crops — J.Otomo/ Y.Taniguchi/ M.Fukunaga
S. Hasegawa
Floriculture — H.Furuya/ T.Katsutani/ Y.Ikeda
Highland Branch — Head: N. Hiraoka
Vegetable crops — T. Ito/ A. Tanaka
Island Branch — Head: T. Funakoshi
Vegetable crops — T. Ushiro/ S. Tanabe/ A. Hirao

11.2 Hiroshima Prefectural Fruit Tree Experiment Station
elevation: 150 m
rainfall : 1,486 mm
sunshine : 2,839 h

Akitsu-machi, Toyota-gun,
Hiroshima-pref. 729-24
Phone: 08464-5-1225/ 08464-5-1227
Dir.: K. Sadai

Subjects for research at this Station are in relation to pomology, soil science and control of insect pests and diseases for citrus and deciduous fruit trees in the Hiroshima prefecture.

Planning and Survey Section — Head: H. Kobayashi/ S. Furui
Cultural Section — Head: K. Sakai
Citrus — T. Akimoto/ M. Nakatani/ T. Yuasa
Deciduous fruit tree — T. Imai/H. Nitta
Soil and Protectional Section — Head: K. Sadai
Soil for citrus orchard — K. Nishida
Soil for deciduous fruit tree orchard — T. Fujiwara/ H. Kimura
Plant pathology — S. Ogasawara
Entomology — K. Matsumoto
Citrus Branch Station — Head: A. Sasaki
Medium and late maturing variety of citrus — K. Ogawa
Soil science — T. Yamazaki
Virus diseases — J. Hiramatsu

11.3 National Fruit Tree Research Station, Akitsu Branch
elevation: 140 m
rainfall : 1,400 mm
sunshine : 2,391 h

Akitsu-machi, Toyota-gun,
Hiroshima-pref. 729-24
Phone: 08464-5-1260/ 08464-5-5370
Dir.: I. Ikeda

Breeding of grape, persimmon and citrus. Of 20 ha, 12 are planted with tree fruits. In addition there is research of the problems concerning fruit tree growing, virus diseases and pest insects of citrus, grape, persimmon and kiwi fruit.

Deciduous fruit Breeding Section	
Grape and Japanese persimmon breeding	Head: Y. Yamane
Grape breeding and tissue culture	T. Hirabayashi
Japanese persimmon breeding	M. Yamada
Citrus Fruit Breeding Section	
Citrus breeding	Head: I. Oiyama/ K. Yoshinaga
Citrus breeding and tissue culture	S. Kobayashi
Fruit Growing Section	
Water stress and evapotranspiration	Head: Y. Hase
Plant physiology	N. Ishikawa
Pathology Section	
Citrus virus and kiwi fruit diseases	Head: Dr. S. Takaya
Virus diseases of citrus and grape	J. Imada
Virus diseases of citrus	H. Sawada
Entomology Section	
Scale insect biology	Head: T. Sakagami
Mites and grape insects	W. Ashihara
Mites and natural enemies	M. Osakabe

11.4 Hiroshima Prefectural Agricultural College
elevation: 220 m
rainfall : 1,555 mm
sunshine : 2,364

Saijo-cho, Higashi-Hiroshima-shi,
Hiroshima-pref. 724
Phone: 0824-22-3184
Dean: Dr. S. Ono

Laboratory of Horticulture - The researches in this Laboratory are as follows: Pomology: factors affecting ripening of grapes and selection of excellent strains of Saijo persimmon; Vegetable growing: effects of plant growth regulators on the growth of vegetables; Floriculture: breeding and tissue culture of carnation, and Fruit and vegetable processing: storage and processing of chestnuts and astringency removal of Saijo persimmon.

Pomology	Prof.Dr. R. Isoda
Vegetable growing	Dr. Y. Hayata
Floriculture	Prof. M. Kakehi
General horticulture	Dr. Y. Niimi
Fruit and vegetable processing	Prof.Dr. T. Manabe

12 Hokkaido

12.1 Hokkaido Central Agricultural Experiment Station
elevation: 16 m
rainfall : 817 mm
sunshine : 1,704 h

Naganuma-cho, Yubari-gun,
Hokkaido 069-13
Phone: 01238-9-2311
Dir.: Dr. T. Baba

This Station serves Hokkaido on all problems concerning the agriculture. Horticultural Division has research programs for breeding, cultural management and storage of apples, grapes, small fruits, vegetables and flowers and also carries out programs for micropropagation and production of virus-free plants by tissue culture.

Horticulture Division	Chief: E. Miki
Fruit Tree Section	
Breeding and growing	Head: T. Minegishi/ F. Matsui
Breeding and tissue culture	H. Muramatsu/ M. Kakizaki
Ebeotsu Apple Branch Station	

Apple breeding and small fruit growing	H. Watanabe
Vegetable and Flower Section	
Growing of vegetables	Head: H.Dohi/ F.Takahashi/ A.Nagao
Breeding and tissue culture of flowers	Y. Shiga
Processing Section	
Storage of fruits and vegetables	Head: T Indo
Tissue culture	T. Kusaka

12.2 Southern Hokkaido Agricultural Experiment Station
elevation: 25 m
rainfall : 1,200 mm
sunshine : 708 h

Ono-cho, Kameda-gun Hokkaido 041-12
Phone: 0138-77-8116
Dir.: Dr. M. Takakuwa

Horticulture under structure: Many plastic and glasshouses. Environment controlled glasshouse (9 rooms). 3 ha are planted with tree fruit (apples, cherries, chestnuts).

Horticultural Section	
Growing of vegetables	Head: K. Sawada
Breeding of strawberry	H. Konno
Growing of tomato, sweetcorn	K. Shiozawa
Growing of turnip, cabbage	S. Katō
Apple and other fruit trees	R. Ogano
Soil and Fertilizer Section	
Soil management	Head: S. Sōma
Strawberry fertilization	S. Kawahara
Organic substance application	T. Meguro
Disease and Insect Section	
Cyclamen-mite	Head: K. Satō
Tomato verticilium wilt	O. Tamura
Prediction and control of Diamondback moth (Plutella xylostella)	S. Mizushima

12.3 Hokkaido National Agricultural Experiment Station
elevation: 70 m
rainfall : 1,196 mm
sunshine : 2,403 h

Hitsujigaoka, Toyohira-Ku Sapporo, Hokkaido 061-01
Phone: 011-851-9141
Dir.: Kyojiro Inoue

This Station, which is located in the northernmost part of Japan, consists of 11 research departments that are concerned with several kinds of crops and livestock. It serves the Hokkaido area for all agricultural problems, and of the 6 ha occupied by the Horticultural Department, 4 ha are planted with fruit trees, and 2 ha with vegetables.

Department of Horticulture and Industrial Crops	Chief: K. Kaneko
1st Horticultural Crop Lab. (Fruit Production)	
Physiology of fruit growth	Head: K. Chiba
Variety test of apple trees	J. Murakami
Breeding of pears	F. Nakajima
Cold resistance of fruit trees and spacing effect of dwarf apple trees	H. Kuroda
2nd Horticultural Crop Lab. (Vegetable Breeding)	
Vegetable breeding	Head: H. Yoshikawa
Physiological and ecological analysis of onion	M. Nagai
Breeding of onion	M. Tanaka
Breeding of asparagus	A. Uragami
3rd Horticultural Crop Lab. (Vegetable Cultivation)	
Vegetable cultivation	Head: N. Matsuo

Physiology of growth — M. Agatsuma

12.4 Hokkaido University, Faculty of Agriculture,
Department of Agronomy, Laboratory of Horticulture
elevation: 16 m
rainfall : 1,135 mm
sunshine : 1,916 h

Nishi 9, Kita 9, Kita-ku
Sapporo 060
Phone: 011-716-2111
Dean: Dr. H. Okajima

Current research activities: Propagation of vegetables and fruit trees, post harvest physiology and storage of horticultural products, breeding of asparagus and some other vegetables, breeding of bulbous plants, design and maintainance of roadside green areas and home gardens, fundamentals in green area planning in metropolitan districts in Hokkaido.

Fruit Tree and Vegetable Crops Laboratory	
Breeding of asparagus	Prof.Dr. T. Yakuwa
Propagation of vegetables and fruit trees	Ass.Prof.Dr. T. Harada
Ornamental Horticulture and Landscape Architecture Laboratory	
Breeding and multiplication of ornamental plants	Prof.Dr. K. Tsutsui
Urban green area planning	Dr. S. Asakawa
Breeding of tulips	H. Chono
Breeding of Lilies	Dr. Y. Asano
Experimental Farm	
Fruit storage	Ass.Prof.Dr. S. Imakawa
Fruit growth	Y. Mino

13 Hyogo

13.1 Hyogo Prefectural Agriculture Center
elevation: 10 m
rainfall : 1,367 mm
sunshine : 2,126 h

Kitaoji-cho, Akashi-shi,
Hyogo-pref. 673
Phone: 078-928-3521
Dir.: T. Fujimura

Breeding and cultivation of horticultural plants. Three ha including 1,500 m^2 of glasshouses and 2,500 m^2 of plastic greenhouses are planted with vegetables and flowers at Akashi. Another 5 ha of the Fruit Trees Experiment Farm are planted with Japanese chestnuts, Japanese pears, grapes, figs and persimmons. Takarazuka and Tajima Branches are growing their local vegetables and fruit trees. Equipment: Plant tissue culture laboratory with glasshouses. Double glazed greenhouse for conserving fuel energy. Waterlevel controlled farm.

Agricultural Experiment Station	Manager: S. Kusaka
Horticultural Division	Chief: T. Fujiwara
Vegetables and Ornamental Plants Section	Head: Dr. M. Fujino*
Environment control for vegetables	Y. Kirimura
Tomato and cucumber	K. Nagai
Water control for vegetables	S. Tokieda
Strawberries	T. Ohnishi
Carnation and rose	A. Uda
Plant tissue culture	T. Iwai
Chrysanthemum	K. Fukushima
Fruit Trees Experiment Farm	Chief: S. Fujiwara
Figs and Japanese pear	T. Kabumoto
Grape	N. Nishitani
Japanese chestnut	Dr. H. Araki
Takarazuka Branch	Chief: Y. Ohmori
Strawberry and Chinese vegetables	T. Kobayashi
Tajima Branch	Chief: T. Imai
Japanese yam	Y. Ikeuchi

Radish and spinach	N. Takada
Japanese Pear Experiment Farm	Chief: K. Ohhashi
Fruit quality	T. Mano
Flower Center	Chief: K. Ohtani
Greenhouse ornamentals	Y. Ikeda
Greenhouse ornamentals	Chief: M. Iwamoto
Orchid	O. Wada
Begonia	M. Yamada
Flower beds	M Horimoto

13.2 Hyogo Prefectural Awaji Agricultural Research Center
elevation: 58 m
rainfall : 1,647 mm
sunshine : 2,139 h

Yagi, Yoginaka, Mihara-cho,
Mihara-gun, Hyogo-pref. 656-04
Phone: 07994-2-4880
Dir.: Y. Haraguchi

Deals with the horticultural problems in the Awaji island. Vegetables, 2 ha, breeding of onions and improving triple cropping system (onion, rice and Chinese cabbage). Measures to control physiological disorders in tree fruits (orange and loquat), 2 ha. Breeding and cultivation of flowers (carnation, chrysanthemum, stock and sweet pea). Equipment: Computorized greenhouses.

Horticultural Division	Head: T. Taniguchi
Vegetables: cultivation, breeding	S. Kobayashi/M. Takegawa
Fruit trees: cultivation, breeding, storage	K. Hamada/Y. Mizuta
Flowers: cultivation, breeding	Y. Takiguchi/Y. Koyama
Plant diseases	J. Nishimura
Insects	M. Fujitomi

13.3 Kobe University, Faculty of Agriculture,
Department of Agriculture and Horticulture
elevation: 125 m
rainfall : 1,385 mm
sunshine : 2,095 h

Rokkodai-cho, Nada-ku, Kobe 657
Phone: 078-881-1212
Telex: 05624-089
Dean: Dr. S. Mizuno

Kobe University is a major national university and is located in uptown Kobe, a cosmopolitan city built around the largest port in the Orient. Horticultural division is composed of the following three laboratories: Pomology, Floriculture and Olericulture, and Preservation Technology. Of 40 ha of the university farm, 4 ha are planted with fruit trees (pears, grapes, peaches, Japanese apricots, persimmons and chestnuts) and 3 ha with vegetables.

Pomology Laboratory	
Physiology of fruit trees	Dr. T. Ichii
Cold resistance of fruit trees	Dr. M. Sawano
Reproductive physiology of fruit trees	Dr. T. Nakanishi
Physiology of grape	T. Ozaki
Polyembryony of citrus fruits	Dr. H. Watanabe
Floriculture and Olericulture Laboratory	
Environmental physiology of vegetables	Dr. M. Terabun
Environmental physiology of ornamentals	Dr. S. Maekawa
Tissue cultures of vegetables	Dr. N. Inagaki
Physiology of vegetables	K. Ishida
Preservation Technology Laboratory	
Transport and storage of fruits and vegetables	Dr. S. Mizuno
Postharvest handling of fruits and vegetables	Dr. T. Hirose
Postharvest physiology of fruits and vegetables	Dr. H. Terai

14 Ibaraki

14.1 Ibaraki Prefectural Horticultural Experiment Station
elevation: 25 m

Ami-machi, Inashiki-gun,
Ibaraki-pref. 300-03

rainfall : 1,250 mm
sunshine : 2,050 h

Phone: 0298-87-1511
Dir.: S. Komori

This Station serves the whole area of Ibaraki prefecture on all problems concerning production, soil and fertilizing, insect pests and plant diseases of horticultural crops.

Tree Fruit Section	Head: Y. Watanabe
Pear and chestnut	M. Yamamoto
Pear and persimmon	T. Ichimura
Chestnut	S. Kasumi
Grape and Japanese apricot	
Vegetable Crops Section	Head: Dr. K. Uchida
Root crops and Chinese cabbage	K. Yamamuro
Watermelon and cucumber	K. Sawahata
Pepper and lettuce	H. Amagai
Tomato	M. Suzuki
Melon and strawberry	Y. Miyagawa
Tissue culture	
Ornamental Crops Section	Head: I. Takatsu
Gladiolus	H. Ohta
Potted plants	Y. Motozu
Tissue culture	
Environment Section	Head: E. Komatsu/N. Fujino
Tree fruit diseases	Dr. S. Yoneyama
Vegetable crop diseases	N. Nakagaki
Tree fruit and vegetable pests	Y. Yanagibashi
Occurrence forecasting of diseases and pests	E. Komatsu
Vegetable crop soil	Y. Matsuzawa
Orchard soil	

14.2 National Fruit Tree Research Station

Yatabe, Ibaraki-pref. 305
Phone: 02975-6-6416
Dir.: Dr. A. Yamaguchi

This Station has 4 Branch Stations. It performs basic research for breeding, cultural practices, control of diseases and insect pests, processing and storage of tree fruits. It consists of 19 ha for Japanese pear and chestnuts in the main station and 9 ha for peach and plums in Chiyoda farm which lies 30 km north of the main station. About 6,000 m^2 of phytotron and greenhouses are for preservation of genetic resources and virus research.

Research Planning and Coordination Division	Head: Dr. Y. Nishiyama
Research planning	Dr. Y. Iba
Fruit physiology	Dr. H. Daito
Fruit breeding	A. Kurihara
Technical training	K. Suzuki
Fruit Breeding Division	Head: T. Shichijo
Breeding methods	Dr. I. Kozaki
Cytology	M. Omura
Tissue culture	N. Matsuta/T. Moriguchi
Japanese pear and chestnut	Dr. Y. Machida/K. Kotobuki Y. Sato/R. Masuda
Stone fruits	Dr. M. Yoshida/Y. Ishizawa M. Yamaguchi
Genetic resources	S. Tsuchiya/J. Soejima
Pomology Division	
Fruit nutrition	Head: Dr. K. Nagai

Fruit culture (Japanese pear, stone fruits)	Dr.T. Yamazaki/S.Murase/H.Watatani
Orchard soil	Y. Sato/Y. Umemiya/M. Yasuda
Orchard meteorology	F. Kamota/H. Honjo/T. Asakura
Plant Protection Division	Head: Dr. A. Otake
Pome fruit virus diseases	Dr. H. Yanase
Grapevine virus diseases	Dr. H. Tanaka
Stone fruit bacterial diseases	Dr. K. Takanashi
Pome fruit fungal diseases	Dr. A. Kudo
Fungicide resistant strain	H. Ishii
Mites and natural enemies	K. Inoue
Fruit insect and biological control	K. Takagi
Fruit insect and viral natural enemies	T. Sato
Fruit insect and fungal natural enemies	K. Yaginuma
Japanese pear insects	S. Moriya
Stone fruit insects	M. Mabuchi

14.3 National Food Research Institute

Kannondai, Yatabe, Tsukuba-gun,
Ibaraki-pref. 305
Phone: 02975-6-8012
Dir.: Dr. N. Tsumura

Processing and distribution technology of fruits and vegetables. Area is 7 ha. Equipment: In addition to the main building there are a food storage laboratory, a food technology laboratory and a radiation laboratory, containing various processing plants.

Division of Food Utilization	Chief: Dr. Y. Tanaka
First Laboratory of Fruit and Vegetable	
Storage technology	Head: Dr. S. Haginuma
Instrumental analysis	H. Yamamoto
Second Laboratory of Fruit and Vegetable	
Biochemistry	Head: M. Kuroki/T. Matsuoka
	H. Hosoda
Third Laboratory of Fruit and Vegetable	
Processing technology	Head: S. Kazumi/N. Muraoka
Distribution technology	T. Isaka/G. Kuroda

14.4 National Institute of Agrobiological Resources, Institute of Radiation
elevation: 80 m
rainfall : 1,400 mm
sunshine : 2,200 h

P.O. Box 3, Ohmiya-machi, Naka-gun,
Ibaraki-pref. 319-22
Phone: 02955-2-1138
Dir.: K. Fujii

Located some 90 km north of NIAR main campus, this Institute has several irradiation facilities, including a large gamma-field (Co-60 2400Ci, 100m radius) and gamma-greenhouse etc. It serves as a center of mutation breeding of all crops from rice and other cereals to fruit trees and forest trees.

Laboratory of Radiation Breeding Technology (I)	
Genetics of mutation	Head: Dr. E. Amano
Induced mutation in cereals	O. Yatou
Induced mutation in vegetables and flowers	S. Iida

14.5 Ibaraki University, Faculty of Agriculture
elevation: 50 m
rainfall : 1,100 mm
sunshine : 2,050 h

Ami-machi, Inashiki-gun,
Ibaraki-pref. 300-03
Phone: 0298-87-1261
Dir.: Dr. T. Akatsuka

Laboratory of Horticulture - Deciduous fruit trees (walnut, chestnut, Japanese persimmon) and vegetable crops (strawberry, melon) are the research objectives.

Asexual propagation of deciduous fruit trees	Prof.Dr. M. Izaki
Ecology of vegetable varieties	Ass. Prof. T. Matsuda
Interaction in plant population	H. Hara

14.6 University of Tsukuba, Institute of Agriculture and Forestry
elevation: 26 m
rainfall : 1,223 mm
sunshine : 2,576 h

Tennodai, Sakura-mura, Niihari-gun, Ibaraki-pref. 305
Phone: 0298-53-4710
Dir.: Dr. T. Akaha

Pomology: Studies on chemotaxonomy, application of tissue culture methods, technical improvement of propagation, meterological condition, hypobaric storage and dwarfing culture systems. Experiments on training methods of grape and productivity of blueberry. Vegetable crop science: Studies on hydroponics, flowering of root and leafy vegetables and environmental conditions on the quality of vegetables. Experiment on heat tolerance of tropical tomatoes. Floriculture: Studies on year-round culture and light conditions of flowers. Collection and reservation of useful flower germplasms.

Pomology Laboratory	
Fruit growing and post-harvest handling	Prof. Dr. C. Oogaki*
Physiology of propagation	Dr. H. Genma
Cultural practice of small fruits	M. Fukushima
Vegetable Crop and Floriculture Laboratory	
Hydroponics and flowering of vegetables	Ass. Prof. Dr. Y. Suzuki*
Vegetable hydroponics	Ass. Prof. M. Shibuya
Floriculture and physiology of ornamental plants	Ass.Prof. T. Abe
Floriculture and chemical control	H. Watanabe
Hydroponics and growing conditions and quality of vegetables	Y. Shinohara*

15 Ishikawa

15.1 Ishikawa Prefectural Agricultural Experimental Station
elevation: 40 m
rainfall : 2,465 mm
sunshine : 1,751 h

Nakabayashi, Nonoichi-machi, Ishikawa-gun, Ishikawa-pref. 921
Phone: 0762-48-1335
Dir.: Dr. I. Wakisaka

This Station is the research center of rice and horticultural crops (apple, pear, chestnut, vegetables and cut flowers) in this prefecture.

Fruit Tree Section	Head: K. Motoda
Production of tree fruits	M. Fujise
Fruit development	K. Tani/Y. Nomura
Vegetable and Flower Section	Head: M. Yamabe
Vegetable growing	K. Fujita
Breeding of radish and local vegetables	H. Hirai
Flower production	E. Kimura

15.2 Ishikawa Prefectural Sand Dune Agricultural Experiment Station

Unoke-machi, Kahoku-gun, Ishikawa-pref. 929-11
Phone: 07628-3-0073
Dir.: H. Nakada
Vice-Dir.: N. Watanabe

Development of cultural techniques to obtain more stable production of fruit trees and vegetables. Establishment of methods for increasing the productivity and quality of fruits and vegetables in sand-dune areas.

Fruit Tree Section

Grape growing	Head: T. Yamakawa
Fruit tree nutrition	Y. Inabe
Environmental control	H. Wakabayashi
Application of growth regulator	S. Yamada

Vegetable Section

Greenhouse-equipment (heat pump)	Head: S. Ohe
Tissue culture and plant protection	T. Odagiri
Watermelon growing	K. Inaba
Vegetable nutrition	N. Fukuoka
Radish and other vegetables growing	Y. Saida

15.3 Ishikawa Prefectural Agricultural College

Suematsu, Nonoichi-machi,
Ishikawa-gun, Ishikawa-pref. 921
Phone: 0762-48-3135
Dir.: Dr. K. Uemura

This College is educating 100 students in a 2-year-course of study. It has about 28 ha of fields; 6 ha of them are planted with fruit trees, 1 ha of glasshouses (1,000 m^2) with vegetable and flower crops.

Laboratory of Horticulture

Fruit development of chestnut	Prof.Dr. K. Shiozawa
Physiological disorder of persimmon fruit	Ass. Prof. S. Tsuchiya
Radish growth and dahlia root development	Ass. Prof.Dr. Y. Kano
Experimental Farm	Dir.: Dr. Y. Nakamura
Fruit development of chestnut	Ass. Prof. S. Mizutani
Production of fruits, vegetables and flowers	C. Kawasaki

16 Iwate

16.1 Iwate Prefectural Horticultural Experiment Station
elevation: 90 m
rainfall : 1,290 mm
sunshine : 1,602 h

Iitoyo-machi, Kitakami-shi,
Iwate-pref. 024
Phone: 0197-68-2331
Dir.: Dr. A. Chiba

Producing of tree fruits, grapes, vegetables and ornamental plants. Of 20 ha, 10.8 are planted with tree fruits (apple, pear, peach and plum), 3.5 ha are vegetables and 5,7 ha are others.

Fruit Tree Division

Testing of different varieties of fruit trees	Head: A. Ito
Production of virus free fruit seedling (apple, grape)	K. Onoda/H. Sasaki

Vegetable ana Ornamental Plant Division

Open cultivation of cucumber, tomato and strawberry	Head: T. Yoshiike
Greenhouse cultivation of strawberry	K. Ishikawa
Production of virus-free seedlings (strawberry, garlic and yam)	O. Fujisawa
Breeding and culture of gentian	

Environmental Division

Ecological study and control of main diseases and pests	Head: T. Hiraragi/R. Takahashi
	T. Chiba

Forecasting methods of diseases and pests	F. Nakatani/K. Muto/S. Sato
Southern Branch	Head: T. Nagabe/T. Abe
Quality and productivity of greenhouse cultivated strawberries and vegetables	T. Kikuchi
All-year-round utilization of plastic greenhouses	
Highland Cool Zone Development Center	Head: T. Sato/Y. Konno
Production of highland cool zone vegetables	M. Nakamura/K. Sugawara
Production of Yamase zone vegetables and plastic greenhouse vegetables	T. Mibuta/T. Fujiwara
Forming of vegetable-producing districts in highland cool zone or Yamase zone	Y. Takahashi
Ohasama Experimental Farm	Head: T. Kawamura
Comparison of different varieties of grapes	T. Kudo
Viticulture for brewing and fresh use	

16.2 National Fruit Tree Research Station, Morioka Branch
elevation: 160 m
rainfall : 1,400 mm
sunshine : 1,930 h

Nabeyashiki, Shimokuriyagawa
Morioka-shi, Iwate-pref. 020-01
Phone: 0195-41-3164
Dir.: Dr. K. Sekiya

Main research projects are apple breeding and improvement of production and fruit quality of apple. This Station plays a role as the centre of research and developmental works of prefectural experiment stations on apple in Japan. 'Fuji', the most extensively distributed apple cultivar in Japan, has been selected in 1958 in this Station.

Laboratory of Fruit Breeding	Head: Dr. Y. Yoshida
Apple and rootstock breeding	Dr. T. Masuda
Micropropagation	H. Bessho
Rootstock trial	
Laboratory of Pomology	Head: Dr. H. Fukuda
Storage and physiological disorder of apple	K. Kudo
Orchard management of apple	Y. Kashimura
Dry matter production of apple tree	F. Takishita
Physiology of apple fruit growth	
Laboratory of Fruit Tree Nutrition	Head: Dr. K. Aoba
Calcium and nitrogen nutrition of apple	Dr. M. Fukumoto
Carbohydrate metabolism of apple	Dr. H. Yoshioka
Laboratory of Plant Pathology	Head: Dr. T. Sakuma/J. Taniuchi
Apple fungal diseases	Dr. H. Koganezawa
Apple viruses	
Laboratory of Entomology	Head: Dr. T. Oku
Fruit borer biology and control	M. Wakou
Apple spider mite control	Y. Ohira
Reproductive ecology of apple leaf rollers	

16.3 National Vegetable and Ornamental Crops Research Station, Morioka Branch Station
elevation: 155 m
rainfall : 1,638 mm
sunshine : 1,930 h

Nabeyashiki, Shimokuriyagawa,
Morioka-shi, Iwate-pref. 020-01
Phone: 0196-41-2031
Dir.: Dr. K. Hozumi

This Station is the main center of tomato breeding and the sub-center of strawberry breeding in Japan. Main objectives of cultivation and plant pathology sections are to develop the desirable cultivation and protection system for cool conditions or for mechanical cultivation of vegetables.

Laboratory of Breeding for Cool-season Vegetables	Head: Dr. Ishiuchi
Breeding of processing and fresh market tomatoes	K. Ito/T. Mochizuki
Laboratory of Breeding Methods	Head: K. Takada

Breeding of strawberry, lettuce and tomato	S. Monma
Laboratory of Vegetable Cultivation	Head: Dr. T. Masaki
Farm mechanization	O. Sakaue
Mechanization of vegetable cultivation	T. Endō/M. Fujino
Cultural system of strawberry	H. Kumakura
Laboratory of Plant Pathology	Head: Dr. T. Sasaki
Ecology and control of vegetable diseases	Dr. Y. Honda

16.4 Iwate University, Faculty of Agriculture
elevation: 133 m
rainfall : 1,299 mm
sunshine : 1,988 h

Ueda, Morioka-shi, Iwate-pref. 020
Phone: 0196-23-5171
Dean: Prof. Dr. M. Yoshida

This Faculty has seven Departments. The following two laboratories belong to the Department of Agronomy, accompanying a small orchard and experimental fields for horticultural studies. In University Farm, there are 40 ha of orchards, of which 30 ha are planted with apples and 10 ha with grapes, pears, chestnuts, cherries, peaches, Japanese apricots and plums. Also a greenhouse for flower production and a processing factory.

Laboratory of Olericulture and Floriculture	
Cytogenetics of Brassica, Allium, and Chrysanthemum	Prof. Dr. S. Iwasa
Breeding of Brassica	Ass. Prof. Dr. M. Endo/I. Inada
Laboratory of Pomology	
Tissue culture of deciduous fruit trees	Prof. Dr. A. Ishihara
Chemical control of apple trees	Ass. Prof. Dr. K. Yokota

17 Kagawa

17.1 Kagawa Prefectural Agricultural Experiment Station
elevation: 35 m
rainfall : 1,245 mm
sunshine : 1,992 h

Takamatsu, Kagawa-pref. 761
Phone: 0878-89-1121
Dir.: K. Morioka

It serves for all areas of Kagawa-pref., on all problems concerning cultivation, insects and plant diseases of main crops (rice, wheat, barley and soyabean), tree fruits, vegetables, ornamental plants and tea trees.

Ornamental Plant Section	Head: T. Yamamoto
Physiology and growing	N. Soichi/Y. Matsumoto
In vitro culture and propagation	H. Jutori
Vegetable Section	Head: K. Miyawaki
Physiology and growing	M. Yamauchi/N. Horikawa
Soil and Fertilizer Section	Head: S. Itose
Field and garden soils	Y. Shirai/H. Noda
Nutrition of vegetables	K. Tanabe/T. Nakao/T. Katamoto
Entomology and Phytopathology Section	Head: T. Kassai
Ecology and control of insects	Y. Sasaki/Y. Nishiyama
Control of fungal and virus diseases	Y. Tsuzaki/K. Sogo
Fuchu Branch Station	Head: Y. Sakai
Physiology and growing of fruit trees	S.Oosawa/S.Doi/Y.Wakabayashi Y.Kagami/M.Ootani/K.Suesawa
Soil management in orchards	M. Ookuma
Control of diseases and pests in orchards	T. Aoki/R. Ooya
Greenhouse culture of fruit trees	Dr. Y. Harada
Bonsai	T. Suekane
Miki Branch Station	Head: K. Tsuyama
Physiology and growing of vegetables	K. Nishitani/K. Nagaoka
Shozu Branch Station	Head: S. Yamaguchi
Olive fruit production	Y. Kawanishi

Physiology and growing of flowers — M. Kawae/H. Tokui
Manno Branch Station — Head: H. Abe
Physiology and genetics of tea trees — K. Tsunekane/K. Yano

17.2 Shikoku National Agricultural Experiment Station
Land Utilization Division
elevation: 90 m
rainfall : 1,100 mm
sunshine : 2,900 h

Ikano-cho, Zentsuji-shi,
Kagawa-pref. 765
Phone: 08776-3-2566
Dir.: S. Nishibe

This Division, one of five divisions of the Shikoku National Agricultural Experiment Station, is the main centre of researches for slope lands utilization in Japan. About 83 ha are used as grassland, citrus orchard and field for vegetables under 6 laboratories and one general affairs section.

Laboratory of Fruit Trees — Head: Dr. F. Ikeda/T. Kihara
Quality improvement of fruits — K. Morinaga
Laboratory of Soil Conservation — Head: A. Ida/K. Yamasaki/T. Ujike
Leaching of elements in slope land
Laboratory of Vegetables — Head: Dr. A. Kotani/Dr. H. Higashio
Laboratory of Agricultural Mechanization — Head: K. Kawasaki/N. Amano
Monorails in slope — N. Itokawa/M. Daikoku
Laboratory of Land Reclamation — Head: S. Nagaishi/Dr. T. Miyazaki
K. Haraguchi
Laboratory of Crop Location — Head: I. Tamaki/F. Fujita
Placement of citrus orchards and vegetable fields

17.3 Kagawa University, Faculty of Agriculture,
Department of Horticulture
elevation: 8.7 m
rainfall : 1,199 mm
sunshine : 2,235 h

Miki-cho, Kida-gun,
Kagawa-pref. 761-07
Phone: 0878-98-1411
Dir.: Dr. T. Okaichi

Kagawa University is a national university and the campus of Faculty of Agriculture is located in a small town, Miki, 6 km from the Takamatsu Airport. Near the campus, there are many fruit, vegetable and flower growers. The department is one of the oldest in horticulture and consists of laboratories of fruit culture, vegetable culture, flower culture, postharvest horticulture, plant pathology, entomology and landscape architecture. Both basic and applied researches are conducted not only to solve local problems, but also outside of Japan.

Laboratory of Pomology
Fruit development and flower bud differentiation — Prof. Dr. H. Inoue
Growth and quality of kaki fruit — Ass. Prof. Dr. T. Chujō
Grape ripening control — Dr. I. Kataoka
Laboratory of Vegetable Crop Science
Vegetable breeding and seed production — Prof. Dr. T. Hirose
Developmental physiology of vegetables — Ass. Prof. Dr. Y. Fujime*
Laboratory of Floriculture
Propagation and forcing of ornamental trees and shrubs — Ass. Prof. Dr. M. Goi*
Propagation and cultivation of Cymbidium and Protea — Ass. Prof. A. Hasegawa*
Propagation and forcing of Phalaenopsis — M. Tanaka*
Laboratory of Postharvest Horticulture
Transportation and storage of fruits and vegetables — Prof. Dr. H. Kitagawa
Postharvest physiology — Dr. K. Kawada

18 Kagoshima

18.1 Kagoshima Prefectural Agricultural Experiment Station

Kamifukumoto-cho, Kagoshima-shi,
Kagoshima-pref. 891-01

elevation: 5 m
rainfall : 2,456 mm
sunshine : 1,926 h

Phone: 0992-68-3231
Dir.: Dr. T. Iwashita

Horticulture Section: Performance tests of vegetable varieties to select well-adaptable ones to Kagoshima prefecture. Studies on the cultivation methods and cropping systems of peas, broad beans and vegetable fruits under plastic film-covered pipehouse and environmental controlled greenhouse. Studies of new cropping system of vegetables. Studies on the protection methods of vegetables from the damages caused by volcanic ashes and gases around Sakurajima Volcano and meteorological disasters. Breeding Section: Breeding of peas. Seed production of taro and sweet potato. Studies on new technology of mass production of seeds. Flower Section: Breeding of lilies, and ornamental plants. Studies on the cultivations of ornamental plants.

Horticulture Section
Vegetable cultivation — Chief: E. Ishida/M. Shimo/K. Aihoshi
H. Togo/M. Kohata/K. Joho

Breeding Section
Vegetable breeding — Chief: M. Homan/K. Ichi/M. Karube
Flower Section
Flower breeding — Chief: M. Kobayashi/T. Iwaki
H. Uehara/M. Himeno/H. Takahashi

Osumi Branch Station — Chief: M. Ebata
Vegetable cultivation — Head: R. Miyaji/K. Tabata/T. Arimura
H. Sakaue

Kumage Branch Station — Chief: C. Takezaki
Vegetable cultivation — Head: S. Miyashita/K. Sameshima
Oshima Branch Station — Chief: H. Obara
Vegetable cultivation — Head: H. Tsukishima/F. Wada
K. Fujita/H. Imafurukawa

18.2 Kagoshima Prefectural Fruit Tree Experiment Station
elevation: 20 m
rainfall : 2,017 mm
sunshine : 2,073 h

Honjo, Tarumizu-shi,
Kagoshima-pref. 891-21
Phone: 09943-2-0179
Dir.: M. Kono

Producing citrus, loquat and deciduous fruit trees. Equipment: Plastic house heated by a soil-air heat exchange system, greenhouse for virus diseases of fruit trees, mass rearing facilities of natural enemy of scales, lysimeter.

Growing Section
Growing of citrus and loquat — Head: M. Fuzisaki
Plant breeding of citrus — R. Kuwahata
Growing of citrus — T. Tokito
Growing of loquat and deciduous fruit trees — H. Okurano
Growing of citrus and plant growth regulators — A. Higashi
Plant Diseases and Pest Control Section
Plant disease and virus — Head: T. Kiku/T. Sakaguchi
Pest control of fruit trees — S. Hashimoto/S. Mizushima
Soil and Fertilizer Section
Soil and fertilizer of fruit trees — Head: T. Tuchimochi/K. Sano
Y. Tatsuda

Nansatsu Branch
Growing of citrus — Head: A. Shikura/T. Niizawa
Hokusatsu Branch
Growing of deciduous fruit trees — Head: M. Suwa/N. Nishimoto
Osumi Branch
Growing of citrus — Head: M. Nagahama/H. Tokudome

18.3 Kagoshima University, Faculty of Agriculture,

Korimoto, Kagoshima-shi,

Department of Horticulture
elevation: 4 m
rainfall : 2,375 mm
sunshine : 2,074 h

Kagoshima-pref. 890
Phone: 0992-54-7145
Dean: Prof. Dr. T. Nishihara

The Department is located at the most south-western part of Japanese mainland, and the research and education are conducted under its meteorologically unique characteristics. The Department consists of four laboratories.

Laboratory of Fruit Science	
Physiology of citrus and subtropical fruits	Prof. Dr. S. Iwahori*
Flowering and fruit quality of citrus	Ass. Prof. S. Tominaga
Growing and propagation of blueberry	T. Kushima
Laboratory of Vegetable Crops	
Male sterility in relation to seed production and breeding Brassica	Prof. Dr. H. Ogura
Genetics and physiology of fruit vegetables	Ass. Prof. T. Johjima
Restoration of fertility in garlic	Dr. T. Etoh
Laboratory of Ornamental Horticulture and Floriculture	
Genetics and breeding of ornamental plants	Prof. Dr. K. Arisumi
Propagation and physiology of bulbous plants	Assoc. Prof. Dr. E. Matsuo
Chemistry and physiology of flower colours	Y. Sakata
Laboratory of Postharvest Physiology and Preservation of Fruits and Vegetables	
Processing and chemistry of citrus and subtropical fruits	Prof. Dr. S. Itoo
Biochemistry of vegetables and substropical fruits	Assoc. Prof. Dr. F. Hashinaga
Postharvest physiology of persimmon and sweet potato	Dr. T. Matsuo
Ibusuki Experimental Botanic Garden	
Acclimatization of tropical and subtropical plants	Assoc. Prof. K. Ishihata
Faculty of Education	
Growth and development of Japanese loquats	Assoc. Prof. Dr. I. Kiyokawa

19 Kanagawa

19.1 Kanagawa Prefectural Agricultural Experiment Station
elevation: 10 m
rainfall : 928 mm
sunshine : 2,357 h

Teradanawa, Hiratsuka-shi,
Kanagawa-pref. 259-12
Phone: 0463-58-0333
Dir.: Dr. T. Itagi

Producing of tuber crops, root crops, cole herbs, salad crops, potherbs and Allium crops in both outdoor field and greenhouse. Breedings of tuber crops, root crops and Allium crops.

Technical Research Division	Chief: Dr. H. Takezawa
Vegetable Section	
Cole herbs and Allium crops	Head: H. Hayashi
Potherbs	M. Mochizuki
Root crops and salad crops	J. Narimatsu
Biotechnological Section: Cell and tissue culture	Head: Dr. Y. Miura/Y. Yamada

19.2 Kanagawa Prefectural Horticultural Experiment Station
elevation: 10 m
rainfall : 1,500 mm
sunshine : 2,240 h

Ninomiya, Naka-gun,
Kanagawa-pref. 259-01
Phone: 0463-71-1052
Dir.: M. Inokuchi

This Station is the main center of horticultural research in Kanagawa prefecture. Breeding and selection, new cropping programs, energy saving, plant protection, control of replant failures, hydroponics of vegetables and flowers in greenhouses.

Cultural Research Division	Chief: Dr. T. Hiraoka
Fruit Vegetable Section	
Beans and eggplant	Head: M. Takahashi
Strawberry and melon	N. Satoh
Energy saving and hydroponics	K. Sasaki
Tomato and cucumber	N. Kita
Deciduous Fruit Section	
Pear and peach	Head: T. Shigeta
Pear and persimmon	T. Akiyama
Grape and blueberry	Y. Katano
Floriculture Section	
Rose breeding and carnation	Head: I. Hayashi
Rose, lily, Lisianthus and Ranunculus	Dr. K. Ohkawa
Pot plant	K. Yamamoto
Environment Section	
Plant protection	Head: K. Ushiyama/N. Aono
	H. Funahashi
Nebukawa Branch Station	
Breeding and selection of mandarin	Chief: S. Futami
Nutrition of mandarin	M. Hirobe
Postharvest of mandarin and kiwi fruit	M. Manago
Plant protection of mandarin	S. Katagi
Sagamihara Branch Station	
Ornamental tree and shrub	Chief: M. Asaoka/M. Okabe
	K. Yamazaki
Miura Branch Station	
New cropping program of vegetables	Chief: I. Seita
Plant protection	N. Ohbayashi
Nutrition of vegetable crops	N. Mizuno
Cultivation of vegetable crops	D. Igarashi
Tsukui Branch Station	
Grape, chestnut and plum	Chief: H. Sakairi
Cultivation of tea	N. Watabe
Cultivation of wild vegetables	T. Ohhashi

19.3 Meiji University, Faculty of Agriculture
elevation: 60 m
rainfall : 1,460 mm
sunshine : 1,940 h

Higashi-mita, Tama-ku, Kawasaki-shi
Kanagawa-pref. 214
Phone: 044-911-8181
Dir.: Dr. Y. Iwamoto

Laboratory of Horticulture	
Biotechnology and breeding of fruit tree	Prof. Dr. T. Akihama
Post-harvest changes in chemical composition of vegetables	Prof. T. Kasukawa
Flower bud differentiation in ornamental plants	M. Hakozaki

19.4 Nihon University, College of Agriculture and Veterinary Medicine, Junior College of Agriculture
elevation: 39.5 m
rainfall : 1,596 mm
sunshine : 2,000 h

Kameino, Fujisawa-shi,
Kanagawa-pref. 252
Phone: 0466-81-6241
Dean: K. Kukita

Area for researching field consists of:
0.5 ha with field of fruit trees, 0.1 ha with vegetable crops, 0.3 ha with ornamental and flowering crops, and five greenhouses (250 m^2).

Laboratory of Pomology and Vegetable Crops	
Physiology of vegetable crops and fruit trees	Dr. B. Takahashi
Flowering and fruiting of fruit trees	H. Inoue
Laboratory of Floriculture	
Growing of rose	H. Sasaki
Tissue culture and flowering of orchids	Dr. K. Yoneda
Laboratory of Horticulture (Junior College)	
Growth and flowering control of ornamental crops	Dr. M. Suzuki
Growing of muskmelon	T. Saito
Physiology of vegetable crops and fruit trees	K. Watanabe

20 Kochi

20.1 Kochi Prefectural Fruit Tree Experiment Station
elevation: 50 m
rainfall : 2,670 mm
sunshine : 2,236 h

Asakura, Kochi-shi, Kochi-pref. 780
Phone: 0888-44-1120
Dir.: K. Yasuoka

Producing and quality keeping of evergreen and deciduous fruits, especially breeding, cultural management systems, chemical regulation, physiological disorder and storage of citrus (pomelo, ponkan, acid citrus and others), loquat, pear, kiwi and other deciduous tree fruits. Of 7.5 ha of hillside field, 5.0 are planted with citrus and loquat, 2.5 ha with pear and other deciduous fruit trees. Equipment: 0.5 ha glass and plastic film greenhouse for study of breeding and PVC film experiments. Two artificial climate equipments for study of virus diseases and physiological disorders of fruit.

Cultural Management Section	
Pomiculture management system	Head: T. Iwakawa
Breeding and physiological disorder of citrus fruit	T. Manabe
Cultural methods of acid citrus and special fruit	T. Aoki
Citrus Section	
Citrus culture under plastic film house	Head: S. Mitsue
Herbicide and plant growth regulator	T. Yoshinaga
Environmental control of citrus orchards	K. Kimura
Deciduous Fruit Section	
Breeding and cultural methods of pear	Head: I. Watanabe
Quality keeping of kiwi and grape fruits	M. Horiuchi

20.2 Kochi Prefectural Horticultural Experiment Station
elevation: 21 m
rainfall : 2,500 mm
sunshine : 2,500 h

Noichi, Kami-gun Kochi-pref. 781-52
Phone: 08875-6-0307
Dir.: T. Yanai

This Station is a centre of flower and vegetable growing research in Kochi-prefecture, and is investigating flowering and physiology of ornamental plants, physiology and ecology of vegetables under structure and in the open, and breeding of vegetables.

Floriculture Section	
Physiology and ecology of ornamental plants	Head: S. Inubushi/H. Nonami
Regulation of flowering of ornamental plants	A. Azuma
Protected Growing Section	
Growth regulation of fruit vegetables	Head: Y. Kubouchi/Y. Fujita
	Y. Shimamura/K. Maeda

Vegetable Growing Section	
Growth regulation of leafy and root vegetables	Head: T. Kanazaya/T. Murakami
Vegetable Breeding Section	
Disease resistant breeding	Head: T. Nakazawa
Breeding of fruit vegetables	S. Saito/N. Sakamoto
Breeding of root vegetables by tissue culture	S. Nakamura

20.3 Kochi Prefectural Highland Agricultural Experiment Station
elevation: 400 m
rainfall : 2,771 mm

Nakamuradaio, Otoyo-cho, Nagaoka-gun, Kochi-pref. 789-03
Phone: 08877-2-0058
Dir.: T. Iwakawa

Efforts are made to introduce favorable crops and establish cultivation techniques in order to advance productivity in intermediate mountain region. The plants tested are *Citrus Junos*, Japanese persimmon, Japanese apricot and maidenhair tree in fruit trees, tomato, pepper, strawberry, lettuce and mioga in vegetables, wasabi, udo salad plant, *Osmunda japonica* and *Aralia elata* in edible wild plants, soybean, barley and wheat in upland crops; and *Panax, Schinseng* and *Senega* in medicinal crops.

Fruit Tree Section	Head: S. Tokuhashi
Deciduous fruit tree growing	
Stability of production and quality improvement of citrus	H. Tachibana
Vegetable and Flower Section	Head: H. Ogasawara
Vegetable growing	H. Kondo/A. Iwasaki
Flower growing and upland farming	N. Yamasaki

20.4 Kochi University, Faculty of Agriculture, Department of Agriculture
elevation: 50 m
rainfall : 2,645 mm
sunshine : 2,273 h

Monobe, Nankoku-shi, Kochi-pref. 873
Phone: 0888-63-4141
Chairman: Dr. Y. Ogura

Laboratory of Pomology	
Physiological disorders of citrus	Prof. Dr. Y. Nakajima
Growth and development of deciduous trees	Ass. Prof. K. Hasegawa
Laboratory of Horticulture	
Growth and development of vegetable crops	Prof. Dr. T. Kato
Propagation and flowering of ornamental plants	Ass. Prof. Y. Sawa
Research Institute of System Horticulture	
Physiology and ecology of vegetables under structure	Ass. Prof. Y. Fukumoto

21 Kumamoto

21.1 Kumamoto Prefectural Agricultural Experiment Station, Horticultural Branch
elevation: 83 m
rainfall : 2,027 mm
sunshine : 1,877 h

Nishigoshi-machi, Kikuchi-gun, Kumamoto-pref. 861-11
Phone: 096-242-0168
Dir.: A. Kitajima

This Station is the center of the research in vegetable crops, flowers and ornamental plants in Kumamoto Pref.

Vegetable Section	
Watermelon growing	Head: H. Kitajima
Tissue culture of vegetables	M. Tanaka
Physiology of vegetables	T. Ishida
Ecology of vegetables	T. Morita
Floricultural Section	
Flower growing	Head: T. Kawakita

Rose growing	T. Morita
Mericlone of flowers	H. Watanabe
Chrysanthemum growing	I. Ogata
Yatsushiro Branch	363, Kagamimura, Kagami-machi, Yatsushiro-gun, Kumamoto-pref. 869-42 Phone: 09655-2-0372 Dir.: K. Chikano
Horticultural Section	
Soil and Fertilizer of vegetables	Head: T. Higashi
Growing and breeding of vegetables	F. Nishimoto
Growing of vegetables	M. Ono
Aso Branch	5896-2, Miyaji, Ichinomiya-machi Aso-gun, Kumamoto-pref. 869-16 Phone: 09672-2-1212 Dir.: T. Matsumoto
Growing of vegetables	F. Nishida
Breeding of flowers	T. Kikuchi
Kuma Agricultural Research Station	2248-16, Kami, Uemura, Kuma-gun, Kumamoto-pref. 868-04 Phone: 09664-5-0470 Dir.: H. Hosono
Crop Section	Head: K. Shibata
Culture of vegetables	Y. Suenaga
Amakusa Agricultural Research Station	636, Hondobaba, Hondo, Hondo-shi, Kumamoto-pref. 863 Phone: 09692-2-4224 Dir.: J. Ota
Crop Section	Head: H. Matsumoto
Culture of vegetables	E. Hayashida
21.2 Kumamoto Prefectural Fruit Tree Experiment Station elevation: 50 m rainfall : 1,646 mm sunshine : 2,668 h	Matsubase-machi, Shimomashiki-gun, Kumamoto-pref. 869-05 Phone: 0964-32-1723 Dir.: S. Yamamoto

This Station is the center of the research in fruit trees in Kumamoto-prefecture.

Evergreen Fruit Culture Section	
Citrus growing	Head: I. Hirata/ H. Shigeoka
	S. Okada/ H. Sakaki
Deciduous Fruit Culture Section	
Pear and plum growing	Head: K. Sakai
Grape and peach growing	H Hirayama
Chestnut growing	S. Sakai
Breeding Section	
Breeding of citrus	Head: M.Matsuda/ A.Isobe/ K.Fujita
Pathology and Entomology Section	
Control of fruit insect pests and mites	Head: M. Uemura/ H. Giyotoku
Ecology and control of fruit diseases	T. Isoda

Soil and Fertilizer Section

Soil and Fertilizer of orchards — Head: M. Hayakami/ M. Nakaji, U. Takahashi

21.3 Kyushu Tokai University, Faculty of Agriculture
elevation: 500 m
rainfall : 2,649 mm
sunshine : 1,489 h

Kawayo, Choyo-mura, Aso-gun
Kumamoto-pref. 869-14
Phone: 09676-7-0611
Dean: Dr. K. Kumazaki

Fruit trees (apple, sweet cherry, Japanese pear, Japanese persimmon and grape) are planted in a 0.7 ha orchard, and vegetable and flower crops are cultivated in a 0.2 ha field, two 240-m^2 plastic houses and a 140-m^2 greenhouse. Camellia vernalis and its allied species are collected from various areas in Japan and from those around the world.

Pomology — Dr. I. Iizuka/ H. Komatsu
Olericulture and Floriculture — Dr. T. Matsumura*/ T. Tanaka

22 Kyoto

22.1 Kyoto Prefectural Institute of Agriculture
elevation: 100 m
rainfall : 1,628 mm
sunshine : 1,632 h

Amarube-cho, Kameoka-shi,
Kyoto-pref. 621
Phone: 07712-2-0424
Dir.: M Matsuda

Development of new cultivation types for year-round leafy vegetable production. Manuring, irrigation on vegetables under non-heating plastic film greenhouses.

Crop Culture Section
Physiology of bamboo shoots — Head: H. Nishino
Physiology of leafy vegetables — O. Mizu
Ecology of leafy vegetables — D. Tanaka
Variety test of vegetables — T. Nishimura
Strawberry culture — K. Inaba
Chutan Branch
Medical crops — A. Takewaka
Wild vegetables — M. Yoshikawa

22.2 Kyoto Prefectural Tango Institute of Agriculture
rainfall : 1,856 mm
sunshine : 1,227 h

Kurobe, Yasaka-cho, Takeno-gun,
Kyoto-pref. 627-01
Phone: 07726-5-2401
Dir.: S. Kakinaka

This Institute is the center of research of agriculture in the Tango district and study for establishment of techniques to improve production of horticultural crops. In addition, it promotes development of the local agriculture.

Vegetable crops — H. Goto
Tree fruits and grapes — J. Nagasawa
Tree fruits — M. Chisaki
Sakyu Branch Laboratory — Head: N. Uemura
Vegetable crops — Y. Himori
Flowers (tulips) — N. Kashimoto

22.3 Kyoto Prefectural Yamashiro Institute of Horticulture

Tanabe-cho, Tsuzuki-gun, Kyoto-shi,
Kyoto-pref. 610-03

elevation: 50 m
rainfall : 1,490 mm
sunshine : 1,605 h

Phone: 07746-2-0048
Dir.: K. Satoh

This Institute, located in the southern part of Kyoto prefecture, is the main center of horticultural research there, and serves on all problems concerning production of horticultural crops under structure or in fields. Main crops are grapes, Japanese persimmon, tomato, cucumber, strawberry, chrysanthemum, and bulbs and tubers.

Fruit Tree Section	
Development and quality of grape	Head: Y. Kobashi
Vegetable Crop Section	
Environmental control of vegetable production	Head: Y. Kanbara
Soilless culture	S. Suzuki
Propagation by tissue culture	T. Ogura
Ornamental Plant Section	
Growth and flowering of chrysanthemum	Head: Y. Nakamura
Growth and flowering of bulbous plants	T. Takeda

22.4 Kyoto Prefectural University, Faculty of Agriculture
elevation: 70 m
rainfall : 1,646 mm
sunshine : 1,460 h

Shimogamo-hangi-cho, Sakyo-ku,
Kyoto-shi, Kyoto-pref. 606
Phone: 075-781-3131
Dean: Dr. T. Hattori

Physiology of the fruit development, particularly of persimmon and peach is studied in the laboratory of pomology. Soilless culture of tomato and other vegetables, and ecology and physiology of the vegetables peculiar to Japan, are studied in the laboratory of vegetable crops.

Laboratory of Pomology	
Physiology and ecology of Japanese persimmon	Prof.Dr. Y. Sobajima
Development and quality of peach fruit	Ass.Prof.Dr. M. Ishida*
Tissue culture and physiology of deciduous fruit trees	A. Kitajima
Laboratory of Vegetable Crops	
Soilless culture of vegetable crops	Prof.Dr. T. Namiki
Taxonomy of Brassica vegetables	Ass.Prof.Dr. S. Yazawa
Dormancy of bulbous crops	S. Terabayashi
Ecology of arrow-head	K. Adachi

22.5 Kyoto University, Faculty of Agriculture,
Department of Agronomy and Horticultural Science
elevation: 41.4 m
rainfall : 1,638 mm
sunshine : 1,919 h

Kitashirakawa, Sakyo-ku, Kyoto-shi,
Kyoto-pref. 606
Phone: 075-751-2111
Dean: Dr. H. Yamagata

The Experimental Farm comprises the Head Farmstead and Kosobe Conservatory which are in Takatsuki city, and Kyoto Farmstead on the main campus. It has collections of 160 varieties of Japanese persimmons (in Kyoto Farmstead) and 3,500 species of cultivars of ornamental plants of tropical or subtropical origin (in Kosobe Conservatory).

Laboratory of Pomology	
Physiology of temperate tree fruits	Dr. T. Tomana
Deastringency of Japanese persimmons	Dr. A. Sugiura*
Mineral nutrition of grape vines	H. Motosugi
Cold hardiness of Citrus spp.	H. Yamada
Laboratory of Vegetable and Ornamental Horticulture	
Physiology and ecology of garden crops	Dr. T. Asahira*
Environmental research of perennial ornamentals	Dr. Y. Takeda*
Physiology and ecology of vegetables	H. Inden

Physiology and ecology of perennials	M. Doi
Laboratory of Weed Science	
Physiology and ecology of weeds	Dr. K. Ueki
Crop-weed competition in orchards	Dr. M. Ito
Experimental Farm	Dir.: Dr. T. Asahira*
General management of vegetable production	Dr. K. Fujimoto
Matter production in Japanese pear	Dr. Y. Furukawa
Clonal propagation of tropical plants	Dr. K. Kawase
Breeding techniques of vegetables	M. Ohi
Fertilizer reaction of fruit vegetables	Y. Yoshida
Dwarfed rootstocks of peach	T. Ogata
Laboratory of Tropical Agriculture	
Agro-ecology of tropical crops	Dr. S. Shigenaga
Environmental physiology of tropical fruit trees	Dr. N. Utsunomiya*
Vegetable production in the tropics	E. Nawata

23 Mie

23.1 Mie Prefectural Agricultural Technical Center	Ureshino-cho, Ichishi-gun,
elevation: 10 m	Mie-pref. 515-22
rainfall : 1,714 mm	Phone: 05984-2-1258
sunshine : 2,248 h	Dir.: K. Kataoka

Department of horticulture studies of variety and cultivation of fruit trees, vegetables and ornamental plants, and distribution processing. Greenhouses 1,500 m^2, orchard 1.5 ha, open field 0.7 ha, and the facilities for production of virus free nursery plants.

Department of Horticulture	Chief: T. Maeda
Fruit Tree Section	
Variety and cultivation of pear, persimmon, citrus	Head: N. Kobayashi
Variety and cultivation of pear and persimmon	Y. Hattori
Vegetable Crop Section	
Variety, cultivation and type of cropping	Head: S. Ito
Mechanization and water culture	I. Nishiguchi
Type of cropping and cultivation of strawberry and asparagus	M. Shyoka
Floriculture Section	
Variety and type of cropping of potted cut flowers	Head: T. Nakano
Physiological disorder of flowering trees	T. Yamabe
Processing Section	
Variety and cultivation of vegetables for processing	Head: Y. Toyotomi
Processing and suitability of vegetables	K. Tanaka
Iga Branch Center	Chief: K. Matsuda
Fruit Tree Section	
Variety and cultivation of citrus	Head: T. Hashimoto
Prevention of disease and pest infesting in citrus fruit	H. Mae
Cultivation of citrus fruit, and soil management	Y. Uragari
Kisozaki Reclaimed Land Research Station	
Cropping season of vegetables	Head: S. Oda
Cultivation and cropping season of vegetables	T. Hosotani

23.2 National Vegetable and Ornamental Crops Research Station	Ano-cho, Age-gun, Mie-pref. 514-23
elevation: 65 m	Phone: 05926-8-1331
rainfall : 1,774 mm	Dir.: Dr. T. Kuriyama
sunshine : 2,121 h	

This Institute, responsible for vegetable and ornamental crop research, is one of the specialized institutes of the Ministry of Agriculture, Forestry and Fisheries in Japan. The main campus is located at Ano, not far from Nagoya city and annexes the phytotron and other facilities. VOCRS has two branch stations at Morioka and Kurume.

23.2.1 Division of Research Planning and Coordination	Chief: Dr. K. Yamakawa
Research Planning Section	Head: T. Obama/ Dr. H. Yoshioka
Coordination Section	Head: S. Tsukada
23.2.2 Department of Plant Breeding	Chief: Dr. S. Komochi
Breeding Methods Section	
Breeding methods and germplasm conservation	Head: Dr. K. Takayanagi
Protoplast manipulation and tissue culture	Dr. T. Nishio
Interspecific hybridization	T. Sato
Cucurbitaceous Crops Section	
Breeding (general)	Head: T. Kanno
Disease resistance in muskmelon and water melon	I. Igarashi
Disease resistance in cucumber	Y. Kuginuki
Solanaceous Crops Section	
Breeding (general)	Head: T. Narikawa
Disease resistance and low temperature tolerance	K. Hida
Interspecific hybridization and tissue culture	Y. Sakata
Cruciferous Crops Section	
Classification of varieties	Head: M. Ashizawa
Interspecific hybridization and tissue culture	H. Yamagishi
Disease and volting resistance	S. Yui
Ornamental Crops Section	
Breeding and germplasm conservation	Head: Dr. M. Amano
Analysis of flowering habit	M. Uda
Protoplast manipulation and tissue culture	M. Shibata
Plant Biotechnology Section	Head: S. Nishimura/ Y. Kuginuki
Farm and Greenhouse Management Section	Head: T. Kawaide
23.2.3 Department of Cultivation and Plant Physiology	Chief: Dr. Y. Ohta
Environmental Physiology Section	
Effect of environment on vegetable growing	Head: K. Arai
Soilless culture for vegetable crops	S. Kagohashi
Development of new cultural techniques	Dr. T. Nakashima
Ecological Analysis Section	
Yield prediction	Head: Dr. K. Hoshino
Ecological analysis	M. Nonaka
Environmental control	M. Oda
Farm Mechanization Section	
Farm mechanization and farm work	Head: Dr. O. Kobori
Ergonomics and sensor technique	H. Kurata
Mechanization systems	Dr. K. Otsuka
Plant Nutrition and Plant Physiology Section	
Plant nutrition	Head: Dr. N. Seyama
Translocation of photosynthates	Dr. Y. Shishido
Photosynthesis of vegetables	S. Imada
Reproductive Physiology Section	
Developmental physiology of vegetable crops	Head: Dr. N. Katsura
Physiology of fruiting	M. Nonaka
Physiology of flowering	K. Gamada
Postharvest Physiology Section	
Biochemistry of vegetables	Head: Dr. R. Saijo
Measuring quality of vegetables	M. Yano
Deterioration control of vegetables	G. Ishii
Ornamental Crop Physiology Section	
Physiology of ornamental crops	Head: Dr. S. Kunishige

Physiology of woody ornamentals	K. Nishio
Physiology of perennial crops	K. Suto
Ornamental Crop Ecology Section	
Ecology of cut-flower	Head: T. Yamaguchi
Environmental stresses	M. Kojima
Environment and flower quality	O. Nakagawa
23.2.4 Department of Plant Protection and Environment	Chief: Dr. T. Shimomura
Plant Pathology Section	
Biology and control of vegetable diseases	Head: Dr. M. Ishii/ D. I. Fujisawa
Diseases of vegetables and ornamental plants	K. Abiko
Soil-born Diseases Section	
Biology and control of soil-born diseases	Head: Dr. S. Takeuchi
Bacterial soil-born diseases	T. Shiomi
Fungal soil-born diseases	S. Horiuchi
Entomology Section	
Biology and control of vegetable pests	Head: T. Okada
Insect migration	K. Kawamoto
Use of attractants and repellents	Y. Shirai
Sucking Pests Section	
Use of plant resistance	Head: K. Tanaka
Biological and integrated control of greenhouse pests	M. Kuwabara
Soil and Fertilizer Section	
Soil physics and irrigation for upland fields	Head: Y. Ohno
Plant nutrition and biochemistry of vegetables	K. Tsuchiya
Inorganic nutrients in upland soil	K. Shibano
23.2.5 Department of Greenhouse Cultivation	Taketoyo-machi, Chita-gun, Aichi-pref. 470-23 Phone: 05697-2-1166 Chief: Dr. Y. Nakagawa
Vegetable Crop Management Section	
Growth control of vegetable crops	Head: Dr. H. Yasui
Growth control of fruit vegetables	K. Tanaka
Vegetable Nutrition Section	
Nutrio-physiology, soilless culture	Head: K. Shimura
Photosynthesis of vegetables	M. Nagaoka
Soil science and plant nutrition	N. Suzuki
Water Management Section	
Water management	Head: Y. Goto
Water relations	Y. Araki
Hydrotechnics	H. Mukai
Ornamental Crop Management Section	
Management of potted plant	Head: M. Hara
Nutrition of potted plant	M. Aoki
Physiology of potted plant	K. Shinoda
Rooting Medium Section	
Soil science and plant nutrition	Head: Y. Nonoyama/ Dr. Y. Uehara
Greenhouse-equipment Section	
Labor saving	Head: T.Atsumi/ S.Itoh/ S.Sakuma
Environmental Control Section	
Environmental control	Head: Dr. F. Naito/ G. Ohara
Farm and Greenhouse Management Section	
Vegetable crop cultivation	Head: M. Kikkawa
23.3 Mie University, Faculty of Agriculture elevation: 5 m rainfall : 1,900 mm sunshine : 2,200 h	Kamihama-cho, Tsu-shi, Mie-pref. 514 Phone: 0592-32-1211 Dean: Dr. Z. Kumazawa

This Faculty is located in the center of Japan. College of Agriculture of Mie, the predecessor of the present University, was founded in 1923 in this place. There are 5 departments in the Faculty: Agronomy, Agricultural Engineering, Forestry, Agricultural Chemistry, and Agricultural Machines, and the Laboratory of Horticulture which is a part of Agronomy.

Laboratory of Horticulture
Air pollutants on fruit trees — Prof.Dr. J. Matsushima
Environmental physiology of vegetable crops under stress — Ass.Prof. S. Tachibana
Removal of astringency in Japanese persimmon fruits — K. Yonemori
Experiment Farm
Flower formation and growth regulators — Prof.Dr. Y. Ogawa
Texture of grape berries — K. Akaura

24 Miyagi

24.1 Miyagi Prefectural Horticulture Experiment Station
elevation: 46 m
rainfall : 1,119 mm
sunshine : 1,854 h

Takadate, Natori-shi,
Miyagi-pref. 981-12
Phone: 02238-2-0121
Dir.: T. Oikawa

This Station, the main center of horticultural research in Miyagi-prefecture, serves on all problems concerning the horticultural industry. Its activities include research on fruit trees (apple, Japanese pear, peach, grape, and Japanese persimmon, strawberry, muskmelon) and vegetables (tomato, cucumber, Chinese vegetables), ornamental shrubs, and annual flowers.

Pomology Section — Head: T. Kawarada
Growth and development physiology — K. Onuma
Growth regulators and herbicides — H. Kikuchi
Cultural practices — Y. Sato
Vegetable Crops Section — Head: S. Takahashi
Chinese cabbage
Strawberry — N. Endo/K. Shoji
Tomato — T. Sasaki
Cucumber — N. Honda
Floriculture Section
Carnation — Head: Y. Yusa
Chrysanthemum — K. Kodama
Rose — Y. Sato
Tissue culture — S. Suzuki
Pathology and Insects Section
Pathology of horticultural crops — Head: M. Maeda
Insects of horticultural crops — M. Odagiri
Soil Science Section
Soil and fertilizer — Head: T. Yokoyama/ M. Saito

24.2 Miyagi Agricultural College, Department of Horticulture
elevation: 126 m
rainfall : 1,064 mm
sunshine : 1,977 h

Hatadate, Yamada, Sendai-shi,
Miyagi-pref. 982-02
Phone: 0222-45-2211
Dir.: T. Abe

This Department consists of 6 laboratories, which are pomology, vegetable crops, floriculture, postharvest horticulture, plant pathology and agricultural economics. It has two purposes of education of two year course students and of research concerning horticultural field.

Laboratory of Pomology
Growth and development physiology — Dr. M. Nakamura

Laboratory of Vegetable Crops	
Vegetable crops management	Ass.Prof. S. Shikano
Laboratory of Floriculture	
Flowering control	Prof. H. Kawai
Laboratory of Postharvest Horticulture	
Quality and storage of fruits and vegetables	Ass.Prof. S. Takahashi

24.3 Tohoku University, Faculty of Agriculture
elevation: 40 m
rainfall : 1,245 mm
sunshine : 1,928 h

Amamiya-machi, Tsutsumi-dori
Sendai-shi, Miyagi-pref. 980
Phone: 0222-72-4321
Dean: Prof.Dr. T. Tsuda

The Faculty has five departments. One of it, the Department of Agronomy, has six laboratories including that of Horticulture. Research activities of Horticulture Laboratory cover the physiological aspects in high density planting-orchards system managements, especially of apple, and physiology of fruit growth, photosynthesis and translocation in fruit vegetables in relation to environmental control of their growth.

Laboratory of Horticulture	
Photosynthesis and translocation in fruit vegetables	Prof. Y. Hori*
Growth control of fruit trees	Ass.Prof.Dr. R. Ogata*
Physiology of fruit growth	Dr. Y. Motomura*
Translocation in relation to growth regulators	Dr. K. Takenō

25 Miyazaki

25.1 Miyazaki Prefectural Agricultural Experiment Station
elevation: 7.3 m
rainfall : 2,300 mm
sunshine : 2,750 h

Sadohara-cho, Shimonaka Miyazaki-gun, Miyazaki-pref. 880-02
Phone: 0985-73-2121
Dir.: N. Okamoto

Producing of tree fruits, vegetables and flowers. 6 ha are planted with fruit trees (citrus, chestnut and persimmon), 2 ha with vegetables and flowers. Equipments: Virus-free plant producing equipment with green- and mesh houses.

25.1.1 Fruit Tree Division	Chief: A. Hidaka
Breeding Section	Head: H. Hatano
Citrus breeding	S. Kushima
Cultural Section	Head: S. Ishikawa
Chestnut and persimmon	E. Nagatomo
Citrus	S. Hiraga
25.1.2 Vegetable and Flower Division	Chief: Y. Goto
Breeding Section	Head: I. Kawahara
Squash breeding	F. Noma
Tomato breeding	S. Kato
Disease resistant breeding	R. Nagata
Cultural Section	Head: H. Takahashi
Cucumber and melon	Y. Uchida/ M. Sugio
Flower Section	Head: Y. Iwakiri
Cut flowers	S. Kawasaki
25.1.3 Subtropical Crop Branch	Chief: K. Kawasaki
Subtropical Fruit Section	Head: Y. Shimogoori/ S. Mutagami T. Yamamoto
Useful Plant Section	Head: K. Kawasaki/ Y. Yoshimura K. Takeuchi

25.2 Miyazaki University, Faculty of Agriculture
Department of Agronomy

Kumano, Miyazaki-shi
Miyazaki-pref. 889-21

elevation: 30 m
rainfall : 2,594 mm
sunshine : 2.276 h

Phone: 0985-58-1116
Dean: Prof.Dr. M. Kimura

Department of Agronomy consists of 8 laboratories: pomology, horticulture under structure, postharvest utilization, crop science, plant breeding, plant pathology, applied entomology and farm economics. The first 3 laboratories serve the warm region of the south area of Kyushu on the problems concerning production and postharvest utilization of fruits, vegetables and flowers.

Laboratory of Pomology	
Physiology of fruit development	Prof.Dr. S. Yamamoto
Pollination physiology of citrus in relation to breeding	Ass.Prof.Dr. K. Yamashita
Laboratory of Horticulture under Structure	
Soil and nutrition of ornamental plants	Prof.Dr. T. Tanaka
Mineral nutrition of vegetable crops	Dr. M. Masuda
Laboratory of Postharvest Utilization	
Postharvest physiology	Ass.Prof.Dr. K. Shimokawa

26 Nagano

26.1 Nagano Prefectural Chushin Agricultural Experiment Station

elevation: 750 m
rainfall : 1,249 mm
sunshine : 1,829 h

Soga, Shioziri-shi,
Nagano-pref. 399-64
Phone: 0263-52-1148
Dir.: Dr. A. Sekiguchi

Located in the center of Nagano prefecture, an area rich in fruits and vegetables growing, this Station consists of Plant Production Section and Plant Breeding Section of new production techniques: Main researches are new rotation systems. The breeding programme aims at high quality and high yield varieties with disease resistance in vegetables, soybean and maize.

Plant Production Section	
Viticulture: table and wine grapes	H. Shiba/ I. Shigehara
Rotation system in vegetables	H. Komatsu
Plant Breeding Section	
Vegetables: tomato, sweet pepper and peas	T.Kobayashi/ H.Baba/ T.Oguchi

26.2 Nagano Prefectural Fruit Tree Experiment Station

elevation: 360 m
rainfall : 1,000 mm
sunshine : 1,660 h

Ogawara, Suzaka-shi,
Nagano-pref. 382
Phone: 02624-6-2411
Dir.: Y Ito

Ten ha are planted with tree fruits (apple, apricot, Japanese pear, sweet cherry, Japanese plum, grape, blueberry, chestnut, walnut and bramble). Main research works are on apple, pear and stone fruit rootstocks, plant virus research, method of insects and plant disease protection.

Cultural Section	Head: Y. Tojyo
Apple	H.Koike/ K.Tsukahara/ A.Usuda
	S. Ito
Stone fruits, grapes and nuts	H. Yamanishi/ H. Kihara
Soil and fertilizer	T. Sakai
Disease and Insect Section	Head: S. Shimizu
Diseases	Dr.S.Komori/ K.Hiroma/ M.Nakamura
Insect pests	Y. Hagiwara/ E. Yoshizawa
Tōbu Branch Station	Head: K. Ozawa

26.3 Nagano Prefectural Nanshin Agricultural Experiment

Shimoichida, Takamori-cho,

Station
elevation: 560 m
rainfall : 1,615 mm
sunshine : 1,849 h

Shimoina-gun, Nagano-pref. 399-31
Phone: 026535-2240
Dir.: M. Iwama

Of 8 ha under sprinkler irrigation, 3 ha are planted with tree fruits (pear, apple, peach, persimmon and walnut) and 1 ha with vegetables (tomato, cucumber, strawberry and melon).

Fruit Growing Section	
Pear and persimmon	J. Shiozawa
Apple, pear and walnut	H. Makita
Pear and peach	T. Shimazu
Vegetable Growing Section	
Tomato and strawberry	T. Nakayama
Cucumber and melon	T. Ozawa

26.4 Nagano Prefectural Vegetable and Ornamental Crops Experiment Station
elevation: 346 m
rainfall : 987 mm
sunshine : 2,435 h

Omuro, Matsushiro-machi,
Nagano-shi, Nagano-pref. 381-12
Phone: 0262-78-6848
Dir.: Y. Miyajima

This Station is the main center of vegetable and ornamental crops research in Nagano prefecture. The principal research themes are as follows: breeding of high quality varieties with disease resistance, improvement of cultivation methods, plant protection by environmental control. Furthermore it is equipped with a mushroom experiment laboratory and a plant tissue culture laboratory.

Vegetable Crops Section	
Vegetable breeding	Head: F. Otani
Breeding of Chinese cabbage and fruit vegetables	M. Fujimori
Breeding of cabbage and lettuce	M. Tsukada
Asparagus growing	S. Matsuyama
Tissue culture and propagation	E. Matsumoto
Growing of leafy vegetables	S. Shimojyo
Ornamental Plants Section	
Floriculture	Head: H. Narusawa
Growing of cut flowers and pot flowers	Y. Miyazawa
Breeding and environmental control of cut flowers	T. Tsukada
Mushroom Section	
Culture of mushrooms	Head: T. Shiratori
Spawn production	Y. Matsubara
Bottle culture of wild mushrooms	T. Yazawa
Breeding of mushrooms	K. Nakamura
Plant Protection and Environment Section	
Plant diseases	Head: T. Matsushita/K. Takeda
Insect pests	H. Nakazawa
Soil and fertilization	B. Nakamura/M. Takahashi

26.5 Shinshu University, Faculty of Agriculture
elevation: 770 m
rainfall : 1,610 mm
sunshine : 2,540 h

Minamiminowa, Kamiina-gun,
Nagano-pref. 399-45
Phone: 02657-2-5255
Dir.: Dr. F. Tokita

This campus has a highland climate. Besides this campus, the Institute for Highland and Cool Zone Agriculture is situated in Nobeyama where elevation is 1,350 m. The major horticultural crops in both campuses are apple, pear, peach, grape, blueberry, tomato, cabbage, lettuce, scarlet runner, cyclamen and chrysanthemum.

Laboratory of Pomology	
Ecology and physiology of apples	Prof. Dr. K. Kumashiro
Desiccation injury in pear leaves	Ass. Prof. Dr. Y. Sato
Injury of leaf mites	Ass. Prof. S. Tateishi
Laboratory of Oleri- and Floriculture	
Coloring of tomato fruits	Prof. Dr. T. Takahashi
Propagation of flowers and ornamental plants	Ass. Prof. Dr. M. Nakayama
Institute for Highland and Cool Zone Agriculture	
Ecology and physiology of scarlet runners	Ass. Prof. Dr. H. Arima

27 Nagasaki

27.1 Nagasaki Prefectural Agricultural and Forestry Experiment Station elevation: 5-10 m rainfall : 2,500 mm sunshine : 1,658 h	Kaizu-cho, Isahaya-shi, Nagasaki-pref. 854 Phone: 09572-6-3330 Dir.: S. Yano

This Station is the main center of agricultural research and technical improvement in Nagasaki prefecture, and has six divisions, i.e. Field crops, Vegetables and flowers, Forest, Environment, Farm management and General affairs.

Vegetables and Flowers Division	Chief: K. Aburaya
Vegetable Section	
Injury by successive cropping	Head: T. Fujiyama
Protected cultivation of fruit vegetables	K. Okano
Tissue culture	N. Mori
Breeding of root crops	Y. Ishibashi
Asparagus cultivation	M. Kobayashi
Chinese vegetable growing	T. Takeda
Flower Section	
Chrysanthemum growing	Head: T. Hayashida
Rose growing	•S. Ishida
Processing Section	
Potato processing	Head: T. Yamamoto
Pickled vegetables	S. Noguchi

27.2 Nagasaki Prefectural Fruit Tree Experiment Station elevation: 70 m rainfall : 1,910 mm sunshine : 2,039 h	Onibashi-cho, Omura-shi, Nagasaki-pref. 856-01 Phone: 09575-5-8740 Dir.: H. Muramatsu

This station is the main centre of fruit tree experiment in Nagasaki prefecture.

Growing Section	
Citriculture	Head: I. Kisino/T. Imamura M. Hashimoto/T. Nakao
Citriculture and loquat growing	K. Oishi/T. Hamaguchi
Growing of deciduous fruit tree	S. Hayashida
Loquat Breeding Section	
Loquat breeding	Head: T.Yoshida/A.Morita/O.Terai
Disease and Pest Control Section	
Bionomics and control of pests	Head: M. Nagano/K. Yokomizo

Bionomics and control of diseases — T. Oota
Integrated control — Dr. N. Ohkubo
Soil Management and Fertilization Section
Improvement of Fertilization — Head: T. Takatsuji
Nutritional Diagnosis — K. Inutsuka

27.3 National Fruit Tree Research Station, Kuchinotsu Branch
elevation: 30 m
rainfall : 1,733 mm
sunshine : 1,908 h

Kuchinotsu, Minamitakaki-gun
Nagasaki-pref. 859-25
Phone: 09578-6-2306
Dir.: N. Ichijima

This Branch undertakes the research on late maturing citrus varieties, growing in warm areas. The subjects of research are breeding, development of techniques for stable production and improvement of fruit quality, chiefly in late maturing citrus varieties. 14 ha are planted mainly with late maturing citrus varieties and partly with other citrus varieties. Equipment: Special greenhouse for low temperature treatment. Low temperature laboratory for studies of storage. Special laboratory and phytotron for studies of plant protection.

Laboratory of Fruit Breeding
Breeding of late maturing citrus — Head: N. Okudai/R. Matsumoto; H. Murata
Breeding of loquat varieties — K. Asada
Laboratory of Pomology
Growing and physiology of late maturing citrus varieties — Head: Dr. I. Iwagaki*/S. Ono; T. Takahara
Laboratory of Fruit Nutrition
Citrus nutrition and manuring — Head: K. Ichiki/M. Uchida
Laboratory of Plant Pathology
Ecology and control of late maturing citrus diseases — Head: S. Kuhara/A. Tanaka
Laboratory of Entomology
Biological control of citrus insect pests and mites — Head: T. Ujiye/T. Kashio/H. Fujii
Field Management Section
Citrus growing — Head: S. Uchihara

28 Nara

28.1 Nara Prefectural Agricultural Experiment Station
elevation: 63 m
rainfall : 1,356 mm
sunshine : 2,318 h

Shijo-cho, Kashiwara-shi,
Nara-pref. 634
Phone: 07442-2-6201
Dir.: Dr. K. Kuroda
Dep.Dir.: Dr. Y. Tatsumi

Established in 1895. The purpose of Horticultural Division is to improve the quality and productivity of crops; mainly of strawberry, eggplant, tomato, spinach, cut and pot flower, perennial flowers, persimmon, grape. Researches are made on variety and culture breeding, protection from pests and diseases, soilless culture, physiology of flowering, flower forcing and climate control of greenhouse and energy-saving, storage, processing and diagnoses of fruit tree. Total number of staff is 105 including 66 researchers. Equipments: Total area of 19 ha, greenhouses with automatic controls 0.5 ha, growth chamber and electron microscope.

Vegetable, Flower, Machinery and Greenhouse Division — Chief: Dr. T. Hisatomi
Vegetable Section
Eggplant breeding and rootstocks — Head: K. Naito
Melon breeding and growing — K. Minamibori
Strawberry breeding and forcing — M. Minegishi
Vegetable culture — T. Nomura
Biotechnology of horticultural crops — M. Hattori
Flower Section
Flowering control of ornamental plants — Head: K. Yokoi

Cut flowers growing and forcing	S. Sasaki
Machinery and Greenhouse Climate Section	
Equipment, structure climate control and energy-saving	Head: N. Kawashima
Covering materials and thermal screens	T. Kurozumi
Protection of diseases by climate control	M. Ohara
Fruit Tree Division	Chief: Dr. K. Kuroda
Persimmon culture	S. Iimuro
Astringency and storage of persimmon	Y. Matsumoto
Diagnosis of nutrient condition of persimmon	Y. Ono
Fruit quality of citrus and peach	Y. Sawamura
Grape and small fruits growing	S. Ueda
Growing, quality and storage of persimmon	S. Takano
Hillside Branch Station	Chief: D. Nishimata
Horticultural Exploitation Section	
Propagation of virus-free strawberry	Head: S. Yasui
Apple, peach and grape culture	A. Tsujii
Tomato, spinach and eatable herbaceous plants	S. Arai
Horticulture Section	
Compost and watering of ornamental pot plants	Head: S. Nagamura*
Strawberry and vegetable culture on sloped land	T. Taimatsu
Forcing of Lilium spp. and strawberry culture	H. Watanabe
Asparagus and spinach culture	K. Tanigawa

29 Niigata

29.1 Niigata Prefectural Horticultural Experiment Station
elevation: 7.1 m
rainfall : 1,421 mm
sunshine : 2,151 h

Mano, Seiro-machi, Kitakanbara-gun,
Niigata-pref. 957-01
Phone: 0254-27-5555
Dir.: T. Tanaka

The farmbelt in Niigata prefecture is distributed among the plain, dune and mountainous areas, and they also have various kinds of soil condition, geographical features and distinctive climates. The horticultural production is managed under such environment. Therefore, there are many difficult problems for its production caused by the unfavorable conditions. This Station is trying to solve these problems and to modernize the management.

Ornamental Crop Section	
Breeding	Head: H. Sakurai*
Cut and pot flower physiology	Y. Nomoto
Propagation	A. Enami
Bulb physiology	T. Nakano
Vegetable Crop Section	
Breeding and physiology	Head: E. Himizu/F. Odagiri
Ecology	N. Ono
Propagation	S. Kurashima
Fruit Tree Section	
Small fruit physiology	Head: K. Shiobara
Physiology	K. Watanabe
Breeding	S. Otake
Propagation	S. Kumaki
Plant Protection, Soil and Fertilizer Section	
Plant pathology	Head: Y. Nakatomi/M. Miyagawa
	Y. Yokoyama
Entomology	S. Sakurai
Fertilizer and soil management	T. Koda/T. Kasahara
Sand Dune Products Section	
Fruit and vegetable physiology	Head: A. Tamada
Propagation	M. Yamamoto

Reproductive physiology — M. Emura

29.2 Niigata University, Faculty of Agriculture
elevation: 1.9 m
rainfall : 1,833 mm
sunshine : 1,866 h

Ikarashi, Niigata-shi,
Niigata-pref. 950-21
Phone: 0252-62-6603
Dir.: Prof. Dr. N. Ogasawara

Camellia japonica var. rusticana growing wild are gathered and kept in the experimental field on the campus. In addition, breaking of dormancy of grape with calcium cyanamide and propagation of horticultural crops by tissue culture have been studied.

Laboratory of Horticulture
Physiology of grapes — Prof. Dr. I. Kuroi
Flower breeding and tissue culture — Dr. Y. Niimi*

30 Ōita

30.1 Ōita Prefectural Citrus Experiment Station
elevation: 20-80 m
rainfall : 1,651 mm
sunshine : 1,752 h

Kunisaki-machi, Higashi-Kunisaki-gun, Ōita-pref. 873-05
Phone: 09787-2-0407
Dir.: T. Akita

This Station serves the citrus areas of Ōita prefecture on all problems concerning citrus cultivation, varieties, storage, soil and fertilizers, diseases and pest control. Citrus unshiu and other kinds of citrus are planted in 10 ha of experimental fields.

Research Division — Head: Y. Watanabe
Cultivation Section — Head: S. Shiba
Growing in plastic greenhouse — N. Kawano
Cultivation — T. Zaizen
Variety and breeding — M. Ohara
Soil and Fertilizer Section — Head: H. Mine
Soil and Fertilizer — S. Koda
Disease and Pest Section — Head: Y. Watanabe
Tsukumi Branch Station — Head: T. Shiraishi
Variety and breeding — T. Mimata
Cultivation and storage — T. Sato
Disease and pest control — I. Kai
Physiological disorder — M. Sato

30.2 Ōita Experimental Station for Floricultural Utilization of Hot-Spring
elevation: 170 m
rainfall : 1,971 mm
sunshine : 1,611 h

Tsurumi, Beppu-shi, Ōita-pref. 874
Phone: 0977-66-0793
Dir.: T. Gotō

In this Station, research is made with the aims of utilization of subterranean heat for breeding, propagation and cultivation of flower plants and ornamental trees. Equipment: a botanical garden (Camellia 400 cultivars, Azalea 40 species, 400 cultivars and Prunus 100 cultivars), a conservatory (about 800 species from tropical and sub-tropical regions).

Research Section — Head: R. Higashi
Breeding and growing of ornamental plants — S.Yoshida/T.Kunimoto/S.Gotō
Y. Morotomi

31 Okayama

31.1 Okayama Prefectural Agricultural Experiment Station

Sanyo-cho, Akaiwa-gun,

elevation: 14 m
rainfall : 1,243 mm
sunshine : 2,133 h

Okayama-pref. 709-08
Phone: 08695-5-0271
Dir.: Dr. S. Fujii
Vice-Dir.: Y. Kimura

Experiments on productive methods and disease and pest control of tree fruits, vegetables and ornamental plants of the locality. Tree fruits are planted with 10 kinds (mainly grape, peach, pear, persimmon), 3 ha with grapes, 2 ha with peaches and others. Vegetables are planted in 2 ha in all seasons, and ornamental flowers are planted in 1 ha.

Fruit Growing Section	Chief: H. Fukai N. Takagi/S. Yoda/T. Kimura F. Tamura/T. Ono/F. Benitani
Vegetable and Flower Section	Chief: M. Akiyama
Vegetable growing	T. Kawai/Y. Naito/T. Fujisawa H. Ichikawa
Floriculture	S. Koono/N. Doi
Chemical Section	Chief: M. Hiraoka
Orchard soil management	H. Kimoto/M. Isoda
Vegetable soil management	Y. Ono/I. Tsuboi
Fruit quality and processing	M. Shigeta/T. Uno
Disease and Pest Control Section	Chief: H. Henmi
Fruit protection	M. Hatamoto/F. Tanaka/H. Nasu H. Date
Vegetable and Flower protection	T. Hiramatsu/S. Kasuyama/K. Nagai
Hokubu Branch Station	Chief: A. Tsuboi
Fruit growing	Head: Y. Okamoto/H. Kagami/ K. Takano
Vegetable growing	Head: T. Mizushima/M. Kaihara

31.2 Okayama University, Faculty of Agriculture
elevation: 3 m
rainfall : 1,223 mm
sunshine : 2,137 h

Tsushima, Okayama-shi,
Okayama-pref. 700
Phone: 0862-52-1127
Dean: Prof. Dr. T. Ogo

The Department of horticulture comprises the four laboratories described below, and the Laboratory of Landscape Architecture. The Department specializes in the field of horticulture from basic study to the practical application of horticultural production, distribution and consumption of the products. The education and research work covers these fields.

Pomology	
Dwarf forms and dwarfing rootstocks of peach	Dr. K. Shimamura
Fertilization and seed development in grape	Dr. G. Okamoto
Growth of tree fruits and environmental conditions	Dr. N. Kubota
Olericulture	
Self-incompatibility and aseptic culture of vegetables	Dr. S. Matsubara
Quality of melon	K. Kinoshita
Floriculture	
Water consumption of greenhouse florist crops	Dr. K. Konishi
Optimum soil nutrient level for cut flower	Dr. Y. Kageyama
Flowering control of perennial florist crops	T. Hayashi
Postharvest Horticulture	
Effect of vibration and impact on respiration of fruit	Dr. R. Nakamura
Mechanism of fruit ripening after harvest	Dr. A. Inaba
Postharvest physiology	Y. Kubo
Research Farm	
Anatomical studies of bulbous plants	Dr. K. Yasui
Disorder of grape berry	M. Nakano

32 Okinawa

32.1 Okinawa Prefectural Agricultural Experiment Station
elevation: 76 m
rainfall : 2,127 mm
sunshine : 2,047 h

Sakiyama, Shuri, Naha-shi,
Okinawa-pref.
Phone: 0988-84-3415
Dir.: T. Matsumoto

Horticultural Branch: Cultivation, disease and pest control of vegetable and ornamental crops.The area of experimental fields is 21 ha including 0.6 ha of greenhouses. Nago Branch: Production and disease control of pineapple and other fruit crops, such as mango, guava,lychee, citrus fruit, with 36 ha of experimental fields. Miyako Branch Station: Production ofsquash and lily tubers. Yaeyama Branch Station: Cultivation of pineapple.

Horticultural Branch	Head: H. Wakugami
Vegetable Crop Laboratory	G. Hiyane/K. Uehara/K. Uechi Y. Nagamine/M. Sakamoto/Y. Iraha
Flower and Ornamental Plant Laboratory	R. Aka/H. Higa/ R. Matsumoto/E. Kinjo R. Higa
Nago Branch Station	Head: E. Tamaki
Fruit Crop Laboratory	M. Miyagi/H. Kinjo/M. Ohshiro
Pineapple Laboratory	A. Onaha/T. Miyasato/M. Higa K. Takaesu
Tropical Fruit Laboratory	G. Tokeshi/N. Yasutomi/H. Ikemiyagi
Miyako Branch Station	Head: Y. Nagamine
Upland Field Crop Laboratory	K. Tana/S. Miyagi
Yaeyama Branch Station	Head: M. Yokota
Horticultural Crop Laboratory	K. Kuramori/T. Shimanaka/K. Deki Y. Shimabukuro/H. Soemori

32.2 University of the Ryukyus, College of Agriculture
elevation: 127 m
rainfall : 2,300 mm
sunshine : 2,061 h

Senbaru, Nishihara-machi,
Nakagami-gun, Okinawa-pref. 903-01
Phone: 09889-5-2221
Dir.: Dr. S. Sunagawa

This University is located in the southern part of Japan. The climate belongs to sub-tropical ones. The main studies are of tropical horticulture.

Pomology	
Physiology of citrus and tropical fruit	Prof. Dr. T. Higa
Tropical fruit growing	Ass. Prof. S. Yonemori
Ornamentals	
Orchid, tissue culture	Ass. Prof. Dr. K. Uesato
Vegetables	
Breeding	Ass. Prof. S. Adaniya
Tropical vegetable growing	Ass. Prof. S. Yonemori

33 Osaka

33.1 Osaka prefectural Agricultural Research Center
elevation: 40 m
rainfall : 1,260 mm
sunshine : 2,214 h

Shakudo, Habikino-shi,
Osaka-pref. 583
Phone: 0729-58-6551
Dir.: N. Nishimura

The Center (33.8 ha) performs its functions as a 'brain' of urban agricultural system in Osaka-pref. and covers protection of nature and development of constant supply of fresh agricultural products. The main works: energy saving in horticultural production, utilization of tissue culture for breeding and raising of virus free plants, production and preservation of horticultural products.

Vegetable Crops Section	
Breeding and growing	Head: K. Yamada/M. Morishita
Soilless culture	H. Tsuji
Ornamental Plants Section	
Propagation of flowering trees	Head: M. Mori/K. Daido
Tissue culture	S. Fukai
Pomology Section	
Forcing of grape fruit production	Head: Y. Okuda/M. Dan
Forcing of citrus fruit production	A. Kato
Preservation and Post-harvest Physiology Section	
Preservation of fruit and vegetable	Head: Y. Okuda
Post-harvest physiology of fruit and vegetable	Dr. K. Abe

33.2 University of Osaka Prefecture, College of Agriculture, Department of Horticultural Science and Agronomy
elevation: 30 m
rainfall : 1,330 mm
sunshine : 2,130 h

Mozuume-machi, Sakai-shi
Osaka-pref. 591
Phone: 0722-52-1161
Dean: Dr. S. Nakagawa*

Founded by Osaka prefecture, one of the largest autonomies. As it is located in densely populated area, suburban horticulture, horticulture under structure and distribution technology of horticultural products are subjects of deep concern. A well-organized staff of 20 researchers work at 5 Laboratories. The University farm for horticulture has 1.6 ha including 20 greenhouses.

Pomology Laboratory	
General viticulture	Prof. Dr. S. Nakagawa*
Plant growth regulators in tree fruits	Ass. Prof. Dr. E. Yuda*
Sugar and protein metabolism in grapes	Dr. H. Matsui
Bud and seed dormancy in fruit trees	Dr. S. Horiuchi
Vegetable Crop Science Laboratory	
Mineral nutrition of vegetable crops	Prof. Dr. T. Osawa
Phylogeny of melons and cucumbers	Ass. Prof. Dr. N. Fujishita
Photosynthesis of vegetable crops	Dr. Y. Oda*
Soilless culture and mineral nutrition of vegetable crops	H. Ikeda*
Floriculture Laboratory	
Regeneration of orchids and bulbous ornamentals	Prof. Dr. Y. Sakanishi
Growth and development of gesneriads and orchids	Ass. Prof. K. Fujihara*
Smoke and ethylene application to bulbs	Dr. H. Imanishi*
In vitro culture of annuals	H. Fukuzumi
Postharvest Physiology and Storage of Horticultural Products	
Chilling injury of fruit	Prof. Dr. T. Iwata
Respiration mechanism of fruit and vegetable	Ass. Prof. Dr. K. Chachin
Enzymatic production of flavor in fruit	Dr. Y. Ueda
Postharvest physiology and keeping quality of mushrooms	Dr. T. Minamide
University Farm	
Fruit growing	Prof. Dr. S. Fukunaga/Dr. H. Kurooka
Vegetable growing	J. Tamura
Regulation of flowering in bulbous plants	Dr. G. Mori

33.3 Kinki University, Faculty of Agriculture, Laboratory of Horticulture
elevation: 35 m

Kowakae, Higashi-Osaka-shi
Osaka-pref. 577
Phone: 06-721-2332

rainfall : 1,300 mm
sunshine : 2,200 h

Chairman: Dr. G. Fuse

Private University. Since a large part of students belonging to the laboratory, are successors to their fathers in farming, education lays importance to actual practices. Alumni are playing an important role in their native places covering the western half of Japan.

Fruit production	Prof. Dr. F. Yoshimura
Greenhouse management	Prof. Dr. J. Fumoto
Plant for landscape	Ass. Prof. N. Suzuki
Flower growing	N. Mizutani

34 Saga

34.1 Saga Prefectural Agricultural Experiment Station
elevation: 3 m
rainfall : 1,800 mm
sunshine : 2,030 h

Kawasoe-cho, Saga-gun,
Saga-pref. 840-23
Phone: 0952-45-2141
Dir.: Y. Yagi

Plant breeding and variety selection and cultivation experiment of vegetables and flowers. 1 ha of glasshouses and vinyl-houses are planted with vegetables (eggplant, strawberry, cucumber, spinach) and flowers (chrysanthemum, carnation, rose etc.). 1 ha of open fields, with vegetables (onion, broccoli, lettuce etc.).

Vegetable Crop and Floriculture Section	Chief: S. Kawasaki
Vegetable Crop Section	
Plant breeding	Head: T. Heguchi
Cultivation	K. Fukuta
Floriculture Section	
Cultivation	Head: M. Tanaka
Plant breeding	M. Tanaka
Mitsuze Branch Station	Dir.: N. Jojima
Cultivation	Head: T. Shimomura
Plant breeding	T. Tanaka
Shiroishi Branch Station	Dir.: M. Tokuyasu
Plant breeding	H. Saito
Cultivation	I. Mori

34.2 Saga Prefectural Fruit Tree Experiment Station
elevation: 30 m
rainfall : 1,800 mm
sunshine : 2,032 h

Ogi-machi, Ogi-gun, Saga-pref. 845
Phone: 0952-73-2275
Dir.: H. Eguchi
Sub. Dir.: T. Ehara

Producing and processing of fruit trees and vines. 6.2 ha are planted with fruit trees (4.6 ha: citrus, 1.0 ha: Japanese pear, 0.6 ha: peach, Japanese apricot, persimmon, loquat, apple and Japanese chestnut), 0.5 ha with vines (grapes and kiwi fruit). Equipment: Greenhouses and nethouse for studies of virus diseases. N-15 emission spectroanalyzer.

Citrus Section	
Cultural practices	Head: T. Nogata
Growth regulators and environmental control	N. Suetsugu
Breeding and propagation	S. Matsuzaki
Packing and transportation	T. Koga
Fruiting, harvesting and storage	H. Iwanaga
Processing and by-products	Y. Shibata
Deciduous Fruit Tree Section	
Growth and chemical control	Head: R. Hirota
Physiology and ecology	Y. Takubo/K. Inadomi

Soil and Fertilizer Section

Plant nutrition	Head: T. Iwakiri
Soil management	S. Matsuse
Fertilizer technology	T. Shindoo

Disease and Pest Section

Citrus diseases	Head: Dr. M. Sadamatsu
Deciduous fruit tree diseases	H. Mikuriya
Pest control	H. Nakamura/M. Muraoka

34.3 Prefectural Upland Cropping Experiment Station
elevation: 110 m
rainfall : 1,747 mm
sunshine : 1,995 h

Karatsu-shi, Saga-pref. 847-01
Phone: 09557-3-3371
Dir.: Y. Koyanagi

This Station, situated in the Uwaba district, the main upland farming area in Saga prefecture, is the center of researches about varieties, cultivations, diseases and insects of upland and vegetable crops, and also the soils in this area.

Upland and Vegetable Crop Section	Head: Y. Matsuo*
Upland and vegetable crop	S. Yamashita
Soil and fertilizer	H. Sumi
Plant protection	N. Tashiro

34.4 Saga University, Faculty of Agriculture,
Department of Horticultural Science
elevation: 3.8 m
rainfall : 1,890 mm
sunshine : 2,033 h

Honjo-machi, Saga-pref. 840
Phone: 0952-24-5191
Dean: Prof. I. Ito

New plants from tissue culture and crosses of Chrysanthemum, Iris, Lycoris, Allium and Colocasia plants, which are used for cytogenetical studies and breeding. Of 7 ha orchard, 5 are planted with citrus and 2 ha with pear, peach, grape, persimmon and loquat. Four hundreds cultivars of citrus and 30 wild species of orange subfamily are planted as genetic resources in the citrus orchard and 2 greenhouses, and used for taxonomy, cytogenetics and breeding. Chemical and physio-mechanical postharvest studies are carried out on the fruit and vegetables to develop new measuring methods of browning, vitamins and other substances, and physical properties.

Laboratory of Olericulture and Floriculture

Tissue culture of vegetables and ornamentals	Ass. Prof. Dr. S. Miyazaki
Genetics and breeding of Allium plants	Dr. Y. Tashiro

Laboratory of Fruit Sciences

Genetics and breeding of citrus fruit	Prof. Dr. M. Iwamasa*
Taxonomy and breeding of citrus fruit	Ass. Prof. Dr. N. Nito*

Laboratory of Food Science and Horticultural Engineering

Postharvest physiology of fruit and vegetables	Prof. Dr. T. Tono
Physico-mechanical properties of fruit and vegetables	Ass. Prof. Dr. T. Kojima
Chemical analyses of fruit and vegetables	Dr. S. Fujita

35 Saitama

35.1 Saitama Prefectural Garden Plants Center
elevation: 81 m
rainfall : 1,123 mm
sunshine : 2,142 h

Kushibiki, Fukaya-shi,
Saitama-pref. 366
Phone: 0485-72-1220
Dir.: C. Ohtsuka
Sub. Dir.: J. Okada

This Station is the main center of garden plant culture and landscape architecture research in Saitama prefecture. Variety preservation: 160 Acer garden varieties, 80 Prunus mume garden varieties, 280 Primura sieboldii garden varieties and 20 Adonis amurensis garden varieties.

Garden Plants Growing Section	
Breeding	Head: K. Itami
Mutiplication and propagation	I. Sakakibara/M. Watanabe
Regulation of flowering	T. Horiguchi
Tissue culture	T. Matsumoto
Landscape Architecture Section	
Planting management	Head: Y. Nakada
Planting technique	Y. Ishii
Planting design	N. Shimizu

35.2 Saitama Prefectural Horticultural Experiment Station
elevation: 12 m
rainfall : 1,235 mm
sunshine : 2,318 h

Rokumanbu, Kuki-shi,
Saitama-pref. 346
Phone: 0480-21-1113
Dir.: O. Moriya

This Station is the main center of horticultural research in Saitama prefecture.

Chemical Section	Head: T. Akimoto
Soil and fertilization	M. Takeda/H. Sato
Weed control	T. Furuya
Floriculture Section	Head: J. Uematsu
	Y. Koyama/H. Tomita/I. Ishizaka
	H. Yajima
Fruit Growing Section	Head: B. Mukai
	T. Okumo/M. Mitobe/S. Asano
	Y. Sakai
Vegetable Section	Head: K. Matsumaru
	S. Watanabe/K. Kogure/Y. Ouchi
	M. Inayamå/K. Watanabe/N. Watanuma
Crop Protection Section	Head: S. Shibukawa
Disease control	R. Zenbayashi/K. Hashimoto
Insect control	K. Takahashi/A. Kubota/H. Nemoto
Tsurugashima Branch	Head: H. Mizumura
Vegetable growing	I. Shiono/M. Sato/F. Uchiyama
	T. Okayasu/T. Takahashi
Crop protection	Y. Shimazaki

36 Shiga

36.1 Shiga Prefectural Agricultural Experiment Station
elevation: 85 m
rainfall : 1,696 mm
sunshine : 1,922 h

Dainaka, Azuchi-cho, Gamo-gun
Shiga-pref. 521-13
Phone: 07486-3081
Dir.: T. Kumagai

Main Station has a Vegetable Section, and is carrying out mainly researches on vegetables under structures. Horticultural Branch Station (Ritto-cho) has a Flower Section, and a Fruit Tree Section. Also 8 greenhouses for carnation, chrysanthemum, rose and other perennial flowers and 4.3 ha orchard of Japanese persimmon, grapevine and Japanese pear.

Vegetable Section	
Leafy vegetables for summer harvest	Head: S. Mori
Irrigation for semi-forcing strawberry	K. Watanabe
Fruit vegetable cultivation	T. Nakada
Kohoku Branch Station	
Strawberry cultivation by nutrient solution	H. Otani

Kosei Branch Station

Breeding of clubroot resistant Japanese turnip — Chief: G. Okeda

Horticultural Branch Station — Chief: R. Ohishi

Fruit Tree Section

High density planting of dwarf type persimmon — Head: Dr. R. Murata

Japanese pear quality — R. Ōishi

Grapevine cultivation — H. Okishima

Flower Section

Control of growth and flowering of perennial flowers — Head: K. Ōsumi

Culture of spray type roses — K. Yoshizawa

Seedling production — H. Kitamura

36.2 Shiga Prefectural Junior College, Faculty of Agriculture, Laboratory of Horticulture

Nishishibukawa-machi, Kusatsu-shi,
Shiga-pref. 525
Phone: 0775-62-1343
Dir.: Dr. T. Sakurai

The College offers a two-year program in agriculture for students to become specialists of medium standing. It also contributes to the community through the studies on horticultural crops.

Pomology — S. Fukuda

Vegetable crops science — . Nishio*

Floricultural science — T. Morita

37 Shimane

37.1 Shimane Prefectural Agricultural Experiment Station
elevation: 20 m
rainfall : 1,910 mm
sunshine : 1,670 h

Ashiwata-cho, Izumo-shi,
Shimane-pref. 693
Phone: 0853-22-6650
Dir.: S. Kitayama

The Station occupies 31.5 ha of clay loam soil field. The research programme is concerned with problems of the main crops of Shimane district i.e. grape, persimmon, tulip, melon and wasabi (Eutrema wasabi Maxim.).

Floriculture Section — Head: Dr. K. Yamada/N. Akimitsu
H. Inamura

Fruit Tree Section — Head: K. Takahashi/Y. Kono/A. Imaoka
H. Azukizawa/K. Yamamoto/Y. Uchida
T. Kurahashi

Vegetable Section — Head: Y. Nakagawa/Y. Fukuyori
K. Haruki/S. Nonomura/M. Kitagawa

Taisha Branch (grapes) — M. Kuranaka

Akana Branch — Head: M. Omoso

Vegetable growing — H. Kono/F. Ishizu

37.2 Shimane University, Faculty of Agriculture
elevation: 15 m
rainfall : 1,957 mm
sunshine : 1,939 h

Nishikawatsu-cho, Matsue-shi,
Shimane-pref. 690
Phone: 0852-21-7100
Dean: Dr. R. Tanaka

Environmental and chemical controls of grape and Japanese persimmon fruit productions, flowering control of ornamentals and taxonomy of tree peony and oriental melon, cropping system of bulbous plants, and mechanical harvest of tomato. Equipment: computor-controlled greenhouse with heating and cooling systems by solar energy, hydroponics for vegetable production and 3 experimental farms, totalling 28 ha.

Laboratory of Pomology
Developmental physiology of grapes — Prof. Dr. R. Naito
Physiological disorders of fruit trees — Ass. Prof. Dr. H. Yamamura
Laboratory of Vegetable and Ornamental Horticulture
Flowering control of ornamentals — Prof. Dr. K. Inaba
Dormancy control and taxonomy of vegetables and ornamentals — Ass. Prof. Dr. T. Hosoki
Experimental Farm
Flower production system of bulbous plants and tree peony — Prof. Dr. S. Yoshino/Dr. N. Aoki
Mechanical properties of tomato for processing — Ass. Prof. Dr. N. Ito
Developmental physiology of grapes — Ass. Prof. H. Ueda

38 Shizuoka

38.1 Shizuoka Prefectural Agricultural Experiment Station
elevation: 30 m
rainfall : 1,786 mm
sunshine : 2,250 h

Tomigaoka, Toyoda-cho,
Shizuoka-pref. 438
Phone: 05383-5-7211
Dir.: T. Shirai

Developing new agricultural technology. Breeding new varieties of vegetables and flowers. Studying new environmental control and plant protection. 7,400 m^2 greenhouses and plastic houses and 4.5 ha of paddy fields.

Horticultural Division — Chief: K. Suzuki
Fruit Vegetable Laboratory
Growing — Head: T. Suzuki
Breeding — S. Nakamura
Leafy and Root Vegetable Laboratory
Growing — Head: M. Sada/Y. Muramatsu
Ornamental Laboratory
Growing — Head: Dr. K. Funakoshi/S. Miwa
M. Matsuda/K. Mito

Resources Laboratory
Plant nutrition — H. Yamashita
Greenhouse Engineering Laboratory
Greenhouse facilities — S. Kimura
Biotechnology Laboratory
Breeding — Head: M. Toda
Tissue culture — T. Ootsuka
Eastern Horticultural Branch
Vegetable growing — T. Takeuchi
Coastal Sand Area Branch
Vegetable growing — M. Kawamura

38.2 Shizuoka Prefectural Citrus Experiment Station
elevation: 30 m
rainfall : 2,114 mm
sunshine : 2,282 h

Komagoenishi, Shimizu-shi,
Shizuoka-pref. 424
Phone: 0543-34-5351
Dir.: Dr. M. Konakahara

This Station was established in 1940 for the citrus industry of Shizuoka-prefecture, one of the coolest but representative citrus growing areas, in Japan. It now has 2 branches in growing areas characterized by climate and soil, and a deciduous fruit tree branch. It is the most important subject to sustain the citrus industry based on satsuma mandarin of the most suitable variety for this prefecture.

Variety Section	Head: S. Hara
Breeding and selection	
Breeding and tissue culture	E. Shikano
Citricultural Section	Head: Y. Takahashi
Soil improvement	T. Suzuki
Controlled environments	M. Okada
Productivity and fruit quality	Y. Makita
Postharvest physiology	
Plant Nutrition Section	Head: N. Okada/K. Yoshikawa
Fertilizer application	H. Hisada/K. Sugiyama
Mineral nutrition	
Pathology and Entomology Section	Head: K. Inoue
Mycology	Dr. K. Furuhashi/A. Tatara
Biological control and population ecology	S. Serizawa
Bacteriology	T. Ichikawa
Virology	
Izu Branch	Head: T. Mita
Cultural practices	H. Suzuki
Growth and postharvest physiology	N. Mitsui
Selection of late varieties	
Seien Branch	Head: S. Kato
Water utilization	I. Iguchi
Selection of early and medium varieties	S. Kikuchi
Cultural practices	
Deciduous Fruit Tree Branch	Head: K. Fukuyo
Cultural practices	T. Ishida
Mineral nutrition	S. Anma
Growth physiology	A. Takahashi
Applied entomology	Y. Ishido
Selection of deciduous fruit trees	

38.3 National Fruit Tree Research Station, Okitsu Branch
elevation: 3-45 m
rainfall : 2,290 mm
sunshine : 2,070 h

Okitsu, Shimizu-shi,
Shizuoka-pref. 424-02
Phone: 0543-69-2111
Dir.: Dr. K. Hirose

This Station is the center of citrus breeding in Japan. About 700 indigenous or cultivated species and varieties, and many satsuma strains have been colleted and preserved. Of 13 ha, 11 ha are planted with about 6,000 crossed seedlings and above-mentioned preserved varieties. In addition, studies on cultivation, fruit storage, processing, citrus diseases and pest control have been conducted.

First Laboratory of Fruit Breeding	Head: Dr. I. Ueno
Genetics and breeding	Y. Yamada
Fruit quality	T. Hidaka
Tissue culture	Y. Ito
Breeding	
Second Laboratory of Fruit Breeding	

38.4 Shizuoka University, Faculty of Agriculture, Department of Horticulture
elevation: 30 m
rainfall : 2,360 mm
sunshine : 2,080 h

Ohya, Shizuoka-shi,
Shizuoka-pref. 420
Phone: 0542-37-1111
Dean: Dr. H. Okazaki

Shizuoka University is located in the coastal area of nearly central part of Honshu island. This district is one of the main citrus and protected crop production areas in Japan. Research programs in the Department are mainly concerned with nutrition of citrus tree, muskmelon and chrysanthemum, propagation of horticultural plants and postharvest physiology of fruit and vegetables.

Laboratory of Citriculture

Nutrition of citrus tree	Prof. Dr. T Suzuki
Morphology of flower development of fruit tree	Ass. Prof. Dr. T. Takagi
Laboratory of Protected Crop Cultivation	
Photosynthesis of vegetable crops in greenhouse	Prof. Dr. K. Takahashi
Nutrition of vegetable crops in greenhouse	Dr. A. Nukaya/Dr. K. Ohkawa
Laboratory of Propagation	
Physiology and morphology in horticultural plant propagation	Prof. Dr. T. Hosoi
Physiology of root formation of cuttings	Prof. Dr. A. Oishi
Tissue culture of fruit tree	Dr. H. Harada
Laboratory of Postharvest Physiology and Preservation	
Postharvest physiology and storage	Prof. Dr. T. Murata*
Fruit ripening and ethylene physiology	Prof. Dr. H. Hyodo*
Chilling injury	Dr. Y. Tatsumi

39 Tochigi

39.1 Tochigi Prefectural Agricultural Experiment Station
elevation: 150-170 m
rainfall : 1,464 mm
sunshine : 2,086 h

Kawaraya-machi, Utsunomiya-shi,
Tochigi-pref. 320
Phone: 0286-65-1241
Dir.: Y. Tsurumi

Selection of good varieties and improved growing techniques for vegetables, fruit trees and ornamental plants. Four ha orchards with a hail prevention net are planted with Japanese pear, grape, chestnut, apple, and peach trees. Also 32 greenhouses for research of growing of vegetables and ornamental plants.

Vegetable Growing Section	Head: S. Miyake
Leafy, root vegetables	O. Chô/M. Kasuya/K. Hiraide Y. Tamura
Fruit vegetables	E. Wada/H. Tochigi/K. Yatabe
Fruit Tree Growing Section	Head: A. Aoki
Japanese pears	T. Kaneko/K. Yamazaki/T. Misaka
Chestnuts	T. Tanaka
Grapes	T. Hayata
Flower Growing Section	Head: A. Yamanaka
Pot plants	N. Minegishi
Cut flowers and pot plants	M. Koguchi/H. Fukazawa
Sano Branch	Head: A. Ôta
Growing of vegetables	T. Shioya
Kuroiso Branch	Head: K. Hanyû
Growing of vegetables	E. Muroi
Kanuma Branch	Head: K. Shibata
Growing of carnations	E. Fukuda
Tochigi Branch	Head: K. Endô
Growing and breeding of strawberry and bottle gourd	H. Kawasato/I. Iwasaki/S. Taguchi H. Akagi/K. Takano/M. Tsuchizawa

39.2 Utsunomiya University, Faculty of Agriculture
elevation: 110 m
rainfall : 1,300 mm
sunshine : 2,100 h

Mine-machi, Utsunomiya-shi,
Tochigi-pref. 321
Phone: 0286-36-1515
Dean: Prof. Dr. T. Muramatsu

The Faculty, originated as Utsunomiya Agricultural College in 1922, covers almost all research fields of agriculture with six departments. Equipment: air-controlled greenhouses (1,500 m^2, laboratory for RI experiment, and institute of weed-control on the faculty campus. Research farm (100 ha) including orchard and vegetable field possesses solar and other saving energy systems.

Laboratory of Horticulture	
Fruit growth and growth substances	Prof. S. Wakabayashi*
Propagation of lilies and Selaginella invelvens	Ass. Prof. H. Sekiya
Flower formation and dormancy in strawberry	Ass. Prof. N. Fujishige
Laboratory of Plant Breeding	
Hybridization and propagation with tissue culture	Ass. Prof. Dr. Y. Matsuzawa
Laboratory of Comparative Agriculture	
Productivity of Japanese pear and tropical fruit	Ass. Prof. Dr. O. Kishimoto
Research Farm	
Growth of bulbous plants	Dr. Y. Iziro
Weed Control Research Institute	
Perennial weed control in orchard	Prof. Dr. T. Takematsu Ass. Prof. Dr. Y. Takeuchi

40 Tokushima

40.1 Tokushima Prefectural Agricultural Experiment Station
elevation: 8 m
rainfall : 1,742 mm
sunshine : 2,090 h

Ishii-cho, Myozai-gun,
Tokushima-pref. 779-32
Phone: 08867-4-1660
Dir.: Y. Nagai

This Station is responsible for Tokushima area's agriculture under a mild climate. It makes experiments on cultural improvement and breeding of strawberry, Japanese radish, Limonium, lily and chrysanthemum.

Vegetable Crop Section	Head: H. Machida
Vegetable Growing	H. Kodo/K. Bando/N. Kita/ H. Kawamura
Vegetable breeding	J. Ogawa
Floricultural Section	Head: A. Sumitomo
Floriculture and breeding	Y. Urakami
Kainan Branch Station	
Vegetable growing	S. Fukuoka/T. Sakaguchi
Ikeda Branch Station	
Vegetable growing	T. Kawashita/K. Nagai

40.2 Tokushima Prefectural Fruit Tree Experiment Station
elevation: 50 m
rainfall : 2,300 mm
sunshine : 1,800 h

Katsuura-cho, Katsuura-gun,
Tokushima-pref. 771-43
Phone: 08854-2-2545
Dir.: M. Nakagawa
Vice-Dir.: Dr. K. Kurokami

Research in this Station is directed to improvement of citricultural methods which consist of productivity, fruit quality, harvesting, storage, nutrition, soil manure and the control of weeds, pests and diseases. There is a 3 ha orchard with 6 greenhouses for studies of virus disease, natural enemies and fruit breeding. The purpose of the branch station is improvement of cultivating methods of several deciduous fruit trees such as Japanese pear, Japanese persimmon, grape, Japanese apricot, kiwifruit and so forth. Additional work is conducted on propagation of citrus budwoods and citriculture under structure.

Growing Section	
Post-harvest physiology and weed control	Head: S. Sagane
Citrus breeding and production	N. Otoi
Growth regulators and fruit quality	H. Hasebe
Protection Section	
Fruit fungus diseases and mycology	Head: H. Yamato
Chemical and biological control of pests	M. Yukinari
Virus diseases and soil-borne diseases	M. Tsuji

Nutrition Section	
Mineral nutrition and soil improvement	Head: H. Wada
Soil chemistry and water relation	M. Yamao
Kenhoku Branch Station	
Fruit pest control	Vice-Dir.: M. Kagawa
Citriculture under structure	A. Jozukuri
Physiological disorders of fruit (mume, kaki)	K. Murakami
Plant hormones and training (grape)	T. Akai
Citrus growing	Y. Shibata
Stone fruits growing	A. Koike
Japanese pear growing	S. Mori
Propagation of budwoods	K. Bando

41 Tokyo

41.1 Tokyo Agricultural Experiment Station
elevation: 76 m
rainfall : 1,591 mm
sunshine : 2,356 h

Fujimi-cho, Tachikawa-shi,
Tokyo 190
Phone: 0425-24-3191
Dir.: K. Ashikawa
Vice-Dir.: N. Sawachi

This Station (main station with 15 ha, Edogawa Branch with 2 ha) conducts researches on agricultural technology and management for encouraging the development of agricultural production in Tokyo area, and also research on air pollution using the plants in air-controlled chambers (200 m^2).

Cultivation Division	Chief: S. Ida
Vegetable Section	
Tissue culture of vegetables	Head: S. Kohno
Native vegetables (spikenard, Japanese horseradish)	S. Kawamura
Breeding (tomato, strawberry)	T. Noro
Improving and testing horticultural implements	Y. Takao
Fruit Tree Section	
Fruit setting on Prunus mume	Head: S. Hijikata
Physiological disorder (Japanese pear)	Dr. S. Kawamata*
Floricultural Section	
Chemical control (growth retardants)	Head: Dr. S. Hashimoto*
Orchids cultivation	O. Kato
Propagation, irrigation system, environmental control	Y. Hamada
Arboricultural Section	
Propagation and growth control	Head: Y. Kato
Weed control	A. Yanagita
Plant Pathology and Entomology Section	
Insect control and nematology	Head: M. Nagasawa
Insecticides	S. Arai
Soil disease and virus control	T. Hirano
Disease control (trees and shrubs)	H. Horie
Air Pollution Section	
Dynamics of air pollution	Head: K. Terakado
Air pollution on plants	H. Kuno
Agricultural Chemistry Division	Chief: Dr. N. Date
Soil and Fertilizer Section	
Analysis of remained pesticide in soil	Head: N. Kanamaru
Soil environmental control	H. Takesako
Soil improvement	H. Miyakoda
Soil conservation	T. Kato
Analysis of harmful substances	T. Ogawa
Processed Food Section	

Improvement of processed food	Head: T. Sato
Quality control (bean sprout)	T. Aoki
Agricultural Management Division	Chief: I. Motohashi
Horticultural Management Section	
Marketing for vegetables	Head: K. Kitajima
Economical production for garden plants	M. Totsuka
Marketing for ornamental plants	M. Takizawa
Edogawa Horticultural Branch Station	Chief: M. Masui
Horticultural Section	
Leafy vegetables	Head: E. Kosuge
Potted flowers	K. Hido
Hydroponics of vegetables	K. Kodera
Soil Nutrition Section	
Phosphate for vegetables	T. Arita
Quality control of vegetables	T. Yoneyama
Plant Pathology and Entomology Section	
Disease control of ornamental plants	Head: S. Sugata
Insect control	M. Iga

41.2 Tamagawa University, Faculty of Agriculture

Machida, Tokyo 194
Phone: 0427-28-3366
Dir.: Dr. F. Yoshida

Studies on growing of flower crops and breeding of chrysanthemums are the main subjects.

Laboratory of Horticulture	
Pot plant production	Head: H. Tanaka*
Flower pigments	K. Inazu
Cell and tissue culture	Y. Tsuyuki

41.3 Tokyo University of Agriculture
elevation: 9 m
rainfall : 1,460 mm
sunshine : 1,492 h

Sakuragaoka, Setagaya-ku, Tokyo 156
Phone: 03-420-2131
Dir.: Dr. T. Suzuki

Faculty of Agriculture	
Laboratory of Pomology	
Deterioration of soil productivity of orchard	Prof. Dr. T. Kawakami
Fruits setting of Prunus mumu	Dr. T. Otsubo
Growth hormones in Citrus spp.	M. Miyata
Laboratory of Vegetable Crops	
Nitrogen nutrition	Prof.Dr. M. Iwata*
Histology and chemistry of herbs	Ass. Prof. Dr. Y. Tomitaka
Nutrition of vegetable crops	M. Ichimura
Laboratory of Floriculture	
Flowering control of ornamental plants	Prof. Dr. H. Higuchi*
Interior quality of foliage plants	Dr. S. Suzuki
Self-incompatibility of flowers	W. Amaki
Junior College of Agriculture	
Laboratory of Horticulture	
Dry matter production of fruit trees	Prof. Dr. S. Hirano
Soil conditions and growth of taro	S. Kuboi
Mineral nutrition of grape	H. Uematsu
Experimental Farm	
Protected cultivation of vegetable crops	Prof. Dr. A. Yoneyasu
Water consumption of vegetable crops	A. Karimata

Soil management of orchard — Ass. Prof. S. Omori
Meteorological factors affecting the quality of Japanese pear — H. Kato
Meteorological factors affecting the quality of satsuma mandarin — N. Ikeda
Asexual propagation of ornamental plants — S. Suzuki
Production system of sweet peas — T. Inoue
Flowering habits of Euphorbia fulugens — Y. Inoue

41.4 Tokyo University of Agriculture and Technology, Faculty of Agriculture
elevation: 58 m
rainfall : 1,501 mm
sunshine : 2,185 h

Saiwai-cho, Fuchu-shi, Tokyo 183
Phone: 0423-64-3311
Dean: Prof. M. Matsumoto*

Production of vegetables, tree fruits and ornamental plants. Of 15 ha under University farm, 4 ha are planted with vegetable crops, tree fruits (Japanese persimmon, pear, chestnut and blueberry), grape, kiwifruit and ornamental plants (Camellia).

Laboratory of Horticulture
Farm works of vegetable production — Prof. M. Matsumoto*
Interspecific hybrids of stone fruits, and of pome fruits — Ass. Prof. Dr. I. Shimura*
Taxonomy of genus Camellia — N. Hakoda
University Farm
Production and cultivars of blueberry — S. Ishikawa*
Cultivation of cyclamen in plastic house — S. Takenaga
Nitrogen-cycling in the vegetable field — S. Matsumura

41.5 University of Tokyo, Faculty of Agriculture
elevation: 22 m
rainfall : 1,503 mm
sunshine : 1,972 h

Yayoi, Bunkyo-ku, Tokyo 113
Phone: 03-812-2111
Telex: UNITOKYO J25510
Dean: Dr. S. Konosu

A 4 ha orchard (grape, Japanese pear, peach, satsuma mandarin, persimmon) and a 20 ha experiment field with 1 ha allotted for vegetable crops, are situated at a distance. Center of Environment Regulation System for Biology is open for researches in plant sciences.

Laboratory of Horticulture
Physiological studies on quality of fruits and vegetables — Dr. R. Sakiyama*
Evaluation of plant potassium status — Dr. N. Sugiyama
Quality of vegetables and environment factors — Dr. H. Miura
Experiment Farm
Labor saving in orchard management — Dr. M. Sato
Effect of arsenicals on citrus enzyme and CoA — Y. Yamaki

42 Tottori

42.1 Tottori Prefectural Fruit Tree Experiment Station
elevation: 35 m
rainfall : 2,018 mm
sunshine : 1,761 h

Daiei-cho Tohaku-gun,
Tottori-pref. 689-22
Phone: 0858-37-4151
ir.: H. Udagawa

Japanese pear, Nijisseiki, is the main object for research because the greatest production in Japan is obtained in this prefecture. Five ha are planted with pears, 4 ha with persimmons, grapes, apples and other deciduous fruits.

Fruit Growing Section
Species of deciduous fruit trees — Head: O. Furuta

Growing of deciduous fruit trees	M. Nagara
Physiological disorder	K. Murata
Plant growth regulation	K. Murao
Protection Section	
Insect pest control	Head: Dr. M. Uchida
Virus	H. Watanabe
Soil and Fertilizer Section	
Soil management	Head: M. Uraki
Diagnosis of nutrient condition	Y. Jinno
Technical Section	
Orchard management	Y. Kadowaki
Kawabara Experiment Field	
Growing of persimmons	K. Urushibara
Hojo Experiment Field	
Growing of grapes	A. Tanaka

42.2 Tottori Prefectural Vegetable and Flower Experiment Station
elevation: 50 m
rainfall : 2,017 mm
sunshine : 1,760 h

Daiei-cho, Tohaku-gun,
Tottori-pref. 689-22
Phone: 0858-37-4211
Dir.: Dr. A. Toyama

Covering problems concerning production of vegetables and flowers in Tottori prefecture, with 10 ha including volcanic ash soil and sandy soil under sprinkler irrigation. Watermelon, netted melon, Chinese yam, Baker's garlic, Welish onion, tulip, and gentian are the main crops.

Vegetable Section	
Vegetable cultivars and herbicides tests	Head: Dr. T. Shimizu
Vegetable growing	T. Saito
Flower Section	
Tissue culture and floriculture	Head: S. Ōmura
Protection Section	
Vegetable pathology	Head: T. Yumoto/I. Sako
Entomology (esp. spider mites)	T. Taniguchi
Soil and Fertilizer Section	
Irrigation and soil management	Head: S. Fujii
Fertilizer	M. Shimonaka
Saihaku Branch	
Vegetable cultivars and growing	Head: M. Nakahara
Strawberry	H. Tokuyama
Sweet potato	Y. Noguchi
Nichinan Branch	
Vegetables at high elevation areas	Head: A. Nitta

42.3 Tottori University, Faculty of Agriculture
elevation: 20 m
rainfall : 2,000 mm
sunshine : 2,100 h

Koyama-cho, Tottori-shi,
Tottori-pref. 680
Phone: 0857-28-0321
Dir.: Prof. Dr. H. Kouno

Laboratory of Horticulture - This Laboratory is playing a central role in the physiological investigation on Japanese pear tree and fruit in western Japan. In addition it serves Japanese pear growers in Tottori region for the education of fundamental physiology of pear growth and the guide of growing techniques.

Japanese pear fruit development and chemical control	Prof. Dr. S. Hayashi*
Physiological disorder of Japanese pear	Ass. Prof. Dr. K. Tanabe*
Flower bud formation of Japanese pear	Dr. K. Banno*
Research Section on Agricultural Use of Sand-dune Area	Hamasaka, Tottori-shi,

This research station is the main centre of investigation on agricultural use of sand-dune fields in Japan

Tottori-pref. 680
Phone: 0857-23-3411

Growing of vegetables in sand-dune fields
Growing of vegetables by brackish water irrigation
Growing of vegetables in arid zone

Prof. Dr. I. Sato
Ass. Prof. M. Yamane
Ass. Prof. Dr. Y. Takeuchi
Ass. Prof. Dr. M. Toyama

43 Toyama

43.1 Toyama Prefectural Research Station for Fruit Growing
elevation: 50 m
rainfall : 2,700 mm
sunshine : 1,800 h

Rokuromaru, Uozu-shi,
Toyama-pref. 937
Phone: 0765-22-0185
Dir.: H. Sawabe

The projects are improvement of fruit growing and selection of cultivars favourable to Toyama prefecture. Of 4.5 ha of the land, 3.5 ha are for orchards, 230 m^2 are under glass, and 50 m^2 for temperature treatment cellars.

Physiology of fruit trees
Fruit cultivars in relation to climate
Climate and quality
Physiological disorder
Disease control

Head: H. Sawabe
M. Oshiro
S. Kawasaki
T. Niiyama
M. Takaguchi

43.2 Toyama Prefectural Vegetable and Ornamental Crops Research Station
elevation: 70 m
rainfall : 2,550 mm
sunshine : 2,090 h

Goromaru, Tonami-shi,
Toyama-pref. 939-13
Phone: 0763-32-2259
Dir.: Dr. J. Kawata*

The projects are breeding and disease control of flower bulbs, and improvement of vegetable growing. Of 6 ha of the land, 4 ha are for experimental fields in the open, 0.1 ha are under glass, 300 m^2 are for bulb storage and 100 m^2 are for temperature treatment cellars.

Vegetable Section
Variety testing
Vegetable cultivars in relation to climate
Nutrient and quality
Vegetative growth and flowering

Head: K. Fukushima
I. Okada
M. Matsumoto
M. Kitada

Plant Breeding Section
Breeding of chrysanthemum
Breeding of tulip and lilies

Head: Dr. J. Kawata*
O. Urashima/Y. Umada

Plant Pathology Section
Virology of flower bulbs
Soil borne diseases of flower bulbs

Head: Dr. T. Yamamoto/K. Nahata
H. Mukobata

44 Wakayama

44.1 Wakayama Prefectural Agricultural Experiment Station
elevation: 50 m
rainfall : 1,519 mm
sunshine : 1,701 h

Kishigawa-cho, Naga-gun,
Wakayama-pref. 640-04
Phone: 0736-64-2300
Dir.: M. Yanagisawa

This Station is the center of vegetable and floricultural research in Wakayama. At present, subjects of studies are pea, strawberry, netted-melon, onion, tomato, rose, carnation, stock, baby's breath and gerbera.

Vegetable and Floricultural Section

Head: H. Nanba

Vegetable growing	A. Sada/N. Wasa/T. Fujioka
Floriculture	M. Fujita*/Y. Ueshima/T. Honda
Integrated Technique Section	Head: H. Sakaguchi
Cultivation technique	J. Hirobe/S. Yano/T. Nishitani
Farming	H. Nishimori
Horticultural facilities	H. Shinto
Insect Pest and Diseases Section	Head: R. Isaka
Disease injury	H. Iemura/M. Tamura/H. Yoshimoto S. Kinoshita
Insect pest	K. Azuma/Y. Jhono
Environmental Conservation Section	Head: Z. Ono
Fertilizer application	S. Hirata/T. Fujii
Soil conservation	N. Ueda/T. Morishita/N. Iwahashi
Soil microbiology	M. Kuriyama
Processed foods	J. Fushihara

44.2 Wakayama Prefectural Fruit Tree Experiment Station
elevation: 55 m
rainfall : 1,756 mm
sunshine : 1,761 h

Kibi-cho, Arita-gun,
Wakayama-pref. 643
Phone: 073752-4320
Dir.: B. Yamamura

This Station was founded for the promotion of production and post-harvest physiology of tree fruits in 1910. Citrus trees 6.2 ha, deciduous fruit trees 1.6 ha. Equipment: Sprinkler irrigation system for multi-purpose utilization. Facilities for producing virus-free plants and test plants. Chambers for sulfur dioxide fumigation. Super-intensive growing under solar house systems. Photo-electric sorting system for harvested fruit.

Cultural Practice Section	
Breeding and propagation of citrus	Head: J. Morimoto/H. Nakaya
Physiology of fruit trees	M. Ogawa/E. Nakao
Physiology of citrus	T. Okunishi
Plant Pathology and Entomology Section	
Plant pathology	Head: S. Yamamoto/K. Natsumi K. Shimazu
Entomology	H. Ohhashi/Y. Yukawa
Soil, Fertilizer and Air Pollution Section	
Fertilization and nutrition	Head: H. Hijikata/Y. Nomi
Air pollution	J. Hayashi/K. Iwao/M. Nakayama
Citricultural Facilities Section	
Plant physiology and post-harvest physiology of citrus	Head: Dr. E. Tomita/S. Yamasaki K. Miyamoto/K. Iwamoto/K. Matsumoto
Kihoku Branch	
Plant physiology and post-harvest physiology of deciduous fruit trees	Head: S. Yamashita/Y. Kitano K. Maesaka/K. Fujimoto/K. Kanaoka
Plant pathology and entomology	H. Komatsu
Experiment Farm	
Plant physiology of citrus and Japanese apricot	Head: H. Harano

45 Yamagata

45.1 Yamagata Prefectural Horticultural Experiment Station
elevation: 100 m
rainfall : 1,217 mm
sunshine : 2,330 h

Shima, Sagae-shi,
Yamagata-pref. 991
Phone: 0237-84-4125
Dir.: T. Sanada

Producing and processing of tree fruits, grapes, vegetables and flowers. Of 14 ha under sprinkler irrigation, 9 are planted with tree fruits (apple, cherry, pear, peach and kaki), 1 ha with grapes and 2 ha are planted with vegetables and flowers. Equipment: special greenhouses and pipe frame greenhouses for studies of virus diseases.

Research Institute of Biotechnology	Head: A. Yoshida
Breeding and genetics	Dr. K. Isawa/Y. Ohnuma/E. Ando K. Noguchi/S. Yamaguchi/K. Uchie
Fruit Tree Section	Head: K. Kido
Growing and storage	S. Matsuda/Y. Takahashi/T. Sato Y. Sato/Y. Suzuki
Vegetable and Flower Section	Head: H. Suzuki
Vegetable crops	M. Oyamada/Y. Kuroda/T. Nunomiya K. Abe
Flowers	K. Ono/H. Sato
Environment Section	Head: W. Ueno
Diseases and pests	H. Shoji/T. Tanaka
Soil and fertilizers	K. Yamaguchi/K. Komabayashi

45.2 Yamagata University, Faculty of Agriculture, Department of Horticulture
elevation: 20 m
rainfall : 1,883 mm
sunshine : 1,787 h

Wakaba-cho, Tsuruoka-shi
Yamagata-pref. 997
Phone: 0235-23-1521
Dean: K. Abe

Laboratory of Pomology	
Histology of flowers and fruits of deciduous fruit trees	Prof. Dr. S. Watanabe
Water-relation and energy exchange	Ass. Prof. Dr. T. Yamamoto*
Physiology of deciduous fruit trees	S. Taira
Laboratory of Vegetable Crop Science	
Flowering and fruiting of vegetable crops	Prof. Dr. T. Saito
Growth and development of bulbous plants	Ass. Prof. Dr. H. Takagi
Mechanism of cucumber fruit curvature	K. Kanahama
Laboratory of Breeding and Propagation	
Propagation and breeding through tissue cultures	Prof. Dr. I. Hiura
Effects of environmental factors on fruit setting in vegetable crops	Ass. Prof. Dr. S. Imanishi
Breeding and propagation of plants	H. Suzuki
Laboratory of Post-harvest Horticulture	
Chilling injury	Prof. Dr. T. Fukushima*
Respiration, ethylene evolution and quality during maturation	Ass. Prof. Dr. T. Kitamura
Post-harvest physiology of persimmon fruits	H. Itamura

46 Yamaguchi

46.1 Yamaguchi Prefectural Agricultural Experiment Station
elevation: 33 m
rainfall : 1,973 mm
sunshine : 1,920 h

Ouchi-mihori, Yamaguchi-shi,
Yamaguchi-pref. 747-13
Phone: 0839-27-0211
Dir.: T. Izu

The performance of this Station includes studies for growing methods, selection of varieties and breeding of horticultural crops.

Horticultural Division	Chief: K. Okamura
Vegetable Crop Section	
Vegetable growing	Head: Y. Nishikawa
Strawberry growing and tissue culture for virus free crops	O. Matsumoto

Vegetable growing	S. Fukuda/Y. Yamamoto
Floriculture Section	Head: I. Kimura
Cutflowers growing	T. Nakamura
Carnation growing and breeding	
Deciduous Fruit Tree Section	Head: K. Nagami
Grape growing	M. Tanaka
Pear growing	T. Kubo
Chestnut and peach growing	
Processing Section	Head: Y. Uchiyama*
Processing technology and quality evaluation of field crops	K. Yoshimatsu
Pre and post harvest physiology	
Tokusa High-and Cold-land Experiment Station	Dir.: Dr. K. Sasaki
Vegetable Crop Section	T. Kawamura/Y. Kimura/S. Tone
Hagi Citrus Experiment Station	Dir.: H. Ito
Citrus (*C. natsudaidai*) growing	A. Miyata
Control of citrus insects	T. Kodama
Control of citrus viruses	Dr. A. Nakada
Soil and fertilizers	M. Nakamura
Oshima Citrus Experiment Station	Dir.: I. Honda
Growing Section	Head: K. Hashimoto/H. Tanaka
Citrus growing	
Environment Section	Head: Dr. T. Kato
Control of citrus insects	M. Anraku/K. Shigeta
Control of citrus diseases	T. Ota
Soil and fertilizers	

46.2 Yamaguchi University, Faculty of Agriculture
elevation: 16 m
rainfall : 1,815 mm
sunshine : 2,116 h

Yoshida, Yamaguchi-shi
Yamaguchi-pref. 753
Phone: 08392-22-6111
Dean: Prof. Dr. A. Sato

Laboratory of Horticulture	
Dormancy of horticultural plants	Prof. Dr. S. Nakamura
Laboratory of Agricultural Environment	
Environmental controlling technique for plant growth	Prof. Dr. Y. Suzuki
Agricultural meteorology	Ass. Prof. Dr. M. Aoki
University Farm	
Soil management in orchard	Prof. Dr. T. Tokimoto

47 Yamanashi

47.1 Yamanashi Prefectural Agricultural Experiment Station
elevation: 317 m
rainfall : 1,092 mm
sunshine : 2,197 h

Futaba-cho, Kitakoma-gun,
Yamanashi-pref. 407-01
Phone: 055128-2496
Dir.: S. Shimizu

Cultivation Research Section	
Vegetables	Head: Dr. S. Takayama
Tissue culture	K. Kinoshita
Growing	Y. Iwade
Soil and fertilizer	
Flowers	Head: A. Niitsu/T. Komori
Growing	K. Amemiya
Tissue culture	Head: S. Itayama
Yatsugatake Branch	

Vegetables	Head: S. Shimizu/S. Kuzukubo
	M. Kobayashi
Flowers	Head: K. Shimizu/M. Shindo
Gakuroku Branch	
Vegetables	Head: Y. Watanabe
Flowers	T. Nishino

47.2 Yamanashi Prefectural Fruit Tree Experiment Station
elevation: 272 m
rainfall : 1,207 mm
sunshine : 2,246 h

Manriki, Yamanashi-shi,
Yamanashi-pref. 405
Phone: 05532-2-1921
Dir.: R. Yano

Production and breeding of tree fruits and grapes.
Eight ha are planted with tree fruits and grapes.

Grape Growing Section	
Growing	Head: A. Harada/M. Aoki/T. Sakurai
	K. Takei
Tissue culture	T. Satō
Tree Fruit Growing Section	
Peaches	Head: T. Yamada/T. Tsuruta
Plums and cherries	Y. Koyaizu/H. Endō
Breeding Section (grapes)	
Breeding and genetics	Head: T. Amemiya/T. Ozawa
	T. Mochizuki
Plant Disease and Insect Section	
Plant insect pest	Head: T. Tsuchiya
Plant diseases	Y. Terai/T. Nishizima
Soil and Fertilizing Section	
Soil and fertilizing	Head: T. Hara/S. Kubokawa/S. Furuya

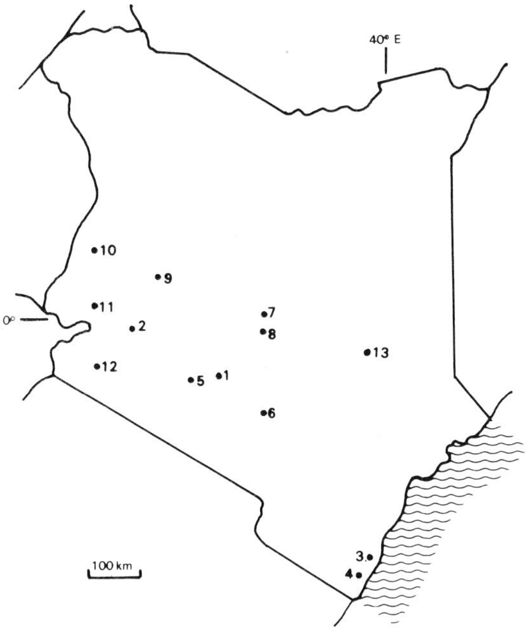
40° E
0°
100 km
•10
•9
•11
•2
•7
•8
•13
•12
•5
•1
•6
3.
4.

1 Thika
2 Molo
3 Mtwapa
4 Matuga
5 Tigoni
6 Katumani
7 Embu
8 Mweya
9 Perkerra
10 Kitale
11 Kakamega
12 Kisii
13 Garissa

Survey

Horticultural research in Kenya is conducted as a section of the Scientific Research Division of the Ministry of Agriculture. It has its own Research Stations as shown on the map, but certain aspects such as soil chemistry, pathology and entomology are dealt with in conjunction with the National Agricultural Laboratories in Nairobi. The headquarters of horticultural research is at the National Horticultural Research Station in Thika. Elevation is 164 m above sea level. The mean annual rainfall is 1100 mm distributed in April-June and October November seasons, the rest of the year being dry and hot.
This station was started in 1957 primarily for pineapple research work, but now deals with a wide range of horticultural crops, including grain legume, mulberry for silk production, sunflower for oil extraction and hard fibres and Kenaf. Also there is training extension. In addition there are three main regional stations situated in different ecological zones of Kenya where various crops are under agronomic experimentation as well as varietal breeding and testing.
The main work at the National Horticultural Research Station in Molo, which lies 2750 m above sea level, is on deciduous fruit trees and cool climate vegetable crops.
At Mtwapa and Matuga, only a few metres above sea level, the work is on the agronomy of coastal crops, mainly coconuts, cashewnuts, mangoes, citrus and vegetables. At Tigoni the Kenya National Potato Research Project has been in existence since July 1970 and is a disease-free centre for the multiplication of healthy stocks of promising cultivars produced through breeding and selection at the nearby National Agricultural Laboratories.
The emphasis has been on breeding for resistance to aphid-borne viruses, bacterial wilt and late blight. Recently floriculture was established.

In addition to these stations a certain amount of field experimentation is done on Regional Agricultural Research Stations at Katumani, Embu, Mwea, Perkerra, Kitale, Kakamega, Kisii and Garissa. The programme at these sites is prepared by the staff at Thika.

1. Thika

1.1 National Horticultural Research Station
elevation: 164 m

P.O. Box 220, Thika
Phone: 21281/2/3/4/5

Prepared by Dr.S.K Njugunah of Nat. Hort. Res. Station, Thika

rainfall : 1100 mm

Horticultural crops, grain legume, mulberries, sunflowers, hard fibres and training extension.

Dir.: S.K. Njugunah, B.Sc., M.Sc. Ph.D.

1.2 Vegetable Production and Vegetable Seed Production

Exotic vegetables; introduction and agronomy
Quality of vegetables and fruits, Food technology

Head: G.G. Madumadu, B.Sc., M.Sc.
D. Mwamba, B.Sc., M.Sc.
E.W. Wambugu, B.Sc., M.Sc.
A.O. Okongo, B.Sc., M.Sc.
E.O. Balah, B.Sc., M.Sc.
E.B. Mwajumwa, B.Sc., M.Sc.

1.3 Fruits and Nut Trees Agronomy
Plant Introduction, tissue culture, gene bank and training banana agronomy
Cultural practices for passion fruit, agronomy of avocado
Agronomy and propagation of macadamia

Agronomy of pineapples
Vegetative propagation of tree crops

Head: C.N. Gathungu, B.Sc, M.Sc.
S.P. Gachanja, B.Sc., M.Sc.
Dr. Z. Worku, B.Sc., M.Sc., Ph.D.
J.M. Mumo, Dip. Hort.
E.N. Khaemba, Dip. Hort.
S. Hirama Iwasaki
J.K. Ruto, Dip. Hort.
Kagiri, Dip. Food Tech.
B. Chege, B.Sc. Hort.
M. Wabule, B.Sc., M.Sc.
B.K. Changwony, Dip. Agric.
W.O. Odenyo, B.Sc. Hort.

1.4 Horticultural Crops Protection
Studies on nematodes of tomatoes onion aphids, thrips

Head: S.T. Kanyagia, B.Sc., M.Sc.
A.A. Seif, B.Sc., M.Sc.
S.J.N Muriuki, B.Sc.
E. Otieno, B.Sc./P. Mwathi, B.Sc.
J.M. Onsando, B.Sc., M.Sc.

1.5 Research Liaison and Extension

Head: M.M. Waiganjo, B.Sc.

1.6 Rice Research

S.O. Sikinyi, B.Sc.

1.7 Mulberrey Agronomy
Cacoon breeding and rearing

Head: J. Bore, Dip. Agric.
M. Kariuki, Dip. Agric.
E.N. Ndoria, B.Sc.
J. Wainaina, Dip. Agric.
D. Menye/J. Onyango, Dip Agric

1.8 Grain Legume Research Project

Bean pathology

S.G.S. Muigai, B.Sc.
A.M.M. Ndegwa, B.Sc.
J. Gathee B.Sc./G. Mbugua, B.Sc.
A. Okoko, B.Sc./J. Muthamia, B.Sc.
G.K. Kinyua, B.Sc., M.Sc.
M. Omunyin, B.Sc., M.Sc.

1.9 Flower Propagation and Agronomy

E.W. Macharia (mrs) B.Sc.

2. Molo

2.1 National Pyrethrum and Horticultural Research Station
elevation : 2750 m
Deciduous fruit trees and high altitude vegetable crops production

P.O. Box 100, Molo
Phone: 41
Off. in Ch.: J.O. Owuor, M.Sc.
E. Kamau (Miss) B.Sc. Agric.
K. Njenga Dip. Hort.

3. Mtwapa

3.1 Coasta Agricultural Research Station	P.O. Kikambala Phone: 485526 Mombasa Off. in Ch.: A.S. Aziz, B.Sc., M.Sc.
Varietal and agronomic research	M. Gethi, B.Sc. K. Mwangi, B.Sc.
Germ plasm	C.B. Chesoli S. Kioko, Dip. Agric.
Vegetable crops for Coast research	A.S. Aziz, B.Sc., M.Sc.

4. Matuga

P.O. Box 96246 Likoni-Mombasa
Phone: 451430 Mombasa
Off. in Ch.: Z.R. Muli, B.Sc. M.Sc.

4.1 Physiological Studies on Flowering of Citrus	P. Kiuru

5. Tigoni

P.O. Box 338, Limuru
Teleph.: 358, 641

5.1 Potato Research Station	Off. in Ch.: I.N. Njoroge, B.Sc., M.Sc.
Potato agronomy	D.N. Njenga, Dip. Agric. C.K. Kamau, B.Sc. Agric. Mrs. L.N. Wambugu, B.Sc. Agric. Miss B.W. Kingori, B.Sc. Agric. P.K.A. Macharia, Dip. Agric.
Potato diseases clonal multiplication	Susan N. Munene, B.Sc., M.Sc. F.M. Wambugu, Dip. Agric. Andrew O. Michieka, M.Sc.
Potato virology	J. Njoroge, B.Sc
Potato certification	D.N. Munene, Dip. Agric.
Potato extension	F.N. Gitungo, B.Sc.
Food technology	N. Kabira, B.Sc.

6. Katumani

Dry Beans

7. Embu

Regional Agricultural Research Station	D. Kitivo, Dip. Agric.

8. Mwea

Regional Agricultural Research Station

9. Perkerra

Phone: 32 Malingat
Off. in Ch.: D. Michieka

10. Kitale

Regional Agricultural Research Station

11. Kakamega

Regional Agricultural Research Station	J.K. Ruto, M.Sc.

12. Kisii

Regional Agricultural Research Station	H.M. Wakhonya, Dip. Agric.

13. Garissa

Garissa Research Station	Off. in Ch.: W.N.P. Mwasya Phone: 100 Garissa

KOREA

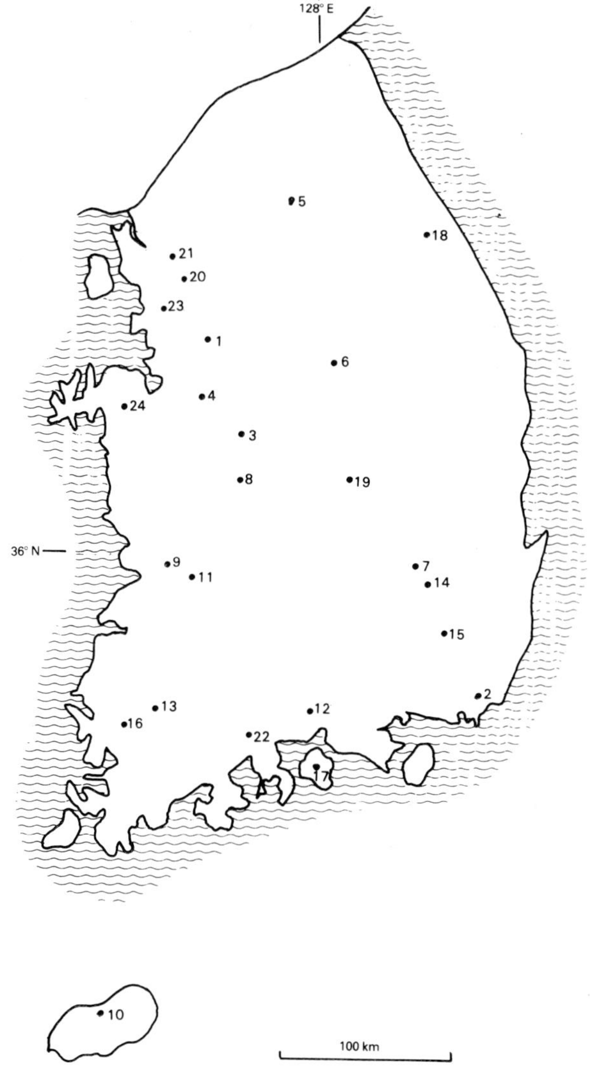
128° E
36° N
5
18
21
20
23
1
6
24
4
3
8
19
9
11
7
14
15
2
13
12
16
22
17
10
100 km

1 Anseong
2 Busan
3 Cheongju
4 Cheonweon
5 Chuncheon
6 Chungju
7 Daegu
8 Daejeon
9 Iri
10 Jeju
11 Jeonju
12 Jinju
13 Kwangju
14 Kyungsan
15 Milyang
16 Naju
17 Namhae
18 Pyungchang
19 Sangju
20 Seongnam
21 Seoul
22 Suncheon
23 Suweon
24 Yesan

Survey

Korea has hilly and mountainous topography and annual precipitation ranging from 900 to 1,300 mm. Summer weather is humid and hot with around 30°C of mean maximum temperature, while winter is severely cold with mean minimum temperature of 5 -7°C in January. It is quite dry in spring and autumn.
Production of Kimchi (salted and fermented leafy and root vegetables with spices) material is the first priority of the vegetable industry. Thus out of 371,000 ha of vegetable production area, 64% or more is covered by Chinese cabbage, radish, hot pepper and garlic that are the most essential items for Kimchi. White potato, cucurbits and onion follow. Recently some exotic western vegetables and processing vegetables began to increase according to the improvement of living standards and the change of food habit.
Glasshouses, polyethylene (PE) covered houses and PE tunnels have rapidly increased to cover 23,000 ha in 1983. Such structures are utilized for supplying various vegetables in cold winter and early spring seasons.
Fruit trees are grown in an acreage of 101,000 ha. In terms of acreage, as well as production amount, apples and mandarin oranges are the most important. Grape, pear, peach and persimmon is the next impor-

Prepared by Dr. Hyo Guen Park, Department of Horticultural Science, Suweon

tant group. Oranges are almost confined to the southern Jeju island which is believed to be the northern most margin of the crop production. Recently processing grapes are rapidly increasing, supported by the strong policy of the government.
Flower and ornamental plant production have been steadily increasing in the last years and the trend is striking since 1981. Acreage of 1983 was 1,892 ha and it is equivalent to 153% of that of 1981. Such increase is closely related to the city decoration prior to the Seoul Olympic Games scheduled in 1988.
Governmental Research Institutions - The Office of Rural Development (ORD) has the responsability of research and extension to support the Ministry of Agriculture and Fishery in agricultural policy and its implementation. ORD has 14 research institutions; 9 of them in different disciplines and the rest for the regional researches.
The Horticultural Experiment Station (HES) is the prime institute in this field. HES was established in 1949 and the main office is now located in Suweon. It has two branch stations, one in Busan mainly for protected cultivation of vegetables and flowers, the other in Naju mainly for persimmon and pear research. The station also has a separate research farm in one of the southern islands, Namhae, for white potato seed production and newly introduced warm-season crops. Jeju Experiment Station and Alpine Experiment Station are partly engaged in the horticultural research, mainly for citrus and white potato, respectively. At provincial level, each of the nine Provincial Offices of Rural Development (PORD) have a horticultural division or section to deal with and to solve the regional problems in horticultural production of each province.
Research in HES is well balanced between half fundamental and half applied fields, while the horticultural part of the PORDs are devoted mostly to the applied field. Major results are published in two different ways: progress reports of ORD (Korean and English) and horticultural research reports.
Universities and Colleges - Education and fundamental research are the responsibility of universities and colleges. Before World War II, the Suweon Agricultural Junior College and Sungsil Junior College (Department of Agriculture) were the only institutions for the advanced education in the field of agriculture. Both colleges opened lectures on olericulture and pomology.
Presently 20 Universities or Colleges and 8 Junior Colleges have horticultural departments and most of the 20 institutions opened graduate courses for the degree of doctrate and/or master majoring in horticultural science.
The Korean Society for Horticultural Science - The Korean Society for Horticultural Science was established in 1963. The address of its office is: Department of Horticulture, College of Agriculture, Seoul National University, Seodun-dong, Suweon 170, Korea. Dr. Young R. Kim is presently serving as President. The society has about 350 members and it publishes the Journal of the Korean Society for Horticultural Science; one volume a year with 500 pages of 4 serial numbers.
Outline of Horticultural Research Vegetables - Self-incompatibility systems and male-sterility have long been a major interest, aiming at the development and utilization of hybrid cultivars in major vegetable crops as radish, Chinese cabbage, red pepper and onions. Technology of off-season production under protected cultivation and by PE mulching is another important field of research. New researches are partly moving their concern to the labour saving technologies, quality improvement and processing of vegetables.
Fruit trees - Besides orchard management techniques including training, pruning and fertilization, mechanization and utilization of chemicals for weed control and thinning are important problems to be studied. Dwarf root stock utilization and its rapid propagation methods are also widely and deeply studied. Extension of harvest period through forcing culture in PE houses has recently been stepped up. Cold hardiness of various fruit trees are presently emphasized. Cultivar selection and varietal breeding is mainly aimed at the high quality, long shelf and storage life and stres and disease resistance.
Flower, ornamentals and turf - Breeding Hibiscus, the national flower, for aphid and cold tolerance is a priority concern. Tissue culture for virus free stock production and rapid propagation are the most popular subjects. Evaluation of Korean native turfs as lawn grass have been a major research subject for last 15 years.

White potato - Clean seed potato production system through apical meristem culture is already established. Early maturity and high yields are major goals in both breeding systems via true seed and seed tuber. Establishment of a fall growing system for table use and introduction of processing cultivars are relatively new subjects.

1 Anseong

1.1 Anseong Agricultural Junior College,
Department of Horticulture
elevation: 24 m

67 Seokjeong-dong, Anseong-eup,
Anseong-gun, Kyunggi-do
Phone: 0334-2703

rainfall : 1104 mm
sunshine : 2558 h

Chairman: Prof. K.B. Yu

Dissease control of fruit trees — Prof. K.H. Park/ J.H. Oh
Ecology of Dodder — Prof.Dr. Y.M. Lee
Vegetative propagation of ornamentals — Assoc.Prof.Dr. M.W. Lee
Plant-water relationship in winter cereals — Ass.Prof. I.Y. Shin
Root growth in radish — Ass.Prof. Y.H. Kim

2 Busan

2.1 Busan Branch Station,
Horticultural Experiment Station
elevation: 69 m
rainfall : 1466 mm
sunshine : 2242 h

Daesa 20 Gangdong-dong, Buk-gu,
Busan
Phone: 051-98-2181-3
Dir.: C.D. Ban

2.1.1 Laboratory of breeding and protection for under-structure — Chief: Sr. Researcher K.Y. Kang
Tomato breeding and *solanaceae* disease — K.Y. Kang/ I.W. Cho
Strawberry breeding — I.C. Yu
Onion breeding — J.K. Suh
2.1.2 Laboratory of cultivation and physiology for under-structure horticultural crops — Chief: Sr. Researcher J.S. Choe
Red pepper physiology — J.S. Choe
Physiology of *solanaceae* vegetables — J.K. An
Melon physiology — H.T. Kim
Environment control in greenhouse cultivation — J.S. Jung
2.1.3 Laboratory of fruit trees and ornamentals — Chief: Sr. Researcher C.J. Yoon
Persimmon cultivation — C.J. Yoon/ M.J. Kim
Flower cultivation in greenhouse — D.W. Lee

2.2 Donga University, College of Agriculture,
Department of Horticulture

749 Goijeong-1-dong, Seo-gu
Busan
Phone: 051-29-6115-8
Chairman: Prof.Dr. S.S. Kwon

3 Cheongju

3.1 Chungbuk Provincial Office of Rural Development,
Research Bureau
elevation: 59 m
rainfall : 1219 mm
sunshine : 2152 h

262 Bokdae-dong, Cheongju
Chungbuk-do
Phone: 0341-4-4101
Dir.: KC. Kwon

Crops Division — Chief: Dr. J.T. Cho
Horticulture Section — G.S. Hong/ Y.J. Song/ K.I. Yoon
Domestication of wild and semi-wild vegetables
Regional problems in horticultural crop production

3.2 Chungbuk National University,
College of Agriculture,
Department of Horticulture

San 5, Gaesin-dong, Cheongju,
Chungbuk-do, Phone: 0341-2-6201-4
Chairman: Ass.Prof.Dr. J.H. Kim

Grape breeding and germplasm evaluation — Assoc.Prof.Dr. S.K. Kim
Photosynthesis and translocation of vegetable crops — Ass.Prof.Dr. J.H. Kim
Anther culture of horticultural crops — Ass.Prof.Dr. J.K. Hwang
In-vitro propagation and breeding of ornamentals — Ass.Prof.Dr. K.Y. Paek

4 Cheonweon

4.1 Yeonam Junior College of Livestock and Horticulture, Department of Horticulture
elevation: 60 m
rainfall : 1162 mm
sunshine : 2796 h

Seonghwan-eup, Cheonweon, Chungnam-do
Phone: 0417-4-3151
Chairman: Prof. K.Y. Kim

Greenhouse cultivation technology of vegetables	Prof. H.J. Yung
Seed germination physiology in radish	Assoc.Prof. B.W. Kim
Tissue culture of ornamentals	Assoc.Prof. E.Y. Kim

5 Chuncheon

5.1 Kangwon Provincial Office of Rural Development, Research Bureau, Crops Division, Horticulture Section
elevation: 74 m
rainfall : 1286 mm
sunshine : 2103 h

402 Udu-dong, Chuncheon, Kangwon-do
Phone: 0361-2-7901-4
Chief: K.K. Lee

Garlic dormancy and ecological classification	K.K. Lee
Domestication of *Codonopsis*	J.H. No
Research on wild garlics	W.B. Kim

5.2 Kangwon National University, College of Agriculture, Department of Horticulture

Hyoja-2-dong, Chuncheon, Kangwon-do
Phone: 0361-2-6011-5
Chairman: Assoc.Prof.Dr. K.E. Lee

Mite control in fruit trees	Prof.Dr. K.P. Han
Insect physiology and control	Dr. O. Bahn
Reproductive physiology of crucifers	Assoc.Prof.Dr. K.C. Yoo
Breeding of peas	Dr. Y.N. Song
Utilization of native plant resources for landscaping	Dr. K.E. Lee

6 Chungju

6.1 Konkuk University-Chungju, College of Natural Sciences, Department of Horticulture
elevation: 70 m
rainfall : 1112 mm
sunshine : 2830 h

Danwol-dong, Chungju, Chungbuk-do
Phone: 0441-3-1881-3
Chairman: Assoc.Prof.Dr. J.Y. Kim

Flowering physiology of vegetables	Prof.Dr. J.Y. Kim
Micro-nutrients in fruit tree growing	Ass.Prof.Dr. Y.J. Yim
Floriculture and landscaping	Dr. B.H. Bae

7 Daegu

7.1 Kyungbuk Provincial Office of Rural Development, Research Bureau, Division of Economical Crops
elevation: 58 m
rainfall : 1005 mm
sunshine : 2368 h

200 Dongho-dong, Buk-gu, Daegu
Phone: 053-30-0631-4
Head: S.B. Lee

Cropping system with garlic and red pepper	S.B. Lee/ J.H. Lim

Apple cultivation and pest control	D.M. Park/ J.S. Kim/ T.M. Yoon
Grape production in PE houses	J.Y. Oh
7.2 Kyungbuk National University, College of Agriculture, Department of Horticulture	1470 Sankyuk-dong, Buk-gu, Daegu Phone: 053-94-5001-45 Chairman: Prof.Dr. W.S. Lee
Eco-classification and chemical assessment of Korean garlic	Prof.Dr. J.K. Joen Assoc.Prof.Dr. Y.G. Suh
Floriculture	Prof.Dr. J.K. Joen Assoc.Prof.Dr. Y.G. Suh
In-vitro propagation of oriental orchids	Ass.Prof.Dr. J.D. Jung
Fertilization in apple tree growing and seed dormancy	Prof.Dr. K.R. Kim Assoc.Prof.Dr. S.T. Jung
Red pepper breeding for resistance to *Phytophthora*	Ass.Prof.Dr. B.S. Kim
7.3 Hyoseong Women's University, Department of Horticulture	1155 Bongdeok-dong, Nam-gu, Daegu Phone: 053-66-2871-5 Chairman: Prof.Dr. S.D. Kim
Sulfuric compounds in *Allium* spp.	Prof.Dr. S.D. Kim
Karyo-classification of *Allium* spp.	Prof.Dr. S.J. Hahn
Genetics and breeding of some ornamentals	Prof.Dr. C.K. Sang
Orchard management in relation with Japanese apple canker	Assoc.Prof.Dr. J.S. Lee
Stabilization of fruit setting in grape	Ass.Prof.Dr. Y.S. Yu
7.4 Kyemyung Junior College, Department of Horticulture	2139 Daemyung-dong, Daegu Phone: 053-67-1321-30 Chairman: Ass.Prof. S.W. Kim
Radish breeding	Ass.Prof. S.W. Kim
Survey on native turf-grasses	Assoc.Prof. B.H. Do
Weed control by chemicals	Ass.Prof. I.K. Kim
Rust control by apple fruits	Y.W. Kim

8 Daejeon

8.1 Chungnam Provincial Office of Rural Developments, Research Bureau, Crops Division, Horticulture Section elevation: 77 m rainfall : 1369 mm sunshine : 2166 h	183 Sangdae-dong, Jung-gu, Daejeon, Chungnam-do Phone: 042-822-0151-5 Chief: D.G. Shin
Regional problems in horticultural crop production	S.W. Ra/ E.R. Im/ B.W. Shin
8.2 Chungnam National University, College of Agriculture, Department of Horticulture	220 Gung-dong, Jung-gu, Daejeon, Chungnam-do Phone: 042-822-0101-9 Chairman: Assoc.Prof.Dr. J.Y. Pyon
Protoplast culture in cruciferous crops	Prof.Dr. Y.R. Kim
Plant growth regulators related to fruit tree physiology	Prof.Dr. J.C. Lee
Action mechanism of herbicides	Prof.Dr. J.Y. Pyon
Post-harvest physiology of cut flowers	Assoc.Prof.Dr. J.S. Lee

Plant tissue culture	Assoc.Prof. Y.B. Lee
Pollution effect on garden plants	Ass.Prof.Dr. J.H. Ku

8.3 Paichai College, Department of Horticulture Science
439-6 Doma-dong, Jung-gu, Daejeon, Chungnam-do
Chairman: Ass.Prof. H.J. Chung

Propagation of ornamental plants	Ass.Prof. H.J. Chung
Design of garden and urban open space	Ass.Prof.Dr. Y.C. Shin
Vegetable crops nutrition	Ass.Prof. J.S. Shim

8.4 Daejeon Junior College, Department of Horticulture Daejeon, Chungnam-do
226-2 Jayang-dong, Dong-gu,
Chairman: Assoc.Prof. J.H. Kim

Agricultural economics	Prof.Dr. J.S. Park
Physiology of fruit trees	Assoc.Prof. J.H. Kim
Floriculture	Assoc.Prof. K.H. Han
Cultivation and physiology	Ass.Prof. S.S. Huh
Vegetable crops	Ass.Prof. U.D. Shin

9 Iri

9.1 Honam Crop Experiment Station, Laboratory of Economic Crops
elevation: 8 m
rainfall : 1246 mm
sunshine : 2715 h

381 Songka-dong, Iri, Jeonbuk-do
Phone: 0653-2-8361-5
Chief: Res. S.J. Yu/ S.K. Suh

Cultivar trials and paddy growing of oilseed rape, onion, and garlic for relay cropping with rice	D.S. Pak/ G.Y. Lee/ S.J. Seon

9.2 Jeonbuk Provincial Office of Rural Development, Research Bureau, Crops Division, Horticulture Section
435-2 Donsan-dong, Iri, Jeonbuk-do
Phone: 0653-2-2307-10
Chief: D.S. Chong/ C.H. Cho
K.D. Choi/ H.C. Lim

Cultivation techniques of ginger and related research
Regional problems in horticultural crop production

10 Jeju

10.1 Jeju Experiment Station, Department of Horticulture
elevation: 22 m
rainfall : 1440 mm
sunshine : 1949 h

1696 Odeung-dong, Jeju, Jeju-do
Phone: 0641-2-2501-4
Head: Sr.Res. D.Y. Moon/ S.H. Kim

Physiology of citrus fruit trees	D.Y. Moon/ S.H. Kim
Breeding of citrus fruit trees	H.Y. Kim/ K.S. Kim
Disease and insect pests in citrus fruit trees	H.M. Kwon
Nutrition and cold hardiness of citrus fruit trees	J.K. Jung/ K.D. Ko

10.2 Jeju Provincial Office of Rural Development, Research Division, Horticulture Section
Yeon-dong, Jeju, Jeju-do
Phone: 0641-2-2521-3
Chief: K.H. Kim

Native ornamentals, collection and domestication	K.H. Kim
Tissue culture with orchids	Jr. Researcher: S.K. Kang
Offseason cultivation technology of garlic, onion, and potato	C.M. Kim/ W.T. Han

Breeding of citrus fruit trees — Y.H. Kim

10.3 Jeju National University,
College of Agriculture,
Department of Horticulture

1 Ara-dong, Jeju, Jeju-do
Phone: 0641-3-2451-5
Chairman: Assoc.Prof.Dr. D.K. Moon

Photosynthesis in horticultural crops	Prof. D.K. Moon
Quality improvement of citrus fruit through cultivation techniques	Prof. H.R. Han
Introduction of tropical fruit trees and related physiology	Assoc.Prof. J.H. Baerk
Garlic ecology and bulbing physiology	J.I. Chang/ Ass.Prof. Y.B. Park
Ecology and in-vitro propagation of orchid landraces	Assoc.Prof.Dr. J.S. Lee Lecturer Dr. I.S. So

11 Jeonju

11.1 Jeonbuk National University, College of Agriculture, Department of Horticulture
elevation: 51 m
rainfall : 1290 mm
sunshine : 2166 h

664-14, Deokjin-dong, Jeonju
Jeonbuk-do
Phone: 0652-3-0031-9
Chairman: Assoc.Prof.Dr. S.D. Oh

Endogenous factors affecting flower bud differentiation in apple tree	Assoc.Prof.Dr. S.D. Oh
Tissue culture with horticultural crops	Prof.Dr. B.K. Lee
Protoplast manipulation and plant cell culture	Assoc.Prof.Dr. J.S. Eun Ass.Prof. H.B. Park
Plant growth regulators	Dr. J.C. Kim

12 Jinju

12.1 Gyungnam Provincial Office of Rural Development, Research Bureau, Crops Division, Horticulture Section
elevation: 22 m
rainfall : 1572 mm
sunshine : 2356 h

1085-1, Chojeon-dong, Jinju,
Gyungnam-do
Phone: 0591-52-4101-3
Chief: K.S. Lee/ K.Y. Hahn

Research Bureau, Crops Division, Horticulture Section
elevation: 22 m
rainfall : 1572 mm
sunshine : 2356 h

Gyungnam-do
Phone: 0591-52-4101-3
Chief: K.S. Lee/ K.Y. Hahn

Regional problems in horticultural crop production

12.2 Gyeongsang National University,
College of Agriculture,
Department of Horticulture

900 Gajoa-dong, Jinju, Gyungnam-do
Phone: 0591-52-4331-5
Chairman: Prof.Dr. J.C. Park

Under-structure cultivation of vegetable crops	Prof.Dr. J.C. Park
Interspecific hybridization of citrus trees	Prof.Dr. H. Kang
Effect of phytohormons on bulb propagation	Prof.Dr. S.K. Um
Nitrogen metabolism in apple tree	Ass.Prof.Dr. S.M. Kang
Dormancy of white potato seed tuber	Prof.Dr. J.L. Cho

12.3 Jinju Junior College of Agriculture and Forestry,

150 Chilam-dong, Jinju, Gyungnam-do

Department of Horticulture

Phone: 0591-52-2378
Chairman: Prof. M.S. Jeong

Flower cultivation	Prof. M.S. Jeong
Growth regulation in grape	K.H. Bae
Temperature effect on mushroom cultivation	B.K. Kang
Flowering and bulbing physiology of onion	H.Y. Kang
Citrus cultivation and propagation	S.H. Park
Factors affecting onion yield	Assoc.Prof. P.S. Park
Critical points of non-heating cultivation in PE houses	Ass.Prof. H.J. Kang

13 Kwangju

13.1 Jeonnam Provincial Office of Rural Development, Research Bureau, Crops Division, Horticulture Section
elevation: 71 m
rainfall : 1316 mm
sunshine : 2284 h

250 Nongseong-dong, Seo-gu,
Kwangju, Jeonnam-do
Phone: 062-33-0151-3
Chief: K.P. Han/ S.K. Choi
T.D. Park

Breeding of watermelon, and regional problems in horticultural crop production

13.2 Jeonnam National University,
College of Agriculture,
Department of Horticulture

Yongbong-dong, Seo-gu, Kwangju,
Jeonnam-do
Phone: 062-55-0011-8
Chairman: Prof.Dr. Heung S. Park*

Growth and development of fruit trees	Prof.Dr. Heung S. Park*
Growth analysis and yield component study	Prof.Dr. Hwa S. Park
Domestication of horticultural plant resources	Assoc.Prof.Dr. C.S. Ahn*
Somatic hybridization of horticultural crops	Ass.Prof. K.S. Kim
Effect of in-ground environment on vegetable growing	Prof.Dr. G.C. Chung
Improvement of photo-environment	S.J. Chung

14 Kyungsan

14.1 Yeongnam University, College of Agriculture and Animal Sciences, Department of Horticulture
elevation: 60 m
rainfall : 1014 mm
sunshine : 3033 h

214-1 Dae-dong, Kyungsan-eup,
Kyungsan-gun, Kyungbuk-do
Phone: 053-8-2331-7
Chairman: Assoc.Prof.Dr. C.U Lee

Disease control of fruit trees	Prof.Dr. C.U. Lee
Growth regulators on apple tree	Prof.Dr. J.K. Byun
Physiology of vegetable crops	Prof.Dr. H.D. Chung*
Tissue culture with ornamental crops	Ass.Prof. K.W. Kim

15 Milyang

15.1 Yeongnam Crop Experiment Station,
Laboratory of Economics Crops
elevation: 13 m
rainfall : 1311 mm
sunshine : 2695 h

1083 Naei-dong, Milyang-eup,
Gyungnam-do
Phone: 0527-2-3051-4

Recently established

16 Naju

16.1 Horticultural Experiment Station, Naju Branch Station
elevation: 20 m
rainfall : 1492 mm
sunshine : 2562 h

1034 Godong-ri, Geumcheon-myeon, Naju, Jeonnam-do
Phone: 0613-2-7278-9

Breeding of Jujube

Dir.: Dr. Y.S. Kim

16.1.1 Laboratory of fruit tree breeding — Chief: W.C. Kim/ W.S. Kim
Pear and jujube breeding and genetic studies

16.1.2 Laboratory of fruit tree physiology — Chief: W.J Lee/ D.S. Son/ J.B. Kim
High-density orchard techniques for persimmon
Under-structure cultivation of grape, pear and jujube
Quality improvement of stored pears

16.1.3 Laboratory of fruit tree protection — Chief: K.H. Hong/ S.B. Jeoung

17 Namhae

17.1 Horticultural Experiment Station, Namhae Research Farm
elevation: 15 m
rainfall : 1772 mm
sunshine : 2717 h

777 Dajeong-ri, Idong-myeon, Namhae-gun, Gyungnam-do
Phone: 0594-2-2357
Dir.: H.Y. Kim/ Y.H. Choi
H.S. Hwang

Cultivation technology of kiwi fruit tree
Survey on the native citrus
White potato seed production and cropping system

18 Pyungchang

18.1 Alpine Experiment Station, Department of Horticulture
elevation: 820 m
rainfall : 1474 mm
sunshine : 2752 h

San 1 Hoengye-ri, Doam-myeon, Pyungchang, Kangwon-do
Phone: 0391-2-4538
Dir.: J.K. Kim

18.1.1 Laboratory of breeding — Chief: Dr. Y.C. Kim/ K.S. Kim
H.Y. Joung/ H.J. Kim

White potato breeding and in-vitro tuberization
Anther and apical meristem culture

18.1.2 Laboratory of pathology — Chief: D.K. Lee/ Y.I. Hahn
S.R. Cheong

Diseases of white potato

18.1.3 Laboratory of cultivation — Chief: G.Y. Shin/ S.I. Kim/ S.Y. Yu
B.D. Kim/ S.J. Hwang/ C.S. Park

Virus free seed potato production
Cultivation technology of potato
Alpine production of vegetable crops — H.K. Chee

19 Sangju

19.1 Sangju Junior College of Agriculture, Department of Horticulture
elevation: 57 m
rainfall : 1116 mm
sunshine : 2519 h

z386 Gajang-ri, Sangju-eup, Gyungbuk-do
Phone: 0582-3-3516-9
Chairman: Prof. M.K. Pak

Control of downy mildew in cucumber	Prof. M.K. Pak
Factors affecting bacterial wilt in cucurbits	S.J. Pak
Rapid propagation of pot ornamentals	S.W. Kang
Survey on Korean persimmon trees	Assoc.Prof. B.I. Pak

20 Seongnam

20.1 Shingu Junior College, Department of Horticulture elevation: 70 m rainfall : 1330 mm sunshine : 2850 h	191 Dandae-dong, Seongnam, Gyunggi-do Phone: 0342-2-4201 Chairman: Prof. S.B. Lee
Rooting of woody ornamentals in pot growing	Prof. S.B. Lee
Physiology and ecology of woody plants used for landscaping	Assoc.Prof. H.B. Leem
Quality improvement of apple	Ass.Prof. K.S. Hwang
Consumption trend of flowers and ornamental plants	J.G. Kim

21 Seoul

21.1 Seoul Municipal College, Department of Environment and Horticulture elevation: 85 m rainfall : 1365 mm sunshine : 2093 h	8-3 Jeonnong-dong, Dongdaemun-gu, Seoul Phone: 02-245-8111-5 Chairman: Prof.Dr. D.W. Han
Environmental control methods in vegetable production under structure	Prof.Dr. D.W. Han
Seedborne disease in vegetable crops	Prof.Dr. D.H. Lee
Somatic hybridization in *solanaceae* vegetables	Ass.Prof.Dr. J.B. Park
21.2 Korea University, College of Agriculture, Department of Horticulture	1 Anam-dong, Seongbuk-gu, Seoul Phone: 02-92-2601-9 Chairman: Prof.Dr. S.B. Kim
Cultural and management practices of fruit trees	Prof.Dr. S.B. Kim
Growing environments for ornamental plants and urban horticulture	Prof.Dr. B.H. Kwack
Physiology of fruit trees	Assoc.Prof.Dr. C.H. Lee
Quality of vegetable crops relating to cultural factors	Prof.Dr. K.W. Park*
21.3 Konkuk University, College of Agriculture, Department of Horticulture	93-1 Mojin-dong, Seongdong-gu, Seoul Phone: 02-445-0061-70 Chairman: Prof.Dr. C.C. Kim
Pomology	Prof.Dr. C.C. Kim
Vegetable production under structure	Prof.Dr. K.H. Lee
Floriculture	Prof.Dr. B.W. Kim
Landscaping	Assoc.Prof. Y.H. Kim
21.4 Dongkuk University, College of Agriculture, Department of Agriculture, Horticulture Major	3-ga, Pil-dong, Jung-gu, Seoul Phone: 02-267-8131-9 Chairman: Ass.Prof.Dr. M.H. Lee
Uptake and translocation of inorganic element in vegetables	Assoc.Prof.Dr. K.J. Kim
Cytogenetics of white potatoes	Prof.Dr. H.Y. Kim

21.5 KyungHee University, College of Industry, Department of Horticulture
1 Hoegui-dong, Dongdaemun-gu, Seoul
Phone: 02-966-0061-5
Chairman: Prof.Dr. J.M. Lee

Storage and self-incompatibility — Prof.Dr. J.M. Lee
Soil and fertilization — Prof.Dr. S.C. Shim
Tissue culture and plant propagation — Lecturer Dr. S.W. Lee

22 Suncheon

22.1 Suncheon College, Department of Horticulture
elevation: 23 m
rainfall : 1765 mm
sunshine : 2486 h
315 Maegok-dong, Suncheon, Jeonnam-do
Phone: 0061-
Chairman: Prof. Y.G. Jeong

Effect of continuous cropping with cucumber
secondary growth and yield in grape — Ass.Prof. Y.O. Jin
Ecology of watercress — S.Y. Yang
Turf and weeds — J.C. Jeong
Self senescence — Dr. I.D. Jin
Breeding of cashcrops — Assoc.Prof. H.J. Kim
Micro-environment in PE houses — Research Ass. W.M. Yang

23 Suweon

23.1 Office of Rural Development, Horticultural Experiment Station
elevation: 34 m
rainfall : 1328 mm
sunshine : 2357 h
475 Imok-dong, Suweon, Gyunggi-do
Phone: 0331-6-8151-6
Dir.Gen.: Dr. J.H. Kim

23.1.1 Division of Vegetable Breeding — Dir.: C.H. Lee
Laboratory of leafy and root vegetable breeding — Chief: Dr. S.S. Lee/ Dr. J.Y. Yoon
Dr. H.M. Yoon/ J.K. Woo

Self-incompatibility in radish and Chinese cabbage
Cytoplasmic male-sterility in *Brassica* and *Raphanus*
Tissue and anther culture in *Brassica* species
Laboratory of fruit vegetable breeding — Chief: Dr. Y.H. Ohm/ Dr. K.S. Choi
D.G. Oh/ D.H. Bae

Breeding of cucurbits with gynoecious lines
Multiple resistance in red pepper
Laboratory of genetic resources — Chief: S.J. Kim*/ H.B. Jung
Dr. H.M. Yoon

Tomato breeding for processing and table use
Amaranthus collection and domestication
Germplasm collection and preservation
Laboratory of seed test — Chief: Dr. K.S. Choi/ D.Y. Park
H.D. Kim

Test of commercial vegetables seeds
23.1.2 Division of Vegetable Cultivation — Dir.: Dr. S.K. Park
Photosynthesis of vegetable crops
Laboratory of vegetable cultivation — Chief: J.H. Chung/ K.Y. Kim/ S.C. Lim
Fertilizer studies for vegetable crops
Hydroponics with vegetables
Seedling raising
Laboratory of vegetable crop physiology — Chief: H.Y. Pak/ J.M.Hwang/ C.I.Lim
Physiological disorders in major vegetables

Environment-plant relation in greenhouse cultivation
Laboratory of cultural environment — Chief: Y.S.Kwon/ H.D. Suh/ Y.B. Lee
Cold and flooding damage in vegetable crops
Development of yield forecast moded in major vegetables

23.1.3 Division of Fruit Tree Breeding — Dir.: Dr. Y.K. Kim
Laboratory of fruit tree breeding — Chief: M.S. Kim*/ Y.U.Shin/S.J.Kang
Cross and introduction breeding of apple, pear, peach and cherries
Breeding of dwarf rootstock in peaches
Nursery-adult correlation of major characters
Laboratory of fruit tree genetics — Chief: Dr.S.B.Hong/ H.M.Jo/ B.W.Yae
Breeding of small fruits and filbert
Anther and tissue culture
Genetic studies with fruit trees
Laboratory of processing and propagation — Chief: Dr.Y.K.Kim/ C.S.Lee/ H.S.Suh
Cultivar screening for processing
Rapid propagation of fruit trees
Packing and processing of fruits
23.1.4 Division of Fruit Cultivation — Dir.: S.B. Kim
Laboratory of fruit tree cultivation — Chief: K.Y. Kim/ M.D. Cho/ J.K. Kim
Labor saving technology in orchard management, mechanization and chemicals
High-density orchard establishment, prunning and training
Laboratory of nutrition and physiology — Chief: Dr. J.Y. Moon/ J.S. Choi K.C. Shin

Cold damage in fruit trees
Water management in fruit trees
Diagnosis through leaf assay
Laboratory of fruit tree protection — Chief: S.B.Kim/ M.S.Yiem/ H.I.Jang
Ecology and physiology of insect and diseases attacking fruit trees and their control
23.1.5 Division of Floriculture — Dir. and Chief: Y.P.Hong/ Y.J.Kim J.H. Jeong

Laboratory of cultivation and physiology
Natural resources development for ornamental purpose
Tissue culture
Rapid propagation of *Rhododendron* and *Azaleas*
Laboratory of breeding — Chief: Dr. J.S. Lee/ J.Y. Kim
Breeding *Hibiscus* for cold hardiness and apid tolerance
Breeding of Chrysanthemum and Azaleas
23.1.6 Division of Potato Research — Dir.: B.H. Hahn
Development of processing varieties
Laboratory of cultivation — Chief: O.H. Ryu/ S.Y. Kim
Improvement of cultural practices of potato
Establishment of fall growing pattern of potato
Laboratory of breeding and genetics — Chief: Dr. I.G. Mok/ H.Y. Kim
Utilization of true potato seed
Tissue culture

23.2 Gyunggi Provincial Office of Rural Development, Research Bureau, Crops Division, Horticulture Section — 315 Gisan-ri, Taean-myun, Hwaseong-gun
Phone: 0331-32-1595-8
Chief: B.W. Choi/ E.Y.Oh/ C.K.Kang

Cultural management study with some western vegetables
Regional problems in horticultural crops production

23.3 Seoul National University, College of Agriculture, Department of Horticulture

103 Seodun-dong, Suweon, Gyunggi-do
Phone: 0331-7-2111-7
Chairman: Prof.Dr. K.C. Ko

Effect of temperature on peach and apple fruits after harvest	Prof.Dr. K.C. Ko
Hastening of ripening of peach fruit	
Physiology of vegetable crops	Prof.Dr. H.K. Pyo
Strawberry dormancy and garlic bulbing	Prof.Dr. B.Y. Lee
Physiology and ecology of turfgrasses	Assoc.Prof.Dr. D.Y. Yeam
Tissue culture and self-incompatibility of crucifers	Assoc.Prof.Dr. H.G. Park

24 Yesan

24.1 Yesan Junior College of Agriculture, Department of Horticulture
elevation: 30 m
rainfall : 1210 mm
sunshine : 2790 h

527 Yesan-ri, Yesan-eup, Chungnam-do
Phone: 0458-2481-3
Dean: Prof. Y.K. Yoon/ Lecturer

Pomology, fertilization of fruit trees	J.H. Jung
Tissue culture	Prof. T.H. Kim
Fertilization application on vegetables	E.P. Shin
Propagation of flowers	C.Y. Lee
Downy mildew disease in greenhouse cucumber	Y.C. Kim

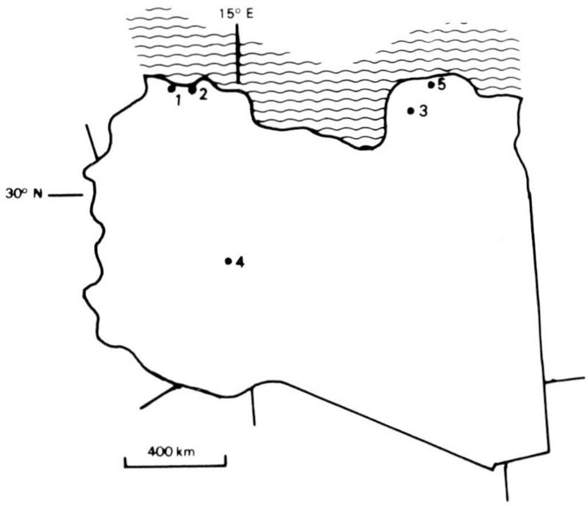
15° E
30° N
1
2
3
4
5
400 km

1 Tripoli
2 Tajoura
3 El-Marj
4 Shebha
5 Al-Beida

Survey

The country is widely distributed with strong variations in soil and climatic conditions. It consists of three main areas, namely western coastal plain with sandy to medium soils having up to 400 mm³ precipitation, eastern highland coast with medium to heavy soils and up to 600 mm³ precipitation, and southern desert having sandy to heavy soils with practically no rainfall. Accordingly three agricultural research stations belonging to the secretariat of agricultural reclamation and land development were established in these regions i.e. in Tajoura, El-Marj and Sebha, respectively and centrally controlled by one head office in Tripoli. These stations serve different agricultural fields including horticulture.
Other organizations interested in horticultural research are the faculty of agriculture in Tripoli (Al-Fatih University) and faculty of agriculture in Al-Beida (Gar-Younis University).

As there is no updated material available, the text of HRI III is repeated.

1 Tripoli

1.1 Agricultural Research Station
Main office:

PO Box 2480, Tripoli
Chrmn.: Dr. A. A. Bin Saad
Dir. of research and studies:
Dr. A.M. Madour
Dir. of public relations:
S.A. Ahmed
Dir. of technical services:
S.S. Al-Gamal

Entomology	Dr. S.A. Hassan
Vegetable crops	Dr. A.I. El-Murabaa
Plant nutrition	Dr. F.M. Chaudhry
Soil chemistry	Dr. R.A. Sakr
Plant physiology	Dr. A.H. Abed

1.2 Al-Fatih University, Faculty of Agriculture — PO Box 13538, Tripoli

Horticulture Department,	Head: Dr. M.M.Ismail
Pomology	Dr. M.M. Ismail/Dr. M.Y. Shurafa, Dr. H.S. Ahmed/M.S. Shaladan, S.A. Naji
Vegetable crops	Dr. W.A. Warid/M.I. Chaudhry
Floriculture	Dr. M.Z. Mehdi/S.H. Sallam
Postharvest physiology	Dr. M.I. El-Yamzini/D.M. Niazi

2 Tajoura

2.1 Agricultural Research Station	Tajoura Dir.: Dr. A.A. Azzy
Vegetable crops	Dr. A.K. Gaafar/A.S. Al-Fallah, A.M. Al-Kiraiw/S.M. Wafi, M.A. Murabi/M.A. Al-Fatisi
Fruit crops	N.A. Farouki
Soil sciences	Dr. A.A. Abdel-Salam, Dr. M.M. Al-Rashid/R.M Al-Khomsi, F. Shandoufi (FAO)/A.H. Abu-Rawi, A.M. Al-Bakouri/A.M. Al-Khraz, I.M. Kalab/M.K. Abdel-Halim, M.A. Ismail/M.A. Ali
Plant protection	Dr. A.A. Azzy/Dr. G.R. Chaudhry, Dr. D.A. Sadiki/Dr. A. Saleh, M.Y. Yossif/M.S Al-Misallaty, H.H. Gulaid/W.M. Dsouki, S.A. Al-Gharbawi/A.H. Abdel-Rahman,
Agricultural economics	A. Abu-Khdair/I.S. Abu-Hifala, A.M. Abdel-Salam

3 El-Marj

3.1 Agricultural Research Station	El-Marj Dir.: M.A. Al-Oushar
Horticulture	F.A. Al-Sheikh/Dr.H. Al-Sawaf(FAO) S.O. Adam/M.A. Al-Ansary
Soil sciences	M.F. Al-Shelwy
Plant protection	Dr. W.A. Ashour/A.F. Azouz, A.A. Atawna
Agricultural economics	A.A. Manah

4 Sebha

4.1 Agricultural Research Station	Sebha Dir.: M.O. Al-Hodairi
Horticulture	M.O. Al-Hodairi
Soil sciences	S. Brashar (FAO)/Dr. A. Kalash, A.A. Ali-Barkouli
Plant protection	M.H. Nehmat-Allah/E.A Zeid

5 Al-Beida

5.1 Gar-Younis University, Faculty of Agriculture, Horticulture Department,	Al-Beida Head: A.M. Serief
Deciduous fruits	

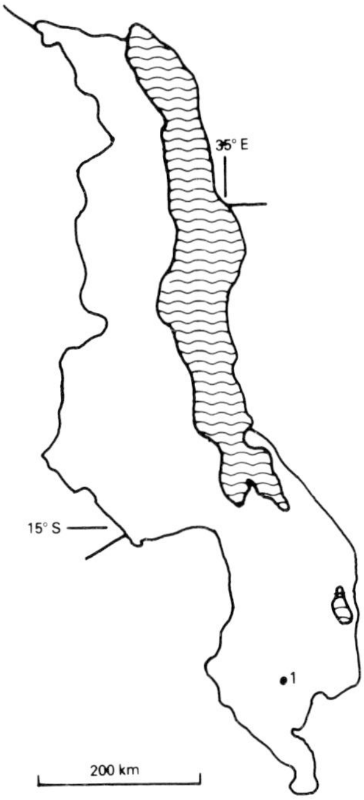
35° E
15° S
1
200 km

1 Bvumbwe

Survey

The Republic of Malawi has approximately 12 million hectares including 9 million hectares of land and 3 million hectares of water. The population was estimated at 8 million in 1980.
Malawi is located approximately 15°S and 35°E with an average rainfall of 1,200 mm annually, most of the rain falls between November and April. Temperatures vary from a very hot region in the Shire Valley to the cool highlands. Frost occurs only at elevations above 2,700 m. Most of the horticultural production is presently in the Southern and Central areas of the country.

1. Bvumbwe

1.1 Bvumbwe Agricultural Research Station
elevation: 1200 m
rainfall : 1200 mm

Limbwe
P.O. Box 5748
Phone: 662-206
Cable: Agrisearch
Officer in Charge: Dr.J.H.A. Maida

Research on production of fruits (temperate, sub-tropical and tropical), nuts, vegetables, potatoes, spices and flowers. Projects include variety evaluations, spacing, fertilization, propagation, irrigation, pruning and growth regulators.

Fruit Crops	E.H.C. Chilembwe*/D.N. Tembwe
Temperate Fruits	D.C. Kalizang'oma
Tropical Fruits	L.E. Nkoloma
Nut Crops	I.M.G. Phiri/W.R.G. Banda
Vegetable Breeding and Spices	C.T. Chizala/Mrs. R. Kumwenda
Potatoes and Vegetables	W.T. Gondwe
(Cultural Practices)	Mrs. R. Nsanjama
Floriculture and Nursery Production	D.N. Tembwe

Prepared by C.E. Arnold, Ministry of Agriculture, Limbwe

MEXICO

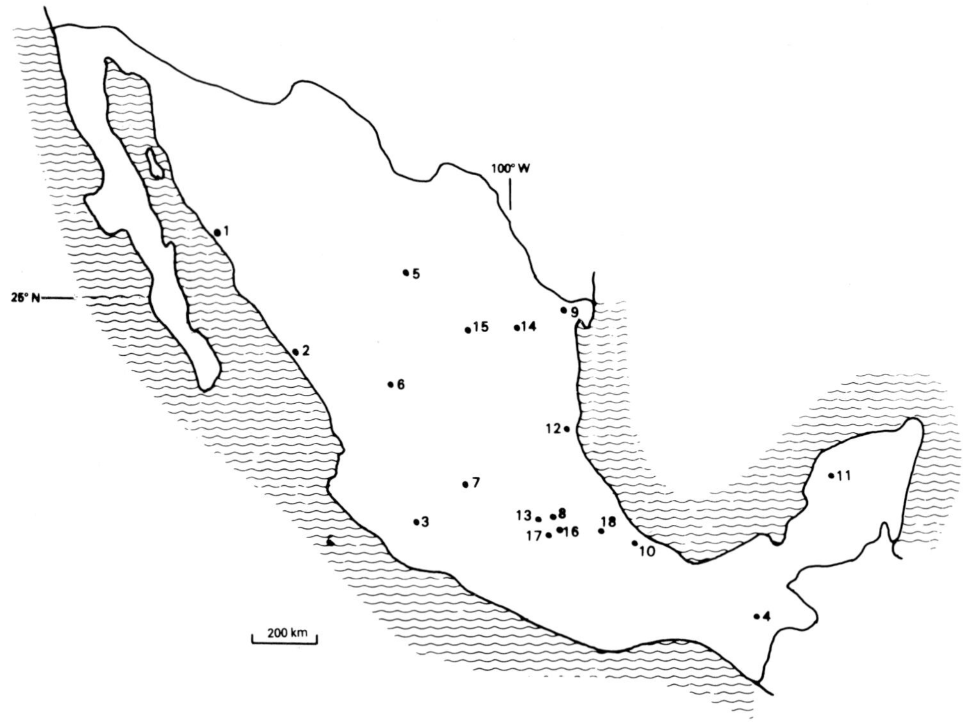
100° W
25° N
•1
•2
•3
•4
•5
•6
•7
•8
•9
•10
•11
12
13
•14
•15
16
17
18
200 km

1 Obregón, 2 Culiacan, 3 Apatzingan, 4 Tuxtla Gtez., 5 Matamoros, 6 Calera, 7 Celaya, 8 Chapingo (CIAMEC), 9 Rio Bravo, 10 Veracruz, 11 Merida, 12 Tampico, 13 Mexico City (Headquarters INIFAP), 14 Monterrey, 15 Saltillo, 16 Chapingo (University), 17 Mexico City (CONAFRUT), 18 Jalapa

Survey

The National Institute of Forestry, Agricultural and Animal Science Research (INIFAP-Instituto Nacional de Investigaciones Forestales, Agrícoles y Pecuarias) created on August 23, 1985, through the union of three institutes (INIF-Instituto Nacional de Investigaciones Forestales, INIA-Instuto Nacional de Investigaciones Agrícolas and INIP-Instituto Nacional de Investigaciones Pecuarias), of the Ministry of Agriculture and Water Resources, has the responsibility of the research activities in Mexico on the basis of its Federal Status. Actually under reorganization. In contrast with more developed countries, the universities offering agricultural degrees, and the Agricultural Colleges in Mexico, currently over 50, perform a very limited amount of research, centered around thesis projects of students. By 1977 only three of the Universities have graduate student programs and only Chapingo offers a master's degree in fruit crops (deciduous).
A great expansion in research activities in horticulture has taken place in the last 15 years in Mexico. The scientific staff at INIA, including all crops and disciplines, has expanded three fold in that period to a current staff of over 900 researchers plus field and administrative support. The country is divided in 12 Regional Centers, each with a Director who is responsible for administrative and technical matters within his center. Every Center has a number of Experiment Stations, and currently 49 of them are distributed in the country. In every Experiment Station a Coordinator of Research is responsible for focusing research activities as well as to assign the proper priority to the local projects within the scope of a National program of Research, which in turn is elaborated by top research people who on a national basis, program and evaluate the overall research activities. In regard to horticultural research, four main divisions and two small ones are considered.

Earlier these crop-oriented divisions were separated Departments, but by late 1977 are only conceptual Units because, operativewise, the Institute now functions with the Experiment Stations as a unit base, and all research should be multidisciplinary oriented with

Prepared by J.A. Garzon Tiznado, CAEB, Celaya

crops as the recurrent subjects, but not as administrative units.

The six divisions are: 1 Vegetable crops - This is one of the original Departments of the former office of Special Studies, and its first results were published prior to 1950. The actual research spreads over 30 different vegetable crops that can be grouped as: a. The most economically important: tomatoes, strawberries, - considered here and not in fruit crops on the basis of the grower's activities - cucumbers, melons, garlic, onion, and other important crops in our export trade. b. Those very important in the internal economy or in the food habits of the people: peppers *(Capsicum annuum)*, which includes more than 20 different and economically important types, potatoes, carrots, snap beans, husk tomato *(Physalis ixocarpa)*, jicama *(Pachyrrizus erosus)*, and other indigenous vegetables. c. Industry oriented crops with moderate internal demand: asparagus, peas, broccoli, cauliflower etc. About 90% of the personnel is working in the first two groups and consequently, very limited research is done in the last category.

2 Citrus - The citrus research was initiated before 1960 mainly in oranges, at the northeast of the country, but lately has been expanded to most of the citrus areas and now includes some of the other commercially important citrus: mexican lime, tangerine, and grapefruit.

3 Temperate of deciduous fruits - By mid 60's the research was initiated with viticulture, and since then, has had a continous expansion to apples, pecans, and peaches, but also to other minor stone fruits.

4 Tropical and sub-tropical fruits - In spite of the great importance of this group, just recently the research has been formally initiated. Most research projects, including the collection and evaluation of creole and native types, have been initiated by 1970. More attention has been placed in the most economically important crops and of wide spread use such as mango, avocado, papaya, banana, and pineapple, but in this section a number of other minor crops are also included, like tamarindus, soursop and guava.

5 Spices and plantation crops - Very limited research is done in pepper *(Piper nigrum)*, vanilla *(Vanilla planifolia)*, clove *(Syzyngium aromaticum)*, and other flavoring spices. Research in cacao *(Theobroma cacao)* and rubber *(Hevea brasilensis)*, is done only at two Experiment Stations and to a limited extent.

6 Ornamentals and flowers - No formal research is done in these crops at the present at INIA or any other Institution in Mexico. A number of commerical enterprises handle this branch of the horticulture, but only from the commercial point of view.

Although the research activities are carried almost completely by INIA, and to a limited extend by the Universities, there are other Federal Organizations which belong to the Ministry of Agriculture and Water Resources, which are also directly involved in horticulture. The National Commission of Fruit Crops (Comisión Nacional de Fruticultura - CONAFRUT), deals exclusively with fruits, in a number of activites such as: commercial nurseries, plant propagation, technical assistance to growers, fruit cooperatives organization and fruit industrialization as well as tissue culture of ornamentals. The National Direction for Agricultural Production and Extension (Dirección General de Extensión y Producción Agrícola - DGEPA), has a Department of Horticulture where fruit and vegetable crops specialists are located, and elaborate and supervises general programs such as home gardens, orchard protection and, most important, disseminate research results, work directly with the grower, and localize and evaluate new problems.

1 Obregón

1.1 Centro de Investigaciones Agrícoles del Noroeste (CIANO) (Agricultural Research Center of the North West
elevation : 40 m
rainfall : 307.1 mm

Apartado 515, Cd. Obregón, Son.
Phone: 503-77, 401-01
Dir.: Dr. J.A. Valencia Vilarreal

Storage facilites for germplasm, greenhouses, laboratories, library

1.2 Campo Agrícola Experimental Valle de Mexicali (CAEMEXI) (Agricultural Experiment Station)
elevation : 4 m
rainfall 60.6 mm

Apartado 1019, Mexicali 3, B.C.
Phone: 671-47, 683-69

Cucurbit, onion, asparagus - agronomy

J.M. González González, Ing.

1.3 Campo Agrícola Experimental Valle del Mayo

Apartado 189, Navojoa, Son.

(CAEMAY) (Agricultural Experiment Station) elevation : 38 m rainfall : 385.9 mm	Phone: 232-44, 238-67
Cucurbits - agronomy	R.G. Alvarez, Mc.
1.4 Campo Agrícola Experimental Costa de Ensenada (CAECOEN) (Agricultural Experiment Station) Potato, tomato - agronomy	Apartado 2777, Ensenada, B.C. Phone: 403-84 B. Cabrera Valle, Pas. Ing.
Apple, peach *vitis vinifera* - agronomy, breeding	J.M. Gutiérrez, Ing.
1.5 Campo Agrícola Experimental Costa de Hermosillo (CAECH) (Agricultural Experiment Station) elevation : 237 m rainfall : 244.9 mm	Apartado 1031, Hermosillo, Son. Phone: 380-55, 319-50
Citrus, apple, peach, pecan - agronomy, breeding	Dr. D.H. Díaz Montenegro, Dr. J.J. Martínez Tellez
1.6 Campo Agrícola Experimental Region de Caborca (CAECAB) (Agricultural Experiment Station) elevation : 292 m rainfall : 185.8 mm	Apartado 125, Caborca, Son. Phone: 206-23
Apple, peach - agronomy	R.J. Gonzalez, Mc.
pecan, citrus, almond - agronomy	Carmen Aída Félix Verdugo, Ing.

2 Culiacán

2.1 Centro de Investigaciones Agrícolas del Pacifico Norte (CIAPAN) (Agriculture Research Center of the Pacific North) elevation : 53 m rainfall : 630.4 mm	Apartado 356, Culiacán, Sin. Phone: 440-50 Dir.: Dr. J.M. Ramírez Díaz
2.2 Campo Agrícola Experimental Valle de Culiacán (CAEVACU) (Agriculture Experiment Station) elevation : 53 m rainfall : 630.4 mm	Apartado 356, Culiacán, Sin. Phone: 440-01, 440-02
Mango, avocado - plant breeding and agronomy	A. Ireta Ojeda, Mc.
Tomato, squash - plant breeding	J. Silva Ríos, Mc.
Tomato, cucumber - phytopathology	P. Manjarrez Morales, Mc.
Tomato - soil fertility	F. Mascareño Castro, Mc.
Tomato - plant breeding	J.L. Rangel Ramírez, Ing.
Tomato, cucumber, sweet pepper - agronomy	A.A. Casado, Ing.
Tomato, cucumber - entomology	Mayra Avilés González, Ing.
Citrus - plant breeding and agronomy	P. Hernández Aguilar, Ing.
2.3 Campo Agrícola Experimental Valle del Fuerte (CAEVAF) (Agriculture Experiment Station) elevation : 10 m rainfall : 320.9 mm	Apartado 342, 81200 Los Mochis, Sin Phone: 601-37, 602-12
Tomato, melon - plant breeding	E. Retes Cázares, Mc.
Tomato - agronomy	A. Rodríquez Escalante, Ing.

Tomato, cucurbits - entomology	S. Armenta Cárdenas, Mc.
Tomato, cucurbits - phytopathology	Beatriz López Ahumada, Biol., S. Silva Vara, Biol.

2.4 Campo Agrícola Experimental Santiago Ixcuintla (CAESIX) (Agriculture Experiment Station)
elevation : 44 m
rainfall : 1178.9 mm

Apartado 100, Santiago Ixcuintla,l Nay.
Phone: 507-10

Pepper - plant breeding	J.M. Cervantes, Ing.

3 Apatzingán

3.1 Centro de Investigaciones Agrícolas del Pacífico Centro (CIAPAC) (Agricultural Research Center of the Central Pacific)
elevation : 500 m
rainfall : 700.1 mm

Apartado 40, Aptzingán, Mich.
Phone: 412-78, 415-84
Dir.: M.C. José de la Luz Sánchez Pérez

3.2 Campo Agricola Experimental Valle de Apatzingán (CAEVA) (Agriculture Experiment Station)
elevation : 500 m
rainfall : 700.1 mm

Apartado 40, Apatzingan, Mich.
Phone: 415-44

Melon - plant breeding	J.F. Arias Suárez, Ms.
Citrus, mango, banana - soil fertility	J. Chávez Contreras, Mc.
Cucurbits - entomology	J. Javier Mercado, Bs.
Melon - weed control	D. Munro Olmos, Mc.
Cucurbits - soil fertility	F. Ordaz Ordaz, Ing.
Cucurbits - phytopathology	A. Vega Piña, Bs.
Melon - phytopathology	J.A. Vidales Fernández, Mc.

3.3 Campo Agricola Experimental Tecomán (CAETECO) (Agriculture Experiment Station)
elevation : 40 m
rainfall : 660.4 mm

Apartado 88, Tecomán, Col.
Phone: 401-33

Citrus - physiology	Dr. S. Becerra Rodríquez
Citrus - phytopathology	J.G. Garza López, Mc.
Citrus - nutrition	V.M. Medina Urrutia, Mc.
Mango - physiology	R. Nuñez Elisea, Ms.
Banana - soil fertility	J. Orozco Romero, Mc.
Citrus - weed control	J.J. Silva Salazar, Ing.
Citrus - horticulture	M. Orozco Santos, Ing.

3.4 Campo Agricola Experimental Costa de Jalisco (CAECJAL) (Agriculture Experiment Station)
elevation : 35 m
rainfall : 905.2 mm

Apartado 2, 48850 La Huerta, Jal.
Phone: 401-05

Cucurbits - production	H. Hurtado Hernández, Ms.
Cucurbits - soil fertililty	J.L.Zepúlveda Torres, Mc.
Cucurbits, mango - phytopathology	J. Velázquez Monrreal, Ing.
Mango - horticulture	G. Diaz González, Ing.

3.5 Campo Agricola Experimental Iguala (CAEIGUA)

Apartado 29, 40000 Iguala, Gro.

(Agriculture Experiment Station)
elevation : 635 m
rainfall : 1086.3 mm

Phone: 210-56, 207-00

Peach - horticulture — M. Payán Jacobo, Ing.

4 Tuxtla Gutiérrez

4.1 Centro de Investigaciones Agrícolas del Pacífico Sur (CIAPAS) (Agricultural Research Center of the Pacific South)
elevation : 864 m
rainfall : 897.5 mm

Apartado 689, Tuxtla Gtz., Chis.
Phone: 276-62, 278-81
Dir.: Dr. E. Betanzos Mendoza

Library, laboratories

4.2 Campo Agrícola Experimental Mixteca Oaxaquena (CAEMOAX) (Agriculture Experiment Station)
Drupaceous, apple, nopal - plant breeding

Apartado 3, Hochixtlán, Oax.
Phone: 201-17
J. Luna Vázquez, Ing.

4.3 Campo Agrícola Experimental Costa de Chiapas (CAECOCHI) (Agriculture Experiment Station)

Apartado 46, Excuintla, Chis.
Phone: 400-04, 400-15

Watermelon - plant breeding, agronomy — E. Colín Castillo, Bs.
Mammee, macadamia, rambután, granadillo, litchi - agronomy — H. Barrios Domínguez, Ing.

4.4 Campo Agrícola Experimental Istmo de Tehuantepec (CAEITE) (Agriculture Experiment Station)

Apartado 51, Juchitán, Oax.
Phone: 203-41

Mango - entomology — P. Márquez Castillo, Bs.

4.5 Campo Agrícola Experimental Valles Centrales de Oaxaca (CAEVOAX) (Agriculture Experiment Station)

Apartado 33 Suc. B., Oaxaca, Oax.
Phone: 661-22

Tomato, pepper - agronomy — M.A. Cano García, Ing.
Nopal - plant breeding — Serafín Miranda Trejo, Bs.

4.6 Campo Agrícola Experimental Costa Oazaquena (CAECOAX)

Apartado 389, Rio Grande, Oax.
Phone: 201-38

Citrus - phytopathology — A. Flores Ricarde, Ing.

4.7 Campo Agrícola Experimental Rosario de Izapa (CAERI)

Apartado 96, Tapachula, Chis.
Phone: 613-44

Banana, mango - phytopathology — J. Hernández Ovalle, Ing.

5 Matamoros

5.1 Centro de Investigaciones Agrícolas del Norte (CIAN) (Agricultural Research Center of the North)
elevation : 1120 m
rainfall : 226.9 mm

Apartado 1, Matamoros, Coah.
Phone: 202-02 al 05

Library, greenhouse, laboratories, collection of vitis

Apartado 247, Torréon, Coah.
Dir.: V.M. Valdez Rodríguez, Mc.

5.2 Campo Agrícola Experimental La Laguna (CAELALA)	Apartado 1, Matamoros, Coah. Phone: 202-02 al 05 Apartado 247, Torreon, Coah
Viticulture	J.M. Castillo Padilla, Ing.
Vitis vinifera - entomology	C. García Salazar, Ms.
Viticulture	Dr. E. Madero Tamargo
Vitis vinifera, pecan - water management	C. Godoy Avila, Ms.
Pecan - phytopathology	Dr. T. Herrera Pérez
Vitis vinifera - phytopathology	Dr. F. Jiménez Díaz
Pecan, apple, pistachio - fruticulture	Dr. A. Lagarda Murrieta
Vitis vinifera, pecan - fruticulture	I. López Montoya, Ing.
Cucurbits - general agronomy	J. de Dios Ruíz de la Rosa, Ing.
5.3 Campo Agrícola Experimental Delicias (CAEDEL) elevation : 1171 m rainfall : 275.2 mm	Apartado 81, Cd. Delicias, Chih. Phone: 219-74, 192-76
Viticulture	Eugenia Montes Martínez, Ing.
Pecan - fruticulture	A. Salas Franco, Mc.
Onion, pepper - agronomy	G.F. Acosta Rodríguez, Ing.
Pepper, onion - water management	J. Martínez Rodriguez, Ing.
5.4 Campo Agrícola Experimental Sierra de Chihuahua (CAESICH) elevation : 2210 m rainfall : 398.8 mm	Apartado 554, Cd. Cuauhtémoc, Chih Phone: 200-86, 231-10
Drupaceous, strawberry, nopal, brambleberry, raspberry, citrus - fruticulture	C. Arguello Mendoza, Mc.
Pear, apple - fruticulture	V.M. Guerrero Prieto, Ms.
Pear, apple - phytopathology	J.L. Jacobo Cuéllar, Ing.
Pear, apple - water management	J.P. Amado Alvarez, Ing.
5.5 Campo Agrícola Experimental Zaragoza (CAEZAR) elevation : 335 m rainfall : 382.6 mm	Apartado 33, Zargoza, Chih. Phone: 604-50
Drupaceous - fruticulture	E.J. Cuéllar Villarreal, Ing.
Pecan - fruticulture	H. Aguilar Pérez, Ing.
5.6 Campo Agrícola Auziliar NUEVO CASAS GRANDES	Apartado 2244, Cd. Juárez, Chih. Phone: 418-08
Drupaceous, apple, pear - phytopathology	M. Ramirez Delgado, Ing.

6 Calera

6.1 Centro de Investigaciones Agrícolas del Norte Centro (Agricultural Research Center of the Central North) elevation : 2612 m rainfall : 330.5 mm	Apartado 18, Calera de V.R., Zac. Phone: 503-63, 503-94 Dir.: Dr. R.A. Martínez Parra
Library, laboratories	
6.2 Campo Agrícola Experimental Zacatecas (CAEZAC) elevation : 2612 m rainfall : 330.5 mm	Apartado 18, Calera de VR, Zac. Phone: 501-98, 501-99

Apple - agronomy	Teresa de Jesús Arellano Ledezma, Ing.
Peach - agronomy	A. Rumayor Rodríguez, Mc., J. Zegbe Domínguez, Ing.
Vitis vinifera - viticulture	Dr. J. Madero Tamargo
Apple, peach - entomology	J. Mena Covarrubias, Ing.
Peach - soil fertility	R. Valdés Zepeda, Ing.
Peach - water management	Dr. J.L. Chan Castañeda, A.G. Bravo Lozano, Mc.
Peach, apple - phytopathology	O. Gutiérrez Treviño, Biol.
Vitis vinifera - enology	Dr. M.I. Cervantes
Pepper - water management	J. Ramos Cano, Mc.

6.3 Campo Agrícola Experimental Pabellón (CAEPAB)
elevation : 2217 m
rainfall : 475.5 mm

Apartado 20, Pabellón, Ags.
Phone: 401-86, 401-67

Peach - agronomy	Dra. Martha Z. Belderas Rodríguez, F. Gitiérrez Acosta, Mc.
Vitis vinifera - viticulture	Dr. J.M. García-Santibañez S.
Garlic, wild potato - agronomy	L.M. Macías Valdés, Ing.
Peach - soil fertility	L.H. Maciel Pérez, Ing.
Peach - phytopathology	P. Valle García, Mc.
Peach - entomology	J. Velázquez Montoya, Ing.

6.4 Campo Agrícola Experimental Valle del Guadiana (CAEVAG)
elevation : 1889 m
rainfall : 440.6 mm

Apartado 186, Durango, Dgo.
Phone: 210-44, 211-55

Apple - agronomy	G. Carbajal Cadena, Ing., A.L. Aguilar, Mc./R.M. Leguizamo, Ing.
Hot pepper (Var. Grossum), potato, onion - agronomy	L. Zamarrón Loredo, Ing.

6.5 Campo Agrícola Experimental San Luis Potosi (CAESAL)

Apartado 2da. Priv., San Luis Potosí S.L.P.
Phone: 379-23, 278-70

Nopal - agronomy	Dr. E. Pimienta Barrios

6.6 Campo Agrícola Experimental Los Canones (CAEDEC)

Apartado Fray José, Jalpa, Zac.
Phone: 522-06, 523-62

Guava - agronomy	Patricia Rangel Jiménez, Ing.
Guava - entomology	E. González Gaona, Ing.

7 Celaya

7.1 Centro de Investigaciones Agrícolas del Bajío (CIAB) (Agricultural Research Center of the Lower Plains)
elevation : 1754 m
rainfall : 597.3 mm

Apartado 112, Celaya, Gto.
Phone: 236-22, 225-99
Dir.: G.A. Longoria Garza, Mc.

Greenhouses, library, laboratories, genetics germplasm, computer center

7.2 Campo Agrícola Experimental Bajío (CAEB)

Apartado 112, Celaya, Gto.

elevation : 1754 m
rainfall : 597.3 mm

Phone: 270-23

Tomato, cucurbits - phytopathology, plant breeding	J.A. Garzón Tiznado
Pepper - plant breeding	Dr. J.G. Salinas González
Onion, garlic, jícama (yam bean)	A. Heredia Zepeda, Mc.
Onion, pepper - plant breeding	C.R. Saray Meza, Mc.
Genetics germplasm	A. Guerrero Moreno, Ms.
Pepper - nutrition	J.A. González Martinez, Ms.
Cucurbits, tomato - entomology	R. Bujanos Muñiz, Mc.
Cucurbits - virology	F. Delgadillo Sánchez, Mc.
Avocado - agronomy	J.D. de la Torre Vizcaíno, Ing.
Drupaceous - plant breeding	Dr. S. Pérez González
Garlic, onion - weed control	A. Arévalo Valenzuela, Bs.
Potato - agronomy	R. Rocha Rodríguez, Ing.
Strawberry - agronomy	P.A. Dávalos González, Ing.

7.3 Campo Agrícola Experimental Sierra Tarasca (CAESIT)
elevation : 2132 m
rainfall : 1041.2 mm

Apartado 32, Pátzcuaro, Mich.
Phone 206-57, 200-73

Atemoya, chirimoya, drupaceous, avocado, apple, pear, macadamia, kiwi - agronomy — M.R. Mendoza López, Mc.

7.4 Campo Agrícola Auziliar Morelia

Apartado 15-G, Morelia, Mich.
Phone: 259-38

Pear, avocado, drupaceous - plant breeding — I. Torres Pacheco, Ing.

7.5 Campo Agrícola Experimental Norte de Guanajuato (CAENGUA)
elevation : 1933 m
rainfall : 418.6 mm

Apartado 25, San José Iturbide, Gto.
Phone: 803-64, 803-71

Pepper, garlic - agronomy — F.H. Najasabadi, Mc.

8 Chapingo

8.1 Centro de Investigaciones Agrícolas de la Mesa Central (CIAMEC) (Agricultural Research Center of the Central Plains)
elevation : 2250 m
rainfall : 644.8 mm

Apartado 10, Chapingo, Méx.
Phone: 585-05-62 (México, D.F.)
422-00 Ext. 5310 (Chapingo)
Dir.: Dr. A. Zuloaga Albarrán

Nopal - agronomy — M.R. Fernández Montes, Mc.

8.3 Campo Agrícola Experimental Zacatepec (CAEZACA)
elevation : 900 m
rainfall : 838.9 mm

Apartado 12, Zacatepec, Mor.

Drupaceous, fig - agronomy	J.A. Alvarez Márquez, Ing.
Tomato - agronomy	J. de Dios Bustamante Orañegui, Ing.
Tomato - plant breeding	M.A. González Hernández, Ing.

8.4 Campo Agrícola Experimental Tecamachalco (CAETECA)
elevation : 2013 m
rainfall : 618.9 mm

Apartado 43, Tecamachalco, Pue.
Phone: 202-12

Nopal, *vitis vinifera* - agronomy — J.C. Martínez González, Ing.
Coffee - phytopathology — R. Roa Durán, Ing.
Coffee - plant breeding — V. Ponce Herrera, Ing.
Husk tomato, tomato, pepper - plant breeding — V. Zamudio Guzmán, Mc.

9 Río Bravo

9.1 Centro de Investigaciones Agrícolas Del Golfo Norte (CIAGON) (Agricultural Research Center of the North Gulf)
elevation : 30 m
rainfall : 517 mm

Apartado 172, Rió Bravo, Tamps.
Phone: 410-46, 407-45
Dir.: Dr. C.A. Rincón Valdés

Greenhouses, laboratories, library

9.2 Campo Agrícola Experimental General Terán (CAEGET)
elevation : 538 m
rainfall : 634.1 mm

Apartado 3, Gral. Terán, N.L.
Phone: 700-25

Apple - fruticulture — D. Cortés Ortega, Mc.
Citrus - virology — M.A. Rocha Peña, Mc. Biol.
Citrus - soil fertility — G.T. de la Cruz, Mc.
Pecan - water management — H. Villarreal Elizondo, Mc.

9.3 Campo Agrícola Experimental Las Adjuntas (CAELAD)
Citrus - plant breeding

Apartado 3, S, Jiménez, Tamps.
J.M. Martínez V., Mc.

9.4 Campo Agrícola Experimental Anahuac (CAEANA)
elevation : 432 m
rainfall : 742.1 mm

Apartado 8, Cd. Anáhuac, N.L.
Phone: 705-95, 701-03

Vitis vinifera, apple - biology — Beatriz Rodríguez Olmos, Biol.

10 Veracruz

10.1 Centro de Investigaciones Agrícolas del Golfo Centro (CIAGOC) (Agricultural Research Center of the Central Gulf
elevation : 16 m
rainfall : 1667.6 mm

Apartado 429, Veracruz, Ver.
Phone: 32-71-04, 36-19-89
Dir.: Dr. S. Acosta Nuñez

10.2 Campo Agrícola Experimental Cotaxtla
elevation : 16 m
rainfall : 1667.6 mm

Apartado 429, Veracruz, Ver.
Phone: 33-55-10, 33-54-21

Mango, papaya - phytopathology — E.N. Becerra León, Mc. Biol.
Mango, papaya - physiology — A. Aguilar Zamora, Ing.
Mango, papaya - plant breeding — F. de los Santos de la Rosa, Bs.

10.3 Campo Agrícola Experimental Cuenca del Papaloapan (CAEPAP)

Apartado 43, Isla, Ver.
Phone: 422-48

Mango, papaya - physiology — M.A. Salazar Valdez, Ing.
Pepper, watermelon - agronomy — H. Guerrero Palomino, Bs.

11 Mérida

11.1 Centro de Investigaciones Agrícolas de la

Apartado 50 Suc. D, Mérida, Yuc.

Peninsula de Yucatan (CIAPY) (Agricultural Research Center of the Yucatan Peninsular) elevation : 9 m rainfall : 940 mm	Phone: 199-46, 393-03 Dir.: Dr. J.S. Martínez González
11.2 Campo Agrícola Experimental Zona Henequenera (CAEZOHE)	Apartado 50 Suc. D, Mérida, Yuc. Phone: 199-46
Pepper (Habanero), tomato, cucurbits - agronomy Citrus, papaya, avocado - agronomy	C. Puente Pérez, Bs. R.S. Garrido Ramírez, Ing.
11.3 Campo Agrícola Experimental Uxmal (CAEUX)	Apartado 50 Suc. D, Mérida, Yuc. Phone: 181-88, 202-03
Citrus, mango, avocado - agronomy	J.J. Argumedo, Ing.

12 Tampico

12.1 Centro de Investigaciones Agrícolas Las Huastecas (CIAHUAS) (Agricultural Research Center of Las Huastecas) elevation : 12 m rainfall : 1079.9 mm	Apartado C-1 Sucl. Aereopuerto, 89000 Tempico, Tamps. Phone: 366-72, 363-74 Dir.: J.M. Reding, Mc.
12.2 Campo Agrícola Experimental sur de Tamaulipas (CAESTAM) elevation : 12 m rainfall : 1079.9 mm	Apartado C-1 Sucl. Aereopuerto, 89000 Tampico, Tamps. Phone: 366-72, 363-74
Pepper - plant breeding Pepper, husk tomato, onion - entomology Pepper - phytopathology	O. Pozo Campodónico, Bs. J. Avila Valdez, Mc. Dr. E. Redondo Juárez
12.3 Campo Agrícola Experimental Ebano (CAEEBA)	Apartado 87, Ebano, S.L.P. Phone: 321-16
Pepper, husk tomato - agronomy	M.R. Meras, Bs.
12.4 Campo Agrícola Experimental Huichihuayán (CAEHUICH)	Apartado 1, Huichihuayán, S.L.P.
Citrus - agronomy	R. Uresti, Cerrillos, Ing.
12.5 Campo Agrícola Auxiliar Papantla	Apartado 41, Papantla, Ver. Phone: 216-37, 203-25
Pepper - agronomy Clitrus - phytopathology	R.R. Martínez, Bs. S.A. Curti Díaz, Bs., R.R. Martínez, Bs., U. Díaz Zorrilla, Bs.

13 Mexico City

13.1 INIFAP (Instituto Nacional de Investigaciones Forestales, Agrícolas y Pecuarias) Central headquaters of INIFAP, Institute recently created	Apartado 6-882 and 6-883, México, Phone: 687-74-21, 687-74-50, 687-74-51, 87-74-16 Dean: Dr. J.M. de la Fuente

Chairman Agricultural Department:	Dr. R. Claverán Alonso
Chairman Forestry Department:	Dr. M. Caballero Deloya
Chairman Animal Science Department:	Dr. C. Arellano Sota

14 Monterrey

14.1 ITESM (Instituto Technológico y de Estudios Superiores de Monterrey)	Suc. de Corréos "J", Monterrey, N.L. 64700 Telex: 0382975 ITESME
Library, computer center, greenhouse, laboratories, offers advanced degrees in various disciplines with horticulture. Pecan, tomatoes, deciduous fruits; irrigation, water consumption	Agric. Dean: P. Reyes Castaneda, Ing.
Horticultural Unit: tomato breeding, fruit set under hot conditions	Dr. I. Flores Reyes

15 Saltillo

15.1 UAAAN (Universidad Autónoma Agraria "Antonio Narro")	Apartado 242, Saltillo, Coah. Phone: 431-00 Dean: J.L. Gutiérrez E., Mc.
Greenhouses, laboratories, library Offers advanced degrees in various disciplines related with horticulture. Viticulture, apple, peaches. Desert flora and its utilization. Efficienty in irrigation systems under arid conditions Horticultural Unit:	
Vegetable crops, cytogenetics	Dr. H. Montelongo
Vegetable crops, breeding, production	Dr. G. Pérez
Vegetable crops	E. Bacopulos Téllez, Ing. Mc., C. Medellín, Ing. Mc.
Fruit crops	S. Rueda, Ing. Mc.

16 Chapingo

16.1 UACH (Universidad Autónoma de Chapingo)	Chapingo, Edo. de México
Computer center, library, laboratories, food technology, greenhouses	
16.2 Under Graduate Division (Escuela Nacional de Agricultura)	Chapingo, Edo. de México Dean: R. Posadas, Ing.
Temperate fruit crops	C.A. Ortega G., Ing. Mc., O.A. Zerecero, Ing., D. Parra G., Ing.
Ornamentals	S. Palacios, Ing. Mc.
16.3 Graduate Division (Colegio de Postgraduados)	Chapingo, Edo. de México Phone: 422-00
Offers advanced degrees in various disciplines related with horticulture	Dean: Dr. M. Villa-Isa
Strawberry breeding and cytogenetics	Dr. F. Barrientos
Avocado cytogenetics	Dr. A. García
Small fruits	J. Rodríquez A., Ing.

elevation: 65 m
rainfall : 523 mm
sunshine : 2944 h
r.h. : 80 %

Phone: 74351
Telex: AGROVET 31873M
Dir.: M. Sédrati

Several departments are concerned with horticultural questions in addition to the activities of the Complex in Agadir.

Plant Protection	
Mycology	M. Besri
Virology	M. Bouhida
Nematology	M. Remah
Weeds science	C. Boulet
Phytopharmacy	M. Ormatella
Horticulture	
Plant physiology	L. Walali
Vegetable crops	A. Skiredj
Culture techniques - citrus, olive tree	R. Loussert
Tissue culture	O. Bel Habib
Food technology	

5.2 Institut National de la Recherche Agronomique (INRA) Directorate

Avenue de la Victorie
B.P. 415, Rabat
Phone: 72642
Dir.: M. Faraj

The horticultural research activity of the INRA is divided into specialized Central Stations. These stations work with 48 regional stations spread over the whole country and where, in particular, culture trials are carried out. (See also 1.2, 3.1, 4.1 and 6.1). This Institution is in the process of being restructured.

Central Station of Fruit Trees	A. Chahbar
Central Station of Olive Trees	A. Chahbar
Central Station of Plant Protection	M. Benani
Station for Crop Breeding	M. Skalli
Horticulture	M. Dafir
Technology Station	

6 Tanger

6.1 Central Station of Radioelements of the Institut National de la Recherche Agronomique (see 5.2)
elevation: 75 m
rainfall : 887 mm
r.h. : 76%

INRA, Bd. de Paris, Tanger
Phone: 38033

Vegetable crops

Mineral nutrition - stress	R. Choukrallah/H. El Attir
Water requirements - greenhouses	M. Sirjacobs/A. El Fadl
Special culture techniques	M. Gérard
Aromatic plants	M. Maréchal
Breeding	A. Hilali

Plant Protection

Mycology	M. Achouri
Bacteriology	J. Colin
Virology	B. Hafidi
Phytopharmacy	L. Pussemier/N. Chtaina
Nematology	H. Ammati
Entomology	A. Benazzoun/A. Hanafi/A. Mazih
Economy	M. Faquir

1.2 Central Station of Market Garden Produce, of the Institut National de la Recherche Agronomique (see 5.2)

B.P. 124, Inezgane (near Agadir)
Dir.: M. Benjamaa

2 Casablanca

2.1 Société Agricole des Services au Maroc
elevation: 50 m
rainfall : 406 mm
sunshine : 2929 h
r.h. : 79%

Allée des Jardins, 206
Ain Sebaa, Casablanca
Phone: 350739
Telex: 26930
Dir.: M. Sqalli

This organisation aims mainly to help with technical management of horticultural farms. It also carries out research, particularly prior to distribution.

3 Kénitra

3.1 Central Station for Research on Citrus Fruit, of the Institut National de la Recherche Agronomique (see 5.2)
elevation: 25 m
rainfall : 596 mm

El Menzeh, km 12 route de Sidi Slimane near Kénitra
Phone: 107 El Assam
Dir.: M. Nador

Several areas of study: breeding, entomology, plant pathology, virology, agronomy.

4 Marrakech

4.1 Central Station of Saharan Agronomy, of the Institut National de la Recherche Agronomique (see 5.2)
elevation: 470 m
rainfall : 242 mm
sunshine : 3094 h
r.h. : 66%

Station de la Ménara, Marrakech
Phone: 31116
Dir.: M. Idrissi

Several areas of study: Prevention of "Bayoud" disease of date palm; multiplication of date palm by in vitro culture; meristematic crops of almond.

5 Rabat

5.1 Institut Agronomique et Vétérinaire Hassan II

B.P. 6202 Rabat

14
18
6
7
1
20
3
8,17
2
16
9
21
11
12
13
10
5
19
14° E
42° N
100 km
15
4

1 Aalsmeer
2 Boskoop
3 's Gravenhage
4 Haren-Groningen
5 Horst
6 Lelystad
7 Lisse
8 Naaldwijk
9 Wageningen
10 Wilhelminadorp
11 Barendrecht
12 Bleiswijk
13 Dordrecht
14 Enkhuizen
15 Haelen
16 De Lier
17 Naaldwijk
18 Noordscharwoude
19 Rilland
20 Voorburg
21 's Gravenzande

Survey

Education, research and extension in a mutual cooperation and close contacts with the growers' community has been the leading principle of the government for promoting horticultural activities in the Netherlands. The governmental start was in 1876 by founding an agricultural school and an experimental station at Wageningen, but in many horticultural districts growers founded stations and experimental gardens afterwards. Although much has changed in hundred years, there is still a strong link between science, education, extension and the growers' community.
Horticultural research, including research on nutrition, storage and processing of horticultural produce is primary conducted by two groups of institutions:
a. Research institutes and research stations under the Ministry of Agriculture and Fisheries
b. Departments of the State Agricultural University at Wageningen.
Research stations. There are seven horticultural stations which are situated in the main areas of specific horticultural crops. They are dealing with all aspects of a horticultural branch to which:- arboriculture, fruit growing, flower culture, vegetable production under glass and in the open, mushroom production and flower bulb culture. Staff and activities are financed in principle for 50% by the growers themselves, and for 50% by the Ministry of Agriculture and Fisheries. The financial contribution by growers gives them a great influence in the programming of research projects. The stations also have frequent contacts with the extension service. In horticultural regions experimental gardens conduct practical experiments coordinated by the horticultural research stations.
Research institutes. The research station can rely on specialized research institutes for long term projects and more specialized aspects such as plant breeding crop protection, horticultural engineering, processing, soil fertility etc. Most of these institutes are situated in Wageningen, a small town in the middle of the country where also the agricultural university of the Netherlands is located. In total there are 24 of these institutes on specific fields of science, but only those who have frequent contact with the horticultural research stations are mentioned in this directory. The management of the institutes is governed by a board in which also growers' (or farmers') organizations take part in order to promote a mental input by the growers community in the selection of research projects. In some cases growers' organizations contribute financially to specific research projects, but in general the budget of institutes is fully financed by the

Prepared by G. Slettenhaar, Directorate for Agricultural Research, Wageningen

Ministry of Agriculture and Fisheries. The staff of research institutes and stations have the status of public servants. The Division for Agricultural Research of the Ministry of Agriculture and Fisheries coordinates the activities of research stations and institutes.
University departments - Departments of the foresaid university are primarily established for education and basic research also in the field of horticulture, but in many cases staff members are dealing with applied research, in which also graduate students take part. There are 70 departments but only those directly related to horticultural aspects are selected for this directory.
The concentration of most of the basic and applied agricultural research in one town facilitates services on documentation, publication, instrument development and production, document delivery (library system which is one of the largest in the world).
In order to provide channels for liaison between departments, institutes and stations, the organization for applied scientific research has established the National Council for Agricultural Research. Apart from the task to intensify the contact between scientists on work level, the Council makes proposals for a national plan for the total agricultural research to the Minister of Agriculture and Fisheries who decides upon in close contact with the Minister for Science Policy. A comprehensive reference book "Agricultural Science in the Netherlands" has been compiled by the International Agricultural Centre, P.O. Box 88, 6700 AB Wageningen.
Private enterprises - Several private enterprises have scientific programmes on horticultural research. It comprises food industry, breeding firms, equipment development etc. Included are a first selection of seed firms with the names of scientists responsible for the breeding programmes.

1 Aalsmeer

1.1 Proefstation voor de Bloemisterij in Nederland	Linnaeuslaan 2a, 1431 JV Aalsmeer
(Research Station for Floriculture)	Phone: 02977-26151
elevation: 4.1 m	Dir.: Dr. T. Reitsma
rainfall : 804 mm	Deputy dir.: Dr. G. Scholten
sunshine: ± 1548 h	Deputy dir. (extension section):
	Ir. C.A.M. Groenewegen
1.1.1 Plant nutrition and soil science	Head: Vacancy
	N.A. Straver
	Ms. M.G. Warmenhoven
	Th. J.M. van den Berg
1.1.2 Economics and management	Head: Dr.Ir. J. van de Vooren*
	Ing. F. Bakema
Coördination	Ir. G. Beers
Economics	Ir. E. van Rijssel (LEI)*
	Ing. L. Oprel
Efficiency	Ms. Ing. M. van der Schilden (IMAG)
Mechanization	Ing. P.A. van Weel (IMAG)
1.1.3 Plant physiology and selection	Head: Ms. Dr.Ir. L. Leffring
Post harvest potplants	Ms. Ir. M.J.H. Rewinkel-Jansen
Growth regulators	Dr.Ir. W. Sytsema*
Post harvest cutflowers	Ms. K.G. Elfering-Koster
	Ing. P. van Leeuwen
	Ms. Ing. E.Ch. Kalkman
Selection and Tissue Culture	Ms. Dr.Ir. L. Leffring
	Ing. H.F. Esendam
	Ms. Ing. Y. Hermes
1.1.4 Horticulture and glasshouse climate	Head:Dr.Ir. C. Vonk Noordegraaf*
Cutflowers carnation	Ing. C.G.T. Uitermark
Different cutflowers	Th.M. van der Krogt
Gerbera, orchids	Ing. P. van Os
Potplants	Ing. M.P. Beuzenberg
	J. Westerhof
New crops	Ir. P.W.M. Lentjes
New crops	Ms. Ing. M.T. van der Zande

	G.J. van den Broek
	Ing. E.M. van Dordrecht
Glasshouse climate	Ir. G.A. van den Berg
	Ms. M. van Schoonhoven
	Ms. M.G. Hoogeveen
	Ing. F. Steinbuch
Reject and waste heat	Ms. Ir. J.V.M. Vogelezang
Variety trials	Ir. A. de Gelder (RIVRO)
	Ms. M.A.C. Reijnders
	Ms. L. Legemaate (RIVRO)
1.1.5 Pests and diseases	Head: Dr. G. Scholten
Mycology	Ir. H. Rattink (IPO)
	Ms. M.C. Dil
	Ms. C. Sanders (NAKS)
Entomology/acarology	M. van de Vrie (IPO)
Spodoptera exigua	Drs. P.H. Smits (LH)
	M. Boogaard
Post harvest treatment	Ir. A.K.H. Wit
Virology and tissue culture	Ir. F.A. Hakkaart (IPO)
	Ms. J.M.A. Versluijs
Chemical control of pests and diseases	I.P. Rietstra
	Ing. J.J. Amsing
1.1.6 General departments	Head: D. Hooijer
Statistics	A.L. Verlind
Informatica	Ing. L.J.G. Aarsen (system analist)
Publication, library, documentation	Head: M.J. 't Hart
Library and documentation	Ms. C.M.M. de Boer
Photography	H. Stephan
Nursery	Head: Ing. J. Bonnyai
Supervisor	H. Kleinhesselink
Visitors: Reception	Ms. M. Los
Guide	G. de Wagt

2 Boskoop

2.1 Proefstation voor de Boomteelt en het Stedelijk Groen (Research Station for nurserystock and urban greenery) elevation: 1.8 m rainfall : 815 mm sunshine: 1520 h	Valkenburglaan 3, 2771 CW Boskoop, Phone: 01727-5220 P.O. Box 118, 2770 AC Boskoop Dir.: Ir. W.J. Bosch Dep. Dir.: Ir. R.J.M. Meijer
Nurserystock	
Propagation and cultivation	Dr.Ir. M.K. Joustra*
	Ing. P.A.W. Verhoeven
Tissue culture	Ir. B.P.A.M. Kunneman
Plant Pathology and weed control	Ir. N.G.M. Dolmans
Soil and fertilizers	Ing. Th.G.L. Aendekerk
Assortment	Ing. G. Fortgens
	H.J. van de Laar
Breeding	Ir. A.S. Bouma
Economics	Ir. A.G. van der Zwaan
	Ing. M. Bouman
Urban Horticulture	
Soil and Water	Ir. A. Maris

Economics

Ing. R. Ligteringen

3 's Gravenhage

3.1 Landbouw-Economisch Instituut
(Agricultural Economics Research Institute)
elevation: 1.2 m
rainfall : 750 mm

Conradkade 175, Den Haag
P.O. Box 29703, 2502 LS Den Haag
Phone: 070-614161
Dir.: Prof. Drs. J. de Veer
Dep. Dir.: Dr. L.C. Zachariasse
Ir. A.L.G.M. Bauwens

The Institute was founded in 1940 by an agricultural organization and soon became a central institute for economic research in agriculture and fisheries, financed by the agricultural industry and Government on a fifty-fifty basis. Since 1 January 1971, it has been under the Ministry of Agriculture and Fisheries. The members of the board are appointed by the Minister of Agriculture and Fisheries.
The institute studies the management and general economics of Dutch agriculture, horticulture, forestry and fisheries. Descriptive and prognostic research is carried out for the Government and for farmers. Research on farms and farm management includes the analysis of factors influencing quantitative ratios in particular technical circumstances. Results are applied in the improvement of farm organization and farm management, improvement of farm structure and the introduction of new crops or types of enterprises. For the Government and professional organizations, the institute carries out policy-orientated research. This includes the following: Analysis of incomes and capital (e.g. costs, returns, profitability, finance, input and productivity); Market research (e.g. supply and demand for products, control of supply the best siting for production and market openings); Analysis of factors optimizing labour, land and durable capital goods.

Horticultural Department	Head: Ir. D. Meijaard*
Ornamentals, open air crops	Dep. Head: Ir. W.G. de Haan*
Glasshouse crops	Drs. A. Boers
Macro-economics, finance	Ir. B.M.M. Kortekaas
Market research	Ir. E.H.J.M. de Kleijn*

4 Haren - Groningen

4.1 Instituut voor Bodemvruchtbaarheid (IB)
(Institute for Soil Fertility)
elevation: 2.5 m
rainfall : 804 mm
sunshine: 1445 h

Oosterweg 92, P.O. Box 30003,
9750 RA Haren (Gr.)
Phone: 050-346541
Dir.: Dr.Ir. K. Harmsen
Dep. Dir.: Dr. P.J. Lont

The institute conducts research on soils and fertilization for the benefit of crop production, grassland farming, and horticulture. Raising production and improving quality of our agricultural and horticultural products are important aspects.
The activities of the Department of Fertilization in Horticulture include the establishment of recommendations for flowers and vegetables on the basis of soil and plant analysis, and the use of various (artificial) substrates (chemical and physical characteristics). In vegetables quality aspects are investigated, specifically the content of nitrate, bromide and heavy metals. Fertilization of tree nursery stock grown in pots receives attention. In flower bulb production, research on nitrogen fertilization is aimed at maximum yield and optimum forcing quality of the bulb. In fruit production, nitrogen and potassium fertilization, the possibility of predicting keeping quality on the basis of fruit analysis, and the efficiency of trickle irrigation are investigated. Research has been started into the soil and fertilizer requirements of trees in urban surroundings.

Soil physics, soil tillage and root functioning	Ir. P. Boekel
Soil chemistry	Dr. A.J. de Groot
Soil biology	Dr. H. van Dijk
Fertilization in horticulture	Dr. Ir. J. van der Boon

5 Horst

5.1 Proefstaton voor de Champignoncultuur (Mushroom Experimental Station) elevation: 23.4 m rainfall : 733 mm sunshine: 1430 h	Peelheideweg 1, 5966 PJ America, P.O. Box 6042, 5960 AA Horst Phone: 04764-1944 Dir.: Dr. L.J.L.D. van Griensven
Mushroom nutrition and composting	Drs. J.P.G. Gerrits
Fructification and casing soil	Drs. H.R. Visscher
Breeding and genetics	Ms. Dr. G. Fritsche
Diseases and pests	Drs. F.P. Geels
Economics	Ir. P.M. Schaper

6 Lelystad

6.1 Proefstation voor de Akkerbouw en de Groenteteelt in de Vollegrond (PAGV) (Research Station for Arable Farming and Field Production of Vegetables) elevation: -4.2 m rainfall : 768 mm sunshine : 1430 h	Visitors: Edelhertweg 1, Lelystad Post: P.O. Box 430, 8200 AK Lelystad Phone: 03200 - 22714 Dir.: vacancy Dep. Dir.: Drs. J. Kamminga
Vegetable crop production	Head: Dr.Ir. G.J.H. Grubben* Dep. Head: Ir. P.H.M. Dekker
Leaf-, stem- and fruit vegetables	Ir. G. van Kruistum/J.T.K. Poll
Cabbages	Ir. R. Booij/C.P de Moel
Bulbous-, tuberous and root plants	Ing. J.A. Schoneveld Ms. Ing. M.H. Roodzant
Leguminous plants	Ir. P.H.M. Dekker/J.J. Neuvel
Research cultural value and new crops	Ing. A.R. Biesheuvel/J. de Kraker Ing. C.A.Ph. van Wijk
Regional research	Ing. F.M.L. Kanters
Farming systems	Head: Ir. C.A.A.A. Maenhout Dep. Head: Ir. H.F.M. Aarts
Soil, fertilizers and mechanization	Ing. J. Alblas/Ir. W.A. Dekkers Ing. L.M. Lumkes Ing. H.H.H. Titulaer
Plant protection	Ir. H.G.M. van den Brand J. Jonkers/Ir. P.M. Spoorenberg Ms. Ir. G. Dijst/A. Ester Ir. C. Kaai/Ms. Ing. R. Meier Ir. P.W.J. Raven
Cropping and farming systems	Ing. O. Hoekstra/Ing. Th. Huiskamp Ir. J.G. Lamers/Ir. H.F.M. Aarts Ing. H. Drenth/Ms. Ing. K. Hindriks Ing. W. Stol/Ir. C.L.M. de Visser Dr. P.H. Vereijken/Ir. F.G. Wijnands
Farm management	Head: Drs. J. Kamminga Dep. Head: Drs. S. Cuperus
Economic research	Drs. S. Cuperus/Ir. C.F.G. Kramer Ing. H. Preuter
Farm management	Ing. K.P. Groot Ing. S.R.M. Janssens Dr. A.T. Krikke/Ir. R. Piepers

Experimental farm	Head: P. ter Steege
Vegetable growing	P.W.V. Bakker
General Advisory Service for arable farming and field production of vegetables	Ir. C.J.J. de Kroon Ing. M. van der Ham/N.J. Snoek

7 Lisse

7.1 Stichting Laboratorium voor Bloembollenonderzoek (Bulb Research Centre)
elevation: -0.3 m
rainfall : 787 mm
sunshine : 1570 h

Vennestraat 22
Postbus 85
2160 AB Lisse
Phone: 02521 - 19104
Dir.: Dr. R.J. Bogers*
Dep. Dir.:Dr. J.C.M. Beijersbergen*
Dep. Dir. and nat. adv. officer for bulb culture: Vacancy

Dept. of Plant Physiology and Biochemistry	Head: Dr. J.C.M. Beijersbergen*
Dept. of Crop Science, Economics and Mechanization	Head: Dr. A.J. Dop
Dept. of Plant Diseases and Crop Protection	Head: Dr. J. van Aartrijk

8 Naaldwijk

8.1 Proefstation voor Tuinbouw onder Glas (Glasshouse Crops Research and Experiment Station)
elevation: 1.2 m
rainfall : 822 mm
sunshine : 1540 h

Zuidweg 38, P.O. Box 8
2670 AA Naaldwijk
Phone: 01740 - 26541
Dir.: Ir. E. Kooistra*
Dep. Dir.: Ir. J.M. Jacobs*
Dr. Ir. G. Weststeijn

Glasshouse crops and glasshouse climate	Ir. C.M.M. van Winden*
Diseases and pests	Dr. Ir. L. Bravenboer*
Plant physiology	Dr. Ir. P.J.A.L. de Lint*
Soils, fertilizers and water management	Ir. J. van den Ende*
Economics, mechanization and rationalization	Ir. J.C.J. Ammerlaan

9 Wageningen

9.1 Centrum voor Agrobiologisch Onderzoek (CABO) (Centre for Agrobiological Research)
elevation: 7 m
rainfall : 780 mm
sunshine : 1440 h

Bornesteeg 65, P.O. Box 14
Phone: 08370 - 19012
6700 AA Wageningen
Dir.: Dr.Ir. J.H.J. Spiertz

Research is focussed on basic processes in plants, crops and eco-systems related to problems of crop husbandry, plant breeding, plant protection, quality of roughage, quality of environment, and regional management of land use.
Fields of research are: plant physiology, biochemistry, ecology, theoretical and experimental crop science, weed science, ecology and management of productive and non-productive vegetation.

Nitrate uptake, metabolism, and accumulation in vegetables	Ms. Ir. M. Blom-Zandstra
Physiological aspects of greenhouse crop production related to energy requirement	Dr. Ir. H. Challa* Drs. J.A. ten Cate/Ir. J.G. Gijzen Ir. P.A.C.M. van de Sanden* Dr. Ir. A.H.C.M. Schapendonk Dr. F.N. Verkleij

Physiological aspects of keeping properties and vase life of cut flowers	Dr. Ir. H.C.M. de Stigter* C.R. Vonk
Root physiology; environmental conditions influencing nitrate accumulatio in vegetative crops	Dr. B.W. Veen
The physiology of plant hormones in higher plants: senescence of plants	Dr. H. Veen
Role of growth hormones in flower development of iris; enzym immuno assays	C.R. Vonk
Growth, development and dry matter production of bulbous crops	Ir. M. Benschop*
Weed ecology and competition	Ir. W. de Groot
Soil covering and weed control	Dr. Ir. M. Hoogerkamp
Physiological research on herbicides: uptake, photosynthesis, resistance	Dr. Ir. J.L.P. van Oorschot
Biological weed control	Dr. Ir. P.C. Scheepens
Weed control in small acreage crops	Ir. P.M. Spoorenberg (PAGV)

9.2 Instituut voor Cultuurtechniek en Waterhuishouding (ICW) (Institute for Land and Water Management Research)
elevation: 7 m
rainfall : 780 mm
sunshine : 1440 h

Marijkeweg 11, P.O. Box 35
Phone: 08370 - 19100
6700 AA Wageningen
Dir.: Ir. G.A. Oosterbaan
Dep. Dir.: Dr. R.A. Feddes

Studies are made in the fields of: hydrology, hydrogeology, evapotranspiration, drainage, salinity problems, sprinkling and sub-irrigation, aeration, land treatment of industrial waste water, pollution of surface and ground water, regional water management schemes, land levelling, changing of soil profiles, subsidence and compaction of soils, requirements of size and shape of fields and holdings, accessibility of fields and farm buildings, site, size and shape of outdoor recreation projects, urbanization aspects of rural regions, physical planning of a region, execution techniques of land consolidation plans, pilot plans of land development schemes, short-and long-term effects of amelioration measures on the economy of farm and region, benefit-cost analyses.

General water quality	Dr. P.E. Rijtema
Water quality modeling	Ir. C.W.J. Roest
Water quality in horticulture	Dr. Ph. Hamaker/Ing. C. Ploegman
General hydrology	Dr. R.A. Feddes
Water management in horticulture	Ir. M. de Graaf
General soil technology	Dr. A.L.M. van Wijk
Soil technology in horticulture	Ir. G.G.M. van der Valk
General land use planning	Ing. Th.J. Linthorst
Layout horticultural areas	Ir. C.G.J. van Oostrom

9.3 Instituut voor Mechanisatie, Arbeid en Gebouwen (IMAG) (Institute of Agricultural Engineering)
elevation: 7 m
rainfall : 780 mm
sunshine : 1440 h

Mansholtlaan 10-12, P.O. Box 43
Phone: 08370 - 94911
6700 AA Wageningen
Dir.: Ir. A. Hagting
Sr. Dep. Dir.: Ir. J.J. Laurs

Research is carried out in order to promote an efficient mechanization and automatization in agriculture and horticulture; an optimal application of labour and an efficient work organization in agriculture and horticulture. The realization of efficient and relative cheap agricultural buildings and greenhouses, including climate control in these spaces; a effective technical contribution of agricultural and rural community to aspects as prevention of pollution, landscape maintenance, open air recreation grounds and liveability.

Engineering division	Head: Ir. U.D. Perdok
Labour and Work Management Division	Head: Drs. K.E. Krolis
Buildings Division	Head: Ir. W.H. de Brabander

Climate Control Division	Head: Ing. J. Maring
Crops Liaison Department	Head: Ir. K. de Koning
Horticulture and Recreation Liaison Department	Head: Ir. C.J. v.d. Post
Livestock and Environment Liaison Department	Head: Ing. G. Postma
Staff Departments:	
Mechanization statistics	Ir. A. Kraai
Research projects developing countries and national and international aspects	Ir. G.J. Poesse

9.4 Instituut voor Plantenziektenkundig Onderzoek (IPO)
(Research Institute for Plant Protection)
Foundation under supervision of the Ministry of Agriculture and Fisheries
elevation: 7 m
rainfall : 780 mm
sunshine : 1440 h

Binnenhaven 12, P.O. Box 9060
6700 GW Wageningen
Phone: 08370 - 19151
Dep. Dir.: Ms. Drs. F. Quak

The Institute's research programme deals with protection of plants against fungi, bacteria, viruses, viroids, insects, mites and nematodes as well as with effects of air pollution on plants. The research findings mainly support the advisory services, the Plant Protection Service, inspection services for propagation material of plants, plant breeding, environmental hygiene and nature management.

9.4.1 Department of Entomology	Head: Dr. Ir. A.K. Minks
Pests of: outdoor vegetables	Ir. H. den Ouden Dr. Ir. J.A.B.M. Theunissen
vegetables in glasshouses	Ir. P.M.J. Ramakers
ornamentals	M. van de Vrie
fruit trees	Dr. L. Blommers
Aphidology	Ms. J.D. Prinsen
Dynamic simulation models	Dr. Ir. J. Booij
9.4.2 Department of Mycology and Bacteriology	Head: Ir. A. Tempel
Fungal diseases of: outdoor vegetables	Ms. Ir. G. Dijst
vegetables in glasshouses	Ir. N.A.M. van Steekelenburg
lettuce and spinach	Ms. Ir. I. Blok
leguminous crops, cucumber, tomato	Dr. Ir. M. Gerlagh
ornamentals	Ir. H. Rattink
potatoes	Dr.Ir. L.J. Turkensteen
Bacterial diseases	Dr. J.M. van Vuurde
Serology	H. Vruggink
9.4.3 Department of Nematology	
Potato cyst nematodes	Ms.Ir. L.J. M. den Nijs/Drs. T.H. Been Ms. Drs. C.H. Schomaker
Stem nematodes, beet cyst nematode	Ir. C. Kaai
Root knot nematodes in tomato and cucumber	Ir. W. Windrich
9.4.4 Department of Virology	Head: Ms. Drs. F. Quak
Virus diseases of: outdoor vegetables	Dr. Ir. L. Bos
vegetables in glasshouses	Dr. Ir. A.T.B. Rast
ornamentals	Ir. F.A. Hakkaart
fruit trees, woody ornamentals	F.A. van der Meer
potatoes	Dr. Ir. J.A. de Bokx
Serology: polyclonal and monoclonal antibodies	D.Z. Maat Dr. P.M. Boonekamp (Agr. Un.)
Chromatography and electrophoresis	W.H.M. Mosch
Biophysics	Dr. Ir. H. Huttinga
Meristem culture	Ms. Drs. F. Quak
9.4.5 Department of Phytotoxicology of Air Pollution	Head: Dr. A.C. Posthumus

Monocotyledonous ornamentals and glasshouse crops	H.G. Wolting
Trees and shrubs	J. Mooi
Diagnosis and damage evaluation	Ir. L.J.M. van der Eerden Drs. A.E.G. Tonneijck
Monitoring of effects	H. Floor/Drs. A.E.G. Tonneijck
Physiology and interactions with pathogens	Drs. C. Kliffen
Dynamic simulation models	Ir. A.M. Leemans (Agr. Un.)

9.5 Stichting ITAL (Association Euratom-ITAL)
Research Institute ITAL
elevation: 25 m
rainfall : 780 mm
sunshine : 1440 h

Keijenbergseweg 6, P.O. Box 48
6700 AA Wageningen
Phone: 08370 - 91911
Dir.: Dr.Ir. A. Ringoet

The Institute's research programme comprises genetic engineering of plant cells, biotechnological production of biocides, soil biology, genetic control of mosquitos and advanced methodology

Plant regeneration from protoplasts	Prof.Dr. B. de Groot Dr. G.M.M. Bredemeijer Dr. Ch.H. Hänisch ten Cate Dr. K. Sree Ramulu/Dr.Ir. S. Roest
Chromosome transfer and somatic hybridization in plants	Prof.Dr. B. de Groot Dr. L.J.W. Gilissen Dr.Ir. A.M.M. de Laat Dr. K.J. Puite/Ir. H.A. Verhoeven Drs. E. de Vries
Molecular biology of gene transformation in the potato	Prof.Dr. B. de Groot Dr. F. Heidekamp/Dr. W.J. Stiekema
Biotechnological production of biocidal compounds (Tagetes spec.)	Prof.Dr. B. de Groot Dr.Ir. H. Breteler/Drs. D.H. Ketel
Modification of toxin production in Bac. thuringiensis	Prof.Dr. B. de Groot Dr.Ir. L. Visser/Dr. C. Waalwijk
Role of microorganisms in the turnover of nutrients in soil	Dr. J.A. van Veen Dr. S.C. van de Geijn Drs. P.J. Kuikman/Dr.Ir. R. Merckx
Dynamics of bacterial species in the rhizosphere and bulk soil	Dr. J.A. van Veen/Dr. J.H. Becking Dr.Ir. J.D. van Elsas Drs. A.F. Dijkstra/Ir. J. Postma
Genetic control of mosquitoes	Dr.Ir. A. Ringoet/Dr. A.M. Feldmann Drs. A. Mans/Dr. A.S. Robinson
Instrumentation, methodology development	Dr. H.A. van Hoof

9.6 Instituut voor Veredeling van Tuinbouwgewassen
(IVT) (Institute for Horticultural Plant Breeding)
elevation: 7 m
rainfall : 780 mm
sunshine : 1440 h

Mansholtlaan 15, P.O. Box 16
6700 AA Wageningen
Phone: 08370 - 19123
Dir.: Dr. N.G. Hogenboom
Dep. Dir.: Ir. M. Nieuwhof

The following four main subjects of research were opted for: resistance to disease and pests, enabling a limited use of chemical control and improving yield stability; efficiency in the use of energy, enabling a saving in energy in glasshouse culture and - by a better use of light and heat - an improvement of quality and profitability; species crosses, enabling a better use of the natural genetic variation for the improvement of various characters; cell techniques, by which the efficiency of breeding programmes can be increased both with respect to selection and to the combination of genetic material.
Besides these main items, objectives receiving attention are the formation of gene banks for some major crops,

quality, nutritional safety, widening of the range of varieties and cooperation with research in developing countries. Development of efficient breeding methods, creation and distribution of populations from which better varieties can be selected by the commercial breeders and, in some cases, the breeding of better varieties by the Institute itself; carrying out basic research in connection with the above tasks; spreading of knowledge (e.g. inheritance of desired characters) useful to the Dutch horticultural industry, particularly with regard to the foregoing items.

9.6.1 Vegetables:	Head: Dr. O.M.B. de Ponti
Cucumber	vacancy
Leaf and stem vegetables	Ir. K. Reinink
Leek, onion, cole crops, new vegetables	Ir. Q.P. van der Meer
Pulses	Dr. E. Drijfhout
Rootcrops	Ir. M. Nieuwhof
Strawberry	L.M. Wassenaar
Tomato, pepper	Dr. W.H. Lindhout
Genetic resources	Ms. Ir. I.W. Boukema
Haploidy, resistance	Ir. R.E. Voorrips
Resistance to insects and mites	Dr. O.M.B. de Ponti

9.6.2 Fruit and Roses	Head: Dr. T. Visser*
Apple, pear	Dr. T. Visser*
Resistance	Ir. W.E. van de Weg
Roses	Ing. D.P. de Vries

9.6.3 Ornamentals	Head: vacancy
Carnation	Dr. L.D. Sparnaaij
Chrysanthemum	Dr. J. de Jong
Freesia, nerine	vacancy
Lily, hyacinth	Dr. J.M. van Tuyl
New ornamentals	G.H. Kroon
Potplants	Ms. Dr. J. Heyting*
Tulip, iris, gladiolus	Ir. J.P. van Eijk
Woody ornamentals	Ms. Ir. A.S. Bouma

9.6.4 Biotechnology	Head: Dr. J.J.M. Dons
Bio- and phytochemistry	Dr. J.J.M. Dons
Cell biology	Ms. Dr. C.M. Colijn-Hooymans
Cytogenetics	Dr. J.N. de Vries
Flower biology	Dr. R.J. Bino
In vitro culture	Ir. J.B.M. Custers

9.6.5 Population Research	Head: Ir. M. Nieuwhof
Biosystematics	Dr. L.W.D. van Raamsdonk
Ecophysiology	Dr. L. Smeets
Resistance testing	A.C. van der Giessen
Statistics, biometrics	Ir. J. Jansen

9.7 Landbouwhogeschool Wageningen
(Wageningen Agricultural University)

Salverdaplein 10, P.O. Box 9101
6700 HB Wageningen

9.7.1 Vakgroep Tuinbouwplantenteelt
(Department of Horticulture)
elevation: 30 m
rainfall : 780 mm
sunshine : 1440 h

Haagsteeg 3, P.O. Box 30
Phone: 08370 - 84096
6700 AA Wageningen
Head: vacancy

The emphasis of research on the effect of the conditions of the environment and growth regulators on plant growth, development and regeneration. The research programme includes aspects of the different types of horticultural crops and can be divided into the following subjects: 1. Propagation (vegetative and generative), 2. Production, 3. Quality.

Growth and development of bulbous and cormous plants	Dr.Ir. J. Berghoef*
Growth regulating substances on horticultural crops; endogenous growth substances	Dr. Ir. J. van Bragt*
Effects of environmental factors on plant-water relations	Ir. G.T. Bruggink
Root stocks of glasshouse roses	Ir. H.W.M. Fuchs
Development of a simulation model describing growth and production of glasshouse tomatoes	Ir. E. Heuvelink
Pomology (apple, pear)	Dr. Ir. P.A.M. Hopmans
Suitability of the Dutch climate for horticultural production	Dr. Ir. H.G. Kronenberg*
Horticultural applications of in vitro culture	Prof. Dr. Ir. R.L.M. Pierik*
Floriculture; flagrance in horticulture	Dr. Ir. P.A. van der Pol
Growth and development of bulbous and cormous plants	Dr. Ir. J. Berghoef*
Ageing of cut flowers	Ir. G. Spikman

9.7.2 Vakgroep Plantenfysiologie
(Department of Plant Physiology)
elevation: 30 m
rainfall : 780 mm
sunshine : 1440 h

Arboretumlaan 4
Phone: 08370 - 82146
6703 BD Wageningen
Head: Prof. Dr. J. Bruinsma

Molecular plant physiology	Dr.Ir. L.C. van Loon
Translocation of assimilates and tuberization	Dr. D. Vreugdenhil
Postharvest physiology of flowers and pollen	Dr.Ir. F.A. Hoekstra
Pre- and postharvest physiology of fruits	Prof.Dr. J. Bruinsma/Dr. E. Knegt Dr.Ir. A. Varga*
Pre- and postharvest physiology of seeds	Prof.Dr. C.M. Karssen
Physiology of cultures in vitro	Dr.Ir. A. Varga*
Methodology of phytohormone determinations	Dr. E. Knegt

9.7.3 Vakgroep Plantenfysiologisch Onderzoek
(Department of Plant Physiological Research)
elevation: 30 m
rainfall : 780 mm
sunshine : 1440 h

Generaal Foulkesweg 72
Phone: 08370 - 82800
6703 BW Wageningen
Head: Prof.Dr. W.J. Vredenberg

The bioenergetic and biocybernetic aspects of light-dependent processes in plants and their regulatory function in growth and development are the main subjects of the research program. Also the effects of NH_3 and SO_4 pollution on photosynthesis are under investigation.

9.7.4 Vakgroep Plantentaxonomie
(Department of Plant Taxonomy)
elevation: 30 m
rainfall : 780 mm
sunshine : 1440 h

Generaal Foulkesweg 37
Postbus 8010, 6700 ED Wageningen
Phone: 08370 - 83160
Head: Prof.Dr.Ir. L.J.G. van der Maesen

Cytotaxonomy of *Apocynaceae*, Begonia, *Orchidaceae*	Ir. J.C. Arends
Cultivated plants, pollen morphology	Dr.Ir. R.G. van den Berg
Dracaena	Dr. Ir. J.J. Bos
Clematis, cultivated plants	Ir. W.A. Brandenburg*
Dichapetalaceae, Connaraceae of Africa	Dr. Ir. F.J. Breteler
Leader Herbarium Staff, Herbarium curator	J. de Bruijn
Identification, editing	Mrs. Drs. F.J.H. van Dilst

Librarian	C.T. de Groot
Cytotaxonomy, database of tropical greenhouses	F.M. van der Laan
Apocynaceae of Africa	Dr. A.J.M. Leeuwenberg
Leguminosae, cultivated plants and relatives	Prof. Dr. Ir. L.J.G. van der Maesen
Identification of European plants	Dr. F.M. Muller
Landolphia (Apocynaceae)	Drs. J.G.M. Persoon
Clematis, Aster, cytotaxonomy, cultivated plants	J.G. van de Vooren
Taxonomy of hardy plants, *Salix*	Mrs. N. Wilders
Aquatic plants, history of biology	Prof. Dr. H.C.D. de Wit
Cultivated plants, history of biology, keeper of botanical gardens	Dr. D.O. Wijnands
Biosystematics of *Lactuca sativa* and related spp.	Mrs. Drs. I.M. de Vries
Begoniaceae of Africa	Dr. Ir. J.J.F.E. de Wilde
PROSEA, documentalist	F.M. Stavast
PROSEA, Plant resources of S.E. Asia, taxonomy	Dr.Ir. P.C.M. Jansen

9.7.5 Botanical Gardens, Agricultural University
elevation: 30 m
rainfall : 780 mm
sunshine : 1440 h

Gen. Foulkesweg 37, Postbus 8010
6700 ED Wageningen
Phone: 08370 - 83160/83461
Keeper: Dr. D.O. Wijnands

Taxonomy of cultivated plants	Dr. D.O. Wijnands
Database of botanical gardens, nursery	J.F. Aleva
Rosa, publicity	J. Belder
Exchange	H.H. de Leeuw
Succlentarium Flevohof, *Faucaria, conophytum*	Ir. L.E. Groen
Educational programmes	Mrs. M. Pott
Curator	W. Keuken
Landscaping	Mrs. J.J.P.M. Segers
Tropical greenhouses	G. Peperkamp
De Dreijen	T. Stoker
Belmonte	R. van Velsen

9.8 Plantenziektenkundige Dienst (PD)
(Plant Protection Service)
elevation: 10 m
rainfall : 780 mm
sunshine : 1440 h

Geertjesweg 15, P.O. Box 9102
6700 HC Wageningen
Phone: 08370 - 19001
Telex: 45163 PD
Dir.: Ir. H.J. de Bruin
Asst. Dir.: Ir. P. Kleijburg
Asst. Dir.: K.F. Scholten

The Plant Protection Service was established in 1899. It reports to the Director-General of Agriculture and Food Supply, Ministry of Agriculture and Fisheries. The head office is located in Wageningen. In addition there are 14 district offices in the country, which are in charge of the field work.

Directorate's advisory staff:

Pesticides and pest control	Ir. H.M. Nollen
Phytosanitary matters	Ir. P. Kleijburg
Import and export matters	Ir. J.A.J. Veenenbos
Operational research and diagnostics	Dr.Ir. P. van Halteren

Technical departments:

Biological laboratory. (Disease and pest control)	Dr. Ir. T. Kooistra
Chemical analysis of pesticides	Drs. A. Martijn
Weed science and weed control	Ir. H. Naber
Field trials. (Disease and pest control)	Dr. W.H. van Eck

Flower bulbs	Ir. C.N. Silver
Vegetables, flower bulbs and public green space	Ir. M.G. Roosjen
Nursery-stock and fruit crops	Ir. C.A.R. Meijneke*
Arable crops	Ir. A. Oldenkamp
Entomology	Dr. S.A. Ulenberg
Mycology	Drs. G.H. Boerema
Bacteriology	Dr. H.J. Miller
Nematology	Ir. P.W.Th. Maas
Import and export	Ir. J.K. Water
Quarantine Matters	Ir. A. Treur

9.9 Rijksinstituut voor het Rassenonderzoek van Cultuurgewassen (RIVRO)
(Government Institute for Research on Varieties of Cultivated Plants)
elevation: 7 m
rainfall : 780 mm
sunshine : 1440 h

Nieuwe Wageningseweg 1
P.O. Box 32, 6700 AA Wageningen
Phone: 08370 - 19056
Dir.: Ir. M.J. Hijink
Dep. Dir.: Ir. H. Vos

The Institute is responsible for research on varieties of cultivated plants. Botanical research on agricultural, horticultural and forest crops on behalf of registration and protection of varieties. Research on the cultural value of varieties of agricultural and horticultural crops with regard to the compilation of descriptive lists of varieties.
Publications: Descriptive lists of varieties of agricultural and horticultural crops; Annual bulletins of varieties; Technical catalogues and reports.

Horticultural Botany	Head: Ir. F. Schneider
Department for ornamental, nursery and fruit crops	Ir. C.J. Barendrecht
Department Vegetables I cabbage, tubers, bulbs, carrots, lettuce and leek	Ir. N.P.A. van Marrewijk
Department Vegetables II other leaf crops, pulses and mushrooms	Ir. H. Koster
Performance trials on horticultural crops	Head: Ir. J.J. Bakker
Department for outdoor vegetables	Ir. F. van der Zweep
Department for glasshouse vegetables	c/o Exp. St., Zuidweg 38 P.O. Box 8, 2670 AA Naaldwijk Ir. J.P. Straatsma
Department for fruit crops	c/o Exp. St., Brugstraat 51 4475 AN Wilhelminadorp Ing. P.D. Goddrie
Department for flowerbulbs	c/o Exp. St., Vennestraat 22 P.O. Box 85, 2160 AB Lisse Ir. H.P. Pasterkamp
Department for ornamental crops	c/o Exp. St., Linnaeuslaan 2a 1431 JV Aalsmeer Ir. A. de Gelder

9.10 Sprenger Instituut (Institute for Research on Storage and Processing of Horticultural Produce)
elevation: 7 m
rainfall : 780 mm
sunshine : 1440 h

Haagsteeg 6, P.O. Box 17
6700 AA Wageningen
Phone: 08370 - 19013
Dir.: Drs. G.J.H. Rijkenbarg
Dep. Dir.: Ing. H.F.Th. Meffert

Grading, packing, transporting and storage of vegetables, flowerbulbs, flowers and fruit; research on the preservation value of new fruit and vegetable types for different processing purposes, technological research and quality research on processed products; farm economic and techno-physical research. The Institute is incorporated in the Foundation for Agricultural Research in which growers, merchants, industrialists and consumers are represented.

Physical engineering	Ir. G. van Beek
Biology	Ms. Chr.E.M. Berkholst
Quality research	Ir. F.J. Westerling
Chemistry	Dr. N. Gorin
Economics	Drs. P. Greidanus
Biochemistry	Ir. W. Klop
Food Chemistry	Drs. M.A. van der Meer
Transport	Ir. G.H. van Nieuwenhuizen
Physics	Ir. J.W. Rudolphij
Physiology	Drs. W.G. van Doorn
Storage	Drs. S.P. Schouten*
Food technology	Ir. E.P.H.M. Schijvens
	Ir. E. Steinbuch
Process technology	Drs. L.M.M. Tijskens

9.11 Stichting voor Bodemkartering (STIBOKA)
(Soil Survey Institute)
elevation: 7 m
rainfall : 780 mm
sunshine : 1440 h

Marijkeweg 11, P.O. Box 98
6700 AA Wageningen
Phone: 08370 - 19100
Telex: 75230VISI
Dir.: Dr. Ir. F. Sonneveld
Dep. Dir: Dr. Ir. J. Bouma

Study of the soil of the Netherlands for classification and mapping; systematic survey of the soils of the Netherlands; interpretations of soil survey data for various types of land use (e.g. crop production, grassland farming, forestry urban and recreational land use); systematic geomorphological survey of the Netherlands (in cooperation with the Geological Survey of the Netherlands); ecological and physiognomic landscape survey of the Netherlands; commissioned soil surveys and soil survey interpretation (e.g. for rural development and reconstruction, development of urban areas, rural and urban zoning); development cooperation.

9.11.1 Soil and Landscape Research and Applications	Dr. J.A. Klijn
Soil Physics and Hydrology	Dr.Ir. J.M.H. Hendrickx
	Ir. J.H.M. Wösten
Land Use	Ir. H.A.J. van Lanen
	Ir. J.G.C. van Dam (horticulture)
Landscape	Drs. J.A.J. Vervloet
	Drs. A.A. de Veer
Soil structure and micropedology	Dr. M.J. Kooistra/Dr. E.B.A. Bisdom
Soil chemistry and clay mineralogy	Dr. Ir. A. Breeuwsma
	Ir. W. de Vries
9.11.2 Soil Survey	Ir. G.J.W. Westerveld
Soil coordination	Ir. G.G.L. Steur
Scientific advisors	Ir. A.F. van Holst/Ir. J. Stolp
	Ir. F.J. Stuurman/Ir. C. v. Wallenburg
Systematic soil survey	Ing. J.J. Vleeshouwer
Soil and landscape surveys on contract	Ir. B.J.A. van der Pouw
Geology, geomorphology and palaeobotany	Drs. J.A.M. ten Cate
	Ir. M.W. van den Berg
Development cooperation	Drs. R.F. van de Weg
	Ir. W. Andriesse
Public relations	Ing. A. Reijmerink
Information system earth sciences-statistics	Ir. A.K. Bregt
	Dr. Ir. J.J. de Gruijter
Cartography	Drs. A.M. van Slobbe
Editing	J.W. Zwolschen
Administration	G.H. Westerbeek

10 Wilhelminadorp

10.1 Proefstation voor de Fruitteelt — Brugstraat 51

(Research Station for Fruit Growing)
elevation: 1.5 m
rainfall : 758 mm
sunshine : 1550 h

4475 AN Wilhelminadorp
Phone: 01100 - 16390
Dir.: Ir. R.K. Elema
Dep. Dir.: Dr. Ir. H.J. v. Oosten

The research station deals with research on top and small fruit (including strawberry) growing and on fruit tree nursery.

Growing techniques top fruit — Dr. Ir. S.J. Wertheim; Ms. Drs. P.S. Wagenmakers; Ing. P.D. Goddrie
Variety testing top fruit — Ir. J. Dijkstra
Growing techniques small fruit — J. Goedegebure
Economics — Dr. J. Tromp
Plant physiology — Dr. P. Delver
Soil sciece and water management — Drs. H.A.T. v.d. Scheer
Phytopathology — Dr. J. Woets
Entomology

Private Enterprises

11 Barendrecht

11.1 Nicherson-Zwaan B.V.
elevation: -0.3 m
rainfall : 809 mm
sunshine : 1530 h

Gebroken Meeldijk 74, P.O. Box 19
2990 AA Barendrecht
Phone: 01806 - 13277
Managing Dir.: Ir. J.H.M. Zwinkels
Dir. R. & D.: J.G. van Hal

Breeding Station Barendrecht — Stat. Manager: Ir. G.P.W. van Bentum

Beans, lettuce, spinach, leek, unions, brood beans, cucumber, gherkins, leek — Ir. G.P.W. van Bentum

Breeding Station "De Hooge Wurft"

Delftweg 13, 1747 GA Tuitjenhorn
Stat. Manager: Ir. B. van Adrichem

Brassica's, radish, beetroot, carrots

Nickerson Zwaan Research Centre
California U.S.A.

Nickerson International Plant
Breeder S.A., P.O. Box 1787
Gilroy CA 95020 U.S.A.
Stat.Manager: P. van den Berg, M.Sc.

Cucumbers, cantaloupes, squash eggplant

12 Bleiswijk

12.1 De Ruiter Zonen (Seed Company)
elevation: -0.5 m
rainfall : 804 mm
sunshine : 1540 h

Hoekeindseweg 39, P.O. Box 4
2665 ZG Bleiswijk
Phone: 01892 - 2741
Dir.: A.C. and H.D. de Ruiter

Department of Research and Plant Breeding — Ir. A.C. de Ruiter
Research — Ir. B.J. v.d. Knaap
Breeding of Tomato — Ing. Th.P. Schotte
Breeding of Cucumber — Ing. J.W. van Weerdenburg
Breeding of Cyclamen — Ir. B.J. v.d. Knaap

13 Dordrecht

13.1 Vreeken Zaden (Seed Company)

P.O. Box 182

elevation: -0.8 m
rainfall : 804 mm
sunshine : 1534 h

3300 AD Dordrecht
Phone: 078 - 135467

Breeding and production — W. Vreeken/A. Vreeken Jr.

14 Enkhuizen

14.1 Enza-Zaden (Seed Company)
elevation: -1.6 m
rainfall : 777 m
sunshine : 1550 h

Haling 1e, P.O. Box 7
1602 DB Enkhuizen
Phone: 02280 - 15844
Head: P. Mazereeuw

Breeding of tomatoes	Ir. F. Herlaar
Breeding of lettuce, endive, spinach	Ir. C.J. de Jong
Breeding of cucumbers, peppers, leek	Ir. R. Kuisten
Breeding of chicory	Ing. R. van Vlimmeren

14.2 Zaadunie B.V.
Sluis and Groot Research

P.O. Box 26, 1600 AA Enkhuizen
Phone: 02280 - 13838
Dir. of Res.:
Ir. J.E. Veldhuyzen van Zanten

Manager Breeding	Dr. Ir. H.A.M. van der Vossen
Manager Technology	Dr. Ir. G.C.A. Bruin
Manager of Resources	Ir. G.W. Hofstede
Development Manager Vegetables	Ir. P. Tjeertes
Development Manager Vegetables	Ir. P.A. Boorsma
Development Manager Flowers	Ir. W.M.N. van Kester

15 Haelen

15.1 Nunhems Zaden B.V. (Seed Company)
elevation: 20 m
rainfall : 708 mm
sunshine : 1450 h

Voort 6, P.O. Box 4005
6080 AA Haelen
Phone: 04759 - 1541
Dir.: Ir. H. Lange/T. Oudenaarde

Department of Plant breeding	Head: Ir. F. Meddens
Breeding of cucumbers	Ir. G. Reuling
Breeding of pea, bean and broadbean	C. Houben
Breeding of pickling cucumbers, chicory witloof and spinach	T. Scheenen
Breeding of tomato, lettuce and leek	Ir. A. Wolters
Breeding of Brussels sprouts, radish and carrots	H. Bongers
Tissue culture	Ir. A. Wolters
Seed technology	Drs. P. v.d. Toorn

16 De Lier

16.1 Zaadunie B.V.
Sluis and Groot Research
elevation: -0.5 m
rainfall : 810 mm
sunshine : 1547 h

Blaker 7, 2678 LW De Lier
Phone: 02380 - 13838
Dir. of Res.:
Ir. J.E. Veldhuyzen van Zanten

Breeding of lettuce, cucumber, tomato and pepper — Head: Ir. J. Sonneveld

16.2 Rijk Zwaan Zaadteelt en Zaadhandel B.V. (Seed Company)	Burg. Crezeelaan 40, P.O. Box 40 2678 ZG De Lier Phone: 01745 - 3941 Dir.: J. Zwaan/H. Beeftink M. Zwaan
Department of Plant Breeding	
Breeding of lettuce	Ir. J.C.J. Velema
Breeding of tomato, cucumber gherkin, egg plant, pepper	Ir. J.W. Koolstra
Breeding of outdoor vegetables	M. Zwaan

17 Naaldwijk

17.1 Bruinsma (Seed Company) elevation: 1.2 m rainfall : 822 mm	Middelbroekweg 67 2675 KG Honselersdijk P.O. Box 24, 2670 AA Naaldwijk Phone: 01740 - 28244, Telex: 33154 Dir.: L. van der Kruk/G. Dannijs C.A. Thoen
Department of Plant Breeding	Head: Ir. L.G. van den Berkmortel
Breeding of tomato and pepper	
Breeding of cucumber	Ir. R. Verhoef
Breeding of tomato	Ir. A.C.M. van de Ven
Breeding of lettuce	Ir. R. van der Laan
17.2 Vandenberg B.V. (Seed Company)	Julianastraat 25, Postbus 25 2671 EH Naaldwijk Phone: 01740 - 27441 Dir.: J.J. van den Berg Jzn J.J. van den Berg Azn
Breeding of paprika, eggplant	Head: Ir. D. Vreugdenhil
Breeding of lettuce	Ing. A. Baelde
Breeding of cauliflower, radish, kohlrabbi	Ing. E. de Gruyter

18 Noordscharwoude

18.1 BEJO-Zaden B.V. (Seed Company) elevation: - 0.8 m rainfall : 780 mm sunshine : 1540 h	Dorpsstraat 612, P.O. Box 9, 1722 ZG Noordscharwoude Phone: 02260-4041, Telex: 57140 bejo Dir.: D. Barten/P.A. Beemsterboer J.A. Jong
Department of Plant Breeding Breeding of: *Allium, Brassica oleracea, Daucus carota, Cichorium and Beta vulgaris (exculenta)*	P. Barten/S. Klaver
18.2 Bakker Brothers of Holland	Oostelijke Randweg 12, P.O. Box 7, 1723 LH Noordscharwoude Phone: 02260-3641 Dir.: Ir. J.O. Bakker J.G.A. Bakker/Ir. R.Ch. Bakker
Plant Breeding Section Breeding of *Phaseolus* beans, black salsify, cornsalad, carrots, onions, leek, radish	Ir. B. Schut

19 Rilland

19.1 D.J. van der Have Co - Plant Breeding Station
Royal Seed Growers & Merchants
elevation: 1.3 m
rainfall : 748 mm
sunshine :1520 h

Vanderhaveweg 2, P.O. Box 1,
4410 AA Rilland
Phone: 01135-2151
Telex: 55265 (breed nl)
Dir.: Ir. D.J. Glas

Department of Onion Breeding — Head: Ing. P.J.A.C. Reijnhoudt
Department of Grass Breeding (Amenity grasses) — Head: Dr.Ir. A.J.P. van Wijk

20 's-Gravenzande

20.1 Leen de Mos Vegetable Seeds
elevation: 0.5 m
rainfall : 808 mm
sunshine : 1546 h

P.O. Box 54, 2690 AB 's-Gravenzande
Phone: 01748-2031
Telex: 34029 De Mos NL
Comm. Dir.: J. de Mos
Technical Dir.: G. de Mos

Breeding of butterhead-lettuce, iceberg-lettuce, tomato, pepper, hot pepper, eggplant, cauliflower, melon, broccoli, radish and cucumbers.

20.2 Leen de Mos Flower Seeds — Comm. Dir.: J. de Mos

Breeding of stocks, primula's and several cutflowers, selling of wide size of beddingplants, cacti and succulents, potplants and cutflower seeds.

20.3 Leen de Mos Tropical Seeds — Comm. Dir.: J. de Mos

Importing and contract-growing of big range of tropical plant-seeds and palmseed, production of young plants.

NEW ZEALAND

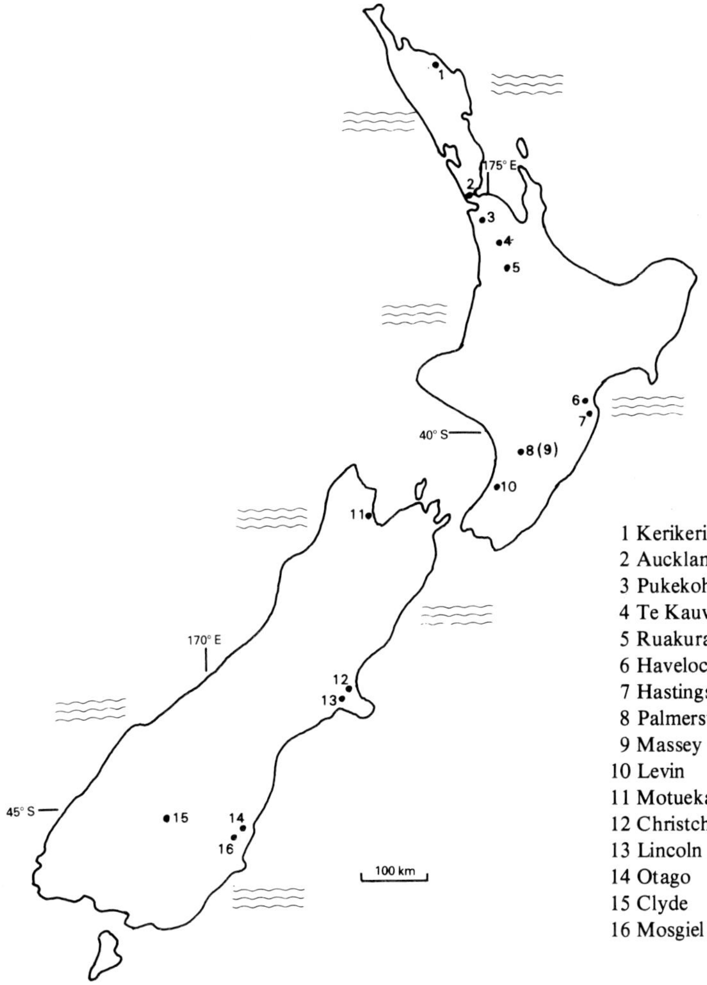
1
175° E
2
3
4
5
6
7
40° S
8 (9)
10
11
170° E
12
13
45° S
15
14
16
100 km

1 Kerikeri
2 Auckland
3 Pukekohe
4 Te Kauwhata
5 Ruakura
6 Havelock North
7 Hastings
8 Palmerston North
9 Massey University
10 Levin
11 Motueka
12 Christchurch
13 Lincoln
14 Otago
15 Clyde
16 Mosgiel

Survey

Horticultural research in New Zealand is carried out by various divisions of the Department of Scientific and Industrial Research, by the Agricultural Research Division of the Ministry of Agriculture and Fisheries, by the Horticulture and other departments of Massey University and Lincoln College and by several research institutes (or associations) which are financed in part by the industries they service and in part by grants from either DSIR or MAF. The research divisions of DSIR are grouped about 3 centres at Auckland, Palmerston North and Lincoln, with substations elsewhere. The Auckland Divisions are primarily concerned with protection of plants from pest and disease and production loss, those at Palmerston North with the quantity and quality of animal feed and pasture, and those at Lincoln with cropping and utilisation of plants. Botany Division provides a reference and information service to all Divisions. The Advisory Services Division of the Ministy of Agriculture and Fisheries (MAF) is involved with horticultural research carried out on growers properties. Frequently this is in association with divisions of DSIR and the Agricultural Research Division of MAF Co-ordination of Research. The National Research Advisory Council established in 1964 reviews research activities and makes recommendations to Government on research planning and administration. NRAC has established various committees which review areas of research including branches of horticulture. Research, advisory and producer interests are represented on these committees.

Prepared by R. Findlay, Ministry of Agriculture and Fisheries, Wellington

1 Kerikeri

1.1 Kerikeri Research Station
Ministry of Agriculture and Fisheries
Agricultural Research Division
elevation: 73 m
sunshine : 1988 h

Phone: 79611
Reg. Dir.: GC Everitt - MSc, PhD, NDA, Dip. Agr., FNZIAS

This division is the focus of the Northland horticultural research programme, investigating all aspects of subtropical fruit production relevant to the region. Priority though is given to crops with export potential.

Soil and Water Management

Consumptive water use for subtropical fruit crops is being assessed based on daily evapotranspiration measurement. Water requirements for tamarillos and kiwifruit are being investigated, as well as optimum scheduling of application and system design. Some modelling of soil water movement is under study.

Environmental Monitoring

Environmental data are being recorded within the sheltered orchard micro-climate. This will allow prediction of heat summation, accumulated winter chilling, and other climatic data for assessing site suitability for a range of locations throughout Northland.

Weed Control

Safe, effective weed control programmes are being formulated for shelter belt establishment on Northland volcanic soils. Suitable chemicals can now be recommended for Cryptomeria, Casuarina, bamboo, and black wattle. Work is continuing with alternative shelter species, such as bluegums.

Cultivar Selection and Evaluation

Improved root stocks and cultivars of subtropical fruit crops at present grown in Northland will be selected by continual evaluation of new plant material. These will include kiwifruit, citrus, tamarillos, feijoas, avocados, macadamia, melons, and other crops with production potential.

Tree Training

New systems of crop architecture are required by improved pruning and training techniques. These will allow greater light penetration, better access into the canopy for pesticides, improved potential for mechanisation (reflecting heavier yields), lower production costs, and improved fruit quality. Such research is specifically required for kiwifruit, persimmons, and feijoas.

Cultural Management

Investigations are commencing on new options for management of subtropical fruit crops, involving plant propagation, planting systems, weed and pest control, nutrition, pruning and training, and harvesting systems.

New Crops

Evaluation of subtropical fruits not previously grown commercially in Northland will identify new crop opportunities, particularly those which show potential for export or import substitution. Examples include persimmons, casimiroa, cherimoya, litchi, pepino, loquat, tropical guavas, and pecan.

Melons, Kiwifruit	R.F. Barber (Scientist-in-charge) BSc, M Agr., Dip. Hort. Sc.
Kiwifruit, Citrus	A.M.D. Rennes - B Hort Sci, PhD.
Kiwifruit, Citrus	A.C. Richardson - B Hort Sci
Kiwifruit, Pepinos	E.F. Walton - B Hort Sci

2 Auckland - DSIR

2.1 Division of Horticulture and Processing
Department of Scientific and Industrial Research, Mt. Albert Research Centre
elevation: 45 m
sunshine : 2067 h

Private Bag, Auckland
Dir.: R.L Bieleski, MSc, PhD F.R.S.N.Z.
Phone: 893660

Liaison:
G.C. Weston, BSc, MA

Undertakes research to improve the yield, quality and diversification of all horticultural crops, particularly pip, stone, citrus, kiwifruit and subtropical fruit.

Investigates the storage and transport, or processing where appropriate, of fruit, ornamentals, wine, vegetables and fish.

2.1.1 Pomology Section	
Plant Growth Substances	R.M. Davison, MSc, PhD, (Asst. Director)
Citrus and subtropical fruits	S.N. Dawes, MSc, PhD
Tissue culture, fruit breeding	L.G. Fraser, MSc, PhD
Stone fruits; persimmons	P.G. Glucina
Chestnuts; louquats; plant rights legislation	K.R.W. Hammett, BSc, PhD
Kiwifruit	M. Lay Yee, B. Hort. Sc
Pomology, subtropicals	R.G. Lowe, BSc
Feijoa, tamarillo	K.J. Patterson*, MSc
Kiwifruit, tamarillo	G.J. Pringle, BSc
Physiology of kiwifruit	W.P. Snelgar, MSc, D. Phil
Pomology, peas, Asian Pears	A.G. White, B. Hort. Sc
2.1.2 Postharvest Physiology and Storage Section	
Fruit physiology, pip and stone fruit storage	E.W. Hewett, B.Sc, PhD (Assistant Director)
Postharvest storage and fruits and vegetables	D.J. Beever, MSc, PhD
Fruit physiology and storage, citrus and subtropicals	J.E. Harman, MSc
Fruit physiology and storage, subtropicals	G. Hopkirk*, M.Hort.Sc, PhD
Fruit physiology, growth, development and storage, subtropicals	N. Lallu*, BSc, PhD
Fruit physiology and storage, chilling injury	E.A. MacRae, BSc, PhD
Fruit physiology and storage, pipfruits	C.B. Watkins, MSc
2.1.3 Food Processing Section	
Food processing	D.J.W. Burns, BSc, PhD
Horticultural processing	
Food industry liaison	J.R. Crossley, MSc, PhD
Chemistry of foods and wines	O.J. Dunbar, MSc, D. Phil
Sensory evaluation of foods	M.G. Hogg, Dip Hs, MS
Fruit and vegetable processing	N. Lodge, MSc, FRIC
Kiwifruit processing	J.A. Venning, B. Food Tech
Fish processing	
Fish processing	D.H. Buisson, MSc, PhD
Microbiology of fish processing	G.G. Fletcher, BSc
Biochemistry of fish processing and fish quality	J.M. Ryder, BSc
Fish quality and storage	D.N. Scott, B. Tech
2.1.4 Physiology and Chemistry Section	
Chemistry of biologically active compounds; gas chromatography and mass spectrometry	H. Young, BSc, PhD
Plant nutrition; biology of the genus Actinidia, history of plant domestication	A.R. Ferguson, MSc, PhD
Nutrient uptake and transport, calcium nutrition	I.B. Ferguson, BSc, PhD
Toxins of phytopathogenic bacteria	R.E. Mitchell, BSc, PhD
Plant lipid biochemistry	P.G. Roughan, MSc, PhD
Mineral nutrition of plants	

2.2 Entomology Division
Department of Scientific and Industrial Research
Mt. Albert Research Centre

Dir.: J.F. Longworth, MSc

Involved in the development of integrated pest management procedures in pip and stone fruit, kiwifruit, citrus, berryfruit, grapes, vegetables and greenhouse crops. Research on post-harvest disinfestation aims to solve quarantine problems - current working on pip and stone fruits, kiwifruit asparagus, flowers and persimmons. Pollination research has focussed on kiwifruit.

Scientific liaison	K.R. Nuzum, MSc
Integrated pest management	C.H. Wearing, D.I.C., PhD
Horticultural entomology, particularly fruit crops	
Postharvest insect disinfestation	T.A. Batchelor, MSc
Insect pheromones	J.R. Clearwater, PhD
Control of lemon tree borer with nematodes	
Lemon tree borer oviposition	
Integrated control	
Sex pheromones of Lepidoptera	S.P. Foster, PhD
Postharvest disinfestation	
Noctuid pest control	M.G. Hill*, PhD
Pesticide-resistant predatory mites	N.R. Markwick, PhD
Greenhouse crop pest control	N.A. Martin, PhD
Integrated control of whitefly	
Earthworms	
Control of insect and mite problems on kiwifruit, with interest in insect dispersal	D. Steven, PhD
Biological control	E.W. Valentine, B.Agr.Sc
Systematics of Chalcidoid Hymenoptera	
Hemipteran systematics	C.F. Butcher, MSc
Coleopteran and Lepidopteran systematics	R.C. Craw, PhD
Lepidopteran systematics	J.S. Dugdale, BSc
Insect rearing dietetics, nutritional physiology	P. Singh, AIARI, PhD
Small RNA viruses of insects	P.D. Scoti, PhD
Virus diagnostic methods, cell structure	
Baculoviruses, molecular biology, cell structure	A.M. Crawford, PhD
Insect protozoa and microsporidia	L.A. Malone, PhD
Ecology of insect pathogens	P.J. Wigley, M Agr Sc, D.Phil
Insect pathology	
Plant parasitic nematodes of fruit and vines including subtropical crops	G.S. Grandison, PhD
Biological control of insect pests by nematodes	W.M. Wouts, Ir. PhD
Chemistry of insect/host plant interaction	R.F.N. Hutchins, PhD

2.3 Plant Diseases Division
Department of Scientific and Industrial Research
Mt. Albert Research Centre
Private Bag Auckland
Dir.: P.J. Brook, MSc, PhD

Involved in studying the nature and control of fungous, bacterial, viral and physiological diseases of all economic plants, including assisting plant breeders to develop disease-resistant plants. Also maintains and develops collections of plant pathogenic micro-organisms as a national resource.

Scientific liaison	A.B. Grace, MSc
2.3.1 Bacteriology Section	
Rhizobium research, bacterial diseases of field crops	C.N. Hale, MSc, PhD
Bacterial diseases of field crops	R.N. Crowhurst, MSc
Bacterial diseases of plants, biological control	D.R.W. Watson, BSc, DIC
Bacterial diseases of orchard crops, bacterial taxonomy	J.M. Young, MSc, PhD
2.3.2 Mycology Section	
Soil-borne diseases, Phycomycetes	G.I. Robertson, MSc, PhD
Fungus taxonomy, Aphyllophorales	P.K. Buchanan, BSc, PhD
Electron microscopy, ultrastructure of Oomycetes	I.C. Hallet, BSc, PhD
Soil-borne diseases	I.J. Horner, MSc
Fungus taxonomy, Ascomycetes and Fungi Imperfecti	P.R. Johnston, BSc

Fungus taxonomy, Hyphomycetes	E.H.C. McKenzie, BSc, PhD
Fungus diseases of fruit crops	S.R. Pennycooke, BSc, PhD
Fungus taxonomy, Ascomycetes and Fungi Imperfecti	G.J. Samuels
2.3.3 Plant Protection Section	
Fungus diseases of vegetables and field crops	B.T. Hawthorne, M.Agr.Sc, PhD
Fungus physiology and genetics	R.E. Beever, MSc, PhD
Soil-borne diseases of field crops	M. Fowler, MSc, PhD
Fungus diseases of field crops, diseases of tropical crops	R.A. Fullerton, B.Agr.Sc, PhD
Epidemiology and disease control	W.F.T. Hartill, BSc, PhD
Soil-borne diseases of vegetable crops	S.A. Menzies, M.Agr.Sc, PhD
Physiology of plant disease	M.D. Templeton, BSc
2.3.4 Virology Section	
Plant virology, viruses of grapevine and subtropical fruits	D.W. Mossop, BSc, PhD
Plant virology, legume viruses	R.L.S. Forster, M.Hort.Sc. PhD
Plant virology, viruses of ornamentals	P.J. Guilford, MSc
Plant virology, citrus viruses	B.A.M. Morris-Krsinich, MSc, PhD

3 Pukekohe

Pukekohe Horticultural Research Station
Ministry of Agriculture and Fisheries
elevation: 82 m
sunshine :1845 h

RD 1, Pukekohe
Phone: 89613

Pukekohe Horticultural Reseach Station research work is grouped in two main areas, research on proven export crops (e.g. kiwifruit, squash, cymbidium orchids) to increase production and/or decrease costs, and research on a range of horticultural crops which are new for N.Z. or internationally. Small areas of the new crops are grown and test-marketed by Pukekohe Horticultural Research Station staff in collaboration with marketing organisations. New crops which are acceptable at market, and have suitable storage life will then be subjected to further agronomic research.

In the outdoor section there is emphasis on nutrition, crop management and weed control of well established export crops such as kiwifruit, squash, onions, potatoes and garlic. New crops recently evaluated include black and white salsify and cardoon.

The new crop evaluation programme includes cut flowers (sandersonia, gloriosa, tuberose, bouvardia), vegetables (Japanese aubergine) and fruit (pepino and table grapes). There are also research programmes on cymbidium orchids with emphasis on nutrition and physiological aspects of vegetative and flowering plants.

Orchids, Matsutake mushrooms	C. Powell (leader)-BSc, PhD
Onions, sandersonia	D.J. Bagyaraj - PhD
Gloriosa, orchids	I.C. Caldwell - BSc
	G. Clark - BSc/R.C. Hutton
Onions, kiwifruit	G.J. Wilson - M.Hort.Sc
Plant and analytical chemistry	I.S. Cornforth (leader)-BSc Phd
	C.J. Asher (visiting scient) PhD
	G.S. Smith - MSc, PhD

4 Te Kauwhata

Division of Horticulture and Processing
Department of Scientific and Industrial Research
elevation: 32 m
sunshine :1906 h

Private Bag, Te Kauwhata
Officer-in-charge:
R.J. Eschenbruch, PhD
Phone: 21

Oenology research yeast physiology, microvinification techniques

5 Ruakura

Horticultural Fruit and Vegetable production
Group Ruakura Soil and Plant Research Station
elevation: 40 m
sunshine :1981 h

Private Bag, Ruakura
Phone: 62839
Dir.: N.A. Cullen

The horticultural group based at Ruakura is involved in research at the following locations:
Ruakura Horticultural Unit
Rukuhia Horticultural Research Area
Te Kauwhata Viticultural Research Station
Manutuke Horticultural Research Station
Matawhero Grape Research Area

Kiwifruit Research
Research on kiwifruit has an emphasis on the future requirements of the New Zealand kiwifruit industry. Involving research on nutritional disorders, plant/water relationships, shelter, artificial pollination, fruit damage, training systems, crop loading and management, pesticide residues and growth regulators.
Grape Research
The Te Kauwhata Viticultural Research Station is the main location of grape research in the region. The research is on variety naming, clonal selection, rootstock evaluation and physiology, resistant phylloxera, rootstock training systems, virus elimination and vine nutrition.
Tree Crops
Tree crop research is investigating nashi (Japanese pear), persimmons, feijoa, cape gooseberry and nut crops, e.g. chestnut.
Berryfruit
Blueberry cultivar evaluations and selection, the use of chemical ripening agents, the test marketing of blueberries - both fresh and dried. Autumn raspberry management and strawberry cultivar evaluations are examples of the studies being run in berryfruit.
Vegetables
There is ongoing research into the production practices of established crops, e.g. asparagus, garlic and buttercup squash. Investigations of new crops with export potential are also conducted, e.g. florence, fennel and chickory.
Shelter
Shelter, its characteristics and effects, has become a major part of the research programme of environmental physicists, engineers and horticulturalists.

Staff list

Blueberries, Tree Crops	F.W. Wood (leader)-M Agr Sc, PhD
	N.S. Brown - M Agr Sc
Kiwifruit pollination	M.E. Hopping - M Hort Sc, PhD
Grapes, Persimmons	D.J. Jordan - B Hort Sc
Kiwifruit, Tamarillos	M.J. Judd - M Sc
Kiwifruit, Gooseberry	D.J Klinac - BSc PhD
Kiwifruit	K.J. McAneyney - BSc PhD
Grapes	R.E. Smart - MSc PhD

6 Havelock North

6.1 Entomology Division Substation
Department of Scientific and Industrial Research, Research Orchard
elevation: 9 m
sunshine :2057 h

Goddards Lane, Havelock North
Phone: 775669
Officer-in-charge: J.G. Charles

Integrated control; pests of berryfruit and grapes; mealybug ecology and control — J.G. Charles, MSc

Integrated control of pip and stone fruit pests, acarology	J.T.S. Walker, M.Hort.Sc

7 Hastings

7.1 Horticultural Research Station
elevation: 9 m
sunshine :2057 h

P.O. Box 1140, Hastings
Phone: 700511

The station which occupies 24 ha is located on the Heretaunga Plains, some 10 km north east of Hastings. It was established to service the extremely important vegetable process industry of the district, and the research programme continues to pay close attention to this sector. Recent district land use changes, have lead to a greater emphasis being placed on perennial crops such as kiwifruit, berryfruit, intensive stone fruit and grapes. The production oriented research programme integrates all aspects of technology including agronomy, cultivar evaluation, and disease, weed and pest control. The recently completed modern station facilities house four scientists, seven technicians, administrative staff and part time support staff.

8 Palmerston North

8.1 Plant Physiology Division
Department of Scientific and Industrial Research
elevation: 34 m
sunshine :1764 h

Private Bag, Palmerston North
Phone: 68019
Dir.: J.P. Kerr, M Agr Sc, PhD

Plant Physiology Division undertakes physiological research on plants of current and potential value to New Zealand. The impact of the environment on plant growth, development, and function is the central theme of research.
The Division operates, as a national research facility, the Climate Laboratory - 24 controlled environment rooms in which environment parameters can be independently controlled for the study of plant responses to changes in climatic variables.

Plant and crop physiology	I.J. Warrington, M Hort Sc
Photosynthetic metabolism	J.T. Christeller, M.Sc, PhD
Photosynthesis	W.A. Laing, B.Sc, PhD
Temperature physiology	H.G. McPherson, M Agr Sc, PhD
Reproductive physiology	I.R. Brooking, B Sc
Lipid biosynthesis	C.R. Slack, M Sc, PhD
Soil physics	B.E. Clothier, B Sc, PhD
Micrometeorology	K.B. McNaughton, B Sc, PhD
Plant growth modelling	P.W. Gandar, M Agr Sc, PhD
Environmental physics	T.W. Spriggs, MS, PhD
Tissue culture, growth regulations	D. Cohen, M Agr Sc, PhD
Molecular biology	B.D. Shaw, B Sc
Microbiology	B.E. Terzaghi, AB, PhD

8.2 Applied Biochemistry Division
Department of Scientific and Industrial Research

Private Bag, Palmerston North
Dir.: C.S.W. Reid, M Sc, PhD

Study plant production in relation to nitrogen fixation and biochemical genetics, plant composition in terms of constituents of importance in animal nutrition and disorders and plant pest resistance. Study and define and improve the nutritive value of New Zealand foods.

Mineral element, physiology	P.F. Reay, B Sc, PhD
Aroma chemistry of fruit	G.J. Shaw, B Sc, PhD
Food quality and dietary fibre	J.A. Monro, B Sc, PhD
Dietary fibre in vegetables	W.D. Holloway, M Sc
Fruit and vegetable composition	F.R. Visser, M Sc, PhD

Crop composition and physiology	J.D. Mann, M Sc, PhD
Crop biochemistry and physiology	J.E. Lancaster, B Sc

8.3 National Plant Materials Centre
Ministry of Works and Development

Private Bag, Palmerston North
Scientist in charge:
C.W.S. van Kraayenoord, Ir Agr

The prime function of this centre is the development and selection of plant materials for erosion control and revegetation on behalf of the National Water and Soil Conservation Organisation. Much of the research is of general horticultural significance. Poplar and willow breeding and selection. Selection of alternative tree species.

Selection of shrub species	R.L. Hathaway, MSc (Hons)
Grasses, legumes and herbs	W.R.N. Edwards, PhD
Tissue culture	A.G. Wilkinson, BSc, (For), Dip Plant Sc
Propagation and establishment, container growing, potting media, fertiliser requirements, and weed control	A.G. Spiers, PhD A.N. Gilchrist, B Agr Sc B.T. Bulloch, BSc (Hons)
Diseases of poplars and willows and other erosion control plants	G.B. Douglas, M Ag Sc (Hons)
Water relations and root systems of land stabilisation trees	
Farm and horticultural shelter	

9 Massey University

9.1 Department of Horticulture and Plant Health
elevation: 34 m
sunshine :1764 h

Private Bag, Palmerston North
Phone: 69099
Head: Prof. K.S. Milne, PhD

Provides courses for horticultural Diplomas and Degrees up to PhD. Conducts producer and ornamental short courses and publishes bulletins.
Facilities include: a range of teaching and research laboratories, automated research and production greenhouses, growth cabinets, conservatory. 22 ha outdoor production/research teaching areas, including 16 ha mixed orchard, landscape development area.

Amenity horticulture	A.V. Boorman, BSc (Hons)
Orchard management	E.A. Cameron, B Hort Sc
Nursery crop production, micropropagation	C.B. Christie, M Hort Sc
Vegetables establishment, seed treatments	D.T. Drost, M Sc
Pest management	P.G. Fenemore, PhD
Vegetables - establishment, physiology, greenhouse	K.J. Fisher*, PhD
Cut flowers - production, post harvest	K.A. Funnell, B Hort Sc (Hons)
Whole plant physiology, seed production	R.J. Hume, PhD
Kiwifruit physiology. Fruit crop rootstocks	G.S. Lawes, PhD
Phytophthora	P.G. Long, PhD
Propagation and physiology of root initiation	B.R. MacKay, B Hort Sc (Hons)
Classification and use of ornamental plants	M.B. MacKay, B Hort Sc (Hons)
Greenhouse white fly. Host plant interactions	J.E. Manley, B Hort Sc
Landscape design	L.M. Maughan, Dip Amen Hort
Bacterial diseases of vegetables	
Phytophthora diseases	S.A. Miller, M Sc
Virus diseases of horticultural crops	K.S. Milne, PhD
Vegetables - plant spacing, physiology	M.A. Nichols*, PhD
Botrytis Fungicide efficacy	T.M. Stewart, B Sc Dip, Pl Sc
Plant physiology. Growth analysis	Prof. J.A. Veale, PhD
Plant growth regulations. Pollination/fruitset	D.E.S. Wood, PhD

Plant growth and development. Source-sink relationships	D.J. Woolley*, PhD
9.2 Department of Agricultural Economics and Farm Management	Head: Prof. R.J. Townsley, PhD
Management of information system	
Kiwifruit management	D.W. Crawford, B Hort Sc
International trade and marketing in horticulture	Prof. A.N. Rae*, PhD
Amenity horticultural management	M.P. Wrigley, B Hort Sc (Hons)
9.3 Department of Agronomy	Head: Prof. B.R. Watkin, PhD
Asparagus mechanical harvesting	S.W. Brown, M Agr Sc
Kiwifruit genetics	I.L. Gordon, PhD
Tractor efficiency, alternate energy sources	R.E.H. Sims, M Sc
9.4 Department of Agricultural Engineering	Head: R.M. Clarke, BE
Electronic sensors in agriculture and horticulture	C.J. Studman, PhD
Modelling and control of the greenhouse environment	C.M. Wells, BE (Hons)
9.5 Department of Soil Science	Head: Prof. J.K. Syers, PhD, DSc
Irrigation of horticultural crops	K.W. McAuliffe, M Agr Sc
Soil water relations	D.R. Scotter, PhD
Foliar uptake of nutrients	Prof. J.K. Syers, PhD, DSc
	R.W. Tillman, BSc (Hons)
Plant analysis	M.A. Turner, PhD
9.6 New Zealand Nursery Research Centre	Dir.: M. Richards, BSc (Hort)
A joint University/Industry/Government venture established to conduct research of interest to the Nursery Industry.	
High health ornamentals	K.R. Everett, MSc
	K.S. Milne, PhD
Propagation Physiology of root initiation	B.R. MacKay, B Hort Sc (Hons)
Plant production studies	M. Richards, BSc (Hort)
19.10 Seed Technology Centre	Dir.: M.J. Hill, PhD
Provides courses designed to promote and improve seed production, processing, quality control and distribution of seeds through a continuing training and research programme in seed technology.	
Seed production, certification, drying and storage	M.J. Hill, PhD
Seed testing and quality control	D.E.M. Meech
Market Research Centre,	Dir.: S.J. West, MBA
Undertakes research into horticultural markets for products and related industries	
9.11 Department of Botany and Zoology	Head: Prof. B.P. Springett, PhD
Role of bumble bees in pollination	N. Pomeroy, PhD
Physiology, growth, development, flowering	Prof. R.G. Thomas, PhD
9.12 Department of Marketing	Head: Prof. J.S. Bridges, MBA
Co-operative marketing organisations.	
Analysing patterns in fresh produce	A.C. Lewis, PhD

10 Levin

10.1 Levin Horticultural Research Centre	Private Bag, Levin
Ministry of Agriculture and Fisheries	Phone: 87059
elevation: 46 m	Dir.: W.M. Kain - M Ag Sc, PhD
sunshine :1854 h	

The centre at Levin is the Ministry's major complex for horticultural research in the Southern North Island Region of New Zealand. It carries out investigations relevant to the production of vegetables for the fresh market and for processing, ornamental crops, berryfruit and other perennial fruit crops in the Southern North Island.

Crop Protection Section

Weed research has accumulated information on control of weeds with a wide range of vegetable crops, and on the biology of important weeds. Current investigations include weed control methods in asparagus, and herbicide tolerance of perennial species. Minimum tillage experiments have been conducted with sweetcorn, peas, broad beans, french beans and cucurbits. Herbicide application methods such as controlled droplet sprays are also investigated. Plant pathology research involves the etiology and control of diseases in tomatoes, brassicas, strawberries, asparagus and stonefruit.

Fruit Research Unit

Berryfruit research is aimed at the development of more economic production techniques. The potential of mechanisation is investigated with research being carried out on pruning and training methods suitable for use on mechanically harvested crops. Research on other fruit crops has commenced more recently; kiwifruit, peaches, nectarines, cherries, apples, pears, persimmons, feijoas and passionfruit.

A special project investigates the use of sand country for fruit crop production to enable intensification of land use from extensive pastoral use. Studies are concentrated on shelter, water and fertiliser requirements.

Ornamentals Research Unit

Cut flower production is major area of study. Selection of proteaceous plants including proteas, waratahs, leucadendrons and leucospermums is a prominent feature and work is also underway on cut flower crops with possible export potential; Limonium species, Zantedeschia, Nerine and Lisianthus. Japan and the USA are the target markets.

Plant Physiology Section

Post harvest storage, transport and subsequent presentation of a wide variety of crops is investigated here. Recent research has included storage of asparagus, berryfruit and cut flowers. Cultural and climatic effects on the quality of berryfruit are under investigation and techniques are being developed to measure quality components by laboratory methods.

The tissue culture laboratory develops techniques for the rapid multiplication of high value crops and of selected high producting lines of other crops such as asparagus, berryfruits and waratah flower crops. Tissue culture is also used to eliminate viruses and other diseases from plant material and to store valuable genetic lines.

Plant Propagation Section

Research is conducted on the propagation and physiology of a range of native and exotic ornamental species, with the aim of producing novel pot plant lines. Research on the propagation of some fruit crops is also undertaken.

Soil and Plant Nutrition Section

Research and development work has led to the wide use of bark as a potting medium by many New Zealand nurseries. Currently the research aims to develop the export potential of New Zealand bark. Simplified techniques and "test kits" have been developed for the analysis of soils and soilless potting mixtures. They are also applied to testing nutrient levels in plant sap. These test kits have now been commercialised.

A major programme is establishing critical plant nutrient levels and associated fertiliser requirements for a range of horticultural crops.

Vegetable Research Unit

This unit investigates the physiology, agronomy and production of outdoor vegetables. Asparagus, cauliflowers, leeks, peas, sweetcorn, potatoes, tomatoes and edible soybeans are studied. Cherry tomatoes and miniature brassicas are also being assessed as possible specialty crops.

Staff List

Floriculture	D.J. Dennis*-B Agr Sc (Hort), PhD
Fruit Production	
Leader Kiwifruit	G.K. Burge - B Hort Sci, MSc, Dip Pl Sci
	C.M. Kingston - B Hort Sc
Strawberries, blackcurrants	L.A. Porter - M Ag Sc (Hort)
	S.J. Shaw - B Hort Sc
Information section	
Leader	D.J. Swain - MSc Agr
Technical Officer	B.S. Eykel - ARPS (Lon), AMPA

Librarian	J.A. Body - BSc, NZLA Cert
Plant physiology	
Leader blackberries, sweetcorn	R.E. Lill - B Ag Sc, PhD
	R.A. Bicknell - BSc
Pot plants, plant propagation	S.M. Butcher - B Ed, MSc
	N.K. Given - MSc
Plant protection	
Leader weed control	T.I. Cox - BSc (Agr)
Stem canker, fungicides	L.H. Cheah - MSc, PhD
	M.J. Esson - BSc
Disease control, tomatoes, asparagus	K.G. Tate - M Agr Sc, PhD
Protected cropping	
Technical officer	N.K. Borst
Soils and plant nutrition	
Leader soil testing and plant analysis	M. Prasad* - BSc, MA, PhD
Irrigation requirements	D.J. Swain - MSc, Agr
Vegetable production section	
Leader vegetable cultivars	W.T. Bussell* - MSc, PhD
Tissue culture	A.R. Robb - B Hort Sc
Biometrics and data processing	
Leader	I.G. Plunkett - MSc
	R.W. Johnston - BSc, Dip ORS
Developmental engineer	S.G.D. McLachlan
Electronics engineer	B.G. Olsen

11 Motueka

11.1 Riwaka Research Station

Division of Horticulture and Processing
Department of Scientific and Industrial Research
elevation: 8 m
sunshine :2423 h

R D 3, Motueka
Phone: 89106
Ass. Dir.: P.E. Smale, M Hort Sc

Agronomy, annual food crops, biometrics	P.A. Alspach, B Sc, B Phil
Plant breeder	R.A. Betson, PhD
Kiwifruit, pests and diseases	N.B. Pyke, M Sc
Processing	J.F. Rohrbach, B. Chem Eng

12 Christchurch

12.1 University of Canterbury

Department of Botany
elevation: 30 m
sunshine :1992 h

Private Bag, Christchurch
Phone: 482009
Head: J.A. McWha, PhD

Plant pathology, fungal taxonomy and physiology	A.L.J. Cole, PhD
Physiology (esp. hormonal) of seed development and germination	J.A. McWha, PhD
Plant and microbial biochemistry	J.R.L. Walker, PhD

12.2 Yates Research Division

Arther Yates & Co. Ltd.

Normandale, Old West Coast Road
Courtenay R.D. 1, Christchurch
Research Manager: C.G. Janson, PhD

Pea and bean breeding	R.J. Casey
Onion cabbage and carrot breeding	J.F.M. Fennell, MSc
Vegetable agronomy	P.B. Bull, M Hort Sc
Vegetable seed production research	W.E. Fell
Agrochemical research	A.J Read

12.3 Wrightson NMA Ltd. - Horticulture

PO Box 6, Auckland
N.Z. Manager - R. Russell

Wrightson NMA have developed a range of co-operative research projects with the Department of Scientific and Industrial Research, the Ministry of Agriculture and Fisheries, and with Massey University in Kiwifruit breeding and pollination. They are investigating new horticultural research projects with Shii-Take Mushroome Ginseng.

Horticultural Research Consultant — H.C. Smith, PhD
P.O. Box 16013, Christchurch

13 Lincoln

13.1 Horticultural Research Station
elevation: 11 m
sunshine :2032 h

P.O. Box 24, Lincoln
Phone: 252511

Horticultural Research in the management, establishment, irrigation requirements, shelter and cultivar comparisons of blackcurrants, asparagus, strawberries, sphagnum moss, blueberries, apples, kiwifruit and pears.

Blackcurrants	P.J. Rhodes, M Agr Sc, Dip Ag
Peas, blackcurrants	R. Stoker, BSc, PhD
Asparagus	R.J. Haynes, B Hort Sc, PhD
Blackberries	P.H. Williams, B Agr Sc
Kiwifruit, irrigation cherries	M.C.T. Trought, BSc, NDA, PhD

13.2 Lincoln Research Centre
Crop Research Division,
Department of Scientific and Industrial Research
Research Centre

Private Bag, Christchurch
Dir.: M.W. Dunbier, M Agr Sc PhD

The Division has a horticultural crop breeding programme for vegetable and berryfruit.
Emphasis is placed on export crops such as asparagus, onions, squash, brambles and strawberries, crops with a high internal value such as tomatoes, potatoes and brassica crops and other vegetables and herb crops not adequately serviced by overseas breeders.
In most cases breeders have specific breeding objectives related to disease resistance and quality aspects.

Scientific liaison	H.T. Bezar, B Agr Sc
Pea breeding	W.A. Jermyn, M Agr Sc, PhD
Tomato, onion and vegetable breeding	M.T. Malone, M Hort Sc
	E.P. McCartney, Dip Hort
Asparagus breeding and agronomy	P.G. Falloon, B Agr Sc (Hons)
	A. Nikoloff, NZCS
Strawberry and apricot breeding	I.K. Lewis, M Sc, PhD
Bramble and raspberry breeding	H.K. Hall, M Sc
	W.F. Braam, Dip Hort, Dip Field Tech
Brassica breeding, herb agronomy and selection	J. Lammerink, Ing
Tree crop selection and establishment	D.J.G. Davies, Dip Ag Sc

13.3 Plant Diseases Division Substation

Private Bag, Christchurch
Officer in charge: J.W. Ashby

Virus diseases of field legumes	J.W. Ashby, B Sc PhD
Fungus diseases of field crops	R.M. Beresford, MSc
Fungus diseases of field crops	M.G. Cromey, M Sc, PhD
Virus diseases of pasture legumes	D.R. Musgrave, B Sc, PhD

13.4 Entomology Division Substation

Private Bag, Christchurch
Officer in charge: W.B. Thomas

Kiwifruit (including pollination), control of vegetable pests — A.M. Ferguson, M Agr Sc

Biological and integrated control in horticulture and tree crops particularly of blackcurrant pests and apple pests, integrated mite control, ecology and control of leafrollers — W.P. Thomas, M Sc

Pesticide resistance — D.M. Suckling, PhD

13.5 Lincoln College
Department of Horticulture, Landscape and Parks

Lincoln College, Canterbury
Head: Prof. R.N. Rowe*, PhD

Vegetable production, biological husbandry — R.A. Crowder, BSc (Hons)
Protected cropping, cut flowers — D.J. Farr, B Hort Sc (Hons)
Fruit and tree physiology, viticulture — D.I. Jackson, PhD
Fruit agronomy and physiology — M.J.S. Morley-Bunker, MSc
Tree crop and root physiology, fruit storage — Prof. R.N. Rowe*, PhD
Plant propagation, container plant nutrition — M.B. Thomas, PhD

13.6 N.Z. Agricultural Engineering Institute
Agronomic approach to mechanised fruit production

Lincoln College, Canterbury
J.S. Dunn, BSc

14 Otago

University of Otago
14.1 Department of Botany
elevation: 2 m
sunshine :1645 h

P.O. Box 56, Dunedin
Phone: 771640
Head: Prof. P. Bannister, PhD

Water relations: resistance to frost, heat, drought — Prof. P. Bannister, PhD

Developmental plant physiology, plant growth substances: tissue culture — P.E. Jameson, PhD

15 Clyde

15.1 Research Station
Division of Horticulture and Processing, Department of Scientific and Industrial Research
elevation: 141 m
sunshine :2034 h

Clyde Research Orchard, R D 6,
Clyde, Phone: 537
Officer in charge: Mr. J. McLaren

Integrated control of stone fruit pests — G.F. McLaren, M Hort Sc, PhD
Mites and integrated mite control — E. Ashley, PhD

16 Mosgiel

16.1 Invermay Agricultural Research Centre
Ministry of Agriculture and Fisheries
elevation: 30 m
sunshine :1685 h

Private Bag, Mosgiel
Phone: 3809

Cultivar evaluations, chemical stimulation, fertilisers and weed control are the areas of horticultural research in the Southern South Island Region of the Agricultural Research Division.
Twelve commercially available cultivars are being compared in Southland on the basis of crop yield, berry size and date of maturity.
In Central Otago and on the Taieri Coast, twenty-two asparagus cultivars are being evaluated under irrigation.
The partial replacement of pruning in the tree training process by using chemical branching agents is being studied.

The effect of nitrogen fertiliser, at various rates and forms on the productivity and survival of nectarine trees is being researched in Central Otago.
Asparagus weed control problems are under investigation, using a range of herbicides.

Blackcurrant	W.H. Risk, M Agr Sc
Asparagus	D. Brash, B Agr Sc
	G.G. Cossens, BSc, Dip AOSM
	W.F.A. Meeklah, NDA, Dip Ag

NIGERIA

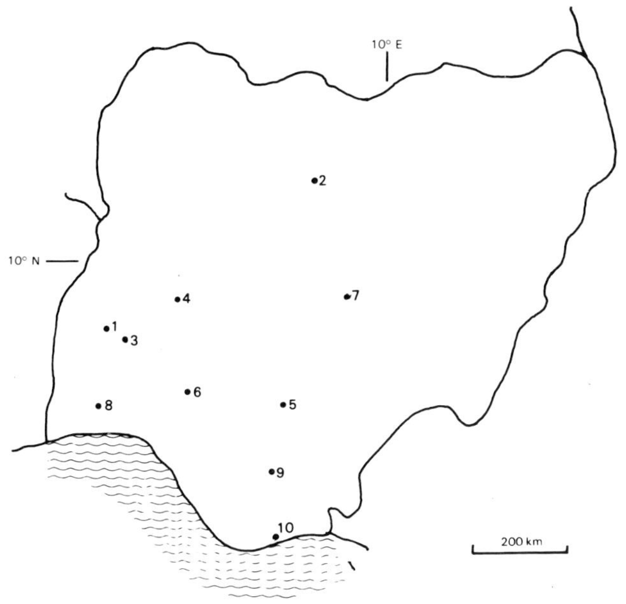
10° E
10° N
1
2
3
4
5
6
7
8
9
10
200 km

1 Ibadan
2 Samaru
3 Ile-Ife
4 Ilorin
5 Nsukka
6 Akure
7 Jos
8 Abeokuta
9 Owerri
10 Port Harcourt

Survey

The climate and soils in many parts of Nigeria are suited to a wide range of fruit and vegetable crops. However, the production of these crops has been mainly cultivated by traditional methods though most recent plantings adopt modern techniques developed by various Research establishments. Prior to the establishment of the Nigerian Institute for Horticultural Research (NIHORT) in 1975 with the assistance of FAO/UNDP some research work on horticultural crops was undertaken by other Institutes/Universities. While some of these Institutions transferred their activities to NIHORT, a large number are still engaged in various aspects of horticultural research. Soon after NIHORT was established, the Horticultural Society of Nigeria was inaugurated in 1977 as a forum for bringing together all classes of horticultural scientists and practitioners.

At the National Horticultural Research Institute, Ibadan, the major Horticultural Research Station in Nigeria, Research is based on five programmes.

1. Citrus Programme, which gives priority to the improvement of sweet orange, tangelo, mandarin, grapefruit, lemon and lime.
2. Fruits Programme, which also gives priority to the improvement of plantain, banana, mango, pineapple, pawpaw, guava, ogbono *(Irvingea gabonensis)* and African breadfruit *(Treculia africana)* in the first instance.
3. Vegetable Programme, with emphasis on tomato, onion, amaranthus, celosia, melon and okra, *Telfairia occidentalis, Corchorus olitorius,* peppers, roselle, *Gnetum africanum and Solanium* spp.
4. Extension Research, Liaison and Training Programme, which aims at transferring the improved horticultural technology to the various grades of users, and
5. Special Programme, which includes projects like germplasm collection and maintenance, storage, marketing, processing and engineering which cut across many crops and activities.

1. Ibadan

1.1 National Horticultural Research Institute.
elevation : 168 m
rainfall : 1450.5 mm
sunshine : 1753.4 h

P. M.B. 5432, Ibadan
Phone: 412490 (022)
Telegrams: NIHORT
Dir: S.A.O. Adeyemi

Prepared by Dr. S.A.O. Adeyemi, National Horticultural Research Institute, Ibadan.

Conducts research into citrus, fruits, vegetable and ornamental plants of economic importance in the country. A total of 50 ha is under cultivation with 18 ha, 20 ha, 12 ha devoted to vegetable programme, citrus and fruits and Extension Research Liaison and Training programmes respectively.

Citrus Division:

Development of improved cultivars of sweet orange, tangelo, mandarin, grape fruits and lemon	C.Amih, B.Sc., M.Sc. T.O. Oseni, M.Sc.,Ph.D.

Chemistry and Utilization Division:

Weed control studies on horticultural crops	N.J. Usoroh, M.Sc., Ph.D. A.O. Akin-Taylor, B.Sc.
Post harvest handling and quality control of horticultural crops	O.A. Taylor (Mrs), B.Sc., Ph.D. G.O. Akhigbe, B.Sc.
Processing and product development	P.O. Ogazi, M.Sc., Ph.D.

Crop Protection Division:

Survey, identification, biology and control of pests and diseases of horticultural crops	M.O. Ogunlana, M.Sc., Ph.D. O. Agunloye, B.Sc., M.Sc. A. Adebanjo, B.Sc., M.Sc. F.O. Anno-Nyako, Ph.D.

Economics and Statistics Division:

Socio-economic studies and marketing of horticultural crops	A. Ogunfidodo, B.Sc., M.Sc. M.S.A. Ishola, (Mrs), B.Sc.

Engineering Division:

Studies on horticultural engineering Design and fabrication of simple horticultural tools	A. Faseun. B.Sc., M.Sc.

Extension Research Liaison and Training Division:

Transfer of improved technology including planting materials.	I.O.A. Emiola (Mrs), B.Sc., M.Sc. S.F. Adedoyin, B.Sc.,MSc.

Floriculture Division:

Survey, identification and collection of economically useful ornamental plants	B. Ajao, B.Sc., M.Sc.

Fruits Division:

Fruit breeding, agronomy and orchard management	O.A. Anyim, B.Sc., M.Sc. B.A. Adelaja, B.Sc., M.Sc.

Genetic Resources Division:

Collection and maintenance of horticultural germplasm	J.O. Oladiran, Ph.D. A.O. Edema, B.Sc., M.Sc.

Vegetable Division:

Vegetable breeding, physiology and agronomy	O.A. Denton, Ph.D. O.A. Olufolaji, B.Sc., M.Sc.

1.2. Institute of Agricultural Research and Training, Horticultural Section	University of Ife, Moor Plantation, P.M.B. 5029, Ibadan Dir.: Prof. Olalokun, Ph.D.
Agronomy of tomato and okra; leaf vegetables Breeding of *Solanum spp* and *capsicum*	B.O. Adelana, Ph.D. J.O.S. Kogbe, Ph.D. M.O. Omidiji, Ph.D.

Nutrition of leaf vegetables	O. Omueti (Mrs) Ph.D.
1.3 Forestry Research Institute of Nigeria	P.M.B. 5054, Ibadan Dir.: Prof. P.R.O. Kio
Flora and *Solanium Linn.*	Z.O. Gbile
1.4 Plant Quarantine Service	P.M.B. 5673, Ibadan Dir.: Dr. M.O. Aluko, Ph.D.
Horticulture Division:	
Glasshouse propagation and environmental physiology	R.O. Esiaba, Ph.D.
Glasshouse plant propagation and culture	O.O. Awosusi (Mrs)
Field plot horticulture	I.O. Olokoshe.
1.5 Department of Agronomy	University of Ibadan, Ibadan Head: Prof. H.R. Chheda, M.Sc., Ph.D.
Water stress in fruits and vegetables, intercropping and allelopathy In vitro culture of vegetables Improvement of vegetables: okra, amaranths, *Corchorus olitorius* and yams	M.O.A. Fawusi, Ph.D. Prof. O. Babalola, Ph.D. C.A. Fatokun/Prof. H.R. Chheda M.E. Aken'ova, Ph.D. M.O. Akoroda, Ph.D.*

2. Samaru

2.1 Department of Agronomy, Faculty of Agriculture Institute for Agricultural Research	Ahmadu Bello University, P.M.B. 1044, Zaria
Weed control in vegetable crops	S.T.O. Lagoke, Ph.D. T.D. Sinha, Ph.D. J.A.Adigun
Agronomy of horticultural crops and post-harvest physiology	S.K. Karikari E.B. Amans., I.S. Majambu

3. Ile-Ife

3.1 Department of Plant Science	University of Ife, Ile-Ife Head: Prof. A.E. Akingbohungbe
Tomato, egg plant, okra, pepper, amaranths, pawpaw, citrus, mango and guava	Profs. A.E. Akingbohungbe, O.A. Adenuga, P.T. Onesirosan, J.L Ladipo, T. Fatunla, Drs. J.I. Olaifa, J.O. Amosu, I.O. Obisesan, C.O. Alofe A.S. Adegoroye, O.A. Akinyemiju

4. Ilorin

4.1 Department of Agricultural Science	University of Ilorin, Ilorin
Nematology of horticultural crops.	J.O. Babatola,Ph.D.

5. Nsukka

5.1 Faculty of Agriculture	University of Nigeria,

Department of Crop Science
elevation : 419 m
rainfall : 135.6 mm/month
sunshine : 61 h/month
r.h. : 60%

Faculty of Agriculture, Nsukka
Head: Dr. J.O. Uzo, Ph.D.*

Vegetable crops breeding and genetics	J.O. Uzo, Ph.D.*
Tropical fruit crops	T.O.C. Ndubizu, Ph.D.
Nutrition of vegetable crops	J.E. Asiegbu, Ph.D.
Genetic variations in okra varieties	A.C.C. Udeogalanya, Ph.D.
Tropical ornamentals and turf grasses	B.N. Mbah, Ph.D.
Breeding of tropical vegetables for processing	M.S.C. Abani
Disease control in melon, bambaragroundnut and pepper	E.U. Okpala, Ph.D.
Nematode/vegetable crops, relationship	R.O. Ogbuji, Ph.D.
Ethiology and control of seed-borne diseases of vegetable crops	C. Iloba, Ph.D.
Weed control in vegetable crops	O.U. Okereke, Ph.D.
Breeding and genetics of vegetable crops	C. Oyolu, Ph.D.

6. Akure

6.1 School of Agriculture and Agricultural Technology

Federal University of Technology,
P.M.B. 704, Akure
Dean: Prof. L.K. Opeke, Ph.D.

Collection and characterisation of indigenous vegetables of Ondo State, (humid tropical area)	Prof. L.K. Opeke, Ph.D.
Studies in the genetic systems of the *Amaranthus viridis*	G.O. Iremiren, Ph.D.
Vegetative propagation studies in *Dacrodes edulis* (African apple)	K. Majasan

7. Jos

7.1 Faculty of Agriculture
Citrus research

University of Jos
Makurdi

8. Abeokuta

8.1 Faculty of Science,
Department of Biological Sciences

Ogun State University
P.M.B. 2002, Ago-Iwoye
Head: Prof. B.A. Oso

Improvement of growth and yield of okra — I.O. Fasidi, Ph.D./O.O. Osonubi, Ph.D

9. Owerri

9.1 School of Agriculture and Agricultural Technology

Federal University of Technology
P.M.B. 1526, Owerri

Fruit agronomy — J.C. Obiefuna, Ph.D.

10. Port Harcourt

10.1 Institute of Agricultural Research and Training

Rivers State University of Science and Technology
P.M.B. 5345, Port Harcourt
Head: N.A. Ndegwe, Ph.D.

Egusi melon agronomy and physiology	N.A. Ndegwe, Ph.D/F.N. Ikpe E.T. Jaja (Mrs.)
Pawpaw growth and production	N.A. Ndegwe, Ph.D./V.J. Chuku
Okra breeding and pathology	L.A. Daniel - Kalio, Ph.D.
Horticultural crops and agroforestry and landscaping	R.M. Dzieciolowski

NORWAY NO

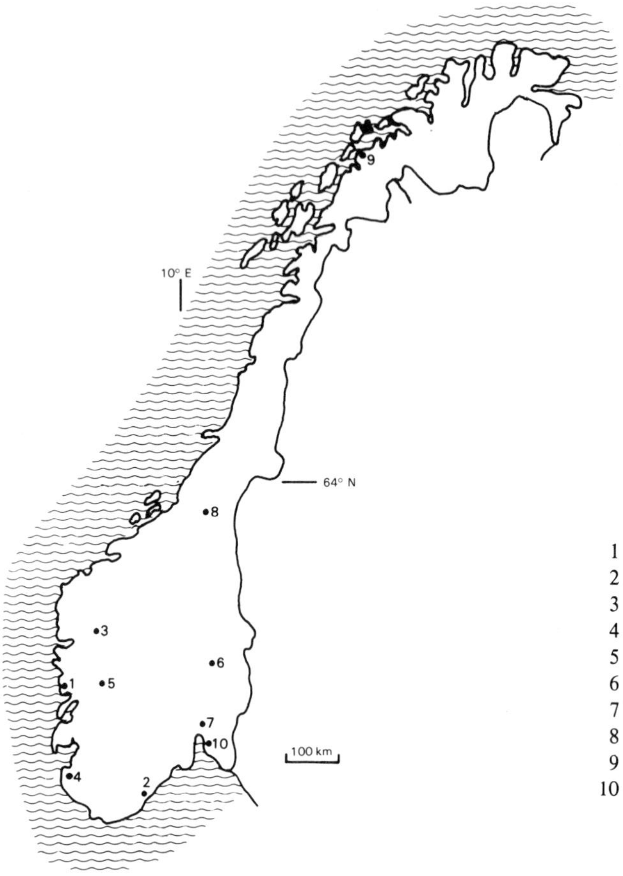
10° E
64° N
100 km

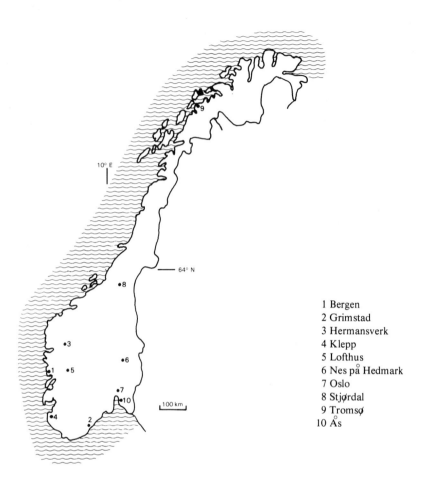

1 Bergen
2 Grimstad
3 Hermansverk
4 Klepp
5 Lofthus
6 Nes på Hedmark
7 Oslo
8 Stjørdal
9 Tromsø
10 Ås

Survey

The Agricultural University of Norway, located at Aas, 30 km south of Oslo, is the centre for academic training and for research in agriculture in Norway. The study is, as at the other academic institutions in Norway, based on the university matriculation examination (examen artium). It is a 5 year program of which the first year is a propedeutical year which consists of introductory courses and practical training. This year is spent at a vocational school in agriculture. However, students who have passed the ordinary curriculum at an agricultural or horticultural vocational school are admitted directly to the second year of study at the university.
The university offers 11 different curricula, one of which is Horticulture. The first part (2nd and 3rd year of study) is mainly composed of courses in basic sciences. But relatively early horticulture is introduced through a compulsory basic course which is a prerequisite for subsequent minor and major courses in the following 4 fields: (1) Floriculture, mainly greenhouse production, (2) Nursery Management with Dendrology and Outdoor Ornamentals, (3) Pomology, and (4) Vegetable Production, Greenhouse and Field Crops. The students must select at least two of these subjects. The last semester is entirely devoted to thesis work usually based on experimental work in one of the four subjects mentioned above. Experimental data are usually collected during the last one or two years of study.
The degree obtained at the University, Cand. Agricultureae, is considered to be on the same level as the Masters degree at an American university. In addition the university offers advanced degrees. The degree Doctor Scientarium is comparable to the Ph.D. The degree Doctor Agriculturae is granted solely on basis of scientific achievement of some substance.

Prepared by: Prof. E. Strømme, Dept. of Floriculture and Greenhouse Crops, Aas

The horticultural courses are given by four institutes or departments corresponding to the four subjects mentioned above. Each institute has grown up around a professorship. The staff consists at present, in addition to the professor, of 3-5 established or permanently engaged scientists and a similar number of graduate students working for advanced degrees. Each institute has its own experimental set-up (field, greenhouses, laboratories etc) and its own technical staff. In 1985 the total number of scientific personnel at the four institutes, including graduate students (research assistant, fellows etc) comprised 29, and the technical staff (including office personnel), 37 persons. The graduates from the Agricultural University become enployed in research, advisory services (state and regional), teaching (vocational schools in agriculture), organizations of different kinds and to some extent in private enterprises.

The Norwegian State Agricultural Research Stations (SFL) consist of 14 stations and 5 substations, located in various parts of the country. The administration is located at Aas (PO Box 100, N-1430 AAS). Most of the stations are devoted to farm crops, but some work also with horticultural crops and 2-3 are specialized in horticulture. The stations are supposed to concentrate on crops and problems important in their area. Vegetable research is carried out at Kise, Landvik, Saerheim, Njøs, Kvithamar and Holt, top fruit research at Kise, Ullensvang and Njøs, small fruit research at Kise, Ullensvang, Njøs, Kvithamar and Holt, and greenhouse crop research at Kvithamar, Saerheim and Landvik.

In addition to the research stations, and often attached to them, a system has been established consisting of local groups of farmers and growers who on basis of certain criteria might obtain financial support and be able to hire a univeristy graduate for conducting local experiments. These experimental "rings" often play a more important role in the advisory service than they do in the way of research.

Horticultural research is also carried out at the following institutions: Norwegian Plant Protection Institute, Norwegian Institute for Food Technology, Norwegian Seed Testing Institute and the Institute for Farm Mechanization. All are located at Aas, close to the Agricultural University.

Publications. Scientific publications in agriculture are : (1) Forskning og forsøk i landbruket (Research in Norwegian Agriculture) which mainly contains reports from the SFL stations. Editorial office at Moerveien 12, N-1430 Aas. Papers are published in English or Norwegian with English summary. (2) Meldinger fra Norges Landbrukshøgskole (Scientific Reports of the Agricultural University of Norway), PO Box 3, N-1432 Aas-NLH. Papers mostly in English. (3) Acta Agriculturae Scandinavica, published by the Scandinavian Association of Agricultural Scientists (NJF). The present editorial address is PO Box 6806, S-11386 Stockholm.

1. BERGEN

1.1 ARBOHA Det Norske Arboret (Norwegian arboretum) and

N-5067 Store Milde
Phone: 05-226972
Curator: P. Søndergaard

Botanisk Hage (University Botanic Gardens)

Box 12, Universitetet
N-5014 Bergen
Phone: 05-212922
Curator: D.O. Øvstedal

ARBOHA is a joint venture of arboretum and botanic gardens. Acclimatization studies, phytogeography and regeneration studies in woody plants. Horticultural and acclimazation studies in herbaceous and woody plants

2. GRIMSTAD

2.1 Statens forskingsstasjon Landvik
(Landvik Agricultural Research Station)
elevation: 6 m
rainfall : 1229 mm
sunshine : 1820 hrs

N-4890 Grimstad
Phone: 041-42266
Head: G. Jonassen

General research on vegetables in the open — J. Vik*
Breeding of onions and other vegetable crops
Physical growth factors in the open and in protected climate — G. Guttormsen*

Seed-crops investigation on vegetables and farm crops	G. Jonassen
Studies of winter hardiness and physical plant-development for bolting in biennial crops	G. Synnevag

3. HERMANSVERK

3.1 Statens Forskingsstasjon Njøs (Njøs Agricultural Research Station)	N-5840 Hermansverk Phone: 056-53611 Head: G. Børtnes
General research on fruit production, cultivars, rootstocks; general small fruit production	P. Husabø
Fruit	S.H. Hjeltnes*
General research on vegetables in the open	G. B/ortnes

4. KLEPP

4.1 Statens Forskingsstasjon Saerheim (Saerheim Agricultural Research Station) elevation : 80 m rainfall : 1100 mm	N-4062 Klepp Phone: 04-420333 Head: G. Taksdal*
General research on vegetables in the open	G. Taksdal*
Greenhouse crops	S.G. Grimstad

5. LOFTHUS

5.1 Statens Forskingsstasjon Ullensvang (Ullensvang Agricultural Research Station) elevation : 50 m rainfall : 1252 mm sunshine : 980 hrs	N-5775 Lofthus Phone: 054-61105 Head: J. Ystaas*
Tree fruit and small-fruit research for Western Norway	
Extension service on fruit production	
General fruit production, fruit-tree nutrition, soil fertility, planting systems, growth regulation, and sweet cherry production	J. Ystaas*
Insect pests	K. Hesjedal
Fruit quality, post-harvest physiology, chemical regulation of fruit crops, rootstocks	L. Sekse
General small fruit production, cultivar and general management trials in small fruit	M. Meland
Extension service, planning and development of Norwegian fruit production	A. Kvaale

6. NES PÅ HEDMARK

6.1 Statens Forskingsstasjon Kise (Kise Agricultural Research Station) elevation : 130 m rainfall : 563 mm sunshine : 1637 hrs	N-2350 Nes på Hedmark Phone: 065-5230, 054-52021 Head: A. Hjeltnes
General pomology. Winter hardiness in fruit crops	A. Nes*
Plant-water relationships in fruit crops	K.L. Kongsrud
Cultivar trials and plant propagation in fruit crops	A. Hjeltnes

Irrigation and soil management trials in vegetable crops	E. Ekeberg/H. Riley
General research on vegetables in the open	S. Dragland

7. OSLO

7.1 Norges Landbruksøkonomiske Institutt (Norwegian Institute of Agricultural Economics)	P.O. Box 8024 - Dep. N-0030 Oslo 1 Phone: 02-424430 Dir.: F. Reisegg
Agricultural economics, farm management	F. Reisegg
Horticultural economics	K. Repstad
7.2 University of Oslo	P.O. Box 1066 Blindern N-0316 Oslo 3 Phone: (02) 455050
Department of Biology	Chairman: S. Nilsen
Division of Botany	Head: S.E. Rognes
Plant physiological studies related to	
plant biochemistry,	S.E. Rognes
plant hormones,	H. Aarnes
photosynthesis	P. Halldal
Ecology	F.-E. Wielgolaski
Phytotron	Head: S. Nilsen
Ecophysiology of cultivated plants, photosynthesis, photorespiration and photoinhibition	S. Nilsen
Greenhouse plant production and cultivation in conditioned climate	S. Nilsen
Plant nutrition especially in tropical climates	A.B. Eriksen
7.3 Universitetets Botaniske Hage og Museum (Botanical Garden and Museum, University of Oslo)	Trondheimsvn. 23 B N-0562 Oslo 5 Phone: 02-686960
Governmental institution. Horticultural studies on a restricted scale	Head: R.Y. Berg
Phytogeography, reproductive biology, embryology, taxonomy	R.Y. Berg
Cytotaxonomy, embryology	L. Borgen
Taxonomy, phytogeography	I. Nordal
Phytosociology, ecology, conservation	E. Marker
Phytosociology, ecology, taxonomy	P. Sunding

8. STJØRDAL

8.1 Statens Forskingsstasjon Kvithamar (Kvithamar Agricultural Research Station) elevation : 35 m rainfall : 817 mm	N-7500 Stjørdal Phone: 07-826211 Head: O.A. Baevre
Cultivation and fertilizing of vegetable crops, extension service on greenhouse crops	
Production of vegetable crops, specially in relation to soil and nutrition, storage of vegetables	M. Flønes
Greenhouse crops	O.A. Baevre
Small fruit production	R. Nestby

Extension service on greenhouse technique	J. Stene

9. TROMSO

9.1 Statens Forskingsstasjon Holt (Holt Agricultural Research Station)	N-9001 Tromsø Phone: 083-84875 Head: R.T. Samuelsen
Vegetable and small fruit crops	R.T. Samuelsen
Insect pests	T.J. Johansen
Cloudberry cultivation	K. Rapp
9.2 Institutt for biologi og geologi, Universitetet i Tromsø (Institute of Biology and Geology, University of Tromsø)	Box 3085 Guleng N-9001 Tromsø Phone: 083-70011 Head: A. Andresen
Physiological studies on cultivated and wild species	O. Junttila/J. Nilsen*
Biochemical plant pathology	B. Robertsen
Biological nitrogen fixation	B. Solheim
Frost resistance	A. Kaurin

10. AAS

10.1 Norges Landbrukshøgskole (Agricultural University of Norway)	N-1432 Aas-NLH Phone: 02-949060
10.1.1 Institutt for Landbruksøkonomi (Department of Agricultural Economics)	Phone: 02-948455 Head: H. Langvatn
General agricultural economics	S. Borgan/H. Giaever
Horticultural economics	H. Langvatn
10.1.2 Institutt for bygningsteknikk (Department of Agricultural Structures)	Phone: 02-949251 Head: A. Nygaard
Glasshouse construction and climate, storage technique for horticultural crops	P. Roer/Z. Sebesta/D. Wenner
10.1.3 Institutt for meieri- og naerings-middelfag (Department of Dairy- and Food Industries	Iphone: 02-948751 Head: R.K. Abrahamsen
Quality, post harvest pphysiology and handling of plant	J. Apeland*/K.Steinsholt/S.Syrrist
10.1.4 Institutt for Blomsterdyrking og Veksthusforsøk (Department of Floriculture)	Phone: 02-948319 Head: R. Moe
General glasshouse crop production	E. Strømme
Growth and flowering of floricultural crops	R. Moe*
Production planning, advisory service	G. Sandved*
Post harvest physiology	T. Fjeld
Propagation technique, growing media, watering and fertilization	H.R. Gislerød
Propagation, growth and flowering	R. Djurhuus*
Mass propagation by tissue culture	T. Borgen*
Tissue culture, disease-free stock plants	M. Appelgren
CO_2, greenhouse climate	L. Mortensen
10.1.5 Institutt for Dendrologi og Planteskoledrift (Department of Dendrology and Nursery Management)	Phone: 02-949890 Head: M. Sandved*
Breeding, ecophysiology, production and management of ornamental trees, shrubs and turfgrasses	A. Haabjørg*
Nursery management, breeding and cultivation of roses	A. Lundstad
Urban horticulture, species and variety trials	M. Sandved*

Ornamental plants in general, phenology	J. Batta
Selection of trees and shrubs for northern regions	S. Horntvedt
Propagation and production of trees and shrubs	O. Billing Hansen
Urban horticulture, air pollution and vegetation	P.A. Pedersen
Establishment and management of landscape vegetation	H. Asheim
10.1.6 Institutt for Fruktdyrking	Phone: 02-948385
(Department of Pomology)	Head: F. Maage*
General pomology, fruit nutrition, growth regulators, dormancy in buds, plum cultivars, soil management	F. Maage*
Fruit storage, post-harvest physiology	R. Landfald
Cherry cultivars, strawberry cultivars, fruit composition fruit quality and time of harvest	S. Vestrheim
Pollination problems, apple cultivars and rootstocks, rubus cultivars, apple and plum breeding	G. Redalen*
Ribes, cultivars, management and quality	N. Heiberg
Fruit transpiration, fruit composition	K. Haffnes
10.1.7 Institutt for Grønnsakdyrking	Phone:02-948402
(Deparment of Vegetable Crops)	Head: H. Hoftun
Genetics and breeding of vegetable crops, vegetable crops production, vegetable trials	A.R. Persson*
Climatic reactions, seed germination. Vegetables for canning and freezing	O. Røeggen
Post-harvest physiology, handling, storage	H. Hoftun
Growth and Development. Controlled atmosphere storage systems	H. Baugerød
Nutrient film technique. Root physiology	J.L.F. van der Vlugt
Post harvest physiology, ethylene	K.J. Willumsen
Post harvest physiology, root crops	M. Knutsen Neergaard
10.1.8 Institutt for Landskapsarkitektur	Phone: 02-949900
(Department of Landscape Architecture)	Head: M. Eggen
Landscape design and landscape planning, parks and open spaces, urban design, residential areas	E. Gabrielsen/O. Bettum
Regional landscape studies, landscape evaluation, roads and industrial landscape	M. Bruun
History of landscape architecture	M. Eggen
Planting design and vegetation management	A.-K. Dyring
Theorectical and applied design	A.-G. Thygesen
Working drawings and perspectives, architectural communication	O.A. Krogness
Town center planning	P. Tveite
Perception of landscape, design theory	K. Jørgensen
10.2 Landbruksteknisk Institutt	Phone: 02-949370
(Institute of Agricultural Engineering)	Head: K. Aas
Farm mechanization	K. Aas
Horticultural engineering	A. Nordby/R. Holmøy
10.3 Statens Frøkontroll (State Seed Testing Station)	P.O. Box 68, N-1432 Aas-NLH
	Phone: 02-949532
Testing agricultural, horticultural and forest-tree seeds and seed potatoes	Head: A. Wold
Seed pathology department	G. Brodal
Purity department	N. Chr. Rogstad
Germination department	P. Overaa
Department of field plot tests	S. Telneset
Cereals	H. Sjøseth
Grasses and clovers	O.S. Buraas

Vegetables	H. Tangeraas
Seed potatoes	S. Telneset
State Seed Testing Station, Branch Station	Kvithamar, 7500 Stjørdal M. Moe
10.4 Statens Plantevern (Norwegian Plant Protection Institute)	H-1432 Aas-NLH Phone: 02-949400 Dir.: J. Fjelddalen
10.4.1 Botanisk Avdeling (Department of Plant Pathology)	Head: H. Røed
General mycology and plant pathology	H. Røed/L. Sundheim
Plant virology	A. Bjørnstad/T. Munthe
Plant bacteriology	A. Sletten
Plant pathology, cereals, grasses	O.N. Elen/H.A. Magnus
Plant pathology, potatoes	E. Førsund
Mycology, plant pathology, fruits, soft berries, ornamentals	H.B. Gjaerum
Plant pathology, vegetables	L. Semb
10.4.2 Ugrasbiologisk Avdeling (Department of Herbology)	Head: A. Bylterud
General weed biology and control	A. Bylterud/O.M. Synnes
Weed control in agricultural crops	R. Skuterud/H. Fykse
Wild oats	J. Netland
Weed control in horticultural crops	H. Fykse
Weed control in forestry and forest nurseries	K. Lund-Høie
Decomposition of herbicides in soil	O. Lode
10.4.3 Zoologisk Avdeling (Department of Entomology)	Head: T. Rygg
General and applied entomology	J. Fjelddalen/T. Hofsvang
Cereal insect pests	A. Andersen
Plant nematology	M. Støen
Vegetable insect pests	T. Rygg
Fruit insects and mites, integrated control	T. Edland
Fruit insects - forecasting	S. Kobro
Greenhouse and small-fruit insect pests, biological control	T. Stenseth
Biochemistry and insecticides	J. Stenersen
10.5 Norsk Institutt for Naeringsmiddelforskning (Norwegian Food Research Institute)	Box 50, N-1432 Aas-NLH Phone: 02-940860 Head: A. Skulberg
Industrial research on meat, poultry, vegetables, fruit and berries	
10.5.1 Department of Food Science	Head: Tore Høyem
Evaluation of methods for determination of internal potato quality	L. Kaaber
Lipid degradation - the importance of flavour development in stored, fresh and frozen vegetables	P. Baardseth
10.5.2 Department of Food Technology	Head: K.I. Hildrum
Quality criteria for vegetables	M. Martens
Vegetable variety trials	H.J. Rosenfeld
Total quality control in the food industry	T. Jarmund
Sensory analysis	M. Martens
Measurement techniques in industrial process control	P.G. Fyhn
Light induced degradation of food	A. Iversen

Near-infrared (NIR) multivariate analysis of the composition of food	H. Martens
Irradiation of food products	E. Risvik
Colour of food products	E. Slinde
Barrier properties of packaging materials	B. Storesund

10.6 Statens Forskingsstasjoner i Landbruk (Norwegian State Agricultural Research Stations)

Drøbakv. 5, N-1432 Aas
Phone: 02-942060
Chairman: A.A. Fretheim

Board and Administration
See No 2.1, 3.1, 4.1, 5.1, 6.1, 8.1, and 9.1.

Head of Administration:
O.B. Olsen

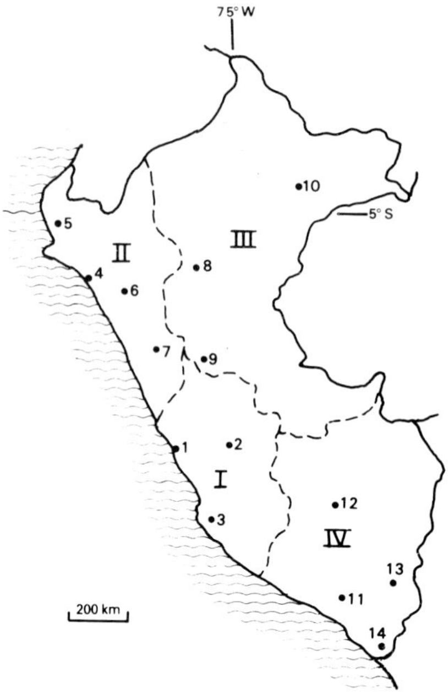
75° W
5° S
I
II
III
IV
1
2
3
4
5
6
7
8
9
10
11
12
13
14
200 km

1 La Molina
2 Chiclayo
3 Tarapoto
4 Arequipa

Survey

The Ministry of Food created in January 1975, is divided into 6 General Directions; one of them is the Dirección General de Investigación. Peru is divided for his Agricultural Research in 4 Centros Regionales de Investigación Agricola (Regional research centers). Each Centro Regional de Investigación Agricola (CRIA), has some Experiment Stations. There are 14 Agricultural Experiment Stations throughout the country.

As there is no updated material available, the text of HRI III is repeated.

1 La Molina (CRIA I)

1.1 Experiment Station La Molina — La Molina - Lima, Apart. 2791
Dir.: Dr. C. Valverde S.
Head: Dr. F. de la Puente

Potatoes	Dr. F. de la Puente
	Ing. C. Vise/Ing. M. Quijandría
Cassava and sweet potatoes	Ing. E. Delgado
Soy and beans	Ing. Rufina Montalv
Fruits and vegetables	Ing. V. Rivadeneyra
Herbicides	Ing. A. Corrales
Cotton	Ing. A. Cruz/Ing. C. Seminario
	Ing. J. Lazo
Wheat	Dr. C. Llosa
Entomology	Ing. C. García
Forages and pastures	Ing. R. Zambrano

1.2 Experiment Station Huancayo — Calle Real 507 - Huancayo
Head: Ing. G. Campos

Agronomy and potatoes; plant breeding	Ing. R. Wisser
Plant pathology	Ing. C. Loayza

Entomology	Ing. P. Alcalá
Wheat and quina	Ing. N. Valencia
Corn	Ing. G. Fukusaki
1.3 Experiment Station Ica	Calle Municipalidad 228 - Ica Head: Ing. D. Farfán
Cotton	Ing. J. Calle
Potatoes	Ing. D. Noriega
Beans	Ing. H. Cano
Forages	Ing. C. Lozano

2 Chiclayo (CRIA II)

2.1 Experiment Station Vista Florida	Apartado 116 - Chiclayo Dir.: Ing. P. Contreras Head: Ing. J. Hernández L.
Rice (breeding)	Ing. J. Hernandez/Ing. H. González
Corn	Ing. J. Sotomayor
Beans	Ing. C. Apolitano
2.2 Experiment Station Chira	Cayetano Heredia 402 - Piura Head: Ing. J. Paredes G.
Cotton	Ing. M. Reyes
Entomology	Ing. J. Senmache
Fruit crops	Ing. M. Guerrero
2.3 Experiment Station Cajamarca	Lima 560 - Cajamarca Head: Ing. T. Fairlie C.
Potatoes	Ing. V. Vásquez
Corn	Ing. L. Narro
Wheat and other cereals	Ing. S. Franco
Beans	Ing. H. de la Cruz
2.4 Experiment Station Huaráz	Ministerio e Alimentación - Huaráz Head: Ing. A. Cueva
Potatoes	Ing. J. Llontop
Wheat	Ing. E. Valencia
Beans	Ing. G. Morales
Corn	Ing. I. Mendoza R.

3 Tarapoto (CRIA III)

3.1 Experiment Station El Porvenir	Apartado 9. Tarapoto - San Martín Dir.: Ing. M. Llaveriá B. Head: M. Lescano Alva
Animal production and forage	Ing. K. Reátegui
Animal health	Dr. M. Ramírez
Tropical fruits	Ing. E. Villalba R.
Cassava	Ing. A. Solórzano
3.2 Experiment Station Tulumayo	Apartado 78 - Tingo María - Huánuco Head: Ing. M. Nurena S.

Rice and beans	Ing. M. López Rafael
Oil palm and tropical crops	Ing. A. Polo
Cassava and corn	Ing. L. Díaz
3.3 Experiment Station Iquitos	Apartado 307 - Iquitos Head: Ing. W. Chávez Flores
Tropical crops	Ing. O. Mendoza
Rice and corn	Ing. R. Beuzeville
Vegetable crops	ing. W. Chávez

4 Arequipa (CRIA IV)

4.1 Experiment Station Arequipa	Luna Pizarro 136 - Arequipa Dir.: Ing. C. Esquivel A. Head: Ing. L. Juárez Galiano
Rice and beans	Ing. F. Carnero
Vegetables	Ing. M. Pinto P.
Potatoes	Ing. E. Almonte
Wheat	Ing. O. Benites
Forages	Ing. R. Velasquez
Animal production	Ing. J. Quintanilla
4.2 Experiment Station Cuzco	Matará 410 - Cuzco Head: Ing. S. Quevedo Willis
Potatoes	Ing. M. Pacheco
Dressing	Ing. J. Pacheco
Entomology	Ing. Frida Gamarra
Corn	Ing. V. de la Colina
Wheat and other cereals	Ing. R. Romero
Tropical crops	Ing. V. Ortiz
Animal production and forage	Dr. D. Perez
4.3 Experimental Station Puno	Av. El Sol 576 - Puno Head: Ing. G. Calderón P.
Potatoes	Ing. V. Huanco
Cereals, quina, cañihua	Ing. G. Calderón P.
Oil crop (colza)	Ing. E. Ormachea
Animal production and forage	Ing. J. Choque
4.4 Experiment Station Tacna	Calle Los Nardos, Nos 46-50 - Urb. Pescasseroli - Tacna Head: Ing. O. Velarde Ruesta
Fruits and vegetables	Ing. O. Velarde R.
Potatoes and others	Ing. J. Reinoso
Corn and wheat	Ing. J. Herna

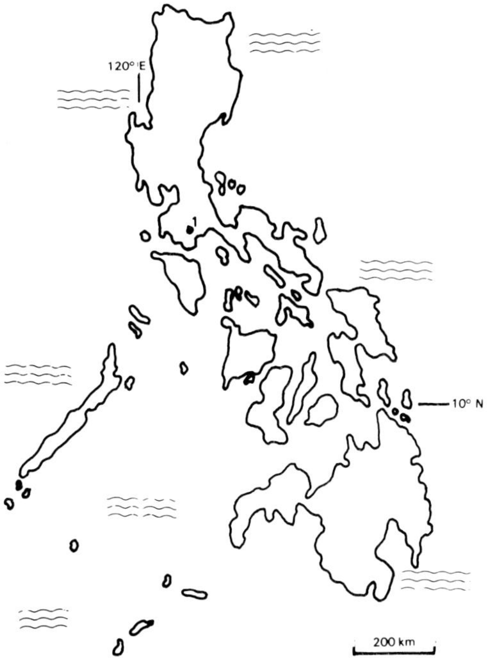
120° E
10° N
1
200 km

1 Los Baños

Survey

The UP at Los Baños (UPLB) complex is located some 63 km. south of Manila. It is composed of the campus, experimental fields, demonstration farms, and research areas. The University also owns 4,244 ha. of forest research in Mt. Makiling and about 4,088 ha. of land grants in Paete, Laguna and La Granja, Negros Occidental. It has 17 colleges, institutes and centres.

A massive 10-year development program, completed in 1973, transformed the University's physical plant into one of the best in Southeast Asia. New physical structures emerged in 1978 and 1980's. The University performs three functions: instruction, research and extension. In pursuing this, the University is guided by the philosophy that this threefold function is interrelated and mututial for building-up high-level manpower, extending the frontiers of knowledge and applying that knowledge to achieve integrated rural development. It is in this sense that the UPLB serves the greatest number of people in Philippine society. The functions of the College are: (1) to turn out its share of manpower requirements for the agricultural and rural development of the country; (2) to carry out research on immediate and long-term problems of Philippine agriculture which will in turn contribute to the advancement of agricultural technology; and (3) to disseminate research findings in a form suitable for adoption by extension workers and farmers; help train extension workers; provide the technical backstop to extension technicians in the field whenever necessary; and conduct pilot action/research projects on agricultural and rural development.

1 Los Baños

1.1 College of Agriculture
University of the Philippines
at Los Baños
elevation: 21.7 m

College, Laguna, Philippines
Phone: 2567, 3585
Dean: R.L. Villareal

Prepared by D.E. Angeles, College of Agriculture, University of the Philippines at Los Baños.

rainfall : 4928 mm
sunshine : 1727 h

1.2 Department of Horticulture — Chairman: J.M. Soriano, M.S.

The Department of Horticulture started out as a mere division under the Department of Agronomy when the College of Agriculture was established in 1909. In 1974, it became a full Department having its own program for instruction, research and extension.
It now boasts for its load of accomplishments which include the chemical induction of mango; year-round mango production using various flower inducers; hormonal assays on mango flower induction; avocado, pineapple and papaya fertilization; evaluation and selection of potato cultivars suited to high and middle elevations; development and release of tomato and cucumber varieties resistant to bacterial wilt and downy mildew, respectively; tissue culture of orchids; embryo culture of anthurium; rose hybridization and mutation; intergeneric hybridization of orchids; evaluation of chrysanthemum cultivars; postharvest handling and storage of papaya, banana and lanzones; mango packaging, maturity indices and CO_2 injury; evaluation of mulberry, coconut and cacao cultivars and their hybrids; studies in pineapple fiber extraction and fabric development, coconut and abaca tissue culture experiments, verification of the medicinal properties of plants and a host of other research.
Research has been greatly facilitated by the establishment of the abaca, coffee, cacao, banana, coconut and medicinal gene banks which serve as a depot for local selections and foreign accessions as well as provide the necessary gene pool for the current and future breeding programs. The fruit crop orchard contains a collection of foreign and local fruit cultivars being used in varietal comparisons. The morphology laboratory and the orchid, macapuno and tissue culture laboratories provide the necessary equipment for both chemical and physiological studies.

1.2.1 Fruit Crops Division	Head: L.O. Namuco, Ph.D.*
Mango research project	N.D. Bondad
Classification of Philippine bananas	R.R.C. Espino*
Cropping systems in banana	D.E. Angeles
Cashew and jackfruit research project	D.E. Angeles
Storage, transport and graftability of scions of fruit trees	E.C. Operio Jr.
Citrus breeding	R.R.C. Espino*
Papaya breeding	R.R.C. Espino*
Screening and development of avocado cultivars for domestic and foreign markets	R.R.C. Espino*
Mutation breeding in Philippine fruit species	R.R.C. Espino*
Screening and development of outstanding rambutan cultivars	R.R.C. Espino*
Adaptability tests on selected grape cultivars	R.R.C. Espino*
Regional variety trials in citrus	R.R.C. Espino*
Plant propagation	F.B. Javier/L.O. Bondoc
1.2.2 Plantation Crops Division	Head: J.B. Sangalang, Ph.D.
Abaca development program	
Development of tissue culture technique as a method of rapid mass propagation for abaca	A.G. del Rosario/M.O. Cedo
Development of on-farm strategies for abaca production technology transfer	I.S. Anunciado
Cacao under coconut research program (cultural management)	I.S. Anunciado
Cacao under coconut research program (field trials)	J.B. Sangalang*
Cacao breeding project	J.B. Sangalang* J.G. Dubouzet
Sericulture research project	I.S. Anunciado
Growth and physiology of macapuno coconut	A.G. del Rosario
Coconut tissue culture	A.G. del Rosario
Seed orchards, seed production areas and	I.S. Anunciado

seed bank of pulping trees	
Locality source and survey of folk uses of potential medicinal crops	E.G. Quintana
Medicinal plants garden and production	R.G. Maghirang
Maintenance of a gene bank for medicinal plants	R.R.C. Espino*
Propagation of medicinal plants	J.D. Saludez
Research on medicinal plants in shade and fertilizer levels	B.C. Felizardo
Coconut breeding project	J.B. Sangalang C.M. Protacio
Tissue culture of coconut inflorescence	A.G. del Rosario*
Field performance of selected coconut cultivars	J.B. Sangalang*
Coconut-based farming system project	I.S. Anunciado*
Selective harvesting and storage of medicinal plant parts	E.G. Quintana*
The possible control of quality and yield of fiber from abaca using plant growth regulators	C.S. de Gutman*
1.2.3 Ornamental Crops Division	Head: Helen L. Valmayor, Ph.D.
Breeding of *Phalaenopsis;* Field trials of Kagawara as cutflower; Rapid growth of *Dendrobium* by tissue culture; *Vanda* and *Phalaenopsis;* Embryology and embryo culture of Philippine orchid species; Tissue culture of selected foliage plants	H.L. Valmayor
Crop improvement in selected foliage plants	T.L. Rosario
Anthurium breeding	T.L. Rosario
Environmental requirements of *Anthurium andreanum* 'Kaumana'	T.J. Rimando
Landscaping	S.R. Bautista
Tissue culture	T.S. Davide
Some cultural studies of production and marketing of foliage plants	H.L. Valmayor*/T.J. Rimando*
1.2.4 Vegetable Crops Division	Head: J.M. Soriano, M.S.
Testing project for vegetable legumes	J.R. Deanon Jr.
Potato improvement program	E.T. Rasco Jr.
Vegetable evaluation trials. Study 27.	E.T. Rasco Jr./T.M. Aurin
Regional trials on crucifers	
Testing program for vegetable legumes	J.M. Soriano
Seed technology	R.C. Mabesa
Vegetable evaluation trials	J.R. Deanon Jr.
Vegetable improvement program	E.T. Rasco Jr./J.R. Deanon Jr. T.L. Rosario
Backyard gardening for improved nutrition	J.M. Soriano*
Socio-economic impact of national cooperative testing for vegetables in selected pilot areas	C.F. Azucena
1.2.5 Postharvest Horticulture Training and Research Center (PHTRC)	Dir.: D.B. Mendoza, Jr.*

The PHTRC conducts basic research and the only training program on postharvest horticulture in the Southeast Asia Region. Its training course is designed to enable graduates to acquire the basic knowledge and skills in postharvest handling of fruits and vegetables.

ASEAN food handling project	Er. B. Pantastico
ASEAN mango and rambutan project	D.B. Mendoza Jr.
ASEAN packinghouse operations and quality control project	O.K. Bautista
ASEAN fruit and vegetable container project	E.V. Labios*
Fruits and vegetable handling trials	E.B. Esguerra

Acceptability and nutritive value of potato and potato products	O.K. Bautista
Postharvest handling and storage of vegetables	E.B. Esguerra
Study II. Transport of cabbage	
Retail handling	M.C.C. Lizada
Wholesale handling	O.K. Bautista*
Bulk handling	M.C.C. Lizada*

1.3 Department of Entomology

Chairman: B.M. Rejesus, Ph.D.*

Research in the Department deals with insect taxonomy, insect ecology, insect anatomy and morphology, insect physiology, insecticide toxicology and pesticide management, insect pathology, veterinary and medical entomology, crop pest management and industrial entomology.

Insect pest management	
Watermelons	I.P. Palacio
Vegetables	V.J. Calilung/E.D. Magallona
Fruits	F.F. Sanchez/E.D. Magallona
Host-plant resistance	C.B. Adalla
Biological control	E.P. Cadapan
Pesticide Toxicology	E.D. Magallona/B.M. Rejesus

1.4 Department of Plant Pathology

Chairman: T.T. Reyes, Ph.D.

The Department is involved in ornamental plant disease research especially on virus diseases of orchids, medicinal plants and sweet potato. Nematode research has been carried out on vegetables, fruits, field legumes and fiber crops.

Nematodes and control	
Vegetables and field legumes	M.B. Castillo
Vegetables and abaca	R.B. Valdez
Vegetables, fruit crops, cotton	R.G. Davide
Sugarcane and rice	T.T. Reyes
Sorghum diseases and control	S.C. Dalmacio
Corn diseases and control	O.R. Exconde
Diseases of ornamental and medicinal plants, root crops and control	G.G. Divinagracia
Legumes and postharvest diseases and control	L.L. Ilag
Mycology	F.T. Orillo
Mycology and mushrooms	T.H. Quimio
Phytobacteriology and fruit crop diseases	A.J. Quimio
Rice and wheat diseases and control	D.B. Lapis
Legume diseases and control	F.C. Quebral/O.S. Opina
	C.J.M. Maroon
Pathogenesis and phytobacteriology	M.P. Natural
Epidemiology	A.C. Molina Jr.
Cotton and abaca diseases	T.T. Reyes

1.5 Department of Soil Science

Chairman: D.A. Carandang, Ph.D.*

The Department has worked on little-leaf malady of Batangas citrus, causes of cadang-cadang, nutrition and fertilizer needs of coffee and coconut and is now active in gathering benchmark information on Philippine soils, in biotechnology and applied microbiology research on various leguminous plants and in action research programs involving perennial based cropping systems.

Soil chemistry	A.M. Briones/E.D. Reyes
	G.O. San Valentin
Soil classification and land use	A.J. Alcantara/R.B. Badayos

	W.C. Cosico
Soil fertility	D.A. Carandang/B.C. Felizardo
	R.N. Nartea/H.P. Samonte
Soil microbiology	I.J. Manguiat/E.S. Paterno
Soil mineralogy	N.C. Fernandez/L.A. Montecillo
Soil physics	A.A. Briones/E.P. Paningbatan Jr.

1.6 National Crop Protection Center — Dir.: F.F. Sanchez, Ph.D.*

This Center serves as the lead agency in the country's crop protection research. It undertakes problem analyses, developmental research and planning required to develop crop protection systems against pests of major economic crops, including horticultural crops.

Other crops pest management team:	
Insect control in mango	S.M. Bato
Rodent control in coconut and pineapple	M.M. Hoque
Vegetable pest management team:	
Disease control	R.G. Davide/R.A. Zorilla
	A.J. Quimio/G.C. Molina
	E.C. Paller
Insect control	E.D. Magallona/V.J. Calilung
Weed control	E.C. Paller/A.M. Baltazar

1.7 Institute of Plant Breeding — Dir.: E.T. Rasco, Jr., Ph.D.

Crop improvement	E.C. Altoveros, M.S. Hort.
White potato improvement project	
Stock seed production of Chinese cabbage and radish	
Plant Biology	N.C. Altoveros, Ph.D.
Collection, evaluation, maintenance of germplasm for vegetables	
Plant Breeding	Renato A. Avenido, B.S. Agr.
Tissue culture of calamansi for seedlessness	
Tissue culture of bamboo for rapid propagation (preliminary stage)	
Plant Breeding/Genetics	Gloria E. Balagtas, BSA/M.S. Gen.
Genetic control of sex expression in selected fruit crops	
Genetic polymorphism in squash and ampalaya	
Seed Production	P.S. Castillo, BSA
Seed production of crucifera, solanaceons legumes and cucurbits	
Comparison of seed field between prostrate and trellisted upo, patola and ampalaya	
Plant Physiology, Genetic Resources Conservation, Plant Propagation	R.E. Coronel, Ph.D.
Evaluation, documentation & Conservation of fruit genetic resources	
Early seedling growth and graftability of different rootstock varieties	
Statistical analysis for determining minimum sample size for propagation experiments	
Tissue Culture	Olivia P. Damasco, M.S. Hort.
Tissue culture	
Tissue culture of banana	
Tissue culture of rambutan and other fruit trees	
Citrus clean-up by meristem culture and shoot	

Tip grafting and thermotherphy	
Plant Physiology Screening vegetable legumes for adaptability to post-rice conditions Screening for drought and waterlogging resistance in tomatoes	Dafrosa A. Del Rosario, Ph.D.
Fruit Crops Breeding: citrus varietal improvement screening and development of outstanding rambutan cultivars mango breeding papaya breeding pollen storage studies of Philippine fruit crops species mutation breeding of Philippine fruits	R.R.C. Espino*, Ph.D.
Crop Physiology Screening for waterlogging and drought resistance in mungbean, and cowpea Genetic studies on stress-inducible gene products in rice and mungbean.	Flordeliza A. Faustino, M.Agr.
Insect Resistance & Crop Improvement (vegetables) Lowland white potato improvement project including TPS evaluation for true seed variety Crop improvement of a ampalaya and upo Insect resistance studies on priority vegetables against thrips and beanfly (new project)	E.C. Fernandez, M.S. Entomology
Soil Science Field screening of mungbean, peanut, soybean and yardlong bean for an enhanced nitrogen fixation	Evelyn C. Fernandez, M.S.
Genetics Interspecific hybridization in squash and bitter gourd	Victoria B. Laurel, M.S.
Food Science Tannins in cowpea: characterization, localization, removal, and effect of in vitro protein digestibility Biochemical and nutritional studies of Philippine indigenous food and forage legumes	A.C. Laurena, M.S.
Horticulture (Ornamentals) Varietal improvement of the ff. crops: okra, squash, amaranth, winged bean, crucifers and lettuce	Rizalina R. Licuanan, B.S.A.
Vegetable Crop Improvement Varietal improvement of pole sitao, bush sitao, snap beans, white beans, garden pea, ?	R.G. Maghirang, M.S. in Plant breeding
Plant Biochemistry Biochemical and nutritional studies of Philippine indigenous food and forage legumes Determination of hybridity and identity of several vegetable crops Regulation of aberrant cell growth of mutant coconut	Evelyn M.T. Mendoza, Ph.D. Biochemistry
Tissue Culture Meristem culture and virus eradication in potato	Cynthia Paet, B.S.A. Agronomy
Varietal screening of mungbean, peanut, soybean and yardlong bean for enhanced nitrogen fixation	Erlinda S. Paterno, Ph.D.
Plant Pathology Varietal screening for resistance to bunchy top and Fusarium wilt	A.R. Pua, M.S.
Genetics Genetic studies on vegetables Genetic control of sex expression in selected	Dolores A. Ramirez, Ph.D.

fruit crops	
Virology, Tissue Culture Cleaning-up of citrus by tissue culture (meristem culture, meristem budding, thermogheraphy)	Claire S. Ramos, B.S. Agriculture
Plant Breeding Varietal improvement of vegetables: breeding for tropical potatoes development of heat tolerant white potato varieties for medium and low elevations improvement of Philippine indigenous vegetables seed stock production of vegetable crops	E.T. Rasco, Jr., Ph.D.
Genetics, Plant Breeding Ornamental crops breeding (foliage plants, Anthuriums, orchids) Bulb crops breeding (onions and garlic) White beans *(Phaseolus vulgaris)* breeding	Teresita L. Rosario, Ph.D.
Crop Production and Management Rapid multiplication of white potato under low elevation conditions	W.F. Sibal, B.S.A. Agronomy
Plant Pathology Resistance of tomato, white potato and other vegetable crops to root-knot and reniform nematode bacterial wilt, major foliar diseases and major virus diseases Studies on diseases affecting Philippine indigenous vegetable crops and screening for resistance against major diseases Screening abaca cultivars and hybrids for resistance to mosaic and bunchy top	R.B. Valdez, Ph.D.
Tissue Culture, Plant Propagation Tissue culture of white potato, banana, citrus with emphasis on multiplication schemes Development of medium term storage protocols and maristem culture	Alfinneta B. Zamora, M.S. Ph.D. (currently enrolled)
1.8 Farming Systems and Soil Resources Institute	Dir.: E.L. Rosario Phone: 2459, 3229
Coconut-based farming systems program Varietal screening of possible intercrops for coconut	Ms. Edna A. Aguilar, M.S. Hort.
Rainfed rice-based farming systems program Lowland rainfed farming systems	Ms. Piedad A. Mendoza, M.S. Hort.

POLAND PL

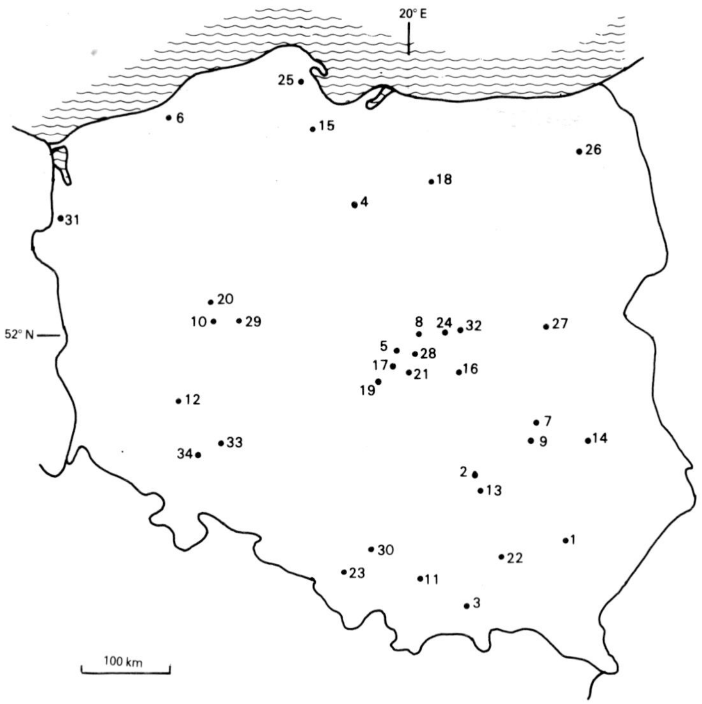
20° E
52° N
100 km
1
2
3
4
5
6
7
8
9
10
11
12
13
14
15
16
17
18
19
20
21
22
23
24
25
26
27
28
29
30
31
32
33
34

1 Albigowa, 2 Bogusławice, 3 Brzezna, 4 Bydgoszcz, 5 Dąbrowice, 6 Dworek, 7 Górna Niwa, 8 Guzów, 9 Końskowola, 10 Kórnik, 11 Kraków, 12 Leszkowice, 13 Lipowa, 14 Lublin, 15 Miłobądz, 16 Nowa Wieś, 17 Nowy Dwór, 18 Olsztyn, 19 Pabianice, 20 Poznań, 21 Prusy, 22 Pryborów, 23 Pszczyna, 24 Reguły, 25 Rekowo, 26 Siejnik, 27 Sinołeka, 28 Skierniewice, 29 Słupia Wielka, 30 Swierklaniec, 31 Szczecin, 32 Warszawa, 33 Wrocław, 34 Wróblowice

Survey

The Polish Academy of Sciences - Since its creation in 1951, the Polish Academy of Sciences (Polska Akademia Nauk) is the highest authority as far as science is concerned in Poland. It is responsible for planning all the research work of the country, especially in basic research. The Polish Academy of Sciences has its own research institutes but only one of them - the Research Institute of Dendrology in Kórnik - is in part concerned with horticulture. In the Department of Agriculture and Forestry of the Polish Academy of Sciences there is a Committee of Horticultural Science. The Committee is responsible for planning and coordinating all horticultural research work in Poland. It is composed of all most prominent Polish horticulturists.

The Research Institute of Pomology and Floriculture - The bulk of agricultural research in Poland is being done at the Institutes under the Ministry of Agriculture and Food Management. The Institute was organized in 1951 at Skierniewice by Prof. S.A. Pieniazek who was its director for 33 years.

The Institute deals with all aspects of pomological and ornamental plants research. The research activities are conducted at the Institute's headquarters in Skierniewice and in fourteen experiment stations scattered all over the country.

Research on woody ornamental plants is being conducted also as one of the activities of the Institute of Dendrology of the Polish Academy of Sciences in Kórnik. Presently research is conducted on the selection of trees and shrubs resistant to low temperatures, to atmospheric pollution and suited for urban and roadside plantings.

The Research Institute of Vegetable Crops was organized in 1964 and directed up to 1973 by late Prof. Dr. E. Chroboczek. The Institute deals with all aspects of vegetable research, edible mushrooms included. There are 10 experiment stations associated

Prepared by Mrs. J. Suska, Research Institute of Pomology, Skierniewice

with the Institute.
Agricultural Universities are subordinated to the Ministry of Science, Higher Education and Technology. A considerable part of horticultural research work is carried out by the horticultural departments of the Agricultural Universities. There are 8 Agricultural Universities in Poland with 16.000 students, namely in Warsaw, Kraków, Lublin, Wrocław, Poznán, Szczecin, Olsztyn and Bydgoszcz. The University of Agriculture in Warsaw with over 4.000 students is the largest. In four of these Universities, Warsaw, Poznán, Kraków and Lublin there are separate Faculties of Horticulture; horticultural students follow, from the "freshman year", a curriculum different from agricultural students.

In addition at the Agricultural University in Warsaw the Faculty of Landscape Architecture is in operation. In the other four Agricultural Universities (Wrocław, Szczecin, Olsztyn and Bydgoszcz there are no faculties of horticulture. However, there are Departments of Horticulture in the Faculties of Agronomy, which work mainly on problems of pomology and vegetable crops.
The Horticultural Faculties of the four Agricultural Universities, although primarily occupied with teaching, have sufficient laboratory and field facilities to conduct research in horticulture on a wider scale. The Departments of Horticulture of the other four Universities are rather small and are mainly occupied with teaching, while research work is conducted on a limited scale.
Publications. The results of horticultural research work in Poland are published in a number of periodicals. There are the Polish Agricultural Annuals (Roczniki Nauk Rolniczych) published by the Polish Academy of Sciences. Each institute publishes its own annual, for instance: The Research Institute of Pomology publishes "Fruit Science Reports" - in English, and "Prace Instytutu Sadownictwa w Skierniewicach" ser. A - concerning pomology, ser. B - concerning ornamental plants. (Experimental Work of the Research Institute of Pomology and Floriculture.) The Institute of Vegetable Crops publishes the Bulletin of Vegetable Crops.
Each Agricultural University has its own Scientific Journal (Zeszyty Naukowe). The Institute of Dendrology publishes "Arboretum Kornickie". The articles in these publications have English and Russian summaries.

1 Albigowa

1.1 Zakład Dóswiadczalny Instytutu Sadownictwa i Kwiaciarstwa (Experimental Station of the Research Institute of Pomology and Floriculture)
elevation: 200 m
rainfall : 600 mm

37-122 Albigowa
Phone: Láncut 30-70, 17-87, 17-97
Telex: 0633148
Dir.: Dr. J. Kleparski

Pomology intensive methods of fruit production; stone and small fruits: nursery; pest and disease control; economics of fruit production
Orchard 142 ha

General pomology	Dr. J. Kleparski
Intensive methods of fruit production	F. Baran Eng.
Blueberries	A. Piekło M.Sc.
Disease and pest control	R. Bachnacki MSc.
Small fruits	Z. Domino Eng./ A. Klimczak M.Sc.
	Z. Radaczynska M.Sc.
Orchard mechanization	R. Rejman M.Sc.
Stone fruits	Dr. M. Stropek

2 Bogusławice

2.1 Warzywniczy Zakład Doswiadczalny Instytutu Warzywnictwa (Experiment Station of the Research Institute of Vegetable Crops)

27-580 Sadowie
Dir.: A. Galicki Eng.
Dep. Dir.: S. Kamiński M.Sc.

Experimental and extension fields (250 ha)
Production of onion, cabbage and carrot seed production.

A. Kamińska M.Sc.

3 Brzezna

3.1 Zakład Doświadczalny Instytutu Sadownictwa i Kwiaciarstwa (Experiment Station of the Research Institute of Pomology and Floriculture)
elevation: 320-360 m
rainfall : 76 (620-1040) mm
sunshine : 877 h

Brzezna, 33-386 Podegrodzie
Phone: Nowy Sacz 22810
Telex: 0322553
Dir.: Prof.Dr. A. Szczygieł*

Problems connected with growing fruit trees and small fruits in mountainous region of South Poland. Main crops: apples, plums, sour cherries, small fruits. Research and extension in the above field. Fruit tree nurseries and strawberry mother plantation of superlite rank.

Research Division	Head: Prof.Dr. A Szczygieł*
Testing varieties and rootstocks	E. Kolbusz Eng.
Soil management and fruit crops protection from damage	P. Kołodziejczak M.Sc.*
Pruning and nutrition	F. Kadzik M.Sc.
Small fruits	K. Pierzga M.Sc./ Z. Pasiut Eng.
Breeding raspberries and blackberries	Dr. J. Danek*
Pest control	Dr. S. Prędki
Disease control	Dr. H. Profic-Alwasiak*
Nematodes (occurence, harmfulness and control)	Prof.Dr. A. Szczygieł*/Dr. A. Zepp

4 Bydgoszcz

4.1 Akademia Rolniczo-Techniczna, Wydział Rolniczy, Zakład Ogrodnictwa (Agricultural University, Faculty of Agriculture. Department of Horticulture)

ul. Bernardyńska 6
85-029 Bydgoszcz
Head: Prof.Dr. M. Jerzy*

Floriculture (chrysanthemums, flower bulbs)	Prof.Dr. M. Jerzy*
Floriculture (chrysanthemums)	Dr. M. Zalewska*
Vegetable genetics and breeding	Dr. P. Nowaczyk
Vegetable physiology	Dr. P. Piszczek

5 Dąbrowice

5.1 Zakład Doświadczalny Instytutu Sadownictwa i Kwiaciarstwa (Experiment Station of the Research Institute of Pomology and Floriculture)
elevation: 128 m
rainfall : 531 mm
sunshine : 1705,9 h

Dąbrowice, 96-100 Skierniewice
Phone: 31-20 Skierniewice
Telex: 885661
Dir.: Dr. K. Dronka

Fruit breeding, orchard soil management, pest and disease control, small fruits. Experimental orchard (215 ha)

General pomology	Dr. K. Dronka
Small fruits, weed control	D. Chlebowska M.Sc.
Plant protection	M. Rejnus M.Sc.
Variety trials	B. Omiecińska M.Sc.
Soil management and nutrition	M. Piątkowski M.Sc.*
General pomology	Z. Szczepanik Eng.

6 Dworek

6.1 Zakład Doświadczalny Instytutu Sadownictwa i

76-036 Borkowice

Kwiaciarstwa (Experiment Station of the Research Institute of Pomology and Floriculture)
elevation: 11 m
rainfall : 600 mm
sunshine : 1080 h

Phone: 81210/81219 Koszalin
Telex: 0532174
Dir.: L. Szczypinski M.Sc.

Variety and rootstock studies, pest and disease control, drip irrigation. Experimental orchard 148 ha

General pomology — M. Gronek M.Sc.
Pest and disease control — W. Szczypinska M.Sc.
Small fruits — B. Pietizak M.Sc./H. Podgorska M.Sc.
Irrigation — M. Bakun M.Sc.

6.2 Radacz
Cranberry and Blueberry Research Station

78-412 Parsęcko
Phone: 47613 Szczecinek
Head: Dr. M. Mackiewicz

General pomology — D. Mackiewicz Eng./ P. Wasiak M.Sc.
Small fruits and ornamentals — Dr. M. Mackiewicz

7 Górna Niwa

7.1 Zakład Doświadczalny Instytutu Sadownictwa i Kwiaciarstwal (Experiment Station of the Research Institute of Pomology and Floriculture)
elevation: 140 m
rainfall : 576 mm
sunshine : 1687 h

Górna Niwa, ul. Dzierżyńskiego 92
24-100 Puławy
Phone: 33-35
Telex: 0643109
Dir.: S. Król M.Sc.

Dwarf rootstocks, pest and disease control, beekeeping. Experimental orchard 20 ha

General pomology — S. Król M.Sc.
Plant protection — Z. Gromisz M.Sc.
Beekeeping — Dr. A. Król

8 Gúzow

8.1 Warzywniczy Zakład Doświadczalny Instytutu Warzywnictwa (Experiment Station of the Research Institute of Vegetable Crops)

Guzów, 05-732 Wiskitki
Dir.: J. Lenkiewicz M.Sc.

Experimental and extension fields (581 ha)
Onion storage for 1200 tons
Organization of production and extension — J. Lenkiewicz MSc.

9 Końskowola

9.1 Dział Prod. Ogrodniczej Wojewódzkiego Ośrodka Postępu Rolniczego (Pomology Section of the Regional Agricultural Experimental Station. Belongs to the Ministry of Agriculture)
elevation: 164 m
rainfall : 554 mm
sunshine : 1687 h

24-130 Końskowola
Head: Dr. K. Tylus
Phone: 16-35 Puławy
Telex: 0642433

Cherry and currant — Dr. K. Tylus

General pomology	Z. Krawiec M.Sc./ H. Osuch M.Sc.
Plant protection	Z. Partyka M.Sc.
Strawberry and ornamental plants	U. Skorupska M.Sc.
Cherry, raspberry and hazelnut	M. Staropiętka
Gooseberry, aronia	M. Dudzic M.Sc.

10 Kórnik

10.1 Instytut Dendrologii (Institute of Dendrology of Polish Academy of Science) elevation: 75 m rainfall : 489 mm	62-035 Kórnik Dir.: Prof.Dr. W. Bugała
10.1.1 Genetics	Prof.Dr. M. Giertych
Population Genetics	Dr. Wl. Chałupka/ Dr. H. Fober G. Kosiński M.Sc.
Biochemical genetics	Ass.Prof. L. Mejnartowicz A. Szmidt M.Sc./A. Lewandowski M.Sc.
Poplar breeding	Ass.Prof. Z. Stecki/ Dr. J. Figaj
10.1.2 Seed biology	Prof.Dr. B. Suszka T. Tylkowski M.Sc.
10.1.3 Biotic and abiotic resistance	Ass.Prof. R. Siwecki
Patogens	Dr. A. Werner/ Dr. K. Przybył L. Rachwał M.Sc.
Frosts	Dr. P. Pukacki
Industrial gases	Dr. G. Lorenc-Plucinska Dr. G. Oleksyn/ Dr. P. Karolewski
10.1.4 Physiology	Ass.Prof. Z. Szczotka Dr. S. Pukacka/ Dr. K. Krawiarz Dr. M. Rudawska Dr. B. Rokicka-Kieliszewska
10.1.5 Introduction and acclimatization	Prof.Dr. W. Bugała Prof.Ass. H. Chylarecki A. Skibinska M.Sc./Dr. K. Bojarczuk Dr. T. Bojarczuk/J. Dolatowski M.Sc. Dr. A.Chodun
10.1.6 Systematics and geography	Prof.Dr. K. Browicz
Chorology	Dr.K.Boratyńska/ Dr.A.Boratyński
Systematics	Dr. J. Zieliński/Dr. J. Hantz

11 Kraków

11.1 Akademia Rolnicza, Wydział Ogrodniczy (Agricultural University, Faculty of Horticultural Science) elevation: 204 m rainfall : 640 mm sunshine : 1850 h	31-425 Kraków Al. 29 Listopada 54 Phone: 11-91-44 Dean: Prof.Dr. E. Pojnar
Institute of Vegetable Crops and Ornamentals	Dir.: Prof.Dr. T. Wojtaszek*
Vegetable crops	Dr. M. Poniedziałek/Dr. A. Libik
Floriculture	Dr. Z. Pindel
Mineral nutrition	Ass.Prof. G. Kozera
Fruit Science	Prof.Dr.K.Kropp/Prof.Dr.M. Łucka Ass.Prof. W. Poniedziałek
Plant Pathology	Ass.Prof. J. Kućmierz
Entomology	Ass.Prof. A. Wnuk

	Ass.Prof. J. Achremowicz
Economics	Ass.Prof. S. Martyna
	Ass.Prof. K. Martyna
Bee research	Ass.Prof. H. Gałuszka
Biochemistry	Prof.Dr. B. Samotus
Botany	Prof.Dr. E. Pojnar
	Prof.Dr. K. Miczynski
Plant physiology	Ass.Prof. J. Myczkowski
	Ass.Prof. Z. Piskornik
Genetics, breeding and seed production	Prof.Dr. W. Kozera*
	Ass.Prof. B. Michalik*

12 Leszkowice

12.1 Warzywniczy Zakład Doświadczalny Instytutu Warzywnictwa (Experiment Station of the Research Institute of Vegetable Crops)	Leszkowice, 67-221 Białołęka Dir.: K. Szczygieł Eng. R. Rzymowska M.Sc.

Experimental and extension fields (890 ha). Organization of vegetable production. Field experiments and extension

13 Lipowa

13.1 Zakład Doświadczalny Instytutu Sadownictwa i Kwiaciarstwa (Experiment Station of the Research Institute of Pomology and Floriculture)	Lipowa, 27-500 Opatów Kielecki Phone: 209 Dir.: H. Zdyb M.Sc.
Variety studies, pest and disease control; orchard soil management. Experimental orchard 152 ha.	
General pomology	H. Zdyb M.Sc.
Orchard mechanization	B. Długowolski M.Sc.
Pomology	A. Kulikowski M.Sc./ M. Wypych Eng.
	J. Zając M.Sc./ L. Zając M.Sc.
Plant protection	M. Sliwiak M.Sc./ K. Wilusz M.Sc.

14 Lublin

14.1 Wydział Roniczy, Akademia Rolnicza (Agricultural University in Lublin, Faculty of Horticulture) elevation: 195 m rainfall : 573 mm sunshine : 1489 h	Akademicka 13 20-934 Lublin Phone: 332-51 Telex: 0643176 AR PL Dir.: Ass.Prof. J. Nurzyński
14.1.1 Institute of Horticultural Production. Pomology	
Raspberries - biology of flowering and fruiting, cultivars, cultural practices	Dr. J. Wieniarska
Sour cherries propagation and training	Dr. J. Selwa
Sour cherries: biology of flowering, fruiting	Dr. St. Wociór
Currants: soil, management, nutrition	Dr. J. Szwedo
Fruit tree propagation and growth	Dr. Z. Kalinowski
Weed control, environmental studies, propagation	Prof.Dr. J. Lipecki*
Strawberries: soil management, nutrition, weed control	E. Zmuda M.Sc.
14.1.2 Ornamentals	
Propagation in vitro	Ass.Prof.J.Dąbrowski*/Dr.M.Dąbski
Growing media for plants grown under covers	T.Baltaziak M.Sc/Ass.Prof.J.Hetma
Cut flowers propagation and cultivation under glass	T.Baltaziak M.Sc/Ass.Prof.J.Hetman
	Dr.H.Laskowska/Dr.E.Pogroszewska
Cultivation and forcing of bulbous plants	Ass.Prof. J. Dąbrowski*

Cut flowers cultivation in unheated plastic tunnels — Ass.Prof.J.Hetman/Dr.H.Laskowska
T.Baltaziak M.Sc/Ass.Prof.J.Hetman
Dr. H. Laskowska

14.1.3 Soil cultivation and nutrition

Nutrition and cultivation methods — Ass.Prof. T. Kesik

Effect of nutrition on the quality of horticultural products — Ass.Prof. J. Nurzyński
Dr.Z.Michałojc/Dr.E.Mokrzecka
J. Kossowski M.Sc.

14.1.4 Horticultural seed and nursery research

Seed physiology, and growth regulators, tissue culture — Ass.Prof. E. Gawroński

Seed physiology and parthenocarpy — Dr. S. Łukasik

Nursery stock propagation improvement — Dr. T. Mitrut

14.1.5 Institute of Horticultural Plant Biology

Plant Physiology

Mineral nutrition — Prof.Dr.Z.Uziak/Dr.E.Borowski
Dr.U.Kruszelnicka/Dr.M.Szymańska
Wl. Michałek M.Sc.

Crop physiology — Dr. Z. Blamowski

Photosynsthesis and plant growth — Dr. R. Stanek

Growth regulators — Dr. E. Borowski/Dr. I. Rukasz
M. Wilkowicz M.Sc.
L. Kozłowska M.Sc.

14.1.6 Botany

Biology of flowering — Prof.Dr. Z. Warakomska

Pollen and nectar production, pollination, anatomy of nectaries, pollen content in honey, pollen and propolis research — Ass.Prof.K.Szklanowska/Dr.B.Dąbska
Dr. Z. Kólasa/Dr. M. Koter
Dr.T.Jaroszyńska/Dr.A.Wróblewska

14.1.7 Genetics and Breeding of Horticultural Plants

Genetics and breeding of garden pea and strawberry, poliploids, interspecific hybridization — Ass.Prof. J. Piech

Genetics and breeding of strawberries — Prof.Dr.T.Hulewicz/Dr.J.Hortyński

Self-and cross - incompatibility and embrology in cherries — Dr. B. Dyś

14.1.8 Department of Phytopathology and Plant Protection

Mycology and disease control — Prof.Dr. B. Łacic/Dr. D. Pięta
Ass.Prof.A. Filipowicz/Dr.A. Wagner

Predators and parasites of insect pests — Prof.Dr. B. Miczulski

Entomology — Ass.Prof. Z. Machowicz-Stefaniak

14.1.9 Department of Entomology

Biology and harmfulness of Aphididae — Dr. B. Jaśkiewicz

Biology and harmfulness of Coccoidae — Dr. B. Łagowska

Biology and harmfulness of Heteroptera — Prof.Dr. T. Ziarkiewicz

Biology and harmfulness of Agrotinae — Dr. J. Napiórkowska-Kowalik

Applied entomology insect pests and economic importance of Apoidae — Prof.Dr. A. Anasiewicz

Instrumentation in plant protection — Dr. W. Huszcza

15 Miłobądz

15.1 Zakład Doświadczany Instytutu Sadownictwa i Kwiaciarstwa (Experiment Station of the Research Institute of Pomology and Floriculture)
elevation: 25 m
rainfall : 521 mm

83-111 Miłobądz
Phone: Miłobądz 22
Telex: 0512654
Dir.: J. Banyś M.Sc.

Variety studies, orchard soil management, plant

protection, storage, floriculture. Experimental orchard 166 ha.

Fruit growing	J. Banyś M.Sc. Dr. K. Chróścicki B. Guzewska-Bogatko Eng.
Plant protection	Dr. A. Burkowicz/ B. Wojtas M.Sc.
Small fruits	T. Biesiada Eng.
Floriculture	C. Wróblewska Eng.
Orchard mechanization	J. Chojnowski M.Sc.
Fruit storage	I. Jakóbczak M.Sc.

16 Nowa Wieś

16.1 Zakład Doświadczalny Instytutu Sadownictwa i Kwiaciarstwa (Experiment Station of the Research Institute of Pomology and Floriculture)
elevation: 130 m
rainfall : 586 mm

Nowa Wieś, 05-600 Warka
Phone: Grójec 123-70
Dir.: E. Gajewski M.Sc.

Variety studies: orchard soil management, fruit production economics, pest and disease control. Experiment orchard 136 ha.

Pest control	J. Dadej M.Sc.
Disease control	Dr. D.K. Olszewski
Soil management and nutrition	M. Mrozowski M.Sc.
General pomology	B. Jarociński Eng.
Economics	E. Gajewski M.Sc.

17 Nowy Dwór

17.1 Kwiaciarski Zakład Doświadczalny (Floricultural Experiment Station of the Research Institute of Pomology and Floriculture)

Nowy Dwór, 96-115 Nowy Kawęczyn
Phone: Skierniewice 36-52
Dir.: A. Tracz M.Sc.

Production of bulbous plants, perennials, ornamental shrubs and trees.

Shrubs and trees	B. Bryk M.Sc.
Perennials	E. Chudziak-Bilska M.Sc.
Bulbous plants	T. Pankiewicz M.Sc.

18 Olsztyn

18.1 Katedra Ogrodnicza Akademii Rolniczo-Technicznej (Department of Horticulture of the Agricultural University)
elevation: 133 m
rainfall : 623 mm
sunshine : 1650 h

10-720 Olsztyn-Kortowo
Phone: 28450
Telex: 0526419
Head: Prof. Z. Kawecki*

Experimental orchard 30 ha
Experimental vegetables 5 ha

General pomology	Prof.Dr. Z. Kawecki*/Dr. W. Kulesza Dr. J. Ważbinska J. Kopytowski M.Sc.
Vegetable crops	Ass.Prof. W. Kryńska Dr. B. Przybyszewska Dr. B. Wierzbicka/Dr. S. Tarkowian

19 Pabianice

19.1 Kwiaciarski Zakład Doświadczalny (Floricultural Experiment Station of the Research Institute of Pomology and Floriculture)
95-200 Pabianice
ul. Ksawerowska 7
Head: W. Witaszek M.Sc.

Introduction of new methods of ornamental plants growing in greenhouses — I. Bieńkowska M.Sc./ J. Jura Eng. H. Poros MSc./E. Rzędzinska M.Sc.

20 Poznań

20.1 Instytut Ochrony Roślin (Institute for Plant Protection)
elevation: 83 m
rainfall : 480 mm
sunshine : 1400 h

ul. Miczurina 20
60-318 Poznań
Phone: 679-021 Poznań
Telex: 0413203
Dir.: Prof.Dr. W. Węgorek

The institute carries out scientific developmental, initiative and extension works. It coordinates the whole research on plant protection in Poland. Belongs to the Ministry of Agriculture. Center in Poznań; branch at Sośnicowice; Experimental Plant Protection Station at Winnogóra (50 km out of Poznań, 640 ha), eight Experiment Stations at Białystok, Człuchow, Opypy, Rzeszów, Sulęcin, Toruń, Trzebnica and Nowy Sącz.

20.1.1 Department of Agricultural Zoology — Head: Prof.Dr. Z. Gołębiowska
Orchard mite biology, control — Ass.Prof.Dr. W. Chmielewski
Codling moth and other orchard pests, control, forecasting, biology — Dr. A. Skorupska/Dr. F. Kagan, Dr. J. Kozłowski
Pests of field vegetables — Dr. L. Korcz
Rodents, population dynamics, control — Ass.Prof.Dr. A. Romankow-Zmudowska

20.1.2 Department of Agricultural Phytopathology — Head: Prof.Dr. A. Zgórkiewicz
Bacterial diseases, tomato, orchards — Prof.Dr. A. Golenia
Virus diseases of vegetables — Dr. A. Jakusz/Dr. H. Pospieszny
Virus diseases of mushrooms — M. Przybylska M.Sc.
Biochemistry and biophysics of virus diseases — Dr. W. Kaniewski
Serology — M. Pruszyńska M.Sc.

20.1.3 Department of Disease and Pest Control Methods — Head: Prof.Dr. J.J. Lipa
Biological control of some pests using biopreparations — Prof.Dr. J.J. Lipa/Dr. J. Ziemnicka, Dr. J. Bartkowski
Mass rearing and release of parasites and predators, integrated control — Ass.Prof. T. Kowalska, Ass.Prof. S. Pruszyński, K. Brzeziński M.Sc.

20.1.4 Department of Research and Control of Pesticides Residues — Head: Dr. J. Dąbrowski
Residues of pesticides in vegetable plants and fruits — Dr. J. Dąbrowski
Use of isotopes in bound residue studies — Dr. E. Czaplicki/J. Dec M.Sc.

20.1.5 Department of Research of Pesticides — Head: Ass.Prof.Dr. W. Witkowski
New insecticides for the control of some orchard and vegetable pests — D. Witkowska M.Sc., K. Szczepańska M.Sc.

20.2 Akademia Rolnicza - Wydział Ogrodniczy (Agricultural University, Faculty of Agriculture)
ul. Wojska Polskiego 28
60-637 Poznań
Phone: 224-581
Telex: 0413322 AR pl

20.2.1 Botany Department
Dendrology, growth, biology, variableness — Prof.Dr. S. Król
Fruit and seeds morphology — Dr. W. Sterna
Chorology and taxonomy of Rosa sp. — Dr. W. Stefanek

20.2.2 Entomology Department

Natural control of orchard pest insects	Prof.Dr. W. Kadłubowski H. Piekarska M.Sc.
Orchard and ornamental pest insects	Ass.Prof.Dr. S. Burdajewicz
Bionomy of aphids	Ass.Prof.Dr. M. Karczewska Dr. B. Wilkaniec
Faunistic and bionomy of leafhoppers	Ass.Prof.Dr. W. Nowacka
20.2.3 Horticultural Plant Nutrition Department	
Plant nutrition, soil science	Prof.Dr. M. Hoffmann
Diagnosis of nutritional requirements	Dr. A. Komosa
Standard levels of microelements	Dr. W. Tyksiński
Critical levels of macroelements for vegetable crops	Dr. A. Golcz
Mineral - organic fertilizers	W. Breś M.Sc.
Mineral nutrition of ornamental and seed plants	E. Kozik M.Sc.
20.2.4 Landscape Agriculture Department	
Agrotechnical problems in landscaping short time decorations	Ass.Prof.Dr. Z. Haber
Introduction of ornamental grasses to the landscape	P. Urbański M.Sc.
History of the gardens, landscape renovation	B. Spikowska M.Sc.
20.2.5 Ornamental Plant Department	
Growing media, nutrition in greenhouse	Dr. A. Lisiecka*
Dormancy and germination of seeds	Dr. K. Grabowski
Bulb crops	Dr. J. Krause*
Propagation by tissue culture	Dr. K. Oszkinis
Introduction and biology of Rhodedendron sp.	Ass.Prof.Dr. M. Czekalski
20.2.6 Plant Pathology Department	
Tomato - soil borne diseases	Prof.Dr. T. Glaser
Carnation disease - rust, wilt and Alternaria spot	Prof.Dr .T.Glaser/Dr.B.Tatarynowicz Dr. M. Werner
20.2.7 Plant Physiology Department	
Pathophysiology of cultivated plants	Prof.Dr.Z.Krzywański/Dr.M.Czech
Application of growth substances to ornamental plants	Ass.Prof.Dr. R. Domański
Phytotoxicity of plant residues	Ass.Prof.Dr. D. Wojtkowiak Dr. B. Politycka
Water economy and water stress of plants	Ass.Prof.Dr.D.Zielińska/Dr.B.Szmal
Protein - metal complexes in plant organelles	Dr. A. Stroński
20.2.8 Plant Protection Methods Department	
Usefulness of new insecticides for protection of some leguminous plants	Prof.Dr. W. Romankow
Biology and chemical control of the chrysanthemum pests	Dr. T. Baranowski
20.2.9 Pomology Department	
General pomology, frost resistance, intensive methods of fruit production	Prof.Dr. T. Hołubowicz*
General pomology, growth substances, fruit cultivars	Ass.Prof.Dr. M. Ugolik
Diseases of small fruits and fruit trees, effects of cultural practices on healthiness of raspberry and strawberry, pomology	Ass.Prof.Dr. Z. Rebandel*
Weed control, frost resistance, irrigation, soil management	Dr. E. Pacholak
Frost resistance, rootstock-interstock relationship, blueberry	Dr. Z. Gruca
Peach cultivation, growth substances in callus tissue culture	Dr. B. Radajewska
Frost resistance of black currants and strawberries, physiology of frost damaged trees	Z. Skowroński M.Sc.
Small fruits, irrigation, soil management	J. Mazur M.Sc.
20.2.10 Seed Science and Nursery Department	
Genetics, ornamental plant breeding	Ass.Prof.Dr. H. Mackiewicz

Fruit trees - breeding and nursery production — W. Legutko M.Sc./ St.Schmidt M.Sc.
Dendrology — Dr. M. Maćkowiak*
Seed physiology — Prof.Dr. B. Sękowski
Prof.Dr. K. Duczmal
R. Hołubowicz M.Sc.
Seed pathology — Dr. K. Tylkowska
Vegetable and flower seed production — Dr.H.Tucholska/Dr.A.Muszyński
Ornamental trees nursery production — S. Korszun M.Sc./ H.Malinowska M.Sc.
E. Borejsza-Wysocka M.Sc.

20.2.11 Vegetable Crops Department — Prof.Dr. T. Pudelski*
Technology of vegetable production under protection, growing media — E. Michalska M.Sc.
Mushroom growing — Ass.Prof. Dr. M. Gapinski*
Mushroom growing, growing media — Dr. M Ziombra
Mushroom growing, evaluation of mushroom varieties — K. Sobieralski M.Sc.
Spawn of Agaricus bisporus and other mushrooms — M. Siwulski M.Sc.
Vegetable field production, irrigation asparagus growing — Dr. M. Knafleski*/Dr. A. Biniek
E. Konys M.Sc.
Glasshouse construction and equipment — Dr. J. Piróg*

21 Prusy

21.1 Zakład Doświadczalny Instytutu Sadownictwa i Kwiaciarstwa (Experiment Station of the Research Institute of Pomology and Floriculture)
elevation: 170 m
rainfall : 450 mm

96-130 Głuchów, Prusy
Phone: 2852 Rawa Maz.
Telex: 884691
Dir.: J. Krzewiński M.Sc.

Variety studies, research in nursery problems and orchard soil management. Experiment orchard 120 ha
Pomology — Dr. A. Jackiewicz/ M. Potocka Eng.
Soil management and nutrition — D. Krzewińska M.Sc.
Plant protection — S. Piotrowski M.Sc./M. Zaggórska
Orchard and nursery mechanization — M. Wilczyński M.Sc.
Nursery production — Z. Zagórski M.Sc.

22 Przyborów

22.1 Warzywniczy Zakład Doświadczalny Instytutu Warzywnictwa (Experiment Station of the Research Institute of Vegetable Crops)

Przyborów, 39-217 Grabiny
Dir.: S. Kilian M.Sc.
Dep.Dir.: J. Flaga M.Sc.

Experiment and extension fields (259 ha), greenhouses (7,5 ha), plastichouses (5 ha)
Organization of vegetable production — K. Bartkowski M.Sc.
Deputy Dir. of research work — S. Kania M.Sc.
Extension — M. Kaczor M.Sc.
Plant breeding — M. Kisiel M.Sc.
Agrotechnique — S. Kubik M.Sc.
Pest and disease control — U. Moczulska M.Sc.
Agrotechnique under glass — J. Pośpiech M.Sc.

23 Pszczyna

23.1 Warzywniczy Zakład Doświadczalny Instytutu Warzywnictwa (Experiment Station of the Research Institute of Vegetable Crops)

43-200 Pszczyna, ul. Cieszyńska 1
Dir.: W. Kurek M.Sc.
Dep. Dir.: K. Czopek M.Sc.

Experimental and extension fields (439 ha) greenhouses (1,5 ha), plastichouses (2 ha).

Organization of vegetable production	K. Czopek M.Sc.
Deputy Dir. for research work	W. Zjawiony M.Sc.
Nematology	Dr. A. Baksik
Agrotechnique	M. Król M.Sc.
Experiments under plastic and glass	A. Polok M.Sc./ M Rozner M.Sc.
	A. Waliczek M.Sc.

24 Reguły

24.1 Warzywniczy Zakład Doświadczalny Instytutu Warzywnictwa (Experiment Station of the Research Institute of Vegetable Crops)

Reguły, 05-820 Piastów
Dir.: A. Kocerba M.Sc.

Experimental and extension fields (342 ha), greenhouses (0,7 ha), plastichouses (2 ha). Chicory production and hydroponic forcing.

Soil management and mineral nutrition	Dr. J. Duch/ L. Jędras M.Sc.
City refuse as organic fertilizer	E. Kostrzewa Eng.
Plant Protection	

25 Rekowo

25.1 Warzywniczy Zakład Doświadczalny Instytutu Warzywnictwa (Experiment Station of the Research Institute of Vegetable Crops)

Rekowo, 84-123 Połchowo
Dir.: S. Kamionka M.Sc.

Experimental and extension fields (602 ha)

Organization of vegetable production in muck soil	
Vegetable production on muck soil	T.Baranowkska M.Sc/S.Kamionka M.Sc.
Cabbage F_1 hybrid experiments	J. Kamionka M.Sc.
Chemical analysis, storage of vegetables	A. Skwiercz M.Sc.
Mechanization	
Nematology	

26 Siejnik

26.1 Sekcja Sadownicza Zakład Doświadczalnego Instytutu Zootechniki (Pomology Section of the Animal Husbandry Experimental Station)
elevation: 183 m
rainfall : 614 mm
sunshine : 1665 h

Siejnik, 19-400 Olecko
Phone: Olecko 22-21
Telex: 852555
Head: Dr. A. Korsak

Belongs to the Ministry of Agriculture.
Experimental orchard 40 ha.

General pomology, variety and rootstock studies, small fruits, winter injury of fruit trees in north eastern Poland	Dr. A. Korsak

27 Sinołęka

27.1 Zakład Doświadczalny Instytutu Sadownictwa i Kwiaciarstwa (Experiment Station of the Research Institute of Pomology and Floriculture)
elevation: 220 m
rainfall : 573 mm

Sinołęka, 08-122 Bojmie
Phone: 124-24, 124-28 Siedlce
Telex: 84669 INSAD
Dir.: M. Cegłowski Eng.

Of 332 ha 182 ha are planted with pomological plants - apples (156 ha), pear, cherries, plums, strawberries, raspberries and currants.	
Rootstocks and interstocks	M. Cegłowski Eng M. Lewandowska M.Sc.
Small fruit breeding	Dr. E. Niezborała
Small fruits	H.Rechnio M.Sc/ B.Niezborała M.Sc

28 Skierniewice

28.1 Instytut Sadownictwa i Kwiaciarstwa (Research Institute of Pomology and Floriculture) elevation: 108 m rainfall : 531 mm sunshine : 1705 h	ul. Pomologiczna 18, Skierniewice Phone: 34-21 Telex: 886659 INSAD Dir.: Prof.Dr. S.W. Zagaja*
Belongs to the Ministry of Agriculture Fruit breeding, fruit growing, control of pests and diseases, fruit storage and fruit suitability for freezing and processing, economics of fruit production, floriculture beekeeping	Dep.Dir.: Prof.Dr. E. Niemczyk* Ass.Prof. W. Skowronek Prof.Dr. R. Rudnicki* Dr. W. Malewski Scientific Secretary: Prof.Dr. Z. Soczek*
Equipment: laboratory buildings, 8 greenhouses (0,24 ha) cold storage, 14 experiment stations, 1800 ha of orchards and nurseries	
28.1.1 Department of Fruit Breeding and Plant Propagation	Head: Ass.Prof. A. Czynczyk*
Rootstocks, interstocks. Variety evaluation Genetics, Breeding	Prof.Dr. S.W. Zagaja*
Rootstocks, interstocks. Nursery production	Ass.Prof. Z. Grzyb P. Radwan-Pytlewski M.Sc.
Rootstocks and fruit tree breeding	T. Jakubowski M.Sc.*
Variety evaluation	Dr. W. Dzięcioł/B. Kowalik M.Sc. D. Kruczynska M.Sc/E. Rozpara M.Sc.
Mutation breeding	Dr. A. Przybyła/S. Pluta M.Sc.
Small fruit breeding	Dr. E. Zurawicz
Tissue culture	Dr. B. Machnik/Dr. T. Orlikowska*
28.1.2 Department of Plant Protection	Head: Prof.Dr. Z.W. Suski
Fruit plants pathology section	Ass.Prof.Dr. J. Cimanowski*
General phytopathology, powdery, mildew	
Bark and wood diseases of fruit trees	Dr. A. Bielenin
Fungus resistance against fungicides	Dr. H. Nowacka*
Fungal diseases of fruit trees	W. Goszczyński M.Sc. M. Olszak M.Sc./ M. Plich M.Sc. O. Salamon Eng.
Bacterial diseases of fruit trees	Dr. P. Sobiczewski/ J.Bogatko M.Sc. St. Berczyński Eng.
Virus diseases of fruit plants	Ass.Prof.Dr. W. Basak Dr.D.Sobczykiewicz*/Dr.B.Zawadzka* H. Antoszewska-Orlińska M.Sc.
Entomology Section. Orchard mites, taxonomy, ecology and control	Prof.Dr. Z.W. Suski
Small fruit insect pests ecology and control	Dr. B. Łabanowska
Tortricidae, taxonomy, ecology and control	Dr. M. Koslińska
Orchard pest control	Dr. T. Badowska-Czubik Dr. Z. Nowakowski/Dr. St. Smolarz A. Maciesiak M.Sc.

Section of biological control of insect and mite pests.	
Predaceous, Heteroptera:	
ecology and taxonomy	Prof.Dr. E. Niemczyk*
Coccinellidae ecology and taxonomy	Dr. R. Olszak
Biological control of insect pest in the orchard	A. Szufa M.Sc
Applied ornithology	Dr. R. Zając
Rodents injuries to fruit and ornamental plants	K. Jaworska M.Sc.
Pesticide residues in fruits	Dr.E. Kępczyńska
28.1.3 Department of Fruit Processing	Head: Dr. W. Lenartowicz
Freezing, Suitability of fruit varieties for	Dr. W. Płocharski*
processing. Canning of fruits. Fruit quality.	J. Zbroszczyk M.Sc.
Freezing of fruits.	
28.1.4 Department of Fruit Storage	
Controlled atmosphere. Fruit storage technology	Head: Dr. E. Lange*
Fruit storage fungal disorders	Ass.Prof. H. Borecka
Orchard and climatic factors in fruit storage	Dr. I. Guzewska
Controlled atmosphere, fruit storage technology	M. Fica M.Sc./Dr. J. Nowacki
28.1.5 Department of Orchard Agrotechnique	Head: Ass.Prof. A. Mika*
Soil science, soil management, watering,	
nutrition	
Plant physiology, growth regulators	Prof.Dr. M. Grochowska*
	J. Maksymiuk M.Sc./ H.Morgaś M.Sc.
Pruning and training of trees, high density orchards	Ass.Prof. A. Mika*/Dr. S. Tymoszuk
Mineral nutrition	Prof.Dr.K. Słowik/Dr.B. Gutyńska
	T. Wąsowski M.Sc./G.Leszczynska M.Sc.
Application of growth regulators	Prof.Dr. Z. Soczek*/Dr. A. Basak
Herbicides	M. Bryk M.Sc.
Management and recultivation	P. Ligocki M.Sc.
Irrigation	W. Treder M.Sc.
Chemistry and nutrition	T. Olszewski M.Sc.
Laboratory of Biochemistry and Radioisotopes	Dr. L. Michalczuk
Growth regulator metabolism. Nutrient accumulation	Dr. E. Lis
Phytosenthesis and distribution of assimilates	
Technique of isotope application	Dr. L. Michalczuk
Phytosenthesis and distribution of assimilates	Dr. U. Dzięcioł
Grafting physiology	
28.1.6 Department of Orchard Mechanization	
Fruit harvesting, grading and packing. Work	
organization	Head: Ass.Prof. Z. Cianciara
Plant protection techniques	Ass.Prof. B. Bera*
	G. Doruchowski M.Sc./ S.Czuba M.Sc.
	S. Popkowicz Eng.
Technique of liquid preparation for plant protection	T. Wojniakiewicz M.Sc.
Techniques of application	Dr. A. Godyń
Fruit harvesting	H. Bożałek M.Sc./ T.Horbal M.Sc.
	P. Wawrzynczak M.Sc.
Pruning mechanization	K. Jagielski M.Sc./ D. Dudek M.Sc.
Horticultural machine design	Z. Salamon M.Sc./ Z.Michalak M.Sc.
	S. Placek M.Sc.
Testing machine with electronic equipment	J. Keller Eng.
Apple harvesting, grading and picking	R. Hołownicki M.Sc.
	J. Rabcewicz M.Sc.
Economics	Dr. P. Zaprzalek*/ W. Zwierz M.Sc.
	Dr. A. Pidek/ L. Zabtocka M.Sc.
28.1.7 Department of Small Fruits	
Variety evaluation. Blueberry, cranberries	Head: Dr. K. Smolarz
Herbicides	Dr. T. Cianciara

Variety evaluation. Breeding	Dr. J. Gwozdecki/Dr. B. Słowik
Mineral nutrition pruning	G. Cieśliński M.Sc.
Grapes	J. Szpunar M.Sc.*
28.1.8 Department of Beekeeping	ul. Kazimierska 2, 24-100 Pulawy
Beekeeping, economics, taxonomy	Ass.Head: Prof. W. Skowronek
Poliination, plant secretion	Prof.Dr. B. Jabłonski
Genetic and breeding	Ass.Prof. M. Gromisz
	L. Koškaa Eng./ J. Wojcik M.Sc.
Bumble-bees and their forage sources	Ass.Prof. A. Ruszkowski
Appiary management	Dr. C. Zmarlicki/ Z. Prychodzien M.Sc.
Bee physiology, queen insemination	Dr. Z. Konopacka
Bee physiology. Nutrition	Dr. J. Muszyńska
Bumble-bee rearing	Dr. M. Biliński
Bee hormones, management	Ass.Prof. W. Skowronek
Mechanization and economics of bee culture	M. Rybak M.Sc.
Appiary management	J. Marcinkowski M.Sc.
Bee product chemistry	Dr. H. Rybak
Plant nectar production	J. Skowronek M.Sc.
28.1.9 Department of Plant Physiology	Head: Dr. H. Plich
Plant physiology. Growth regulators in fruit bearing	Dr. H. Plich/ J.Sobolewska M.Sc.
Growth regulators. Dormancy of fruit trees	Ass.Prof. B. Borkowska*
Growth regulators. Frost resistance	B. Zrały M.Sc.
28.1.10 Department of Extension	
Extension service. Plant protection	Head: Dr. A. Holewiński*
Extension service. Plant protection	Ass.Prof. J. Włodek/ B.Iwanek MSc
	E. Jesiotr M.Sc./ W.Karolczak M.Sc.
Experimental design. Statistical methods	Prof.Dr. K. Szczepański*
	Dr. S. Rejman
Computer programming	J. Łys M.Sc./ J.Zwierz M.Sc.
	Cz. Smolarek Eng.
Library	E. Marcinkowska M.Sc.
	Z. Zurawicz M.Sc.
Publications	J. Suska M.Sc./ K. Durska M.Sc.
	T. Ligocka M.Sc./ T.Figiel Eng.
Documentation	A. Wojniakiewicz M.Sc.
	E. Strojna M.Sc./ M.Sobiczewska MSc
Research planning	E. Czynczyk Eng./ M. Gwozdecka Eng.
	M. Matyjaszczyk Eng.
28.1.11 Horticulture Research Division	Dir.: Prof.Dr. R.M. Rudnicki*
Equipment: experimental fields (10 ha), experimental greenhouses and plastic tunnels (0.3 ha), laboratories	
28.1.12 Department of Plant Breeding and Seed Production	Head: Ass.Prof. K. Mynett*
	Dr. B. Grabowska
Breeding of Alstromeria, Dahlia, Gerbera, Gladiolus, Iris, Lilium, Narcissus and Tulipa	Dr. J. Kepczyński/ M.Osiecki M.Sc.
	A. Wilkońska Eng.
	A. Pobudkiewicz M.Sc.
28.1.13 Department of Cultivation	Head:Dr.H.Wisniewska-Grzeszkiewicz
Cold storage and cuttings	R. Dobrowolska M.Sc.
Chemical weed control	J. Maszkowska M.Sc.
Propagation of various pot plants	T. Hempel M.Sc./ J. Rak M.Sc.
Plant nutrition and media cultivation methods of carnations, chrysanthemums, alstromerias, gerberas and roses	M. Jagusz M.Sc./ W.Kowalczyk M.Sc.
	E. Michalak M.Sc/Cz.Pytlewski M.Sc
	Dr. Z. Strojny/ B. Matysiak M.Sc.
28.1.14 Department of Ornamental Nursery Crops	Head: Dr. J. Marcinkowski
Generative propagation	Dr. M. Grzesik/Dr. W. Kamiński

Vegetative propagation	A. Lenartowicz M.Sc.
Container culture	M. Robak M.Sc.
Herbaceous perennials-variety and propagation	J. Tracz M.Sc.
Growth regulators	
28.1.15 Department of Plant Protection	
Laboratory of Entomology of Ornamental Plants	Head: Dr. G. Łabanowski
Nematodes control. Control of shrubs and tree pests in greenhouses	W. Bogatko M.Sc./ E. Pala Eng.
	M. Wojtowicz M.Sc.
28.1.16 Department of Plant Protection	
Laboratory of Phytopathology of Ornamental Plants	Head: Ass.Prof. L.B. Orlikowski*
Etiology and control of soil-borne pathogens	Ass.Prof. M Kamińska
Control of bulbous plants diseases	A. Saniewska Eng.
Registration and identification of various diseases	Cz. Skrzypczak M.Sc.
Epitemiology and control of bacterial diseases	A. Wojdyła M.Sc./J.Bogatko M.Sc.
28.1.17 Department of Culture and Storage of Floricultural Plants	Head: Dr. M. Hempel
Laboratory of tissue culture	K. Cywinska-Smoter M.Sc.
	E. Gabryszewska M.Sc.
	M. Podwyszynska M.Sc.
	U. Soczek M.Sc./ J.Tymoszuk M.Sc.
Laboratory storage of horticultural plants	Head: Ass.Prof.Dr. M. Saniewski*
	L. Banasik M.Sc.
	Dr. D. Goszczyńska*/ Dr. J. Halaba
	B. Michalczuk M.Sc.
	ASss.Prof. J. Nowak
28.1.18 Department of Extension	Z. Miziołek M.Sc./ Z. Rolewska M.Sc.
Extension Service	E. Taczanowska Eng.
28.2 Instytut Warzywnictwa (Research Institute of Vegetable Crops)	ul.22 Lipca 1/3 96-100 Skierniewice Phone: 22-11, Telex: 886444 Dir.: Ass.Prof. J. Skierkowski* Dep.Dir.: J. Izdebski M.Sc. Scient.Secr.: Dr. K. Viscardi

Belongs to the Ministry of Agriculture. Vegetable breeding, control of pests and diseases, mechanization, storage, nutrition, processing and freezing, economics, physiology and biology, agrotechnique, seed biology.
Experimental fields (60 ha) experimental greenhouses and plastic tunnels (3 ha), laboratories: mashroom laboratory, cold storage, 10 experimental stations (3605 ha), greenhouses (21 ha) and 10 ha plastic tunnels.

28.2.1 Department of Vegetable Plant Breeding	
F_1 hybrids of cucumbers, tomatoes, spinach	Head: Ass.Prof. I. Paszkowska
Cabbage, radish, kohlrabi	J. Borowiak M.Sc/ E. Horodecka M.Sc.
	Dr. I. Wasilewska
	B. Kuszlejko M.Sc/ M. Hałas M.Sc.
F_1 hybrids of onions, monogerm beets, cauliflower	Ass.Prof. R.W. Doruchowski*
F_1 hybrids io early cabbage and cauliflower	Ass.Prof. J. Hoser-Krause
Varieties and F_1 hybrids of greenhouse cucumbers	Dr. M. Rengwalska*
Field and greenhouse tomatoes	Dr. H. Potaczek
Onion disease resistance	T. Kotlińska M.Sc.
Tomato disease resistance	E. Kozik M.Sc/ A. Marianowski M.Sc.
28.2.2 Department of Seed Production	
Pollination entomophilous vegetable seed production	Head: Ass.Prof. H. Woyke*
Mineral nutrition in seed cultures	A. Sokołowska M.Sc.
Method of production F_1 tomatoes hybrids	B. Pinchinat M.Sc.
F_1 hybrids of cabbage, cauliflower and onions	M. Dudek M.Sc.
Carrot seed production	A. Szafirowska M.Sc.

Seed borne diseases	Dr. S. Czyżewska*
Seed vigor and emergence	
28.2.3 Department of Outdoor Vegetables Production	
General problems in outdoor vegetable production	Head: Dr. J. Rumpel*
Irrigation	Dr. S. Kaniszewski
Variety trials	I. Babik M.Sc./Dr. J. Robak
	A. Grajewska-Wieczorek M.Sc.
	K. Grudzień M.Sc.
	K. Felczyński M.Sc./ E.Sikora M.Sc.
28.2.4 Department of Weed Control	
General problems on the selectivity and phytotoxity of herbicides	Head: Ass.Prof. A. Dobrzański*
	Z. Anyszka M.Sc/ J.Palczyński M.Sc
28.2.5 Department of Glasshouse Vegetable Production	Head: M. Owczarek M.Sc*
Method of young plant production	Dr. J. Dobrzańska*
Vegetables under plastics and glass	Dr. C. Slusarski/ W.Rozpara M.Sc.
	T. Glapś M.Sc./ J.Babik M.Sc.
	J. Dyśko M.Sc.
28.2.6 Department of Nutrition	Head: Prof.Dr. O. Nowosielski*
General problems of vegetable crops nutrition	Dr. B. Szmidt*/Dr. E. Szwonek
	A. Bereśniewicz M.Sc.
Organic and mineral nutrition	H. Wiśniewska M.Sc./ M.Mijas M.Sc.
28.2.7 Department of Plant Protection	
Pest control	Prof.Dr. J. Narkiewicz-Jodko*
Nematodes	Prof.Dr. M. Brzeski
Fungus diseases	Dr. W. Rondomański*/Dr. M. Macias
Aphid biology and control	Dr. J. Szwejda/Dr. S. Kotliński
Phytopathology	J. Sobolewski M.Sc.
28.2.8 Department of Mechanization	
Mechanization in vegetable crop growing and harvesting	Head: A. Krokowski M.Sc.
Mechanization in vegetable crop growing and harvesting	H. Charzewska M.Sc./ D. Viscardi M.Sc.
	B. Kuc M.Sc./ P. Murczek M.Sc.
	K. Michałowski M.Sc.
28.2.9 Department of Vegetable Storage	Head: Dr. F. Adamicki*
General problems of vegetable storage	Dr. L. Umięcka
Prepacking, short-term storage	
Onion storage	M. Perłowska M.Sc.
Storage of vegetables in controlled atmosphere	M. Grzegorzewska M.Sc.
28.2.10 Department of Vegetable Processing and Freezing	
Chemical composition of vegetables	Head:Dr.J. Bąkowski/Dr.H. Michalik
Problems in cucumber fermentation	Dr. K. Elkner
Pesticide residue in vegetables	M. Horbowicz M.Sc.
Biochemistry	Dr. J. Czapski
Proteins	Dr. R. Kosson/ W. Dobrzanski M.Sc.
28.2.11 Department of Economics and Organization of Production	Al. Jerozolimskie 55, 00-697 Warszawa
Economics of vegetable crops	Head: Ass.Prof. J. Szklarska
	Dr.K. Ozarowska/ Dr.M.Przedpełska
	B.Głogowska MSc/ J.Mierwinska MSc
Cost calculation in vegetable production	Dr. J. Reinschmidt/ A.Gembicka M.Sc
	Dr. W. Cieślak-Wojtaszek
	B. Bibro M.Sc./ D. Domanski M.Scv.
Organization of horticultural production	Z. Barańska M.Sc.*
	G. Sokołowski M.Sc.
Mathematical methods in horticulture	A. Radzikowska M.Sc./ H. Pielat Msc
28.2.12 Department of Biology	
Tissue cultures and growth substances	Head: Dr. K. Górecki

Physiological diseases	Ass.Prof. J. Borkowski* Ass.Prof. K. Smierzchalska Dr. K. Górecka/ M.Habdas M.Sc.
Plant anatomy	B. Dyki M.Sc.
Biochemistry food irradition	Dr. J. Ostrzycka E. Wojniakiewicz M.Sc.
28.2.13 Department of Extension Service	Head: Dr. M. Plucinska
Editorial secretary	B. Narkiewicz M.Sc.
Extension	E. Sikora M.Sc./ E.Komorowska M.Sc. A. Podczaska M.Sc.
Research Documentation	R. Rumpel Eng.
Librarian	G. Barańska M.Sc./ L.Krokowska
28.2.14 Department of Planning and Organization	
Planning and organization of research	Head: M Skierkowska M.Sc.
Section of Foreign Relations	J. Brzeska M.Sc./W.Zygadlewicz M.Sc.

28.3 Warzywniczy Zakład Doświadczalny Instytutu Warzywnictwa (Experiment Station of the Research Institute of Vegetable Crops)

ul. Sobieskiego 16
96-100 Skierniewice
Dir.: J. Melon M.Sc.

Experimental and extension fields (230 ha) greenhouses (0,4 ha), plastichouses (1 ha). Organization of vegetable production. Research work organization	M. Romanowska M.Sc. Z. Różański M.Sc. S. Trzak

28.4 Szklarniowy Zakład Doświadczalny Instytutu Warzywnictwa (Greenhouse Experiment Station of the Research Institute of Vegetable Crops)

ul. Sobieskiego 89
96-100 Skierniewice
Dir. M. Owczarek M.Sc.*

Experimental and extension fields (12 ha) greenhouses (2,6 ha)	M. Owczarek M.Sc.*
Organization of production under glass	J. Brzozowska M.Sc.
Agrotechnique and extension	B. Ciereszko M.Sc./G. Kapusta M.Sc.
Agrotechnique under glass	B. Koceba Eng. A. Małachowski Eng/ F.Łuczak M.Sc. E. Pisulewska Eng/T. Pisulewski Eng K. Salamończyk M.Sc. B. Sikorska M.Sc.

28.5 Zakład Doświadczalny-Produkcyjny Grzybow Jadalnych Instytutu Warzywnictwa (Experiment Station of Edible Fungi, of the Research Institute of Vegetable Crops)

ul. Sobieskiego 20
96-100 Skierniewice
Dir.: K. Szudyga M.Sc.
Dep.Dir.: M. Stamborowski M.Sc.

Experimental and productive mushroom plant 1200 tons per year of champignons	J. Siański M.Sc/ K. Laudan M.Sc.
Method of mushroom growing	E. Kozłowska M.Sc.
Mushroom mycology	E. Kołodziejek M.Sc.
Mushroom production	Dr. J. Szymański
E. Nagórska M.Sc.	Dr. J. Maszkiewicz

29 Słupia Wielka

29.1 Sekcja Ogrodnicza Centralnej Stacji Doświadczalnej Oceny Odmian (Central Experiment Station for Variety Evaluation)

Słupia Wielka, 63-000 Sroda Wielkp.
Head: P. Rudy Eng.

Belongs to the Ministry of Agriculture. Evaluation of productivity, preparation of description, control of purity and identify of cultivated varieties. The Station is conducting this work in several stations all over Poland, under different climates and soil conditions. The results of the experiments are the basis for inclusion in the "State Register of Original Plant Varieties" and in the "List of Selected Varieties" of the varieties introduced by Polish breeding stations.

Variety evaluation of vegetable crops — T. Feliński M.Sc.
Variety evaluation of ornamental plants — T. Hrycak M.Sc.

30 Świerklaniec

30.1 Zakład Doświadczalny Instytutu Sadownictwa i Kwiaciarstwa (Experiment Station of the Research Institute of Pomology and Floriculture)

42-622 Świerklaniec
Phone: 85-22-77
Dir.: T. Piskor M.Sc.

Small fruit variety studies, orchard soil management, nursery, pest and disease control. Experimental orchard (240 ha).

General pomology. Disease control — K. Bystydzieńska M.Sc.
Pest control — B. Cisek M.Sc./ B. Kobiela
Orchard mechanization
General pomology — M. Lisowski M.Sc./ E.Piskor M.Sc.
Fruit storage — S. Saba M.Sc./ J. Sikora
Small fruits

31 Szczecin

31.1 Instytut Ogrodnictwa Akademii Rolniczej (Institute of Horticulture of the Agricultural University)

ul. Janosika 8, 70-965 Szczecin
Dir.: Ass.Prof. M. Orłowski

31.1.1 Section of Pomology

General pomology, fruit physiology and fruit storage

Head: Prof.Dr. W. Ostrowski
Dr. K. Ostrowski
P. Chełpiński M.Sc.

31.1.2 Section of Vegetable Crops

Field vegetables. Weed control. Seed production

Head: AssProf. M. Orłowski
E. Rekowska M.Sc.
R. Dobromilska M.Sc.
M. Abramowicz M.Sc.

32 Warszawa

32.1 Wydział Ogrodniczy, Szkoła Główna Gospodarstwa Wiejskiego-Akademia Rolnicza (Faculty of Horticulture, Warsaw Agricultural University)

Nowoursynowska 166
02-766 Warszawa
Dean: Assoc.Prof.Dr. S. Kryczyński

32.1.1 Katedra Sadownictwa (Department of Pomology)

Head: Prof.Dr. S. Sadowski

Breeding and cultivar testing of apples and blueberries; rootstock selection for stone fruits; propagation of blueberries, stone fruits and apples;

mineral nutrition of sour cherries; apples, red currants and lingonberries; growth retardants in apples and pears; factors affecting storage quality of apples

Prof.emer.Dr. A. Rejman*
Ass.Prof.Dr. F. Jaumień*
Dr. J. Ciesielska-Piper
Dr. E. Jadczuk
Dr. F. Kempski/Dr. E. Pitera*
Dr. K. Pliszka*/Dr. K. Ścibisz*
K. Tomala M.Sc/A. Leszczyński M.Sc
J. Maciak/ J. Zuchowska M.Sc.

32.1.2 Katedra Warzywnictwa i Roślin Leczniczych (Dep. of Vegetable Crops and Medicinal Plants)

Head: Prof.Dr. H. Skąpski*

32.1.2.1 Zakład Warzywnictwa (Section of Vegetable Crops)

"Hydro-peat" method of greenhouse tomato and cucumber

Ass.Prof.Dr. W. Gosiewski
Dr. B. Dąbrowska

production; energy-saving methods of greenhouse crop production; cultivar testing of tomatoes, cauliflowers and multiflora bean; methods of culture and seed production of Polish tomato cvs; internal browning of Brussels sprouts; methods of growing of greenhouse tomato cvs in open field

Dr. S. Gawroński/Dr. G. Janowski
Dr. J. Kobryń/Dr. A. Lewandowska
Dr. Z. Lipiński/ M.Podkoński M.Sc
G. Jaroszewicz M.Sc.

32.1.2.2 Zakład Roslin Leczniczych i Specjalnych (Section of Medicinal and Spice Plants)

Chemical composition and processing of spice plants; variability in populations of Matricaria chamomilla, Papaver somniferum and Patinaca sativa. Application of growth regulators in Atropa belladonna, Datura innoxia, Capsicum annuum, Solanum laciniatum. Seed production and vegetative propagation of selected medicinal plants.

Head: Prof.Dr. K. Karwowska
Prof.emer.Dr. A. Rumińska*
Ass.Prof.Dr. K. Suchorska
Dr. Z. Węglarz

32.1.3 Katedra Roślin Ozdobnych (Department of Ornamental Plants)

Flower morphogenesis of bulbous, cormous and rhizome plants; physiology of flowering of cormous plants; prolongation of vaselife of cut flowers; chemotaxonomy; breeding of pelargonium, peony and iris. Vegetative propagation of selected herbaceous plants, woody ornamental trees and shrubs; micropropagation; methods of growing of woody ornamentals in containers; frost resistance of perennials and woody ornamentals; reproduction and forcing of bulbous plants; herbicides in nursery and in seed production of ornamentals.

Head: Ass.Prof.Dr. H. Chmiel
Ass.Prof.Dr. W. Szlachetka*
Ass.Prof.Dr. W. Seneta
Dr. B. Chlebowski/Dr. J. Gołos
Dr. A. Łukaszewska
Dr. Sz. Marcyński/Dr. J. Rokosza
Dr. M. Solecka/Dr. J. Tonecki
Dr. M. Witomska
W. Zelechowski M.Sc.

32.1.4 Katedra Uprawy Roli i Nawożenia Roślin Ogrodniczych (Department of Soil Management and Fertilization of Horticultural Crops)

Nutrient film technique for tomatoes; pine bark as substrate for vegetable and ornamental plants; urban and industrial wastes as organic fertilizers; fertilization of gerbera with microelements; foliar nutrition of cuttings of woody ornamentals

Head: Prof.Dr. J.R. Starck
Ass.Prof.Dr. A. Kropisz*
Dr. I. Kacperska
Dr. W. Oswięcimski
Dr. J. Stojanowska
D. Przeradzki M.Sc.

32.1.5 Katedra Organizacji i Ekonomiki Ogrodnictwa (Department of Horticultural Economics)

Consumption and prices of fruits, vegetables and flowers; salaries and prices of means of production in horticulture; export of horticultural products; economics and organization of horticultural production; horticultural farm management; economics of landscape architecture

Head: Ass.Prof.Dr. T. Dąbrowski*
Prof.emer.Dr. N. Krusze*
Ass.Prof.Dr. W. Ciechomski*
Dr. E. Matejek/Dr. L. Jabłonska
E. Połaski M.Sc./ K.Gorecka M.Sc.

32.1.6 Katedra Fitopatologii (Dep. of Plant Pathology)

Identification and properties of plant viruses and viroids; Virus and viroid diseases of potato, chrysanthemum and soybean; and Phytophora spp.; resistance of apple cvs to bark cancer fungi; susceptibility of peach cvs and seedlings to Cytospora cancer; diseases of vegetable crops caused by fungi, breeding of vegetables for diseases resistance; fungous diseases of ornamental plants; mycytoxins and ochratoxins

Head: Prof.Dr. Z. Borecki
Prof.emer.Dr. J. Kochman
Ass.Prof.Dr. S. Kryczyński
Ass.Prof.Dr. J. Chełkowski
Dr. C. Zamorski/Dr. B. Leski
Dr. J. Bednarek/Dr. R. Dzięcioł
Dr. J Marcinkowska
Ass.Prof.Dr. B. Nowicki
Dr. M. Schollenberger
M. Szyndel M.Sc./ L. Wakulinski MSc

32.1.7 Katedra Entomologii Stosowanej (Department of Applied Entomology)

Head: Prof.Dr. J. Boczek
Prof.Dr. J. Dmoch
Prof.Dr. Z.,T. Dąbrowski

Factors affecting the resistance of small grains to aphids; ecology of aphids in horticultural crops; ecology of urban communities of plant feeding and predatory mites; physiology of plants infested by spider mites; biological bases for control of cruciferous pests and of carrots; energy budget of stored product mites; chemical messengers of insects on cabbage plants; economics of protection of greenhouse crops	Ass.Prof.Dr. D. Kropczyńska Ass.Prof.Dr. E. Cichocka Dr. W. Goszczyński Dr. S. Ignatowicz Dr. M Kozłowski Dr. A. Tomczyk A. Starzyński M.Sc.
32.1.8 Katedra Genetyki i Hodowli Roślin Ogrodniczych (Dep. of Genetics and Breeding of Horticultural Plants)	Head: Ass.Prof.Dr. S. Malepszy Ass.Prof.Dr. J. Pala
Determination and evolution of sex in higher plants; Heterosis and its use in plant breeding; genetics, breeding and seed production of tomato, Cucurbitaceae, spinach, strawberry and rye; methods of culture in vitro for obtaining new cultivated plants; biochemical and genetical analysis of variability in proteins and enzymatic systems of lines of rye and cucumber; intergenetic hybridization between Fragaria and Potentilla species	Dr. K. Niemirowicz-Sczytt Dr. Z. Przybecki M. Śmiech M.Sc. M. Ratynski M.Sc.
32.1.9 Zakład Praktyk Ogrodniczych i Upowszechnienia (Section of Practical Training and Extension Service in Horticulture)	Head: Dr. T. Pyzik
Practical training of students in horticulture, extension of research achievements of the faculty.	
Vegetable crops	Dr. T. Pyzik/J. Gajc
Plant Pathology	Dr. K. Szopa
Pomology	J. Marciak
Genetics and Plant Breeding	L. Drużyński
Ornamental Plants	W. Wałęza/ A. Tytkowski
Medicinal Plants	Dr. P. Metera
32.1.10 Pracownia Roślin Antyrakowych (Laboratory of Anticancer Plants)	Head: Ass.Prof.Dr. A. Sadowska*
Selection and propagation of Catharanthus roseus; determination of optimum harvesting date; application of growth regulators; methods of culture under plastic and in open field; poliploidyzation by means of colchinine	M. Racka M.Sc.
32.1.11 Katedra Projektowania w Architekturze Krajobrazu (Dep. of Landscape Planning and Design)	Head: Prof.Dr. E. Bartman Prof.emer.Dr. W. Niemirski
Information system for landscape planning; radio-aesthetic survey of the influence of plants upon human organism; socionatural criteria of determination, furnishing and design of recreational areas; development of green area system and structure in cities and suburbs; methodology of landscape management in the industrial buffer zone	Dr. M. Kiciński Dr. S. Rutkowski Dr. J. Rylke/Dr. P. Wolski Dr. Z. Sobotkowski/ J.Jaszczak M.Sc. Dr. J. Skalski M.Sc. M. Szumański M.Sc. D. Dobiech MSC/ M. Pawłowska MSc G. Sokołowska
32.1.12 Katedra Urzadzania i Pielegnowania Krajobrazu (Dep. of Landscape Construction and Maintainance)	Head: Ass.Prof.Dr. L Majdecki Ass.Prof.Dr. M. Siewniak
Green area system; open landscape in relation to health and recreation needs of urban population; improvement of construction and maintainance methods for urban and open landscape; preservation and revalorization of historical gardens according to contemporary	Dr. M. Mędrzycki Dr. M. Kosmala Dr. J. Wojtatowicz P. Głowacki A. Haber M.Sc./ M. Kulus M.Sc.

expectations; land improvement in industrial areas — Z. Suski M.Sc./ K. Piechna M.Sc.

32.1.13 Katedra Ochrony Środowiska
(Department of Environment Protection)
Air pollution by industry with special reference to petroleum industry; urban ecology in relation to its biocenotic degradation; landscape protection with the use of traditional preservation techniques and modern methods in landscape parks and protected landscape areas

Head: Prof.Dr. H. Zimny
Ass.Prof.Dr. W. Nowakowski
As s.Prof.Dr. J. Janecki
Dr. C. Wysocki/ Dr. L. Indeka
E. Sawczuk M.Sc.
M. Zakrzewska M.Sc.

32.2 Wydzial Rolniczy, Szkoła Główna Gospodarswa Wiejskiego Akademia Rolnicza (Faculty of Agronomy, Warsaw Agricultural University)

Rakowiecka 26/30
02-528 Warszawa
Dean: Prof.Dr. Z. Brogowski

32.2.1 Katedra Gleboznawstwa (Dep. of Soil Science)
Chemical and physico-chemical properties of orchards soils as affected by fertilization; effects of salinity in green urban areas and means of recultivation of salinized soils along the streets

Head: Ass.Prof.Dr. Czerwiński
Ass.Prof.Dr. M. Kępka
T. Kozanecka M.Sc.

32.3 Centralny Ośrodek Badawczo-Rozwojowy (Research and Development Center of Horticultural Market Economy and Technology)

ul. Chodakowska 53/57
03-816 Warszaw
Phone: 10-62-88
Dir.: Ass.Prof. S. Rosowski

Economics and organization of horticultural products marketing. Technology of horticultural products storage and processing. Economics and organization of cooperatives. Main economic problems of horticultural production. Economic policy in horticulture

32.3.1 Economic and organization of marketing
Costs, prices and profits in marketing and storage
Harvest forecasting and market deliveries of fruits and vegetables

Ass.Prof. K. Kubiak
D. Prus-Głowacka M.Sc.
M. Czckanski M.Sc.

32.3.2 Economics and organization of horticultural cooperatives Optimalization of organized structure and management
Work organization

Dr. R. Marynowski/ D.Sikora M.Sc.
A. Pociecha M.Sc./ W.Petlic M.Sc.
A. Okreglicki M.Sc.

32.3.3 Horticultural production development
Allocation costs and profitability of fruits and vegetable production
Tax policies

Ass.Prof. S. Rosowski
J. Świetlik M.Sc./ J. Nowak M.Sc.
D. Migalska M.Sc.
M. Kościelak MSc/ G.Szumańska MSc
E. Anuszewska Eng/ A.Figura M.Sc.
J. Felski M.Sc./ J. Sysiak M.Sc.
A. Chojnowksa M.Sc.

32.3.4 Marketing and storage technology of horticultural procuts

Dr. A. Golisz/Dr. T. Cąderek
M. Gajewski M.Sc./ Z.Kosmalski M.Sc
H. Wardzyńska M.Sc./ E. Kozikowska MSc

32.3.5 Processing technology of horticultural products

M. Szemiel MSc/ A. Krajewski MSc
M. Ochorowicz M.Sc.

32.3.6 Engineering development in marketing of horticultural products

J. Baran Eng./ D. Maciszewski M.Sc.
M. Strzeszewski Eng.

33 Wrocław

33.1 Zakład Ogrodnictwa Akademia Rolnicza (Department of Horticulture, Agricultural University)
elevation: 120 m
rainfall : 592 mm
sunshine : 1848 h

ul. Cybulskiego 39,50-205 Wrocław
Phone: 22-25-63
Telex: 0715327 AR W PL

Teaching and research in horticulture — Head: Ass.Prof.Dr. E. Kołota; A. Szewczuk M.Sc.
Cultivation, mineral nutrition, irrigation, pomology, variety studies. Experimental orchard (20 ha)
Cultivation, irrigation, nutrition of vegetable, greenhouse tomatoes. Experimental fields (30 ha) — Ass.Prof.Dr. E. Kołota; Dr. M. Osińska/ A. Wojłkiewicz; A. Biesiada M.Sc./ A. Szewczuk MSc; J. Krèzel M.Sc.

33.2 Katedra Entomologii Rolniczej, Akademia Rolnicza Dep. of Agriculture Entomology, Agricultural University)

ul. Cybulskiego 29,50-205 Wrocław
Head: Prof.Dr. C. Kania

Insect resistance in crops plants — Prof.Dr. C. Kania
Insects attacking bean, pea and broad bean — Ass.Prof.Dr. P. Niezgodzinski
Effect of insecticidal treatments on the environment — Ass.Prof.Dr. M. Goos
Side effect of herbicides on insects

33.3 Ogród Botaniczny Instytutu Botaniki i Biochemii Uniwersytetu Wrocławskiego (Botanical Garden of the Institute of Botany and Biochemistry of the Wrocław University)

ul. Sienkiewicza 23
50-335 Wrocław
Phone: 22-59-57
Head: Dr. T. Nowak

Tissue culture of decorative plants, orchid propagation — Prof. K. Kukułczanka*
Frost resistance of trees and shrubs — Dr. M Tokarski

34 Wróblowice

34.1 Sadowniczy Zakład Doświadczalny Instytutu Sadownictwa i Kwiaciarstwa (Experiment Station of the Research Institute of Pomology and Floriculture)
rainfall : 579 mm
sunshine: 1360 h

Wróblowice, 55-023 Lutynia
Phone: Wrocław 49-38-53
Dr.: Dr. J. Caputa

Pomology intensive methods of fruit production, stone and small fruits, nursery, pest and disease control. Experimental orchard (166 ha).
Soil management, plant protection — E. Bureć M.Sc.
Pomology, general pomology — J. Szszurek M.Sc.
Small fruits — Z. Hoffman M.Sc.
Orchard mechanization — J. Karczewski M.Sc./A. Rejman M.Sc.*

PORTUGAL

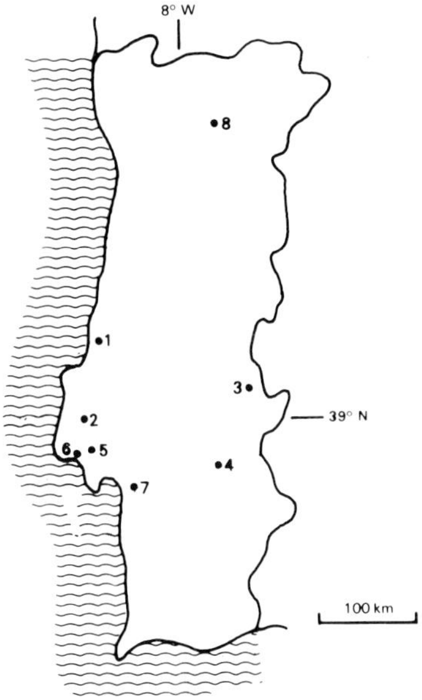
8° W
•8
•1
3•
•2
— 39° N
6• •5
•4
•7
100 km

1 Alcobaça
2 Dois Portos
3 Elvas
4 Evora
5 Lisboa
6 Oeiras
7 Setúbal
8 Vila Real

Survey

Two different public agencies are in charge of horticultural research in Portugal: the Ministry of Education and Science through the Universities and the Ministry of Agriculture and Fisheries through the National Institute for Agrarian Research. There is no tradition of research financied by private companies or even conducted by private agencies or enterprises.
Horticulture is an ancient agricultural subsector and a traditional job in Portugal but one must go back to the late century to find a scientific approach of the vegetable and fruit crop studies; in fact only viticulture began to be carefully studied at that time, the first scientific book being published in the beginning of the century (1900) by L.C. Costa with a detailed description of the grape varieties of the country.
During this century a special name must be mentioned: the research work done by the late Prof. J.V. Natividade on fruit and vegetable breeding and who founded the first national fruit crops research center at Alcobaca around 1950.
The contribution of the horticultural industry to the agricultural gross income in Portugal reaches about one third of the total production. 6 per cent is from potatoes and dry beans (100×10^3 ha.), and 5 per cent from the remaining vegetables (40×10^3 ha.), 9 per cent is the contribution of viticulture (380×10^3 ha.), 4 per cent comes from olives (750×10^3 ha.) and 9 per cent from other fruit crops (180×10^3 ha.). The ornamental production is quite new as an industry in the country.

Plastic tunnel greenhouses and plastic mulch row crops have developed from the early 70's mainly in the south and the total area is now near 3000 ha. (1100 ha. of greenhouses).
The most important horticultural processing industry is the tomato. Tomato paste and whole tomatoes. Canneries are modern with an up-to-date technology and Portugal as the third world exporter plays an important role in the world market of these products.

1 Alcobaça

1.1 Estação Nacional de Fruticultura Vieira Natividade
(National Research Station for Fruit Crops)
2460 Alcobaça
Phone: 062.42027

Prepared by Dr. A.A. Monteiro, Ph.D., Lisboa

elevation: 50 m
rainfall : 840 mm
sunshine : 2480 h

Dir.: Eng. I. Saraiva

It is the research station for pomology located in the center of a very important and traditional fruit industry area. The departments of olives and citrus are situated in Elvas and Setúbal (see 3.1 and 7.1)

Section of pommes and prunes

Propagation and cultural practices; apple breeding	Head: Eng. J.T. Ferreira Eng. O. Pereira

Section of Nuts

Climate adaptation; cultural practices	Head: Eng. I. Saraiva Eng. J. Gomes Pereira

Section of Miscealaneous research

Fertilization; leaf analysis	Head: Eng. Ma. Luisa Duarte
Fruit storage	Eng. D. Marques
Micropropagation	Eng. Teixeira de Sousa
Photosynthesis	Eng. Ma. do Céu Matos
Apricot citology	Eng. Clara Medeira
Integrated pest control	Eng. C. Matias

Section of growth regulators

Apple and pear physiology	Head: Eng. A. Avelar do Couto

2 Dois Portos

2.1 Estação Vitivinicola Nacional (National Research Station for Viticulture and Oenology)

Dois Portos, 2575 Runa
Phone: 061.72106
Dir.: Eng. A. Curvelo Garcia

This national research station includes oenology and viticulture research and is situated in one of the most important areas for table wine production. It has 13 ha of experimental grapeyards.

Pruning, training and varieties selection	Eng. P.C. Pereira
Morphology and ampelography	Eng. J.F. Dias
Physiology of growth and development	Eng. J. Serralheiro

3 Elvas

3.1 Estação Nacional de Fruticultura
Department of olive crops
elevation: 200 m
rainfall : 604 mm
sunshine : 3030 h

7350 Elvas
Phone: 068.62858

Bioclimatology; Vegetative propagation	Head: Eng. J. F. Serrano
Mechanical harvest	Eng. L. Santos

4 Evora

4.1 University of Evora
elevation: 75 m
rainfall : 764 mm
sunshine : 2960 h

Apartado 94
7001 Evora Codex
Phone: 066.25398

The university is situated in an important area for extensive production of cereals and cattle. It runs directly 3 large farms which are an important aspect of its integration in regional agriculture.

Department of crop science — Head: Prof.Dr. Dargent de Albuquerque

Viticulture — Dr. J. Araújo; Eng. J. Mota Barroso

Vegetables for processing; crucifera; melon — Eng. A. Calado*

5 Lisboa

5.1 Instituto Superior de Agronomia
elevation: 60 m
rainfall : 694 mm
sunshine : 2790 h

Tapada da Ajuda
1399 Lisboa Codex
Phone: 01. 638161/2

It is the College of Agriculture and Forestry of the Technical University of Lisbon, located in a 100 ha park.

Section of horticulture
Potatoes; vegetables for processing — Chairman: Prof.Dr. C. Portas*
Protected cultivation of vegetables — Dr. A.A. Monteiro*
Fruit growing — Dr. J.M. Matos Silva
Fruit pruning and training — Dr. R. de Castro; Eng. Cristina Oliveira
Viticulture — Eng. A. Costa

6 Oeiras

6.1 Estação Agronomica Nacional
elevation: 50 m
rainfall : 690 mm
sunshine : 2790 h

Quinta do Marquês
2780 Oeiras
Phone: 01.2430442
Dir.: Dr. C. Cardoso

This is the main research station of I.N.I.A. (National Institute for Agrarian Research) which belongs to the Ministry of Agriculture. It includes not only horticulture but many different aspects of basic and applied agriculture research.

Department of Genetics and plant breeding — Head: Eng. Viana e Silva

It deals with various genetical aspects and breeding of field and horticultural crops.

Section of olive breeding, morphological and physiological study of cultivars; fat content of fruits; rooting of herbaceous cuttings — Head: Eng. F. Leitão; Eng. Ma. de Fátima Potes; Eng. Ma. L. Calado

Department of Phytopathology

General aspects of basic and crop phytopathology including some pests and diseases of horticultural crops.

Virus diseases of citrus, grapes and vegetables for processing — Head: Prof.Dr. O. Sequeira; Dr. J.C. Sequeira; Eng. Amarilis Mendonça; Lic. Ma. de Lurdes Borges; Lic. Ma. Eugénia Pinto; Eng. Rosália Freitas

Fungi and bacteria diseases of vegetables, fruittrees and grapes	Dr. J.F. Ferraz Lic. J.M. Martins Eng. Leopoldina Silva Eng. A.M. Fernandes Lic. Elvira Melo
Vegetable breeding for diseases resistance	Eng. Sara Loureiro Eng. Marta Sequeira Lic. Luisa Freitas
Nematodes	Dr. M.B. Lima/ Eng. Gerson Reis
6.2 Department of Vegetables Crops	Quinta do Marquês 2780 Oeiras Phone: 01.2431505

It is the research center for vegetable crops in the National Institute for Agrarian Research (I.N.I.A.).

Vitro culture; propagation	Head: Dr. A. Gardé
Cultural practices for open and protected cultivation	Eng. A. Abrantes
Breeding and varieties	Eng. Ma. de Lurdes Matos

7 Setúbal

7.1 Estação Nacional de Fruticultura- Department of citrus crops elevation: 40 m rainfall : 600 mm sunshine : 2900 h	Estrada de Palmela 2900 Setúbal
Viruses; rootstocks	Head: Eng. Teixeira Duarte Eng. Joana Abrantes

8 Vila Real

8.1 Instituto Universitário de Trás-os-Montes e Alto Douro elevation: 450 m rainfall : 1019 mm sunshine : 2614 h	5000 Vila Real Phone: 059.236688

It is a 10 years old University with a College of Agriculture. Most of its research and development projects are related with the Northeast of Portugal and the Valley of Douro where port wine is produced

Department of crop science	Head: Eng. A. Machado
Grapes, varieties and growing	Eng. N. Magalhães
Garlic and cole crops	Eng. E. Rosa
Fruit growing and propagation	Eng. A. Santos
Potato growing and varieties	Eng. F. Martins

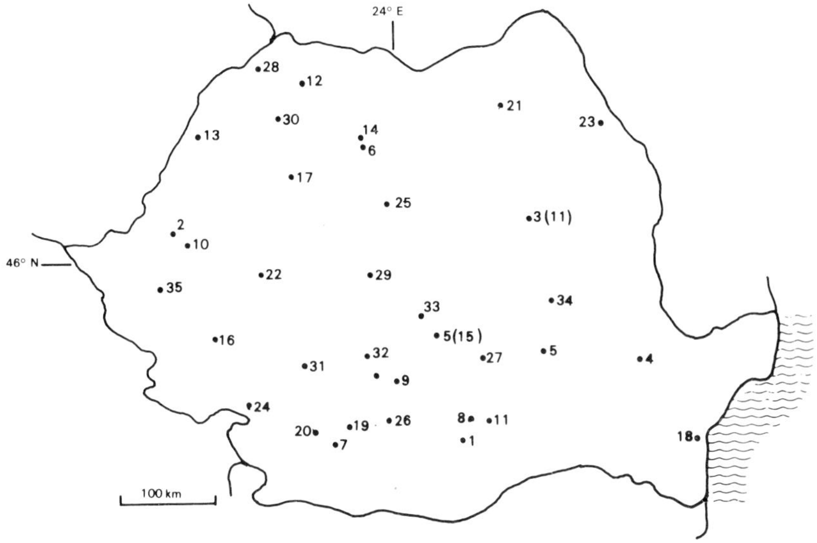
24° E
46° N
100 km
28
12
21
30
13
14
6
23
17
25
3(11)
2
10
22
29
35
33
34
5(15)
16
32
27
5
4
31
9
24
8
11
20
19
26
7
1
18

1 Vidra, 2 Arad, 3 (11) Bacău, 4 Brăila, 5 (15) Buzau, 6 Iernut, 7 Işalnita, 8 Bucarest, 9 Măracineni-Pitesti, 10 Lipova, 12 Baia Mare, 13 Oradea, 14 Bistrita, 16 Caransebes,, 17 Cluj, 18 Valul lui Traian, 19 Craiova, 20 Cîrcea, 21 Fălticeni, 22 Geoagiu, 23 Jasi, 24 Mehedint,i, 25 Tirgu Mureş, 26 Strejeşti, 27 Măgurele, 28 Satu Mare, 29 Sibiu, 30 Zalău, 31 Tîrgu Jiu, 32 Vîlcea, 33 Voineşti, 34 Focşani, 35 Timişoara

Introduction

En Roumanie la recherche scientifique dans le domaine de la horticulture est entreprise dans des unités spécialisées et par les chaires horticoles de l'enseignement supérieur. La recherche agricole est coordonnée par l'Académie de Sciences Agricoles et Forestières - présidée par le Prof.Dr.Doc. I. Muresan - amélioration des plantes - à travers la Section de horticulture et usinage des produits horticoles. Au cadre du Ministère de l'Agriculture et de l'Industrie Alimentaire le coordinateur de toute l'activité dans le domaine de la horticulture est Prof.Dr. I. Ceauşescu - horticulteur. La recherche scientifique est organisée par programmes, par produits ou groupes de produits. Chaque programme comprend: économie, organisation, création de nouvelles variétés, perfectionnement de la technologie de culture et valorisation, etc. A l'activité de recherche contribuent des collectifs formés de chercheurs scientifiques, cadres de l'enseignement supérieur, spécialistes de la grande production et d'autres domaine d'activité. Cela permet une meilleure orientation des thèmes de recherche, la coopération entre la recherche, la production et l'enseignement, de même que la vulgarisation de résultats de la recherche dans la production le plus rapidement possible. Schéma de l'Organisation de la Recherche Horticole en Roumanie:

-Institut de Recherches pour Culture Maraîchères et Floriculture
-Institut de Recherches pour Arboriculture
-Institut de Recherches et Projets pour la Valorisation des Légumes et des Fruits
-Laboratoires de Recherches pour Culture Protégées

Les instituts coordonnent des stations de recherches ou de laboratoires à spécialisation particulière dans de différents domaines.

1 Vidra

1.1 Institut de Recherches pour Cultures Maraîchères et Florales - Vidra
altitude : 55 m
pluviométrie : 535 mm
ensoleillement: 2180 h

8268 Vidra, Dép. de Giurgiu
Tél.: 13.92.82
Dir.: L. Stoian
Dir.adj.sc.: V. Poli

Préparé par Dr.Ing. C. Iordăchescu, Bucarest

Superficie totale = 432 ha, les champs expérimentaux et de vérification des résultats occupant 64 ha. Coordination de toute l'activité de recherche scientifique concernant les légumes et les fleurs cultivées en plein champ ou en culture protégée. Activité de création de nouvelles variétés de légumes et fleurs et de production de semences de base. Essais concernant le perfectionnement et le développement de la technologie de culture des plantes maraîchères, la lutte contre les mauvaises herbes, l'irrigation des cultures, l'utilisation des substances régulatrices de croîssance, la nutrition et la fertilisation des plantes, la prévention et la lutte contre les maladies et les nuisibles des cultures en plein champ et protégées, la mécanisation et l'automatisation des procésus, organisation de la production maraîchère, etc. Conseils et assistance à l'activité d'application des nouvelles variétés et des résultats des recherches dans les unités de production.

1.1.1 Laboratoire de Génétique et Amélioration	Chef: Maria Tânâsescu
Tomates	J. Poncu/ C. Tânâsescu
Piments et aubergines	Maria Tânâsescu/ Eugenia Şuţa
	M. Tudor
Oignon	N. Popandron
Carottes	Alexandrina Budişan
Concombres et cornichons	Milica Scurtu
Melon, courgettes et pastèque	N. Gherman/ Maria Dumitru
Pois	I. Scurtu
Haricots	C. Ionescu
Chou	Maria Sveatchevici
Haploïdie et cultures de tissues	Aurelia Ionescu/ Elisabeta Mirghiş
1.1.2 Laboratoire de Production des Semences	Chef: C. Posaşcâ
Sélection conservatrice des:	
Tomates et poivrons	F. Gapşa/ A. Savu
Oignon et radis	Olimpia Lascu
Carottes, céléry, persil	Bogdana Bardosi
Pois et haricots	Lucia Ionescu/ Elena Ciocan
Melon et pastèque	M. Leca
Chou	E. Grosulescu
Technologie des cultures de semences	C. Poaşcâ
Purité variétale	C.Poaşcâ/ F. Gapşa
	Florica Nacu
1.1.3 Laboratoire de Technologie des Cultures Maraîchères	Chef: M. Dumitrescu
Technologie et mécanisation des cultures maraîchères	Al. Conea/ T. Ilie/ Elena Iancu
	Virginia Teodorescu
	Râdiţa Roşulescu/ M. Râşcanu
Irrigations	D. Buzescu
Herbicides	M. Dumitrescu/ V. Râdoi/ V Miron
Biostimulants et régulateurs de croîssance	Floarea Croitoru/ M. Dumitrescu
1.1.4 Laboratoire de Cultures Protégées	Chef: S. Dragomir
Cultures sous abris en plastique	Florica Cîrstoiu/ S. Dragomir
	Florica Constantinescu
	Adriana Gheorghe
Sources d'énergies non-conventionnelles	I. Lascu
Bioclimatologie	R. Mirghiş
1.1.5 Floriculture	Chef: Rodica Ţepordei
Cultures de fleurs en plein champ	Alexandrina Hâţişi/ G. Manolache
Culture de roses	Constanţa Argatu
Cultures de fleurs sous abris en plastique	Rodica Ţepordei/ Maria Buciu
1.1.6 Laboratoire d'Agrochimie	Chef: V. Lâcâtuş
Nutrition et fertilisation des légumes cultivées - en plein champ	V. Lâcâtuş/ M. Podoleanu
	T. Munteanu/ M. Gavriluc
et sous abris en plastique	Liliana Roman
Biochimie	Carmen Botez/ P. Negulescu
Physiologie végetale	Aurora Ciupalâ

1.1.7 Laboratoire de Protection des Plantes	Chef: M. Costache
Mycoses	M. Costache/ Florica Trandaf
	N. Dragomir
Bactérioses	Gh. Marinescu
Viroses	A. Stoenescu
Entomologie	E. Cîndea/ T. Roman
La résistance des plantes maraîchères aux:	
Trachéomycoses	M. Costache/ Ana Tomescu
Maladies bactériennes	Gh. Marinescu
Maladies à virus	A. Stoenescu
1.1.8 Laboratoire d'Economie et Calcul	Chef: L. Jidav
Organisation, rotation des cultures, écologie	L. Jidav
Economie et calcul	Maria Gheorghe/ Gherghina Marinescu
	Gh. Băjenaru

2 Arad

2.1 Station de Recherches pour Cultures Maraîchères - Aradul Nou
altitude : 101 m
pluviométrie : 575 mm
ensoleillement: 2030 h

212, Rue Karl Marx, 2900 Arad
Dép. de Arad
Tél.: 966-16691
Dir.: H. Popescu

Superficie totale = 433 ha; les champs expérimentaux et de vérification des résultats = 10 ha. Recherches d'amélioration du piment rouge destiné à la transformation en poudre et de l'oignon, de production de semences, de technologie de culture des légumes précoces en plein champ, de lutte contre les mauvaises herbes, d'irrigation et de cultures protégées dans la partie occidentale de la Roumanie. Production de semences de base de tomates, piments, melons, oignon, concombres, carottes, etc. Conseils et assistance à l'activité d'application des résultats des recherches dans les unités de production.

Amélioration: piment	I. Frenţ
Amélioration: oignon	Gh. Neamţu
Sélection conservatrice	Brîndusa Şipa/ D. Milaş
	Adriana Milaş
Technologie des cultures maraîchères	V. Beldea
Herbicides	L. Hălmăgean
Irrigations	H. Popescu
Cultures protégées	Doina Popescu
Protection des plantes	P. Vărădie

3 Bacău

3.1 Station de Recherches pour Cultures Maraîchères - Bacău
altitude : 91 m
pluviométrie : 550 mm
ensoleillement: 1992 h

220, Route de Bîrlad
5500 Bacau, Dep. de Bacău
Tel.: 931/42452
Dir.: D. Davidescu

Superficie totale = 568 ha; les champs expérimentaux et de vérification des résultats = 39 ha. Recherches concernant l'amélioration et la production de semences de légumes, la technologie de culture des plantes maraîchères, la lutte contre les mauvaises herbes et la culture des fleurs. Conseils et assistance à l'activité d'application des résultats des recherches dans les unités de production.

Amélioration: haricot	M. Tomescu
Amélioration: chou et chou-fleur	N. Munteanu
Amélioration: chicorée witloof et salade	P. Bulzan
Sélection conservatrice	Elena Gherghi

Technologie des cultures maraîchères	Larisa Luca/Camelia Bălan
Herbicides	V. Timofte
Cultures protégées	D. Davidescu/Silvia Ambăruş
Floriculture	Marcela Fălticeanu
Nutrition et fertilisation	L. Stoian
Protection des plantes	N. Manole/Maria Chitic

4 Brăila

4.1 Station de Recherches pour Cultures Maraîchères - Brăila
altitude : 15 m
pluviométrie : 450 mm
ensoleillement: 2144 h

Route de Foçsani, km. 5
6100 Brăila, Dép. de Brăila
Tel.: 966/12.860
Dir.: P. Mălcică

Superficie totale = 1084 ha; les champs expérimentaux et de vérification des résultats = 15 ha. Recherches sur la production de semences de base, la technologie et la mécanisation des cultures maraîchères, la lutte contre les mauvaises herbes et les maladies, la culture des fleurs. Conseils et assistance à l'activité d'application des résultats des recherches dans les unités de production.

Essai variétal	Adriana Marin
Sélection conservatrice	Veronica Agrigoroaie/Magda Diaconu
Technologie des cultures maraîchères	P. Mălcica/I. Giugă
	Maria Iorga
Herbicides	Sevastita Găzdaru
Mécanisation des cultures maraîchères	C. Vîju
Floriculture	Maria Ardeleanu
Protection des plantes	Gh. Paraschiv

5 Buzău

5.1 Station de Recherches pour Cultures Maraîchères - Buzău
altitude : 90 m
pluviométrie : 500 mm
ensoleillement: 2150 h

23, Rue Mesteacănului
5100 Buzău, Dép. de Buzău
Tél: 974/11.601
Dir.: Lucia Luncă

Superficie totale = 492 ha; les champs expérimentaux et de vérification des résultats = 12 ha. Activité de recherche et production des cultures maraîchères en champs protégées; amélioration de l'oignon, utilisation des herbicides et production de semences. Conseils et assistance a l'activité d'application des résultats des recherches dans les unités de production.

Essai variétal	V. Gogoci/P. Andrei
Sélection conservatrice	I. Stan/Elena Ioan/Iulia Gogoci
Technologies des cultures maraîchères - en plein champ	Maria Stoicesu/Gh. Drăguţ
et sous abris en plastique	Lucia Luncă
Irrigations	G. Teodorescu
Mécanisation	C. Vlad
Cultures protégées	G. Mogoş/Petruţa Bebea
Nutrition et fertilisation	Florica Gheorghe
Protection des plantes	Marcela Iosifescu

6 Iernut

6.1 Station de Recherches pour Cultures Maraîchères - Iernut
altitude : 309 m

4351 Iernut, Dép. de Mures
Tél.: 954/26.792
Dir.: Al. Gyorgy

pluviométrie : 675 mm
ensoleillement: 2004 h

Superficie totale = 319 ha; les champs expérimentaux et de vérification des résultats = 16 ha. Recherches concernant la production de semences et la technologie de culture des plantes maraîchères dans la plaine de la Transylvanie. Conseils et assistance à l'activité d'application des résultats des recherches dans les unités de production.

Sélection conservatrice	Varvara Moldovan/Corina Simedrea
Technologie des cultures maraîchères	Al. Gyorgy/Clara Butu/V. Eisler

7 Işalniţa

7.1 Station de Recherches pour Cultures Maraîchères - Işalniţa
altitude : 90 m
pluviométrie : 550 mm
ensoleillement: 2057 h

1126 Işalniţa, Dép. de Dolj
Tél.: 941/15.636
Dir.: D. Diţu

Superficie totale = 935 ha; les champs expérimentaux et de vérification des résultats = 16 ha. Recherches concernant l'amélioration et la technologie des cultures maraîchères en plein champ et sous abris en plastique, recherches d'amélioration du pois, des haricots, des tomates, des poivrons et des concombres, études de production de semences de base. Conseils et assistance a l'activité d'application des résultats des recherches dans les unités de production.

Amélioration - tomates	V. Poli/Nicoleta Roşu
- poivrons	I. Pintilie
- cornichons	V. Poli
- pois et haricots	N. Lascu
Sélection conservatrice	Dorina Burileanu/Dorina Jilcu
	Lorelei Glăvan/Cornelia Buse
	Ecaterina Stancu
Technologie des cultures maraîchères	D. Ditu/M. Jilcu
Herbicides	M. Giorgota
Floriculture	Mariana Giorgota
Nutrition et fertilisation	Domnica Mitrache
Protection des plantes	V. Lemeni/M. Echert

8 Bucarest

8.1 Chaire de Culture Maraîchère et Floriculture
altitude : 92 m
pluviométrie : 555.1 mm
ensoleillement: 2177 h

Institut Agronomique
"N. Balcescu", 59, Boulevard
Mărăşti, Bucarest
Tel.: 90/182230

Cours de culture maraîchère générale, recherches concernant la technologie des cultures maraîchères en plein champ et dans espaces protégés; recherches sur l'écologie légumière; l'effet de la lumière artificielle sur les cultures maraîchères	I. Ceauşescu
Cours de culture maraîchère spéciale, recherches sur l'égologie et la physiologie des cultures de tomates en serre, la technologie des cultures maraîchères en plein champ et serres, l'effet de la lumière sur la croissance et la développement des tomates	V. Voican
Cours de culture maraîchère, recherches sur les cultures en serre, l'effet des radiations sur les légumes, la	Elena Florescu

culture des concombres sous abri en matériel plastique	
Recherches sur la nutrition des tomates de serre, la technique de production de semences de légumes annuelles, la technologie de culture des tomates pour industrialisation	M. Mihalache
Recherches concernant la création de nouvelles variétés de tomates précoces, tomates de serre et concombres à résistance génétique aux maladies	C. Petrescu
Technologie du pois de jardin, espèces de légumes pour la diversification de l'assortiment	Gabriela Doru
Biologie et technologie du piment cultivé en serre, recherches sur l'écologie des cultures en serre	V. Popescu
Effet du paillage sur les cultures maraîchères, détermination de la valeur d'utilisation des légumes	Ruxandra Ciofu
Technologie de culture des concombres pour industrialisation, problèmes de mécanisation des cultures maraîchères	N. Atanasiu
Cours de floriculture, la biologie et la nutrition minérale des glaïeuls, gerbéras, tubéreuses, l'utilisation des substances stimulatrices de la croissance	Amelia Milițiu*
Cours d'arrangement des espaces verts, la biologie et la technologie des tulipes, l'effet des substances stimulatrices de la croissance sur *Pelargonium,* la production de semences de *Petunia* et *Viola*	Elena Selaru
Cours d'arboriculture ornementale et d'architecture paysagiste, la multiplication des plantes ornementales, l'arrangement des parcs, la création de variétés de *Antirhinum magus*	Ana-Felicia Iliescu
Recherches sur la résistance aux températures basses	Ioana Molea
Techniques nucléaires appliquées à l'étude de la nutrition des tomates, piments et concombres	C. Caramete
Culture intensive de l'abricotier. Recherches sur l'amélioration et la physiologie de l'abricotier	N. Cepoiu
Recherches sur l'amélioration du pommier et griottier	Caliope Rițiu
Technologie de culture du prunier	Maria Tertecel
Recherches concernant la biologie et l'amélioration du nectarinier et pêcher	Monica Murvai
Recherches concernant la biologie et la technologie de culture du framboisier	T. Tudor
Recherches sur la technologie de culture et la valorisation du fraisier, des légumes et des fruits	Maria Elena Ceaușescu

8.2 Station Fruitière Baneasa

71592 Bucarest, I-e secteur
4, Boulevard Ion Ionescu de la Brad
Tél.: 90/332127
Dir.: T. Stancu

Etude des variétés et création de nouvelles variétés d'abricotier, y compris porte-greffe. Recherches sur la prévention de la mort prématurée de l'abricotier. Technologie de culture du fraisier en serre.

Création de variétés d'abricotier. Cytogénétique	Antonia Ivașcu/Viorica Balan
Technologie de culture de l'abricotier	Niculina Burloi
Technologie de culture du fraisier en serre	Evelina Zambrowicz
Floriculture - plantes d'appartement	Viorica Canarache
Technologie de culture des oeillets	M. Mares
Maladies des plantes	Elisabeta Stoian

8.3 Chaire d'Arboriculture

Institut Agronomique
"N. Bălcescu", Faculté de

Cours d'arboriculture générale. Culture intensive et superintensive. Recherches d'écologie, physologie et nutrition minérale des arbres. Recherches d'embryoculture et culture de tissues d'arbres et arbustes fruitiers	Horticulture, 59, Boulevard Mărăsti, Bucarest Tél.: 90/182230
Cours d'arboriculture. Culture intensive et superintensive du pommier	N. Cepoiu A. Negrilă
Cours de horticulture. Culture intensive du pêcher et cerisier	N. Stefan
Cours d'arboriculture spéciale. Technologie de culture du pêcher	Gr. Mihăescu
Cours de horticulture. Technologie de culture du poirier	Fl. Lupescu
Cours de technologie des fruits et légumes. Recherches sur la physiologie et biochimie des fruits et légumes	
Recherches sur la biologie et technologie de culture du pêcher	Creola Mănescu

8.4 Institut de Recherches et Projets pour la Valorisation des Légumes et des Fruits

1A, Rue Intrarea Binelui
75614 Bucarest, IV-e secteur
B.P. 93, Tél.: 823440
Dir.: A. Teodorescu
Dir.adj.sc. S. Gîrbu

Modernisation des technologies de triage-calibrage et conservation des légumes, pommes de terre, fruits frais et raisins, ainsi que la technologie de culture et la valorisation des champignons de couche. Technologies d'usinage des légumes et fruits pour l'obtention des produits deshydratés, concentrés, stérilisés, fermentés, diététiques et des jus.
Détermination des types d'emballages et des technologies d'emballage des légumes et des fruits frais et preparés. Etablissement des moyens et des meilleurs systèmes de transport, ainsi que des méthodes et techniques de commercialisation des produits. Etablissement de l'efficience économique des divers moyens de conservation des produits frais et de transformation industrielle des légumes et des fruits; études de marketing. Etablissement du schéma des machines et outillages pour la mécanisation des procédés de triage-calibrage, transport, entreposage, transformation industrielle, etc. Amélioration des types constructifs d'entrepôts et de sections d'usinage. Elaboration de la documentation pour les entrepôts à ventilation mécanique, pour les entrepôts frigorifiques et à athmosphère controlée, pour les fabriques de produits conservés et semi-conservés, pour la modernisation des sections d'usinage et des champignonnières, etc. L'activité de l'institut se dévéloppe en six compartiments:

8.4.1 Laboratoire de conservation des légumes et des fruits	Chef: A. Gherghi
Organisation de l'activité dans les entrepôts modernes	S. Gîrbu
Conservation de l'oignon, de l'ail et des poireaux	C. Iordăchescu
Valorisation des légumes de serre, des legumes racineuses et des choux	Olga Iordăchescu/ Nicoleta Rusu Constanta Alexe
Maintien de la qualité des légumes perissables: tomates, poivrons, aubergines	Mihaelă Ciurel
Conservation des pommes de terre	F. Niculescu
Conservation des pommes et des poires	A. Gherghi/ K. Millim
Conservation temporaire des pêches, abricots, néctarines, cerises, griottes et des petits fruits	Stefania Fugel/ Margareta Oancea
Technologie de valorisation des raisins	Elena Popa
Problèmes phytopathologiques de la valorisation des légumes et des fruits	Gh. Ţaşca/ Margareta Giurea Nicoleta Mihăilescu
Problèmes physiologiques après la récolte des légumes et des fruits	I. Burzo
Conservation de la qualité des fleurs coupées et les traîtements thermiques des bulbes	Alexandrina Amăriuţei
8.4.2 Laboratoire de recherches concernant l'usinage des produits horticoles	Chef: O. Burtea

Usinage des produits horticoles par stérilisation	Stela Bàbiceaunu/ V. Micu C. Popescu
Deshydratation des légumes, des fruits et des pommes de terre	Rodica Cricoveanu
Technologie des produits obtenus par fermentation lactique	Doina Popa
Technologie des pâtes des fruits et des légumes et des produits du type "instant"	O. Burtea
Congélation des légumes et des fruits	Gh. Mihalca
Aliments diététiques et pour enfants	Elena Bacheş/Stela Babiceanu
8.4.3 Laboratoire des jus, des concentres, des aromes et des boissons non-alcooliques	Chef: J. Jurubiţă
Technologie de l'obtention des enzymes	Mariana Pele
Technologie de l'emploi industriel des membranes	
Obtention des concentres et aromes naturelles	Roxandra Glâman
Diversification des sortes de jus et de boissons non-alcooliques	Adina Constantinescu
8.4.4 Laboratoire d'emballage, transport, organisation et économie de la valorisation	Chef: M. Irimia
Technologie d'emballage pour les légumes et les fruits frais et diversement préparés	I. Mircea/ Ir. Popescu
Pré-emballage et presentation des produits	Doina Tâtaru
Transport des produits horticoles	M. Irimia/ A. Jalea
Présentation et commercialisation des légumes et des fruits	Rodica Cernat
Economie de la valorisation, le marketing, l'organisation du travail dans les entrepôts modernes, l'étude des tendances de consommation	Delia Bogdan
Organisation de l'activité de valorisation par les methodes d'optimisation mathématique	R Toma/ G. Matei
8.4.5 Laboratoire de microbiologie et biochemie	Chef: Maria Olaru
Microbiologie des produits usinés des légumes et des fruits	Mariana Henter
Analyse des résidus des pesticides sur les légumes et les fruits frais et usinés	Miruna Bibicu
Dinamique des indices de qualité pour les légumes et les fruits frais et usinés	Maria Flueraru/ M. Dobreanu
Perfectionnement des méthodes d'analyse physique-chimique et microbiologique	Elena Ionescu/ Dorina Pîrvulescu
8.4.6 Bureau technique pour les projets d'investissements dans le domaine de la valorisation des légumes, des fruits des raisins et des pommes de terre	Chef: L. Ionescu

Projets de construction des objectifs spécifiques à l'activité de production et de valorisation-entrepôts de légumes et fruits, congélateurs, fabriques et sections d'usinage, champignonnières, serres, centres de vinification, fabriques de champagne, de jus et de boissons non-alcooliques. Etudes et recherches concernant l'assimilation des outillages, la mécanisation des procédés de valorisation, la croissance des indices d'exploitation des machines, outillages, instalations, etc. Organisation de l'activité de production et de valorisation dans les entreprises horticoles spécialisées.

8.5 Laboratoire de Recherches pour Cultures Protégées	110, Rue Moşoaia, 75616 Bucarest IV-e secteur, Tél.: 83.31.18 Chefs: P.Focşâneanu/ O.Mânucâ
Amélioration des légumes de serre et obtention des semences hybrides	Narcisa Sindile
Amélioration des espèces des fleurs cultivées dans	Viorica Canarache

les serres et obtention du matériel bon à planter	
Perfectionnement des technologies de culture des légumes de serre avec une consommation réduit d'énergie thérmique	Gh. Ilie
Perfectionnement des technologies de culture des fleurs dans les serres	M. Mareş
Etude des maladies produites par les fungi phytopathogènes; méthodes pour les combattre dans les cultures de serre	Gabriela Stan
Combat des bactérioses aux cultures de serre	I. Tănase
Combat des insectes nuisibles aux cultures de serre	S. Mihăilescu
Nutrition et fertilisation des cultures de serre	O. Mănucă

9 Mărăcineni

9.1 Institut de Recherches pour Arboriculture
altitude : 287 m
pluviométrie : 700 mm
ensoleillement: 2074.2 h

0312 Mărăcineni, Dép. de Arges
Tél.: 976/34292
Dir.: P. Parnia

Superficie totale d'expérimentation et de vérification des résultats (y compris les 26 stations) = 42 000 ha. Organisation, conseil et control de toute l'activité de recherche et production dans le domaine de l'arboriculture de Roumanie. Création de nouvelles variétés de prunier, griottier, poirier, fraisier, noisetier; technologies modernes de production de matériel de plantation - arbres, fraisiers et arbustes; multiplication vegétative dans des espaces protégés, dévirosation. Technologie de culture intensive et superintensive des pommiers, poiriers, pruniers, cerisiers et griottiers. La biologie et la lutte contre les pests, testes de pesticides et insectofungicides, problèmes de fertilisation et d'agronomie. Le perfectionnement du schéma de machines pour arboriculture. Irrigations, économie et marketing des arbres et arbustes fruitiers.

Création de variétés de prunier et griottier	V. Cociu
Technologie de culture du pommier	A. Suta
Création de variétés de poirier	N. Braniste
Etude des variétés. Creation de varietes de cerisier et griottier	Chiriacula Micu/ S. Budan
Technologie de culture des arbustes fruitiers	Paulina Mladir
Amélioration du prunier	R. Roman
Technologie de culture du fraisier	M. Coman
Cultures de méristèmes, thermo-thérapie, virologie	Tatiana Coman/ Maria Isac Luminita Neculae
Bactérioses des arbres à pépins	Valentina Amzăr/ A. Richiţearu
Virologie des arbres à pépins	
Nématodes - biologie et lutte	Georgeta Teodorescu
Entomologie, protection phytosanitaire complèxe. Tested de pesticides et insectofungicide	Victoria Suta/ Adina Gava
Biostimulateurs, retardants, réglementation chimique de la production	Sabina Stan
Technologie de culture du griottier	Cornelia Parnia
Utilisation des herbicides dans l'arboriculture	S. Coman
Physiologie des arbres, photosynthèse	A. Paul-Bădescu
Cytologie - génétique des arbres à pépins	Irina Popescu/ A. Popescu
Génétique, biochimie	Evelina Rudi
Technologie de culture du pommier, coupages	M. Cotorobai
Techniques de production du matériel de plantation	I.Onea/ N.Stanciu/ Steliana Popescu
Création et sélection de nouveaux port-greffes pour pommier, poirier, prunier, cerisier et griottier	Gh. Mladin/ I. Dutu/ N. Stanciu
Mécanisation	Gh. Stan/ I. Bebu/ D. Moiceanu
Amélioration. Ecologie	I. Neamţu

Amendements. Irrigations	M. Iancu
Economie de l'arboriculture. Marketing	A. Otteanu
Floriculture - dendrologie	N. Puşcăsu
Technologie de culture de noyer	V. Vasilescu

9.2 Station Fruitière Argeş	Commune de Mârăcineni, 0312 Argeşel, Dép. de Argeş Tél.: 976/32179 Dir.: Gh. Badescu
Etude et amélioration du pommier pour la zone sous-montane (800-860 m). Sélection clonale du pommier et myrtille.	Adina Perianu
Collection nationale de myrtille et roses pour confiture	
Aménagement des terrains en pente pour la culture des arbustes fruitiers	I. Cirstea
Amélioration et technologie de culture du pommier.	Gh. Bădescu
Multiplication du myrtille	
Agrotechnique du pommier	S. Diaconu
Protection phitosanitaire	Valeria Copâescu

10 Lipova

10.1 Station Fruitière - Lipova altitude : 130 m pluviométrie : 577 mm ensoleillement: 2122.5 h	6 Rue Lugojului, Lipova - 2875 Dép. Arad Tél.: 960/61514 Dir.: N. Voştinar
Technologie de culture du pommier et prunier; culture intensive et superintensive; amélioration	N. Ionescu
Etude des variétés des arbustes fruitiers	

11 Bacău

11.1 Station Fruitière Bacău altitude : 167 m pluviométrie : 544.3 mm ensoleillement: 1992 h	27 Rue Romanului, Bacău - 5500 Dép. Bacău Tél.: 931/34514 Dir.: Gh. Onea
Etude des variétés du pommier; technologie de culture.	
Drainage-assèchement, dessiccation	Gh. Onea
Etude des variétés; amelioration	V. Raţi
Drainage, dessiccation	C. Florescu

12 Baia Mare

12.1 Station Fruitière Baia Mare altitude : 194 m pluviométrie : 976 mm ensoleillement: 2110.5 h	232, Rue Victoriei, 4800 Baia Mare Dép. de Maramureş Tél.: 994/34991 Dir.: A. Lazăr

Amélioration du marronnier, collection national de marronnier. Technologie de culture intensive et superintensive des espèces: marronnier, pommier, prunier, groseillier noir. Schéma de machines pour terrains en pente.

Etudes des variétés, technologie de multiplication du marronnier	I. Popa
Drainage - assèchement, dessiccation, organisation de la production	A. Pop

Lutte contre les maladies et les pests du marronnier	Smaranda-Rodica Florea
Aménagement des terrains pour la culture	P. Lăzăroiu/ A. Gallov
Technologie de culture des arbustes fruitiers	O. Roşca

13 Oradea

13.1 Station Fruitière Bihor altitude : 137 m pluviométrie : 635 mm ensoleillement: 2092.2 h	1, Rue Bolintineanu, 3700 Oradea Dép. de Bihor Tél.: 991/32986 Dir.: O. Petre

Etude des variétés et création de nouveaux variétés de pêcher. Détermination de la technologie de multiplication de l'abricotier, du pêcher et de l'amandier, porte-greffes. Valorisation des terrains sablonneux. Irrigations.	Monica Ştefan
Création de variétés de pêcher et amandier	V. Scheau
Culture de l'abricotier et du pêcher sur terrains sablonneux, irrigations	A. Bunea
Technologie de production de matériel de plantation pour l'abricotier et le pêcher	I. Stefan
Protection phytosanitaire	Silvia Murg

14 Bistriţa

14.1 Station Fruitière Bistriţa altitude : 358 m pluviométrie : 680 mm ensoleillement: 2004 h	3, Rue Drumul Dumitrei Nou 4400 Bistriţa Năsăud Dép. de Bistriţa Năsăud Tél.: 990/17895 Dir.: N. Drăgan

Technologie de culture superintensive du pommier sur terrains en pente. Agrotechnique dans les vergers de pommiers, cerisiers et griottiers.

Création de variétés de cerisier	I. Ivan
Agrotechnique du sol des vergers sur terrains en pente	I. Dumitrache
Production de matériel de plantation libre de viroses	N. Minoiu/ Doina Vlădianu
Aménagement des terrains en pente	
Biochimie	K. Pattantyus

15 Buzău

15.1 Station de Recherches d'Arboriculture Cîndeşti - Vernesti altitude : 122 m pluvimetrie : 580 mm ensoleillement: 2030 h	5120, Cîndeşti, Dép. de Buzău Tél.: 974/16296 Dir.: C. Solzea

Technologie de la culture du prunier et du pommier sur les collines	C. Solzea
Amélioration du prunier	
Organisation et économie de l'arboriculture	

16 Caransebeş

16.1 Station Fruitière Caransebeş altitude : 201 m	94, Rue Godeanu, 1650 Caransebeş Dép. de Caraş-Severin

pluviométrie : 737 mm	Tél.: 965/12302
ensoleillement: 2032.3 h	Dir.: N. Vulpeş

Sélection du cerisier, prunier, griottier et pommier.	
Amélioration et technologie de culture mécanisée du prunier.	
Sélection des arbres à pépins	N. Mutaşcu
Technologie de culture intensive et superintensive du cerisier et griottier	N. Feneşan
Protection phytosanitaire complèxe	Gh. Simeria

17 Cluj

17.1 Station Fruitiere Cluj	5, Rue Horticultorilor
altitude : 363 m	3400 Cluj-Napoca, Dép. de Cluj
pluviométrie : 613 mm	Tél. 951/41038
ensoleillement: 1987.8 h	Dir.: St. Oprea

Création de variétés d'arbustes fruitiers et fraisier.
Sélection du pommier, poirier, cerisier, griottier et fraisier. Création de porte-greffes nains pour cerisier et griottier.
Lutte contre les pests du fraisier et des arbustes fruitiers.

Amélioration et création de variétés de poirier	Mircea Străulea
Etude et amélioration du fraisier, création de variétés d'arbustes fruitiers	A. Botar
Création de porte-greffes nains pour cerisier et griottier	St. Wagner

Laboratoire de Floriculture

Amélioration des oeillets, gerberas et glaïeuls	Lucia Litan
Amélioration des roses	St. Wagner
Arbres et arbustes ornementaux. Culture des fleurs en pots	St. Oprea
Technologie de culture des fleurs en serres	D. Zaharia

Laboratoire de Cultures Maraîchères

Amelioration des tomates et du piment et production de semences	Z. Otvös

17.2 Chaire de Horticulture, Institut Agronomique "Dr. Petru Groza", Cluj

Cours de culture maraîchère, recherches concernant les cultures précoces en plein champ, la culture des légumes en serre, la création de hybrides de concombres et chou	D. Indrea
Ecologie et technologie de culture de l'oignon semis directe	Ana Radu
Influence des facteurs génétiques et écologiques sur la composition chimique des légumes	Gh. Marca
Cours d'arboriculture. Amélioration du pommier	St. Oprea
Le système radiculaire des arbres	E. Rusu

18 Valul lui Traian

18.1 Station Fruitière Valul lui Traian	1, Rue Pepinierei, 8763 Valul lui
altitude : 32 m	Traian, Dép. de Constantza
pluviometrie : 378.8 mm	Tél.: 912/34349
ensoleillement: 2281.7 h	Dir.: V. Zglobiu

Collection nationale de cerisier et pêcher, création de nouvelles variétés d'abricotier, pêcher et amandier. Cultures comparatives de variétés de pommier d'été, abricotier, pêcher, nectarinier, amandier et cerisier. Technologie de culture intensive et superintensive du pêcher et abricotier. Irrigations. Economie de l'arboriculture.

Technologie de culture du pêcher et abricotier. Irrigations	Preda Ionescu
Etude des variétés et amélioration du pêcher et abricotier	Elena Topor

Laboratoire de Floriculture

Maladies des arbres fruitiers	Mărioara Trandafirescu
Amélioration et production de semences, culture de tulipes et glaïeuls	El. Bîtlan
Dendrologie	Joana Velicu

19 Craiova

19.1 Chaire de Culture Maraîchère et Floriculture, Université de Craiova — 13, Rue Cuza Vodă, 1100 Craiova, Tél.: 941/11711

Cours de culture maraîchère générale, recherches concernant l'effet des facteurs de végétation sur les plantes légumières	Chef: Gr. Radu
Cours de culture maraîchère spéciale, recherches sur la culture des tomates sur sables	Gr. Radu
Cours de culture maraîchère, recherches sur la production des pommes de terre sur sables	Gh. Dina
La nutrition minérale extraradiculaire et l'effet des substances retardantes sur les tomates cultivées en serre	Pelaghia Chilom
La nutrition minérale extraradiculaire et l'effet des substances retardantes sur les concombres cultivés en serre	Adriana Duţa
Cours de floriculture, études d'écologie des fleurs	Afrodita Pavel
Technologie des fleurs à bulbes	Cornelia Lemeni

19.2 Chaire de Horticulture, Université "Al. I. Cruza" — Faculté de Horticulture, Craiova

Cours d'arboriculture spéciale. Technologie de culture du pêcher et abricotier sur sables	M. Popescu*
Cours d'arboriculture générale. Technologie de culture intensive et superintensive du pommier	I. Miliţiu
Cours de technologie des produits horticoles. Technologie de valorisation des fruits et legumés	L. Roşu
Biologie et culture du noyer	I. Godeanu
Ecologie du pêcher sur sables	Elena Voica
Culture du framboisier sur sables	Tatiana Pădureanu
Technologie des fruits et légumes	Doina Anton

20 Cîrcea

20.1 Station Fruitière Dolj — 1112 Cîrcea, Dép. de Dolj

altitude : 105 m — Tél.: 941/45490
pluviometrie : 523 mm — Dir.: C. Magherescu
ensoleillement: 2181.7 h

Sélection du cerisier, abricotier, pêcher, griottier, framboisier et fraisier. Perfectionnement des méthodes de production de matériel de plantation - arbres à pépins.

Technologie de culture annuelle du fraisier	M. Diaconeasa
Etude des variétés d'abricotier et amandier, problèmes de résistance au gel	Sabina Iliescu/ T. Slămnoiu
Protection phytosanitaire de l'abricotier et amandier	Virginia Popescu

21 Fălticeni

21.1 Station Fruitière Fălticeni
altitude : 335 m
pluviométrie : 538 mm
ensoleillement: 1921.9 h

10 Plut. Ghimită
5750 Fălticeni, Dép. de Suceava
Tél.: 988/42284
Dir.: Gh. Lazăr

Sélection du pommier, griottier et groseillier noir. Création de port-greffes pour cerisier et griottier. Technologie de culture du framboisier. Aménagements anti-érosion.

Etude en cultures comparatives du groseillier noir, framboisier, groseillier à maquereau et myrtille (y compris la multiplication)	Maricica Silochi/ O. Petru
Problèmes des porte-greffes de pommier, cerisier et griottier	Mircea Movileanu
Technologie de culture intensive et superintensive des variétés d'arbres à pépins	V. Moroşanu
Etude des variétés des arbustes fruitiers	D. Iftime

22 Geoagiu

22.1 Station Fruitière Geoagiu
altitude : 230 m
pluviométrie : 613.3 mm
ensoleillement: 2032.3 h

141, Rue Romanilor, 2616 Geoagiu
Dép. de Hunedoara
Tél.: 956/41593
Dir.: Elisie Pasc

Collection nationale de noyer, création de variétés de noyer, pommier, poirier, prunier. Création de portegreffes végétatifs d'arbres à pépins. Biologie, écologie et lutte contre les araigniers nuisibles des arbres.

Diagnostic, aspects de la nutrition des arbres	Elisie Pasc
Création de porte-greffes de prunier et pommier	St. Casavela
Lutte contre lea araigniers nuisibles	Gh. Lefter
Utilisation des herbicides dans les vergers et pépinières	D. Prică

23 Jassy

23.1 Chaire de Horticulture
altitude : 100 m
pluviométrie : 517.8 mm
ensoleillement: 1995.3 h

Institut Agronomique "Ion Ionescu de la Brad" 3, Allée M. Sadoveanu
Jassy
Tél.: 981/40798

Cours de culture maraîchère générale, étude des hybrides simples et doubles de tomates, cultures maraîchères protégées par máteriaux plastiques	P. Savitzchi
Cours de culture maraîchère spéciale, la biologie et la technologie des tomates cultivées sous abris et matériaux plastiques	N. Stan
Cours de floriculture, recherches de physiologie des espèces légumières, la production de semences	Natalia Ailincăi
Cours d'arboriculture ornementale et d'architecture paysagiste, la valorisation de la flore ligneuse spon-	L. Palade

tanée de Roumanie par l'arboriculture ornementale
Chaire de Physiologie
Etude des biostimulateurs et des retardants sur les arbres et les légumes — C. Milicá

23.2 Station Fruitière Jassy — 175, Rue Voinesti, 6600 Jassy
Office postale 11, c.p. 60
Dép. de Jassy
Tél.: 981/42820
Dir.: Gh. Dumitrescu

Collection nationale de cerisier et création de variétés de pommier, poirier, cerisier, griottier et abricotier.
Sélection de porte-greffes de cerisier et griottier.
Technologie de culture intensive et superintensive du cerisier et griottier.
Etude des variétés, selection clonale du cerisier et griottier; création de variétés autofertilies. Technologie de culture — I. Bodi/ P. Ludovic

23.3 Faculte de Horticulture — Institut Agronomique "Ion Ionescu de la Brad", Jassy

Cours de technologie des produits horticoles. Valorisation des fruits et légumes — I. Potec
Cours d'arboriculture générale. Culture du cerisier et griottier — V. Cireașá
Cours d'arboriculture spéciale. Coupage du pêcher — Gh. Drobotá
Cours d'arboriculture. Coupage du cerisier — Florica Roșu
Biologie et culture du prunier — Mariani Drobotá
Valorisation des fruits et légumes — A. Cotrău

24 Mehedinți

24.1 Station Fruitière Mehedinți — 277, Rue Traian, 1500 Drobeta Turnu Severin, Dép. de Mehedinți
altitude : 70 m
pluviométrie : 661 mm
ensoleillement: 2176.7 h
Tél.: 978/12368
Dir.: Marian Pufan

Collection nationale d'amandier et figuier. Sélection de l'abricotier, amandier, pommier, griottier et marronnier. Sélection de porte-greffes pour l'amandier. Technologie de culture intensive et superintensive de l'amandier. Economie de l'arboriculture.

Amélioration des arbres à pépins — Ioana Zaharia
Sélection de porte-greffes, technologie de production du matériel de plantation - abricotier et amandier
Technologie de culture des arbres fruitiers — Paula Giura
Protection phytosanitaire — C. Zaharia

25 Tîrgu Mureș

25.1 Station Fruitière Mureș — 1, Rue Pásunii, 4300 Tîrgu Mureș
altitude : 309 m
pluviométrie : 636 mm
ensoleillement: 2004 h
Dép. de Mures
Tél.: 954/36506
Dir.: T. Habor

Collection nationale de cornouiller, argousier, frambosier. Sélection comparative de prunier, pommier, griottier, noyer. Multiplication du cornouiller et de l'argousier, technologie de culture.

Sélection du noyer, technologie du culture du noyer
Sélection du prunier et griottier — I. Micu

Drainage - assèchement dessication — J. Samuilă
Protection phytosanitaire — Florentina Togănel

26 Strejeşti

26.1 Strejeşti-Olt
altitude : 182 m
pluviométrie : 578.8 mm
ensoleillement: 2211.5 h

0572 - Strejeşti, Dép. Olt
Tél.: 944/13441
Dir.: I. Nicolaescu

Technologie de culture du prunier; culture intensive.
Techniques de production du matériel de plantation.
Bactérioses et vinologie des arbres.

27 Măgurele

27.1 Station Fruitière Prahova
altitude : 164 m
pluviométrie : 588 mm
ensoleillement: 2115.8 h

2129 Măgurele, Dép. de Prahova
Tél.: 971/26859
Dir.: I. Neagu

Technologie de culture très dense du pommier. Lutte contre les maladies et les pests. Mécanisation des travaux dans les vergers.

Etude des variétés et technologie de culture du pommier — I. Neagu
Modernisation des méthodes de traitement contre les maladies — D. Popovici
Mécanisation des travaux dans les vergers — Lidia Rasa

28 Satu-Mare

28.1 Station Fruitière Viile Satu-Mare
altitude : 129 m
pluviométrie : 667.9 mm
ensoleillement: 2110.5 h

3, Rue Republicii
Dép. Satu Mare - 3958
Tél.: 997/36847
Dir.: I. Moldovan

Technologie de culture. Irrigations. Culture de framboisier.

29 Sibiu

29.1 Station Fruitière Sibiu
altitude : 416 m
pluviométrie : 662 mm
ensoleillement: 1924.1 h

97, Rue Măgurii, 2437 Cisnădie
Dép. de Sibiu
Tél.: 925/61026
Dir.: V. Sandru

Etude en cultures comparatives des arbres fruitiers — R. Thiesz
Etude des variétés de myrtille, multiplication des arbustes fruitiers. — Didina Sihleanu
Sélection du noyer, technologie — N. Oprean
Lutte contre les maladies et pests des arbustes fruitiers — Nicoleta Thiesz

30 Zalău

30.1 Station Fruitière Zalău
altitude : 288 m
pluviométrie : 705.3 mm
ensoleillement: 1987.8 h

24, Rue Cerbului, Zalău - 4700
Dép. Sălaj
Tél.: 996/12942
Dir.: Cosma Dorel

Aménagement des terrains en pente pour la culture des arbustes fruitiers.

Etude des variétés des arbustes fruitiers; protection phytosanitaire complèxe — V. Cantor/ Maria Cantor

31 Tîrgu Jiu

31.1 Station Fruitière Tîrgu Jiu
altitude : 210 m
pluviométrie : 753 mm
ensoleillement: 2040.6 h

72, Route de Bucarest
1400 Tîrgu Jiu, Dép. de Gorj
Tél.: 929/12471
Dir.: I. Tomescu

Collection nationale de cognassier. Sélection du noyer, marronnier, prunier, griottier et cognassier. Sélection de porte-greffes de noyer, perfectionnement de la technologie de multiplication du noyer. Amélioration des variétés.

Création de nouvelles variétés d'arbres à pépins — R. Roman
Technologie de culture du noyer, perfectionnement de la technique de multiplication, sélection de nouveaux porte-greffes — Cristea Tetileanu, Valeria Petricà
Traitement contre les maladies — Teodora Tetileanu

32 Vîlcea

32.1 Station Fruitière Vîlcea
altitude : 242 m
pluviométrie : 707.3 mm
ensoleillement: 2028.8 h

Càzànesti 1008
Dép. de Vîlcea
Tél.: 957/10940
Dir.: A. Duțà

Collection nationale de myrtille. Sélection du pommier, prunier et noyer. Variétés de pommier, prunier, fraisier et arbustes fruitiers. Sélection clonale de porte-greffes de prunier.

Sélection de pommier et prunier; lutte contre les maladies — Elena Turcu/ Elena Ioachim
Perfectionnement de la technique de production du matériel de plantation et sélection de nouveaux porte-greffes — I. Botu
Technologie de culture — O. Mihàescu/ C. Càtusanu

33 Voinești

33.1 Station Fruitière Voinești
altitude : 380 m
pluviométrie : 560 mm
ensoleillement: 2115.8 h

0262 Voinești Dép. de Dîmbovitza
Tél.: 926/79323, 926/34106
Dir.: I. Achimescu

Collection nationale de pommier. Sélection du pommier, purnier, cerisier et noyer. Porte-greffes pour pommier, poirier et prunier. Technologie de culture intensive et superintensive des variétés d'arbres à pépins.

Technologie de culture très dense du pommier et poirier — Gh. Petre
Porte-greffes pour arbres à pépins
Sélection du pommier — Luca Șerboiu
Protection phytosanitaire — Albertina Șerboiu

34 Focșani

34.1 Station Fruitière Vrancea
altitude : 60 m
pluviométrie : 503.1 mm
ensoleillement: 2120 h

75, Rue Cuza Vodà, 5300 Focșani
Dép. de Vrancea
Dir.: S. Scàrlàtescu

Sélection clonale et cultures comparatives de griottier, pommier, abricotier et prunier. Porte-greffes de griottier. Technologie de culture superintensive du pommier et griottier.

Méthodes de traitement contre les maladies — Maria Herţug

35 Timişoara

35.1 Chaire de Horticulture — Institute Agronomique, Timişoara

Cours d'arboriculture. Technologie du cerisier — Gh. Predescu

Génétique - amelioration du poirier — E. Drăgănescu

Cours de culture maraîchère. Technologie des cultures précoces de légumes. Biologie, écologie — H. Butnaru

Génétique et amélioration de l'oignon — I. Suciu

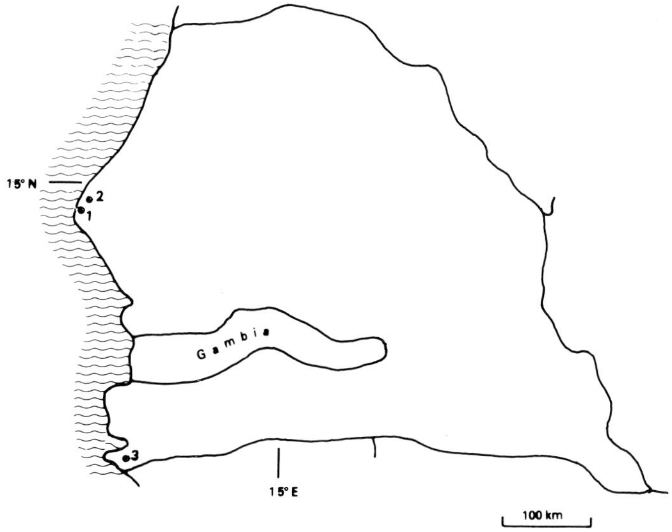
15° N
2
1
Gambia
3
15° E
100 km

1 Dakar
2 Cambérène
3 Singher

1 Dakar

1.1 Centre pour le Développement de l'Horticulture I.S.R.A.

C.D.H., B.P. 2619 Dakar
Tél.: 21.22.05
Chef du Projet FAO: H. Van der Veken
Dir.: M. Ly

Crée en 1972 avec l'asistance de la FAO, le CDH fait actuellement partie de l'Institut Sénégalais des Recherches Agricoles (ISRA) et se spécialise dans le domaine de la recherche sur les cultures maraîchères. Au début la recherche s'était orienté vers l'étalement de la production, l'introduction variétale, l'application des techniques culturales, la protection phytosanitaire ainsi que l'amélioration d'un large éventail de cultures légumières (chou, oignon, tomate, pomme de terre, piment, manioc, jaxatu, gombo, carotte, laitue, haricot, concombre, melon, pasteque, pois, courgette, bissap et aubergines). Par la suite un programme prioritaire a été développé pour les cultures de tomate, oignon, pomme de terre, manioc, gombo, jaxatu, piment et patate douce. C'est ainsi qu'à partir des années 82-83, le Centre est composé des Services de Recherches suivants:

Les Services de Recherche "Piment-patate douce", "Manioc-oignon", "Tomate-gombo", et "Pomme de terre-Jaxatu" dont les programmes consistent dans l'introduction de variétés et le trivarietal de ces espèces maraîchères prioritaires en vue de l'étalement et l'amélioration de leur production, l'amélioration génétique de certaines espèces pour une meilleure adaption aux conditions climatiques et aux problèmes phytosanitaires, létude des problèmes dans le domaine des techniques culturales et de la conservation des récoltes, ainsi que la production de semences et d'autre matériel de multiplication végétative. Le conditionnement des semences et du matériel végétatif est sous la responsabilité du Service de Recherche "Pomme de terre-Jaxatu".

Le Service de Recherche "Protection" régroupe les activités de recherche dans les domaines d'entomologie, pathologie, virologie et nématologie. Il identifie les problèmes phytosanitaire et étudie les moyens de lutte. Il supporte aussi les autres Services de Recherche.

Le Service de Recherche "Commercialisation" mène des enquêtes et des études dans les domaines de la production maraîchère, les circuits de commercialisation, les facteurs socio-économiques qui régissent les exploitations maraîchères ainsi que sur l'évolution de la consommation des légumes et l'évolution des prix de vente.

Le Service de Recherche "Prévulgarisation et Formation" assure la liaison entre la recherche maraîchère et le développement de ce secteur à travers les organismes de dévelopment. Ceci ce fait par la participation à la formation de l'encadrement, la démonstration des acquis de la recherche au Centre et sur le terrain, la publication sous forme simple des résultats de recherche.

Préparé par M. Diédhiou, Projet Fruitier, Singher

Pomme de terre-Jaxatu	G. Van den Plas/A. Seck
Manioc-oignon	G. Delanoy
Piment-patate douce	A. Ba Diallo
Tomate-gombo	A. Thiam/T. Ba
Protection phytosanitaire:	
Pathologie - nématologie (virologie)	L. De Maeyer/A. Mbaye
Entomologie	V. Coly
Comercialisation	P. Seck
Prévulgarisation et formation	J. Beniest*/O. Seck
Sous-station Ndiol (fleuve)	R. Verdonck

2 Cambérène

2.1 Direction des Parcs et Jardins
altitude : 10 m
pluviométrie: 518 mm
hygrométrie : 70 à 85%

B.P. 5.109 - Dakar
Région du Cap Vert
Dir.: M. Danso

3 Singher

3.1 Station de Singher
pluviométrie : 1085 mm
ensoleillement: 28°C-32°C
hygrométrie : 70%

B.P. 132, Ziguinchor
Dir.: M. Diédhiou

Projet fruitier (bananier, ananas)

55° 30′ W
4° 30′ N
1
25 km

1 Grand' Anse

Survey

The Seychelles with a population of 65000 lies 4°-8°S of the Equator and has a tropical climate which is warm wet and humid during the N.W. Monsoon (28 - 31°C from October to March) and rather dry and cool in the S.E. Monsoon (24 - 28°C from April to September). The yearly average rainfall is 2000 mm which consists of light showers during the South East and very heavy downpours during the North West Monsoon. The Seychelles is made up of 106 islands, 52 being granitic and the rest coralline. Horticulture is carried out mostly on the largest granitic island Mahe of 145 km^2 in an area whose highest elevation is nearly 950 m above sea level and 400 m. Being granitic, it is relatively poor of major nutrients and trace elements. It consists mainly of three main soil types - calcareous (derived from coral), colluvial and red earth (lateritic). Research on Horticultural crops at the Grand'Anse Experimental Centre started in 1964. The Station is directly involved in experimentation whereby more emphasis is attached to vegetables, tropical fruits and root crops growing. It must be stressed however, that the policy of the Centre is that of an applied research set-up.
One of the objectives at the Station is to find out by means of continuous trials the cultivars suitable to our tropical conditions and especially off-season varieties, with good yield potential, quality and resistance to pests and diseases. The introduction, trials and evaluation of new cultivars form a major part of our work. The other section of our research work deals with manurial trials and cultural techniques.
Fourteen years of innovative work by the research staff of the Grand'Anse Experimental Centre has resulted in a packag of technical and practical recommendations which have played a major role in the increased production of vegetables, and promotion of root crops and tropical fruits.

The new Soil and Plant Diagnostic Laboratory, opened in July 1983, has provided an added boost towards improving the facilities vital to providing better services to all agricultural producers in Seychelles. The Laboratory is sufficiently equipped to enable extensive analysis of soil and plant tissue, pest and disease problems associated with the wide range of soil and crop types cultivated. Presently a redevelopment project is being implemented to modernise the research facilities. As a result of the improvement, the Centre will be better equipped to provide a wide range of services to all producers.

1 Grand'Anse

1.1 Grand'Anse Experimental Centre
elevation: 25 m (Main Station)
400 m (Annex Station)
rainfall : 2000 mm
sunshine : 87.7 h
r.h. : 80%

P.O. Box 166, Grand'Anse, Mahe
Seychelles
Phone: 78252, 78428, 78429
Telex: 2312 MINDEV SZ
Dir.: C.S. Adam*

Continuous testing of improved fruit, root crop and vegetable varieties.

Prepared by C.S. Adam, Grand'Anse, Mahe

Multiplication of fruit, and vegetable and root crop varieties for distribution.
Screening of improved varieties with resistance to diseases and insects.
Evaluation of cultural practices and production methods applicable to Seychelles conditions.
Testing of improved irrigation systems suitable to small farmers and larger producers.
Providing up-to-date information to farmers.
Providing advice and information as needed to other national organizations including Coopera-ives, Pilot farms, parastatals, Seychelles Agricultural Development Company, Island Development Company, National Agro Industries and Marketing Board.
Strengthening the close contacts established with National Youth Service and Seychelles Polytechnic.

Horticulture and Promotion	Head: C.S. Adam*
Soil Science	A.M. Moustache
Plant Pathology	Y. Johnson

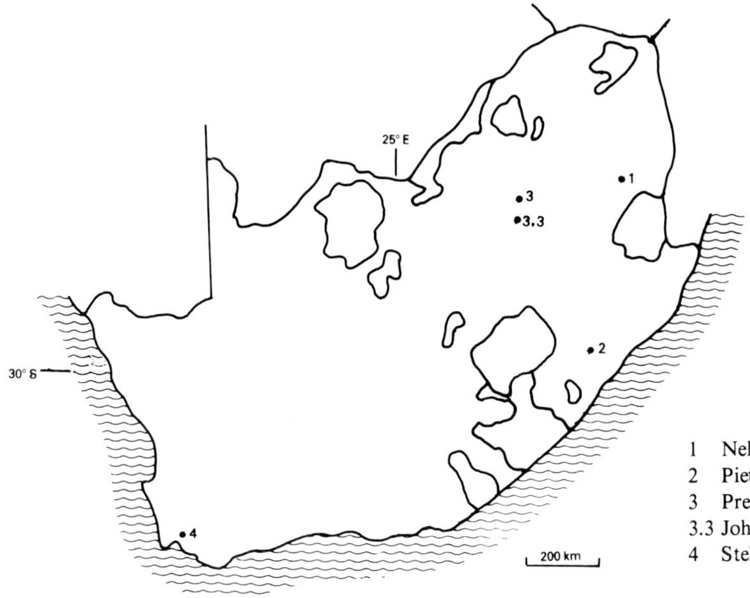
25° E
30° S
1
2
3
3.3
4
200 km

1 Nelspruit
2 Pietermaritzburg
3 Pretoria
3.3 Johannesburg
4 Stellenbosch

Survey

Organized horticultural research in South Africa is of relatively recent origin, dating from approximately 1928. The first horticultural research institutions of any significance to be established in South Africa were the Citrus and Subtropical Horticultural Research Station (now named the Citrus and Subtropical Fruit Research Institute) circa 1928, and the Western Province Fruit Research Station at Stellenbosch (now named the Fruit and Fruit Technology Research Institute) circa 1937.
Since then the Horticultural Research Institute at Elsenburg near Stellenbosch, has also been established. Horticultural activities are carried out in South Africa under a very large diversity of climatic conditions, on widely different soil types and with a very large variety of horticultural crops. Consequently numerous practical problems, requiring urgent attention, have been encountered. Although in horticultural research work, applied research so far has been mostly emphasized, nevertheless many practical problems still remain to be solved. Horticultural research in South Africa mainly concern the following: deciduous fruit and table and wine grapes, suptropical fruits including citrus, vegetables and ornamental horticulture. Horticultural research on a national basis is at present carried out mainly at four research institutes, viz.:
1. Citrus and Subtropical Fruit Research Institute at Nelspruit researching citrus and subtropical crops.
2. Horticultural Research Institute near Pretoria, working mainly on vegetables, potatoes, floriculture and ornamentals.
3. Fruit and Fruit Technology Research Institute at Stellenbosch, working mainly on deciduous fruit under summer and winter-rainfall conditions, on packing, transport and storage and fruit technology problems.
4. Viticultural and Oenological Research Institute at Stellenbosch.
In addition horticultural research is being carried out by agricultural faculties of Universities. All agricultural including horticultural, research work is co-ordinated by the Management: Agriculture and Water Supply which has its offices in Pretoria.
The Management consist of five full-time Chief Directors, one of which represents horticulture. The chairman of the Management is directly responsible to the Head of the Department. The initiative for new horticultural research work comes to a large extent from the research institutions but Advisory Committees on which growers are represented can also initiate research projects. Records are kept of all agricultural research projects in progress. Annual progress reports are submitted to the responsible Chief Director on all registered research projects. Final summaries of reports on completed research projects are published annually in "Agricultural Research" an official publication of the Department of Agriculture and Water Supply.
Research work on deciduous fruit table grapes and

Prepared by Dr.J.J.du Toit, Dept. of Agriculture and Water Supply, Pretoria

wine grapes and on citrus and subtropical fruits is mainly concerned with problems of immediate practical importance. These include cultivar and rootstock studies, breeding and selection of new improved cultivars, propagation methods, cultural practices including soil cultivation, fertilization, irrigation, pruning, packing, transport and storage problems and disease and pest control.
Research work on vegetables involves mainly cultivar tests and the breeding of new, improved cultivars and hybrids, some cultural practices, water requirements, and fertilizer experiments. Problems dealing with disease and pest control are also being investigated. Research work on ornamental horticulture concerns mainly the collection and study of ornamental trees and shrubs, lawn grasses and rose varieties, breeding work with indigenous flowering plants with the object to develop suitable cut flower cultivars and the production, storage, etc. of flower bulbs. The inspection and certification of horticultural and other agricultural seeds produced under the Government's Seed Certification Scheme, and the inspection of nursery plants and general plant inspection is the responsibility of the Division of Plant and Seed Control. Other phytosanitary services regarding the importation and exportation of plant propagating material is jointly rendered by the Plant Protection Research Institute and the Directorate of Plant and Seed Control.

Through the joint efforts of the Agricultural Departments' four horticultural research institutes, the Plant Protection Research Institute, the Directorate of Plant and Seed Control, the South African Plant Improvement Organization (SAPO), which was constituted by the three South African fruit boards and the South African Nurseryman's Association, the stage has now been reached where the South African fruit industry can be provided with elite propagating material known as South African Super Plants. These plants will be free of all known viruses and other diseases and pests. Furthermore they will be genetically the very best cultivars of clones presently available. Control over marketable quality and processing of horticultural products fall under the Directorate of Agricultural Product Standards of the Department of Agricultural Economics and Marketing.

1. Nelspruit

1.1 Citrus and Subtropical Fruit Research Institute
elevation: 716 m
rainfall : 820 mm
sunshine : 7,2 h
r.h.: 45-76%

Private Bag X11208
Nelspruit 1200
Phone: 01311-24241
Dir.: Dr.J.H. Terblanche
Deputy: A. van Oostrum

Research into the most effective nursery and cultural practices, cultivar improvement, research into plant nutrition, irrigation and biochemical problems, leaf and soil analysis, plant pathological, research, microbiological and entomological research (with the emphasis on biological studies and the long-term biological control of pests), research into post-harvest handling and storage of subtropical crops, research into the effects of weather on crop production, meristem, tissue culture and advanced biotechnical studies. The 24 crops that are being researched, include citrus, banana, avocado, pineapple, mango, litchi, guava, papaya, pecan and macadamia nuts, ginger, coffee and tea as well as spices.

Horticulture (citrus)	P.J. Müller
Horticulture (subtropical crops)	A.J. Joubert (Asst. Dir.)
Plant nutrition and irrigation	Dr.S.F. du Plessis (Asst. Dir.)
Plant pathology and virology	Dr.J.N. Moll (Asst. Dir.)
Entomology, nematology and acarology	Dr.M.A. van den Berg (Asst. Dir.)
Meteorology	N.B. Human

2. Pietermaritzburg

2.1 Faculty of Agriculture, University of Natal
elevation: 742 m
rainfall : 834 mm
sunshine : 6,6 h
r.h. : 53 - 79%

P.O. Box 375
Pietermaritzburg 3200
Phone: 0331-63320

Academic education and research	Prof.P. Allan*
	Prof.B.N. Wolstenholme

(Horticultural Science: Ecophysiology of tropical and subtropical fruit) — Prof.F.H.J. Rijkenberg

3. Pretoria

3.1 Horticultural Research Institute
elevation: 1400 m
rainfall : 705 mm
sunshine : 8,9 h
r.h. : 39 - 69%

Private Bag X293
Pretoria 0001
Phone: 012-821647
Dir.: Dr.J.T. Meynhardt*
Deputy Dir.: Dr.W.P. Burger

Research on all aspects of the production of vegetables, potatoes and flowers for the whole of South Africa. Thus including breeding and selection of suitable cultivars for the various production areas, nutrition and cultural practices, diagnoses and control of diseases, including virus diseases, control of pests, weed control, water requirements and climatic adaptation of crops, shrubs and ornamental gardening. Meristem, tissue culture and other advanced biotechnical methods are explored.

Vegetable crops production breeding etc.	A.F. Coertze/J.J.B. van Zÿl
Floriculture-Ornamentals	Dr.D.I. Ferreira*/G.J. Brits
Plant Pathology	Dr.B.H. Boelema/B.W. Young
Entomology	Dr.C.C. Daiber
Virology	G.J. Thompson (Miss)
Biotechnology	J.G Jansen van Rensburg (Miss)
Potato Research (leader)	F.I. du Plooy
Breeding	A.F. Visser
Production	B.J. Pieterse

3.2 Faculty of Agriculture, University of Pretoria
elevation: 1400 m
rainfall : 705 mm
sunshine : 8,9 h
r.h. : 39 - 69%

University of Pretoria
Pretoria 0002
Phone: 012-436381

Academic education and research. Horticulture: citrus and subtropical fruit — Prof.L.C. Holtzhausen, Prof.P.J. Robertse

3.3 Randse Afrikaanse University (RAU)

P.O. Box 524
Johannesburg 2000

Academic education and research. Dept. of Biochemistry: Physiological abnormalities in avocado — Prof.J.C. Schabort

3.4 University of the Witwatersrand
elevation: 1759 m
rainfall : 843 mm
sunshine : 8,7 h
r.h. : 41 - 70%

P.O. Box 1176
Johannesburg 2000
Phone: 011-7163189

Academic education and research. Dept. of Microbiology: Research into the isolation, taxonomy, morphology and in vitro control of the greening organism in citrus. — Prof. Helen M. Garnet

4. Stellenbosch

4.1 Fruit and Fruit Technology Research Institute
elevation: 257 m
rainfall : 733 mm

Private Bag X5013
Stellenbosch 7600
Phone: 02231-2001

sunshine : 8,2 h
r.h. : 48 - 93%

Dir.: Dr.P.G. Marias
Deputy: Dr.J.H. Terblanche

Research on all aspects of deciduous fruit production, including cultural practices, evaluation and breeding of cultivars, tissue culture and biotechnical studies, nutrition, control of pests and diseases, cold storage, post-harvest physiology and the processing of fruit including biochemical, microbiological and technological studies. Determination of quality standards for fruit and fruit products.

Deciduous fruit propagation and production	O. Bergh/ P. van Huysteun C.W.J. Bester/ J.D. Stadler
Deciduous fruit plant improvement by breeding and selection.	J.H. Acker
Chemistry and soilscience	Dr.W.A.G. Gotzé/M. du Preez
Radio isotope techniques and applications.	J.H.M. Karsten
Biochemistry and plant physiology	Dr.HJ.J van Zyl/Dr.J. Steenkamp L. Gouws
Tissue culture	L.J. von Mollendorff
Entomology	Dr.P.L. Swart
Plant pathology	Dr.W.F.S. Schwabe Dr.H.J. du Plessis
Fruit processing	Dr.B.K. Nortje/C.F. Hansmann
Post-harvest fruit handling	Dr.G.J. Eksteen/Dr.J.C. Combrink

4.2 Viticultural and Oenological Research Institute
elevation: 257 m
rainfall 733 mm
sunshine : 8,2 h
r.h. : 48 - 93%

Private Bag X5026
Stellenbosch 7600
Phone: 02231-70110
Dir.: Dr.J. Deist
Deputy: C. Kok

Research on all aspects of the production of wine, table and dried grapes including cultural practices such as fertilization, soil cultivation, irrigation, trellising, pruning and mechanization. Studies on aspects of selection and breeding plant physiology and the control of pests and diseases are also carried out, as well as the evaluation of rootstock cultivars. In addition, wine chemistry, wine technology and wine microbiology receive attention, with emphasis on wine flavour compounds, phenolic substances in wine and the selection, evaluation and study of wine yeast strains. The main research centre is situated at Nietvoorbij, near Stellenbosch, with several experimental farms in various viticultural regions. Research is carried out by about 40 graduated researchers and 100 support personnel.

Viticulture	E. Archer
Table and Dried Grapes	D. Saayman
Plant Protection	Dr.C.A. de Klerk
Soil Science	J.L. van Zyl
Wine and Brandy Research	O.P.H. Augustyn

4.3 Faculty of Agriculture, University of Stellenbosch
Academic education and research

University of Stellenbosch
Stellenbosch 7600

4.3.1 Department of Horticultural Science Physiology of growth, deciduous and protea-species	Prof.D.K. Strydom*/Dr.S.P. Erasmus J. de V. Lötter*
4.3.2 Department of Viticulture. Propagation and breeding	Prof.C.J. Orffer P.G. Goussard
4.3.3 Department of Oenology. Flavour compounds and Malolactic fermentation	Prof.C.J. van Wyk T.J. van Rooyen
4.3.4 Department of Food Science	Prof.G. van Noort

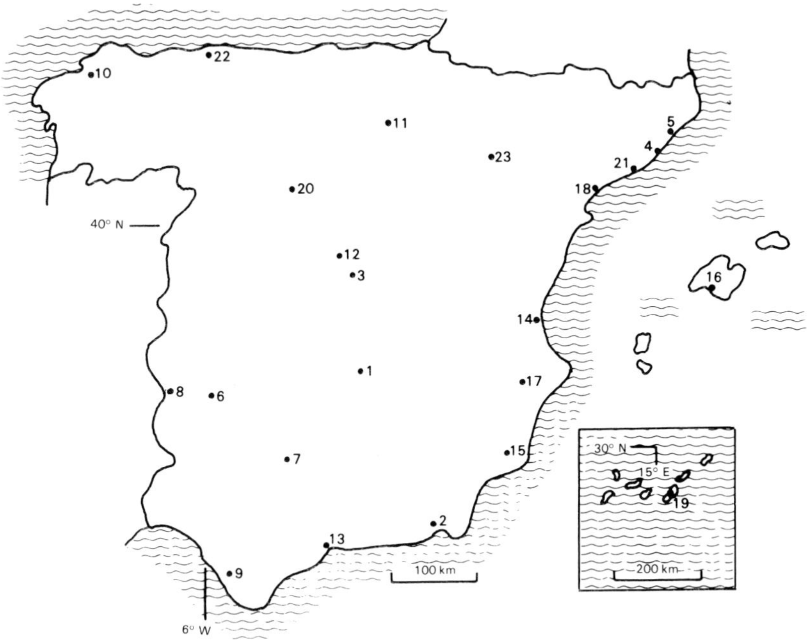
10
22
11
23
5
4
21
18
20
40° N
12
3
16
14
1
8
6
17
15
30° N
15° E
19
7
2
13
9
100 km
200 km
6° W

1 Alcázar de San Juan, 2 Almeria, 3 Aranjuez, 4 Barcelona, 5 Cabrils, 6 Almendralejo, 7 Córdoba, 8 Guadajira, 9 Jeréz de la Frontera, 10 La Coruña, 11 Logroño, 12 Madrid, 13 Málaga, 14 Moncada, 15 Murcia, 16 Palma de Mallorca, 17 Requena, 18 Reus, 19 Sta. Cruz de Tenerife, 20 Valladolid, 21 Vilafranca del Panadés, 22 Villaviciosa, 23 Zaragoza

Survey

Agricultural research in Spain recently underwent a far-reaching administrative change within the framework of the new autonomous stucture of the Kingdom of Spain. This particularly affected the National Institute for Agricultural Research (INIA).
This Institute is dependent on the Ministry of Agriculture, Fisheries and Food (MAPA), which transferred the Centres and Agricultural Research Departments to the autonomous regional Communities. Formerly they were directly dependent of INIA.
These centres and Research Departments are attributed to the Council of Agriculture in the autonomous regions in vue of the administrative function, and also research and experiments of regional character.
Yet, there will be a close bond with the Central Services of INIA, in case of big research projects, which will be brought together in the National Plan of Agricultural Research, represented by MAPA. It will be coordinated by the Services of Planning and Co-ordination, which will also be responsible for the continuation. All Centres and Agricultural Research Departments which are linked with the Central Services of INIA for all contacts and relations of international character are also linked with the Technical Board of Scientific Relations of INIA. The address of this Service of International Relation is José Abascal 56, 28003 - Madrid, where interested people can obtain more information.

The rest of the centres of research in agriculture, forestry and animal husbandry dependent of the High Council of Scientific Research of the Ministry of Education and Science, or of the Universities did not undergo any alterations.

Publications: A list is available by request directed to the chief of the Section of Publications, - José Abascal, 56, 28003-Madrid -Anales de la Estación Experimental de Aula Dei: is obtainable at the Estación Experimental de Aula Dei, Biblioteca, Apartado 202, Zaragoza.

Prepared by Dr.E. Prieto Heraud, INIA, Madrid

1 Alcázar de San Juan

1.1 Estación de Viticultura y Enología (Station of Viticulture and Oenology)

Problems of vine and wine in relation with the region La Mancha

. Alcázar de San Juan (Ciudad Real), Comunidad Autónoma Castilla-La Mancha
Phone: 926-540537

2 Almería

2.1 Centro de Investigación y Desarrollo Hortícola (Centre of Horticultural Research and Development)
elevation: 150 m
rainfall : 224.7 mm
sunshine : 3050 h

. Communidad Autónoma de Andalucía
Phone: 951-341514
Dr.Ing. Agr. F. Capdevila Gómez

3 Aranjuez

3.1 Estación de Horticultura (Station of Horticulture)

. Finca la Pavera, Carretera de Chinchón, km. 3, Aranjuez (Madrid)
Communidad Autónoma de Madrid
Phone: 91-8910088

General problems of horticulture, especially *Frangaria* and asparagus, physiology and genetics of horticultural plants

4 Barcelona

4.1 Unidad de Horticultura y frutos secos de los Servicios de Investigación Agraria de la Generalidad

. Consejeria de Agricultura Urgel, 187, Barcelona 08011
Ing.Agr.P. Arús

4.2 Instituto de Agrobiología de Cataluna (Agrobiological Institute of Cataluna)
Ministry of Science and Education
Edafology and applied physiology for horticultural plants

Faculty of Pharmacy, University of Barcelona, Barcelona - 08014
Phone: 93-3309061
Dr.J. Blanco

5 Cabrils

5.1 Servicios de Investigación Agraria (Agricultural Research Institute)
elevation: 85 m
rainfall : 600 mm
sunshine : 4.458 h

. Cabrils, Barcelona
Generalidad de Cataluña
Phone: 93-7532511

Department of Ornamental Plants
Problems of national floriculture, particularly studies on carnations, roses, bulbous plants
Physiology of ornamental plants improvement

Dr.Ing. de Montes
A. Nadal

6 Almendralejo

6.1 Estación de Viticultura y Enología (Station of Viticulture and Oenology)

Problems of vine and wine in relation with the Extremadura Region

. Carretera de Sevilla s/n
Comunidad Autónoma de Extremadura
Almendralejo (Badajoz)
Phone: 924-660532
Dr.Ing. Agr. P. Vidal

7 Córdoba

7.1 Centro Regional de Investigación y Desarrollo (Regional Centre of Agricultural Research and Development)
elevation: 110 m
rainfall : 652 mm
sunshine : 2.766 h

. Alameda del Obispo, Apartado 240
Córdoba
Phone: 957-293766
Telex: 76686-INIA-E

7.2 Department of Oliveculture

Comunidad Autónoma de Andalucía
Phone: 957-292288
Dr.Ing.Agr. J.M. Fernández

Studies on olive tree: agronomy, plant pathology, industrialization etc.

Dr.Ing. Agr. J. Humanes

8 Guadajira

8.1 Centro Regional de Investigaciones Agrarias (Regional Centre of Agricultural Research)
elevation: 237 m
rainfall : 437 mm
sunshine : 2.767 h

. Finca, "La Orden" Guadajira
(Badajoz)
Comunidad Autónoma de Extremadura
Phone: 924-440150
Telex: 28738-INIA-E

Problems of regional horticulture especially pome fruits and vegetables

Dr.Ing. Agr. M.M. Martin

9 Jerez de la Frontera

9.1 Department of Viticulture and Enologie
Estación Experimental Rancho "La Merced"
(Experimental Station Rancho "La Merced")
elevation: 55 m
rainfall : 629 mm
sunshine : 3.000 h

. Apartado 76 - Jerez de la Frontera
Comunidad Autónoma de Andalucía
Phone: 956-336570
Dr.Ing. Agr.A. García Gil de Bernabé

10 La Coruña

10.1 Centro Regional de Investigación y Desarrollo Agrario (Regional Centre of Agricultural Research and Development)
elevation: 97 m
rainfall : 986 mm
sunshine : 1.600 h

. Apartado 10 - La Coruña
Comunidad Autónoma de Galicia
Phone: 981-673000
Dr.Ing. Agr. J. Zea

11 Logroño

11.1 Estación de Rioja-Navarra* (Research Station of Horticulture, Viticulture and Oenology)
elevation: 360 m
rainfall : 361 mm
sunshine : 1.956 h

. Logroño 26080
Apartado de Correos 1056
Comunidad Autónoma de La Rioja
Phone: 948-645126
Dr.Ing. Agr. J. Provedo

General problems on vegetables. Selection and improvement techniques

Department of Viticulture and Enology

Head: Dr.J. Provedo

Sanitary selection of vines — J. Provedo
Aroma and polyphenolic compounds in grapes and wines — T. López
Variety determination of local vines — F. Martínez-Zaporta
Morphology and physiology of flowering and fruit set of vines — A. Blasco
Department of Horticulture (Vegetables) — Head: Isabel Trigo
Plant breeding. Pepper and artichoke — Ana Simón
Isoenzymatic determination of asparagus cultivars
Direct sowing of vegetables. Seed and seedling establishment — Luisa Suso
Soil/plant water relations. Soil conditions and seedlings emergence — A. Pardo
Cultural practices in *Cucurbitaceas* crops — F. Jorge/J. Arbeloa
Microbiology of fermented vegetables
Field experiments. Cereals and leguminous crops — A. López-Para

11.2 Estación de Viticultura y Enología de Haro (Station of Viticulture and Oenology of Haro) — Haro (Logroño)

Problems of vine and wine in La Rioja — Dr.Ing. Agr. A.J. Baró

12 Madrid

12.1 Instituto Nacional de Investigaciones Agrarias Servicios Centrales (National Insitute of Agricultural Research)
elevation: 595 m
rainfall : 438 mm
sunshine : 2.723 h

C/ José Abascal, 56, 28003-Madrid
Phone: 91-4420109
Telex: 48989-INIA-E
Dr.Ing. Agr. A. Gimeno

General problems of national agriculture, central services, coordination and control of work plans international relationships

12.2 Departamentos centrales de Investigación básica y aplicada (Central Research Departments of Basic and Applied Horticulture)
Regional Horticultural problems, particularly fruits, small fruits, vegetables and ornamental plants. Plant protection

Avda. Puerta de Hierro,
Carretera de La Coruña
Km. 7, 28040-Madrid
Phone: 91-2078240
Telex: 47308-INIA-E
Dr.Ing. Agr. F. Lázaro

12.3 Departamento Nacional de Viticultura y Enologia (National Department of Viticulture and Enologie)

Finca "El Encin", Km. 38.200
Carretera Madrid-Barcelona
Alcalá de Henares, Madrid
Comunidad Autónoma de Madrid
Phone: 91-8860302

General studies on the problems of vine and wine — Dr.Ing. Agr. L. Hidalgo

13 Málaga

13.1 Centro Experimental Económico Agrario (Experimental Centre for Agronomical Economics)
elevation: 20 m
rainfall : 460 mm
sunshine : 3.000 h

"La Mayora" - El Algarrobo Málaga

Superior Council of Scientific Research, Ministry of — Dr. A. Aguilar

Science and Education
Technical and economical problems of horticulture in the subtropical region

13.2 Departamento de Hortofruticultura y Cultivos Tropicales (Department of Horticulture and Tropical Fruit)
Problems of sub-tropical horticulture, tropical fruits, ornamental plants, early vegetables

Málaga
Comunidad Autónoma de Andalucía

Cortijo de la Cruz - Churriana, Málaga
Phone: 952-511000
Dr.Ing. Agr. E. Grana

14 Moncada

14.1 Instituto Valenciano de Investigaciones Agrarias (Agricultural Research Institute of Valencia)
elevation: 62 m
rainfall : 382.5 mm
sunshine : 2.412 h

Moncada, Carretera de Moncada, Km. 5 (Valencia)
Comunidad Autónoma Valenciana
Phone: 96-1391000
Dir.: Dr.Ing. Agr. L. Navarro Lucas

Department of Ecology
Soils and climate, fertilizers, irrigation systems, weed control, soil-water relationships
Head: Dr.D.G. de Barreda

Department of Plant Protection
Entomology, biological control, plant pathology, disease prospecting, bacteriology, virology, serology
Head: Dr.F. Martí-Fabregat

Department of Citriculture
Citrus cultural methods, stock-scion behaviour, varieties selection, rootstocks, virus-free citrus plants, in vitro micrografting, nucellar plants cultured in vitro (nucelle and ovule), virus indexing, citrus cultivars characterization, mutants and hybrids
Head: Dr.E.P. Millo

Department of Processing Technology
Cold and modified atmosphere storage; citrus degreening, post-harvest physiology and diseases
Head: Dr.A. Albert

Department of Development
Citrus marketing, economics, statistics, biometry, domestic citrus market, citrus maturity process, vegetable breeding
Head: P. Caballero

Horticultural Group
Head: Dr.V. Castell

15 Murcia

15.1 Servicio de Investigaciones Agrarias (Agricultural Research Service)
elevation: 60 m
rainfall : 255 mm
sunshine : 2900 h

Comunidad Autónoma de Murcia
Apartado Oficial - La Alberca, Murcia 30071
Phone: 968-840150
Dir.: Dr.Ing. Agr. J. Costa

Department of Horticulture and Fruits
Vegetable culture under glass, pepper breeding for paprika production, rootstock varieties for fruit trees, introduction of avocado and pistacchio culture, sugar cane trials, farm production increase, cotton and silk, lemons
Head: Dr.P. Florián

15.2 Centro de Edafología y Biología Aplicada del

Universidad de Murcia

Segura (Centre of Edafology and Applied Biology of Segura)

Phone: 968-239050

Superior Council of Scientific Research. Ministry of Science and Education
Physiology and edafology of horticultural plants

16 Palma de Mallorca

16.1 Granja Experimental (Experimental Farm)

Comunidad Autónoma Balear
Palma de Mallorca

General problems of horticulture on the Balear Islands

17 Requena

17.1 Estación de Viticultura y Enología (Station of Viticulture and Oenology)
elevation: 692 m
rainfall : 430 mm
Regional studies on vine and wine

Comunidad Autónoma de Valencia
Pl. Garcia Morato, 1, Requena (Valencia)
Phone: 96-2301415
Dr.Ing. Agr. M. Haba y Jarque

18 Reus

18.1 Estación de Viticultura y Enología (Station of Viticulture and Oenology)

Problems of regional oenology and viticulture

P° de Nuñer s/n, Reus (Tarragona)
Generalidad de Cataluña
Phone: 977-310450

19 Santa Cruz de Tenerife

19.1 Instituto Canario de Investigaciones Agrarias (Agricultural Research Institute of the Canary Islands)
elevation: 295 m
rainfall : 462 mm
sunshine : 2.035 h

Apartado 60 - La Laguna, Canarias
Comunidad Autónoma de Canarias
Phone: 922-540150
Telex: 92069-INIA-E

Regional horticulture of the Canary Islands, especially banana, tomato, potato, avocado, and tropical fruit, melon and other vegetables, ornamental plants

Dr.Ing. Agr. A. Arroyo

19.2 Jardín de Aclimatacion de la Orotava (Garden of Acclimatization of la Orotava)

La Orotava (Santa Cruz de Tenerife

Studies of the acclimatization of the species

Dr.Ing. Agr. A. Arroyo

20 Valladolid

20.1 Servicio de Investigación Agraria (Agricultural Research Service)
elevation: 700 m
rainfall : 450 mm
sunshine : 2.670 h

Comunidad Autónoma de Castilla-Léon, Finca Zamadueñas, Valladolid 47080, Apartado 733
Phone: 983-333166
Telex: 26566-INIA-E

Study of regional horticulture

Dr.Ing. Agr. J.L. Montoya

21 Vilafranca des Panadés

21.1 Estación de Viticultura y Enología (Station of Viticulture and Oenology

Problems of vine and wine of the Cataluna región

. Amalia 21, Vilafranca des Panadés (Barcelona)
Comunidad Autónoma de Cataluña
Phone: 93-8900211

22 Villaviciosa

22.1 Centro de Experimentación Agraria (Agricultural Experiment Centre)
rainfall : 98.6 mm
sunshine: 138.5 h

Studies on pome and stone fruits, nursery technique, industrialization

. Villaviciosa (Asturias)
Comunidad Autónoma de Asturias

Dr.Ing. Agr. P. Castro Alonso

23 Zaragoza

23.1 Servicio de Investigación Agraria* (Agricultural Research Service)
elevation: 225 m
rainfall : 413.5 mm
sunshine : 2.475 h

. Carretera de Montañana, 117
Apartado 202 - Campus "Aula Dei", Zaragoza
Comunidad Autónoma de Aragón
Phone: 976-709311
Telex: 58194-IDAE-E
Dir.: Dr.Ing. Agr. J. Alibés Rovira

Department of Fruit Crops
Studies on fruit and vegetables, specially deciduous fruit. Prunus and Pyrus

Head: Dr.Ing. Agr. A.F. Mansergas

23.2 Estación Experimental de "Aula Dei" (Experimental Station of Aula Dei)

Apartado 202, Zaragoza
Phone: 976-709511

Superior Council of Scientific Research, Ministry of Science and Educations
Studies on pome and stone fruits. Selection and improvement, technique of cultivation and propagation
Selection and breeding of poliploids in vegetables

Dr.Ing. Agr. L. Heras

23.3 Instituto Agronómico Mediterráneo de Zaragoza (International Centre of High Mediterranean Study (CIHEAM)

Apartado 202, Zaragoza
Phone: 976-709311
Telex: 58672-IAMZ-E

Post-graduate courses in vegetable and fruit culture

Dr.Ing. Agr. M. Mut

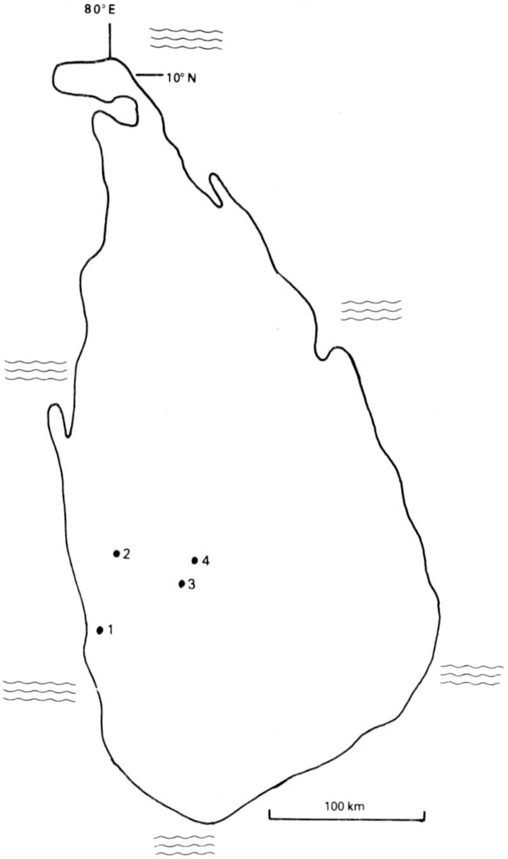
80°E
10°N
2
4
3
1
100 km

1 Colombo
2 Lunuwila
3 Peradeniya
4 Kandy

Survey

Agricultural research is done mainly by the Government of Sri Lanka. The Government's Department of Agriculture carries out research on fruit, flowers and vegetables. Most of the work is done by several research officers in nine Regional Research Centres and seven sub-stations spread throughout the country. Each of these Regional Centres are under a Regional Deputy Director Agriculture (Research).

All of whom are under the Deputy Director Agriculture (Research) based at Peradeniya. The other part of the work is undertaken by the Division of Horticulture in Peradeniya, which is manned by a Deputy Director Agriculture and by two additional Deputy Director Agriculture. The Royal Botanic Gardens, Peradeniya and the Botanic Gardens at Hakgala and Gampaha is administered by the Superintendent, Royal Botanic Gardens, Peradeniya. Research on plantation crops such as tea, rubber and coconut are carried out under the Director of the Research Institutes for these crops which are situated at Talawakele, Agalawatta and Lunuwila respectively.
Research and development of spice crops are carried out under the Director of Minor Export Crops, stationed at Peradeniya, with the research station based at Matale.
The Ceylon Institute of Scientific and Industrial Research (C.I.S.I.R.) is another Government institution which does research in utilization of fruits and vegetables and also in in vitro culture of papaya.
The Agricultural Faculty as well as the Post Graduate Institute of Agriculture of the University of Peradeniya carries out some research on horticultural and plantation crops and their products.

1 Colombo

1.1 Ceylon Institute of Scientific and Industrial 363 Bauddhaloka Mawatha,

Prepared by M.E.R. Pinto, Division of Horticulture, Peradeniya

Research	Colombo 7 Phone: 01-93807

Utilization of fruits, vegetables and spices

Post harvest biology, processing of fruits and vegetables and tissue culture	Head: Dr. K.G. Gunatileke
Natural products, spice processing	Head: Dr. U. Senanayake
Fruit and vegetable processing	R.G. Cutis (Mrs)
Post harvest biology of fruits and vegetables	Dr. S.W. Wijeratnam

2 Lunuwila

2.1 Coconut Research Institute (Quasi Government Organization) elevation : 30 m rainfall : 1800 mm r.h. : 70-80% sunshine : 2400 h	Bandirippuwa Estate, Lunuwila, N.W. Province Phone: 030-3795 Dir.: Dr. D.T. Wettasinghe Dep.Dir. (Research): Dr. R. Mahindapala

The coconut Research Institute, established in 1929, is presently managed by the coconut Research Board and is financed by a grant from the Government. The laboratories of the Coconut Research Institute are located at Bandirippuwa Estate, Lunuwila and are well equipped to conduct research in the relevant disciplines. The Institute has six outlying stations and three seed gardens.

The primary function of the institute is research and development in the growth and the cultivation of the coconut palm, coconut based cropping systems and the processing of coconut products and by-products. Although the Coconut Research Institute is not directly responsible for extension activities, it advises the extension personnel and other related authorities regarding developments in research and train their staff to enable them to transfer the new technologies to the appropriate end-users.

The institute acts as a centre for the collation and dissemination of technical information on coconut to research workers and others engaged in the coconut industry in Sri Lanka and abroad.

The institute produces nearly all the advisory literature used in extension activities and several other technical as well as non-technical publications. The institute also provides high quality seed nuts to the industry.

Genetics and Plant Breeding Division:	
Plant breeding, hybridization and selection, pollen studies, drought tolerance	Dr. M.R.T. Wickramaratne
Drought tolerance, seed selection	R.R.A. Peiris/U. Fernando
Soils and Plant Nutrition:	
Foliar analysis	M. Jeganathan
Soil physics	K.S. Jayasekera
Analytical chemistry	L.L.W. Somasiri
Fertilizers	M.B.M.N. Dias (Miss)
Soil microbiology	A. Tennakoon
Agronomy Division:	
Soil moisture conservation, pasture planting densities	L.V.K. Liyanage (Mrs)
Agricultural economics	M.A. Tilakasiri
Cropping systems	H.A.J. Gunatilake/M. de S. Liyanage
Crop ecology	D.N.S. Fernando
Soil moisture conservation	T.G.L.G. Gunasekera, H.P.S. Jayesundera
Coconut Processing Research Division:	
Fibre, energy	P.A.N. Ratnayake
Crop Protection Division:	
Insect pathology, pesticide residues	Dr. P. Kanagaratnam
Insect ecology, biological control	P.A.C.R. Perera
Insect ecology, pesticides	L.C.P. de Silva/C. Dissanayake (Mrs)

Tissue Culture Unit:	
Plant tissue culture	S.M. Karunaratne (Mrs)
Biometry Unit:	
Design of experiments, crop weather relationships, statistics, biometry	D.T. Mathes, T.S.G. Peiris
Plant Physiology Unit:	
Plant physiology	C.W. Jayesekera (Mrs)

3 Peradeniya

3.1 Department of Agriculture (Government Department)	Peradeniya Phone: 08-88331 Dir.: Dr. S.D.I. Gunawardene Dep.Dir.: Dr. H.M.E. Herath
3.2 Central Agricultural Research Institute elevation : 520 m rainfall : 2500 mm	Gannoruwa Phone: 08-88011 Reg.Dep.Dir.(Research): Dr. M.H.J.P. Fernando
Wet zone fruit, vegetables and yams	
Vegetable breeding and agronomy	I.S. Padmasiri (Mrs)
Fruit breeding and agronomy	H.M.S. Heenkenda
Vegetable breeding	K.D.A. Perera (Miss)/I. Medagoda (Mrs)
Fruit agronomy	S. Radhakrishna
Food technology	D.B.T. Wijeratne
Yam and tuber crop research	S.D.G. Jayawardene/K.P.U.D. Silva
In-vitro culture propagation	K.K.S. Fernando (Mrs)
Bio-techniques	H.M. Mendis
3.3 Divison of Horticulture	Gatambe
Dep. Dir. Agriculture (Horticulture)	Dr. H.M.E. Herath
Add. Dep. Dir. Agriculture (Horticulture)	Dr. C.R. de Vaz
Add. Dep. Dir. Agriculture (Horticulture) and Project Coord., FAO Horticulture Project	M.E.R. Pinto
All Island fruit, soybean, roots and tuber crops and export oriented vegetables	
Fruit breeding	A.O.C.de Zoysa/Dr. V.A. Yoigaratnam
Soybean co-ordinator	C.D. Dharmasena
Soybean food processing	T.D.W. Siriwardena
Soybean breeding	V. Arulnandi
Soybean food processing	F. Hewavitharana (Mrs)
Soya microbiology	P. Thirukumaran/C. Thirukumaran (Mrs)
3.4 Royal Botanic Gardens	Peradeniya Phone: 08-88238 Supt.: B. Sumithraarachchi
Flowers, mainly anthuriams and orchids and systematic botany	
Systematic botanist	M. Jayasuriya
Tissue culture	W.M. Abeyratne
3.5 Angunakolapelessa, Regional Research Centre	Phone: Angunakolapelessa 804

elevation : 40 m rainfall : 900 mm	Reg.Dep.Dir. (Research): Dr. J. Handawala
Dry zone fruits and vegetables	
Vegetables and fruits breeding and agronomy Soybean agronomy	J.A. Sirisena P.B. Jayamana
3.6 Bandarawela, Regional Research Centre elevation : 1200 m rainfall : 1300 mm	Hope Estate Phone: 057-2520 Reg.Dep.Dir. (Research): Dr. S.L. Amarasiri
Sub tropical fruits and vegetables	
Fruit agronomy Vegetable breeding Vegetable agronomy	S.K. Tharmarajah S.D. Abeyratne (Mis) P. Alahakoon
3.7 Bombuwela, Regional Research Centre elevation : 15 m rainfall : 280 mm	Phone: 034-22673 Reg.Dep.Dir. (Research): G.A. Gunatileke
Wet zone fruits and vegetables	
Fruits breeding and agronomy Vegetable breeding and agronomy	E.M. Dissanayake (Mrs) L. Senanayake/K. Mirihagalla, G.S. Ratnapala (Miss)
Roots and tuber crop research	J.B.D.S. Kahandawala, S.N. Harischandra (Mrs)
3.8 Girandurukotte, Regional Research Centre Affiliated Station elevation : 120 m rainfall : 1500 mm	Reg.Dep.Dir.(Research): Dr. S.H. Upasena
Dry zone fruits and vegetables	
Vegetable breeding and agronomy Fruit agronomy and vegetables Roots and tuber crops research Roots and tuber crops and fruit research	A. Bentota (Miss) G.A.S.S. Jeevananda G.W.K. Weerasinghe S. Peiris
3.9 Kalpitiya, Agricultural Research Station elevation : 2 m rainfall : 700 mm	Res.Off.in.Ch: D.S.P. Kuruppuarachchi
Dry zone fruits, vegetables and tubers in regesol soil type	
3.10 Karadian Aru, Regional Research Centre elevation : 15 m rainfall : 1000 mm	Res.Off.in.Ch.: S. Kandasamy
Dry zone fruits, vegetables and yams	
3.11 Kilinochchi, Regional Research Centre elevation : 20 m rainfall : 1200 mm	Phone: Kilinochchi 327 (Act.) Reg.Dep.Dir. (Research): B.N. Emerson

Dry zone fruits and vegetables

Fruit breeding and agronomy — S. Sundaralingam (Miss)
Horticulture agronomy — V. Vellupillai (Mrs)

3.12 Maha Illuppallama, Regional Research Centre — Phone: 025-2050 Ext. 02
elevation : 120 m — Reg.Dep.Dir.(Research):
rainfall : 1500 mm — Dr. M. Sikurajapathy

Dry zone fruits, vegetables and soybean

Vegetable and fruit research — H. Samaratunga
Soybean breeding and agronomy — A. Pathirana (Mrs)
Onion agronomy — M.H.D. Fonseka

3.13 Makandura, Regional Research Centre — Gonawila, N.W. P.
elevation : 20 m — Reg.Dep.Dir. (Research):
rainfall : 1800 mm — Dr. H. Somapala

Intermediate zone fruits, vegetables and yams

Fruit breeding and agronomy — B.S. Jeyaratnam
Yams and vegetable research — W.L. Stanley

3.14 Moneragala, Agriculture Research Station — Res.Of.in.Ch.: W.M.S.M. Bandara
elevation : 30 m
rainfall : 1600 mm

Dry and intermediate zone fruits
Fruit research — E.R.U.W. Nilmalgoda

3.15 Rahangala, Agricultural Research Station — Phone: 057-5235
elevation : 330 m — Res.Of.in.Ch.: L.G. Herath
rainfall : 1700 mm

Sub-tropical fruits, vegetables and tubers

3.16 Sita Eliya, Nuwara Eliya, Agricultural Research Station — Phone: 052-2615
elevation : 2000 m — Res.Of.in.Ch.: K.G.W. Abetunge
rainfall : 2200 mm

Sub-tropical fruits, vegetables and tubers

Potato breeding and agronomy — B.S. Rafael
Fruit and vegetable research — G. Rafael (Mrs)
Potato research — S. Abetunge (Mrs)

3.17 Tirunelvely, Jaffna, Agricultural Research Station — Phone: 021-24177
elevation : 3 m — Res.Of.in.Ch.: T. Kugadasan
rainfall : 1100 mm

Dry zone fruits, vegetables and tubers

Potato research — K. Devasabai
Onion research — M. Thavabalachandran (Mrs)
Vegetable agronomy — S. Jayapathy

3.18 Vanathavillu, Agricultural Research Station
elevation : 3 m
rainfall : 1000 mm

Res.Of.in.Ch.: K.H. Sarananda

Dry zone fruits, vegetables and yams in latosol type of soil

4 Kandy

4.1 Department of Minor Export Crops
elevation : 500 m
rainfall : 2200 mm

Gatambe, Peradeniya
Phone: 08-88363
Dir.: Dr. S.T.W. Kirinde
Dep.Dir.(Research):
Dr. S. Kathirgamathyar
S.Ass.Dir.of Res.: P. Wickramasinghe

Research, extension and production of spice crops such as pepper, cinnamon, cardamon, nutmeg, cloves etc.

Breeding	Dr. P. Gurusinghe
Agronomy	P. Wickramasinghe
Entomology	H.A.S. Perera
Pathology	S. Kularatne
In-vitro culture	Y.M. Heenbanda
Post-harvest technology	M. Senanayake (Mrs)
Cropping patterns	W.H.E. Premaratne

SUDAN

SD

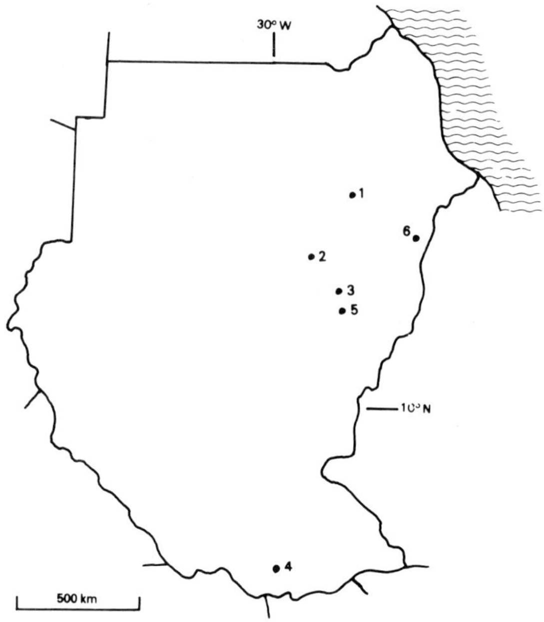
30° W
1
6
2
3
5
10° N
4
500 km

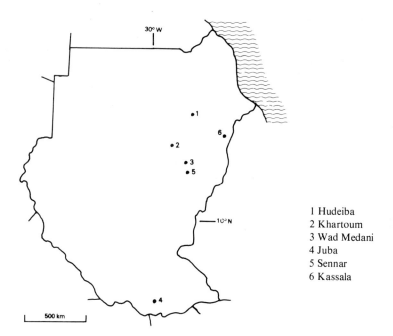

1 Hudeiba
2 Khartoum
3 Wad Medani
4 Juba
5 Sennar
6 Kassala

Survey

The Sudan is the largest country in Africa covering an area of one million square miles, extending approximately between 22°-38° East Longitude and 4°-23° North Latitude. The climatic conditions spectrum ranges from the hot dry desert type in the north to the humid tropics in the south. The rainfall is almost zero in the northern region amounting to 1500 mm/annum in the south. Also the soils vary very much from desert type (sand) in the north to heavy clays in the centre to the lateritic soils in the south. The irrigation water for horticultural crops production is mainly from the River Nile water and its several tributaries. Some people use the underground water.
These variations in the natural resources i.e. climatic, soils, irrigation water etc. avail good chances for the Sudan to produce a large number of horticultural crops, for example the most important fruits produced successfully in the Sudan are citrus spp. (sweet orange, grapefruit, mandarin, lemon and lime). The Sudan grapefruit especially "foster variety" is known to be one of the best all over the world. The Sudan also produces mango *(Mangifera indica)*, date palm *(Phoenix dacty lifera)*, banana *(Musa Sapientum)* and guava *(Psidium gua-java)*. Principal vegetables include onion *(Allium cepa)*, okra *(Abolimiscous esculentus)*, jewsmallow *(Corchorus olitorius)*, purslane *(Portulaca olevacea)*, tomato *(Lycopersicon esculentum)*, eggplant *(Solanum melongena)*, pepper *(Capsicum spp.)*, potato *(Solanum tuberosum)*, water melon *(Citrullus vulgaris)*, pumpkin *(Cucurbita sp.)*, muskmelon *(Cucumis melo)*, squash *(Cucurbita pepo)*, cucumber *(Cucumis vulgaris)*, sweet potato *(Ipomoea batatas)*, cassava *(Manihot utilissima)*, roquette *(Eruca sativa)* and radish *(Raphanus sativus)*. In addition to these many other temperatezone vegetables are successfully produced outdoor during the winter time. These include carrot, cabbage, cauliflower, table beet, peas and lettuce. The horticultural education is offered in the Departments of Horticulture in the Agricultural Faculties in Khartoum, Gezira and Juba Universities. Also some institutes offer courses for the diploma e.g. Shambat, Abu Naama and Abu Haraz institutes.
The University Departments together with the several research stations in Hudeiba, Shambat, Wad Medani, Sennar, Kassala and Kashm El Girba etc. contribute very much for the progress of research in the field of horticulture.
The experiments carried out mainly include, variety trials, sowing dates, population density, irrigation, pest and disease control in both vegetables and fruits. Meagre work is done on the ornamental plants, and the Sudan Horticultural Society is working hard to find a place under the sun for Sudan floriculture and therefore it is keen in holding the annual flower show that attracts a lot of people. With the available natural resources - previously mentioned - the Sudan should have been one of the leading African countries in the field of horticulture. But many problems have contributed to this and hampered the horticultural

Prepared by Prof. A.T. Abdel Hafeez, University of Khartoum, Sudan

development. Some of these problems are: Lack of certified seeds, inefficient means of transport, lack of capital, lack of spare parts, lack of technical know-how, pest and diseases, lack of storage facilities and absence of proper marketing system.
Inspite of all this it can be said that compared with the previous decade the horticultural education, research, understanding and progress have developed very much in Sudan.
Both Government and private sector people have realized the role played by the horticultural crops in the Sudan food and economy. The beginning of exporting some vegetables and fruits from the Sudan to some Arab, African and European countries is a step towards achieving the goal that says "The Sudan is the World Bread Basket".

1 Hudeiba

1.1 Hudeiba Experiment Station of the Research Division of the Ministry of Agriculture

Hudeiba

It is situated in the Northern Province for research in relation to the crops of the northern region. The section has established modern orchard planting to be used for research and demonstration and serve as a varietal collection. These include citrus, mango, date palm and grapevines. The horticultural research programme includes pomology, olericulture, plantation crops and vegetable and fruit post-harvest physiology.
Pomology: Research activities include 1. Survey for virus diseases of fruits as a basis for future virus work on citrus, 2. Introduction of virus-free budwood of commercial citrus, 3. Introduction of citrus rootstocks and establishment of citrus-rootstock nursery, 4. Variety trials.
Because of the general nature of the preliminary ground work it is not easy to indicate precisely responsibilities associated with each research worker.

2 Khartoum

2.1 Department of Horticulture, Faculty of Agriculture University of Khartoum

Shambat, Khartoum
North Sudan
Phone: 611171

Created in 1960 the Department of Horticulture is one of nine departments in the Faculty of Agriculture. Horticulture is one of eight options offered to the final-year students for specialization.

2.1.1 Food Research Centre
This is presently a joint project between FAO and the Horticultural Division established for processing research in fruits, vegetables and cereals, training of personnel, use of modern techniques and improvement of local traditional processing methods of local food.

Shambat, c/o UNDP, P.O. Box 913,
Khartoum
Project Manager: M.S. Joshi

3 Wad Medani

3.1 Agricultural Research Corporation

Dir. Gen.: Dr. O. Gameel

3.2 Gezira University

Wad Medani

4 Juba

4.1 Faculty of Agriculture and Natural Resources

Juba

5 Sennar

5.1 Horticultural Research Section

Sennar

6 Kassala

6.1 Horticultural Research Section

Kassala

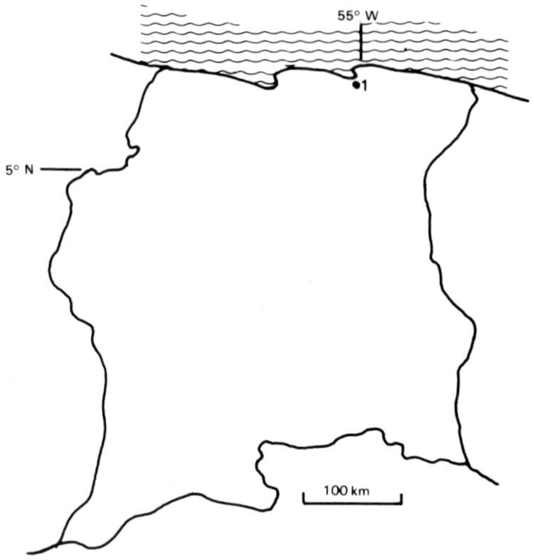
55° W
5° N
1
100 km

1 Paramaribo

1 Paramaribo

1.1 Landbouwproefstation Suriname
(Surinam Agricultural Experiment Station)
elevation: sea level
rainfall : 2500 mm
sunshine : 58% per month

Cultuurtuinlaan, P.O. Box 160
Paramaribo
Dir.: Ing. R.R. Huiswoud

Soil fertility	Vacant
Soil physics and agrohydrology	Vacant
Entomology	Mrs. A. van Sauers-Muller
Nematology	Mrs. M. Dipotaroeno-Sakrama
Virology	Ir. F.E. Klas
Mycology/ Bacteriology	Mrs.Drs. H. van de Lande
Weed control	Ing. S. Ausan
Citrus	Ing. R. Mansour
Musa and vegetables	Ing. F.W. Soerodimedjo
Grassland and animal nutrition	Ir. P. Kerkhoff
Cereals	Ing. S. Ausan
Pulses and pomology	Ing. F.W. Soerodimedjo
Coconut and oilpalm	Dr. V. Thomas Alexander
Miscellaneous crops	Vacant
Contacts with the Extension Service	Vacant
Plant protection and production	H.I. Te Vrede
Experimental farms	R. Santokhi

Prepared by Tj. F. Soekhai, Ministerie van Landbouw, Veeteelt en Visserij, Paramaribo

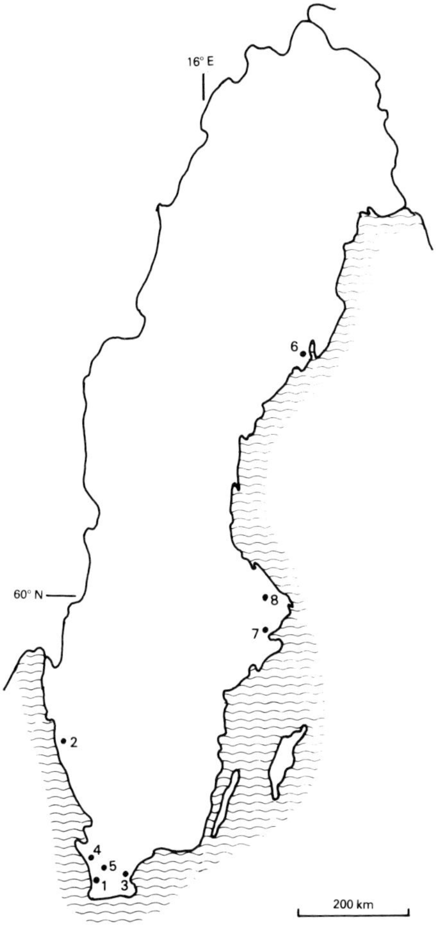
16° E
60° N
6
8
7
2
4
5
1
3
200 km
1 Alnarp
2 Göteborg
3 Hammenhög
4 Landskrona
5 Lund
6 Röbäcksdalen
7 Stockholm
8 Uppsala

The Swedish University of Agricultural Sciences was reorganized in 1977 and consists now of the Faculty of Agriculture (with Horticulture), and the Veterinary and Forestry Faculties. The center of the university is located at Ultuna, near Uppsala, where Agriculture, Veterinary and Forestry are situated. The College of Horticulture of the university is located at Alnarp, in the south of Sweden between Malmoe and Lund, being part of the Faculty of Agriculture and having a local administration at Alnarp.

The higher education in horticulture at the university consists of a five-year program. After high school there is one propeudeutic year at Alnarp, 2 years at Ultuna (basic subjects) and the final 2 years at Alnarp (applied subjects). 25 students are admitted to horticulture each year. For landscape architecture 30 students are admitted, also on a five-year programme with 2 years at Ultuna and 3 years at Alnarp. It is also possible to continue for a doctor's degree. The research work of the Faculty of Agriculture covers most aspects of agriculture. It is carried out in different departments. Usually the departments are built up around a professorship, with the professor as head of the department. Some of the departments are engaged in experimental work, having for this purpose one or more experimental divisions.

Applied research work on an experimental basis is carried out according to a general programme for the

Prepared by Prof. L. Ottosson, Swedish University of Agric. Sciences, Alnarp whole country. This programme is worked out in the divisions metioned above, which are in very close connection with practical agriculture and horticulture. At the Faculty of Agriculture there is a committee for experimental work which functions as a central authority for applied research. Finally, the programme is approved by the Board of the University.

The experimental work outside Alnarp is under the supervision of the "field-trial organization". The country is divided into three experimental districts for horticulture. In each district there are local experimental stations (Alnarp, Ultuna, Röbäcksdalen) with district experimental officers as leaders and coordinators of the work. Investigations are, however, not only carried out at these experimental stations but also at a number of local experimental fields. It should be mentioned that sometimes growers are cooperating as hosts for such experiments, which are, for instant, performed in their orchards or glasshouses.

1 Alnarp

1.1 The Swedish University of Agricultural Sciences*, College of Horticulture elevation: 10 m rainfall : 542 mm sunshine : 1912 h	S-230 53 Alnarp Phone: 040 - 41 50 00
1.1.1 Avdelningen för frukt- och bärodling (Department of Pomology)	Head: Prof. Dr. I. Fernqvist*
Division of research and teaching	Prof. Dr. I. Fernqvist*
Fruit tree physiology and pomology	
Domestication of *Vaccinium vitis-idaea* and *V. augustifolium*	Mrs. I. Hjalmarsson
Cytogenetical studies in *Ribes*	Prof. emer. F. Nilsson
Experimental division of fruit growing	Head: Dr. B. Bjurman*
Propagation, tree fruits	Dr. E. Goldschmidt*
Cultivar and rootstock experiments	
Crop management, pollination, pruning	
Fruit quality and storage	Dr. N.-A. Ericsson*
Crop management, irrigation, planting systems	
Propagation, small fruits	Dr. B. Bjurman*
Crop managements, strawberries	
Cultivar experiments, small fruits	Dr. K. Sakshaug
Crop managements, bush fruits	
Experimental division of fruit breeding	Bålsgard, Fjälkestadsvägen 123-1, S-291 94 Kristianstad Head: Dr. V. Trajkovski
Disease-resistance, methods	
Disease-resistance, Gleosporium	Mrs. K. Ralsgård
Breeding of stone fruits and rootstocks	Dr. V. Trajkovski
Breeding of apple and pears	P.-O. Bergendal
Breeding of strawberries	Mrs. K. Trajkovski
Breeding of bush fruits	Dr. V. Trajkovski/ B.Sjöstedt
Breeding of cowberries and lowbush blueberries	Dr. I. Fernqvist*
	Mrs. I. Hjalmarsson
1.1.2 Avdelningen för köksvaxtödling (Department of Vegetable Crops)	Head: Prof. Dr. L. Ottosson*
Division of research and teaching	Dr. L. Ottosson*
Maturation of vining peas	
Timing, prognostication and production of vegetable crops	
Cytogenetic studies in *Pisum*	Prof. emer. R. Lamm*
Post-harvest physiology of vegetables	Dr. T. Nillson*
Food value and chemical composition of vegetables	
Chemical analyses as a basis for quality and	P. Kristiansson

compositional studies in vegetable	
Irrigation and fertilization of greenhouse tomatoes	Mrs. M. Magnusson
Optimal climate of greenhouse tomatoes	A. Waldberg
Seed production of vegetable crops	Miss. B. Ottosson
Seed physiology and seed preparations	G. Granqvist*
Experimental division of vegetable crops	Head: Dr. S. Lindfors*
New varieties and species of vegetables for greenhouses and on open land	
Effects of environmental factors in greenhouse cucumbers	Miss. I. Hansson
Production systems in greenhouse in relation to energy consumption	Miss. I. Hannson G. Erlandsson
Pollination and setting in tomatoes	G. Erlandsson
Improved establishment of vegetables in the field	Dr. A. Wredin*
Irrigation and nutrient supply of field vegetables	
Official herbicid test	H. Åvall
Growing techniques and varieties - chinese cabbage, onions, cauliflower, lettuce	
Optimum harvest-period of vegetables in the open	
1.1.3 Avdelningen för prydnadsväxtodling (Department of Floriculture and Ornamental Techniques)	Head: Prof. Dr. T. Kristoffersen*
Division of research and teaching	Dr. T. Kristoffersen*
Flower-crop production, greenhouse climate, growth and development	
Photosynthesis and growth in relation to environ-factors in *Pelargonium* and *Croton*	Dr. O. Hellgren
Meristem culture of *Pelargonium* and *Cordyline*	Dr. T. Welander*
Tissue culture of *Senecio*	Mrs. U. Gertsson
Tissue culture of woody ornamentals	Dr. M. Welander*
Teaching	Miss. C. Alsved
Flowering in *Pelargonium*	Dr. T. Welander*
Experimental division of ornamental crops	Head: Dr. J. Johansson*
Experimental work with glasshouse and field crops	
Propagation and testing of collected woody ornamental plants	R. Bengtsson
Cultivation trials with ground covers	
Container growing of nursery plants	Mrs. R. Gajdos
Growth regulators for floricultural crops	Dr. R. Larsen*
Cultivation of ornamental foliage plants	
Protected cultivation of perennials	
Improved growing technique in order to decrease the production cost for pot plants	
Production program for *Begonia* x elatior	Mrs. E. Löfvenberg*
Production program for *Begonia* x lorraine	
New pot plants	H. Schüssler
Variety trials with lawn-grass	R. Svensson*
Testing of herbicides	
Trials with evergreen ornamental shrubs	
Taxus under different cultivation conditions	
Hardiness trials with woody ornamental bushes of various provenence	
Problems connected with the planting of woody nursery plants	
Weed control through covering with bark and other substrates	
Extensive turfgrass - plant material and management	

1.1.4 Avdelningen för trädgårdsekonomi (Department of Horticultural Economics)	Head: Prof. M. Carlsson*
Research and teaching in production economics and marketing	Miss. L. Ekelund*
Research: Methods for management and entrepreneurship (in connection with the horticultural consult bureau TEU)	
The use of computers in planning of horticultural enterprises	B. Håkansson
Competition and marketing conditions for horticultural products	Miss. L. Ekelund*
1.2 Institutionen för landskapsplanering (Department of Landscape Planning)	Head: Prof. O.R. Skage
Section of Landscape Architecture	Head: Vacant
	Prof. emer. P. Friberg
Teaching and research in landscape and urban environmental design	
Design principles for planning gardens, parks and landscapes, stressing human experience of visual qualities	G. Sorte
Functional aspects of the outdoor environment, stressing studies of human requirements for and exploitation of land and areas for leisure time and recreation	
Historical garden and park constructions in Sweden, stressing historical reference material and restoration problems	
Section of Plants and their Environment	Head: R. Gustavsson
Teaching and research in design, establishment and management of vegetation types for landscape, public and private green areas.	
Construction, change and functional suitability of the urban vegetation	R. Gustavsson
Structure and structural changes in rural vegetation, studies of structural qualities and types and their practical importance	
The plant material's establishment and development, stressing management models concerning construction, restoration and regeneration	
Section of landscape planning	Head: Prof. O.R. Skage
Teaching and research on policies in local open space planning and local and regional land use planning based on ecological principles.	
Physical planning and landscape development	
Physical planning - theory and method development	
Open space policy	
1.3 Södra Trädgårdsförsöksdistriktet (South Swedish Horticultural Experiment District)	Distr.Off.: Dr. A. Wredin*
Coordination of experiment work	
1.4 Institutionen för växt- och skogsskydd (Department of Plant and Forestry Protection)	
Experimental division of pest control	Head: Dr. H. Rosen
Biological control of greenhouse pests	Mrs. B. Nedstam
Pests of fieldgrown vegetables - forecasting under	C. Tornéus

control

Experimental division of nematology — Head: Dr. S. Andersson

Migratory plant-parasitic nematodes: Investigations in horticultural crops — Miss. A. Bank*

Experimental division of virus diseases — Head: Dr. B. Nilsson*

Diagnosis and investigation of virus diseases in ornamentals, fruits and vegetables — Mrs. G. Åman*/Mrs. E. Gripwall*

Virus elimination in totally infected cultivars of vegetatively propagated horticultural plants — Dr. M. Akius*

Experimental division of fungal and bacterial diseases — Head: Dr. B. Olofsson

Fungal diseases of horticultural plants — G. Svedelius*

Desinfection in greenhouses

Control of Red-core disease on strawberry, *Phytophthora fragariae*

Biological control of fungal diseases in greenhouses and fruit orchards

Control of Stem blight disease on cucumber, caused by *Didymella bryoniae*

Experimental division of plant resistance — Head: Dr. B. Leijerstam

Pests of horticultural crops: — Mrs. I. Gustafsson*

Resistance against lettuce downy mildew. Test and selection methods for non-specific resistance in lettuce

Research information centre/division of plant protection

Horticultural crops — Mrs. I. Åkesson*

2 Göteborg

2.1 Göteborgs Botaniska Trädgård
(Göteborg Botanical Garden)
elevation: 25-120 m
rainfall : 670 mm
sunshine : 1,600 h

Carl Skottsbergs Gata 22
S-413 19 Göteborg
Phone: 031 - 41 37 50/41 81 12
Dir.: Prof. G. Weimarck

Garden proper (20 hectares) with plantings for various horticultural and educational purposes, especially Rhododendron and E. Asiatic collections; extensive rock-garden; bulb-frames with mainly C. and SW. Asiatic material; hothouses and alpine houses with particularly rich collections of orchids, succulents, Begonia etc. Nature Park (20 hectares), nature reserve. Outer Area (135 hectares), nature reserve incl. Arboretum (15 hectares) with stands of American, European and (mainly) Asiatic trees and shrubs in semi-natural surroundings. Horticultural activity: mainly introduction of new material for gardens and parks in W. Sweden. Status: municipal.

Taxonomy of Liliaceae — Prof. G. Weimarck

Education, demonstrations etc — Dr. I. Nordin

Herbaceous plants — J. Persson

Woody plants — B. Aldén

Hothouses — M. Neuendorf

Park etc — K. Rundqvist

Administration — E. Hansson

2.2 Svenska Livsmedelsinstitutet (SIK)
(The Swedish Food Institute)

Box 5401, S-402 29 Göteborg
Phone: 031 - 40 01 20
Telex: 21651 SIK S
Dir.: Prof. N. Bengtsson

Preservation and storage of food

The Swedish Food Institute covers research and development, documentation and education on processing, preservation and storage of foods and food products, and related matters concerning food raw materials, packaging, distribution and consumption.

3 Hammenhög

3.1 Svalöf AB/Hammenhögs, Växtförädlings- och konsulentavdelning (Plant-Breeding and Advisory Department)

S-270 50 Hammenhög
Phone: 0414 - 404 00
Telex: 32272 HSEEDS S
Dir.: G. Åkesson

Plant-breeding of vegetables, turf-grass and ornamental flowers	
Section for vegetables	S. Leijon/ L. Jeppsson
Section for turf-grass	T. Månsson
Section for ornamental flowers	vacant
Section for resistance	Mrs. C. Larsson

4 Landskrona

4.1 Weibullsholms Växtförädlingsanstalt*
(Plant Breeding Institute Weibullsholm)
elevation: 5 m
rainfall : 650 mm
sunshine : 1611 h

S-261 24 Landskrona
Phone: 0418 - 78000
Telex: 72388
Dir.: G. Ewertson

The Institute is owned by a private enterprise, W. Weibull Ltd., but is given a yearly subsidy by the government; this covers 8-10% of the breeding costs. Plant-breeding of tomatoes and greenhouse cucumbers and plant husbandry of vegetables and ornamental flowers.
Laboratories: Cytological and pathological studies, chemistry, analyse services.

Breeding work on carrots, outdoor cucumber, cauliflower, onion, red beets, beans	Dr. L. Fors
Breeding work on tomatoes and greenhouse cucumbers	Mrs. M. Lundin*
Breeding work on flowers: *Callistephus, Matthiola, Chrysanthemum, Tagetes, Ageratum, Antirrhinum, Calendula, Cosmos*	E. Schutz
Plant pathology	Mrs. M. Lundin*/Dr. P.N. Lundin
Cytogenetics, mutations	Dr. S.G. Blixt
Breeding work on lawn grasses	H.A. Jönsson/ P. Weibull
Publ.: Agri Hortique Genetica (once a year)	

5 Lund

5.1 Institutionen för Lantbrukets Byggnadsteknik (Department of Farm Buildings)

S-220 06 Lund
Phone: 046 - 11 75 10
Head: Prof. R. Henriksson

Division of horticultural engineering

S-230 53 Alnarp
Head: B. Landgren*

Horticultural buildings, designing and testing different types of glasshouse and stores construction, climate control in greenhouses and stores

5.2 Statens utsädeskontroll (Swedish Seed Testing and Certification Institute)
elevation: 25 m
rainfall : 630 mm
sunshine : 1700 h

Box 33, S-221 00 Lund
Phone: 046 - 12 45 20
Dir.: Prof. L. Kåhre

Varietal control unit — Head: Dr. G. Andersson
Vegetable section (laboratories, greenhouses and fields) — P. Fredriksson

6 Röbäcksdalen

6.1 Norra trädgårdsförsöksdistriktet och försöksavdelningen för norrländsk tradgårdsodling (North Swedish Experimental Horticultural District and Experimental Division of North Swedish Horticulture)
elevation: 10 m
rainfall : 601 mm
sunshine : 2013 h

S-900 05 Umeå
Phone: 090 - 13 53 10
Head: Dr. B. Nilsson*

Applied research on experimental basis with fruit and small fruit, vegetables, ornamental plants, breeding and research work on northern species of Rubus, Ribes, Fragaria etc.

6.1.1 Öjebyns försöksstation (Experimental Horticultural Station, Öjebyn)
elevation: 5 m
rainfall : 512 mm
sunshine : 2230 h

S-943 00 Ojebyn
Head: Dr. B. Nillson*

7 Stockholm

7.1 Bergianska Botaniska Trädgården (Bergius Botanical Garden)
elevation: 0 - 19 m
rainfall : 540 mm
sunshine : 1760 h

Frescati, Box 50017
S-104 05 Stockholm
Phone: 08 - 16 28 53
Dir.: Prof. B. Jonsell

Since July 1, 1969, the Bergius Botanic Garden belongs to the Swedish State. From July 1, 1977, it is managed by the University of Stockholm in cooperation with the former owner, the Bergius Foundation of the Royal Swedish Academy of Sciences.
Arboretum, temperate plants, tropical houses.
Ecology and morphology of temperate plants.

8 Uppsala

8.1 Distriktsförsöksstation, Mellersta trädgårdsförsöksdistriktet (District Experimental Horticultural Station, Ultuna)
elevation: 10 m
rainfall : 554 mm
sunshine : 1641 h

Box 7003, S-750 07 Uppsala
Phone: 018 - 171730
Distr. Off.: Mrs. L. Gäredal

Main station of the Middle Swedish Experimental Horticultural District
Applied research on experimental basis with fruit crops, vegetables and ornamental plants — A. Gustafsson

8.1.1 Rånna försöksstation (Experimental Horticultural Station)
elevation: 150 m
rainfall : 707 mm
sunshine : 1746 h

S-541 91 Skövde
Phone: 0500 - 364 39
Distr. Off.: Mrs. L. Gäredal

Substation of the Middle Swedish Experimental Horticultural District
Applied research on experimental basis with fruit crops and ornamental plants — C-G Einarson

8.2 Statens Maskinprovningar (The National Machinery Testing Institute)

S-750 07 Uppsala
Phone: 018 - 30 19 00
Telex: 76126
Head: E. Johansson

Testing of agricultural, forestal and horticultural machinery.

Compulsory Testing — Head: M. Linder
General Testing — Head: Y Jonsson
Testing Stations:
S-900 05 Umeå
S-230 53 Alnarp
S-750 07 Uppsala

SWITZERLAND

CH

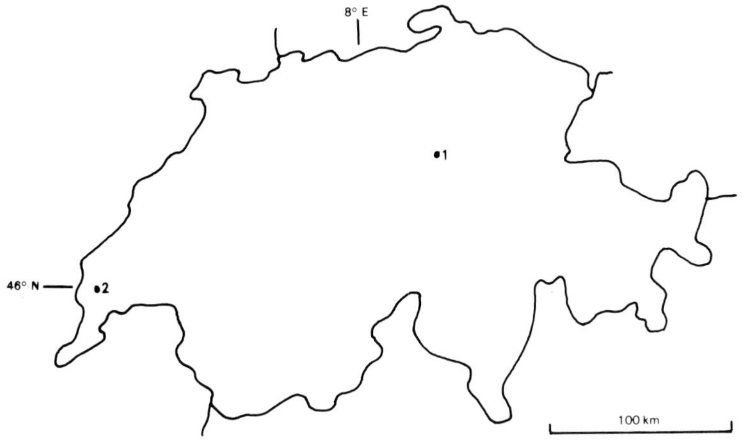
8° E
•1
46° N
•2
100 km

1 Wädenswil, 2 Changins

Survey

In Switzerland horticultural research is primarely done at the Federal Research Stations of Wädenswil and Changins. Theirs activities are, together with 5 more agricultural research stations, coordinated by the directory of the Swiss Ministery of Agriculture. Wädenswil is responsible for horticultural research in the German speaking part (mainly the fruit growing area of the North-East-Region of Switzerland) and Changins for the French and Italian speaking part of Switzerland (including the Valais- and the Ticino-areas). Both stations are organising their own extension services, which is added by local extension services paid by the cantons. Horticultural education is offered on three levels: Apprenticeship with a duration of 3 years, at the schools of engineering specialised in horticulture and on the University level at the Swiss School of Technology in Zürich (ETHZ). In addition, some basic biological disciplines are located at the Universities of Basel, Bern, Fribourg, Genève, Lausanne, Neuchâtel and Zürich. Several staff members of the Research Stations are participating in giving lectures on all three levels.

1 Wädenswil

1.1 Eidg. Forschungsanstalt für Obst-, Wein- und Gartenbau (Swiss Federal Research Station of Fruit Growing, Viticulture and Horticulture)
elevation: 438 m
rainfall : 1365 mm
sunshine : 1411 h

CH - 8820 Wädenswil
Phone: 01 780 13 33
Dir.: Dr. W. Müller

Integrated production of fruits, plant physiology	Dr. R. Schumacher*/Dr. W. Koblet
Studies of physiological disorders	Dr. R. Schumacher*/P. Perret
	R. Theiler*
Development of new production methods in fruit growing,	A. Widmer/Dr. W. Koblet
viticulture, vegetables and ornamental plant production	Dr. F. Kobel/F. Keller
Genetical research (tests), pollination, breeding of new fruit varieties, evaluation of fruit varieties	Dr. M. Kellerhals

Prepared by Dr. W. Müller, Swiss Federal Research Station of Fruit Growing, Viticulture and Horticulture, Wädenswil

Farm management and economical aspects	T. Meli
Evaluation of alternative and wild fruits (black elderberries, quince, walnut, filbert, kiwi, etc.)	U. Gremminger
Rootstocks and nursery problems	W. Riesen
Computer management and statistics	F. Fankhauser
Optimization of climatic conditions in glasshouse crops	Dr. F. Kobel
Breeding of vegetables and small fruits, production of seeds and disease free propagation material	Dr. F. Kobel/H.P. Buser/M. Lutz
Variety testing and production techniques of vegetables and small fruits	F. Keller/M. Lutz
Research in fruit tree physiology: control of reproductive and vegetative development	Dr. J. Berüter
Cultivation of mushrooms	Dr. P. Kalberer
In vitro cultures of woody and ornamental plants	R. Theiler*
Physiological studies on grapewines, clonal selection, varietal studies, rootstock studies	Dr. W. Koblet/Dr. P. Basler P. Perret
Development of integrated pest management (IPM) methods, pest forecast systems, pest tresholds, resistant cultivars, non pesticidal control methods. Pesticides: efficacy, ecological side effects, risk assessment, registration	Dr. Th. Wildbolz et al.
Phytopathology: orchard and grape diseases, soil biology	Dr. H. Schüepp Dr. Elisabeth Bosshard
Soil cultivating methods, herbology	Dr. U. Niggli
Diseases of vegetables and soft fruit	Dr. H.P. Lauber
Bacterial diseases	R. Grimm
Virus diseases of fruit trees	G. Schmid
Zoology: orchard pests	Dr. Th. Wildbolz/Dr. E. Mani
Grape pests, fruit flies	Dr. E. Boller
Vegetable pests, insect-plant relations	Dr. E. Städler
Plant and insect nematodes	Dr. J. Klingler
Insect sex pheromones	Dr. H. Arn
Pesticides: Registration	Dr. H.P. Bosshardt
Analysis, environmental contamination	Dr. H.R. Buser
Mutagenicity	Dr. J. Seiler
Plant quarantine measures	Dr. E. Mani
Fruit juice and wine technology	Dr. U. Schobinger
Improvement of instrumental and sensory methods to assess the quality of juices, concentrates, and essences. Management of a sensory test panel	Dr. P. Dürr
Investigations on the malolactic fermentation of wine and the occurrence of biogenic amines in wines produced by bacteria	Dr. K. Mayer/D. Pulfer
Evaluation of the chemical composition of raw material in fruit juices, wines and spirits to improve the quality	H. Tanner
Development of new analytical methods for fruit juices, wines and spirits	H.R. Brunner/H. Limacher
Irradiation by gamma rays for the conservation and longer preservation of foods, for pest control and breeding	H.J. Zehnder
Chilling and refrigerated storage of fruits and vegetables: effects on quality and nutritive value. Processing and preservation of fruits and vegetables: freezing, drying and heating	Dr. E. Höhn
Phytohormons (metabolism, application, effect on fruit quality)	Dr. J. Hurter
Soil (analysis, migration of nutrients eg. nitrate)	Dr. Chr. Gysi

Plant nutrition (influence of fertilizer on yield and quality of plants)	
Product quality (vegetable quality and its control, development of rapid methods for quality control, nutritional studies eg. bio-availability of minerals)	Dr. A. Temperli/U. Künsch
Preservation (lactic acid fermentation for preservation and processing of vegetables)	

1.2 Experimental Farm for Fruit Growing
elevation: 415 m
rainfall : 1096 mm
sunshine : 1446 h

CH - 4451 Breitenhof, Wintersingen BL
Foreman: W. Zbinden

1.3 Experimental Farm for Fruit Growing
elevation: 410 m
rainfall : 930 mm
sunshine : 1503 h

CH - 8594 Güttingen TG
Foreman: Chr. Krebs

1.4 Experimental Farm for Viticulture
elevation: 414 m

CH - 8712 Stäfa ZH
Foreman: W. Fürer

2 Changins

2.1 Station Fédérale de Recherches Agronomiques de Changins (Swiss Federal Agricultural Research Station)

CH - 1260 Nyon
Phone: 022 61 54 51
Telex: 22 785
Dir.: Dr. A. Vez

Soil research, physical and chemical properties, organic matter, problems of long term fertility	Dr. J.A. Neyroud/Dr. J.P. Quinche
Soil analysis and advice for fertilization	Dr. J.P. Ryser
Research on fertilization, optimization of the use of farm manure	Dr. J.A. Neyroud
Bioclimatology and plant ecology	F. Calame
Soil tillage, crop management, control of erosion	Dr. A. Maillard
Plant physiology	Dr. G.F. Collet
Weed research	Dr. E. Beuret
Tissue culture	L. Lê
Research in plant protection (integrated protection, biology and biotechnical control)	Dr. A. Bolay/Dr. A. Stäubli Dr. D. Gindrat
Research on virus diseases, development of testing methods	Dr. P. Gugerli
Production and release of virus-free rootstocks and grafts	Dr. F. Pelet
Wine technology, development of wine-based beverages with low alcool content	J. Crettenand/P. Cuénat Dr. J. Aerny
Research on fermentation	O. Cazelles
Breeding of vegetables	Dr. S. Badoux

2.2 Station Fédérale de Recherches Agronomiques de Changins, Centre d'Arboriculture et d'Horticulture des Fougères

CH - 1964 Conthey
Phone: 027 36 27 22
Head: Dr. Ch. Darbellay

Fruit production techniques, variety testing, crop management (including berries)	W. Pfammatter
Production techniques for vegetables (open air and under glass or plastic), variety testing	A. Granges

Research on storage of fruits and vegetables	A. Schwarz
Testing of energy-saving devices for greenhouses	A. Reist*
2.3 Station Fédérale de Recherches Agronomiques de Changins, Domaine du Caudoz	CH - 1009 Pully Phone: 021 28 28 66 Head: J.-L. Simon
Viticulture, crop management and soil tillage	F. Murisier
Studies on propagation material of vines (rootstocks and vines), production techniques	J.-L. Simon
Breeding of vine varieties	A. Jaquinet
2.4 Sottostazione Fédérale di Ricerche Agronomiche	CH - 6593 Cadenazzo Phone: 092 62 18 31 Head: Dr. G. Jelmini
Studies of specific problems for the southern part of Switzerland in horticulture and viticulture, research on acid soils	Dr. G. Jelmini

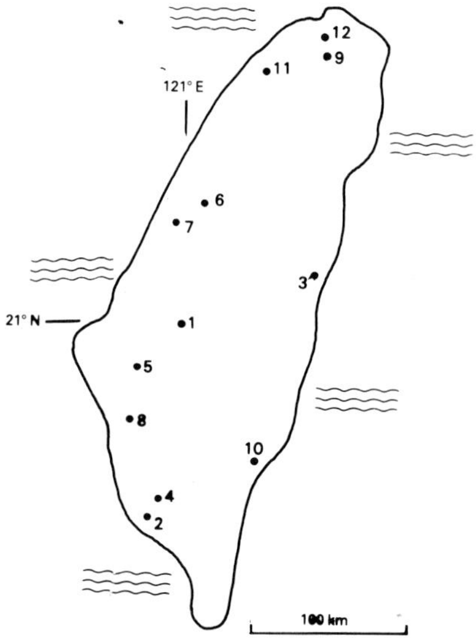

121° E
21° N
12
9
11
6
7
3
1
5
8
10
4
2
100 km

1 Chiayi
2 Fengshan
3 Hwalien
4 Pingtung
5 Shanhua
6 Shinshieh
7 Taichung
8 Tainan
9 Taipei
10 Taitung
11 Taoyuan
12 Yungmingshan

1 Chiayi

1.1 Chiayi Junior Agricultural College Department of Horticulture elevation : 50 m rainfall : 1,998 mm sunshine : 1,500 h	Chiayi (600) Phone: (05) 2277148
Citrus culture and plant propagation practices	Prof. C.C. Huang
1.2 Chiayi Branch Station of Taiwan Agricultural Research Institute (TARI), Department of Horticulture-	Chiayi (600) Phone: (05) 2271340 2 Head: C.K. Chu
Banana and coffee culture	C.K. Chu
Citrus breeding and rootstock	C.C. Lin/ A.S. Hwang
Pineapple breeding and culture	C.C. Chang
Lychee, longan; passion fruit, mango and tropical fruits breeding and culture	C.R. Yen
Avocado, canistel, other tropic and subtropical rare fruit breeding, preservation and tissue culture	T.M. Zong

2 Fengshan

2.1 Fengshan Tropical Horticultural Experiment Station, TARI elevation : 50 m rainfall : 1,923 mm sunshine : 2,427 h	Fengshan (830) Phone: (07) 7310191 3 Dir.: T.F. Sheen, M.Sc.
2.1.1 Department of Vegetable	Head: C.H. Lin, M.Sc.
Watermelon, melon, cucumber	C.H. Lin, M.Sc.
Beans, water vegetables	T.D. Liou, MSc.
Cabbage, radish	T.F. Sheen, M.Sc.
Onion, garlic	T.C. Teng

Prepared by Prof. Kang Yeou-der, Dept. of Horticulture, National Taiwan University, Taipei.

Hydroponic vegetables	T.F. Sheen, M.Sc.
2.1.2 Department of Tropical Fruit Tree	Head: Dr. M.S. Chen
Papaw, waxapple	D.N. Wang
Pineapple	Y.K. Lin
Mango	Z.H. Shu, M.Sc./ Dr. M.S. Chen
Passionfruit	Y.T. Lin
Litchi, longan, grape	Y.H. Teng
Carambola	U.C. Wang
Guava	Dr. M.S. Chen/ U.C. Wang
Breeding for seedless fruits	Dr. M.S. Chen
Nutritional and reproductive physiology of tropical fruits	Z.H. Shu, M.Sc.
2.1.3 Department of Plant Protection Fruit and	Head: H.S. Lee
Vegetable Pests	H.S. Lee/ H.C. Wen/ F.M. Lu
Fruit and vegetable diseases	Dr. C.C. Lin/ H.L. Wang/ H.S. Chien
2.1.4 Department of Management and Utilization	Head: C.T. Liou, M.Sc.
Agricultural machinery	L.C. Lee
Fruit and vegetable processing	R.C. Yu,M.Sc/ B.T. Hwang/ R.S. Hwang

3 Hualien

3.1 Hualien District Agricultural Improvement Station, Provincial Dept. Agriculture and Forestry (PDAF)	Hualien (935) Phone: (038) 521108 110
elevation : 50 m rainfall : 1,981 mm sunshine : 1,648 h	
Yam, day-lily, vegetable varietal, cultural tests	D.C. Perng
Watermelon, tomato, blueberry and kiwifruit varietal, cultural tests	P. Hwang
Strawberry and floral varietal and cultural tests	Y.S. Tsai

4 Pingtung

4.1 Taiwan Banana Research Institute elevation : 50 m rainfall : 2,438 mm sunshine : 2,398 h	Chiuju (901) Phone: (087) 392111 3 Dir.: H.H. Lai
4.2 Pingtung Institute of Agriculture, Department of Horticulture	Pingtung (900) Phone: (087) 227121, 228041 Head: Assoc.Prof. H.H. Liu
Promotion of tissue culture of herbaceous flowers	Assoc.Prof. H.H. Liu
4.3 Kaohsiung District Agricultural Improvement Station	Pingtung (900) Phone: (087) 229466
Orchid breeding	S.Y. Chen
Vegetable breeding and culture	S.Y. Chen/ YR. Chen/ C.M. Han Z.N. Wang
Orchard culture	Y.M. Hsu
Floriculture	S.Y. Chen/ Y.M. Hsu

5 Shanhua

5.1 Asian Vegetable Research and Development Center	PO Box 42, Shanhua (741)

(AVRDC)
elevation : 50 m
rainfall : 2,026 mm
sunshine : 2,618 h

Phone: 06-5837801
Telex: 73560 AVRDC
Cable: ASVEG, SHANHUA
Dir.: Dr. G.W. Selleck*

AVRDC has selected six vegetable crops for research and improvement: soybean, mungbean, tomato, chinese cabbage, sweet potato and white potato.

6 Shinshieh

6.1 Taiwan Seed Service, PDAF
elevation : 200-500 m
rainfall : 2,000 mm
sunshine : 2,460 h

Shinshieh (426)
Phone: (045) 811311 3

Tomato disease resistance breeding	S.Y. Chen
Cabbage, cauliflower and broccoli breeding	C.I. Liao
Plant tissue culture and orchid culture	Dr. W.J. Ho
Bulbs and carnation culture	F.W. Hou MS
Tomato breeding	S.Y. Chen
Watermelon breeding	P.K. Huang
Spinach breeding	W.J. Ho/ P.K. Huang
Chinese cabbage breeding	S.H. Deng

7 Taichung

7.1 National Chung Hsing University, Department of
Horticulture
elevation : 100-200 m
rainfall : 1,784 mm
sunshine : 2,461 h

Taichung (400)
Phone: (04) 2873181
Head: Prof. N.-T. Fan

Citrology, fruit-tree physiology	Prof. N.-T. Fan
Ornamental plants, landscape design	Prof. C.-H. Peng
Floricultural science	Prof. M.-C. Huang
Vegetable breeding	Prof. W.-N. Chang*
Mineral nutrition of horticultural crops	Prof. K.-C. Lee
Vegetable crops, plant propagation,	Assoc.Prof. S.-J. Luo
Tropic and subtropic fruits	Assoc.Prof. S.-W. Weng
Viticulture	Assoc.Prof. Y.-S. Yang
Postharvest physiology of horticultural crops	Assoc.Prof. H.-J. Chiu
Handling of horticultural products	Instructor D.-T. Horng
Flowering bulbs, herbaceous flowers	Instructor T.-Y. Wang
Temperate fruits	Instructor C.-C. Nee

7.2 Taichung District Agricultural Improvement Station PDAF
elevation : 50 m
rainfall : 1,565 mm
sunshine : 2,460 h

Changhua (515)
Phone: (048) 523101 8

Pea breeding	J.Y. Kuo
Breeding of rust-resistant kidney bean varieties	J.W. Guu
Purification of chinese leek	W.J. Chung
Environmental control in horticulture	F.Y. Kwo
High-land vegetable culture	T.C. Lin

Viticulture and pear culture	J.H. Lin
Forcing culture of peach and plum	W.J. Liaw
Flower differentiation and pomology physiology	L.R. Chang

7.3 Taiwan Agricultural Research Institute (TARI)
Department of Horticulture
elevation : 100-200 m
rainfall : 1,784 mm
sunshine : 2,461 h

Wu-feng (431)
Phone: (043) 302301 5
Head: S.C. Lin, Ph.D.*

Citrus breeding and culture improvement	C.Y. Yeh/ P.C. Liu
Fruit and vegetable storage	S.C. Lin*/CC Huang
Pear and plum breeding	H.T. Hsu
Peach and kaki breeding	Y.C. Wen, MSc.
Table grape breeding and culture improvement	W.Y. Wang, M.Sc.
Physiology and biochemistry of fruit crops	M.T. Wu, Ph.D.
Tropical fruits breeding and culture improvement	S.K. Ou, M.Sc.
Cucumber, melon and cabbage breeding	C.H. Hsiao, Ph.D.
Potato and pea breeding	S.J. Tsao, Ph.D.*
Cucumber and luffa spp. breeding	W.Z. Yang
Vegetable crops germplasm collection and breeding	Y.M. Chang
Chrysanthemum and orchid improvement	R.S. Lin, M.Sc.
Cultivation of flower crops	C.L.L.

8 Tainan

8.1 Tainan District Agricultural Improvement Station
elevation : 50 m
rainfall : 1,839 mm
sunshine : 2,618 h

Tainan (700)
Phone: (06) 2679526 9
Dir.: Dr. C.C. Tu

Asparagus breeding, culture and seed production	Dr. C.C. Tu/ Y.W. Chen/ Y.F. Yen
Muskmelon and processing tomato breeding and culture	S.L. Hwang
Breeding and culture improvement of onion and garlic	S.R. Yang
Culture improvement of strawberry and vegetables	C.F. Wang
Flower culture and seeds production	C.Y. Wang
Subtropic grape and passion fruit culture	M.T. Chang
Mulching plant culture on slopeland orchard	C.L. Lai
Bamboo shoot production and tropic fruit culture	G.J. Juang
Mango breeding and culture, soil and water conservation	M.F. Liou
Pineapple production	Y.K. Wang

9 Taipei

9.1 National Taiwan University,
Department of Horticulture
elevation : 50 m
rainfall : 2,100 mm
sunshine : 1,646 h

Taipei (107)
Phone: (02) 3510231
Head: Prof. TT Fang

Tissue culture propagation of horticultural plants	Prof. S.S. Ma, M.Sc.*
Citrus varieties and citrus viruses	Prof. P. Lin, M.Sc.
Physiology and culture of deciduous fruit trees	Prof. Y.D. Kang Ph.D.
Mineral nutrition of fruit trees and radiotracer methodology	Prof. Y.Y. Yeh
Fruits culture on slopeland	Prof. M.C. Liao, Ph.D.
Plant growth and development	Prof. W.C. Chang, Ph.D.

Tissue and cell culture	Prof. W.C. Chang, Ph.D.
Physiology of fruit trees	Prof. C.Y. Cheng, Ph.D.
Vegetable and asparagus breeding	Prof. L. Hung, Ph.D.*
Physiology and culture of vegetables (tomato, garlic and pepper)	Prof. H. Huang, M.Sc.
Mushroom culture	Prof. S. Cheng
Effects of ethylene and storage conditions on the toughening of post-harvest asparagus spears	Prof. C.N. Chang, Ph.D.
Physiology and nutrition of floricultural crops and ornamental horticulture	Prof. N. Lee, M.Sc.
Photosynthesis and nutrient of vegetables	Assoc.Prof. H.H. Lai, Ph.D.
Crop genetics	Assoc.Prof. C.T. Shii*
General horticulture	Prof. M.N. Chiang
Controlling pectinesterase activity on the texture of horticultural products	Prof. Y.C. Huang
Fruit juice and frozen vegetables	Prof. T.T. Fang
Chemistry of natural pigments and flavor, hypobarid storage of fruits and vegetables	Prof. P.L. Tsai, Ph.D.
Postharvest physiology	Assoc.Prof. T.S. Lin, Ph.D.*
Studies on chinese gardens and landscape	Prof. D.L. Ling
Studies on the relationship between ornamental plants and landscape gardening	Prof. D.W. Cheng
Studies on landscape ecology in Taiwan	Prof. C. Liu

10 Taitung

10.1 Taitung District Agricultural Improvement Station PDAF elevation : 30-100 m rainfall : 1,844 mm sunshine : 1,877 h	Taitung (930) Phone: (089) 325110
Citrus culture	S.S.Chang/ S. Wang
Pineapple	S.C.Lee/ J.R. Lin
Ginger culture	S. Wang
Day-lily variety trial	S.C. Lee
Fern	S.A. Pai

11 Taoyuan

11.1 Taoyuan District Agricultural Improvement Station PDAF elevation: 50 m rainfall : 1,765 mm sunshine : 1,925 h	Hsinwu (327) Phone: (034) 775116
11.1.1 Department of Crop Improvement, Section of Horticulture	Head: D.H. Sheu Chief: C.M. Lee
Strawberry and kiwifruit	C.M. Lee
Vegetable production	S.W. Shieh/ T.L. Chen
Potato and vegetables	W.T. Ni
Postharvest and processing	L.F. Lin
Chicory, sweet fennel and leaf mustard	C. Cheng
Radish, bean and water convolvulus	S.J. Fan
Citrus and litchi	T.J. Tsay
Vegetable seed production on high-land	W.C. Tai/ C.C. Sheu
11.1.2 San-chung Branch Station	Taipei (241)

	Phone: (02) 9712585
Cabbage and Chinese kale breeding	F.S. Liao, M.Sc.
Cabbage and Chinese cabbage culture	Y.H. Chen
Green onion breeding	Y.P. Hsu
Amaranth breeding	T.R. Chang, M.Sc./ S.C. Wang
Studies on the rotation system for the cultivation of summer vegetables on upland	F.S. Liao/ T.R. Chang
Bamboo shoot culture	C.I. Chang
Chinese tea rose, jasmin and patchouli culture and the processing of the natural perfumes	T.C. Soong/ T.R. Chang
Studies on the package and storage of harvested vegetables	T.R. Chang/ F.S. Liao
Improvement of culture technique on leatherleaf fern	W.S. Lee, M.Sc.
Seasonal occurrence of the diamondback moth *(Plutella Xylostella L.)*	S.S. Wang
Observation and mass rearing of the cabbage webworm	S.S. Wang/ S.H. Jean
Studies of sclerotinia blight of green onion	T.I. Tu, M.Sc.

12 Yangmingshan

12.1 Chinese Culture University,
Department of Horticulture
elevation: 600 m
rainfall : 3,297 mm
sunshine : 1,600 h

Yangmingshan (113)
Phone: (02) 8610511
Head: D.W. Cheng

Floriculture and marketing	Prof. D.W. Cheng, Ph.D.
Ancient Chinese gardens	Prof. C.S. Chen, Ph.D.
Special culture on vegetable crops	Prof. P.N. Lee
Camellia and azalea breeding	Prof. S.P. Yang
Horticultural insects	Prof. C.C. Tao
Horticultural plant taxonomy	Prof. S.Y. Lee, Ph.D.
Processing of horticultural products	Assoc.Prof. K.T. Yang, Ph.D.
Advanced pomology	Assoc.Prof. S. Wang
Ornamental trees	M.F. Lee M.Sc.
Garden construction	L.U. Hwang, M.Sc.
Genetics and biostatistics	Y.T. Teng, Ph.D.
Horticultural pathology	T.S. Yang, M.Sc.
Uses of horticultural plants	C.H. Lin, M.Sc.
Tissue culture	C.I. Hwang, M.Sc.

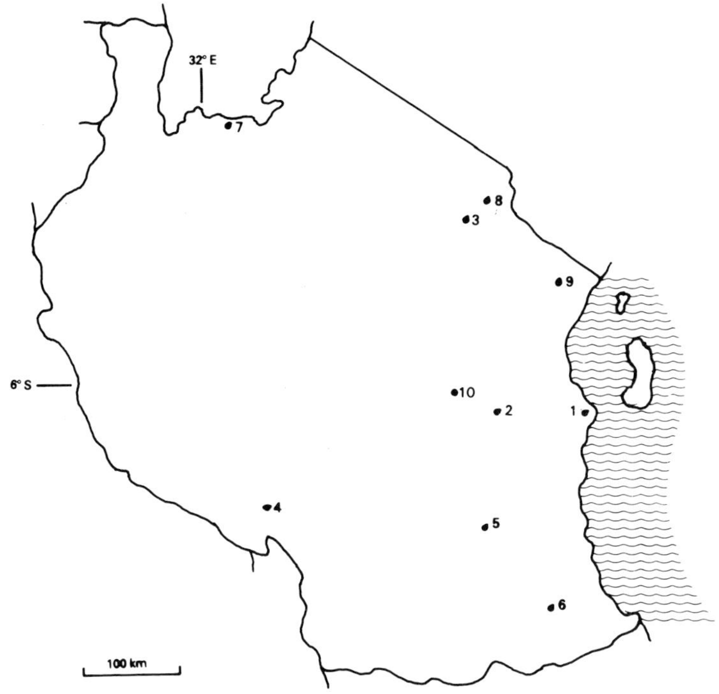
32° E
6° S
7
8
3
9
10
2
1
4
5
6
100 km

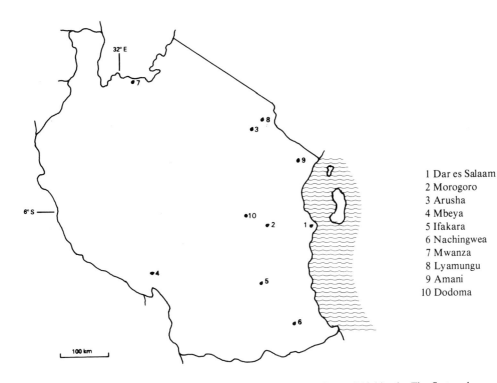

1 Dar es Salaam
2 Morogoro
3 Arusha
4 Mbeya
5 Ifakara
6 Nachingwea
7 Mwanza
8 Lyamungu
9 Amani
10 Dodoma

Survey

Lying just south of the equator in Eastern Africa, the United Republic of Tanzania comprises the mainland of 362,000 m^2 (including 20.000 m^2 of inland waters) and the islands of Zanzibar (640 m^2) and Pemba (380 m^2). The mainland population rates about 18 million and the islands 500.000.
Because of the large variations in altitude the climate displays a great range from tropical warmth of the coast to the cool highlands. The first or long rains occur from April to May, while the short rains come between November and December.
Tanzania has four major horticultural districts: The Coastal Belt (Mangoes, citrus pineapples), Morogoro (vegetables, citrus), the Usambaras (vegetables), Southern Highlands (temperate fruits, potatoes).
All institutes dealing with agriculture and horticulture are placed under the Ministry of Agriculture and Livestock Development.

1 Dar es Salaam

1.1 Research and Training Division — PO Box 9192, Dar es Salaam

1.2 Crop Development Division — PO Box 9071, Dar es Salaam

Extension and technical services — Coord: R.D. Kyamba, R.A.D. Mreta, M.Sc.

2 Morogoro

2.1 Sokoine University of Agriculture — PO Box 3000, Morogoro

Vegetable variety trials
Growth and developmental physiology of indigenous vegetable species — F.J. Teri M.Sc.

Prepared by N.A. Mnzava, Sokoine University of Agriculture, Morogoro

Fruit and seed setting, growth regulation, cultural methods in vegetable production	N.A. Mnzava, Ph.D./F.J. Teri, M.Sc.

3 Arusha

3.1 Horticultural Training and Research Institute, Tengeru elevation : 1235 m rainfall : 850 mm	PO Box 1253, Arusha
Indigenous vegetables, exotic vegetables and potatoes Citrus, bananas, and most of the tropical fruits	
Citrus rootstock trial	R.E.A. Swai, M.Sc.
Citrus disease control	I.S. Swai, B.Sc.
Vegetable disease control	N.A. Kaaya, M.Sc.
Temperate zone fruits	R. Nchanjala, B.Sc.
Vegetable variety trials	
Diploma (Horticulture) training	N. Maliwa
Project leader	J.R. Butler

4 Mbeya

4.1 Uyole Agricultural Centre elevation : 1700 m rainfall : 1200 mm sunshine : 6.9 h p.day	PO Box 400, Mbeya
Temperate zone fruits	A.M.S. Nyomora, M.Sc.
Cv testing, rooting of cuttings	
Vegetable variety trials	A. Meela, B.Sc.
Vegetable diseases control	F. Swai, M.Sc.
Round potato variety trial	J. Nsekela, Dip.Hort.

5 Ifakara

5.1 Kilombero Agricultural Training and Research Institute (KATRIN)	Private Bag, Ifakara, Morogoro Region
Irrigated horticulture	
Citrus rootstock trials	O. Mchau, Dip.Hort.

6 Nachingwea

6.1 Naliendele Research Station	Private Bag, Nachingwea
Casewnut research	M.E.R. Sijaona, B.Sc.

7 Mwanza

7.1 Ukiriguru Research Station	Private Bag, Mwanza
Cassava and sweet potato research	M. Msabaha, M.Sc.

Other commodities of horticultural interest:
Coconut development is administered from Dar Es Salaam and has stations in Tanga and Zanzibar.

8 Lyamungu

8.1 Lyamungu Research Center

PO Box 3004, Moshi
Kilimanjaro Region

Sub-tropical horticulture including essential oils
Coffee researach

9 Amani

9.1 Tea research is centered in Marikitanda near Amani

9.2 Spice collection is maintained at Zigi and Amani

10 Dodoma

10.1 Grapes Research Station (DODEP)
elevation : 1100 m
rainfall : 500 mm

PO Box 1676, Dodoma

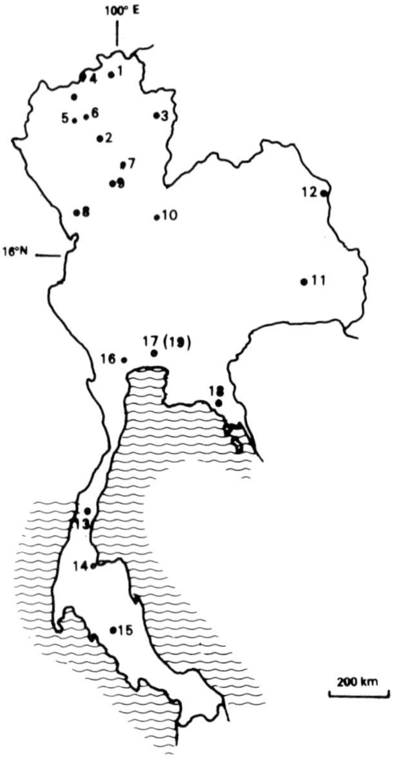
100° E
16°N
200 km

1 Chieng Rai
2 Hang Chat
3 Nan
4 Fang
5 Mae John Luang
6 Khun Wang
7 Pichit
8 Doi Muser
9 Tha Chai
10 Koa Kor
11 Sri Sa Ket
12 Nakorn Phanom
13 Chumporn
14 Surajthani
15 Trang
16 Rajburi
17 Bangkok Noi
18 Plew
19 Bangkok

Survey

The former Horticulture Division, Department of Agriculture, Ministry of Agriculture and Cooperatives was established in 1972. It was entrusted to promote national actions in increasing food security and horticultural production.

In order to cope with the need and demand of horticultural products to meet population's needs, the Horticulture Research Institute was established in 1984. The structure of this organization has been changed to fit to the new DOA policies but the subjects matter falling within the former organization's task is carried out within them.

Horticulture Research Institutes (HRI) - The Institute mainly obtained national policy for planning and monitoring research projects in order to produce priority economic horticultural plants in sufficient level for local consumption and for exporting.

The HRI network consists of 6 Horticulture Research Centers and 12 Horticulture Experiment Stations across the country. The Research Centers have the objectives to serve the priority in each region which is varied by topography and climatic conditions. And also, the Horticulture Experiment Stations are the satellites of each center to do research in a specific area in order to complete the network covering the whole horticultural plantation area of the nation.

Structures - The lay-out of Horticulture Research Institutes consists of the Central Scientist Board, Central Executive Sections, Horticulture Research Centers and Horticulture Experiment Stations.

Objectives - The objectives of HRI are emphasized in particular on implement tasks in line with the policies of the Ministry of Agriculture and Cooperatives; intensification of research according to the National Agricultural Policies; monitoring and evaluation of research projects, in collaboration with government organizations and the private sector, and joined research projects with related departments.

Framework - The Horticulture Research Center Framework consists of an Administration Section, Plant Breeding Section, Seed Production Section,

Prepared by Mr. Ampol Senanarong, Department of Agriculture, Bangkok

Cultivation Practice Section, Post Harvest Section, Soil Science Section, Agri-Chemical Section and a Plant Protection Section.
As the difference in topographic and climatic conditions have influenced the various kinds of plants, the following data are grouped withing geographic different regions of Thailand, i.e. Northern, North-Eastern, Southern and Central region, respectively.

Norther Region

1 Chieng Rai

1.1 Horticulture Research Center
elevation: 412 m
rainfall : 1669,5 mm
sunshine : 2494 h

Muang District, Chieng Rai Province
Dir.: Somsak Chaisilrapin

Research activities for regional and commercial crops:
Vegetables: Mustard green, garlic, tomato, potato, shallot
Fruit trees: Lychee, longan, mango, citrus, grape, strawberry, banana, macadamia nut, avocado, pineapple
Ornamental: Orchids, Gladiolus, Carnation, Roses, Chrysanthemum, Gerbera, Aster, Anthurium, Statice, Snapdragon, Straw flower, Bird of Paradise and local ornamentals
Others: Coffee, tea, pepper, spices

2 Lampang

2.1 Horticulture Experiment Station
elevation: 300 m
rainfall : 1071.8 mm
sunshine : 2412 h

Hang Chat District Lampang Province
Dir.: Plean Wangchareoan

Research activities for regional and commercial crops:
Vegetables: Tomato, Mustard green, pepper, yard-long bean
Fruit trees: Mango, longan, grape, tamarind, pineapple, citrus
Ornamental: Rose, Aster, local ornamentals

3 Nan

3.1 Horticulture Experiment Station
elevation: 264.03 m
rainfall : 1386.5 mm
sunshine : 2160 h

Muang District, Nan Province
Dir.: Tawin Kaisuwan

Research activities for regional and commercial crops:
Vegetables: Tomato, asparagus, Mustard green cabbage
Fruit trees: Citrus, mango, longan, lychee
Ornamental: Rose, local ornamentals
Others: Tea

4 Fang

4.1 Horticulture Experiment Station
elevation: 520-580 m
rainfall : 1,436.5 mm

Fang District, Chieng Mai Province
Dir.: Surasak Intharakamheang

Research activities for regional and commercial crops:
Vegetables: Potato, Mustard green, cabbage, Pe-tsai cabbage, Chinese mustard, Chinese kale, sweet potato
Fruit trees: Mácadamia nut, lychee, longan, avocado, strawberry
Ornamental: Anthurium, Orchid, Chrysanthemum, Snapdragon, local ornamentals
Others: Tea, spices

5 Mae John Luang

5.1 Highland Horticulture Research Station
elevation: 1,200-1,300 m
rainfall : 1,982.46 mm
sunshine : 2167 h

Mae Jam District
Chieng Mai Province
Dir.: Vithoon Rattana

Research activities for regional and commercial crops:
Vegetables: Cauliflower, broccoli, Brussels sprouts, kohlrabi
Fruit trees: Peach, apple, pear, *Myrica esculenta,* jam, persimmon
Ornamental: Semi-tropical flowers
Others: Tea, coffee, spices

6 Khun Wang

6.1 Highland Horticulture Experiment Station
elevation: 1,300-1,400 m
rainfall : 1,800 mm
sunshine : 3780 h

San Pha Thong District
Chieng Mai Province
Dir.: Sathit Khewkeaw

Research activities for regional and commercial crops:
Vegetables: Cauliflower, broccoli, Brussels sprouts, kohlrabi
Fruit trees: Peach, apple, pear, persimmon
Ornamental: Semi-tropical flowers
Others: Tea, coffee, spices

7 Pichit

7.1 Horticulture Research Station
elevation: 35 m
rainfall : 1204.6 mm

Muang District, Pichit Province
Dir.: Chamnarn Thongprab

Research activities for regional and commercial crops:
Vegetables: Tomato, Chinese convolvulus, cucumber, Chinese radish
Fruit trees: Pomelo, banana, papaya, mango, citrus, grape, tamarind, guava, jackfruit
Ornamental: Jasmine, Chrysanthemum, Rose, Aster, local ornamentals
Others: Coconut

8 Doi Muser

8.1 Horticulture Experiment Station
elevation: 800-1,000 m
rainfall : 1,715.8 mm

Muang District, Tak Province
Dir.: Tawatchai Sasiplarin

Research activities for regional and commercial crops:
Vegetables: Tomato, garden pea, cauliflower, broccoli, Brussels sprouts, kohlrabi, cabbage
Fruit trees: Avocado, macadamia, strawberry
Ornamental: Gladiolus, Rose, Carnation, Gerbera, Statice, Bird of Paradise, local ornamentals
Others: Tea, coffee, spices

9 Tha-Chai

9.1 Horticulture Experiment Station
elevation: 50 m
rainfall : 1,500 mm
sunshine : 2340 h

Sri-Satchanarai District
Sukhothai Province
Dir.: Suthat Arunpairoj

Research activities for regional and commercial crops:
Vegetables: Chilli, Chinese convolvulus, egg plant
Fruit trees: banana, grape, mango, papaya, tamarind, sapodila
Ornamental: Gladiolus, Jasmine, Gerbera, Chrysanthemum, Aster, local ornamentals
Other: Cashew nut

10 Kao Kor

10.1 Highland Horticulture Research Station
elevation: 750 m
rainfall : 1,800 mm

Muang District, Petchaboon Province
Dir.: Chaiwat Wattanachat

Research activities for regional and commercial crops:
Vegetables: Cauliflower, broccoli, Brussels sprouts, kohlrabi, asparagus, potato
Fruit trees: Peach, apple, pear, *Myrica esculenta,* persimmon
Ornamental: Semi-tropical flowers
Others: Tea, coffee

North-Eastern Region

11 Sri-Sa-Ket

11.1 Horticulture Research Center
elevation: 125 m
rainfall : 1483.7 mm
sunshine : 2628 h

Muang District, Sri-Sa-Ket Province
Dir.: Prasert Anupand

Research activities for regional and commercial crops:
Vegetables: Shallot, tomato, garlic
Fruit trees: Mango, papaya
Ornamental: Jasmine, Gladiolus, Chrysanthemum, Gerbera, Aster, local ornamentals
Main crop: Cashew nut

12 Nakorn Phanom

12.1 Horticulture Experiment Station
elevation: 144 m
rainfall : 2,096.7 mm

Muang District
Nakorn Phanom Province
Dir.: Preecha Cherchum

Research activities for regional and commercial crops:
Vegetables: Chilli, egg plant, Chinese radish, Chinese kale
Fruit trees: Mango, tamarind, durian, lychee, longan, rambutan, jackfruit, custard apple
Ornamental: Rose, local ornamentals

Southern Region

13 Chumporn

13.1 Horticulture Research Center
elevation: 12.5 m
rainfall : 1887.2 mm
sunshine : 1944 h

Sawee District, Chumporn Province
Dir.: Anupab Theerakul

Research activities for regional and commercial crops:
Fruit trees: Durian
Others: Coconut, cocoa, coffee, spices
Main crop: Coconut

14 Surajthani

14.1 Horticulture Research Center
elevation: 19 m
rainfall : 1962 mm

Karnchanadit District
Surajthani Province
Dir.: Nakorn Sarakun

Research activities for regional and commercial crops:
Vegetables: Sweet potato
Fruit trees: Durian, rambutan, mangosteen, banana, pineapple, lansium, citrus
Ornamental: Anthurium, Rose, local ornamentals
Others: Coffee, cocoa, oil palm, cashew nut, spices
Main crop: Oil palm

15 Trang

15.1 Horticulture Experiment Station
elevation: 35 m
rainfall : 2,064 mm

Sikao District, Trang Province
Dir.: Prayoon Patthong

Research activities for regional and commercial crops:
Vegetables: *Parkia speciosa* Hassk.
Fruit trees: Durian, rambutan, mangosteen, banana, pineapple, lansium, citrus
Ornamental: Anthurium, Rose, local ornamentals
Others: Coconut, coffee, cocoa, oil palm, cashew nut, spices
Main crop: Coconut, cocoa, oil palm

Central Region

16 Rajburi

16.1 Horticulture Research Center

Rajburi Province
Phone: 579-5582
Dir.: Hiran Hiranpradit

Research activities for regional and commercial crops:
Vegetables: Yard-long bean, tomato, sweet potato, cilli, egg plant
Fruit trees: Papaya, citrus, mango, rambutan, durian, mangosteen, banana, grape, jackfruit, sapodila, tamarind, guave
Ornamental: Orchid, Rose, Chrysanthemum, Jasmine, Gerbera, Anthurium, Aster, local ornamentals
Others: Coconut, spices

17 Bangkok Noi

17.1 Horticultural Experiment Station
Elevation : 25 m
rainfall : 1,162 mm

Bangkok Noi District, Bangkok
Phone: 424-0201
Dir.: Suchep Chayantrakom

Research activities for regional and commercial crops:
Vegetables: Yard-long bean
Fruit trees: Mango, tamarind, lychee, guava, jackfruit
Ornamental: Anthurium, Jasmine, Gerbera, Aster, local ornamentals
Others: Spices

18 Plew

18.1 Horticulture Experiment Station
elevation: 15 m
rainfall : 2,828.5 mm

Plew District, Chanthaburi Province
Dir.: Thong Yoo Pinpiban

Research activities for regional and commercial crops:
Vegetables: Tomato, yard-long bean, Chinese convolvulus
Fruit trees: Rambutan, durian, mangosteen, citrus, lansium, salacca, jackfruit, avocado
Ornamental: Jasmine, Chrysanthemum, Gerbera, Anthurium, local ornamentals
Others: Coffee, cocoa, spices

19 Bangkok

19.1 Headquarters Horticulture Research Institute, Department of Agriculture, Ministry of Agriculture and Cooperatives

DOA Building Bangkhen,
Bangkok 10900
Phone: 579-0583/579-0554
Telex: 84478 INTERAG TH
Dir.: Banchong Sikkhamondhol

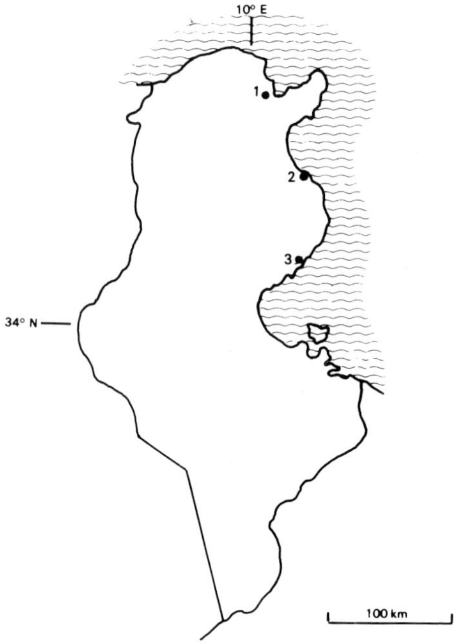
10° E
1
2
3
34° N
100 km

1 Ariana
2 Sousse
3 Sfax

Survey

Research and experimentation work in the field of horticulture started in the early forties first on fruit tree improvement and then on vegetables. G. Valdeyron and then P. Crossa Raynaud initiated the first breeding programs on apricot and citrus mainly, at I.N.R.A.T. Since the independance the activities have been expanded and diversified. In 1962 I.N.R.A.T. has intensified and extended the horticultural research to most important fruit and vegetable crops in the country including breeding, propagation, agronomy, physiology and plant protection.
Other existing education and research institutions became involved in horticultural research e.g. I.N.A.T. (crop production and protection) and C.R.G.R. (irrigation). New institutions have been created: E.S.H.C.M. (Horticultural School), and I.O. (Olive tree Institute), which contributed to strengthen the national research system on horticulture.
The main research and education institutions involved today in horticulture are: I.N.R.A.T., I.N.A.T., E.S.H.C.M., C.R.G.R. and I.O. - All these institutions work closely together with different kinds of development and extension agencies (Groupements - Offices des Périmètres Irrigues - Regional Extension Services - Nurseries - Seed Companies).

Experimental farms are located at: La Soukra (citrus grapes), Koubba (citrus), Mornag (all crops), Sbikha (apricot), Le Krib (pome fruit), Sbiba (pome fruits, apricot), Bembla (apricot, almond, peach), Sfax (olive, almond, pistachio), Tozeur (dates), Metline (potato), Sahline (greenhouses), Teboulba (vegetables).

1 Ariana

1.1 Institut National de la Recherche Agronomique de Tunisie (I.N.R.A.T.) (National Agricultural Research Institute of Tunisia)
elevation: 10 m
rainfall : 474 mm
sunshine: 3049 h

Ave de l'Indépendance, 2080 Ariana
Phone: 230 024/ 231 023/ 231 693
Dir.: M. Lasram

Fruit breeding and production:
Apricot, pistachio — Head of Lab.: M. M'lika
Citrus, subtropical fruits — M. Lasram

Prepared by Mr. M. Lasram, Institut National de la Recherche Agronomique de Tunisie, Ariana

Almond, peach and rootstocks	B. Jraïdi
Pome fruits	S. El Aouini
Pistachio (pollination and cytology)	Z. Belfakh
Date palm	A. Ben Abdallah
Tissue culture and propagation	N. Elloumi
Viticulture:	
Breeding	Head of Lab.: F. Askri
Physiology	H. Nehdi
Tissue culture	M. Ben Slimane
Vegetables breeding and production:	
Tomato, pepper, plastic greenhouses	Head of Lab.: N. Hamza
Melon, watermelon, cucumber	H. Jebbari
Potato	N. Khammassi
Breeding for virus resistance	A. Djemmali
Plant protection:	
Entomology: med. fruit fly	Head of Lab.: M. Cheikh
Black scale, aphids, carpocaps	H. Ben Salah
Potato tuber moth	J. Gouider
Virology: tomato, potato	Head of Lab.: C. Cherif
Grapewine	N. Chebbouh
Pathologie: fruit diseases	Head of Lab.: A. M'laïki
Vegetable diseases	A. Guermach
Sfax Station: plant pathology (fruit and vegetables)	Head of Lab.: A. Trigui
Fruit tree entomology	R. Cherif
Irrigation and fertilization	
Water and irrigation	Head of Lab.: J. Benzarti
Plant nutrition	Head of Lab.: T. Tnani
Fertilization (soil and leaf analysis)	S. Riahi
Ornamental	
Breeding and propagation	Head of Lab.: Y. Chatty
Botany	N. Ben Brahim

1.2 Centre de Recherche de Génie Rurale (C.R.G.R.)
(Research Centre oxf Urbanisation)
rainfall: 474 mm

Av. de l'Indépendance, B.P. 10
2080 Ariana
Phone: 231 624/ 321 628

1.3 Institut National Agronomique de Tunisie (I.N.A.T.)
(National Agricultural Institute of Tunisia)

13 Av. Charles Nicolle, 1002 Tunis
Phone: 283 124

2 Sousse

2.1 Ecole Supérieure d'Horticulture (E.S.H.)
(School of Higher Horticultural Studies)
elevation: 25 m
rainfall : 272 mm
sunshine: 2821 h

4042 Chott Mariem
Phone: 03-48 134

3 Sfax

3.1 Institut de l'Olivier (I.O.) (Olive Tree Institute)
elevation: 25 m
rainfall : 212 mm
sunshine: 3146 h

Rue Arrafat, 3029 Sfax
Phone: 04-41 240

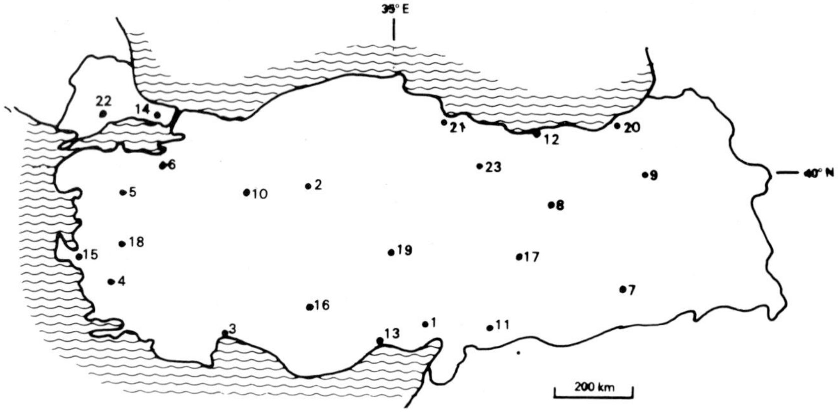
35° E
40° N
200 km
22
14
6
5
10
2
21
12
20
23
9
8
18
15
19
17
4
7
16
3
13
1
11

1 Adana, 2 Ankara, 3 Antalya, 4 Aydın, 5 Balıkesir, 6 Bursa, 7 Diyarbakir, 8 Erzincan, 9 Erzurum, 10 Eskisehir, 11 Gaziantep, 12 Giresun, 13 Içel, 14 Istanbul, 15 Izmir, 16 Konya, 17 Malatya, 18 Manisa, 19 Nevsehir, 20 Rize, 21 Samsun, 22 Tekirdag, 23 Tokat

Survey

Research services in Turkey are supplied by the Government. Agricultural research activities are carried out either by the universities or by the Ministry of Agriculture, Forestry and Rural Affairs. Some of the activities are also supported by the Scientific and Technical Research Council of Turkey. Twelve of the Universities (eight of them have been established within the last 2-3 years) in the country, have faculties of agriculture in which both research and instruction are carried out. Postgraduate training to doctorate level is also given. The General Directorate of Project and Implementation is concerned with agricultural research in the Ministry (Milli Mudafa Caddesi, No: 20, Kizilay-Ankara/Turkey). Under this General Directorate there are 36 research institutes, of which some deal only with horticulture while others deal with different disciplines of agriculture besides horticulture. Besides these research institutes and stations, there are 31 nurseries distributed all over the country. Their task is to produce fruit seedlings, vegetable seed, flower seed and seedlings, and rooted and grafted vine twigs for the growers.

1 Adana

1.1 University of Çukurova, Faculty of Agriculture, Department of Horticulture
elevation: 23 m
rainfall : 700 mm
sunshine : 3150 h

Adana
Phone: 24571/2151
Univ.Dir.: Prof.Dr. M. Ozsan
Head: Prof.Dr. N. Kaska

The department has got nearly all facilities for growing all deciduous tree fruits and subtropical fruits including citrus, olive, avocado, pecan, loquat, pomogranate and fig. Early grapes, early vegetables such as tomato,

Prepared by M. Sürek, Ministry of Agriculture, Forestry and Rural Affairs, Ankara cucumber, eggplant, watermelon, melon and onion are grown as well. In addition there is about 0,3 ha in the department with modern greenhouses, laboratories, cold storage rooms of 40 tons capacity and common stores of 500 tons capacity. In the department research teaching and training are being carried out together. The whole orchard is about 123 ha.

Fruit Section	Prof.Dr. M. Ozsan/Prof.Dr. N. Kaska
	Prof.Dr. M. Yılmaz
	Ass.Prof.Dr. O. Gezerel
	Ass.Prof.Dr. O. Tuzcu
	Ass.Prof.Dr. F. Ergenoglu
Viticulture Section	Dr. M. Akilli
Vegetable Section	Prof.Dr. N. Kaska
Storage Section	

2.1 University of Ankara, Faculty of Agriculture, Department of Horticulture
elevation: 850 m
rainfall : 377 mm
sunshine : 2600 h

Ankara
Phone: 472100/298
Head: Prof.Dr. A Gunay

Producing, breeding and cold storage of horticultural crops (vegetables, fruits and grapes). Ten ha are planted with tree fruits, vegetables and grapes. Equipment: cold stores, phytotron, callusing room for propagation of grafted cuttings and four greenhouses (200 m^2).

Vegetable Section	Prof.Dr.A.Günay/Ass.Prof.Dr.K.Abak
Fruit Section	Prof.Dr. M. Ayfer
	Ass.Prof.Dr. I. Köksal
Viticulture Section	Prof.Dr. Y. Fidan
	Ass.Prof.Dr. H. Çelik

2.2 Central Anatolia Regional Agricultural Research Institute
elevation: 1055 m
rainfall : 377 mm
sunshine : 2650 h

P.O. Box 226, Ankara
Phone: 153244
Telex: 42994 Cimy Tr.
Attn. Dr. S Anil
Dir.: Dr. B. Yılmaz
Ass.Dir.: Dr. S. Anil

This department works on growing and breeding of vegetables especially winter types, temperate tree fruits, grapes and ornamental plants. Three ha out of 30 ha are under sprinkler irrigation. 14 ha of the total acreage are planted with tree fruits: apple, pear, apricot, cherrie, almond, and prune; 8 ha are under vegetables: carrot, celleries, leek, radish, tomato, and pepper; 6 ha with grapes, and 2 ha with ornamental plants.

Horticulture Department	Head: Dr. A. Talay
Pomology Section	Dr. S Anil
Vegetables Section	Dr. A. Talay
Viticulture Section	C. Sarifakioglu
Ornamental Plants Section	F. Kedici

3 Antalya

3.1 Citrus Research Institute
elevation: 39 m
rainfall : 1068 mm
sunshine : 3100 h

P.O. Box 35, Antalya
Phone: 19232, 11465
Dir.: K. Morali

Producing and processing of subtropical tree fruits, mainly citrus, avocado, pecan, and loquat. The Institute also works on coffee, banana and pomogranate. In addition, it serves to the regional area of South-West Turkey on the problems concerning fertilizer application of horticultural crops.

Horticultural Section — Head: Dr. A.Y. Hızal
T. Goral/ Y. Apaydin

Food Processing Section — Head: E.N. Bagriyanik

Soil and Leaf Analysis Laboratory — Head: Dr. A.T. Koseoglu

3.2 Vegetable Research Institute
elevation: sea level
rainfall : 1065 mm
sunshine : 3100 h

P.O. Box 130, Antalya
Phone: 11468
Ass.Dir: U. Ertekin

This institute is the centre of the Greenhouse Research Project in the country. The institute has got 100 ha land. Tomato, pepper, cucumber, and eggplant are the main interests.

Breeding Section — Head: E. Genc
E. Durceylan/ O. Yalcın

Plant Pathology Section — Head: K. Yelboga, M.Sc.

Greenhouse Technology Section — Head: C. Cetin
U. Ertekin/ H. Karatas

4 Aydın

4.1 Agricultural Research Institute
elevation: 65 m
rainfall : 642 mm
sunshine : 2900 h

P.O. Box 10, Incirliova
Erbeyli-Aydın
Phone: Incirliova 3
Dir.: M. Nalbant

The Institute works on breeding and processing of figs. In the orchard there are planted 4 ha of Sarilop fig plantation, 3 ha of Caprifig collection (58 varieties) and 5,6 ha of fig collection (271 varieties). The institute also has got 3 ha of fig seedling nursery.

Breeding Section — Head: A.S. Eroglu
Dr. A. Kabasakal/ G. Koyuncu
S. Ozgen/ Z. Eroglu

5 Balıksir

5.1 Oliveculture Research Station
elevation: 10 m
rainfall : 739 mm
sunshine : 2900 h

P.O. Box 40, Edremit-Balıksir
Phone: 1557, 1371
Dir.: E. Atalay

The station has been working on olive breeding, propagation and processing since 1959. Total area is about 11 ha. 4 ha of the total area are under sprinkler irrigation. There are 2 propagation and 1 research greenhouses, 1 processing laboratory and 1 tissue culture laboratory.

Plant Breeding Section — Head: H.M. Dincer
Y. Luma/ D. Saralp/ Y. Özen

Technology Section — Head: H. Ozen

Sapling Propagation Section — Head: C. Corbacioglu

5.2 Vegetable Seed Experiment and Production Station
elevation: 139 m
rainfall : 603 mm
sunshine : 2525 h

P.O. Box 68, Balikesir
Phone: 11702, 16663
Dir.: I. Dogan
Ass.Dir.: N. Gurel

Producing and processing of vegetable seeds are the

main subjects of the Station. Total area is 100 ha. 70 ha are cultivated for seed production, 30 ha for seed storage, seed processing, research work, and greenhouse.

Seed Production and Research Section — Head: I. Hasdagli, M. Evyapan

Plant Protection Section — Head: G. Aslan

6 Bursa

6.1 University of Uludag, Faculty of Agriculture, Department of Horticulture
elevation: 155 m
rainfall : 713 mm
sunshine : 2600 h

P.O. Box 343, Bursa
Phone: 31244, 31053
Head: Prof.Dr. A. Eris*

This department conducts courses for undergraduates and graduates and research on horticultural plant growing. Also breeding in relation to physiology, morphology, citology, breeding, storage, and post harvest physiology of fruits, vine, vegetables and flowers. At the same time, it is also interested in mushroom production, forcing vegetable growing, standardization, marketing, and exporting of horticultural crops.

Fruit and Viticulture Section — Prof.Dr. A. Eris*, Prof.Dr. Y.S. Agaoglu*, Ass.Prof.Dr. R. Turk/ Dr. A. Soylu

Vegetable Growing and Seed Production Section — Prof.Dr. A. Eris*, Ass.Prof.Dr. V. Seniz

Floriculture and Landscape Architecture — Ass.Prof.Dr. I.V. Alptekin, Dr. A. Menguc

6.2 Sericulture Research Institute
elevation: 100 m
rainfall : 713 mm
sunshine : 2600 h

P.O. Box 1, Bursa
Phone: 31020
Dir.: S.F. Un

Moriculture is the main subject (seed and sapling production). Three ha of mulberry orchard consists of 1 ha of local and exotic variety, 1 ha of production, and 1 ha of nursery stock-unit. The Institute has got a bottom-heat greenhouse.

Moriculture Section — Head: O. Sipahioglu

7 Diyarbakir

7.1 South-East Anatolia Regional Agricultural Research Institute
elevation: 660 m
rainfall : 600 mm
sunshine : 3150 h

P.O. Box 72, Diyarbakir
Phone: 2931
Dir.: Dr. A.E. Firat

The subjects are researches on viticulture and vegetables

Horticulture Section — Head: C. Yaman

8 Erzincan

8.1 Horticultural Research, Production and Training Centre
elevation: 1200 m
rainfall : 400 mm
sunshine : 2590 h

P.O.Box 18, Erzincan
Phone: 1080, 1884
Dir.: Dr. M. Eltez

The centre has got a 405 ha area. 250 ha are used for seedling production and 155 ha for research work. The centre is mainly interested in apple, pear, mulberry, apricot, peach, sour and sweet cherries, plum, grape, and vegetables.

Fruit Section — Dr. M. Eltez/ T. Yalcin/ T. Ceran
Viticulture Section — K. Yılmaz
Vegetable Section — M. Pamir/ G. Dabanlioglu

9 Erzurum

9.1 University of Ataturk, Faculty of Agriculture, Department of Horticulture
elevation: 1950 m
rainfall : 800 mm
sunshine : 2700 h

Erzurum
Phone: 13714, 13689
Head: Prof.Dr. A. Istar

Producing and processing of tree fruits and grapes.

Viticulture Section — Prof.Dr. A. Istar
Landscape Architecture Section — Prof.Dr. F. Tanriverdi
Fruit Growing Section — Ass.Prof.Dr. M. Guleryuz
Greenhouse and Vegetables Section — Ass.Prof.Dr. R. Alan
Ornamental Plants — Dr. K. Guclu

10 Eskisehir

10.1 Agricultural Research Institute
elevation: 790 m
rainfall : 600 mm
sunshine : 2500 h

P.O. Box 18, Eskisehir
Phone: 11030
Dir.: Dr. F. Altay

The Institute works on vegetable researches, and produces vegetable seeds.

Vegetable Section — Head: T. Cetinel
H. Turac

11 Gaziantep

11.1 Agricultural Research Institute
elevation: 850 m
rainfall : 550 mm
sunshine : 3100 h

P.O. Box 32, Gaziantep
Phone: 11057, 20363
Dir.: A.M. Bilgen

This Institute is the centre of the Pistachio Research and Development Project. Other subjects are olive, viticulture and temperate fruits. An area of 60 ha is under sprinkler irrigation.

Pistachio Section — N. Uygur/ Dr. C Kuru/ H. Tekin
Viticulture Section — S. Yurdakul
Oliveculture Section — A. Ulusarac
Economy Section — C. Cakmur
Technology Section — R. Karaca

12 Giresun

12.1 Hazelnut Research and Training Centre
elevation: 10 m

P.O. Box 46, Giresun
Phone: 1136

rainfall : 1500 mm
sunshine : 1800 h

Dir.: K. Altay

The institute only works on hazelnut research: selection, adaptation, and fertilizer.

Breeding Section	A.N. Okay/ N. Koc
Agronomy Section	Y. Kucuk
Physiology Section	A. Kucuk
Economy Section	Y. Yakut/ A. Kaya
Technology Section	Dr. A. Uzun

13 Icel

13.1 Horticultural Research and Training Center
elevation: 10 m
rainfall : 613 mm
sunshine : 3800 h

P.O. Box 27, Erdemli-Içel
Phone: 1084
Dir.: Dr. E. Cetiner

This Institute is located in the area where citrus trees and subtropical fruits are grown. There is 15 ha cultivated land for researches. The main research subjects are selection, adaptation and cultural practices on tree fruits, mainly citrus, other subtropical fruits, peaches, and fig, grapes, vegetables, ornamentals and medicinal plants. There are 5 greenhouses, several plastic houses in which tomatoes, eggplants, cucumber, and green pepper adaptation experiments have been carried out. In addition, rose, banana trials and vegetative propagation experiments are in progress in the greenhouses.

Fruit Section	Head: R. Aydin K. Clikel/ H. Ayanoglu/ E. Ozdemir M. Saglamer/ B. Sen
Vegetable Section	Head: S. Kesici C. Celikel/ M. Cakir
Viticulture Section	Head: A. Inal
Ornamental and Medicinal Plants Section	Head: K.S. Cetiner O. Karaguzel/ I. Aspir
Food Technology Section	Head: O. Yunculer A. Nizamoglu/ G. Yunculer
Leaf and Soil Analysis Laboratory Section	Head: N. Bayram

14 Istanbul

14.1 Ataturk Horticultural Research Institute
elevation: sea level
rainfall : 1100 mm
sunshine : 2600 h

P.O. Box 15, Yalova-Istanbul
Phone: 1005, 2520
Dir.: R. Ergin

Main subjects of studies are breeding new varieties for the region and producing breeder-foundation seed, seedling and nursery stock of these new varieties; increasing the yields by cultural methods, finding better methods for horticultural crops, transferring new technology to the extension service.

Fruit Section	Head: S. Demiroren Dr. O. Konarli/ M. Buyukyilmaz
Vegetable Section	Head: Dr. T. Turkes Dr.N. Turkes/ G. Simsek/ N. Surmeli
Mushroom Section	Head: Dr. S.E. Isik I. Erkel
Viticulture Section	Head: I. Uslu H. Samancı
Olive Section	Head: A.E. Fidan

	A.R. Sutcu
Ornamentals Section	Head: N. Ertan K. Gursan
Post-Harvest Physiology Section	Head: Dr. U. Ertan Dr. Z. Ozelkok/ K. Kaynas
Plant Health Section	Head: K. Akar M. Yureturk/ Dr. T. Muftuoglu
Food Technology Section	Head: O.M. Baykal H.H. Cetin/ F. Fidan
Agricultural Economy Section	Head: A. Yucel A. Safak/ S. Erkal

15 Izmir

15.1 University of Aegean, Faculty of Agriculture, Department of Horticulture
elevation: 10 m
rainfall : 850 mm
sunshine : 2850 h

Bornova-Izmir
Phone: 180110
Dir.: Prof.Dr. M. Dokuzoguz

The subjects are citrus, other subtropical fruits, temperate fruits like apple, pear, plum, sweet and sour cherries; almond, grapes and vegetables.

Fruit Section	Prof.Dr. R. Gulcan Prof.Dr. R. Ozcagiran Ass.Prof.Dr. K. Mendilcioglu
Viticulture Section	Prof.Dr. E. Ilter
Vegetables Section	Prof.Dr.H. Vural/Prof.Dr.F. Macit*
Landscape Architecture Section	Prof.Dr. A. Bayraktar Prof.Dr. A. Hatipoglu

15.2 Aegean Regional Agricultural Research Institute
elevation: 10 m
rainfall : 850 mm
sunshine : 2850 h

P.O. Box 9, Menemen-Izmir
Phone: 149131
Dir.: Dr. K. Temiz

Main subjects of studies are breeding and adaptation in vegetables: eggplant, pepper, musk-melon, cucumber, tomatoes, cauliflower, lettuce and carrot; mistpropagation in cherries, nursery selection in pear, selection in citrus, plum, and sour cherry; hybrid variety development in tulip, and improvement of reproduction techniques of *Lilium spp.*, collection evaluation, storage and rejuvenation of plant genetic material.

Fruit Section	Head: Dr. N. Gonulsen Dr. N. Karabiyik/ M. Ulubelde, M.Sc
Vegetable Section	Head: E. Uraz M.N. Alan/ T. Bas
Ornamental Plant Section	Head: D. Ozkahya S. Uraz

15.3 Oliveculture Research Station
elevation: 28 m
rainfall : 700 mm
sunshine : 2600 h

Bornova-Izmir
Phone: 161036, 161035
Dir.: I. Dikmen

The Institute works on olive breeding and processing in collaboration with the International Olive Oil Council and UNDP/FAO. Activities are researches on olive breeding, fertilization, processing table olive, olive oil and sapling propagation (100 000 saplings/year). Equipments: 2 greenhouses each 120 m^2, and plastic tunnels each 30 m^2 for sapling propagation, olive oil mill, international olive oil laboratory, table olive processing room, leaf and soil analysis laboratory. Total area of the institute is 78 ha.

Olive Propagation Section — Head: D. Usanmaz
A. Uluskan/ A. Cavusoglu/ H. Arsel
E. Ozahci/ B. Yuce/ A. Sayin

Olive Oil Technology Section — Head: A. Oktar
B. Ersoy/ H. Acar/ T. Isikli

Table Olive Technology Section — Head: A. Erol
D. Tetik/ N. Ozyılmaz

Leaf Analysis Laboratory Section — Head: O. Canozer
M. Cakir/ G. Puskulcu

Economy and Marketing Section — Head: B. Akman
Mechanization Section — Head: D. Caran

16 Konya

16.1 University of Selcuk, Faculty of Agriculture, Department of Horticulture
elevation: 1030 m
rainfall : 380 mm
sunshine : 2850 h

Konya
Phone: 18510
Head: Ass.Prof.Dr. MF Ecevit

The department works on viticulture.

17 Malatya

17.1 Agricultural Research Station
elevation: 900 m
rainfall : 400 mm
sunshine : 2700 h

PO Box 43, Malatya
Phone: 11297, 17406
Dir.: -
Ass.Dir.: N. Demirel

Main subjects of studies are selection of apricot varieties and determination of their characteristics as genetic stock, adaptation for sweet cherry, peach, and pear; selection and training system for vine. The station has got 350 ha land. 150 apricot cultivars are available in the collection.

Fruit Section — Head: T. Pektekin
Viticulture Section — Head: N. Demirel

18 Manisa

18.1 Viticulture Research Institute
elevation: 71 m
rainfall : 747 mm
sunshine : 2850 h

P.O. Box 12, Manisa
Phone: 1293
Dir.: Dr. A. Caliskan

The main function of the Institute is raisin production, processing and marketing. Besides raisins, the Institute works also on phylloxera resistant rootstock production. The Institute has got 15,5 ha nursery and 5,5 ha vineyard. There are 200 regional grape cultivars in the collection.

Genetic and Breeding Section — I. Ilhan
Protection Section — A. Ertem
Raisin Technology Section — E. Karagozoglu

19 Nevsehir

19.1 Viticulture Research Station
elevation: 1260 m
rainfall : 388 mm

P.O. Box 12, Nevsehir
Phone: 1281
Dir.: Y. Demirbuker

The main subject of the station is viticulture (research, propagation and processing).

Viticulture Section — I. Yuksel
Agronomy Section — H. Dabanli
Technology Section — H. Sarioglu
Economy and Marketing Section — S. Gencturk

20 Rize

20.1 Tea Research Institute — Rize
elevation: sea level
rainfall : 1750 mm
sunshine : 1800 h

This Institute is under the Ministry of Finance and Customs. The main subject of study is variety selection, variety preservation and cultural methods.

21 Samsun

21.1 University of May 19, Faculty of Agriculture, Department of Horticulture
elevation: 10 m
rainfall : 735 mm
sunshine : 3000 h

Samsun
Phone: 19688
Dir.: Prof.Dr. F. Tosun
Head: Prof.Dr. H. Apan

A new horticultural orchard is being set up on the University Campus for demonstration and research studies. In the campus area, about 60 ha of land is available for horticultural purposes. All temperate zone fruits like apple, pear, apricot, peach, plum, cherry, hazelnut, walnut, fig, etc will be grown in this area.

Vegetable Section — Prof.Dr. H. Apan
Viticulture Section — Ass.Prof.Dr. F. Odabas
Fruit Section — Ass.Prof.Dr. S.M. Sen

22 Tekirdag

22.1 Viticulture Research Institute
elevation: 10 m
rainfall : 600 mm
sunshine : 2500 h

P.O. Box 7, Tekirdag
Phone: 4893, 2042
Dir.: C. Baris

The Institute mainly works on viticulture. It is the centre of the viticulture research projects carried out in Turkey. It has got 100 ha land of which 60 ha are under irrigation, 20 ha of grapes planted, 20 ha of rootstocks mother plantation, 40 ha for sapling propagation. 600.000 saplings are propagated annually.

Breeding and Amphelografic Studies Section — Head: K. Gurnil
Vine Physiology Section — Head: Dr. E. Gokcay
Cultural Technics and Propagation Section — Head: H. Eryildiz
Vine Technology — Head: F. Yayla

23 Tokat

23.1 University of Republic, Faculty of Agriculture, Department of Horticulture
elevation: 650 m
rainfall : 460 mm
sunshine : 2150 h

Tokat
Phone: 1870
Head: Prof.Dr. A. Yazgan

This is a new faculty. In the Department of Horticulture only vegetables studies are carried out.

USSR

Union of Soviet Socialist Republics

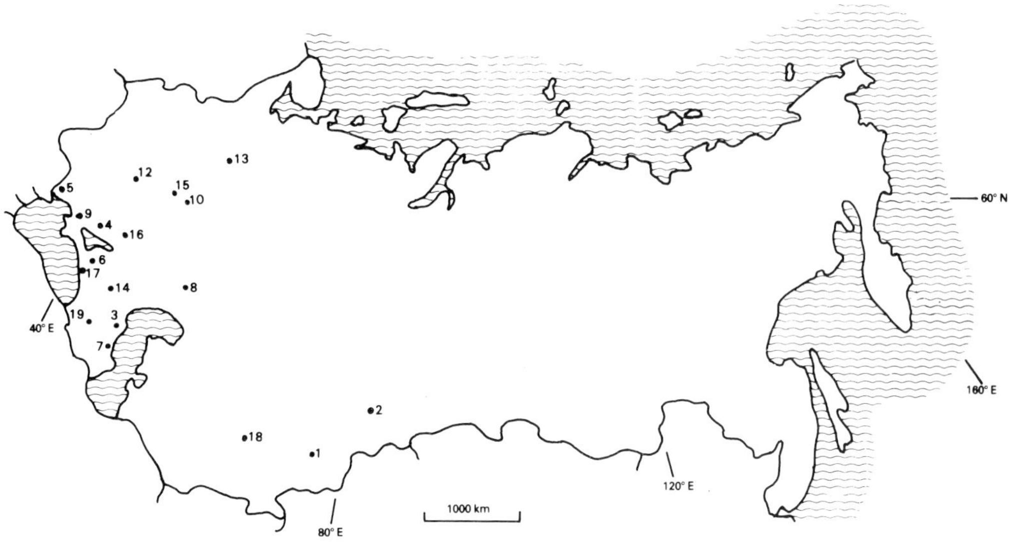
1
2
3
4
5
6
7
8
9
10
12
13
14
15
16
17
18
19
60° N
40° E
80° E
120° E
160° E
1000 km

1 Alma-Ata, 2 Barnaul, 3 Dushanbe, 4 Kiev, 5 Kishinëv, 6 Krasnodar Territory, 7 Kuba, 8 Kuibyshev, 9 Melitopol', 10 Michurinsk, 11 Vashnil, 12 Minsk, 13 Moscow, 14 Nal'chik, 15 Orël, 16 Rossosh', 17 Sochi, 18 Tashkent, 19 Tbilisi

1 Alma Ata

1.1 Kazan Research Institute of Production, Storage and Processing of Fruit and Vegetables (Ministry of Agriculture of Kazachstan)

. 480032 Kazachstan SSR, Alma Ata
Gagarin Avenue 238a
Phone: 458.575
Dir.: K.N. Oetarakov, Cand.Agr.Sc.

Selection, introduction and multiplication of fruit- grapes and berry-cultures, development of technology for growing, harvesting, storage and processing of fruit, berries and grapes, studies of biochemical characteristics of fruits, berries and grapes, development of systems of soil maintenance, irrigation and fertilizers, plant protection against pests and diseases in crops. Selection of equipment for complex mechanization of production processes in horticulture.

2 Barnaul

2.1 Lisavenko Siberia Institute of Fruit Production (Siberian Section of Vaskhnil)

. 656020 Altai Territory, Barnaul 20,
Zmeinogorodsky, Trakt 49
Phone: 319.86
Dir.: Mrs. Dr. I.P. Kalinina,
Memb. V.IL., Acad. Agr. Sc.

Development of technology for horticultural production under the climate of the Altai and Siberia, studying of varieties and selection of fruits, berries and ornamental plants, mechanization of labour consuming processes in berry growing.

3 Dushanbe

3.1 Tadzik Research Institute of Horticulture and

. 735022 Tadzhik SSR, a/j 191,

Prepared by V.L. Simonov, Ministry of Horticulture, Dept. Foreign Relations (Gosagroprom) Moscow.

Viticulture (Ministry of Horticulture of Tadzjikistan)

Phone: 311.653
Dir.: Dr. U.E. Eshankulov, Cand. Agr. Sc.

Scientific research of production technology for high density growing in gardens on irrigated soil, development of production technology for viticulture on irrigated and reclaimed soils, development of technology for citrus growing and vegetable growing in the open air and in glasshouses, development of technology for seedlings of grapes, citrus fruits and nut-cultures, the introduction of new varieties of grapes, fruit and subtropical fruits.

Departments: Citric and subtropical cultures, horticulture, viticulture, fruit, melons and potatoes, plant protection, scientific technical information.

4 Kiev

4.1 Ukrainian Research Institute of Fruit Production (Ministry of Food Industry of the Ukranian SSR)

. 252027 Kiev 27, Teremki
Phone: 666.548, 666.117
Dir.:Dr.V.I.Maidebura, Dr.Agr.Sc.

Development of technology for production of planting material, intensive horticulture and berry-culture under the climate of the Ukraine, introduction and variety studies and the selection of new varieties, mechanization of labour consuming processes, technology for storage and processing of fruit and berries, economics and organization of horticultural production.

Departments: Agrotechnics, nurseries, mechanization, plant protection, selection and variety studies, agrochemics, soil sciences, physiology and cold resistance, storage technology and processing of fruit and berries, department for reclamation of sandy and low productivity soils, economics and organization of agrocultural production.

5 Kishinev

5.1 Kodru Moldavian Research Institute of Fruit Production (Ministry of Horticulture of Moldavia)

. 277019 Kishinev 19,
ul. Fruktovaya 14,
Phone: 526.418
Dir.: A.V. Vylku, Cand. Econ.Sc.

Development of intensive technology for horticulture and planting material for the production of seed and cloned rootstocks for fruit

6 Krasnodar Territory

6.1 North Caucasion Zonal Research Institute of Fruit Production and viticulture (All Russian Department of Vasknhil)

. 350029 Krasnodar, PO 29,
Shosseinaya Street 39
Phone: 412.06
Dir.:Dr.I.N.Pereverzev,Dr.Ec.Sc.

Development of technology for production of fruit-, grape- and berry- cultures, selection and variety studying of seeds and kernels, development of means of mechanization of labour in horticulture

7 Kuba

7.1 Azerbaidzjan Research Institute for Horticulture and Subtropical Cultures (Ministry of Agriculture of Azerbaidzjan)

. Kuba, Kuba district, Pos. Zardabi
Phone: 3717
Dir.: F.M. Mamedov, Cand. Biol. Sc.

Selection and variety studies, improving production

technology for fruit cultures, subtropical cultures and tea

8 Kuibyshev

8.1 Kuybishev Zonal Experiment Station of Horticulture (Ministry of Horticulture RSFSR)

. 443072 Kuibyshev, PO 72,
Phone: 530.912
Dir.: F.N. Rykalin, Cand. Agr. Sc.

Development of intensive technology for horticulture, introduction of new varieties of fruits, berries and grapes

9 Melitopol

9.1 Ukrainian Research Institute of Fruit Production on Irrigated Lands (Ministry of Food Production of the UKR.SSR)

. Zaporozhsky Region,
332311 Melitopol 11,
Vakulinchuk Street 99
Phone: 31120
Dir.: Dr. V.I. Senin, Cand.Agr.Sc.

Improving technology of fruit production in highly productive nurseries, under the climate of the southern and central steppe of the Ukraine, improving plant protection against diseases and pests, technology and processing of fruit and berries, introduction variety studies and selection of seed and kernel cultures.

Departments: Agrotechnics, irrigation-regimes, sprinkling technics, selection and variety studies, mechanization of labour consuming processes, storage and processing, plant protection, economics information and introduction of advanced technology.

10 Michurinsk

10.1 Michurin All Union Research Institute of Fruit Production (Ministry of Horticulture of the USSR)
elevation : -155 m
rainfall : 502 mm
sunshine : 4484 hrs

. 393740 Tambov Region,
Michurinsk 14, Michurin st. 30
Phone: 42161
Dir.: M.I. Boldirev, Cand. Agr. Sc.

Coordination of scientific research for horticulture, research for more sophisticated ways of intensification of horticulture via mechanization of production processes, development of advanced technology of growing, harvesting and storage, developing of new varieties and rootstocks, improvement of labour organization and labour payment, introduction into practice of scientific recommendations, equipment, varieties, rootstocks and standards, selection of ornamental plants: asters, gladioli and lilies.

Departments: Coordination, agrotechnics, selection, multiplication, mechanization, economics, storage, plant protection, introduction of new technologies, extension service.

11 Vaskhnil

11.1 Michurin Central Genetical Laboratory

. 393740 Tambov Region,
Michurinsk 10
Phone: 425.52, 92220
Dir.: Dr.G.A. Kursakov, Cand.Biol.Sc.

Development of new methods for breeding winter hardiness and high quality cultivars of fruit and berries, with high consumer qualities. Diagnostic methods for winter hardiness, studies of the effect of physical and chemical agents treatment on the fertilization process, development of heredity of fruit and berry plants.

12 Minsk

12.1 Byelorussian Research Institute of Horticulture and Potato Growing (Ministry of Horticulture BSSR)

223013 Minsk Region, Samokhvalovichi
ul. Kovalev 2,
Phone: 363441, 941145
Dir.: A.V. Krugliakov, Cand.Agr.Sc.

Development of technology for growing fruit and berry cultures in the climate of the BSSR.

13 Moscow

13.1 Zonal Research Institute of Fruit Production of the Non-Black Soil Belt of the RSFSR

115547 Moscow M547, Biriulevo,
Phone: 384.8496, 384.8668
Dir.: Dr.V.G.Trushechkin, Dr.Agr.Sc.

Development of technology for fruit- and berry-cultures under the conditions of the non-black soil belt of the RSFSR, variety studies and developing of new varieties of fruits and berries

14 Nalchik

14.1 Kabardino-Balkar Experiment Station of Horticulture (Ministry of Horticulture of the RSFSR)

360004 Kabardino-Balkar Autonomous Republic, Nalchik, Zatish-E 2
Phone: 25294, 27590
Dir.: Dr. A.K. Kairov, Cand.Agr.Sc.

Development of advanced technology for fruit and grape growing on a commercial scale, improvement of low productive sloping lands and swamps

15 Orel

15.1 Orel Zonal Experimental Station of Fruit Production (Ministry of Horticulture RSFSR)
elevation : 203 m
rainfall : 515 mm
sunshine : 1964 hrs

303130 Orel, P.O. Zjilina,
Phone: 444.79, 447.79
Dir.: J.V. Osipov, Cand.Agr.Sc.

Development of technology for intensive production of orchards and berry-gardens, selection and variety studies of new resistent species of seeds and kernel cultures

16 Rossosh

16.1 Rossosh Zonal Experiment Station of Horticulture and Berry Growing (Ministry of Horticulture RSFSR)
elevation : 147 m
rainfall : 402 mm
sunshine : 1924 hrs

Rossosh, Veronezh Region
Phone: 91110, 98136
Dir.: Dr.Y.E. Fomenko, Cand.Agr.Sc.

Breeding of planting material for fruit, berry-cultures and ornamental plants, improving the technology of growing them on a commerical scale

17 Sochi

17.1 Research Institute of Mountain Fruit Production

354002 Sochi,

(Ministry of Agriculture of the USSR)

Fabritsius Street 2/28,
Phone: 927.361, 996.821
Dir.:Dr. V.A. Grjazev, Dr.Agr.Sc.

Development of technology for production of fruit, grapes and berries, tea, subtropical and ornamental plants under the climate of the Black Sea coast on a commerical scale

18 Tasjkent

18.1 Shreder Uzbek Research Institute of Fruit Production, Viticulture and Winemaking
(Middle-Asian Department of Vaskhnil)

700000 Tashkent, PO 16,
Ordzhonikidze District
Phone: 445.077, 249.730
Dir.: Dr. M.M. Mirzayev,
Corres.Memb.V.I. Lenin Acad.Agr.Sc.

Selection, agrotechnics, mechanization, biochemics, physiology, plant protection, economics of fruit-, berries and grape-cultures, breeding of planting material

19 Tbilisi

19.1 Georgian Research Institute of Fruit Production, Viticulture and Winemaking
(State Committee for Agricultural Production)

380015 Tbilisi, 15 Vashladzhvara,
Phone: 514.611
Dir.: Dr. N.S. Chkartishvili,
Cand. Agr. Sc.

Development of technology for fruit and grape production on a commerical scale, storage and processing of fruit

UNITED KINGDOM GB

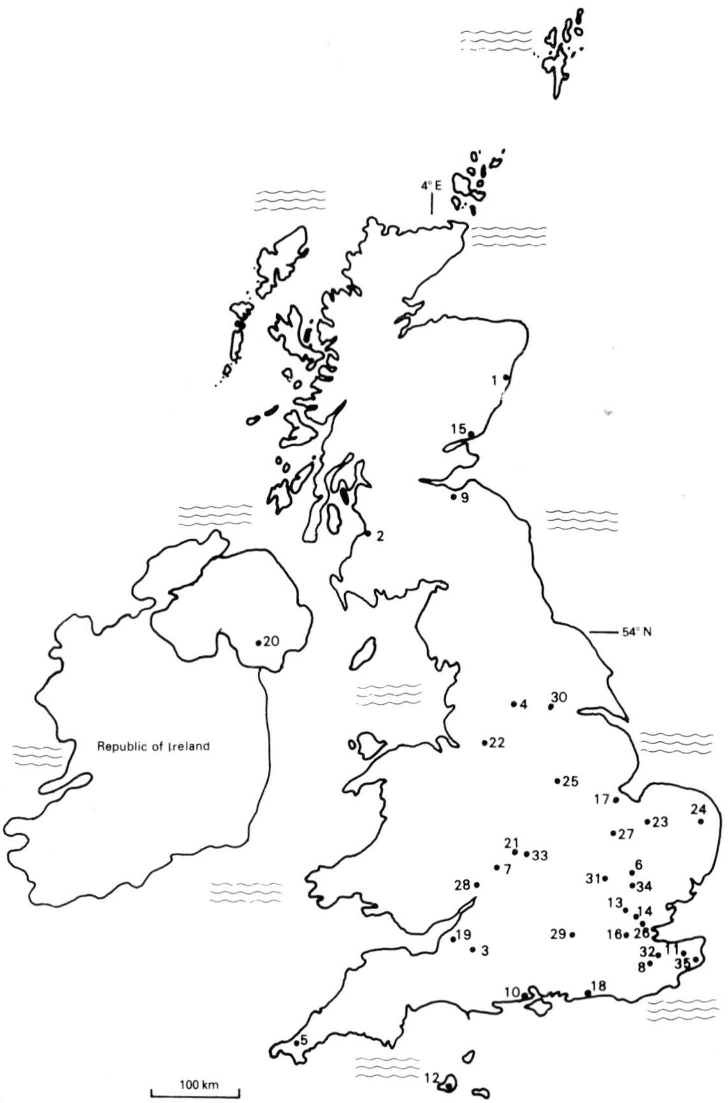
4° E
54° N
Republic of Ireland
100 km
1
2
3
4
5
6
7
8
9
10
11
12
13
14
15
16
17
18
19
20
21
22
23
24
25
26
27
28
29
30
31
32
33
34
35

1 Aberdeen, 2 Auchincruive, 3 Bath, 4 Bingley, 5 Camborne, 6 Cambridge, 7 Chipping Campden, 8 East Malling, 9 Edingburgh, 10 Efford, 11 Faversham, 12 Guernsey, 13 Harpenden, 14 Hoddesdon, 15 Invergowrie, 16 Kew, 17 Kirton, 18 Littlehampton, 19 Long Ashton, 20 Loughgall, 21 Luddington, 22 Manchester, 23 Mepal, 24 Norwich, 25 Nottingham, 26 Ongar, 27 Peterborough, 28 Preston Wynne, 29 Reading, 30 Selby, 31 Silsoe, 32 Sittingbourne, 33 Wellesbourne, 34 Whittlesford, 35 Wye

Survey

Organized research in agricultural subjects in Great Britain began in 1843 with the founding of Rothamsted Experimental Station at Harpenden by Sir John Lawes. Experiments there soon attracted the attention of scientists throughout the world, but it was not until the first decade of the present century that similar institutions began to be set up to investigate the problems of horticultural crops.

Each research station has had a different history and is governed in a different way, although with the institutes founded since 1944 a similar pattern of administration has emerged.

The Research Stations - Long Ashton Research Station originated in 1903 as the National Fruit and

Prepared by B.F. Self, East Malling Research Station, Maidstone, Kent.

Cider Institute and in 1912 it became associated with the University of Bristol. East Malling Research Station started in 1913 as a field station of Wye College. East Malling, now independent of Wye, specializes in work on fruit crops, hardy ornamental nursery stock, and pathology of hops. The tradition of field research in hops is continued at Wye College in a department which studies all aspects of the production and growing of the crop. In 1969 Ditton Laboratory, which opened in 1929 and was noted for its research on controlled atmosphere storage of apples, was combined with the adjacent East Malling Research Station. Other research institutes have been established more recently. At the National Vegetable Research Station, Warwickshire, work in progress includes breeding of new varieties of outdoor vegetables and watercress, weed studies, and irrigation. The Glasshouse Crops Research Institute, Littlehampton, Sussex, was established at its present site in 1953, but was originally founded in 1914 as the Experimental Research Station, Cheshunt, Hertfordshire. It is concerned with all aspects of growing glasshouse crops, hardy nursery stock, bulbs and mushrooms.

The Scottish Crop Research Institute, Invergowrie, devotes particular attention to horticultural problems in Scotland with special emphasis on the culture of soft fruits and vegetables and the study of plant diseases.

In addition to the institutes already mentioned, two other Agricultural and Food Research Council (AFRC) sponsored bodies have horticultural departments. The National Institute of Agricultural Engineering at Silsoe has engineering departments concerned with the design of sprayers and other horticultural machinery and an environmental control division which investigates problems of glasshouse heating and watering. The Food Research Institute was established at Norwich adjacent to the University of East Anglia. The State-aided institutes in England and Wales now draw their grants from the AFRC. Those in Scotland are financed by the Department of Agriculture and Fisheries for Scotland on the advice of the Council, which has responsibility for agricultural, horticultural and food research in Great Britain. The AFRC is responsible to the Secretary of State for Education and Science.

In 1972 a body was set up to advise the Agricultural Research Council, the Department of Agriculture and Fisheries for Scotland, and the Ministry of Agriculture, Fisheries and Food on research and development needs and priorities. This is the Joint Consultative Organisation for Research and Development in Agriculture and Food (JCO). The major objective of the re-organisation was to secure a closer integration of research and development work supported from public funds by the AFRC, MAFF, DAFS, and the universities. Since 1975 about one half of the funds for AFRC are provided by the Ministry of Agriculture, Fisheries and Food as payment for work commissioned with the Council and carried out at AFRC institutes.

The Universities - Five universities have horticultural courses, viz. the Universities of Bath, London (Wye College), Nottingham, Reading, and Strathclyde (in association with the West of Scotland College of Agriculture). All engage in research as an adjunct to their teaching activities. Work done in the botany departments of universities includes biochemistry, physiology, genetics and ecology. The money to support this work is mostly derived from the Government grant to the universities, and may be supplemented by research grants from the AFRC to cover particular projects. The Council gives extended support to certain leading scientific workers, providing them with additional staff in small teams known as research units. The AFRC Unit of Nitrogen Fixation at the University of Sussex at Brighton is an example.

The Agricultural Development and Advisory Service - A wide range of practical horticultural problems are studied on the Experimental Horticulture Stations which form part of the ADAS of the Ministry of Agriculture, Fisheries and Food. These stations are intended to provide facilities for the testing of research findings on a wider scale than can be done at the research stations under conditions approximating to those found on commercial farms and nurseries. They provide facilities for the study of important local problems and in addition conduct variety trials. The stations are located in the main horticultural areas of England. The Directors are senior members of ADAS and are assisted by horticultural officers and trained crop recorders. The first Experimental Horticultural Station to be established was at Luddington near Stratford-upon-Avon, where apples, plums, soft fruits, hardy ornamental nursery stock, and a range of vegetable crops are studied. At Stockbridge House in Yorkshire a similar station is concerned particularly with vegetables and glasshouse crops suitable for northern English conditions, and rhubarb for forcing. Rosewarne in Cornwall caters for the needs of the growers of early vegetables and potatoes, and flowers in the south west and the Isles of Scilly. At Efford in Hampshire work covers soft fruit, vines, vegetables, and crops grow in structures and glasshouses. Experiments with cucumbers, lettuce, and other protected crops, including mushrooms, are undertaken at the Lee Valley EHS at Hoddesdon north of London.

Variety Trials - There are two important organisations concerned with variety testing: the National Institute of Agricultural Botany (NIAB) at Cambridge, and the National Fruit Trials at Brogdale EHS.

The NIAB, sponsored by the Ministry of Agriculture, exists for the classification and testing of varieties of crop plants and for the maintainance of mother stocks of varieties which official breeders have raised. In addition to its own farm at Cambridge it makes use of farms, agricultural colleges, and EHS's as regional trial centres. For fruit crops similar functions are performed by Brogdale EHS National Fruit Trials near Faversham, Kent. Trials of apple, pear, plum, cherry, black currant, strawberry, raspberry, gooseberry and red currant are conducted. At Brogdale a living collection of over 2,000 named varieties of apple is maintained. National collections of plants include the Royal Botanic Gardens at Kew and Edinburgh. At Kew there is a herbarium with a large collection of preserved plants from many parts of the world. In addition, there is the collection of living plants comprising some 25,000 species and varieties. Scientific work is concerned with the identification, classification of plants, study of their anatomy and cytology, and their economic uses. The Royal Botanic Gardens, Edinburgh, was established in 1670. Several universities of the United Kingdom have botanic gardens providing material for plant investigations. The Royal Horticultural Society's garden at Wisley, Surrey, performs a valuable service in testing varieties of garden plants. Pests and diseases of plants are also studied. The national rose collection is maintained at St. Albans, Hertfordshire, by the Royal National Rose Society. Many horticultural firms have collections of plants for selection and breeding.

Northern Ireland - Research on horticultural crops is one of the responsibilities of the Ministry of Agriculture for Northern Ireland, which maintains a Horticultural Department at Loughgall in County Armagh, where work covers apples, black currants, strawberries, mushrooms, and a number of pest and disease and weed control problems.

Scotland - The Scottish Crop Research Institute is situated in Invergowrie, near Dundee. Three agricultural colleges, the North of Scotland College of Agriculture at Aberdeen, the West of Scotland at Auchincruive, and the Edinburgh School of Agriculture, are all engaged in advisory work, experiments and variety trials.

Other Research - Most of the organisations mentioned are maintained by funds provided directly or indirectly by the Government. There are two further groups which conduct research of a more specialized kind. The first are trade associations for research purposes, for example the Campden Food Preservation Research Association, Chipping Campden. Secondly, there are the manufacturers of agricultural chemicals, who are concerned mainly with the development and field testing of their company's products.

Publications - The best known among the scientific journals is the 'Journal of Horticultural Science'. It is sponsored by East Malling and Long Ashton research stations, but receives contributions from all institutes in the UK conducting research and from many overseas centres. The journal entitled 'Crops Research' covers research and experimental work on all horticultural crops in the UK and from overseas countries. Each of the main research stations issues an annual report which contains full details of their publications. Annual reviews are issued by the ADAS EHS's. 'Horticultural Abstracts' is prepared by the Commonwealth Bureau of Horticulture and Plantation Crops, East Malling, and appears monthly. The index and the volumes of abstracts are a key to the horticultural literature of the world.

Although the Bureau producing the Abstracts is specially linked with the countries of the British Commonwealth, the journals and publications include those of almost every country with an interest in horticulture.

In addition, the Bureau publishes Technical Communications, for example TC22 'Sand and water culture, methods used in the study of plant nutrition' by E.J. Hewitt, is a standard reference work. Occasional reviews are also produced. An extensive series of bulletins and advisory leaflets are published by the Ministry of Agriculture, Fisheries and Food; by the Department of Agriculture for Scotland; and by the Department of Agriculture for Northern Ireland. Many of the manufacturers of spray chemicals, fertilizers and other technical products also issue leaflets or bulletins.

List of publications - 1. Abstracting journals: 'Horticultural Abstracts'. Other abstracting journals in related fields: 'Forestry Abstracts', 'Pesticides Abstracts', 'Soils and Fertilizers', 'Weed Abstracts'.

2. Scientific or technical journals: 'Agriculture in Northern Ireland', 'Annals of Applied Biology', 'Annals of Botany', 'Forestry', 'Crops Research', 'Journal of Agricultural Engineering Research', 'Journal of Experimental Botany', 'Journal of Horticultural Science', 'Journal of the Science of Food and Agriculture', 'Journal of the Sports Turf Research Institute', 'Plant Pathology', 'Tropical Agriculture', 'Weeds Research', 'World Crops'.

3. The horticultural press: 'Gardeners Chronicle and Horticulture Trade Journal', 'Nurseryman and Garden Centre', 'The Grower' (weekly).

4. Reports of official research stations and similar bodies: ADAS Research and Development Reports issued in a number of parts e.g.: 'Protected Crops Vegetables', 'Protected Crops Ornamentals', 'Field Vegetables', 'Fruit and Hops', 'Hardy Ornamental Nursery Stock', 'Bulbs and Allied Flower Crops'; Agricultural and Food Research Council; Food Research Institute; East Malling Research Station; Edinburgh School of Agriculture; Efford EHS; Forestry Commission Report on Forest Research; Glass-

house Crops Research Institute; Wye College Department of Hop Research; John Innes Institute; Kirton EHS; Lee Valley EHS; Long Ashton Research Station; Luddington EHS; Department of Agriculture for Northern Ireland (Research and Technical Work); National Institute of Agricultural Botany; National Institute of Agricultural Engineering; National Vegetable Research Station; School of Agriculture, Aberdeen; Processors and Growers Research Organisation; Arthur Rickwood Experimental Husbandry Farm (vegetables); Rosemaund Experimental Husbandry Farm (for hops); Rosewarne EHS; Scottish Crop Research Institute; Stockbridge House EHS; University of Nottingham School of Agriculture; West of Scotland Agricultural College.
5. Yearbooks issued by the Royal Horticultural Society: 'Daffodils', 'Lilies and other Liliaceae', 'Rhododendron and Camellias'.
6. Other yearbooks and annuals: The specialised societies concerned with particular plants issue yearbooks.
7. Reports on economics of commercial horticulture issued by the University Departments of Agricultural Economics. Reports available include items from: University of Bath, University of Bristol, University of London (Wye College), University of Nottingham, University of Reading, Edinburgh School of Agriculture
8. Surveys of agricultural research in the United Kingdom: 'List of Research Workers in the Agricultural Sciences in the Commonwealth' published by the Commonwealth Agricultural Bureaux 1981, includes a list of research workers in the UK and their main lines of study.

1 Aberdeen

1.1 The North of Scotland College of Agriculture and Aberdeen University Department of Agriculture School of Agriculture
Research laboratories and controlled environment facilities are established for postgraduates horticultural studies.

581 King Street, Aberdeen AB9 1UD
Scotland
Phone: 0224 40291
Telex: 73538
Principal: Prof. G.A. Lodge

Horticulture Division	Head: Dr. G.R. Dixon*
Hardy nursery stock specialist	A.Q.M. Blain
Ornamental bulbs specialist	M.W. Sutton

1.2 Experimental Horticulture Unit
elevation: 80 m
rainfall : 816 mm
sunshine : 1310 h

Craibstone Estate, Bucksburn,
Aberdeen, Scotland
Phone: 0224 712616

Facilities include 6 ha of land, heated glass, cold and heated polythene structures, insect proofed structure, netting tunnels, laboratories and controlled environment chambers. Research concentrates on the problems of hardy nursery stock, ornamental bulbs and protected vegetable crops.

Unit manager	F. Wilson
Trials officer	C.P. Britt

1.3 Experimental Horticulture Unit
elevation: 30 m
rainfall : 620 mm
sunshine : 1350 h

Aldroughty Farm, Elgin,
Morayshire, Scotland
Phone: 0343 45026

Research is concerned with field vegetables and soft fruit crops, experiments are rotated around 50 ha of arable farmland. Facilities include field laboratories, unheated plastic tunnels and a cold store.

Field vegetables specialist	J.W. Berridge
Soft fruit specialist	J.P. Sutherland
Trials officer	J.G. Fraser

2 Auchincruive

2.1 The West of Scotland Agricultural College,
Department of Horticulture and Beekeeping

Auchincruive - Ayr - Scotland
Phone: 0292 520331
Head: Prof. H.J. Gooding*

Senior Lecturers: P.J. Dudney/ F.S. Hardy
I.C Maxwell (Horticulturist and Apiarist)

Lecturers: Miss. L.W. Dick/ I.C.G. Dougall
G.D. Watson/ C.M. Taylor

Demonstrators: Miss. F.J. Newton/ A.M. Stirrat (Apiarist)

Courses Offered:
A 4-year Honours BSc degree course in Horticulture, taught in collaboration with the University of Strathclyde (Glasgow) and the science, engineering and economics Divisions/Departments of the College. (Introduced 1972.)
A 3-year SCOTEC Diploma in Horticulture (DH), with final year options in Amenity or Commercial Horticulture. Introducted in 1982. It is planned to introduce modules in 1985: which will correspond in most aspects with the DH. Success in defined groups of modules will be validated as a College Diploma in Horticulture or Higher Diploma in Horticulture.
A 1-year College Certificate course in Beekeeping.
Postgraduate research facilities for MSc and PhD degree courses are available for field and protected crops. Project area topics are available on application.

2.2 The West of Scotland Agricultural College,
Glasshouse Investigational Unit for Scotland
elevation: 45 m
rainfall : 927 mm
sunshine : 1347 h

Auchincruive - Ayr - Scotland
Phone: 0292 520419
Telex: 77740
Head: R.A.K. Szmidt
Exp.Off.: G.M. Hitchon*/J.H.F. Smith

An experimental horticultural Unit established in 1970 to serve the needs of glasshouse growers in Scotland. Administered by the West of Scotland Agricultural College. Station Technical Committee includes representatives of the three Agricultural Colleges, the Department of Agriculture for Scotland, the Scottish Crop Research Institute and the National Farmers' Union for Scotland.

3 Bath

3.1 University of Bath*, School of Biological Sciences
Plant Biology Group
elevation: 180 m
rainfall : 850 mm
sunshine : 1460 h (estimate)

Bath University, Claverton Down
Bath BA2 7AY England
Phone: 0225 61244
Telex: 449097
Head: Prof. G.G. Henshaw

Horticultural section
Entomology — Head: V.W. Fowler
Vegetable production and nutrition — D.C. Cull
Tissue culture — Dr. A.W. Flegmann
Seed production and protected cropping — R.A.T. George*
Horticultural economics — Dr. M.J. Sargent*
Weed biology — R.J. Stephens
Management
Fruit production — H. Wainwright
Crop protection section
Physiology and fine structure in wilt diseases — Head: Dr. R.M. Cooper
Mushroom physiology and pathology. Biomass in composts. — To be appointed
Diseases of ornamentals
Mode of action of insecticides; gut flora of insects — Dr. AK Charnley

Plant sciences section

Seed development, germination and seedling physiology	Head: Dr. K.G. Moore
Flower longevity. Root initiation.	
The environmental control of bolting and flowering	T.D.H. Catchpool
Mode of action of herbicides	Dr. A.D. Dodge
Watercress physiology	Dr. L.W. Robinson

4 Bingley

4.1 The Sports Turf Research Institute
elevation: 198 m
rainfall : 940 mm

Bingley, West Yorkshire,
BD16 1AU England
Phone: 0274 565131
Dir.: Dr. P. Hayes

The institute is the UK's national centre for sport and amenity turf. Independent and non-commercial, it conducts sponsored research for Government agencies etc. in addition to its own programme of research, and operates an advisory service.

Research section	Ass.Dir.: J.P. Shildrick
Soil physics	Dr. S.W. Baker
Plant pathology	A.R. Woolhouse
Soil chemistry	Dr. D. Lawson
Research officers	P.M. Canaway/ C.H. Peel/ M.J. Bell
	Dr. G. Holmes

5 Camborne

5.1 Rosewarne Experimental Horticulture Station
ADAS, Ministry of Agriculture
elevation: 43-79 m
rainfall : 1092 mm
sunshine : 1609 h

Camborne, Cornwall TR14 0AB England
Phone: 0209 716 673
Dir.: M.R. Pollock
Dep.Dir.: A.A. Tompsett

Area 52 ha. Variety improvement and field experiments in vegetable crops, especially winter heading cauliflower, spring cabbage and early potatoes. Stock improvement, field experiments and control of flowering in narcissus, anemones and allied flower crops. Crop production under glass and plastic structures. Trees and shrubs for shelter and cut foliage production. A sub-station on the Isles of Scilly is working on early flowering narcissus and early potatoes. Area 1.1 ha.

B.H. Houghton/ J.R. Smith
C.R. Treble/ C.R. Tapsell (Plant breeder on secondment from Nat. Vegetable Research Station, Wellesbourne)

6 Cambridge

6.1 National Institute of Agricultural Botany*

Huntingdon Road, Cambridge CB3 OLE England
Phone: 0223 276381
Telex: 817455
Dir.: Dr. G.M. Milbourn
Dep.Dirs.: Mrs. V. Silvey
Dr. J.K. Doodson

Vegetable and seed testing of general agricultural crops including vegetables. Official Seed Testing Station for England and Wales.

Vegetable Branch	Head: J.W. Chowings
Trials section	M.J. Day
Ornamental plants	A.J. George
Systematic botany	J.L. Evans
Glasshouse crops	Dr. R.W.K. Holland

Pathology	Dr. C Knight
Regional Trials: Stockbridge House EHS	R.F. Cowin
Regional Trials: Kirton EHS	B.J. Withers
Seed Production Branch	Head: J.D.C. Bowring
Vegetable seed certification	Dr. A. Bould

7 Chipping Campden

7.1 The Campden Food Preservation Research Association
elevation: 122 m
rainfall : 705 mm

Chipping Campden, Gloucester
GL55 6LD England
Phone: 0386 840319
Telex: 337017
Dir.: K. Dudley

Industrial research association concerned with all aspects of food production, processing, packaging and quality.

Division of Agriculture and Quality	V.D. Arthey
Division of Food Science, incorporating food chemistry, biochemistry, maths and computing, and electron microscopy	J.D. Henshall
Division of Food Technology, incorporating food engineering, processing and packaging	CD. .Dennis
Co-ordinator of Research, including Library and Information Service	S.D. Holdsworth

8 East Malling

8.1 East Malling Research Station*
elevation: 32 m
rainfall : 660 mm
sunshine : 1554 h

East Malling, Maidstone, Kent,
ME19 6BJ, England
Phone: 0732 843833
Telex: 957251
Acting Dir.: Dr. J.E. Jackson*
Ass. to Dir.: D.A. Holland

The station, dealing with deciduous fruit crops and hardy ornamental nursery stock, belongs to the Kent Incorporated Society for Promoting Experiments in Horticulture, a body representing universities, scientific societies, fruit growers, hop growers, and administrative bodies. It is grant-aided from public funds through the AFRC. Area over 235 ha.

8.1.1 Crop production and plant science division	
Environmental effects, yield variation, light interception, orchard systems	Head: Dr. J.E. Jackson*
Plant Propagation department	
Growth regulators	Head: Dr. J.D. Quinlan*
Apple and pear thinning, fruit set, growth regulators	Dr. J.N. Knight*
Growth regulators, rootstocks, clonal selection scheme	A.D. Webster
Clonal selection of hardy ornamentals	G.C. Thomas
Histology and electron microscope	Dr. D.S.Skene/ Dr. K.A.D.MacKenzie
Orchard agronomy	G.C. White
Orchard system productivity, light utilisation	Dr. J.W. Palmer
Pears, fruit set, growth regulators	Dr. G. Browning
Raspberry agronomy, training, recording	R.I.C. Holloway
Selection of clones of Cox and Bramley	T. Sparks
Strawberry agronomy	M. Beech
Irrigation, strawberry agronomy, soil management	Miss. C.M.S. Thomas
Plant Propagation Department	

Physiology of vegetative propagation - technique development	Head: Dr. B.H. Howard
Physiology of vegetative propagation - environmental aspects	Dr. R.S. Harrison-Murray
Selection for vegetative propagation	H.R. Shepherd
Micropropagation of hardy ornamentals	Dr. T. Marks
Plant physiology department	
Plant hormone action: calcium biochemistry	Head: Dr. M.A. Venis
Mechanism of rootstock effect, *in vitro* culture	Dr. O.P. Jones
Genetic manipulation	Dr. D.J. James
Plant nutrient analysis	Dr. T.J. Samuelson
Uptake of nutrients	Dr. K.K.S. Bhat
Stress physiology; hormone immunoassay	Dr. H.G. Jones
8.1.2 Crop protection division	Head: Vacant
Plant pathology department	Head: Vacant
	Dep.Head: Dr. J.M. Thresh
Fireblight: bacterial canker of cherry and plum; crown gall	Dr. C.M.E. Garrett
Airborne fungal diseases; post-harvest diseases	D.J. Butt/ A.A.J. Swait
Biochemistry of host/parasite relations	Dr. R.C. Hignett/ J. H. Carder
Hop verticilium wilt, resistance and control	D.A. Chambers
Replant diseases	Dr. G.W.F. Sewell
Strawberry mycology, epidemiology and control	Dr. D.C. Harris
Phytophthora spp.	
Virus and MLO diseases of fruit, hops and ornamentals,	Dr. J.M. Thresh/ Dr. M.F. Clark
detection, identification and epidemiology	Dr. A.N. Adams/ D.L. Davies
Plant protective chemistry department	Head: Dr. D.J. Austin
Metabolic profiling	Dr. D.J. Austin
	Dr. R.A. Murray
Pesticide application	J.G. Allen/ Dr. D.J. Austin
	T.M. Warman/ L.D. Hunter
Electrophoresis, detection of predators	Dr. R.A. Murray
Plant growth regulators	Dr. D.J. Austin
	T.M. Warman/ L.D. Hunter
Rooting of hardwood cuttings	Dr. D.J. Austin/ P.S. Blake
Damson-hop aphid, resistance to pesticides	Dr. R.A. Murray
Zoology department	Head: Dr. J.J.M. Flegg
Damson-hop aphid, pear sucker	Dr. C.A.M. Campbell
Integrated control of pests	Dr. M.G. Solomon/ M.A. Easterbook
Nematodes	Dr. J.J.M. Flegg/ Dr. D.G. McNamara
Fruit breeding department	
Fruit, genetic resources	Head: Dr. R. Watkins/ R.A. Smith
Apple, pear (including pear and quince rootstocks)	Dr. F.H. Alston/ R.A. Smith
Apple, cherry, and woody ornamentals	K.R. Tobutt/ R.A. Smith
Plums (scions and rootstocks)	R.P. Jones/ R.A. Smith
Rubes	V.H. Knight
Strawberries	Dr. D.W. Simpson
8.1.3 Fruit storage division	Head: Dr. R.O. Sharples
Quality of stored fresh fruits	Dep.Head: Dr. M. Knee
Prestorage factors and storage behaviour	D.S. Johnson
Storage environment, new varieties	Dr. J.R. Stow
Fruit maturity; fruit quality; extension of shelf life	Dr. S.M. Smith
Fruit composition and storage quality	M.A. Perring
Store construction and operation	J. Jameson
Ethylene removal, control of weight loss	C.J. Dover
Postharvest biochemistry of fruit	Dr. M. Knee/ Dr. I.M. Bartley
	S.G.S. Hatfield

Statistics department
Design and analysis of experiments — Head: Dr. D.A. Preece; Dep.Head: J. Tamsett
Biometry and documentation — C.S. Moore
Computing — J. Tamsett
Liaison department — Head: B.F. Self*
Liaison and information — B.F. Self*/ Dr. A.C. McVittie

9 Edinburgh

9.1 Agricultural Scientific Services, Department of Agriculture and Fisheries for Scotland
elevation: 61 m

East Craigs, Edinburgh EH12 8NJ
Scotland
Phone: 031-339 2355
Telex: 727348 DAFASS
Dir.: Dr. D.C. Graham

There are three Divisions, all of which contribute incidentally to horticultural research: Plant Varieties and Seeds; Potato and Plant Health; and Pest Control and Pesticides

9.1.1 Plant varieties and seeds — Dep.Dir.: R.D. Seaton

Official Seed Testing Station for Scotland: testing methods, physiology, taxonomy and pathology of seeds — S.R. Cooper/ W.J. Rennie

Vegetable and herbage crops: varietal taxonomy and differentiation — W.G. Sutton

9.1.2 Potato and plant health — M.J. Richardson

Potato: production of healthy clonal stock, seed potato certification, disease monitoring physiology, and varietal taxonomy and differentiation — T.D. Hall/ D.M. MacDonald; C.J. Jeffries

Potato and disease control: fungi, bacteria and viruses — C.E. Quinn/ W.M.R. Laidlaw

Plant health: phytosanitary considerations generally, seed potato quarantine, bulb and soft fruit pathology — Dr. P.J. Howell/ P.S. Harris

Cropp loss assessment: including seed pathology and the effect of seed treatment (now amalgamated with Pesticide Usage) — G. Hosie

Nematology: potato cyst eelworms and other plant parasitic nematodes — T.W. Mabbott

9.1.3 Pest control and pesticides — J.R. Cutler

Crop entomology: potato aphid epidemiology, phytosanitary considerations generally — Mrs. L.A.D. Turl

Pesticide usage/Crop loss assessment — G. Hosie/ G.G. Tucker

Chemistry: pesticide residues and potato tuber disease control by fumigation — G.A. Hamilton/ Miss. E. Findlay; Dr. K. Hunter/ A.D. Ruthven; C. Griffiths

10 Efford

10.1 Efford Experimental Horticulture Station
ADAS, Ministry of Agriculture
elevation: 6-18 m
rainfall : 779 mm
sunshine : 1722 h

Lymington, Hampshire
SO4 0LZ England
Phone: 0590 73341
Dir.: R.F. Clements

Area 142 ha with 1.1 ha of glasshouses and other protection. Variety, cultural and environmental trials with glasshouse vegetable and flower crops, particularly early tomatoes, chrysanthemums and carnations. Nursery stock production including propagation, nutrition and irrigation of container grown shrubs. Soft fruits, particularly strawberries, protected and field grown crops. Outdoor vegetable production, varieties and cultural methods.

11 Faversham

11.1 Brogdale Experimental Horticulture Station
National Fruit Trials, ADAS, Min. of Agriculture
elevation: 43-61 m
rainfall : 622 mm
sunshine : 1577 h

Faversham, Kent ME13 8XZ England
Phone: 0795 535 462
Dir.: R.R. Stapleton

Area 75 ha. National centre for the maintainance of collections of hardy tree fruit varieties of apples, pears, plums and cherries also soft fruit crops. Variety and cultural trials of the above fruits including pollination and fruit set, rootstocks, orchard systems, irrigation, fruit storage, Micropropagation of strawberries and other fruits. Testing for distinctness of new varieties on behalf of the Plant Variety Rights Office.

12 Guernsey

12.1 States of Guernsey Horticultural Advisory Service
elevation: 104 m
rainfall : 855 mm
sunshine : 1820 h

Experimental Station, Burnt Lane,
St Martin's, Guernsey, CI
Phone: 0481 35741
Dir.: R.D. Pollock*

This station is the headquarters of the Guernsey Horticultural Advisory Service and its function is to support the Island's glasshouse industry. Major crops are tomatoes, cut flowers and ornamental plants.

13 Harpenden

13.1 Harpenden Laboratory, Ministry of Agriculture, Fisheries and Food
elevation: 128 m
rainfall : 723 mm
sunshine : 1502 h

Hatching Green, Harpenden
Herts AL5 2BD, England
Phone: 05827 5241
Telex: 826363 PPLHAR G
Dir: R.A. Lelliott

The Laboratory forms the scientific and technical headquarters responsible for preventing the introduction into England and Wales of plant diseases and pests that attack agricultural and horticultural crops; for containing and eliminating harmful organisms that gain entry and for preventing or limiting the spread of established diseases and pests by statutory means. It has responsibilities for many aspects of health standards for planting material, including seeds, for the home and export markets; for national aspects of plant diseases and pests; and for actively promoting international co-operation in phytosanitary matters. It is also responsible for the registration of pesticides under the Pesticides Safety Precautions Scheme and the Agricultural Chemicals Approval Scheme; for the investigation of residues of pesticides and other toxic substances in crops; for the assessment of hazards to operators from pesticide use and to the environment from spray drift; and for analysis and determination of pesticides. In phytosanitary and in pesticide questions, it is responsible for guiding the administrative divisions of the Ministry as a basis for action.
Apart from the bench work associated with these responsibilities, it does research and development work on problems that arise from its functions and particularly on: methods for the analysis and determination of pests and diseases; methods for the analysis and determination of pesticides; risks to operators using pesticides; methods for surveying pests, diseases, pesticide residues and pesticid usage; conducting such surveys; forecasting the pest and disease incidence and assessing associated loss; resistance of insect pests and fungal pathogens to pesticides; assessment of risks from non-indigenous pests and diseases; and virus disease problems in crop certification and in Regional advisory work.

13.1.1 Entomology	Head: H.J. Gould
Plant health	C.R.B. Baker
Pest intelligence	K.S. George
Pest control	Dr. J. Cotten
Pesticide surveys	J.M.A. Sly
Identification and nematology	J.F. Southey

13.1.2 Pest Control Chemistry	Head: D.F. Lee
Crop protection chemistry	D.F. Lee
Pesticides residues	Dr. N.A. Smart
Operator protection	G.A. Lloyd
Pesticides formulations analysis	B. Crozier
Pesticides methods of analysis and specifications	Dr. G.R. Raw
Mass spectrometry	J.P.G. Wilkins
13.1.3 Pesticides Registration	Head: J.A.R. Bates
Risk evaluation (data appraisal)	T.E. Tooby
Risk evaluation (toxicology and environment)	Dr. A.D. Martin
Efficacy evaluations	G. Stell
Technical services	Mrs. G.A. Lloyd
13.1.4 Plant Pathology	Head: H.J. Wilcox
Alien diseases	A.W. Pemberton
Domestic diseases	Dr. D.L. Ebbels
Disease assessment	Dr. J.E. King
Mycology	Dr. J.S.W. Dickens
Bacteriology	Dr. D.E. Stead
Virology	S.A. Hill

14 Hoddesdon

14.1 Lee Valley Experimental Horticulture Station
ADAS, Ministry of Agriculture
elevation: 41 m
rainfall : 648 mm
sunshine : 1409 h

Ware Road, Hoddesdon, Hertfordshire
EN11 9AQ England
Phone: 0992 463623
Dir.: A.J. Dyke

Area 10 ha of which 1.7 ha are used for protected cropping with glasshouses or plastic clad structures. A mushroom production unit is fitted with insulated film plastic cladding. The experimental programme includes variety, cultural, environmental and storage trials on mushrooms, cucumbers, lettuce and minor vegetable crops. Ornamental crops include pot and bedding plants.

15 Invergowrie

15.1 Scottish Crops Research Institute
elevation: 30 m
rainfall : 683 mm
sunshine : 1396 h

Invergowrie, Dundee DD2 5DA
Scotland
Phone: ~~08267 731~~ 0382 562731
Dir.: Dr. C.E. Taylor

The Scottish Crop Research Institute, which is financed by the Department of Agriculture and Fisheries for Scotland, was formed in 1981 by the amalgamation of the Scottish Horticultural Research Institute and the Scottish Plant Breeding Station. The SCRI has some 400 ha of land for trials, and research staff are provided with laboratories, controlled environment, glasshouse and field experimentation facilities, supported by technical and administrative services.

The work of the SCRI is to improve the productivity and quality of crops, by studying their breeding, culture and protection from diseases and pests; fundamental research is also done which contributes to the establishment of scientific principles. Potato, barley (especially for malting), forage brassica (especially swede, rape and kale), raspberry and black currant are the major crops in the research programme.

15.1.1 Plant Protection Division	Head: R.A. Fox
Mycology and bacteriology department	
Fungal diseases of potato	Head: R.A. Fox Dr. J.M. Duncan/ Dr. J.G. Harrison Dr. J.F. Malcolmson
Bacterial diseases of potato	R.A. Fox/ Dr. M.C.M. Perombelon Dr. D.A. Perry

Fungal diseases of cane fruit	R.A. Fox/ Dr. J.M. Duncan Dr. B. Williamson
Bacterial diseases of cane fruit	Dr. M.C.M. Perombelon
Fungal diseases of strawberry	Dr. J.M. Duncan
Zoology department	Head: Dr. D.L. Trudgill
Tolerance of potato cyst nematode	Dr. D.L. Trudgill
Nematode problems of soft fruit	Dr. B. Boag
Nematode pests of vegetables	Dr. D.J.F. Brown
Nematode transmitted virus in potato	Dr. J.M.S. Forrest
Mechanisms of resistance to potato cyst nematode	M.S. Phillips
Control of potato cyst nematode with resistant cultivars	S.C. Gordon
Insect and mite pests of soft fruit	Dr. J.A.T. Woodford
Aphid transmitted virus in potato	Head: Dr. N.L. Innes
15.1.2 Plant Breeding Division	Head: G.R. Mackay
Potato breeding division	Dr. P.D.S. Caligari
Commercial breeding: production of new cultivars and associated research	Dr. R.L. Wastie
Pathology: screening for resistance to diseases and pests and associated research	D.R. Glendinning
Novel germ plasms: exploration and exploitation of non-tuberosum germ plasm, maintainance of Commonwealth Potato Collection	
Soft fruit breeding	Head: Dr. D.L. Jennings
Plant breeding raspberries and blackberries	R.J. McNicol
Plant breeding black currants	M.M. Anderson
Tissue culture and cytology unit	Head: Dr. I.H. McNaughton
Potato micropropagation, in vitro storage, compex explant culture and anther culture	J.E. Middlefell Williams E.M. Borrino
15.1.3 Virology Division	Head: Dr. B.D. Harrison
Virology Department	Head: Dr. B.D. Harrison
Potato viruses	Dr. H. Barker
Virus resistance in potato	Dr. P.R. Massalski
Monoclonal antibodies to potato viruses	Dr. A.T. Jones
Raspberry and rubus viruses	W.P. Mowat
Narcissus virus diseases	Dr. A.M. Lennon
Cassava viruses	
15.1.4 Crop Science Division	Head: Dr. P.D. Waister
Physiology and crop production	Head: Dr. P.D. Waister
Crop Physiology	Dr. H.V. Davies/ Dr. K.J. Oparka
Physiology of potato growth and development	H.A. Ross
Environmental physiology of potato growth	Dr.R.A. Jeffries/Dr.D.K.L. MacKerron
Mathematical modelling of potato development and growth	Dr. B. Marshall
Vegetable agronomy	R. Thompson/ H. Taylor
Weeds and herbicides	H.M. Lawson
Herbicide evaluation	J.S. Wiseman
Soft fruit physiology	Dr. D.T. Mason
Soft fruit culture and mechanical harvesting	M.R. Cormack*
Competition effects on growth of potato plant	P.A. Gill
Information officer	R.J.A. Exley

16 Kew

16.1 Royal Botanic Gardens* — Kew, Richmond, Surrey, TW9 3AB

elevation: 5-11 m
rainfall : 635 mm
sunshine : 1560 h

England
Phone: 01 940 1171
Telex: 296694 KEWGAR
Dir.: Prof. E.A. Bell
Dep.Dir.: Prof. K. Jones

Also at: Wakehurst Place, Ardingley, Haywards Heath, West Sussex

Research into taxonomy and distribution of plants, especially in tropical regions; anatomy and cytogenetics; research into economic botany including biochemistry. Duration of comprehensive collections of plants (living and preserved) and relevant literature for use by researchers. Micropropagation development unit. Quarantine service for tropical crops, e.g. cocoa. Seed Bank and research into seed physiology at Wakehurst Place. Education and training in botanical horticulture through Kew Diploma Course (3 years) and international trainee scheme. Museum exhibits and provision of public access to the living collections. Preparation of publications and indexes on botany.

Herbarium
Classification and naming of plants from all parts of the world, preparation of monographs, revisions and floras. Collection of nearly 5 millions sheets of dried specimens, supplemented by intensive series of dried fruits and seeds, specimens in liquid preservation and by a collection of pollen slides.
Keeper: G.Ll. Lucas
Dep. Keeper: Dr. W.D. Clayton

Jodrell Laboratory
Systematic anatomy of flowering plants and microscopical identification of vegetable material, physiology and biochemistry of seed germination response (Seed Bank), chemotaxonomy, cytotaxonomy.
Keeper: Prof. K. Jones

Systematic anatomy — Dr. D.F. Cutler
Cytogenetics — Dr. P.E. Brandham
Physiology and Seed Bank — R.D. Smith
Biochemistry — T. Reynolds

Living Collections
Within historic landscaped gardens at Kew (120 ha), Wakehurst Place (200 ha), extensive documented collection of living plants (50,000 taxa). Major greenhouse collections of ferns, orchids, succulents, cycads and palms, and extensive collections of hardy plants. Botanical reserve at Wakehurst. Micropropagation unit at Kew.
Curator: J.B.E. Simmons
Dep.Curator (Kew): R.I. Beyer
Dep.Cur. (Wakehurst): A.D. Schilling

School of Horticulture
Supervision of Studies:
L.A. Pemberton

Quarantine Unit
A.G. Bailey

Library
Contains over 100,000 bound volumes, 140,000 separates, 175,000 plant illustrations and 10,000 maps.
Chief Libr. and Archivist:
Miss S.M.D. FitzGerald

Museums
General Museum with economic botany exhibits and Wood Museum. Also large reference collections not open to public but may be consulted by bona fide students on application. New museum building currently under construction
Officer in charge: Miss.R.C.R.Angel

17 Kirton

17.1 Kirton Experimental Horticulture Station — Boston, Lincolnshire PE20 1EJ

ADAS, Ministry of Agriculture
elevation: 4 m
rainfall : 556 mm
sunshine : 1480 h

England
Phone: 0205 722 391
Dir.: I. Sandwell

Area 42 ha. Outdoor vegetable crops especially brassicas and dry bulb onions, variety and cultural trials including propagation and transplanting methods; vegetable storage under controlled conditions. Bulb flower crops, tulips and narcissus, bulb production, handling and storage, health of bulb stocks, flower forcing. Strawberry production for processing.

18 Littlehampton

18.1 Glasshouse Crops Research Institute*
elevation: 5 m
rainfall : 735 mm
sunshine : 1800 h

Worthing Road, Littlehampton
West Sussex BN17 6LP, England
Phone: 0903 716123
Dir.: Dr. D. Rudd-Jones
Chief Scient.Liaison Officer:
Dr. D. Price

An independent research institute under the supervision of the Agricultural and Food Research Council and largely funded by state grants. Initially founded in 1914 as the Experimental Research Station, Cheshunt. Established on present site as the Glasshouse Crops Research Institute 1953. Responsible for research on the cultivation of crops under protection and mushrooms and of bulbs, flowers and hardy ornamental nursery stock grown in the open. Three research divisions with associated laboratories and glasshouses for research and field trials, including three glasshouse complexes providing facilities for multifactorial investigation of the glasshouse environment. Extensive controlled environment facilities including artificially lit and naturally illuminated growth cabinets. Total staff 245 of which 91 are graduate scientists.

18.1.1 Crop Science Division — Head: Dr. A.R. Rees

Crop Science Department

Growth and development of bulbous ornamentals and allied crops	Head: Dr. A.R. Rees
Physiology and propagation of bulbous ornamentals and allied crops	G.R. Hanks
Flower bulb agronomy; plant growth regulators	Dr. R. Menhenett
Chrysanthemum: growth, development and breeding; narcissus breeding genetics and micropropagation	Dr. F.A. Langton
Hardy ornamental nursery stock: propagation from leafy and dormant cuttings	Dr. K. Loach
Hardy ornamental nursery stock: field nutrition, field establishment from containers, clonal selection and new introductions	D.N. Whalley
Hardy ornamental nursery stock: container production, substrates for containers	R.L. Jinks*
CO_2 assimilation measurement, crop responses to environment, air pollution in glasshouses	Dr. D.W. Hand*
CO_2 measurement and control; operation of daylit controlled environment cabinets	M.A. Hannah
Effects of environment on growth and yield of glasshouse vegetable crops	G. Slack
Energy saving and nutrient film culture of glasshouse vegetable crops	Dr. C.J. Graves
Lettuce breeding and genetics	J.W. Maxon-Smith
Mushroom Culture: effects of environment	D.J. Humphries
Mushroom Culture: physical, chemical and microbiological aspects of compost preparation	J.F. Smith/ Miss. P.E. Randle
Energy studies and flowering	Dr. K.E. Cockshull/ J.S. Horridge

Biomathematics Department

Mathematical models applied to glasshouse crops; computing	Head: Dr. D.P. Aikman
Applied statistics	J. Fenlon
Statistical design and analysis of horticultural experiments	R.N. Edmondson
Statistical design, analysis of horticultural experiments and computing	Miss. H.D. Robinson
Management of computer services	R.D. Preece
Computer data acquisition and control; computer programming and system engineering	D.J. Fitter
18.1.2 Crop Protection and Microbiology Division	Head: Dr. D.R. Rudd-Jones
Plant Pathology and Microbiology Department	Head: Dr. J.M. Lynch
Mycology and Bacteriology Section	
Diseases of ornamental bulbs and their control	Head: Dr. D. Price
Diseases of protected crops and their control	Miss. M. Ebben
Biological control of horticultural pathogens	Dr. J.M. Whipps
Diseases of hardy ornamental nursery stock	Vacant
Microbial Technology Section	
Composting microbiology, mushroom physiology, lignocellulolysis	Head: Dr. D.A. Wood
Biological control, lignocellulolysis	Dr. N. Claydon
Mushroom genetics	T.J. Elliott
Mushroom diseases, compost microbiology	Dr. T.R. Fermor
Straw degradation	Dr. N. Magan/ Dr. P. Hand
Plant Virology Section	
Viruses of glasshouse crops and tropical food crops	Head: Dr. A.A. Brunt
Viruses and virus diseases of mushrooms	Dr. R.J. Barton
Viruses of glasshouse vegetable crops and hardy nursery stock plants	Dr. B.J. Thomas
Electron microscopy of plant and fungal viruses	R.T. Atkey
Propagation of virus-free horticultural crops	R.R. Pawley
Entomology and Insect Pathology Department	
Biological and integrated pest control	Head: Dr. C.C. Payne
Integrated control of pests of glasshouse crops	M.S. Ledieu
Evaluation of ULV techniques for pesticide applicatic	Dr. N.E.A. Scopes
Effects of pest incidence on crop yield	Dr. I.J. Wyatt
Biological control of aphids by polyphagous predators	Dr. K.D. Sunderland
Biological control of aphids by aphid-specific predators	Dr. R.J. Chambers
Insect pathology: bacterial pathogens of insects	Dr. H.D. Burges
Genetic manipulation of *Bacillus thuringiensis*	P. Jarrett
Fungal pathogens of insects	Dr. R.A. Hall
Viral pathogens of insects	Dr. N.E. Crook
Insect parasitic nematodes	P.N. Richardson
Chemical control of mushroom flies	P.F. White
18.1.3 Physiology and Chemistry Division	Head: Dr. D. Vince-Prue
Biochemistry Group	
Tomato fruit quality and genetics	Head: Dr. G.E. Hobson
Plant and substrate analysis	A.R. Brown
Photosynthesis Group	
Photosynthesis and respiration	Head: Dr. J.T. Ludwig/ A.C. Withers
Photosysthetic enzymes	Dr. R.J. Besford
Photobiology Group	
Photoperiodism and photomorphogenesis	Head: Dr. D. Vince-Prue
Molecular biology of photosynthesis and photomorphogenesis	Dr. B.J. Jordan

Biochemistry of photomorphogenesis	Dr. M.D. Partis
Development of photobiology	Dr. B. Thomas
Perception of light; construction of specialised equipment	Dr. P.H. Knapp
Physiology and Senescence Group	
Post-harvest physiology of horticultural produce	Head: Dr. R. Nichols*
Ethylene metabolism	K. Manning
Plant Nutrition Group	
Nutrition and quality of glasshouse crops	Head: P. Adams
Plant and soil analysis	M.H. Adatia

19 Long Ashton

19.1 Long Ashton Research Station
elevation: 50 m
rainfall : 900 mm
sunshine : 1508 h

Long Ashton, Bristol BS18 9AF
England
Phone: 0272 392181
Dir.: Prof. K.J. Treharne
Dep.Dir.: Dr. F.W. Beech

This Station, which is the Department of Agriculture and Horticulture of the University of Bristol, is a grant-aided institute mostly financed by the Agricultural and Food Research Council, but receives additional funds from other official bodies and from industry. The Station serves the agricultural and food industries by improving crop production and utilization. It has special responsibilities for basic and applied research on plant growth regulation and on the control of weeds, pests and diseases and for the integration of environmentally acceptable crop production practices into agricultural and related systems. Interests relate to the pest and disease problems of arable crops, disease forecasting, fungicide-resistance, spray application, plant-pathogen relations, nitrogen assimilation, plant growth regulation and environmental physiology. Food science studies aim to quantify the factors that give improved quality to foods and beverages, and to this end research covers quality assessment, natural plant pigments, spoilage and domestic food handling.

19.1 Crop Protection Division	Head: Dr. K.J. Brent
Crop Pathology Group	Head: Dr. K.J. Brent
Resistance of pathogens to fungicides	Dr. G.A. Carter/ T. Hunter
Biology and control of cereal diseases	Dr. V.W.L. Jordan
Disease forecasting	Dr. D.J. Royle/ Dr. M.W. Shaw
Crop Protection Chemistry Group	Head: Dr. N.H. Anderson
Physical chemistry of sprays	Dr. N.H. Anderson
Redistribution of pesticide on leaves	Dr. E.A. Baker
Effects of azole plant growth regulators and fungicides on plant metabolism	Dr. R.S. Burden
Improved pesticides delivery. Metabolism of pesticides	Dr. D.R. Clifford
Pesticides and nitrogen fixation	Dr. D.J. Fisher
Surfactants and pesticide behaviour	Dr. P.J. Holloway
Photochemical degradation of pesticides	Dr. D.A.M. Watkins
Mechanisms of fungicide action	Dr. D.J. Fisher/ Dr. R.S.T. Loeffler
Metabolism of azole fungicides	AHB Deas/ Dr. T. Clark
Pesticide residue analysis	J.A. Pickard
Tree diseases and phytoalexins	M.S. Kemp
Physiological Plant Pathology Group	Head: Dr. R.J.W. Byrde
Enzymes and pathogenesis	A.H. Fielding
Disease resistance mechanisms in French bean	Dr. J.A. Bailey
Disease resistance mechanisms in cereals	Dr. J.A. Hargreaves
Spray application group	Head: Dr. E.C. Hislop
Biological efficiency of crop spraying techniques, including ULV and electrostatic studies	Dr. B.K. Cooke
Zoology Group	Head: Dr. B.D. Smith
Pests and virus vectors in cereals and grass crops	Dr. D.A. Kendall

and control in cropping systems	
19.1.2 Plant Science Division	Acting Head: Dr. G.V. Hoad
Biochemistry Group	Head: Dr. B.A. Notton
Nitrogen assimilation - nitrate reductase	Dr. B.A. Notton
Inorganic nitrogen metabolism	Dr. D.P. Hucklesby
Biochemistry of di- and polyamines	Dr. T.A. Smith
Developmental Physiology Group	Head: Dr. G.V. Hoad
Phytohormones	Dr. G.V. Hoad
Use of plant growth regulators on arable crops	R.D. Child
Biochemistry and physiology of plant growth regulators	Dr. J.R. Lenton/ Dr. P. Hedden
Movement of growth regulators in plants	Dr. D.N. Butcher
Plant growth regulators and legume proteins	Dr. D.H.P. Barratt
Plant tissue culture	Dr. A.J. Abbott
Components of yield and protein of field bean	Dr. D.G. Hill-Cottingham
	C.P. Lloyd-Jones
Environmental Physiology Group	Acting Head: Dr. D.R. Butler
Microclimate, weather and disease epidemics. Effects of environment on cereal physiology	Dr. D.R. Butler
Modelling disease effects on physiology and yield	Dr. N.D.S. Huband
Consequence of disease to cereal growth and yield	Miss. J.M. Bennett
Modelling growth and yield formation in plants	Dr. J.R. Porter
19.1.3 Pomology Division	Head: Dr. A.I. Campbell
Pome fruit viruses; induction and assessment of mutants	Dr. A.I. Campbell
Factors affecting fruit set; cider pomology	R.R. Williams
Improvement of trees and shrubs	C.S. Gundry
Plum breeding and biomass studies	K.G. Stott
Strawberry breeding and agronomy	H.M. Anderson
19.1.4 Food and Beverages Division	Head: Dr. F.W. Beech
Analytical Chemistry Group	
Chemistry of fruit juices and cider	Dr. L.F. Burroughs
Composition of fresh and processed foods	Miss. S.J. Smith
Domestic Food Handling Group	
Microbial hazards in domestic food handling UK Total Diet Study	Miss. G.A. Fine
Microbiology Group	
Ecology of spoilage yeasts and bacteria	Dr. R.R. Davenport
Wine spoilage organisms	Dr. D.S. Thomas
Ecology of non-acid tolerant bacteria	K.A. Goverd
Wine spoilage organisms	Dr. D.S. Thomas
Ecology of non-acid tolerant bacteria	K.A. Goverd
Natural Products Chemistry Group	
Non-toxic plant food colours	Dr. C.F. Timberlake
Phenolic components of foods and beverages	Dr. A.G.H. Lea
Wine oxidation and tartrate instability	Dr. N.W. Preston
Sensory Analysis Group	
Analytical and sensory components of food quality	Dr. A.A. Williams
Taints in fermented beverages. Microbial flavour formation	Dr. O.G. Tucknott
Biometrics and Computing Section	Head: Dr. P. Brain
Experimental design, biological scaling	Miss. M.E. Holgate
Models of crop production processes, population dynamics	Ms. G.M. Arnold
Multivariate analysis of crop processes	Ms. S.F. Todd
Computing	S. Lucey/ P.D. Moody
Plantations and Glasshouses	Head: E.G. Gilbert

Management of experiments, agronomy herbicides, pesticides

Scientific Liaison — Head: Dr. R.K. Atkin

Organization of scientific meetings - reception of scientific visitors from overseas and UK - liaison with user industries

20 Loughgall

20.1 The Horticultural Centre
elevation: 50 m
rainfall : 802 mm
sunshine : 1285 h

Loughgall, Co. Armagh, BT61 8JA
N. Ireland
Phone: 076289 208
Dir.: G.H. McElroy

The Centre was established in 1951 by the Ministry of Agriculture to improve production efficiency of horticultural enterprises on Northern Ireland farms through a programme of applied research and development. It extends to a total area of 40 ha with 0.3 ha of glass and polythene, and work covers all the main horticultural diciplines - mushrooms, fruit, vegetables, hardy nursery stock and glasshouse crops.

Mushrooms

Development of synthetic composts and the evaluation and development of reduced input production systems. — Miss. D. Moore

Fruit

The control of the major crop fruit diseases of apple canker, mildew and scab. Growth control and the regulation of cropping in Bramley's Seedling, fruit storage and marketing. — B.S. Watters

Vegetables

Continuity of cropping including production techniques and varietal assessment in the major vegetable crops. Cooling storage and marketing of produce. — Vacant

Nursery Stock

Post propagation techniques including containers, compost nutrition, irrigation and herbicides also tree production techniques. — Miss. S. O'Neill

Glasshouse Crops

Varietal evaluation in lettuce, tomatoes and cucumbers. Alternative crops with particular reference to those with low energy requirements. — Vacant

Biomass

Emergy from biomass including the production of purpose grown *Salix* as an alternative source of energy for heating and as a supplement for animal feedstuffs. — W.M. Dawson

21 Luddington

21.1 Luddington Experimental Horticulture Station
ADAS, Ministry of Agriculture
elevation: 34-58 m
rainfall : 610 mm
sunshine : 1418 h

Stratford-upon-Avon, Warwickshire
CV37 9SJ, England
Phone: 0789 750 601
Dir.: M.R. Shipway

Area 93 ha. Fruit crops include apples, plums, black currants, strawberries and cane fruits and the trials cover varieties, rootstocks, cultural systems and pests and disease control. Outdoor vegetable crop production of brassicas, alliums, lettuce, asparagus with variety, soil management, irrigation and weed control trials. Hardy ornamental field-grown nursery stock, propagation and production. Protected crops mainly lettuce and newer vegetables. Shelter studies. Unit for investigational and advisory work with hive bees.

22 Manchester

22.1 University of Manchester, Faculty of Science, Department of Botany
elevation: 75 m
rainfall : 750 mm
sunshine : 1268 h

Manchester M13 9Pl, England
Phone: 061 273 7121
Experimental Grounds, Jodrell Bank, Cheshire
Phone: 0477 71213

5 ha for field experiments at Jodrell Bank.
Miscellaneous glasshouses, growth rooms and cabinets on University Campus.

Weed science	Dr. R.A. Benton
Crop physiology	Dr. P. Newton*
Plant pathology (fungi and bacteria)	Dr. H.A.S. Epton
Plant pathology (viruses)	Dr. R.R. Frost

23 Mepal

23.1 Arthur Rickwood Experimental Husbandry Farm ADAS, Ministry of Agriculture
elevation: -1.5 m
rainfall : 579 mm
sunshine : 1437 h

Mepal, Ely, Cambridgeshire CB6 2BA, England
Phone: 03543 2531
Dir.: J. MacLeod

Area 76 ha. Located in a black fen peat area, cultural and other investigations on vegetable root crops, celery and onions. Basic studies on peat wastage and mineralisation.

24 Norwich

24.1 John Innes Institute

Colney Lane - Norwich - NR4 7UH
Norfolk - England
Dir.: Prof. H.W. Woolhouse

The Institute founded in 1910 by a bequest of the late John Innes of Merton is supported by a grant-in-aid from the AFRC. It works on a wide range of problems concerned with the fundamental and applied genetics of plants and bacteria, as well as plant viruses

24.1.1 Department of Applied Genetics — Head: Prof. D.R. Davies

The use of conventional and novel techniques to develop new methods of crop improvement.

Plant tissue culture	Dr. G. Hussey
Anther and pollen culture	Dr. J. Dunwell
Molecular biology of pea storage proteins	Dr. R. Casey
Pea seed development	Dr. C. Hedley/ Dr. T. Wang
Pea pathology	Dr. P. Matthews
Population genetics of Rhizobium	Dr. P. Young
Molecular biology of the pea genome	Dr. C. Cullis/ Dr. N. Ellis
cms in sugar beet	Prof. D.R. Davies
Microinjection of plant cells	Prof. D.R. Davies

24.1.2 Department of Genetics — Head: Prof. D.A. Hopwood FRS

Fundamental genetical research on micro-organisms and flowering plants and of relationships between them.

Molecular biology of bacterial plant pathogens.	Dr. M.J. Daniels
Genetics and biochemistry of Rhizobium-legume	Dr. A.W.B. Johnston/ Dr. N. Brewin

symbioses.	Dr. A. Downie
Transposable elements in Antirrhinum	Dr. C. Martin/ Dr. E. Coen Miss. R. Carpenter
Genetics and molecular biology of actiomycetes	Dr. K. Chater/ Dr. M. Bibb Mrs. H.M. Wright
Genetics and biochemistry of crown gall disease and its use in transferring foreign DNA to plants	Dr. J.L. Firmin/ Dr. A.G. Hepburn
24.1.3 Department of Cell Biology	Head: Dr. K. Roberts
Plant cell structure and developmental biology	
Plant cell protoplasts	Dr. J.W. Watts
Structure and composition of biological macromolecules	Dr. P. Shaw/ G.J. Hills
Cell wall and cytoskeleton in relation to morphogenesis	Dr. K. Roberts/ Dr. J. Burgess Dr. C. Lloyd
24.1.4 Department of Virus Research	Head: Dr. J. Davies
The structure, multiplication and manipulation of plant viruses	
Molecular biology of plant viruses	Dr. R. Hull/ Dr. J. Stanley
Develop gene vectors from virus nucleic acid sequences	Dr. G. Lomonossoff/ Dr. R. Townsend Dr. T.M.A. Wilson/ Dr. S. Covey
Plant cell protoplasts	Dr. A. Maule
Virus-insect relationships	Dr. P.G. Markham
24.1.5 Photosynthesis group	Head: Prof. H.W. Woolhouse
The study of the carbon economy of crop plants in terms of photosynthesis, photorespiration and dark respiration	Prof. H.W. Woolhouse/ Dr. A. Smith

25 Nottingham

25.1 University of Nottingham School of Agriculture	Sutton Bonington, Loughborough Leicestershire LE12 5RD, England Phone: 0602 506101 Telex: 37346 UNINOTG
25.1.1 Department of Agriculture and Horticulture	Head: Prof. J.D. Ivins
Ornamental plants, vegetative propagation, plant health	Dr. P.G. Alderson
Glasshouse crops, physiology of flowering	Dr. J.G. Atherton
Fruit, vegetables, intra-plant competition	Dr. C.J. Wright
Horticultural mechanisation	B. Wilton
25.1.2 Department of Physiology and Environmental Studies	Head: Prof. W.J. Whittington
Environmental physics, micro-meteorology	Prof. J.L. Monteith/ Dr. J.A. Clark Dr. J. Colls
Soil processes	Dr. M. McGowan
Genetics	Dr. W.J. Whittington/Dr.I.B. Taylor
Agricultural zoology	J.Y. Ritchie
Plant physiology	Dr. C.R. Black/ Dr. D. Grierson
Plant pathology	Dr. T.F. Hering/ Dr. S. Rossall
25.1.3 Department of Biochemistry and Nutrition	Head: Prof. R.A. Lawrie
Plant biochemistry	Dr. G. Norton/ G. Tucker

26 Ongar

26.1 May & Baker Agrochemicals	Fyfield Road, Ongar, Essex CM5 0HW England Phone: 0277 362127 Telex: 28691

Evaluation of new chemicals for possible herbicidal insecticidal and fungicidal activity.

General Manager: B.M. Drew
Ag. Res Manager: Dr. B.M. Savory

27 Peterborough

27.1 Processors & Growers Research Organisation
elevation: 40 m
rainfall : 580 mm
sunshine : 1383 h

Thornhaugh, Peterborough
Cambridgeshire PE8 6HJ, England
Phone: 0780 782585
Dir.: G.P. Gent

Applied research on the production and harvesting of peas and beans for dry harvest or vegetable production plus the provision of a wide range of technical services for these crops. Headquarters trial ground of 8 ha at Thornhaugh and regional trials extending from southern England to eastern Scotland. Full canning and quick-freezing, seed testing and pathology facilities available.

Biology section
Rotational investigations — Head: A.J. Biddle*
New seed treatments — Mrs. F. Herbert
Pest and disease control — P. Jackson
Seed testing
Seed vigour investigations

Agronomy and Botany sections — Head: Ms. C.M. Knott
Evaluation of new varieties — D.J. Eagle
Evaluation of new herbicides and fungicides — M.E. Tait
Plant population studies — S.P. Goater
Quality aspects for processing
Pea and bean husbandry

28 Preston Wynne

28.1 Rosemaund Experimental Husbandry Farm
ADAS, Ministry of Agriculture
elevation: 84 m
rainfall : 664 mm
sunshine : 1432 h

Preston Wynne, Hereford HR1 3PG
England
Phone: 0432 78 444
Dir.: S.C. Meadowcroft

Total area 176 ha with 17 ha of hops. Husbandry and pest and disease control investigations with hops, evaluation of wirework systems and spray machinery, drying, storage and quality of hops.

29 Reading

29.1 University of Reading*, Faculty of Agriculture and Food, Sub-Department of Horticulture

Earley Gate, Reading RG6 2AU
England
Phone: 0734 875123 Ext. 6319
Telex: 847813
Head: Prof. G.F. Pegg

Landscape design and amenity horticulture: lawn and tree management — R.J. Bisgrove
Glasshouse technology: artificial lighting — A.E. Canham*
Plant-insect relations: biological control — Prof. H.F. van Emden
Plant pathology: crop protection — Dr. R.T.V. Fox
Vegetable production: crop physiology — Dr. P. Hadley
Protected cropping: flowering physiology — Dr. G.P. Harris
Fruit production: plant physiology — Dr. J.K. Wall

30 Selby

30.1 Stockbridge House Experimental Horticulture Station, ADAS, Ministry of Agriculture
elevation: 8 m
rainfall : 610 mm
sunshine : 1330 h

Cawood, Selby, N. Yorkshire
YO8 0TZ, England
Phone: 0757 86 275
Dir.: J.D. Whitwell

Area 65 ha with 1.4 ha of glasshouses and other protection. Outdoor vegetable crops, brassicas, root vegetables, alliums, lettuce, variety and cultural trials including weed control and pest and disease control; production for the vegetable processing industry. Rhubarb for outdoor cropping and forcing. Soft fruit varieties and production for local conditions. Glasshouse crop production including cucumbers, lettuce and tomatoes, evaluation of varieties and new cultural systems, environmental studies, pest and disease control.

31 Silsoe

31.1 Silsoe College*
elevation: 59 m
rainfall : 557 mm
sunshine : 1470 h

Silsoe, Bedford MK45 4DT, England
Phone: 0525 60428
Telex: 825072
Head: Prof. B.A. May

The College is a constituent part of the Cranfield Institute of Technology. The College's programme consists of taught courses at undergraduate (B.Sc and B.Eng) and postgraduate (M.Sc and Diploma) levels; research study (M.Sc and Ph.D): short courses and professional development programmes; consultancy and sponsored research.

Agricultural engineering

Agricultural production technolgy	Prof. R.W. Radley
Low-cost cultivation equipment for developing countries	C.P. Crossley
Farm machinery use and crop processing; crop protection	R.T. Lewis
Assessment of damage to horticultural crops	R.F.A. Murfitt
Tropical crop production and processing	Prof. H.D. Tindall*
Design of agricultural machine systems	Dr. R.J. Godwin
Dr. B. Clarke	
Design of agricultural and horticultural machinery for crop production	J. Dyson
Design and development of harvesting machinery for horticultural produce	J. Kilgour
Environmental control and instrumentation	B.C. Stenning*/ G.T. Stone
Soil and water engineering	M.G. Kay
Climatology studies related to crop production	Dr. R.P.C. Morgan
Soil conservation	Prof. G. Spoor
Monitoring growth of crops by remote sensing	Dr. J.C Taylor/ A. Belward
Marketing and management	Prof. R.W. Hill
Marketing of horticultural produce	I.M. Crawford
Food engineering	Prof. E. Morris

31.2 National Institute of Agricultural Engineering*

Wrest Park, Silsoe, Bedford
MK45 4HS, England
Phone: 0525 60000
Telex: 825808
Dir.: J. Matthews

A research institute grant-aided by the Agricultural and Food Research Council. Research and development concerned with the mechanisation and control of agricultural processes. OECD and EEC tractor and tractor safety cab testing. 110 ha of land. Facilities for commercial contract work.

31.2.1 Agricultural vehicles

Design of tractors for heavy draught operations	Acting head: Dr. M.J. Dyer
Design of low ground pressure vehicles	Dr. M.J. Dyer

Farm materials transport	W.T.B. Marchant
Ride vibration	R.M. Stayner
Tractor and machinery noise	J.C.D. Talamo
Ergonomics of information presentation	D.H. O'Neill
Machine dynamic performance and overload protection	Dr. C.J. Chisholm
31.2.2 Crop engineering division	Head: Dr. D.S. Boyce
Spray physics - electrostatics	Dr. P.C.H. Miller
Field sprayer technology - boom suspensions and spray rate controls	Dr. A.R. Frost
Drying of agricultural crops	Dr. M.E. Nellist
Operational research	E. Audsley
31.2.3 Engineering Design and Development Division	Acting head: R.M. Stayner
Design	M.J. Dyer
Wear	Dr. A.G. Foley
31.2.4 Farm buildings division	Head: Dr. J.M. Randall
Environmental engineering	G.A. Carpenter/ C.R. Boon Dr. J.M. Randall
Waste engineering	Dr. V.R. Phillips/ T.R. Cumby C.P. Schofield/ Dr. A.G. Williams
Wind loading	Dr. R.P. Hoxey
Retaining walls for agricultural materials	P. Moran
31.2.5 Field machinery division	Head: W.E. Klinner
Soil dynamics and the design of cultivation implements	Dr. J.V. Stafford
Cultivation implements	D.E. Patterson
Research and development in the treatment and collection of forage crops	O.D. Hale
Research and development in crop mowing	Dr. M.J. O'Dogherty
31.2.6 Horticultural engineering division	Head: Dr. B.J. Legg*
Seed drills to give high uniform seedling emergence	F.R. Brown
Root handling	W.P. Billington
Harvesting and preparation of vegetables	J.B. Holt
Hardy ornamental nursery stock	F.R. Brown
Efficient use of energy in greenhouses	Dr. B.J. Bailey
Design of greenhouses to give maximum light transmission	D.L. Critten
Monitoring and control of greenhouse environment	Dr. B.J. Bailey
31.2.7 Instrumentation and control division	Head: M.E. Moncaster
Field machinery monitoring and control	G.O. Harries
Moisture measurement	G.E. Bowman
Instrumentation	W.R. Wignall
Time series analysis	Dr. P.F. Davis
31.2.8 Overseas division	Head: R.D. Bell
Agricultural engineering for the developing countries	
Engineering services department	Head: L.E. Osborne
Information services department	Head: Dr. G.F. Forster

32 Sittingbourne

32.1 Sittingbourne Research Centre
Shell Research Limited
elevation: 74 m
rainfall : 600 mm

Broad Oak Road, Sittingbourne
Kent ME9 8AG, England
Phone: 0795 24444
Telex: 96181 SHELL G
Dir.: Dr. J.D. Shimmin

Sittingbourne Research Centre has the role of discovering and developing products new to Biosciences of relevance to the Group's interest. This includes: 1. the discovery and development of agrochemicals; 2. the discovery and development of microbiological products and processes of value to the oil and related industries. The laboratories plus associated support services are located on a 200 ha farm. Screening of compounds is carried out in glasshouses and on the farm prior to more extensive testing by Shell in many parts of the world.

Director of Agrichemicals R & D	T. Chapman
Director of Biotechnology and Toxicology	Dr. E. Thorpe
Head of Biological Evaluation	M.A. Pinnegar
Insecticides	Dr. J.P. Fisher
Herbicides	Dr. R. Fenton
Fungicides	D.P. Highwood
Plant Growth Regulants	Dr. L. Phillips

33 Wellesbourne

33.1 National Vegetable Research Station*
elevation: 47 m
rainfall : 598 mm
sunshine : 1372 h

Wellesbourne, Warwickshire CV35 9EF
England
Phone: 0789 840382
Dir.: Prof. J.K.A. Bleasdale
Dep.Dir.: Dr. P.J. Salter
Scientific Liaison Officer:
Dr. C.C. Wood

Supported by government funds through the Agricultural and Food Research Council. The only research station to concentrate on all aspects of British outdoor vegetable production. It was set up in 1949 to co-ordinate research on vegetable crops. The station occupies some 191 ha of which approximately 45 ha are used annually in rotation for field experiments. In addition there are laboratories, growth rooms, electron microscopes and 1 ha of protected growing area under glass and polythene.

33.1.1 Biochemistry

Plant disease resistance, biochemical interactions between host and virus; genetic engineering	Head: Dr. R.S. Fraser Dr. C.J.R. Thomas
Plant growth regulators; new assay methods, and role in virus-infected plants	Dr. R.J. Whenham

33.1.2 Soil Science

Control of soil fertility; remote sensing	Head: Dr. D.J. Greenwood
Major element nutrition: computer simulation	Dr. I.G. Burns
Seed bed preparation: seedling emergence	Dr. H.R. Rowse
Prediction of N-fertilizer requirements: Micro-nutrient disorders	Dr. M.A. Scaife
Yield variability: soil/plant water relations	Dr. P.A. Costigan
Methods of chemical analysis	J. Hunt
Soil cultivation: remote sensing	D.A. Stone

33.1.3 Entomology

General control of vegetable pests	Head: G.A. Wheatley
Plant resistance to pests of crucifers, carrots, onions and lettuce	Dr. P.R. Ellis
Biology and behaviour of insect pests of vegetables	Dr. S. Finch
Control of insect pests of vegetables: mainly cabbage root fly, carrot fly, cabbage aphid	Dr. A.R. Thompson
Biochemistry of plant resistance to aphids	Dr. R.A. Cole
Insecticide residues; behaviour in soil and uptake by plants	D.L. Suett

33.1.4 Pathology

Diseases of vegetables. Epidemiology of air-borne diseases	Head: Dr. R.T. Burchill

Downy mildew disease of lettuce; clubroot disease, resistance studies	Dr. I.R. Crute
White rot disease of onion	Dr. A.R. Entwistle
Seed-borne fungal diseases of vegetable crops: neck rot of onions; *Botrytis* diseases of salad (green) onions	Dr. R.B. Maude
Bacterial diseases of vegetables	Dr. J.D. Taylor
Virus diseases, purification and serology of viruses *Brassica* virus diseases and virus chemotherapy	Dr. J.A. Tomlinson
Virus diseases, elimination by meristem culture Necrotic disorders of white cabbage. Virus resistance in *Phaseolus* beans	Dr. D.G.A. Walkey
Use of soil partial sterilants in field vegetable production. *Brassica* seedling establishment diseases. Cavity spot of carrots	Dr. J.G. White
Seed-borne diseases of horticultural brassicas - oilseed rape	Dr. F.M. Humpherson-Jones

33.1.5 Physiology

Crop physiology, crop production systems	Head: Dr. P.J. Salter
Sources of plant-to-plant variability in vegetable crops	Dr. L.R. Benjamin
Stress-induced affects in vegetable seedlings	Dr. N.L. Biddington
Physiology of the onion crop	Dr. J.L. Brewster
Physiology of seed germination and seedling establishment	Dr. P.A. Brocklehurst
Fluid drilling affects on vegetable crops	Dr. W.E. Finch-Savage
Crop physiology, seed production, seed quality and seedling establishment	Dr. D. Gray*
Distribution of assimilates in metamerically segmented vegetable crops	Dr. R.C. Hardwick
Dry matter distribution in vegetable crops	Dr. C.C. Hole
Hormonal control of growth and development of vegetable species, particularly root crops	Dr. T.H. Thomas
Crop physiology, environmental factors at early stages of growth	Dr. D.C.E. Wurr*

33.1.6 Plant Breeding

Lettuce - resistance to downy mildew, lettuce mosaic virus, lettuce root aphid	Head: Dr. T.J. Riggs
Autumn cauliflower, broccoli including calabrese	Dr. P.C. Crisp
Onions and carrots, biometric approach to selection	B.D. Dowker
Seed production and testing of new varieties	J.C. Jackson
Anther culture in Brassicae, interspecific crosses in onions and beans	Dr. D.J. Ockendon
Genetics of plant resistance to disease	Dr. D.A.C. Pink
Brussels sprouts, biometric approach	B.M. Smith
Winter cauliflower, biometric approach	Dr. C.R. Tapsell
Gene Bank	Dr. D. Astley

33.1.7 Statistics

Design of experiments, methods of data analysis Methods of data analysis	G.E.L. Morris/ Mrs. K. Phelps
	Dr. R.A. Sutherland

33.1.8 Weeds

Cultural and chemical control of weeds	Head: H.A. Roberts
Behaviour of herbicides and fungicides in soil	Dr. A. Walker

34 Whittlesford

34.1 Ciba-Geigy Agrochemicals (private institution)

Hill Farm, Whittlesford
Cambridge CB2 4QT, England
Phone: 0223 833621
Telex: 81325

Agrochemicals research and development on horticultural crops	Head: T.C. Breese
Technical development manager	T.R. Barnes

35 Wye

35.1 Wye College (University of London)
elevation: 56 m
rainfall : 715 mm
sunshine : 1567 h

Wye College, Ashford, Kent TN25 5AH
England
Phone: 0233 812401
Telex: 96118 ANZEEC G
Principal: I.A.M. Lucas

The college has a 400 ha estate including 35 ha devoted to commercial horticulture. Two thirds of this area is occupied by orchards and soft fruit and protected cropping is carried out in 1 ha of automated glasshouses. Other facilities include gardens laid out with a range of flowering shrubs, annual and perennial plants especially suited to calcareous soils. Research laboratories, 0.25 ha of experimental glasshouses, 15 environment controlled growth rooms and 32 daylit growth cabinets.

35.1.1 School of Rural Economics and Related Studies

Agricultural Economics	Head: Prof. A.E. Buckwell
Economic aspects of fruit production	D. Ray
Farm Business	Head: Prof. J.S. Nix
Management applied to horticultural holdings particularly tomato and apple crops; price forecasting	J.A.H. Nicholson*
Marketing: the structure of horticultural marketing systems throughout the world and their evolution in response to changing economic and social conditions; the development of international trade; aspects of co-operative marketing	

35.1.2 Department of Horticulture

Reproduction physiology, dormancy, natural growth hormones, photoperiodism, vernalization, phyllotaxis, porthemocarpic fruit set and juvenility	Head: Prof. W.W. Schwabe*
Establishment of forest trees, weed competition	Dr. G.P. Buckley
Microclimatology, environmental control, nutrient film	Dr. S.W. Burrage
Fruit propagation and production systems	P.B. Dodd
Environmental physiology of vegetable crops	Dr. R. Fordham
Pathenocarpy; hormone sprays in relation to fruit set, tissue culture	Dr. G.K. Goldwin*
Hormonal and environmental control of cereal plant development and stress resistance	Dr. P.D. Hutley-Bull
Water relations and nutrition of plants	Dr. S. McCain
Amenity horticulture, gardens and parks management	T.W.J. Wright
Vegetation establisment and management	Mrs. Shirley E. Wright
Unit for Advanced Propagation Systems. Vegetative propagation, especially of coconut palm, cocoa and potato, using tissue culture	Dr. Jennet Blake
Vegetative propagation of coconut palm by tissue culture	Dr. R.L. Branton/ Dr. F.J. Paton
Factors controlling differentiation in oil palm callus	Dr. Avril L. Brackpool
Micropropagation; roles of stress metabolism in tissue differentiation	Dr. S.H. Mantell

35.1.3 Department of Biological Sciences

	Head: Prof. D.A. Baker
Plant Pathology: mechanisms of disease resistance and fungal and bacterial pathogenicity	Dr. J.M. Mansfield
Physiological basis of virus-induced stunting in plants and general studies of viruses of legumes	Dr. K.W. Bailiss

Genetics: Aspects of crop plant chromosome structure and function	Dr. G.P. Chapman
Genetics and physiology of legumes	Dr. W.E. Peat
Plant Physiology: Source/sink interactions and the control of long-distance transport of assimilates	Prof. D.A. Baker
Physiological aspects of weed/crop interactions	Dr. T.A. Hill
Microbiology: The recycling of organic wastes including high-temperature composting	Dr. J. Lopez-Real
Entomology: Physiology, anatomy, behaviour and ecology of parasitic *Hymenoptera,* particularly *Chalcidoidea* with special reference to biological control of insect pests in heated glasshouses	Dr. M.J.W. Copland
Ecological factors affecting population levels and inter-plant movement within and between crops by aphids, phyllids and mites with reference to food quality, antibiosis, population density effects and predation	Dr. C.J. Hodgson
Aspects of honey production in honey bees	Dr. D.S. Madge

35.2 Department of Hop Research

Wye College, Ashford, Kent, England
Phone: 0233 812401
Telex: 96118 (ANZEEC G)
Dir.: R.E. Gunn

13 ha of hop gardens, chemical and biological laboratories, glasshouse, kilns for experimental drying of hops, picking machines, oast house.

Plant breeding section

Resistance to verticillium wilt	Head: R.E. Gunn
Resistance to downy mildew	P. Darby
Resistance to powdery mildew	
High yield and brewing value	
Resistance to hop aphid	

Field section

Irrigation	Head: Dr. G.G. Thomas
Harvest date	
Reduced labour requirements	
Weed control	

Crop physiology section

Environmental effects of growth, yield and quality	Head: Dr. G.G. Thomas
Hormones and growth regulators	

Chemistry section

Chemistry of hop constituents	Head: Dr. C.P. Green
Storage losses of resins	

Pathology section

Biology, aetiology and control of Fusarium canker	Head: Dr. P. Darby

UNITED STATES OF AMERICA US

100° W
Canada
Washington
J
Oregon
Montana
Idaho
North Dakota
Minnesota
South Dakota
G
Wisconsin
Michigan
F
New Hampshire
Vernont
Maine
A
B
New York
Massachusetts
Rhode Island
Connecticut
Pennsyl-
vania
Maryland
New Jersey
40° N
Nevada
I
Utah
Wyoming
Nebraska
Iowa
Illinois
Indiana
Ohio
West
Virgi-
nia
C
Virginia
Californie
K
Colorado
Kansas
Missouri
Kentucky
Arizona
New Mexico
Oklahoma
Arkansas
E
Tennessee
North Carolina
South
Carolina
H
Texas
Mississippi
D
Alabama
Georgia
Louisiana
Florida
Mexico
500 km

In the following review of horticultural research in the USA the States have been grouped as follows:

A **New England States**: Maine, New Hampshire, Vermont, Massachusetts, Connecticut, Rhode Island
B **North Atlantic States**: New York, Pennsylvania, New Jersey
C **Middle Atlantic States**: Delaware, Maryland, Virginia, West Virginia
D **South Atlantic and Gulf Coast States**: North Carolina, South Carolina, Georgia, Florida, Alabama, Mississippi, Louisiana
E **South Central States**: Tennessee, Arkansas, Kentucky, Missouri
F **North Central States**: Ohio, Indiana, Illinois, Michigan, Wisconsin
G **Great Plains Region**: Minnesota, North Dakota, South Dakota, Iowa, Nebraska, Kansas
H **Southwestern States**: Oklahoma, Texas, New Mexico, Arizona
I **Mountain States**: Colorado, Utah, Nevada, Wyoming, Idaho, Montana
J **Pacific Northwest States**: Washington, Oregon
K **California**
L **Alaska**
M **Hawaii**
N **Puerto Rico**

In the alphabetical key at the end of this book the letters A-N marking the groups of States, have been added to the indication "US" behind the names of the research workers.
The names of the heads of departments who collected the data on horticultural research in the groups of States are mentioned at the end of each group.

A. NEW ENGLAND STATES

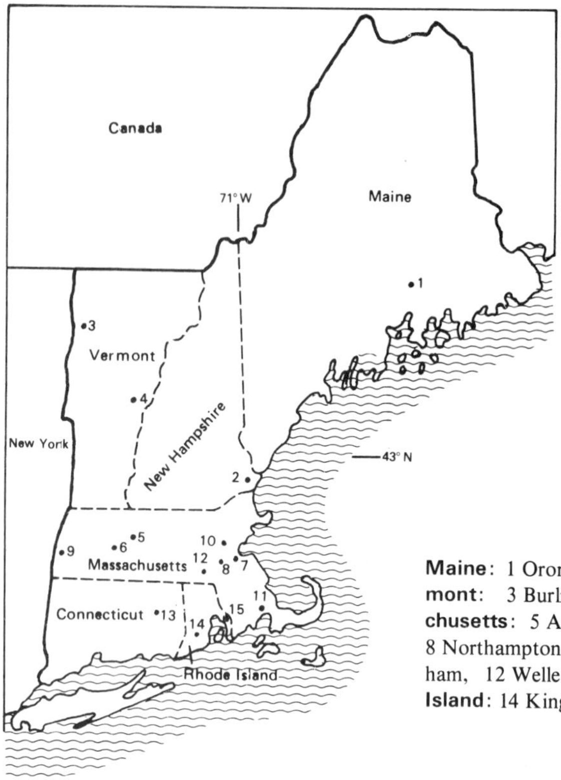
Canada
71° W
Maine
1
3
Vermont
4
New Hampshire
New York
43° N
2
5
6
9
10
Massachusetts
12
8
7
11
15
Connecticut
13
14
Rhode Island

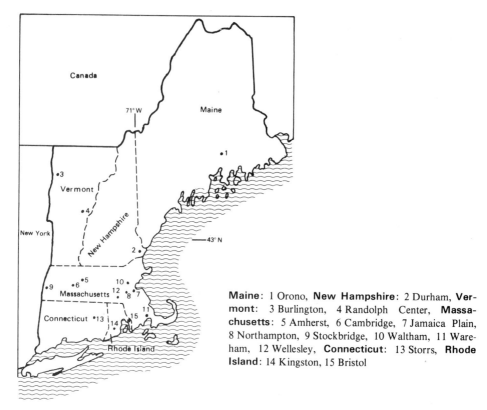

Maine: 1 Orono, **New Hampshire**: 2 Durham, **Vermont**: 3 Burlington, 4 Randolph Center, **Massachusetts**: 5 Amherst, 6 Cambridge, 7 Jamaica Plain, 8 Northampton, 9 Stockbridge, 10 Waltham, 11 Wareham, 12 Wellesley, **Connecticut**: 13 Storrs, **Rhode Island**: 14 Kingston, 15 Bristol

Survey

The northeastern states represent one of the oldest horticultural areas in eastern United States, and one of the first organized groups of people interested in horticultural in the country. The Massachusetts Horticultural Society was founded February 24, 1829. That organization has done much to increase interest in the development and to stimulate research. It has received world-wide recognition for its sponsorship of annual exhibits, horticultural fairs, innovations and research. Horticulture is that segment of New England agriculture which is growing and which appears to have a great deal of potential. Although coastal sections are highly populated and industrialized there remain inland many rural areas in which large acreages are adapted to apple growing, the principal tree-fruit crop, high- and low-bush blueberries *(Vaccinium)* and, in the southern New England states, peaches and grapes. Much of the entire area, with the exception of the mountainous sections, is adapted to growing strawberries and red raspberries.

The apple industry of New England presently centres around the McIntosh variety for which these states are famous, although other varieties are gaining rapidly in prominence. The industry is exceptionally well organized by alert, ambitious growers who are fully aware of their problems and are constantly on the alert to meet the challenge of them. Privately-owned, controlled-atmosphere storage plants and innovations in bulk handling, storage, grading and transportation are to be observed frequently through the six-state area. Nursery trees, shrubs and ornamental plants are grown extensively in each of the states. This too represents another segment of horticulture which is growing. While hardy species of deciduous and coniferous trees and shrubs make up the greater part of nursery-grown stock, added emphasis is placed on annuals and potted plants growing in increasing areas under glass. State and National Park and highway-beautification programs are under way and receiving greater attention as a stimulus to development of a rapidly growing recreational industry. Research in horticulture and plant sciences is centered in the colleges and universitites of the six states and in other public and privately endowed institutions which will be mentioned later. In recent years investigations have placed less emphasis on cultural practice, and more on problems of genetics and breeding, nutrition, plant growth and development, auxins and other growth substances and their composition and effects. Increasing emphasis is being placed on studies of light effects, the uses of herbicides, and on the interactions of plant nutrients. In these states the problems of hardiness of plants will always receive a great deal of attention. These, and many other projects of importance, all empahize the need for close cooperation of the physical and biological sciences in order to resolve answers to increasingly basic problems.

Maine

Aroostook County, Maine, has long been known for its tremendous potato production and a research farm for that work is located just south of Presque Isle, Maine. There is now also considerable interest in development of alternate crops for the same area.

In the southern part of the state there is a growing apple industry with a research farm at Monmouth, Maine. In the same region there is a concentration of vegetables for roadside stands. The low-bush blueberry industry is also located principally in the southern and south-eastern parts of the state, and a great deal of emphasis is being placed on research on this crop with a research farm at Jonesboro, Maine.

New Hampshire

Much of New Hampshire is covered with coarse and medium-textured soils of glacial till. The northern half of the state includes steep mountains with shallow soils to bed rock. This land is mainly in forest and some potatoes and is in an area with a short cool growing season. While soils in the south half of the state are derived from glacial till of granatic rock and are therefore fairly shallow, they do respond well to fertilizer and water. Apples are the most important orchard fruit, grown with high and low-bush berries, strawberries and red raspberries, the leading small-fruit corps. Turfgrass, greenhouse corps, particularly ornamentals, and nursery crops are all increasing rapidly in production, and are quite uniformly distributed through the area south of the White Mountains. Truck-crop produciton centered in the Merrimack Valley includes a diversity of vegetables produced under irrigation.

Vermont

The important horticultural areas of Vermont are located in the Champlain Valley, the Connecticut River Valley and in Grand Isle County. The principal horticultural crops grown in the state are apples, strawberries, blueberries and truck-crops for roadside stands and for market gardens, also woody ornamental and annual bedding plants.

Massachusetts

Massachusetts, centrally located among the New England states, has traditionally been recognized for its leadership in the horticulture of this country. Although the state has become heavily industrialized it still has the most widely diversified horticulture of any of the six-state group, and productive areas are developing further from centres of population. The principal horticultural industries (63% of cash farm receipts) in Massachusetts and their major locations are: 1. Floricultural and nursery products (about 27% of the value of all agricultural products in Mass.), located principally in Middlesex, Essex, Norfolk, Worcester and Plymouth counties. 2. Vegetable crops (about 10% of the value of all agricultural products in the state), located in the Connecticut Valley, northeastern Mass. and north-central Bristol County. 3. Tree fruits: apple, peach, pear (about 5% of the value of all agricultural products in the state), located primarily in Worcester and Middlesex counties and parts of western Mass. 4. Cranberries (about 18% of the value of all agricultural products of Mass.), located in Plymouth and Barnstable counties.

Connecticut

Connecticut, the most southern state in New England has a widely diversified florist crop industry of 6,300,000 sq. ft. in protected cultivation. Flowering pot plants such as poinsettias, chrysanthemums, etc., bedding plants, and cut roses are the principle crops. The production nursery industry combined with garden centers and other retail outlets of nursery products and landscape services play an increasing role in the horticulture of this heavily industralized and urbanized area. The Connecticut River Valley is an important area for field grown nursery stock and combined with an increased emphasis on container production represents a sizable industry. A diversified vegetable industry exists with crops such as tomato, sweet corn, lettuce, cabbage, peppers, carrots and snapbeans grown for fresh market. Connecticut grown, fresh, high quality, direct marketing by the producer to the consumer at roadside farm stands dominates the industry. Fruit orchards exist throughout the state at the higher elevations. Small fruits such as strawberries are often widely grown and marketed as pick-your-own. Acreage in wine grapes has increased recently making Connecticut a leader in the production of grapes and premium wine in New England. Farm wineries and vineyards are located throughout the state. The University of Connecticut at Storrs has teaching, research, and extension program to support all areas of horticulture in Connecticut.

Rhode Island

Much of the state of Rhode Island is in range of the modifying effects of the ocean. Therefore, many horticultural species that will not stand the more rigorous climate of northern New England will develop well here. Ornamental and nursery crops are grown to supply the requirements of a densely populated industrial area and a very large outstate trade. This industry is developed on both sides of Narragansett Bay. Vegetables are grown principally to supply the requirements of a thriving roadside-stand business. Apples, peaches and small fruits are grown on higher land with better soils and air drainage in the central and north western parts of the state.

MAINE

1 Orono

1.1 University of Maine, College of Life Sciences and Agricultural Experiment Station . Orono, Maine 04469

Department of Plant and Soil Sciences	Head: Dr. V.H. Holyoke
Apple culture and post-harvest physiology	Dr. W.C. Olien*
Blueberry nutrition physiology	Dr. J.M. Smagula
Potato breeding	Dr. A. Reeves
Potato culture and fertility, weed control	Dr. G. Porter
Potato tuberization	Dr. A.R. Langille
Small fruit and woody ornamental germplasm	Dr. P.R. Hepler
Ornamentals	Dr. W. Mitchell
Extension specialist in tree fruit	Dr. H. Wave
Extension specialist in blueberries	T. DeGomez
Extension specialist in ornamentals	Prof. L. Littlefield
Extension specialist in vegetables	Dr. W. Erhardt
Experimental Farms:	
Arootstock Farm	Preesque Isle 04769
Potatoes	Superintend.: J.A. Lloyd
Blueberry Hill Farm	RFD - Addison 04606
Blueberries	D. Emerson
Highmoor Farm	Monmouth 04259
Tree fruits	J. Harker

NEW HAMPSHIRE

2 Durham

2.1 Unversity of New Hampshire, College of Life, Sciences and Agriculture . Durham, New Hampshire 03824
Phone: (603) 862-3205

rainfall : 990 mm
sunshine : 55%

Department of Plant Science	Chrm: Dr. O.M. Rogers
Cell genetics	Dr. T.M. Davis
Ornamental genetics	Dr. O.M. Rogers
Population genetics	Dr. Y.T. Kiang
Vegetable genetics and breeding	Dr. L.C. Pierce/Dr. J.B. Loy
Post harvest physiology	Dr. J.E. Pollard
Plant biochemistry	Dr. D.G. Routley
Extension specialist in fruit	W. Lord
Extension specialist in turf	Dr. J. Roberts
Extension specialist in vegetables	Dr. O.S. Wells
Area agent in ornamentals	Dr. C.H. Williams

VERMONT

3 Burlington

3.1 University of Vermont, College of Agriculture, Agricultural Experiment Station — Burlington, Vermont 05405

Department of Plant and Soil Science	Chrm.: Dr. F.R. Magdoff
Ornamentals	Drs. N. Pellet/L.P. Perry*
Integrated pest management	Dr. L. Berkett

Apples — J.F. Constante
Strawberries, apples, small fruit — Dr. B. Boyce
Research emphasis on cold acclimation, winter hardiness, rootstocks

4 Randolph Center

4.1 Vermont Technical College . Randolph Center, Vermont 05061

Agricultural Department — R.C. Brown
Horticulture — K. Wigrem

MASSACHUSETTS

5 Amherst

5.1 University of Massachusetts . Amherst, Massachusetts 01003
Phone: (413) 545-2243

5.1.1 Department of Plant and Soil Sciences — Head: Dr. J.H. Baker

The department is concered with those sciences which deal with economic plants, soils and land use.

Floriculture:	
Mineral nutrition	Dr. W.J. Rosenau
Morphology	Dr. G.B. Goddard
Physiology, tissue culture	Dr. T. Boyle
Taxonomy	N.L. Gamabrants, B.S.
Nursery and Ornamentals:	
Water stress	Dr. J.R. Havis
Low temperature effects	
Pomology:	
Post-harvest physiology	Dr. W.J. Bramlage/Dr. W.R. Autio
Growth regulators	Dr. D.W. Greene
Size-controlling rootstocks	Dr. W.R. Autio/Dr. D.W. Greene
Varieties	J.F. Anderson
Weed-control herbicides	Dr. W.R. Autio
Mineral nutrition	Dr. J.H. Baker/Dr. W.J. Bramlage
Turf: pathology, nematology, weed control, culture	Dr. W.A. Torello
Vegetables:	
Plant breeding and genetics	
Mineral nutrition	Dr. A.V. Barker
Weed control, varieties, culture	Dr. R.J. Cooper
Physiological stress	H.V. Marsh/Dr. L.E. Craker

6 Cambridge

6.1 New England Botanical Club . 22 Divinity Avenue, Cambridge, Massachusetts 02138
Pres. Dr. Alice Tryon

Incorporated for the study of New England botany, Maintains an herbarium of plant specimens and records the distribution of plants in the New England states. Phanerogamic herbarium. Publ.: Rhodora (quarterly).

7 Jamaica Plain

7.1 Arnold Arboretum . Jamaica Plain, Massachusetts 02130
Dir.: Dr. P. Shaw Ashton

A trust administered by Harvard University operating in Jamaica Plain, Weston and Cambridge, Massachusetts. Founded in 1872 and functions to increase the knowledge of the use of woody plants, display plantings of special nature and research greenhouses, laboratories and offices. An additional 110 acres in Weston, Massachusetts, serve as a nursery and research area for plant introduction and development and as a display area for collections of special purposes, i.e. ground covers, small-street trees, etc. Laboratories and offices in Cambridge are associated with herbaria and libraries for teaching and research. An herbarium of nearly 800,000 specimens and a library of 53,000 books are divided into horticultural and floristic sections.

8 Northampton

8.1 Botanical Garden of Smith College . Northampton, Massachusetts 01060
Phone: (413) 584-2700

An area of 122 acres planted and maintained as a center of student instruction in botany and horticulture, and as a background for academic buildings. Rock gardens, perennial beds, display greenhouses, shrubs and tree plantings contain about 3,500 species.

9 Stockbridge

9.1 Berkshire Garden Center . Stockbridge, Massachusetts 01262
Dir.: C.R. Boutard

An area of 8.5 acres established in 1934 to provide a meeting place for those interested in horticulture and trial grounds for quality perennials, annuals, herbs, shrubs and trees hardy in Western Massachusetts.

9.2 New England Wild Flower Society Framington, Massachusetts 01701
Dir.: J. Shaw

There are eight sanctuaries. The New England Garden-In-The-Woods is the largest.

9.3 Heritage Plantation Sandwich, Massachusetts 02563

Collection of rhododendrons and azaelas — Jean Gillis, B.Sc.
Horticulture

9.4 Old Sturbridge Village Sturbridge, Massachusetts 01566
Dir.: L. Crawford

Features a herb collection

10 Waltham

10.1 University of Massachusetts, Agricultural Experiment Station, Suburban Experiment Station . Waltham, Massachusetts 02154
Phone: (617) 891-0650
Head: Dr. G. Fadoul

10.1.1 Department of Environmental Sciences

Breeding of commercial varieties of carnations for New England. New annual variety evaluation under New England conditions, nutritional and physiological studies on potted plants. Vegetable breeding for improvement of quality and adaptability. Sources and requirement of magnesium and carbondioxide for greenhouse tomatoes, seed improvement.

Floriculture — F.J. Campbell, B.Sc.
Vegetable crops — Dr. C.W. Nicklow

11 Wareham

11.1 Unversity of Massachusetts, Agricultural . East Wareham, Massachusetts 02538

Experiment Station,
Cranberry Station

Phone: (617) 295-2212
Head: I.E. Demoranville, M.Sc.

Influence of pesticide residues on environmental conditions. Uptake of herbicides, study of some destructive insects, mechanical harvesting, cultural systems, growth regulators physiology, plant pathology

Dr. K.H. Deubert/Dr. R. Devlin, Dr. B.M.Zuckermann/Prof. J.S.Morton, M.Sc./I.E. Demoranville, M.Sc.

12 Wellesley

12.1 Alexandra Botanic Garden and Hunnewell Arboretum of Wellesley College

Wellesley College, Wellesley, Massachusetts 02181
Dir.: Dr. G. Sanford

An area of 24 acres established in 1923 in association with the Department of Biological Sciences. Used in general teaching and educational programs of the College.

W.J. Jennings

12.2 Hunnewell Pinetum

845 Washington Street, Wellesley, Massachusetts 02181

Founded in 1852 as the H.H. Hunnewell Pinetum, the 40 acres estate will remain open to visitors requesting permission. The collection of coniferous plants and rhododendrons is one of the finest collections in New England.

CONNECTICUT

13 Storrs

13.1 Storrs Agricultural Experiment Station,

Storrs, Connecticut 06268

Department of Plant Science	Head: Dr. E.R. Emino
Integrated pest management	Dr. R.G. Adams
Park restoration, urban park ecology	J. Alexopoulos
Crop physiology	Dr. D.W. Allinson
Herbicide physiology and biological competition, vegetables	Dr. R.A. Ashley
Stress physiology and biological competition, vegetables	Dr. B.B. Bible
Biotechnology, tissue culture	Dr. M.P. Bridgen
Enology (organic chemistry)	Dr. J.M. Bobbitt
Horticultural marketing of retailing	Dr. E.D. Carpenter*
Nursery crop culture, propagation and plant selection	Dr. E.G. Corbett/Dr. S. Waxman
Turf nutrition and culture	Dr. W.M. Dest
Plant nutrition	Dr. F.H. Emmert
Historic landscape preservation	R. Favretti
Plant fertility, soil testing lab	Dr. G.F. Griffin
Ecology and mycorrhiza	Dr. A.J.R. Guttay
Floristry, design of interior landscapes	W.L. Harper
Phenology and physiology	Dr. W.C. Kennard
Tree fruit training and cultural systems	Dr. D.A. Kollas
Florist crop cultural system, computerized controlled environment	Dr. J.S. Koths*
Bartlett Arboretum	Stamford, Cennecticut
Living museum of dwarf conifers, ericaceous plants and other ornamental plants	T. Lockwood
Plant breeding and genetics	Dr. R.D. Parker/Dr. I. Greenblatt, Dr. E.G. Corbett

Conservation tillage systems	Dr. R.A. Peters
Horticultural entomology	Dr. M.G. Savos
Plant pathology	Dr. D.B. Schroeder
Department of Agricultural Engineering	Head: Dr. R.A. Aldrich
Greenhouse engineering	Dr. R.A. Aldrich
Greenhouse mechanization	J.W. Bartok

RHODE ISLAND

14 Kingston

14.1 University of Rhode Island, College of Resource Development and Agricultural Experiment Station . Kingston, Rhode Island 02881
Phone: (401) 792-2791
rainfall : 1066 mm

Department of Plant Sciences	Chrm.: Dr. W.C. Mueller
Floriculture: culture and management	Asst.Chrm.: Dr. R.J. Shaw
Fruit: physiology and post-harvest physiology	Dr. R.E. Gough
Turfgrass: culture and management	Dr. C.R. Skogley
Weed control and management	J.A. Jagschitz, M.Sc.
Physiology and stress environment	Dr. D.T. Duff
Physiology and biochemistry - general	D.R.J. Hull
Breeding and genetics	
Agronomy	Dr. R.C. Wakefield
Woody ornamentals:	
Plant classification	
Physiology and propagtion	Dr. J.J. McGuire
Weed control and management	J.L. Pearson, M.Sc.
Tissue culture: embryogenesis	Dr. W.R. Krul
Landscape design	
Construction	J. Dunnington, M.L.A.
Landscape design	
Genetic engineering	Dr. B.C. Kim
Landscape architecture	Prof. R. Hanson, M.L.A.
Floral art	Ms Kathleen Maison
Virology	Dr. C. Moeller
Entomology - integrated pest management	Dr. R. Casagronde
Plant pathology	Dr. N. Jackson/Dr. C. Beckman
Entomology	Dr. P. Logan
Plant pathology	Dr. L. Englander
Entomology	Dr. R. Lebrun

15 Bristol

15.1 Blithewold Gardens and Arboretum . Ferry Rd., Bristol,
Rhode Island 02809
Dir.: Dr. D. Buma

Prepared by Everett R. Ermino, Dept. of Plant Science, Univ. of Connecticut, Storrs, Conn.
Allen V. Barker, Dept. of Plant and Soil Sciences, Univ. of Massachusetts, Amherst, Mass.
Owen M. Rogers, Dept. of Plant Science, Univ. of New Hampshire, Durham, New Hampshire
John J. Mc.Guire, Dept. of Plant Science, Univ. of Rhode Island, Kingston, Rhode Island

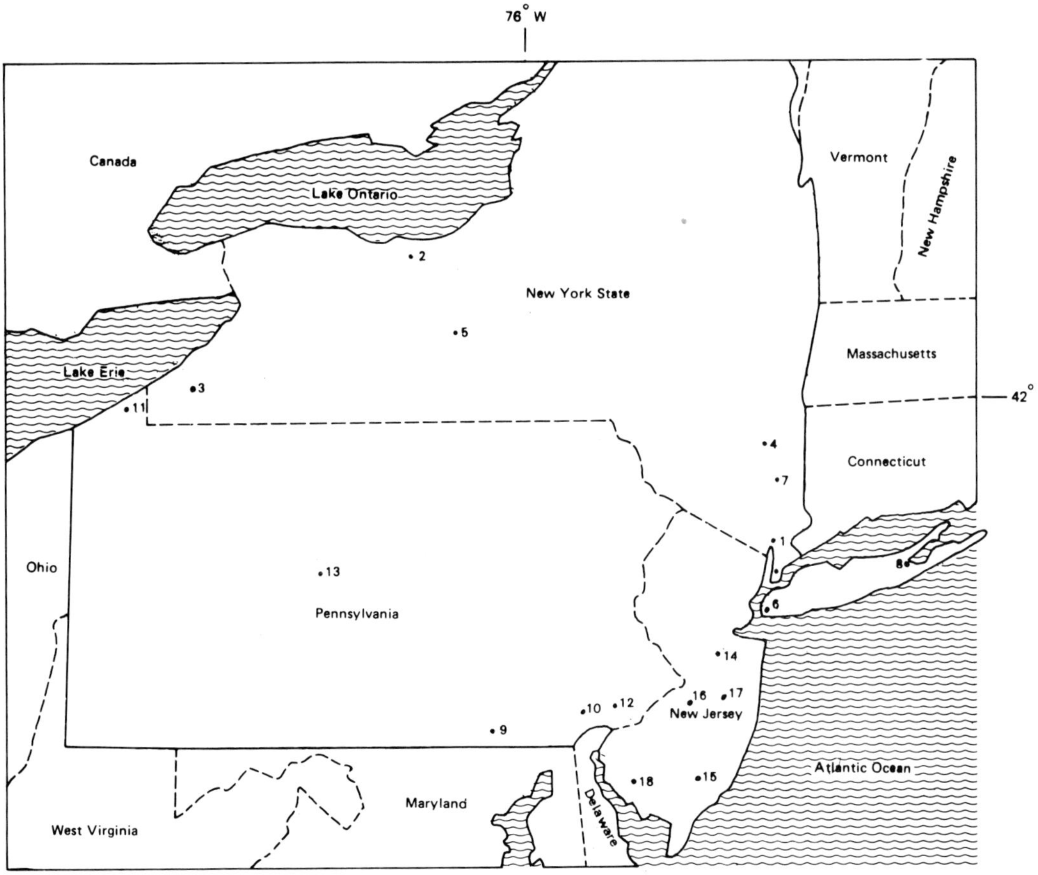
76° W
Canada
Lake Ontario
New York State
Lake Erie
Vermont
New Hampshire
Massachusetts
42°
Connecticut
Ohio
Pennsylvania
New Jersey
Atlantic Ocean
Maryland
West Virginia
Delaware
1
2
3
4
5
6
7
8
9
10
11
12
13
14
15
16
17
18

New York State: 1 Bronx, 2 Geneva, 3 Fredonia, 4 Highland, 5 Ithaca, 6 Brooklyn, 7 Ossining, 8 Riverhead - Long Island, **Pennsylvania**: 9 Biglerville, 10 Kennett, 11 North East, 12 Philadelphia (Wyndmoor), 13 University Park, **New Jersey**: 14 New Brunswick, 15 New Lisbon, 16 Cream Ridge, 17 Adelphia, 18 Bridgeton

Survey

New York State

Horticulture as an industry did not begin in the State until after the American Revolution, although fruits were grown by the Indians and early settlers. In New York State, the horticultural industry began with the nursery industry. Almost every family grew fruit of one kind or another. Grape growing was known on Long Island and in the Hudson River Valley but the important grape area in western New York started about 1830 in the Keuka Lake Region and became a commercial industry about 1853. The small-fruit industry in the State came into being about the same time. The commercial vegetable industry began in the late nineteenth century. Before this period, vegetables were grown about the home or in some truck patches. Seed farming started somewhat earlier, about 1800. Commercial flower production began in the 1830's and initially developed primarily near or in larger cities, a pattern which continues today. In the 1860's the horticultural industry began to take on a new look with the establishment of agricultural colleges and experiment stations. Important old fruit varieties originating in New York State were: apples: Green Newton, Jonathan, Twenty Ounce and Northern Spy; pears: Lawrence and Sheldon; European plums: Imperial Gage and Yellow Gage: grapes: Clinton and peaches: Chili

Institutions:

Cornell University was founded by Ezra Cornell and Andrew D. White in 1865 on farm land overlooking Cayuga Lake in the Finger Lake Region of New York State at Ithaca. Cornell provided the land and funds, while White gave direction and leadership and became its first president. The State Legislature gave

New York State's portion of the Morrill Land Grant Act of 1862 for the founding of colleges for teaching agriculture and mechanic arts. It was Cornell's idea to found an institution 'where any person can find instruction in any study'. The University today is one of the great institutions of the nation, offering courses in engineering, medicine, arts and sciences, in hotel administration, architecture, industrial and labor relations, human ecology, veterinary medicine, business management and agriculture. Under-graduate and graduate enrollment is about 18,000 students.

Probably no people had as much to do with the development of horticulture and agriculture at Cornell University as Dr. Liberty Hyde Bailey and Isaac Roberts. Bailey came to Cornell University in 1888 and together with Roberts developed practical and experimental horticulture. The College of Agriculture was founded in 1896. The first appropriation from the State Legislature in 1904 for the construction of a new agricultural building, established the State College of Agriculture at Cornell University. L.H. Bailey was its Dean and much of its growth and philosophy was due to this international figure.

The New York State Agricultural Experiment Station at Geneva has been a part of the State College of Agriculture at Cornell University since 1923. The Station was founded in 1882 by an act of the State Legislature 'for the purpose of promoting agriculture in its various branches by scientific investigations and experiment'. It has had a long and outstanding history in fruit research which has served the state, the nation and the world.

Cornell has field research stations at the following sites: Hudson Valley Laboratory: Fruit (3). Long Island Horticultural Research Laboratory, Riverhead: Vegetables, Potatoes, Florist, Nursery and Turfgrass Crops, Grapes (8). Vineyard Laboratory, Fredonia: Grapes (9).

Bronx, The New York Botanical Garden was established in 1891 by an Act of the State Legislature authorizing the Commissioner of Parks to set aside land in Bronx Park, New York City. The Garden was created to establish and maintain a botanical garden, museum and arboretum for plants, flowers, shrubs and trees; to conduct research and instruction in the same; to exhibit ornamental and decorative horticulture and gardening and to provide entertainment, recreation and instruction. A Board of Managers governs the New York Botanical Garden.

The Brooklyn Botanic Garden is one of the four divisions of the Brooklyn Institute of Arts and Science which is affiliated with the City of New York. The Garden was established in 1911 by Prof. Franklin W. Hooper, Director of the Institute and Alfred T. White, an esteemed citizen of Brooklyn, to conduct popular education and scientific research.

The Boyce Thompson Institute for Plant Research, formerly at Yonkers and now on the Cornell University campus at Ithaca, is a privately endowed, tax-exempt, non-profit organization. It was founded by Col. W. B. Thompson to acquire basic knowledge on plant and animal life and develop sound principles for solving problems of plant and animal culture and use.

Pennsylvania

Pioneers coming to Pennsylvania found an abundance of fruit growing wild and grown by the Indians: apples, peaches, plums, grapes, raspberries, blackberries, dewberries, cranberries, gooseberries and blueberries. It is believed that the peach was brought to this country by the Spaniards when they visited Florida in 1565. The Indian was evidently responsible for its wide spread since the peach was a prized fruit. Early settlers found many wild growing peach trees. Most of American fruit production was founded on varieties discovered growing wild.

In early times fruits were grown primarily for drink (wines, brandy, cider) and frequently to feed livestock. Only occasionally were they eaten. The beginnings of American grape growing were in Lancaster County in Southeastern Pennsylvania. From 1740 to 1840, Philadelphia was the center of scientific learning in America and held undisputed leadership in horticulture. Three men are largely responsible for Pennsylvania's horticulture: Benjamin Franklin, John Bartram and Richard Peters. Franklin was instrumental in the founding of the Philadelphia Society for the Promotion of Agriculture in March, 1785, the first agricultural organization in America. From it, the first horticultural society was organized in 1827, the Pennsylvania Horticultural Society. Bartram was the first American Botanist and established the first botanical garden in America in 1728. Richard Peters was a judge and Philadelphia lawyer. He was primarily responsible in bringing about government recognition of agriculture and recognizing the importance of instruction and research in Agriculture.

Institutions:

Longwood Gardens, located 3 miles east from Kennett Square, is one of the most outstanding horticultural gardens in the United States. It compromises nearly 1,000 acres of natural rolling countryside with about 350 acres which have been made into many gardens of various kinds. One of the most famous fountain systems of the world, covering 5 acres and illuminated at night with coloured lights, is a special evening attraction.

The land at Longwood Gardens dates back to William Penn who gave a grant of land to George Peirce in 1700. His property was purchased by Pierre S. du Pont in 1906 for his personal use as a country estate.

Pennsylvania State University, College of Agriculture at University Park. This university was founded in 1855 as the Farmer's High School. In 1863, after the passage of the Morrill Act, it was designated as the 'land-grant college' of the Commonwealth. The university consists of 16 campuses with the main campus located at University Park. Total student enrollment is about 50,000 with about 25,000 on the main campus. The College of Agriculture provides instruction in the sciences which underlie agriculture, conducts research under the Agricultural Experiment Station and extension through the Agricultural Extension Service.

New Jersey

New Jersey is called 'the Garden State' which reflects readily the importance of the horticultural industry in the state. New Jersey is situated in the center of the belt of heavy population and industry between Washington D.C. and Boston, Mass. The total value of agriculture in the state is about $ 400 million, with fruits and vegetables combined amounting to about $ 175 million. Vegetables, both for fresh market and processing outlets, are about $ 70 million and $ 30 million, respectively. Fruits amount to about $ 40 million and flowers and ornamentals about $ 50 million.

Institutions:

Rutgers', the State University, was founded as Queens College in 1766 by members of the Dutch Reformed Church. In 1825 the name of the college was changed to Rutgers in honour of Col. Henry Rutgers, a former trustee. In 1864, Rutgers' Scientific School was designated as the 'land-grant college' of New Jersey, becoming the College of Agriculture and later Cook College, at New Brunswick. In 1945 the New Jersey Agricultural Experiment Station became a part of the University. The university has campuses at Newark and Camden and offers courses in engineering, science, liberal arts, medicine, law and agriculture. Total enrollment is about 30,000 students. Horticultural research is centered at New Brunswick in the Department of Horticulture and Forestry, which has responsibility for research, teaching and extension in floriculture, forestry, landscape design, ornamentals, pomology and vegetable crops. Research is both basic and applied, primarily serving the industry of the state. Two important research areas are breeding and nutrition including trace elements. Many vegetable and fruit varieties introduced from this station are internationally grown.

Rutgers has field stations at the following locations: Cranberry and Blueberry Research Center, New Lisbon, Fruit Research and Development Center, Cream Ridge, Soils and Crops Research Center, Adelphia and Rutgers Research and Development Center in Bridgeton.

NEW YORK STATE

1 Bronx

1.1 The New York Botanical Garden
elevation : 3 m
rainfall : 1088 mm
sunshine : 2600 h

Phone: (212) 220-8777
Dir. Dr. J.M. Hester

The garden covers 250 acres including the Bronx River Gorge and Falls, the Hemlock Forest, 40 acres of original forest, high boulders of the glacial age, meadows and rocky knolls. Plantings include evergreen and deciduous trees; the Pinetum, the Montgomery Conifer Collection of over 200 species of evergreens, a large collection of azaleas, rhododendrons, lilacs and magnolias, a rose garden of over 7,000 plants of 80 natural species and over 400 cultivars, the Jane Watson Irwin Perennial Garden, the Thompson Memorial Rock Garden, the Herb Garden, a Native Plant Garden and a seasonal outdoor display of many flowering plants.

The Botanical Garden has a large conservatory, the Enid A. Haupt Conservatory, which was completely restored and reopened in 1978, and a herbarium containing over 4,500,000 plant specimens from all parts of the world but with an emphasis on the Americas, and one of the foremost research libraries* on botany and horticulture in the United States with over 1,000,000 holdings. It has undertaken or sponsored over 500 plant expeditions to various parts of the world as part of its taxonomy research program. Research on the taxonomy of the Flora of the Americas with an emphasis on the Intermountain Region and Tropical South America is the main emphasis of the herbarium staff. The Garden's Institute of Economic Botany is concerned with identification of new food and fuel plants especially in tropical rain forests. The education program includes cooperative baccalaureate and graduate programs with Herbert H. Lehman College of the City University of New York, a cooperative associate degree (A.A.S.) program in ornamental horticulture with Bronx Community College, an adult continuing education program and a special program for children. The associated Mary Flagler Cary Arboretum of the New York Botanical Garden was founded in 1971 and covers an area of over 2,000 acres in Millbrook, New York. It is the site of the Garden's Institute of Ecosystems Studies which focuses its research on the ecology of disturbance and recovery in the North Eastern United States.

Botanical research	G. T. Prance, D.Phil.
Herbarium	Patricia K. Holmgren, Ph.D., M.J. Balick, Ph.D./R.C. Barneby D.Sc.
Graduate fellowships	Admin.: P.M. Richardson, Ph.D.
Biochemical systematics and tissue culture unit	S. Mante, Ph.D./P.M. Richardson Ph.D.
Scientific publications	Maria L. Lebron-Luteyn, Ph.D., Elsie T. Doherty
Institute of economic botany	Dir.: G.T. Prance, Ph.D., Ass.Dir.: M.J. Balick, Ph.D.
Horticulture	Vice Pres.Hort.: C.A. Totemeier,Jr
Education services	J.F. Reed, AMLS/C.R. Long, MA,SMLS
Library and plant information service	L. Lynas, MA,MLS/Rose Li,AMLS
Institute of ecosystems studies	Dir.: G.E. Likens, Ph.D./ Admin: J.S. Warner, MA

2 Geneva

2.1 New York State Agricultural Experiment Station*	Geneva, New York
Cornell University	Phone: (315) 787-2211
elevation : 218.8 m	Telex: 937478
rainfall : 815 mm	Dir.: Dr. L.F. Hood
sunshine : 2377 h	

Primarily fruit and vegetable research for the processing industry. Investigations in food science and technology, plant breeding and genetics, seed testing and improvement, rootstocks fruit and vegetable culture and nutrition, integrated pest management, and biotechnology. Equipment: vegetable and orchard plots, laboratories, greenhouses, growth chambers, library, omputer center.

2.1.1 Administration, Jordan Hall	Dir.: Dr. L.F. Hood
2.1.2 Department of Horticultural Sciences	Chairman: Dr. G.E. Harman
Breeding (resistance, stone fruits)	Dr. R.C. Lamb*
Plant nutrition, soil fertility, mechanization	Dr. N.H. Peck
Rootstocks, propagation	Dr. J.N. Cummins
Orchard management - Hudson Valley	Dr. C.G. Forshey
Genetics, plant breeding (peas, beets)	Dr. G.A. Marx
Genetics, plant breeding (tomatoes, cucumbers, lettuce, spinach)	Dr. R.W. Robinson
Genetics, plant breeding (snap beans, cabbage, broccoli)	Dr. M.H. Dickson
Seed physiology	Dr. A.A. Khan
Vineyard management and physiology (grape)	Dr. R.M. Pool*
Seed microbiology	Dr. G.E. Harman
Grape and apple physiology	Dr. A.N. Lakso
Plant biochemistry	Dr. J.C. Stanford
Breeding (strawberry, raspberry)	Dr. N.F. Weeden
Seed science	Dr. A.G. Taylor
Orchard management/tree training	Dr. T.L. Robinson
Breeding and genetics (grape)	Dr. B.I. Reisch
Plant Introduction Station (USDA)	Dr. D.D. Dolan
Plant Introduction Station (USDA)	Dr. R. Alconero
2.1.3 Department of Entomology	Chairman: Dr. G.A. Schaefers
Biology and control of insects on small fruits and host resistance	Dr. G.A. Schaefers
Insect attractant research	Dr. W.L. Roelofs
Biology and control of vegetable insect pests in Eastern New York	Dr. R.W. Straub

Biology and control of vegetable insect pests	Dr. C.J. Eckenrode
Biology and control of fruit arthropod pests in Eastern New York	Dr. R.W. Weires, Jr.
Biology and control of arthropod pests in pome fruits	Dr. W.H. Reissig
Biology and control of vegetable insects	Dr. A.M. Shelton
Insecticide toxicology, metabolism and mode of action of pesticides	Dr. D.M. Soderlund
Soil insect ecology	Dr. M.G. Villani
Insect molecular genetics	Dr. D.C. Knipple
Biology and control of grapes	Dr. T.J. Dennehy
Quantitative insect ecology	Dr. J.P. Nyrop
Plant Introduction Station (USDA)	Dr. B.J. Fiori
2.1.4 Department of Plant Pathology	Chairman: Dr. H.S. Aldwinckle
Biology and control of airborne bacterial and fungal diseases of vegetable crops	Dr. J.E. Hunter
Biology and control of soilborne diseases of vegetable crops	Dr. G.A. Abawi
Development of disease resistant fruit crops; tissue culture	Dr. H.S. Aldwinckle
Virus diseases of fruit crops; basic virology	Dr. D. Gonslaves
Biological control; cytology of host-parasite interactions	Dr. H.C. Hoch
Biology and control of airborne fungal diseases of grapes and small fruits	Dr. R. C. Pearson
Pest management of fruit crops; quantitative epidemiology and modeling	Dr. R.C. Seem
Biology and control of tree fruit diseases	Dr. T.J. Burr
Fruit disease extension, research on soil borne diseases of fruit crops	Dr. W.F. Wilcox
Diseases of processing vegetables in Western New York	Dr. H.R. Dillard
Virus diseases of vegetables	Dr. R. Provvidenti
Biology and control of airborne fungal diseases of fruit	Dr. D.A. Rosenberger
2.1.5 Department of Food Science and Technology	Chairman: Dr. D.F. Splittstoesser
Fermented foods, microbiology	Dr. K.H. Steinkraus
Post-harvest physiology and biochemistry	Dr. L.M. Massey
Microbiology of fermentation	Dr. J.R. Stamer
Carbohydrate chemistry; enzymology	Dr. R.S. Shallenberger
Texture measurements; vegetable processing	Dr. M.C. Bourne
Plant chemistry; instrumentation; enology	Dr. L.R. Mattick
Biochemistry of fruits and vegetables	Dr. J.P. VanBuren
Microbiology of spoilage organisms; fermentation	Dr. D.F. Splittstoesser
Food processing extenstion	Dr. D.L. Downing
Toxicology of naturally occurring components, additives, pesticides	Dr. G.S. Stoewsand
Processing waste chemistry and utilization	Dr. R.H. Walter
Pesticide residues, feed and fertilizer control	Dr. J.B. Bourke
Food chemistry; fruit and vegetable processing, nutritional quality	Dr. C.Y. Lee
Enology	Dr. T.H.E. Cottrell
Food processing, engineering, energy, physical chemistry	Dr. M.A. Rao
Plant chemistry; phenolics pigments	Dr. G. Hrazdina
Microbiology; processing waste treatment	Dr. Y.D. Hang
Flavor chemistry, instrumentation, food biochemistry	Dr. T.E. Acree
Fruit vegetable processing; computerization	Dr. M.R. McLellan
Microbiology of spoilage organisms	Dr. S.B. Rodriguez

2.1.6 Germplasm programs, cooperative ventures of USDA and State Agricultural Experiment Stations

Northeast Regional Plant Introduction Station — Dr. D.D. Dolan/Dr.R. Alconero, Dr. B. Fiori

National Clonal Repository for Apple and Eastern Grapes — P.L. Forsline*

2.2 Integrated pest Management Support Group, Cornell University — Leader: Dr. J.P. Tette

3 Fredonia

3.1 Vineyard Research Laboratory
elevation : 232 m
rainfall : 965 mm
sunshine : 2200 h

Fredonia, NY 14063
Phone: (716) 672-7336

4 Highland

4.1 Hudson Valley Laboratory
elevation : 47 m
rainfall : 1020 mm
sunshine : 2400 h

Highland, Box 727, NY 12528
Phone: (914) 691-7231

Orchard management — Dr. C.G. Forshey

Biology and control of fruit arthropod pests in Eastern New York — Dr. R.W. Weires, Jr.

Biology and control of airborne fungal diseases of fruit crops in the Hudson Valley; fruit crops extenstion pathology — Dr. D.A. Rosenberger

Biology and control of vegetable insect pests in Eastern New York — R.W. Straub

5 Ithaca

5.1 Cornell University, College of Agriculture, and Agricultural Experiment Station
elevation : 293 mm
rainfall : 896 mm
sunshine : 2200 h

Ithaca, NY 14853
Phone: (607) 256-6530
Telex: 937478
Dir.: N.R. Scott

Horticulture is divided into three departments and areas for basic and applied research,teaching and extension: Floriculture and Ornamental Horticulture, Pomology, and Vegetable Crops. These are supported by several departments such as Agronomy, Plant Pathology, Entomology, Agricultural Engineering, Plant Breeding, Agricultural Economics, Bailey Hortorium and the Cornell Plantation, the University's arboretum, botanical garden and natural area. Equipment: Laboratories of various types, radioisotope facilities, elaborate greenhouses with automatic controls, specially designed growth chambers, controlled atmosphere storages and research facilties, gardens and orchards.

5.1.1 Department of Floriculture and Ornamental Horticulture — Chairman: Dr. C.F. Gortzig*

Floriculture

Florist crop production and physiology, soils and nutrition, interior horticulture, operations management — Dr. T.C. Weiler*

Florist crop production and physiology, environmental factors, greenhouse structures and energy, growth chamber technology — Dr. R.W. Langhans*

Retail floriculture, horticultural therapy — Dr. R.T. Fox

Floral design, interior horticulture — C.C. Fischer, M.Sc.

Dr. J.W.Boodley/Dr. J.G.Seeley*, Prof.Em.

Landscape Horticulture	
Nursery crop production and physiology, landscape horticulture, soils and nutrition	Dr. G.L. Good
Nursery crop production and physiology, propagation physiology, mycorrhizal relations of horticultural crops	Dr. K.W. Mudge
Plant materials and landscape horticulture	Dr. R.G. Mower*
Homes and grounds	R.E. Kozlowski, B.Sc.
Turfgrass Science and Management	
Turfgrass physiology, soils and nutrition	Dr. A.M. Petrovic
Turfgrass physiology and cultural management	Dr. N.W. Hummel Jr.
	Dr. J.F. Cornman, Prof.Em.
Horticultural Weed Science	
Physiology of herbicide action, weed control in florist, nursery and turfgrass crops	Dr. J.C. Neal
	Dr. A. Bing*, Prof.Em.
Urban Horticulture (Institute)	
Physiology of plants in the urban environment	Prog. Coord.: Dr. N.L. Bassuk*, Dr. T.H. Whitlow
Post-harvest and Horticultural Physiology	
Post-harvest physiology of florist, nursery and turfgrass crops, carbohydrate metabolism, enzyme biochemistry	Dr. F.B. Negm
Department Extension Programs	Dept.Ext.Leader: Dr. G.L. Good
	J. Gruttadaurio, M.P.S.(Agric.)
Commercial Programs Coordinator	R.E. Kozlowski, B.Sc.
Homes and Grounds Program Coordinator	E.F. Schaufler, M.Sc.,Prof.Em.
Landscape Architecture Program	
Program Coordinator, teaching, extension	P.J. Trowbridge, M.L.A.
Teaching, extension	M.I. Adleman, M.L.A.,
	T.H. Johnson, M.L.A.
Teaching	D.W. Krall, M.L.A./L.J. Mirin, M.L.A.
Teaching, extension, research in ecology-based regional landscape planning, agroforestry, landscape ecology	A.S. Lieberman, M.Sc.
Research	L.Y. Mudrak, M.S.L.A.
Teaching, extension	R.T. Trancik, M.L.A.
	R.J. Scannell, Prof.Em.
Freehand Drawing	
Program Coordinator, teaching	R.J. Lambert, M.Sc.
Teaching	A.E. Elliot, B.F.A.
5.1.2 Department of Pomology	Chairman: Dr. G.H. Oberly
Auxins, biochemistry of growth and plant behaviour	Dr. L.E. Powell
Biochemistry of flavours in horticultural crops	Dr. L.L. Creasy
Chemical analysis of nutrients	M.A. Rutzke, B.Sc.
Growth regulators	Dr. L.J. Edgerton
Nutrition, soil and leaf analysis service	Dr. G.H. Oberly
Post-harvest physiology, controlled atmosphere storage, apple scald, colour formation, environment and keeping quality	Dr. G.D. Blanpied*
Storage physiology	Dr. F.W. Liu
Extension specialist in small fruits and fresh grapes	Dr. M.P. Pritts*
5.1.3 Department of Vegetable Crops	Chairman: Dr. E.E. Ewing
Anatomy and morphology	Dr. L.D. Topoleski
Bean and cabbage breeding	Dr. D.H. Wallace
Culture, variety evaluation	Drs. D.W. Wolfe/D.A. Wilcox
Culture, legume crops	Dr. D.E. Halseth

Culture, nutrition	Dr. P.L. Minotti
Home gardening	R.A. Kline, M.Sc.
Muck crops	Dr. R.E. Ellerbrock
Potato production	J.B. Sieczka, M.Sc./Dr. D.E. Halseth
Potato physiology	Dr. E.E. Ewing
Crop physiology	Dr. H.C. Wien*
Post-harvest physiology, controlled-atmosphere storage	Drs. J.R. Hicks/P.M. Ludford
Tomato, onion, celery and cucumber breeding	Drs. H.M. Munger*/L.D. Topoleski
Weed control	Dr. R.R. Bellinder/D.T. Warholic, M.Sc.
5.1.4 Department of Agricultural Engineering	Chairman: Dr. G.E. Rehkugler
	E.S. Shepardson, M.Sc.
5.1.5 Department of Agronomy	Chairman: Dr. R.F. Lucey
Agricultural climatology, bioclimatology	Dr. B.E. Dethier
Micro-climate studies	Dr. R.W. Zobel
5.1.6 Department of Entomology	Chairman: M.J. Tauber
5.1.7 Department of Plant Breeding	Chairman: Dr. W.D. Pardee
Potato breeding	Dr. R.L. Plaisted
Tomato and brassica breeding	Dr. M.A. Mutschler
Cabbage and dry bean breeding	Dr. D.H. Wallace
Tomato molecular genetics	Dr. S.D. Tanksley
Brassica tissue culture	Dr. E.D. Earle
Insect resistance in potatoes, tomatoes	Dr. D.A. Ave/Dr. P. Gregory
5.1.8 Department of Plant Pathology	Chairman: Dr. W.E. Fry
Culture of disease free plants, designing of new growth rooms	
Nematodes, viruses	Dr. W.F. Mai
Tissue culture	Dr. R.K. Horst*

Liberty Hyde Bailey Hortorium — Dir.: Dr. D.A. Young

A vast collection of pressed and dried plants from all parts of the world.

Wiegand Herbarium — Dir.: Dr. D.A. Young

A collection of 300,000 dried specimens of selected flowering plants

A.R. Mann Library* — Dir.: J. Olsen, M.L.S.

The second-largest agricultural library in the U.S. (250,000 volumes)

5.2 Boyce Thompson Institute — Man.Dir.: Dr. R.A. Young

Non-profit research institution affiliated with the College of Agriculture and Life Sciences. The Institute is divided into four program areas: environmental biology, plant stress, nitrogen and crop yields, and biological control. Research is conducted on plant growth, seed physiology, response of plants to environmental stresses such as air pollution and salinity, biochemistry of parasitism, insect physiology and pathology, the nature and control of plant pests, nitrogen fixation, plant molecular genetics, and plant molecular biology.

6 Brooklyn

6.1 Brooklyn Botanic Garden and Arboretum
elevation : 5 m
rainfall : 1136 mm
sunshine : 2600 h

1000 Washington Ave
Brooklyn, New York 11225
Phone: (718) 622-4433
Pres.: D.E. Moore, B.Sc.

Private institution. Land belongs to New York City which supports it in part. The Botanic Garden includes a greenhouse with many outstanding displays: economic plants, cycads, ferns, cacti and succulents, orchids, bromeliads, and the largest bonsai collection in the U.S. There is a Japanese garden, a rock garden, a children's garden, a Shakespeare garden, lily pools, a rose garden, a herb garden, a garden of fragrance for the blind, a local flora section, and an arboretum.

Vice Presidents	Miss Elisabeth Scholtz, D.H.L,D.Sc
Operations	R.S. Tomson, B.Sc.
Finance	L.M. Keegan, B.Sc.
Horticulture Director	E.O. Moulin, M.S.
Scientific Affairs Director	Dr. S. K-M Tim
Education Director	Miss L.E. Jones, M.S.
Clark Garden Director	Dr. H. Irwin
Education Coordinator	Mrs M.A. Castellana

7 Ossining

7.1 Brooklyn Botanic Garden Research Center
(formerly Kitchawan Research Station)
elevation : 73 m
rainfall : 1279 mm
sunshine : 2400 h

712 Kitchawan Rd, Ossining NY 10562
Phone: (914) 941-8886
Chairman: Dr. S.S. Hagar

Research Department of Brooklyn Botanic Garden. The general objective of the research department is to acquire basic knowledge of plant life, especially on the interrelationships between plants, their diseases and their environments, and how these relate to human welfare.

Plant pathology, air pollution, virology	Dr. C.R. Hibben
Plant pathology	Dr. S.S. Hagar
Plant breeding	Dr. L. Koerting
Plant physiology	Dr. D. Wright
Tissue culture	Mrs I. Biedermann
Horticulturist	Miss M.B. Anderson
Education coordinator	Mrs D.W. Purdy

8 Riverhead

8.1 Long Island Horticultural Research Laboratory
elevation : 30 m
rainfall : 1151 mm
sunshine : 2600 h

39 Sound Ave, Riverhead NY 11901
Phone: (516) 727-3595
Supt.: J.B. Sieczka, M.Sc.

The laboratory is a field station of Cornell University and has as its mission research and extension in horticultural crops (vegetables, potatoes, florist and nursery crops, turfgrass, grapes) with emphasis on production and marketing under Long Island conditions.

Potato culture	J.B. Sieczka, M.Sc.
Potato diseases	Dr. R. Loria
Insects of potatoes, vegetables, florist and nursery crops	Dr. M. Semel
Pest management of potatoes	R. Wright, M.Sc.
Weed control in florist and nursery crops and turfgrass	Dr. A. Bing*, Prof. Em.
Diseases of florist and nursery crops	M.L. Daughtrey, M.Sc.

PENNSYLVANIA

9 Biglerville

9.1 Pennsylvania State University
Fruit Research Laboratory
elevation : 152 m
rainfall : 1047 mm
sunshine : 2400 h

Biglerville, PA 17307
Phone: (717) 677-6116
Scientist-in-Charge: Dr. K.D. Hickey

Part of the Pennsylvania Agricultural Experiment Station. Concerned primarily with field research to serve the fruit industries of Pennsylvania.

Management of fruit diseases, fungicide evaluation	Dr. L.A. Hull
Management of fruit insects and mites, insecticide evaluation	
Control of diseases caused by viruses and nematodes	Dr. B.A. Jaffee
Fruit tree culture, rootstock, cultivar evaluation and tree nutrition	Dr. G.M. Greene*

10 Kennett Square

10.1 Longwood Gardens*
elevation : 52 m
rainfall : 1154 mm
sunshine : 2600 h

Kennett Square, PA 19348
Phone: (215) 388-6741

Private institution. One of the most outstanding horticultural gardens in the United States. In the conservatory (4 acres under glass) are displays of begonias, lilies, caladiums, browallias, chrysanthemums, camellias, acacias, rhododendrons, and azaleas. Tropical Terrace Garden, Rose House, Fern Passage, Desert House, Economic House of Plants valuable to man as food, fiber and medicine, displays of espalier trained fruit trees and orchids, Rock Garden, Herb Garden, pools of tropical waterlilies, several rose gardens, Topiary Garden, Heather Garden, Italian Water Garden, Wild Flower Garden, Open Air Theater and illuminated fountain displays.

Design and horticultural displays	F.E. Roberts/F.J.Carstens, D.R. Gregg/Mrs L. Landon Scarlett
Ornamental-plant taxonomy	Dr. D.G. Huttleston
Ornamental-plant breeding with emphasis on cold-hardy camellias, penstemons, cannas, New Guinea impatiens, asarums	Dr. R.J. Armstrong
Education	Dr. D.A. Apps

11 North East

11.1 Erie County Research Laboratory
elevation : 223 m
rainfall : 1001 mm
sunshine : 2200 h

462 N. Cemetery Road,
North East, PA 16428

Part of the Pennsylvania Agricultural Experiment Station. The staff are members of the university faculty. Staff members located at University Park also conduct research at this laboratory. The laboratory mainly serves the important concord grape industry in Pennsylvania's section along Lake Erie.

Grape culture Concord, wine and table varieties	Dr. C.W. Haeseler
Grape entomology	Dr. G.L. Jubb, Sr.

12 Philadelphia (Wyndmoor)

12.1 Eastern Regional Research Center
Agricultural Research Service, United States
Department of Agriculture (ARS, USDA)
elevation : 2 m
rainfall : 1052 mm
sunshine : 2600 h

600 East Mermaid Lane
Philadelphia, PA 19118
Phone: (215) 247-5800
Dir.: J.P. Cherry

One of the four Regional Research Laboratories in the U.S. authorized by the federal government in 1938 for conducting basic and applied, pre- and post-harvest research on agricultural commodities including milk, meat, animal hides, wool, animal fats, fruits and vegetables. The Center consists of six laboratories whose mission is to develop through basic, applied, and developmental research, new knowledge and technology that will ensure an abundance of high quality agricultural commodities and products at reasonable prices to meet the increasing needs and to provide a continued improvement in the standard of living of all Americans. Research objectives are to: improve productivity of animals and crop plants and reduce pre- and post-harvest losses; develop new and improved products and processing technology; upgrade nutritional value; open new and expand existing domestic and foreign markets; reduce marketing costs; eliminate health-related problems; utilize waste products, particularly potentional pollutants; minimize energy consumption; reduce soil erosion; improve plant/soil nutrition; and improve quality and economy in respect to consumer interests. Center Laboratories engaged in this research include:

Animal Biomaterials Laboratory	Chief: Dr. S.H. Feairheller
Engineering Science Laboratory	Chief: J.C. Craig, Jr.
Food Safety laboratory	Chief: Dr. D.W. Thayer
Food Science laboratory	Chief: Dr. J.H. Woychik
Physical Chemistry and Instrumentation Laboratory	Chief: Dr. J.R. Cavanaugh
Plant Science Laboratory	Dr. D.D. Bills

13 University Park

13.1 Pennsylvania State University, College of Agriculture, and Pennsylvania Agricultural Experiment Station
elevation : 358 m
rainfall : 956 mm
sunshine : 2200 h

University Park (State College)
PA 16802
Phone: (814) 865-2571
Agric. Admin. Bdg.
Acting Dean: W.W. Hinish

Land-Grant Institution for Pennsylvania. Founded in 1855.

13.1.1 Department of Horticulture

Tyson Building
Head: Dr. F.H. Witham

Responsibility of research, instruction and extension in floriculture, ornamental horticulture, plant breeding, plant nutrition, pomology, vegetable and fruit technology. Research is of a basic and applied nature oriented to the industry. A field station for grapes is located in Erie County (see 10.1) and for tree fruits in Adams County (see 8.1). Cooperative research is conducted with several Departments

Agricultural climatology	
Floriculture	
Plant physiology, greenhouse environment, growth regulators	Dr. J.W. Mastalerz
Soils, nutrition, water relations	Dr. J.W. White
Extension specialist	Dr. D.J. Wolnick
Flower crop production growth regulators	Dr. E. J. Holcomb
Extension specialist	Dr. R.G. Brumfield
Soils, nutrition, fertilizers	Dr. G.C. Elliott
Extension specialist	Dr. R.F. Fortney
Ornamental horticulture	Dr. C.W. Heuser
Propagation, plant materials, flower gardens	
Plant propagation, weed control	Dr. C. Haramaki*
Nutrition, pruning, crabapple varieties	Dr. D.J. Beattie
Flower gardens	Dr. R.F. Fortney
Extension specialists	Dr. J.R. Nuss/Dr. L.J. Kuhns
Plant breeding	
Sweet corn breeding, corn biochemical genetics	Dr. C.D. Boyer
Ornamentals and flower breeding and genetics, seed-propagated geraniums	Dr. R. Craig

Cytology and cytogenetics in solanum species	Dr. P. Grun
Plant nutrition	
Nutrition of horticultural crops	Dr. C.B. Smith
Vegetable nutrition, virus-nutrition interrelationship	Dr. E.L. Bergman*
Plant physiology	
Transport and utilization of photosynthates, tissue culture	Dr. J.C. Shannon
Photosynthesis, growth regulators	Dr. R.N. Arteca
Post-harvest physiology	Dr. K.B. Evensen
Pomology	
Growth and development, growth regulators, tree training, plant-environmental relationships, mechanical harvesting	Dr. L.D. Tukey*
Small-fruit culture	Dr. R.R. Daniels
Post-harvest physiology	
Grape culture, wine varieties (North East: see 11.1)	Dr. C.W. Haeseler
Weed control, fruit tree culture, rootstocks, variety evaluation (Biglerville: see 9.1)	Dr. G.M. Greene*
Extension specialist in tree fruits	Dr. R.M. Crassweller
Vegetables	
Tomato production, mechanical harvesting, variety trials	Dr. M.D. Orzolek
Fluid drilling	
Extension specialist in major fresh and processing crops and greenhouse culture	Dr. P.A. Ferretti
Extension specialist in potatoes	Dr. R.H. Cole
Extension specialist in home horticulture	Dr. J.R. Nuss
With cooperation from Food Science Department:	
Processing of fruits and vegetables, table-wine evaluation	Dr. R.B. Beelman
Potato chipping quality	Dr. G.D. Kuhn
Department of Agronomy	Acting Head: Dr. D.D. Fritton
Turf, breeding and management	Dr. J.M. Duich
Department of Landscape Architecture	Head: N.H. Porterfield M.L.A.
Department of Entomology	Head: C.W. Pitts
Department of Plant Pathology	Head: Dr. J.M. Skelly
Mushrooms	Dr. P.J. Wuest

NEW JERSEY

14 New Brunswick

14.1 Rutgers, The State University of New Jersey, Cook College and Agricultural Experiment Station
elevation : 38 m
rainfall : 1156 mm
sunshine : 2600 h.

New Brunswick, NJ 08903
Phone: (201) 932-9447
Dean & Dir.: Dr. S.J. Kleinschuster

Research is directed at a solution to problems existing in the industry. These efforts include plant breeding for improved quality, productivity, pest resistance, and adaptability to mechanization. In addition, parallel programs in physiology, culture, and management have similar objectives. Lastly, special emphasis is placed on ornamental horticulture research designed to improve the range of plant materials available to the general public.

14.1.1 Department of Horticulture and Forestry	Chairman: Dr. M.R. Henninger
Natural resources - wildlife	Dr. J.E. Applegate
Vegetable physiologist	Dr. G. Berkowitz
Vegetable breeding and genetics	Dr. C.D. Carter
Tissue culture	Dr. Chee-Kok Chin

Landscape architecture	R.H. De Boer, M.L.A.
Floriculture	Dr. D.J. Durkin
Cranberry and blueberry nutrition	Dr. P. Eck
Biology/ecology	Dr. D.W. Ehrenfeld
Asparagus breeding	Dr. J.H. Ellison
Post-harvest physiology	Dr. C. Frenkel
Extension specialist - vegetable culture and physiology	Dr. S.A. Garrison
Pomology plant physiology - growth regulator	Dr. T.J. Gianfagna
Extension specialist - small fruits	Dr. B. Goulart
Forest mensuration	Dr. E.J. Green
Ornamental horticulture	Dr. B.A. Hamilton
Extension specialist - potatoes	Dr. M.R. Henninger
Extension specialist - tree fruits	Dr. J.T. Hopfinger
Vegetable physiology - greenhouse crops	Dr. H. Janes
Plant breeding and genetics (strawberries)	Dr. G.L. Jelenkovic
Extension specialist - vegetable crops	W.B. Johnson, M.Sc.
Forest Biology and genetics	Dr. J.E. Kuser
Extension specialist - home horticulture	D.B. Lacey, M.Sc.
Extension specialist - nursery management	L.D. Little, Jr., M.Sc.
Pomology - breeding and genetics	Dr. S.A. Mehlenbacher
Ornamental horticulture	Dr. R.H. Merritt
Extension specialist - arboriculture and pesticide coordinator	W.R. Oberholtzer, BS
Ornamental breeding	Dr. E.R. Orton, Jr.
Floriculture physiology	Dr. J.N. Sacalis*
Landscape architecture	W.F. Scerbo, M.L.A./S. Strom, M.L.A.
Urban forestry	Dr. R.L. Tate
Extension specialist - forestry	Dr. M. Vodak
Breeding and genetics - blueberries and cranberries	Dr. N. Vorsa
Nursery crops	Dr. A. Vrecenak
Landscape architecture	C.A. and J.F. Webster, M.L.A.
Wildlife biology	Dr. L.J. Wolgast
Extension specialist - floriculture	Dr. G.J. Wulster
14.1.2 Department of Agricultural economics and marketing	Chairman: Dr. A.R. Koch
Direct marketing	Dr. M. Fabian
Electronic marketing	Dr. F. Perkins
14.1.3 Department of Biological and Agricultural Engineering	Chairman: W.J. Roberts, M.Sc.
Greenhouse design	
Extension specialist - trickle irrigation	H.E. Carpenter
14.1.4 Department of Entomology and Economic Zoology	Chairman: Dr. H.T. Streu
Extension specialist - pest management	Dr. G. Ghidiu
Insect ecology	Dr. J.H. Lashomb
Fruit insects	Dr. F.C. Swift
Extension specialist - entomology	Dr. L.M. Vasvary
Extension specialist - fruit	Dr. S.R. Race
14.1.5 Department of Meteorology and Physical Oceanography	Chairman: Dr. M.D. Shulman
Agricultural meteorologist	K.R. Arnesen
14.1.6 Department of Plant Pathology	Chairman: Dr. R.A. Cappellini
Extension specialist - vegetable diseases	Dr. S.A. Johnston
Vegetable diseases	Dr. G.D. Lewis
Cranberry and blueberry diseases	Dr. A.W. Stretch
Turf diseases	Dr. P.M. Halisky
Ornamental - forest diseases	Dr. J.L. Peterson
Extension specialist - fruit diseases	Dr. J. Springer

Extension specialist - ornamentals	Dr. B. Clark
14.1.7 Department of Soils and Crops	Chairman: Dr. L.A. Douglas
Turf management	Dr. R.E. Engel
Extension specialist - soil fertilization agronomic crop	Dr. R.L. Flannery
Turf breeding	Dr. C.R. Funk, Jr.
Weed science	Dr. R.D. Ilnicki
Extension specialist - turf management	Dr. H.W. Indyk
Extension specialist - weed control - vegetables	Dr. B.A. Majek
Extension specialist - weed control - agronomic and fruit	Dr. J.A. Meade
Extension specialist - soil fertilization - vegetable crop	Dr. J.W. Paterson

15 New Lisbon

15.1 Cranberry and Blueberry Research Center
elevation : 24 m
rainfall : 1163 mm
sunshine : 2600 h

. New Lisbon, NJ 080-9
Phone: (609) 894-8740
Faculty Member in Charge:
Dr. A.W. Stretch

Part of the New Jersey Agricultural Experiment Station in New Brunswick. The staff are members of the university faculty who conduct basic, applied and development research on blueberries and cranberries to aid and support the large industry in New Jersey.

16 Cream Ridge

16.1 Rutgers Fruit Research and Development Center
elevation : 17 m
rainfall : 1077 mm
sunshine : 2400 h

. Cream Ridge, NJ 08514
Phone: (201) 932-9711
Faculty Member in Charge:
Dr. J.A. Hopfinger

The Researach Center, laboratories and orchards and part of the New Jersey Agricultural Experiment Station in New Brunswick. Concerned primarily with field and laboratory research to serve the fruit industry of New Jersey.

17 Adelphia

17.1 Soils and Crops Research Center
elevation : 59 m
rainfall : 1166 mm
sunshine : 2600 h

. Aldelphia, NJ 07728
Phone: (201) 492-9120
Faculty Member in Charge:
Dr. L.A. Douglas

Part of the New Jersey Agricultural Experiment Station in New Brunswick. The staff are members of the university faculty who conduct their field research at the outlying station. Field plots are set up in turf, grains, aquatic weed science, soil fertilization, etc.

18 Bridgeton

18.1 Rutgers Research and Development Center
elevation : 37 m
rainfall : 1042 mm
sunshine : 2600 h

. Bridgeton, NJ 08302
Phone: (609) 455-3100
Dir.: Dr. J.W. Paterson

The Vegetable Research Center is part of the New Jersey Agricultural Experiment Station in New Brunswick. Vegetable research field trials are conducted to support the intensive fresh market vegetable industry in New Jersey.

Prepared by Carl F. Gortzig, Dept. of Floriculture and Ornamental Horticulture, Cornell Univ., Ithaca, New York
Ms. Elisabeth Scholtz, Brooklyn Botanical Garden, Brooklyn, New York
Michael J. Balick, The New York Botanical Garden, Bronx, New York

C. MIDDLE ATLANTIC STATES

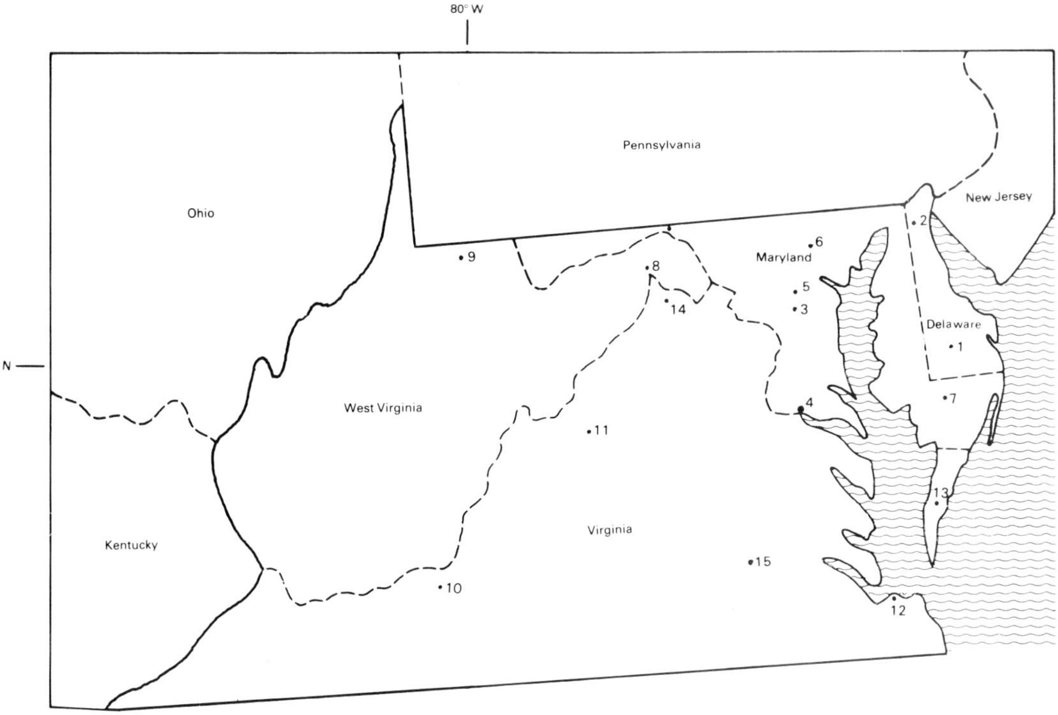
80° W
N
Pennsylvania
Ohio
New Jersey
Maryland
Delaware
West Virginia
Kentucky
Virginia
1
2
3
4
5
6
7
8
9
10
11
12
13
14
15

Delaware: 1 Georgetown, 2 Newark, **Maryland**: 3 Beltsville, 4 Washington (DC), 5 College Park, 6 Keedysville, 7 Salisbury, **West Virginia**: 8 Kearneysville, 9 Morgantown, **Virginia**: 10 Blacksburg, 11 Steeles Tavern, 12 Virginia Beach, 13 Painter, 14 Winchester, 15 Blackstone

Survey

Horticulturally speaking these Middle Atlantic States are best known for large mountain orchards and for truck farming on the Delmarva Peninsula (Delaware plus the eastern shore of Maryland and Virginia). The apple leads all horticultural crops in value in the region. Peaches, potatoes, tomatoes, peas, snap beans, lima beans, and strawberries are grown in large quantities also. Agricultural research in the area dates back to the plant introduction work of the US Patent Office before the Department of Agriculture was organized. The major research institutions at present are the Agricultural Research Center at Beltsville, Maryland, the Applachian Fruit Research Station (USDA) at Kearneysville, West Virginia, and the Agricultural Experiment Stations of the four states that were authorized by the Hatch Act. This act was passed March 2, 1887. All four states established their stations in the year 1888. All four stations were established at agricultural colleges which had been designated as Land Grant Colleges, and all have continued at their original locations.

Agricultural research at Beltsville began with the purchase of the Beltsville Farm in 1910 by the United States Department of Agriculture (USDA). Facilities were expanded and many units of the Department began doing research there. By the 1930's it became the logical point for consolidating all research being done at Beltsville, the Arlington Farm, greenhouses on the Capitol Mall, and other locations in the general vicinity of Washington, D.C. This consolidation was accomplished officially by the appointment on August 28, 1934 of the first Director of the Beltsville Research Center. It was renamed the Agricultural Research Center following the establishment of the Agricultural Research Service in 1953 and the Appalachian Fruit Research Station in 1979. The specific research activities of all these stations has followed much the same pattern. From the beginning, plant materials were assembled, evaluated, and disseminated. Better cultural practices were an early target. More recent objectives have been improvement through selection and breeding, improved handling and storage methods, and efficient marketing procedures. In recent years, much emphasis is being given to basic research.

DELAWARE

1 Georgetown

1.1 University of Delaware, College of Agricultural Sciences and Delaware Agricultural Experiment Station, Substation
elevation : 15.3 m
rainfall : 1108 mm

Georgetown, Delaware 19947
Phone: (302) 856-5254
Dir.: G.W. Chaloupka

This 100 ha substation located in South Delaware on primarily loamy sands is the primary site of field research on horticultural crops (mainly vegetable) in Delaware. Most research/demonstration is concerned with cultural practices (fertilization, irrigation, pest control) and cultivar testing of primarily sweet corn, cucurbits, lima beans, and white potatoes.

2 Newark

2.1 University of Delaware, College of Agricultural Sciences and Delaware Agricultural Experiment Station, Main Station
elevation : 30.7 m
rainfall : 1108 mm

Newark, Delaware 19717-1303
Phone: (302) 451-2501
Dir.: Dr. D.F. Crossan

The four-year undergraduate program in Plant Science allows students to concentrate in one of 4 areas: Plant Science, Ornamental Horticulture, Pathology, or Agronomy. At the (post)-graduate level, two curricula are available: Plant Protection and Improvement; and Soil Chemistry and Plant Nutrition. Research of faculty and graduate students ranges from basic to applied and is conducted in laboratories, growth chambers, greenhouses, and the field.

Department of Plant Science	Chairman: Dr. A.L. Morehart
Agronomy, weed science	Dr. W.H. Ahrens
Horticulture, ornamentals - extension specialist	S.S. Barton
Plant pathology	Dr. R.B. Carroll
Soil scientist (soil testing)	L.J. Cotnoir
Plant pathology	Dr. D.F. Crossan
Horticulture, vegetables culture and pathology	Dr. D.J. Fieldhouse*
Horticulture, ornamental horticulture	Dr. D.R. Frey
Plant physiology, ion uptake	Dr. H. Frick
Agronomy, plant genetics and breeding	Dr. J.A. Hawk
Horticulture, vegetables - extension specialist	W.E. Kee
Horticulture, plant cell and tissue culture	Dr. S.L. Kitto
Horticulture, landscape design	C. Lydon
Plant pathology, mycology	Dr. A.L. Morehart
Plant pathology, extension specialist	R.P. Mulrooney
Horticulture, vegetables physiology - fluid drilling	Dr. W.G. Pill
Plant anatomy/morphology	Dr. T.D. Pizzolato
Plant pathology, bacteriology	Dr. M. Sasser
Agronomy, soil fertility	Dr. T.J. Sims
Agronomy, soil chemistry	Dr. D.L. Sparks
Horticulture, ornamentals	Dr. J.E. Swasey
Agronomy, crop management - extension specialist	Dr. R.W. Taylor
Agronomy, plant nutrition	Dr. M.R. Teel

MARYLAND

3 Beltsville

3.1 Beltsville Agricultural Research Center,

Beltsville, Maryland 20705

U.S. Department of Agriculture, Agriculture Research Service	Phone: (301) 344-3338
Federal national program specialist - horticulture	Dr. H.J. Brooks
Weed investigations, horticultural crops	Dr. J.L. Hilton
Herbicides	W.J. Gentner
Physiology of fruit - growth regulators, deciduous fruit culture	Dr. M. Faust*/Dr. G. Steffens
Small fruit breeding	Dr. G. Galletta
Fruit nutrition	Dr. R.F. Korcak
Fruit physiology - juvenility studies and tissue culture research	Dr. R.H. Zimmerman
Deciduous fruit viruses, bacteriophage	Dr. E.L. Civerolo
Blueberry and strawberry breeding	Dr. A. Draper
Strawberry diseases	Dr. J.L. Maas
Post-harvest research	Dr. C.Y. Wang
Horticultural crops, quality evaluation	Dr. A.E. Watada
Post-harvest diseases, vegetables	Dr. H.E. Moline
Post-harvest diseases, fruit	Dr. W. Conway
Post-harvest physiology	Dr. B. Whitaker/Dr. J. Anderson, Dr. J.E. Baker
Ripening, senescence	Dr. K. Gross
Quality evaluation, fruit and vegetables	Dr. Judy Abbott
General horticulture	Dr. A.A. Piringer
Physiology of ornamentals	Dr. R. Griesbach/Dr. M. Roh
Diseases of ornamentals	Dr. R.H. Lawson/Suzanne Hurtt
Genetics and improvement of ornamentals	P. Semeniuk/Dr. T. Arisumi
Insects on ornamentals	Dr. R.E. Webb/Dr. J. Neal/Dr. H. Laren
Controlled environment for plant growth	Dr. D.T. Krizek
Breeding and diseases, potatoes and vegetables	Dr. R.E. Webb
Tomatoes and cucurbits breeding	Dr. A.K. Stoner
Potato breeding for insect resistance	Dr. L.L. Sanford
Cytogenetics of *Solanums*	Dr. G.D. McCollum, Jr.
Potato pysiology	Dr. S.L. Sinden
Potato diseases	Dr. R.W. Goth
Tomato diseases	Dr. T.H. Barksdale
Vegetable physiology, disease resistance	Dr. J.P. San Antonio/Dr. K. Deahl
Vegetable entomology	Dr. W.W. Cantelo
Plant introduction	Dr. G.A. White
Glenn Dale Plant Introduction Station	Dr. B. Parliman
Plant virology laboratory	Dr. J.M. Kaper/Dr. T.O. Diener, Dr. R.E. Davis/Dr. A. Hadidi, Dr. R.A. Owens
Biological control of soil-borne diseases	Dr. G.C. Papavizas/Dr. J.A. Lewis, Dr. R.D. Lumsden/Dr. P.B. Adams

4 Washington, D.C.

4.1 National Arboretum, Agricultural Research Service U.S. Department of Agriculture	Washington, D.C. 20002 Phone: (202) 475-4815 Dir.: Dr. H.M. Cathey*
Ornamental culture: physiology	Dr. H.M. Cathey*
Ornamental shrub breeding	Dr. D.R. Egolf
Shade tree breeding	Dr. A. Townsend/Dr. F.S. Santamour
Nomenclature: taxonomy	Dr. F.G. Meyer/Dr. T.R. Dudley

5 College Park

5.1 Maryland Agricultural Experiment Station	College Park, Maryland 20742

State Institution	Phone: (301) 454-3707 Dir.: Dr. W.L. Harris
5.2 University of Maryland, Department of Horticulture State Institution	College Park, Maryland 20742 Phone: (301) 454-3614 Chairman: Prof. B. Quebedeaux
Food science	Dr. D.V. Schlimme/Dr. B.A. Twig, Dr. R.C. Wiley
Forestry extenstion	Dr. J.F. Kundt
Landscape architecture	W. Gould, M.LA/D.G. Pitt, M.LA
Ornamentals and floriculture	Dr. F.R. Gouin/Dr. W.F. Healy, Dr. L. LaSota/Dr. H.G. Mityga, Dr. D.P. Stimart
Pomology	Dr. G.W. Stutte/Dr. H.J. Swartz, Dr. C.S. Walsh
Plant physiology	Dr. B. Quebedeaux
Post harvest physiology	Dr. T. Solomos/Dr. D.V. Schlimme
Vegetable breeding	Dr. T.J. Ng/Dr. J.C. Bouwkamp
Vegetable extension	Dr. C.A. McClurg

6 Keedysville

6.1 Western Maryland Research and Education Center, Maryland Agricultural Experiment Station State Institution Fruit extension and research	Route 1, Box 49B, Keedysville, Maryland 21756 Phone: (301) 791-2298 Head: Dr. G.L. Jubb, Jr.

7 Salisbury

7.1 Vegetable Research Farm Maryland Agricultural Experiment Station State Institution	Route 5, Quantico Road, Salisbury, Maryland 21801 Phone: (301) 742-8788
Farm superintendence and extension programs	Dr. F.D. Schales
Vegetable herbicides	Dr. C.E. Beste

WEST VIRGINIA

8 Kearneysville

8.1 West Virginia University Experiment Sub-Station for Tree Fruit elevation : 171 m rainfall : 965 mm	Kearneysville, West Virginia 25430 Res.Coord.-in-Ch.: Dr. S. Blizzard
Extension pathologist	Dr. J.G. Barrat
Extension entomologist	Dr. H. Hogmiere
Extension horticulturist	Dr. Tara Auxt Baugher
Horticulture and ward control	Dr. R. Young
Meteorologist	J.R. Holtz, M.Sc.
8.2 Appalachian Fruit Research Station U.S. Department of Agriculture Agricultural Research Service	Route 2, Box 45, Kearneysville, West Virginia 25430 Phone: (304) 725-3451 Dir.: B.A. Butt

Pathology - pome fruit diseases	Dr. T. Van der Zwet
Pathology - stone fruit diseases	Dr. C. Wilson
Post-harvest pathology	Dr. W. Janisiewicz
Horticulture - cultural studies of tree fruits	Dr. S. Miller
Post-harvest physiology	Dr. F. Abeles
Plant physiology	Dr. E. Ashworth
Pome fruit breeding/genetics	Dr. R.L. Bell*
Soil sciences	Dr. D.M. Glenn
Stone fruit breeding/genetics	Dr. R. Scorza
Small fruit culture/physiology	Dr. F. Takeda*
Stone fruit culture/physiology	Dr. B. Horton
Agricultural engineering - production and harvesting mechanization	Dr. D. Peterson
Fruit sorting instrumentation and controls	Dr. B. Upchurch
Fruit insect research - tree fruit entomology	B. Butt
Fruit insect systems	Dr. M. Brown
Weed and vaccinium research - weed science	Dr. W. Welker
Blueberry and cranberry pathology	Dr. A. Stretch

9 Morgantown

9.1 West Virginia University, Agricultural Experiment Station elevation : 200 m rainfall : 1000 mm	Main Station, Morgantown, West Virginia 26505 Dir.& Dean: Dr. R. Maxwell Dir. Div. Plant Sc.: Dr. M. Gallegly Ass.Dir.: Dr. S. Blizzard
Horticulture section	Chairman: Dr. B. Bearce
Floriculture and greenhouse management	Dr. B. Bearce
Physiology and biochemistry	Dr. D. Stelzig/Dr. J. Brooks
Tree fruits	Dr. S. Singha
Small fruits	C. Hickman, M.Sc.
Extension small fruits and vegetables	N.C. Hardin, M.Sc.
Post-harvest physiology	Dr. M. Ingle

VIRGINIA

10 Blacksburg

10.1 Agricultural Experiment Station Virginia Polytechnic Institute and State University, Main Station	Blacksburg, Virginia 24061 Dir.: Dr. J.R. Nichols Ass.Dir.: Dr. E.N. Boyd
State Institution	
Department of Horticulture	Head: Dr. T.A. Fretz
Developmental plant physiology	Dr. L.H. Aung
Tree-fruit physiology	Dr. J.A. Barden*
Post-harvest physiology - tree fruits	Dr. S.K. Lee
Tree-fruit production	Dr. R.E. Marini
Floriculture production	Dr. R.S. Lindstrom
Physiology of flowering	Dr. R.E. Lyons
Small-fruit physiology	Dr. J.M. Williams
Viticulture and enology	B.K. Zoecklein, B.Sc.
Small-fruit production	Dr. C.L. McCombs*
Nutrition	Dr. R.D. Wright
Nursery crop production	Dr. P.L. Smeal
Landscape architecture	R.F. McDuffie, M.LA
Landscape horticulture	Dr. J.S. Coartney

Landscape design	J.A. Faiszt, M.LA.
Environmental horticulture	Dr. R.T. Johnson*
Vocational-technical horticulture	Dr. A.R. McDaniel
Vegetable crop physiology	Dr. R.D. Morse
Vegetable breeding and genetics	Dr. R.E. Veilleux
Vegetable crop production	C.R. O'Dell, M.Sc.
International horticulture program	Dr. J.S. Caldwell
Urban-consumer horticulture	Dr. P.D. Relf

11 Steeles Tavern

11.1 Shenandoah Valley Research Station, part of the Agricultural Experiment Station at Blacksburg — Steeles Tavern, Virginia 24476

Peach, apple and grape production	E.L. Phillips, M.Sc.

12 Virginia Beach

12.1 Virginia Truck and Ornamentals Research Station, part of the Agricultural Experiment Station at Blacksburg — Virginia Beach, Virginia 23458
Dir.: Dr. E.A. Borchers

Physiology of nursery crops	Dr. D.C. Milbocker
Plant propagation and tissue culture	Dr. T.J. Banko
Insect pests of ornamentals	Dr. P. Shultz
Plant nutrition	R. Walden, M.Sc.

13 Painter

13.1 Virginia Truck and Ornamentals Research Station, Eastern Shore Branch, part of the Agricultural Experiment Station at Blacksburg — Painter, Virginia 23420
Sc.-in-ch.: Dr. R.E. Baldwin

Vegetable crop nutrition	Dr. H.E. Hohlt
Vegetable crop physiology	Dr. S.B. Sterrett
Herbicide physiology	Dr. H.P. Wilson
Soils	Dr. G. Evanylo
Insecticide physiology	Dr. G. Zehnder

14 Winchester

14.1 Winchester Fruit Research Laboratory, part of the Agricultural Experiment Station at Blacksburg — Winchester, Virginia 22601
Supt.: Dr. R.L. Horsburgh

Tree-fruit physiology	Dr. R.E. Byers
Tree-fruit production	Dr. H.A. Rollins, Jr.*
Plant pathology-fruit diseases	Dr. K.S. Yoder
Viticulture-grape production	Dr. T.K. Wolf

15 Blackstone

15.1 Southern Piedmont Research and Education Center, part of the Agricultural Experiment Station at Blacksburg — Blackstone, Virginia
Dir.: Dr. J.L. Tramel

Small fruit production	Dr. H.D. Stiles*

MIDDLE ATLANTIC STATES

Prepared by Wallace Pill, Dept. of Plant Science, Univ. of Delaware, Newark, Delaware
B.A. Butt, Appalachian Fruit Research Station, Kearneysville, West Virginia
Thomas A. Fretz, College of Agriculture and Life Sciences, Blacksburg, Virginia

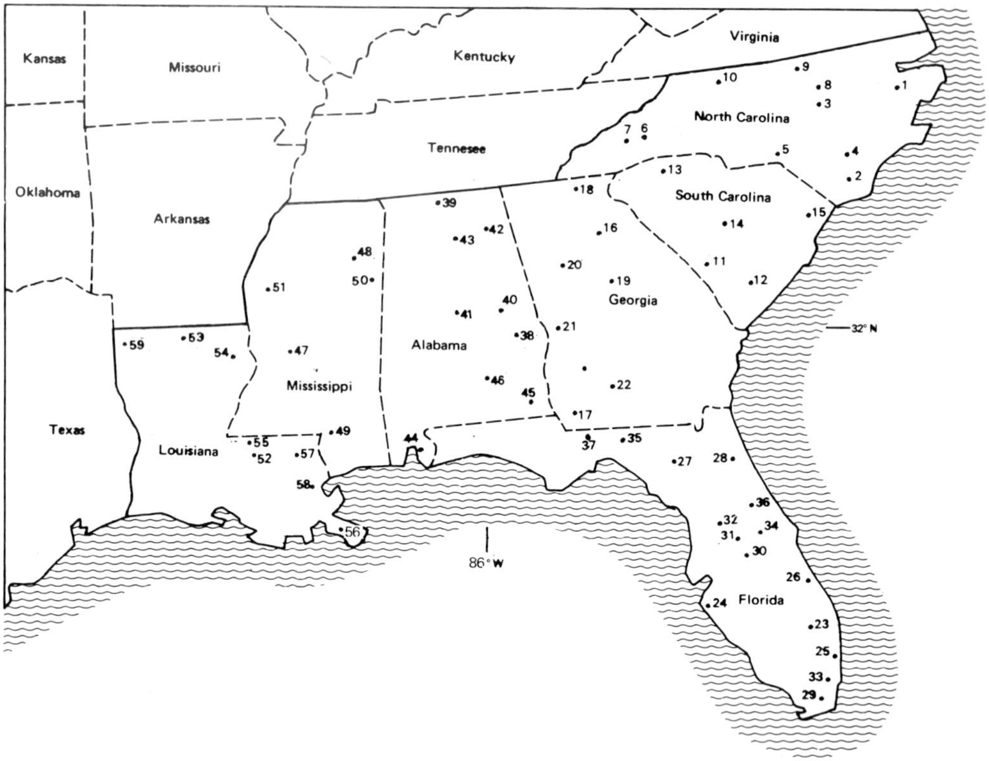
Kansas
Missouri
Kentucky
Virginia
Oklahoma
Arkansas
Tennessee
North Carolina
South Carolina
Georgia
Alabama
Mississippi
Louisiana
Texas
Florida
32° N
86° W

North Carolina: 1 Plymouth, 2 Castle Hayne, 3 Clayton, 4 Clinton, 5 Jackson Springs, 6 Fletcher, 7 Waynesville, 8 Raleigh, 9 Oxford, 10 Reidsville, **South Carolina**: 11 Blackville, 12 Charleston, 13 Clemson, 14 Columbia, 15 Florence, **Georgia**: 16 Athens, 17 Attapulgus, 18 Blairsville, 19 Byron, 20 Experiment, 21 Pine Mountain, 22 Tifton, **Florida**: 23 Belle Glade, 24 Bradenton, 25 Fort Lauderdale, 26 Fort Pierce, 27 Gainesville, 28 Hastings, 29 Homestead, 30 Lake Alfred, 31 Apopka, 32 Leesburg, 33 Miami, 34 Orlando, 35 Monticello, 36 Sanford, 37 Quincy, **Alabama**: 38 Auburn, 39 Belle Mina, 40 Camp Hill, 41 Clanton, 42 Crossville, 43 Cullman, 44 Fairhope, 45 Headland, 46 Spring Hill, **Mississippi**: 47 Crystal Springs, 48 Pontotoc, 49 Poplarville, 50 Mississippi State, 51 Lorman, **Louisiana**: 52 Baton Rouge, 53 Calhoun, 54 Chase, 55 Clinton, 56 Port Sulphur, 57 Hammond, 58 New Orleans, 59 Shreveport

Survey

The States of North Carolina, South Carolina, Georgia, Alabama, Mississippi and Louisiana comprise an area of 290,491 square miles and, together with the State of Florida, constitute the South Atlantic and Gulf Coast Region of the United States. From the Standpoint of climate these six contiguous states have factors in common and a factor in distinction. Factors in common are the amount and distribution of the rainfall, the amount of sunshine, and the proximity to large bodies of water. The mean annual rainfall varies from 1200 mm for North Carolina and South Carolina to 1375 mm for Louisiana.

The amount of sunshine varies with the length of the light period and the degree of cloudiness. The number of sunny days is greater than the number of partially cloudy or wholly cloudy days. The proximity to either the Atlantic Ocean or the Gulf of Mexico moderates the temperature level, particularly within a rather narrow belt on the coast. This is particularly the case during the winter and spring months.

The factor in distinction is the differences in elevation at the eastern or northern borders. In these areas North Carolina and South Carolina have elevations of 1,200 m or more. Alabama has a maximum elevation of about 600 m and Mississippi and Louisiana have maximum elevations of about 240 and 150 m, respectively. These differences in elevation account for the differences in temperature level and the length of the frost-free growing season. In general, the mountainous districts of North Carolina, South Carolina, Georgia and Alabama have a relatively short frost-free period (170-220 days), whereas the lower elevated areas of Mississippi and Louisiana have a relatively high temperature level and a relatively long frost-free period (215-250 days).

Organization of Horticultural Research.

Passage of the Hatch Act provided for the establishment of an experiment station in conjuction with, but not always at, the same location of the Land-Grant college. Further, with the passage of the Smith-Lever Act in 1914, the state and federal governments became partners not only in the education of all the people, and in the development of basic and applied research in agriculture and horticulture, but also in the dissemination of research results to all the people. At first only one station with its horticultural staff was established in each state. However, because of the differences in climates and soils and the differences in climate and soil requirements of the crops, a comparatively large number of research stations have become established within each state.

Florida

The state of Florida is pre-eminently a horticultural state, with two-thirds of its agricultural income produced by horticultural crops. No other state has so high a proportion of horticulture to total agriculture, and only California has a higher value of horticultural crops. Citrus fruits dominate Florida horticulture, accounting for about 60 percent of its value, with vegetables a good second at 35 percent, and ornamental crops third. Florida soils are mostly rather infertile, and it is climate rather than soil which makes Florida the leading state in production of both citrus fruits and winter vegetables. Florida's climate is subtropical, not tropical except for Key West, perhaps. It ranges from warm subtropical in the southern areas to cool subtropical in the northern ones. Thus North Florida is too warm in winter for most deciduous fruits to get adequate chilling, while South Florida has freezing temperatures too often for tropical fruits to be secure. Since oranges and grapefruit are subtropical in their requirements, they have proved better adapted to Florida than any other fruits. For most of the length of the peninsula there is a central ridge of sand hills with good air and water drainage, while the coastal areas and the whole width of the southern third of the state are low and flat, with drainage problems. In the past the citrus industry has been located mostly along the Ridge, but in recent years extensive plantings have been made on poorly drained, coastal soils, with confidence in huge pumps to keep the watertable from rising too high even in the rainy season. Such tropical fruits as mango, avocado, and lime are mostly grown south of Miami, centered in Homestead. Organized horticultural research in Florida began in 1888, when the Florida Agricultural Experiment Station was established in connection with the existing College of Agriculture at Lake City. In 1906 the College was merged with three other state institutions to form the University of Florida at Gainesville, and in 1907 the Experiment Station moved to Gainesville also.

The present system of branch stations dates from 1921, when the citrus station was inaugurated. Today there are 20 branch stations scattered over the state. Not all of these are concerned with horticulture, but most of them are.

The United States Department of Agriculture began horticultural research in Florida in 1893, at Eustis. Today it has a large group at Orlando investigating citrus problems and a small one near Miami working on tropical fruits and ornamentals. The Florida State Department of Agriculture has only been involved in research since the former State Plant Board became its Plant Industry Division in 1961, and only in very recent years had this agency carried on horticultural research. The Florida Citrus Commission supports horticultural research conducted cooperatively at the Citrus Experiment Station, providing funds for salaries and equipment. The Commission was established by law in 1935, and has supported research since 1943. Two private institutions now carry on horticultural research, both having been established in 1938-39. The Fairchild Tropical Garden in Miami investigates problems of ornamental plant culture, especially of palms, while the Soil Science Foundation in Lakeland is concerned with citrus nutrition.

NORTH CAROLINA

All outlying Research Stations are jointly administered by the North Carolina Agricultural Research Service of North Carolina State University (NCSU) and the North Carolina Department of Agriculture (NCDA). The station ownership is indicated in the ( ).

1 Plymouth

1.1 Tidewater Research Station (NCDA) — Plymouth, North Carolina 27962

elevation : 6 m
rainfall : 1278 mm
sunshine : 2800 h

Phone: (919) 793-4118
Supt:. J.W. Smith

Cultural and breeding research on vegetables. Main crops are potatoes, cucumbers, peppers and cole crops.

2 Castle Hayne

2.1 Horticultural Crops Research Station (NCSU)
elevation : 12 m
rainfall : 1350 mm
sunshine : 2900 h

Castle Hayne, North Carolina 28429
Phone: (919) 675-2314
Supt.: T.L. Blake

Main station for blueberry research. Cultural and breeding research on strawberries, grapes, brambles, woody ornamentals, flower-bulbs, sweet potatoes, carrots, cucurbits, beets and cole crops.

Dr. C.M. Mainland*

3 Clayton

3.1 Central Crops Research Station (NCSU)
elevation : 101 m
rainfall : 1200 mm
sunshine : 2800 h

Clayton, North Carolina 27520
Phone: (919) 553-6468
Supt.: W.R. Baker, Jr.

Cultural and breeding research on grapes, apples, peaches, cucurbits, sweet potatoes, tomatoes, cole crops, brambles, peppers and carrots.

4 Clinton

4.1 Horticultural Crops Research Station (NCDA)
elevation : 48 m
rainfall : 1235 mm
sunshine : 2840 h

Clinton, North Carolina 28320
Phone: (919) 592-7939
Supt.: F.E. Cumbo

Cultural breeding research on blueberries, cucurbits, sweet potatoes strawberries, grapes, tomatoes, peppers, asparagus, snapbeans, carrots, eggplant and okra.

5 Jackson Springs

5.1 Sandhills Research Station (NCSU)
elevation : 223 m
rainfall : 1245 mm
sunshine : 2870 h

Jackson Springs,
North Carolina 27281
Phone: (919) 974-4673
Supt.: C.S. Black

Cultural and breeding research on peaches, grapes, nectarines and asparagus.

6 Fletcher

6.1 Mountain Horticultural Crops Research Station (NCSU)
elevation : 631 m
rainfall : 1220 mm
sunshine : 2650 h

Fletcher, North Carolina 28732
Phone: (704) 684-7197
Supt.: H.E. Blackwell

Cultural and breeding research is conducted on Strawberries, brambles, grapes, potatoes, carrots, cole crops, blueberries, ginseng, apples, woody ornamentals and Christmas trees.

Department of Horticultural Science	
Culture of vegetable crops	Dr. T.R. Konsler
Tomato breeding	Dr. R.G. Gardner
Apple pre-harvest physiology	Dr. C.R. Unrath
Culture of woody ornamentals	R.E. Bir
Department of Plant Pathology	
Diseases of vegetables	Dr. P.B. Shoemaker
Department of Soil Science	
Vegetable nutrition	Dr. G.D. Hoyt
Ornamental and fruit nutrition	Dr. J.E. Shelton
Department of Zoology	
Vole research	W.T. Sullivan

7 Waynesville

7.1 Mountain Research Station (NCDA)
elevation : 810 m
rainfall : 1190 mm
sunshine : 2550 h

Waynesville, North Carolina 28786
Phone: (704) 456-3943
Supt.: J.R. Edwards

Cultural and breeding research on potatoes, blueberries, tomato, woody ornamentals and Christmas trees.

8 Raleigh

8.1 North Carolina State University
elevation : 122 m
rainfall : 1190 mm
sunshine : 2680 h

Raleigh, North Carolina 27695-7609
Phone: (919) 737-3131

The N.C. Agricultural Research Service was established at NCSU in 1887. There are six university research units in the Raleigh area. Horticultural programs are conducted at two of them.

8.1.1 Department of Horticultural Science	Head: Dr. A.A. De Hertogh*
Genetics - sweet potatoes, potatoes	Dr. W.W. Collins/Dr. F.L. Haynes, Jr
Tomatoes, watermelons and cucumbers	Dr. W.R. Henderson*/Dr. T.C. Wehner
Culture of vegetables	Dr. M.M. Peet*/Dr. C.H. Miller
	Dr. D.C. Sanders/Dr. W.J. Lamont
Physiology	Dr. L.K. Hammett/Dr. D.M. Pharr
Genetics - blueberries, strawberries, grapes and peaches	Dr. J.R. Ballington/Dr. R.G. Goldy
	Dr. D.J. Werner
Apples, culture	Dr. C.R. Unrath/Dr. E. Young
Post-harvest physiology of fruit	Dr. W.E. Ballinger
	Dr. S.M. Blankenship
Nutrition, floricultures	Dr. P.V. Nelson
Cultural practices, floriculture	Dr. W.C. Fonteno/Dr. R.A. Larson
Culture of ornamentals including Christmas trees and propagation	Dr. T.E. Bilderback/Dr. F.A. Blazich
	Dr. L.E. Hinesley/Dr. J.C. Raulston*
	Dr. V.P. Bonaminio
Cultures of small fruits	Dr. E.B. Poling
Plant taxanomy	Dr. P.R. Fantz
Weed control	Dr. T.J. Monaco/Dr. W.A. Skroch
	Dr. A.R. Bonanno

Agricultural meteorology	Dr. K.B. Perry
8.1.2 Department of Plant Pathology	Head: Dr. W.L. Klarman
Diseases of fruits	Dr. R.D. Milholland Dr. D.F. Ritchie/Dr. T.B. Sutton
Diseases of vegetables	Dr. E. Echandi/Dr. S.F. Jenkins, Jr. Dr. P.B. Shoemaker/Dr. J.W. Moyer Dr. G.V. Gooding
Diseases of ornamentals and floriculture	Dr. D.M. Benson/Dr. R.K. Jones Dr. D.L. Strider
Air pollution	Dr. R.A. Reinert/Dr. A.S. Heagle Dr. S.R. Shafer
8.1.3 Department of Entomology	Head: Dr. R.J. Kuhr
Bees	Dr. J.T. Ambrose
Cultural practices	Dr. G.G. Kennedy/Dr. J.R. Meyer Dr. F.P. Hain/Dr. G.C. Rock

9 Oxford

9.1 Oxford Tobacco Research Station (NCDA)
elevation : 152 m
rainfall : 1125 mm
sunshine : 2670 h

Oxford, North Carolina 27565
Phone: (919) 693-2483
Supt.: W.C. Clements

Breeding research on tomatoes and cucumbers

10 Reidsville

10.1 Upper Piedmont Research Station (NCSU)
elevation : 271 m
rainfall : 1090 mm
sunshine : 2750 h

Reidsville, North Carolina 27320
Phone: (919) 349-8347
Supt.: H.O. Gentry, Jr

Breeding research on cucumbers

SOUTH CAROLINA

All outlying research/extension/education sites of Clemson University are adminstered by the College of Agricultural Sciences. Ownership of the land and facilities is public. The names of all five sites have been changed to be Research and Education Centers to acknowledge their continuing education (extension) and formal degree program teaching roles.

11 Blackville

11.1 Edisto Research and Education Center
elevation : 99 m
rainfall : 1219 mm
sunshine : 3000 h

PO Box 247, Blackville,
South Carolina 29817
Phone: (803) 284-3343
Res.Dir.: Dr. J.R. Hill, Jr.

Breeding cantaloupes, cucumber, watermelons	Dr. B.B. Rhodes*
Breeding of sweet potatoes	Dr. M.G. Hamilton

12 Charleston

12.1 Coastal Research and Education Center
elevation : 2-2 m

2865 Savannah Highway, Charleston,
South Carolina 29407

rainfall : 1500 mm sunshine : 3000 h	Phone: (803) 766-3761 Res.Dir.: Dr. W.R. Sitterly
Plant pathology and squash breeding	Dr. W.R. Sitterly
Vegetable post-harvest physiology	Dr. J.W. Rushing
Vegetable culture	Dr. W.P. Cook
Tomato breeding	Dr. W.H. Courtney III
12.2 United States Vegetable Laboratory	2875 Savannah Highway, Charleston, South Carolina 29407 Phone: (803) 566-0840 Act.Dir.: Dr. G. Fassuliotis
Entomology	Dr. C.S. Creighton/Dr. J.M. Shalk, Dr. K.D. Elsey
Plant pathology	Dr. P.D. Dukes/Dr. C.E. Thomas
Nematology	Dr. G. Fassuliotis
Horticulture	Dr. P.E. Nugent
Research chemist	Dr. J.A. Wells/Dr. J.K. Peterson
Genetics	Dr. R.L. Fery*/Dr. A. Jones
Weed science	Dr. H.F. Harrison

13 Clemson

13.1 College of Agricultural Sciences, Department of Horticulture elevation : 250 m rainfall : 1370 mm sunshine : 3900 h	Clemson University, Poole Agricultural Center, Clemson South Carolina 29634-0375 Phone: (803) 656-3403
Fruit breeding	Head: Dr. R.L. Andersen*
Post-harvest physiology	Dr. R.A. Baumgardner
Ornamental horticulture	Dr. D.W. Bradshaw
Integrated pest management	Dr. J.A. Brittain
Peach breeding	Dr. D.W. Cain*
Small fruit culture	Dr. Judith D. Caldwell
Pomology	Dr. D.C. Coston*
Ornamental horticulture	Prof. J.P. Fulmer
Landscape design	Dr. R.G. Halfacre, Prof. Mary T. Haque
Floriculture	Dr. J.W. Kelly
Small fruit culture	Dr. G.A. King, Jr.
Ornamental horticulture	Dr. Alta R. Kingman
Turfgrass	Dr. A.R. Mazur/Dr. L.C. Miller
Olericulture	Dr. W.L. Ogle
Floriculture	Dr. A.J. Pertuit, Jr.
Postharvest physiology	Dr. E.T. Sims, Jr.*
Woody ornamentals	Dr. D.F. Wagner
Weed science	Dr. T. Whitwell

14 Columbia

14.1 Sandhill Research and Education Center elevation : 137 m rainfall : 1270 mm sunshine : 3000 h	PO Box 280, Elgin, South Carolina 29045 Phone: (803) 788-5700 Res.Dir.: Dr. J.K. Golden

Nematology	Dr. J.K. Golden
Pecans	Dr. J.B. Aitken
Pomology	Dr. G.L. Reighard
Vegetables	Dr. J.R. Johnson
Virology	Dr. S. Scott

15 Florence

15.1 Pee Dee Research and Education Center	PO Box 271, Florence,
elevation : 43 m	South Carolina 29503
rainfall : 1215 mm	Phone: (803) 662-3526
sunshine : 2900 h	Res.Dir.: Dr. J.B. Pitner
Vegetable culture	Dr. D.R. Decoteau

GEORGIA

The Georgia Agricultural Experiment Station has three main stations, viz. at Athens, Experiment and Tifton

Dean & Coord.: Dr. W. Flatt
Dir.Exp.Stations: Dr. C. Donoho
Dir. Coop. Ext.: Dr. T. DuVall
Dir.Res.Instr.: Dr. C.J.B. Smit

16 Athens

16.1 College Station	Athens, Georgia 30602
elevation : 271 m	Phone: (404) 542-2471
rainfall : 1145 mm	
sunshine : 2800 h	

The administration of a discipline state wide is handled by a committee of department heads.

Department of Horticulture	Head: Dr. C.H. Hendershott
Physiology and culture of tree fruits	Dr. G.A. Couvillon
Physiology and culture of ornamentals	Dr. F.A. Pokorny/Dr. J.H. Tinga, Vivian Munday/Dr. M.Dirr, Dr. A. Armitage
Nutrition and physiology of pecans	Dr. H. Wetzstein/Dr. D. Sparks
Post-harvest physiology	Dr. H.M. Vines/Dr. S. Kays
Vegetable nutrition	Dr. H. Mills/Dr. J.B. Jones, Jr.
Department of Plant Pathology	Head: Dr. W. Garrett
Fruit diseases	Dr. F. Hendrix
Nematodes of fruit and nut trees	Dr. W. Powell
Vegetable diseases	Dr. S. McCarter
Department of Agricultural Engineering	Head: Dr. R. Brown
Fruit tree irrigation	Dr. J. Chesness
Extension pomology	M.E. Feree, Ph.D.
Extension floriculture	J. Lewis, Ph.D.
Extension ornamentals	G.E. Smith, M.Sc.
Extension ground maintainence	G. Wade, Ph.D.
Extension floriculture, ornamentals (Statesboro)	H. Clay, M.Sc.

16.2 Russel Research Center (USDA-SEA-AR)	Athens, Georgia 30604
	Phone: (404) 546-3320

Horticulture crops laboratory	Chief: Dr. J. Dull
Quality assessment	Dr. J. McGee/P. Merdith, M.Sc.
Horticultural crops composition	Dr. S. Senter

Cell wall chemistry	Dr. R. Pressey
Post-harvest physiology	Dr. E. Dekazos/Dr. A. Williams
Post-harvest engineering	Dr. J. Birth/Dr. S. Nelson, Dr. W. Tyson/Dr. R. Leffler, Dr. R. Horvat/R. Forbus, M.Sc., A. Bennett, M.Sc.

17 Attapulgus

17.1 Attapulgus Extension Research Center elevation : 85 m rainfall : 1283 mm sunshine : 2900 h	Attapulgus, Georgia 31715 Phone: (912) 465-3421 Supt.: Dr. D. Granberry
State institution	
Vegetable culture	Dr. W. McLaurin

18 Blairsville

18.1 Georgia Mountain Station elevation : 642 m rainfall : 1338 mm sunshine : 2700 h	Blairsville, Georgia 30512 Phone: (404) 745-2655 Supt.: J. Dobson, M.Sc.
State institution	
Vegetable culture and extension	P. Colditz, M.Sc.

19 Byron

19.1 Southeastern Tree and Nut Research Laboratory (USDA-SEA-AR) elevation : 163 m rainfall : 1195 mm sunshine : 3000 h	Box 351, Byron, Georgia 31008 Phone: (912) 956-5656
Genetics and breeding of tree fruits	Dr. J.M. Thompson
Stone fruit diseases and nematodes	Dr. E.J. Wehunt/Dr. D.J. Weaver, Dr. J. Wells
Nutrition, physiology and production of fruit trees	Dr. B. Horton/Dr. J. Edwards
Fruit insects	Dr. J. Payne

20 Experiment

20.1 Georgia Experiment Station elevation : 310 m rainfall : 1268 mm sunshine : 2800 h	Experiment, Georgia 30212 Phone: (404) 228-7263 Dir.: Dr. C. Laughlin
Department of Horticulture	Head: Dr. C. Johnson
Vegetable production and breeding	Dr. R. Beverly
Pomology	Dr. J. Daniell
Christmas trees	T.S. Davis, M.Sc.
Small fruit breeding and culture	Dr. R. Lane
Ornamentals	W.L. Corley, M.Sc.
Department of Agricultural Engineering	Head: Dr. B. Verma
Ornamental mechanization	Dr. B. Verma
Department of Food Science	Head: Dr. T. Nakayama

Fruit food products	Dr. L.F. Flora
Maintaining pecan quality	Dr. L.R. Beuchat/Dr. E.K. Heaton
Utilization of fruit and vegetable waste	Dr. N.P. Moon
Increasing shelf life of fruits and vegetables	Dr. T. Nakayama
Department of Plant Pathology	Head: Dr. T. Walker
Ornamental diseases	Dr. T. Walker
Vegetable diseases	Dr. J.W. Demski
Vegetable transplant diseases	Dr. C. Chang

21 Pine Mountain

21.1 Ida Cason Calloway Gardens
elevation : 272 m
rainfall : 1275 mm
sunshine : 2850 h

Pine Mountain, Georgia 31822
Phone: (404) 663-2281
Head: Dr. W. Barrick

Private institution, located between Columbus and Atlanta. Native ornamental trees and shrubs in naturalistic surroundings. Large collection of hollies and azaleas, home-fruit and vegetable garden with many varieties of common and uncommon kinds of tree fruits, small fruits and vegetables. Taxomony and woody ornamentals. Floriculture and woody ornamentals, vegetable-crop varieties.

22 Tifton

22.1 Georgia Costal Plains Experiment Station
elevation : 123 m
rainfall : 1207 mm
sunshine : 2850 h

Tifton, Georgia 31794
Phone: (912) 386-3338
Dir.: Dr. W. McCormick

Department of Horticulture	Head: Dr. M. Austin
Small fruit culture	Dr. M. Austin
Peach culture	Dr. D. Evert
Vegetable breeding and culture	Dr. M. Hall
Pecan nutrition and culture	Dr. R. Worley
Vegetable culture	Dr. S.C. Phatak/Dr. S. Smittle, Dr. D. Batal
Vegetable weed control	Dr. N. Glaze
Vegetable transplants	Dr. C. Jaworksi*
Department of Plant Pathology	Head: Dr. R. Stouffer
Pecan diseases	Dr. P. Bertrand
Vegetable diseases	Dr. D.R. Summer/R. Gitiatis
Extension nut crops	Dr. T. Crocker
Extension fruits and vegetables	I.M. Barber, M.Sc.
Extension pomology	Dr. G. Krewer
Turf diseases	Dr. H.D. Wells
Vegetable and turf nematodes	Dr. A.W. Johnson

FLORIDA

23 Belle Glade

23.1 Everglades Research and Education Center

PO Drawer A, Belle Glade,
Florida 33430
Phone: (305) 996-3062

State institution. Established in 1921, for solution of problems of crop production on the muck soils around Lake Okechobee. Horticultural research began in 1925 and expanded rapidly as discovery of the importance of Cu in plant nutrition on this soil enabled commercial crops to be grown. Vegetables, sugar cane, fiber crops, beef cattle.

Vegetable varieties	Dr. V.L. Guzman
Weed science	Dr. J.A. Dusky
Breeding beans, celery and sweet corn	Dr. E.A. Wolf
Vegetable crops	Dr. R. Subramanya*

24 Bradenton

24.1 Gulf Coast Research and Education Center

5007 60th Street East,
Bradenton, Florida 34203
Phone: (813) 755-1568
Dir.: Dr. E.W. Waters

State institution. Established in 1925 for study of tomato diseases and since 1951 chiefly working on problems of growing vegetables and ornamentals

Production of agronomic crops	Dr. C.G. Chambliss
Systems for vegetable crops	Dr. A.A. Csizinsky
Nutritional problems of vegetables	Dr. C.M. Geraldson
Systems for ornamental crops	Dr. B.K. Harbaugh
Physiological disorders of vegetable and ornamental crops	Dr. T.K. Howe
Nematology	A.J. Overman, M.Sc.
Economist for horticultural crops	Dr. J.W. Prevatt
Entomology for cut flowers and ornamentals	Dr. J.F. Price
Diseases of vegetable crops	Dr. J.P. Jones
Entomologist for vegetable crops	Dr. D.J. Schuster
Water requirements of ornamentals and vegetables	Dr. C.D. Stanley
Breeding of new ornamental crops	Dr. G.J. Wilfret
Physiological diseases of vegetables and ornamentals	Dr. S.S. Woltz
Vegetable breeding	Dr. J.W. Scott
Weed science	Dr. J.P. Gilreath
Extension vegetable specialist	Dr. D.N. Maynard*

25 Fort Lauderdale

25.1 Agricultural Research Center

3205 SW, 70th Avenue,
Fort Lauderdale, Florida 33314
Phone: (305) 475-8990
Dir.: W.B. Ennis, Jr., Ph.D.

Turf and ornamental diseases	R.A. Atilano, Ph.D.
Turfgrass extension	B.J. Augustin, Ph.D.
Turfbreeder	P. Busey, Ph.D.*
Turf and ornamentals	J.A. Reinert, Ph.D.
Tropical ornamentals	T.K. Broschat, Ph.D.
Extension	D.G. Burch, Ph.D.
Palm culture	H.M. Donselman, Ph.D., F.W. Howard, Ph.D/R.E. McCoy, Ph.D. J.H. Tsai, Ph.D.
Aquatic weeds	J.K. Balciunas, Ph.D/T.D. Center, Ph.D. D.L. Sutton, Ph.D./V.V. Vandiver, Ph.D. K.K. Steward, Ph.D.
Environmental quality	G.E. Fitzpatrick, Ph.D.

26 Fort Pierce

26.1 Agricultural Research Center

PO Box 248, Fort Pierce,
Florida 33454
Phone: (305) 461-4371

Soil fertility and water management	Head: Dr. D.V. Calvert
Citrus, rootstocks and diseases	Dr. M. Cohen
Scales and mites, pesticides equipment	Dr. R.C. Bullock
Vegetables: variety trials, tomato breeding, fertilizers, crop spacing, herbicides	
Tomato breeding and diseases	Dr. R.M. Sonoda
Vegetable crops	Dr. P.J. Stoffella*

27 Gainesville

27.1 Institute of Food and Agricultural Sciences	Gainesville, Florida 32611 Phone: (904) 392-2971 Vice.Pres.: Dr. K.R. Tefertiller Dean: Dr. F.A. Wood Ass.Dean: Dr. J.M. Davidson, Dr. V.G. Perry
27.2 Department of Entomology and Nematology	Chairman: Dr. D.L. Shankland
Insects and mites on vegetables	Dr. D.H. Habeck
Turf insects	Dr. S.H. Kerr
Nematodes affecting horticulture crops	Dr. D.T. Kaplan/Dr. G.C. Smart, Dr. A.C. Tarjan
27.3 Department of Food Science and Human Nutrition	Chairman: Dr. J. Busta
Biochemistry and nutrition of foods	Dr. E.M. Ahmed/Dr. L.B. Bailey, Dr. J.S. Dinning/Dr. J.F. Gregory, Dr. R.C. Robbins/Dr. R.B. Shireman, Dr. H.S. Sitren
Technology of foods	Dr. J.P. Adams/Dr. R.P. Bates
Microbiology of foods	Dr. J.A. Koburger/Dr. J.L. Oblinger
Pesticide residues	Dr. H.A. Moye/Dr. N.P. Thompson, Dr. W.B. Wheeler
27.4 Department of Fruit Crops	Act.Chairman: Dr. W.J. Wiltbank*
Incompatibility in tangelo, physiology of citrus and deciduous fruits	Dr. A.H. Krezdorn
Breeding peaches and blueberries	Dr. R.H. Sharpe
Breeding blackberries and plums	Dr. W.B. Sherman
Precooling citrus fruits, tea culture	Dr. J. Soule*
Cultural problems with peaches	Dr. R.H. Biggs
27.5 Department of Ornamental Horticulture	Chairman: Dr. W.J. Carpenter
Nursery culture of woody ornamentals	Dr. J.E. Barrett/Dr. R.J. Black, Dr. B. Dehgan/Dr. H.M. Donselman, Dr. D.L. Ingram/Dr. C.R. Johnson
Foliage plant culture	Dr. D.G. Burch/Dr. R.W. Henley, Dr. D.B. McConnell
Flower culture and nutrition	Dr. J.N Joiner/Dr. T.A. Nell*, Dr. T.J. Sheehan/Dr. B.O. Tjia
Turfgrass culture	Dr. B.J. Augustin/Dr. A.E. Dudeck
27.6 Department of Plant Pathology	Chairman: Dr. C.L. Niblett
Diseases of turfgrasses	Dr. T.E. Freeman
Viruses of ornamentals	Dr. D.E Purcifull/Dr. F.W. Zettler
Diseases of ornamentals	Dr. G.W. Simone
27.7 Department of Vegetable Crops	Chairman: Dr. D.J. Cantliffe*
Vegetable breeding	Dr. M.J. Bassett
Postharvest physiology	Dr. J.K. Brecht
Molecular genetics	Dr. C.D. Chase
Post-harvest physiology/vegetable composition	Dr. D.D. Gull
Vegetable physiology	Dr. C.B. Hall*

Biochemical genetics	Dr. L.C. Hannah
Extension vegetable specialist	Dr. G.J. Hochmuth
Post-harvest physiology	Dr. D.J. Huber
Vegetable physiology	Dr. T.E. Humphreys
Crop production	Dr. S.R. Kostewicz/Dr. M.B. Lazin
Herbicide nutrition	Dr. S.J. Locascio*
Post-harvest vegetable crops specialist	Dr. M. Sherman*
Vegetable crops specialists	Dr. W.M. Stall/J.M. Stephens
Physiological genetics	Dr. C.E. Vallejos
27.8 Department of Soil Science	Chairman: Dr. C.F. Eno
27.9 Department of Agricultural Engineering	Chairman: Dr. G.L. Zachariah
Irrigation of citrus and vegetables	J.M. Myers
Mechanization of tomato harvesting	Dr. L.S. Shaw

28 Hastings

28.1 Agricultural Research Center

PO Box 728, Hastings,
Florida 32045
Phone: (904) 692-1792

State institution, started in 1923 for the study of potato diseases and now dealing with all problems of growing cabbages and potatoes

Soil chemistry: fertilizing	Off.-in-Ch.: Dr. D.R. Hensel
Culture and variety trials - IPM	Dr. J.R. Shumaker
Diseases of cabbage and potato	Dr. D.P. Weingartner
Insects on cabbage and potatoes Drainage and irrigation	Dr. R.B. Workman

29 Homestead

29.1 Tropical Research and Education Center

18905 SW, 280th Street,
Homestead, Florida 33031
Phone: (305) 247-4624

State institution started in 1929 for studies of fruits and vegetables of the humid tropics and subtropics. Emphasis on variety improvement, propagation, crop establishment, pest management, irrigation, fertilization and harvesting.

Fruit fly research; biological control; taxonomy-hemiptera	Dir.: Dr. R.M. Baronowski
Vegetable crops ; cultural systems including crop establishment and mechanical harvesting	Dr. H.H. Bryan*
Vegetable root crops - such as aroids, cassava, sweet potato and potato	Dr. S.K. O'Hair
Fruit crops - germplasm collection and culture of avocado, lime, mango, lychee	Dr. C.W. Campbell*
Fruit crops - physiology of flowering and abscission growth and development	Dr. T.L. Davenport
Fruit crops ; tissue culture, cloning and para-sexual hybridization	Dr. R.E. Litz
Plant pathology - tomato, cucurbit, and aroid breeding	Dr. R.B. Volin
Plant pathology - field control of diseases of fruit and vegetables	Dr. R.T. McMillan, Jr.
Plant pathology - integrated pest management	Dr. K.L. Pohronezny
Entomology - control of insects on vegetable crops	Dr. V.H. Waddill
Soils and climatology - irrigation, fertilization, climatic and environmental effects	Dr. P.G. Orth
Vegetable crops production	Dr. A.A. Duncan*

30 Lake Alfred

30.1 Citrus Research and Education Center	700 Experiment Station Road Lake Alfred, Florida 33850 Phone: (813) 956-1151 Dir.: Dr. W.J. Kender* Ass.Dir.: Dr. W.S. Castle
Fruit harvesting mechanization (DOC)	G.E. Coppock
Fruit harvesting mechanization (USDA)	S.L. Hedden
Harvesting; spraying	Dr. J.D. Whitney
Fruit physiology	Dr. C.R. Barmore
Decay control (DOC)	Dr. G.E. Brown
Fruit physiology ; pollution abatement (DOC)	M.A. Ismail
Fresh fruit handling; coordinator pilot plants	Dr. W.M. Miller
Fungicide residue analysis; post-harvest methodologies (DOC)	Dr. S. Nagy
Post-harvest fruit physiology. Packinghouse extension	Dr. A.C. Purvis
Fruit abscission; fruit color; acidity regulation; cold-hardiness (DOC)	Dr. W.F. Wardowski/Dr. W.C. Wilson
30.2 DOC Scientific Research Department	Dir.: Dr. J.A. Attaway (chemist)
Coord. processing and by-products research; juice definition (DOC)	Dr. S.V. Ting
Product analysis (DOC)	
Specialty products	R.W. Barron
Juice definition; acidity regulation (DOC)	Dr. R.J. Braddock
Processing; energy studies (DOC)	Dr. B.S. Buslig
Processing; pectin	Dr. Chin Shu Chen
New food products and sensory evaluation (DOC)	Dr. P.G. Crandall/Dr. P.J. Fellers
Organic constituents of citrus juices and by-products; analytical methods (DOC)	Dr. J.F. Fisher
Bacteriology; new food products; folic acid studies (DOC)	E.C. Hill
Microbiology of citrus products	
Citrus juice color measurement (DOC)	Dr. T.R. Graumlich
Citrus specialty products	R.L. Huggart
Mico-analysis of trace minerals in juices (DOC)	J.W. Kesterton/Dr. J.A. McHard (in Gainesville)
Product analysis and new food products (DOC)	M.D. Maraulja
New food products	Dr. J.E. Marcy
Enzymes and pectic substances; new food products; citrus nutrition (DOC)	Dr. E.L. Moore
New analytical techniques for juice analysis (DOC)	D.R. Petrus
Vitamin analysis;nutrition;new methods of analysis (DOC)	Dr. R.L. Rouseff
Fresh fruit quality; blight research	Dr. L.G. Albrigo
Pest management	Dr. J.C. Allen
Soils and fertilizer program	Dr. C.A. Anderson
Remote sensing	Dr. C.H. Blazquez
Electron microscopy	Dr. R.H. Brlansky
Insect control and spray equipment	Dr. R.F. Brooks
Rootstocks	Dr. W.S. Castle*
Insect and mite control	Dr. C.C. Childers
Instrumental analysis	G.J. Edwards
Pest management	T.R. Fasulo
Disease physiology	Dr. A.W. Feldman
Irrigation; drainage; rootstocks	Dr. H.W. Ford

Disease physiology	Dr. R.W. Hanks
Pest management, IPM specialist	Dr. J.L. Knapp
Irrigation and fertilizers	Dr. R.C.J. Koo
Diseases of citrus	Dr. R.F. Lee
Insect and mite control	Dr. C.W. McCoy
Extenstion economist	R.P. Muraro
Insect toxicology	Dr. H.N. Nigg
Water resources	Dr. L.R. Parsons
Variety improvement	Dr. A.P. Pieringer
Fruit color; plant growth regulators	Dr. I. Steward
Water relations	Dr. J.P. Syvertsen
Tree decline diseases	Dr. L.W. Timmer
Extension in citrus production	Dr. D.P.H. Tucker
Growth regulators	Dr. T.A. Wheaton*
Fungus diseases	Dr. J.O. Whiteside
Worker exposure to pesticides	Dr. Geraldine Wicker
Soil microbiology	Dr. R.M. Zablotowicz

31 Apopka

31.1 Agricultural Research Center

Rt.3, Box 580, Apopka,
Florida 32703
Phone: (305) 889-4161
Off.-in-Ch.: Dr. C.A. Conover*

State institution. Established in 1968 to provide information on production of foliage plants and cut ferns.

Nutrition - soils	Dr. C.A. Conover*
Physiology - foliage	Dr. R.T. Poole
Plant pathology	Dr. A.R. Chase
Foliage plant breeding	Dr. R.J. Henny
Extension horticulture	Dr. R.W. Henley

32 Leesburg

32.1 Agricultural Research Center

PO Box 388, Leesburg, Florida 32748
Phone: (904) 787-3423
Dir.: Dr. G.W. Elmstrom

State institution for study of water melon and grape diseases

Dr. W.C. Adlerz/Dr. J.M. Crall,
Dr. D.L. Hopkins/Dr. J.A. Mortensen

Plant pathology : watermelon breeding, variety trials of watermelon and cantaloupe, fertilizers and mulches for watermelons. Insecticide evaluation, pollination of watermelons by bees. Breeding and culture of grapes and peaches.

33 Miami

33.1 Fairchild Tropical Garden

10901 Old Cutler Road, Miami,
Florida 33156
Dir.: Dr. J. Popenoe

Private institution. Founded by friends of the late David Fairchild and supported by memberships, admission fees and gifts. 85 acres, 5,000 species of plants.

Woody-plant introduction	Dr. J. Popenoe
Taxonomy	Dr. D.S. Correll
Morphology	Dr. J.B. Fisher
Tissue culture	Dr. K. Norstog

34 Orlando

34.1 U.S. Horticultural Research Laboratory (USDA)	2120 Camden Road, Orlando, Florida 32803 Dir.: Dr. R.H. Young
Horticulture-breeding research unit	Res.Leader: Dr. R.H. Young
Genetics	Dr. H.C. Barrett/Dr. C.J. Hearn, Dr. D.J. Hutchison
Plant pathology	Dr. S.M. Garnsey/Dr. G.R. Grimm, Dr. S. Nemec, Jr./Dr. G. Yelenosky
Horticulture	Dr. K. Wutscher
Physiology research unit	Res.Leader: Dr. G.K. Rasmussen, Dr. M.G. Bausher/P.L. Davis
Entomology-nematology research unit	Res.Leader: A.G. Selhime/B. Beavers, Dr. D.T. Kaplan/Dr. W.J. Schroeder
Market quality and transportation research unit	Res.Leader: Dr. T.T. Hatton*
Agricultural marketing	P.W. Hale/L.A. Risse
Research plant pathologist	Dr. J.J. Smoot

35 Monticello

35.1 Agricultural Research Center	Rt.4, Box 63, Monticello, Florida 32344 Phone: (904) 997-2596
Deciduous fruits	Act.Dir.: Dr. W.J. French
Diseases, fruits	Dr. W.J. French
Woody ornamentals	F.J. Regulski

36 Sanford

36.1 Central Florida Research and Education Center	PO Box 909, Sanford, Florida 32911 Phone: (305) 322-4134 Dir.: Dr. J.F. Darby
Vegetable diseases	Dr. J.F. Darby
Vegetable varieties, fertilizers	Dr. R.B. Forbes
Insects	Dr. A. Ali
Nematodes	Dr. H.L. Rhoades
Herbicides on vegetables	Dr. W.T. Scudder
Vegetable crops, insects	Dr. G.L. Leibee
Environmental program	Dr. K.R. Reddy
Vegetable crops, culture	Dr. J.M White

37 Quincy

37.1 North Florida Research and Education Center	Route 3, Box 638, Quincy, Florida 32351 Phone: (904) 627-9236 Dir.: Dr. D.C. Herzog
Entomology, soybeans	Dr. D.C. Herzog
Small grains breeding	Dr. R.D. Barnett
Field crops	Dr. J.E. Funderburk
Vegetable crops	Dr. S.M. Olson
Soil science and water relations	Dr. F.M. Rhoads
Field crops	Dr. F.M. Shokes
Pest management	Dr. R.K. Sprenkel
Forage crops	Dr. R.L. Stanley, Jr.

Peanuts, tobacco — W.B. Tappen
Extension agronommist — Dr. D.L. Wright

ALABAMA

38 Auburn

38.1 Alabama Agricultural Experiment Station

Auburn, Auburn University,
Alabama 36830
Act.Dir.: Dr. D.H. Teem
Ass.Dir.: C.W. Bruce, M.Sc.
Act.Ass.Dir.: L.A. Smith, B.Sc.

State institution. Established in 1883. In Alabama all research at sub-stations and field stations is under the direction of all project leaders at Auburn.

Department of Horticulture — Head: Dr. D.Y. Perkins
Physiology of fruit and nut crops — Dr. H.A. Amling/Dr. W.A. Dozier
Genetics and breeding of pepper and tomato — Dr. O.L. Chambliss
Physiology and culture of vegetables — J.W. Turner, M.Sc./Dr. J.E. Brown
Physiology and culture of ornamentals — Dr. D.A. Cox/Dr. G.J. Keever, W.C. Martin, B.Sc./K.C. Sanderson, Dr. F.B. Perry/Dr. H.G. Ponder, Dr. C.H. Gilliam
Genetics and breeding of cantaloupe, watermelon, cucumber, southern pea and plums — Dr. J.D. Norton/Dr. O.L. Chambliss
Processing of fruits and vegetable products — Dr. K.S. Rymal/Dr. D.A. Smith

39 Belle Mina

39.1 Tennessee Valley Substation

Belle Mina, Alabama 35615
Supt.: W.B. Webster, M.Sc.
V.H. Calvert II, M.Sc.

State institution. Although most research is on field crops and livestock, experiments on varieties of vegetables for processing have been initiated.

40 Camp Hill

40.1 Piedmont Substation

Camp Hill, Alabama 36850
Supt.: W.A. Griffey, M.Sc.
E. Burgess, B.Sc.

State institution. Experiments on fruit crops especially apples.

41 Clanton

41.1 Chilton Horticulture Substation

Clanton, Alabama 35045
Supt.: J.A. Pitts, M.Sc.
K.C. Short

State institution. Research on fruit and vegetables grown in the area.

42 Crossville

42.1 Sand Mountain Substation

Crossville, Alabama 35962
Supt.: J.T. Eason, M.Sc.
M.E. Ruf, M.Sc.

State institution. Experiments for the benefit of the numerous small farms in the area. Recent research includes nutrition and varietal studies with potatoes which have become major crop.

43 Cullman

43.1 North Alabama Horticulture Substation

Cullman, Alabama 35055
Supt.: M.H. Hollingsworth

State institution. Fruits and vegetables grown in the area.

44 Fairhope

44.1 Gulf Coast Substation

Fairhope, Alabama 36532
Supt.: E.L. Carden, M.Sc.
N.R. McDaniel, M.Sc.

State institution. Pecans, potatoes and other crops grown in the area.

45 Headland

45.1 Wiregrass Substation

Headland, Alabama 36345
Supt.: H.W. Ivey, B.Sc.

State institution. Field crops and livestock. Varietal studies of peaches, plums and vegetables are under way.

L.W. Wells, M.Sc.
L.T. Solomon, B.Sc.

46 Spring Hill

46.1 Ornamental Horticulture Substation

Spring Hill, Alabama 36608
Act.Supt.: J.C. Stephenson, M.Sc.

State institution. Greenhouse and ornamental crops grown in the district.

MISSISSIPPI

All outlying research stations are administered by the Mississippi Agricultural and Forestry Experiment Station (MAFES) of Mississippi State University.

47 Crystal Springs

47.1 Truck Crops Branch Station

Crystal Springs, Mississippi 39059
Supt.: Dr. S.L. Windham

Located in the one time intensive Truck Crops District of Crystal Springs and Hazelhurst

Small farms research - culture of fruits and vegetables including garden demonstration plots and tomato, pepper, cabbage, watermelon and other vegetable variety trials and trials of grapes, peaches, apples and blueberries

Dr. S.L. Windham/Dr. C.P. Hegwood

48 Pontotoc

48.1 Pontotoc Ridge-Flatwood

Pontotoc, Mississippi 38863
Supt.: Dr. H. Palmertree

Culture of peach, apple and plum fruits and sweet potato culture trials. Foundation seed program with sweet potatoes.

49 Poplarville

49.1 South Mississippi Branch — Poplarville, Mississippi 39470
Supt.: Dr. F. Tyner

Culture of fruits, vegetables and ornamentals	Dr. A.J. Laiche, Jr/Dr. B. Graves, M.Sc.

50 Mississippi State

50.1 Mississippi Agricultural and Forestry Experiment Station, Main Station — Mississippi State, Mississippi 39762-5519

Department of Horticulture	Head: Dr. C.C. Singletary
Culture of ornamental crops	L. Estes, M.Sc./Dr. S. Newman
Culture of vegetable crops	Dr. J. Garner/Dr. P. Thompson
Culture of Fruit Crops	
Department of Agricultural Economics	Head: Dr. V. Hurt
Economics and marketing of ornamental crops	Dr. T. Phillips
Economics and marketing fruit and vegetables crops	Dr. L. Bateman
Department of Plant Pathology and Weed Science	Head: Dr. J. McGuire
Diseases of pecans and peaches	Dr. C.H. Graves, Jr.
Diseases of ornamentals	Dr. J.H. Spencer
Weed control in vegetables	Dr. A.W. Cole
Weed control in turf	Dr. G.E. Coats
Agricultural and bioligical engineering	Head: Dr. W.R. Fox
Development of machinery to improve the culture and harvest of fruit and vegetables	Dr. L.H. Chen

51 Lorman

51.1 Alcorn Branch Experiment Station, Alcorn State University — Lorman, Mississippi 39096
Supt.: Dr. J. Collins

Variety and culture studies on peaches, plums, and apples	Dr. O.P. Vadhaw
Variety and nutrition studies on vegetables, grapes and brambles	Dr. S. Tiwari

LOUISIANA

52 Baton Rouge

52.1 Louisiana Agricultural Experiment Station, Main Station — Baton Rouge, Louisiana 70893
Phone: (504) 388-4181
Dir.: Dr. D. Chambers

elevation : 19 m
rainfall : 1416 mm
sunshine : 5.9 h

State institution. Research projects in horticulture began in 1889, particularly with strawberries, potatoes and sweet potatoes. Investigations with these important crops have continued throughout the years and research with other crops has been initiated and vigorously pursued.

Department of Agricultural Economic and Agribusiness	Head: Dr. L.J. Guedry
Marketing of fruits and vegetables	Dr. R.A. Hinson
Department of Agricultural Engineering	Head: Dr. M.E. Wright
Mechanical harvesting of sweet potatoes	
Department of Botany and Plant Pathology	Head: Dr. D.R. Mackenzie

Diseases of vegetables	Dr. L.L. Black
Diseases of sweet potatoes	C.A. Clark, Ph.D.
Diseases of ornamentals	Dr. G.E. Holcomb
Department of Horticulture	Head: A.C. Purvis
Physiology and culture of vegetables	Dr. J.F. Fontenot/Dr. D.H. Picha, Dr. W.M. Randle/Dr. L.C. Standifer, Dr. F.J. Sundstrom/Dr. D.W. Walker, Dr. P.W. Wilson
Breeding of vegetables	Dr. J.F. Fonetenot/Dr. W.M. Randle
Physiology and culture of flowers and ornamental crops	Dr. E.P. Barrios/Dr. W.A. Meadows, Dr. E.N. O'Rourke, Jr/Dr. K.C. Torres, Dr. W.R. Woodson
Physiology and culture of tree and small fruits	Dr. J.E. Boudreaux/Dr. C.A. Lundergan
Food technology	Dr. D.H. Picha/Dr. P.W. Wilson
Department of Entomology	Head: Dr. J.B. Graves
Insects on vegetables	Dr. L.H. Rolston/Dr. R.N. Story
Insects on ornamentals	Dr. A. Oliver

53 Calhoun

53.1 Calhoun Research Station
elevation : 55 m
rainfall : 1279 mm

PO Box 10, Calhoun, Louisiana 71225
Phone: (318) 644-2662
Res.Dir.: Dr. W.A. Young

State institution established in 1885. Emphasis is placed on projects with peaches, watermelons, sweet potatoes and vegetable peas.

Physiology and culture of peach	Dr. W.A. Young
Production and cultural studies on southern peas, sweet cantaloupe and other vegetables	Dr. O.M. Lancaster
Peach and watermelon breeding and culture	Dr. C.E. Johnson

54 Chase

54.1 Sweet Potato Research Station
elevation : 25 m
rainfall : 1324 mm

PO Box 120, Chase, Louisiana 71324
Phone: (318) 435-4584
Res.Dir.: Dr. L.M. Robbins

State institution, established in 1948 primarily for the production of high-quality productive varieties of sweet potatoes. Later the program was enlarged to conduct research on other vegetables.

Breeding and physiology of sweet potatoes	Dr. L.M. Robbins
Foundation seedstock and physiology of sweet potatoes	Dr. W.C. Porter
Vegetable production and outfield studies	Dr. W. Mulkey

55 Clinton

55.1 Idlewild Research Station
elevation : 61 m
rainfall : 1514 mm

Drawer 985, Clinton,
Louisiana 70722
Phone: (504) 683-5848
Res.Dir: Dr. F.J. Peterson

State institution, located 35 miles north of Baton Rouge. Estalished in 1957, 200 acres for research on fruits and vegetables.

Physiology and culture of fruits and vegetables Breeding	Dr. J.E. Boudreaux

56 Port Sulphur

56.1 Citrus Research Station
elevation : 1 m
rainfall : 1690 mm

Rt. 1, Box 628, Port Sulphur,
Louisiana 71301
Phone: (504) 564-2467
Res.Dir.: A.J. Adams, M.Sc.

State institution. Located within the citrus-industry region of Louisiana. Established in 1949. Research on citrus, fruit, potato, cabbage, tomato, cantaloupe and ornamentals.

Breeding and production of citrus and peaches	A.J. Adams, M.Sc.
Production of vegetables	Dr. H.V. Hanna
Breeding and production of citrus	Dr. W.J. Bourgeois

57 Hammond

57.1 Hammond Research Station
elevation : 14 m
rainfall : 1606 mm

5925 Old Corington Highway
Hammond, Louisiana 70401
Phone: (504) 345-4110
Res.Dir.: Dr. R.J. Constantin

State institution. Established in 1921. Since the station is located within the world-famous Hammond strawberry district, most of the research has been centered on this important crop. However other crops grown in the area are not neglected.

Breeding and production of strawberries, culture of ornamentals	Dr. R.J. Constantin
Culture of ornamentals	Dr. W.L. Brown/B.R. Williams, M.Sc.
Genetics and breeding of muscadine-bunch grapes	R.P. Bracy, M.Sc.
Production problems of strawberries, vegetables	D.W. Wells, M.Sc.

58 New Orleans

58.1 United States Southern Regional Research Laboratory
elevation : 3 m
rainfall : 1517 mm

New Orleans, Louisiana 70150

Federal institution. Established in 1936 to develop new products from crops grown in southern U.S. A recent development is sweet potato flakes, a dehydrated product of the fleshy roots. Consumer-acceptance tests have been encouraging. Engineering phases of new products.

59 Shreveport

59.1 Pecan Research-Extension Station
elevation : 77 m
rainfall : 1114 mm
sunshine : 6.3 h

PO Box 5519, Shreveport,
Louisiana 71135
Phone: (318) 797-8034
Res.Dir.: Dr. R. O'Barr

Federal institution. Problems in pecan industry in Louisiana and adjacent states.

Diseases of pecan	Dr. R.S. Sanderlin
Pecan insects	Dr. M.T. Hall
Physiology of pecan	Dr. L.J. Grauke

Pepared by A.A. De Hertogh, Dept. of Horticultural Science, North Carolina State Univ., Raleigh, North Carolina
Clyde C. Singletary. Dept. of Horticulture, Mississippi, Mississippi State
C.H Hendershott, Division of Horticulture, College Station, Athens, Georgia
Daniel J. Cantliffe, Vegetable Crops Department, Inst. of Food & Agric. Sciences, Gainesville, Florida
J.F. Fontenot, Dept. of Horticulture, Louisiana Agric. Exp. Station, Baton Rouge, Louisiana
R.D. Rouse, Agricultural Experiment Station, Auburn Univ., Auburn, Alabama

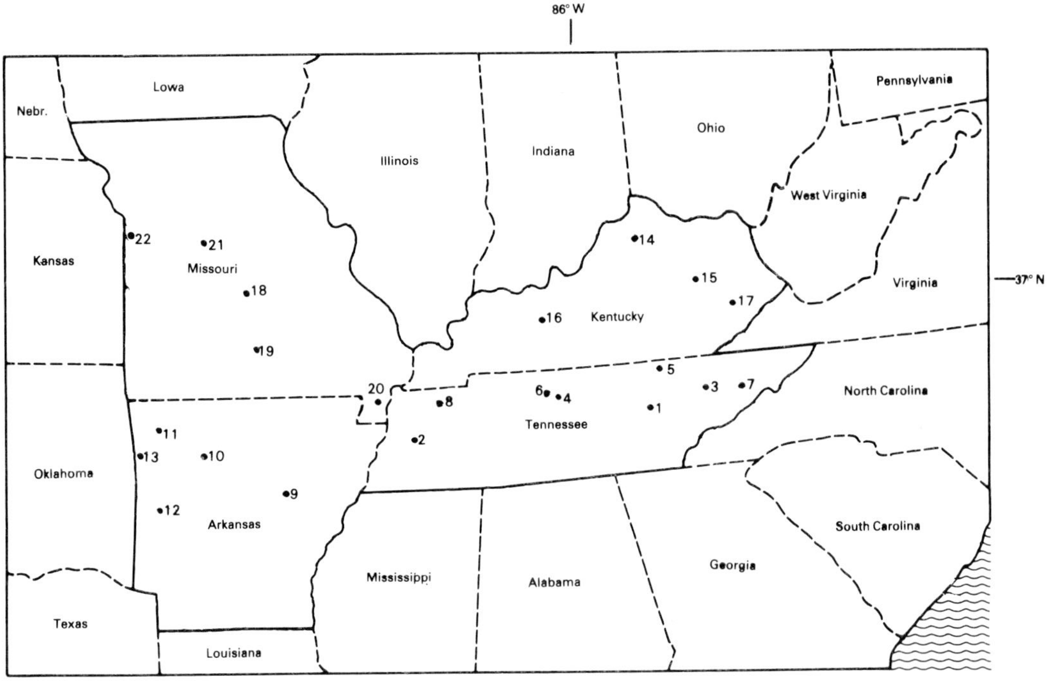

86° W
37° N
Lowa
Nebr.
Illinois
Indiana
Ohio
Pennsylvania
West Virginia
Kansas
Missouri
Kentucky
Virginia
North Carolina
Tennessee
Oklahoma
Arkansas
South Carolina
Georgia
Mississippi
Alabama
Texas
Louisiana

Tennessee: 1 Crossville, 2 Jackson, 3 Knoxville, 4 Nashville, 5 Springfield, 6 Spring Hill, 7 Greeneville, 8 Martin, **Arkansas**: 9 Baldknob, 10 Clarksville, 11 Fayetteville, 12 Nashville, 13 Alma, **Kentucky**: 14 Eden Shale, 15 Lexington, 16 Princeton, 17 Quicksand, **Missouri**: 18 Columbia, 19 Mountain Grove, 20 Portageville, 21 New Franklin, 22 Kingsville

Survey

The four-state area, including Arkansas, Kentucky, Missouri and Tennessee, is divided by the Mississippi River with Kentucky and Tennessee east of the river and south of the Ohio River, Missouri and Arkansas west of the river, and for the most part, south of the Missouri River. The Appalachian mountains form the eastern boundary of Tennessee. The Allegheny Plateau extends across eastern Kentucky through Tennessee west of the southwest flowing Tennessee River.

The Bluegrass area of Kentucky lies at the western edge of the Allegheny Plateau. With various interruptions, this area extends into the Central Basin of Tennessee which is drained by the Cumberland River into the Ohio. These areas lie at elevation of 153 to 457 m above sea level. From the area lying west of the north- flowing Tennessee River, the slope of the land is toward to Mississippi descending from about 244 m elevation to about 107 m at Memphis. The western-most part of this area and the eastern- most part of the states of Arkansas and Missouri as far north as St. Louis, are part of the Mississippi delta, low lying plains, often flooded, offering much for cotton, rice, corn and vegetable production.

Westward, south of the Missouri, the land rises to a height of more than 610 m forming the Ozark and Ouachita Mountains. The northern portion of Missouri, lying north of the Missouri River, is an extension of the dissected plain making up Iowa and southern Minnesota.

Considerable areas of wind-blown soils occur along the Missouri River. In western Kentucky and Tennessee, areas of loess occur on the Mississippi-river highlands.

The climate of these four states is essentially continental except for areas modified by mountain terrain in the eastern Tennessee. Arboral climate is found in the high elevations of the Appalachians. The dominate weather structures influencing temperature, rainfall, and storm development are the artic highs and lows that sweep southeast, east of the Rocky Mountain cordillera, moving toward the Gulf of Mexico, and the maritime highs and lows that move north ward from the Gulf of Mexico. Rainfall generally increases from northwest to southwest and varies from less than 889 mm per year in north-western

Missouri to more than 1270 mm in the Ouachita mountains in southwestern Arkansas and 2032 mm in parts of the Appalachians of East Tennessee.
The generally mild climate of the area is occasionally interupted with short periods of hot or cold weather that provide pleasant variety. The early spring, coupled with long nights, results in some uncertainty about frost during the blooming season for most fruits. However, frost damage is usually slight. Spring is earliest along the Mississippi River and lays somewhat along the eastern and western boudaries of the area. In each of the four states, horticultural departments were established in the Land-Grant Colleges shortly after the Morrill Act of 1862 was passed.

Four horticultural substations were established in Arkansas to handle specific problems. These are located at Nashville, Alma, Clarksville and Baldknob. In Missouri, the State Department of Agriculture established an experiment station at Mountain Grove and the Missouri Agricultural Experiment Station has a field station at Portageville dealing with vegetables. Tennessee has no seperate horticultural substations, but carries on work at Jackson, Springfield, Crossville and Knoxville. Kentucky has field work at Princeton, Quicksand and Eden Shale Farm in Owen County. The State Departments of Agriculture carry out the regulatory work with regard to nursery inspection, fruit and vegetable inspection, the collection of statistical information and control of pesticide use.

TENNESSEE

1 Crossville

1.1 Plateau Experiment Station, Branch Station of Tennessee Agricultural Experiment Station	. Crossville, Tennessee 38555 Phone: (615) 484-0034
General horticulture	Supt.:Dr. R.D. Freeland Dr. C.A. Mullins

2 Jackson

2.1 West Tennessee Agricultural Experiment Station, Branch Station of Tennesse Agricultural Experiment Station	. Jackson, Tennessee 38301 Phone: (901) 424-1643 Supt.: Dr. J.F. Brown
General horticulture and extension	Dr. J. Wyatt/Dr. W.D. Sams

3 Knoxville

3.1 Tennessee Agricultural Experiment Station Main Station, state institution elevation : 888 m rainfall : 1173 mm sunshine : 56 %	. Knoxville, Tennessee 37901 Phone: (615) 974-7201 Supt.: J. Hodges
Department of Ornamental Horticulture	Head: Dr. G.D. Crater
Vegetable production	Dr. D.L. Coffey/Dr. C. Sams
Fruit production	Dr. D.E. Deyton*
Fruit extension	Dr. D.W. Lockwood
Vegetable and small fruits extension	Dr. A.D. Rutledge
Turf	Dr. L.M. Callahan
Ornamental morphology	Dr. E.T. Graham
Floriculture	D.G.L. McDaniels
Ornamentals - nursery	Dr. H. van de Werken/Dr. J.W. Day/ Dr. D.B. Williams/Dr. W.T. Witte
Landscape design	S. Rogers
Ornamental extension	Dr. J.L. Pointer/Dr. K. Tilt

4 Nashville

4.1 Tennessee State University . Nashville, Tennessee 37203
Phone: (615) 320-3650

Vegetable production S. Hayslett/Dr. H.W. Carter

4.2 University of Tennessee Extension Nashville, Tennessee 37222
Phone: (615) 832-6550

Vegetable extension specialist R. Winston/Dr. K. Johnson

5 Springfield

5.1 Highland Rim Experiment Station, Branch Station of Tennessee Agricultural Experiment Station . Springfield, Tennessee 37172
Phone: (615) 384-5292

General horticulture Supt.: L. Safley

6 Spring Hill

6.1 Middle Tennessee Experiment Station, Branch Station of Tennessee Agricultural Experiment Station . Spring Hill, Tennessee 37174
Phone: (615) 486-2129
Supt.: J. High

7 Greeneville

7.1 Tobacco Experiment Station . Greeneville, Tennessee 37743
Phone: (615) 638-6532
Supt.: P. Hunter

8 Martin

8.1 University of Tennessee . Martin, Tennessee 38237
Phone: (901) 587-7263

General horticulture A. Smith

ARKANSAS

9 Baldknob

9.1 Strawberry Substation, Substation of Arkansas Agricultural Experiment Station . Baldknob, Arkansas 72010
Phone: (501) 724-3368
Off.in Ch.: E.C. Baker
elevation : 185 m
rainfall : 1290 mm

10 Clarksville

10.1 Fruit Substation, Substation of Arkansas Agricultural Experiment Station . Clarksville, Arkansas 72830
Phone: (501) 754-2406
Off.in Ch.: Dr. J.R. Clark
elevation : 245 m
sunshine : 2775 h

11 Fayetteville

11.1 Arkansas Agricultural Experiment Station, . Fayetteville, Arkansas 72701

Main Station
elevation : 450 m
rainfall : 1115 mm
sunshine : 2825 h

Phone: (501) 575-4446
Dir.: Dr. P.E. La Ferney

State institution
Horticulture and Forestry

Culture and physiology of vegetables	Head: Dr. G.A. Bradley*/ Dr. C.R.Anderson
Breeding of Southern peas, spinach, beans	Dr. T.E. Morelock
Breeding of tomatoes, okra	Dr. S.J. Scott
Breeding of brambles, grapes, strawberries, apples, peaches	Dr. J.N. Moore*
Tree-fruit physiology, breeding, culture	Dr. R.C. Rom
Ornamental horticulture	Dr. A.E. Einert/Dr. G.L. Klingaman
Food Science	Head: Dr. A.A. Kattan/Dr. D.R. Davis
Post-harvest physiology	Dr. R.W. Buescher/Dr. A.R. Gonzalez
Grapes, mechanization studies	Dr. J.R. Morris
Plant pathology	
Fruit pathology	Head: Dr. D.A. Slack
Vegetable pathology	Dr. M.J. Goode
Viruses	Dr. Rose Gergerich
Entomology	Dr. T.M. Johnson/Dr. P. McLoud
Herbicides	Dr. R.E. Talbert
Marketing horticultural products	Dr. C. Price

12 Nashville

12.1 Peach Substation, Substation of Arkansas Agricultural Experiment Station
elevation : 168 m
rainfall : 1330 mm

. Nashville, Arkansas 71852
Phone: (501) 845-1764
Off.in Ch.: E.J. Arrington

13 Alma

13.1 Vegetable Substation, Substation of Arkansas Agricultural Experiment Station
elevation : 123 m
rainfall : 1080 m

. Alma, Arkansas 72921
Phone: (501) 474-0475
Off.in Ch.: D.R. Motes

KENTUCKY

14 Eden Shale

14.1 Eden shale Experiment Farm Substation, Kentucky Agricultural Experiment Station
elevation : 280 m
rainfall : 993.75 mm
sunshine : 2630 h

. Owenton, Kentucky 40359
Phone: (502) 484-5531
Manager: J.W. Wyles

Production of tree fruits, greenhouse crops. No formal horticultural programs at Eden Shale.

15 Lexington

15.1 Kentucky Agricultural Experiment Station
Main Station, State Institution

. Lexington, Kentucky 40546
Phone: (606) 257-4772

elevation : 270 m
rainfall : 1112.25 mm
sunshine : 2550 h

Dir.: Dr. C.E. Barnhart, Dean

This station is the main center for agricultural research. Horticulture research consists of approximately 40 acres of vegetable crops, 20 acres of fruit crops and 17 acres for ornamentals. Greenhouse space is also available for the below mentioned sections.

Horticulture Section	Head: Dr. A.S. Williams
Vegetable breeding	Dr. H.C. Mohr/Dr. D. Knavel
Vegetables, physiological genetics	Dr. J. Snyder
Vegetables, physiology	Dr. R. Houtz
Vegetables, commercial, culture	Dr. C.R. Roberts
Vegetables, fruit, fresh market	Dr. J. Strang
Vegetables, fruit, ornamentals, biochemistry	Dr. T. Kemp
Fruit, physiology	Dr. D. Archbold
Floriculture	Dr. J. Buxton/Dr. R. Anderson, S. Bale
Home horticulture	Dr. M. Witt
Ornamental horticulture	Dr. R. McNiel/Dr. W. Fountain
Plant propagation	Dr. L. Stoltz
4-H horticulture	Dr. W. Fountain
Landscape Architecture Section	Head: Dr. A.S. Williams
Landscape architecture	J. Cervelli/H. Schach/R. Southerland
Landscape architecture - computer land use planning	Dr. T. Nieman

16 Princeton

16.1 Kentucky Research and Education Center
Substation of Kentucky Agriculture Experiment Station
elevation : 135 m
rainfall : 1135 mm
sunshine : 2800 h

. PO Box 469, Princeton,
Kentucky, 42445
Phone: (502) 365-7541
Manager: D. Davis

This is a major substation for research and education applicable to Western Kentuckly. Horticultural research is aimed primarily at fruit (10 acre orchard), specialty vegetables and ornamentals.

Horticulture Section	Head: Dr. A.S. Williams
Fruit	Dr. G.R. Brown
Ornamentals, vegetables	Dr. W. Dunwell
Culture, horticulture crops	M. Hurley

17 Quicksand

17.1 Robinson Substation, Kentucky Agricultural Experiment Station
elevation : 275 m
rainfall : 1132 mm
sunshine : 2420 h

. Quicksand, Kentucky 41363
Phone: (606) 666-2438
Manager: G. Armstrong

This station is responsible for research and education programs for Eastern Kentucky. Horticultural programs are available in general horticulture and demonstrates applicable crops for the area. The overall station also includes 15,000 acres of forest research land.

Horticulture Section	Head: Dr. A.S. Williams
General horticulture	Dr. T. Jones

MISSOURI

18 Columbia

18.1 Missouri Agricultural Experiment Station . Columbia, Missouri 65211
Phone: (314) 882-7511

Horticulture	Head: Dr. R.R. Rothenberger
Pomology	Dr. M.R. Warmund/Dr. A.E. Gaus
Turfgrass	Dr. J.H. Dunn/Dr. D.D. Minner
Nutrition, growth regulators	Dr. D.D. Hemphill
Floriculture	Dr. N.R. Natarella/Dr. D.N. Trinklein
	Dr. M.N. Rogers
Ornamental	Dr. C.S. Starbuck
Vegetable breeding	Dr. R.A. Schroeder*/Dr. V.N. Lambeth*
Landscape design	Dr. L.C. Snyder, Jr./Dr. R.E. Taven

19 Mountain Grove

19.1 Mountain Grove State, Fruit Experiment Station of the State Department of Agriculture . Mountain Grove, Missouri 65711
Phone: (417) 926-4105
Dir.: Dr. J. Moore

Fruit culture	P. Andersen/H. Townsend/M. Haag

20 Portageville

20.1 Missouri Agricultural Experiment Station, Field Station . Portageville, Missouri 63873
Phone: (314) 379-5431
Supt.: J. Scott

Vegetables	H.F. DiCarlo

21 New Franklin

21.1 New Franklin Horticultural Research Station . New Franklin, MO
Phone: (816) 848-2268
Supt.: J. Shopland

22 Kingsville

22.1 Powell Horticultural and Natural Resources Center . Route no. 1, Box 90, Kingsville, MO 64061
Phone: (816) 566-2600
Supt.: D. Vismara

Prepared by George Bradley, Dept. of Horticulture and Forestry, Univ. of Arkansas, Fayetteville, Arkansas
A.S. Williams, Dept. of Horticulture and Landscape Architecture, Univ. of Kentucky, Lexington, Kentucky
G. Douglas Crater, Dept. of Ornamental Horticulture and Landscape Design, Univ. of Tennessee, Knoxville, Tennessee
R.R. Rothenberger, Dept. of Horticulture, Univ. of Missouri, Columbia, Missouri

85° W
40° N
Lake Superior
Canada
Minnesota
Wisconsin
Lake Michigan
Lake Huron
Michigan
Lake Ontario
New York
Lake Erie
Iowa
Pennsylvania
Illinois
Ohio
Indiana
ML
West Virginia
Missouri
Kentucky
Virginia

Ohio: 1 Canfield, 2 Celeryville, 3 Columbus, 4 Fremont, 5 Jackson, 6 Ripley, 7 Wooster, **Indiana**: 8 Vincennes, 9 Lafayette, **Illinois**: 10 Simpson, 11 St. Charles, 12 Urbana, 13 Wichert, **Michigan**: 14 East Lansing, 15 Clarksville, 16 Fennvile, 17 Grand Rapids, 18 Sodus, 19 Traverse City, 20 Tipton, **Wisconsin**: 21 Ashland, 22 Hancock, 23 Madison (Arlington), 24 Spooner, 25 Sturgeon Bay, 26 Lancaster, 27 Rhinelander

Survey

This region, which can be referred to as the North-Central states, includes Illinois, Indiana, Michigan, Ohio and Wisconsin, with an area of 248,283 square miles. It has a north latitude range from 37°0' at Cairo, Illinois, to 47°30' at the northern tip of Michigan's Upper Peninsula; and a west longitude range from 80°42' at the Ohio-Pennsylvania border to 92°54' at the most western part of Wisconsin. As can be seen on the map a part of the boundary line of each of the 5 states is contiguous to one or more of the Great Lakes, all 5 of which comprise that most extensive body of fresh water in the world; Lake Ontario lies to the east of this region. The Ohio River forms all of the southern boundary of the Region and the Mississippi River most of the western boundary. The elevation above sea level varies from 98 m at

Cairo, Illinois, to 592 m at the top of Rib Mountain in Wisconsin. However, much of the cropland of the region lies between 152 and 304 m, except in the river bottom of the Ohio, Mississippi and other rivers where it is frequently lower. In the vicinity of the four contiguous Great Lakes the elevation is usually approximately 183 m, since the elevation of Lake Superior is 200 m and that of Lake Michigan, Huron and Erie 6 to 9 m lower. The elevation increases with distance back from the shore line.

The wide range in latitude provides a suitable climate within the region for the commerical production of each of the important deciduous fruits, more than 30 kinds of vegetables, and many species of ornamentals. A major factor in this crop adaptability is the broad spectrum of soil types, ranging from muck, which generally occurs near the Great Lakes, through sandy loam, loam and clay loam, to the loess deposits of the hills along the Mississippi, Illinois and Ohio rivers.

The annual precipitation throughout the region is adequate for the crops grown if distrubuted relatively uniformly during the year, particularly during the growing season. However, yield-reducing droughts may occur during any part of the growing season in each of the 5 states, but are most frequent during July and August. Irrigation is being used increasingly and will probably come to be considered a standard production practice for some crops, especially on the less water-retentive soils and in the most drought-prone areas.

The effect of large bodies of water on climate relative to fruit adaptation is illustrated by the extensive fruit area along the eastern shore of Lake Michigan. For example, peaches are grown extensively in Michigan, whereas they cannot be grown along the western shore of the lake in Illinois and Wisconsin; the temperature-stabilizing effect of the prevailing southwest, west and northwest winds blowing across the lake makes this possible.

Ohio, one of the most important horticultural states, is characterized by a particularly wide diversification of agricultural and horticultural crops and likewise is an outstanding leader industrially. Horticultural produce represents from 10 to 15 percent of the farmers gross income, in terms of processed products, the value involves a much higher percentage of total agri cultural income. The most important environmental factors responsible for this pronounced horticultural diversification are: 1. wide range of, not extreme, temperatures from south to north, 2. considerable diversity of soil, and 3. generally favourable moisture (rainfall) conditions. From an economic viewpoint, the strategic geographical location and large population are very important. Ohio has a very large number of cities with a population over 25,000 well distributed over the whole state except the southeast. Over 30 vegetable crops are grown commercially of which tomatoes, potatoes, sweet corn, cabbage, celery, onions, peppers, cucumbers, and radish are of greatest importance. Tomatoes grown for processing rank second in importance in the USA while the state ranks first in greenhouse vegetable production.

Ohio's annual apple crop is in excess of 3 million bushels. It also produces significant crops of strawberries and grapes and is an important wine producing state.

Florist crops such as roses and chrysanthemum are widely grown in greenhouses throughout the state. Potplants, likewise, comprise a significant part of the florist industry. The largest nursery area in the USA is centered in northeastern Ohio. A wide range of herbaceous and woody ornamentals form the basis for this area of horticulture. Furthermore, nurseries producing specialized crops, such as rhododendrons and yews comprise an outstanding segment of the industry. The Ohio Agricultural Experiment Station was established at Columbus in 1882 as an institute separate financially from the Ohio State University and has continued as such since that time. Upon removal to Wooster, in 1893, the great proportion of research was conducted at Wooster but since 1929 a significant amount has been carried on by staff members located at Columbus. Since 1929 the Department of Horticulture and Forestry at the Ohio State University and the Department of Horticulture at the Experiment Station at Wooster have been administered as a single unit. Members of the Department of Horticulture and Forestry have been part time Center personnel since 1929 and since 1964 Research Center personnel have also held University appointments as well. This arrangement has resulted in research in all horticultural areas being conducted at six Center Branches under the direction of Center faculty. The name of the Ohio Agricultural Experiment Station was changed to Ohio Agricultural Research and Development Center in 1865. The name of the department at the Ohio State University was changed in 1970 to Horticulture Department when a new Forestry Department was established.

Indiana. The topography of Indiana is hilly in the southern one-third back from the Ohio River bottoms, and gently rolling over most of the remainder of the state, except the north-west corner adjacent to Lake Michigan, where it is relatively flat. Soils from a loess type in the southern hills through sandy loam, loam, clay loam to muck in the north adjacent to Lake Michigan.

Apples are grown in all parts of the state. Most of the young trees are dwarfs or spur types. Peaches are grown in the northwest corner near the south end of Lake Michigan, and in the southern one-third of the state, where it is warmest. Leading fresh market vegetables are cabbage, cantaloupes, cucumbers, onions, potatoes, tomatoes and watermelons. The horticultural crops include an extensive greenhouse crop and ornamental industry. A unique crop in Indiana is mint, grown extensively in the muck area of the northern part of the state.

The Horticultural Department was established at Purdue University in 1884 with the appointment of James Troop as Professor of Horticulture and Entomology. In 1912, Professor Troop was made head of the newly-created Department of Entomology and C.G. Woodbury was made head of the Department of Horticulture thus establishing the basis for the present department.

Early investigations centered on the apple with variety adaptation receiving major attention. The establishement of the Agricultural Experiment Station following the passage of the Federal Hatch Act in 1887, and subsequent Federal and State acts provided the fund and stimulus for the beginning of research in all phases of horticulture. From this basis the present extensive research program has developed.

Illinois is called the 'Prairie State' because a large percentage of its topography is relatively flat to gently rolling. Paralleling the Mississippi River, which forms the western boundary of the state, and the Illinois River, and between them, are high hills of loess soil. The southern part of the state is covered by this same general type of hills. The northeast corner, extending back from Lake Michigan has areas of peaty or muck soil. The prairie soils are generally suitable for fruit growing, but since they are well suited to grain crops, corn, soybeans, oats and wheat are the major crops. An exception is the southcentral, relatively-flat, slowly-drained area, where apples, peaches and strawberries are grown but not as extensively as in former years.

The commercially-grown fruits are apples, peaches, grapes and strawberries. Apples are grown in the south-central area, in the hilly areas of the state. Peaches are grown in the south of the state with a few small plantings along the Mississippi River as far north as Moline. Minimum winter temperature prevents commercial production in most of the northern two-thirds of the state. Strawberries are grown for pick-your-own markets throughout the state, and blueberries are grown mostly in a trial way for the same markets.

As to vegetables, leading fresh market crops are asparagus, snap beans, cabbage, carrots, cantaloupes, sweet corn, cucumbers, horseradish, onions, tomatoes, and watermelons.

Illinois produces about 80 percent of the pumpkins processed in the United States. The major acreage of vegetables grown for processing is in northern Illinois. Illinois is a leading state in the production of greenhouse crops and woody ornamental plants. This industry is located primarily in the counties near Chicago, but also to a lesser extent in other areas of the state. The present University of Illinois was established as the Illinois Industrial University in 1868. The importance of horticulture was recognized by the appointment of T.J. Burrill as Professor of Botany and Horticulture in 1869; he retained this appointment until his retirement in 1912. Technically, research in horticulture began with his appointment to the staff of the newly-created Illinois Agricultural Station in 1887. However, he conducted experiments in horticulture from his first appointment and his world-famous report that fireblight was caused by bacteria was published in 1880 and he described and named the causal organism in 1882.

Horticulture was organized as a separate department in 1899, with J.C. Blair as Head, a position in which he served until his retirement in 1939. At present the department includes pomology, vegetable crops and ornamentals (floriculture, turf, and nursery crops).

Michigan fruit production is concentrated along the western side of the lower peninsula because of the tempering effect of Lake Michigan on the climate. Among tree fruits, apples lead both in terms of acreage and farm value, followed by tart cherries, sweet cherries, peaches, and plums. Michigan leads all states in production of tart cherries; only Washington and New York produce more apples. Grapes and blueberries are also important crops, the farm value of blueberries surpassing that of all other fruits except apples in some years. Large portions of these crops are processed.

Vegetable crops, in order of acreage planted, are cucumbers, asparagus, green beans, and tomatoes; however, farm values of onions, celery, and carrots all exceed that of green beans. Nurseries produce approximately $ 50 million worth of herbaceous and woody perennials, and bedding plants are also an important commodity. Only California produces more. Michigan soils vary from sandy and gravelly loams (eastern Michigan) to mucks (south-central), the latter being well suited for crops such as celery and onions. Sandy soils predominate in the northern portion of the lower peninsula, an area once covered by forests of white pine.

Horticulture has been taught at Michigan State University from its founding as Michigan Agricultural College in 1857, but was not established as a separate department until 1883. L.H. Bailey, a native of Michigan and a graduate of M.A.C., was Department Chairman from 1885 to 1888, and drafted plans for the first building (Eustace Hall) specifically designed for the teaching of horticulture in the United States. He was succeeded by L.R. Taft, noted for his pioneering work in the use of chemical fungicides. The staff and facilities continued to develop under later Department chairmen, including V.R. Gardner, H.B. Tukey, and H.J. Carew. Some notable contributions of staff members include the introduction of the "Haven" peaches and numerous vegetable cultivars, as well as pioneering experiments with apple rootstocks, radioisotopes, plant growth regulators, fruit storage, and forcing of bulbs.

Wisconsin. The production of horticultural crops for sale and many enterprises closely related to these activities constitute important aspects of Wisconsin's agriculture. Fruits, flowers, vegetables and woody ornamentals are grown extensively in the state. Vegetable crops for canning are the largest segment of the horticultural industry.

Wisconsin's climate is a typical north temperate continental climate exhibiting wide extremes of summer and winter temperatures. Lakes Superior and Michigan and the Mississippi River modify conditions in areas adjacent to them and provide restricted regions for specialized horticultural development, particulary tree fruits.

Much of Wisconsin's topography is gently rolling. The soils range from extensive muck and peat areas in the central and southeastern parts of the state to sand areas in the central, northeast and northwest areas. Along the shores of Lake Superior and Michigan are large areas of red clay that are generally unsuited for horticultural crops. The Door Peninsula in northeast Wisconsin is limestone derived.

Horticultural production is centered in areas uniquely favored by soil, climate or markets. Cranberries are located in the marsh areas of central and northern Wisconsin, which is the only state in this group that produces them commercially. Tree fruits are concentrated in the Door Peninsula, southeastern counties near Lake Michigan, Mississippi River Valley and the Bayfield Peninsula on Lake Superior. Canning crops are extensively grown in the southern and eastern section of the state with extensive acreage of potatoes, green beans and cucumbers in the central sand plain. Ornamental crops are produced near large urban centers with a few scattered nurseries in other areas. Temperate-zone fruit production began in the middle of the 19th century as farm orchards. Large commerical plantings occurred early in the 20th century. The first cranberry bogs were developed in the 1870's and new ones are still being developed. Canning crops production started after the turn of the century and has continued to flourish to the present. Organized horticultural research began as early as 1868 through the cooperation of the then struggling University of Wisconsin and the Wisconsin Horticultural Society. This initial effort was chiefly variety testing and was destined to be short lived. The first true research program in horticultural began in 1889 with the appointment of Emmett S. Goff as Professor of Horticulture. The research program has continued to grow at Madison since that time. Branch stations for horticultural, as well as other agricultural research were established in several areas of the state in the period after 1910.

In 1959 the main research plantings of the Department were moved from Madison to a new farm 25 miles north of Madison (recently named Arlington Horticulture Research Farm).

OHIO

1 Canfield

1.1 Mahoning County Branch - OARDC . Canfield, Ohio
Dir.: C.A. Morrison

2 Celeryville

2.1 Muck Crops Branch - OARDC . Celeryville, Ohio
R. Nassell

3 Columbus

3.1 The Ohio State University, College of Agriculture, Home Economics and Natural Sources . 2001 Fyffe Court
Columbus, Ohio
43210-1096

Department of Plant Pathology
Floral crop diseases

Chairman: Dr. C. Curtis
Assoc. Chairman: Dr. L.E. Williams
Dr. C.C. Powell
Phone: 614-422-1800

Department of Horticulture

Floriculture:	
Growing media, nutrition	Dr. J.C. Peterson
Economics of production and marketing	Dr. W.T. Rhodus*
Post harvest physiology, hypo-storage	Dr. T.A. Prince*
Cultural practices	Dr. H.K. Tayama
Ornamental Horticulture:	
Propagation, tissue culture	Dr. R.D. Lineberger
Nutrition and fertilizer practices	Dr. E.M. Smith
Propagation, plant materials	Dr. S.M. Still
Shade tree evaluation	Dr. T.D. Sydnor
Pomology:	
Cultivar evaluations, cultural practices	Dr. R.C. Funt*
Stone fruit management and rootstocks	Dr. Diane Miller
Orgard management	Dr. H.A. Rollins, Jr.
Vegetable Crops:	
Culture and weed control	Dr. S.F. Gorske
Food Processing and Technology:	
Food chemistry, flavor	Dr. A. Teng
High protein foods, lipids	Dr. A.C. Peng
Department of Entonology	Act. Chairman: Dr. D. Horn Assoc. Chairman: Dr. R.E. Treece
Integrated control	Dr. H. Wilson

4 Fremont

4.1 Vegetable Crops Branch - OARDC .	Fremont, Ohio C.C. Willer

5 Jackson

5.1 Jackson Branch - OARDC .	Jackson, Ohio R.M. McConnell

6 Ripley

6.1 Southern Branch - OARDC .	Ripley, Ohio J.D. Wells

7 Wooster

7.1 Ohio Agricultural Research and Development Center .	Wooster, Ohio 44691 Dir.: F. Hutchinson
7.1.1 Department of Plant Pathology	Chairman: Dr. I.W. Deep Assoc. Chairman: Dr. C. Curtis
Fruit crops diseases	Dr. M.A. Ellis
Diseases of woody ornamentals	Dr. H.A.J. Hoitink
Vegetable diseases	Dr. R.C. Rowe
7.1.2 Department of Horticulture	Phone: 216-263-3818 Chairman: Dr. R.A. Kennedy Assoc. Chairman: Dr. G. Cahoon
Ornamental Horticulture:	
Plant physiology	Dr. A. Miller
Food Processing and Technology:	
Enology fruit processing	Dr. J.F. Gallander
Vegetable Crops:	
Greenhouse vegetable culture and energy conservation	Dr. W.L. Bauerle, Jr.*

Breeding processing tomatoes	Dr. S.Z. Berry
Cultural practices processing vegetables	Dr. D.W. Kretchman*
Potatoes and fresh market vegetables	Dr. R. Precheur
Pomology:	
Pear Breeding	Dr. C. Chandler
Viticulture and nutrition	Dr. G.A. Cahoon
Rootstocks, high-density orchards, photosynthesis	Dr. D. Ferree*
7.1.3 Department of Entomology	Act. Chairman: Dr. D. Horn Assoc. Chairman: Dr. R.E. Treece
Insects of fruit crops	Dr. F.R. Hall
Fruit insects	Dr. R.N. Williams
Greenhouse pests and ornamentals	Dr. R.K. Lindquist
7.1.4 Department of Agricultural Engineering	Chairman: Dr. W.I. Roller
Greenhouse design and energy conservation	Dr. T.H. Short

INDIANA

8 Vincennes

8.1 Southwestern Indiana Agricultural Research Center State Institution elevation: 121 m rainfall : 1087 mm	. PO Vincennes, Indiana Supr.: M. Lang

Research on fruit and vegetable crops

9 Lafayette

9.1 Indiana Agricultural Experiment Station, Main Station elevation: 187 m rainfall : 936 mm State Institution	. Lafayette, Indiana 47907 Phone (317)494-8360 Dir.: Dr. B.R.. Baumgardt
9.1.1 Department of Botany and Plant Pathology	Head: Dr. G.E. Shaner
Breeding of scab resistant apples	J. Crosby
Diseases of fruit crops	Dr. P.C. Pecknold
Diseases of mint crops	Dr. R.J. Green
Diseases of turf	Dr. D.H. Scott
9.1.2 Department of Food Science	Head: Dr. P.E. Nelson
Food chemistry and quality of processed foods	Dr. J.E. Hoff
9.1.3 Department of Entomology	Head: Dr. E.E. Ortman
Fruit insects	Dr. D.L. Mathew
Vegetable insects	Dr. A.C. York
Ornamental insects	Dr. D.L. Schuder
Breeding apples for insect resistance (USDA)	Dr. H.F. Goonewardene
9.1.4 Department of Horticulture	Head: Dr. B.C. Moser
Water stress physiology	Dr. R. Bressan
Landscape horticulture	Dr. P.L. Carpenter
Physiology of senescence and hormone action	Dr. J.H. Cherry
Urban horticulture	Dr. M.N. Dana
Physiology and production of tree fruits	Dr. F.H. Emerson
Breeding and genetics of potato, salpiglossis and streptocarpus	Dr. H.T. Erickson
Phenology of woody ornamentals	Dr. H.L. Flint*
Molecular biology of horticultural crops	Dr. P. Goldsborough
Breeding and genetics of greenhouse tomatoes	Dr. L. Hafen

Floriculture crop production	Dr. P.A. Hammer*
Post harvest physiology	Dr. A.K. Handa
Fruit tree rootstocks, pruning and training; small fruits	Dr. R.A. Hayden*
Plant cell, tissue and organ culture	Dr. P.M. Hasegawa
Breeding and genetics of fruit crops	Dr. J. Janick*
Woody plant physiology	Dr. R.J. Joly
Stress physiology and seed dormancy	Dr. C.A. Mitchell
Plant physiology/biochemistry	Dr. D. Rhodes
Weed control and culture of vegetable crops	Dr. J. Simon*
Economics and marketing of horticulture crops	Dr. G.H. Sullivan
Breeding and genetics of vegetable crops	Dr. E.C. Tigchelaar
Weed control in fruit and vegetable crops	Dr. S.C. Weller
Nutrition of vegetable crops	Dr. G.E. Wilcox
Physiology of floriculture crops	Dr. W.R. Woodson

ILLINOIS

10 Simpson

10.1 Dixon Springs Agricultural Center . Simpson, Illinois 62985
Supt.: Dr. J.W. Courter

Small fruit and vegetable production, greenhouse production, horticulture marketing

11 St. Charles

11.1 St. Charles, Horticulture Research Center . St. Charles, Illinois 60174

Cultural research in vegetables and small fruits — Dr. J.M. Gerber/W.H. Shoemaker

12 Urbana

12.1 State Natural History Survey . Urbana, Illinois 61801
Chief: Dr. P. Risser
elevation: 225 m
rainfall : 881 mm
sunshine : 52%

This state-supported organization is housed at the University of Illinois but has a separate budget and administrative board. Some staff members hold joint appointments with the Illinois Agricultural Experiment Station. Incidence of plants by insects and diseases, control methods.

12.1.1 Applied Botany and Plant pathology:	Head: Dr. C. Grunwald
Diseases of shade and forest trees	Dr. D.F. Schoeneweiss/Dr. E.B. Himelich,
Diseases of woody ornamentals	Dr. D. Neely
12.1.2 Economic Entomology:	Head: Dr. W.G. Ruesink
Insects of horticultural crops	Dr. R. Randell

12.2 University of Illinois College of Agriculture — Urbana, Illinois 61801
Dean: J.R. Campbell

Experiment Station, Main Station — Dir.: D.A. Holt

12.2.1 Department of Food Science:	Head: Dr. A.J. Siedler
Processing Vegetables	Dr. W.E. Artz
12.2.2 Department of Horticulture:	Acting Head: Dr. D.B. Dickinson
Growth, development, flowering of floriculture crops	Dr. M.C. Carbonneau

Horticultural crops physiology and biochemistry	Dr. D.B. Dickinson
Post harvest physiology of horticultural crops	Dr. B.A. Eisenberg
Commercial production of turf	Dr. T.W. Fermanian
Landscape construction and design	F.A. Giles, M.Ed.
Vegetable genetics and breeding	Dr. J.A. Juvik
Rooting physiology of woody plant	Dr. G.J. Kling
Genetics and breeding of tree fruits	Dr. S.S. Korban
Commercial fruit production	Dr. D.B. Meador*
Tissue culture, growth, development and propagation of herbaceous and woody plants	Dr. M.M. Meyer, Jr.*
Landscape planning and design	W.R. Nelson
Flower arrangement and retailing	D.A. Noland
Horticultural physiology and biochemistry, chlorophyll biosynthesis, cell-free culture, pesticide development	Dr. C.A. Rebeiz
Information for home gardeners	J.C. Schmidt
Greenhouse structures, floriculture production	W.J. Sherry
Anatomy and morphology of deciduous fruits, rootstock performance culture	Dr. R.K. Simons*
Small fruit breeding, tissue culture	Dr. R.M. Skirvin
Woody landscape plants, tissue culture of woody plants	Dr. M.A.L. Smith
Physiology and production of vegetable crops; growth regulators	Dr. W.E. Splittstoesser
Horticulture crop atmosphere-plant-water relations, crop ecophysiology	Dr. L.A. Spomer
Genetics and breeding of vegetable crops	Dr. J.G. Sullivan
Fruit crop nutrition and physiology	Dr. J.S. Titus
Turfgrass community dynamics and ecology, cultural systems	Dr. D.J. Wehner
Production and management of nursery crops	Dr. D.J. Williams
Vegetable crops culture and production	Dr. J.M. Gerber
Vegetable crops fertility, soils	Dr. J.M. Swiader
12.2.3 Department of Plant Pathology:	Head: Dr. R.E. Ford
Turfgrass disease	Dr. H.T. Wilkinson
Vegetable crop diseases	Dr. B.J. Jacobsen
Bacterial and fungus diseases of fruits	Dr. S. Ries
Viruses of horticultural plants	Dr. C. D'Arcy

13 Wichert

13.1 Kankakee River Valley Sand Field	St. Anne, Illinois 60964
Use of sandy soils for vegetable production	R.K. Lindstrom/Dr. J.M. Swiader

MICHIGAN

14 East Lansing

14.1 Michigan Agricultural Experimental Station	East Lansing, Michigan 48824 Dean: Dr. J.H. Anderson Dir.: Dr. R.G. Gast
14.1.1 Department of Horticulture	Chairman: Dr. J.F. Kelly*
International horticulture	Dr. H.C. Bittenbender*
Floriculture: physiology/taxonomy	Dr. W. Carlson/Dr. R.D. Heins*, Dr. L. Taylor/J.A. Birnbaum
Breeding, genetics	Dr. L. Ewart
Design, marketing	Dr. B. Fails
Landscape Horticulture: physiology/taxonomy/	Dr. A. Cameron*/Dr. C. Peterson,

maintenance/design	J. Saylor/ R. Schutzki
Fruit Crops: culture and physiology of tree fruits	Dr. M.J. Bukovac*, Dr. F.G. Dennis*,Jr./ Dr. J. Hull, Dr. J.A. Flore*/Dr. R.L. Perry*, Dr. E. Hanson
Genetics of tree fruits	Dr. A. Iezzoni
Physiology of small fruits	Dr. G.S. Howell
Genetics of small fruits	Dr. J. Hancock
Post harvest handling and physiology	Dr. D. Dilley
Vegetable Crops: culture and physiology	Dr. H.C. Price*/Dr. S. Ries, Dr. B. Zandstra/Dr. I. Widders
Breeding and genetics	Dr. K. Sink/Dr. L. Ewart
Post harvest physiology	Dr. R. Herner
Weed science	Dr. A. Putnam
14.1.2 Department of Entomology	Chairman: Dr. J. Bath
Nematodes	Dr. G. Bird
Fruit insects	Dr. M. Whalon/Dr. A. Howitt
Vegetable insects	Dr. E. Grafius
Ornamental insects	Dr. D. Smitley
14.1.3 Department of Botany and Plant Pathology	Chairman: Dr. E.J. Klos
Vegetable diseases	Dr. C. Stephens/Dr. M. Lacy, Dr. R. Hammerschmitt
Fruit diseases	Dr. E.J. Klos/Dr. A. Jones, Dr. D. Ramsdell
Ornamental diseases	Dr. G. Adams
Turf diseases	Dr. J. Vargas
14.1.4 Horticulture Research Center	Supt.: F. Richey, M.S.

15 Clarksville

15.1 Clarksville Horticultural Experiment Station	Clarksville 48815
Fruits, vegetables, woody ornamentals	Supt.: J. Skeltis, M.S.

16 Fennville

16.1 Trevor Nichol Experimental Farm	Fennville 49408
Fruit pests	Supt.: M. Dally, B.Sc.

17 Grand Rapids

17.1 Graham Horticultural Experiment Station	N.W. Grand Rapids, Lake Michigan District 49504
Tree fruits	Supt.: J. Gilmore, M.Sc.

18 Sodus

18.1 Sodus Experiment Station	Sodus 49126
Vegetables and small fruits	Supt.: Dr. H. Price

19 Traverse City

19.1 Northwest Michigan Horticultural Research Station	Traverse City 49684, Dir.: Dr. C.D. Kessner
Stone fruits, strawberries, brambles	Mgr.: W. Klein

20 Tipton

20.1 Hidden Lake Gardens	Tipton 49278
Native botanical collection, conservatory	Cur.: F.W. Freeman, M.Sc.

WISCONSIN

21 Ashland

21.1 University Experimental Farm — Ashland, Wisconsin 54806
elevation: 198 m — Phone: (715) 682-6844
rainfall : 783 mm — Supt.: T.D. Syverud, M.S.
sunshine : 2350 h
State institution. Research plots including strawberries, small fruit and home horticulture.

22 Hancock

22.1 Agricultural Research Station — Hancock, Wisconsin 54943
elevation: 327 m — Phone: (715) 249-5961
rainfall : 752 mm — Supt.: G.G. Weis, M.S.
sunshine : 2505 h
Varietal testing cultural practices, production of vegetables and small fruits under irrigation, potato research and breeding

23 Madison (Arlington)

23.1 Main Agricultural Research Station — Madison, Wisconsin 53706
elevation: 329 m — Dir.: M. Finner, M.S.
rainfall : 783 mm
sunshine : 2498 h
Small fruit, orchard, vegetable crops, nursery, and ornamental research projects. — Supt.: Dr. R. Newman

23.1.1 Department of agronomy:	Chairman: Dr. J. Forsberg
Sweet corn breeding	Dr. T. Tracy
Pea breeding	Dr. E.T. Gritton
Weed control in peas and corn	Dr. R. Doersch
23.1.2 Department of Entomology:	Chairman: Dr. J. Wyman
Corn rootworm and other soil insects	Dr. J. Wedberg
Biology of apple maggot	Dr. G.M. Boush
Evaluation of insecticides for vegetables	Dr. R.K. Chapman
Fruit crop insecticides	Dr. D. Mahr
23.1.3 Department of Horticulture:	Chairman: Dr. H.J. Hopen
Culture of geraniums and poinsettia	Dr. G. Beck
Adaptation of woody ornamentals, selection of strains in Junipers	Dr. E.R. Hasselkus
Pomology:	
Weed control, nutrition and culture of cranberries and strawberries	Dr. M.N. Dana
Apple rootstock evaluation	Dr. E. Stang*
Breeding systems for self-pollinated vegetables *Phaseolus sp.*	Dr. F. Bliss
Techniques for breeding onions, red beets, green beans; genetics of carotene synthesis; mineral-nutrition	Dr. W.H. Gabelman
Genetics and breeding of *Cucumis sativus*	Dr. R. Lower
Breeding and genetics of ornamentals	Dr. R.W. Hougas
Carrot, cucumber and onion improvement	Dr. C. Peterson
Potato culture and varietal trials	Dr. J.A. Schoeneman
Cytogenetics of *Solanum* sp.	Dr. S.J. Peloquin
Genetics and cytogenetics of *Solanum* sp.	Dr. R.E. Hanneman

Chemical control of weeds	Dr. L.K. Binning/Dr. H.J. Hopen
Cultivar evaluation and weed control in turf and tobacco	Dr. R. Newman/Dr. H.J. Hopen
Vegetable crop and tobacco nutrition	Dr. L. Peterson
Effects of environment and pollution on horticultural crops	Dr. T.W. Tibbitts*
Home garden horticulture	Dr. H. Harrison
Cucumber research and evaluation	Dr. J. Staub
Biochemistry of plant growth regulators	Dr. M. Sussman
Molecular biology of plant development	Dr. R. Vierstra
Statistical research applied to horticulture	Dr. B. Yandell
Irrigation research, vegetable crops	Dr. D. Curwen
23.1.4 Department of Plant Pathology	Chairman: Dr. D. Maxwell
Diseases of cranberry, strawberry, raspberry and genetics of *Venturia inequalis*	Dr. S. Jeffers
Nematodes in crops, potato virus studies	Dr. S. Slack
Bean diseases	Dr. D. Hagadorn
Lettuce diseases	Dr. L. Sequeira
Evaluation of fruit and vegetable fungicides	Dr. W. Stevenson
Cabbage and cucumber diseases	Dr. P. Williams
Evaluation of fungicides for ornamentals	Dr. G. Worf
Bacterial diseases and storage disorders	Dr. A. Kelman
Soil borne disease of vegetable crops	Dr. D. Rouse
Nematodes and related pathological problems	Dr. A. MacGuidwin

23.2 Arlington Agricultural Research Station
elevation: 329 m
rainfall : 783 mm
sunshine : 2498 h

Arlington, Wisconsin 53511
Phone: (608) 846-3761
Supt.: D.A. Schlough, M.S.

24 Spooner

24.1 Agricultural Research Station
elevation: 334.4 m
rainfall : 735 mm
sunshine : 2460 h

Spooner, Wisconsin 54801
Phone: (715) 635-3735
Supt.: R.E. Rand, M.S.

State institution. Research plots, small fruit, potato weed control, varietal testing, vegetable crops, cultural studies.

25 Sturgeon Bay

25.1 Agricultural Research Station
elevation: 199 m
rainfall : 795 mm
sunshine : 1900 h

Sturgeon, Bay, Wisconsin 54235
Phone: (414) 743-5406
Supt.: Dr. R.W. Weidman, M.S.

State institution. Fruit research, apple rootstocks, strawberry breeding and culture, ornamental and USDA Potato Introduction Station.

26 Lancaster

26.1 Agricultural Research Station
elevation: 317 m
rainfall : 835 mm
sunshine : 2600 h

Lancaster, Wisconsin 53813
Phone: (608) 723-2580
Supt.: Dr. W.H. Paulson

Apple evaluation research and fruit culture

27 Rhinelander

27.1 Wisconsin Potato Research Station
elevation: 487.4 m
rainfall : 777,5 mm
sunshine : 2130 h

Rhinelander, Wisconsin 54501
Phone: (715) 362-5719
Supt.: D. Kichefski, B.S.

Potato breeding and genetics research and potato varietal evaluation

Prepared by Herbert J. Hopen, College of Agricultural and Life Science, Univ. of Wisconsin, Madison, Wisconsin
V.L. Lechtenberger, Agricultural Experiment Station, Purdue Univ., Lafayette, Indiana
D.B. Dickinson, Dept. of Horticulture, Univ. of Illinois, Urbana, Illinois
Rosemary Crassweller, Dept. of Horticulture, Ohio State Univ., Columbus, Ohio
J.D. Carlson, Dept. of Entomology, Michigan State Univ., East Lansing, Michigan

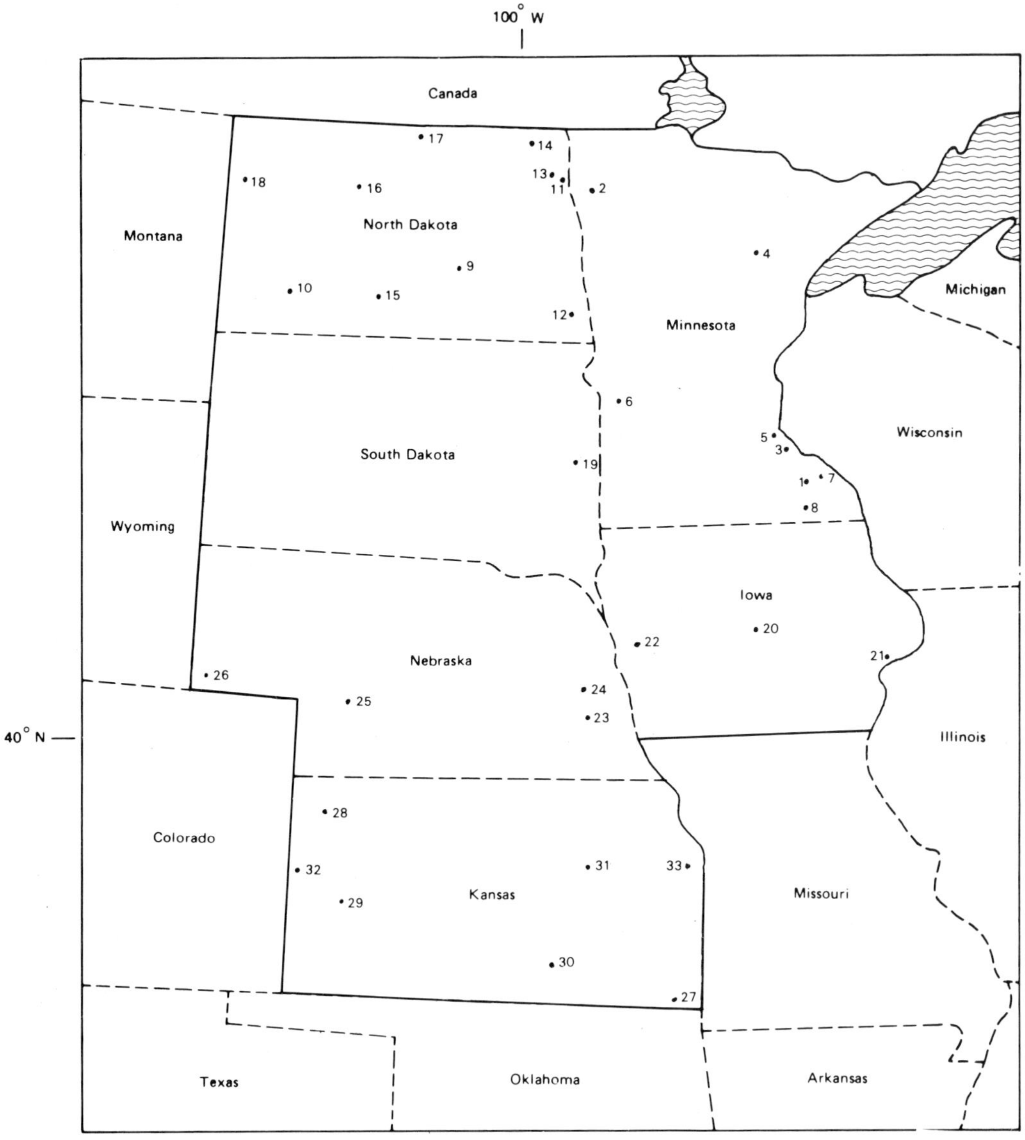
100° W
40° N
Canada
Montana
North Dakota
South Dakota
Wyoming
Nebraska
Colorado
Kansas
Texas
Oklahoma
Minnesota
Michigan
Wisconsin
Iowa
Illinois
Missouri
Arkansas
1
2
3
4
5
6
7
8
9
10
11
12
13
14
15
16
17
18
19
20
21
22
23
24
25
26
27
28
29
30
31
32
33

Minnesota: 1 Chaska, 2 Crookston, 3 Lamberton, 4 Grand Rapids, 5 Becker, 6 Morris, 7 Saint Paul, 8 Waseca, **North Dakota**: 9 Carrington, 10 Dickinson, 11 East Grand Forks, 12 Fargo, 13 Grand Forks, 14 Langdon, 15 Mandan, 16 Minot, 17 Rolla, 18 Williston, **South Dakota**: 19 Brookings, **Iowa**: 20 Ames, 21 Muscatine, 22 Whiting, **Nebraska**: 23 Lincoln, 24 Mead, 25 North Platte, 26 Scottsbluff, **Kansas**: 27 Chepota, 28 Colby, 29 Garden City, 30 Wichita, 31 Manhattan, 32 Tribune, 33 DeSoto

Survey

MINNESOTA

Minnesota extends from latitude 49°23'N. at the Canadian border 358 miles south to latitude 40°30'N. at the northern boundary of Iowa. Topography is generally flat to gently rolling. Highest elevation is 694 m and the lowest 184 m. The Red River Valley which forms a common boundary with the Dakotas is the largest area of nearly level land in the world and is almost free from microclimate variations.

In the Twin Cities (St. Paul-Minneapolis) area where the University is located, mean temperatures range from 12.4°F for January to 72.3°F for July. Summers are generally pleasant and temperatures of 100°F or above are uncommon. Rainfall averages 636 mm in the Twin Cities area (eastern border) and decreases to less that 508 mm at the western border.

Horticulture - The state contains a wide range of soils and, in most areas, ample water for supplemental irrigation. The horticultural industry base is comprised of potatoes and cool season vegetables for processing. In addition, the state hosts large commercial floriculture, landscape nursery, and turf industries. Around population centers, the pick-your-own fresh fruit and vegetable business has and contiues to enjoy significant growth.

Native to the state are raspberry (*Rubus spp),* blueberry *(Vaccinium spp),* cranberry *(Vaccinum* spp), highbush cranberry *(Viburnum trilobum),* elderberry *(Sambucus* spp), crab apple *(Malus ionensis),* choke cherries, wild plums *(Prunus* spp), service berry *(Amelanchier* spp), and wild grapes *(Vitus riparia).*

Apples are grown in commercial orchards throughout the southern third of the state along the Mississippi and St. Croix River valleys east of the Twin Cities.

Horticultural Research - Research facilities include a central facility (Alderman hall) and a greenhouse range located on the St. Paul campus and a 870 acre landscape arboretum combined with a horticultural research center located at Chaska. These latter two facilities combine to form a unique field laboratory for all perennial horticultural plants.

A major component of the research programs has been and continues to be the study of cold hardiness and the development of hardy cultivars. Begun in 1890, more than 70 cultivars of fruit such as apple, plum, cherry, blueberry, pear and apricot have been released to the public. Development of improved and hardy landscape materials has resulted in the release of more than 40 cultivars of azaleas, shrubs, and shade and ornamental trees.

In support of the breeding programs, the department's Cold Hardiness Laboratory has been active in the elucidation of mechanisms involved in cold acclimation and hardiness to extremely low temperatures. Current research has expanded into the study of all plant stresses at all levels of structure from whole-plant to subcellular. Research programs in other aspects of plant physiology have and continue to be very productive. Included are the physiological bases of self incompability, juvenility, cell membrane phenomena, hormonal growth regulation, starchsugar conversion, in vitro and in vivo propagation, and protein synthesis in potato.

Vegetable, including potato, research centers around cultivar improvement and stand establishment. Major programs include sweet corn breeding for corn borer resistance and potato breeding for disease tolerance and processing quality. Floriculture research enjoys a large and respected history of service to the industry. Current emphases are in reducing energy requirements through cultural manipulation and in chrysanthemum breeding for day neutrality. Turf research activities include maintenance levels, cultivar evaluation, sod transport and *Poa annua* breeding.

The landscape architecture program is jointly sponsored by the Department of Horticulture and the School of Architecture. Its design-centered curricula and computer-aided regional analysis are nationally-recognized. The program recently was awarded membership in the Graduate School and has developed a research component through which students may be awarded an M.L.A. degree.

Being located in a large metropolitan area, the demands of and opportunities for teaching and research in various aspects of urban horticulture are virtually limitless. A major effort is underway to meet the needs of amateur horticulturists, both urban and rural, through the training of para-professionals (Master Gardeners).

NORTH DAKOTA

Agriculture and Horticulture. North Dakota is an agricultural state characterized by mixed farming with wheat, barley, flax, soybeans and corn the major cultivated crops, and because of the large livestock industry comprising cattle, hogs, sheep, and poultry, the additional establishment of natural and cultivated grasslands are important.

The potato is the most important of the vegetable crops averaging 120.000 acres with a value of $ 60 million. The entire range of garden vegetables except those confined to semi-tropical and tropical zones are cultivated. For example, well-adapted cultivars of tomatoes, cucumbers, melons, peppers and sweet corn are grown with considerable success. Tree fruits are slowly becoming a commercial venture particularly in more populous areas and in areas of considerable natural protection (valleys, etc). The apple is most common along with plums which appear in home gardens in every area of the state. The strawberry is the most common small fruit and has commercial possibilities as well as common home garden use.

Horticultural Research. This has been in continuous existence since the founding of the North Dakota Agricultural Experiment Station in 1890. Research problems over the years dealt largely with the development of fruits, flowers, vegetables, and ornamentals adapted to the climate. The North Dakota Agricultural Experiment Station of North Dakota State

University at Fargo, North Dakota, in the department of Horticulture and Forestry has been largely responsible for the bulk of the horticultural research and development in the state.

The U.S.D.A. horticulture field station located at Mandan, North Dakota, to its closing in 1965 was largely responsible for fruit breeding and shelterbelt studies. Publications from the Mandan station on work in the above named areas were classics of their time. The U.S.D.A. station and experiment station worked closely together in comparable research problems. At the present time the Agricultural Research Station (U.S. Department of Agriculture) is only investigating shelterbelt influences on crop (wheat) production.

The Agricultural Experiment Station at Fargo operates several branch stations and horticultural work is carried on at these. The largest single area of research is with the potato crops. Since 1956 nine potato varieties have been introduced. Of these nine, three - Norland, Norchip and Norgold - are among the top 10 varieties in total production in the United States in 1970 and comprise about 60 percent of the total potato acreage in North Dakota. Cultural research is the other facet of potato research.

The ornamental work has largely been concerned with hardiness and establishment studies and of late, the testing of new species and varieties of various materials as poplar and potentilla for identity of types and their potential use. Evaluation and establishment of deciduous and evergreen species is underway to develop recreation areas on the prairie. An arboretum has been established to aid in hardiness studies and teaching.

The fruit work is largely cultural and has emphasized the use of hardy rootstocks, interstocks and determination of hardy dwarfing materials.

Cultural studies with vegetable crops under irrigation is an important area of research as this phase of agriculture is developed. Studies on the use of plastic mulches to aid in the establishment of early producing crops is proving to be helpful in the development of the vegetable industry in the state.

SOUTH DAKOTA

South Dakota occupies the area between latitudes 42°29' and 45°57'N and longitudes 96°26' and 10°43'W. It is 380 miles long from east to west and 200 miles wide from north to south. Its area is 77,047 square miles. The Missouri River bisects the state north and south and this river and its tributaries form the principle drainage system of the state. Four dams on the Missouri River create large reserviors covering most of the river valley in South Dakota. Numerous small lakes are found in the northeast corner of the state. The greater part of South Dakota consists of plains, but a small mountainous area, the Black Hills, occurs in western South Dakota. The climate is continental. Temperatures of all seasons are variable but extremes in temperature are moderated as one travels from the northwest to the southeast. January temperatures averaged from 10°F to 20°F in different zones.

The average number of frost-free days ranges from 100 in the Black Hills to 160 in south-eastern South Dakota.

Horticulture. The low temperature extremes in winter and dry, often semi-arid climate in summer limit the range of horticultural crops which can be grown and determine one of the primary directions of research in horticulture, namely the securing of cultivars of flowers, fruits, vegetables, and ornamental plants adapted to the climate. The commercial value of horticultural products is not high compared with livestock and field crops which are the chief source of income in the state. Flower crops and bedding plants account for most greenhouse production. Vegetable crops and strawberries are produced for local consumption. Potatoes are grown for both local and interstate markets. Nursery crop production and garden centers to serve horticulture in local areas are an important phase of horticulture.

Horticultural Research. Horticultural research in South Dakota centers on the South Dakota Agricultural Experiment Station which is a part of South Dakota State University located at Brookings. Primary attention has been given to breeding and selecting cultivars of horticultural crops and trees for windbreaks. Numerous cultivars have been introduced in apples, plums, pears, grapes, sand cherries, ornamental trees and shrubs, tomatoes, and peppers. The department also pioneered the use of polyethylene greenhouses in cold climates. Other fruitful research has concerned the culture of vegetables, fruit, and ornamentals, determination of factors influencing flowering in Amazon lilies, research on tree spacing and design of windbreaks, and cultural practices for horticultural crops. The department has 214 acres for research on ornamentals, fruit, vegetables, and shelterbelts.

IOWA

A complex of geography climate, economy and population historically and presently determines the character of horticulture in Iowa. Iowa is located near the center of the continental United States from east to west, somewhat nearer the northern boundary than the southern, and approximately in the center of the North American continent. It lies between latitudes 40'24" and 43'30" North and longitudes 90'5" and 96'40" West. The climate, continental in character, is subject to considerable variations in temperatures in short periods of time and to unpredictable droughts with considerable variations in time and distribution of precipitation.

In general, the climate is favorable for forage crops,

small grains, Indian corn maize, and soy beans, a wide range of vegetables, and a very satisfactory range of temperate-zone ornamental trees, shrubs, and flowers. Both fruit trees and small fruit call for adjustments of variety, protection, and cultural practices to overcome the vicissitudes of climatic extremes, particularly of unseasonable low temperature. In central Iowa, snowfall and rainfall together average 762 mm annually. The average length of the growing season is 156 days. Northwest Iowa has an average of 140 frost-free days, the south border of the state has 170. Twenty-five percent of the soil of the United States, rated grade 1 on the basis of productiveness, is located in Iowa, and the state contributes about 10 percent of the total national food supply. Soil is not a limiting factor in the production of horticultural crops.

Horticulture. Commercial vegetable production in Iowa centers in potatoes, onions, cabbage and melons for the fresh market; sweet corn, tomatoes, potatoes, and carrots for processing. Almost the whole range of vegetables is produced for local markets and, except possibly for melons, in all parts of the state. Home gardening is practiced everywhere in both town and country. The commercial crops are consumed both within and beyond the borders of the state. The largest production of potatoes, onions, and carrots occurs in drained lake beds of peat or organic soils in the north central counties, especially Kossuth, Hancock, Winnebago, Cerro, Gordo, and Mitchell. Sweet corn for canning is scattered throughout the state. Tomatoes and cucumbers for canning and pickling are largely grown in the Mississippi Valley, especially in Scott, Louisia, and Muscatine counties. Melons for local markets appear in isolated small plantings over a wide area of the state but those grown for shipment are produced mostly in Muscatine and Lee counties in the southeast corner of the state on the sandy soils of the Mississippi Valley.

Apples are the major fruit grown commercially in Iowa and most of them are marketed in the state. The majority of orchards grown on the loess soils of the Mississippi Valley from Clinton to Keokuk and in the Mississippi Valley from Council Bluffs to the Missouri line. Many are also located near cities and large towns in the central counties such as Cedar and Polk. Equally favourable sites and locations may be found in most of the southern two-thirds of the state but planting is limited by economic factors. Iowa is noted for the wholesale production of nursery plants covering a very wide range of hardy and semi-hardy perennial flowers, shrubs, shade trees, ornamentals, and fruit trees. The largest wholesale nursery operators in the country are located here largely because of excellent soils, partly because of a combination of other fortuitous circumstances. The largest acreages are found in Fremont and Page counties in the southwest corner of the state and in Polk, Franklin, Floyd, and Lynn counties farther east. Local nurseries are found in every county. In 1970, there were 257 producing nurseries with approximately 6300 acres in actual production. Commercial floriculture is represented chiefly by the glasshouse culture of cut flowers and pot plants. Nearly every city and town has a greehouse range. There are approximately 3 million square feet of glass devoted to florists' crops in Iowa.

Horticultural Research. In Iowa, horticultural research is determined principally by two factors: 1 - the annual range and distribution of its temperatures over long periods of years, and 2 - the needs and desires of its citizens. The main soil types found in the state, nearly all ranking above average in fertility, afford no serious problems in the production of those vegetables, fruits, or ornamentals adapted to the cimate, which cannot be solved by modifications and adaptations of such standard cultural practices as selection of adapted localities and sites, drainage, irrigation, fertilization, and pest control. Major topographical determinants are also absent in a state where elevations above sea level range between 510 m at the highest to 146 m at the lowest. Adjustments to local topography, however, are always desirable and often necessary as they affect exposures to wind, light, flooding, and erosing on individual sites. The human needs and desires as related to horticultural research is naturally complex. It has varied from pioneer times 100 to 150 years ago, when home food production of edible vegetables and fruits was the prime horticultural objective, to a modern civilization which depends chiefly on commercial production for its horticultural foods. The search for climatically adapted varieties of flowers, vegetables, ornamental shrubs, shade, and shelter trees, and particularly for fruit trees, was the first and greatest horticultural research undertaking in Iowa. These continue to rate high in importance, but differ from early-variety testing techniques by the introduction of plant breeding to produce new varieties having better adaptations to the climate of the state and the marketable and ornamental qualities of horticultural plants and their products. The uncertain distribution of summer rainfall requires investigations in irrigation, and have since 1923. Problems of soil management and fertilization for specific crops are in constant study. Storage for the preservation of apples, cut flowers, potatoes, and nursery stock to prolong their seasons of usefulness has a large place in horticultural research in Iowa. Growing in importance in its research efforts are investigations in marketing, involving packaging, quality control, and determination of standards of excellence for nursery, fruit, vegetable, and flower products.

Historically, fruitful fields of horticultural research

which have resulted in notable contributions include: importations and evaluation of varieties of apples from abroad, especially from Russia; effects of low winter temperatures on injury to and killing of many species of fruits and ornamental plants; introduction of new varieties of apples, pears, peaches, and plums by plant breeding; introduction of new varieties of strawberries, raspberries, junipers, chrysanthemums, and geraniums; extensive investigations on the effects of various storage temperatures and conditions, harvest maturity, and variety on the keeping of apples in common and controlled storages; adaptation of understocks to the propagation of apples and roses; hydroponic techniques applied to greenhouse plants; introduction of new cultivars of sweet corn, potatoes, onions, pumpkins and melons; quick freezing for home and market storage of fruits and vegetables; fertilization and soil management of apple orchards. Literature relating to the above investigations is contained in bulletins of the Iowa Agricultural and Home Economics Experiment Station; proceeding of the Iowa State Horticultural Society, State House, Des Moines, Iowa; Proceedings of the American Society for Horticultural Science; and various horticultural journals of the time. Active research in horticulture in Iowa proceeds under public, semipublic, and private organizations and institutions. Consistent, continuing, long-range research sustained by scientifically trained personnel is chiefly developed in the Iowa Agricultural Experiment Station, Iowa State University, Ames, Iowa, under the guidance of the Department of Horticulture. A horticulture Farm, established in 1968-1969, consisting of 2276 acres, 5 research buildings, and extensive irrigation facilities is located 9 miles north of the Ames campus. New facilities on the campus were occupied during the summer of 1980. In addition, a 40 acre research station is located at Muscatine and a 20 acre station at Whiting.

NEBRASKA

Nebraska is the most westerly of the North Central group of states. Its total area is 77,227 m^2 of which only 615 are water covered. The Missouri river bounds the entire northeast corner and the east side of the state and its tributaries drain the northern quarter of the area. Nebraska has become the largest irrigated region closest to the metropolitan East. Extensive areas of productive soils and a climate similar to major producing areas suggest a potential for increased horticultural crops production in the State. Vegetable yield results in the state and economic analyses of vegetable and principal agronomic crop production indicate that, under a high level of management, vegetable production in irrigated land can be competitive with other uses of resources. The climate is distinctly continental and is characterized by a wide temperature difference between winter and summer. Temperatures in January and July respectively, average 23.5 and 76.0°F. Lower average temperature, an increasing range in daily temperatures and a shorter growing season accompany the increase in elevation from about 305 m in eastern Nebraska to 1220 m in the west. Annual precipitation averages over 762 mm in eastern Nebraska and decreases to less than 407 m in the west.

Horticulture research centers in the Department of Horticulture at the university of Nebraska in Lincoln. The horticultural research activities of the Department may be summarized as follows: 1 - Routine but necessary variety testing to evaluate the many new introductions under local conditions. National turfgrass and flowering crab trials are included. 2 - Agricultural climatology, plant physiology, plant breeding and genetics, biochemistry and post-harvest physiology, and the propagation and production of turfgrass, vegetables and ornamental plant materials are involved in the present-day research projects and programs of the Department. Research is conducted at : Agricultural Research Division, Institute of Agriculture and Natural Resources, Lincoln, Agricultural Research and Development Center, Mead, UNL West Central Research and Extension Center, North Platte, UNL Panhandle Research and Extension Center, Scottsbluff, Northwest Agricultural Laboratory, Alliance, all in the State of Nebraska.

KANSAS

The State of Kansas is located very near the geographical center of the United States. Its area is 82,270 square miles. It is rectangular in shape, 410 miles from east to west and 208 miles from north to south. It lies between latitudes 37° and 40° North and longitudes 94°30' and 102° West. The highest altitude is 1260 m and the lowest 213 m above sea level. Its climate is continental in character, subject to rapid changes from high to low temperatures in all seasons of the year. Extremes may vary from 100°F or more in summer to 10°F to 20°F on rare occasions in winter. The average annual temperature is 550°F ranging from 58°F in the southeast to 52°F in the northwest.

Agriculture. Agriculturally, Kansas is a wheat and beef state with important productions also of corn (maize), soya beans, forage grasses, sheep, hogs and poultry. Horticulturally, the Kaw Valley produces commercial crops of potatoes and nursery plants. Apples are grown along the Missouri river and its tributaries in the north-east corner of the state and in the Arkansas River Valley where peaches are also produced on certain sandy soils. Market gardens are found in the vicinity of all cities in the eastern third. Dry beans are produced on about 25,000 acres in Western Kansas.

Kansas is rather on the northern edge of pecan-growing territory but the pecan is native and excellent hardy varieties are grown. In spite of the rigors of its continental climate, Kansas with an earlier spring and a longer summer than its neighbours to the north and northeast, is far enough south to admit a somewhat wider range of ornamental trees and shrubs. Crepe myrtle *(Lagerstroemia indica)* and *Pyrocantha* sp, for example, appear in parks and ornamental plantings everywhere along the east side of the state and flowering dogwood *(Cornus florida)* is native to the woodlands. The florist industry is concentrated around the population centres in Eastern Kansas and the Wichita area and includes the production of pot plants, cut flowers and bedding plants. The nursery industry also is located chiefly in the eastern area with individual nurseries ranging from large wholesale to small retail establishments.

Historically, horticultural research in Kansas was directed to testing the adaptation of the varieties of fruits, vegetables and ornamentals of the Atlantic seaboard, New York, Pennsylvania and other established states. Gradually, the production of new varieties by breeding and adaptations of irrigations to offset summer droughts and inadequate rainfall have been added. Refinements of techniques such as grafting of hardy pecans on desirable stocks, planting of semi-dwarf apple types, and the use of plastics in greenhouses, are examples of numerous individual problems which constantly require attention in such a field as horticulture. In common with the other north-central states horticultural research is carried on chiefly through the agency of the agricultural experiment station as part of Kansas State University at Manhattan, Kansas. Besides the horticultural farm at the University, the department operates research farms at Wichita in south-central Kansas which experiments with ornamentals, fruits and vegetables. The southeast Kansas experimental farm at Chepota deals chiefly with pecans. There are also branch stations at Colby, Garden City and Tribune. Organizations within the state which are principally concerned with horticulture are: The Kansas State Horticultural Society, and the Federated Garden Clubs of Kansas.

MINNESOTA

1 Chaska

1.1 Landscape Arboretum and Horticultural Research Center
elevation : 219 m
rainfall : 726 mm

Chaska, Minnesota 55318
Phone: (612) 443-2460
Dir.: P. Olin, M.L.A.

2 Crookston

2.1 North West Experiment Station
elevation : 271 m
rainfall : 509 mm

Crookston, Minnesota 56716
Phone: (218) 281-6510
Dir.: Dr. L. Smith

3 Lamberton

3.1 South West Experiment Station
elevation : 349 m
rainfall : 635 mm

Lamberton, Minnesota 56152
Phone: (507) 752-7372
Dir.: Dr. W.W. Nelson

4 Grand Rapids

4.1 North Central Experiment Station
elevation : 399 m
rainfall : 670 mm

Grand Rapids, Minnesota 55744
Phone: (218) 327-1790
Dir.: Dr. B. Nyvall

5 Becker

5.1 Sand Plains Experimental Field
elevation : 316 m
rainfall : 704 mm

Becker, Minnesota 55330
Phone: (612) 261-4063
Dir.: Dr. R. Thompson

6 Morris

6.1 West Central Experiment Station
elevation : 347 m
rainfall : 607 mm

Morris, Minnesota 56267
Phone: (612) 589-1711
Dir.: Dr. R. Vatthaner

7 Saint Paul

7.1 University of Minnesota, College of Agriculture, Department of Horiticultural Science and Landscape Architecture
elevation : 295 m
rainfall : 669 mm

St. Paul, Minnesota 55101
Phone: (612) 373-1026
Head: Dr. J.F. Bartz

Fruit breeding and physiology	Dr. J. Luby/Dr. E. Hoover*
Weed ecology	Dr. L. Hertz
Vegetable pre- and post-harvest physiology and stand establishment	Dr. L. Waters
Vegetable breeding	Dr. D.W. Davis
Stress physiology	Dr. J.V. Carter
Plant propagation	Dr. P.E. Read*
Landscape architecture	R. Martin, M.L.A.
Landscape horticulture	Dr. C.G. Hard/M. Eisel, M.Sc.
Potato breeding	Dr. F.I. Lauer
Potato genetics	Dr. S.L. Desborough
Hardiness physiology	Dr. P.H. Li/Dr. A. Markhart
Genetics, incompatibility	Dr. P.D. Ascher
Landscape horticulture	Dr. R. Mullin
Sensory evaluation	Shirley T. Munson, M.Sc.
Ornamental breeding and landscape management	Dr. B.T. Swanson/Dr. H. Pellet
Turf management	Dr. D.B. White
Protoplasmatology	Dr. E. Stadelmann
Ornamentals, urban horticulture	J. McKinnon, M.Sc./Deborah L. Brown, M.Sc.
Nutrition	Dr. C. Rosen
Floriculture	Dr. R.E. Widmer*/Dr. H.F. Wilkins*
Juvenility	Dr. W. Hackett
Potato physiology, carbohydrate metabolism	Dr. J.R. Sowokinos
Fruit hormone phypsiology	Dr. M.L. Brenner

8 Waseca

8.1 Southern Experiment Station
elevation : 351 m
rainfall : 778 mm

Waseca, Minnesota 56093
Phone: (507) 835-3620
Dir.: Dr. R.H. Anderson

NORTH DAKOTA

9 Carrington

9.1 Carrington Experiment Irrigation Station

Carrington, North Dakota 58421

Branch station, state institution
Irrigation research, potatoes, vegetables, ornamentals and fruits — H. Olson

10 Dickinson

10.1 Dickinson Experiment Station — Dickinson, North Dakota 58601

Branch station, state institution
Vegetables, fruits and ornamentals — T.J. Conlon

11 East Grand Forks

11.1 Potato Processing Laboratory — East Grand Forks, Minnesota 56721
Dir.: P. Orr

Cooperative United States Department of Agriculture, States of Minnesota, North Dakota, Red River Valley Potato Growers Association, Federal state, private basic and applied studies on processing. — Drs. J. Varns/J. Sowokinos, E. Lulai

12 Fargo

12.1 North Dakota State University, Agricultural Experiment Station — Fargo, North Dakota 58105

Development and selection of fruits, vegetables and ornamental woody plants adapted to state environment. Culture of horticultural crops - irrigation, plastic mulches, hardiness - shelter belt studies.

Vegetable breeding — Head: Prof. Dr. A.A. Boe*
Evaluation of ornamentals
Potato breeding — Prof. Dr. R.H. Johansen
Potato-production studies — Prof. Dr. D.C. Nelson
Cultural studies with fruits, vegetables, ornamentals — Prof. N.S. Holland
Ass. Prof. Dr. E.W. Scholz
Herbicides, propagation, tissue culture — Prof. R.H. Heintz
Assoc. Prof. J.B. Barley
Shelterbelt studies, tree improvement — Ass. Prof. G.A. Tusken

13 Grand Forks

13.1 Research Farm - Red River Valley — East Grand Forks, Minnesota 58201
Exec. Secr.: L. Schmidt

Potato Growers Association, private institution — L. Schmidt
Breeding and cultural problems — D. Preston

14 Langdon

14.1 Langdon Experiment Station — Langdon, North Dakota 58249

Branch station, state institution
Fruits, potatoes and ornamentals — J. Lukach

15 Mandan

15.1 United States Northern Great Plains Field Station — Mandan, North Dakota 58554

Federal institution (U.S. Deparment of Agriculture). Terminated in 1965.
Influence on crop production

16 Minot

16.1 Minot Experiment Station — Minot, North Dakota 58701

Branch station, state institution
Potatoes, fruits, ornamentals — B.K. Hoag

17 Rolla

17.1 International Peach Garden — Rolla c/o Horticultural Department, NDSV, North Dakota 58367

Cooperative, operated by Canada and United States — O. Hagen
Ornamentals

18 Williston

18.1 Williston Experiment Station — Williston, North Dakota 58801

Branch station, state institution
Potatoes, vegetables, fruits and ornamentals — E.W. French

SOUTH DAKOTA

19 Brookings

19.1 University of South Dakota, South Dakota Agricultural Experiment Station — Brookings, South Dakota 57006

As there is no updated material available, the text of HRI III is repeated.

State institution. Apple and grape breeding. Forestry: shelterbelt spacing, hybrid trees. Landscape design, turf and woody plants. Cultural practices in forestry.

Derpartment of fruit breeding for hardiness	Head: Prof R.M. Peterson, Ph.D.
Landscape design	Assoc. Prof. L.C. Johsnon, M.Sc.
Ornamental horticulture, nutrition, cold hardiness	Ass. Prof. J.E. Klett
Park management	Assoc. Prof. P.E. Nordstrom, Ph.D.
Vegetable crops: tomato breeding	Assoc. Prof. D.P. Prashar,Ph.D.

IOWA

20 Ames

20.1 Iowa State University,
Agricultural Experiment Station
elevation : 291 m
rainfall : 776 mm
sunshine : 60%

Ames, Iowa 50011
Phone: (515) 294-2751
Telex: 283-359 IASU-UR

State institution, coordinated with the U.S. Department of Agriculture and with industrial organizations and others who sustain research fellowships. Advisory relationships with numerous organized horticultural bodies of the State Department of Horticulture. Development of safe methods for the disposal of excess pesticides used by farmers and agricultural applicators.

Iowa vegetable crop production - research and breeding	Head: Prof. C.V. Hall
Breeding and culture of bedding plants	
Breeding, culture, and propagation of herbaceous ornamental plants and roses	Prof. G.J. Buck
Small fruits breeding, physiology, and mechanization	Prof. E.L. Denisen*
Environmental and physiological interactions and the pathogenesis of grass disease	Prof. C.F. Hodges
Development of impatiens and flowering shrub cultivars with greater stress tolerance	Prof. J.L. Weigle
Feasibility study of potential nut crops for Iowa	Ass. Prof. C.D. Fear
Tree fruit adaptation and culture	
Scion/rootstock and interstem effects on apple tree growth and fruiting	Prof. P.A. Domoto*
Packaging, marketing, and propagation of nursery plants	Assoc. Prof. J.D. Kelley
Cultural and localized problems of vegetable crop production	Prof. H.G. Taber
Plant introduction	Ass. Prof. M.P. Widrlechner
Turfgrass cultural practices and improvement	Assoc. Prof. N.E. Christians Ass. Prof. M.L. Agnew
Floriculture production	Assoc. Prof. D.S. Koranski
The effects of pre- and post-harvest practices on the quality of horticultural products	Assoc. Prof. R.J. Gladon
Landscape ornamentals	Ass. Prof. M.W. Hefley
Environmental factors: influence on the distribution and culture of ornamentals and related crops	Assoc. Prof. L.T. Midcap
Breeding and genetics of vegetable crops to improve disease resistance, nutrient content, and lower production risk	
Quality and nutritive value of processed potatoes	Assoc. Prof. W.L. Summers
Tissue culture methods and applications	Ass. Prof. L.C. Stephens

21 Muscatine

21.1 Muscatine Station
elevation : 170 m
rainfall : 855 mm
sunshine : 60%

Muscatine, Iowa 52761
Phone:(319) 2672-8787
Supt.: V.F. Lawson

Outlying station, state institution
Culture of potatoes, tomatoes, sweet corn, muskmelons, watermelons, and other vegetables involving irrigation in sandy soil in the Mississippi Valley.

22 Whiting

22.1 Whiting Station
elevation : 342 m
rainfall : 629 mm
sunshine : 60%

Whiting, Iowa 51063
Phone: (712) 458-2660
Supt.: D.A. DeBuchananne

Outlying station (Mo. Valley Hort, Res. Sta)
Culture of vegetable and small fruit crops suitable for western Iowa to be grown with irrigation.

NEBRASKA

23 Lincoln

23.1 University of Nebraska

Lincoln, Nebraska 68583

elevation : 350 m
rainfall : 730 mm
sunshine : 5606 h

Phone: (402) 472-2854
Dir.: R.G. Arnold, Vice Chancellor

Department of Horticulture	
Ornamentals	Head: Prof. R.D. Uhlinger
Floriculture	Assoc. Prof. J.B. Fitzgerald
Turf breeding	Assoc. Prof. T.P. Riordan
Turf management research	Assoc. Prof. R.C. Shearman
Native woody plants	Assoc. Prof. R.K. Sutton
Plant breeding and genetics, vegetable crops (beans, squash)	Prof. D.P. Coyne*/R.L. Austin
Floriculture	Ass. Prof. E.T. Paparozzi
Plant pysiology, plant-stress and turf research	Prof. E.J. Kinbacher
Ornamentals research	Assoc. Prof. S.S. Salac
Extension horticulturist	Prof. D.H. Steinegger
Climatology	Prof. R.E. Neild
Pecan evaluation	Assoc. Prof. W.A. Gustafson

24 Mead

24.1 University of Nebraska
Research and Development Center
elevation : 373 m
rainfall : 764 mm
sunshine : 5431 h

Mead, Nebraska 68045
Phone: (420) 624-5935
Dir.: W.W. Sahs, Supt.

25 North Platte

25.1 University of Nebraska
West Central Research and Extension Center
elevation : 846 m
rainfall : 498 mm
sunshine : 5782 h

North Platte, Nebraska 69101
Phone: (308) 531-3611
Dir.: L.J. Sumption

Ornamentals breeding propagation	Assoc. Prof. D.T. Lindgren

26 Scottsbluff

26.1 University of Nebraska
Panhandle Research and Extension Center
elevation : 2206 m
rainfall : 369 mm
sunshine : 5606 h

Scottsbluff, Nebraska 69361
Phone: (308) 632-1230
Dir.: R.D. Fritschen

Horticulture, research and extension service	Ass. Prof. D.S. Nuland
Potato, dry beans	Prof. R.B. O'Keefe

KANSAS

27 Chetopa

27.1 Pecan Experimental Field,
Horticulture Department Field
elevation : 274 m
rainfall : 889 mm

Chetopa, Kansas 67336
Phone: (316) 597-2972
Supt.: W.R. Reid

28 Colby

28.1 Colby Experiment Station
(KSU Branch Station - General Horticulture)
elevation : 275 m
rainfall : 473 mm

Colby, Kansas 67701
Phone: (913) 462-7575
Head: Dr. L. Robertson

29 Garden City

29.1 Garden City Experiment Station
(KSU Branch Station - General Horticulture)
elevation : 876 m
rainfall : 508 mm

Garden City, Kansas 67846
Phone: (316) 276-8286
Head: Dr. G. Herron

30 Wichita

30.1 Horticulture Research Center
(Horticulture Department Field)
elevation : 415 m
rainfall : 737 mm

Weichita, Kansas 672330
Phone: (316) 788-0492
Supt.: Dr. J.C. Pair
Dr. M.L. Allison*

31 Manhattan

31.1 KSU, Agricultural Experiment Station
Ashland Horticulture Farm - general horticulture
Rocky Ford Turfgrass Field
elevation : 314 m
rainfall : 762 mm

Manhattan, Kansas 66506
Phone: (193) 539-3991
Head: Dr. P.H. Jennings

State institution. Department of Horticulture

Fruit crops	Asst. Prof. Dr. E.W. Hellman
Growth regulators and herbicides	Prof. Dr. R.W. Campbell
Turfgrass science	Asst. Prof. Dr. J.L. Nus
Plant breeding and genetics, beans	Prof. Dr. C.D. Clayberg
Ornamental horticulture physiology	Asst. Prof. F.D. Gibbons
Vegetable crops physiology	Prof. Dr. J.K. Greig*
Stress physiology - ornamentals	Assoc. Prof. Dr. S.C. Wiest
Woody ornamentals	Assoc. Prof. Dr. H. Khatamian
Floriculture	Assoc. Prof. Dr. R.K. Kimmins
Horticultural therapy	Prof. Dr. R.H. Mattson
Fruit crop physiology	Asst. Prof. Dr. C.B. Rajashekar
Ornamental horticulure	
Extension horticulturist, ornamental horticulture	Assoc. Prof. L.D. Leuthold
Extension horticulturist, crop protection	Assoc. Prof. Dr. C.E. Long
Extension horticulturist, vegetable crops	Prof. Dr. C.W. Marr
Extension horticulturist, fruit and nut	Prof. Dr. F. D. Morrison
Extension horticulturist, landscape horticulture	Prof. Dr. G.A. Van der Hoeven

32 Tribune

32.1 Tribune Experiment Station
(KSU Branch Station - General Horticulture)
elevation : 1100 m
rainfall : 457 mm

Tribune, Kansas 67879
Phone: (316) 376-4761
Head: Dr. R. Gwin

33 DeSoto

33.1 Eastern Kansas Horticulture Field
- Vegetable Crops
elevation : 256 m
rainfall : 889 mm

DeSoto, Kansas 66018
Supt.: T. Schaplowsky

Prepared by A.A. Boe, Dept. of Horticulture, N.S.D.U., Fargo, North Dakota
Roger D. Uhlinger, Dept. of Horticulture, Univ. of Nebraska, Lincoln, Nebraska
Charles V. Hall, Dept. of Horticulture, Iowa State Univ., Ames, Iowa
H.F. Wilkins, Dept. of Horticultural Science and Landscape Architecture, Univ of Minnesota, St. Paul, Minnesota
Paul H. Jennings, Dept. of Horticulture, Kansas State Univ., Manhattan, Kansas

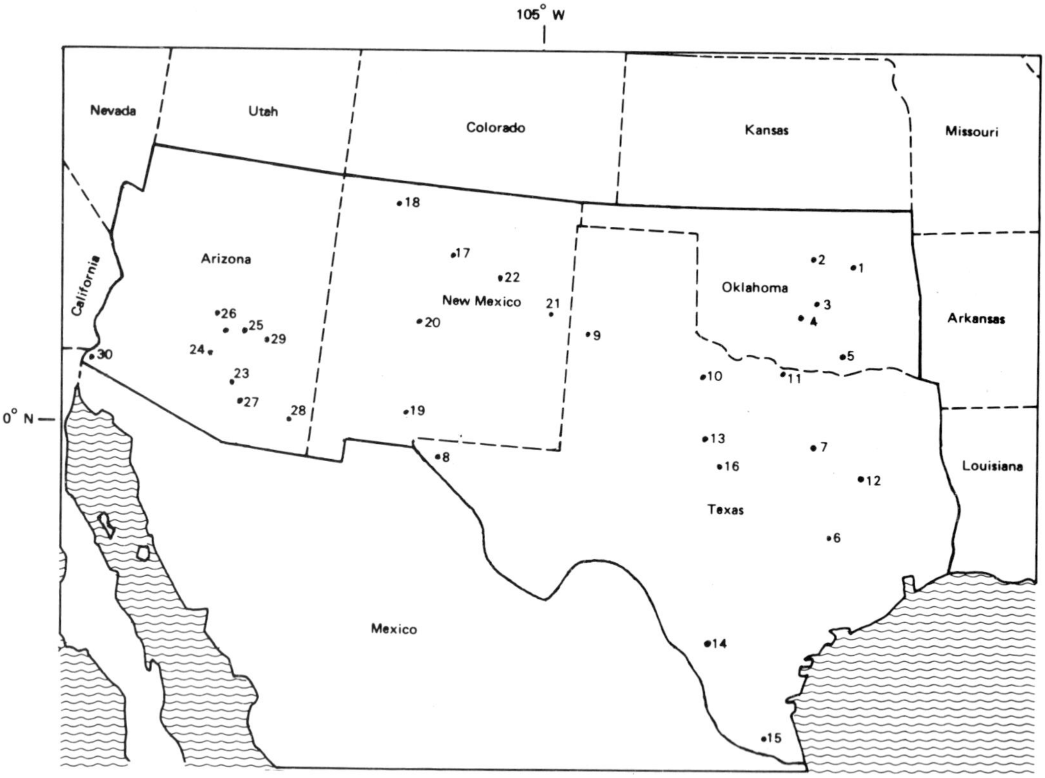
105° W
0° N
Nevada
Utah
Colorado
Kansas
Missouri
Arizona
California
New Mexico
Oklahoma
Arkansas
Louisiana
Texas
Mexico
1
2
3
4
5
6
7
8
9
10
11
12
13
14
15
16
17
18
19
20
21
22
23
24
25
26
27
28
29
30

Oklahoma: 1 Bixby, 2 Sparks, 3 Stillwater, 4 Perkins, 5 Lane, **Texas**: 6 College Station, 7 Dallas, 8 El Paso, 9 Lubbock, 10 Munday, 11 Montague, 12 Overton, 13 Stephenville, 14 Uvalde, 15 Weslaco, 16 Brownwood, **New Mexico**: 17 Alcalde, 18 Farmington, 19 Las Cruces, 20 Los Lunas, 21 Tucumcari, 22 Mora, **Arizona**: 23 Marana, 24 Maricopa, 25 Mesa, 26 Phoenix, 27 Tucson, 28 Safford, 29 Superior, 30 Yuma

Survey

OKLAHOMA

Although Oklahoma's agriculture is mainly animal and wheat production, horticulture does play an important role in the lives of the people of the state and makes up a significant part of the state's agriculture. The potential for horticulture production has not yet begun to be realized, but energy availability and cost may dictate an increase in the production of horticulture crops in Oklahoma in the future.

Ornamental horticulture is continuing to increase in the state and many nurseries are being started. Flower crop production is also expanding but not at the same rate as nursery stock production. The high incidence of winter sun, mild winter conditions, availability of energy sources and increasing population in this part of the "Sun Belt" will probably encourage even greater expansion of these industries.

Vegetable crops are also expanding and the future for new crops as well as those now grown looks very bright. Most vegetable crops grow well in Oklahoma and our climate allows for two plantings each year of many of these crops. The principal crops grown are watermelon, cantaloupe, green beans, okra, radishes, spinach, potatoes, sweet potato and squash. Fruit crops such as peaches, apples and strawberries are well established and acreage is increasing and there is interest in other fruit crops such as blueberries and blackberries. The transistion to "you-pick" operations has resulted in an increase in the acreages of most fruit crops.

The pecan is native to Oklahoma's creek and river bottoms. Many of these native groves have come under management and new pecan orchards are being planted. With the emphasis on proper tree management and the increase of pecan acreage, it appears that the future for the pecan industry is very good.

Variation in climate, topography and ecology of the state is great. Rainfall varies from 1524 mm in the Southeast to 380 mm in the Northwest. Elevations vary from 152 - 1520 m. Eastern sections of the state are hilly with much timber, creeks and river valleys. The western area is more flat, barren and arid while the middle portion of the state reflects the transistion from the East to the West.

Irrigation is used with most horticultural crops and there is plenty of irrigation available to soils that are well suited to the production of horticulture crops. We believe that there will be significant increases in production in the future.

TEXAS

Texas ranks third in agricultural production in the U.S., producing large quantities of agronomic crops. The state ranks third in all horticultural crops with vegetables being the most valuable at about $ 400 million per year. Onions, cabbage carrots, potatoes, watermelons and cantaloupes are the most valuable vegetable crops, respectively, returning over $ 175 million to growers annually. Major production areas include the lower Rio Grande Valley, the Winter Garden, the San Antonio area and the High Plains. The El Paso Valley is increasing in importance.

Total vegetable acreage has stabilized, trending toward crops that can be mechanically harvested. Citrus production with red fleshed grapefruit as a base leads other fruit crops. Pecans are the second most important fruit crop with rapid expansion of improved cultivar acreage. Total production may exceed 75 million pounds in favourable crop years.

Total peach acreage is increasing with high income per acre resulting from retail sales by the grower. Blueberry acreage is also increasing. Several hundred acres of wine grapes have been planted in recent years with the construction of fourteen wineries. There is limited production of apples, pears, plums, strawberries, blackberries, and figs.

Grower returns for ornamentals and turf is estimated at over $ 300 million and is increasing. There has been an influx of greenhouse and nursery producers into Texas, along with increased acreage by long time producers. Nursery stocks continue to shift from the field to containers. Rose bush production remains viable with annual shipments of 10 - 15 million bushes.

The climate and soils of Texas are extremely varied. Rainfall ranges from less than 153 mm annually in west Texas to 1397 mm in east Texas. Temperature ranges from areas of no frost to 23°F below zero. Vegetables or fruits are available in every month of the year from one or more areas of the state and some 30 different vegetables are produced commercially.

Water for irrigation is plentiful in the eastern half of the state and is sufficient in large areas of west Texas for use on high-cash horticultural crops.

NEW MEXICO

New Mexico is primarily a livestock-production state. Cotton and alfalfa hay are important economic crops. Horticultural income in 1983 amounted to $ 137 million which is about 14% of total agricultural income to this state.

New Mexico contains 315,115 km^2. The topography is varied with elevations ranging from 840 in the lower Sonoran Desert to 3900 m in the mountains. Some valleys are large enough to make agricultural operations possible with irrigation. New Mexico is semi-arid with an annual precipitation of about 250 mm. Dona Ana County has the largest income from horticultural enterprises. The largest pecan orchard (1600 ha) is located here. Chile and onions are the important vegetable crops in the state. Lettuce acreage continues to decline because of competition. Generally, trends in the state are toward dry land agriculture in eastern NM and more intensive agriculture such as horticulture crop production. Recently, approximately 1600 ha of grapes were planted and a wine industry is developing. Apple production continues to decline in northern mountain valleys. However, potential in southern valleys is very promising.

Organized horticultural research in Mexico began in 1889. New Mexico State University has its origin in 1888, 23 years before New Mexico became the 47th state. In addition to the main station located in Las Cruces, the University operates 7 Agricultural Science Centers throughout the state. Four of these science centers at Alcalde, Los Lunas, Farmington and Tucumcari have horticulturists on staff. The Mora Research Station investigates the potential for Christmas tree production and other woody ornamental crops for northern New Mexico. State Department of Agriculture established under the Board of Regents at New Mexico State University has the function of protecting and maintaining the agriculture state of New Mexico.

ARIZONA

The southeastern one-third and two-thirds of the upper half of the state consist of undulating plains, steep ridges and high peaks. Virtually the entire state is in the drainage basin of the Colorado River. The principal gorge is the world famed Grand Canyon. The temperature range between day and night is extreme, especially at elevations above 1524 m where the nights are cool and comfortable. The southern section of Arizona has a delightful winter season which has made the area a Mecca for winter tourists,

retired persons, and health seekers. At lower elevations in the Gila and Colorado River Valleys, a few locations often escape killing frosts for two to three years in succession. Most of the vegetable crops are grown in the southeastern part of the state. In this area, rainfall generally averages about 127 m a year with a few areas receiving as much as 254 m.
Early spring lettuce, cantaloupes, green onions, cabbage and miscellaneous vegetables are grown. More labor-saving equipment will be in the fields soon for all vegetable crops. The principal vegetable producing areas are Yuma and Maricopa Counties, with smaller plantings in the Cochise, Pima, Pinal and LaPaz Counties. Citrus (grapefruit, oranges and lemons) is produced only in Maricopa and Yuma Counties. Ornamental horticulture is increasing in importance and contributes to the state's agricultural income. Nursery crops account for 91 percent of all income attributed to ornamental horticulture. Cut flowers is an important segment of ornamental horticulture. The ornamental industry is centered in Maricopa, Yuma, and Pima Counties.

OKLAHOMA

1 Bixby

1.1 Vegetable Research Station, Substation
State Institution

Bixby, Oklahoma 74008
Supt.: B. Bostian

2 Sparks

2.1 Pecan Research Station, Substation
State Institution

Sparks, Oklahoma 74869
Supt.: H.L. Davis

3 Stillwater

3.1 Oklahoma Agricultural Experiment Station
Main Station, State Institution

Stillwater, Oklahoma 74074
Dir.: Dr. C.B. Browning

Department of Plant Pathology	Head: Dr. L. Littlefield
Ornamental diseases	
Fruit and vegetable diseases	Dr. K. Conway
Department of Entomology	Dr. L. Crowder
Pecan insects	Dr. R.D. Eikenbary
Department of Agricultural Engineering	Dr. D. Thompson
Horticulture crop production	Dr. L.F. Roth/Dr. M.D. Paine,
	Dr. D.P. Schwab/Dr. A.D. Barefoot
Department of Horticulture	Head: Dr. D. Buchanan
Floriculture	Dr. R.N. Payne
Woody ornamentals	Dr. C.E. Whitcomb
Vegetable evaluation, culture	Dr. J.E. Motes/Dr. R.E. Campbell
Tree fruits and pecans, valuation, production, culture	Dr. M.W. Smith

4 Perkins

4.1 Perkins Fruit Research Station, Substation

Perkins, Oklahoma 74059
Foreman: K. Karner

5 Lane

5.1 Southeastern Oklahoma Agricultural Research and
Extension Center
Fruits and vegetables

Lane, Oklahoma

TEXAS

6 College Station

6.1 Texas Agricultural Experiment Station
elevation : 96 m
rainfall : 889 mm
sunshine : 3550 h

College Station, Texas 77843
Phone: (409) 845-8484
Dir.: Dr. N.P. Clarke

Located on the campus of Texas A&M University, this facility consists of greenhouses, field plots and orchards where production, breeding and physiology research on ornamentals, fruits and vegetables takes place.

Department of Horticulture	Head: H.G. Vest, Ph.D
Post-harvest physiology	C. Andersen, Ph.D
Horticulture business management	C. Townsend, Ph.D
Socio-horticulture	J. Novak, Ph.D
Pomology:	
Fruit breeding, physiology and irrigation	D.H. Byrne, Ph.D
Pecan nutrition, growth regulators, irrigation	J.B. Storey, Ph.D*
Floriculture:	
Propagation, physiology of nursery crops	F.T. Davies, Ph.D
Nutrition, physiology and tissue culture	J. Frett, Ph.D
Floral design	J. Johnson, M.Sc.
Foliage plants, native species, physiology and taxonomy	E. McWilliams, Ph.D
Growth regulators, nutrition, greenhouse crop production	A.E. Nightingale, Ph.D
Physiology, permeability and nutrition	D.W. Reed, Ph.D
Teaching greenhouses	Mngr: M. Sweatt, M.Sc.
	Ass.Mngr.: Lisa Lipscomb
Vegetables:	
Breeding of onions, carrots, cucumbers	L. Pike, Ph.D
Breeding of potatoes and legumes	J.C. Miller, Jr., Ph.D*
Processing of fruits and vegetables	E.E. Burns, Ph.D
Stress physiology	G. Cobb, Ph.D

6.2 Texas Agricultural Extension Service

College Station, Texas 77843
Phone: (409) 845-7898
Dir.: Dr. Z. Carpenter

Project Leader	B.G. Hancock, M.Sc.
State wide extension: pecan specialist,	G.R. McEachern, Ph.D*
fruit specialist,	C. Lyons, Ph.D
vegetables specialists,	S. Cotner, Ph.D/T. Longbrake, M.Sc.
	D. McCraw, Ph.D
landscape specialist,	W.C. Welch, Ph.D
food technology specialist,	A. Wagner, Ph.D
greenhouse specialist	D. Wilkerson,Ph.D

7 Dallas

7.1 Texas A&M University Research and Extension Center
elevation : 214 m
rainfall : 748 mm

17360 Coit Rd., Dallas, Texas 75252
Phone: (214) 231-5362
Dir.: Dr. J. Reinert

This station located in the urban area of Dallas, Texas provides the opportunity for urban horticulture research primarily on landscape plant materials both native and domesticated.

Landscape plants: propagation, physiology,	D. Morgan, Ph.D
Native ornamental plants	B. Simpson, M.Sc.

Ornamentals: area extension specialist — M.L. Baker, M.Sc.*

8 El Paso

8.1 Texas A&M University Research Center
elevation : 1109 m
rainfall : 196 mm

1380 A&M Circle, El Paso,
Texas 79927
Phone: (915) 859-9111
Dir.: H. Malstrom

At this center, research on production of vegetables and pecan crops with saline water is studied. The use of native plants for landscapes is researched to reduce water use.

Pomology: nutrition, growth regulators, irrigation	R. Marquard, Ph.D
Vegetables: physiology and cultural practices	R.M. Taylor, Ph.D
Landscape plants: native ornamentals	J. Tipton, Ph.D
Pomology: area extension specialist (pecans and grapes)	S. Helmers, Ph.D
Vegetables: area extension specialist	W. Peavy, Ph.D

9 Lubbock

9.1 Texas A&M Agricultural Research and Extension Center
elevation : 1026 m
rainfall : 457 mm

Rt. 3, Lubbock, Texas 79401
Phone: (806) 746-7101
Dir.: Dr. J. Abernathy

Located in the High Plains of Texas, this center is the site for vegetable and grape research with an effort to conserve diminishing water supplies while maintaining high yields of quality produce.

Pomology: grape culture	W. Lipe, Ph.D
Vegetables: culture and physiology	D. Bender, Ph.D
Area extension specialist	R. Roberts, Ph.D

10 Munday

10.1 Texas A&M University Vegetable Research Station
elevation : 442 m
rainfall : 607 mm

Rt. 2 Box 2E, Munday, Texas 76371
Phone: (817) 442-4531
Dir.: Dr. D. Nagel

Vegetable: culture and physiology — P. Richwine, M.Sc.

11 Montague

11.1 Texas A&M University Res-Demonstrating Center
elevation : 355 m
rainfall : 850 mm

Montague, Texas 76251
Phone: (817) 894-2906

This station functions for fruit production and breeding research

12 Overton

12.1 Texas A&M University Research and Extension Center
elevation : 122 m
rainfall : 1130 mm

Drawer E, Overton, Texas 75684
Phone: (214) 834-6191
Dir.: Dr. C. Long

Located in East Texas, this station is used for vegetable, rose and fruit research.

Pomology: breeding, physiology	K. Patten, Ph.D
Vegetables: physiology, post-harvest physiology, breeding	D. Paterson, Ph.D
Rose culture: production, propagation	B. Pemberton, Ph.D
Vegetables: area extension specialist	T. Menges, Ph.D
Ornamentals: area landscape horticulturist	D.S. Hall, Ph.D

13 Stephenville

13.1 Texas A&M University of Agricultural Research and Extension Center
elevation : 402 m
rainfall : 711 mm

Box 292, Stephenville, Texas 76401
Phone: (817) 968-4144
Dir.: J.S. Newman, M.Sc.

Located in the rolling hills of Texas, research at this center concentrates on peach and pecan culture, and physiology.

Pomology : culture and physiology	J. Worthington, Ph.D
Horticulture extension specialist	L. Stein, Ph.D

14 Uvalde

14.1 Texas A&M University of Agricultural Research and Extension Center
elevation : 279 m
rainfall : 607 mm

PO Drawer 1050, Uvalde, Texas 78801
Phone: (512) 278-9151
Dir.: Dr. W. Holloway

This station is located in the Winter Garden area of Texas, an area of significant vegetable production. Research at the center concentrates on vegetable production and disease control.

Vegetable: culture and physiology	F. Dainello, Ph.D
Pomology: area extension specialist	L.W. Shreve, Ph.D

15 Weslaco

15.1 Texas A&M University of Agricultural Research and Extension Center
elevation : 160 m
rainfall : 559 m

2415 East Hwy. 83 Weslaco, Texas 78596
Phone: (512) 968-5585
Dir.: Dr. C. Connolly

Located in the Lower Rio Grande Valley, a rich horticulture production area, research includes vegetable breeding and physiology, ornamental production and citrus rootstocks. All aspects of research include insect and disease control.

Ornamentals: foliage plant culture and physiology	Y.T. Wang, Ph.D
Vegetables: breeding of lettuce, onions and tomatoes	P. Leeper, M.Sc.
Vegetables: culture and physiology	R. Dufault, Ph.D
Vegetables: plant pathology	M. Miller, Ph.D
Vegetables: breeding of disease resistant pepper	B. Villalon, Ph.D
Pomology: citrus production, rootstocks	R. Rouse, Ph.D
Vegetable extension specialist	T. Hartz, Ph.D

Citrus and ornamentals, extension specialist — J. Sauls, Ph.D

16 Brownwood

16.1 U.S. Department of Agriculture, Pecan Field Station
elevation : 409 m
rainfall : 655 mm

Box 579, Brownwood, Texas 76801
Phone: (915) 646-0593
Off-in-ch.: Dr. R. Hunter

Plant pathology — R. Hunter, Ph.D
Genetics — T. Thompson, Ph.D
N-pacts — F. Young, Ph.D

NEW MEXICO

17 Alcalde

17.1 Agricultural Science Center
elevation : 1704 m
rainfall : 234 mm

Alcade, New Mexico 87511
Phone: (505) 852-4241
Supt.: F.B. Matta

State institution for research on agronomic, fruit, vegetable, and Christmas tree crops for small farms. Crops grown under irrigation on 7 ha include apples, grapes, beans, dry flowers and Christmas trees.

18 Farmington

18.1 Agricultural Science Center
elevation : 1615 m
rainfall : 190 m

Farmington, New Mexico 87401
Phone: (505) 327-7757
Supt: J. Gregory

State Institution with 54.6 ha under irrigation. Research focuses on nutrition and varietal adaptability of fruit and vegetables.

19 Las Cruces

19.1 New Mexico State University, New Mexico Agricultural Experiment Station, Main Station
elevation : 1050 m
rainfall : 200 mm

University Park, Las Cruces
New Mexico 88003
Phone: (505) 646-1999

State institution on the campus of New Mexico State University. Research is conducted on two experimental farms devoted to traditional agriculture (80 ha) and ornamental horticulture (20 ha). Greenhouse facilities include a 1080 m^2 geothermally heated facility. Research focuses on onions, chile, tissue culture, Christmas trees and new plant introductions.

Department of Crop and Soil Sciences — Head: open
Soils — Dr. B. McCaslin
Turf and turf genetics — Dr. A.A. Baltensperger
Department of Entomology and Plant Pathology — Head: J. Owens
General entomology — Dr. M. English
Fruit and ornamental insects — Dr. E. Shannon
Fruit and ornamental diseases — Dr. D. Lindsey
Department of Horticulture — Head: J.G. Mexal
Ornamental horticulture, propagation and morphology — Dr. F.B. Widomyer

Vegetable physiology and culture plant environments, greenhouse-vegetables, production, plant-water requirements	Dr. D.J. Cotter
Post-harvest physiology, maturity, fruit pecan culture	Dr. E. Herrera
Vegetable physiology, vegetable-variety trials, dormancy and rest in woody plants, growth regulators	Dr. J.N. Corgan
Tissue culture	Dr. G. Phillips
Plant breeding, chile	
Nursery management	
Landscape design	
Christmas tree physiology	Dr. J.T. Fisher

20 Los Lunas

20.1 Agricultural Science Center
elevation : 1475 m
rainfall : 207 m

Route 1, Box 28, Los Lunas
New Mexico 87031
Supt: Dr. R. Hooks

State institution with 80 ha under irrigation. Research focuses on native plant introductions, peanut breeding, salt tolerance, and variety adaptability.

21 Tucumcari

21.1 Agricultural Science Center
elevation : 1249 m
rainfall : 396 mm

Northeastern Branch Station,
PO Box 689, Tucumcari,
New Mexico 88401
Supt.: R. Kirksey

State institution with research focused on vegetable production. Center consists of 188 ha with 53 ha under irrigation.

22 Mora

22.1 Mora Research Center
elevation : 2225 m
rainfall : 584 mm

Box 357, Mora, New Mexico 87732
Supt.: G. Fancher

State institution with 21 ha under irrigation	
Christmas tree production and management	G. Fancher
Tree improvement	R. Neumann

ARIZONA

23 Marana

23.1 Marana Agricultural Center
elevation : 597.4 m
rainfall : 279 mm

Marana, Arizona 85238
Phone: (602) 682-38722

24 Maricopa

24.1 Maricopa Agricultural Center
elevation : 353.6 m
rainfall : 187 mm
humidity : 48, 27%

Maricopa, Arizona 85329
Phone: (602) 437-1366

25 Mesa

25.1 Mesa Agricultural Center, State Institution
Mesa Branch Station
elevation : 374.9 m
rainfall : 191 mm
sunshine : 86%
humidity : 49, 29%

Mesa, Arizona 85201
Phone: (602) 964-1725

Post harvest physiology	R.E. Foster II, Ph.D
Vegetable breeding	J.M. Nelson

26 Phoenix

26.1 Desert Botanical Garden of Arizona

Box 547, Tempe, Arizona 85281

Private institution, founded in 1928, 150 acres. Temperature range 20-125°F, cacti and leaf succulents. Cacti, desert trees and shrubs.

27 Tucson

27.1 Tucson Campus Agricultural Center
University of Arizona, Arizona Agricultural
Experiment Station, Main Station
elevation : 710.2 m
rainfall : 266 mm
sunshine : 86%
humidity : 48, 32%

Tucson, Arizona 85721
Phone: (602) 621-7201
Dir.: L.W. Dewhirst, Ph.D

State institution	
Department of Plant Sciences	Head: B.B. Taylor, Ph.D
Extension: vegetables	P.M. Bessey, Ph.D*
Horticultural research	M.H. Jensen, Ph.D
Landscape horticulture	D.A. Palzkill, Ph.D*
Physiology of vegetables	J.M. Kobriger, Ph.D
Extension: vegetables	N.F. Oebker, Ph.D*
Extension: fruit/nut crops	M.W. Kilby, Ph.D
Nursery and greenhouse management	L. Hogan, Ph.D*
Physiology fruit and nut crops	J.W. Moon, Ph.D
Extension landscape	C.M. Sacamano, Ph.D
Weed control	K.C. Hamilton, Ph.D/S. Heathman, M.Sc
Department of Entomology	Head: W.S. Bowers
Taxonomy of insects	
Department of Plant Pathology	Head: M. Nelson, Ph.D
Diseases of ornamental crops	S.M. Alcorn, Ph.D
Soil-borne diseases	
Forest-tree diseases	
Virus diseases	

28 Safford

28.1 Safford Agricultural Center
elevation : 883.9 m
rainfall : 214 m
humidity : 48, 36%

Safford, Arizona 85546
Phone: (602) 428-2432

29 Superior

29.1 Boyce Thompson Southwestern Arboretum

Superior, Arizona 85273
Phone: (602) 689-2811
Dir.: W.R. Feldman, Ph.D

Private institution, founded in 1928, consists of 900 acres of which 30 are in the arboretum. Introduction of plants to the more arid parts of the Southwest: *acacia, atriplex, callistemon, condalia, eucalyptus, melalecua, pistacia, phus, anacardiacae cactacae, leguminoseae and myrtacea.*

30 Yuma

30.1 Yuma Valley and Yuma Mesa
Branch Station
elevation : 58.2 m
rainfall : 70.4 mm
sunshine : 91%
humidity : 54, 26%

Yuma, Arizona 85364
Phone: (602) 782-3836

Citrus diseases

E.L. Nigh, Ph.D

30.2 Yuma Valley Agricultural Center
elevation : 36.6 m
rainfall : 65.3 mm
sunshine : 91%
humidity : 54, 26%

State instution. Vegetables and citrus research. Vegetable production.

Prepared by Grant Vest, Dept. of Horticulture, Texas A & M Univ., College Station, Texas
J.G. Mexal, Dept. of Horticulture, New Mexico State Univ., Las Cruces, New Mexico
Dave W. Buchanan, Dept. of Horticulture and Landscape Architecture, Oklahoma State Univ., Stillwater, Oklahoma
B.B. Taylor, Dept. of Plant Sciences, Univ. of Arizona, Tucson, Arizona

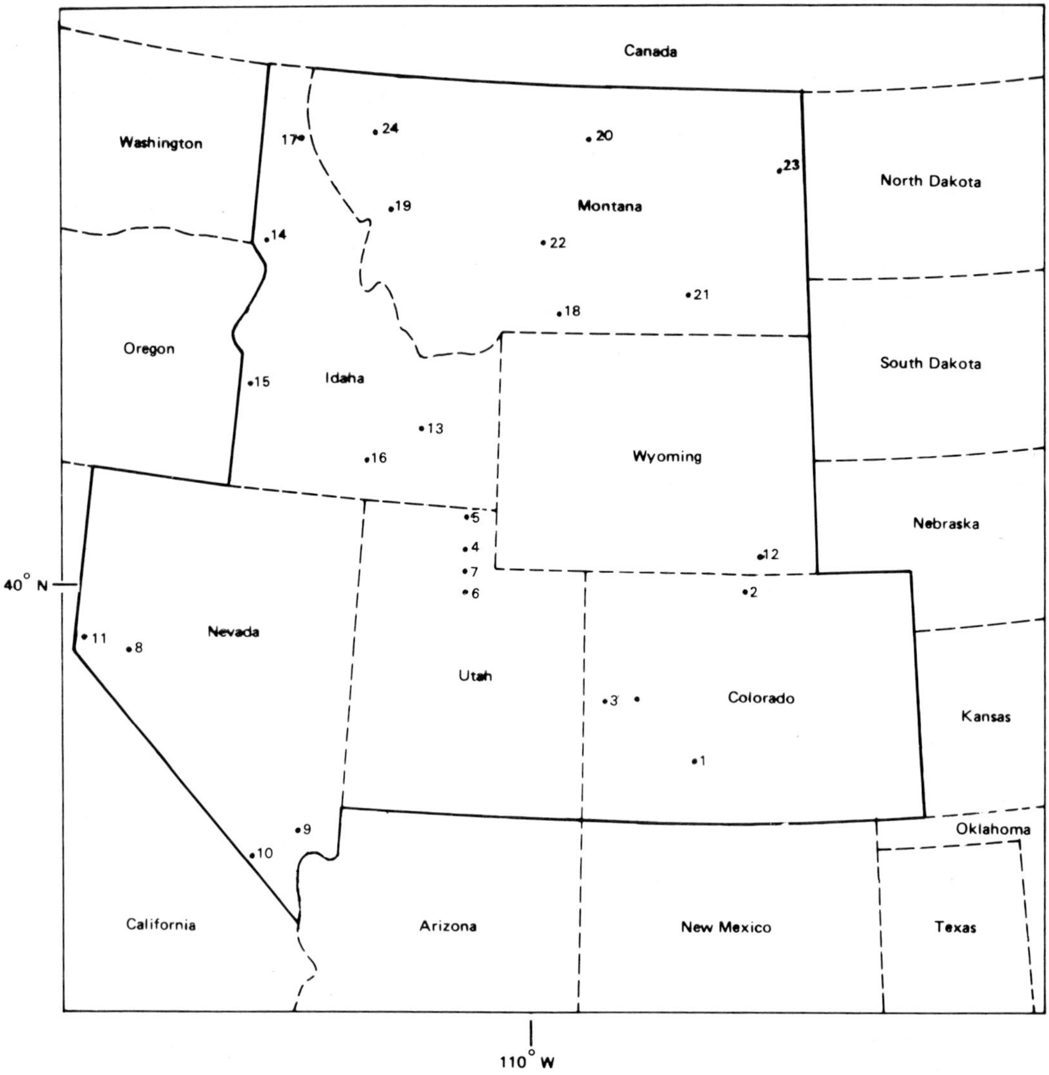
Canada
Washington
Montana
North Dakota
Oregon
Idaho
South Dakota
Wyoming
Nebraska
40° N
Nevada
Utah
Colorado
Kansas
Oklahoma
California
Arizona
New Mexico
Texas
110° W

Colorado: 1 Center, 2 Fort Collins, 3 Grand Junction, **Utah**: 4 Farmington, 5 Logan, 6 Provo, 7 Salt Lake City, **Nevada**: 8 Fallon, 9 Logandale, 10 Pahrump, 11 Reno, **Wyoming**: 12 Laramie, **Idaho**: 13 Aberdeen, 14 Moscow, 15 Parma, 16 Kimberley, 17 Sandpoint, **Montana**: 18 Bozeman, 19 Corvallis, 20 Havre, 21 Huntley, 22 Moccasin, 23 Sidney, 24 Creston

Survey

COLORADO

Horticultural crops produced commercially in Colorado, return about 145 million dollars annually to growers. It is of direct importance to the ecomonical life of Colorado by the contribution to transportation, chemical supplies, containers, power equipment and labor used in the growing and marketing of the crops. The potato crop dominates the horticultural crops accounting for about 50 percent, fresh and processed vegetable crops 29 percent, fruit crops (apples, peaches, pears, cherries, etc.) 8 percent, under-glass flower crops 12 percent, flower and vegetable seeds, 1 percent of the total dollar income to the state. There is expansion in the field of ornamental horticultural plants including trees, shrubs and outdoor flowering plants and in turf grass. Colorado's climate is modified by the Rocky Mountain range which cuts down through the center of the state. It has the highest mean elevation of any state and crops are grown at elevations which vary from 1400 to 3070 m above sea level. The major crops are grown in areas which in general have low average precipitation, low humidities, intense sunlight and few cloudy periods. The major fruit, potato and fresh vegetable crops are produced in districts with an annual total percipitation of 203 - 457 mm. The apple crop is grown at elevations from 1435 to 2070 m above sea level while the major potato growing area is located at an elevation of 2540 m above sea level.

The development of irrigation in Colorado and the southwest goes back to the Pueblo Indians and irrigation works probably were built between 1200 and 1400 A.D. The Spanish colonists, according to authentic records, were issued court decrees for irrigation rights on Colorado streams in 1852, near the town of San Luis, Colorado. Since that time Colorado has developed a network of irrigation canals and storage reservoirs which has set not only the legal pattern on water usage from rivers but determined the history and development of horticulture in Colorado. The history of research in horticulutre goes back to date of the establishment of Colorado State University in 1870 although the first official research report was not made until 1883.

The agricultural experiment station was officially established in 1888 following the passage of the Federal Hatch Act. Colorado Agricultural Experiment Station now operates 10 research centers. The Arkansas Valley Research Center has been in continuous operation since 1888 and the research work there pioneered in the development of the western cantaloupe industry. Most of the horticulture research conducted there now involves the onion crop.

UTAH

The horticulture industry in Utah was started the first day the Mormon pioneers reached what is now Salt Lake City in 1847. Five acres of potatoes were planted the day of arrival. Fruit-tree seed and scions were also brought by these pioneers with standard varieties being planted in 1850. Most of the cultivated lands of the state are at an elevation of between 1400 and 1850 m. Only the Dixie area in the southwestern corner of the state is lower in elevation, averaging under 1000 m.

The most important agricultural areas for crop production have a growing season of from 120 to 150 days. The climate is dry with relatively little fog and wind and not many cloudy days. The days are clear, bright and warm, and the nights are relatively cool. Only 7 percent of the land area receives more than 500 mm of precipitation a year. About 41 percent receives from 250 to 500 mm, and 6 percent less than 150 mm annually. Although the seasonal distribution of precipitation within the state varies, in the major crop producing area, which is the north-central part of the state, the maximum percipitation comes in the winter and spring and the minimum in the summer months. The fruit and vegetable plantings are concentrated in this area which includes Utah, Box Elder, Davis, Weber, Salt Lake and Cache counties. Irrigation water is usually plentiful in this area. The water is impounded in mountain reservoirs or diverted from streams to the fields in canals. Principal vegetable crops grown in Utah are tomatoes, potatoes, snap beans, dry onions, bunching onions, peppers, cucumbers, sweet corn, lettuce, squash, peas, celery, eggplant, carrots, beets, parsley, cauliflower and cabbage. Principal fruit crops in Utah are apples, peaches, sweet and sour cherries and pears. There are also some apricots and raspberries.

NEVADA

Around the turn of the century the state of Nevada boasted a thriving pomological industry. Shortly therafter, with the onset of a mining depression, the industry gradually failed and horticultural research was dormant until 1960 when interest revived. Small horticultural greenhouse businesses struggled for several years but likewise have been unsuccessful. Locally, strong interests in sod production and retail ornamentals have developed. The major horticultural crops in western Nevada are potatoes, cantaloupe, onions and garlic; in southern Nevada is fall grown lettuce at Pahrump, with experimental plantings of grapes and nuts. Production in the remainder of the state is negligible. Nevada is mostly high desert with numerous mountain ranges. Western Nevada valleys have elevations of 1370 m, an average annual rainfall of 200 mm and an average growing season of 130 days. Daytime temperatures in winter are in the mid forties (°F) and summer average 90°F. Winter lows may reach -16°F, but more commonly are in the low 20's (°F). North central and eastern desert valleys are higher in elevation, have shorter growing seasons and harsher climates.

Southern Nevada valleys are 365 to 550 m in elevation, average 100 mm annual precipitation, and are extremely hot in summer (100°F/80°F day-night) and cold in winter (50°F/30°F day-night). The growing season averages about 250 days.

Soils range from sandy loam to silty clay loams throughout the state with extremes of sandy to heavy clay. Most have a high pH (7.4 to 8.2) and many may be sodic and/or saline.

WYOMING

Horticultural enterprises in Wyoming are small in number because of the low population density and because most of the state is grassland and forest. Vegetables and small fruits for local market are grown on a few farms throughout the state. Greenhouse, nursery and sod production operations are scattered and small in number. Twenty-eight thousand acres of dry beans are grown annually on irrigated land. Yield is about 500,000 cwt. valued at $ 9 million. Potatoes are grown on 3,000 irrigated acres with a yield of about 850,000 cwt.

IDAHO

The principal horticultural crop in Idaho is the potato. Potatoes are grown principally in the Snake River and areas all across southern Idaho. Other vegetable crops of importance include onions, mostly sweet Spanish, sweet corn and lima beans for processing, green peas and spinach for freezing. Food processing is Idaho's most rapidly growing industry. Recently there has been an increased interest in the production of snap beans for processing. The production of greenhouse tomatoes is increasing. A very active vegetable seed industry has been established in Idaho for a long time. Idaho produces the major proportion of the following vegetable seed for the United States: green beans, onions, leaf lettuce, carrots, sweet corn and peas. Lesser amounts of radish, turnip, parsnip and other minor vegetables are also grown for seed-production purposes. Apples, plums, sweet cherries and peaches are the leading fruit crops. These are grown principally in the Boise and Payette River Valleys in southwestern Idaho and are sold for fresh market. There is a limited production of pears, apricots, cane fruits and strawberries. A flower seed industry is developing and a large nursery stock industry in northern Idaho.

MONTANA

Since Montana is pre-eminently a cereal grain and livestock producing state, traditional horticultural enterprises accounts for only about 2.5 percent of the total agricultural income. The best estimate of the contribution of horticultural enterprise to the economy of Montana is that the annual value of horticultural endeavor is between $ 12 million and $ 15 million.

Horticulture in Montana is characterized by a modest assemblage of small business, engaged in intensive agriculture with varying degrees of specilization. Some horticultural enterprises, such as commercial floriculture and home gardening, are rather generally distributed throughout the state; others, such as sweet and sour cherry production, canning crops, potato growing, and Christmas-tree production are concentrated in relatively limited areas as a result of climatic factors, availability of irrigation, or association with centers of population, as with retail florists' shops and garden centers. Recently, increased emphasis is being placed on environmental horticulture, especially as it is related to realization of the potential inherent in outdoor recreation in sparsely populated, but space-, water-, and scenery-rich Montana.

Tourism ranks high among Montana enterprises as a source of income. Both public and private agencies are augmenting their efforts in this area. Increasing attention is being given also to environmental aspects of urban horticulture. Potatoes are the most extensively grown horticultural crop in Montana, with about 8000 acres devoted to this crop. The greatest concentration of production is in western Montana. Somewhat more than half the acreage is devoted to production of certified seedstock, with 'Russett Burbank' being the most important cultivar. Vegetables are grown mostly in home gardens, or occasionally for roadside sales, with sweet corn and cucumbers popular items for the latter purpose. One vegetable cannery, producing peas and carrots, is in operation in Carbon County.

Commercial floriculture is dispersed in comparatively small units throughout the state, but the greatest concentration of production is in Cascade County. Bedding plants are important in the profit picture, but pot plants are grown both year-around and seasonally, depending on the species. Production of such cut flowers as roses and carnations has declined. A start has been made in production of greenhouse tomatoes, encouraged by the poor quality of greenwraps shipped in during the winter. Most nursery products merchandized in Montana are produced elsewhere, but there is a growing demand for nursery stock of native plants for use on natural sites, highway embankments, and for revegetation of disturbed areas such as those resulting from strip-mining activities. A modest amount of turf is being produced. Christmas trees, principally Scots pine and Douglas-fir, are being produced primarily in Western Montana. Wild-harvested Christmas trees are declining in importance. Planning, planting and maintenance of outdoor recreational areas is assuming increasing importance. Sweet-cherry production is limited essentially to the vicinity of Flathead Lake in northwestern Montana, where 1200-1400 acres are devoted principally to production of the cultivar 'Lambert'. Principal market outlet is as fresh fruit maturing in late July and early August. Pollenizer varieties are commonly brined for production of maraschinos and other processed products. Low winter temperatures and rain near harvest are the principal climatic hazards. Commercial sour-cherry production in Montana is restricterd essentially to the Bitterroot Valley of western Montana, where a vertically integrated cannery is in operation. About 300 acres are devoted to production of 'Montmorency' as the principal cultivar.

COLORADO

1 Center

1.1 San Luis Valley Research Center
elevation : 1951 m
rainfall : 175 mm
sunshine : 3339 h

Center, Colorado 81125
Phone: (303) 754-3594
Supt.: D.G. Holm, Ph.D.

State institution

Potato cultivar development, fertility, irrigation management, seed improvement and certification	R.D. Davidson, M.S.
High-altitude vegetable crops physiology	M.K. Thornton, M.S.

2 Fort Collins

2.1 Colorado State University, Agricultural Experiment Station, Main Station
State Institution

Fort Collins, Colorado 80523
Phone: (303) 491-5371
Dir.: Dr. R.D. Heil
Dep.Dir.: Dean M.H. Niehaus

2.2 Colorado State University, Main Station
elevation : 1270 m
rainfall : 372 mm
sunshine : 2987 h

Fort Collins, Colorado 80523
Phone: (303) 491-7019

State institution

Department of Horticulture	Head: K.M. Brink, Ph.D.
Research problems of turf grasses and winter hardiness	Dr. J.D. Butler
Vegetable-crop, computer application research	Dr. C.W. Basham
Plant-physiology research on problems of vegetable crops, cultural problems related to mechanical production and harvesting, variety evaluation	Dr. J.E. Ells
Genetics and plant breeding: small fruits	Dr. H.G. Hughes
Floriculture: carnations, bedding and pot plants, environmental research, CO_2, light and temperature effects	Dr. K.L. Goldsberry
Floriculture: physiological research with carnations, roses, foliage plants	
Floriculture: carnation, rose research irrigation regimes and soil-factor problems, physiology, computerization	Dr. J.J. Hanan
Post-harvest physiology: potatoes and fruit	Dr. M. Workman
Ornamentals and nursery management	Dr. J.E. Klett
Potato research and certification	Dr. K.W. Knutson
Vegetables, high altitude research, computer modeling	Dr. F.D. Moore
Landscape design and construction	Prof. G.W. Reid/Prof. A.E. Simeoni
Plant physiology	Dr. E.E. Roos/Dr. L.E. Towill, USDA
Stress physiology	Dr. S.J. Wallner
Landscape plants	Dr. J.R. Feucht (Denver)

3 Grand Junction

3.1 Orchard Mesa Research Center
elevation : 1209 m
rainfall : 210 mm
sunshine : 3066 h

Grand Junction, Colorado 81501
Phone: (303) 434-3264
Supt.: A.R. Renquist, Ph.D.

State institution. Fruit, rootstock research, production management and environmental control.

Fruit insects	A.D. Bulla, B.S.
Fruit diseases	

3.2 Hotchkiss, Rogers Mesa Research Center . Hotchkiss, Colorado 81419
elevation : 1359 m Phone: (303) 872-3387
rainfall : 279 mm Supt.: K.S. Yu, Ph.D.
sunshine : 2935 h

State institution
Fruit performance; testing rootstocks and production management. M.K. Rogoyski, Ph.D.

UTAH

4 Farmington

4.1 Horticultural Research Station . Farmington, Utah 84025
elevation : 1450 m Phone: (801) 451-2763
rainfall : 508 mm Supt.: C. Thompson
growing season : 161 days

State institution. Vegetables, ornamentals, small-fruit culture and stone-fruit virus research.

5 Logan

5.1 Utah Agricultural Experiment Station, Main Station . Logan, Utah 84322
elevation : 1500 m Phone: (801) 750-2215
rainfall : 440 mm Dir.: D.J. Matthews
growing season : 159 days

State institution

Department of Agricultural Economics	
Fruit marketing	Dr. L.H. Davis
Department of soils and Meteorology	
Meteorology, temperature and frost resistance on stone fruit	Prof. A. Richardson/Dr. G.L. Ashcroft
Department of Biology	Head: J.A. MacMahon
Fluorine damage and iron chlorosis (enzyme studies)	Dr. G.W. Miller
Insects on fruit trees	Dr. D.W. Davis
Fruit, vegetable and ornamental insects and safe use of pesticides	Dr. J.B. Karren
Pollination of vegetable and seed crops	Dr. G.E. Bohart
Vegetable seed physiology	Dr. W.F. Campbell
Nematodes with small and stone fruit	Dr. G.D. Griffin
Fruit, vegetable and ornamental diseases	Dr. S.V. Thomson
Bee management and biology	W.P. Nye, M.Sc.
Department of Plant Science	Head: Dr. K.R. Allred
Tree- and small-fruit culture, growth regulators and weed control	Dr. J.L. Anderson*
Fruit, physiology, growth	Dr. S.D. Seeley*
Tomato culture and weed control	Dr. A.R. Hamson
Nutrition, irrigation and rest of fruit trees	Dr. D.R. Walker
Culture of ornamental shrubs	

6 Provo

6.1 Brigham Young University, Private Institution
elevation : 1400 m
rainfall : 377 mm
growing season: 142 days

Provo, Utah 84601
Phone: (801) 378-1211

Floriculture and ornamentals culture — Dr. T.D. Davis
Vegetable culture — Dr. F. Williams*
Fruit culture — Dr. R.H. Walser

7 Salt Lake City

7.1 University of Utah, State Institution
elevation : 1500 m
rainfall : 389 mm
growing season : 161 days

1600 E. 2nd South Street
Salt Lake City, Utah 84112
Phone: (801) 581-7200

Department of Biology — Head: Dr. W.J. Dickinson
Plant-water relations — Dr. J. Ehleringer
Air pollution of ornamentals — Dr. M. Treshow
Plant genetics — Dr. R.K. Vickery

NEVADA

8 Fallon

8.1 Fallon Branch Station, State Institution
elevation : 1210 m
rainfall : 200 mm
growing season : 132 days

111 Scheckler Road
Fallon, Nevada 89406
Phone: (702) 423-2844
E.L. Chouinard

9 Logandale

9.1 Logandale Branch Station
elevation : 385 m
rainfall : 100 mm
growing season : 250 days

Box 126, Logandale, Nevada 89021
Phone: (702) 423-2844
J.A. Anderson

Greenhouse, grapes — Dr. D.G. Robison
Water use efficiency — Dr. D.A. Devitt

10 Pahrump

10.1 Pahrump Branch Station, State Institution
elevation : 825 m
rainfall : 100 m
growing season : 250 days

Box 1090, Pahrump, Nevada 89041
Phone: (702) 727-5532
R.W. Hammond, M.Sc.

General horticulture

11 Reno

11.1 Nevada Agricultural Experiment Station, Main Station, State Institution
elevation : 1340 m
rainfall : 200 mm
growing season : 103 days

Reno, Nevada 89557
Phone: (702) 784-4910
Dir.: Dr. B.M. Jones

Kimlick & Boynston Ln.

11.2 Valley Road Branch Station, State Institution — 910 Valley Road, Reno, Nevada 89557
elevation : 1370 m — Phone: (702) 784-6600
rainfall : 200 mm — P. Robles
growing season : 133 days

Ornamental crops — Dr. W.S. Johnson
Greenhouse crops — R.L. Post, M.Sc.

WYOMING

12 Laramie

12.1 Wyoming Agriculture Experiment Station — Laramie, Wyoming 82071
Main Station, State Institution — Phone: (307) 766-3667
elevation : 2184 m — Dir.: Dr. C.C. Kaltenbach
rainfall : 283 mm
growing season : 105 days

Division of plant science — Head: Dr. J.E. Lloyd
General horticulture — Dr. J.A. Cook
Potatoes — Dr. K.E. Bohnenblust
General horticulture — L.C. Ayres, M.Sc.

IDAHO

13 Aberdeen

13.1 Aberdeen Branch Experiment Station — Aberdeen, Idaho 83210
State Institution
elevation : 1341 m — Phone: (208) 397-4181
rainfall : 200 mm
sunshine : 2850 hrs

Potatoes — Prof. W.C. Sparks/Dr. R.B. Dwelle
Crop diseases — Dr. J.R. Davis/Dr. D. Corsini (USDA)
Potato breeding — Dr. J.J. Pavek (USDA)
Weeds — Dr. L. Haderlie
Insects — Dr. Sandvol
Plant breeding potatoes — Dr. S. Love

14 Moscow

14.1 Idaho Agricultural Experiment Station, — Moscow, Idaho 83843
Main Station, State Institution — Phone: (208) 885-6276
elevation : 811 m — Dir.: R.J. Miller
rainfall : 564 mm
sunshine : 2650 hrs

Control of vegetable insect pests — Dr. L.E. O'Keeffe
Post harvest — Dr. B.S. Kiles
Plant pathology, fruit tree diseases — Dr. A.W. Helton
Physiology of horticultural crops — Dr. W. Kochan
Ornamental horticulture and landscape gardening — Dr. R. Tripepi

15 Parma

15.1 Southwest Idaho Research and Extension Center — Parma, Idaho 83660

elevation : 675 m
rainfall : 278 mm
sunshine : 3000 h

Phone: (208) 722-5186

State institution	
Soils	Dr. L. Ewing/Dr. R. Mahlert
Seed pathology	Dr. N. Schaad
Tree fruits/horticulture	Dr. W. Kochan/Dr. W.M. Colt*
Vegetable seed crops	
Diseases	Prof. W.R. Simpson
Potatoes and onions	Dr. G.R. Beaver

16 Kimberly

16.1 Research and Extension Center, State Institution . Kimberly, Idaho 83341
Phone: (208) 743-3600

elevation : 1149 m
rainfall : 222 mm
sunshine : 2950 hrs

Crops diseases	Dr. R.L. Forster
Crop physiology	Dr. G. Kleinkopf

17 Sandpoint

17.1 Research and Extension Center . Sandpoint, Idaho 83864

elevation : 640 m
rainfall : 831 mm
sunshine : 2450 hrs

Cole crop research	
Breeding beans	Dr. J.J. Kolar
Insects of beans and peas	Dr. R.L. Stoltz
Weeds	Dr. S. Dewey

MONTANA

18 Bozeman

18.1 Montana Agricultural Experiment Station, Main Station . Bozeman, Montana 59717
Phone: (406) 994-3681
Dir.: Dr. J.R. Welsh

elevation : 1480 m
rainfall : 3927 mm
sunshine : 61%

State institution, founded 1893	
Department of Plant Pathology	Head: Dr. E.L. Sharp
Virus diseases of potatoes	Dr. T.W. Carroll
Physiological bases for disease resistance, potatoes, field crops	Dr. G.A. Strobel
Potato-seedstock certification	Dr. M.K. Sun
Air-borne diseases of field crops, genetic bases for disease resistance, genetics of pathogenicity	Dr. E.L. Sharp
Department of Plant and Soil Science	Head: Dr. D.G. Miller
Genetics	Dr. J.M. Martin/Dr. T.K. Blake, Dr. J.R. Schaeffer
Weed control	Dr. P.K. Fay/Dr. M.E. Foley
Crop physiology	Dr. J.H. Brown

Agricultural climatology, phenology — Dr. J.M. Caprio
Environmental horticulture:
Turf, woody ornamentals, propagation — Dr. G.E. Evans
Herbaceous perennials — Dr. C.W. Smith
Annuals, vegetable varieties, and culture — Dr. R.H. Lockerman
Extension specialist:
Weeds — M.J. Jackson, M.Sc./Dr. J.E. Nelson
Soils
Soil surveys, land-use planning — Dr. G.A. Nielsen
Soil chemistry — Dr. R.A. Olsen
Soil fertility — Dr. E.O. Skogley/Dr. J.R. Sims
Seed physiology — Dr. L.E. Wiesner/L.I. Hart, M.Sc.
Landscape design — R.K. Pohl, M.L.A.

19 Corvallis

19.1 Western Montana Branch Station . Corvallis, Montana 59828
elevation : 1105 m — Phone: (406) 961-3025
rainfall : 2997 mm — Supt.: D.R. Graham, M.Sc.
sunshine : 53%

State institution, started in 1907 to serve the research needs of Bitterroot Valley orchardists and ranchers. Biology and control of insects affecting fruit, vegetables, ornamentals; fruit-, vegetable-, and ornamental-variety evaluation.

Plant nutrition — D.R. Graham, M.Sc.
Tree fruits — Dr. N.W. Callan

20 Havre

20.1 Northern Montana Branch Station . Havre, Montana 59501
elevation : 796 m — Phone: (406) 265-6115
rainfall : 4933 mm — Supt.: D.C. Anderson, M.Sc.
sunshine : 62%

State institution, started in 1913 to serve the research needs of norther Montana ranchers. Dryland shelterbelts, dryland fruit culture, and variety evaluation.

21 Huntley

21.1 Southern Montana Branch Station . Huntley, Montana 59037
elevation : 911 m — Phone: (406) 348-3400
rainfall : 3381 mm — Supt.: Dr. G.F. Stallknecht
sunshine : 62%

State institution, started in 1910 to assist ranchers in solving problems related to irrigated crop production on the Huntley Irrigation Project. Environmental horticulture, new crops.

22 Moccasin

22.1 Central Montana Branch Station . Moccasin, Montana 59462

elevation : 1311 m
rainfall : 3889 mm
sunshine : 63%

Phone: (406) 423-5227
Supt.: Dr. G. Jackson

Dryland shelterbelts, woody ornamental testing

23 Sidney

23.1 Eastern Montana Branch Station
elevation : 585 m
rainfall : 3452 mm
sunshine : 57%

Sidney, Montana 59270
Phone: (406) 482-2208
Supt.: Dr. J.W. Bergman

State institution, started in 1947 to assist eastern Montana ranchers in solving problems in irrigated and dryland agriculture. Environmental horticulture and soils.

24 Creston

24.1 Northwestern Montana Branch Station
elevation : 896 m
rainfall : 4933 m
sunshine : 53%

Kalispell, Montana 59901
Phone: (406) 755-4303
Supt.: V.R. Stewart, M.Sc.

State institution, started in 1949 to serve the research needs of ranchers in the Flathead District of northwestern Montana. Weed control, plant breeding, potato variety evaluation.

Prepared by David R. Walker, Plant Science Dept., Utah State Univ., Logan, Utah
K.M. Brink, Dept. of Horticulture, Colorado State Univ., Fort Collins, Colorado
Robert R. Tripepi, Dept. of Plant, Soil and Entomological Sciences, Univ. of Idaho, Moscow, Idaho
J.A. Cook, College of Agriculture, Univ. of Wyoming, Laramie, Wyoming

J. PACIFIC NORTHWEST STATES

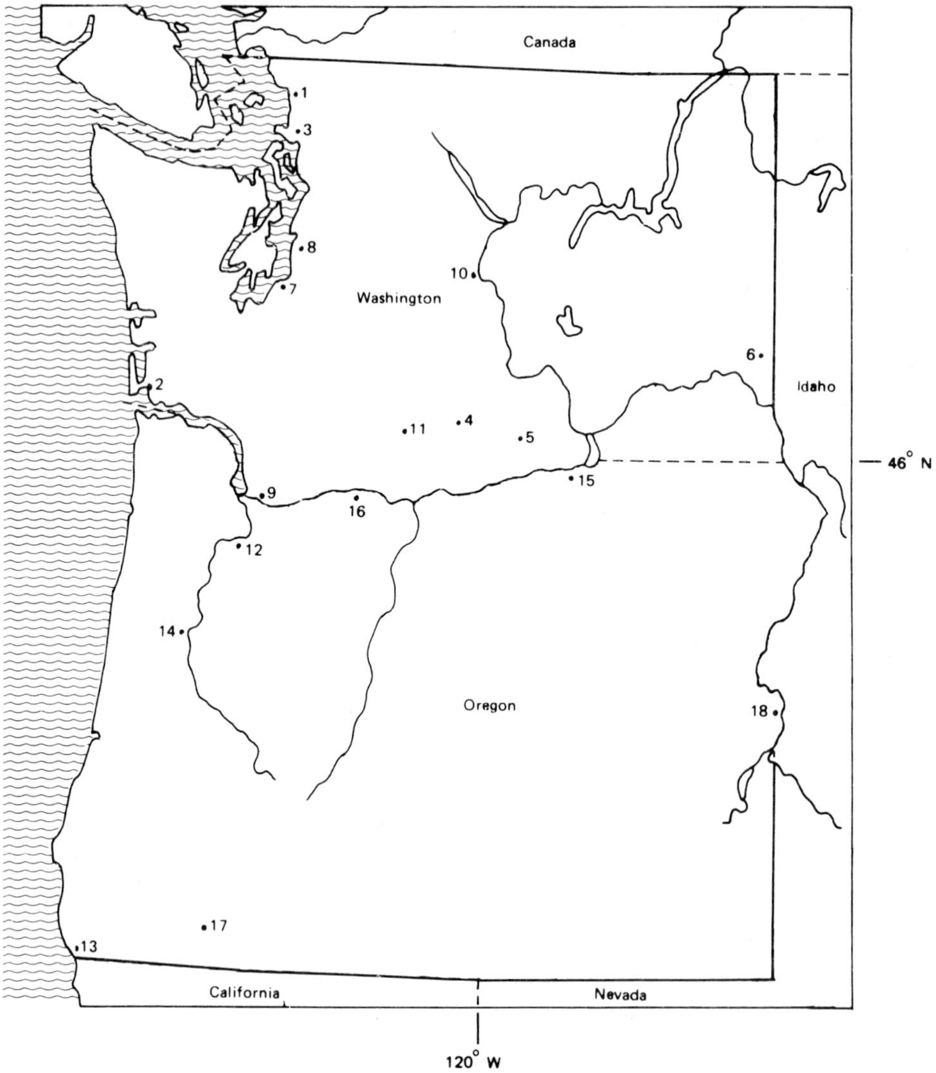
Canada
Washington
Idaho
Oregon
California
Nevada
46° N
120° W
1
2
3
4
5
6
7
8
9
10
11
12
13
14
15
16
17
18

Washington: 1 Bellingham, 2 Long Beach, 3 Mount Vernon, 4 Moxee, 5 Prosser, 6 Pullman, 7 Puyallup, 8 Seattle, 9 Vancouver, 10 Wenatchee, 11 Yakima, **Oregon**: 12 Aurora, 13 Brookings, 14 Corvallis, 15 Pendleton, 16 Hood River, 17 Medford, 18 Ontario

Survey

Several states have a greater total production of fruits, nuts, vegetables and speciality horticultural crops than the states of the Pacific Northwest, but certain of these crops are produced in abundance in this region. The Pacific Northwest states lead all states in the production of apples, winter pears, red raspberries, trailing blackberries for processing, filberts, sweet cherries, and Irish potatoes. They are second in the production of Bartlett pears, prunes, and walnuts. Although ranking third after fruits and vegetables in this region, ornamental crops are increasing at a faster pace. The region leads the nation in the production of bulbs, shade trees, evergreen trees, shrubs and Christmas trees and is the sole commercial producer of English holly. The Pacific Northwest produces approximately 55% of the U.S. production of apples and 70% of the pears, with Washington having the greatest production. The region also produces roughly 80% of its sweet cherries. California leads in prune and plum production. Oregon is second. Washington leads in sweet cherry production and Oregon is second. Washington is a substantial producer of peaches and apricots. The two states produce about one eighth of the country's berries, and Washington is a leading producer of juice grapes. The two states produce all the nation's filberts and approximately 6% of the vegetables grown in the U.S. come from Northwest states. These are principally for processing. Oregon is second in the nation in snap bean production for processing and approximately 20% of the national production is from Oregon and Washington. The Northwest states, including Idaho, also produce 22% of the nation's sweet corn and 30% of the peas for processing. The Northwest has a variety of climatic settings as a result of variations in surface form and elevations. The influence of the Pacific Ocean is modified by two parallel mountain ranges as distance increases inland from the coast line. The Cascades dominate the region and serves to divide both Oregon and Washington into two climate subregions: the mild, comparatively wet west and the more extreme, drier east.

Tree-fruit research in the region during the past 50 years has been directed principally at finding varieties suited to climate and soils of the region and the demands of an export market. The latter has also required considerable research in handling and storage. The selection of virus-free varieties and rootstocks resistant to certain disease problems and low winter temperatures has received considerable attention. Orchard-management research has concerned itself with soil management, irrigation, frost control, pest and disease control, pollination, chemical control of fruit set and rootstocks. The breeding of small fruits and vegetables for processing and for resistance to certain serious disease problems has received considerable attention in recent years. Management problems in these crops, as in the fruits and ornamentals, are the same the world over. The general trend in recent years has been toward more basic inquiry into the physiology and genetics of these horticultural crops, as both growers and researchers have become more sophisticated and refined in their understanding of crop production.

Organized horticultural research in the Northwest has its beginning with the establishment of the Oregon Agricultural Experiment Station in 1888 and the Washington Agricultural College and Experiment Station in 1892. There was a Horticultural Department at Oregon State Agricultural College as early as 1890, working on such problems as the use of paris green in controlling 'codlin' moth. Branch Stations to study area problems soon followed. The Hood River Station was established in 1912 and the Medford Station in 1911 to help solve problems in this rapidly expanding apple center. Although at first a station for field demonstrations only, the present Western Washington Research and Extension Center at Puyallup has been interested in horticultural problems since 1894. At present there are six branch stations in Washington and five at Oregon that have established the research services of the State Universities in most of the Northwest's leading horticultural districts.

The University of Washington Arboretum at Seattle was established in 1935 to introduce, test and distribute new or rare trees and shrubs. This activity is under the administration of the University's College of Forest Resources. The Pacific Bulb Grower's Research and Development Station at Harbor, Oregon, for lily bulb research is supported with funds supplied by the bulb growers and jobbers, but is under the administration of Oregon State University with the aid of bulb grower's advisory committee.

The United States Department of Agriculture has been active in horticulture research at both the central and branch stations in Oregon and Washington from the beginning. The Federal State research activities have been on a co-operative basis and the USDA workers were considered as members of the staff of these stations. The Washington State Department of Agriculture's Horticultural Division co-operating with the State University manages a Plant Introduction and Quarantine Station (Moxee) for the purpose of providing certified, true-to-name, pathogen free nursery stock to Washington fruit growers. The Oregon State Department of Agriculture, established in 1931, works in a similar manner with Oregon State University. The USDA Ornamental Plants Research laboratory was established at Oregon State University in 1973. The Northwest Plant Germplasm Repository was established in 1980 at Oregon State University to collect, maintain and distribute genetic material of pears, strawberries, blueberries, and other small fruits, filberts, hops and mint. These selections are available to plant breeders and other scientists throughout the country. A number of commissions and councils have been established to encourage and financially support horticultural research in the Northwest. The Washington State Apple Commission, Fruit Commission, Potato Commission, and Wine and Grape Growers Council are examples in Washington, while Oregon has its Filbert Commission, Nursery Advisory Council and Oregon Strawberry Commission.

WASHINGTON

1 Bellingham

1.1 Washington State Nursery
elevation : 45.4 m
rainfall : 897 mm

Bellingham, Washington 98225
Phone: (510) 774-1099,
Answer: COLL.AG.PMAN for all WSU outlying stations
Dept.Sup: B. Cole
Supt.: R.O. Holland

Established in 1912. Since 1971 operated under the Washington State Department of Natural Resources. Forest seedling research, potato seed certification, potato variety evaluation, and field testing of tree fruits.

2 Long Beach

2.1 Coastal Washington Research and Extenstion Unit
elevation : 7.6 m
rainfall : 2588 mm
sunshine : 1783 h

Long Beach, Washington 98631
Supt.: A.Y. Shawa*

State institution, established in 1923 as a branch station for cranberry and blueberry research with major emphasis on control measures for insects, diseases, and weed, soil management and frost control.

3 Mount Vernon

3.1 Northwestern Washington Research and Extension Unit
elevation : 4.3 m
rainfall : 871 mm

Mount Vernon, Washington 98273
Supt.: Dr. R.A. Norton*

Established in 1947 as a branch station of the Washington State Agricultural Research Center. Peas, cabbage, sweet corn, vegetable seed crops, small fruit crops, and ornamental greenhouse crops.

4 Moxee

4.1 Plant Introduction and Quarantine Station
elevation : 472.4 m
rainfall : 216 mm

Moxee, Washington 98936
Dep.Sup.: C. Neilson
Dir.: M.D. Aichele

Established in 1944, cooperation of the Plant Industries Division of the State Department of Agriculture and the Washington State Agricultural Research Center for executing the Nursery Improvement Program designed to supply certified fruit trees to the state's fruit tree industry.

5 Prosser

5.1 Irrigated Agriculture Research and Extension Center
elevation : 275.2 m
rainfall : 196 m

Prosser, Washington 99350
Supt.: Dr. L.R. Faulkner

Part of the Washington State Agricultural Research Center system, established in 1919. Research on all phases of growth and management of the major irrigate crops in Washington.

6 Pullman

6.1 Washington State Agricultural Research Center
elevation : 775.7 m
rainfall : 541 mm
sunshine : 2605 h

Pullman, Washington 99164-6414
Dean: Dr. J.L. Ozbun
Dir.: Dr. L.L. Boyd

State institution, established in 1892, cooperating with the USDA and the Division of Plant Industry of the Washington State Department of Agriculture.

Department of Entomology — Act.Chrm.: Dr. E.P. Catts
Most of the entomologists working on horticultural crops are at the branch research centers and units
Pollination and bee poisoning

Department of Horticulture and Landscape Architecture — Chrm.: Dr. H.P. Rasmussen
Plant introduction in cooperation with Plant Germplasm Introduction and Testing, USDA. Landscape architecture and land use planning. Culture and breeding of greenhouse florist crops. Culture and maintenance of fruit and vegetable crops, including Irish potatoes. Post-harvest physiology of horticultural crops

Department of Plant Pathology — Chrm.: Dr. A.D. Davison
Most of the plant pathologists working on horticultural crops are at the branch research centers and units

7 Puyallup

7.1 Western Washington Research and Extension Center
elevation : 15.2 m
rainfall : 1036 mm

Puyallup, Washington 98371
Supt.: Dr. E.C. Bay

State institution, started in 1894 as a demonstration station to provide farmers with information of their farming prolems. Research on greenhouse and nursery crops, berry crops and commercial vegetable crops.

8 Seattle

8.1 University of Washington Arboretum
elevation : 18.3 m
rainfall : 980 mm
sunshine : 2019 h

Seattle, Washington 98195
Dir.: Dr. H.B. Tukey, Jr.*

Established in 1935 and operated by the College of Forest Resources of the University of Washington: (1) to form and maintain a living museum of woody plants suited to the area; (2) to introduce, test and distribute new and rare trees and shrubs; and, (3) to supply information on and to form a library and herbarium dealing with such plants.

9 Vancouver

9.1 Southwestern Washington Research Unit
elevation : 64 m
rainfall : 1044 mm

Vancouver, Washington 98665
Supt.: Dr. C.H. Shanks

Branch station of the Washington Agricultural Research Center. Problems peculiar to the area; berry fruits, vegetable crops, tree fruits and Christmas trees. Mechanical harvesting of raspberries.

10 Wenatchee

10.1 Tree Fruit Research Center
Washington State University
elevation : 243.8
rainfall : 406 mm

Wenatchee, Washington 98801
Supt.: S.C. Hoyt

Management of indirect insect and mite pests of apple — S.C. Hoyt

Horticulture, orchard systems, tree design, and rootstocks	B.H. Barritt*
Management of direct insect pest of tree fruits	J.F. Brunner
Management of insect and mite pests of pear	E.C. Burts
Epidemiology and control of bacterial and fungal diseases of tree fruits	R.P. Covey
Horticulture, evaluation of Red Delicious strains, winter hardiness and cold acclimation	D.O. Ketchie*

10.2 Tree Fruit Research Laboratory, USDA, Agricultural Research Service — Wenatchee, Washington

Post harvest physiology - fruit maturity and storage	K.L. Olsen*
Post harvest physiology - effects of growth regulators on fruit quality	E.A Curry*
Post harvest physiology - packaging and transport fruit	Proj.Leader: J. Fountain
Post harvest physiology - engineer - energy conservation in cold storage	G.E. Yost
Post harvest physiology - biochemical changes in fruit maturity	S. Drake
Production physiology - mineral nutrition and fruit quality	T. Raese
Production physiology - endogenous growth regulation in tree fruit	E. Stahly
Production physiology - bioregulation of tree growth and fruiting	Loc.Leader: Dr. M.W. Williams*
Virology - stone and pome fruit diseases	L. Parish

11 Yakima

11.1 Yakima Agricultural Research Laboratory, USDA, SEA/ARS — Yakima, Washington 98902
elevation : 324.3 m
rainfall : 203 mm
sunshine : 2909 h

Fruit and vegetable insect control of insect pests, utilization of pheromones in insect control research.

OREGON

12 Aurora

12.1 North Willamette Experiment Station — Aurora, Oregon 97002
Phone: (503) 678-1264
Supt.: Dr. L.W. Martin*
elevation : 46 m
rainfall : 1050 mm

Branch station of Oregon Agricultural Experiment Station (Corvallis). Small fruits, vegetables and nursery crops.

13 Brookings

13.1 Pacific Bulb Growers Research and Development Station — Brookings, Oregon 97415
Phone: (503) 469-2215
elevation : 21 m
rainfall : 2056 mm

Cooperative venture between Pacific Bulb Growers Association, Oregon State University, University of California and USDA, administered by a joint growers Advisory committee and a Technical Research Committee appointed by the Dean of Agriculture of Oregon State University. Lily bulb production problems, breeding and variety testing, soil management, insect and disease control, virus diseases and vectors, storage, handling, subsequent greenhouse forcing, basic physiology.

14 Corvallis

14.1 Oregon Agricultural Experiment Station
Main Station
elevation : 69 m
rainfall : 1008 mm

Corvallis, Oregon 97331
Phone: (503) 754-2331
Dean: E.J. Briskey
Dir.: R.E. Witters

State institution cooperating with USDA

Department of Agricultural Engineering — Head: J.R. Miner
Design and development of new or improved equipment for mechanizing harvesting and handling operations
Department of Botany and Plant Pathology — Head: Dr. R. Scanlan
Department of Entomology — Head: Dr. B.F. Eldridge
Department of Food Science and Technology — Head: Dr. R.E. Wrolstad
Department of Horticulture — Head: Dr. C.J. Weiser*

15 Pendleton

15.1 Umatilla Station, Branch Station, Columbia Basin Agricultural Center
elevation : 453 m
rainfall : 375 mm

Pendleton, Oregon 97801
Phone: (503) 276-5721
Supt.: S. Land

State institution in cooperation with USDA. Research on crops suitable for the newly developed Hermiston Irrigation District. Testing of new horticultural crops for the surrounding irrigated area. Fruit- and vegetable production problems including variety testing weed, insect and disease control and winter hardiness.

16 Hood River

16.1 Mid Columbia Station, Branch Station
elevation : 152 m
rainfall : 779 mm

Hood River, Oregon 97031
Phone: (503) 386-2030
Supt.: Dr. E.A. Mielke*

State institution. Control of entomological orchard pests, control of tree fruit diseases, soil fertility, winter hardiness and pollination, cultural practices. The main crops under study are pears, apples, cherries and peaches.

17 Medford

17.1 Southern Oregon Station, Branch Station
elevation : 444 m
rainfall : 540 mm

Medford, Oregon 97502
Phone:(503) 772-5165
Supt.: J.A. Yangen

State institution. Improvement of cultural practices, control of pests, introduction of superior varieties of pears, peaches, apples and sweet corn, pear rootstock research, pear breeding.

18 Ontario

18.1 Malheur Station, Branch Station
elevation : 678 m
rainfall : 252 mm

Ontario, Oregon 97914
Phone:(503) 889-2174
Supt.: Dr. C.C. Stock

State institution. Research on livestock feeding, irrigation, soil management, variety testing of forage crops and to a limited extent horticultural crops (potatoes, onions, tree fruits).

Prepared by H. Paul Rasmussen, Dept. of Horticulture and Landscape Architecture, Washington State Univ., Pullman, Washington
C.J. Weiser, Dept. of Horticulture, Oregon State Univ., Corvallis, Oregon

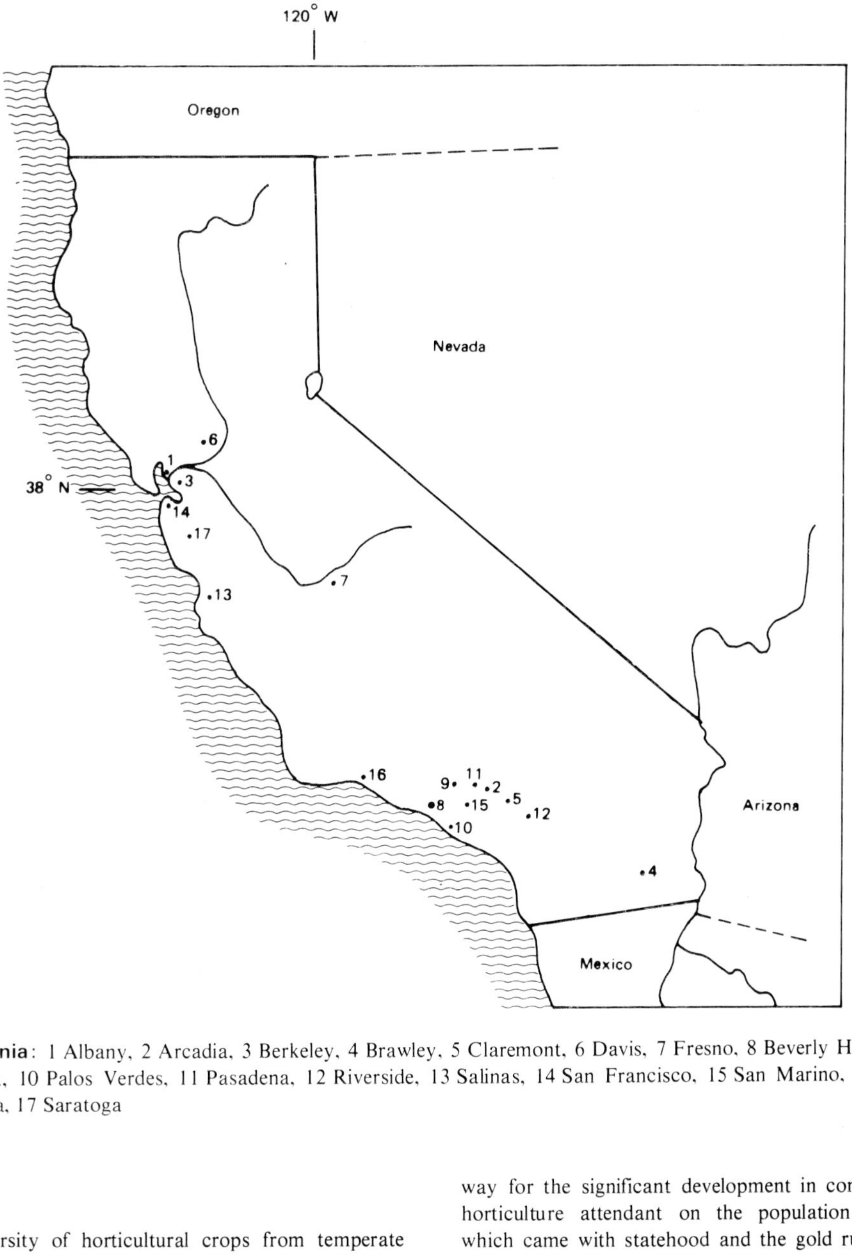
120° W
Oregon
Nevada
38° N
Arizona
Mexico
1
2
3
4
5
6
7
8
9
10
11
12
13
14
15
16
17

California: 1 Albany, 2 Arcadia, 3 Berkeley, 4 Brawley, 5 Claremont, 6 Davis, 7 Fresno, 8 Beverly Hills, 9 La Canada, 10 Palos Verdes, 11 Pasadena, 12 Riverside, 13 Salinas, 14 San Francisco, 15 San Marino, 16 Santa Barbara, 17 Saratoga

Survey

A diversity of horticultural crops from temperate zone to subtropical species are adaptable to California's varied climate and topography and irrigated agriculture. Many are so well adapted that California ranks at or near the top in the production of them. California's horticulture began in the late 1700's with the Spanish colonization and the agricultural development associated with the Jesuit missions. In the early 1800's, pioneer fruit growers, with American settlers prominent among them, established vineyards and orchards of citrus and deciduous fruits, preparing the way for the significant development in commercial horticulture attendant on the population growth which came with statehood and the gold rush. The opening of the transcontinental railroad and the advent of refrigerated transportation later in the century were a major stimulus to the expansion of the fruit and vegetable industries. Ornamental or landscape horticulture also has grown in importance with the increasing population and urbanization in the state. The eminent position of California horticulture today is also due in part to contributions from horticultural research. Over the past 60 years especially, the University of California through its College of Agriculture and Agricultural Experiment Station has carried a major responsibility for such research with the efforts divided among several departments appropriate to the diversity and economic importance of the state's horticultural industries. Today horticultural research in the University is centered primarily on two of its campuses, Davis and Riverside.

CALIFORNIA

1 Albany

1.1 Western Regional Research Center
USDA, Agricultural Research Center

800 Buchanan Street, Albany,
California 94710
Dir.: S.H. Feairheller

The Center at Albany is one of the four original Regional Research Laboratories in the USA authorized by the Federal Government in 1938 for conducting basic and applied research to find new and wider use for American farm commodities, better products and wider markets. The Center consists of seven research units. These conduct basic and applied research on composition, senescence, hormonal and energy-use activity, control of metabolic processes, N-fixation, photosynthesis, chemical reactions related to quality changes, and the control of micro-organisms in food products.

2 Arcadia

2.1 Los Angeles State and County Arboretum
Operated by the Department of Arboreta and
Botanic Gardens of Los Angeles County
elevation : 166-195 m
rainfall : 450 mm

Arcadia, California 91006
Phone: (818) 446-8251
Dir.: F. Ching
Asst.Dir.: R. Ito
Supt.: J. Provine

Testing and introduction of ornamental plants from other parts of the world that are suitable and desirable for Southern California. Principal plant collections include *Eucalyptus, Callistemon, Melaleuca, Cassia, Tabebuia, Erythrina, Cycadaceae, Zamiaceae, Palmae, Orchidaceae, Magnolia, Ficus, Pinus, Zingiberaceae, Bamboo.* To date, over 100 plants have been introduced to the Southern California landscape. Principal geographic plantings include Australia, Asia/North America, Africa, Latin America, South America, Southwestern United States.

J. Baum/S. Granger

Herbarium consists of 16,000 specimens, primarily of Arboretum's living collection and ornamental plants of Southern California; permanent collection mapped and computerized. Annual Index Seminum available for exchange, plant identification and cultural information available, liaison with Los Angeles County Poison Control Center with identification specimens available.

Education and Public Services Programs for general public include following displays: turfgrasses, groundcovers, herb garden, garden of old roses, synopic juniper collection, prehistoric and jungle garden, aquatic garden, meadowbrook, native oak woodland, tropical greenhouse, begonia greenhouse, demonstration home gardens, garden for all seasons, bird sanctuary. Programs for information include horticultural and botanical information brochures, mini courses, adult classes, lectures, walks, Arbor Day program with 1,000

T. Niiya

schools participating, guided tours.
Historical significance including four buildings, the earliest constructed in 1840, with all buildings restored. Gabrielino Indians made their home here because of the abundance of water, and the area subsequently became Rancho Santa Anita.

3 Berkeley

3.1 University of California
elevation : 8.8 m
rainfall : 460 mm

Berkeley, California 94720
Phone: (415) 642-7171

3.1.1 Department of Entomological Sciences

Management of floricultural and berry pests	W.W. Allen
Manipulation of naturally occurring biological control agents in agroecosystems through vegetation management	M.A. Altieri
Medical and veterinary entomology	J.R. Anderson
Plant-herbivore interactions. Behaviour and physiology of phytophagous insects with special reference to feeding and food selection	E.A. Bernays
Biological control of arthropod pests of pome and stone fruits. Taxonomy of Copidosomatini (Encyrtidae).	zL.E. Caltagirone
Toxicology and insecticide chemistry	J.E. Casida
Alimentary physiology, feeding behavior and nutrition	R.H. Dadd
Insect ecology, forest entomology, avian insectivores, and urban insect problems	D.L. Dahlsten
Biosystematics and evolutionary biology, Hymenoptera, especially Appoidea, pollination	H.V. Daly
Biosystematics of Coleoptera (Tenebrionidae), experimental methods in systematics	J.T. Doyen
Integration of pathogens into pest management programs	L.A. Falcon
Urban pest management	G.W. Frankie
Biological control of mosquitoes	R. Garcia
Mathematical ecology and population dynamics, renewable resource management, optimization models and optimum control	W.M. Getz
Biochemistry, nutrition, toxicology	H.T. Gordon
Biological control of pest arthropods in alfalfa, cotton and grapes	A.P. Gutierrez
Nutrition and ecophysiology of Entomophagous insects, systematics of Cocinellidae and Anthicidae	K.S. Hagen
Acarology	M.A. Hoy
Arthropod pest mangement on ornamental plants	C.S. Koehler
Natural product chemistry and arthropod control	I. Kubo
Ecology of ticks and tick borne diseases Biosystematics of Tabanidae	R.S. Lane
Circadian control of behaviour patterns	W.J. Loher
Developmental biology, morphogenesis, evolution	G.F. Oster
Insect morphology	R.L. Pipa
Nematode parasites and associates of invertebrates	G.O. Poinar
Biosystematics of Lepidoptera	J.A. Powell
Insect vectors of plant viruses	A.H. Purcell
Aquatic entomology	V.H. Resh
Ecology of generalist predators. Ecological theory and pest management. Ecology of sustainable agriculture in the tropical temperate areas	S.J. Risch

Mosquito biology, biochemistry and control	G.H. Schaefer
Systematics of flies and spiders, data management of collections, ecology of wild areas	E.I. Schlinger
Pest management, field and vegetable insects	C.G. Summers
Insect vectors of plant viruses, statistics	E.S. Sylvester
Insect pathology, epizootiology, immunity, microbial control	Y. Tanada
Studies of insect baculoviruses with emphasis on specificity of infection and mechanisms or replication	L.E. Volkman
Forest pest management, insect population dynamics, biometrics	W.E. Waters
Hormonal causes and functional consequences of neuronal regression in insects	J.C. Weeks
Helminthology, immunology, helminth-vector relations	C.J. Weinmann
Arthropod interactions and arthropod pest management with emphasis on deciduous fruit and nut crops	S.C. Welter
Forest pest management, pheromones, biology of bark beetles	D.L. Wood
Botanical substances with anti-feeding effects on insects	I. Kubo
Animal and insect virology, comparative virology	L.E. Volkman

3.1.2 Department of Plant Pathology

Control of diseases caused by soil-borne fungi, root diseases	L.J. Ashworth, Jr.
Forest tree diseases, root diseases, rusts, integrated control of forest pests	F.W. Cobb, Jr.
Ecology of soil-borne pathogens, soil biology, physiology and biochemistry of plant diseases	J.G. Hancock, Jr.
Biochemistry of plant pathogens and plant disease, microbiology of soil-borne pathogens	O.C. Huisman
Ecology and physiology of leaf surface microorganisms, bacteria incited frost damage, biological control of weeds and brush	S.E. Lindow
Diseases of ornamentals and trees, control programs	A.H. McCain
Plant virology, comparative virology, viroid diseases	T.J. Morris
Genetics of plant pathogens, pathogen-host interactions, bacterial diseases	N.J. Panapoulos
Forest tree diseases, root disease fungi, mistletoes, insect-disease relationships	J.R. Parmeter, Jr.
Management of diseases of ornamentals, Armillaria root rot, integrated control	R.D. Raabe
Plant virology, comparative virology, physiology and cytology of virus diseases	D.E. Schlegel
Ecology and physiology of bacterial diseases, biological control, plants and soils as reservoirs for human bacterial pathogens	M.N. Schroth
Molecular genetics of host-parasite interactions	B.J. Staskawics
Somatic plant genetics, genetic analysis of host-parasite interactions	Z.R. Sung
Disease and pathogen physiology, potato diseases, biology of soil-borne pathogens	A.R. Weinhold

3.1.3 Department of Plant and Soil Biology

Pedology - specifically the study of pedogenic processes and the environmental and anthropogenic factors which affect them	R.G. Amundson

Trace element distribution and mobility in soils — H.E. Doner

Ecology of microbes in soil as influenced by the physiological capabilities of microbes and the physical/chemical characteristics of the soil — M. Firestone

Biogeochemical cycling in terrestrial ecosystems with emphasis on forest ecosystems — J.G. McColl

Introduction of radioactive nuclides into the food chain of man via the plant uptake pathway — R.K. Schulz

Study of limiting factors in photosynthesis, the influence of environmental factors on plant growth, the physiology of heavy metal toxicity in crop plants and the foles of trace elements in mineral nutrition — N. Terry

Mechanics of root permeated soil in relation to slope stability, soil erosion and compaction, the mechanics of unsaturated soil — L.J. Waldron

4 Brawley

4.1 Irrigated Desert Research Station
USDA, Agricultural Research Service

4151 Highway 85, Brawley,
California 92227
Off.-in-Ch.: C.M. Brown

Worksite for breeding of improved stress-resistant varieties of lettuce, cucurbits (melons and Cucurbita) and cole crops. Five greenhouses, two seed-storage rooms and 150 acres of growing area (for vegetables and agronomic crops). The work is under the direction of Dr. E.J. Ryder and Dr. J.D. McCreight, who are located at the U.S. Agricultural Research Station, Salinas.

5 Claremont

5.1 Rancho Santa Ana Botanic Garden

1500 North College Avenue,
Claremont, California 91711
Phone: (714) 628-8767
Dir.: Dr. T.S. Elias
Hort.: J. Dourley
Supt.: G. Steenhuizen

Private institution, 100 acres, located in Eastern Los Angeles County at an elevation of 1200 feet. Mediterranean climate. Library, scientific laboratories, herbarium (750,000 specimens), greenhouses, propagating room, walk-in refrigerator, lath house, seed-storage room, 3 acres experimental growing grounds. Special emphasis to native groundcover plants and ornamentals developed through hybridization of native California species.

Mycology: Mucorales and Laboulbeniales — Dr. R.K. Benjamin

Anatomy and morphology: wood anatomy, biota of long dispersal, insular floras — Dr. S. Carlquist

Taxonomy, Onagraceae, phylogeny, flora of California — Taxonomist and curator herbarium: Dr. R.F. Throne

Primitive angiosperms, amphitropical disjuncts in Western North and South America

Experimental systematice, biochemical systematics — Dr. R. Scogin

Systematic and evolutionary relationships of the Poaceae — Dr. K. Tomlinson

Anatomy of secretory tissues, systematics of woody species and floristic and biogeographic relationships between North America and Eurasia — Dr. T. Elias

6 Davis

6.1 University of California — Davis, California 95616

6.1.1 Department of Plant Pathology

Epidemiology and control of fungal and bacterial diseases of deciduous fruits and nuts	W.D. Gubler
Grape virus diseases	A.C. Goheen
Soil-borne diseases of deciduous fruit and nut trees	S.M. Mircetich
Fruit tree virus diseases; mycoplasmas	B. Kirkpatrick
Deciduous fruit and nut tree diseases; postharvest	J.M. Ogawa
Physiology host-parasite interactions in deciduous fruit and nut tree diseases	R.M. Bostock
Grape diseases; epidemiology, mathematical modeling	J.J. Marois

6.1.2 Department of Environmental Horticulture
University of Davis

Davis, California 95616
Phone: (916) 752-3071
Chrm.: Prof. R.M. Sachs

Physiology of flowering in horticultural and agronomic species, biomass production and culture systems	Prof. R.M. Sachs
College of Agricultural and Environmental Sciences	Dean: C.E. Hess*
Assessment of nitrogen-fixing landscape plants, root and root nodule development	Asst.Prof.: Alison M. Berry
Woody plant developmental physiology, cell and tissue culture, plant propagation, nursery management	Asst.Prof.: D.W. Burger
Management of commercial greenhouse flower evaluation of plant environment relations	Specialist: T.G. Byrne*
Genetics and breeding of flower crops with emphasis on Gerbera, evolutionary biology of California native plants	Prof. J.A. Harding
Arboriculture, park management, woody plant physiology	Prof. R.W. Harris*
Floriculture and greenhouse production, post-harvest technology and physiology	Prof. A.M. Kofranek*
Floriculture, greenhouse production systems, crop productivity	Prof. Emer. H.C. Kohl
Landscape plant introduction and selection, taxonomy and ecology or ornamental plants, revegetation of disturbed areas. Teaching taxonomy of woody ornamental plants	Prof. A.T. Leiser
Crop modeling, greenhouse nursery crop systems analysis	Asst. Prof. J.H. Lieth
Management of container media and landscape soils, soil-plant relations	Prof. J.L. Paul
Postharvest technology, handling and marketing of environmental plants, postharvest physiology	Prof. M.S. Reid
Turfgrass breeding and management, ecological genetics and stress physiology	Ass.Prof. L.L. Wu

6.1.3 Department of Pomology

Chrm.: Prof. D.J. Durzan*

Genetics of small fruits, especially strawberry breeding, sytogenetics and evolution	Prof. R.S. Bringhurst*
Orchard soil fertility, mineral nutrition aspects of rootstocks, analytical chemistry of soils, plants, waters	R.M. Carlson
Responses of deciduous fruit trees to salinity and to soil waterlogging. Walnut flower abscission	P.B. Catlin*
Applications of molecular genetics for germplasm development of fruit and nut crops. Application of recombinant DNA techniques for germplasm analysis, structure and function of useful genes, development of gene transfer systems	A.M. Dandekar

Environmental plant physiology: photosynthetic efficiency of tree crops relative to the utilization of nitrogen, water, and solar radiation	Ass.Prof. T.M. De Jong*
Whole tree field physiology, water, nitrogen and amino acid metabolism, propagation and genetic improvement through cell and tissue culture, laser photobiology, pistachio and cherry production problems, biotechnology in agroforestry systems	Prof. D.J. Durzan*
Quantitative genetics, plant breeding	Prof. P.E. Hansche
Postharvest biology and technology of horticultural crops, regulation of fruit ripening, quality of fruits and nuts in relation to pre- and post-harvest factors	Prof. A.A. Kader*
Almond variety and rootstock improvement, clonal variations in perennial fruit crops, plant propagation, tissue culture and mircropropagation, development of phenology models	Prof. D.E. Kester*
Postharvest biology of fruit and nuts, emphasis on cell wall polysaccaride metabolism and biochemistry of host-pathogen interactions	Ass.Prof. J.M. Labavitch
Tree fruit physiology, emphasis on hormones and applied growth regulators. Control of rest, flowering, fruit set, fruit growth and abscission	Prof. G.C. Martin
Methods for maintenance and storage of germplasm repository, curator of National Plant Germplasm Repository at Davis	D.E. Parfitt
Reproductive biology of flowering plants, especially tree species. Interacellular calcium dynamics in plant cells	Ass.Prof. V.S. Polito
Cellular senescence, with emphasis on the use of cultured plant cells, mitochondrial metabolism and omeostatic, postharvest physiology and biochemistry	Prof. R.J. Romani
Synthesis, transport and roles of synthetic and natural compounds, especially growth regulators in plant and fruit development. High density planting, training systems, their effects on flower initiation, fruit set and growth	Prof. K. Ryugo*
Postharvest physiology, pathology, handling, storage and transportation of fruits. Mycotoxins in fruit, vegetables and nuts	N.F. Sommer*
Clonal propagation by tissue culture, physiology of tissues in culture and of regenerated plants	Asst.Prof. Ellen Sutter
Mineral nutrition and water relations of deciduous fruit trees	Prof. K. Uriu*
Small fruit research, especially strawberries	V. Voth
Utilization of fertilizer nitrogen by deciduous fruit trees. Phenology of deciduous fruit trees. Fruit set phenomenology and pollination biology in fruit tree	Ass.Prof. S.A. Weinbaum

6.1.4 Department of Vegetable Crops — Chrm.: Prof. L. Rappaport*

elevation : 13 m
rainfall : 150 mm
sunshine : 10-12 hrs/day

Postharvest physiology, membrane behavior in relation to senescence and quality loss of harvested vegetables	Asst.Prof. A.B. Bennett
Nutrient acquisition and carbon assimilation in relation to environmental stress	Asst.Prof. A.J. Bloom

Seed physiology, stress physiology, hormonal physiology	Asst.Prof. K.J. Bradford
Tomato genetics and breeding, genetic basis for photosynthate allocation	Asst.Prof. J.D. Hewitt
Biochemical genetics, tomato breeding, genetic basis of tomato quality characteristics and stress tolerance	Ass.Prof. R.A. Jones
Physiology and biochemistry of low temperature (chilling) injury in crop plants	Prof. J.M. Lyons*
Lettuce genetics, tissue culture and breeding, host-pathogen interactions	Asst.Prof. R.W. Michelmore
Cell wall metabolism, regulation of cell elongation	Prof. D.J. Nevins
Genetics and breeding of celery and Brassicas	Asst.Prof. C. Quiros
Biochemistry and physiology of gibberellins and other plant growth regulators. Cell and tissue culture studies related to disease resistance	Prof. L. Rappaport
Cytogenetics of tomato and related species. Plant exploration and germplasm development. Maintenance of Tomato Genetics Stock Center	Prof. C.M. Rick
Postharvest physiology, physiology and environmental control of ripening, ethylene physiology and plant responses to mechanical stress, control of senescence and deterioration of leafy crops	Asst.Prof. M.E. Saltveit, Jr.
Soil, plant, water relations, cultural practices	Asst.Prof. Carol Shennan
Cultural practices related to production and quality of potatoes. Environmental and air pollution effects on bean seed production	Dr. H.H. Timm*
Postharvest biochemistry and physiology. Regulation of biosynthesis and mode of action of ethylene in senescent or stressed tissues	Prof. Shang Fa Yang
Agricultural labor efficiency studies, seed production in cucumbers, onions and lettuce	M. Zahara, B.S.
Melon breeding	F.W. Zink, M.S.*

6.1.5 Department of Viticulture and Enology
University of California

Davis, California 95616
Chrm.: Prof. C.S. Ough

Chemical engineering aspects of fermentation and wine processing. Wine making equipment selection, winery design and layout, and the economics of investment and operation. Fermentation kinetics, mathematical modeling, computer simulation and control of enological operations. Stability and non-microbial disorders of wines	Ass.Prof. R.B. Boulton
Influence of environment (temperature, light intensity, humidity, soil moisture, nutrients, wind) on stomatal behavior, photosynthesis, growth, budbreak, bud fruitfulness, fruit set, coloration and composition of grapevines and their fruit, metabolism of malic and tartaric acids, translocation, storage, and utilization of photosynthate, nitrogen metabolism, especially amino acids, evaluation of trellising systems, long term effects of defoliation, especially as it relates to mechanical harvesting, canopy management and pest damage, control of vegetative growth	Prof. W.M. Kliewer
Growth of and fermentation by malolactic and other lactic acid bacteria, wine yeast metabolism including ethanol production and tolerance and other control mechanisms (fusel oil and H_2S), wine spoilage, yeast strain selection	Prof. R.E. Kunkee

Environmental control of growth and productivity emphasizing photosynthesis, biophysics of cell expansion, determination of the physiological mechanisms involved in responses to water deficits, interactions of nutrient levels and water deficits, varietal improvement	Asst.Prof. M.A. Matthews
Somatic cell genetics and tissue culture of grape and other higher plants, resistance to environmental and biological stress, improvement of grape varieties	Asst.Prof. Carole P. Meredith
Growth and development of grape vines and fruit. Climatic control of bud development, flower initiation, fruit development and fruit composition. Cell wall factors controlling plant growth. Grape vine anatomy	Asst.Prof. Janice C. Morrison
Evaluation of effect of viticultural and enological treatments and of individual components on wine sensory properties. Correlation of instrumental and sensory evaluation of wine flavor	Ass.Prof. Ann C. Noble
Wine analysis techniques, development and evaluation, testing of new additives and applications to winery use, fermentation factors as they affect must and wine composition and quality, and evaluation of grape variety with regard to wine composion and quality	Prof. C.S. Ough
Yeast metabolism and quantitative physiology related to wine and ethanol fermentation. Optimization of fermentation variables related to the quality improvement of wine. Application to immobilized enzymes and immobilized whole cell systems to the development of continuous wine fermentation and malolactic fermentation. Research effort directed toward a development of more energy efficient material saving process technoloy for wine making. Recovery and utilization of grape by-products and winery wastes	Prof. D.D.Y. Ryu
Chemistry of wine aging, accelerated maturation of wines, processing-quality relationships, tannins and polyphenolics of grapes and wine, oxidation reactions in grapes and wines, and raisin chemistry	Prof. V.L. Singleton
Plant hormones and regulators, factors improving quality of wine and table grapes including girdling, pruning, thinning, hormones, cropping, water, etc., dormancy, abscission, translocation, photosynthesis	Prof. R.J. Weaver

7 Fresno

7.1 Horticultural Crops Research Laboratory,
Agricultural Research Service, USDA
elevation : 100 m
rainfall : 265 mm
sunshine : 4450 h

2021 Soutpeach Avenue
Fresno, California 93727
Phone: (209) 487-5334
Lab.Dir.: P.V. Vail, Ph.D.

7.2 Protection and Quarantine Research Unit

Development of knowledge and procedures that help reduce losses caused by plant diseases, insect pests and physiological deterioration of horticultural crops during domestic and export marketing

Plant diseases and storage of tree fruit, grapes and berries; air and surface transport of fruits; fumigation and handling of grapes and tree fruit for	J.M. Harvey, Ph.D.

export; controlled atmospheres; development of quarantine procedures for various types of fruits and vegetables

Transportation of fruits; development of quarantine treatments for fruits — C.M. Harris

Postharvest diseases of citrus and subtropical crops and their control; citrus fruits: controlled atmosphere effects, phytotoxicity of quarantine treatments; fumigation, heat, cold treatments, and chilling injury — L.G. Houck, Ph.D.

7.3 Quality Maintenance, Genetics and Transportation Research Unit

Research on maintaining the culinary, nutritional and economic value of fruits and vegetables; genetic manipulation and breeding to improve quality and pest resistance of stone fruits and grapes; development of genetic techniques to increase the efficiency of breeding programs

Effects of pre- and post-harvest environments, including controlled atmosphere storage, on quality retention of vegetables and melons — W.J. Lipton, Ph.D.

Physiological and biochemical factors involved in development of chilling injury of melons and citrus fruits compositional factors and enzyme behavior involved in differential susceptibility of vegetables to injury induced by low 0_2 or high $C0_2$ levels in the storage atmosphere — C.F. Forney

Factors influencing inoculum potential of postharvest pathogens of horticultural crops; influence of temperature and modified atmospheres on pathogen-host interaction; elucidation of factors involved in development of various physiological disorders of stone fruits — D.J. Phillips, Ph.D.

Development of new and improved cultivars of grapes and stone fruits for the fresh market, including resistance to insects and diseases and improved culinary quality; development of root stocks with resistance to soil pests; development of genetic procedures for increasing efficiency of plant breeding — D.R. Ramming, Ph.D.

Influence of packaging using polymeric films on quality retention in fruits and vegetables; influence of variables in the storage environment on gas permeability of polymeric films — R.E. Rij

8 Beverly Hills

8.1 Virginia Robinson Gardens
Operated by the Department of Arboreta and Botanic Gardens of Los Angeles County
elevation : 152.4 m

Beverly Hills, California
Phone: (818) 446-8251
Dir.: F. Ching
Asst.Dir.: R. Ito

Testing of tropical and subtropicals in microclimate that does not exist at Arboretum in Arcadia. Original plantings of *Archontophoenix Cunninghamiana* has naturalized. Other plants including *Zingiberaceae, Heliconia* and flowering tropical trees.

9 La Canada-Flintridge

9.1 Descanso Gardens
Operated by the Department of Arboreta and Botanic

La Canada-Flintridge,
California 91011

Gardens of Los Angeles County
elevation : 381 m

Phone: (818) 790-5571
Dir.: F. Ching
Supt.: G. Lewis

Testing of plants that require cooler temperatures than normally occur at the Arboretum in Arcadia (see 2.1). Camellias, roses, azaleas, rhododendrons, lilacs, iris, naturalized daffodils, California native plants. Testing of ornamental trees and shrubs for resistance to oak root fungus under natural growing conditions.

10 Palos Verdes

10.1 South Coast Botanic Garden
Operated by the Department of Arboreta and Botanic Gardens of Los Angeles County
elevation : 118.8 m

Palos Verdes Peninsula,
California 90274
Phone: (213) 377-0468
Dir.: F. Ching
Asst.Dir.: R. Ito

First landfill botanic garden developed (1961) on a site that was once a diatomaceous earth mine, then a refuse dump. As a consequence, has become an outdoor laboratory studied by botanists, horticulturists, biologists, ecologists, and environmentalists from all parts of the world.
A pinetum, shrubs, ground covers, flowering fruit and other ornamentals trees totaling 500 genera, 1,000 species, and 150,000 plants, plus a man-made lake and stream, providing a haven for 175 observed species of birds, testify to triumph over high soil temperatures, subsidence, methane and carbon dioxide gases, and other landfill-related problems.

11 Pasadena

11.1 Fruit and Vegetable Chemistry Laboratory
Agricultural Research Service, USDA
elevation : 264 m
rainfall : 450 mm

263 South Chester Avenue, Pasadena
California 91106
Phone: (818) 796-0239
Dir.: Dr. V.P. Maier

The mission of this unit is to carry out basic research on the chemistry, biochemistry and physiology of subtropical and arid region plants and plant products. Emphasis is on the biosynthetic, metabolic, regulatory and developmental systems of plants, the biological activities of plant constituents and their derivatives, and the mechanisms and responses of plants to environmental stress. Results are applied to the solution of agricultural problems relating to the following:

Product quality, product yield and development of valuable by-products. — Dr. R.M. Horowitz/Dr. H. Yokoyama,

Bioregulation of plant and plant constituents to control processes such as growth, development, maturition and senescence for improved crop composition, performance, yield and postharvest stability. — Dr. S. Hasegawa/Dr. R.E. Bennett, C.F. Vandercook/Dr. M.A. Berhow, Dr. E. Hayman/Dr. W.J. Hsu, S.M. Norman/Dr. B.H. Tisserat

Characterizing aspects of stress amenible to biochemical and genetic manipulation for increasing crops resistance to environmental stresses.

Substation

Aiea-Hawaii 96701
Phone: (808) 487-5561
Dr. P.H. Moore/Dr. B.J. Wood,
Dr. D.A. Grantz

12 Riverside

12.1 University of California

Riverside, California 92521
Phone: (714) 787-3101
Dean: I.W. Sherman

12.1.1 Department of Botany and Plant Sciences	Chrms: I.P. Ting
Genetics and breeding of avocado, peppers, and tomatoes	B.O. Bergh
Tree spacing; growth control of subtropical trees and shrubs by mechanical, chemical, or other methods	S.B. Boswell
Function and genetics of plant hormones	E.A. Bray
Reducing environmental heat and cold stress on citrus and other crops; air pollution studies with important crops in the San Joaquin Valley	R.F. Brewer
Plant genetics; molecular structure and evolution of the chloroplast genome; genetics of plant populations	M.T. Clegg
Plant physiologist - influence of plant growth on horticultural crops, and analytical methods for plant growth regulators	C.W. Coggins, Jr.
Botany; developmental morphology	D.A. DeMason
Carbohydrate metabolism and mineral element metabolism	W.M. Dugger, Jr.
Population ecology and evolutionary genetics	N.C. Ellstrand
Mineral nutrition and fertilization of citrus and avocado; related nitrate pollution of ground waters	T.W. Embleton*
Environmental plant physiology and crop ecology	A.E. Hall
Plant and cell physiology, with particular respect to investigation of photosynthetic mechanisms and oxidative damage to membranes	R.L. Heath
Weed science: herbicide resistance, physiological and icological studies in the area of weed-crop competition, ecological approaches to weed control	J.S. Holt
Factors controlling cell growth in yeast and cultured plant cell lines, role of asparagine in cell growth; mutations in cultured plant cell lines	G.E. Jones
Plant physiologist - weed science; herbicide physiology, metabolism	L.S. Jordan
Research program aimed at elucidating chemistry, biosynthesis, and mode of action of growth-regulating compounds; metabolism determination of residues of herbicides	J. Kumamoto
Aspects of mineral nutrition in plants	C.K. Labanauskas*
Root physiology; structure and function of plant cell membranes as related to ion transport	R.T. Leonard
Plant physiology; genetics and plant breeding	L.F. Lippert
Botany; developmental morphology; reproductive biology	E.M. Lord
Regulation of nucleotide and arginine metabolism, biochemistry of mineral nutrition, citrus physiology	C.J. Lovatt
Plant tissue culture studies	T. Murashige
Factors associated with the propagation and maintenance of primary citrus budwood sources	E.M. Nauer
Plant physiology; interactions of cytoskeleton, membrane, and cell wall; light emission as a probe of senescence	E.A. Nothnagel
Plant breeding especially citrus; biochemical genetics plant evolution	M.L. Roose
Evolution, bio- and chemosystematics of various plant taxa, with emphasis on the *Autantioidea*	R.W. Scora
Genetics and nucellar embroyony of citrus	R.K. Soost*
Effects of air pollutants on agricultural crops and natural ecosystems (also in Statewide Air Pollution Research Center)	O.C. Taylor
Structure and function in cell biology; ultrastructure,	W.W. Thomson

structural, developmental and functional correlations in plant and animal systems	
Plant metabolism and gas transfer between plants and environment	I.P. Ting
Systematics and evolution; breeding systems in natural populations; phytogeography; vegetation ecology; vegetative reproduction	F.C. Vasek
Cytogenetics, plant breeding, and crop evolution-emphasis on wheat and beans	J.G. Waines
Genetics and molecular genetics of pathogen infection and protection from infection	L. Walling
Genetics and plant breeding	D.M. Yermanos

12.1.2 UCR Botanic Garden

Riverside, California 92521
Phone: (714) 787-3706
Dir.: Dr. J.G. Waines

Greenhouse, lathhouse, information center, shop, office and herbarium, and it encompasses a total of 37 acres

13 Salinas

13.1 U.S. Agricultural Research Station
USDA-ARS
elevation : 16.8 m
rainfall : 350.5 mm

1636 E. Alisal Street
Salinas, California 93905
Phone: (408) 443-2253
Loc.Leader: Dr. E.J. Ryder

Equipment: greenhouses, growth chambers, laboratories, 170 acres for field projects, electron microscope.

Lettuce - breeding for resistance to lettuce mosaic, big vein, infectious yellows, tipburn, corky root rot, lettuce root aphid; stress in growth and development; breeding methods; genetic studies	Dr. E.J. Ryder
Crucifers - genetics and breeding of broccoli and cauliflower; male sterility and incompatibility; disease resistance	Dr. J.D. McCreight
Cucurbits - breeding for resistance to watermelon mosaic, powdery mildew, squash leaf curl, infectious yellows; studies on male sterility, "bird's nest" plant type	Dr. J.D. McCreight
Artichoke - breeding for yield, uniformity, resistance to plume moth; insect monitoring, pheromone, parasitic control studies	Dr. N.E. De Vos*/Dr. M.A. Bari

14 San Francisco

14.1 Strybing Arboretum and Botanical Gardens*

Golden Gate Park, 9th Avenue,
Lincoln Way, San Francisco,
California 94122
Phone: (415) 558-3622
Dir.: W.R. Valen
Asst.Dir.: G. Georgeades

Owned by the City and the County of San Francisco and operated by the Recreation and Park Department. 64 Acres, 5500 species of plants from all over the world. Rhododendrons, magnolias, succulents, plant accession, education, library.

14.2 Strybing Arboretum Society

Phone: (415) 661-1316

15 San Marino

15.1 Huntington Botanical Gardens

1151 Oxford Road, San Marino

elevation : 183 m
rainfall : 380 mm
sunshine : 3000 h

California 91108
Phone: (818) 405-2160
Cur.: M. Kimnach

Private institution, 207 acres, library, herbarium.

Taxonomy of Cactaceae and Mexican Crassulaceae; transmission of virus in *Camellia;* taxonomy of orchaea (Orchidaceae). Breeding of new hybrids of succulents. Importation, testing and distribution of plants new to California horticulture, with emphasis on succulents, palms, cycads, bamboos, conifers, herbs, old rose cultivars and subtropical plants — M. Kimnach/F. Moore/J. Folsom

16 Santa Barbara

16.1 Santa Barbara Botanical Garden
elevation : 230 m
rainfall : 583 mm

1212 Mission Canyon Road,
Santa Barbara, California 93105
Dir.: Dr. R.N. Philbrick

Private institution, 65 acres. Native plants, *Ceanothus,* chaparral plants, *Mahonia, Rhamnus, Ribes and Arctostapylos,* herbarium (80,000 sheets)

Plants of the California Islands, Pacific Coast Cacti	Dr. R.N. Philbrick
Breeding and selection of native California plants	D. Emery
Taxonomy of *Polemoniaceae and Ribes*	Dr. S.L. Timbrook
Ecology of California island plants	M.C. Hochberg
Taxonomoy of California native plants	S.A. Junak
Propagation and culture of native California species	C.J. Bornstein

17 Saratoga

17.1 Saratoga Horticultural Foundation
elevation : 120 m
rainfall : 625 m
sunshine : 70 %

15185 Murphy Avenue, San Martin,
California 95046
Phone: (408) 779-3033
Dir.: P. McMillan-Browse

Selection and evaluation of superior forms of ornamental plants for Mediterranean-type climates - principally California. Develops procedures for propagation and production of these selections for the horticultural trade. Emphasizes shrubs and ornamental shade trees suitable for California.

Prepared by R.K. Webster, Dept. of Plant Pathology, Univ. of California, Davis, CA.
Sandra Fielden, Dept. of Environmental Horticulture, Davis, CA.
Don J. Durzan, Dept. of Pomology, Davis, CA.
Anne Rundstrom, Dept. of Vegetables, Davis, CA.
C.S. Ough, Dept. of Viticulture and Enology, Davis, CA.
W.J. Lipton, Protection and Quarantine Research Unit, Fresno, CA.
Philip McMillan-Browse, Saratogo Horticultural Foundation, San Martin, CA.
Edward J. Ryder, Agricultural Research Station, Salinas, CA.
Margaret D. Good, Dept. of Botany and Plant Sciences, Riverside, CA.
Barbara M. Pitschel, Strybing Arboretum and Botanical Gardens, San Francisco, CA.

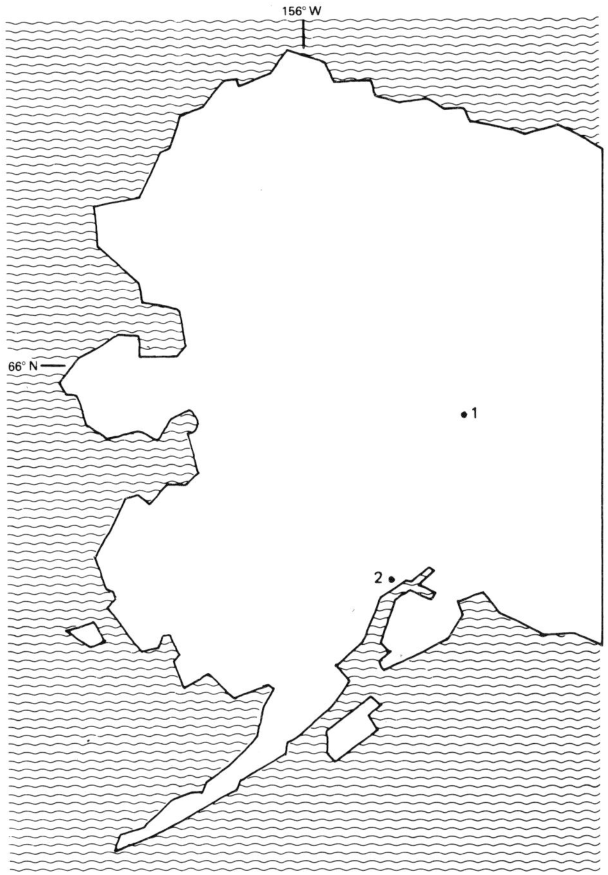
156° W
66° N
1
2

1 Fairbanks
2 Palmer

Survey

The Alaska Agricultural Experiment Station is a part of the Land Grant College with objectives similar to Land Grant Colleges of the other states. The approaches to solving problems confronting humanity in Alaska are through teaching, research and extension. The Experiment Station Center is 7 miles west of Fairbanks at College, Alaska, the site of the University of Alaska, Fairbanks. At College, Alaska the University has staff persons for work in Horticulture, Agronomy, Soils, Agricultural Economics, Animal Sciences and lesser related subjects.
The Experiment Station Research Center is at Palmer and Matunuska, 7 miles apart on the north shore of Cook Inlet and 40 miles northwest of the University of Alaska, Anchorage. At Palmer, the University has workers in Agronomy, Agricultural Engineering and Animal Science.

The U.S. Department of Agriculture invested heavily in structures and personnel in 1948 to further northern agricultural research. In 1967 the facilities and a few workers were transferred to the state. U.S. Department of Agriculture through Science and Education Administration maintains research workers in Horticulture, Agronomy and Soils, all of which are stationed at Palmer, Alaska. They are attached for Administration and logistics to the North Central Region at Fargo, North Dakota.

1 Fairbanks

1.1 Agricultural and Forestry Experiment Station
elevation : 145 m
rainfall : 313 mm

Afes 309, O'Neill Bldg, Univ. Alaska
Fairbanks, Alaska 99701
Phone: (907) 474-7188

sunshine : 3816 hrs

Telex: 35414 (Geophysical Institute refer to AFES)
Dir.: J.V. Drew

The station conducts variety trials and develops cultural practices for agronomic and horticultural crops for interior Alaska. Projects include conservation tillage of small grains in Delta, Alaska; horticultural crop production using simulated waste heat; development of forages and alternative feed sources for reindeer and livestock production; nutrient cycling in forest soils; study of tree and small grain diseases.

Range management	W.B. Collins
Weed science	J. Conn
Agricultural engineering	R.F. Cullum
Plant physiology, horticulture	M. Griffith
Horticulture	P. Holloway
Animal science	F.M. Husby
Agricultural education	C.A. Kirts
Resource management	C.E. Lewis
Plant pathology	J.H. McBeath
Forest management	E.C. Packee
Agronomy, soil microbiology	S.D. Sparrow
Agricultural economics	W.C. Thomas
Forest soils	K. Van Cleve
Agronomy	F. Wooding

2 Palmer

2.1 Agricultural Research Center
elevation : 46 m
rainfall : 398 mm
sunshine : 3715 hrs

AFES, University of Alaska,
533 E. Fireweed, Palmer,
Alaska 99645
Phone: (907) 745-3257
Telex: 35414 (Geophysical Institute refer to AFES)
Assis.Dir: S.H. Restad

The station at Palmer conducts variety trials and develops cultural practices for horticultural, agronomic, and forage crops. Other projects include breeding of small grains and forage crops, soil and tissue nutrient analysis, soil classification, study of potato diseases, and beef and dairy herd managment.

Agricultural engineer	L.D. Allen
Animal scientist	B. Bruce
Horticulture, plant pathology	D.E. Carling
Agronomy, forage crops	L.J. Klebasadel
Agronomy, range management	J.D. McKendrick
Agronomy, forage crops	W.W. Mitchell
Soils, soil classification	C.-L. Ping
Agronomy, plant breeding	R.L. Taylor

2.2 Alaska Plant Materials Center

elevation : 15 m
rainfall : 398 mm
sunshine : 3715 h

Palmer, Alaska 99645
State Div.of Agriculture, SRB 7440
Phone: (907) 745-4469
Mng.: R.H. Parkerson

Development, propagation, and seed production of hardy plant materials for agriculture, horticulture, and revegetation in Alaska. Trials are conducted at 11 sites throughout the state.

Potato disease control	B. Cambell
Seed analyst	B.J. Hall
Foundation seed project	E.J. Heyward
Conservation plants	N.J. Moore
Plant propagation	D. Ross
Horticulturist	C.I. Wright*
Plant materials specialist	S.J. Wright

Prepared by Marilyn Griffith, Agricultural and Forestry Experiment Station, Fairbanks, Alaska

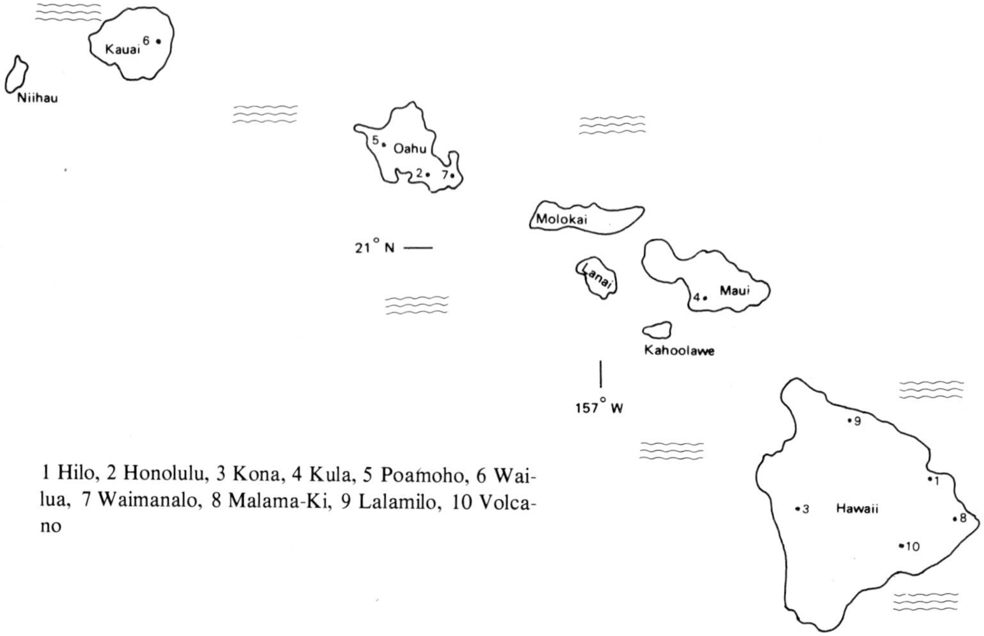
Kauai 6
Niihau
5 Oahu 2 7
Molokai
21° N
Lanai
4 Maui
Kahoolawe
157° W
9
1
3 Hawaii
8
10

1 Hilo, 2 Honolulu, 3 Kona, 4 Kula, 5 Poamoho, 6 Wailua, 7 Waimanalo, 8 Malama-Ki, 9 Lalamilo, 10 Volcano

Survey

In the last 60 years the University of Hawaii's College of Tropical Agriculture and Human Resources has grown from the humble beginnings to an internationally known research center dealing with special problems of tropical and subtropical areas.
Its facilities are found on every major island in the Hawaiian chain. Its staff has been called for consultation on projects in Asia, Australia, New Zealand, Pacific Islands, Africa, Central and South America. Its staff and advanced students are among the leading agriculturists in the state and in the Pacific Basin. Charged with broad responsibility for educating the people of the state as well as conducting research for the growers of tropical crops, the College maintains its research and educational conduct through branch stations of the Hawaii Institute of Tropical Agriculture and Human Resources and is in a large measure responsible for the steady improvement in agricultural methods and products.
Although most of the growing regions in the state may be considered to have a tropical climate, variations in termperature and rainfall varying with the altitude and extremely different micro-climate over relatively small areas provide an opportunity for growing a great variety of horticultural crops. The expansion of the horticultural industry consisting of macadamia, papaya, orchids, proteas, anthuriums, bananas, coffee, guava, avocado, foliage plants and others, is steadily increasing. Bananas and cherimoyas are increasing in importance. Taro, watercress, lettuce, cabbage and tomatoes are of high quality and available throughout the year. Sweet corn, sweet potatoes, and watermelons are common. Among the ornamentals, few are highly prized as orchids, anthuriums, plumerias (frangipani), giner, heliconias, bird of paradise, carnations, roses and hibiscus.

1 Hilo

1.1 Substation
elevation : 161 m
rainfall : 4064 mm

Branch Station of the Hawaii Institute of Tropical Agriculture and Human Resources.

461 W. Lanikaula Street, Hilo,
Hawaii, Hawaii 96720
Phone: (808) 935-2885
Admin: Dr. T. Higaki

Floricultural crop management and culture	Dr. T. Higaki
Breeding fruit and nut crops	Dr. P.J. Ito*
Foliage plant management and culture	Dr. R.Y. Iwata
Vegetable crop management and culture	Dr. B.A. Kratky
Fruit crop physiology and culture	Dr. M.A. Nagao

Malama-Ki Field Station for Fruit and Nut Crops — (see 8 on Map)
elevation : 92 m
rainfall : 2286 mm

Lalamilo Field Station for Vegetable Crops — (see 9 on Map)
elevation : 810 m
rainfall : 635 mm

Volcano Field Station for Vegetable and Fruit Crops — (see 10 on Map)
elevation : 1223 m
rainfall : 3810 mm

2 Honolulu

2.1 Bishop Museum

Intensive research in taxonomy of horticultural crops

1355 Kalihi Street, Honolulu,
Oahu, Hawaii 96819
Phone: (808) 847-3511
Dir.: Dr. E.C. Creutz

2.2 Hawaii Institute of Tropical Agriculture and Human Resources, University of Hawaii, Main Station

3050 Maile Way, Honolulu,
Oahu, Hawaii 96822
Phone (808) 948-8234
Dean: Dr. N.P. Kefford
Dir.: Dr. C.T.K. Ching
Ass.Directors: Dr. A. Demb,
Y. Kitagawa/Dr. K. Rohrbach

State Institution

Department of Horticulture

3190 Maile Way, Honolulu,
Oahu, Hawaii 96822
Phone: (808) 948-8389
Chairman: Dr. R.K. Nishimoto

The Department of Horticulture is part of the State Unversity supported largely by state funds with some federally supported projects. University of Hawaii is the only U.S. land-grant university situated in the tropics. The Department specializes in such crops as macadamia, papaya, coffee, guava, anthurium, orchids, ornamentals, and vegetables. A major interest has been genetics and breeding of horticultural crops. In 1970, the Department moved into a 5 million dollar Plant Science Building on Central Campus. This building, supported by the National Science foundation and the State of Hawaii, houses the Departments of Horticulture, Botany and Plant Pathology.

Genetics and breeding of corn, leguminous trees	Dr. J.L. Brewbaker
Food quality and techonology	Ms C.G. Cavaletto, M.Sc.*
Fruit crop improvement	Dr. C.L. Chia
Ornamental crop physiology and culture	Dr. R.A. Criley*
Vegetable crop physiology and culture	Dr. R.R. Coltman
Weed science	Dr. J. DeFrank
Genetics and breeding of bean, lettuce	Dr. R.W. Hartmann
Genetics and breeding of orchids, anthuriums	Dr. H. Kamemoto*
Crop modelling, fruit crop physiology	Dr. K.D. Kobayashi
Floricultural crop physiology, tissue culture	J.T. Kunisaki, M.Sc.
Ornamental crop improvement	Dr. K.L. Leonhardt
Genetics and breeding of papaya and guava	Dr. R.M. Manshardt*
Turfgrass management and culture	Dr. C.L. Murdoch

Weed science — Dr. R.K. Nishimoto
Ornamental crop management and culture — Dr. F.D. Rauch
Tissue culture — Dr. Y. Sagawa*
Vegetable crop improvement — Dr. K.Y. Takeda
Genetics and breeding of tomato, eggplant, sweet potato — J.S. Tanaka, M.Sc.

2.3 Harold L. Lyon Arboretum
University of Hawaii

3860 Manoa Road, Honolulu,
Oahu, Hawaii 96822
Phone: (808) 988-3177
Dir.: Y. Sagawa*

Intensive program in collection and exchange of horticultural plants.

2.4 Honolulu Bontanic Garden

Foster Botanical Garden,
50 N. Vineyard Boulevard,
Honolulu, Oahu, Hawaii 96817
Phone: (808) 533-3406
Dir.: P. Weissich

Intensive program in the introduction and collection of ornamental plants is conducted on Oahu.

3 Kona

3.1 Substation
elevation : 545 m
rainfall : 1270 mm

PO Box 208, Kealakekua,
Hawaii, Hawaii 96750
Phone: (808) 322-2718
Admin.: Dr. T. Higaki

Field research station of Hawaii Institute of Tropical Agriculture and Human Resources. Culture and improvement of coffee, macadamia and other fruit trees

4 Kula

4.1 Substation
elevation : 856 m
rainfall : 711 mm

PO Box 269, Kula, Maui,
Hawaii 96790
Phone: (808) 878-1213
Admin.: D. Shigeta

Field research station of the Hawaii Institute of Tropical Agriculture and Human Resources.
Protea improvement, culture and management — Dr. P.E. Parvin*

5 Poamoho

5.1 Substation
elevation : 266 m
rainfall : 1143 mm

Poamoho, Oahu, Hawaii 96791
Phone: (808) 948-7138
Admin.: Ms B. Arakawa

Field research station of the Hawaii Institute of Tropical Agriculture and Human Resources. Culture and improvement of vegetable, fruit and nut crops.

6 Wailua

6.1 Substation
elevation : 162 m
rainfall : 2286 mm

PO Box 278-A
Wailua, Kauai, Hawaii 96746
Phone: (808) 822-4984
Admin.: D. Ikehara

Branch station of Hawaii Institute of Tropical Agriculture and Human Resources.

Culture and breeding of horticultural crops — Dr. T. T. Sekioka

7 Waimanalo

7.1 Substation
elevation : 21 m
rainfall : 1143 mm

Waimanalo, Oahu, Hawaii 96795
Phone: (808) 948-7138
Admin.: Ms B. Arakawa

Field research station to the Hawaii Institute of Tropical Agriculture and Human Resources. Culture and improvement of horticultural crops.

Prepared by Roy K. Nishimoto, College of Tropical Agriculture and Human Resources, Univ. of Hawaii, Manoa, Hawaii

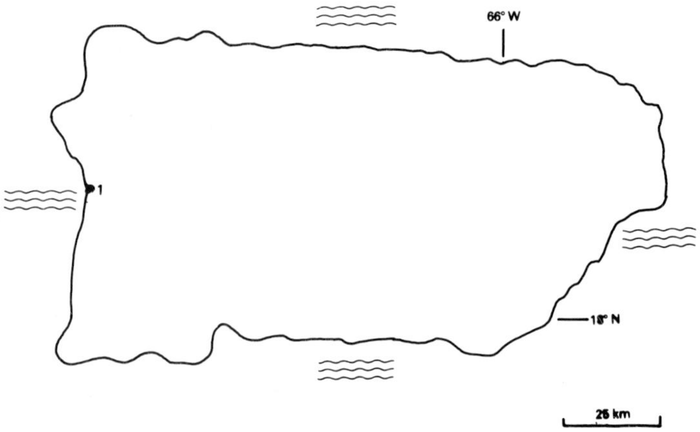
66° W
1
18° N
25 km

1 Mayaguez

Survey

The island of Puerto Rico is a Commonwealth of the U.S.A. It has a population of 3.2 million people; the climate is tropical, hot and humid with 1524 - 2540 mm rain and temperatures of 60 - 90°F.
The University of Puerto Rico was established in 1911. The Department of Horticulture is the academic and administrative division of the College of Agricultural Sciences dealing mainly with teaching, research and technical divulgations related to fruits, vegetables, ornamentals, starchy crops and coffee.
Research in the department is divided in six commodities, each one with a leader. Commodities are: coffee, fruits, vegetables, bananas and plantains, ornamentals, and starchy crops. Research priorities are discussed in annual meetings where teaching, research, extension and representatives of the Department of Agriculture and private industry are present. The Caribbean Basin Advisory Group, also sponsors some research projects in the Department. The Department maintains a summer exchange program with Rutgers University at New Brunswick, USA.
Extension activities in horticulture are carried on by specialists on the following commodities: coffee, starchy crops, vegetables, ornamentals and fruits.

1 Mayaguez

1.1 University of Puerto Rico,
College of Agricultural Sciences

Mayaguez Campux, College Station
Mayaguez, 00708

Acting Chancellor	L. Martinez Picó, Ph.D.
Acting Dean	J. Bird, Ph.D.
Acting Associate Dean	J. Román, Ph.D.
Acting Associate Dean	S. Hernández Gayá, M.S.

1.2 Department of Horticulture

Vegetables	E. Caraballo, M.S./G. Fornaris, M.S. Brunilda Luciano, M.S., Sonia Martínez, M.S., F.W. Martin, Ph.D./A. Morales, Ph.D. E. Orengo, M.S./J.A. Quiñones, M.s. A. Roque, M.S/D. Unander, Ph.D. Felícita Varela, M.S.
Fruits	A. Cedeño, Ph.D./C.A. Fierro, Ph.D. P. Márquez, M.S./F.H. Ortiz, B.S. O.D. Ramirez, Ph.D./I. Reyes, M.S. E. Toro, M.S.
Starchy crops	M. Cordero, M.S./J.J. Green, M.S.

Coffee
Ornamentals

H. Irizarry, Ph.D./M. Santiago, M.S.
E. Boneta, B.S.A./M. Monroig, M.S.
L. Flores, M.S./C. Mántaras, M.S.

Prepared by Jose L. Martinez Picó, College of Agricultural Sciences, Univ. of Puerto Rico, Mayaguez, Puerto Rico

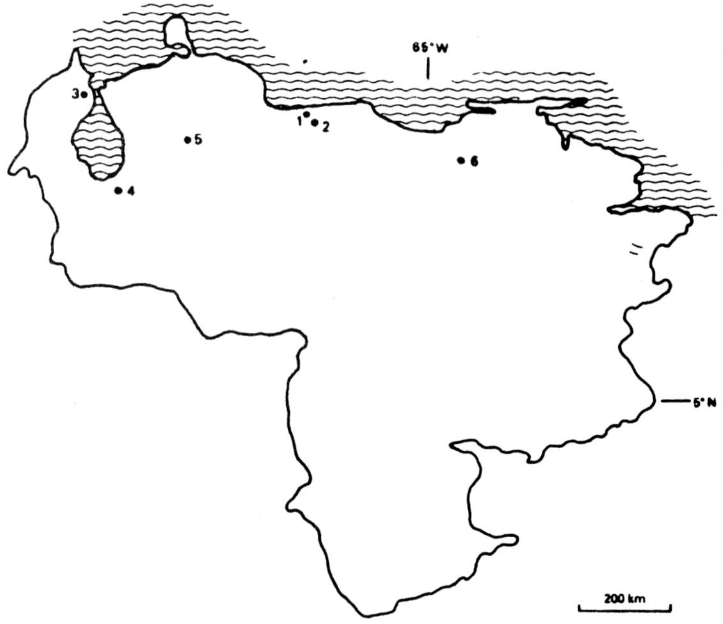
65°W
5°N
200 km
3
5
1
2
6
4

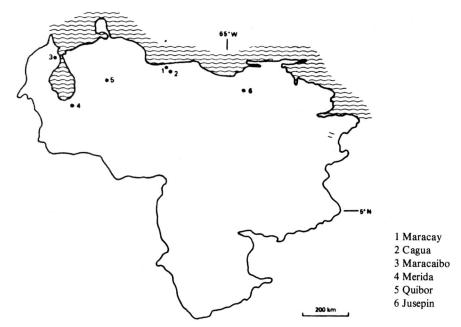

1 Maracay
2 Cagua
3 Maracaibo
4 Merida
5 Quibor
6 Jusepin

Survey

Venezuela is a country with climatic and soil conditions favourable for growing different horticultural plants. Vegetables (cool and warm season crops), fruits (tropical and temperate) and flowers are produced mostly for local consumption and some exports.
FONAIAP (1) has been working for about 30 years on research in fruits (tropical) and some vegetable crops (tomato, potato). It has several research stations around the country. FUSAGRI (2), a private non-profit organization was founded in 1952. Since then it has been working on vegetables (tomato, cucurbits, pepper, cabbage, onion, carrot, beet and garlic) and fruit crops (peach, citrus). FUSAGRI (3) at Zulia State was founded in 1973 and the center for viticultural development was started in 1977. They work on table grape production and wine. This program also makes some research on sapodilla and guava. The Institute for Agricultural research (4) makes some research on temperate fruits and vegetables.
FONAIAP (5) investigates on some aspects of vegetable crops (tomato, onion, pepper). Universidad de Oriente (6) investigates about tropical fruits and vegetables.
Publications - Agronomia Tropical, Noticias Agricolas and some technical reports.

1 Maracay

1.1 FONAIAP-CENIAP
elevation : 400 m

El Limón, Maracay, Estado Aragua

Agronomic research on tropical fruits (mango, guava, sapodilla, soursop) citrus and some vegetables (potato, tomato)

Ing. M. Figueroa
Ing. E. Debrot

2 Cagua

2.1 FUSAGRI, Vegetable Crops Program
elevation : 400 m

PO Box 162, Cagua,
2122 Aragua

Prepared by Dr. A. Flores G., FUSAGRI, Cagua

Cultural practices, varietal evaluation and handling practices of tropical vegetables, tomato, onion, pepper, cucumber, melons, cabbage, garlic	Ing. A. Flores G.*
Fruit Program Cultural practices on citrus, rootstock evaluations and agronomic research on bananas	Ing. R. Mendt

3 Maracaibo

3.1 FUSAGRI, Fruit Crops Program elevation : 4 m	Calle 76 N° 46-21, Frente Stadium, Maracaibo, Estado Zulia
Agronomic research on sapodilla, guava, tablegrapes, wine making	Ing. P. Corzo

4 Merida

4.1 Instituto de Investigaciones (Institute for Agricultural Research) elevation : 1600 m	Universidad de los Andes, Merida
Agronomic research on temperate fruits and vegetables	Ing. A. Rincón

5 Quibor

5.1 FONAIAP elevation : 300 m	Estación Experimental, El Cují, Quibor, Estado Lara
Agronomic practices on tomato, bell pepper, onion, eggplant	Ing. D. Delgado

6 Jusepin

6.1 Universidad de Oriente elevation : 80 m	Jusepín, Estado Monagas
Agronomic research on tropical fruits and vegetables	Ing. G. León

YUGOSLAVIA

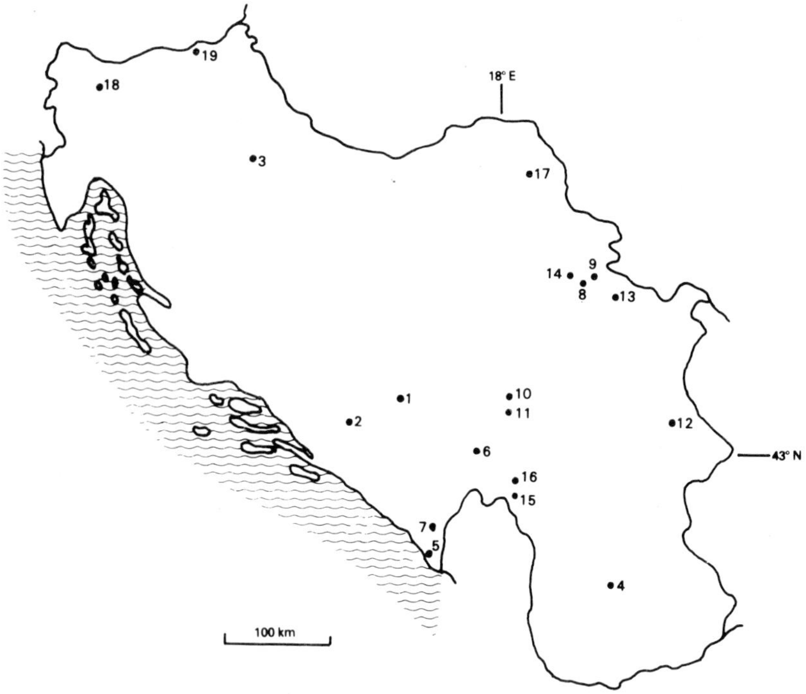

19
18
18° E
3
17
9
14
8
13
1
10
11
2
12
43° N
6
16
15
7
5
4
100 km

SR Bosnia and Hercegovina: 1 Sarajevo, 2 Mostar, **SR Croatia**: 3 Zagreb, **SR Macedonia**: 4 Skopje, **SRv Montenegro**: 5 Bar, 6 Bijelo Polje, 7 Titograd, **SR Serbia**: 8 Beograd, 9 Bolec, 10 Čačak, 11 Guča, 12 Niš, 13 Smederevska Palanka, 14 Zemun, **SAP Kosovo**; 15 Peć, 16 Priština, **SAP Vojvodina**: 17 Novi Sad, **SR Slovenia**: 18 Ljubljana, 198 Maribor

Survey

The Socialist Federal Republic (SFR*) Yugoslavia consists of 6 republics and 2 socialist autonomous regions, (in alphabetic order): SR Bosnia and Hercegovina, SR Croatia, SR Macedonia, SR Montenegro, SR Serbia including SAP Vojvodina, SAP Kosovo and SR Slovenia. The countries described extend between latitudes 40° and 47° N and between longtitudes 13° and 23° E. The total area amounts to 255,804 km^2. About 85% of the country is situated in the moderate continental climate, 10% is classed as continental and another 5% as littoral climate. Maximum elevation is about 2,842 m but commercial growing is below 600 m, except for some fruit like blueberry, etc. The population is about 22.5 million, agriculture population is 19.9% (1981).

Horticulture growing region and climate.

According to the climate conditions, the horticulture production in Yugoslavia is located in three main regions. In littoral type of climate, the horticulture growing region covers all islands, a narrow belt of Adriatic coast and river estuaries. In these areas the principal fruit crops are olives, figs, marasca, cherries, almonds, peaches, citrus fruits (mandarine, orange, and lemon), *Ceratonia siliqua, Punica granatum* and *Diospyros kaki.* The intensive orchards and other crops are mostly irrigated.

In the continental climate horticultural growing regions are located on the Pannonian lowlands (mainly SAP Vojvodina) and nearby. In these areas the main horticultural crops are grapes, apples, apricots, peaches, pears and sour cherries, and ornamental plants. The intensive horticultural fields are mostly irrigated. In the moderate continental climate which spreads throughout the rest of Yugoslavia, the horticultural growing regions are to be found in the mountainous areas and some in the plain areas. In

* Abbreviations: SFR Sociajalistička Federativna Republika (Socialist Federal Republic), SR Socijalistička Republika (Socialist Republic), SAP Socijalistička Autonomna Pokrajina (Socialist Autonomous Province).

Prepared by Prof. Dr. S.A. Paunović, Faculty of Agronomy, Horticulture Department, Čačak.

mountainous areas the main fruit crops are plums and prunes, small fruits, quince, cherries, apricots, nuts, grapes, potato and ornamental plants and in plain areas there are apples, pears, peaches, vegetables, grapes and flowers. The intensive horticultural fields are mostly irrigated, in the plain areas and some of them in the mountainous areas.
The most frequent types of soil in production areas are: chernozem, degraded chernozem, various podzolic, brown, chesnut and sandy soils, light and heavy black soils, sierosem, terra rossa and alluvial soils.

Development of horticultural industries.

Yugoslavia's horticultural industries can be divided into two periods: before and after World War II. The first period is generally characterized by extensive horticultural production, except in some areas where there was very intensive fruit and grape growing. In orchards, trees as standards were planted far apart, management was inadequate, e.g. spraying programmes, manuring, irrigation etc. were often absent. Plum cultivation was most developed. Other fruits had no cultivars of a high commercial value. Plantations were small, like gardens. But in this period a start was made to introduce different kinds of fruits and cultivars from abroad, mainly from USA and France e.g. apples, pears, plums, peaches, sweet and sour cherries. Grapes, vegetables and floriculture, like fruit trees were not grown well, except grapes. There were mainly wine grapes and a few table cultivars, vegetables were almost only grown in season; there were not enough good cultivars in floriculture. Research and technical staff were few. There were only two, very new, horticulture research stations (Gorazde and Maribor), many small state nurseries, two Departments of Horticulture at the Agricultural Faculty (Beograd, Zagreb), three technical agricultural schools and a few others. During the Second World War, the horticultural industry was almost ruined and much was destroyed. The second period, after World War II, can be divided into six phases.

1. New farms (cooperative and state) were formed from the lands which were already planted with horticultural plants following agrarian reform and other state laws. These orchards and vineyards were too small and scattered, and consisted of different kinds of fruit with many cultivars.
2. This period (1946-1955) can be marked as follows: the planting of new orchards and vineyards took place mainly on collective farms and to a small extent on state farms; orchards and vineyards were planted in small plots, up to 10 ha. Tall standard fruit trees, many kinds and cultivars and small numbers of high-quality cultivars were planted; management of the young and bearing orchards and vineyards was better than on private farms, but still not good enough to get high production.
3. In this period (1956-1959), which followed the reorganization of the cooperative farm and some stabilization of the socialist farms, a faster development occurred. New plantations were planted by the state and by cooperative farms in larger sized plots (10 to 30 up to 50 ha), with not many kinds of fruit and only few cultivars. Tall standard fruit trees were rarely planted and better preparation of the soil occurred.
4. In the next period (1960-1968) was the application of modern principles of intensive fruits, grapes, and vegetables production. The main characteristics are: large orchards and vineyards, up to 1,000 ha; deep ploughing, up to 1 m of all orchards and vineyard plots; very large quantities of manure, intensive planting in orchards, wider spacing in vineyards (3-4 m), a new commercial system of growing (espailiers), bush trees in orchards; management was such that mechanization could be used for cultivation, spraying and picking. Everything was done to get rapid growth in the first few years and earlier fruiting and to obtain maximum yields. This period was the beginning of concentration and specialization in orchards, vineyards, vegetables and floriculture fields.
5. This period (1969-1978) is similar to the previous ones, except that they started to plant new fruit plantations much more densely (pillar system) up to 3,000 trees per ha, and to control yields by chemical means and weeds by herbicides and in the summer to apply foliar fertilization together with fungicides; also more attention was paid to cultivars and rootstocks, e.g. more to quality than to quantity of fruits, and generally to profitability of fruit production.
6. This period is from 1979 up to the present. Characteristics of this phase are producing, more or less in mass virus free plant materials. Then export of nursery fruit and grape plants, and in the last two years the production by tissue culture was started on some rootstocks and small fruits, as well as flowers several years ago. In the most intensive orchards approximative yields range from 35,000 to 45,000 kg/ha.

There are now 4 specialized Fruit Research Institutes, Čačak, Ljubljana, Sarajevo, Skopje; 8 Fruit Research Stations, Bijelo Polje, Maribor, Mostar, Peć, Samobor for temperature fruits and Bar, Split-Kastel Stari for subtropical fruits; 1 specialized Vineyard Research Institute, Niš; Center for Genetics and Breeding of Grapes, Radmilovac-Vinca, near Beograd; 2 specialized Vegetable Research Institutes, Smederevska Palanka, Novi Sad; 1 Potato Research Station, Guča, Institute for Adriatic Culture, Split; PKB "Agroekonomic"; Fruit Research Station, Radmilovac-Vinca, near Beograd; 8 Horticultural Departments of Agriculture Faculties, Zemun-Beograd, Zagreb, Ljubljana, Novi Sad, Sarajevo, Skopje, Priština and Cacak; 6 Departments of Floriculture and Landscape Architecture, Beograd, Zagreb, Ljubljana, Sarajevo,

Novi Sad, Skopje; 2 commercial laboratories for virus free materials, Čačak, Sarajevo; 2 commercial laboratories for propagation of fruit plants by tissue culture, Knjazevac, Split. In total about 435 researchers and professors are employed. No private organization undertakes any research on fruits, grapes and vegetables.

Present conditions in horticulture.

Today there are about 14,254,000 ha agricultural lands of which 3.51% is under orchards, vineyards 1.67% and vegetables 4.48%, and nurseries 0.034%. In 1984 the total number of fruit trees, including subtropical plants (6,694,000 plants) amounts to about 173,343,000 of which 142,525,000 are in bearing with an average (1980-1984) production of 2,077,317 tons. In this production temperate fruit takes part of 85-61%, subtropical 1,58%, small fruits 8,22% and wild not-cultivated fruits 4,59% (in the forest as strawberries, raspberries, blackberries, blueberries, nuts, chestnuts, dogwoods, dogroses). The principal temperate fruit crops in Yugoslavia are plums with 43.52%, apples 30.32%, pears 7.79%, peaches 4.97%, sour cherries 4.97%, and all other kinds of temperate fruits about 8.43%, with increase of nuts, apricots and quince. Olive 0.8%, figs 0.6% and mandarine 0.41% are the most important subtropical fruits, as well as strawberries 5.27% and raspberries 2.37% as small fruits, but with rapid increase-blackberries 1.05% and blueberries too, and beginning to be planted more and more: *Actinidia Chinensis*. Home canning of dried fruits amounts to 26,820 tons, of which dry prunes account for 22,140 tons, jam 13,369 tons, olive oil 24,310 hl, plum brandy (slivovits) 662,280 hl of 25° and 204,100 hl of 45°, brandy from other fruits 34,000 hl, and 22 other products.

SR Serbia ranks first in fruit production among the Republics. More than 62% of the total Yugoslavia fruit production is there. (SR Serbia 85.5%, SAP Vojvodina 11%, SAP Kosovo 3.5%) followed by SR Bosnia and Hercegovina 14%, SR Croatia 12%, SR Macedonia 7%, SR Slovenia 3.5% and SR Montenegro 1.5%. Total area under *vineyards* is 238,000 ha. On the American rootstocks 212,000 ha, land under the domestic *Vitis vinifera* and hybrids 26,000 ha. The mentioned areas cover 1,329 million of grape plants from of which 1,251 million are in fruiting stage. Varieties which are grafted into American rootstocks produce grapes from 1,313,000 to 1,780,000 tons (1980-1984). There is very modern winemaking which produces from 6,291,000 hl to 8,576,000 hl of wine and grape brandy (komovitz) 340,000 hl, as well as other products (vinjak= cognac). In Yugoslavia there are 21 vineyards regions, 31 subregions, and 190 vineyard areas (large locations), and special state law relating to growing areas of grape cultivars and wine types production. Grape-wine cultivars are grown in relation to table-grape cultivars as 90:10% and coloured wine (black wine) in relation to white wine as 60:40%.

Total harvested area under commercial production of principal *vegetables* amounts to about 567,411 ha with a total production in 1984 of about 5,006,000 tons. The principal vegetable crop is potato with 49.08%, followed by cabbage and kale 15.20%, tomato 9,58%, cantaloupe and watermelon 9.04%, onion and garlic 7.6%, pepper 6.57%, but other kinds of vegetables are grown in a small percentage 2.98%. Beside the registrated, also produced are many thousands of hectares under non-registered kinds of vegetables: lettuce, spinach, carrots, greens, beet, hren, radish, leek, string beans, zucchini, mushroom, sweet corn etc., for local market and also 350 ha under greenhouse, plastic materials (mostly tomato, cucumber, peppers, etc.). Canned vegetables amount to about 208.113 tons - final production. The tendency in outdoor vegetable production is for the principal cultivars to be grown throughout the year, since Yugoslavia has excellent climatic conditions. Cool-season and warm-season varieties are produced in the most suitable climate (littoral, continental and moderate continental) according to the season.

Ornamental floriculture.

This industry comprises two distinct parts: commercial floriculture and nursery plant production. Commercial floriculture is mainly the production of bulbs and seed outdoors in glasshouses and plastic houses. With gladiolus, petunia and roses there are active breeding projects. Nursery plant production consists of growing woody ornamentals in the field or in containers. Principal crops are mainly cultivars of holly, camellia, ligustrum, barberry and others.

Each year Yugoslavia exports about 165,000 tons of fruit, 130,000 tons of vegetables, brandy and wine about 80,000 to 90,000 tons, 5,000 tons of table grapes etc. Within the fruit sector the sour cherries take first place in the export, followed by raspberries, prunes and plums, dried prunes, strawberries and within the vegetable sector peppers, followed by potatoes and mushrooms (5,000 tons). Yugoslavia imports every year large quantities of citrus fruits, banana and other subtropical and tropical fruits and spends about $ USA 80-100 million a year on it.

Specialization in fruit, grape and vegetable growing.

Fruit growing within the socialist sector is about 9.98%, pure fruit or agriculture industrial plants is about 7.91% and agricultural farms 2.07%. Private sector (individual farms) is 90.02%. Fruit industrial plants are fruit organization for production, storage and marketing. Agricultural industrial plants have a special fruit section which is organized as fruit industrial plants. Both fruit and agricultural industrial plants are located over all Republics in the best ecological and economical conditions. These organizations represent a powerful factor which can and should carry out the modernization and reconstruction of the fruit industry, provided that these organizations operate profitably. The fruit organizations mentioned have orchards mostly in one complex of 100 ha to 1,000 ha or have concentrated fruit production in small areas, up to 6,000 ha ("Beograd", "Opuzen", "Smederevo").

Fruit growing in agricultural farms is a part of mixed farming. Private individual farms are dominant in Yugoslav fruit production, with maximum size of arable lands 10 ha. On these mixed farms the sizes of orchards range from 0.5 to 2 ha, and on pure fruit farms orchard size ranges from 5 to 8 ha. Also, many non-agricultural persons grow fruits, grapes or vegetables in small gardens up to 0.5 to 1 ha (many kinds and cultivars).

In the last 10 years already established and going on in the progress to establish more and more new ones e.g. orchards or vineyard farmers united lands which range from 25 to 200 ha, or more (one kind of fruit) in one complex on the basis of economic and social collaboration between the enterprise and united farmers.

Wild fruit as possible source of genotypes.

In mountainous regions of Yugoslavia, besides the forest trees, there can be found many wild fruit trees which grow very densely and frequently in the form of regular forests. Usually fruit forests are composed of wild fruit trees of different species and in some cases almost pure stands of one species. The fruit forests of pears *(Pyrus communis)* grow in East Serbia, *Pyrus amygdaliformis* in Macedonia, walnuts *(Juglans regia)* in Hercegovina, filbert *(Corylus avellana)* in East of Serbia and Hercegovina, blueberries *(Vaccinim myrtillus, V. uliginosum, V. vitis idaes),* strawberries *(Fragaria vesca),* red raspberries *(Rubus idaeus subsp. vulgatus),* blackberries (Rubus 14 species), *Cornus mas, Crataegus* (5 species) and *Sorbus* 4 species, nearly in all mountainous areas of Yugoslavia, peaches *(Prunus persica* Stock) locally called "domestic vineyards peach", mostly used for rootstocks, plums and prunes *(Prunus domestica, Prunus insititia, Prunus spinosae, Prunus cerasifera, Prunus cocomilia, Prunus prostrata, Prunus avium, Prunus mahaleb)* mostly in Serbia, Bosnia and Hercegovina and Macedonia, *Amygdalua nana, Amygdalus webbii* mainly in Macedonia and Hercegovina, *Malus pumila, Malus florentina, Corylus colurna, Vitis sylvestris* can be found in Macedonia, Serbia, Hercegovina. Also, other wild fruit species can be seen in the forest.

Principal horticultural industries - Temperate Fruits.

There is a commercial industry of various sorts of temperate fruits in the following (alphabetical) list of Republics: SR Bosnia and Hercegovina: plum, prune, pear, peach, sweet and sour cherries, walnuts, figs and quince; SR Croatia: apple, pear, sweet and sour cherries, peach, olive, fig, citrus and walnuts; SR Macedonia: apple, pear, peach, sour cherries, apricot, walnuts and *Actinidia;* SR Montenegro: peach, fig, olive and citrus; SR Serbia: plum, prune, apple, sweet and sour cherries, pear, peach, apricot, walnuts, filbert, raspberry, strawberries, blackberry, currants and quince; SR Slovenia: apple, pear, peach, sweet cherries and currants. Strawberries are grown in all Republics and commercial industry of olive and figs is near the Adriatic Coast.

Viticulture

The commercial grape industry is in SR Serbia, south and south-west of SR Macedonia, north and near the Adriatic of SR Croatia, north-east and near the Adriatic coast of SR Slovenia, the regions of Hercegovina and the central and near Adriatic coast of SR Montenegro.

Vegetables

Several major kinds of vegetables are grown on a commercial scale in Yugoslavia, but mostly in SR Serbia and SR Croatia, then in SR Bosnia and Hercegovina and SR Macedonia. Other kinds are mainly grown for local markets. The commercial vegetables industry consists of the following species: potato, cabbage, tomato, onion, pepper, beans, peas, watermelon, carrot with a variety of domestic and imported cultivars.

Ornamental floriculture

More than 40 different flower cultivars are grown in various regions.

SR BOSNIA and HERCEGOVINA

1 Sarajevo

ul. Butmirska 222
71210 Ilidża, Sarajevo
Phone: (071) 621-314

elevation : 630 m
rainfall : 920 mm
sunshine : 160 days/year

Area 135 ha, comprising experimental plots 50 ha, glasshouses 50 m^2 and screen house 75 m^2.

Fruit Section:

Clone selection of plums and sour cherries, growing plums possibility on poor productive soils and products of plum by applied low temperature	Dr. D. Jarebica/Ing.Agr. N.Stajić
Cultivar trials of apple and small fruits, clonal selection of *Prunus cerasifera*	Dr. A. Šoškić/ Ing.Agr. U. Karahasanović
Hazelnut and walnuts growing and selection	Dr. B. Manušev/K. Efendić
Cultivar trials of peach, apricot and sweet cherry	Ing.Agr. M. Pirnat
Foliar nutrition of plums and trials of small fruits	Ing.Agr. T. Čarkić
Cultivar trials of pear	Ing.Agr. S. Lukić
Morphology of cultivars, application of growth regulators, storage of fruits	Dr. A.S. Muratović

Virus Section:

Spreading, host, races and vectors of virus sharka (plum pox), virus of apple	Dr. H. Festić
Vectors of virus sharka and viruses of *Vitis Vinifera,* integral plant protection	Ing.Agr. S. Djurić

1.2 Faculty of Agriculture, Institute of Fruit Growing and Viticulture

ul. Zagrevačka 18
71000 Sarajevo
Phone: (071) 613-033

Established in 1947. Lecturing and research on fruit growing, viticulture processing of fruits, grapes and vegetables. The experimental work of this institute is mainly done on fields of estate agricultural properties and of Fruit and Viticulture Station in Sarajevo.

Development of root systems of stone fruits on different rootstocks, nutrition, pome fruit trials	Prof.Dr. V. Lučić
Clone selection of rootstocks for plums and apricots, cultivar trials in plum and apricots, application of growth regulators	Ing.Agr. N. Mićić
Air pollution in fruit growing, selection of clone rootstocks for plum and apricots, fruit nursery	Prof.Dr. V. Prica
Selection of grape cultivars, development of root system, processing of grapes and the use of by-products	Prof.Dr. P.T. Vuksanović
Selection of grape cultivars, processing of grape	Ing.Agr. D. Mijatović
Canning of fruits and vegetables by technics of drying, cooling, freezing and concentrating	Dr. M. Savić

2 Mostar

2.1 APRO Research Development Institute
elevation : 60 m
rainfall : 1300 mm
sunshine : 2,400 h

ul. Put za Glavisu 10
79000 Mostar
Phone: (079) 55-220, 480-107
Telex: 46272

Research and development on fruit growing, viticulture, wine, vegetables, floriculture, landscape gardening, forestry, plant protection, soils, nutrition and economics. Areas of several thousand ha, comprising experimental plots 15 ha, glasshouses several hectares, plastic tunnels. Laboratory for pedology and plant nutrition, vegetative propagation by tissue culture, plant protection and enology.

Viticulture, fruit growing and enology	Ing.Agr. R. Kovačina, M. Sc.
Floriculture, landscape architecture and vegetables	Dr. J. Pehar
Tobacco	Dr. J. Beljo
Plant protection	Ing.Agr. V. Nadaždin
Soils and nutrition	Dr. E. Hanić
Agroeconomic	Dr. V. Trninić

SR CROATIA

3 Zagreb

3.1 Faculty of Agriculture, Institute for Fruit Growing, Viticulture, Enology and Vegetables elevation : 123 m rainfall : 870 m sunshine : 1363 h	ul. Šimunska cesta 25 41000 Zagreb Phone: (041) 216-777

Established in 1933. Area 25 ha state farm land. Physiology, selection, cultivar trials of fruit trees, grapes, vegetables ornamental plants and vines.

Fruit Section:	
Morphology and distribution of root system of peaches and apples, iron deficiency chlorosis of pear, apples and peaches on calcareous soils, interrootstocks in pears	Prof.Dr. I. Miljković
Pear cultivars, structure of pear fruits, selection of sour cherries	Ass.Prof.Dr. I. Dubravec
Hazelnut fertilization and selection of indigenous hazelnuts	Prof.Dr. I. Miljković, Ass.Prof.Dr. I. Dubravec
Plum rootstock trials	Ing.Agr. N. Paulić
Foliar analysis	Ing.Agr. V. Iveković
Viticulture:	
Ecology, training, pruning and nutrition of grapevines	Ass.Prof.Dr. R. Bišof
Genebank of grape cultivars	Ing.Agr. N. Mirošević, M.Sc.
Selection and pollination in grapevine cultivars	Ing.Agr. N. Mirošević M.Sc., Ing.Agr. M. Vičić
Foliar diagnoses	Ing.Agr. I. Gagro
Introduction and selection of grape cultivars	Dr. M. Fazinić/Ing.Agr. Z. Ivanković
Vegetable Section:	
Growth and development of vegetables, method of production	Prof.Dr. R. Lešić, J. Borošić, M.Sc./Ing.Agr. B. Novak
Peas, beans, spinach, carrot, cucumber, cabbage, onions, cauliflower and mushrooms cultivar trials	Ing.Agr. D. Tomas/Ing.Agr. I. Zutić
Selection and technics of seeds production of vegetable cultivars	Ing.Agr. J. Lovoković, Ing.Agr. L. Vrbanc
Floriculture:	
Landscape gardening and growing of ornamental plants	Prof.Dr. V. Jurčic
Growing of ornamental plants	Ing.Agr. M. Toplak, M.Sc., Ing.Agr. I. Vršek
Landscape gardening, planning, planting, growing	Ing.Agr. M. Kurtela

3.2 Faculty of Agriculture, Institute for Plant Protection	ul. Šimunska cesta 25 41000 Zagreb Phone: (041) 216-777

Established 1933. Area 9 ha (Zagreb, Rim Street 98). There are situated vineyards, orchards, vegetables and greenhouses intended exclusively for experimental aims.

Zoology Section:	
Entomology problems in orchards, vineyards, vegetables and ornamental plants, integrated control	Dr. M. Maceljski
Nematodes in various crops	Ing.Agr. Lj. Oštrec, M.Sc.
Aphids on various crops, head of forecast service for SR Croatia	ng.Agr. I. Igrc, M.Sc.

Insect and other pests of vegetables and ornamentals in fields and glasshouses, acarology	Ing.Agr. N. Pagliarini, M.Sc.
Phytopathology Section:	
Virus diseases of fruit trees, grapes and vegetables	Dr. A. Šarić
Fungus diseases of fruit trees, grapes and vegetables, official testing fungicides in these crops	Dr. B. Cvjetković
Weed control	Ing.Agr. V. Lodeta

3.3 Faculty of Forestry, Department of Forestry and Ornamental Plants

Šimunska cesta 25
41000 Zagreb

Established in 1922. Area 6 ha. Ecology, phytocenology, technics of forest growing, amelioration of forest soils, parks and landscape gardening.

Dendrology of forest trees, ornamental trees and shrubs, phytocenology and ecology	Prof. Dr. M. Antić
Technics of forest growing	Prof.Dr. I. Dekanić/Ing. S. Matić
Amelioration of forest soils	Ing. A. Horvat/Ing. A. Tomašivić
Ornamental trees and shrubs, parks and landscape gardening	Ing. Z. Badovinac
Biology of forest trees	Dr. B. Prpić
Phytocenology in forestry	Ing. D. Rauš

3.4 Faculty of Agriculture, Research Institute for Mechanization, Technology and Construction in Agriculture

ul. Šimunska cesta 25
41000 Zagreb
Phone: (041) 216-777

Research in soils management, plant protection and picking by machines, fruit, grapes and vegetables

Machinery in vegetable production	Prof. Dr. J. Brčić
Machinery in fruits and grape production	Prof. Dr. M. Dujmović
Machinery in vegetables and plant protections	Ing. B. Antonić, M.Sc.
Machinery in potato production	Ing. J. Barcić

SR MACEDONIA

4 Skopje

4.1 Faculty of Agriculture,
Fruit Research Institute
elevation : 245 m
rainfall : 489 mm
sunshine : 180 days

ul. Prvomajska 5
91000 Skopje
Phone: (091) 230-557, 232-257

Established in 1947. Area 90 ha and nursery production. Biology, physiology, selection, pomology, agrotechnics.

Introduction and pomology of fruit species and cultivars	Prof. Dr. B. Ristevski
Walnuts, hazelnuts, almonds	Ing.Agr. I. Kuzmanovski, M.Sc.
Apricots, peaches and citrus	Ass.Prof. Dr. Z. Mitevski
Biology of pears and *Actinidia Chinensis*	Dr. R. Spirovska
Plums, dogrose, *A. Chinensis*	Ing.Agr. D. Georgiev, M.Sc.
Apples and citrus	Ing.Agr. H. Popovski, M.Sc.
Small fruits	Ing.Agr. D. Veleva
Sweet and sour cherries and irrigation	Ing.Agr. O. Kosevska

Canning fruits	Dr. L. Sivakov
4.2 Faculty of Agriculture, Department of Plant Pathology	ul. Autokomanda bb 91000 Skopje
Diseases of stone fruits, infection, degeneration, powdery mildew of grapes, rootrot in tomato and peppers	Prof.Dr. K. Minev/Ing. F. Pejčinovski
Taxonomical and applied entomology	Prof. Dr. P. Atanasov
4.3 Department of Vegetables	ul. Autokomanda bb 91000 Skopje

Area 30 ha and experimental fields of the Agricultural Research Institute (500 ha), glasshouses 1,500 m^2.

Selection in tomato, peppers and melons, tomato growing and agrotechnics	Dr. S. Popov
Morphological and qualitative properties of vegetables and growing of peppers	Prof. Dr. M. Cirkova-Georgievska
Collection and research of local cultivars of onion, production of vegetables under polyethylene film	D. Simonov/T. Tudžakov, V. Demirovska/D. Jankulovski

SR MONTENEGRO

5 Bar

5.1 University of "Veljko Vlahović", Agricultural Research Institute, Subtropical Research Station — 81350 Bar
elevation : 10 m
rainfall : 1527 mm
sunshine : 2525 h

Established 1937. Area 14 ha and state farm lands.
Citrus, pomegranate, olive, figs, *Actinidia Chinensis*

Citrus, Pomegrante, figs, *A. Chinensis*	Ing.Agr. M. Plamenac
Olive	Dr. K. Miranović, Ing.Agr. S. Vujanović
Pesticides	Ing.Agr. Č. Gojnić
Chemical and technological properties of fruits	Dipl.Chem. V. Nikčević

6 Bijelo Polje

6.1 University of "Veljko Vlahović", Agricultural Research Institute, Fruit Research Station — 8400 Bijelo Polje
elevation : 550 m
rainfall : 750 mm
sunshine : 1600 h

Established in 1950. Area 40 ha and state farm lands. Apples, pears, plums, sweet and sour cherries and small fruits under the ecological conditions of the northern parts of Montenegro and Sandžak; nursery plants production.

Selection of apples and pears, cultivar trails of apples, pears and small fruits	Dr. R. Jovančević
Biological properties and cultivar trails of plums, cherries and small fruits	Dr. Lj. Krgović

Chemical and technological properties of fruits — Ing.Agr. J. Zečević

7 Titograd

7.1 University of "Veljko Vlahović", Agricultural Research Institute
elevation : 52 m
rainfall : 1500 mm
sunshine : 2519 h

Bul. Revolucije 9
81000 Titograd
Phone: (081) 41-940, 41-764
Dir.: Dr. Z. Kalezić

Established in 1946. Area 35 ha and state farm lands, 40 m^2 under glass. There are 7 departments: vegetables, viticulture, fruit growing, floriculture, shrubs trees and other ornamental plants, plant protection, pedology and agroeconomics, and 2 research stations in Bijelo Polje and the Subtropical Research Station in Bar. Investigations of grapes, peaches, some less of sweet and sour cherries and almonds, of tomato, peppers, cantaloupe, watermelon, onions, peas, beans in the open fields and under protected cultivation, diseases, pests and weeds.

Viticulture	Dr. M. Ulicević/Dr. Lj. Pejović
Fruit culture	Prof. Dr. S. Bulatović
Vegetables	Ing.Agr. M. Golović
Phytopathology	Dr. M. Mijušković, Ing.Agr. Z. Vučinić, Ing.Agr. J. Tiodorović
Entomology	Ing.Agr. V. Velimirović, M.Sc.
Agroeconomics	Dr. Ž. Kalezić/Ing. Agr. B. Bulavić, Dipl.ecc R. Djukić

SR SERBIA

8 Beograd

8.1 Plant Protection Institute
elevation : 110 m
rainfall : 657 mm
sunshine : 1930 h

ul. Teodora Drajzera 7
11000 Beograd

Established in 1945. Diseases, pests, nematodes, mites, control by chemical and biological methods, pesticides and their value.

Diseases Department:	
Virus diseases of fruit trees and grape	Dr. M. Jordović
Fungus diseases of fruit trees	B. Borić
Pest Department:	
Nematodes in agricultural plants	Dr. G. Gruičic
Mites in agricultural plants	Dr. T. Stamenković
Aphids and coccides	Dr. N. Mitić-Mužina
Phytopharmacological Department:	
Pestidices and raw materials for formulation of pesticides	Dr. N. Ostojić/A. Zabel, D. Matijević/Lj. Stojanović
Biological Control Department:	
Biological control of *Hyphantria cunea* and *Limantria cunea*	Dr. M. Maksimović
Integrated plant protection	Dr. M. Injac
Influence of insecticides on entomophages, biological control of *Pyrausta nubilalis*	B. Manojlović
Rodents in agriculture	Dr. A. Ružic
Weed Department:	

Weed control in agriculture — Dr. S. Čuturilo/Dr. K. Mijatović

8.2 Faculty of Forestry, Institute for Landscape Architecture
elevation : 110 m
rainfall : 657 mm
sunshine : 1930 h

ul. Kneza Višeslava 1
11000 Beograd
Phone: (011) 553-122
Dir.: Dr. V. Avdalović

Established in 1960. Landscape planning and organization, designing green areas in and out of urban settlement, etablishing and management of all landscape categories, production of ornamental trees, shrubs, flowers, physiology and bioecology.

Planning and designing green areas in and out of urban settlement, recreation	Prof. S. Milinković
Planning, designing and organization of landscape	Prof. Dr. V. Macura
Functionality of urban and suburban landscape, investigation of bioecological phenomena, planting and management of green areas	Ing. N. Anastasijević, M.Sc.
Taxonomy, biological and aesthetical investigation of dendroflora in and out of urban settlement	Prof. Dr. E. Vukićević
Investigation of present and potential vegetation, phytocenology	Prof. Dr. Z. Tomić
Systematics, morphology, ecology and technology of flower production, flower application	Prof. Dr. O. Mijanović
Genetics and breeding of ornamental trees, shrubs, climbers and flowers	Prof. Dr. Tucović
Nursery of ornamental plants and production	Prof. Dr. S. Stilinović
Harmful insects of ornamental plants and protection	Prof. Dr. Tomić
Diseases of ornamental plants and protection	Prof. Dr. Marinković
Environmental protection and improvement	Prof. Dr. V. Avdalović

9 Bolec

9.1 Fruit and Grape Research Station, PKB "Institute Agro-economic", Agricultural Combination "Beograd"
elevation : 92 m
rainfall : 657 mm
sunshine : 1930 h

ul. Smederevski put bb
11307 Bolec (kod Beograda)
Phone: (011) 4884-833, 4885-971, 4885-705
Dir.: J. Medigović

Established in 1971. Area 300 ha, nursery 30 ha, quarantine plots 10 ha. Produce 2 million plants. Breeding of fruit and grapes, fruit and grape cultivar trials, systems of growing, application of retardant, physiology, soil research, plant protection, mechanical engineering, technology and economy, harvesting, packing, transport of fruits and publication of "Nauka u praksi" (Science in practice).

Fruit and grape breeding	Dr. P.D. Mišic/R. Todorović, M. Mirković
Pome fruits	Lj. Pavlović/D. Janković, S. Marić
Stone fruits	Prof. Dr. S.A. Paunović, V. Pavlović/M. Tisma/A. Obradović
Grapes	I. Viotošević/M. Plavsić
Hormones in fruit and grapes	Dr. I. Ninkovski/D. Petrović
Plant protection	M. Stojanović/B. Ristić, J. Zelenović
Soils	V. Komnenić, M.Sc.
Nursery plants research	J. Medigović/M. Djaković

Micropropogation of fruit trees	Dr. D. Vinterhalter
Harvesting and fruit storage, technology of fruits	Lj. Vuković
Fruit and grape machinery	T. Jocić

10 Čačak

10.1 Faculty of Agronomy,
Department of Fruit Growing, Viticulture and Vegetables
elevation : 242 m
rainfall : 777 mm
sunshine : 1740 h

ul. Cara Dušana 34
32000 Čačak
Phone: (032) 43-452

Established in 1978. Lecturing and research on fruit growing, viticulture, vegetable, plant protection, biochemistry, agrochemistry, podology, land reclamation, mechanization, economy, organization of work, marketing. The experimental plots are mainly on the state farm lands and on the private farms.

Breeding and selection of stone fruits, cultivar and rootstock trials of stone fruits, genebank of plums, apoplexy of apricots	Prof. Dr. S.A. Paunović, T. Milošević
Viticulture	Prof. Dr. V. Četković/R. Kojović
Vegetables, meristem culture	Prof. Dr. P. Maksimović
Plant physiology	Prof. Dr. D. Djokić/M. Djurić
Fruits, grapes and vegetable diseases	M. Cvetić, M.Sc.
Chemistry	Prof. Dr. D. Stojanovski, Dr. M. Spasojević
Agrochemistry, biochemistry	Ass. Prof. Dr. M. Bojić
Microbiology	M. Djukić, M.Sc.
Pedology, water-physical properties of hard soils	Prof. Dr. Babović/V. Turudić
Horticulture and agriculture land reclamation	Prof. Dr. M. Mićović, G. Šekularac
Meteorology, climatology and hydrology	Prof. Dr. D. Labus
Machinery in horticulture	Prof. Dr. D. Komarčević, R. Koprivica
Genetics	Prof.Dr. Ž. Jestrović
Botany	Prof.Dr. P. Veljović/S. Simović
Plant selection in horticulture	Prof. Dr. Č. Rakočević
Agro-ecology	Prof. Dr. Dj. Perić
Economy	Ass.Prof. Dr. Dj. Djukić
Organization of horticulture and agriculture work at the state and united lands of private farms	Prof. Dr. R. Djordjević, M. Jovanović, M.Sc.
Fruits, grapes, vegetables and agriculture marketing	Prof. Dr. M. Nikšić, Prof. Dr. S. Mitić
Agrarian politics in horticulture	Prof. Dr. S. Mandić/Z. Maslać
Sociology aspects of the protection of human surroundings in and out of urban settlement	Prof. Dr. T. Prodanović

10.2 Fruit Research Institute

ul. Vojvode Stepe 9
32000 Čačak
Phone: (032) 47-411, 47-4022

Established in 1946. Area 165 ha and state farm lands, 330 m² glasshouses and building for walnuts propagation. Breeding and selection, pomology, cultural practices, plant protection, biochemical and technological fruit properties, nutrition, physiology, tissue culture.

Pomology and breeding of apples and small fruits	Ž. Tešović, M.Sc.
Pomology and breeding of plums and peaches	Dr. D. Ogašanović
Pomology and breeding of apricots	R. Plazinić

Pomology and breeding of pears, quince and cherries	Dr. M. Nikolić/Z. Kostadinović
Pomology and breeding of small fruits	M. Stanisavljević
Breeding and grafting of walnut and hazelnuts	M. Mitrović, M.Sc.
Cytogenetics and biology of fertilization in stone fruits	R. Cerović
Fruit nutrition and weed control	R. Janković, M.Sc.
Soils and irrigation	M. Rakićević
Plant propagation	Dj. Ružić
Virus diseases of fruits	Dr. M. Ranković/Svetlana Paunović
Pests of fruit trees	Dr. S. Stamenković
Technological value of stone and small fruits, drying prunes	Dr. Lj. Janda
Chemistry	J. Gavrilović

11 Guča

11.1 Potato Research Station
elevation : 330 m
rainfall : 888 mm
sunshine : 1660 h

ul. Albanske Spomenice 21
32230 Guča
Phone: (032) 854-220, 854-226

Established in 1954. Area 80 ha and state farm lands, under glass 117 m^2. Biology, physiology, ecology, system of growing, plant protection, storage, breeding, selection, seed production of elite and marketable potatoes.

Genetics, breeding, selection	Ing.Agr. Z. Vasiljević/S. Savić, R. Djekić
Agrotechnics	Ing.Agr. S. Sušić
Physiology and agrochemistry	Dr. B. Stoiljković/S. Babić
Plant protection and seeds control	Ing.Agr. M. Domanović, D. Milošević
Seeds Department	Ing.Agr. S. Slavković

12 Niš

12.1 Viticulture and Enology Research Station
elevation : 195 m
rainfall : 571 mm
sunshine : 1670 h

ul. Kolonija EI 6
18000 Niš
Phone: (018) 33-738

Established in 1953. Area 40 ha and state farms lands. Selection, biology, ecology, system of growing, grape protection, storage, processing, use, transportation, economy of production and processing of wine.

Decrease of SO_2 - excess in wine, must and wine distillate, separation of iron as excess from wine, industrial processing and standardization of black wine, rootstock influence on contents and quality of wine, economic and technology characteristics of the Prokupac cultivar under SR Serbia conditions	Prof. Dr. D. Zirojević
Breeding and selection of grapes	Dr. S. Ilić
Nursery, nutrition, irrigation	Ing.Agr. S. Stanković
Grape protection	Ing.Agr. G. Petrović
Pruning and properties of wine	Ing.Agr. B. Stojanović

13 Smederevska Palanka

13.1 Vegetable Research Institute "Palanka"
elevation : 120 m
rainfall : 620 mm
sunshine : 2135 h

ul. Karadjordjeva 73
11420 Smederevska Palanka
Phone: (026) 33-762
Dir.: Ing.Agr. Ž. Nikosavić

Established in 1946. Genetics, breeding, selection, agrotechnics, plant protection, growing, seeds production, élite and other categories. Area 300 ha, hydroponic glasshouses, glasshouses and plastic tunnels, storage house and manipulative spaces, centre for processing of seeds and special destination for vegetable seeds.

Breeding and selection of peppers	Ž. Miladinović/D. Stevanović
Breeding and selection of tomato	Prof. Dr. M. Popović/Ž. Marković
Breeding and selection of cucumbers and watermelon	Ing.Agr. Lj. Djunisijević
Breeding and selection of cabbages	B. Kandić/Ing.Agr. T. Bertolino
Breeding and selection of peas	I. Djinović
Breeding and selection of carrot	Ing.Agr. M. Zdravković
Breeding and selection of French beans and common beans	D. Čorokalo
Breeding and selection of onions	J. Panajotović
Agrotechnics	M. Damjanović
Fungus diseases on vegetables	Dr. Ž. Aleksić/N. Marinković
Virus diseases on vegetables	Ing.Agr. M. Mijatović

14 Zemun

14.1 Faculty of Agriculture,
Horticulture Research Institute
elevation : 70 m
rainfall : 578 mm
sunshine : 1921 h

ul. Nemanjina 6
11080 Zemun
Phone: (011) 215-315

Established in 1919. Area 100 ha and glasshouses at the Faculty Experimental Farm under the name: Center for Viticulture and Enology "Radmilovac", Vinca-Belgrade (12 km from Beograd along the road Beograd-Smederevo.) Lecturing and research on fruit growing, viticulture, vegetables, processing. Laboratory for biology, enology, beekeeping, meteorology, climatology and other Departments and Laboratories of Agriculture.

Fruit Section:	
Genetics, breeding and selection	Prof. Dr. B. Pejkić, Prof. Dr. M. Milutinović
Pomology of fruit trees, cultivars and rootstocks	Prof. Dr. D. Rahović, Dr. M. Mijačika
Physiology of fruit nutrition	Prof. Dr. M. Jovanović, S. Veličković
Irrigation and retardants	Prof. Dr. S. Savić
Problems in bee-culture	Prof. Dr. B. Konstatinović
Viticulture Section:	
Breeding, selection and ampelography	Prof. Dr. L. Avramov
Physiology of grapes	Prof. Dr. M. Milosavljević, A. Nakalemić
Nutrition and agrotechnics of grapes	Prof. Dr. R. Lović/B. Ristić
Vegetable Section:	
Breeding and selection	Prof. Dr. M. Popović/V. Bijelić
Genetics and seeds	Prof. Dr. M. Marić/R. Sabovljević
Agroecology	Prof. Dr. M. Stojanović, R. Cvetković
Weeds control	Prof. Dr. B. Šinžar
Meteorology, Climatology Section:	
Ecology of grapes	Prof. Dr. S. Stanojević, Ass.Prof. N. Todorović
Floriculture	Lj. Prijić
Medicinal plants	Ass.Prof. N. Krstić-Pavlović
Technology Section:	
Fruit juice, processing of fruits and vegetables	Prof. Dr. G. Niketić-Aleksić, B. Bukvić

Technology of wine	Prof. Dr. M. Daničić/S. Jović
Technology of alcohol	Prof. Dr. R. Paunović/B. Djurisić
Biochemistry	Prof. Dr. M. Džamić, Prof.Dr. D. Veličković/D. Vajagić
Economic Section:	
Biometrics	Prof. Dr. V. Erdeljan, Prof. Dr. Ljesov
Economy	Prof. Dr. P. Marković, Prof. Dr. V. Randjelović, Prof. Dr. B. Radovanović
Organization of fruit growing and viticulture	Prof. Dr. M. Furundžic, M. Radoičić
Marketing	Prof. Dr. A. Tomin/M. Djorović
Financial analysis	Prof. Dr. M. Petrović/J. Rodić
Plant Protection Section:	
Phytopathology	Prof. Dr. M. Babović, Prof.Dr. M. Panić/Prof.Dr. M. Tošić
Entomology	Prof. Dr. N. Tanasijević, Prof. Dr. B. Ilić
Problems in plant protection	Prof. Dr. N. Ostojić
Phytopharmacia	Prof. Dr. R. Kljajić, Ass.Prof. M. Šestović
Physiology and Chemistry Section:	
Plant physiology and agrochemistry	Prof. Dr. Djurdje Jelenić, Prof. Dr. R. Džamić
Irrigation	Prof. Dr. D. Stojićević, J. Milivojević
Microbiology	Prof. Dr. M. Todorović, V. Bogdanović
Pedology	Prof. Dr. M. Živković, Prof. Dr. R. Korunović
Land reclamation	Prof. Dr. R. Čorović, Ass.Prof. M. Pejković
Hydrology and hydraulics	Prof. Dr. R. Marković
Machinery Section:	
Machinery in fruit growing and grapes	Prof. Dr. P. Nenić/M. Urošević
Mechanization of production and draying	Prof. Dr. J. Lukić/D. Raičević
Thermodynamics and termotechnicque	Dr. Ć. Todorović, C. Džodžo
Power machinery and equipment	Prof. Dr. C. Stegešek
Tractors and supply units	S. Božić

14.2 Research Institute for Mechanization of Agriculture
elevation : 88 m
rainfall : 620 mm
sunshine : 1990 h

Zemun Polje
11080 Zemun, PO Box 41
Phone: (011) 212-403

Established in 1947. There are five laboratories. Problems of agricultural mechanization in Yugoslavia, investigation and testing of all types of agricultural machinery, new agricultural organization and outfit. The Institute has its own bulletin "Poljoprivredna tehnika".

Machinery in fruit and grape production	Ing.Agr. B. Jovanović, Ing.Agr. I. Suković
Machinery in vegetable production	Ing.Agr. R. Dabetić
Machinery for soil management	Dr. I. Milinković, Ing.Agr. V. Popović
Machinery for wheat harvesting and forage crops	Ing.Agr. S.Z. Živkovic

Economy of machinery	Ing.Agr. S. Lazarević
Sowing and methods of field production of tomato and pepper	Ing.Agr. R. Savić
Application of pneumatics in agriculture	Ing.Mach. N. Milovanović
Exploitation and standardization of machines (tractors) in agriculture	Ing.Agr. D. Obradović
Workshops for service and reparation	Ing. M. Vasić

SAP KOSOVO

15 Peć

15.1 Biotechnical Institute . ul. Maršala Tita 128
38300 Peć
elevation : 523 m
rainfall : 876 mm
sunshine : 162 days p.y.

Established in 1949. Area 55 ha and state farm lands. Apples, plums, pears, sweet and sour cherries, training, pruning, nutrition, irrigation, technological value of fruits, plant protection.

Agrotechnics, manuring, system of growing, pruning	Prof. Dr. V. Djorović, Ing.Agr. E. Taljati/R. Muhadjeri
Ampelography, cultivars and rootstocks, system of growing grapes, manuring	A. Kojić, M.Sc./Ing.Agr. R. Muhadja
Plant protection	D. Vulević, M.Sc./R. Garić
Agrochemical research	Dr. M. Brković

16 Priština

16.1 Faculty of Agriculture, Fruit Growing, Viticulture and Vegetable Research Station . ul. Lenjinova bb
38000 Priština
elevation : 573 m
rainfall : 580 mm
sunshine : 157 days p.y.

Established in 1974. Botany, biology, ecology, fruit growing, viticulture, vegetables, economics, machinery, lecturing and research.

Fruit species, cultivars and rootstocks	Prof. Dr. H. Hadrović/A. Zajmi, S. Ferhatović
Viticulture and ampelography	A. Kojić, M.Sc./Ing.Agr. B. Korenica
Vegetable growing in- and out-doors	Prof. Dr. V. Savić
Genetics	Prof. Dr. D.N. Stolić
Plant physiology	Prof. Dr. M. Jablanović
Phytopathology and entomology	Prof.Dr. B. Brkić/Lj. Susuri, M.Sc., M. Pireva, M.Sc.
Meteorology and climatology	Ing.Agr. F. Šalja
Agrochemistry and plant nutrition	Prof. Dr. B. Dobrodoljani
Pedology	Prof. Dr. M. Dauti/Ing.Agr. S. Marić
Agricultural land reclamation	Prof. Dr. B. Kabaši, Prof. Dr. U. Lugonja
Agricultural machineries	Prof.Dr. B. Jačinac/Ing.Agr. F. Jaha

SAP VOJVODINA

17 Novi Sad

17.1 Faculty of Agriculture, Institute for Viticulture, . ul. Veljka Vlahovića 2

Fruit Growing and Floriculture
elevation : 80 m
rainfall : 613 mm
sunshine : 192 days p.y.

21000 Novi Sad
Phone: (021) 58-366

Established in 1947. Lecturing and research of viticulture, fruit growing and floriculture. Area 20 ha and state farm lands.

Breeding, selection and ampelography	Ing.Agr. N. Bajkan/R. Vidojković
Irrigation, manuring, rootstocks and physiology of grape cultivars	Prof. Dr. D. Burić, Ing.Agr. Dj. Paprić
Breeding of resistant early table- and wine-grape cultivars, breeding of new rootstocks for grapes	Prof. Dr. P. Cindrić
Rootstock trial for grapes, management of soils and manuring, biological and technological proterties of grape berries	Prof. Dr. M. Zorzić
Rootstock trial, fruit nursery, apricot apoplexy	Ass.Prof. Dr. B. Djurić
Breeding and selection of sour cherries and cultivar trials	Prof. Dr. M. Nilovankić
Walnut selection and nurseries	Prof. Dr. M. Korać
Strawberry cultivars, bud formation in pears and sweet cherries	Ing.Agr. O. Grbić
Irrigation and manuring in apple, pear and peach trees	Prof. Dr. V. Janjić
Bud formation in apple, pear and cherry trees	Dr. V. Varga
Rootstock trials, training and pruning for apples, postharvest physiology of apple and pear	Prof. Dr. D. Gvozdenović
Floriculture and ecology protection	Prof. Dr. M. Sapundžić, Ing.Agr. D. Sikoparia

SR SLOVENIA

18 Ljubljana

18.1 Agricultural Research Institute of Slovenia
elevation : 298 m
rainfall : 1471 mm
sunshine : 1531 h

ul. Hacquetova 2
61000 Ljubljana
Phone: (061) 323-064, 323-192

Established in 1898. Area (horticulture) 14 ha and state farm lands. Agriculture, seed production and control (elite and original), animal and milk husbandry, fruit growing, viticulture, vegetables, plant production, mechanization, economics, plastics, central laboratory and information bureau.

Selection of apples, pears and peaches, cultivar trials in pears and peaches, rootstock trials in apples and peaches	Ing.Agr. M. Lekšan
Fruit nutrition and soil management in orchards and vineyards	Ing.Agr. M. Jazbec, Ing.Agr. L. Briški
Selection, cultivar trials, rootstocks, ampelography, nutrition of grapes	Ing.Agr. T. Zafošnik
Plant protection in orchards and vineyards, degeneration of grape cultivars	Ing.Agr. V. Šišaković
Cultivar trials, selection of french beans and cultural technics in vegetable production	Ing.Agr. S. Avšić
Effect of plastics on vegetable production and cultural technics in vegetables for processing industry	Ing.Agr. M. Černe
Economy in fruit and wine production	Ing.Agr. O. Štefula

18.2 Biotechnical Faculty, University Edvard Kardelj

ul. Jamnikarjeva 101

of Ljubljana, Department of Fruit Growing and Viticulture

61000 Ljubljana
Phone: (061) 264-761
Dir.: Prof. Dr. D. Modić

Established in 1951. Area 3 ha planted with fruit trees of apples, pears and hazelnut cultivars. Lecturing and research.

18.2.1 Fruit Research Station
elevation : 153 m
rainfall : 1505 mm
sunshine : 1972 h

Post send to Biotechnical Faculty:
ul. Jamnikarjeva 101
61000 Ljubljana
Phone: (016) 264-761
Head: Zora Korošec, M.Sc.

Area 4 ha under vineyards, greenhouse and heatchamber for studies of virus diseases of *Vitus vinifera* cvs.

Selection of hazelnuts, olives, apples, pears, pollination of hazelnuts, application of plant growth regulators, technology, projecting of orchards	Prof. Dr. D. Modić
Breeding of peach, selection of apricots, cherries, plums and peaches, pollination of cherries, rootstock trials	Prof. Dr. J. Smole
Breeding and clonal selection of apples, apple rootstock trials, technology value of apples for processing, chemicals for apple thinning	Dr. J. Črnko/M. Marn
Influence of pruning time on crop and fruit of apples	Dr. A. Štruklec
Selection and vegetative propagation of walnuts and chesnuts	Ing.Agr. T. Hlišč
Technology of pruning on the various vegetative rootstocks	Ing. Agr. M. Babnik
Selection and testing of grape cvs, testing and propagation of virus free *Vitis* plant materials, technology	Ing.Agr. Zora Korošec, M.Sc.

18.3 Biotechnical Faculty, Department of Horticulture and Landscape Architecture-Chair of Field and Forage Crops and Vegetable Growing

ul. Krekov trg 1
61000 Ljubljana

Chair of Vegetable Growing was established in 1951.

Research of indigenous vegetable cvs, research on onion, tomato, cauliflower, kohlrabi, physiology and technics	Prof. Dr. E. Leskovec
Acclimatization of ornamental woody plants, nursery cultivation techniques	Sp.Agr. A. Šiftar

19 Maribor

19.1 Fruit Research Station
elevation : 270 m
rainfall : 1039 mm
sunshine : 1684 h

ul. Vinarska 14/I
62000 Maribor
Phone: (062) 21-755
Head: Dr. J. Črnko

This Station belongs to the Biotechnical Faculty of Ljubljana. Area 2 ha of apples, hazelnuts and walnuts, rootstocks. Established in 1892.

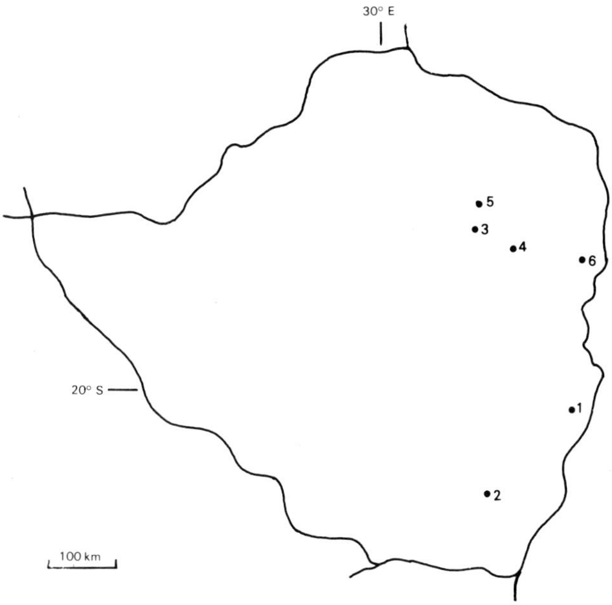
30° E
20° S
100 km
5
3
4
6
1
2

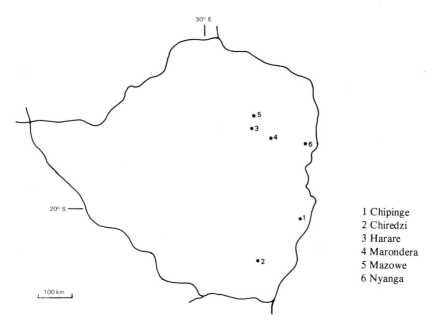

1 Chipinge
2 Chiredzi
3 Harare
4 Marondera
5 Mazowe
6 Nyanga

Survey

With the exception of work on sugar and tobacco, most agricultural research is the responsibility of Research and Specialist Services (R & SS), a Department within the Ministry of Lands, Agriculture and Rural Resettlement. Similarly, most advisory work is done by a sister Department of Agricultural, Technical and Extension Services (Agritex), which employs horticultural advisers in the main centres. In the past, such personnel have sometimes undertaken research projects related to their main advisory interests, and with the introduction of horticulture as an option in the Crop Science degree at the University of Zimbabwe, more projects of this nature can be expected. Some investigational work is undertaken by commercial enterprises, notably at Mazowe, see entry 5.1. Zimbabwe lies well within the tropics but it has a wide altitudinal range, enabling an equally wide range of crops to be grown - from tropical in the SE Lowveld and Zambezi Valley to temperate in the E Highlands. To illustrate this point, mean altitude is shown for each Station. To avoid giving an inflated picture of the personnel situation, names of technical and support staff are omitted from individual entries, although their role is perhaps greater in developing than in developed countries. In contrast, programmes are given in greater detail than will be found elsewhere in HRI IV.

1 Chipinge (formerly Chipinga)

1.1 Coffee Research Station
elevation : 1130 m
rainfall : 1030 mm
sunshine : 2780 hrs
r.h. : 73%

PO Box 61, Chipinge
Phone: 137 2476

Agronomy: *Coffea arabica:* short cycle intensive management systems involving plant populations and spatial arrangements; NPK, cultivar, legume live mulch, minimum nutrient input and establishment trials; fertilizer/irrigation/population trial; investigation of nursery techniques and root excavations.
Miscellaneous crops including *Vernonia galamensis*.

Officer in Charge: Dr. D. Kumar
Agronomist: position vacant

Prepared by C.B. Payne, Rhodes Inyanga Experimental Station, Rusape.

Entomology: development of scouting procedures for *Leucoptera meyricki* (leaf miner), and monitoring coffee pests in general through light trap; control with contact and systemic insecticides.

Entomologist: position vacant

Pathology: chemical control of rust *(Hemileia vastatrix)* and Fusarium bark disease and epidemiology of FBD.

Pathologist: position vacant

2 Chiredzi

2.1 Lowveld Research Stations

Chiredzi Research Station (headquarters of the complex)
elevation : 430 m
rainfall : 600 mm
sunshine : 2960 hrs
r.h. : 63%

PO Box 97, Chiredzi
Phone: 133-8 2397/8
Head: E. Jones
Horticulturist: M.D.S. Nzima
Entomologist: position vacant

Cultivar evaluation of litchee, mango, pineapple, pecan and coffee. Introduction of oil palm. Grapefruit rootstock trial. Investigations into improvement of pecan bud break. Studies on banana sucker development. Introduction and testing of jojoba, buffalo gourd and marama bean. Extension of cropping season for cabbage, forage rape, okra and tomato. Selection of heat-tolerant tomato and Chinese cabbage lines and cvs. Screening and evaluation of virus-resistant sweet potato lines, onion cultivar evaluation. Nursery techniques for fruits, nuts and vegetables, and investigations into cropping systems for small scale irrigation schemes.

3 Harare (formerly Salisbury)

3.1 Department of Research and Specialist Services (headquarters)
elevation : 1500 m
rainfall : 821 mm
sunshine : 2920 hrs
r.h. : 61%

PO Box 8108, Causeway, Harare
Phone: 10 704531 (all offices exc. where otherwise specified)
Dir.: Dr. P.R.N. Chigaru

Responsible for conducting all agricultural research except that indicated in the Survey. Publications: "Zimbabwe Agricultural Journal" (bi-monthly) and "The Zimbabwe Journal of Agricultural Research" (bi-annually)

3.1.1 Crop Research Division

Controls individual Stations and other units where most of the horticultural (and agronomic) research is carried out.

Asst.Dir.: R.J. Fenner

Crop Breeding Institute

PO Box 8108, Causeway, Harare
Head: R.C. Olver
M.J. Joyce

Breeding of high yielding cultivars of locally important agricultural crops including potatoes - mainly for improved quality and resistance to *Phytophthora infestans* (late blight). (See also entry 6.1)

3.1.2 Research Services Division

Asst. Dir.: Mrs. N.R. Gata

Chemistry and Soil Research Institute

Head: T.J.T. Madziva

Collaborates with all Research and Experiment Stations in fertilizer trials; carries out routine and research soil and leaf analyses; investigates methods and defines standards for leaf analysis of deciduous and subtropical fruits and coffee.

Crop Nutrition: F. Tagwira

National Herbarium and Botanic Garden

Phone: 10 725313
Head: Th. Muller
Keeper National Herbarium: R.B. Drummond
Taxonomists: Mrs. K.E. Bennett
E.B.Best(Hon.)/Miss B.M.Browning
Asst. Curator Botanic Garden: M.J. Leppard

The Herbarium maintains a comprehensive collection of 250,000 dried specimens of indigenous plants and cultivated ornamental species, and provides an identification service for workers in agricultural and related fields. Taxonomy of flora of South-Central Africa including groups of horticultural interest, eg. *Euphorbia, Orchidaceae and Stapeliae.* Publication: Kirkia (irregular) and contributions to *Flora Zambesiaca*. The 58 ha garden specializes in shrubs and trees of the region and of climatically similar areas.

Plant Protection Research Institute

Head: Dr. S.S. Mlambo

Entomology: control of coffee leaf miner *Leucoptera meyricki* and stem borer *Stephanoderes hampei;* control of diamond back moth *Plutella xylostella* and several spp. of aphid in brassicas.

R. Tanyongana/E. Zitsanza

Nematology: studies of *Pratylenchus* spp. as pests of agricultural and horticultural crops; effect of nematode control on apple tree performance; trials with systemic nematicides on apple; monitoring miscellaneous deciduous fruits for nematode infestation.

B.Dube

Pathology: control of coffee diseases (see also entry 1.1); control of early and tale blight on potato, and virus diseases of potato; control of *Alternaria alternata* on apple.

C. Mguni/C.N. Mzira

Seed Services

Phone: 10 720370
Head: Dr. T.M. Musa
Seeds Off.: Mrs. B. Asueri
Mrs. J.L. Chigogora/D.A.Duncombe
V.E.E. Gwarazimba/P.Maramba
C.N. Ntungakwa
Snr Seeds Analysts: Mrs.S. Madyara
Mrs. N. Ruwende

Administration of official seed certification scheme involving field inspections and routine testing of a wide range of field crop and vegetable seeds; ad hoc investigation of fruit, vegetable and tree seed germination for Government research institutions and commercial seed growers.

3.2 Harare Municipality Amenities Division (Department of Works)

PO Box 1583, Harare
Phone: 10 727875

Investigation into provision of modern urban amenities including parks, and embracing diverse activities from refuse disposal to street tree species selection.

City Amenities Manager: G.F. Munyoro
Hortic: B.J. Maclaurin

3.3 University of Zimbabwe, Faculty of Agriculture - Department of Crop Science

P. Bag MP 167, Mount Pleasant, Harare, Phone: 10 303211
Chrmn:Prof. M.A. Schweppenhauser

Field plots, laboratory and constant temperature facilities are available for post-graduate research; although no formal course work is presented at graduate level, undergraduate horticultural courses have been introduced.

Dr. R.C. Herner

4 Marondera (formerly Marandellas)

4.1 Horticultural Research Centre
elevation : 1630 m

P. Bag 3748, Marondera
Phone: 128 4122

rainfall : 885 mm
sunshine : 2960 hrs
r.h. : 71%

Off. in Ch.: M.L Vogel
R.L. Msika
Adv. Hort.(Agritex): position vacant

Cultivar screening trials for apricot, nectarine, peach, grape, kiwifruit, rabbiteye blueberry, strawberry, macadamia and *Protea* species. Intensive systems for peach, and peach nutrition.Trellis systems for kiwifruit, and pruning systems for Chenin Blanc grape. Cultivar and cultural trials of amaranth, cabbage, pea and tomato. Pesticide screening for grape and vegetables. Operation of an elite scheme for strawberry and sweet potato. Field testing of *Vernonia* species.

5 Mazowe (formerly Mazoe)

5.1 Mazoe Citrus Estates
elevation : 1170 m
rainfall : 915 mm
sunshine : 2880 hrs
r.h. : 65%

PO Mazowe
Phone: 175 481/2
Gen.Manager: A.F. Heberden

Citrus: rootstock/scion combinations, high density plantings (orange), pH problems caused by fertilizer placement, NPK trial (orange), leaf analysis, control of greening, juice processing investigations and quality control, soft citrus cultivar appraisal. Survey of weed species. Control of *Trioza erytraeae, Scirtothrips aurantii* and *Phoma citricarpa*. Cultivar screening and economic evaluation of avocado, macadamia and pecan. Testing of under-tree micro-irrigation systems.

Citrus Products Manager:
R.G.P Medley
Hort.: L.H. Kirton
Entomologist: position vacant
Consult. Entomologist: D. Miller

6 Nyanga (formerly Inyanga)

6.1 Rhodes Inyanga Experiment Station
elevation : 1870 m
rainfall : 1120 mm
sunshine : 2760 hrs
r.h. : 76%

P.Bag 8044, Rusape
Phone: 129-8 28125
Off. in Ch.: C.B. Payne

Cultivar trials for apple, apricot, pear, plum, brambles and kiwifruit; evaluation of apple, apricot, pear and plum rootstocks, and improvement of propagation techniques for rootstocks and own-rooted trees. Intensive systems for apple, apricot and plum. Vegetable cultivar screening trials for peasant farmers. Miscellaneous investigations with *Vernonia* spp., *Trigonella foenum-graecum* and *Plectranthus esculentus*. Potato virus-testing and cultivar breeding/selection by Crop Breeding Institute staff - see entry 3.1.1

INDEX OF NAMES OF PLACES

Aalsmeer NL 1
Aalst BE 1
Aarslev DK 1
Aas NO 10
Abeokuta NG 8
Aberdeen GB 1
Aberdeen US-I 13
Acireale IT 1
Adana TR 1
Addis Ababa ET 10
Adelaide AU 36
Adelphia US-B 17
Agadir MA 1
Ag. Anargyroi GR 1
Agassiz CA 1
Ag. Deca GR 2
Ag. Paraskevi GR 3
Ahrensburg DE 1
Aichi JP 1
Akhelia CY 3
Akita JP 2
Akola IN 13
Akure NG 6
Alagoinha BR 9
Albany AU 50
Albany US-K 1
Al-Beida LY 5
Albigowa PL 1
Alcalde US-H 17
Alcázar de San Juan ES 1
Alcobaça PT 1
Aliartos GR 4
Alice Springs AU 64
Allahabad IN 23
Alma US-E 13
Alma Ata SU 1
Almendralejo ES 6
Almeria ES 2
Alnarp SE 1
Alstonville AU 1
Altamira BR 3
Amani TZ 9
Ames US-G 20
Amherst US-A 5
Angers FR 1, 19
Ankara TR 2
Anseong KR 1
Antalya TR 3
Antibes FR 2, 20
Antwerpen BE 2
Aomori JP 3
Apatzingán MX 3
Apopka US-D 31
Aracaju BR 11
Arad RO 2
Aranjuez ES 3
Arcadia US-K 2
Arequipa PE 4
Århus DK 9
Ariana TN 1
Armidale AU 2
Arta GR 5
Arusha TZ 3
Ashland US-F 21
Ath BE 3
Athalassa CY 1
Athens GR 6
Athens US-D 16
Attapulgus US-D 17
Auburn US-D 38
Auchincruive GB 2
Auckland NZ 2
Aurora US-J 12
Avignon FR 3
Awassa ET 6
Aydın TR 4
Ayr AU 13

Bacău RO 3, 11
Bad Neuenahr-Ahrweiler DE 2
Bad Zwischenahn DE 3
Baia Mare RO 12
Bako ET 5
Balcarce AR 1
Baldknob US-E 9
Balıkesir TR 5
Bandung ID 1.2
Bangalore IN 1, 11
Bangkok TH 19
Bangkok Noi TH 17
Bar YU 5
Barcelona ES 4
Barendrecht NL 11
Bari IT 2
Barnaul SU 2
Bath GB 3
Bathurst AU 3
Baton Rouge US-D 52
Bavendorf DE 4
Beaverlodge CA 5
Becker US-G 5
Beer-Sheva IL 4
Belém BR 2
Bella Vista AR 2
Bellegarde FR 21
Belle Glade US-D 23
Belle Mina US-D 39
Bellingham US-J 1
Bello CO 5
Belo Horizonte BR 15
Beltsville US-C 3
Bento Gonçalves BR 31
Beograd YU 8
Bergen NO 1
Berkeley US-K 3
Berlin DD 1
Berlin DE 5
Bersée FR 22
Bet-Dagan IL 1
Beverly Hills US-K 8
Bhubaneswar IN 16
Biglerville US-B 9
Bijelo Polje YU 6
Bingley GB 4
Bistrita RO 14
Bixby US-H 1
Blacksburg US-C 10
Blackstone US-C 15
Blackville US-D 11
Blairsville US-D 18
Bleiswijk NL 12
Bogotá CO 7
Bogusławice PL 2
Bolec YU 9
Bologna IT 3
Bonn DE 6
Bonn-Bad Godesberg DE 7
Bordeaux FR 4
Boskoop NL 2
Botucatu BR 22
Bouctouche CA 39
Bowen AU 14
Bozeman US-I 18
Bradenton US-D 24
Brăila RO 4
Bramminge DK 12
Brasília BR 25
Brastagi ID 2
Bratislava CS 1
Braunschweig DE 8
Brawley US-K 4
Brest FR 23
Brétigny sur Orge FR 24
Bridgeton US-B 18
Brisbane AU 15
Bristol US-A 15
Brno CS 2
Bronx US-B 1
Brookings US-G 19
Brookings US-J 13
Brooklyn US-B 6
Brooks CA 6
Brownwood US-H 16
Brussel - Bruxelles BE 4
Brzezna PL 3
Bucaramanga CO 3
Bucarest RO 8
Budakalász HU 1
Budapest HU 2
Buenos Aires AR 3
Bunbury AU 51
Bundaberg AU 16
Burlington US-A 3
Burnie AU 49

Burnley AU 27
Bursa TR 6
Busan KR 2
Busselton AU 52
Buzău RO 5, 15
Bvumbwe ML 1
Bydgoszcz PL 4
Byron US-D 19

Cabrils ES 5
Čačak YU 10
Cagliari IT 4
Cagua VE 2
Cairns AU 17
Caïro EG 1
Calera MX 6
Calhoun US-D 53
Cambérène SN 2
Camborne GB 5
Cambrai FR 25
Cambridge GB 6
Cambridge US-A 6
Camp Hill US-D 40
Campinas BR 20
Canberra AU 65
Canfield US-F 1
Caransebes RO 16
Carlow IE 1
Carnarvon AU 53
Carquefou FR 26
Carrington US-G 9
Casablanca MA 2
Castelar AR 4
Castle Hayne US-D 2
Catamarca AR 5
Catania IT 5
Celaya MX 7
Celeryville US-F 2
Center US-I 1
Cerexhe-Heuseux BE 5
Česká Skalice CS 3
Chaidari GR 7
Chambourcy FR 27
Chanea GR 8
Changins CH 2
Chapingo MX 8, 16
Charleston US-D 12
Charlottetown CA 43
Chase US-D 54
Chaska US-G 1
Chaubattia IN 24
Cheongju KR 3
Cheonweon KR 4
Chetopa US-G 27
Chiayi TW 1
Chiba JP 4
Chiclayo PE 2
Chieng Rai TH 1
Cruz das Almas BR 12
Crystal Springs US-D 47
Culiacán MX 2
Cullman US-D 43

Dabrowice PL 5
Daegu KR 7
Daejeon KR 8
Chipinge ZI 1
Chipping Campden GB 7
Chiredzi ZI 2
Christchurch NZ 12
Chumporn TH 13
Chuncheon KR 5
Chungju KR 6
Chûtes-aux-Outardes CA 26
Cienaga CO 1
Cipanas ID 1.4
Cîrcea RO 20
Clanton US-D 41
Claremont US-K 5
Clarksville US-E 10
Clarksville US-F 15
Clayton US-D 3
Clemson US-D 13
Clermont Ferrand FR 5, 28
Cleveland AU 18
Clinton US-D 4
Clinton US-D 55
Clonroche IE 2
Cluj RO 17
Clyde NZ 15
Colby US-G 38
College Park US-C 5
College Station US-H 6
Colmar FR 6
Colombo LK 1
Columbia US-D 14
Columbia US-E 18
Columbus US-F 3
Como IT 6
Concordia AR 6
Conegliano IT 7
Cordoba AR 7
Córdoba ES 7
Corfu GR 9
Corrientes AR 8
Corvallis US-I 19
Corvallis US-J 14
Cosenza IT 8
Craiova RO 19
Cream Ridge US-B 16
Creston US-I 24
Crookston US-G 2
Crossville US-D 42
Crossville US-E 1

Dakar SN 1
Dallas US-H 7
Dapoli IN 15
Dar es Salaam TZ 1
Dareton AU 4
Darwin AU 62
Davis US-K 6
Debre Zeit ET 9
Debrecen HU 3
De Lier NL 16
Delta del Parana AR 9
Deschambault CA 27
DeSoto US-G 33
Devonport AU 43
Dickinson US-G 10
Dijon FR 7
Diyarbakir TR 7
Dobřichovice CS 4
Dodoma TZ 10
Doi Muser TH 8
Dois Portos PT 2
Dordrecht NL 13
Dossenheim DE 9
Dresden DD 2
Dryanovo BG 1
Dublin IE 3
Durham US-A 2
Dushanbe SU 3
Dworek PL 6

East Grand Forks US-G 11
East Lansing US-F 14
East Malling GB 8
East Melbourne AU 26
Eden Shale US-E 14.1
Edinburgh GB 9
Edmonton CA 7
Efford GB 10
Ehime JP 5
El-Kanater EG 3
El-Kassasein EG 4
El-Marj LY 3.1
El-Nobaria EG 6
El Paso US-H 8
El-Sabahia EG 2
Elvas PT 3
Embu KE 7
Enkhuizen NL 14
Erzincan TR 8
Erzurum TR 9
Eskisehir TR 10
Espinal CO 2
Essen DE 10
Evora PT 4
Experiment US-D 20

Fairbanks US-L 1
Fairhope US-D 44
Faizabad IN 21

Fallon US-I 8
Fâlticeni RO 21
Fang TH 4
Fargo US-G 12
Farmington US-I 4
Farmington US-H 18
Farnham CA 28
Faversham GB 11
Fayetteville US-E 11
Fengshan TW 2
Fennville US-F 16
Firenze IT 9
Flakkebjerg DK 5
Fletcher US-D 6
Florence US-D 15
Florianópolis BR 28
Foçsani RO 34
Fortaleza BR 8
Fort Collins US-I 2
Forth AU 44
Fort Lauderdale US-D 25
Fort Pierce US-D 26
Frankston AU 28
Fredericton CA 40
Fredonia US-B 3
Freising-Weihenstephan DE 11
Fremont US-F 4
Fresno US-K 7
Fukui JP 6
Fukuoka JP 7
Fukushima JP 8

Gainesville US-D 27
Garden City US-G 29
Garissa KE 13
Gastouni GR 11
Gatton AU 19
Gaziantep TR 11
Geisenheim DE 12
Gembloux BE 6
General Roca AR 10
Geneva US-B 2
Gent BE 7
Geoagiu RO 22
Georgetown US-C 1
Geraldton AU 54
Giessen DE 13
Gif sur Yvette FR 30
Gifu JP 9
Giresun TR 12
Glen Osmond AU 37
Gode ET 8
Gödöllö HU 4
Göteborg SE 2
Goiânia BR 26
Górna Niwa PL 7
Gorna Oryachovitsa BG 2
Gosford AU 5
Grand'Anse SC 1
Grand Forks US-G 13
Grand Junction US-I 3
Grand Rapids US-F 17
Grand Rapids US-G 4
's Gravenhage NL 3
's-Gravenzande NL 20
Graz AT 1, 10
Greeneville US-E 7
Grenoble FR 31
Griffith AU 6
Grignon FR 8
Grimstad NO 2
Grossbeeren DD 3
Gross Enzersdorf AT 8.4
Grove AU 45
Guadajira ES 8
Guča YU 11
Guelph CA 15
Guernsey GB 12
Gunma JP 10
Guzów PL 8

Haelen NL 15
Haidegg AT 11
Hamburg DE 14
Hamilton CA 16
Hammenhög SE 3
Hammond US-D 57
Hancock US-F 22
Hang Chat TH 2
Hannover DE 15
Harare ZI 3
Haren NL 4
Harpenden GB 13
Harrow CA 17
Hastings NZ 7
Hastings US-D 28
Havelock North NZ 6
Havre US-I 20
Headland US-D 45
Healesville AU 35
Heidelberg DE 16
Hellerup DK 6
Helsinki FI 10
Heraclion GR 10
Hermansverk NO 3
Heřmanův Městec CS 5
Highland US-B 4
Hilario Ascasubi AR 11
Hilo US-M 1
Hiroshima JP 11
Hisar I 9
Hobart AU 46
Hoddesdon GB 14
Hokkaido JP 12
Holetta ET 2
Holice CS 6
Holovousy CS 7
Homestead US-D 29
Honolulu US-M 2
Hood River US-J 16
Hornum DK 2
Horst NL 5
Hualien TW 3
Hudeiba SD 1
Huntley US-I 21
Hurbanovo-Sesíleš CS 8
Hveragerdi IC 1
Hyderabad IN 6
Hyogo JP 13

Ibadan NG 1
Ibaraki JP 14
Içel TR 13
Ierapetra GR 12
Iernut RO 6
Ifakara TZ 5
Ile-Ife NG 3
Ilorin NG 4
Indian Head CA 9
Innisfail AU 20
Innsbruck AT 2, 12
Invergowrie GB 15
Iri KR 9
Irymple AU 31
Işalniţa RO 7
Ishikawa JP 15
Istanbul TR 14
Itabuna BR 14
Ithaca US-B 5
Ivanka pri Dunaji CS 9
Iwate JP 16
Izmir TR 15

Jaboticabal BR 23
Jackson US-E 2
Jackson US-F 5
Jackson Springs US-D 5
Jakarta ID 1
Jalapa MX 18
Jamaica Plain US-A 7
Jaroměř CS 10
Jassy RO 23
Jeju KR 10
Jeneponto ID 4
Jeonju KR 11
Jerez de la Frontera ES 9
Jimma ET 4
Jinju KR 12
Jodhpur IN 18
Joensuu FI 14
Jokioinen FI 2
Jomala FI 17
Jorhat IN 7
Jork DE 17

Jos NG 7
Juba SD 4
Junin AR 12
Jusepín VE 6
Jusepin VE 6
Jyväskylä FI 13

Kagawa JP 17
Kagoshima JP 18
Kakamega KE 11
Kalamata GR 13
Kalyani IN 28
Kanagawa JP 19
Kandy LK 4
Kanpur IN 22
Kao Kor TH 10
Karlsruhe DE 18
Karlštejn CS 11
Kasaragod IN 3
Kassala SD 6
Kassel DE 19
Katherine AU 63
Katumani KE 6
Kazanl"k BG 4
Kearneysville US-C 8
Kecskemét HU 5
Keedysville US-C 6
Kematen AT 3
Kénitra MA 3
Kennett Square US-B 10
Kentville CA 41
Kerikeri NZ 1
Keszthely HU 6
Kew GB 16
Khartoum SD 2
Khun Wang TH 6
Kiev SU 4
Kifissia GR 14
Kimberly US-I 16
Kingston US-A 14
Kingsville US-E 22
Kinsealy IE 4
Kirton GB 17
Kiryat Shemona IL 3
Kishinev SU 5
Kisii KE 12
Kitale KE 10
Klagenfurt AT 13
Klčov CS 12
Klepp NO 4
Klosterneuburg AT 4
Knoxfield AU 30
Knoxville US-E 3
København DK 3
Kochi JP 20
Kolding DK 10
Köln DE 20
Kona US-M 3
Końskowola PL 9
Konya TR 16
Kórnik PL 10
Kostinbrod BG 6
Kraków PL 11
Krasnodar Territory SU 6
K"rdzahli BG 3
Krefeld DE 21
Krems AT 14
Kuba SU 7
Kuibyshev SU 8
Kula US-M 4
Kumamoto JP 21
Kumasi GA 2
Kununurra AU 55
Kuopio FI 15
Kvetoslavov CS 13
Kwadaso GA 3
Kwangju KR 13
Kyoto JP 22
Kyungsan KR 14
Kyustendil BG 5

La Canada-Flintridge US-K 9
La Consulta AR 13
La Coruña ES 10
Lafayette US-F 9
La Force FR 32
La Gaude FR 29
La Hulpe BE 8
Lake Alfred US-D 30
Lalamilo US-M 9
La Londe FR 33
Lamberton US-G 3
La Ménitré FR 34
Lamia GR 15
La Molina PE 1
Lancaster US-F 26
Landernau FR 9
Landskrona SE 4
Lane US-H 5
Langdon US-G 14
Langenlois AT 15
La Plata AR 14
La Pocatière CA 30
Laramie US-I 12
Larissa GR 16
Las Cruces US-H 19
l'Assomption CA 29
Laukaa FI 9
Launceston AU 47
Lavras BR 16
Lédenon FR 35
Lednice CS 14
Leesburg US-D 32
Legon GA 1
Lelystad NL 6
Lesbos GR 17
Les Ponts de Ce FR 42
Leszkowice PL 12
Lethbridge CA 8
Leuven BE 9
Levin NZ 10
Lexington US-E 15.1
Libramont BE 10
Liége BE 11
Lincoln NZ 13
Lincoln US-G 23
Linz AT 23
Lipova RO 10
Lipowa PL 13
Lisboa PT 5
Lisse NL 7
Littlehampton GB 18
Ljubljana YU 18
Lofthus NO 5
Logan US-I 5
Logandale US-I 9
Logroño ES 11
London CA 18
Londrina BR 27
Long Ashton GB 19
Long Beach US-J 2
Lorman US-D 51
Los Baños PH 1
Los Lunas US-H 20
Loughgall GB 20
Louvain-la-Neuve BE 12
Loxton AU 38
Lubbock US-H 9
Lublin PL 14
Lucknow IN 26
Luddington GB 21
Ludhiana IN 17
Lullymore IE 5
Lund SE 5
Lunderskov DK 8
Lunuwila LK 2
Lyamungu TZ 8
Lycovrissi GR 18
Lyngby DK 4
Lysá CS 15

Maaninka FI 6
Madison US-F 23
Madrid ES 12
Mae John Luang TH 5
Măgurele RO 27
Málaga ES 13
Malama-Ki US-M 8
Malang ID 1.5
Malatya TR 17
Malda IN 29
Malemort sur Correze FR 36
Malicorne FR 37
Manaus BR 1

Manchester GB 22
Mandan US-G 15
Manhattan US-G 31
Manisa TR 18
Manjimup AU 56
Maracaibo VE 3
Maracay VE 1
Mărăcineni RO 9
Marana US-H 23
Mareeba AU 21
Maribor YU 19
Maricopa US-H 24
Marondera ZI 4
Marrakech MA 4
Martin US-E 8
Maryborough AU 22
Massey NZ 9
Matamoros MX 5
Matuga KE 4
Mayaguez US-N 1
Mazowe ZI 5
Mbeya TZ 4
Mead US-G 24
Medford US-J 17
Medina AU 57
Mehedinti RO 24
Meise BE 13
Mekele ET 7
Melitopol SU 9
Melka Werer ET 3
Melle BE 14
Mendoza AR 15
Mepal GB 23
Merelbeke BE 15
Mérida MX 11
Merida VE 4
Mesa US-H 25
Meudon Bellevue FR 38
Mexico City MX 13,17
Miami US-D 33
Michurinsk SU 10
Midland AU 58
Mie JP 23
Mikkeli FI 5
Milano IT 10
Mildura AU 33
Milly La Foret FR 39
Miłobadz PL 15
Milyang KR 15
Minot US-G 16
Minsk SU 12
Mississauga CA 19
Mississippi State US-D 50
Miyagi JP 24
Miyazaki JP 25
Mlyňany CS 16
Moccasin US-I 22
Modra CS 17
Molo KE 2
Moncada ES 14
Montague US-H 11
Monterrey MX 14
Monticello US-D 35
Montpellier FR 10
Montreal CA 31
Mora US-H 22
Morden CA 13
Morgantown US-C 9
Morogoro TZ 2
Morphou CY 2
Morris US-G 6
Moscow SU 13
Moscow US-I 14
Mosgiel NZ 16
Mosonmagyaróvár HU 7
Mostar YU 2
Motueka NZ 11
Mount Vernon US-J 3
Mountain Grove US-E 19
Moxee US-J 4
Mtwapa KE 3
Muddusjärvi FI 11
Munday US-H 10
Murcia ES 15
Muscatine US-G 21
Mutěnice CS 18
Mwanza TZ 7
Mwea KE 8

Naaldwijk NL 8, 17
Nachingwea TZ 6
Nagano JP 26
Nagasaki JP 27
Naju KR 16
Nakorn Phanom TH 12
Nalchik SU 14
Nambour AU 23
Namhae KR 17
Nan TH 3
Nancy FR 11
Naoussa GR 19
Nara JP 28
Narbonne FR 12
Nashville US-E 12
Nashville US-E 4
Nazareth ET 1
Nea Moudania GR 20
Nea Philothei GR 21
Nelspruit ZA 1
Nes på Hedmark NO 6
Neustadt DE 22
Nevsehir TR 19
Newark US-C 2
New Brunswick US-B 14
New Delhi IN 2
New Franklin US-E 21
New Lisbon US-B 15
New Orleans US-D 58
Niigata JP 29
Niš YU 12
Niterói BR 19
Nitra CS 19
Noordscharwoude NL 18
Normandin CA 32
Northampton US-A 8
North East US-B 11
North Platte US-G 25
Norwich GB 24
Nossen DD 4
Nottingham GB 25
Novi Sad YU 17
Nowa Wieś PL 16
Nowy Dwór PL 17
Nsukka NG 5
Nuriootpa AU 40
Nyanga ZI 6

Obregón MX 1
Odder DK 11
Odense DK 7
Oeiras PT 6
Oita JP 30
Okayama JP 31
Okinawa JP 32
Olomouc CS 20
Olsztyn PL 18
Ongar GB 26
Ontario US-J 18
Oppenheim DE 23
Ora IT 11
Oradea RO 13
Orange AU 7
Orechová CS 21
Orel SU 15
Orlando US-D 34
Orléans FR 13, 40
Ormeignies BE 16
Orono US-A 1
Osaka JP 33
Oslo NO 7
Osnabrück DE 24
Ossining US-B 7
Otago NZ 14
Ottawa CA 20
Oulu FI 16
Övermark FI 3
Overton US-H 12
Owerri NG 9
Oxford US-D 9

Pabianice PL 19
Padova IT 12
Pahrump US-I 10

Painter US-C 13
Palermo IT 13
Pälkäne FE 4
Palma de Mallorca ES 16
Palmer US-L 2
Palmerston North NZ 8
Palmira CO 4
Palos Verdes US-K 10
Pantnagar IN 20
Paramaribo SR 1
Parbhani IN 12
Paris FR 41
Parma IT 14
Parma US-I 15
Pasadena US-K 11
Pasto-Obonuco CO 8
Patra GR 22
Peć, YU 15
Pelotas BR 30
Pendleton US-J 15
Peradeniya LK 3
Perkerra KE 9
Perkins US-H 4
Perth AU 59
Perugia IT 15
Peterborough GB 27
Petit Bourg (Guadeloupe) FR 18
Philadelphia US-B 12
Phoenix US-H 26
Piacenza IT 16
Pichit TH 7
Pietermaritzburg ZA 2
Piikkiö FI 1
Pine Mountain US-D 21
Pingtung TW 4
Piracicaba BR 24
Pisa IT 17
Pitzelstätten AT 5
Pleven BG 7
Plew TH 18
Plovdiv BG 8
Plymouth US-D 1
Poamoho US-M 5
Pomorie BG 9
Pontotoc US-D 48
Poplarville US-D 49
Portageville US-E 20
Port Harcourt NG 10
Portici IT 18
Porto Alegre BR 29
Porto Velho BR 4
Port Sulphur US-D 56
Poznań PL 20
Praha CS 22
Preston Wynne GB 28
Pretoria ZA 3
Prievidza CS 23
Princeton US-E 16
Priština YU 16
Prosser US-J 5
Provo US-I 6
Průhonice CS 24
Prusy PL 21
Przyborów PL 22
Pszczyna PL 23
Pukekohe NZ 3
Pullman US-J 6
Puyallup US-J 7
Pyrgos GR 23
Pyungchang KR 18

Quedlinburg DD 5
Quibor VE 5
Quicksand US-E 17
Quincy US-D 37

Rabat MA 5
Rahuri IN 14
Rajburi TH 16
Raleigh US-D 8
Randolph Center US-A 4
Reading GB 29
Recife BR 10
Regina CA 10
Reguły PL 24
Rehovot IL 2
Reidsville US-D 10
Rekowo PL 25
Rennes FR 14
Reno US-I 11
Requena ES 17
Retz AT 16
Reus ES 18
Rhinelander US-F 27
Rhodes GR 24
Richmond AU 8
Ridgetown CA 21
Rilland NL 19
Rinn AT 17
Rio Branco BR 5
Río Bravo MX 9
Rionegro CO 6
Ripley US-F 6
Ritzlhof AT 18
Riverhead US-B 8
Riverside US-K 12
Rize TR 20
Röbäcksdalen SE 6
Rockhampton AU 24
Rolla US-G 17
Roma IT 19
Rosario AR 16
Roseworthy AU 41
Rossosh SU 16
Rovaniemi FI 8
Ruakura NZ 5
Rubmeke BE 17
Ruukki FI 7

Sabour IN 8
Safford US-H 28
Saga JP 34
Saharanpur IN 25
Saint Marcellin FR 43
Saint Paul US-G 7
Saint-Paul-lez-Durance FR 44
Saint-Remy-de-Provence FR 45
Saitama JP 35
Saittas CY 44
Salerno IT 20
Salinas US-K 13
Salisbury US-C 7
Salta AR 17
Saltillo MX 15
Salt Lake City US-I 7
Salvador BR 13
Salzburg AT 6
Samaru NG 2
Samokov BG 10
Samsun TR 21
Sandanski BG 11
Sandpoint US-I 17
Sanford US-D 36
San Francisco US-K 14
San Giuliano (Corse) FR 15
Sangju KR 19
San José CR 1
San Juan AR 18
San Marino US-K 15
San Pedro AR 19
San Rafael AR 20
Sanremo IT 22
Santa Barbara US-K 16
Santa Cruz de Tenerife ES 19
Santiago del Estero AR 21
São Luis BR 6
São Paulo City BR 21
Sarajevo YU 1
Saratoga US-K 17
Sarrians FR 46
Sarstedt DE 25
Saskatoon CA 11
Sassari IT 23
Satu-Mare RO 28
Schlachters DE 26
Scottsbluff US-G 26
Scottsdale AU 48
Seattle US-J 8
Sebha LY 4
Sede Boqer IL 5
Segunung ID 1.3
Seibersdorf AT 24
Selby GB 30
Sennar SD 5

Seongnam KR 20
Seoul KR 21
Setúbal PT 7
Sfax TN 3
Shanhua TW 5
Shepparton AU 29
Shiga JP 36
Shimane JP 37
Shinshieh TW 6
Shizuoka JP 38
Shreveport US-D 59
Sibiu RO 29
Sidney CA 2
Sidney US-I 23
Sids EG 7
Siejnik PL 26
Silberberg AT 19
Silistra BG 12
Silsoe GB 31
Simcoe CA 23
Simla IN 4
Simpson US-F 10
Singher SN 3
Sinołeka PL 27
Sint-Katelijne-Waver
(Mechelen) BE 18
Sint-Truiden BE 19
Sittingbourne GB 32
Sitzenberg AT 7
Skierniewice PL 28
Skopje YU 4
Słupia Wielka PL 29
Smederevska Palanka YU 13
S. Michele all'Adige IT 21
Smržice CS 25
Sochi SU 17
Sodus US-F 18
Sofia BG 13
Solan IN 10
Solok ID 3
Sousse TN 2
South Tahrir EG 5
Sparks US-H 2
Spooner US-F 24
Springfield US-E 5
Spring Hill US-D 46
Spring Hill US-E 6
Sri-Sa-Ket TH 11
St. Andrä AT 20
Stanthorpe AU 25
St-Augustin CA 33
St. Catherines CA 22
St. Charles US-F 11
Ste-Anne-de-Bellevue CA 37
Steeles Tavern US-C 11
Ste-Foy CA 38
Stellenbosch ZA 4
Stephenville US-H 13
St-Hyacinthe CA 34
Stillwater US-H 3
St-Jean-sur-Richelieu CA 35
St. John's CA 44
Stjordal NO 8
St-Lambert-de-Levis CA 36
Stockbridge US-A 9
Stockholm SE 7
Stoneville AU 60
Storrs US-A 13
Straelen DE 27
Strejesti RO 26
Struan AU 39
Sturgeon Bay US-F 25
Stuttgart DE 28
Summerland CA 3
Suncheon KR 22
Superior US-H 29
Surajthani TH 14
Suweon KR 23
Swan AU 61
Swan Hill AU 34
Swierklaniec PL 30
Swift Current CA 12
Sydney AU 9
Szczecin PL 31
Szentes HU 8

Tafo GA 4
Taichung TW 7
Tainan TW 8
Taipei TW 9
Taitung TW 10
Tajoura LY 2
Tampico MX 12
Tanger MA 6
Taoyuan TW 11
Tarapoto PE 3
Tasjkent SU 18
Tatura AU 32
Tbilisi SU 19
Těchobuzice CS 26
Te Kauwhata NZ 4
Tekirdag TR 22
Teresina BR 7
Tervuren BE 20
Tha-Chai TH 9
Thessaloniki GR 25
Thika KE 1
Tifton US-D 22
Tigoni KE 5
Timişoara RO 35
Tipton US-F 20
Tîrgu Jiu RO 31
Tîrgu Mureş RO 25
Titograd YU 7
Tlekung ID 1.6
Tochigi JP 39
Tokat TR 23
Tokushima JP 40
Tokyo JP 41
Torino IT 24
Tottori JP 42
Toulouse FR 16
Toyama JP 43
Trang TH 15
Traverse City US-F 19.1
Trenton CA 24
Tribune US-G 32
Trier DE 29
Tripoli LY 1
Trivandrum IN 5
Tromso NO 9
Troyan BG 14
Truro CA 42
Tsar Krum BG 15
Tucson US-H 27
Tucuman AR 22
Tucumcari US-H 21
Turku FI 12
Turnov CS 27
Turrialba CR 2
Tuxtla Gutierrez MX 4

Udaipur IN 19
Udine IT 25
University Park US-B 13
Uppsala SE 8
Urbana US-F 12
Uvalde US-H 14

Valladolid ES 20
Valul lui Traian RO 18
Vancouver CA 4
Vancouver US-J 9
Varanasi IN 27
Varna BG 16
Vaskhnil SU 11
Velké Losiny CS 28
Velký Krtíš CS 30
Veltrusy CS 29
Veracruz MX 10
Verona IT 26
Versailles FR 17
Veselé CS 31
Vicosa BR 17
Vidra RO 1
Viedma AR 23
Vilafranca des Panadés ES 21
Vila Real PT 8
Vîlcea RO 32
Villaviciosa ES 22
Vilvoorde BE 21
Vincennes US-F 8
Vineland Station CA 25
Virginia Beach US-C 12

Vitória BR 18
Viterbo IT 27
Voineşti RO 33
Volcano US-M 10
Volos GR 26

Wädenswil CH 1
Wad Medani SD 3
Wageningen NL 9
Wagga Wagga AU 10
Wailua US-M 6
Waimanalo US-M 7
Waite AU 42
Wakayama JP 44
Waltham US-A 10
Wareham US-A 11
Warszawa PL 32
Waseca US-G 8
Washington DC US-C 4
Waynesville US-D 7
Weinsberg DE 30
Wellesbourne GB 33
Wellesley US-A 12
Wenatchee US-J 10
Weslaco US-H 15
Wetteren BE 22
Wetzawinkel AT 21
Wexford IE 6
Whiting US-G 22
Whittlesford GB 34
Wichert US-F 13
Wichita US-G 30
Wien AT 8, 25
Wies AT 22
Wieselburg AT 9
Wilhelminadorp NL 10
Williston US-G 18
Winchester US-C 14
Winnipeg CA 14
Witzenhausen DE 31
Wolbeck DE 32
Wollongbar AU 11
Wooster US-F 7
Wróblowice PL 34
Wrocław PL 33
Würzburg-Veitshöchheim DE 33
Wye GB 35

Yakima US-J 11
Yamagata JP 45
Yamaguchi JP 46
Yamanashi JP 47
Yanco AU 12
Yangmingshan TW 12
Yesan KR 24
Yuma US-H 30

Zagreb YU 3
Zalău RO 30
Zaragoza ES 23
Želešice CS 32
Zemun YU 14
Zyghi CY 5

INDEX OF NAMES OF RESEARCH WORKERS

Aaouine, M. MA 1.1
Aarnes, H. NO 7.2
Aarsen, L.J.G. NL 1.1.6
Aarts, H.F.M. NL 6.1
Aas, K. NO 10.2
Abadjieva, M. BG 13.1
Abak, K. TR 2.1
Abaka-Gyenin, A.K GA 1.2
Abani, M.S.C. NG 5.1
Abawi, G.A. US-B 2.1.4
Abbas, S.R. IN 26.1
Abbate, V. IT 5.3
Abbott, A.J GB 19.1.2
Abbott, J. US-C 3.1
Abbott, T.S. AU 9.1.2
Abdel-Halim, M.K. LY 2.1
Abdel-Rahman, A.H. LY 2.1
Abdel-Salam, A.A LY 2.1
Abdel-Salam, AM LY 2.1
Abdul-Karim, A. GA 4.5
Abe, H. JP 10.1, 17.1
Abe, K. JP 8.2 33.1, 45.1, 45.2
Abe, N. JP 8.2
Abe, T. JP 14.6, 16.1, 24.2
Abebe, B ET 5.1
Abebe, Y. ET 2.1
Abed, A.H. LY 1.1
Abel, J. DK 3.2
Abeles, F. US-C 8.2
Abernathy, J US-H 9.1
Abetunge, K.G.W. LK 3.16
Abetunge, S. LK 3.16
Abeyratne, S.D. LK 3.6
Abeyratne, W.M. LK 3.4
Abiko, K. JP 23.2.4
Abo-Zeid, M.H. EG 7.1
Abraham, C.C. IN 5.3
Abrahamsen, R.K. NO 10.1.3
Abrahão, E. BR 15.1
Abramof, L. BR 10.1
Abramowicz, M. PL 31.1.2
Abrantes, A. PT 6.2
Abrantes, J. PT 7.1
Abreu, J.M. de BR 14.1
Abreu Choairy, S. BR 9.1
Abu-Hifala, I.S LY 2.1
Abu-Khdair, A. LY 2.1
Abu-Rawi, A.H. LY 2.1
Aburaya, K. JP 27.1
Abutiate, W.SY. GA 3.1
Acar, H. TR 15.3
Accati Garibaldi, E. IT 24.3
Acciarri, N. IT 20.1
Acharya, M.M. IN 19.2
Achimescu, I. RO 33.1
Achituv, M IL 1.2.3
Achkova, Z. BG 13.1
Achouri, M. MA 1.1
Achremowicz, J. PL 11.1.4
Acker, J.H. ZA 4.1
Ackerl, I. HU 5.3.1
Acosta, A.R. AR 18.1
Acosta, J. CR 1.1
Acosta Nuñez, S. MX 10.1
Acosta Rodriquez, G.F. MX 5.3
Acree, T.E. US-B 2.1.5
Acs, A. HU 3.1.1
Acs, E. HU 2.1.5
Adachi, K. JP 22.4
Adalla, C.B. PH 1.3
Adam*, C.S. SC 1.1
Adam, R. CS 3.1
Adam, S.O. LY 3.1
Adamicki, F. PL 28.2.9
Adamowicz, S. FR 2.1.1
Adams, A.J. US-D 56.1
Adams, A.M. GB 8.1.2
Adams, G. US-F 14.1.3
Adams, J.P. US-D 27.3
Adams, K. DE 22.1
Adams, P. GB 18.1.3
Adams, P.B. US-C 3.1
Adams, R.G US-A 13.1
Adaniya, S. JP 32.2
Adansi, M.A. GA 3.4
Adatia, M.H. GB 18.1.3
Adatto, Y. IL 1.2.5
Adda, J. FR 7.1.4
Addae-Kagyah, K.A. GA 3.1
Addison, E.A. GA 3.2
Adebanjo, A. NG 1.1
Adedoyin, S.F. NG 1.1
Adegoroye, A.S. NG 3.1
Adelaja, B.A. NG 1.1
Adelana, B.O. NG 1.2
Adenuga, O.A. NG 3.1
Adeyemi, S.A.O. NG 1.1
Adhikari, K.S. IN 24.1.3
Adigun, J.A. NG 2.1
Adiri, N. IL 1.2.4
Adleman, M.I. US-B 5.1.1
Adomako, B. GA 4.5
Adomako, D. GA 4.2
Adriaensens, W. BE 18.1
Adriansen, E. DK 1.2
Adu-Ampomah, Y. GA 4.5
Aendekerk, Th.G.L. NL 2.1
Aerny, J. CH 2.1
Agami, M. IL 5.1.2, 5.1.4
Agaoglu*, Y.S. TR 6.1
Agarwal, P.K. IN 1.1.1
Agatsuma, M. JP 12.3
Agble*, F. GA 3.1
Aggarwal, R.K. IN 18.4
Agnew, M.L. US-G 20.1
Agnihotri, R.P. IN 10.3
Agrigoroaie, V. RO 4.1
Aguilar, A. ES 13.1
Aguilar, A.L. MX 6.4
Aguilar, E.A. PH 1.8
Aguilar Perez, H. MX 5.5
Aguilar Zamora, A. MX 10.2
Aguiree, J.A. CR 1.1
Agunloye, O. NG 1.1
Aharoni, N. IL 1.3.1
Aharoni, Y. IL 1.3.1
Ahenkorah, Y. GA 4.3
Ahlawat, V.P. IN 9.1
Ahmed, E.M. US-D 27.3
Ahmed, H.S. LY 1.2
Ahmed, S. IN 19.1
Ahmed, S.A. LY 1.1
Ahn*, C.S. KR 13.2
Ahonen, S. FI 10.1.1
Ahrens, W.H. US-C 2.1
Ahuja, A.K. IN 1.1.8
Aiazzi, M. AR 7.1
Aichele, M.D. US-J 4.1
Aidar, H. BR 26.1
Aihoshi, K. JP 18.1
Aikman, D.P. GB 18.1.1
Ailincăi, N. RO 23.1
Aitken, J.B. US-D 14.1
Ait Oubahou, H. MA 1.1
Aizu, H. JP 3.2.3
Ajao, B. NG 1.1
Aka, R. JP 32.1
Akagi, H. JP 39.1
Akaha, T. JP 14.6
Akai, T. JP 40.2
Akamatsu, S. JP 5.2
Akar, K. TR 14.1
Akatsuka, T. JP 14.5
Akaura, K. JP 23.3
Aken'ova, M.E. NG 1.5
Åkesson, G. SE 3.1
Åkesson*, I SE 1.4
Akhade, M.N. IN 4.1
Akhigbe, G.O. NG 1.1
Akiba, F. BR 19.2
Akihama, T. JP 19.3
Akilli, M. TR 1.1
Akimitsu, N. JP 37.1
Akimoto, T. JP 11.2, 35.2
Akin-Taylor, A.O. NG 1.1
Akingbohungbe, A.E. NG 3.1

Akinyemiju, O.A. NG 3.1
Akita, T. JP 30.1
Akius*, M SE 1.4
Akiyama, M. JP 31.1
Akiyama, T. JP 19.2
Akiyoshi, H. JP 5.3
Akman, B. TR 15.3
Akoroda*, M.O. NG 1.5
Akoumianakis, C. GR 6.1.3
Akoumianakis, N. GR 6.1.4
Alabouvette, C. FR 7.1.3
Alahakoon, P. LK 3.7
Alan, M.N. TR 15.2
Alan, R. TR 9.1
Al-Ansary, M.A. LY 3.1
Al-Bakouri, A.M. LY 2.1
Alberghina, O. IT 5.1
Albério, A.C. BR 2.3
Albert, A. ES 14.1
Albertini, A. IT 19.1.1
Alblas, J. NL 6.1
Albouy, J. FR 17.1.2
Albrigo, L.G. US-D 30.2
Albuquerque Regina, M. BR 15.1
Albu-Yaron, A. IL 1.3.2
Alcalá, P. PE 1.2
Alcantara, A.J. PH 1.5
Alconero, R. US-B 2.1.2, 2.1.6
Alcorn, S.M. US-H 27.1
Aldegretti, J.L.D.H. BR 29.1
Aldén, B. SE 1.4
Alderson, P.G. GB 25.1.1
Alderz, W.C. US-D 32.1
Aldrich, R.A. US-A 13.1.
Aldwinckle, H.S. US-B 2.1.4
Aleksić, Z. YU 13.1
Aleva, J.F. NL 9.7.5
Alexander, D.McE. AU 65.2
Alexander, M.P. IN 1.1.9
Alexander, V.T. SR 1.1
Alexandrakis, B. GR 8.1.1
Alexandrova, M. BG 8.1
Alexe, C. RO 8.4.1
Alexopoulos, J. US-A 13.1
Alexopoulou, P. GR 14.1.2
Al-Fallah, A.S. LY 2.1
Al-Fatisi, M.A. LY 2.1
Al-Gamal, S.S. LY 1.1
Al-Gharbawi, S.A. LY 2.1
Algie, J.E. AU 65.4
Al-Hodairi, M.O. LY 4.1
Ali, A. ET 6.1
Ali, A. US-D 36.1
Ali, M.A. LY 2.1
Ali-Barkouli, A.A LY 4.1
Alibés Rovira, J. ES 23.1
Alishev, H. BG 15.1
Alivizatos, A.S. GR 14.1.1
Al-Khomsi, R.M. LY 2.1
Al-Khraz, A.M. LY 2.1
Al-Kiraiw, A.M. LY 2.1
Allan*, P. ZA 2.1
Allard, C. FR 17.1.2
Allavena, A. IT 20.1
Allemand, P. FR 2.1.3
Allen, J.C. US-D 30.2
Allen, J.G. GB 8.1.2
Allen, L.D. US-L 2.1
Allen, R.N. AU 11.1
Allen, W.R. CA 25.2
Allen, W.W. US-K 3.1.1
Allerup, S. DK 3.4
Allinson, D.W. US-A 13.1
Allison*, M.L. US-G 30.1
Allred, K.R. US-I 5.1
Almeida, A.M.R. BR 27.1
Almeida, J.L. Jr. BR 10.1
Almeida Burity, H. BR 10.1
Almeida Camargo, U. BR 31.1
Almeida Drummond, O. BR 15.1
Almeida Oliveira, M. BR 12.1
Almeida da Silva, G. BR 31.1
Al-Misallaty, M.S. LY 2.1
Almonte, E. PE 4.1
Alofe, C.O. NG 3.1
Aloni, B. IL 1.1.1
Alonso, J.M. AR 22.3
Alonzo, F. CR 2.2
Al-Oushar, M.A. LY 3.1
Alper, M. IL 1.1.2
Alper, Y. IL 1.4, 1.4.1
Alpi, A. IT 17.2
Alptekin, I.V. TR 6.1
Al-Rashid, M.M. LY 2.1
Al-Sawaf, H. LY 3.1
Al-Sheikh, F.A. LY 3.1
Al-Shelwy, M.F. LY 3.1
Alspach, P.A. NZ 11.1
Alston, F.H. GB 8.1.2
Alsved, C. SE 1.1.3
Altay, F. TR 10.1
Altay, K. TR 12.1
Altieri, M.A. US-K 3.1.1
Altman, A. IL 2.1
Altoveros, E.C. PH 1.7
Altoveros, N.C. PH 1.7
Altwell, K. DK 3.2
Aluko, M.O. NG 1.4
Alvare Márquez, J.A. MX 8.3
Alvarez, R.G. MX 1.3
Alvarez, S. AR 22.3
Alves Aguiar, F.F. BR 21.1
Alvarez Costa, E. AR 23.1
Alves, E.J. BR 12.1
Alves, R.J.P. BR 19.1
Alves Costa, J.T. BR 8.2
Alves da Silva, J. BR 25.3
Alves de Menezes Sobrinho, J. BR 25.5
Alvim, P. de T. BR 14.1
Alvim, R. BR 14.1
Amado Alvarez, J.P. MX 5.4
Amagai, H. JP 14.1
Åmaki, W. JP 41.3
Åman*, G. SE 1.4
Amano, E. JP 14.4
Amano, M. JP 23.2.2
Amano, N. JP 17.2
Amano, S. JP 5.3
Amans, E.B. NG 2.1
Amaral, D.I. BR 29.2
Amarasiri, S.L. LK 3.6
Amăriutei, A. RO 8.4.1
Amazonas A. da Silva, M.A. BR 31.1
Ambărus, S. RO 3.1
Amberger, A. DE 11.1.3
Ambrose, J.T. US-D 8.1.3
Amemiya, K. JP 47.1
Amemiya, T. JP 47.2
Amici, J.F. IT 19.2
Amico Roxas, U. IT 13.2
Amih, C. NG 1.1
Amiot, M-J. FR 3.1.8
Amiraux, A. FR 35.1
Amling, H.A. US-D 38.1
Amma, A. AR 19.1
Ammati, H. MA 1.1
Ammerlaan, J.C.J. NL 8.1
Amos, T. AU 27.1
Amosu, J.O. NG 3.1
Ampe, G. BE 17.1
Amponsah, J.D. GA 4.5
Amsen*, M.G. DK 1.2
Amsing, J.J. NL 1.1.5
Amundson, R.G. US-K 3.1.3
Amzăr, V. RO 9.1
An, J.K. KR 2.1.2
Anadoliev, G. BG 3.1
Anadoliev, Z. BG 3.1
Anagnou, M. GR 14.1.2
Anais, G. FR 18.1
Anajaneyulu, K. IN 1.1.1
Anand, N. IN 1.1.2
Ananda, S.A. IN 10.3
Anasiewicz, A. PL 14.1.9
Anastasijević, N. YU 8.2

Anastasiou, G. GR 23.1
Anders, A. DE 15.3.1
Andersen, A. NO 10.4.3
Andersen*, Aa. DK 3.1
Andersen*, A.S. DK 3.1
Andersen, C. US-H 6.1
Andersen, G.P. DK 3.7
Andersen, H. DK 1.2
Andersen, M. DK 3.7
Andersen, P. US-E 19.1
Andersen, R.L. US-D 13.1
Anderson, C.A. US-D 30.2
Anderson, C.R. US-E 11.1
Anderson, D.C. US-I 20.1
Anderson, H.M. GB 19.1.3
Anderson, J. US-C 3.1
Anderson, J.A. US-I 9.1
Anderson, J.F. US-A 5.1.1
Anderson, J.H. US-F 14.1
Anderson, J.L. US-I 5.1
Anderson, J.R. US-K 3.1.1
Anderson, M.B. US-B 7.1
Anderson, M.G. CA 41.1
Anderson, M.M. GB 15.1.2
Anderson, N.H. GB 19.1.1
Anderson, R. US-E 15.1
Anderson, R.H. US-G 8.1
Andersson, G. SE 5.2
Andersson, N.E. DK 1.2
Andersson, S. SE 1.4
Ando, E. JP 45.1
Ando*, T. JP 4.4
Andor, D. HU 2.1.2
Andrásfalvy, A. HU 5.3.2
André, J.P. FR 2.1.2
André, P. BE 12.1
Andrada, H.N. AR 18.1
Andrade, J.G.M. BR 27.1
Andrade Kato, M. do S. BR 3.1
Andrei, P. RO 5.1
Andréoli, C. FR 2.1.3
Andreoli, Cl. BR 25.5
Andreotti, R. IT 14.1
Andresen, A. NO 9.2
Andrew*, W.T. CA 7.1
Andriesse, W. NL 9.11.2
Androulakis, J. GR 8.1.1
Angel, R.C.R. GB 16.1
Angel, S. IL 1.3.2
Angeles, D.E. PH 1.2.1
Angeliev, V. BG 8.3
Angelini, S. IT 19.1.1
Angelov, T. BG 8.3
Angerilli, N.P.D. CA 3.1
Anglade, P. FR 4.1.7
Anil, S. TR 2.2
Anjaneyulu, K. IN 1.1.9
Anma, S. JP 38.2
Anno-Nyako, F.O. NG 1.1
Anraku, M. JP 46.1
Ansótegui, N. AR 5.1
Anserwadekar*, K.W. IN 12.1
Anstett, A. FR 17.2
Antero Neto, J.F. BR 8.1
Antić, M. YU 3.3
Antognozzi, A. IT 15.1
Anton, D. RO 19.2
Antonacci, D. IT 7.4
Antonelli, M IT 19.1
Antonić, B. YU 3.4
Antoniani, C. IT 3.2
Antônio, H. BR 27.1
Antonioletti, R. FR 3.1.4
Antoniou, M. GR 14.1.2
Antoszewska-Orlinska, H. PL 28.1.2
Anunciaçaõ de Andrade, V. BR 30.2
Anunciado, I.S. PH 1.2.2
Anupand, P. TH 11.1
Anuszewska, E. PL 32.3.3
Anvari, S.F DE 28.1.1
Anyim, O.A. NG 1.1
Anyszka, Z. PL 28.2.4
Aoba, K. JP 16.2
Aoki, A. JP 39.1
Aoki, H. JP 4.1
Aoki, K. JP 4.1
Aoki, M. JP 1.1.1, 23.2.5, 46.2, 47.2
Aoki, N. JP 37.2
Aoki, T. JP 17.1, 20.1, 41.1
Aono, N. JP 19.2
Aoyagi, S. JP 4.1
Apan, H. TR 21.1
Apaydin, Y. TR 3.1
Apeland*, J. NO 10.1.3
Apelbaum, A. IL 1.3.1
Apolitano, C. PE 2.1
Apostol, E. HU 2.1.2
Apostol, J. HU 2.1.2
Apostolakis, C. GR 3.1.2
Appelgren, M. NO 10.1.4
Appiah, M.R. GA 4.3
Applegate, J.E. US-B 14.1.1
Apps, D.A. US-B 10.1
Aquilizan, F.A. AU 24.1
Arad, S. IL 4.1
Arafa, A.I. EG 1.3
Arai, K. JP 23.2.3
Arai, S. JP 28.1, 3.2.3, 41.1
Arakawa, B. US-M 5.1, 7.1
Araki, C. JP 38.3
Araki, H. JP 13.1
Araki, Y. JP 23.2.5
Aranda, D.E. AR 15.3
Aranha, O. BR 20.1
Araújo, C.M. BR 19.2
Araújo, J. PT 4.1
Araujo, L.C.S. BR 13.2
Araujo, M. BR 15.1
Araujo, P.E.S. BR 10.1
Araya, J. JP 3.3.1
Arbeloa, J. ES 11.1
Arce, J. CR 2.3
Archambault, J. CA 30.1
Archbold, D. US-E 15.1
Archer, E. ZA 4.2
Ardeleanu, M. RO 4.1
Arellano Sota, C. MX 13.1
Arends, J.C. NL 9.7.4
Arévalo Valenzuela, A. MX 7.2
Argatu, C. RO 1.1.5
Argerich, C. AR 10.1
Arguello, J. AR 7.1
Arguello Mendoza, C. MX 5.4
Argumedo, J.J. MX 11.3
Argyrakis, E. GR 10.1
Argyriou, L.C. GR 14.1.2
Arias, J.A. AR 17.1
Arias Suárez, J.F. MX 3.2
Arihara, K. JP 4.1
Arima, H. JP 26.5
Arimura, T. JP 18.1
Arino, A. JP 2.1
Arisumi, K. JP 18.3
Arisumi, T. US-C 3.1
Arita, T. JP 41.1
Ariza, U. CO 7.2
Arjona, C. AR 15.2.1
Arkoudilos, J. GR 18.1.2
Arleu, R.J. BR 18.1
Armenta Cárdenas, S. MX 2.3
Armitage, A. US-D 16.1
Armstrong, D.W. AU 47.1
Armstrong, G. US-E 17.1
Armstrong, R.J. US-B 10.1
Arn, H. CH 1.1
Arnesen, K.R. US-B 14.1.5
Arnold, G.M. GB 19.1.4
Arnold, R.G. US-G 23.1
Arnoux, M. FR 3.1.12
Aronson, A. IL 4.1
Arora, R.L. IN 20.1
Arrington, E.J. US-E 12.1
Arroyo, A. ES 19.1
Arruda, G.P. BR 10.1

Arsel, H. TR 15.3
Arsenault, C. CA 31.1
Arteca, R.N. US-B 13.1.1
Arthey, V.D. GB 7.1
Artz, W.E. US-F 12.2.1
Arulnandi, V. LK 3.3
Arunpairoj, S. TH 9.1
Arús, P. ES 4.1
Arvieu, J.C. FR 2.1.2
Arya, R.S. IN 2.2
Arze, J. CR 2.2
Asada, K. JP 27.3
Asada, T. JP 3.4
Asahira*, T. JP 22.5
Asakawa, S. JP 12.4
Asakura, T. JP 14.2
Asano, S. JP 35.2
Asano, Y. JP 12.4
Asante, S.K. GA 4.4
Asaoka, M. JP 19.2
Asare-Nyako, A. GA 4.6
Asari, A. JP 2.1
Asari, M. JP 2.2
Ascher, P.D. US-D 7.1
Asgrimsson, O. IC 1.1
Ashbell, G. IL 1.3.3
Ashby, J.W. NZ 13.3
Ashcroft, G.L. US-I 5.1
Ashcroft, W. AU 32.1
Asheim, H. NO 10.1.5
Asher, C.J. AU 15.5
Ashihara, W. JP 11.3
Ashikawa, K. JP 41.1
Ashizawa, M. JP 23.2.2
Ashley, R.A. US-A 13.1
Ashok, T.H. IN 11.2
Ashour, W.A. LY 3.1
Ashtaputre, J.U. IN 12.2
Ashworth, E. US-C 8.2
Ashworth, Jr., L.J. US-K 3.1.2
Asiegbu, J.E. NG 5.1
Askri, F. TN 1.1
Aslan, G. TR 5.2
Aso, P.J. AR 22.3
Aspir, I. TR 13.1
Assaf, R. IL 1.2.2
Assefa, I. ET 1.1
Assefa, W/t. H. ET 4.1
Asselin, A. CA 38.1
Asselin, C. FR 1.1.5
Assis, M. BR 30.1
Assis Paiva, F. BR 15.1
Astié, M. FR 19.4
Astley, D. GB 33.1.6
Astorga, C. CR 2.3
Asueri, B. ZI 3.1.2
Atalay, E. TR 5.1
Atanasiu, N. RO 8.1
Atanasov, P. YU 4.2
Atanassov, A. BG 13.1
Atanassov, Y. BG 11.1
Atawna, A.A. LY 3.1
Atger, P. FR 3.1.12
Athanasopoulos, Th. GR 23.1
Athayde, A. de O. BR 18.1
Atherton, J.G. GB 25.1.1
Atilano, R.A. US-D 25.1
Atkey, R.T. GB 18.1.2
Atkin, R.K. GB 19.1.4
Atsumi, T. JP 23.2.5
Attaway, J.A. US-D 30.2
Attene, G. IT 23.1
Atubra, O.K. GA 2.1
Atzemis, A. GR 25.2.2
Aubé, C.B. CA 35.1
Aubert, S. FR 3.1.8
Auclair, D. FR 13.1.3
Audemard, H. FR 3.1.9
Audergon, J-M. FR 3.1.3
Audsley, E. GB 31.2.2
Augé, R. FR 19.5
Augé, P. FR 2.1.3
Augustin, B.J. US-D 25.1, 27.5
Augustin Oliveira, E. BR 30.1
Augustyn, O.P.H. ZA 4.2
Auling, G. DE 15.3.4
Aung, L.H. US-C 10.1
Aurich, M. IT 11.1
Aurin, T.M. PH 1.2.4
Ausan, S. SR 1.1
Aussenac, G. FR 11.1.2
Austin, D.J. GB 8.1.2
Austin, M. US-D 22.1
Austin, R.L. US-G 23.1
Autio, W.R. US-A 5.1.1
Avagnina de del Monte, M.S. AR 15.1
Avall, H. SE 1.1.2
Avanzato, D. IT 19.1
Avdalović, V. YU 8.2
Ave, D.A. US-B 5.1.7
Avelar do Couto, A. PT 1.1
Avellaneda, M.O. AR 15.2.2
Avendano, L. CR 2.2
Avenido, R.A. PH 1.7
Avgelis, A. GR 10.2
Avigdori, H. IL 1.1.2
Avila Valdez, J. MX 12.2
Avilés González, M. MX 2.2
Avramov, L. YU 14.1
Avšić, S. YU 18.1
Awad*, G. DE 5.1.1
Awamura, M. JP 7.1
Awasthi, M.D. IN 1.1.8
Awasthi, R.P. IN 10.2
Awosusi, O.O. NG 1.4
Ayala, H.G. AR 22.1
Ayanoglu, H. TR 13.1
Aydin, R. TR 13.1
Ayele, A.Z. ET 10.1
Ayfer, M. TR 2.1
Ayres, L.C. US-I 12.1
Azambuja Centeno, G. BR 30.3
Azirin, A. ID 1.2
Aziz, A.S. KE 3.1
Aziz, I. CY 1.1
Azouz, A.F. LY 3.1
Azukizawa, H. JP 37.1
Azuma, A. JP 20.2
Azuma, K. JP 44.1
Azzy, A.A LY 2.1

Ba, T. SN 1.1
Baardseth, P. NO 10.5.1
Baba, H. JP 26.1
Baba, T. JP 12.1
Babalola, O. NG 1.5
Babatola, J.O. NG 4.1
Babel, Y.S. IN 19.4
Babić, S. YU 11.1
Băbiceaunu, S. RO 8.4.2
Babik, I. PL 28.2.3
Babik, J. PL 28.2.5
Babnik, M. YU 18.2.1
Babós, L. HU 2.2.4
Baboth, E. HU 5.3.2
Babović, M. YU 14.1
Babović, YU 10.1
Babrikov, D. BG 8.3
Babu, H. IN 6.1
Bacalacos, C. GR 11.1
Bacelar, W. BR 26.2
Bach, I. HU 2.7
Bachariev, D. BG 8.1
Baches, E. RO 8.4.2
Bachmann, H. AT 3.1
Bachmann, H. IT 11.1
Bachnacki, R. PL 1.1
Bachthaler, E. DE 15.1.8
Bachvarov, St. BG 2.1
Bäcker, G. DE 12.1.14
Bacogiannis, A. GR 26.1
Bacopulos Tellez, E. MX 15.1
Badayos, R.B. PH 1.5
Baden, C. DK 3.6
Bădescu, Gh. RO 9.2

Ba Diallo, A. SN 1.1
Badiyal, J. IN 10.3
Badoux, S. CH 2.1
Badovinac, Z. YU 3.3
Badowska-Czubik, T. PL 28.1.2
Bae, B.H. KR 6.1
Bae, D.H. KR 23.1.1
Bae, K.H. KR 12.3
Baelde, A. NL 17.2
Baerenholdt-Jensen, O. DK 7.1
Baerk, JH KR 10.3
Baev, Ch. BG 8.2
Baevre, O.A. NO 8.1
Baez, M.J. AR 21.1
Bagade, T.R. IN 13.6
Bagnall, R.H. CA 40.3
Bagnara, D. CA 35.1
Bagnaresi, U. IT 3.1
Bagriyanik, E.N. TR 3.1
Bagtzoglou, Z. GR 25.2
Bahn, O. KR 5.2
Bahnmüller, H. DE 16.1
Baigent, D.R. AU 11.1
Baigorria, G. AR 4.1.4
Baille, A. FR 3.1.4
Bailey, A.G. GB 16.1
Bailey, B.J. GB 31.2.6
Bailey, J.A. GB 19.1.1
Bailey, L.B. US-D 27.3
Bailiss, K.W. GB 35.1.3
Bajenaru, Gh. RO 1.1.8
Bajkan, N. YU 17.1
Bajpai, P.N. IN 22.1
Bajpai, S.K. IN 22.1
Bajtay, I. HU 5.3.1
Bajwa, G.S. IN 17.1
Bakalounakis, B. GR 10.2
Bakarcic, M. AR 9.1
Bakcsa, F. HU 2.1.3
Bakema, F. NL 1.1.2
Baker, C.R.B. GB 13.1.1
Baker, D.A. GB 35.1.3
Baker, E.A. GB 19.1.1
Baker, E.C. US-E 9.1
Baker, G.J. AU 36.2
Baker, H. AU 9.1.2
Baker, I. AU 62.1
Baker, J.E. US-C 3.1
Baker, J.H. US-A 5.1, 5.1.1
Baker*, M.L. US-H 7.1
Baker, S.W. GB 4.1
Baker Jr., W.R. US-D 3.1
Bakhchevanova, S. BG 11.1
Bakker, J.G.A. NL 18.2
Bakker, J.J. NL 9.9
Bakker, J.O. NL 18.2
Bakker, P.W.V. NL 6.1
Bakker, R.Ch. NL 18.2
Bakonyi, K. HU 6.1
Bakos, F. HU 4.2
Bakos, I. HU 4.2
Bakowski, J. PL 28.2.10
Baksik, A. PL 23.1
Bakun, M. PL 6.1
Bal, E. BE 19.1
Bal, J.S. IN 17.1
Balagopalan, C. IN 5.1
Balagtas, G.E. PH 1.7
Balah, E.O. KE 1.2
Bålan, C. RO 3.1
Balan, V. RO 8.2
Balázs, K. HU 2.5
Balázs, S. HU 2.2.12, 5.3, 5.3.1
Balciunas, J.K. US-D 25.1
Balderas Rodriquez, M.Z. MX 6.3
Baldini, E. IT 3.1
Baldini*, P. IT 14.1
Baldrati, G. IT 14.1
Baldwin, R.E. US-C 13.1
Baldy, C. FR 3.1.4
Bale, S. US-E 15.1
Balick, M.J. US-B 1.1
Balinov, I. BG 6.1
Balinova, A. BG 4.1
Bálint, J. HU 2.2.9
Bálint, K.E. HU 2.2.7
Balla, Cs. HU 2.2.1
Balla, I. HU 2.1.2
Balla, V. HU 2.2.9
Ballade, P. FR 28.1
Ballal, A.L. IN 12.2
Ballinger, W.E. US-D 8.1.1
Ballington, J.R. US-D 8.1.1
Bálo, B. HU 5.2
Balogh, I. HU 2.2.10
Balogh, P. HU 5.3.2
Balogh-Ormos, I. HU 2.2.4
Balsay, J. HU 2.6
Baltazar, A.M. PH 1.6
Baltaziak, T. PL 14.1.2
Baltensperger, A.A. US-H 19.1
Balthazard, J. FR 6.1.4
Baltzakis, N. GR 10.2
Ban, C.D. KR 2.1
Banasik, L. PL 28.1.17
Banda, W.R.G. ML 1.1
Bandara, W.M.S.M. LK 3.14
Bando, K. JP 40.1, 40.2
Bandu, R.J. BR 28.1
Banfi, G.N. AR 6.1
Bangard, W. BR 29.1
Bangerth*, F. DE 28.1.1
Banham, F.L. CA 3.1
Bánházi, Gy. HU 4.2
Banik, B. IN 28.1
Banin, A. IL 2.4
Bank*, A. SE 1.4
Bankapur, V.M. IN 11.3
Bankl, R. AT 8.10
Banko, T.J. US-C 12.1
Bannerot, H. FR 17.1.1
Bannister, P. NZ 14.1
Banno, K. JP 42.3
Banville, G. CA 26.1
Bányai, A. HU 5.2
Banyś, J. PL 15.1
Barańska, G. PL 28.2.13
Barańska, Z. PL 28.2.11
Barabé, D. CA 31.1
Bar-Akiva, A. IL 1.2.3
Baran, F. PL 1.1
Baran, J. PL 32.3.6
Baranowkska, T. PL 25.1
Baranowski, T. PL 20.2.8
Baranyi, J. HU 5.1.4
Baratta, B. IT 13.1
Barba, M. IT 19.2
Barbe, J.-P. FR 37.1
Barber, I.M. US-D 22.1
Barber, R.F. NZ 1.1
Barbieri, G. IT 14.1, 18.2
Barbosa Cabral, J. BR 10.1
Barbosa de Oliveira, M.V.L. BR 21.2
Barcelos, A.P. BR 19.2
Barcelos Kuhn, G. BR 31.1
Barcić, J. YU 3.4
Barclay, G.M. CA 40.2
Barden*, J.A. US-C 10.1
Bardosi, B. RO 1.1.2
Barefoot, A.D. US-H 3.1
Barendrecht, C.J. NL 9.9
Bareš, I. CS 22.4
Bargioni, G. IT 26.1
Bari, M.A. US-K 13.1
Barilari, C.T. AR 3.1.2
Baris, C. TR 22.1
Barkai-Golan, R. IL 1.3.1
Barke, R.E. AU 15.1
Barker, A.V. US-A 5.1.1
Barker, H. GB 15.1.3
Barkley, P. AU 9.1.1
Barksdale, T.H. US-C 3.1
Barlass*, M. AU 65.2
Barley, J.B. US-G 12.1
Barlow, E.W.R. AU 9.4

Barmore, C.R. US-D 30.1
Barneby R.C. US-B 1.1
Barnes, J.A. AU 16.1
Barnes, T.R. GB 34.1
Barnett, D. AU 65.4
Barnett, R.D. US-D 37.1
Barnhart, C.E. US-E 15.1
Barnóczki, A. HU 5.3.4
Barnola, P. FR 28.1
Baró, A.J. ES 11.2
Barón, C.G. AR 3.1.1
Baron, I. IL 1.2.1
Baroni, G. IT 26.1
Baronowski, R.M. US-D 29.1
Barooah, S. IN 7.1
Barralis, G. FR 7.1.8
Barrass, I. AU 32.1
Barrat, J.G. US-C 8.1
Barratt, D.H.P. GB 19.1.2
Barreira Bilhalva, A. BR 30.3
Barreto, H.B.F. BR 21.2
Barreto, J.A.E. BR 8.2
Barreto, P.D. BR 8.1
Barreto Figueiredo, M. BR 21.2
Barrett, H.C. US-D 34.1
Barrett, J.E. US-D 27.5
Barrick, W. US-D 21.1
Barrientos, F. MX 16.3
Barrios, E.P. US-D 52.1
Barrios Dominguez, H. MX 4.3
Barritt*, B.H. US-J 10.1
Barron, R.W. US-D 30.2
Barros, B.C. BR 21.2
Barta, A. HU 8.1
Barta, J. HU 2.2.1
Bárta, M. CS 32.1
Bartell, R.J. AU 65.5
Barten, D. NL 18.1
Barten, P. NL 18.1
Barth, G. AU 36.2
Barth, H.-G. DE 15.2.3
Bartha, Cs. HU 2.7
Bartha, J. HU 2.1.4
Bartkowski, J. PL 20.1.3
Bartkowski, K. PL 22.1
Bartley, I.M. GB 8.1.3
Bartman, E. PL 32.1.11
Bartok, J.W. US-A 13.1
Bartolini, G. IT 9.2
Barton, R.J. GB 18.1.2
Barton, S.S. US-C 2.1
Bartosik*, M.-L. FI 3.1
Bartuchotto, C.V. AR 15.2.4
Bartz, J.F. US-G 7.1
Barzic, M.R. FR 1.1.3
Bas, T. TR 15.2
Basak, A. PL 28.1.5
Basak, W. PL 28.1.2
Basco, H.J. AR 22.3
Bash, W. US-F 3.1
Basham, C.W. US-I 2.2
Basker, D. IL 1.3.2
Basky, Zs. HU 5.3.1
Basler, P. CH 1.1
Basoccu, L. IT 24.3
Bassème, F. FR 22.1
Bassett, M.J. US-D 27.7
Bassi, D. IT 3.1
Basso, C. BR 28.1
Basso, M. IT 17.1
Bassuk*, N.L. US-B 5.1.1
Bastos, C.R. BR 20.1
Batal, D. US-D 22.1
Batchelor, T.A. NZ 2.2
Bateman, L. US-D 50.1
Bateman, M.A. AU 65.5.1
Bates, J.A.R. GB 13.1.3
Bates, R.P. US-D 27.3
Bath, E. BR 21.2
Bath, J. US-F 14.1.2
Batista da Silva, J. BR 25.1
Bato, S.M. PH 1.6
Batra, R.C. IN 17.1
Batta, J. NO 10.1.5
Batten, D.J. AU 1.1
Bauckmann, M. DE 12.1.7
Bauer, F. HU 5.3.1
Bauerle*, Jr., W.L. US-F 7.1.2
Baugerød, H. NO 10.1.7
Baugher, T.A. US-C 8.1
Baum, J. US-K 2.1
Baumann, H. DE 6.1
Baumgardner, R.A. US-D 13.1
Baumgardt, B.R. US-F 9.1
Baumgärtel, G. DE 15.1.5
Bausher, M.G. US-D 34.1
Bautista, O.K. PH 1.2.5
Bautista, S.R. PH 1.2.3
Bauwens, A.L.G.M. NL 3.1
Bavappa, K.V.A. IN 3.1
Bavaresco, L. IT 16.1
Baxter, L.B. AU 25.1
Bay, E.C. US-J 7.1
Bay, L. DK 9.1
Baykal, O.M. TR 14.1
Bayraktar, A. TR 15.1
Bayram, N. TR 13.1
Baytorun, A. DE 15.1.7
Bazand, L. AT 3.1
Bazzocchi, R. IT 3.1
Beal, P.R. AU 18.1
Beale, A. CR 2.2
Bearce, B. US-C 9.1
Beardsell, D. AU 30.1
Beattie*, B.M. AU 43.1
Beattie, D.J. US-B 13.1.1
Beattie, G.A.G. AU 9.1.3
Beaver, G.R. US-I 15.1
Beavers, B. US-D 34.1
Bebea, P. RO 5.1
Bebeli, P. GR 6.1.5
Bebu, I. RO 9.1
Becerra León, E.N. MX 10.2
Becerra Rodriquez, S. MX 3.3
Bech, K. DK 3.1
Beck, G. US-F 23.1.3
Beck, W. AT 8.10
Becker, H. DE 12.1.2
Becker, W.F. BR 28.1
Becking, J.H. NL 9.5
Beckman, C. US-A 14.1
Beczner, L. HU 2.5
Bédard, R. CA 38.1
Bedding, R. AU 46.6
Bede, P. HU 2.6
Bednarek, J. PL 32.1.6
Beech, F.W. GB 19.1, 19.1.4
Beech, M. GB 8.1.1
Beeftink, H. NL 16.2
Beek, M.A. BR 25.5
Beel, E. BE 22.1
Beelman, R.B. US-B 13.1.1
Beemsterboer, P.A. NL 18.1
Been, T.H. NL 9.4.3
Beers, G. NL 1.1.2
Beever, D.J. NZ 2.1.2
Beever, R.E. NZ 2.3.3
Bégin, S. CA 27.1
Begtrup, J. DK 4.1
Behrens, V. DE 15.1.4
Beijersbergen*, J.C.M. NL 7.1
Bejarano, W. CR 2.2
Bekiari, S. GR 8.1.1
Bélair, G. CA 35.1
Bélanger, A. CA 35.1
Belcourt, J. CA 35.1
Beldea, V. RO 2.1
Belder, J. NL 9.7.5
Belehu, T. ET 1.1
Belfakh, Z. TN 1.1.
Bel Habib, O. MA 5.1
Belichki, I. BG 11.1
Beljo, J. YU 2.1
Bell, E.A. GB 16.1
Bell, M.J. GB 4.1
Bell, R.D. GB 31.2.8

Bell*, RL. US-C 8.2
Bellas, T.E. AU 65.5
Belleli, E. AR 15.2.1
Bellert, M.R. AU 11.1
Belletti, P. IT 24.4
Bellinder, R.R. US-B 5.1.3
Bellini*, E. IT 9.1
Bellion, Cl. FR 19.3
Belward, A. GB 31.1
Belyakov, V. BG 8.2
Belzunces, L. FR 3.1.6
Bem, F.P. GR 14.1.1
Benani, M. MA 5.2
Ben Abdallah, A. TN 1.1.
Bénard, P. FR 12.1
Ben-Arie, R. IL 1.3.1
Ben-Asher, J. IL 5.1.2
Beñatena, H.N. AR 6.1
Benavent, J.M. AR 17.1
Benazzoun, A. MA 1.1
Ben Brahim, N. TN 1.1
Benčať, F. CS 16.1, 16.1.1
Bender, D. US-H 9.1
Bendezu, J.M. BR 15.1
Bendjouya, B. BR 29.1
Benetti, M.P. IT 19.2
Bengoa, R.E. AR 7.1
Bengtsson, N. SE 2.2
Bengtsson, R. SE 1.1.3
Ben-Haim, G. IL 1.2.1
Beniest*, J. SN 1.1
Benincá de Salles, L.A. BR 30.1
Ben-Ismail, M.C. MA 1.1
Benitani, F. JP 31.1
Benites, O. PE 4.1
Benitez, C. AR 10.1
Benito, M.R. AR 18.1
Benjamaa, M. MA 1.2
Benjamin, L.R. GB 33.1.5
Benjamin, R.K. US-K 5.1
Bennett, A. US-D 16.2
Bennett, A.B. US-K 6.1.4
Bennett, J.M. GB 19.1.2
Bennett, K.E. ZI 3.1.2
Bennett, P. IE 2.1
Bennett, R.E. US-K 11.1
Benoit, D. CA 35.1
Benoit*, F. BE 18.1
Ben Salah, H. TN 1.1
Benschop*, M. NL 9.1
Ben-Shalom, N. IL 1.3.2
Ben Slimane, M. TN 1.1
Benson, D.M. US-D 8.1.2
Bental, Y. IL 1.2.4
Benton, F.P. BR 14.1
Benton, R.A. GB 22.1
Bentota, A. LK 3.8
Ben Yaacov*, A. IL 1.2.5
Ben-Yacov, Y. IL 1.1.2
Ben-Yehoshua, S. IL 1.3.1
Bényei, F. HU 2.2.10
Bényei-Himmer, M. HU 2.2.7
Benzarti, J. TN 1.1
Benzioni, A. IL 4.1
Beppu, E. JP 5.2
Bera*, B. PL 28.1.6
Beran, N. DE 22.1
Bérard*, L. CA 35.1
Berasategui, L. AR 3.1.2
Berbert, P.R.F. BR 14.1
Berczi, L. HU 4.2
Berczyński, St. PL 28.1.2
Berényi, M. HU 5.3.3
Beresford, R.M. NZ 13.3
Bereśniewicz, A. PL 28.2.6
Beretta, D. IT 6.1
Berg, R.Y. NO 7.3
Bergamaschi, H. BR 29.1
Bergamini*, A. IT 19.1.1
Bergé, J.B. FR 2.1.5
Bergendal, P-O. SE 1.1.1
Berger, H. AT 8.11
Bergh, B.O. US-K 12.1.1
Bergh, O. ZA 4.1
Berghoef*, J. NL 9.7.1
Bergman*, E.L. US-B 13.1.1
Bergman, J.W. US-I 23.1
Bergna, D. AR 10.1
Berhow, M.A. US-K 11.1
Berkett, L. US-A 3.1
Berkholst, Chr.E.M. NL 9.10
Berkowitz, G. US-B 14.1.1
Berman, J. AR 15.3
Bernard, G.E. CA 40.2
Bernard, J. BE 6.4
Bernard, V. CS 27.1
Bernáth, J. HU 1.1
Bernays, E.A. US-K 3.1.1
Bernhard, R. FR 4.1.1
Bernhardt, L.W. BR 20.2
Bernier, G. BE 11.1
Berninger*, E. FR 2.1.6
Berric, G. FR 35.1
Berridge, J.W. GB 1.3
Berry, A.M. US-K 6.1.2
Berry, S.Z. US-F 7.1.2
Bertolini, P. IT 3.3
Bertolino, T. YU 13.1
Bertram, A. DK 11.1
Bertrand, H. FR 19.5
Bertrand, P. US-D 22.1
Bertrandy, J. FR 17.1.2
Bertus, A.L. AU 9.1.1
Bertuzzi, P. FR 3.1.7
Berüter, J. CH 1.1
Berville, A. FR 7.1.6
Besford, R.J. GB 18.1.3
Besold, D. DE 5.1.1
Besri, M. MA 5.1
Bessard, A. FR 17.1.3
Bessey*, P.M. US-H 27.1
Bessho, H. JP 16.2
Best, E.B. ZI 3.1.2
Beste, C.E. US-C 7.1
Bester, C.W.J. ZA 4.1
Bestvater, C.R. AR 10.1
Betanzos Mendoza, E. MX 4.1
Betlach, J. CS 20.1
Betson, R.A. NZ 11.1
Bet-Tal, Y. IL 1.2
Bettner, W. DE 12.1.1
Bettum, O. NO 10.1.8
Beuchat, L.R. US-D 20.1
Beuret, E. CH 2.1
Beuzenberg, M.P. NL 1.1.4
Beuzeville, R. PE 3.3
Beveridge, H.J.T. CA 3.1
Beverly, R. US-D 20.1
Bevington, K.B. AU 4.1
Beyer, R.I. GB 16.1
Bezar, H.T. NZ 13.2
Bezerra, J.E.F. BR 10.1
Bezerra, J.L. BR 14.1
Bezerra Mendonça, J.F. BR 4.1
Bhagat, K.N. IN 24.1.3
Bhagawati, K.N. IN 7.1
Bhandari, T.P.S. IN 24.1.6
Bhansali, R.R. IN 18.2
Bhardwaj, S.P. IN 10.2
Bhargava, B.S. IN 1.1.9
Bhargava, J.N. IN 10.1
Bhartiya, S.P. IN 10.2
Bhat, K.K.S. GB 8.1.1
Bhat, N.R. IN 14.1
Bhat, R.N. IN 1.1.3
Bhattacharayya, S.K. IN 4.1
Bhattacharyya, A.R. IN 3.1
Bhore, D.P. IN 14.6
Bhumannavar, B.A. IN 1.1.7
Bhumannavar, B.S. IN 1.1.7
Bhutani, V.P. IN 10.3
Biache, G. FR 17.1.6
Biacs, P. HU 2.4
Bianchi, A. AR 17.1
Bianchini, P. AR 19.1
Bianco*, V.V. IT 2.2, 2.3
Bibb, M. GB 24.1.2
Bibicu, M. RO 8.4.5
Bible, B.B. US-A 13.1

Bibro, B. PL 28.2.11
Bicknell, R.A. NZ 10.1
Bidaud, A. FR 34.1
Biddappa, C.C. IN 3.1
Biddington, N.L. GB 33.1.5
Biddle*, A.J. GB 27.1
Biderbost, E. AR 7.1
Biedermann, I. US-B 7.1
Bielenin, A. PL 28.1.2
Bieleski, R.L. NZ 2.1
Bielig, H.-J. DE 5.3.1
Bieńkowska, I. PL 19.1
Biernbaum, J.A. US-F 14.1.1
Biesheuvel, A.R. NL 6.1
Biesiada, A. PL 33.1
Biesiada, T. PL 15.1
Biggs, A.G. AU 8.1
Biggs, A.R. CA 25.2
Biggs, R.H. US-D 27.4
Biglia, J. AR 19.1
Bignami, C. IT 27.1
Bigre, J.P. FR 19.5
Bijelić, V. YU 14.1
Bilderback, T.E. US-D 8.1.1
Bilgen, A.M. TR 11.1
Biliński, M. PL 28.1.8
Billette, C. FR 2.1.3
Billiet, F. BE 13.1
Billington, W.P. GB 31.2.6
Billot, C. FR 3.1.12
Bills, D.D. US-B 12.1
Bimboni, H. AR 19.1
Binder, A. IC 1.1
Bindra, A.S. IN 17.1
Bing*, A. US-B 5.1.1, 8.1
Bini, G. IT 9.1
Biniek, A. PL 20.2.11
Binning, L.K. US-F 23.1.3
Bino, R.J. NL 9.6.4
Bin Saad, A.A. LY 1.1
Biondi, G. IT 3.3
Bir, R.E. US-D 6.1
Biran, I. IL 2.2
Bird, G. US-F 14.1.2
Bird, J. US-N 1.1
Biris, D. GR 26.1
Birnbaum, E. IL 4.1
Biró, F. HU 4.1
Biró, P. HU 5.3.1
Birot, Y. FR 3.2.1
Birth, J. US-D 16.2
Bisdom, E.B.A. NL 9.11.1
Bisgrove, R.J. GB 29.1
Bishop*, C.J. CA 20.1
Bishop, D.G. AU 65.4
Bisht, I.S. IN 24.1.6
Bišof, R. YU 3.1
Bist, L.D. IN 20.1
Biston, R. BE 10.1
Biswas, B. IN 28.1
Biswas, S.R. IN 1.1.10
Bitencourt, A.A. BR 21.2
Bitlan, El. RO 18.1
Bittencourt da Silva, C. BR 25.5
Bittencourt Salazar V. Pessoa, H. BR 25.5
Bittsánszky, J. HU 5.3.1
Bjarnason, E. AU 27.1
Bjerregaard, G. DK 3.9
Bjørnstad, A. NO 10.4.1
Bjurman*, B. SE 1.1.1
Blabjerg, M. DK 3.7
Black, C.R. GB 25.1.2
Black, C.S. US-D 5.1
Black, J.D.F. AU 26.2
Black, L.L. US-D 52.1
Black, R.J. US-D 27.5
Blackburn, K. AU 62.1
Blackstock, J.M. AU 26.3
Blackwell, H.E. US-D 6.1
Blagden, P. IE 6.1
Blain, A.Q.M. GB 1.1
Blaisinger, P. FR 6.1.3
Blake, A. AU 59.2
Blake, J. GB 35.1.2
Blake*, J.R. AU 15.1
Blake, P.S. GB 8.1.2
Blake, T.K. US-I 18.1
Blake, T.L. US-D 2.1
Blamowski, Z. PL 14.1.5
Blanc*, D. FR 2.1.1
Blancard, D. FR 3.1.5
Blanco, J. ES 4.2
Blanco, L.E. AR 17.1
Blanco, M.P. AR 7.1
Blank, H.-G. DE 17.1
Blank, W. DE 17.1
Blankenship, S.M. US-D 8.1.1
Blanpied*, G.D. US-B 5.1.2
Blasco, A. ES 11.1
Bläsing, D. DE 15.1.4
Blatt, C.R. CA 41.1
Blattny, C. CS 26.1
Blay, E. GA 1.1
Blažek, J. CS 7.1
Blazich, F.A. US-D 8.1.1
Blazquez, C.H. US-D 30.2
Bleasdale, J.K.A. GB 33.1
Bleinroth, E.W. BR 20.2
Blenkhorn, M.J. CA 42.1
Blennerhassett, R.M. AU 31.1
Blesing, M.A. AU 65.1
Bleyaert, P. BE 17.1
Bliss, F. US-F 23.1.3
Blixt, S.G. SE 4.1
Blizzard, S. US-C 8.1, 9.1
Bloch, N. DK 9.1
Blok, I. NL 9.4.2
Blom, T.J. CA 25.1
Blomme, R. BE 22.1
Blommers, L. NL 9.4.1
Blom-Zandstra, M. NL 9.1
Blondon, F. FR 30.1.1
Bloom, A.J. US-K 6.1.4
Blumenfeld*, A. IL 1.2.5
Blundell, D.R. AU 6.1
Boag, B. GB 15.1.1
Board, P.W. AU 65.4
Boateng, P.Y. GA 2.1
Bobbitt, J.M. US-A 13.1
Boccon-Gibod, J. FR 19.5
Bocelli, J.O. AR 18.1
Bock, H.-H. DE 12.1.13
Bocklet, M.F. AR 15.2.5
Bockstaele, L. BE 17.1
Bockstedte, W. DE 17.1
Boczek, J. PL 32.1.7
Bödecs, I. HU 2.7
Bodi, I. RO 23.2
Bodson, M. BE 11.1
Boe*, A.A. US-G 12.1
Boekel, P. NL 4.1
Boelema, B.H. ZA 3.1
Boerema, G.H. NL 9.8
Boers, A. NL 3.1
Boesman, G. BE 7.1
Bogatko, J. PL 28.1.16, 28.1.2
Bogatko, W. PL 28.1.15
Bogdan, D. RO 8.4.4
Bogdán, K. HU 2.7
Bogdanović, V. YU 14.1
Bogers*, R.J. NL 7.1
Boggio, A. FR 34.1
Bognár, V. HU 2.2.1
Bohart, G.E. US-I 5.1
Bohling*, H. DE 18.1.1
Bohnenblust, K.E. US-I 12.1
Boiler, J. AT 8.10
Boiteau, G. CA 40.3
Boivin, G. CA 35.1
Bojappa, K.M. IN 11.2
Bojarczuk, K. PL 10.1.5
Bojarczuk, T. PL 10.1.5
Bojić, M. YU 10.1
Bolay, A. CH 2.1
Boldirev, M.I. SU 10.1
Bolhar-Nordenkampf*, H. AT 8.6
Bollen-Vandergeten, J. BE 19.1
Boller, E. CH 1.1
Bolton, A.T. CA 20.2

Boman, U. FI 17.1
Bonaminio, V.P. US-D 8.1.1
Bonanno, A.R. US-D 8.1.1
Bonatsos, D. GR 22.1.1
Bond, E.J. CA 18.1
Bondad, N.D. PH 1.2.1
Bondoc, L.O. PH 1.2.1
Bondoux, P. FR 1.1, 1.1.3
Boneta, E. US-N 1.2
Bonetti, J.I.S. BR 28.1
Bongen, M. FR 28.1
Bongers, H. NL 15.1
Bongi, G. IT 15.1
Bongiorno, A. AR 10.1
Bonhomme, M. FR 5.1
Bonicel, A. FR 31.1
Bonin, V. BR 28.1
Bonino, R.F. AR 15.3
Bonn, W.G. CA 17.1
Bonneau, M. FR 11.1
Bonnet, A. FR 3.1.2
Bonnet, Ph. FR 2.1.3
Bonnyai, J. NL 1.1.6
Bontemps, J. FR 29.1
Bontovits, L. HU 5.3.1
Boodley, J.W. US-B 5.1.1
Boogaard, M. NL 1.1.5
Booij, J. NL 9.4.1
Booij, R. NL 6.1
Boon, C.R. GB 31.2.4
Boon, R. BE 9.2
Boon, W. BE 9.2
Boonekamp, P.M. NL 9.4.4
Boorman, A.V. NZ 9.1
Boorsma, P.A. NL 14.2
Boppaiah, M.G. IN 26.1
Boratyńska, K. PL 10.1.6
Boratyński, A. PL 10.1.6
Borchers, E.A. US-C 12.1
Borcheva, N. BG 6.2
Bore, J. KE 1.7
Borecka, H. PL 28.1.4
Borecki, Z. PL 32.1.6
Borejsza-Wysocka, E. PL 20.2.10
Borg, G. DK 7.2
Borgan, S. NO 10.1.1
Borgel, A. FR 35.1
Borgen, L. NO 7.3
Borgen*, T. NO 10.1.4
Borgo, M. IT 7.1
Borgo, R. AR 13.1
Borić, B. YU 8.1
Borka, Gy. HU 6.1
Borkowska*, B. PL 28.1.9
Borkowski*, J. PL 28.2.12
Bormans, H. BE 19.1
Bornstein, C.J. US-K 16.1
Borochov, A. IL 2.2
Borochov, U. IL 2.2
Borošić, J. YU 3.1
Boross, I. HU 2.2.9
Borowiak, J. PL 28.2.1
Borowski, E. PL 14.1.5
Borrino, E.M. GB 15.1.2
Borsani, L.G. AR 15.2.3
Borst, N.K. NZ 10.1
Børtnes, G. NO 3.1
Bos, J.J. NL 9.7.4
Bos, L. NL 9.4.4
Bosc, B. FR 34.1
Bosch, W.J. NL 2.1
Boscheri, S. IT 11.1
Bose, P.C. IN 2.1.2, 2.1.4
Bose, T.K. IN 28.1
Boselli, M. IT 16.1
Bosman, P. BE 15.3
Bosshard, E. CH 1.1
Bosshardt, H.P. CH 1.1
Bossis, M. FR 14.1.3
Bostanian, N.J. CA 35.1
Bostian, B. US-H 1.1
Bostock, R.M. US-K 6.1.1
Boswell, S.B. US-K 12.1.1
Botar, A. RO 17.1
Botelho, W. BR 15.1
Botelho de Andrade, E. BR 2.1
Botez, C. RO 1.1.6
Botos, Gy. HU 5.3.2
Böttcher, H. IC 1.1
Botu, I. RO 32.1
Bouček, Z. CS 24.1
Bouchard, A. CA 31.1
Bouchard, C. CA 33.1
Bouchard, C.J. CA 33.1
Boucher, W. AU 45.2
Boudreaux, J.E. US-D 52.1, 55.1
Bouharmont, J. BE 12.2
Bouhida, M. MA 5.1
Bouhot, D. FR 7.1.3
Bouillon, G. DE 7.1
Boukema, I.W. NL 9.6.1
Boulard, Th. FR 3.1.4
Bould, A. GB 6.1
Boulet, C. MA 5.1
Boulet, M. CA 38.1
Boulidard, L. FR 17.1.1
Boulton, R.B. US-K 6.1.5
Bouma, A.S. NL 2.1, 9.6.3
Bouma, J. NL 9.11
Bouman, M. NL 2.1
Bounias, M. FR 3.1.6
Bounous, G. IT 24.1
Bouquet, A. FR 10.1.3
Bourbos, E. GR 8.1.1
Bourgeois, M. FR 1.1.3
Bourgeois, W.J. US-D 56.1
Bourgin, J.-P. FR 17.1.4
Bourke, J.B. US-B 2.1.5
Bourne, M.C. US-B 2.1.5
Bourque, P. CA 31.1
Bourrigault, M.F. FR 34.1
Boush, G.M. US-F 23.1.2
Boutard, C.R. US-A 9.1
Bouvarel, L. FR 13.1.3
Bouwkamp, J.D. US-C 5.2
Bové, J.M. FR 4.1.5
Bovio, M. IT 24.1, 24.2
Bøvre, O. DK 2.1
Bowden, D.M. CA 3.1
Bowden, R.P. AU 15.2
Bowen, B. AU 59.2
Bower, C.C. AU 3.1, 7.1
Bowers, W.S. US-H 27.1
Bowman, G.E. GB 31.2.7
Bowring, J.D.C. GB 6.1
Boxus*, P. BE 6.1
Boy, A. AR 19.1
Boy, E. DE 23.1
Boyadzhiev, B. BG 4.1
Boyce, B. US-A 3.1
Boyce, D.S. GB 31.2.2
Boychev, A. BG 9.1
Boyd, E.N. US-C 10.1
Boyd, L.L. US-J 6.1
Boyer, C.D. US-B 13.1.1
Boyer, N. FR 28.1
Boykov, A. BG 13.2
Boyle, T. US-A 5.1.1
Bozałek, H. PL 28.1.6
Božić, S. YU 14.1
Braam, W.F. NZ 13.2
Brackpool, A.L. GB 35.1.2
Bracy, R.P. US-D 57.1
Braddock, R.J. US-D 30.2
Bradfield, S. CR 2.2
Bradford, K.J. US-K 6.1.4
Bradley, B.F. AU 15.2
Bradley*, G.A. US-E 11.1
Bradshaw, D.W. US-D 13.1
Brady, C.J. AU 65.4
Braga Bastos, J. BR 1.3
Bragantini, C. BR 25.3
Braga Paiva, J. BR 8.2
Brahmachari, V.S. IN 8.1
Brain, P. GB 19.1.4
Bramlage, W.J. US-A 5.1.1
Brand, R. FR 3.1.11
Brandan, E. AR 22.2
Brandenburg*, W.A. NL 9.7.4
Brander, P.E. DK 2.1
Brandham, P.E. GB 16.1
Braniste, N. RO 9.1

Branton, R.L. GB 35.1.2
Brar, S.S. IN 17.1
Brashar, S. LY 4.1
Bravenboer*, L. NL 8.1
Bravo, B. IL 2.1
Bravo, F.B. AR 17.1
Bravo Lozano, A.G. MX 6.2
Bray, E.A. US-K 12.1.1
Braz Tinôco, P. BR 1.1
Brčak, J. CS 22.2
Brčić, J. YU 3.4
Brecht, J.K. US-D 27.7
Bredemeijer, G.M.M. NL 9.5
Bredmose*, N. DK 1.2
Breese, T.C. GB 34.1
Breeuwsma, A. NL 9.11.1
Bregt, A.K. NL 9.11.2
Brendel, G. DE 12.1.12
Brennan, P. IE 3.1
Brenner, M.L. US-G 7.1
Brenner, N.L. BR 27.2
Brent, K.J. GB 19.1.1
Breś, W. PL 20.2.3
Brescia, V.P. AR 17.1
Bressan, R. US-F 9.1.4
Breteler, F.J. NL 9.7.4
Breteler, H. NL 9.5
Breuils, G. FR 3.1.11
Breuils, L. FR 3.1.8
Brew, A.H. GA 4.4
Brewbaker, J.L. US-M 2.2
Brewer, R.F. US-K 12.1.1
Brewin, N. GB 24.1.2
Brewster, J.L. GB 33.1.5
Bridgen, M.P. US-A 13.1
Bridges, J.S. NZ 9.9
Brielmaier, U. DE 5.4
Brigati, S. IT 3.3
Brighenti, E. BR 28.1
Bringhurst*, R.S. US-K 6.1.3
Brink, K.M. US-I 2.2
Briones, A.A. PH 1.5
Briones, A.M. PH 1.5
Briquet, M. BE 12.1
Briskey, E.J. US-J 14.1
Briški, L. YU 18.1
Brisson, J.D. CA 33.1
Brits, G.J. ZA 3.1
Britt, C.P. GB 1.2
Brittain, J.A. US-D 13.1
Brkić, B. YU 16.1
Brković, M. YU 15.1
Brlansky, R.H. US-D 30.2
Broadbent, A.B. CA 25.2
Brockhus, M.A. AU 31.1
Brocklehurst, P.A. GB 33.1.5
Brodal, G. NO 10.3
Broener, G.L. BR 30.2
Brogowski, Z. PL 32.2
Brohier, L. AU 30.1
Bromand, B. DK 4.2
Bronner, A. FR 6.1.4
Brook, P.J. NZ 2.3
Brooking, I.R. NZ 8.1
Brooks, H.J. US-C 3.1
Brooks, J. US-C 9.1
Brooks, R.F. US-D 30.2
Broschat, T.K. US-D 25.1
Brouillette, L. CA 31.1
Browicz, K. PL 10.1.6
Brown, A.R. GB 18.1.3
Brown, B.I. AU 15.2
Brown, C.M. US-K 4.1
Brown, D.J.F. GB 15.1.1
Brown, D.L. US-G 7.1
Brown, F.R. GB 31.2.6
Brown, G.E. US-D 30.1
Brown, G.R. US-E 16.1
Brown, J.E. US-D 38.1
Brown, J.F. US-E 2.1
Brown, J.H. US-I 18.1
Brown, M. US-C 8.2
Brown, N.S. NZ 5.1
Brown, R. AU 27.1
Brown, R. US-D 16.1
Brown, R.C. US-A 4.1
Brown, R.H. CA 21.1
Brown, S.W. NZ 9.3
Brown, W.L. US-D 57.1
Brown, W.V. AU 65.5
Browning, B.M. ZI 3.1.2
Browning, C.G. US-H 3.1
Browning, G. GB 8.1.1
Browning, R. GB 12.1
Bruce, B. US-L 2.1
Bruce, C.W. US-D 38.1
Bruckler, L. FR 3.1.7
Brückner, U. DE 12.1.11
Bruggink, G.T. NL 9.7.1
Brugmans, W. BE 19.1
Bruin, G.C.A. NL 14.2
Bruins, H. IL 5.1.3
Bruinsma, J. NL 9.7.2
Brulfert, J. FR 30.1.2
Brumfield, R.G. US-B 13.1.1
Brun, A. FR 13.1.5
Brun, P. FR 15.1.3
Brun, R. FR 2.1.1, 3.1.14
Brundell, D.J. NZ 3.1
Brune, S. BR 7.1
Brunel, E. FR 14.1.3
Brunet, Y. FR 3.1.4
Brunner, H.R. CH 1.1
Brunner, J.F. US-J 10.1
Brunner, T. HU 2.1.2
Brunt, A.A. GB 18.1.2
Bruun, M. NO 10.1.8
Bruzau, F. FR 4.1.4
Bryan*, H.H. US-D 29.1
Bryk, B. PL 17.1
Bryk, M. PL 28.1.5
Brzeska, J. PL 28.2.14
Brzeski, M. PL 28.2.7
Brzeziński, K. PL 20.1.4
Brzozowska, J. PL 28.4
Bubán, T. HU 2.1.4
Buchanan, D. US-H 3.1
Buchanan, G.A. AU 31.1
Buchanan, P.K. NZ 2.3.2
Buchenauer, H. DE 15.1.6
Buchloh, G. DE 28.1.1
Buchwaldt, L. DK 4.1
Buciu, M. RO 1.1.5
Buck, G.J. US-G 20.1
Buckland, R. CA 37.1
Buckley, G.P. GB 35.1.2
Buckwell, A.E. GB 35.1.1
Budan, S. RO 9.1
Budisan, A. RO 1.1.1
Budowski, G. CR 2.1
Budrukkar, N.D. IN 12.2
Bueno, A. BR 12.1
Bueno de Paula, M. BR 15.1
Bues, R. FR 3.1.9
Buescher, R.W. US-E 11.1
Bufler, G. DE 28.1.1
Bugała, W. PL 10.1
Buisson, D.H. NZ 2.1.3
Bujanos Muñiz, R. MX 7.2
Bujdosó, G. HU 5.3.1
Bukovac*, M.J. US-F 14.1.1
Bukvić, B. YU 14.1
Bulatović, S. YU 7.1
Bulavić, B. YU 7.1
Bulblitz, E.O. BR 28.1
Bulcke, R. BE 7.9
Bulfin, M. IE 4.1
Bulíř, P. CS 24.1
Bulit, J. FR 4.1.6
Bull, P.B. NZ 12.2
Bulla, A.D. US-I 3.1
Bullitta, P. IT 23.1
Bulloch, B.T. NZ 8.3
Bullock, R.C. US-D 26.1
Bulnes Mendoza, J. AR 14.1
Bulzan, P. RO 3.1
Buma, D. US-A 15.1
Bunea, A. RO 13.1
Bünemann*, G. DE 15.1.4
Buraas, O.S. NO 10.3
Burba, J.L. AR 7.1
Burch, D.G. US-D 25.1, 27.5
Burchill, R.T. GB 33.1.4
Burdajewicz, S. PL 20.2.2

Burden, R.S. GB 19.1.1
Bureau, C. FR 19.2
Bureć, E. PL 34.1
Buret, M. FR 3.1.8
Burgaard, E. DK 7.2
Burge, G.K. NZ 10.1
Burger, D.W. US-K 6.1.2
Burger, E. DE 15.3.1
Burger, W.P. ZA 3.1
Burgerjon, A. FR 3.1.9
Burges, H.D. GB 18.1.2
Burgess, E. US-D 40.1
Burgess, J. GB 24.1.3
Burgos, C. CR 2.2
Burgyán, J. HU 2.5
Burić, D. YU 17.1
Burileanu, D. RO 7.1
Burke, W. IE 4.1
Burkowicz, A. PL 15.1
Burloi, N. RO 8.2
Burns, D.J.W. NZ 2.1.3
Burns, E.E. US-H 6.1
Burns, I.G. GB 33.1.2
Burr, T.J. US-B 2.1.4
Burrage, S.W. GB 35.1.2
Burroughs, L.F. GB 19.1.4
Burrows, J.M. CA 41.1
Burt*, J.R. AU 53.1
Burtea, O. RO 8.4.2
Burts, E.C. US-J 10.1
Burzo, I. RO 8.4.1
Buse, C. RO 7.1
Buser, H.P. CH 1.1
Buser, H.R. CH 1.1
Busey, P. US-D 25.1
Buslig, B.S. US-D 30.2
Buso, J.A. BR 25.5
Buss, A. BR 28.1
Bussell, W.T. NZ 10.1
Bussi, C. FR 3.1.12
Bussières, Ph. FR 3.1.1
Busta, J. US-D 27.3
Bustamante, A. AR 15.2.1
Bustamante, E. CR 2.2
Bustos, J.E. AR 18.1
Buszard*, D.J. CA 37.1
Butcher, C.F. NZ 2.2
Butcher, D.N. GB 19.1.2
Butcher, S.M. NZ 10.1
Buteler, M.J. AR 7.1
Butler, D.R. GB 19.1.2
Butler, J.D. US-I 2.2
Butler, J.R. TZ 3.1
Butnaru, H. RO 35.1
Butt, B. US-C 8.2
Butt, B.A. US-C 8.2
Butt, D.J. GB 8.1.2
Butu, C. RO 6.1
Buwalda, J.G. NZ 3.1
Buxton, J. US-E 15.1
Buyukyilmaz, M. TR 14.1
Buzescu, D. RO 1.1.3
Byers, R.E. US-C 14.1
Byett, A. NZ 10.1
Bylterud, A. NO 10.4.2
Byrde, R.J.W. GB 19.1.1
Byrne*, D.H. US-H 6.1
Byrne*, T.G. US-K 6.1.2
Bystydzieńska, K. PL 30.1
Byun, J.K. KR 14.1

Caballero, P. ES 14.1
Caballero Deloya, M. MX 13.1
Cabannes, M. FR 46.1
Cabibel, B. FR 3.1.7
Cabibel, M. FR 3.1.8
Cabidoche, Y.M. FR 18.1
Cabral de Miranda, J.E. BR 25.5
Cabral do Vale, E. BR 8.2
Cabrera Valle, B. MX 1.4
Cabus Maaze, U. BR 10.1
Cáceres, E.M. AR 18.1
Cadapan, E.P. PH 1.3
Caderek, T. PL 32.3.4
Cadic, A. FR 1.1.2
Cadman, R. AU 33.1
Cahill, R. AU 30.2
Cahoon, G. US-F 3.1, 7.1.2
Cain*, D.W. US-D 13.1
Cain, N. CA 15.1
Cain, P.A. AU 65.1
Caixeta, T.J. BR 15.1
Cakir, M. TR 13.1, 15.3
Cakmur, C. TR 11.1
Calabrese, F. IT 13.1
Calado*, A. PT 4.1
Calado, M.L. PT 6.1
Calame, F. CH 2.1
Calcagnolo, G. BR 21.2
Caldas, L.S. BR 25.4
Calderón, G. CO 1.1
Calderon, M. IL 1.3.3
Calderon P., G. PE 4.3
Caldwell, J. US-F 3.1
Caldwell, J.D. US-D 13.1
Caldwell, J.S. US-C 10.1
Caliandro, A. IT 2.2
Caligari, P.D.S. GB 15.1.2
Calilung, V.J. PH 1.3, 1.6
Caliskan, A. TR 18.1
Callahan, L.M. US-E 3.1
Callan, N.W. US-I 19.1
Calle, J. PE 1.3
Callesen, O. DK 1.3
Callupil, N. AR 23.1
Calò, A. IT 7.1
Calot, L. AR 3.1.1
Caltagirone, L.E. US-K 3.1.1
Calvar, D.J. AR 10.1
Calvar, J. AR 3.1.2
Calvert, D.V. US-D 26.1
Calvert II, V.H. US-D 39.1
Calvo, G. CR 2.2
Calzavara, B.B.G. BR 2.3
Camba, J.J. AR 15.1
Cambell, B. US-L 2.2
Cameron*, A.C. US-F 14.1.1
Cameron, E.A. NZ 9.1
Cameron, I.J. AU 58.1
Camilo, A.P. BR 28.1
Camorani, F. IT 3.1
Camos, J.P. BR 17.1
Campacci, A.A. BR 21.2
Campbell, A.I. GB 19.1.3
Campbell, C.A.M. GB 8.1.2
Campbell*, C.W. US-D 29.1
Campbell, F.J. US-A 10.1.1
Campbell, J.E. AU 7.1
Campbell, J.R. US-F 12.2
Campbell, M. AU 27.1
Campbell, R.E. US-H 3.1
Campbell, R.W. US-G 31.1
Campbell, W.F. US-I 5.1
Campeglia, O.G. AR 13.1
Campelo, A.M.F.L. BR 14.1
Campion, B. IT 20.1
Campo, R.J. BR 27.1
Camporota, P. FR 7.1.3
Campos, G. PE 1.2
Campos, J. BR 17.1
Campos, O. CR 2.1
Campos, S. AR 22.2
Canarache, V. RO 8.2, 8.5
Canaway, P.M. GB 4.1
Cancellier, S. IT 7.1
Candelá, C. FR 34.1
Candal Neto, J.F. BR 18.1
Candresse, Th. FR 4.1.6
Canham*, A.E. GB 29.1
Cano, H. PE 1.3
Cano Garcia, M.A. MX 4.5
Canozer, O. TR 15.3
Cantarutti, R.B. BR 14.1
Cantelo, W.W. US-C 3.1
Canteros, B.J. AR 2.1
Cantliffe*, D.J. US-D 27.7
Cantor, M. RO 30.1
Cantor, V. RO 30.1
Cantos, F. AR 21.1
Capdevila Gomez, F. ES 2.1
Cappadocia, M. CA 31.1, 31.3
Cappelleri, G. IT 7.1
Cappellini, P. IT 19.1

Cappellini, R.A. US-B 14.1.6
Caprio, J.M. US-I 18.1
Caputa, J. PL 34.1
Caraballo, E. US-N 1.2
Caracotche, O. AR 11.1
Caramete, C. RO 8.1
Caran, D. TR 15.3
Carandang, D.A. PH 1.5
Carbajal Cadena, G. MX 6.4
Carballo, M. CR 2.2
Carbonari, R. BR 22.1
Carbonneau, M.C. US-F 12.2.2
Carbonneau, R. CA 31.3
Carden, E.L. US-D 44.1
Carder, J.H. GB 8.1.2
Cardin, L. FR 2.1.3
Cardin, M.C. FR 2.1.5
Cardinaels, C. BE 9.1
Cardoso, C. PT 6.1
Cardoso, J.E. BR 5.1
Cardoso, V.T.M. BR 28.1
Caredda, A. IT 23.1
Čarkić, T. YU 1.1
Carletti, G. BE 6.8.12
Carletti, M.G. IT 4.1
Carletto, G.A. BR 14.1
Carling, D.E. US-L 2.1
Carlone, R. IT 24.2
Carlquist, S. US-K 5.1
Carlson, R.M. US-K 6.1.3
Carlson, W.H. US-F 14.1.1
Carlsson*, M. SE 1.1.4
Carmi, Y. IL 1.3.3
Carmine Dianese, J. BR 25.4
Carnero, F. PE 4.1
Caron, M. CA 29.2
Carpenter*, E.D. US-A 13.1
Carpenter, G.A. GB 31.2.4
Carpenter, H.E. US-B 14.1.3
Carpenter, P.L. US-F 9.1.4
Carpenter, R. GB 24.1.2
Carpenter, W.J. US-D 27.5
Carpenter, Z. US-H 6.2
Carreras, J. AR 7.1
Carroll, E.T. AU 15.1
Carroll, R.B. US-C 2.1
Carroll, T.W. US-I 18.1
Carseldine, M.L. AU 18.1
Carstens, F.J. US-B 10.1
Cartagena, J.R. CO 2.1
Cartechini, A. IT 15.1
Carter, C.D. US-B 14.1.1
Carter, E.M. AU 59.1
Carter, G.A. GB 19.1.1
Carter, H.W. US-E 4.1
Carter, J.V. US-G 7.1
Carter, M.V. AU 42.1
Cartwright, D. AU 36.2
Caruso, A. IT 1.1
Caruso, P. IT 13.2
Carvalho, R.A. BR 3.1
Carvalho, R.P.L. BR 30.1
Carvalho, S.L.C. BR 27.2
Carvalho Carelli, M.L.
BR 20.1
Carvalho Guedes, A.L. BR 1.3
Cary*, P.R. AU 65.6
Cary, R. AU 59.2
Casado, A.A. MX 2.2
Casafús, C.M. AR 6.1
Casagronde, R. US-A 14.1
Casali, V.W.D. BR 17.1
Casamiquela, C. AR 10.1
Casamiquela, L. AR 23.1
Casano, F.J. IT 19.2
Casanova, M.R. AR 22.3
Casares, J.M. AR 4.1.2
Casavela, St. RO 22.1
Cascarano, A. IT 2.3
Casela, C.R. BR 30.2
Casella, G. IT 2.1
Casement, E.B. CA 6.1
Casen, M.L. AR 22.1
Cáseres, S. AR 2.1
Casey, R. GB 24.1.1
Casey, R.J. NZ 12.2
Casida, J.E. US-K 3.1.1
Casimir, D.J. AU 65.4
Casimir, M. AU 9.1.3
Casini*, E. IT 9.1
Casolari, A. IT 14.1
Cassaniti, S. IT 5.3, 5.4
Cassetari, D. BR 18.1
Cassiano da Rocha, A. BR 18.1
Cassidy, J.C. IE 4.1
Cassin, J. FR 15.1.4
Cassino, A. AR 10.1
Cassino, R. BR 19.2
Cassol, E.A. BR 29.2
Castaigne, F. CA 38.1
Castelain, C. FR 3.1.5
Castell, V. ES 14.1
Castellana, M.A. US-B 6.1
Castellane, P.D. BR 23.1
Castelli, S. IT 9.2
Castelli, V. IT 14.1
Castillo, M.B. PH 1.4
Castillo, P.S. PH 1.7
Castillo Padilla, J.M. MX 5.2
Castle*, W.S. US-D 30.2
Castle, W.S. US-D 30.1
Castro, C. BR 30.1
Castro, H. AR 10.1
Castro, T.S. AR 18.1
Castro Alonso, P. ES 22.1
Castro Gamarra, J. AR 14.1
Catania, C.D. AR 15.1
Catchpool, T.D.H. GB 3.1
Cathey*, H.M. US-C 4.1
Catizone, P. IT 3.2
Catlin*, P.B. US-K 6.1.3
Catts, E.P. US-J 6.1
Cătusanu, C. RO 32.1
Caubel, G. FR 14.1.3
Causse, R. FR 3.1.9
Caussin, R. BE 6.8.10
Cavalcante, P.B. BR 2.2
Cavaletto*, C.G. US-M 2.2
Cavallo, A. AR 7.1
Cavanaugh, J.R. US-B 12.1
Cavazza, L. IT 3.2
Cavusoglu, A. TR 15.3
Cayrol, J.C. FR 2.1.5
Cayrol, R. FR 3.1.9
Cazelles, O. CH 2.1
Ceausescu, I. RO 8.1
Ceausescu, M.E. RO 8.1
Ceballos, H. AR 7.1
Ceccarelli, N. IT 17.2
Cedeño, A. US-N 1.2
Cedo, M.O. PH 1.2.2
Cegłowski, M. PL 27.1
Celestino, P. BR 1.1
Celik, H. TR 2.1
Celikel, C. TR 13.1
Center, T.D. US-D 25.1
Cepoiu, N. RO 8.1, 8.3
Ceran, T. TR 8.1
Cerchiai, E. AR 15.2.5
Cereda, E. BR 22.1
Cerezo, C.E. AR 18.1
Cerkauskas, R.F. CA 25.2
Černá, K. CS 8.1
Cernat, R. RO 8.4.4
Černe, M. YU 8.1
Černý, J. CS 10.1
Černý, L. CS 22.2
Cerović, R. YU 10.2
Cerqueira Gomes, C. BR 12.1
Cersosimo, A. IT 7.1
Cervantes, J.M. MX 2.4
Cervantes, M.I. MX 6.2
Cervelli, J. US-E 15.1
Cesar, P. BR 19.2
Cetin, C. TR 3.2
Cetin, H.H. TR 14.1
Cetinel, T. TR 10.1
Cetiner, E. TR 13.1
Cetiner, K.S. TR 13.1
Cetković, V. YU 10.1
Ceulemans*, R. BE 2.1
Ceustermans, N. BE 18.1
Cezar, J. de O. BR 14.1
Chachin, K. JP 33.2

Chacko, C.I. IN 1.1.6
Chadha*, K.L. IN 1.1
Chadha, T.R. IN 10.1
Chagnon, R. CA 35.1
Chahbar, A. MA 5.2
Chaisilrapin, S. TH 1.1
Chakarvarty, B. IN 7.1
Chakrabarti, U. IN 29.1
Chakrawar, V.R. IN 12.1
Chalfoun de Souza, S.M. BR 15.1
Chalfun, N.N.J. BR 16.1
Challa*, H. NL 9.1
Challa, P. IN 1.1.8
Chalmers, D.J. AU 32.1
Chaloupka, G.W. US-C 1.1
Chałupka, Wł. PL 10.1.1
Chalutz, E. IL 1.3, 1.3.1
Chambers, D. US-D 52.1
Chambers, D.A. GB 8.1.2
Chambers, R.J. GB 18.1.2
Chambliss, C.G. US-D 24.1
Chambliss, O.L. US-D 38.1
Chambon, J.-P. FR 17.1.3
Chambonnet, D. FR 3.1.2
Chambroy, Y. FR 3.1.8
Chamel*, A. FR 31.1
Chamorro, H. CO 4.2
Champagnat, M. FR 28.1
Champagnat, P. FR 28.1
Champéroux, A. FR 2.1.1
Chanana, Y.R. IN 17.1
Chan Castañeda, J.L. MX 6.2
Chand, I. IN 25.1
Chander, R. IN 10.2
Chander, S. IN 1.1.4
Chandler, B.V. AU 65.4
Chandler, C. US-F 7.1.2
Chandrasekhar Rao, K. IN 6.1
Chandravadana, M.V. IN 1.1.8
Chang, C. US-D 20.1
Chang, C.C. TW 1.2
Chang, C.I. TW 11.1.2
Chang, C.N. TW 9.1
Chang, J.I. KR 10.3
Chang, L.R. TW 7.2
Chang, M.T. TW 8.1
Chang, S.S. TW 10.1
Chang, T.R. TW 11.1.2
Chang, W.C. TW 9.1
Chang*, W.N. TW 7.1
Chang, Y.M. TW 7.3
Changwony, B.K. KE 1.3
Chanzy, A. FR 3.1.7
Chapa, J. FR 4.1.1
Chaplin, G.R. AU 65.4
Chapman, G.P. GB 35.1.3
Chapman, J.C. AU 22.1
Chapman, K.R. AU 23.1
Chapman, K.S.R. AU 43.1
Chapman, R.K. US-F 12.1.2
Chapman, T. GB 32.1
Chapon, J.F. FR 45.1
Charan, R. IN 19.4
Charchar, J.M. BR 25.5
Chardas, G. GR 18.1.3
Charles, J.G. NZ 6.1
Charnay, D. FR 28.1
Charnley, A.K. GB 3.1
Charpentier, S. FR 19.5
Chartier, A. FR 32.1
Charzewska, H. PL 28.2.8
Chase, A.R. US-D 31.1
Chase, C.D. US-D 27.7
Chasseriaux, G. FR 19.5
Chater, K. GB 24.1.2
Chattopadhyay, P.K. IN 28.1
Chattopadhyay, T.K. IN 28.1
Chatty, Y. TN 1.1
Chaturvedi, O.P. IN 22.1
Chaudhary, A.S. IN 22.1
Chaudhary, S.K. IN 17.1
Chaudhry, F.M. LY 1.1
Chaudhry, G.R. LY 2.1
Chaudhry, M.I. LY 1.2
Chaufaux, J. FR 17.1.6
Chauham, H.S. IN 4.1
Chauhan*, J.S. IN 10.1
Chauhan, P.S. IN 10.2
Chavarria, A. AR 22.1
Chavarria, H. CR 2.2
Chavdarov, I. BG 13.2
Chávez, W. PE 3.3
Chávez Contreras, J. MX 3.2
Chávez Flores, W. PE 3.3
Chayantrakom, S. TH 17.1
Cheah, L.H. NZ 10.1
Chebbouh, N. TN 1.1
Chee, H.K. KR 18.1.3
Cheema, S.S. IN 17.1
Chege, B. KE 1.3
Cheikh, M. TN 1.1
Chełkowski, J. PL 32.1.6
Chełpinski, P. PL 31.1.1
Chen, C.S. TW 12.1
Chen, L.H. US-D 50.1
Chen, M.S. TW 2.1.2
Chen, S.Y. TW 4.3, 6.1
Chen, T.L. TW 11.1.1
Chen, Y.H. TW 11.1.2
Chen, Y.R. TW 4.3
Chen, Y.W. TW 8.1
Cheng, B.T. CA 36.1
Cheng, C. TW 11.1.1
Cheng, C.Y. TW 9.1
Cheng, D.W. TW 9.1, 12.1
Cheng, S. BR 2.1
Cheng, S. TW 9.1
Cheong, S.R. KR 18.1.2
Cherchum, P. TH 12.1
Cherif, C. TN 1.1
Cherif, R. TN 1.1
Cherry, J.H. US-F 9.1.4
Cherry, J.P. US-B 12.1
Chesnel, A. FR 34.1
Chesness, J. US-D 16.1
Chesoli, C.B. KE 3.1
Chevreau, E. FR 1.1.1
Chheda, H.R. NG 1.5
Chia, C.L. US-M 2.2
Chiang, M.N. TW 9.1
Chiang, M.S. CA 35.1
Chiba, A. JP 16.1
Chiba, K. JP 12.3
Chiba, M. CA 25.2
Chiba, S. JP 3.1
Chiba, T. JP 16.1
Chibbaro, A. CR 2.1
Chien, H.S. TW 2.1.3
Chiesa, A. AR 3.1.1
Chigaru, P.R.N. ZI 3.1
Chigogora, J.L. ZI 3.1.2
Chikaizumi, S. JP 5.3
Chikano, K. JP 21.1
Chiko, A.W. CA 2.1
Child, R.D. GB 19.1.2
Childers, C.C. US-D 30.2
Childiyal, P.C. IN 24.1.2
Chilembwe, E.H.C. ML 1.1
Chilom, P. RO 19.1
Chin, Chee-Kok, US-B 14.1.1
Ching, C.T.K. US-M 2.2
Ching, F. US-K 2.1, 8.1, 9.1, 10.1
Chin Shu Chen, US-D 30.2
Chisaki, M. JP 22.2
Chisholm, C.J. GB 31.2.1
Chitic, M. RO 3.1
Chittirai Chelvan, R.C. IN 1.1.1
Chitzanidis, A. GR 14.1.1
Chiu, H.J. TW 7.1
Chiusoli, A. IT 3.1
Chizala, C.T. ML 1.1
Chkartishvili, N.S. USSR 19.1
Chlebowska, D. PL 5.1
Chlebowski, B. PL 32.1.3
Chmela, C. CS 20.1
Chmiel, H. PL 32.1.3
Chmielewski, W. PL 20.1.1
Cho, C.H. KR 9.2
Cho, I.W. KR 2.1.1
Cho, J.L. KR 12.2
Cho, J.T. KR 3.1

Cho, M.D. KR 23.1.4
Chô, O. JP 39.1
Chodun, A. PL 10.1.5
Choe, J.S. KR 2.1.2
Choer Moraes, E. BR 30.1
Chohan, G.S. IN 17.2
Choi, B.W. KR 23.2
Choi, J.S. KR 23.1.4
Choi, K.D. KR 9.2
Choi, K.S. KR 23.1.1
Choi, S.K. KR 13.1
Choi, Y.H. KR 17.1
Chojnowksa, A. PL 32.3.3
Chojnowski, J. PL 15.1
Cholakov, T. BG 7.1
Cholevas, C.D. GR 14.1
Chong, C. CA 25.1
Chong, D.S. KR 9.2
Chono, H. JP 12.4
Chopra, H.R. IN 17.1
Chopra, S.K. IN 10.1
Choque, J. PE 4.3
Choudhari, K.B. IN 14.1
Choudhari, K.G. IN 14.1
Choudhari, S.M. IN 14.1
Chouinard, E.L. US-I 8.1
Choukrallah, R. MA 1.1
Chourdaki, A. GR 14.1.4
Chow, T.L. CA 40.3
Chowings, J.W. GB 6.1
Chrétien, L. CA 31.2
Christ, E. DE 11.1.1
Christakis, P. GR 11.1
Christeller, J.T. NZ 8.1
Christensen, F.G. DK 3.3
Christensen*, J.V. DK 1.3
Christensen*, O.V. DK 1.2
Christensen, P.E. DK 1.3
Christian, J.H.B. AU 65.4
Christians, N.E. US-G 20.1
Christiansen, O. DK 7.1
Christias, Ch. GR 3.1.3
Christie, C.B. NZ 9.1
Christoffers, D. DE 15.3.1
Christou, M. GR 18.1.3
Christoulas, C. GR 4.1
Christov, B. BG 2.1
Christov, L. BG 6.2
Chronopoulos, I. GR 6.1.4
Chróścicki, K. PL 15.1
Chryssagi, M. GR 14.1.4
Chtaina, N. MA 1.1
Chu, C.K. TW 1.2
Chu, C.L. CA 2.1
Chubey, B.B. CA 13.1
Chudyk, R.V. CA 25.1
Chudziak-Bilska, E. PL 17.1
Chujo, T. JP 17.3
Chuku, V.J. NG 10.1
Chung*, B. AU 43.1
Chung, G.C. KR 13.2
Chung, H.D. KR 14.1
Chung, H.J. KR 8.3
Chung, J.H. KR 23.1.2
Chung, S.J. KR 13.2
Chung, W.J. TW 7.2
Chupeau, Y. FR 17.1.4
Chuu, E. JP 3.3.2
Chvojka, L. CS 22.2
Chylarecki, H. PL 10.1.5
Ciancaglini, N. AR 15.2.2
Cianciara, T. PL 28.1.7
Cianciara, Z. PL 28.1.6
Cicala, A. IT 5.1
Cichocka, E. PL 32.1.7
Ciechomski, W. PL 32.1.5
Ciereszko, B. PL 28.4
Ciesielska-Piper, J. PL 32.1.1
Cieślak-Wojtaszek, W. PL 28.2.11
Cieslinski, G. PL 28.1.7
Cifranič, J. CS 31.1
Cimanowski*, J. PL 28.1.2
Cimato, A. IT 9.2
Cimino, A. IT 3.3
Cîndea, E. RO 1.1.7
Cindrić, P. YU 17.1
Cinta, W. AR 20.1
Cintra, F.L.O. BR 12.1
Ciocan, E. RO 1.1.2
Ciociola, A.I. BR 16.1
Ciofu, R. RO 8.1
Cirami, R.M. AU 40.1
Cireasă, V. RO 23.3
Cirkova-Georgievska, M. YU 4.3
Cirrito, M. IT 22.1.1
Cirstea, I. RO 9.2
Cîrstoiu, F. RO 1.1.4
Cisek, B. PL 30.1
Ciupală, A. RO 1.1.6
Ciurel, M. RO 8.4.1
Civerolo, E.L. US-C 3.1
Clark, B. US-B 14.1.6
Clark, C.A. US-D 52.1
Clark, C.J. NZ 5.1
Clark, J.A. GB 25.1.2
Clark, J.R. US-E 10.1
Clark, M.C. CA 40.3
Clark, M.F. GB 8.1.2
Clark, R.J.T. GB 12.1
Clark, T. GB 19.1.1
Clarke, B. GB 31.1
Clarke, N.P. US-H 6.1
Clarke, R. AU 27.1
Clarke, R.M. NZ 9.4
Claverán Alonso, R. MX 13.1
Claxton, R.A. AU 9.2
Clay, H. US-D 16.1
Clayberg, C.D. US-G 31.1
Claydon, N. GB 18.1.2
Clayton, W.D. GB 16.1
Clayton-Greene*, K. AU 30.1
Clearwater, J.R. NZ 2.2
Clegg, M.T. US-K 12.1.1
Clemens, J. AU 9.3
Clement, Ch.R. BR 1.2
Clements, R.F. GB 10.1
Clements, W.C. US-D 9.1
Clerjeau, M. FR 4.1.6
Clifford, D.R. GB 19.1.1
Clignez, M. BE 6.9
Clikel, K. TR 13.1
Cline, R.A. CA 25.1
Clingeleffer, P.R. AU 65.2
Cloquemain, G. FR 6.1.5
Clothier, B.E. NZ 8.1
Coartney, J.S. US-C 10.1
Coats, G.E. US-D 50.1
Cobb, G. US-H 6.1
Cobb Jr., F.W. US-K 3.1.2
Cobbe, R.V. BR 25.5
Cobianchi*, D. IT 19.1.2
Cociu, V. RO 9.1
Cockshull, K.E. GB 18.1.1
Cocozza Talia, M. IT 2.2
Coelho, R. BR 15.1
Coelho, R.G. BR 19.1
Coelho e Silva, W. BR 25.5
Coelho Pedrosa, A. BR 10.1
Coen, E. GB 24.1.2
Coertze, A.F. ZA 3.1
Coffey, D.L. US-E 3.1
Coffin, R. CA 15.1, 40.3
Coggins Jr., C.W. US-K 12.1.1
Cohat, J. FR 9.1
Cohen, A. IL 1.1.2, 1.2.3
Cohen, D. NZ 8.1
Cohen, E. IL 1.3.1
Cohen*, M. US-D 26.1
Cohen, Z. IL 5.1.6
Colbran, R.C. AU 15.4
Coldewey-zum Eschenhoff, H. DE 15.1.5
Colditz, P. US-D 18.2
Cole, A. IE 5.1
Cole, A.L.J. NZ 12.1
Cole, A.W. US-D 50.1
Cole, B. US-J 1.1
Cole, K.S. AU 59.1
Cole, P. AU 38.2
Cole, R.A. GB 33.1.3

Cole, R.H. US-B 13.1.1
Cole, S. AU 65.4
Cole, T.J. CA 20.2
Coleman S.E. CA 40.2
Coleman, W.K. CA 40.3
Colijn-Hooymans, C.M. NL 9.6.4
Colin, J. MA 1.1
Colin Castillo, E. MX 4.3
Collapietra, M. IT 7.4
Collet, G.F. CH 2.1
Collicutt, L.M. CA 13.1
Collino, D. AR 7.1
Collins, J. US-D 51.1
Collins, W.B. US-L 1.1
Collins, W.W. US-D 8.1.1
Colls, J. GB 25.1.2
Colombo, N. AR 17.1
Colorio, G. IT 19.1
Colt*, W.M. US-I 15.1
Coltman, R.R. US-M 2.2
Coly, V. SN 1.1
Comai, M. IT 21.1
Coman, M. RO 9.1
Coman, S. RO 9.1
Coman, T. RO 9.1
Combrink, J.C. ZA 4.1
Côme, D. FR 38.1
Comeau, J.E. CA 30.2
Concaret, J. FR 7.1.2
Conceição, F.A.D. BR 22.1
Conea, Al. RO 1.1.3
Conei, V. AR 7.1
Conicella, C. IT 18.1
Conlon, T.J. US-G 10.1
Conn, J. US-L 1.1
Connolly, C. US-H 15.1
Conover*, C.A. US-D 31.1
Conroy, R.J. AU 9.1.1
Constante, J.F. US-A 3.1
Constantin, R.J. US-D 57.1
Constantinescu, A. RO 8.4.3
Constantinescu, F. RO 1.1.4
Conte, E. IT 19.2
Conteas, Th. GR 6.1.1
Conti, A. CA 35.1
Conti, L. IT 6.1
Conti, S. IT 3.2
Contin, H. AR 7.1
Continella, G. IT 5.1
Contreras, P. PE 2.1
Conway, A. IE 6.1
Conway, K. US-H 3.1
Conway, W. US-C 3.1
Cook, A.V. AU 38.1
Cook, J.A. US-I 12.1
Cook, W.P. US-D 12.1
Cooke, B.K. GB 19.1.1
Coolen, W.A. BE 15.2
Coombe*, B.G. AU 42.1
Cooper, D.W. AU 9.4
Cooper, I.M. AU 41.1
Cooper, R.J. US-A 5.1.1
Cooper, R.M. GB 3.1
Cooper, S.R. GB 9.1.1
Coosemans, J. BE 9.1
Copâescu, V. RO 9.2
Copani, V. IT 5.4
Copeman, R.J. CA 4.2
Copland, M.J.W. GB 35.1.3
Coppock, G.E. US-D 30.1
Coquen, C. FR 19.4
Coradin, L. BR 25.3
Corazza, L. IT 19.2
Corbacioglu, C. TR 5.1
Corbett, E.G. US-A 13.1
Corbin, E.J. AU 12.1
Corbineau, F. FR 38.1
Cordeiro, Z.J.M. BR 12.1
Cordero, M. US-N 1.2
Corgan, J.N. US-H 19.1
Corino, L. IT 7.2
Corley, W.L. US-D 20.1
Cormack*, M.R. GB 15.1.4
Cornejo, R. AR 18.1
Cornellier, N. CA 31.1
Cornforth, I.S. NZ 5.1
Cornic, J-F. FR 3.2.2
Cornillon, P. FR 3.1.1
Cornman, J.F. US-B 5.1.1
Cornu, A. FR 7.1.6
Cornuet, J-M. FR 3.1.9
Corokalo, D. YU 13.1
Coronel, R.E. PH 1.7
Corović, R. YU 14.1
Corrales, A. PE 1.1
Correa, G. BR 6.1
Correa de Sal, M.A. AR 5.1
Correia, E. BR 29.2
Correia de Faria, J. BR 26.1
Correia Nunes, M.U. BR 5.1
Correll, D.S. US-D 33.1
Correo, E.B. BR 10.1
Corriols-Thevenin, L. FR 17.1.1
Corsini, D. US-I 13.1
Cortes Ortega, D. MX 9.2
Cortez Bezerra, J.R. BR 7.1
Corzo, P. VE 3.1
Cosico, W.C. PH 1.5
Cossens, G.G. NZ 16.1
Cossentine, J. CA 3.1
Cossio*, F. IT 26.1
Costa, A. PT 5.1
Costa, A.V. BR 26.2
Costa, G. IT 3.1, 25.1
Costa, J. ES 15.1
Costa, N.B. AR 6.1
Costa, R.G. BR 11.1
Costa Allem, A. BR 25.3
Costache, M. RO 1.1.7
Costacurta, A. IT 7.1, 7.2
Costamagna, O. AR 1.1
Costa Pinto de Araujo, L. BR 12.2
Costa Poltronieri, M. BR 3.1
Costa Veloso, C.A. BR 6.1
Costes*, C. FR 8.1
Costigan, P.A GB 33.1.2
Costilla, M. AR 22.3
Coston*, D.C. US-D 13.1
Côté, J.C. CA 35.1
Cother, E. AU 12.1
Cotner, S. US-H 6.2
Cotnoir, L.J. US-C 2.1
Coto, T. CR 2.2
Cotorobai, M. RO 9.1
Coträu, A. RO 23.3
Cotten, J. GB 13.1.1
Cottenie, A. BE 7.8
Cotter, D.J. US-H 19.1
Cottrell, T.H.E. US-B 2.1.5
Coulombe, L.J. CA 35.1
Couranjou, J. FR 4.1.1
Courduroux, J.C. FR 27.1
Courter, J.W. US-F 10.1
Courtney III., W.H. US-D 12.1
Courtney, W.H. CA 23.1
Cousin, M-T. FR 17.1.2
Cousin, R. FR 17.1.1
Couteaudier, Y. FR 7.1.3
Coutisse, S. BE 6.8.5
Couvillon, G.A. US-D 16.1
Couvreur, L. BE 10.1
Covassi, M. IT 9.3
Covatta, F. AR 3.1.2
Covey, R.P. US-J 10.1
Covey, S. GB 24.1.4
Cowan, C. IE 3.2
Cowin, R.F. GB 6.1
Cowley, K.R. AU 41.1
Cox, D.A. US-D 38.1
Cox, J.I. AU 43.1
Cox, N.R. NZ 5.1
Cox, T.I. NZ 10.1
Coyne*, D.P. US-G 23.1
Crabbé*, D. BE 9.1
Crabbé*, J. BE 4.2, 6.8.5
Cragneolini, C. AR 7.1
Craig, Jr., J.C. US-B 12.1
Craig, R. US-B 13.1.1
Craig, W.E. CA 42.1
Craig-Brown, H. AU 30.1

Craker, L.E. US-A 5.1.1
Crall, J.M. US-D 32.1
Crandall, P.G. US-D 30.2
Crassweller, R.M. US-B 13.1.1
Crater, G.D. US-E 3.1
Craw, R.C. NZ 2.2
Crawford, A.M. NZ 2.2
Crawford, D.W. NZ 9.2
Crawford, H.R. CA 15.1
Crawford, I.M. GB 31.1
Crawford, L. AU 27.1
Crawford, L. US-A 9.4
Creasy, L.L. US-B 5.1.2
Creemers, P. BE 19.1
Creighton, C.S. US-D 12.2
Crescimanno, F.G. IT 13.1
Crête, R. CA 35.1
Crettenand, J. CH 2.1
Creutz, E.C. US-M 2.1
Crew, P.S. AU 25.1
Cricoveanu, R. RO 8.4.2
Criley*, R.A. US-M 2.2
Cripps, J.E.L. AU 59.1
Crisp, P.C. GB 33.1.6
Cristoferi, G. IT 3.1
Critten, D.L. GB 31.2.6
Crnko*, J. AR 15.1
Crnko, J. YU 18.2.1, 19.1
Crocci, D.M. AR 8.1
Crocker, C. AU 59.2
Crocker, T. US-D 22.1
Croise, R. FR 34.1
Croitoru, F. RO 1.1.3
Cromey, M.G. NZ 13.3
Crook, N.E. GB 18.1.2
Crosby, J. US-F 9.1.1
Crossan, D.F. US-C 2.1
Crossa-Raynaud, P. FR 3.1.3
Crossley, C.P. GB 31.1
Crossley, J.R. NZ 2.1.3
Crouzel, I.S. AR 4.1.4
Crowder, L. US-H 3.1
Crowder, R.A. NZ 13.5
Crowe*, A.D. CA 41.1
Crowhurst, R.N. NZ 2.3.1
Crozier, B. GB 13.1.2
Crozier, L.M. CA 42.1
Crüger, G. DE 8.1
Cruiziat, P. FR 5.1
Crute, I.R. GB 33.1.4
Cruz, A. PE 1.1
Csáki, Gy. HU 2.2.12
Csáky, A. HU 4.1.1
Csapai, M. HU 2.7
Cselőtei, L. HU 4.1.1
Csemez, A. HU 2.2.4
Csepregi, P. HU 2.2.10
Cserni, I. HU 5.3.1
Csillèry, G. HU 5.3.2
Csillik, M. HU 2.2.4
Csima, P. HU 2.2.4
Csizinsky, A.A. US-D 24.1
Csizmadia, L. HU 5.3.2
Csőke, K. HU 2.2.9
Csontos, Gy. HU 3.1.1
Csorbai, F. HU 5.1.2, 5.1.3
Csukás, L. HU 4.2
Cuany, A FR 2.1.5
Cucchi, N.J.A. AR 15.1
Cuéllar Villarreal, E.J. MX 5.5
Cuénat, P. CH 2.1
Cueva, A. PE 2.4
Cull, B.W. AU 23.1
Cull, D.C. GB 3.1
Cullen, N.A. NZ 5.1
Cullis, C. GB 24.1.1
Cullum, R.F. US-L 1.1
Cumbo, F.E. US-D 4.1
Cumby, T.R. GB 31.2.4
Cumming, B. AU 30.1
Cumming, D.B. CA 3.1
Cummins, J.N. US-B 2.1.2
Cunha, R.J.P. BR 22.1
Cunningham, I.C. AU 21.1
Cuocolo, L. IT 18.2
Cupertino, F.P. BR 25.4
Cuperus, S. NL 6.1
Cuppels, D.A. CA 18.1
Curatolo, G. IT 13.2
Curi, S.M. BR 21.2
Curir, P. IT 22.1
Curran, P.L. IE 3.1
Currie, A.W. CA 40.1
Curry*, E.A. US-J 10.2
Curti Díaz, S.A. MX 12.5
Curtis, C. US-F 3.1, 7.1.1
Curvelo Garcia, A. PT 2.1
Curwen, D. US-F 23.1.3
Custers, J.B.M. NL 9.6.4
Cutcliffe*, J.A. CA 43.1
Cutis, R.G. LK 1.1
Cutler, D.F. GB 16.1
Cutler, J.R. GB 9.1.3
Cuturilo, S. YU 8.1
Cvetić, M. YU 10.1
Cvetković, R. YU 14.1
Cvjetković, B YU 3.2
Cywinska-Smoter, K. PL 28.1.17
Czaplicki, E. PL 20.1.4
Czapski, J. PL 28.2.10
Czckanski, M. PL 32.3.1
Czech, M. PL 20.2.7
Czekalski, M. PL 20.2.5
Czerwiński, PL 32.2.1
Czopek, K. PL 23.1
Czuba, S. PL 28.1.6
Czukor, B. HU 2.4
Czynczyk*, A. PL 28.1.1
Czynczyk, E. PL 28.1.10
Czyzewska, S. PL 28.2.2

Dabanli, H. TR 19.1
Dabanlioglu, G. TR 8.1
Dabbs*, D.H. CA 11.1
Dabetić, R. YU 14.2
Dabin, P. BE 12.2
Dabov, St. BG 8.1
Dąbrowska, B. PL 32.1.2.1
Dąbrowski*, J. PL 14.1.2
Dąbrowski, J. PL 20.1.4
Dąbrowski*, T. PL 32.1.5
Dąbrowski, Z.T. PL 32.1.7
Dabska, B. PL 14.1.6
Dabski, M. PL 14.1.2
Da Conceição, A.J. BR 12.2
Da Costa, A.N. BR 18.1
Da Costa, C.P. BR 24.1
Da Costa, J.G.C. BR 26.1
Da Costa, S.I. BR 20.2
Da Cruz, C. de A. BR 19.1
Da Cruz, N.D. BR 20.1
Da Cruz, P.F.N. BR 14.1
Da Cunha, G.M. BR 29.1
Da Cunha Gastal, M.F. BR 30.2
Dadd, R.H. US-K 3.1.1
Dadej, J. PL 16.1
Dadson, R.B. GA 1.1
Da Eira, P.A. BR 19.1
Dafir, M. MA 5.2
Da Gama Wanderley, L.J. BR 10.1
Dagnelie, P. BE 6.8.12
Daguenet, J. FR 19.1
Daguin, F. FR 19.2
Dahlenburg*, A.P. AU 36.2
Dahlsten, D.L. US-K 3.1.1
Daiber, C.C. ZA 3.1
Daido, K. JP 33.1
Daikoku, M. JP 17.2
Dainello, F. US-H 14.1
Daito, H. JP 14.2
Dakwa, J.T. GA 4.6
Dale, A. CA 23.1
Dale, J.L. AU 15.4
Dale, M.O. AU 31.1
Dall'Aglio, G. IT 14.1
Dall'Agnol, A. BR 27.1
Dalla Guda, C. IT 22.1
Dall'Armellina, A. AR 23.1
Dally, M. US-F 16.1
Dalmacio, S.C. PH 1.4
Dalman*, P. FI 5.1

Dalmasso, A. FR 2.1.5
Dalmasso, J. US-F 3.1
Daly, H.V. US-K 3.1.1
Damasco, O.P. PH 1.7
Damato, G. IT 2.2
Dambre, P. BE 22.1
Damiano*, C. IT 19.1.3, 22.1
Damigella, P. IT 5.1
Damjanovic, M. YU 13.1
Damke, M.M. IN 13.4
D'Amore, R. IT 20.1
Dan, A. IN 1.1.5
Dan, E. BR 18.1
Dan, M. JP 33.1
Dana, M.N. US-F 23.1.3
Dana, M.N. US-F 9.1.4
Danailov, B. BG 16.1
Dandekar, A.M. US-K 6.1.3
Danek*, J. PL 3.1
Danek-Jezik, K. AT 8.1
Dangale, K.M. IN 15.1
D'Angelo, G. IT 6.1
Daničić, M. YU 14.1
Daniel, L. HU 5.3.1
Daniel-Kalio, L.A. NG 10.1
Daniell, J. US-D 20.1
Daniells*, J.W. AU 20.1
Daniels, J. BR 30.1
Daniels, M.J. GB 24.1.2
Daniels, R.R. US-B 13.1.1
Daniels, R.W. CA 42.2
Danilatos, N. GR 18.1.4
Dann, I. AU 32.1
D'Anna, F. IT 13.2
Dannijs, G. NL 17.1
Danós, E. AR 6.1
Dansereau, B. CA 38.1
Danso, M. SN 2.1
Danu, N.S. IN 24.1.2
Daorden, M.E. AR 3.1.1
Da Ponte Filho, J.J. BR 8.2
D'Araujo Couto, F.A. BR 15.1
Darbellay, Ch. CH 2.2
Darbind, IN 12.2
Darbonne, H. FR 39.1
D'Arcy, C. US-F 12.2.3
Darby, J.F. US-D 36.1
Darby, P. GB 35.2
Dardenne, P. BE 10.1
Darfeld, H.J. DE 6.1.2
Daris, B. GR 18.1.1
Darisse, F. CA 29.2
Da Rocha, A.C.P.N. BR 3.1
Da Silva, A.C.F. BR 28.1
Da Silva, E.M.N. BR 19.2
Da Silva, J. BR 1.3
Da Silva, J.F. BR 17.1
Da Silva, J.S.M. BR 18.1
Da Silva, L.A. BR 8.1
Da Silva, M.C.L. BR 10.1
Da Silva, N. BR 25.5
Da Silva, N.M. BR 13.1
Da Silva, S.D. BR 20.2
Da Silva Coelho, Y. BR 12.1
Da Silva Cordeiro, D. BR 30.2
Da Silva Freire, C.J. BR 30.1
Da Silva Fulko, W. BR 29.1
Da Silva Gomes, A. BR 30.2
Da Silva Gomes, J.E. BR 29.1
Da Silva Martins, C. BR 1.3
Da Silva Moreira, C. BR 26.2
Da Silva Souza, L. BR 12.1
D'Artagnan de Almeida, L. BR 20.1
Dartigues, A. FR 1.1.4
Das, A.K. IN 16.2
Das, B. IN 28.1
Das, B.B. IN 4.1
Das, B.K. IN 16.2
Das, D.K. IN 16.2
Das, G.C. IN 16.2
Das, J. IN 16.2
Das, P. IN 16.2
Das, P.K. IN 3.1
Das, R. IN 7.1
Das, R.C. IN 16.2
Das, T.K IN 16.2
Das Chagas, E.F. BR 6.1
Das Chagas A. Paz, F. BR 5.1
Daskalov, St. BG 13.1
Dass, H.C. IN 18.2, 18.3, 18.4
Date, H. JP 31.1
Date, N. JP 41.1
Daubeny*, H.A. CA 4.1
Daudet, F.A. FR 5.1
Daughtrey, M.L. US-B 8.1
Daulay, H.S. IN 18.2
Daulta, B.S. IN 9.1
Daunay, M-C. FR 3.1.2
Daunicht*, H.J. DE 5.1.1
Dauplé, P. FR 3.1.10
Daussant, J. FR 38.1
Dauti, M. YU 16.1
Dautrebande, S. BE 6.8.10
Dávalos González, P.A. MX 7.2
Dave, B.V. IN 19.1
Davenport, R.R. GB 19.1.4
Davenport, T.L. US-D 29.1
David, J. CA 37.1
Davide, R.G. PH 1.4, 1.6
Davide, T.S. PH 1.2.3
Davidescu, D. RO 3.1
Davidson, C.G. CA 13.1
Davidson, J.G.N. CA 5.1
Davidson, H.R. CA 44.1
Davidson, J.M. US-D 27.1
Davidson, R.D. US-I 1.1
Davies, D.J.G. NZ 13.2
Davies, D.L. GB 8.1.2
Davies, D.R. GB 24.1.1
Davies, F.T. US-H 6.1
Davies, H.V. GB 15.1.4
Davies, J. GB 24.1.4
Dávilla, J. AR 15.2.5
Davis, D. US-E 16.1
Davis, D.R. US-E 11.1
Davis, D.W. US-G 7.1
Davis, D.W. US-I 5.1
Davis, H.L. US-H 2.1
Davis, J.R. US-I 13.1
Davis, L.H. CR 1.1
Davis, L.H. US-I 5.1
Davis, P.F. GB 31.2.7
Davis, P.L. US-D 34.1
Davis, R.E. US-C 3.1
Davis, T.D. US-I 6.1
Davis, T.M. US-A 2.1
Davis, T.S. US-D 20.1
Davison, A.D. US-J 6.1
Davison, R.M. NZ 2.1.1
Davous, M.P. FR 15.1.4
Dawes, S.N. NZ 2.1.1
Dawson, W.M. GB 20.1
Day, J.W. US-E 3.1
Day, M.J. GB 6.1
Dayal, K. IN 25.1
De Abreu, J.M. BR 14.1
De Agudelo, F. CO 4.1
Deahl, K. US-C 3.1
Deák-Karácsonyi, S. HU 2.2.5
De Alba, A.P.C. BR 21.2
De Albuquerque, D. PT 4.1
De Albuquerque, F.C. BR 2.1
De Almeida, J.I.L. BR 8.1
De Almeida, L.C.C. BR 14.1
De Almeida, P.A. BR 12.1
De Almeida Conrado, P.R. BR 21.2
De Almeida Lima, J. BR 25.5
Dean, P. AU 59.2
De Andrade Neto, A.P.M. BR 18.1
Deanon, Jr. J.R. PH 1.2.4
De Aquino, C.R.N. BR 13.2
De Aquino P. Castro, T. BR 26.1
De Araújo, F.E. BR 8.1
De Araújo, J.P.P. BR 26.1
De Araújo, M.L. BR 19.1
De Araújo Filho, G.C. BR 8.1

De Araújo Torres, C. BR 10.1
De Arruda Veiga, R.A. BR 22.1
Deas, A.H.B. GB 19.1.1
De Assis Lemos de Souza, F. BR 27.2
De Athayde, J.T. BR 19.1
De Avila, A.C. BR 25.5
De Azevedo, J.N. BR 7.1
De Azevedo Campelo, G.J. BR 7.1
De Baldini, A.M.D. AR 15.2.5
De Barreda, D.G. ES 14.1
De Barros, J.C.M. BR 19.1
De Barros, L.F. BR 21.1
Debergh, P. BE 7.3
De Boer, C.M.M. NL 1.1.6
De Boer, R.H. US-B 14.1.1
De Boer, S.H. CA 4.1
De Bokx, J.A. NL 9.4.4
De Bondt, K. BE 1.1
De Bont, G. BE 13.1
De Boodt, M. BE 7.4
De Brabander, W.H. NL 9.3
De Brito Giordano, L. BR 25.5
Debritu, W/o. ET 5.1
Debrot, E. VE 1.1
De Bruijn, J. NL 9.7.4
De Bruin, H.J. NL 9.8
DeBuchananne, D.A. US-G 22.1
Dec, J. PL 20.1.4
De Cannière, BE 12.1
De Caro, A. IT 2.2
De Carvalho, A.R. BR 14.1
De Carvalho, J.E.B. BR 12.1
De Carvalho, P.C.I.L. BR 12.1
De Carvalho, V.D. BR 15.1
De Carvalho Genu, P. BR 25.2
De Carvalho Neto, J.S. BR 13.2
De Castro, J.V. BR 20.2
De Castro, L.A.B. BR 25.3
De Castro, R. PT 5.1
Deckers, T. BE 19.1
De Cleene, M. BE 7.12
De Clercq, R. BE 15.2
De Cockborne, A-M. FR 3.1.7
De Cockborne, P. FR 3.1.2
De Cormis, L. FR 3.1.6
Decoteau, D.R. US-D 15.1
Decourtye*, L. FR 1.1.1, 1.1.2
De Cruz Machado, J. BR 16.1
De Cunha, R. BR 19.1
De Dias, R.M. AR 15.2.5
De Diaz Botto, B. AR 22.2
De Dios Bustamante Orañegui, J. MX 8.3
De Dios Ruiz de la Rosa, J. MX 5.2
De Donato, M. IT 24.3
Dedonder, A. BE 7.12
Deep, I.W. US-F 71.1
Deer, T.W. AU 31.1
De Fatima, M. BR 8.1
De Fátima Batista, M. BR 25.3
De Fatima Potes, Ma. PT 6.1
De Figueirebo, J.M. BR 14.1
De Filippis, L.F. AU 10.1
De Franchi, S. IT 2.3
DeFrank, J. US-M 2.2
Degani, H. IL 1.2.5
De Gelder, A. NL 1.1.4, 9.9
Degeyter, L. BE 22.1
De Goes, A. BR 19.1
DeGomez, T. US-A 1.1
De Graaf, M. NL 9.2
De Grisse, A. BE 7.6
De Groot, A.J. NL 4.1
De Groot, B. NL 9.5
De Groot, C.T. NL 9.7.4
De Groot, W. NL 9.1
De Gruijter, J.J. NL 9.11.2
De Gruyter, E. NL 17.2
De Guiran, G. FR 2.1.5
De Gutierrez, B. CO 4.1
De Haan, W.G. NL 3.1
De Hertogh*, A.A. US-D 8.1.1
Dehgan, B. US-D 27.5
Dehne, J. DD 5.1
Deidda, M. IT 4.1, 23.1
Deist, J. ZA 4.2
De Jesus Arellano Ledezma, T. MX 6.2
De Jesus Soria Vasco, S. BR 31.1
De Jesus Vernetti, F. BR 30.2
De Jong, C.J. NL 14.1
De Jong, H. CA 40.3
De Jong, J. NL 9.6.3
De Jong, R.W. AU 35.1
De Jong*, T.M. US-K 6.1.3
Dekanić, I. YU 3.3
Dekasoz, E. US-D 16.2
Dekazos, H. GR 6.1.1
De Kekan, V. CO 7.1
Deki, K. JP 32.1
Dekker, P.H.M. NL 6.1
Dekkers, W.A. NL 6.1
De Kleijn*, E.H.J.M. NL 3.1
De Klerk, C.A. ZA 4.2
De Koning, K. NL 9.3
De Kraker, J. NL 6.1
De Kroon, C.J.J. NL 6.1
De Laat, A.M.M. NL 9.5
De la Colina, V. PE 4.2
De la Cruz, G.T. MX 9.2
De la Cruz, H. PE 2.3
De la Cruz, R. CR 2.2
De la Fosse, R. AR 4.1.2
De la Fuente, J.M. MX 13.1
De la Iglesia, F.J. AR 15.2.3
De la Luz Sánchez Pérez, J. MX 3.1
De Landeros y P., P.T. MX 17.1
Delanoy, G. SN 1.1
De la Puente, F. PE 1.1
De la Puente Ciudad, F. BR 25.5
De la Roche, A.I. CA 20.2
Delas, J. FR 4.1.2
De Lasson*, A. DK 9.1
De la Torre Vizcaino, J.D. MX 7.2
Delatour, C. FR 11.1.4
Delaunay, M. FR 28.1
Delbridge, R.W. CA 42.1
Delecolle, B. FR 3.1.5
Delecolle, R. FR 3.1.4
De Leeuw, H.H. NL 9.7.5
De Ley, J. BE 7.12
Delgadillo Sánchez, F. MX 7.2
Delgado, D. VE 5.1
Delgado, E. PE 1.1
Delhaye, J-P. BE 4.2
De Lima Alves, F. BR 18.1
De Lima Mesquita, J.E. **BR 5.1**
De Lint*, P.J.A.L. NL 8.1
De Lizarraga, B.E.S. AR 22.3
Dellacecca, V. IT 2.2
Della Strada, G. IT 19.1
Delmon, D. FR 29.1
Del Monte, R.F. AR 15.1
Delmotte, C. BE 6.2
Delmotte, R. BE 8.1
De Loose, R. BE 14.1
De los Santos de la Rosa, F. MX 10.2
Del Rosario, A.G. PH 1.2.2
Del Rosario, D.A. PH 1.7
Delroy, N.D. AU 51.1
Deltour, J. BE 6.8.5, 6.8.11
De l'U, D. AU 65.4
De Lucena Duarte Ribeiro, R. BR 19.2
De Luque, C. CO 7.1
De Lurdes Borges, Ma. PT 6.1
De Lurdes Matos, Ma. PT 6.2
Delvaux, J. BE 9.1
Delver, P. NL 10.1
De Maeyer, L. SN 1.1
De Magalhaes, J.R. BR 25.5

De Manzino, M.E.B. AR 15.3
De Marcó, C.M.A. AR 6.1
Demarly, Y. FR 19.2
De Martin, Z.J. BR 20.2
De Martínez, N. AR 22.3
DeMason, D.A. US-K 12.1.1
De Matos, P.L.P. BR 12.1
De Matos Lopes, A. BR 26.1
De Matos Malavasi, M. BR 19.2
Dematté, M.E. Soares P. BR 23.1
Demb, A. US-M 2.2
De Medeiros, A.R.M. BR 30.1
Demeke, T. ET 9.1
De Mello Freire, L.M. BR 31.1
De Melo, V.J. BR 23.1
De Melo Amaral, R.E. BR 21.2
De Melo Freire, J. BR 31.1
De Melo Lisboa, S. BR 4.1
De Melo Moura, G. BR 5.1
De Menezes Pereira, I.T. BR 21.1
De Menezes Porto, O. BR 29.1
Demere, A.A. ET 10.1
De Mesquita Filho, M.V. BR 25.5
De Meyere, BE 13.1
De Michele, A. IT 13.1
De Millán, N.J.G. AR 13.1
Demirbuker, Y. TR 19.1
De Miranda, A.R. BR 25.3
De Miranda Flor, H. BR 25.4
De Miranda Santos, M. BR 1.1
Demirel, N. TR 17.1
Demiroren, S. TR 14.1
Demirovska, V. YU 4.3
Demjén, I. HU 2.2.4
De Moel, C.P. NL 6.1
Demolin, G. FR 3.2.2
De Montes, ES 5.1
De Moraes, E. BR 1.3
De Moraes, V.H. BR 1.1
Demoranville, I.E. US-A 11.1
De Mos, G. NL 20.1
De Mos, J. NL 20.1, 20.2, 20.3
De Moura Barros, L. BR 8.1
Dempsey, P. IE 3.1
Demski, J.W. US-D 20.1
Demuro, A. IT 4.1
Denardi, F. BR 28.1
Denby, L.G. CA 3.1
Deng, S.H. TW 6.1
Denisen*, E.L. US-G 20.1
Dennehy, T.J. US-B 2.1.3
Dennett, J.M. AR 5.1
Den Nijs, L.J.M. NL 9.4.3
Dennis, C.D. GB 7.1
Dennis, D.J. NZ 10.1
Dennis*, Jr., F.G. US-F 14.1.1
Den Ouden, H. NL 9.4.1
Denton, O.A. NG 1.1
De Nunez, N.S. AR 4.1.3
Deol, I.S. IN 17.1
De Oliveira, A.F.F. BR 2.1
De Oliveira, A.M. BR 19.1
De Oliveira, D.P. BR 14.1
De Oliveira, H.G. BR 8.2
De Oliveira, M.L. BR 14.1
De Oliveira, S. BR 12.1
De Oliveira, S.L. BR 12.1
De Oliveira, V.H. BR 5.1
De Oliveira, W.S. BR 1.2
De Oliveira Lima, T.S. BR 5.1
De Pacheco, M.E.B. AR 22.1
De Pádua, J.G. BR 15.1
De Pádua, T. BR 16.1
De Paiva, J.R. BR 1.1
De Paula Godinho, F. BR 15.1
De Paula Lima, L.A. BR 16.1
De Petter, E. BE 7.12
De Pinto Pereira, R. BR 13.1
De Ponti, O.M.B. NL 9.6.1
De Prest, G. BE 7.7
De Queiroz, G.M. BR 8.1
De Queiroz Pinto, A.C. BR 25.2
De Ramalo, N.V. AR 22.3
De Ramirez, M. del H.C. AR 2.1
De Ranieri, M. IT 22.1.1
Dereuddre, J. FR 38.1
De Ricci, M.T.D. AR 22.2
Derolez, J. BE 17.1
De Rosa, T. IT 7.1
De Ruiter, A.C. NL 12.1
De Ruiter, H.D. NL 12.1
Desai, A.G. IN 15.1
Desai, M.N. IN 14.4
Desai, U.T. IN 14.1
De Salvador, R. IT 19.1
De Sancho, J. AR 3.1.1
De Sanctis, F. IT 19.2
De Santana, C.J.L. BR 14.1
De Saravia, E.M.G. AR 17.1
Desbiez, M.O. FR 28.1
Desborough, S.L. US-G 7.1
Descamps, A. BE 8.1
Desclaux, J.L. FR 42.1, 46.1
Deshmukh, P.A. IN 12.1
Deshmukh, P.P. IN 12.3
Deshmukh, S.S. IN 14.7
Deshmukh, U.G. IN 12.1
Deshmukh, Z.V. IN 14.1
Deshpande, A.A. IN 1.1.2
De Silva, J.B. BR 15.1
De Silva, L.C.P. LK 2.1
De Silvestri, M.P. AR 15.3
De Siqueira, J.O. BR 16.1
Desjardins, Y. CA 38.1
Desle, G.Y. IN 14.2, 14.5
Desle, J.S. IN 14.1
De S. Liyanage, M. LK 2.1
De Sosa Gonzalez, G. AR 9.1
De Souza, B.H. BR 29.1
De Souza, F.R.S. BR 3.1
De Souza, M. BR 16.1
De Souza, P. BR 16.1
De Sousa, T. PT 1.1
De Souza Tanaka, M.A. BR 15.1
De Spiegeleer, F. BE 7.8
Dest, W.M. US-A 13.1
De Stanchina, G. IT 21.1
De Stigter*, H.C.M. NL 9.1
Desvignes, J.C FR 32.1
De Targa Araújo, M. BR 25.5
Detcheva, R. BG 4.1
De Terraciano, L.C. AR 4.1.1
Dethier, B.E. US-B 5.1.5
De Toloza, S.C. AR 12.1
Detroux, L. BE 6.3, 6.7
Deubert, K.H. US-A 11.1
Devasabai, K. LK 3.17
De Vaz, C.R. LK 3.3
Devcic, J. AR 4.1.1
De Veer, A.A. NL 9.11.1
De Veer, J. NL 3.1
De Venter, J. BE 9.2
Devergne, J.C. FR 2.1.3
Devgire, IN 14.3
De Villèle, O. FR 3.1.4
De Visser, C.L.M. NL 6.1
De Vita, M. IT 22.1.2
Devitt, A.C. AU 59.1
Devitt, D.A. US-I 9.1
De Vleeschauwer, A. BE 7.7
Devlin, R. US-A 11.1
De V. Lötter, J. ZA 4.3.1
De Vos*, N.E. US-K 13.1
De Vries, D.P. NL 9.6.2
De Vries, E. NL 9.5
De Vries, I.M. NL 9.7.4
De Vries, J.N. NL 9.6.4
De Vries, W. NL 9.11.1
De Wael, L. BE 15.2
De Wagt, G. NL 1.1.6
De Wandeleer, V. BE 13.1
Dewey, S. US-I 17.1
Dewhirst, L.W. US-H 27.1
De Wilde, J.J.F.E. NL 9.7.4
De Wit, H.C.D. NL 9.7.4
Deyton*, D.E. US-E 3.1
De Zoysa, A.O.C. LK 3.3

Dhaka, R.S. IN 19.1
Dhaliwal, G.S. IN 17.1
Dhanvantari, B.N. CA 17.1
Dharmasena, C.D. LK 3.3
Dhatt*, A.S. IN 17.1
Dhawan, S.S. IN 9.1
Dhillon, B.S. IN 17.1
Dhillon, D.S. IN 17.1
Dhillon, R.S. IN 17.1
Dhua, R.S. IN 28.1
Dhumal, S.A. IN 14.3
Diaconeasa, M. RO 20.1
Diaconu, M. RO 4.1
Diaconu, S. RO 9.2
Diamond, L. CA 4.2
Dias Nogueira, F. BR 15.1
Dias, J.F. PT 2.1
Dias, L.A. dos S. BR 14.1
Dias, M.B.M.N. LK 2.1
Diaz, D. AR 3.1.1
DiCarlo, H.F. US-E 20.1
Di Césare, L.N. AR 20.1
Dick, L.W. GB 2.1
Dickens, J.S.W. GB 13.1.4
Dickinson, D.B. US-F 12.2.2
Dickinson, W.J. US-I 7.1
Dickler, E. DE 9.1
Dickson, M.H. US-B 2.1.2
Di Collalto, G. IT 9.1
Diédhiou, M. SN 3.1
Dietrich, R.A.E. AR 17.1
Dieguez, R. AR 4.1.3
Diehl, J.F. DE 18.1
Diekmann, H. DE 15.3.4
Diener, T.O. US-C 3.1
Dienes, H. HU 5.3.2
Dietrich, H. DE 12.1.4
Digat, B. FR 1.1.3
Dijkstra, A.F. NL 9.5
Dijkstra, J. NL 10.1
Dijst, G. NL 6.1, 9.4.2
Dikmen, I. TR 15.3
Dil, M.C. NL 1.1.5
Dillard, H.R. US-B 2.1.4
Dilley, D.R., US-F 14.1.1
Diltrich, R.C. BR 28.1
Di Marco, L. IT 13.1
Di Martino Aleppo, E. IT 1.1
Di Martino, E. IT 1.1
Dimasi-Theriou, K. GR 25.1.1
Dimchev, B., BG 16.1
Dimény, I., HU 2.2, 2.2.9
Dimény, J., HU 2.2.12
Dimitrievits, Gy. HU 4.2
Dimitrov, G. BG 8.1
Dimitrov, St. BG 10.1
Dimov, I. BG 8.1
Dimsey, R.T. AU 31.1
Dina, Gh. RO 19.1
Dincer, H.M. TR 5.1
Dinesen, Ib.G. DK 4.1
Dinning, J.S. US-D 27.3
Dinoor, A. IL 2.6
Dinopoulos, A. GR 19.1
Dinova, M. BG 13.2
Diófási, L. HU 5.2
Dipotaroeno-Sakrama, M. SR 1.1
Dirr, M. US-D 16.1
Dissanayake, C. LK 2.1
Dissanayake, E.M. LK 3.7
Dittrich, H.H. DE 12.1
Ditu, D. RO 7.1
Divakar, B.L. IN 24.1.3
Divinagracia, G.G. PHC1.4
Dixit, V.S. IN 19.2
Dixon*, G.R. GB 1.1
Dixon, K.W. AU 59.3
Dixon, M. CA 15.1
Djaković, M. YU 9.1
Djekic, R. YU 11.1
Djelianov, D. BG 6.2
Djemmali, A. TN 1.1
Djeneva, A. BG 9.1
Djinović, I. YU 13.1
Djokic, D. YU 10.1
Djordjević, R. YU 10.1
Djorović, M. YU 14.1
Djorović, V. YU 15.1
Djukić, Dj. YU 10.1
Djukic, M. YU 10.1
Djukic, R. YU 7.1
Djunisijević, Lj. YU 13.1
Djurhuus*, R. NO 10.1.4
Djurić, B. YU 17.1
Djurić, M. YU 10.1
Djurić, S. YU 1.1
Djurisić, B. YU 14.1
Djurović, D. YU 10.1
Djuvinov, V. BG 8.2
Dlouhá, J. CS 4.1
Długowolski, B. PL 13.1
Dmoch, J. PL 32.1.7
Do, B.H. KR 7.4
Do Amaral, S.F. BR 21.2
Doazan, J.P. FR 4.1.8
Dobiech, D. PL 32.1.11
Dobos, L. HU 5.1.3
Dobreanu, M. RO 8.4.5
Dobritz, A. DE 15.1.5
Dobrodoljani, B. YU 16.1
Dobromilska, R. PL 31.1.2
Dobrovoda, K. CS 1.1
Dobrowolska, R. PL 28.1.13
Dobrzánska*, J. PL 28.2.5
Dobrzánski*, A. PL 28.2.4
Dobrzánski, W. PL 28.2.10
Dobson, B.G. NZ 10.1
Dobson, J. US-D 18.1
Docampo, D. AR 7.1
Do Carmo, C.A.S. BR 18.1
Do Ceu Matos, Ma. PT 1.1
Dochev, D. BG 8.2
Dodd, B.C. AU 25.1
Dodd, P.B. GB 35.1.2
Dodge, A.D. GB 3.1
Doepel, R.F. AU 59.1
Doersch, R. US-F 23.1.1
Dogan, I. TR 5.2
Dogras, C. GR 25.1.2
Doherty, E.T. US-B 1.1
Dohi, H. JP 12.1
Doi, M. JP 22.5
Doi, N. JP 31.1
Doi, S. JP 17.1
Doijode, S.D. IN 1.1.9
Doku, E.V. GA 1.1
Dokuzoguz, M. TR 15.1
Dolan, D.D. US-B 2.1.2, 2.1.6
Dolatowski, J. PL 10.1.5
Dolmans, N.G.M. NL 2.1
Domanović, M. YU 11.1
Domanski, D. PL 28.2.11
Dománski, R. PL 20.2.7
Domiciano, N.L. BR 27.2
Dominguez, J.A. AR 22.1
Domino, Z. PL 1.1
Domoto*, P.A. US-G 20.1
Donadio, L.C. BR 23.1
Donahaye, E. IL 1.3.3
Donald, E.R.C. BR 11.1
Donchev, A. BG 7.1
Donchev, P. BG 14.1
Doner, H.E. US-K 3.1.3
Donoho, C. US-D 16.1
Dons, J.J.M. NL 9.6.4
Donselman, H.M. US-D 25.1, 27.5
Doodson, J.K. GB 6.1
Doohan, D.J. CA 40.1
Dooley*, L. AU 30.1
Dop, A.J. NL 7.1
Do Prado Sobral, J. BR 11.1
Dorel, C. RO 30.1
Doreswamy, R. IN 1.1.8
Döring, J. DE 5.1.1
Dorotea, G. IT 16.1
Dorrell, D.G. CA 8.1
Doru, G. RO 8.1
Doruchowski, G. PL 28.1.6
Doruchowsk*i, R.W. PL 28.2.1
Do Sacramento, C.K. BR 14.1
Dosba, F. FR 4.1.1

Dos Santos, C.A. BR 6.1
Dos Santos, L.C. BR 20.2
Dos Santos, W.J. BR 27.2
Dos Santos Vieira, R. BR 1.2
Doubková, M. CS 29.1
Doucet, R. CA 34.1
Dougall, I.C.G. GB 2.1
Douglas, G. IE 4.1
Douglas, G.B. NZ 8.3
Douglas, J.A. NZ 5.1
Douglas, L.A. US-B 14.1.7, 17.1
Douglas, T.J. AU 65.2
Dourley, J. US-K 5.1
Dover, C.J. GB 8.1.3
Dovrat, A. IL 2.3
Dowker, B.D. GB 33.1.6
Downey, G. IE 4.1
Downie, A. GB 24.1.2
Downing, D.L. US-B 2.1.5
Downs, C. NZ 10.1
Downton, W.J.S. AU 65.1
Doyen, J.T. US-K 3.1.1
Doyon, D. CA 33.1
Doytchev, K. BG 5.1
Dozier, W.A. US-D 38.1
Drågan, N. RO 14.1
Drågånescu, E. RO 35.1
Draganov, G. BG 9.1
Dragland, S. NO 6.1
Dragomir, N. RO 1.1.7
Dragomir, S. RO 1.1.4
Drågut, Gh. RO 5.1
Drake, S. US-J 10.2
Drapeau, R. CA 32.1
Draper, A. US-C 3.1
Dreatta, I. dos S. BR 20.2
Drenth, H. NL 6.1
Dresher, R.W. 6.1
Drew, B.M. GB 26.1
Drew, J.V. US-L 1.1
Drew, R.A. AU 18.1
Dreze, P. BE 6.8.11
Driessen, A.C. BR 28.1
Drijfhout, E. NL 9.6.1
Drilleau, J.F. FR 14.1.4
Drimon, J. FR 31.1
Drobný, J. CS 7.1
Drobotå, Gh. RO 23.3
Drobotå, M. RO 23.3
Dronka, K. PL 5.1
Drosopoulos, A.S. GR 14.1.2
Drosos, N. GR 12.1
Drost, D.T. NZ 9.1
Druart*, P. BE 6.6
Drummond, R.B. ZI 3.1.2
Drużyński, L. PL 32.1.9
Dry, P.R. AU 41.1
Dsouki, W.M. LY 2.1
Duarte, M.L. PT 1.1
Duarte, T. PT 7.1
Duarte Costa, N. BR 13.1
Duarte Cruz, A. BR 30.3
Dubé, A. CA 36.1
Dube, B. ZI 3.1.2
Dubey*, K.C. IN 1.1.1
Dubey, P.S. IN 22.1
Dubos, B. FR 4.1.6
Dubouzet, J.G. PH 1.2.2
Dubravec, I. YU 3.1
Dubuc-Lebreux, M.A. CA 31.2
Ducasse, D. AR 7.1
Duch, J. PL 24.1
Ducloux, F. FR 33.1
Ducrey, M. FR 3.2.1
Duczmal, K. PL 20.2.10
Dudeck, A.E. US-D 27.5
Dudek, D. PL 28.1.6
Dudek, M. PL 28.2.2
Dudley, K. GB 7.1
Dudley, T.R. US-C 4.1
Dudney, P.J. GB 2.1
Dudzic, M. PL 9.1
Dufault, R. US-H 15.1
Duff, D.T. US-A 14.1
Duffek, J. CS 22.1
Dugdale, J.S. NZ 2.2
Dugger, Jr., W.M. US-K 12.1.1
Duich, J.M. US-B 13.1.1
Dujmovic, M. YU 3.4
Dukes, P.D. US-D 12.2
Dulieu, H. FR 7.1.6
Dull, J. US-D 16.2
Dumas, Y. FR 3.1.1
Dumas de Vaulx, R. FR 3.1.2
Du Merle, P. FR 3.2.2
Dumitrache, I. RO 14.1
Dumitrescu, Gh. RO 23.2
Dumitrescu, M. RO 1.1.3
Dumitru, M. RO 1.1.1
Dunbar, O.J. NZ 2.1.3
Dunbier, M.W. NZ 13.2
Duncan*, A.A. US-D 29.1
Duncan, J.M. GB 15.1.1
Duncombe, D.A. ZI 3.1.2
Dunez, J. FR 4.1.6
Dungate, L. AU 59.2
Dunn, J.H. US-E 18.1
Dunn, J.S. NZ 13.6
Dunne, R.M. IE 4.1
Dunnington, J. US-A 14.1
Dunwell, J. GB 24.1.1
Dunwell, W. US-E 16.1
Duodo, Y.A. GA 3.3
Du Plessis, H.J. ZA 4.1
Du Plessis, S.F. ZA 1.1
Du Plooy, F.I. ZA 3.1
Duprat, F. FR 3.1.8
Du Preez, M. ZA 4.1
Durand, R. FR 2.1.2
Duranti, A. IT 18.2
Durceylan, E. TR 3.2
Durkin, D.J. US-B 14.1.1
Duron, M. FR 1.1.2, 19.2
Dürr, P. CH 1.1
Durska, K. PL 28.1.10
Durzan*, D.J. US-K 6.1.3
Dusek, J. CS 20.1
Dusky, J.A. US-D 23.1
Dutå, A. RO 19.1, 32.1
Dutra, L.G. BR 26.2
Dutrecq, A. BE 6.2
Dutta*, C.P. IN 2.2
Dutta, O.P. IN 1.1.2
Dutu, I. RO 9.1
DuVall, T. US-D 16.1
Dvivedi, R.M. IN 22.1
Dvořák, A. CS 26.1
Dwelle, R.B. US-I 13.1
Dwivedi, M.P. IN 10.3
Dwyer, B. IE 4.1
Dwyer, M.J. GB 31.2.1, 31.2.3
Dyck, V.A. CA 3.1
Dyke, A.J. GB 14.1
Dykeman*, B.W. CA 40.1
Dyki, B. PL 28.2.12
Dykstra, M. CA 6.1
Dynia, J.F. BR 30.1
Dyring, A.-K. NO 10.1.8
Dyś, B. PL 14.1.7
Dyśko, J. PL 28.2.5
Dyson, J. GB 31.1
Dzięciol, R. PL 32.1.6
Dzięciol, U. PL 28.1.5
Dzięciol, W. PL 28.1.1
Dzięciolowski, R.M. NG 10.1

Eady*, F.C. CA 25.1
Eagle, D.J. GB 27.1
Earle, E.D. US-B 5.1.7
Eason, J.T. US-D 42.1
East, R. NZ 5.1
Easterbook, M.A. GB 8.1.2
Eaton*, G.W. CA 4.2
Ebata, M. JP 18.1
Ebbels, D.L. GB 13.1.4
Ebben, M. GB 18.1.2
Ebert, A. BR 28.1
Ecevit, M.F. TR 16.1
Echandi, E. US-D 8.1.2
Echert, M. RO 7.1
Echeverri, J. CR 2.2
Echeverria, E.M. AR 18.1

Eck, P. US-B 14.1.1
Eckenrode, C.J. US-B 2.1.3
Economakis, C. GR 8.1.1
Economides, C.V. CY 1.1
Economopoulos, A. GR 3.1.1
Economou*, A. GR 25.1.3
Edema, A.O. NG 1.1
Eder, R. AT 19.1
Edge, V.E. AU 9.1.3
Edgerton, L.J. US-B 5.1.2
Edinger, W. CA 25.1
Edland, T. NO 10.4.3
Edmondson, R.N. GB 18.1.1
Edwards, D.G. AU 15.5
Edwards, G.J. US-D 30.2
Edwards, G.R. AU 42.1
Edwards, J. US-D 19.1
Edwards, J.R. US-D 7.1
Edwards, M.E. AU 31.1
Edwards, W.R.N. NZ 8.3
Efendić, K. YU 1.1
Efthimiadou, E. GR 6.1.1
Eggen, M. NO 10.1.8
Eggens, J.L. CA 15.1
Egger, L. IT 7.3
Egolf, D.R. US-C 4.1
Eguchi, H. JP 34.2
Ehara, T. JP 34.2
Ehleringer, J. US-I 7.1
Ehrendorfer, F. AT 8.6
Ehrenfeld, D.W. US-B 14.1.1
Ehrenthaler, F. AT 13.1
Eichhorn, K.-W. DE 22.1
Eikenbary, R.D. US-H 3.1
Einarson, C.-G. SE 8.1.1
Einert, A.E. US-E 11.1
Eisel, M. US-G 7.1
Eisenberg, B.A. US-F 12.2.2
Eisler, V. RO 6.1
Ekeberg, E. NO 6.1
Ekelund*, L. SE 1.1.4
Eksteen, G.J. ZA 4.1
El-Abdein, A.Z. EG 1.3
El-Ahwah, F. EG 1.4
El Aouini, S. TN 1.1
El Attir, H. MA 1.1
El-Aziz, B.A. EG 1.2
Elber, Y. IL 1.1.2
El-Daslerity, S. EG 1.4
El-Deen, A.-B.S. EG 1.2
El-Deen, S. EG 1.2
El-Din Awad, K. EG 1.3
El-Din*, S.B. EG 1.1
Eldridge, B.F. US-J 14.1
Elema, R.K. NL 10.1
Elen, O.N. NO 10.4.1
Elena, M.A. AR 17.1
El Fadl, A. MA 1.1
Elfering-Koster, K.G. NL 1.1.3
Elfving, D.C. CA 23.1
Elia, A. IT 2.3
Elias, T.S. US-K 5.1
Elkner, K. PL 28.2.10
Ellerbrock, R.E US-B 5.1.3
Elliot, A.E. US-B 5.1.1
Elliott, G.C. US-B 13.1.1
Elliott, J.F. AU 59.1
Elliott, T.J. GB 18.1.2
Ellis, M.A. US-F 7.1.1
Ellis, N. GB 24.1.1
Ellis, P.J. CA 4.1
Ellis, P.R. GB 33.1.3
Ellison, J.H. US-B 14.1.1
Ellison, P. AU 12.1
Ellissèche, D. FR 9.1
Elloumi, N. TN 1.1.
Ells, J.E. US-I 2.2
Ellstrand, N.C. US-K 12.1.1
El Mahjoub, M. FR 23.1
Elmdust, M. DE 15.5
El-Moneim, S.A. EG 1.1, 1.3
Elmstrom, G.W. US-D 32.1
El-Murabaa, A.I. LY 1.1
El-Nokrashi, M.A. EG 1.2
El Otmani, M. MA 1.1
Elsey, K.D. US-D 12.2
El-Sherbini, M. EG 1.3
El-Shiaty, M. EG 1.4
Eltez, M. TR 8.1
El-Yamzini, M.I. LY 1.2
El-Zeftawi, B. AU 30.1
Ema, S. JP 1.1.6
Emanouilidis, E. GR 18.1.2
Emanuel, N. GR 6.1.6
Embleton*, T.W. US-K 12.1.1
Embree*, C.G. CA 41.1
Emerson, B.N. LK 3.11
Emerson, D. US-A 1.1
Emerson, F.H. US-F 9.1.4
Emery, D. US-K 16.1
Emino, E.R. US-A 13.1
Emiola, I.O.A. NG 1.1
Emmanuil, V. GR 6.1.7
Emmert, F.H. US-A 13.1
Emmett, R.W. AU 31.1
Emond, G. CA 29.1
Emura, M. JP 29.1
Ena, S. IT 6.1
Enami, A. JP 29.1
Endô, H. JP 47.2
Endo, K. JP 39.1
Endo, M. JP 4.1, 16.4
Endo, N. JP 24.1
Endo, T. JP 16.3
Engel, G. DE 6.1.1
Engel, R.E. US-B 14.1.7
Engels, H. BE 14.1
Englander, L. US-A 14.1
Engler, J.C. BR 24.1
English, M . US-H 19.1
Engsbro, B. DK 4.1
Ennis, Jr., W.B. US-D 25.1
Eno, C.F. US-D 27.8
Enomoto, M. JP 8.1
Enriquez, G. CR 2.2
Entwistle, A.R. GB 33.1.4
Epp, F. AT 14.1
Epstein, E. IL 1.2.4
Epton, H.A.S. GB 22.1
Erasmus, S.P. ZA 4.3.1
Erbenová, M. CS 7.1
Erdeljan, V. YU 14.1
Erdélyi, L. HU 2.2.1
Erdős, Z. HU 2.1.1
Erez*, A. IL 1.2.2
Ergenoglu, F. TR 1.1
Ergin, R. TR 14.1
Erhardt, W. US-A 1.1
Erickson, H.T. US-F 9.1.4
Ericsson*, N-A. SE 1.1.1
Eriksen, A.B. NO 7.2
Eriksen, E.N. DK 3.1
Eris*, A. TR 6.1
Erkal, S. TR 14.1
Erkel, I. TR 14.1
Erlandsson, G. SE 1.1.2
Erner, Y. IL 1.2.3
Ernst, D. DE 15.3.1
Eroglu, A.S. TR 4.1
Eroglu, Z. TR 4.1
Erol, A. TR 15.3
Errasti, M. AR 7.1
Errico, A. IT 18.1
Ersoy, B. TR 15.3
Ertan, N. TR 14.1
Ertan, U. TR 14.1
Ertekin, U. TR 3.2
Ertem, A. TR 18.1
Ertsey, K. HU 2.7
Eryildiz, H. TR 22.1
Esbjerg, P. DK 4.1
Eschenbruch, R.J. NZ 4.1
Esendam, H.F. NL 1.1.3
Esguerra, E.B. PH 1.2.5
Eshankulov, U.E. SU 3.1
Eshed, N. IL 2.6
Esiaba, R.O. NG 1.4
E. Silva, V. PT 6.1
Eskesen, J.G. DK 7.1
Esler, W.J. AU 4.1
Esmenjaud, D. FR 29.1
Espinal Aguilar, J.A. BR 25.5
Espinillo, M. AR 1.1

Espino*, R.R.C. PH 1.2.1, 1.7
Espul, J.C. AR 15.1
Esquivel A., C. PE 4.1
Essau, R. CA 6.1
Estabrooks*, E.N. CA 40.3
Ester, A. NL 6.1
Estes, L. US-D 50.1
Esteves, M. do C.F. BR 19.1
Estrella, H. AR 15.2.5
Ethiraj, S. IN 1.1.5
Etling, D. DE 15.5
Etoh, T. JP 18.3
Ettenauer, E. AT 15.1, 15.2
Eun, J.S. KR 11.1
Eurola, S. FI 16.1
Euverte, G. FR 25.1
Evagelopoulos, I. GR 25.2.2
Evans, G.E. US-I 18.1
Evans, J.L. GB 6.1
Evanylo, G. US-C 13.1
Evenari, M. IL 5.1.3
Evensen, K.B. US-B 13.1.1
Everett, K.R. NZ 9.6
Everitt, G.C. NZ 1.1
Evert, D. US-D 22.1
Evyapan, M. TR 5.2
Ewart, A.J.W. AU 41.1
Ewart, L.C. US-F 14.1.1
Ewertson, G. SE 4.1
Ewing, E.E. US-B 5.1.3
Ewing, L. US-I 15.1
Exconde, O.R. PH 1.4
Exley, R.J.A. GB 15.1.4
Eykel, B.S. NZ 10.1
Eynard, I. IT 24.1, 24.2

Faas, K. DE 29.1
Fabbri, A. IT 9.1
Fabian, M. US-B 14.1.2
Fabre, J-P. FR 3.2.2
Fachinello, J.C. BR 30.3
Facsar, G. HU 2.2.7
Fader, W. DE 22.1
Fadhly, M. ID 4.1
Fadoul, G. US-A 10.1
Faedi, W. IT 19.1.2
Fahy, P.C. AU 9.1.1
Fails, B.S. US-F 14.1.1
Fairlie C., T. PE 2.3
Faiszt, J.A. US-C 10.1
Faita, E. AR 3.1.2
Faivre, A. FR 17.1.2
Faivre-Amiot, A. FR 17.1.2
Falavigna, A. IT 20.1
Falch, E. AT 12.2
Falcon, L.A. US-K 3.1.1
Falesi, I.C. BR 2.1
Falloon, P.G. NZ 13.2
Fâlticeanu, M. RO 3.1
Faluba, Z. HU 2.7
ludi, I. HU 2.2.1
Famelart, M. CA 31.1
Fan, N.T. TW 7.1
Fan, S.J. TW 11.1.1
Fancher, G. US-H 22.1
Fang, T.T. TW 9.1
Fankhauser, F. CH 1.1
Fanourakis, N. GR 10.1
Fantom, C.J. BR 18.1
Fantz, P.R. US-D 8.1.1
Faoro, I.D. BR 28.1
Faquir, M. MA 1.1
Faragher, J. AU 30.1
Faragó, L. HU 8.1
Faraj, M. MA 5.2
Farestveit, B. DK 3.1
Farfán, D. PE 1.3
Fargas, J. CR 2.2
Fargues, J. FR 17.1.6
Fári, M. HU 5.3.2
Farias Bezerril, E. BR 8.1
Faria Vieira, R. BR 26.1
Farina, E. IT 22.1
Farkas, J. HU 5.3.1, 6.1
Farkas, K. HU 2.2.8
Farlow, P.J. AU 18.1
Farmahan, H.L. IN 10.3
Farooqui, A.A. IN 11.2
Farouki, N.A. LY 2.1
Farr, D.J. NZ 13.5
Faseun, A. NG 1.1
Fasidi, I.O. NG 8.1
Fasih, M. IN 26.1
Fassotte, C. BE 6.4
Fassuliotis, G. US-D 12.2
Fasulo, T.R. US-D 30.2
Fatokun, C.A. NG 1.5
Fatta, G. IT 13.1
Fatunla, T. NG 3.1
Faulkner, L.R. US-J 5.1
Faure, A. FR 3.1.7
Faust*, M. US-C 3.1
Faustino, F.A. PH 1.7
Fauvel, G. FR 10.1.2
Fauzy, F. EG 1.2
Favret, E.A. 4.1.1
Favretti, R. US-A 13.1
Fawusi, M.O.A. NG 1.5
Fay, P.K. US-I 18.1
Fayza, E.G. 1.3
Fazekas, I. HU 2.7
Fazinic, M. YU 3.1
Fazuoli, L.C. BR 20.1
Feairheller, S.H. US-B 12.1
Feairheller, S.H. US-K 1.1
Fear, C.D. US-G 20.1
Fedalto, A.A. BR 25.5
Feddes, R.A. NL 9.2
Fehér, A. HU 2.7
Fehér, B. HU 5.1.4
Fehér, T. HU 5.1.4, 5.3.1
Fehér-Ravasz, M. HU 2.2.3
Fehrmann, W. DD 2.1
Feiehtenberger, E. BR 21.2
Feilberg, L. DK 3.3
Feitosa, M.I. BR 21.2
Fejes, J. HU 5.1.3
Felbinger, B. DE 11.1.2
Felczyński, K. PL 28.2.3
Feldmann, A.M. NL 9.5
Feldman, A.W. US-D 30.2
Feldman, W.R. US-H 29.1
Felfoldi, E.M. AU 27.2
Felhős-Vaczi, E. HU 2.2.7
Feliński, T. PL 29.1
Félix Verdugo, C.A. MX 1.6
Felizardo, B.C. PH 1.2.2, 1.5
Feliziani, M. FR 37.1
Fell, W.E. NZ 12.2
Feller, R. IL 1.4.3
Fellers, P.J. US-D 30.2
Fellin, L.R. AR 17.1
Felpeto, C. AR 3.1.1
Felsenstein, G. IL 1.4.5
Felski, J. PL 32.3.3
Fendrik, I. DE 15.3.2
Fenemore, P.G. NZ 9.1
Fenesan, N. RO 16.1
Fenlon, J. GB 18.1.1
Fennell, J.F.M. NZ 12.2
Fenner, R.J. ZI 3.1.1
Fenton, R. GB 32.1
Ferauge, M-Th. BE 6.5
Feree, M.E. US-D 16.1
Ferenczy, A. HU 2.2.9
Ferenczy, T. HU 5.1.1
Ferguson, A.M. NZ 13.4
Ferguson, A.R. NZ 2.1.4
Ferguson, I.B. NZ 2.1.4
Ferhatović, S. YU 16.1
Fermanian, T.W. US-F 12.2.2
Fermor, T.R. GB 18.1.2
Fernandes, A.M. PT 6.1
Fernández, C.E. CR 1.1
Fernández, E.C. PH 1.7
Fernández, F. AR 22.1
Fernández, F.M. AR 21.1
Fernández, J.M. ES 7.2
Fernández, N.C. PH 1.5
Fernández, R. AR 22.2
Fernández, R.V. AR 22.2
Fernández Lozano, J. AR 3.1.1
Fernańdez Montes, M.R. MX 8.1

Fernando, D.N.S. LK 2.1
Fernando, K.K.S. LK 3.2
Fernando, M.H.J.P. LK 3.2
Fernando, U. LK 2.1
Fernqvist*, I. SE 1.1.1
Ferrandes, P. FR 3.2.3
Ferrara, E.T. IT 2.1
Ferrari, P. IT 21.1
Ferrari, R. IT 9.3
Ferrari, V. IT 20.1
Ferrario, S. FR 2.1.1
Ferrato, J. AR 16.1
Ferraz, E. BR 10.1
Ferraz, E.C. de A. BR 14.1
Ferraz, J.F. PT 6.1
Ferree*, D. US-F 7.1.2
Ferreira, A.A.F. BR 30.3
Ferreira, B.S.C. BR 27.1
Ferreira*, D.I. ZA 3.1
Ferreira, E. BR 26.1
Ferreira, F.A. BR 15.1
Ferreira, F.R. BR 25.3
Ferreira, J.Ms. BR 11.1
Ferreira, J.T. PT 1.1
Ferreira, L.R. BR 18.1
Ferreira, M.T.T. BR 29.2
Ferreira, V.L.P. BR 20.2
Ferreira Antunes, I. BR 26.1
Ferreira Cintra, A. BR 21.2
Ferreira da Silva, R. BR 17.1
Ferreira de Lima, M. BR 11.1
Ferreira Mota, I. BR 13.1
Ferrer de Almeida, H. BR 7.1
Ferrèro, F. FR 2.1.7
Ferretti, P.A. US-B 13.1.1
Ferriera, J.F. AR 3.1.2
Ferro Frazão, J.M. BR 7.1
Ferron, P. FR 17.1.6
Ferronatto, E.M. de O. BR 14.1
Ferry, P. FR 3.1.8
Fery*, R.L. US-D 12.2
Festić, H. YU 1.1
Feucht, J.R. US-I 2.2
Feucht*, W. DE 11.1.1
Feutry, A. FR 34.1
Feyen, J. BE 9.1
Fiala, F. AT 8.7
Fiala, V. FR 17.1.5
Fic, M. CS 1.1
Fic, V. CS 14.1.7, 18.1
Fica, M. PL 28.1.4
Fida, P. AT 8.11
Fidan, A.E. TR 14.1
Fidan, F. TR 14.1
Fidan, Y. TR 2.1
Fideghelli, C. IT 19.1
Fieldhouse*, D.J. US-C 2.1
Fielding, A.H. GB 19.1.1
Fierro, C.A. US-N 1.2
Fies, J-C. FR 3.1.7
Figaj, J. PL 10.1.1
Figiel, T. PL 28.1.10
Figueira, A. dos R. BR 16.1
Figueiredo, P. BR 21.2
Figueiredo Fortes, J. BR 30.1
Figueiredo Pereira, J. BR 30.1
Figueroa, M. VE 1.1
Figueroa, P.N. AR 17.1
Figura, A. PL 32.3.3
Filgueira, F.A.R. BR 26.2
Filho, A. de C.V. BR 14.1
Filho, E.L.R. BR 25.3
Filho, H.P.S. BR 12.1
Filho, J.G. BR 31.1
Filho, O.P. BR 20.1
Filho, R.V. BR 24.1
Filho, W. dos Santos s. BR 12.1
Filipowicz, A. PL 14.1.8
Filippone, E. IT 18.1
Filiti*, N. IT 3.1
Filius, I. HU 5.1, 5.1.3
Fils, B. FR 3.1.8
Finardi, N.L. BR 30.1
Finch, S. GB 33.1.3
Finch-Savage, W.E. GB 33.1.5
Findlay, E. GB 9.1.3
Findlay, W.I. CA 17.1
Fine, G.A. GB 19.1.4
Finger, H. DE 23.1
Finnamore, D.B. CA 40.1
Finner, M. US-F 23.1
Fiori, B. US-B 2.1.6
Fiori, B.J. US-B 2.1.3
Fiorino, P. IT 9.2
Fırat, A.E. TR 7.1
Firestone, M. US-K 3.1.3
Firmin, J.L. GB 24.1.2
Firpo, J.T. AR 16.1
Fischer, C.C. US-B 5.1.1
Fischer, I. HU 5.3.2
Fischer, M. DE 12.1.8
Fischer, P. DE 11.2.1
Fischer-Kolbri, P. AT 8.11
Fisher, D.J. GB 19.1.1
Fisher, J.F. US-D 30.2
Fisher, J.B. US-D 33.1
Fisher, J.M. AU 42.1
Fisher, J.P. GB 32.1
Fisher, J.T. US-H 19.1
Fisher*, K.H. CA 25.1
Fisher*, K.J. NZ 9.1
Fitter, D.J. GB 18.1.1
Fitzell, R.D. AU 11.1
Fitzgerald, J.B. US-G 23.1
FitzGerald, S.M.D. GB 16.1
Fitzpatrick, G.E. US-D 25.1
Fiume, F. IT 20.1
Fjeld, T. NO 10.1.4
Fjelddalen, J. NO 10.4, 10.4.3
Flaga, J. PL 22.1
Flannery, R.L. US-B 14.1.7
Flatt, W. US-D 16.1
Flegg, J.J.M. GB 8.1.2
Flegmann, A.W. GB 3.1
Flehr, J.B. AU 31.1
Fletcher, B.S. AU 65.5.1
Fletcher, G.C. AU 31.1
Fletcher, G.G. NZ 2.1.3
Flett, S. AU 32.1
Flint*, H.L. US-F 9.1.4
Flønes, M. NO 8.1
Floor, H. NL 9.4.5
Flora, L.F. US-D 20.1
Flore*, J.A. US-F 14.1.1
Florea, S.-R. RO 12.1
Flores*, A. VE 2.1
Flores, L. US-N 1.2
Flores, M. AR 7.1
Flores Cantillano, R.F. BR 30.1
Flores G.*, A. VE 2.1
Flores Reyes, I. MX 14.1
Flores Ricarde, A. MX 4.6
Florescu, C. RO 11.1
Florescu, E. RO 8.1
Florián, P. ES 15.1
Floyd, R.M. AU 59.1
Flueraru, M. RO 8.4.5
Fober, H. PL 10.1.1
Focsǎneanu, P. RO 8.5
Foda, S. EG 1.3
Fodor, L. HU 8.1
Fodor, P. HU 2.2.3
Foguet, J.L. AR 22.3
Folchi, A. IT 3.3
Földesi, D. HU 1.1
Foley, A.G. GB 31.2.3
Foley, M.E. US-I 18.1
Folk, Gy. HU 2.2.8
Folsom, J. US-K 15.1
Fölster, E. DE 15.1.3
Fomenko, Y.E. SU 16.1
Fonnesbech*, M. DK 3.1
Fonseca, M.A. de C. BR 10.1
Fonseka, M.H.D. LK 3.12
Fontan, H.M. AR 7.1
Fontanazza, G. IT 15.1
Fontemachi, E.C. AR 18.1
Fonteno, W.C. US-D 8.1.1

Fontenot, J.F. US-D 52.1
Fontes Urben, A. BR 25.3
Fontes Vieira, R. BR 25.3
Forbes, A.R. CA 4.1
Forbes, J.J. AU 44.1
Forbes, R.B. US-D 36.1
Forbus, R. US-D 16.2
Forchthammer, L. DE 11.2.1
Ford, H.W. US-D 30.2
Ford, R.E. US-F 12.2.3
Fordham, R. GB 35.1.2
Fores de Castro, L.L. BR 18.1
Fornaris, G. US-N 1.2
Fornasier, J.B. BR 20.1
Fornavier, M. BR 18.1
Forney, C.F. US-K 7.3
Forrest, J.M.S. GB 15.1.1
Forró, E. HU 2.2.11
Fors, L. SE 4.1
Forsberg, J. US-F 23.1.1
Forshey, C.G. US-B 2.1.2
Forshey, C.G. US-B 4.1
Forsline*, P.L. US-B 2.1.6
Forster, G.F. GB 31.2.8
Forster, R.L. US-I 16.1
Forster, R.L.S. NZ 2.3.4
Førsund, E. NO 10.4.1
Fortes, J.M. BR 17.1
Fortgens, G. NL 2.1
Forti, M. IL 4.1
Fortney, R.F. US-B 13.1.1
Fortugno, C. AR 4.1.4
Foster, S.P. NZ 2.2
Foster II, R.E. US-H 25.1
Foti*, S. IT 5.3, 5.4
Fouarge, G. BE 10.1
Foucard, J.C. FR 27.1
Fouilloux, G. FR 17.1.1
Fountain, J. US-J 10.2
Fountain, W. US-E 15.1
Fourcy, A. FR 31.1
Fournet, J. FR 18.1
Fournier, D. FR 2.1.5
Foury, C. FR 17.2
Fowler, M. NZ 2.3.3
Fowler, V.W. GB 3.1
Fox, H. CA 9.1
Fox, R.A. GB 15.1.1
Fox, R.T. US-B 5.1.1
Fox, R.T.V. GB 29.1
Fox, W.R. US-D 50.1
Fráguas, J.C. BR 31.1
França, F.H. BR 25.5
França-Dantas, M.S. BR 11.1
Franca Neto, J.B. BR 27.1
Francile, S. AR 10.1
Francki, R.I.B. AU 42.1
Franco, C.F.O. BR 9.1
Franco, S. PE 2.3
Frangi, H. AR 19.1
Frank, H.K. DE 18.1.1
Frankel, H. IL 1.4.6
Frankel, R. IL 1.1.3
Frankie, G.W. US-K 3.1.1
Franz*, Ch. DE 11.1.1
Frappell, B.D. AU 43.1
Fraselle, J. BE 6.8.4
Fraser, J.G. GB 1.3
Fraser, L.G. NZ 2.1.1
Fraser, R.S. GB 33.1.1
Frazer, D.B. CA 4.1
Fredericq, H. BE 7.12
Fredriksson, P. SE 5.2
Freeland, R.D. US-E 1.1
Freeman*, B.M. AU 6.1
Freeman, F.W. US-F 20.1
Freeman*, J.A. CA 1.1
Freeman, T.E. US-D 27.6
Fregoni, M. IT 16.1
Freire, A.B. BR 26.1
Freire, L.B. BR 14.1
Freitas, L. PT 6.1
Freitas, R. PT 6.1
French, C.J. CA 2.1
French, E.W. US-G 18.1
French, J. CR 2.2
French, W.J. US-D 35.1
Frenkel, C. US-B 14.1.1
Frent, I. RO 2.1
Frenz*, F-W. DE 11.2.5
Frenzel, A. DE 15.5
Fresnaye, J. FR 3.1.9
Fretheim, A.A. NO 10.6
Frett, J. US-H 6.1
Fretz, T.A. US-C 10.1
Freund, H. DE 15.1.4
Freve, A. CA 30.2
Frey, D.R. US-C 2.1
Friberg, P. SE 1.2
Frick, H. US-C 2.1
Fridrich, J. CS 20.1
Friess, H. DE 23.1
Frigyesy, F. HU 5.1.4
Friis, K. DK 1.2
Frith*, G. AU 30.1
Fritsche, G. NL 5.1
Fritschen, R.D. US-G 26.1
Fritton, D.D. US-B 13.1.1
Fritz*, P.D. DE 11.1.1
Frochot, H. FR 11.1.2
Fromberg, H. CR 2.3
Frosi, J.F. BR 28.1
Frossard, J.S. FR 5.1
Frost, A.R. GB 31.2.2
Frost, R.R. GB 22.1
Fruhlich, G. AT 7.1
Frusciante, L. IT 18.1
Frutal, M. AR 22.1
Fry, W.E. US-B 5.1.8
Fuchs, H.W.M. NL 9.7.1
Fuchs*, Y. IL 1.3.1
Fuentes, J.C. AR 20.1
Fugel, S. RO 8.4.1
Füglistaller, P. IC 1.1
Fujieda, K. JP 7.4
Fujihara*, K. JP 33.2
Fujii, H. JP 27.3
Fujii, K. JP 14.4
Fujii, S. JP 31.1, 42.2
Fujii, T. JP 44.1
Fujii, Y. JP 2.2
Fujikawa, Y. JP 1.1.5
Fujime*, Y. JP 17.3
Fujimori, M. JP 26.4
Fujimoto, J. JP 2.1
Fujimoto, K. JP 22.5, 44.2
Fujimoto, T. JP 9.2
Fujimura, T. JP 13.1
Fujino*, M. JP 13.1
Fujino, M. JP 16.3
Fujino, N. JP 14.1
Fujioka, T. JP 44.1
Fujisawa, I. JP 23.2.4
Fujisawa, O. JP 16.1
Fujisawa, T. JP 31.1
Fujise, M. JP 15.1
Fujishige, N. JP 39.2
Fujishita, N. JP 33.2
Fujita, B. JP 6.2
Fujita, F. JP 17.2
Fujita, K. JP 3.3.1, 15.1, 18.1, 21.2
Fujita*, M. JP 44.1
Fujita, S. JP 34.4
Fujita, Y. JP 20.2
Fujitomi, M. JP 13.2
Fujiwara, S. JP 13.1
Fujiwara, T. JP 11.2, 13.1 16.1
Fujiyama, T. JP 27.1
Fukai, H. JP 31.1
Fukai, S. JP 33.1
Fukaya, M. JP 1.1.1, 2.2
Fukazawa, H. JP 39.1
Fukuda, C. BR 12.1
Fukuda, E. JP 39.1
Fukuda, H. JP 16.2
Fukuda, S. JP 36.2, 46.1
Fukuda, W.M.G. BR 12.1
Fukumoto, M. JP 16.2
Fukumoto, Y. JP 20.4
Fukunaga, M. JP 1.1.4, 11.1
Fukunaga, S. JP 33.2
Fukuoka, N. JP 15.2

Fukuoka, S. JP 40.1
Fukusaki, G. PE 1.2
Fukushima, C. JP 3.2.3
Fukushima, K. JP 13.1, 43.2
Fukushima, M. JP 14.6
Fukushima*, T. JP 45.2
Fukuta, K. JP 34.1
Fukuta, M. JP 1.1.3
Fukutomi, F. JP 9.3
Fukuyama, T. JP 5.3
Fukuyo, K. JP 38.2
Fukuyori, Y. JP 37.1
Fukuzumi, H. JP 33.2
Fuleki, T. CA 25.1
Fullerton, R.A. NZ 2.3.3
Fulmer, J.P. US-D 13.1
Fumoto, J. JP 33.3
Funagami, K. JP 5.2
Funahashi, H. JP 19.2
Funakoshi, K. JP 38.1
Funakoshi, T. JP 11.1
Funderburk, J.E. US-D 37.1
Funes, N.P. AR 23.1
Funk, Jr. C.R. US-B 14.1.7
Funnell, K.A. NZ 9.1
Funt*, R.C. US-F 3.1
Fürer, W. CH 1.4
Füri, J. HU 5.2
Furness, G.O. AU 38.2
Fürst, D. DE 15.2.3
Furtado, M.J. BR 18.1
Furuhashi, K. JP 38.2
Furui, S. JP 11.2
Furukawa, Y. JP 22.5
Furumoto, O. BR 25.5
Furundžić, M. YU 14.1
Furuta, O. JP 42.1
Furuya, H. JP 11.1
Furuya, S. JP 7.3, 47.2
Furuya, T. JP 35.2
Fuse, G. JP 33.3
Fushihara, H. JP 7.1
Fushihara, J. JP 44.1
Fushtey, S.G. CA 1.1
Füstös, Zs. HU 5.3.2
Futami, S. JP 19.2
Futatsudera, T. JP 9.1
Fuzisaki, M. JP 18.2
Fyhn, P.G. NO 10.5.2
Fykse, H. NO 10.4.2
Fytizas, E. GR 14.1.2
Fytizas, R. GR 14.1.4

Gaafar, A.K. LY 2.1
Gaash*, D. IL 1.2.2
Gaaziero, D.L.P. BR 27.1
Gabelman, W.H. US-F 23.1.3
Gabillard, D. FR 35.1
Gabriels, R. BE 14.1
Gabrielsen, E. NO 10.1.8
Gabryszewska, E. PL 28.1.17
Gachanja, S.P. KE 1.3
Gad, A. IL 2.1
Gadelha, R.S.S. BR 19.1
Gadewar, A.B. IN 4.1
Gage*, J.F. AU 18.1
Gagen, S. AU 27.1
Gagnaire, J. FR 31.1
Gagro, I. YU 3.1
Gaikwad, M.R. IN 14.2
Gaioso Cardoso, R.M. BR 21.2
Gajaraja, C.P. IN 4.1
Gajc, J. PL 32.1.9
Gajdos, R. SE 1.1.3
Gajewski, E. PL 16.1
Gajewski, M. PL 32.3.4
Galambos, J. HU 2.2.4
Galardo, V.R.B. BR 29.2
Galarraga, I.M. AR 15.2.3
Gale, J. IL 5.1, 5.1.1
Galerani, P.R. BR 27.1
Galichet, P. FR 3.1.9
Galicki, A. PL 2.1
Galindo, J.J. CR 2.2
Gallander, J.F. US-F 7.1.2
Gallardo, G.S. AR 13.1
Gallasch, P.T. AU 38.2
Gallegly, M. US-C 9.1
Galletta, G. US-C 3.1
Gallov, A. RO 12.1
Galmarini, H.R. AR 15.1
Galoux, M. BE 6.7
Gałuszka, H. PL 11.1.6
Gälzer, R. AT 8.5
Gamabrants, N.L. US-A 5.1.1
Gamada, K. JP 23.2.3
Gamarra, F. PE 4.2
Gamboa, B.S. AR 14.1
Gameel, O. SD 3.1
Gamietea, R. AR 19.1
Ganapathy, M.M. IN 1.1.1
Ganczaugh, L. HU 2.1.5
Gandar, P.W. NZ 8.1
Ganeshan, S. IN 1.1.9
Ganeson, S. AU 59.2
Gapinski*, M. PL 20.2.11
Gapsa, F. RO 1.1.2
Gapska, F. RO 1.1.2
Garami, M. HU 2.2.11
Garbaye, J. FR 11.1.1
Garcia, A. BR 27.1, 30.1
Garcia, A. MX 16.3
Garcia, A.F. AR 9.1
Garcia, C. PE 1.1
Garcia, C.D. AR 11.1
Garcia, J.L.M. BR 20.2
García, J.R. BR 14.1
García, M.A. AR 22.1
García, M.F. AR 15.1
García, R. US-K 3.1.1
García, S.M. AR 16.1
García Bessa, J.M. BR 10.1
García Gil de Bernabe, A. ES 9.1
García Salazar, C. MX 5.2
García-Santibañez S., J.M. MX 6.3
Gardan, L. FR 1.1.3
Gardé, A. PT 6.2
Gardner, R.G. US-D 6.1
Gäredal, L. SE 8.1, 8.1.1
Gargiulo, A.A. AR 20.1
Gargiulo, C.A. AR 22.3
Garibaldi, N.L. BR 30.3
Garić, R. YU 15.1
Garner, J. US-D 50.1
Garnet, H.M. ZA 3.4
Garnsey, S.M. US-D 34.1
Garrán, S.M. AR 6.1
Garrec, J.P. FR 31.1
Garrett, C.M.E. GB 8.1.2
Garrett, R. AU 27.1
Garrett, W. US-D 16.1
Garrido Ramírez, R.S. MX 11.2
Garrison, S.A. US-B 14.1.1
Gärtner, G. AT 2.1
Gartner, H. AT 20.1
Garton, R.W. CA 17.1
Gary, C. FR 3.1.4
Garza López, J.G. MX 3.3
Garzoli, K.V. AU 65.6
Garzón Tiznado, J.A. MX 7.2
Gasia, M.C. BE 6.8.11
Gaspard, J. BE 6.9
Gasparotto, L. BR 1.1
Gast, R.G. US-F 14.1
Gata, N.R. ZI 3.1.2
Gately, T. IE 6.1
Gates, L.F. CA 17.1
Gathee, J. KE 1.8
Gathercole, F.J. AU 38.2
Gathungu, C.N. KE 1.3
Gatica de Mathey, M.E. AR 15.1
Gaudiel, R. CA 6.1
Gaul, S.O. CA 41.1
Gaunce, A.P. CA 3.1
Gaur, G.N. IN 22.1
Gaus, A.E. US-E 18.1
Gava, A. RO 9.1
Gaviola, J.C. AR 13.1
Gavrilović, J. YU 10.2
Gavriluc, M. RO 1.1.6

Gawish, R. DE 5.1.1
Gawroński, E. PL 14.1.4
Gawroński, S. PL 32.1.2.1
Gay, G. IT 24.2
Gaynor, J.D. CA 17.1
Gàzdaru, S. RO 4.1
Gazelas, C. GR 14.1.2
Gazit*, S. IL 2.1
Gazitt*, S. IL 1.2.5
Gbile, Z.O. NG 1.3
G"bova, R. BG 8.2
Gé, J.P. FR 34.1
Geard, I.D. AU 46.3
Gebhardt, DE 11.1.1
Gebre Selassie, K. FR 3.1.5
Géczi, L. HU 3.1.1
Geels, F.P. NL 5.1
Geenen, J. BE 15.3
Geertsen, V. DK 1.2
Geier, T. DE 12.1.10
Gelencsér, É. HU 2.4
Gellatley, J.G. AU 9.1.3
Gembicka, A. PL 28.2.11
Genc, E. TR 3.2
Genchev, S. BG 8.1
Gencturk, S. TR 19.1
Gendraud, M. FR 28.1
Genest, J. CA 27.1
Genma, H. JP 14.6
Gent, G.P. GB 27.1
Gentner, W.J. US-C 3.1
Gentry, Jr., H.O. US-D 10.1
Georgakopoulou-Voyiatzi, Chr. GR 25.1.3
George, A.J. GB 6.1
George, A.P. AU 23.1
George, K.S. GB 13.1.1
George, M.K. IN 5.3
George*, R.A.T. GB 3.1
Georgeades, G. US-K 14.1
Georgiadis, C. GR 4.1
Georgiadis, S. GR 25.2.1
Georgiev, A. BG 7.1
Georgiev, Ch. BG 13.1
Georgiev, D. YU 4.1
Georgiev, V. BG 5.1
Georgopoulos, S.G. GR 6.1.7
Geraldson, C.M. US-D 24.1
Gérard, M. MA 1.1
Gerber, J.M. US-F 11.1, 12.2.2
Gergely, I. HU 2.1.2
Gergely, K. HU 2.6
Gergerich, R. US-E 11.1
Gerlach, R. DE 19.1
Gerlagh, M. NL 9.4.2
Germain, E. FR 4.1.1
Germain, R. FR 34.1
Germanà, G. IT 5.1
Gerrits, J.P.G. NL 5.1
Gerson, U. IL 2.5
Gertsson, U. SE 1.1.3
Gethi, M. KE 3.1
Getz, W.M. US-K 3.1.1
Geypens, M. BE 9.2
Gezerel, O. TR 1.1
Ghai, S.K. IN 3.1
Ghanem, I.B. CA 40.1
Gheorghe, A. RO 1.1.4
Gheorghe, F. RO 5.1
Gheorghe, M. RO 1.1.8
Gherardi, P.B. AU 52.1
Gherardi, S. IT 14.1
Gherghi, A. RO 8.4.1
Gherghi, E. RO 3.1
Gherman, N. RO 1.1.1
Ghidiu, G. US-B 14.1.4
Ghisleni, P.L. IT 10.1
Ghosh, B. IN 28.1
Ghosh, D. IN 28.1
Ghosh, S. IN 28.1
Ghosh, S.N. IN 28.1
Ghosh, S.P. IN 5.1
Ghunke, S.M. IN 14.3
Giacomelli, E.J. BR 20.1
Giaever, H. NO 10.1.1
Giallelis, G. GR 23.1
Gianfagna, T.J. US-B 14.1.1
Gianinazzi, S. FR 7.1.7
Gianinazzi-Pearson, V. FR 7.1.7
Giannopolitis, C.N. GR 14.1.3
Giarrizzo, L. AR 15.3
Gibbons, F.D. US-G 31.1
Gidrol, X. FR 4.1.4
Giertych, M. PL 10.1.1
Gijzen, J.G. NL 9.1
Gil, P. AR 18.1
Gil, V.L. BR 2.1.1
Gilbert, E.G. GB 19.1.4
Gilbert, F.A. US-F 25.1
Gilbert, W.S. AU 9.1.2
Gilbey, D.J. AU 59.1
Gilchrist, A.N. NZ 8.3
Giles, F.A. US-F 12.2.2
Gilet, R. FR 31.1
Gilissen, L.J.W. NL 9.5
Gill, P.A. GB 15.1.4
Gill, R.P.S. IN 17.1
Gill, S.S. IN 17.1
Gillard, A. BE 7.5
Gilles*, G. BE 19.1
Gillespie, D. CA 2.1
Gillet, J. BE 6.9
Gilliam, C.H. US-D 38.1
Gillis, J. US-A 9.3
Gilly, G. FR 2.1.1
Gilmore, J. US-F 17.1
Gilreath, J.P. US-D 24.1
Gimenes Calbo, A. BR 25.5
Gimeno, A. ES 12.1
Gindrat, D. CH 2.1
Ginoux, G. FR 3.1.10
Ginsburg, H. IL 1.1.2
Giokas, G. GR 23.1
Giorgessi, F. IT 7.1
Giorgota, M. RO 7.1
Girard, E. CO 6.2
Girard, J.-M. CA 38.1
Girardin, P. FR 6.1.1
Gîrbu, S. RO 8.4, 8.4.1
Girija, V. IN 1.1.6
Giroto, R. AR 17.1
Giroux, M. CA 29.1, 36.1
Gislerød, H.R. NO 10.1.4
Gitiatis, R. US-D 22.1
Gitungo, F.N. KE 5.1
Giugà, I. RO 4.1
Giulovo, C. IT 12.1
Giura, P. RO 24.1
Giurea, M. RO 8.4.1
Given, N.K. NZ 10.1
Giyotoku, H. JP 21.2
Gjaerum, H.B. NO 10.4.1
Gladon, R.J. US-G 20.1
Glàman, R. RO 8.4.3
Glandard, A. FR 3.1.5
Glapś, T. PL 28.2.5
Glas, D.J. NL 19.1
Glaser, T. PL 20.2.6
Glàvan, L. RO 7.1
Glaze, N. US-D 22.1
Glen, D.M. GB 19.1.1
Glendinning, D.R. GB 15.1.2
Glenn, D.M. US-C 8.2
Glenn, T. AU 38.2
Glennie, J.D. AU 23.1
Glits, M. HU 2.2.8
Glofke, E. AT 8.11
Glogowska, B. PL 28.2.11
Glorie, K. BE 18.1
Głowacki, P. PL 32.1.12
Glubrecht, H. DE 15.3.1
Glucina, P.G. NZ 2.1.1
Go, T. JP 9.1
Goater, S.P. GB 27.1
Goblet, J.P. BE 12.1
Göblös, G. HU 4.2
Goddard, G.B. US-A 5.1.1
Godden*, G.D. AU 30.2
Goddrie, P.D. NL 9.9, 10.1
Godeanu, I. RO 19.2
Godini*, A. IT 2.1
Godley, G.L. AU 51.1

Godoy Avila, C. MX 5.2
Godwin, R.J. GB 31.1
Godyń, A. PL 28.1.6
Goedegebure, J. NL 10.1
Goell*, A. IL 1.2.3
Goffinet, R. BE 4.1
Goffings, G. BE 21.1
Gogoci, I. RO 5.1
Gogoci, V. RO 5.1
Gogova, K. BG 8.2
Goheen, A.C. US-K 6.1.1
Goi*, M. JP 17.3
Goidanich, G. IT 3.3
Gojnić, C. YU 5.1
Gokcay, E. TR 22.1
Golan, A. IL 5.1.2
Golcz, A. PL 20.2.3
Goldbach, H. DE 11.1.3
Golden, J.K. US-D 14.1
Goldsberry, K.L. US-I 2.2
Goldsborough, P. US-F 9.1.4
Goldschmidt*, E. SE 1.1.1
Goldschmidt, E.E. IL 2.1
Goldwin, G.K. GB 35.1.2
Goldy, R.G. US-D 8.1.1
Gołebiowska, Z. PL 20.1.1
Golenia, A. PL 20.1.2
Goliáš, J. CS 14.1.6
Golisz, A. PL 32.3.4
Gołos, J. PL 32.1.3
Golović, M. YU 7.1
Gólya, E. HU 5.1.3
Gombköto, G. HU 2.2.3
Gomes, G.C. BR 25.5
Gomes, J.A. BR 18.1
Gomes dos Santos, Z. BR 11.1
Gomes dos Santos Filho, B. BR 30.3
Gomes Marques, M. BR 19.1
Gomes Pereira, J. PT 1.1
Gomez Riera, P. AR 15.1
Gomi, K. JP 4.4
Gonçalves, A.L. BR 21.1
Gonçalves Barreira, J. BR 8.1
Gonçalves de Assis, V.L. BR 30.3
Gonçalves Neto, L. BR 10.1
Gönczy, J. HU 2.4
Gonda, I. HU 2.1.4
Gondwe, W.T. ML 1.1
Gonnet, M. FR 3.1.9
Gonsalves, D. US-B 2.1.4
Gonulsen, N. TR 15.2
Gonzálex Hernandez, M.A. MX 8.3
González, A.R. US-E 11.1
González, H. PE 2.1
González, J. AR 19.1
González, J.L. AR 22.3
González, M. AR 16.1
González, M.A. AR 22.1
González, M.L. AR 15.1
González, R.J. MX 1.6
González, T. AR 7.1
González, W. CR 2.2
Gonzalez Caldwell, S.N. AR 15.1
González Gaona, E. MX 6.6
González González, J.M. MX 1.2
González Martínez, J.A. MX 7.2
Gonzalez Terán, C. AR 22.3
Good, G.L. US-B 5.1.1
Goode, M.J. US-E 11.1
Gooding, G.V. US-D 8.1.2
Gooding*, H.J. GB 2.1
Goodwin, P.B. AU 9.3
Goodwin, S. AU 5.1
Goonewardene, H.F. US-F 9.1.3
Goos, M. PL 33.2
Gopalakrishnan, P.K. IN 5.3.1
Gopalakrishna Rao, K.P. IN 1.1.5
Goppel, Ch. DE 11.12
Goral, T. TR 3.1
Gorda, J.L. FR 20.1
Gordon, H.T. US-K 3.1.1
Gordon, I.L. NZ 9.3
Gordon, S.C. GB 15.1.1
Górecka, K. PL 28.2.12, 32.1.5
Górecki, K. PL 28.2.12
Goren, R. IL 2.1
Goren*, M. IL 1.2.5
Gorin, N. NL 9.10
Gormley, T.R. IE 4.1
Gorrochetegui, M. AR 23.1
Gorske, S.F. US-F 3.1
Gortzig*, C.F. US-B 5.1.1
Gosiewski, W. PL 32.1.2.1
Gospodinova, M. BG 8.2
Gosse, G. FR 17.1.7
Gosselin*, A. CA 38.1
Goswami, A.M. IN 2.1.2, 2.1.4
Goszczyńska*, D. PL 28.1.17
Goszczyński, W. PL 28.1.2, 32.1.7
Goth, R.W. US-C 3.1
Goto, A. JP 38.3
Goto, H. JP 22.2
Goto*, M. JP 9.3
Goto, R. BR 13.1
Goto, S. JP 30.2
Goto, T. JP 30.2
Goto, Y. JP 23.2.5, 25.1.2
Götz, G. DE 30.1
Gotzé, W.A.G. ZA 4.1
Goubran, F. AU 30.1
Gougas, E. GR 18.1.3
Gough, N. AU 15.3
Gough, R.E. US-A 14.1
Gouider, J. TN 1.1
Gouin, F.R. US-C 5.2
Goulart, B. US-B 14.1.1
Goulas, J.P. FR 4.1.3
Gould, H.J. GB 13.1.1
Gould, W. US-C 5.2
Goulpaud, J. FR 35.1
Gouramanis, G. GR 26.1
Goussard, P.G. ZA 4.3.2
Gouws, L. ZA 4.1
Goverd, K.A. GB 19.1.4
Goyal, R.S. IN 18.3
Grabowska, B. PL 28.1.12
Grabowski, K. PL 20.2.5
Graca, J. BR 19.1
Grace, A.B. NZ 2.3
Gracia, O. AR 15.1
Graf, H. DE 17.1
Grafiadellis*, M. GR 25.2.1
Grafius, E. US-F 14.1.2
Graft, A. AT 8.7
Graham, C. CA 16.1
Graham, D. AU 65.4
Graham, D.C. GB 9.1
Graham, D.R. US-I 19.1
Graham, E.T. US-E 3.1
Graham, P. AU 59.2
Graifenberg, A. IT 17.2
Grajewska-Wieczorek, A. PL 28.2.3
Grana, E. ES 13.2
Granada, G.A. CO 4.1
Granberry, D. US-D 17.1
Grande Teixeira, M. BR 26.1
Grandi, M. IT 3.1
Grandison, G.S. NZ 2.2
Granger, J. FR 19.1
Granger, R.L. CA 35.1
Granger, S. US-K 2.1
Granges, A. CH 2.2
Granqvist*, G. SE 1.1.2
Grant, G. CA 6.1
Grant, W.J.R. AU 65.1
Grantz, D.A. US-K 11.1
Gras, R. FR 2.1.1
Grasselli, L. IT 7.3
Grassely, Ch. FR 3.1.3
Grassi, G. IT 19.1
Grassotti, A. IT 22.1.1
Gratte, H.V. AU 51.1
Grauke, L.J. US-D 59.1
Graumlich, T.R. US-D 30.2

Grauslund, J. DK 1.3
Graves, B. US-D 49.1
Graves Jr., C.H. US-D 50.1
Graves, C.J. GB 18.1.1
Graves, J.B. US-D 52.1
Gray*, D. GB 33.1.5
Gray, V.P. CA 25.1
Grbic, O. YU 17.1
Greber, R.S. AU 15.4
Green, C.P. GB 35.2
Green, E.J. US-B 14.1.1
Green, J.J. US-N 1.2
Green, R.J. US-F 9.1.1
Greenblatt, I. US-A 13.1
Greene, D.W. US-A 5.1.1
Greene*, G.M. US-B 9.1, 13.1.1
Greenhalgh, F. AU 27.1
Greenhalgh*, W.J. AU 9.3
Greenwood, D.J. GB 33.1.2
Greffet, A. FR 34.1
Gregg, D.R. US-B 10.1
Gregoriou, C. CY 1.1
Gregory, J. US-H 18.1
Gregory, J.F. US-D 27.3
Gregory, P. US-B 5.1.7
Greidanus, P. NL 9.10
Greig*, J.K. US-G 31.1
Grellier, B. FR 19.2
Grelon, J. FR 19.5
Gremminger, U. CH 1.1
Grewal, J.S. IN 4.1
Grewal*, S.S. IN 17.1
Grice, M.S. AU 47.4
Grierson, D. GB 25.1.2
Griesbach, R. US-C 3.1
Grieve, A.M. AU 4.1
Griffey, W.A. US-D 40.1
Griffin, G.D. US-I 5.1
Griffin, G.F. US-A 13.1
Griffith, M. US-L 1.1
Griffiths, C. GB 9.1.3
Grigoletti Jr., A. BR 31.1
Grigoriu, A. GR 6.1.7
Grigorov, Y. BG 8.2
Grillo, S. IT 18.1
Grimm, G.R. US-D 34.1
Grimm, R. CH 1.1
Grimstad, S.G. NO 4.1
Gripp, A. BR 25.3
Gripwall, E. SE 1.4
Gritton, E.T. US-F 23.1.1
Grjazev, V.A. SU 17.1
Grochowska*, M. PL 28.1.5
Grodzinski, B. CA 15.1
Grodzki, L. BR 27.2
Groen, L.E. NL 9.7.5
Groenewegen, C.A.M. NL 1.1
Gromisz, M. PL 28.1.8
Gromisz, Z. PL 7.1
Gronek, M. PL 6.1
Groot, K.P. NL 6.1
Grosclaude, C. FR 3.1.5
Groshkov, I. BG 13.2
Grospierre, P. FR 43.1
Gross, K. US-C 3.1
Grosulescu, E. RO 1.1.2
Grouet, D. FR 17.1.2
Groven*, I. DK 2.1
Grozdin, A. BG 8.2
Grubben*, G.J.H. NL 6.1
Gruber, D.J. CA 9.1
Gruca, Z. PL 20.2.9
Grudzień, K. PL 28.2.3
Gruičić, G. YU 8.1
Grun, P. US-B 13.1.1
Grunewaldt, J. DE 15.1.1, 15.3.2
Grunwald, C. US-F 12.1.1
Gruppe, D.W. DE 13.1
Gruttadaurio, J. US-B 5.1.1
Grzegorzewska, M. PL 28.2.9
Grzesik, M. PL 28.1.14
Grzyb, Z. PL 28.1.1
Guadalupe, A. CR 2.3
Gubicza, I. HU 5.3.2
Gubler, W.D. US-K 6.1.1
Guclu, K. TR 9.1
Guedes, A.C. BR 25.5
Guedry, L.J. US-D 52.1
Guennelon, R. FR 3.1.7
Guerif, M. FR 3.1.4
Guermach, A. TN 1.1
Guerrero, M. PE 2.2
Guerrero Moreno, A. MX 7.2
Guerrero Palomino, H. MX 10.3
Guerrero Prieto, V.M. MX 5.4
Guerriero*, R. IT 17.2
Gugerli, P. CH 2.1
Guichard, E. FR 7.1.4
Guiheneuf, Y. FR 19.5
Guilford, P.J. NZ 2.3.4
Guinot, J-P. FR 3.1.4
Gulaid, H.H. LY 2.1
Gulcan, R. TR 15.1
Guldager, S. DK 3.2
Guleryuz, M. TR 9.1
Gull, D.D. US-D 27.7
Gunasekera, T.G.L.G. LK 2.1
Gunatilake, H.A.J. LK 2.1
Gunatileke, G.A. LK 3.7
Gunatileke, K.G. LK 1.1
Gunawardene, S.D.I. LK 3.1
Gunay, A. TR 2.1
Gundacker, E. AT 7.1
Gundry, C.S. GB 19.1.3
Gunjate, R.T. IN 15.1
Gunjkar, S.N. IN 12.1
Gunn, R.E. GB 35.2
Gupta, A. IN 18.2
Gupta, A.K. IN 9.1
Gupta, B.P. IN 24.1.4
Gupta, G.K. IN 10.2, 26.1
Gupta, J.H. IN 25.1
Gupta, L.K. IN 24.1.8
Gupta, O.P. IN 9.1
Gupta, S.G. IN 22.1
Gupta, U.C. CA 43.1
Gupta, V.C. IN 2.2
Gur*, A. IL 2.1
Gurav, S.B. IN 14.2
Gurcheva, Y. BG 9.1
Gurel, N. TR 5.2
Gurnil, K. TR 22.1
Gurov, Al. BG 13.2
Gursan, K. TR 14.1
Gurusinghe, P. LK 4.1
Gusenleitner, J. AT 8.10
Gustafson, W.A. US-G 23.1
Gustafsson, A. SE 8.1
Gustafsson*, I. SE 1.4
Gustavsson, R. SE 1.2
Gutierrez, A.P. US-K 3.1.1
Gutierrez, J. FR 10.1.2
Gutiérrez, J.M. MX 1.4
Gutiérrez Acosta, F. MX 6.3
Gutiérrez E., J.L. MX 15.1
Gutiérrez Jarquin, A. MX 17.1
Gutiérrez Treviño, O. MX 6.2
Gutser, D.E. 11.1.3
Guttay, A.J.R. US-A 13.1
Gutterman, Y. IL 5.1.4
Guttormsen, G. NO 2.1
Gutyńska, B. PL 28.1.5
Guu, J.W. TW 7.2
Guy, G. AU 27.1
Guy, M. IL 5.1.1
Guyon, J-P. FR 3.2.1
Guyot, G. FR 3.1.4
Guzewska, I. PL 28.1.3
Guzewska-Bogatko B. PL 15.1
Guzman, N. CR 2.2
Guzman, V. US-D 23.1
Gvozdenović, D. YU 17.1
Gwaraimba, V.E.E. ZI 3.1.2
Gwatkin, P. AU 65.4
Gwin, R. US-G 32.1
Gwozdecka, M. PL 28.1.10
Gwozdecki, J. PL 28.1.7
Gwynne, R.T. AU 59.1
Gyenes, M. HU 5.3.1
Györffy, S. HU 5.3.1
Gyorgy, Al. RO 6.1

Győri, E. HU 2.7
Gysi, Chr. CH 1.1
Gyulai, B. HU 2.2.5
Gyuris, K. HU 5.3.1
Gyuró, F. HU 2.2.2
Gyurós, J. HU 2.2.12

Haabjørg*, A. NO 10.1.5
Haag, H.P. BR 24.1
Haag, M. US-E 19.1
Haapala, T. FI 10.1.1
Haas, J.C. BR 31.1
Haba y Jarque, M. ES 17.1
Habdas, M. PL 28.2.12
Habeck, D.H. US-D 27.2
Habegger, R. DE 15.1.3
Haber, A. PL 32.1.12
Haber, W. DE 11.1.5
Haber, Z. PL 20.2.4
Habib, R. FR 3.1.1
Habor, T. RO 25.1
Hache, V. ET 10.1
Häckel, H. DE 11.4
Hackett*, C. AU 65.7
Hackett, W. US-G 7.1
Haderlie, L. US-I 13.1
Hadidi, A. US-C 3.1
Hadincova, M. CS 27.1
Hadley, P. GB 29.1
Hadrović, H. YU 16.1
Hadziyev, D. CA 7.2
Haegeman, J. BE 14.1
Haeseler, C.W. US-B 11.1, 13.1.1
Hafen, L. US-F 9.1.4
Haffnes, K. NO 10.1.6
Hafidi, B. MA 1.1
Hagadorn, D. US-F 23.1.4
Hagar, S.S. US-B 7.1
Hagen, K.S. US-K 3.1.1
Hagen, O. US-G 17.1
Hagiladi, A. IL 1.1.2
Haginuma, S. JP 14.3
Hagiwara, H. JP 1.1.3
Hagiwara, Y. JP 26.2
Hagley, E.A.C. CA 25.2
Hagting, A. NL 9.3
Hahn, B.H. KR 23.1.6
Hahn, K.Y. KR 12.1
Hahn, S.J. KR 7.3
Hahn, Y.I. KR 18.1.2
Haile, B. ET 3.1
Haile-Mariam*, S-N. ET 10.1
Hain, F.P. US-D 8.1.3
Hajdu, E. HU 5.2
Hajnal, J. CS 19.1
Hakansson, B. SE 1.1.4
Hakkaart, F.A. NL 1.1.5, 9.4.4
Hakkola, H. FI 7.1
Hakoda, N. JP 41.4
Hakozaki, M. JP 19.3
Halaba, J. PL 28.1.17
Halász, A. HU 2.4
Halász, G. HU 3.1
Halász-Zelnik, K. HU 2.2.12
Halbmayr, H. AT 8.1
Hale, C.N. NZ 2.3.1
Hale, O.D. GB 31.2.5
Hale, P.W. US-D 34.1
Haleem, S.A. IN 1.1.1
Hales, D.R. AU 9.4
Halevy*, A.H. IL 2.2
Halfacre, R.G. US-D 13.1
Haliburton, T.H. CA 42.2
Halisky, P.M. US-B 14.1.6
Hall, A.E. US-K 12.1.1
Hall, B.J. US-L 2.2
Hall*, C.B. US-D 27.7
Hall, C.V. US-G 20.1
Hall, D.S. US-H 12.1
Hall, F.R. US-F 7.1.3
Hall, H.K. NZ 13.2
Hall*, I.V. CA 41.1
Hall, M. US-D 22.1
Hall, M.T. US-D 59.1
Hall, R.A. GB 18.1.2
Hall, T.D. GB 9.1.2
Hallaire, V. FR 3.1.7
Hałas, M. PL 28.2.1
Halldal, P. NO 7.2
Hallet, A. BE 5.1
Hallet, I.C. NZ 2.3.2
Hallinan, M. IE 3.1
Halm, B.J. GA 4.3
Hălmăgean, L. RO 2.1
Halseth, D.E. US-B 5.1.3
Hälvä, S. FI 10.1.1
Hamachi, F. JP 7.1
Hamada, K. JP 13.2
Hamada, Y. JP 41.1
Hamaguchi, T. JP 27.2
Hamaker, Ph. NL 9.2
Hamar, N. HU 5.1.1, 5.3.1
Hämet-Ahti, L. FI 10.1.3
Hamid, M.T. IN.24.1.4
Hamill, A.S. CA 17.1
Hamilton, B.A. US-B 14.1.1
Hamilton, G.A. GB 9.1.3
Hamilton, J.T. AU 9.1.3
Hamilton, K.C. US-H 27.1
Hamilton, M.G. US-D 11.1
Hamilton, R.I. CA 4.1
Hammer*, P.A. US-F 9.1.4
Hammerschmitt, R. US-F 14.1.3
Hammett, K.R.W. NZ 2.1.1
Hammett, L.K. US-D 8.1.1
Hammond, R.W. US-I 10.1
Hampson, M.C. CA 44.1
Hamson, A.R. US-I 5.1
Hamza, N. TN 1.1
Han, C.M. TW 4.3
Han, D.W. KR 21.1
Han, H.P. KR 5.2
Han, H.R. KR 10.3
Han, K.H. KR 8.4
Han, K.P. KR 13.1
Han, W.T. KR 10.2
Hanada, K. JP 7.4
Hanafi, A. MA 1.1
Hanan, J.J. US-I 2.2
Hanaoka, Y. JP 10.1
Hancock, B.G. US-H 6.2
Hancock, J.F. US-F 14.1.1
Hancock Jr., J.G. US-K 3.1.2
Hand*, D.W. GB 18.1.1
Hand, P. GB 18.1.2
Handa, A.K. US-F 9.1.4
Handawala, J. LK 3.5
Hang, Y.D. US-B 2.1.5
Hanger*, B. AU 30.1
Hanić, E. YU 2.1
Hanin, H.H. EG 1.2
Haniotakis GR 3.1.1
Hänisch ten Cate, Ch.H. NL 9.5
Hanks, G.R. GB 18.1.1
Hanks, R.W. US-D 30.2
Hanna, H.V. US-D 56.1
Hannah, L.C. US-D 27.7
Hannah, M.A. GB 18.1.1
Hanneman, R.E. US-F 23.1.3
Hanniford, G.G. US-F 3.1
Hansche, P.E. US-K 6.1.3
Hansen, A.J. CA 3.1
Hansen, B. DK 2.1
Hansen, H. DE 18.1.1
Hansen, J. DK 1.2
Hansen, K. DK 3.6
Hansen, L.M. DK 4.1
Hansen, L.S. DK 4.1
Hansen*, M. DK 7.1
Hansen, O.B. NO 10.1.5
Hansen*, P. DK 3.1
Hansmann, C.F ZA 4.1
Hanson, B.R. AU 36.4
Hanson, E.J. US-F 14.1.1
Hanson, R. US-A 14.1
Hansson, E. SE 2.1
Hansson, I. SE 1.1.2
Hantz, J. PL 10.1.6
Hanyû, K. JP 39.1
Haque, M.T. US-D 13.1
Haquenne, W. BE 6.7
Hara, H. JP 14.5

Hara, M. JP 23.2.5
Hara, S. JP 38.2
Hara, T. BR 19.2
Hara, T. JP 47.2
Harada, A. JP 47.2
Harada, H. JP 38.4
Harada, T. JP 12.4
Harada, Y. JP 17.1
Haraguchi, K. JP 17.2
Haraguchi, Y. JP 13.2
Haramaki*, C. US-B 13.1.1
Harano, H. JP 44.2
Haraszthy, J. HU 3.1.1
Harbaugh, B.K. US-D 24.1
Hard, C.G. US-G 7.1
Hardin, N.C. US-C 9.1
Harding, J.A. US-K 6.1.2
Hardman, J.M. CA 41.1
Hardwick, R.C. GB 33.1.5
Hardy, F.S. GB 2.1
Hardy, R.J. AU 46.2
Harford, B.A. AU 33.1
Hargitai, L. HU 2.2.11
Hargreaves, J.A. GB 19.1.1
Hargreaves, J.R. AU 18.1
Härig, R. DE 3.1
Harischandra, S.N. LK 3.7
Haritou, A. GR 24.1
Harker, J. US-A 1.1
Harman, G.E. US-B 2.1.2
Harman, J.E. NZ 2.1.2
Harmat, I. HU 3.1.1
Harmat, L. HU 2.1.4
Harmsen, K. NL 4.1
Harney, P.M. CA 15.1
Harpaz, I. IL 2.5
Harper, F.R. CA 8.1
Harper, W.L. US-A 13.1
Harradine, A.R. AU 46.7
Harries, G.O. GB 31.2.7
Harris, C.H. CA 18.1
Harris, C.M. US-K 7.2
Harris, D.C. GB 8.1.2
Harris, G.P. GB 29.1
Harris, P.S. GB 9.1.2
Harris*, R.W. US-K 6.1.2
Harris, W.L. US-C 5.1
Harrison, B.D. GB 15.1.3
Harrison, D. AU 27.1
Harrison, H. US-F 23.1.3
Harrison, H.F. US-D 12.2
Harrison, J.G. GB 15.1.1
Harrison-Murray, R.S. GB 8.1.1
Harsányi, J. HU 2.7
Hart, L.I. US-I 18.1
Hart, M. AU 47.3
Hartge, K.-H. DE 15.4
Hartill, W.F.T. NZ 2.3.3
Hartmann, H. CA 2.1
Hartmann*, H.D. DE 12.1.6
Hartmann*, W. DE 28.1.1
Hartmann, R.W. US-M 2.2
Hartz, T. US-H 15.1
Haruki, K. JP 37.1
Harvey, J.M. US-K 7.2
Hasdagli, I. TR 5.2
Hase, Y. JP 11.3
Hasebe, H. JP 40.2
Hasegawa*, A. JP 17.3
Hasegawa, H. JP 3.1
Hasegawa, K. JP 20.4
Hasegawa, P.M. US-F 9.1.4
Hasegawa, S. JP 11.1
Hasegawa, S. US-K 11.1
Hasegawa, Y. JP 38.3
Hashimoto, K. JP 35.2, 46.1
Hashimoto, K.-I. BR 2.1
Hashimoto, M. JP 27.2
Hashimoto, N. JP 8.2
Hashimoto*, S. JP 18.2
Hashimoto*, S. JP 41.1
Hashimoto, T. JP 23.1
Hashinaga, F. JP 18.3
Hashizume, T. BR 20.2
Haskel, A. IL 1.2.4
Hass, I. DE 12.1.8
Hassan, S.A. LY 1.1
Hasselkus, E.R. US-F 23.1.3
Hässler, J. CS 9.1
Hatai, S. JP 3.3.2
Hatakeyama, J. JP 2.1
Hatamoto, M. JP 31.1
Hatano, H. JP 25.1.1
Hatfield, S.G.S. GB 8.1.3
Hathaway, R.L. NZ 8.3
Hatipoglu, A. TR 15.1
Hâtisi, A. RO 1.1.5
Hatton*, W.J. US-D 34.1
Hattori, M. JP 28.1
Hattori, T. JP 22.4
Hattori, Y. JP 23.1
Hatziharissis, J. GR 19.1
Hatzinikolis, E. GR 21.1
Haubs, H. DE 12.1.3
Haunold, E. AT 24.1
Haupt, G. DK 1.2
Hauschildt, H. DE 17.1
Hausher, L.G. CA 6.1
Haushofer, J. AT 4.1
Hauxner, M. DK 3.2
Havis, J.R. US-A 5.1.1
Hawk, J.A. US-C 2.1
Hawlitzky, N. FR 17.1.3
Hawson, M.G. AU 59.1
Hawthorne, B.T. NZ 2.3.3
Hay, S. CA 31.1
Hayakami, M. JP 21.2
Hayakawa, Y. JP 6.1
Hayashi, G. JP 1.1.3
Hayashi, H. JP 3.3.2, 19.1
Hayashi, I. JP 19.2
Hayashi, J. JP 44.2
Hayashi*, M. JP 4.4
Hayashi, S. JP 1.2, 8.2
Hayashi, S*. JP 42.3
Hayashi, T. JP 31.2
Hayashi, Y. JP 7.1
Hayashida, E. JP 21.1
Hayashida, S. JP 27.2
Hayashida, T. JP 27.1
Hayata, T. JP 39.1
Hayata, Y. JP 11.4
Hayden*, R.A. US-F 9.1.4
Hayes, P. GB 4.1
Hayman, E. US-K 11.1
Haynes Jr., F.L. US-D 8.1.1
Haynes, R.J. NZ 13.1
Hayslett, S. US-E 4.1
Heagle, A.S. US-D 8.1.2
Healy, W.F. US-C 5.2
Hearn, C.J. US-D 34.1
Heath, R.L. US-K 12.1.1
Heathman, S. US-H 27.1
Heaton, E.K. US-D 20.1
Heaton, J.B. AU 25.1
Heberden, A.F. ZI 5.1
Hédde, L. FR 46.1
Hedden, P. GB 19.1.2
Hedden, S.L. US-D 30.1
Hedding, R.R. AU 26.3
Hedley, C. GB 24.1.1
Heegaard, C. DK 7.2
Heenbanda, Y.M. LK 4.1
Heenkenda, H.M.S. LK 3.2
Hefley, M.W. US-G 20.1
Hegde, D.M. IN 1.1.9
Heguchi, T. JP 34.1
Hegwood, C.P. US-D 47.1
Heiberg, N. NO 10.1.6
Heidekamp, F. NL 9.5
Heikal, I. EG 1.5
Heil, R.D. US-I 2.1
Heimer, Y.M. IL 5.1.2
Hein, K. DE 23.1.
Heins, B. DE 15.1.5
Heins*, R.D. US-F 14.1.1
Heintz, R.H. US-G 12.1
Heinze, W. DE 5.2.1
Hejný, S. CS 24.2
Hellgren, O. SE 1.1.3
Hellman, E.W. US-G 31.1
Helmers, S. US-H 8.1
Heltmann-Tulok, M. HU 2.2.12

Helton, A.W. US-I 14.1
Hemeng, B. GA 2.2.1
Hemerly, F.X. BR 18.1
Hemle-Jánosi, M. HU 2.2.6
Hempel, M. PL 28.1.17
Hempel, T. PL 28.1.13
Hemphill, D.D. US-E 18.1
Hemsy, V. AR 22.3
Hendershott, C.H. US-D 16.1
Henderson, D. AU 38.1
Henderson*, W.R. US-D 8.1.1
Hendrickx, J.M.H. NL 9.11.1
Hendrickx, M. BE 9.1
Hendriks, L. DE 15.6
Hendrix, F. US-D 16.1
Henis, Y. IL 2.6
Henley, R.W. US-D 27.5, 31.1
Henmi, H. JP 31.1
Henmi, S. JP 8.1
Hennart, J.W. FR 35.1
Hennebo, D. DE 15.2.2
Hennerty*, M.J. IE 3.1
Henning, A. BR 27.1
Henninger, M.R. US-B 14.1.1
Henny, R.J. US-D 31.1
Henriet, J. BE 6.3
Henriksen, K. DK 1.1
Henriksson, R. SE 5.1
Hensel, D.R. US-D 28.1
Henshall, J.D. GB 7.1
Henshaw, G.G. GB 3.1
Henter, M. RO 8.4.5
Henze, J. DE 6.1
Henzell, R.F. NZ 5.1
Hepburn, A.G. GB 24.1.2
Hepler, P.R. US-A 1.1
Heras, L. ES 23.2
Herath, H.M.E. LK 3.1, 3.3
Herath, L.G. LK 3.15
Herbert, F. GB 27.1
Herbst, R. ET 10.1
Herde, M. DK 4.1
Heredia Zepeda, A. MX 7.2
Hergár, E. HU 2.2.1
Hering, T.F. GB 25.1.2
Herlaar, F. NL 14.1
Herlyn, U. DE 15.2.1
Hermann, L. FR 1.1.1
Hermann, M. DE 5.1.1
Hermes, Y. NL 1.1.3
Hermus, R. AU 12.1
Hernandez, J. PE 2.1
Hernández Aguilar, P. MX 2.2
Hernández Gayá, S. US-N 1.1
Hernández L., J. PE 2.1
Hernández Ovalle, J. MX 4.7
Herna, J. PE 4.4
Herner, R.C. ZI 3.3
Herner, R.C. US-F 14.1.1
Herregods, M. BE 21.1
Herrera, E. US-H 19.1
Herrera, F. CR 2.2
Herrera Perez, T. MX 5.2
Herrington, M.E. AU 18.1
Herrmann, Ch. DE 11.1.2
Herron, G. US-G 29.1
Hertug, M. RO 34.1
Hertz, L. US-G 7.1
Hervé, Y. FR 14.1.1
Herzfeld, F. DE 15.3.3
Herzog, D.C. US-D 37.1
Hesjedal, K. NO 5.1
Heslop, A. AU 34.1
Hess*, C.E. US-K 6.1.2
Hester, J.M. US-B 1.1
Hetman, J. PL 14.1.2
Heungens, A. BE 7.5
Heursel, J. BE 14.1
Heuser, C.W. US-B 13.1.1
Heuvelink, E. NL 9.7.1
Hevesi, M. HU 2.5
Hévizi, I. HU 5.1.1
Hewavitharana, F. LK 3.3
Hewett, E.W. NZ 2.1.2
Hewitt, J.D. US-K 6.1.4
Heylen, J. BE 13.1
Heyting*, J. NL 9.6.3
Heyward, E.J. US-L 2.2
Hibben, C.R. US-B 7.1
Hibberd, A.M. AU 18.1
Hibino, Y. JP 9.1
Hickey, K.D. US-B 9.1
Hicklenton, P.R. CA 41.1
Hickman, C. US-C 9.1
Hicks, J.R. US-B 5.1.3
Hida, K. JP 23.2.2
Hidaka, A. JP 25.1.1
Hidaka, T. JP 38.3
Hidalgo, L. ES 12.3
Hido, K. JP 41.1
Hieke, K. CS 24.1
Higa, H. JP 32.1
Higa, M. JP 32.1
Higa, R. JP 32.1
Higa, T. JP 32.2
Higaki, T. US-M 1.1, 3.1
Higashi, A. JP 18.2
Higashi, R. JP 30.2
Higashi, T. JP 21.1
Higashio, H. JP 17.2
Higgins, K.B. AU 30.2
Higgins, W.J. CA 42.2
Higginson, F.R. AU 9.1.2
High, J. US-E 6.1
Highwood, D.P. GB 32.1
Hignett, R.C. GB 8.1.2
Higuchi*, H. JP 41.3
Higuita, F. CO 7.2
Hiirsalmi*, H. FI 1.1
Hijikata, H. JP 44.2
Hijikata, S. JP 41.1
Hijink, M.J. NL 9.9
Hilali, A. MA 1.1
Hildebrand, P.D. CA 41.1
Hildebrandt*, W. DE 15.1.10
Hildrum, K.I. NO 10.5.2
Hilggarg, E.R. BR 29.1
Hill, E.C. US-D 30.2
Hill, G.N. AU 1.1
Hill Jr., J.R. US-D 11.1
Hill*, M.G. NZ 2.2
Hill, M.J. NZ 9.7
Hill, R.W. GB 31.1
Hill, S.A. GB 13.1.4
Hill, S.B. CA 37.1
Hill, T.A. GB 35.1.3
Hill, T.R. AU 56.1
Hill-Cottingham, D.G. GB 19.1.2
Hiller, R.G. AU 9.4
Hills, G.J. GB 24.1.3
Hilton, J.L. US-C 3.1
Himelich, E.B. US-F 12.1.1
Himeno, M. JP 18.1
Himeno, S. JP 7.1
Himizu, E. JP 29.1
Himori, Y. JP 22.2
Hinde, R.W. AU 9.4
Hindriks, K. NL 6.1
Hinds, R.P. CA 40.2
Hinesley, L.E. US-D 8.1.1
Hingand, L. FR 14.1.2
Hinish, W.W. US-B 13.1
Hino, A. JP 5.3
Hinson, R.A. US-D 52.1
Hinterholzer, J. AT 8.7
Hirabayashi, T. JP 11.3
Hiraga, S. JP 25.1.1
Hirai, H. JP 15.1
Hirai, M. JP 38.3
Hirai, T. JP 3.3.2
Hiraide, K. JP 39.1
Hirakawa, N. JP 7.1
Hiramatsu, J. JP 11.2
Hiramatsu, T. JP 31.1
Hirano, S. JP 41.3
Hirano, T. JP 7.2, 41.1
Hiranpradit, H. TH 16.1
Hirao, A. JP 11.1
Hiraoka, M. JP 31.1
Hiraoka, N. JP 11.1
Hiraoka, T. JP 19.2
Hiraragi, T. JP 16.1
Hirata, I. JP 21.2

Hirata*, N. JP 4.4
Hirata, S. JP 44.1
Hirata, Y. JP 7.3
Hiratsuka, S. JP 4.4
Hirayama, H. JP 21.2
Hiremath, I.G. IN 11.7
Hirobe, J. JP 44.1
Hirobe, M. JP 19.2
Hiroma, K. JP 26.2
Hirose, K. JP 38.3
Hirose, T. JP 13.3, 17.3
Hirota, K. JP 1.1.1
Hirota, R. JP 34.2
Hiruki, C. CA 7.1
Hisada, H. JP 38.2
Hisatomi, T. JP 28.1
Hishida, A. JP 1.1.2
Hislop, E.C. GB 19.1.1
Hitchon*, G.M. GB 2.2
Hitokuwada, W. JP 4.1
Hiuane, G. JP 32.1
Hiura, I. JP 45.2
Hiwale, S.S. IN 1.1.1
Hiyane, G. JP 32.1
Hızal, A.Y. TR 3.1
Hjalmarsson, I. SE 1.1.1
Hjartarson, O. IC 1.1
Hjeltnes, A. NO 6.1
Hjeltnes*, S.H. NO 3.1
Hlavička, A. CS 1.2
Hlišć, T. YU 18.2.1
Hnízdil, F. CS 23.1
Hnízdil, M. CS 4.1
Ho, W.J. TW 6.1
Hoad, G.V. GB 19.1.2
Hoag, B.K. US-G 16.1
Hobolth, L.A. DK 4.1
Hobson, G.E. GB 18.1.3
Hoch, H.C. US-B 2.1.4
Hochberg, M.C. US-K 16.1
Hochmuth, G.J. US-D 27.7
Hock, B. DE 11.1.5
Hockenhull*, J. DK 3.5
Hocking, A.D. AU 65.4
Hockley, D.G. AU 65.4
Hoda, M.N. IN 8.1
Hodge, D. AU 40.1
Hodges, C.F. US-G 20.1
Hodges, J. US-E 3.1
Hodgson, C.J. GB 35.1.3
Hodossi, S. HU 5.3.1
Hódosy, S.A. HU 5.3.1
Hoekstra, F.A. NL 9.7.2
Hoekstra, O. NL 6.1
Hoepfner, M.A. BR 27.2
Hoeppe, C. DE 31.1
Hoff, J.E. US-F 9.1.2
Hoffman, Z. PL 34.1
Hoffmann, D. DE 12.1.12
Hoffmann*, G.M. DE 11.1.3
Hoffmann, M. PL 20.2.3
Hofman, E. DE 12.1.12
Hofstede, G.W. NL 14.2
Hofsvang, T. NO 10.4.3
Hoftun, H. NO 10.1.7
Hogan*, L. US-H 27.1
Hogenboom, N.G. NL 9.6
Hogg, D.R. AR 6.1
Hogg, M.G. NZ 2.1.3
Hogmiere, H. US-C 8.1
Hogue, E.J. CA 3.1
Hohlt, H.E. US-C 13.1
Hohmann, G. DE 20.1
Höhn, E. CH 1.1
Hoitink, H.A.J. US-F 7.1.1
Holanda, L. BR 8.2
Holcomb, E.J. US-B 13.1.1
Holcomb, G.E. US-D 52.1
Holdsworth, S.D. GB 7.1
Hole, C.C. GB 33.1.5
Holewiński*, A. PL 28.1.10
Holgate, M.E. GB 19.1.4
Holland, D.A. GB 8.1
Holland, N.S. US-G 12.1
Holland, P.T. NZ 5.1
Holland, R.O. US-J 1.1
Holland, R.W.K. GB 6.1
Hollett, J.M. CA 42.1
Holley, J. CA 40.3
Hollingsworth, M.H. US-D 43.1
Holloway, P. US-L 1.1
Holloway, P.J. GB 19.1.1
Holloway, R.I.C. GB 8.1.1
Holloway, W. US-H 14.1
Holloway, W.D. NZ 8.2
Holm, D.G. US-I 1.1
Holmann, M. CS 25.1
Holmenlund, N.P. DK 3.7
Holmes, G. GB 4.1
Holmes, R.J. AU 30.2
Holmgren, P.K. US-B 1.1
Holmøy, R. NO 10.2
Hołownicki, R. PL 28.1.6
Holst, H. DE 12.1.12
Holt, D.A. US-F 12.2
Holt, J.B. GB 31.2.6
Holt, J.S. US-K 12.1.1
Hölters, J. DE 15.1.8
Holtz, J.R. US-C 8.1
Holtzhausen, L.C. ZA 3.2
Holub, J. CS 22.1
Holubová, M. CS 28.1
Hołubowicz, R. PL 20.2.10
Hołubowicz*, T. PL 20.2.9
Holyoke, V.H. US-A 1.1
Homan, M. JP 18.1
Hompanera, N. AR 19.1
Honda, F. JP 7.3
Honda, I. JP 46.1
Honda, N. JP 24.1
Honda, R. JP 6.1
Honda, T. JP 44.1
Honda, Y. JP 16.3
Hong, G.S. KR 3.1
Hong, K.H. KR 16.1.3
Hong, S.B. KR 23.1.3
Hong, Y.P. KR 23.1.5
Honjo, H. JP 14.2
Hood, L.F. US-B 2.1
Hood, P.M. AU 49.1
Hoogerkamp, M. NL 9.1
Hoogeveen, M.G. NL 1.1.4
Hooijer, D. NL 1.1.6
Hooks, R. US-H 20.1
Hoover*, E. US-G 7.1
Hopen, H.J. US-F 23.1.3
Hopfinger, J.A. US-B 16.1
Hopfinger, J.T. US-B 14.1.1
Hopkins, D.L. US-D 32.1
Hopkirk*, G. NZ 2.1.2
Hopmans, P.A.M. NL 9.7.1
Hopping, M.E. NZ 5.1
Hoppmann, D. DE 12.2
Hopwood, D.A. GB 24.1.2
Hoque, M.M. PH 1.6
Horánszky, Zs. HU 2.2.9
Horbal, T. PL 28.1.6
Horbowicz, M. PL 28.2.10
Hore, J. IN 28.1
Hori*, Y. JP 24.3
Horie, H. JP 41.1
Horiguchi, T. JP 35.1
Horikawa, N. JP 4.3, 17.1
Horimoto, M. JP 13.1
Horino, Y. BR 25.5
Horiuchi, M. JP 20.1
Horiuchi, S. JP 23.2.4, 33.2
Hörmann, D.M. DE 15.1.2
Horn, D. US-F 3.1, 7.1.3
Horn*, W. DE 11.1.1
Horner, I.J. NZ 2.3.2
Horng, D.T. TW 7.1
Hornok, L. HU 2.2.12
Horntvedt, S. NO 10.1.5
Horodecka, E. PL 28.2.1
Horowitz, M. IL 1.1.2
Horowitz, O. IL 1.1.2
Horowitz, R.M. US-K 11.1
Horridge, J.S. GB 18.1.1
Horsburgh, R.L. US-C 14.1
Horst*, R.K. US-B 5.1.8
Horton, B. US-C 8.2
Horton, B. US-D 19.1
Hortyński, J. PL 14.1.7

Horvat, A. YU 3.3
Horvat, R. US-D 16.2
Horváth, A.Z. HU 2.2.5
Horváth, E. HU 2.2.5
Horváth, G. HU 2.7
Horváth, Gy. HU 8.1
Horváth, I. HU 5.1.2
Horváth-Kerkai, E. HU 2.2.1
Hoschke, A. HU 2.4
Hoser-Krause, J. PL 28.2.1
Hoshikawa, S. JP 10.1
Hoshino, K. JP 23.2.3
Hosie, G. GB 9.1.2, 9.1.3
Hosking, D.C. AU 59.1
Hosoda, H. JP 14.3
Hosoi, T. JP 38.4
Hosoki, T. JP 37.2
Hosono, H. JP 21.1
Hosotani, T. JP 23.1
Hosoya*, M. JP 4.3
Hossain, M. IN 28.1
Hossain, T. IN 28.1
Høst, O. DK 3.6
Höster, H.-R. DE 15.2.4
Hostetter*, G.W.B. CA 22.1
Hotta, Y. JP 1.1.1
Hou, F.W. TW 6.1
Houben, C. NL 15.1
Houck, L.G. US-K 7.2
Hougas, R.W. US-F 23.1.3
Houghton, B.H. GB 5.1
Houtz, R. US-E 15.1
Hovadik, A. CS 20.1
Howard, B.H. GB 8.1.1
Howard, F.W. US-D 25.1
Howard, R.J. CA 6.1
Howe, J.A.G. CA 9.1
Howe, T.K. US-D 24.1
Howell, G.S. US-F 14.1.1
Howell, P.J. GB 9.1.2
Howitt, A. US-F 14.1.2
Hoxey, R.P. GB 31.2.4
Hoy, M.A. US-K 3.1.1
Høyem, T. NO 10.5.1
Høyer, L. DK 1.2
Hoyoux, J-M. BE 6.4
Hoyt, G.D. US-D 6.1
Hoyt, P.B. CA 3.1
Hoyt, S.C. US-J 10.1
Hozumi, K. JP 16.3
Hraskó, I. HU 5.3.1
Hrazdina, G. US-B 2.1.5
Hrdlička, O. CS 22.3
Hričovský, I. CS 23.1
Hričovský, J. CS 19.1
Hron, R. AT 8.7
Hrycak, T. PL 29.1
Hsiao, C.H. TW 7.3
Hsu, H.T. TW 7.3
Hsu, W.J. US-K 11.1
Hsu, Y.M. TW 4.3
Hsu, Y.P. TW 11.1.2
Huanco, V. PE 4.3
Huang, C.C. TW 1.1, 7.1
Huang, C.S. BR 25.4
Huang, H. TW 9.1
Huang, M.C. TW 7.1
Huang, P.K. TW 6.1
Huang, S.P. BR 25.5
Huang, Y.C. TW 9.1
Huba, A. CS 9.1
Hubac, C. FR 30.1.2
Hubáček, V. CS 11.1
Hubáčkova, M. CS 22.4
Huband, N.D.S. GB 19.1.2
Huber, D.J. US-D 27.7
Huber, W. DE 11.1.5
Huber, W. IT 11.1
Hucklesby, D.P. GB 19.1.2
Huddar, A.G. IN 11.2
Hudec, P. CS 23.1
Hudon, M. CA 35.1
Hudska, G. CS 7.1
Hudson, R.R. AR 15.1
Huett, D.O. AU 1.1
Hugard*, J. FR 10.1.1
Huggart, R.L. US-D 30.2
Hughes, H.G. US-I 2.2
Hughes, I.K. AU 15.4
Hughes, J. CA 15.1
Hugon, E. FR 28.1
Huguet, Cl. FR 3.1.1
Huguet, J-G. FR 3.1.1
Huh, S.S. KR 8.4
Huiskamp, Th. NL 6.1
Huisman, O.C. US-K 3.1.2
Huiswoud, R.R. SR 1.1
Hulamani, N.C. IN 11.3
Hulewicz, T. PL 14.1.7
Hull, D.R.J. US-A 14.1
Hull Jr., J. US-F 14.1.1
Hull, L.A. US-B 9.1
Hull, R. GB 24.1.4
Human, N.B. ZA 1.1
Humanes, J. ES 7.2
Hume, R.J. NZ 9.1
Hummel Jr., N.W. US-B 5.1.1
Humpherson-Jones, F.M. GB 33.1.4
Humphreys, T.E. US-D 27.7
Humphries, D.J. GB 18.1.1
Hung, L. TW 9.1
Hunt, J. GB 33.1.2
Hunter*, A. IE 3.1
Hunter, D.M. CA 17.1
Hunter, J.E. US-B 2.1.4
Hunter, K. GB 9.1.3
Hunter, L.D. GB 8.1.2
Hunter, P. US-E 7.1
Hunter, R. US-H 16.1
Hunter, T. GB 19.1.1
Huokuna, E. FI 5.1
Hurka, W. DE 19.1
Hurley, M. US-E 16.1
Hurt, V. US-D 50.1
Hurtado Hernández, H. MX 3.4
Hurter, J. CH 1.1
Hurtt, S. US-C 3.1
Husabø, P. NO 3.1
Husby, F.M. US-L 1.1
Hussey, G. GB 24.1.1
Huszcza, W. PL 14.1.9
Hutchins, R.F.N. NZ 2.2
Hutchinson, F. US-F 7.1
Hutchinson, J. AU 30.1
Hutchison, D.J. US-D 34.1
Hutley-Bull, P.D. GB 35.1.2
Huttinga, H. NL 9.4.4
Huttleston, D.G. US-B 10.1
Hutton*, R.J. AU 12.1
Huuhtanen, P. FI 1.1
Huus-Bruun*, T. DK 9.2
Hvalsøe, E. DK 3.7
Hwang, A.S. TW 1.2
Hwang, B.T. TW 2.1.4
Hwang, C.I. TW 12.1
Hwang, H.S. KR 17.1
Hwang, J.K. KR 3.2
Hwang, J.M. KR 23.1.2
Hwang, K.S. KR 20.1
Hwang, L.U. TW 12.1
Hwang, P. TW 3.1
Hwang, R.S. TW 2.1.4
Hwang, S.L. TW 8.1
Hwang, S.J. KR 18.1.3
Hyodo*, H. JP 38.4

Ialongo, M.T. IT 19.2
Iancu, E. RO 1.1.3
Iancu, M. RO 9.1
Iannini, B. IT 7.1
Iapichino, G. IT 13.2
Iba, Y. JP 14.2
Ichi, K. JP 18.1
Ichihashi, S. JP 1.4
Ichii, T. JP 13.3
Ichijima, N. JP 27.3
Ichikawa, H. JP 31.1
Ichikawa, T. JP 38.2
Ichiki, K. JP 27.3
Ichiki, S. JP 3.2.4
Ichimura, M. JP 41.3
Ichimura, T. JP 14.1
Ichinohe, H. JP 3.2.1

Ichita, T. JP 3.1
Ida, A. JP 17.2
Ida, S. JP 41.1
Idrissi, M. MA 4.1
Ieki, H. JP 38.3
Iemura, H. JP 44.1
Iezzoni, A. US-F 14.1.1
Iftime, D. RO 21.1
Iga, M. JP 41.1
Igarashi, D. JP 19.2
Igarashi, I. JP 23.2.2
Igarzabal, D. AR 7.1
Ignatowicz, S. PL 32.1.7
Igrc, I. YU 3.2
Iguchi, I. JP 38.2
Iida, S. JP 14.4
Iida, T. JP 1.1.2
Iida, W. JP 4.4
Iimuro, S. JP 28.1
Iino, K. JP 6.1
Iizuka, I. JP 21.3
Iizuka*, M. JP 4.4
Ikeda, F. JP 17.2
Ikeda, H. JP 7.2, 7.3
Ikeda*, H. JP 33.2
Ikeda, I. JP 11.3
Ikeda, N. JP 41.3
Ikeda, Y. JP 11.1, 13.1
Ikehara, D. US-M 6.1
Ikemiyagi, H. JP 32.1
Ikeuchi, Y. JP 13.1
Ikpe, F.N. NG 10.1
Ilag, L.L. PH 1.4
Iland, P.G. AU 41.1
Ilhan, I. TR 18.1
Ilić, B. YU 14.1
Ilić, S. YU 12.1
Ilie, Gh. RO 8.5
Ilie, T. RO 1.1.3
Iliescu, A.-F. RO 8.1
Iliescu, S. RO 20.1
Iliev, I. BG 9.1
Iliev, P. BG 1.1
Ilnicki, R.D. US-B 14.1.7
Iloba, C. NG 5.1
Ilter, E. TR 15.1
Im, E.R. KR 8.1
Imabori, K. JP 6.3
Imada, J. JP 11.3
Imada, S. JP 23.2.3
Imafurukawa, H. JP 18.1
Imagawa, H. JP 1.1.3
Imai, K. JP 3.2.1
Imai, T. JP 11.2, 13.1
Imakawa, S. JP 12.4
Imamura, T. JP 3.3.1, 27.2
Imanishi*, H. JP 33.2
Imanishi, S. JP 45.2
Imaoka, A. JP 37.1
Imbroglini, G. IT 19.2
Impens, I. BE 2.1
Imre, K. HU 2.2.8
In, K. BE 6.8.12
Inaba, A. JP 31.2
Inaba, K. JP 15.2, 22.1, 37.2
Inabe, Y. JP 15.2
Inada, I. JP 16.4
Inadomi, K. JP 34.2
Inagaki, N. JP 13.3
Inal, A. TR 13.1
Inamura, H. JP 37.1
Inayama, M. JP 35.2
Inazu, K. JP 41.2
Incalcaterra, G. IT 13.2
Incze, F. HU 2.2.4
Indeka, L. PL 32.1.13
Inden, H. JP 22.5
Indo, T. JP 12.1
Indrea, D. RO 17.2
Indyk, H.W. US-B 14.1.7
Ingle, A. NZ 10.1
Ingle, M. US-C 9.1
Ingram, B.F. AU 25.1
Ingram, D.L. US-D 27.5
Ingratta*, F.J. CA 25.1
Injac, M. YU 8.1
Innes, N.L. GB 15.1.2
Inokuchi, M. JP 19.2
Inose, O. JP 5.3
Inoue, H. JP 17.3, 19.4
Inoue, K. JP 12.3, 14.2, 38.2
Inoue, M. JP 4.1
Inoue, S. JP 8.2
Inoue, T. JP 41.3
Inoue, Y. JP 41.3
Insero, O. IT 19.1.3
Intharakamheang, S. TH 4.1
Intrieri*, C. IT 3.1
Intrigliolo, F. IT 1.1
Inubushi, S. JP 20.2
Inutsuka, K. JP 27.2
Ioachim, E. RO 32.1
Ioan, E. RO 5.1
Ioannidis, I. GR 26.1
Ionescu, A. RO 1.1.1
Ionescu, C. RO 1.1.1
Ionescu, E. RO 8.4.5
Ionescu, L. RO 1.1.2, 8.4.6
Ionescu, N. RO 10.1
Ionescu, P. RO 18.1
Iordăchescu, C. RO 8.4.1
Iordăchescu, O. RO 8.4.1
Iorga, M. RO 4.1
Iosifescu, M. RO 5.1
Iraha, Y. JP 32.1
Iremiren, G.O. NG 6.1
Ireta Ojeda, A. MX 2.2
Irimia, M. RO 8.4.4
Irizarry, H. US-N 1.2
Ironside, D.A. AU 23.1
Irving, A.R. AU 65.4
Irving, D.E. NZ 10.1
Irwin, H. US-B 6.1
Isaacs, A.R. AU 15.2
Isac, M. RO 9.1
Isaka, R. JP 44.1
Isaka, T. JP 14.3
Isawa, K. JP 45.1
Ishibashi, H. JP 4.2
Ishibashi, Y. JP 27.1
Ishida, E. JP 18.1
Ishida, K. JP 13.3
Ishida*, M. JP 22.4
Ishida, S. JP 27.1
Ishida, T. JP 4.1, 21.1, 38.2
Ishida, Y. JP 5.2
Ishido, Y. JP 38.2
Ishihara, A. JP 16.4
Ishihata, K. JP 18.3
Ishii, G. JP 23.2.3
Ishii, H. JP 14.2
Ishii, M. JP 23.2.4
Ishii, Y. JP 35.1
Ishikawa, H. AR 2.1
Ishikawa, K. JP 5.2, 16.1
Ishikawa, M. JP 4.1
Ishikawa, N. JP 11.3
Ishikawa, S. JP 25.1.1
Ishikawa*, S. JP 41.4
Ishitani, M. JP 3.3.1
Ishiuchi, JP 16.3
Ishiwata, H. JP 4.1
Ishiyama, M. JP 3.2.2
Ishizaka, I. JP 35.2
Ishizawa, Y. JP 14.2
Ishizu, F. JP 37.1
Ishola, M.S.A. NG 1.1
Isik, S.E. TR 14.1
Isikli, T. TR 15.3
Ismail, M.A. LY 2.1
Ismail, M.A. US-D 30.1
Ismail, M.M. LY 1.2
Isman, M.B. CA 4.2
Isobe, A. JP 21.2
Isoda, M. JP 31.1
Isoda, R. JP 11.4
Isoda, T. JP 21.2
Isogai, I. JP 9.4
Israilides, K. GR 18.1.2
Issa, E. BR 21.2

Issar, A. IL 5.1.2
Issell, L. AU 32.1
Istar, A. TR 9.1
Istas, J. BE 20.1
Istas, W. BE 14.1
Itagi, T. JP 19.1
Itami, K. JP 35.1
Itamura, H. JP 45.2
Itayama, S. JP 47.1
Itier, B. FR 17.1.7
Ito, A. JP 16.1
Ito, H. JP 4.2, 46.1
Ito, I. JP 34.4
Ito, K. JP 1.1.3, 16.3
Ito, M. JP 22.5
Ito, N. JP 37.2
Ito*, P.J. US-M 1.1
Ito, R. US-K 2.1, 8.1
Ito, S. JP 23.1, 26.2
Ito*, T. JP 4.4
Ito, T. JP 11.1
Ito, Y. JP 7.2, 26.2, 38.3
Itoh, S. JP 23.2.5
Itokawa, N. JP 17.2
Itoo, S. JP 18.3
Itose, S. JP 17.1
Iuchi, T. BR 30.1
Ivan, I. RO 14.1
Ivanković, Z. YU 3.1
Ivanov, Sp. BG 8.2
Ivanov, V. BG 6.2
Ivany, J.A. CA 43.1
Ivascu, A. RO 8.2
Ive, J.R. AU 65.7
Iveković, V. YU 3.1
Iversen, A. NO 10.5.2
Ivey, H.W. US-D 45.1
Ivins, J.D. GB 25.1.1
Ivker, S. IL 1.1.2
Iwade, Y. JP 47.1
Iwagaki*, I. JP 27.3
Iwahashi, N. JP 44.1
Iwahori*, S. JP 18.3
Iwai, T. JP 13.1
Iwakawa, T. JP 20.1, 20.3
Iwaki, T. JP 18.1
Iwakiri, T. JP 34.2
Iwakiri, Y. JP 25.1.2
Iwama, M. JP 26.3
Iwamasa*, M. JP 34.4
Iwamoto, K. JP 44.2
Iwamoto, M. JP 13.1
Iwamoto, Y. JP 19.3
Iwanaga, H. JP 34.2
Iwanek, B. PL 28.1.10
Iwao, K. JP 44.2
Iwasa, S. JP 16.4
Iwasaki, A. JP 20.3
Iwasaki, I. JP 39.1
Iwasaki, S.H. KE 1.3
Iwase, T. JP 3.3.2
Iwashita, T. JP 18.1
Iwata*, M. JP 41.3
Iwata, R.Y. US-M 1.1
Iwata, T. JP 33.2
Iwaya, A. JP 3.3.1
Iyengar, B.R.V. IN 1.1.9
Iyer, C.P.A. IN 1.1.1
Iyer, R.D. IN 3.1
Izaki, M. JP 14.5
Izdebski, J. PL 28.2
Izhar, S. IL 1.1
Iziro, Y. JP 39.2
Izsak*, E. IL 1.1.1
Izu, T. JP 46.1

Jablanović, M. YU 16.1
Jabłonska, L. PL 32.1.5
Jabłonski, B. PL 28.1.8
Jačinac, B. YU 16.1
Jackiewicz, A. PL 21.1
Jackson, D.I. NZ 13.5
Jackson, E.D. CA 41.1
Jackson, G. US-I 22.1
Jackson, J.C. GB 33.1.6
Jackson*, J.E. GB 8.1, 8.1.1
Jackson, M.J. US-I 18.1
Jackson, N. US-A 14.1
Jackson, P. GB 27.1
Jackson*, T.H. ET 10.1
Jacob, H. DE 12.1.7
Jacobo Cuéllar, J.L. MX 5.4
Jacoboni, A. IT 27.1
Jacoboni, N. IT 15.1
Jacobs, J.M. NL 8.1
Jacobsen, B.J. US-F 12.2.3
Jacobsen, J. DK 4.1
Jacobsen, N. DK 3.6
Jacobsohn, R. IL 1.1.1
Jacoli, G.G. CA 4.1
Jacqomain, E. CA 31.1
Jacquemond, C. FR 15.1.1
Jacquemond, M. FR 3.1.5
Jacquemont, R. FR 29.1
Jacques, M. FR 30.1.1
Jacques, R. FR 30.1, 30.1.1
Jacquet, M. FR 35.1
Jadczuk, E. PL 32.1.1
Jadhav, R.B. IN 14.3
Jaffee, B.A. US-B 9.1
Jaganath, S. IN 11.3
Jaganmohan, N. IN 1.1.7
Jagdale, IN 14.11
Jager, L. AU 30.1
Jagielski, K. PL 28.1.6
Jagschitz, J.A. US-A 14.1
Jagusz, M. PL 28.1.13
Jaha, F. YU 16.1
Jailloux, F. FR 4.1.6
Jain, B.L. IN 18.3
Jain*, B.P. IN 8.1
Jain, M.C. IN 19.1
Jain, V.B. IN 24.1.5
Jaivenois, A. BE 16.1
Jaja, E.T. NG 10.1
Jákó, N. HU 5.2
Jakob, L. DE 22.1
Jakóbczak, I. PL 15.1
Jakobsson, T. IC 1.1
Jakovác, F. HU 4.2
Jakubowski*, T. PL 28.1.1
Jakusz, A. PL 20.1.2
Jalea, A. RO 8.4.4
Jalikop, S.H. IN 1.1.1
Jamagne, M. FR 13.1.5
Jámbor, I. HU 2.2.4
Jámbor-Benczur, E. HU 2.2.4
James, D.J. GB 8.1.1
James, L. AU 30.1
James, P.A. AU 40.1
Jameson, J. GB 8.1.3
Jameson, P.E. NZ 14.1
Jamieson, A.R. CA 41.1
Jamieson, G.I. AU 15.1
Janda, Lj. YU 10.2
Janecki, J. PL 32.1.13
Janes, H. US-B 14.1.1
Jang, H.I. KR 23.1.4
Jáni, Gy. HU 2.2.9
Janick*, J. US-F 9.1.4
Janisiewicz, W. US-C 8.2
Janjić, V. YU 17.1
Janković, D. YU 9.1
Janković, R. YU 10.2
Jankulovski, D. YU 4.3
Jannamico, L. AR 23.1
Jannotti, O. BR 21.2
Janowski, G. PL 32.1.2.1
Jansen, H. DE 15.3.3
Jansen, J. NL 9.6.5
Jansen, P.C.M. NL 9.7.4
Jansen van Rensburg, J.G. ZA 3.1
Janson, C.G. NZ 12.2
Janssen*, H. DE 28.1.3
Janssens, S.R.M. NL 6.1
Janýška, A. CS 20.1
Janzein, FR 17.2
Janzsó, J. HU 2.6
Japhar Berniz, J.M. BR 6.1
Jaques, R.P. CA 17.1
Jaquinet, A. CH 2.3
Jaramillo, J. CO 4.2
Jarebica, D. YU 1.1

Járfás, J. HU 5.1.4
Jarmund, T. NO 10.5.2
Jarociński, B. PL 16.1
Jarošová, M. CS 20.1
Jaroszewica, G. PL 32.1.2.1
Jarowzyńska, T. PL 14.1.6
Jarrett, P. GB 18.1.2
Jarvis, W.R. CA 17.1
Jaša, B. CS 2.1
Jasić, J. CS 9.1
Jáskiewicz, B. PL 14.1.9
Jaszczak, J. PL 32.1.11
Jaumién*, F. PL 32.1.1
Jauregui, M.A. AR 13.1
Javier, F.B. PH 1.2.1
Javier Mercado, J. MX 3.2
Jaworksi*, C. US-D 22.1
Jaworska, K. PL 28.1.2
Jayamana, P.B. LK 3.5
Jayanth, K.P. IN 1.1.7
Jayapathy, S. LK 3.17
Jayasankar, N.P. IN 3.1
Jayaseelan, R. IN 1.1.3
Jayasekera, K.S. LK 2.1
Jayasuriya, M. LK 3.4
Jayawardene, S.D.G. LK 3.2
Jayesundera, H.P.S. LK 2.1
Jazbec, M. YU 18.1
Jean, S.H. TW 11.1.2
Jeannequin, B. FR 3.1.14
Jebbari, H. TN 1.1
Jędras, L. PL 24.1
Jeevananda, G.A.S.S. LK 3.8
Jeffers, S. US-F 23.1.4
Jeffries, C.J. GB 9.1.2
Jeffries, R.A. GB 15.1.4
Jeganathan, M. LK 2.1
Jelenić, Dj. YU 14.1
Jelenkovic, G.L. US-B 14.1.1
Jelev, I. BG 8.2
Jelley, R.M. IE 4.1
Jelmini, G. CH 2.4
Jenčo, M. CS 21.1
Jenkins, A.C. AU 41.1
Jenkins, I.B. AU 54.1
Jenkins, P.T. AU 27.1
Jenkins Jr., S.F. US-D 8.1.2
Jennings, D.L. GB 15.1.2
Jennings, P.H. US-G 31.1
Jennings, W.J. US-A 12.1
Jensen, A. DK 4.1
Jensen*, E. DK 7.1
Jensen, G. DK 3.4
Jensen, H.A. DK 10.1
Jensen, H.E.K. DK 1.2
Jensen, J. DK 1.1, 3.6
Jensen, J.B. DK 3.8
Jensen, J.H. DK 7.1
Jensen, K.I.N. CA 41.1
Jensen, M.H. US-H 27.1
Jenser, G. HU 2.5
Jeong, J.C. KR 22.1
Jeong, J.H. KR 23.1.5
Jeong, M.S. KR 12.3
Jeong, Y.G. KR 22.1
Jeoung, S.B. KR 16.1.3
Jeppsson, L. SE 3.1
Jerie, P.H. AU 32.1
Jermyn, W.A. NZ 13.2
Jerzy*, M. PL 4.1
Jesiotr, E. PL 28.1.10
Jessop, R.S. AU 2.1
Jestin, Y. FR 36.1
Jestrović, Ž. YU 10.1
Jewett, T.J. CA 17.1
Jeyaratnam, B.S. LK 3.13
Jhono, Y. JP 44.1
Jidav, L. RO 1.1.8
Jilcu, D. RO 7.1
Jilcu, M. RO 7.1
Jiménez, F. CR 2.2
Jiménez, J. CR 2.2
Jiménez Díaz, F. MX 5.2
Jin, I.D. KR 22.1
Jin, Y.O. KR 22.1
Jinbo, N. JP 4.1
Jindal, K.K. IN 10.1
Jindal, P.C. IN 2.1.1
Jinks*, R.L. GB 18.1.1
Jinno, Y. JP 42.1
Jinturkar, S.P. IN 12.1
Jiřík, J. CS 20.1
Jitsukawa, S. JP 4.2
Jo, H.M. KR 23.1.3
Jocić, T. YU 9.1
Jodo, S. JP 5.3
Joen, J.K. KR 7.2
Jogdande, N.D. IN 14.4
Johansen, R.H. US-G 12.1
Johansen, T.J. NO 9.1
Johansson, E. SE 8.2
Johansson, J. SE 1.1.3
Johjima, T. JP 18.3
Johns, G.G. AU 1.1
Johns, R. AU 33.2
Johnson, A.W. US-D 22.1
Johnson, C. US-D 20.1
Johnson, C.E. US-D 53.1
Johnson, C.R. US-D 27.5
Johnson, D.S. GB 8.1.3
Johnson, J. US-H 6.1
Johnson, J.R. US-D 14.1
Johnson, K. US-E 4.2
Johnson, R.C. CA 2.1
Johnson*, R.T. US-C 10.1
Johnson, T.H. US-B 5.1.1
Johnson, T.J. CA 40.1
Johnson, T.M. US-E 11.1
Johnson, W.B. US-B 14.1.1
Johnson, W.S. US-I 11.2
Johnson, Y. SC 1.1
Johnston, A.W.B. GB 24.1.2
Johnston, D. AU 60.1
Johnston, P.R. NZ 2.3.2
Johnston, R.L. AU 65.4
Johnston, S.A. US-B 14.1.6
Johnstone, G.R. AU 46.5
Joho, K. JP 18.1
Johsnon, L.C. US-G 19.1
Joiner, J.N. US-D 27.5
Jojima, N. JP 34.1
Jolivet, E. FR 17.1.5
Jolliffe*, P.A. CA 4.2
Joly, R.J. US-F 9.1.4
Jona, R. IT 24.1, 24.2
Jonassen, G. NO 2.1
Jones, A. US-D 12.2
Jones, A. US-F 14.1.3
Jones, A.R.C. CA 37.1
Jones, A.T. GB 15.1.3
Jones, B.M. US-I 11.1
Jones, E. AU 12.1
Jones, E. ZI 2.1
Jones, G.E. US-K 12.1.1
Jones, H.G. GB 8.1.1
Jones Jr., J.B. US-D 16.1
Jones, J.P. US-D 24.1
Jones, K. GB 16.1
Jones, K.M. AU 46.2
Jones, L.E. US-B 6.1
Jones, O.P. GB 8.1.1
Jones, R.A. US-K 6.1.4
Jones, R.K. US-D 8.1.2
Jones, R.P. GB 8.1.2
Jones, T. US-E 17.1
Jong, J.A. NL 18.1
Jonkers, J. NL 6.1
Jonsell, B. SE 7.1
Jönsson, H.A. SE 4.1
Jonsson, Y. SE 8.2
Joolka, N.K. IN 10.3
Jordan, B.J. GB 18.1.3
Jordan, D.T. NZ 5.1
Jordan, L.S. US-K 12.1.1
Jordan, R.A. AU 15.2
Jordan, R.B. NZ 5.1
Jordan, V.W.L. GB 19.1.1
Jordović, M. YU 8.1
Jorge, F. ES 11.1
Jørgensen, H.A. DK 4.1
Jørgensen, K. NO 10.1.8
Jørgensen, K.B. DK 4.1
Jørgensen, K.R. AU 15.1
Jørgensen*, M.B. DK 1.1

Jørgensen, L.N. DK 4.2
Jørgensen, P.E. DK 7.2
Joseph, J. IN 3.1
Joshi, G.D. IN 15.1
Joshi, Km.R. IN 24.1.4
Joshi, K.R. IN 24.1.1
Joshi, M.S. SD 2.1.1
Joshi, R.P. IN 24.1.2
Joshi, S.M. IN 24.1.3
Jošt, J. CS 5.1
Jotic, P. AU 46.2
Jouan, B. FR 14.1.2
Jouannic, F. FR 33.1
Joubert, A.J. ZA 1.1
Joubert, G. FR 21.1
Joung, H.Y. KR 18.1.1
Jourdheuil, P. FR 2.1.4
Jouret, M. FR 16.1
Joustra*, M.K. NL 2.1
Jovančević, R. YU 6.1
Jovanović, B. YU 14.2
Jovanović, M. YU 10.1, 14.1
Jović, S. YU 14.1
Joyce, M.J. ZI 3.1.1
Jozukuri, A. JP 40.2
Jraïdi, B. TN 1.1
Ju, H-Y. CA 42.2
Juang, G.J. TW 8.1
Juárez Galiano, L. PE 4.1
Jubb Jr., G.L. US-C 6.1
Jubb Sr., G.L. US-B 11.1
Judd, G.J.R. CA 17.1
Judd, M.J. NZ 5.1
Juhanoja, S. FI 1.1
Juhász, O. HU 2.2.10
Jullian, P. FR 3.1.10
Junak, S.A. US-K 16.1
Jung, H.B. KR 23.1.1
Jung, H.J. KR 4.1
Jung, J.D. KR 7.2
Jung, J.H. KR 24.1
Jung, J.K. KR 10.1
Jung, J.S. KR 2.1.2
Jung, S.T. KR 7.2
Junker, K. DK 4.2
Junnila, S. FI 1.1
Junttila, O. NO 9.2
Jura, J. PL 19.1
Jurčić, V. YU 3.1
Jürgensen, C. DE 15.1.9
Jurubità, J. RO 8.4.3
Juste, C. FR 4.1.2
Jutori, H. JP 17.1
Juven, B. IL 1.3.2
Juvik, J.A. US-F 12.2.2

Kaaber, L. NO 10.5.1
Kaack, K. DK 1.3
Kaai, C. NL 6.1, 9.4.3
Kaaya, N.A. TZ 3.1
Kabasakal, A. TR 4.1
Kabasawa, F. JP 10.1
Kabaši, B. YU 16.1
Kabira, N. KE 5.1
Kable, P.F. AU 9.1.1
Kabumoto, T. JP 13.1
Kackar, N.L. IN 18.2
Kacperska, I. PL 32.1.4
Kaczor, M. PL 22.1
Kadam, B.A. IN 12.1
Kadam, P.Y. IN 14.3
Kaddous, F.G. AU 28.1
Kader*, A.A. US-K 6.1.3
Kadłubowski, W. PL 20.2.2
Kadman*, A. IL 1.2.5
Kadman-Zahavi*, A. IL 1.1.2
Kadowaki, Y. JP 42.1
Kadoya*, K. JP 5.3
Kadrekar, S.B. IN 15.1
Kadzik, F. PL 3.1
Kafetzakis, N. GR 1.1
Kaffka, K. HU 2.4
Kagami, H. JP 31.1
Kagami, Y. JP 17.1
Kagan, F. PL 20.1.1
Kagan-Zur, V. IL 4.1
Kagawa, M. JP 40.2
Kagaya, H. JP 2.1
Kageyama, Y. JP 31.2
Kagiri, KE 1.3
Kagohashi, S. JP 23.2.3
Kahandawala, J.B.D.S. LK 3.7
Kahn, V. IL 1.3.2
Kåhre, L. SE 5.2
Kai, I. JP 30.1
Kaihara, M. JP 31.1
Kainrath, D. AT 8.2
Kairov, A.K. SU 14.1
Kaisuwan, T. TH 3.1
Kajiura*, I. JP 38.3
Kakehi, M. JP 11.4
Kakihara, F. JP 5.3
Kakinaka, S. JP 22.2
Kakizaki, M. JP 12.1
Kako, S. JP 1.3
Kalab, I.M. LY 2.1
Kalášek, J. CS 32.1
Kalash, A. LY 4.1
Kállay, E. HU 2.1.5
Kállay, T. HU 2.1.2
Kalberer, P. CH 1.1
Kale, P.N. IN 14.1
Kalezić, Ž. YU 7.1
Kalfov, St. BG 8.1
Kalinina, I.P. SU 2.1
Kalinke*, H. DE 12.1.13
Kalinowski, Z. PL 14.1.1
Kalizang'oma, D.C. ML 1.1
Kalkman, E.Ch. NL 1.1.3
Kalmoukos, P.E. GR 14.1.4
Kalofonos, P. GR 23.1
Kalra, S.K. IN 17.1, 26.1
Kaltenbach, C.C. US-I 12.1
Kaltsikis, P. GR 6.1.5
Kalyvas, G. GR 5.1
Kamada, C. JP 3.2.1
Kamakura, J. JP 3.2.4
Kamau, C.K. KE 5.1
Kamau, E. KE 2.1
Kamel, H.A. EG 1.1, 1.5
Kamemoto*, H. US-M 2.2
Kamińska, A. PL 2.1
Kamińska, M. PL 28.1.16
Kamiński, S. PL 2.1
Kamiński, W. PL 18.1.14
Kaminsky, M. CR 1.1
Kamionka, S. PL 25.1
Kamiya, T. JP 9.2
Kamminga, J. NL 6.1
Kamoen, O. BE 15.3
Kamota, F. JP 14.2
Kamura, T. JP 1.1
Kanafusa, A. JP 7.1
Kanagaratnam, P. LK 2.1
Kanahama, K. JP 45.2
Kanai, U. JP 10.1
Kanamaru, N. JP 41.1
Kanaoka, K. JP 44.2
Kanasirska, V. BG 8.1
Kanaujia, J.P. IN 24.1.6
Kanazaya, T. JP 20.2
Kanbara, Y. JP 22.3
Kanbe, K. JP 2.3
Kanda, M. JP 4.3
Kandasamy, S. LK 3.10
Kandić, B. YU 13.1
Kaneko, J. JP 2.1
Kaneko, K. JP 12.3
Kaneko, M. JP 1.1.1
Kaneko, T. JP 39.1
Kang, B.K. KR 12.3
Kang, C.K. KR 23.2
Kang, H. KR 12.2
Kang, H.J. KR 12.3
Kang, H.Y. KR 12.3
Kang, K.Y. KR 2.1.1
Kang, S.J. KR 23.1.3
Kang, S.K. KR 10.2
Kang, S.M. KR 12.2
Kang, S.W. KR 19.1
Kang, Y.D. TW 9.1
Kania, C. PL 33.2
Kania, S. PL 22.1
Kaniewski, W. PL 20.1.2

Kaniszewski, S. PL 28.2.3
Kankam, J.S. GA 2.2.1
Kannan, K. IN 5.6
Kanner, J. IL 1.3.2
Kanno, T. JP 23.2.2
Kano, T. JP 38.3
Kano, Y. JP 15.3
Kantartzis, N. GR 6.1.4
Kantas, A. GR 9.1
Kanters, F.M.L. NL 6.1
Kanwar*, J.S. IN 17.1
Kanyagia, S.T. KE 1.4
Kapatos, E. GR 9.1
Kapeller, K. HU 5.3.3
Kaper, J.M. US-C 3.1
Kapetanakis, E.G. GR 14.1.4
Kapitány, J. HU 5.3.3
Kaplan, D.T. US-D 27.2, 34.1
Kapotis, G. GR 6.1.3
Kappel, F. CA 17.1
Kapur, S.P. IN 17.1
Kapusta, G. PL 28.4
Karabetsos, J. GR 13.1
Karabiyik, N. TR 15.2
Karaca, R. TR 11.1
Karagianni, E. GR 19.1
Karagozoglu, E. TR 18.1
Karaguzel, O. TR 13.1
Karahasonovic, U. YU 1.1
Karai, J. HU 2.2.5
Karaivanov, V. BG 8.1
Karalakis, M. 12.1
Karale, A.R. IN 14.1
Karantonis, N. GR 18.1
Karatas, H. TR 3.2
Karavitis, P. GR 13.1
Karchi, Z. IL 1.1.1
Karczewska, M. PL 20.2.2
Karczewski, J. PL 34.1
Kardos, Z. HU 6.1
Kärelampi, L. FI 15.1
Karihaloo, J.L. IN 1.1.3
Karikari, S.K. NG 2.1
Karikari, S.K. GA 1.2
Karimata, A. JP 41.3
Kariuki, M. KE 1.7
Karkara, B.K. IN 10.2
Karlsen, P. DK 3.1
Karmakar, P.G. IN 1.1.4
Karnatz*, A. DE 5.1.1
Karner, K. US-H 4.1
Karnik, A.R. IN 15.1
Karolczak, W. PL 28.1.10
Karolewski, P. PL 10.1.3
Karren, J.B. US-I 5.1
Karssen, C.M. NL 9.7.2
Karsten, J.H.M. ZA 4.1
Kartalov, P. BG 8.3
Karube, M. JP 18.1
Karunaratne, S.M. LK 2.1
Karvounis, G.C. GR 9.1
Karwowska, K. PL 32.1.2.2
Karydes, J. GR 13.1
Kasahara, T. JP 29.1
Kashimoto, N. JP 22.2
Kashimura, Y. JP 16.2
Kashio, T. JP 27.3
Kashyap, A.S. IN 10.3
Kasijadi, F. ID 1.5
Kaska, N. TR 1.1
Kass, D. CR 2.2
Kassai, T. JP 17.1
Kassimis, D. GR 9.1
Kasukawa, T. JP 19.3
Kasumi, S. JP 14.1
Kasuya, M. JP 1.1.1, 39.1
Kasuyama, S. JP 31.1
Kasza-Kerék, G. HU 4.2
Katagi, S. JP 19.2
Katamoto, T. JP 17.1
Katan, Y. IL 2.6
Katano, Y. JP 1.1.1, 19.2
Kataoka, I. JP 17.3
Kataoka, K. JP 23.1
Katerji, N. FR 17.1.7
Katerov, K. BG 7.1
Kathirgamathyar, S. LK 4.1
Katiyar, B.C. IN 22.1
Katiyar, R.S. IN 22.1
Kato, A. JP 33.1
Kato, H. JP 9.2, 41.3
Kato, K. JP 8.2
Kato, M. JP 1.1.1, 5.3
Kato, O. JP 41.1
Kato, S. JP 1.1.3, 2.2, 12.2, 25.1.2, 38.2
Kato, T. JP 1.1.1, 3.2.4, 20.4, 41.1, 46.1
Kato, Y. JP 41.1
Katranis, N. GR 15.1
Katsaboxakis, K. GR 18.1.2
Katsoyiannos, V. GR 25.1.6
Katsuda, H. JP 6.3
Katsumine, M. JP 1.1.1
Katsura, H. JP 4.2
Katsura, N. JP 23.2.3
Katsuragawa, N. JP 9.3
Katsurayama, Y. BR 28.1
Katsutani, T. JP 11.1
Kattan, A.A. US-E 11.1
Kaukovirta*, E. FI 10.1.1
Kaul, G.L. IN 2.2
Kaul, J.L. IN 10.1
Kaulgud, S.N. IN 14.8
Kaurin, A. NO 9.2
Kavanagh, E. AU 65.4
Kavanagh, J.A. IE 3.1
Kavanagh, T. IE 4.1
Kaven, E. AT 8.2
Kawabuchi, A. JP 1.1.1
Kawada, K. JP 17.3
Kawae, M. JP 17.1
Kawahara, I. JP 6.2, 25.1.2,
Kawahara, S. JP 12.2
Kawai, A. JP 7.3
Kawai, H. JP 24.2
Kawai, K. JP 3.1
Kawai, T. JP 31.1
Kawaide, T. JP 23.2.2
Kawakami, R. JP 5.2
Kawakami, T. JP 41.3
Kawakita, T. JP 21.1
Kawamata*, S. JP 41.1
Kawamoto, K. CP 23.2.4
Kawamura, H. JP 40.1
Kawamura, M. JP 38.1
Kawamura, S. JP 41.1
Kawamura, T. JP 16.1, 46.1
Kawanishi, Y. JP 17.1
Kawano, N. JP 30.1
Kawarada, T. JP 24.1
Kawasaki, C. JP 15.3
Kawasaki, K. JP 17.2, 25.1.3
Kawasaki, S. JP 25.1.2, 34.1, 43.1
Kawasaki, Y. JP 4.1
Kawasato, H. JP 39.1
Kawase, K. JP 22.5, 38.3
Kawase, S. JP 4.1
Kawashima, K. JP 3.2.3
Kawashima, M. JP 9.3
Kawashima, N. JP 28.1
Kawashita, T. JP 40.1
Kawata*, J. JP 43.2
Kawathalkar, M.P. IN 13.5
Kawauchi, H. JP 5.1
Kawazoe, F. JP 1.2
Kawecki*, Z. PL 18.1
Kay, I. AU 13.1
Kay, M.G. GB 31.1
Kaya, A. TR 12.1
Kaynas, K. TR 14.1
Kays, S. US-D 16.1
Kazuma, T. JP 6.1
Kazumi, S. JP 14.3
Kearney, B. IE 3.2
Keblovský, L. CS 31.1
Kecskeméti, L. HU 5.3.1
Kecskés-Szabó, I. HU 2.2.4
Kedar, N. IL 2.3
Kedici, F. TR 2.2
Kee, W.E. US-C 2.1
Keegan, L.M. US-B 6.1
Keever, G.J. US-D 38.1

Kefford, N.P. US-M 2.2
Keiding, H. DK 3.3
Keller, F. CH 1.1
Keller, J. PL 28.1.6
Kellerhals, M. CH 1.1
Kelley, J.D. US-G 20.1
Kellock, A.W. AU 35.1
Kelly, J.C. IE 4.1
Kelly*, J.F. US-F 14.1.1
Kelly, J.W. US-D 13.1
Kelman, A. US-F 23.1.4
Kemenes, M. HU 2.1.2
Kemp, M.S. GB 19.1.1
Kemp, T. US-E 15.1
Kempenaers, A. BE 4.1
Kempler, C. CA 2.1
Kempski, F. PL 32.1.1
Kendall, D.A. GB 19.1.1
Kender*, W.J. US-D 30.1
Keng, J.C.W. CA 1.1
Kennard, W.C. US-A 13.1
Kennedy, G.G. US-D 8.1.3
Kennedy, R.A. US-F 3.1, 7.1.2
Kennel, W. DE 28.1.3
Kenneth, R. IL 2.6
Kepczyńska, E. PL 28.1.2
Kepczyński, J. PL 28.1.12
Kępka, M. PL 32.2.1
Kerek, M.M. HU 2.1.1
Kerényi-Nemestóthy, K. HU 2.2.8
Kerkhoff, P. SR 1.1
Kerlan, C. FR 14.1.2
Kermarrec, A. FR 18.1
Kern, H. DE 12.1.4
Kerr, A. AU 42.1
Kerr, J.P. NZ 8.1
Kerr, R.M. NZ 10.1
Kerr, S.H. US-D 27.2
Kerr, W.E. BR 6.2
Kerridge, G.H. AU 65.2
Kersten, E. BR 30.3
Kesava Murthy, S.V. IN 1.1.9
Kesici, S. TR 13.1
Kesik, T. PL 14.1.3
Keskar, B.G. IN 14.1
Kesselring, H. AT 10.2
Kessler, B. IL 1.2.2
Kessner, C.D. US-F 19.1
Kester*, D.E. US-K 6.1.3
Kesterton, J.W. US-D 30.2
Ketchie, D.O. US-J 10.1
Ketel, D.H. NL 9.5
Kettner, H. DE 11.2.6
Keuken, W. NL 9.7.5
Khader, S.A.E. IN 26.1
Khaemba, E.N. KE 1.3
Khajuria, H.N. IN 17.1
Khalifa, A. EG 3.1
Khammassi, N. TN 1.1
Khan, A.A. US-B 2.1.2
Khan, M.M. IN 11.7
Khandelwal, R.C. IN 19.1
Khangarot, K.S. IN 19.1
Khangura, J.S. IN 17.1
Khanna, R.C. IN 2.1.3, 2.1.4
Khatamian, H. US-G 31.1
Khedkar, D.M. IN 12.1
Khewkeaw, S. TH 6.1
Khokhar, U.U. IN 10.3
Khurdiya, D.S. IN 2.1.6
Kiang, Y.T. US-A 2.1
Kichefski, D. US-F 27.1
Kiciński, M. PL 32.1.11
Kido, K. JP 45.1
Kiefer, W. DE 12.1.1
Kiemayer, AT 6.1
Kiemstedt, H. DE 15.2.4
Kienzel, H. AT 8.6
Kiermeier, F. DE 11.2.8
Kihara, H. JP 26.2
Kihara, T. JP 17.2
Kiihl, R.A.S. BR 27.1
Kikkawa, M. JP 23.2.5
Kiku, T. JP 18.2
Kikuchi, H. JP 24.1
Kikuchi, S. JP 38.2
Kikuchi, T. JP 3.4, 16.1, 21.1
Kilby, M.W. US-H 27.1
Kiles, B.S. US-I 14.1
Kilgour, J. GB 31.1
Kim, B.C. US-A 14.1
Kim, B.D. KR 18.1.3
Kim, B.S. KR 7.2
Kim, B.W. KR 4.1, 21.3
Kim, C.C. KR 21.3
Kim, C.M. KR 10.2
Kim, E.Y. KR 4.1
Kim, H.D. KR 23.1.1
Kim, H.J. KR 18.1.1, 22.1
Kim, H.T. KR 2.1.2
Kim, H.Y. KR 10.1, 17.1, 21.4, 23.1.6
Kim, I.K. KR 7.4
Kim, J.B. KR 16.1.2
Kim, J.C. KR 11.1
Kim, J.G. KR 20.1
Kim, J.H. KR 3.2, 8.4, 23.1
Kim, J.K. KR 18.1, 23.1.4
Kim, J.S. KR 7.1
Kim, J.Y. KR 6.1, 23.1.5
Kim, K.H. KR 10.2
Kim, K.J. KR 21.4
Kim, K.R. KR 7.2
Kim, K.S. KR 10.1, 13.2, 18.1.1
Kim, K.W. KR 14.1
Kim, K.Y. KR 4.1, 23.1.2, 23.1.4
Kim, M.J. KR 2.1.3
Kim*, M.S. KR 23.1.3
Kim, S.B. KR 21.2, 23.1.4
Kim, S.D. KR 7.3
Kim, S.H. KR 10.1
Kim, S.I. KR 18.1.3
Kim*, S.J. KR 23.1.1
Kim, S.K. KR 3.2
Kim, S.W. KR 7.4
Kim, S.Y. KR 23.1.6
Kim, T.H. KR 24.1
Kim, W.B. KR 5.1
Kim, W.C. KR 16.1.1
Kim, W.S. KR 16.1.1
Kim, Y.C. KR 18.1.1, 24.1
Kim, Y.H. KR 1.1, 10.2, 21.3
Kim, Y.J. KR 23.1.5
Kim, Y.K. KR 23.1.3
Kim, Y.R. KR 8.2
Kim, Y.S. KR 16.1
Kim, Y.W. KR 7.4
Kimball, E.R. CA 41.1
Kimmins, R.K. US-G 31.1
Kimnach, M. US-K 15.1
Kimoto, H. JP 31.1
Kimotó, T. BR 22.1
Kimpinski, J. CA 43.1
Kimura, E. JP 8.1, 15.1
Kimura, H. JP 11.2
Kimura, I. JP 46.1
Kimura, K. JP 20.1
Kimura, M. JP 25.2
Kimura, N. JP 1.1.1
Kimura, O. BR 19.2
Kimura, S. JP 38.1
Kimura, T. JP 31.1
Kimura, Y. JP 10.1, 31.1, 46.1
Kinbacher, E.J. US-G 23.1
Kinbara, T. JP 1.1.3
Kinet*, J-M. BE 11.1
King, G. NZ 10.1
King Jr., GA US-D 13.1
King, J.E. GB 13.1.4
King, P.D. NZ 3.1, 5.1
King, R.R. CA 40.3
Kingman, A.R. US-D 13.1
Kingori, B.W. KE 5.1
Kingston, C.M. NZ 10.1
Kinjo, E. JP 32.1
Kinjo, H. JP 32.1
Kinoda, M. JP 3.2.3
Kinoshita, K. JP 31.2, 47.1
Kinoshita, S. JP 44.1

Kinoshita, T. JP 1.1.1
Kinsella, M.N. AU 26.3
Kinyua, G.K. KE 1.8
Kio, P.R.O. NG 1.3
Kioko, S. KE 3.1
Király, L. HU 5.1.1
Király, Z. HU 2.5
Kirchgatter, G. DE 11.2.4
Kirilov, D. BG 13.2
Kirimura, Y. JP 13.1
Kirinde, S.T.W. LK 4.1
Kirkham, R.P. AU 35.1
Kirkpatrick, B. US-K 6.1.1
Kirksey, R. US-H 21.1
Kirton, L.H. ZI 5.1
Kirts, C.A. US-L 1.1
Kiselnic, L. DE 15.3.2
Kishimoto, O. JP 39.2
Kishino, A.Y. BR 27.2
Kishore*, D.K. IN 2.1.2
Kishun, R. IN 1.1.6
Kisiel, M. PL 22.1
Kisino, I. JP 27.2
Kiss, A. HU 2.1.3
Kiss, F.H. HU 5.3.1
Kiss, I. HU 2.4
Kiss, L.Z. HU 2.2.9
Kita, K. JP 5.2
Kita, N. JP 19.2, 40.1
Kitada, M. JP 43.2
Kitagawa, H. JP 17.3
Kitagawa, M. JP 37.1
Kitagawa, Y. US-M 2.2
Kitaguchi, M. JP 4.1
Kitahara, H. JP 8.1
Kitajima, A. JP 21.1, 22.4
Kitajima, E.W. BR 25.4
Kitajima, H. JP 21.1
Kitajima, K. JP 41.1
Kitamura, H. JP 36.1
Kitamura, T. JP 45.2
Kitano, Y. JP 44.2
Kitayama, H. JP 3.2.2
Kitayama, S. JP 37.1
Kitivo, D. KE 7
Kitsos, G. GR 22.1
Kitto, S.L. US-C 2.1
Kiuru, P. KE 4.1
Kiyasu, K. JP 5.1
Kiyokawa, I. JP 18.3
Kiyosawa, Y. JP 7.3
Klapzubova, H. CS 4.1
Klarman, W.L. US-D 8.1.2
Klas, F.E. SR 1.1
Klaus, R. AT 8.2
Klaver, S. NL 18.1
Klebasadel, L.J. US-L 2.1
Kleber de Abreu Mattos, J. BR 25.4
Kleijburg, P. NL 9.8
Klein, I. IL 1.2.4
Klein, W. US-F 19.1
Kleinhesselink, H. NL 1.1.6
Kleinkopf, G. US-I 16.1
Kleinschuster, S.J. US-B 14.1
Klement, Z. HU 2.5
Klenczner, A. HU 2.2.9
Klenert, M. DE 15.3.2
Klenkhart, E. AT 8.2
Kleparski, J. PL 1.1
Klett, J.E. US-G 19.1
Klett, J.E. US-I 2.2
Kliewer, W.M. US-K 6.1.5
Kliffen, C. NL 9.4.5
Klijn, J.A. NL 9.11.1
Klimaj, P. HU 5.3.1
Klimczak, A. PL 1.1
Klimpel, J. CS 7.1
Klinac, D.J. NZ 5.1
Klinck, H.R. CA 37.1
Kline, R.A. US-B 5.1.3
Kling, G.J. US-F 12.2.2
Klingaman, G.L. US-E 11.1
Klinger, F. AT 8.1
Klingler, J. CH 1.1
Klinner, W.E. GB 31.2.5
Kljajić, R. YU 14.1
Klop, W. NL 9.10
Kloppstech, K. DE 15.3.3
Klos, E.J. US-F 14.1.3
Klug, L. AT 18.1
Knafleski*, M. PL 20.2.11
Knapp, J.L. US-D 30.2
Knapp, P.H. GB 18.1.3
Knavel, D. US-E 15.1
Knee, M. GB 8.1.3
Kneen, D.A. AU 34.1
Knegt, E. NL 9.7.2
Knight, C. GB 6.1
Knight*, J.N. GB 8.1.1
Knight, V.H. GB 8.1.2
Knipple, D.C. US-B 2.1.3
Knoblauch, F. DK 2.1
Knott, C.M. GB 27.1
Knowles*, N.R. CA 4.2
Knutson, K.W. US-I 2.2
Knälmann, M. DE 15.3.2
Ko, K.C. KR 23.3
Ko, K.D. KR 10.1
Kobashi, Y. JP 22.3
Kobayashi, H. JP 11.2
Kobayashi, K.D. US-M 2.2
Kobayashi, M. JP 18.1, 27.1, 47.1
Kobayashi, N. JP 23.1
Kobayashi, S. JP 11.3, 13.2
Kobayashi, T. JP 13.1, 26.1
Kobayashi, Y. JP 7.1
Kobel, F. CH 1.1
Kobiela, B. PL 30.1
Koblet, W. CH 1.1
Kobori, O. JP 23.2.3
Kobriger, J.M. US-H 27.1
Kobro, S. NO 10.4.3
Kobryń, J. PL 32.1.2.1
Koburger, J.A. US-D 27.3
Koc, N. TR 12.1
Kocęba, B. PL 28.4
Kocerba, A. PL 24.1
Koch, A.R. US-B 14.1.2
Koch, J. DK 3.5
Koch, N. HU 2.1.5
Koch, N.E. DK 3.2
Kochan, W. US-I 14.1, 15.1
Kochman, J. PL 32.1.6
Köck, L. AT 17.1
Köck, O. HU 2.7
Kocsis, F. HU 2.6
Kocsis, L. HU 2.7
Koda, S. JP 30.1
Koda, T. JP 29.1
Kodachi, K. JP 4.2
Kodama, K. JP 24.1
Kodama, T. JP 46.1
Kodera, K. JP 41.1
Kodo, H. JP 40.1
Koehler, C.S. US-K 3.1.1
Koerting, L. US-B 7.1
Koffmann, W. AU 3.1
Kofranek*, A.M. US-K 6.1.2
Koga, T. JP 34.2
Koganezawa, H. JP 16.2
Kogbe, J.O.S. NG 1.2
Koguchi, M. JP 39.1
Kogure, K. JP 35.2
Kohata, M. JP 18.1
Kohl, H.C. US-K 6.1.2
Kohli, R.R. IN 1.1.1
Kohlmann, H. AT 8.11
Kohno, S. JP 41.1
Kohta, K. JP 4.1
Koide, H. JP 1.1.1
Koike, A. JP 40.2
Koike, H. JP 26.2
Koizumi, M. JP 38.3
Kojić, A. YU 15.1, 16.1
Kojima, H. JP 1.1.4
Kojima, M. JP 23.2.3
Kojima, T. JP 34.4
Kojović, R. YU 10.1
Kok, C. ZA 4.2
Kokate, A.S. IN 14.1
Kokkalos*, T. CY 1.1
Köksal, I. TR 2.1
Kolar, J.J. US-I 17.1

Kołasa, Z. PL 14.1.6
Kolb, W. DE 33.1
Kolbusz, E. PL 3.1
Koleda, I. HU 2.2.6
Kolev, K. BG 15.1
Koll, H. AT 5.1
Kollányi, L. HU 2.1.3
Kollar, G. HU 2.1.2
Kollas, D.A. US-A 13.1
Koller*, O.L. BR 28.1
Kollias, A. GR 18.1.3
Kolling, J. BR 29.1
Köllner, V. DE 5.4
Kołodziejczak*, P. PL 3.1
Kołodziejek, E. PL 28.5
Kołota, E. PL 33.1
Kolozsvari, B. HU 2.2.9
Komabayashi, K. JP 45.1
Komamura, K. JP 8.2
Komarčević, D. YU 10.1
Komatsu, E. JP 14.1
Komatsu, H. JP 21.3, 26.1, 44.2
Komazaki, S. JP 38.3
Komiszar, L. HU 2.2.4
Komlósi, Cs. HU 5.3.1
Komnenić, V. YU 9.1
Komochi, S. JP 23.2.2
Komori, S. JP 14.1, 26.2
Komori, T. JP 47.1
Komorowska, E. PL 28.2.13
Komosa, A. PL 20.2.3
Kon, T. JP 3.2.4
Konahar, T. IN 16.2
Konakahara, M. JP 38.2
Konarli, O. TR 14.1
Koncz, K. HU 2.2.1
Kondarev, M. BG 8.2
Kondilis, Th. GR 11.1
Kondo, H. JP 7.1, 20.3
Kondo, S. JP 2.2
Konečný, J. CS 27.1
Kongsrud, K.L. NO 6.1
Konishi, K. JP 31.2
Konno, H. JP 12.2
Konno, Y. JP 16.1
Kono, H. JP 37.1
Kono, M. JP 18.2
Kono, T. JP 5.1
Kono, Y. JP 37.1
Konomi, K. JP 7.2
Konopacka, Z. PL 28.1.8
Konosu, S. JP 41.5
Konradsen, P. DK 1.2
Konsler, T.R. US-D 6.1
Konstantakis, A. GR 24.1
Konstantinidou, E. GR 22.1.1
Konstatinović, B. YU 14.1
Kontonikolaou, D. GR 18.1.2
Konys, E. PL 20.2.11
Koo, R.C.J. US-D 30.2
Kooistra*, E. NL 8.1
Kooistra, M.J. NL 9.11.1
Kooistra, T. NL 9.8
Koolstra, J.W. NL 16.2
Koono, S. JP 31.1
Kopec, K. CS 8.1
Kopel, R. IL 5.1.1
Koppensteiner, O. AT 8.2
Koprivica, R. YU 10.1
Kopytowski, J. PL 18.1
Korać, M. YU 17.1
Koranski, D.S. US-G 20.1
Korban, S.S. US-F 12.2.2
Korbuly, J. HU 2.2.6
Korcak, R.F. US-C 3.1
Korcz, L. PL 20.1.1
Korematsu, H. JP 11.1
Korenaga, R. JP 38.3
Korenica, B. YU 16.1
Körmendi, D.A. HU 6.1
Kornova, K. BG 8.2
Környei, B. HU 5.2
Kőrös, L. HU 5.3.1
Korošec, Z. YU 18.2.1
Koroxenidis, N. GR 25.2.3
Korsak, A. PL 26.1
Korszun, S. PL 20.2.10
Kort, J. CA 9.1
Kortekaas, B.M.M. NL 3.1
Korunović, R. YU 14.1
Kościelak, M. PL 32.3.3
Košek, J. CS 27.1
Koseoglu, A.T. TR 3.1
Kosevska, O. YU 4.1
Koshikawa, K. JP 9.1
Koshy, P.K. IN 3.1
Kosi, F. YU 14.1
Kosina, V. CS 7.1
Kosiński, G. PL 10.1.1
Kośka, L. PL 28.1.8
Koslinska, M. PL 28.1.2
Kosmala, M. PL 32.1.12
Kosmalski, Z. PL 32.3.4
Kosseva, D. BG 4.1
Kosson, R. PL 28.2.10
Kossowski, J. PL 14.1.3
Kostadinović, Z. YU 10.2
Koster, H. NL 9.9
Kostewicz, S.R. US-D 27.7
Kostov, D. BG 8.1
Kostova, D. BG 8.1
Kostrzewa, E. PL 24.1
Kosuge, E. JP 41.1
Kotani, A. JP 17.2
Koter, M. PL 14.1.6
Koths*, J.S. US-A 13.1
Kothwade, D.V. IN 14.4
Kotlińska, T. PL 28.2.1
Kotliński, S. PL 28.2.7
Kotmajer, Gy. HU 2.7
Kotobuki, K. JP 14.2
Kotoula-Syka, E. GR 25.2.2
Kótun, K. HU 6.1
Koufakis, E. GR 25.1.2
Koukourikou-Petridou, M. GR 25.1.4
Koukoylakis, P. GR 25.2.3
Kouno, H. JP 42.3
Koutepas*, N.G. GR 21.1
Koutroubas, A. GR 26.1
Koutsaftakis, A. GR 8.1.1
Kouvaraki, K. GR 6.1.7
Kouyama, R. JP 7.3
Kouzo Kato, A. BR 2.1
Kováčik, A. CS 22.4
Kovačina, R. YU 2.1
Kovács, A. HU 2.6, 5.1.3
Kovács, G. DK 3.5
Kovács, Gy. HU 5.1.4
Kovács, J. HU 6.1
Kovács, S. HU 2.2.2
Kovaleski, A. BR 30.1
Kováts, Z. HU 2.1.5
Kowalczyk, W. PL 28.1.13
Kowalenko, C.G. CA 1.1
Kowalik, B. PL 28.1.1
Kowalska, T. PL 20.1.3
Koyaizu, Y. JP 47.2
Koyama, H. JP 1.2
Koyama, Y. JP 13.2, 35.2
Koyanagi, Y. JP 34.3
Koyuncu, G. TR 4.1
Kozai*, T. JP 4.4
Kozaki, I. JP 14.2
Kozanecka, T. PL 32.2.1
Kozár, F. HU 2.5
Kozera, G. PL 11.1.2
Kozera*, W. PL 11.1.10
Kozik, E. PL 20.2.3, 28.2.1
Kozikowska, E. PL 32.3.4
Kozłowska, E. PL 28.5
Kozłowska, L. PL 14.1.5
Kozłowski, J. PL 20.1.1
Kozłowski, M. PL 32.1.7
Kozlowski, R.E. US-B 5.1.1
Kozma, Jr. P. HU 5.2
Kozma, P. HU 2.2.10
Kraai, A. NL 9.3
Krahn, T.R. CA 6.1
Krajewski, A. PL 32.3.5
Krall, D.W. US-B 5.1.1
Kramarovsky, E. AR 3.1.1
Kramer, C.F.G. NL 6.1

Kratky, B.A. US-M 1.1
Kratochvílová, H. CS 20.1
Kraus, A. DE 11.2.7
Kraus, Lj. IC 1.1
Kraus, V. CS 14.1.5
Krause*, J. PL 20.2.5
Krause, M. AU 38.1
Krawiarz, K. PL 10.1.4
Krawiec, Z. PL 9.1
Krczal, H. DE 9.1
Krebs, Chr. CH 1.3
Kreiter, S. FR 10.1.2
Krekule, J. CS 22.2
Kretchman*, D.W. US-F 7.1.2
Kretschmer*, M. DE 12.1.6
Krewer, G. US-D 22.1
Krexner, R. AT 8.11
Krezdorn, A.H. US-D 27.4
Krèzel, J. PL 33.1
Krgović, Lj. YU 6.1
Kricnar, K. CS 7.1
Kriedemann*, P.E. AU 42.1
Kriegl, J. AT 10.1
Krienke, Ch. DE 23.1
Křikava, K. CS 2.1
Křikava, M. CS 14.1.4
Krikke, A.T. NL 6.1
Krishnakumar, N.K. IN 1.1.7
Krishnamurthy, A. IN 1.1.7
Krishna Murthy, P.N. IN 1.1.7
Krishnamurthy, S. IN 1.1.5
Krishnan, R. IN 1.1.4
Krishnappa, K.S. IN 11.2
Krishna Prasa, K.S. IN 4.1
Krishna Prasad, V.S.R. IN 1.1.2
Krishna Rao, D.S. IN 1.1.7
Kristensen*, H.R. DK 4.1
Kristensen*, L.N. DK 1.2
Kristiansen, K. DK 1.2
Kristiansson, P. SE 1.1.2
Kristoffersen*, T. SE 1.1.3
Kristóf-Kégl, E. HU 2.2.12
Kriszta, E. HU 2.2.1
Křivský, K. CS 5.1
Krizek, D.T. US-C 3.1
K"rjin, Chr. BG 6.1
Krogness, O.A. NO 10.1.8
Krokowska, L. PL 28.2.13
Krokowski, A. PL 28.2.8
Król, A. PL 7.1
Król, M. PL 2.3.1
Król, S. PL 7.1, 20.2.1
Krolis, K.E. NL 9.3
Kronenberg*, H.G. NL 9.7.1
Kronmann, J. DK 7.1
Kroon, G.H. NL 9.6.3
Kroon, K.H. AU 30.2
Kropáč, Z. CS 24.2
Kropczyńska, D. PL 32.1.7
Kropisz*, A. PL 32.1.4
Kropp, K. PL 11.1.2
Krpeš, J. CS 11.1
Krstić-Pavlović, N. YU 14.1
Kruczynska, D. PL 28.1.1
Krug, G. BE 6.2
Krug*, H. DE 15.1.3
Kruger*, N.S. AU 15.1
Krugliakov, A.V. SU 12.1
Krul, W.R. US-A 14.1
Krusze*, N. PL 32.1.5
Kruszelnicka, U. PL 14.1.5
Kryczyński, S. PL 32.1, 32.1.6
Kryńska, W. PL 18.1
Krzewińska, D. PL 21.1
Krzewiński, J. PL 21.1
Krzywański, Z. PL 20.2.7
Ku, J.H. KR 8.2
Kubečka, D. CS 1.1
Kubiak, K. PL 32.3.1
Kubik, S. PL 22.1
Kubo, I. US-K 3.1.1
Kubo, T. JP 46.1
Kubo, Y. JP 31.2
Kuboi, S. JP 41.3
Kubokawa, S. JP 47.2
Kubota, A. JP 35.2
Kubota, N. JP 31.2
Kubota, S. JP 5.2
Kubouchi, Y. JP 20.2
Kuc, B. PL 28.2.8
Kučera, V. CS 22.4
Kucmierz, J. PL 11.1.4
Kucuk, A. TR 12.1
Kucuk, Y. TR 12.1
Kudo, A. JP 14.2
Kudo, K. JP 16.2
Kudo, N. JP 3.2.1
Kudo, S. JP 3.2
Kudo, T. JP 3.2.1, 16.1
Kugadasan, T. LK 3.17
Kuginuki, Y. JP 23.2.2
Kuhara, S. JP 27.3
Kuhlmann, H. DE 15.1.5
Kuhn, G.D. US-B 13.1.1
Kuhns, L.J. US-B 13.1.1
Kuhr, R.J. US-D 8.1.3
Kuikman, P.J. NL 9.5
Kuisten, R. NL 14.1
Kukita, K. JP 19.4
Kukułczanka*, K. PL 33.3
Kularatne, S. LK 4.1
Kulesza, W. PL 18.1
Kulikowski, A. PL 13.1
Kulkarni, R.M. IN 12.1
Kulkarni, V. IN 6.1
Kulus, M. PL 32.1.12
Kumagai, M. JP 2.2
Kumagai, T. JP 36.1
Kumaki, S. JP 29.1
Kumakura, H. JP 16.3
Kumakura, M. JP 8.2
Kumamoto, J US-K 12.1.1
Kumanov, B. BG 8.1
Kumar Agarwal, R. IN 24.1.5
Kumar, A. IN 22.1
Kumar, D. ZI 1.1
Kumar, G. IN 20.1
Kumar, O.V. IN 20.1
Kumar, R. IN 8.1
Kumar, S. IN 1.1.2, 1.1.6, 22.1
Kumar, T.V. IN 1.1.4
Kumashiro, K. JP 26.5
Kumazaki, K. JP 21.3
Kumazawa, Z. JP 23.3
Kume, Y. JP 2.2
Kummert, J. BE 6.8.8
Kumwenda, R. ML 1.1
Kundt, J.F. US-C 5.2
Kunev, K. BG 15.1
Kunimoto, T. JP 30.2
Kunisaki, J.T. US-M 2.2
Kunisawa, T. JP 8.2
Kunishige, S. JP 23.2.3
Kunkee, R.E. US-K 6.1.5
Kunneman, B.P.A.M. NL 2.1
Kuno, H. JP 41.1
Künsch, U. CH 1.1
Kunwar, R. IN 24.1.3
Kunze, L. DE 9.1
Kuo, J.Y. TW 7.2
Kurahashi, T. JP 37.1
Kuramori, K. JP 32.1
Kuranaka, M. JP 37.1
Kurashima, S. JP 29.1
Kurata, H. JP 23.2.3
Kurek, W. PL 23.1
Kuriakose, T.F. IN 5.3.2
Kurian, C. IN 3.1
Kurihara, A. JP 14.2
Kurihara, K. JP 10.1
Kurihara, N. JP 10.1
Kuriu, K. JP 3.3.1
Kuriyama, M. JP 44.1
Kuriyama, T. JP 23.2
Kurki*, L. FI 3.1
Kuroda, G. JP 14.3
Kuroda, H. JP 12.3
Kuroda, K. JP 28.1
Kuroda, Y. JP 45.1
Kuroi, I. JP 29.2
Kurokami, K. JP 40.2

Kuroki, M. JP 14.3
Kurooka, H. JP 33.2
Kurosawa, S. JP 2.2
Kurozumi, T. JP 28.1
Kursakov, G.A. SU 11.1
Kurtán, S. HU 2.2.5
Kurtela, M. YU 3.1
Kurtz, J.O. BR 28.1
Kuru, C. TR 11.1
Kuruppuarachchi, D.S.P. LK 3.9
Kusaka, S. JP 13.1
Kusaka, T. JP 12.1
Kusano, S. JP 7.1
Kuser, J.E. US-B 14.1.1
Kushima, S. JP 25.1.1
Kushima, T. JP 18.3
Kushwaha, P.K.S. IN 19.1
Kuszlejko, B. PL 28.2.1
Kuwabara, M. JP 23.2.4
Kuwabara, T. JP 9.1
Kuwahata, R. JP 18.2
Kuzmanovski, I. YU 4.1
Kuzukubo, S. JP 47.1
Kvaale, A. NO 5.1
Kwack, B.H. KR 21.2
Kweta, G. AT 8.7
Kwo, F.Y. TW 7.2
Kwon, H.M. KR 10.1
Kwon, K.C. KR 3.1
Kwon, S.S. KR 2.2
Kwon, Y.S. KR 23.1.2
Kyamba, R.D. TZ 1.2
Kyriakopoulos, I. GR 1.1
Kyriakopoulou, P.E. GR 14.1.1
Kyritsis, S. GR 6.1.8

Labanauskas*, C.K.
US-K 12.1.1
Łabanowska, B. PL 28.1.2
Łabanowski, G. PL 28.1.15
Labarère, J. FR 4.1.3
Labavitch, J.M. US-K 6.1.3
Labergère, M. FR 32.1
Labios, E.V. PH 1.2.5
Labonne, G. FR 10.1.2
Laborde, J. FR 4.1.3
Laborde, L.J. AR 15.1
Laborde, M. FR 41.2
Labrecque, M. CA 31.1
Labus, D. YU 10.1
Lăcătus, V. RO 1.1.6
Lacaze, J.F. FR 13.1.1
Lacey, D.B. US-B 14.1.1
Lacey, M.J. AU 65.5
Lacey, S.A. AU 18.1
Lachet, B. FR 31.1
Łacic, B. PL 14.1.8
Lacointe, A. FR 5.1
LaCroix, L.J. CA 14.1
Lacy, M. US-F 14.1.3
Lad, B.L. IN 15.1
Ladipo, J.L. NG 3.1
La Ferney, P.E. US-E 11.1
Lafolie, F. FR 3.1.7
Lafon, R. FR 4.1.6
Lagarda Murrieta, A. MX 5.2
Lagoke, S.T.O. NG 2.1
Lagouarde, J-P. FR 3.1.4
Łagowska, B. PL 14.1.9
Laguna, G. AR 7.1
Lahav, E. IL 1.2.5
Lahav, N. IL 2.4
Lahitte, J.M. AR 1.1
Lai, C.L. TW 8.1
Lai, H.H. TW 4.1, 9.1
Laiche, A.J. US-D 49.1
Laidlaw, W.M.R. GB 9.1.2
Laing, I.A.F. AU 59.1
Laing, W.A. NZ 8.1
Lajoux, Cl. FR 19.1
Lakhanpal, S.C. IN 10.3
Lakos, L. HU 2.6
Laksmi, ID 1.1
Lakso, A.N. US-B 2.1.2
Lal, B.B. IN 10.1
Lal, G. IN 20.1
Lal, H. IN 22.1, 24.1.8
Lal, J. IN 20.1
Lal, R.L. IN 20.1
Lal, S. IN 20.1
Lal, S.D. IN 24.1.2
Lalev, N. BG 9.1
Lallu*, N. NZ 2.1.2
La Malfa*, G. IT 5.2, 5.4
Lamb*, R.C. US-B 2.1.2
Lambert, J. BE 12.1
Lambert, R.J. US-B 5.1.1
Lambeth*, V.N. US-E 18.1
Lamers, J.G. NL 6.1
Lamm*, R. SE 1.1.2
Lammens, E. BE 13.1
Lammerink, J. NZ 13.2
Lamond, D. FR 28.1
Lamond, M. FR 28.1
Lamont*, G.P. AU 5.1
Lamont, W.J. US-D 8.1.1
Lamptey, P. GA 1.1
Lanaridis, P. GR 18.1.4
Lancaster, J.E. NZ 8.2
Lancaster, O.M. US-D 53.1
Land, S. US-J 15.1
Landfald, R. NO 10.1.6
Landgren*, B. SE 5.1
Landon Scarlett, L. US-B 10.1
Landry, B. CA 35.1
Lane, A.G. AU 65.4
Lane, R. US-D 20.1
Lane, R.S. US-K 3.1.1
Lane*, W.D. CA 3.1
Lang, M. US-F 8.1
Lang, Z. HU 2.2.5
Lange, E. PL 28.1.4
Lange, H. NL 15.1
Lange*, P. DE 11.1.1
Langer, H. DE 15.2.4
Langer, V. DK 7.2
Langhans*, R.W. US-B 5.1.1
Langille, A.R. US-A 1.1
Langschwager, L. DK 12.1
Langton, F.A. GB 18.1.1
Langvatn, H. NO 10.1.1
Lanini Neto, J.A. BR 28.1
Lannoye, R. BE 4.2
Lansari, A. MA 1.1
Lanteri, S. IT 24.4
Lanterman, W. CA 2.1
Lantin, B. FR 1.1.1
Lantmann, A.F. BR 27.1
Lanza, G. IT 1.1
Lapierre, H. FR 17.1.2
Lapis, D.B. PH 1.4
Lapushner, D. IL 1.1.3
Lara, J.C.C. BR 20.2
Lareau, M.J. CA 35.1
Laren, H. US-C 3.1
Larocca, L.H. AR 6.1
Laroche, M. BE 12.1
Larroque, O. AR 14.1
La Rosa, G. IT 5.1
Larsen, A.B. DK 7.1
Larsen, E. DK 7.2
Larsen, F. DK 3.1
Larsen, J. DK 1.1
Larsen, O.N. DK 2.1
Larsen*, R. SE 1.1.3
Larson, R.A. US-D 8.1.1
Larsson, C. SE 3.1
Lasca, C.C. BR 21.2
Lascu, I. RO 1.1.4
Lascu, N. RO 7.1
Lascu, O. RO 1.1.2
Lasheen, A. MA 1.1
Lashomb, J.H. US-B 14.1.4
Láska, P. CS 20.1
Laskowska, H. PL 14.1.2
LaSota, L. US-C 5.2
Lasram, M. TN 1.1
Lastra, J. CR 2.2
Lászay, Gy. HU 2.2.4
Lászlo, M. HU 2.1.3
Lászlóffi, A. HU 5.3.2
Laterrot, H. FR 3.1.2
Latrasse, A. FR 7.1.4
Lattanzio, V. IT 2.3

Latz, P. DE 11.1.5
Laubenheimer, K.H. DE 23.1
Lauber, H.P. CH 1.1
Laudan, K. PL 28.5
Lauer, F.I. US-G 7.1
Laughlin, C. US-D 20.1
Laughlin*, J.C. AU 43.1
Laumond, C. FR 2.1.5
Laurel, V.B. PH 1.7
Laurena, A.C. PH 1.7
Laurière, C. FR 38.1
Laurinen*, E. FI 1.1
Lauro, E.M. CA 25.1
Laurs, J.J. NL 9.3
Laury, J.C. FR 19.1
Lavania, M.L. IN 20.1
Lavarenne, A. FR 28.1
Lavarenne, S. FR 28.1
Lavee*, S. IL 1.2.4
Lavergne, F. FR 19.2
Lavezzi, A. IT 7.1
Lavi, U. IL 1.2.1
Lavigne, P. CA 34.1
Lavon, R. IL 1.2.3
Lavsen, E.R. DK 9.1
Lawande, K.E. IN 14.2.1
Lawes, G.S. NZ 9.1
Lawrence, M.E. AU 65.1
Lawrence, R.A. CA 41.1
Lawrie, R.A. GB 25.1.3
Lawson, D. GB 4.1
Lawson, H.M. GB 15.1.4
Lawson, R.H. US-C 3.1
Lawson, V.F. US-G 21.1
Lay Yee, M. NZ 2.1.1
Layne, R.E.C. CA 17.1
Lazăr, A. RO 12.1
Lazăr, Gh. RO 21.1
Lázár, J. HU 5.2
Lazarević, S. YU 14.2
Lázaro, F. ES 12.2
Lăzăroiu, P. RO 12.1
Lazarovits, G. CA 18.1
Lazin, M.B. US-D 27.7
Lazo, J. PE 1.1
Lê, L. CH 2.1
Lea, A.G.H. GB 19.1.4
Leal, N.R. BR 19.1
Leamon, K. AU 33.1
Leandri, A. IT 19.2
Leão, A.C. BR 14.1
Lebeda, A. CS 25.1
Le Bohec, J. FR 26.1.2
Leblanc, P.V. CA 39.1
Leblond, C. FR 38.1
Lebron-Luteyn, M.L. US-B 1.1
Lebrun, R. US-A 14.1
Leca, M. RO 1.1.2
Lecharny, A. FR 30.1.1
Leclant, F. FR 10.1.2
Leclercq, G. FR 22.1
LeComte, B.E. 8.1
Lecoq, H. FR 3.1.5
Ledent, J. BE 12.1
Lederle, E. DE 15.1.3
Lederman, I.E. BR 10.1
Ledesma, A. AR 7.1
Ledezma, E. CR 2.2
Ledger, S.N. AU 15.2
Ledieu, M.S. GB 18.1.2
Lee, B.Y. KR 23.3
Lee, B.K. KR 11.1
Lee, C.H. KR 21.2, 23.1.1
Lee, C.M. TW 11.1.1
Lee, C.S. KR 23.1.3
Lee, C.U. KR 14.1
Lee, C.Y. KR 24.1
Lee, C.Y. US-B 2.1.5
Lee, D.F. GB 13.1.2
Lee, D.H. KR 21.1
Lee, D.K. KR 18.1.2
Lee, D.W. KR 2.1.3
Lee, E. AU 65.4
Lee, G.Y. KR 9.1
Lee, H.S. TW 2.1.3
Lee, J. IE 6.1
Lee, J.C. KR 8.2
Lee, J.M. KR 21.5
Lee, J.S. KR 7.3, 8.2, 10.3, 23.1.5
Lee, K.C. TW 7.1
Lee, K.E. KR 5.2
Lee, K.H. KR 21.3
Lee, K.K. KR 5.1
Lee, K.S. KR 12.1
Lee, L.C. TW 2.1.4
Lee, L.S. AU 22.1
Lee, M.F. TW 12.1
Lee, M.H. KR 21.4
Lee, M.W. KR 1.1
Lee, N. TW 9.1
Lee, P.N. TW 12.1
Lee, R.F. US-D 30.2
Lee, S.B. KR 7.1, 20.1
Lee, S.C. TW 10.1
Lee, S.K. US-C 10.1
Lee, S.Y. TW 12.1
Lee, S.S. KR 23.1.1
Lee, S.W. KR 21.5
Lee, T.H. AU 37.1
Lee, T.T. CA 18.1
Lee, W.J. KR 16.1.2
Lee, W.S. KR 7.2
Lee, W.S. TW 11.1.2
Lee, Y.B. KR 8.2, 23.1.2
Lee, Y.M. KR 1.1
Leela, D. IN 1.1.8
Leem, H.B. KR 20.1
Leemans, A.M. NL 9.4.5
Leeper, P. US-H 15.1
Leeuwenberg, A.J.M. NL 9.7.4
Lefèbre, J.M. FR 7.1.2
Lefèvre, B. FR 34.1
Lefèvre, R. FR 23.1
Leffler, R. US-D 16.2
Leffring, L. NL 1.1.3
Lefort, P-L. FR 4.1.1
Lefter, Gh. RO 22.1
Legave, J.M. FR 10.1.1
Legemaate, L. NL 1.1.4
Legg*, B.J. GB 31.2.6
Leggo, D. AU 11.1
Legrand, G. BE 12.1
Leguizamo, R.M. MX 6.4
Legutko, W. PL 20.2.10
Lehmann, M. HU 2.1.5
Lehmushovi*, A. FI 1.1
Lehoczky, J. HU 2.5, 5.2
Leibee, G.L. US-D 36.1
Leidenfrost, E. DE 12.1.11
Leidermann, L. BR 21.2
Leijerstam, B. SE 1.4
Leijon, S. SE 3.1
Leippert, H. DE 15.3.5
Leippert, S. DE 15.3.5
Leiser, A.T. US-K 6.1.2
Leitão, F. PT 6.1
Leitão, M.F de F. BR 20.2
Leite, H.F. BR 6.1
Leite, J. de O. BR 14.1
Leite, L.A. BR 19.1
Leite, R.M. de O. BR 14.1
Leite Meirelles, M. BR 15.1
Lekšan, M. YU 18.1
Lelakis, P. GR 6.1.2
Le Lezec, M. FR 1.1.1
Lelièvre, F. FR 15.1
Lelley, J. DE 21.1
Lelliott, R.A. GB 13.1
Lellis, W.T. BR 14.1
LeMaguer, M. CA 7.2
Lemaire*, F. FR 1.1.4
Lemaire, J-M. FR 3.1.5
Lemanceau, P. FR 19.5
Lemattre, M. FR 17.1.2
Lemattre*, P. FR 17.2
Lemcoff, J.H. AR 3.1
Lemeni, C. RO 19.1
Lemeni, V. RO 7.1
Lemoine, J. FR 1.1.1
Lemoine, M. FR 13.1.2
Lemos Fonseca, J.N. BR 25.3
Le Nard, M. FR 9.1
Lenartowicz, A. PL 28.1.14

Lenartowicz, W. PL 28.1.3
Lenka, P.C. IN 16.2
Lenkiewicz, J. PL 8.1
Lennon, A.M. GB 15.1.3
Lenormand, M. FR 14.1.2
Lentjes, P.W.M. NL 1.1.4
Lenton, J.R. GB 19.1.2
Lenz*, F. DE 6.1
Lenz, LM.. CA 14.1
Léon, H. FR 1.1.5
Léon, G. VE 6.1
Leonard, R.T. US-K 12.1.1
Leone, A. IT 18.1
Leonhardt, K.L. US-M 2.2
Leoni, C. IT 14.1
Leoni, S. IT 4.1
Le Picard, D. FR 23.1
Lépine, M. FR 40.1
Lepoivre, P. BE 6.8.3
Lepori*, G. IT 24.4
Leppard, M.J. ZI 3.1.2
Lercari, B. IT 17.2
Lerman, S. IL 4.1
LeRoux, E.J. CA 20.1
Lescano Alva, M. PE 3.1
Leshem, B. IL 1.1.2
Lešić, R. YU 3.1
Leski, B. PL 32.1.6
Leskovec, E. YU 18.3
Lespinasse, J-M. FR 4.1.1
Lespinasse, Y. FR 1.1.1
Lessaint, C. FR 17.1.5
Leszczyński, A. PL 32.1.1
Leszczyńska, G. PL 28.1.5
Le Tacon, F. FR 11.1.1
Letal, J. CA 6.1
Létard, M. FR 26.1.1
Letham, D.B. AU 9.1.1
Letouzé, R. FR 19.2
Leuthold, L.D. US-G 31.1
Leva, A.R. IT 9.2
Lévai, P. HU 5.1.3
Levan, N. IL 1.1.2
Levi, A. IL 1.3.2
Levi, D. IL 1.1.1
Levieux, J. FR 13.1.4
Lévy, G. FR 11.1.1
Levy, Y. IL 1.2.3
Lewandowska, A. PL 32.1.2.1
Lewandowska, M. PL 27.1
Lewandowski, A. PL 10.1.1
Lewis*, I.U. ET 10.1
Lewis, A.C. NZ 9.9
Lewis, C.E. US-L 1.1
Lewis, G.D. US-B 14.1.6
Lewis, G. US-K 9.1
Lewis, I.K. NZ 13.2
Lewis, J. US-D 16.1
Lewis, J.A. US-C 3.1
Lewis, R.T. GB 31.1
Leyman, V. BE 13.1
Li, P.H. US-G 7.1
Li, R. US-B 1.1
Liao, C.I. TW 6.1
Liao, F.S. TW 11.1.2
Liao, M.C. TW 9.1
Liaw, W.J. TW 7.2
Liberal, O.H.T. BR 19.1
Libik, A. PL 11.1.1
Lichtemberg, L.A. BR 28.1
Licuanan, R.R. PH 1.7
Lidster, P.D. CA 41.1
Lieberman, A.S. US-B 5.1.1
Liebeswar, K. AT 8.6
Liebig*, H.-P. DE 15.1.3
Liebricht, H. IC 1.1
Liesecke, H.J. DE 15.2.2
Lieth, J.H. US-K 6.1.2
Ligocka, T. PL 28.1.10
Ligocki, P. PL 28.1.5
Ligteringen, R. NL 2.1
Likens, G.E. US-B 1.1
Lilien-Kipnis*, H. IL 1.1.2
Lill, R.E. NZ 10.1
Lim, C.I. KR 23.1.2
Lim, H.C. KR 9.2
Lim, J.H. KR 7.1
Lim, S.C. KR 23.1.2
Lima Pereira, A. BR 30.1
Lima, A.B.A. BR 27.1
Lima, C.C.A. BR 10.1
Lima, G.R. do A. BR 10.1
Lima, M.B. PT 6.1
Limacher, H. CH 1.1
Limongelli, F. IT 19.1.3
Limongelli, J.C. AR 3.1.1
Lin, C.C. TW 1.2, 2.1.3
Lin, C.H. TW 12.1, 2.1.1
Lin, J.R. TW 10.1
Lin, J.H. TW 7.2
Lin, L.F. TW 11.1.1
Lin, P. TW 9.1
Lin, R.S. TW 7.3
Lin, S.C. TW 7.3
Lin, T.S. TW 9.1
Lin, T.C. TW 7.2
Lin, W.C. CA 2.1
Lin, Y.K. TW 2.1.2
Lin, Y.T. TW 2.1.2
Linardakis, D. GR 10.1
Lind, F. DK 4.1
Linden, R. BE 6.1
Linder, M. SE 8.2
Lindfors*, S. SE 1.1.2
Lindgren, D.T. US-G 25.1
Lindhout, W.H. NL 9.6.1
Lindner, P. IL 1.3.2
Lindow, S.E. US-K 3.1.2
Lindquist, R.K. US-F 7.1.3
Lindsey, D. US-H 19.1
Lindstrom, R.K. US-F 13.1
Lindstrom, R.S. US-C 10.1
Lineberger, R.D. US-F 3.1
Ling, D.L. TW 9.1
Lingaiah, B. IN 11.8
Link*, H. DE 18.1.3
Linsalata, D. IT 2.3
Linthorst, Th.J. NL 9.2
Lionakis, S. GR 8.1.1
Liotta, M.A. AR 18.1
Liou, C.T. TW 2.1.4
Liou, M.F. TW 8.1
Liou, T.D. TW 2.1.1
Lipa, J.J. PL 20.1.3
Lipari*, V. IT 5.2
Lipe, W. US-H 9.1
Lipecki*, J. PL 14.1.1
Lipiński, Z. PL 32.1.2.1
Lippert, L.F. US-K 12.1.1
Lippert, T. DK 3.7
Lips, H. IL 5.1.2
Lips, J. BE 15.1
Lipscomb, L. US-H 6.1
Liptay, A. CA 17.1
Lipton, W.J. US-K 7.3
Lis, E. PL 28.1.5
Lisbao, R.S. BR 20.1
Lisiecka*, A. PL 20.2.5
Liska, J. CS 30.1
Lisker, N. IL 1.3.3
Lisowski, M. PL 30.1
List, D. DE 5.3.1
Litan, L. RO 17.1
Little, C. AU 30.1
Little, Jr., L.D. US-B 14.1.1
Littlefield, L. US-A 1.1
Littlefield, L. US-H 3.1
Litz, R.E. US-D 29.1
Liu, C. TW 9.1
Liu, F.W. US-B 5.1.2
Liu, H.H. TW 4.2
Liu, P.C. TW 7.1
Liuni, C. IT 7.4
Liverani, A. IT 19.1.2
Liyanage, L.V.K. LK 2.1
Ljesov YU 14.1
Llano, J. CO 6.2
Llaveriá B., M. PE 3.1
Lleras Perez, E. BR 25.3
Llontop, J. PE 2.4
Llorente, A. AR 10.1
Llosa, C. PE 1.1
Lloyd, C. GB 24.1.3
Lloyd, G.A. GB 13.1.2, 13.1.3

Lloyd, J.A. US-A 1.1
Lloyd, J.E. US-I 12.1
Lloyd-Jones, C.P. GB 19.1.2
Loach, K. GB 18.1.1
Loayza, C. PE 1.2
Lobo, A. AU 59.2
Locascio*, S.J. US-D 27.7
Locher, P. FR 27.1
Lockerman, R.H. US-I 18.1
Lockhart, C.L. CA 41.1
Lockwood, D.W. US-E 3.1
Lockwood, T. US-A 13.1
Lode, O. NO 10.4.2
Loder, M.A. AU 10.1
Lodeta, V. YU 3.2
Lodge, G.A. GB 1.1
Lodge, N. NZ 2.1.3
Loeffler, R.S.T. GB 19.1.1
Loeser, H. DE 16.1
Löfvenberg*, E. SE 1.1.3
Logan, B. AU 12.1
Logan, P. US-A 14.1
Loginova, E. BG 8.1
Lo Giudice, V. IT 1.1
Loher, W.J. US-K 3.1.1
Lohweg, E. DE 11.2.2
Loidl, H. DE 5.2.1
Loiselle, R. CA 20.2
Lombardo, N. IT 8.1
Lomman, G.J. AU 36.5
Lomonossoff, G. GB 24.1.4
Loneragan, P. AU 8.1
Long, C. US-H 12.1
Long, C.E. US-G 31.1
Long, C.R. US-B 1.1
Long*, J.K. AU 9.2
Long, P.G. NZ 9.1
Longbrake, T. US-H 6.2
Longly, B. BE 12.1
Longo, G. IT 5.3
Longoria Garza, G.A. MX 7.1
Longuenesse*, J-J. FR 3.1.4
Longworth, J.F. NZ 2.2
Lont, P.J. NL 4.1
Looney*, N.E. CA 3.1
Lopes, C.A. BR 19.2, 25.5
Lopes, E.B. BR 9.1
Lopes, J.F. BR 25.5
Lopes, L.C. BR 17.1
Lopes Ban, E. BR 18.1
Lopes Teixeira, S. BR 17.1
Lopes Vilar, R.R. BR 3.1
López, T. ES 11.1
López Ahumada, B. MX 2.3
Lopez Camelo, A. AR 11.1
López Montoya, I. MX 5.2
López-Para, A. ES 11.1
López Rafael, M. PE 3.2
López-Real, J. GB 35.1.3
Lord, E.M. US-K 12.1.1
Lord, J. CA 16.1
Lord, L. CA 30.1
Lord, W. US-A 2.1
Lorenc-Plucinska, G. PL 10.1.3
Lorenz, H.-P. DE 22.1
Lorenzi, R. IT 17.2
Loreti*, F. IT 17.1
Loria, R. US-B 8.1
Loria, R.S. AR 22.3
Lőrincz, A. HU 2.2.10
Lőrincz, B. HU 5.1.3
Lorusso, A.O. AR 3.1.2
Los, M. NL 1.1.6
Løschenkohl, B. DK 4.1
Lošonský, L. CS 17.1
Losso, M. BR 28.1
Lot, H. FR 3.1.5
Lott, H. DE 23.1
Louant, B. BE 12.1
Loublier, Y. FR 3.1.9
Lougheed*, E.C. CA 15.1
Loughton*, A. CA 23.1
Louis, L. ET 4.1
Loupassaki, M. GR 8.1.1
Loureiro, S. PT 6.1
Loussert, R. MA 5.1
Louvet, J. FR 7.1.3
Lovato, A. IT 3.2
Lovatt, C.J. US-K 12.1.1
Love, S. US-I 13.1
Lővei, G. HU 2.5
Lovenstein, H. IL 5.1.3
Loveys, B.R. AU 65.1
Lović, R. YU 14.1
Lovoković, J. YU 3.1
Lowe, R.G. NZ 2.1.1
Lower, R. US-F 23.1.3
Loy, J.B. US-A 2.1
Loyola Dantas, J.L. BR 12.1
Lozano, C. PE 1.3
Lozano Cruzado, A.C. AR 21.1
Lozanov, P. BG 2.1
Lu, F.M. TW 2.1.3
Lubenov, I. BG 6.1
Lubet, E. FR 4.1.2
Luby, J. US-G 7.1
Luca, L. RO 3.1
Lucas, G.Ll. GB 16.1
Lucas, I.A.M. GB 35.1
Lúčanska, H. CS 31.1
Lucey, R.F. US-B 5.1.5
Lucey, S. GB 19.1.4
Lucia, A. BR 12.1
Luciano, B. US-N 1.2
Lučić, V. YU 1.2
Łucka, M. PL 11.1.2
Luczak, F. PL 28.4
Lüdders*, P. DE 5.1.1
Ludford, P.M. US-B 5.1.3
Ludovic, P. RO 23.2
Ludvik, K. CS 7.1
Ludwig, J.T. GB 18.1.3
Ludwig, K. DE 5.2.1
Ludwig, W. AT 8.2
Luffman, M. CA 39.1
Lugonja, U. YU 16.1
Luisetti, J. FR 1.1.3
Lukach, J. US-G 14.1
Lukács, J.Sz. HU 2.2.5
Lukács, P. HU 5.3.2
Łukasik, S. PL 14.1.4
Łukaszewska, A. PL 32.1.3
Lukić, J. YU 14.1
Lukić, S. YU 1.1
Lukins, R.G. AU 65.5.1
Lulai, E. US-G 11.1
Lullfitz*, R. AU 59.2
Luma, Y. TR 5.1
Lumis, G.P. CA 15.1
Lumkes, L.M. NL 6.1
Lumsden, R.D. US-C 3.1
Luna, R. AR 7.1
Luna Vázquez, J. MX 4.2
Luncă, L. RO 5.1
Lund, A. DK 3.2
Lund-Høie, K. NO 10.4.2
Lundergan, C.A. US-D 52.1
Lundgaard, Sv. DK 7.1
Lundin*, M. SE 4.1
Lundin, P.N. SE 4.1
Lundsgaard, T. DK 3.5
Lundstad, A. NO 10.1.5
Lundsten, K. DK 3.1
Luntz, O. HU 5.2
Luo, S.J. TW 7.1
Lupescu, Fl. RO 8.3
Luppi, G. IT 24.3
Luque, G. AR 7.1
Luque, J.L. AR 15.1
Luttringer, M. FR 3.1.6
Lutz, M. CH 1.1
Luz, E.D.M.N. BR 14.1
Lužny, J. CS 14.1.4
Luzzardi, G.C. BR 30.3
Ly, M. SN 1.1
Łyś, J. PL 28.1.10
Lydon, C. US-C 2.1
Lynas, L. US-B 1.1
Lynch, D.R. CA 8.1
Lynch, J.M. GB 18.1.2
Lynch, K.V. CA 40.1
Lyon, J.P. FR 2.1.4
Lyons, C. US-H 6.2

Lyons*, J.M. US-K 6.1.4
Lyons, R.E. US-C 10.1
Lystlund, K. DK 9.1

Ma, S.S. TW 9.1
Maage*, F. NO 10.1.6
Maas, J.L. US-C 3.1
Maas, P.W.Th. NL 9.8
Maat, D.Z. NL 9.4.4
Mabbott, T.W. GB 9.1.2
Mabesa, T.C. PH 1.2.4
Mabuchi, M. JP 14.2
Maccaferri, M. IT 3.3
MacCanna, C. IE 4.1
MacCarthaigh, D. DE 15.1.4
MacDonald, D.M. GB 9.1.2
Maceljski, M. YU 3.2
MacFarlane, J. AU 27.1
MacFarlane, J. CA 22.1
MacGiolla Ri, P. IE 2.1
MacGuidwin, A. US-F 23.1.4
Machado, A. PT 8.1
Machado, E.L. BR 29.1
Machado dos Santos, A. BR 30.1
Macharia, E.W. KE 1.9
Macharia, P.K.A. KE 5.1
Machida, H. JP 40.1
Machida, I. JP 3.2.3
Machida, Y. JP 14.2
Machnik, B. PL 28.1.1
Machovec, J. CS 14.1.3
Machowicz-Stefaniak, Z. PL 14.1.8
Machurek, J. CS 13.1
Maciak, J. PL 32.1.1
Macias, M. PL 28.2.7
Macias Valdés, L.M. MX 6.3
Maciel Pérez, L.H. MX 6.3
Maciesiak, A. PL 28.1.2
Maciszewski, D. PL 32.3.6
Macit*, F. TR 15.1
MacKay, B.R. NZ 9.1, 9.6
Mackay, G.R. GB 15.1.2
MacKay, M.B. NZ 9.1
MacKenzie, A.F. CA 37.1
Mackenzie, D.R. US-D 52.1
MacKenzie, K.A.D. GB 8.1.1
MacKerron, D.K.L. GB 15.1.4
Mackiewicz, D. PL 6.2
Mackiewicz, H. PL 20.2.10
Mackiewicz, M. PL 6.2
Maćkowiak*, M. PL 20.2.10
Mackroth*, K. DE 12.1.14
MacLachlan, J.B. IE 2.1
Maclaurin, B.J. ZI 3.2
MacLeod, J. GB 23.1
MacLeod, L.B. CA 43.1
MacMahon, J.A. US-I 5.1
MacPahil, B. CA 42.1
Macquaire, G. FR 4.1.6
MacRae, E.A. NZ 2.1.2
MacRae, H.F. CA 42.2
Macura, V. YU 8.2
Maczák, B. HU 8.1
Madalageri, B.B. IN 11.2
Maddens, K. BE 17.1
Madero Tamargo, E. MX 5.2
Madero Tamargo, J. MX 6.2
Madge, D.G. AU 31.1
Madge, D.S. GB 35.1.3
Madge, P.E. AU 36.2
Madhavachari, S. IN 6.1
Madhosingh, C. CA 18.1
Madi, L.F.C. BR 20.2
Madinelli, C. IT 26.1
Madne, N.S. IN 14.11
Madour, A.M. LY 1.1
Madsen, A.C. DK 3.4
Madsen, F.B. DK 3.7
Madsen, K. DK 9.1
Madumadu, G.G. KE 1.2
Madyara, S. ZI 3.1.2
Madziva, T.J.T. ZI 3.1.2
Mae, H. JP 23.1
Maeda, K. JP 20.2
Maeda, M. JP 3.2.4, 24.1
Maeda, S. BR 4.1
Maeda, T. JP 23.1
Maekawa, S. JP 13.3
Maelzer, D.A. AU 42.1
Maenhout, C.A.A.A. NL 6.1
Maesaka, K. JP 44.2
Maestri, M. BR 17.1
Mafla, A. CO 5.1, 6.1
Magalhães, A.F. de J. BR 12.1
Magalhães, N. PT 8.1
Magallona, E.D. PH 1.3, 1.6
Magan, N. GB 18.1.2
Maganaris, A. GR 19.1
Magarey, P.A. AU 38.2
Magdoff, F.R. US-A 3.1
Magdum, M.B. IN 15.1
Magein, H. BE 6.8.1
Maggini, F. IT 27.1
Maghami*, P. FR 39.2
Magherescu, C. RO 20.1
Magherini, R IT 9.1
Maghirang, R.G. PH 1.2.2, 1.7
Maginnes*, E.A. CA 11.1
Magnani, M. BR 30.1
Magnanini, E. IT 3.1
Magnifico*, V. IT 2.3
Magnus, H.A. NO 10.4.1
Magnusson, M. SE 1.1.2
Magrisso, J. BG 7.1
Magyar, K. HU 2.1.2
Magyar-Bándi, J. HU 2.2.5
Mahadeva, S. CA 6.1
Maher, M.J. IE 4.1
Mahindapala, R. LK 2.1
Mahlert, R. US-I 15.1
Maholay, M.N. IN 1.1.6
Mahr, D. US-F 23.1.2
Mahur, K.P. IN 25.1
Mai, W.F. US-B 5.1.8
Maia, A. de S. BR 14.1
Maïa, N. FR 2.1.3
Maida, J.H.A. ML 1.1
Maidebura, V.I. SU 4.1
Maier, N.A. AU 36.2
Maier, V.P. US-K 11.1
Maillard, A. CH 2.1
Maillet, J.P. FR 19.1
Mailloux, G. CA 29.1
Mailloux, M. CA 28.1
Maini*, S.E. IN 2.1.6
Mainié, Ph. FR 17.2
Mainland*, C.M. US-D 2.1
Mainou, A. GR 19.1
Maison, K. US-A 14.1
Maisonneuve, B. FR 17.1.1
Maiti, S.C. IN 28.1
Maity, R.G. IN 28.1
Majambu, I.S. NG 2.1
Majasan, K. NG 6.1
Majdecki, L. PL 32.1.12
Majek, B.A. US-B 14.1.7
Majgaonkar, S.V. IN 15.1
Majoros, L. HU 2.7
Majumder, P.K. IN 2.1.1, 2.1.2
Makhija, M. IN 9.1
Makishima, N. BR 25.5
Makita, H. JP 26.3
Makita, Y. JP 38.2
Maklouf Carvalho, E.J. BR 3.1
Makropodi, M. GR 9.1
Maksimović, M. YU 8.1
Maksimović, P. YU 10.1
Maksymiuk, J. PL 28.1.5
Makuch, M.A. AR 13.1
Małachowski, A. PL 28.4
Malathrakis, N. GR 10.2
Malatinszki, Gy. HU 5.3.1
Malausa, J.C. FR 2.1.4
Malazzi, P.R. BR 21.2
Malburg, J.L. BR 28.1
Mălcică, P. RO 4.1
Malcolmson, J.F. GB 15.1.1
Maldiney, R. FR 30.1.1
Maldonado, J.F.M. BR 19.1
Malenin, I. BG 7.1
Malepszy, S. PL 32.1.8

Malet, Ph. FR 3.1.4
Malewski, W. PL 28.1
Malhi*, C.S. IN 17.1
Malhotra, V.P. IN 4.1
Malinowska, H. PL 20.2.10
Maliwa, N. TZ 3.1
Malla Reddy, K. IN 6.1
Mallett, K. CA 6.1
Mallidis, K. GR 18.1.2
Mallik, B. IN 11.7
Mallik, S. IN 28.1
Malone, L.A. NZ 2.2
Malone, M. NZ 11.2
Malone, M.T. NZ 13.2
Malstrom, H. US-H 8.1
Maltais, B. CA 33.1
Mamane, S.A. IN 14.1
Mamarov, P. BG 7.1
Mamedov, F.M. SU 7.1
Mamezuka, S. JP 7.1
Manabe, T. JP 11.4, 20.1
Manago, M. JP 19.2
Manago, N. JP 1.1.1
Manah, A.A. LY 3.1
Manceau, C. FR 1.1.3
Mancini, L. IT 2.2
Manciot, R.R. BR 11.1
Mandal, B. IN 8.1
Mandal, R.C. IN 3.1
Mandal, S.C. IN 8.1
Mandelli, M.A. BR 28.1
Mandge, A.S. IN 12.3
Mandic, S. YU 10.1
Mǎnescu, C. RO 8.3
Manfredini, S. BR 31.1
Mangave, K.K. IN 14.3
Manguiat, I.J. PH 1.5
Mani, E. CH 1.1
Manios, B. GR 10.1
Manjarrez Morales, P. MX 2.2
Manjunath, K.L. IN 1.1.6
Manley, J.E. NZ 9.1
Mann, J.D. NZ 8.2
Mann, W. DE 15.1.3
Manning, K. GB 18.1.3
Mannini, F. IT 24.2
Mano, T. JP 13.1
Manohar*, M.S. IN 19.1
Manojlović, B. YU 8.1
Manolache, G. RO 1.1.5
Manole, N. RO 3.1
Manolopoulou, E. GR 18.1.2
Manoukas GR 3.1.1
Mans, A. NL 9.5
Mansergas, A.F. ES 23.1
Mansfield, J.M. GB 35.1.3
Manshardt, R.M. US-M 2.2
Mansour, R. SR 1.1
Månsson, T. SE 3.1
Mantaka, Aik. GR 14.1.2
Mántaras, C. US-N 1.2
Mante, S. US-B 1.1
Mantell, S.H. GB 35.1.2
Mantinger, H. IT 11.1
Mǎnuča, O. RO 8.5
Manuelyan, Kh. BG 8.1
Manušev, B. YU 1.1
Manzo, P. IT 19.1
Mapplebeck, L. CA 42.2
Maramba, P. ZI 3.1.2
Marangoni*, B. IT 3.1, 25.1
Maranhao, E.H. de A. BR 10.1
Maraulja, M.D. US-D 30.2
Marca, Gh. RO 17.2
Marcelle*, R. BE 19.1
Marcellin, P. FR 38.1
Marchant, W.T.B. GB 31.2.1
Marchoux, G. FR 3.1.5
Marcinkowska, E. PL 28.1.10
Marcinkowska, J. PL 32.1.6
Marcinkowski, J. PL 28.1.8, 28.1.14
Marcó, G.M. AR 6.1
Marcó, M.A. AR 6.1
Marcy, J.E. US-D 30.2
Marczyński, Sz. PL 32.1.3
Mareček, F. CS 22.4
Mareček, J. CS 24.1
Maréchal, M. MA 1.1
Mares, M. RO 8.2, 8.5
Mari, M. IT 3.3
Mariano, A.H. BR 14.1
Marianowski, A. PL 28.2.1
Marias, P.G. ZA 4.1
Marić, M. YU 14.1
Marić, S. YU 9.1, 16.1
Marin, A. RO 4.1
Marinari, A. IT 9.3
Marinescu, Gh. RO 1.1.7, 1.1.8
Maring, J. NL 9.3
Marini, R.E. US-C 10.1
Marinković, N. YU 13.1
Marinkovic. YU 8.2
Marino, G. IT 3.1
Marinov, P. BG 12.1
Maris, A. NL 2.1
Marker, E. NO 7.3
Markham, P.G. GB 24.1.4
Markhart, A. US-G 7.1
Markides, A.J. AU 10.1
Markkula, M. FI 2.1.2
Marković, P. YU 14.1
Marković, R. YU 14.1
Marković, Z. YU 13.1
Marks, C.F. CA 17.1
Marks, T. GB 8.1.1
Márkus, F. HU 5.3.3
Markwick, N.R. NZ 2.2
Marn, M. YU 18.2.1
Marocke, R. FR 6.1.1
Marogiannopoulos, G. GR 6.1.8
Marois, J.J. US-K 6.1.1
Maroon, C.J.M. PH 1.4
Maroquin, C. BE 6.2
Marquard, R. US-H 8.1
Marques, D. PT 1.1
Marques, J.R.B. BR 10.1
Marques da Silveira, P. BR 26.1
Marques Porto, M.c. BR 12.1
Márquez, P. US-N 1.2
Márquez Castillo, P. MX 4.4
Marr, C.W. US-G 31.1
Marrapodi, J.L. AR 1.1
Marras, G.F. IT 23.1
Marriage, P.B. CA 17.1
Mars, S. FR 2.1.1
Marsh, H.V. US-A 5.1.1
Marshall, B. GB 15.1.4
Márta, K. HU 2.2.4
Martel, P. CA 35.1
Matel, Y. CA 40.3
Martens, H. NO 10.5.2
Martens, L. BE 7.11
Martens, M. NO 10.5.2
Marti-Fabregat, F. ES 14.1
Martijn, A. NL 9.8
Martin, A.D. GB 13.1.3
Martin, C. GB 24.1.2
Martin, Cl. FR 7.1.10
Martin, F. FR 11.1.1
Martin, F.W. US-N 1.2
Martin, G.C. US-K 6.1.3
Martin, G.M. AR 15.1
Martin, J.M. US-I 18.1
Martin*, L.W. US-J 12.1
Martin, M.M. ES 8.1
Martin, N.A. NZ 2.2
Martin, R. US-G 7.1
Martin, R.R. CA 4.1
Martin, S.L.D. BR 18.1
Martin, W.C. US-D 38.1
Martinengo de Mitidieri, J. AR 19.1
Martinetti, L. IT 10.1
Martinez, C. AR 15.3
Martinez, G. AR 22.1
Martinez, J.A. BR 21.2
Martinez, O. AR 14.1
Martinez, O.R. CO 3.1
Martinez, R. AR 23.1
Martinez, R. CR 2.1, 2.2
Martinez, R.R. MX 12.5

Martínez, S. US-N 1.2
Martínez González, J.C. MX 8.4
Martínez González, J.S. MX 11.1
Martínez Parra, R.A. MX 6.1
Martínez Pelaez, H. AR 12.1
Martínez Picó, L. US-N 1.1
Martínez Quintana, O. AR 14.1
Martínez Rodríquez, J. MX 5.3
Martínez Tellez, J.J. MX 1.5
Martínez V., J.M. MX 9.3
Martinez-Zaporta, F. ES 11.1
Martinovich, V. HU 5.3.2
Martins, F. PT 8.1
Martins, J.M. PT 6.1
Martins Chaves, G. BR 17.1
Martins e Silva, H. BR 1.1
Martinson, V.A. GA 4.5
Martins Pereira, J.F. BR 30.1
Martins Silva, D. BR 24.2
Martin-Tanguy, J. FR 7.1.10
Mártonffy, B. HU 2.7
Martouret, D. FR 17.1.6
Marty, J. FR 2.1.6
Martyn, E.J. AU 46.1
Martyna, S. PL 11.1.5
Martyna, K. PL 11.1.5
Maruo, T. JP 4.4
Marushima, Y. JP 4.2
Maruyama, Y. JP 9.1
Marx, E. FR 42.1
Marx, G.A. US-B 2.1.2
Maryowski, R. PL 32.3.2
Marzi, V. IT 2.2
Mas, P. FR 3.1.5
Masaki, T. JP 16.3
Masca, M.G.C.C. BR 23.1
Mascareño Castro, F. MX 2.2
Maschio, L. BR 27.2
Maser, R. AT 8.11
Masia, A. IT 12.1
Maslac, Z. YU 10.1
Mason, D.T. GB 15.1.4
Massai, R. IT 17.2
Massalski, P.R. GB 15.1.3
Massey, L.M. US-B 2.1.5
Massig, D. IL 5.1.3
Massini, R. IT 14.1
Massonie, G. FR 4.1.7
Mastalerz, J.W. US-B 13.1.1
Masuda, M. JP 25.2
Masuda, R. JP 14.2
Masua, T. JP 16.2
Masuda, Y. JP 3.2.3
Masui, M. JP 41.1
Maszkiewicz, J. PL 28.5
Maszkowska, J. PL 28.1.13
Matei, G. RO 8.4.4
Matejek, E. PL 32.1.5
Mathai, C.K. IN 1.1.8
Mathe, A. HU 2.2.6
Mathes, D.T. LK 2.1
Mathew, C. IN 3.1
Mathew, D.L. US-F 9.1.3
Mathew, J. IN 3.1
Mathiassen*, H.P. DK 7.1
Mathiassen, O.H. DK 9.1
Matias, C. PT 1.1
Matić, S. YU 3.3
Matijević, D. YU 8.1
Maton, A. BE 15.1
Matos, C.S. BR 28.1
Matos Cavalcante, E. BR 8.1
Matos Silva, J.M. PT 5.1
Matouš, J. CS 24.1
Matsubara, S. JP 31.2
Matsubara, Y. JP 26.4
Matsuda, K. JP 23.1
Matsuda, M. JP 3.3.2, 21.2, 22.1, 38.1
Matsuda, S. JP 45.1
Matsuda, T. JP 14.5
Matsui, F. JP 12.1
Matsui, H. JP 33.2
Matsui, I. JP 2.2
Matsui, M. JP 7.2
Matsui, S. JP 9.4
Matsukawa, T. JP 7.1
Matsukawa, Y. JP 8.2
Matsumaru, K. JP 35.2
Matsumoto, A. JP 7.2
Matsumoto, E. JP 26.4
Matsumoto, H. JP 21.1
Matsumoto*, K. BR 25.3
Matsumoto, K. JP 5.3
Matsumoto, K. JP 11.2, 44.2
Matsumoto*, M. JP 41.4
Matsumoto, M. JP 43.2
Matsumoto, N. JP 8.2
Matsumoto, O. JP 46.1
Matsumoto, R. JP 27.3, 32.1
Matsumoto, T. JP 21.1, 32.1, 35.1
Matsumoto, Y. JP 17.1, 28.1
Matsumura, H. JP 9.1
Matsumura, S. JP 41.4
Matsumura*, T. JP 21.3
Matsunaka, K. JP 3.3.1
Matsunami, T. JP 10.1
Matsuo, E. JP 18.3
Matsuo, N. JP 12.3
Matsuo, T. JP 4.3, 18.3
Matsuo*, Y. JP 34.3
Matsuoka, T. JP 14.3
Matsuse, S. JP 34.2
Matsushima, J. JP 23.3
Matsushita, T. JP 5.2, 26.4
Matsuta, N. JP 14.2
Matsuura, M. JP 6.1
Matsuyama, M. JP 6.1
Matsuyama, S. JP 26.4
Matsuzaki, S. JP 34.2
Matsuzawa, Y. JP 14.1, 39.2
Matta, F.B. US-H 17.1
Matteoni, J.A. CA 25.2
Matthes, L.A. BR 20.1
Matthews, D.J. US-I 5.1
Matthews, J. GB 31.2
Matthews, M.A. US-K 6.1.5
Matthews, P. GB 24.1.1
Mattick, L.R. US-B 2.1.5
Mattos, J.R. BR 29.2
Mattson, R.H. US-G 31.1
Matus, M.S. AR 15.2.3
Matusoka, K. BR 17.1
Matyjaszczyk, M. PL 28.1.10
Matysiak, B. PL 28.1.13
Maude, R.B. GB 33.1.4
Mauget, J.C. FR 5.1
Maughan, L.M. NZ 9.1
Maule, A. GB 24.1.4
Maurer, A.R. CA 1.1
Mauromicale, G. IT 5.4
Maurya, A.N. IN 27.1
Maurya, V.N. IN 22.1
Maximov, N. BG 8.2
Maxon-Smith, J.W. GB 18.1.1
Maxwell, D. US-F 23.1.4
Maxwell, I.C. GB 2.1
Maxwell, R. US-C 9.1
May, B.A. GB 31.1
Mayak, S. IL 2.2
Mayclair, R.M. NZ 10.1
Mayer, K. CH 1.1
Mayers, P. AU 23.1
Maynard*, D.N. US-D 24.1
Maynard, J.R. AU 43.1
Mazereeuw, P. NL 14.1
Mazière, Y. FR 37.1
Mazih, A. MA 1.1
Mazomenos, B. GR 3.1.1
Mazumder, B.C. IN 28.1
Mazur, A.R. US-D 13.1
Mazur, J. PL 20.2.9
Mazza, G. CA 13.1
Mbah, B.N. NG 5.1
Mbaye, A. SN 1.1
Mbugua, G. KE 1.8
McAleese, D.M. IE 3.1
McAneney, K.J. NZ 5.1
McAuliffe, K.W. NZ 9.5

McBeath, D.K. CA 13.1
McBeath. J.H. US-L 1.1
McCain, A.H. US-K 3.1.2
McCain, S. GB 35.1.2
McCarter, S. US-D 16.1
McCarthy, M.G. AU 40.1
McCartney, E.P. NZ 13.2
McCaslin, B. US-H 19.1
McClanahan, R.J. CA 17.1
McClean, G.D. AU 36.1
McClurg, C.A. US-C 5.2
McColl, C. AU 64.1
McColl, J.G. US-K 3.1.3
McCollum, Jr., G.D. US-C 3.1
McCombs*, C.L. US-C 10.1
McConchie, M.R. AU 8.1
McConnell, D.B. US-D 27.5
McConnell, R.M. US-F 5.1
McCormick, S.J. NZ 5.1
McCormick, W. US-D 22.1
McCoy, C.W. US-D 30.2
McCoy, R.E. US-D 25.1
McCraw, D. US-H 6.2
McCreight, J.D. US-K 13.1
McDaniel, A.R. US-C 10.1
McDaniel, N.R. US-D 44.1
McDaniels, D.G.L. US-E 3.1
McDonald, J.G. CA 43.1
McDuffie, R.F. US-C 10.1
McEachern*, G.R. US-H 6.2
McElgunn, J.D. CA 5.1
McEllister, F. AU 64.1
McElroy, G.H. GB 20.1
McGarvey, B.D. CA 25.2
McGeary, D.J. AU 28.1
McGee, J. US-D 16.2
McGlasson*, W.B. AU 65.4
McGowan, M. GB 25.1.2
McGrath, D.J. AU 14.1
McGregor*, G.R. AU 35.1
McGuire, J. US-D 50.1
McGuire, J.J. US-A 14.1
McHard, J.A. US-D 30.2
Mchau, O. TZ 5.1
McIsaac, D.W. CA 42.1
McKay, A. AU 56.1
McKendrick, J.D. US-L 2.1
McKenzie, A.R. CA 40.3
McKenzie, E.H.C. NZ 2.3.2
McKenzie, R.C. CA 6.1
McKeown, A.W. CA 23.1
McKevan, D.K.E. CA 37.1
McKinnon, J. US-G 7.1
McLachlan, S.G.D. NZ 10.1
McLaren, G.F. NZ 15.1
McLauchlan, R.L. AU 15.2
McLaurin, W. US-D 17.1
McLellan, M.R. US-B 2.1.5
McLeod, C.D. CA 40.3
McLeod, D.G.R. CA 18.1
McLeod, J.A. AU 8.1
McLeod, R.W. AU 9.1.1
McLoud, P. US-E 11.1
McMahon, C.R. AU 19.1
McMaster, L.C. AU 36.5
McMillan, Jr., R.T. US-D 29.1
McMillan-Browse, P. US-K 17.1
McMullen, R.D. CA 3.1
McNamara, D.G. GB 8.1.2
McNaughton, I.H. GB 15.1.2
McNaughton, K.B. NZ 8.1
McNicol, R.J. GB 15.1.2
McNiel, R. US-E 15.1
McNulty, P. IE 3.1
McPherson, H.G. NZ 8.1
McVittie, A.C. GB 8.1.3
McWha, J.A. NZ 12.1
McWilliams, E. US-H 6.1
Meade, J.A. US-B 14.1.7
Meador*, D.B. US-F 12.2.2
Meadowcroft, S.C. GB 28.1
Meadows, W.A. US-D 52.1
Meagher, J.W. AU 26.1
Mecenas, L.M.R. BR 27.1
Mécs, J. HU 5.3.3
Meddens, F. NL 15.1
Medeira, C. PT 1.1
Medellín, C. MX 15.1
Medgoda, I LK 3.2
Medhi, G. IN 7.1
Medigović, J. YU 9.1
Medina, V. BR 12.1
Medina Urrutia, V.M. MX 3.3
Medley, R.G.P. ZI 5.1
Medrado da Silva, E. BR 25.2
Medrzycki, M. PL 32.1.12
Meech, D.E.M. NZ 9.7
Meeklah, W.F.A. NZ 16.1
Meela, A. TZ 4.1
Meeus, P. BE 6.3
Meffert, H.F.Th. NL 9.10
Meguro, T. JP 12.2
Megyeri, A. HU 5.1.2
Megyeri, L. HU 2.1.2
Mehdi, M.Z. LY 1.2
Meheriuk*, M. CA 3.1
Mehlenbacher, S.A. US-B 14.1.1
Mehrete, T. ET 2.1
Mehta, N.S. IN 24.1.3
Mehta, P.K. IN 9.1
Mei, A. AR 15.3
Mei, A.O. AR 15.2.5
Meier, R. NL 6.1
Meijaard, D. NL 3.1
Meijer, R.J.M. NL 2.1
Meijneke*, C.A.R. NL 9.8
Meilland, A. FR 20.1
Meinx, R. AT 8.7
Meiss, W. DE 19.1
Mejnartowicz, L. PL 10.1.1
Mekers, O. BE 14.1
Mekhamer, M. AU 27.1
Meland, M. NO 5.1
Melanta, K.R. IN 11.2
Melbert, A. DE 15.3.6
Meldrum, S.K. AU 65.4
Meleegy, M. EG 1.5
Meli, T. CH 1.1
Melichar, M. CS 4.1
Melis, L.O. AR 15.2.4
Melo, A.A. de O. BR 14.1
Melo, E. PT 6.1
Melo, Q.M.s. BR 8.1
Melo de Moura, R.J. BR 10.1
Melon, J. PL 28.3
Melo Passos, E.E. BR 11.1
Membier, A.T. 8.8
Mena, A. AR 22.2
Mena Covarrubias, J. MX 6.2
Menary, R.C. AU 46.8
Mendes, R.A. BR 25.3
Mendes Rodrigues, F. BR 1.1
Mendilcioglu, K. TR 15.1
Mendis, H.M. LK 3.2
Mendlinger, S. IL 4.1
Mendonça, A. PT 6.1
Mendonça, N.P. BR 21.2
Mendoza, Jr. D.B. PH 1.2.5
Mendoza, E.M.T. PH 1.7
Mendoza, O. PE 3.3
Mendoza, P.A. PH 1.8
Mendoza López, M.R. MX 7.3
Mendoza R., I. PE 2.4
Mendt, R. VE 2.1
Meneses, R. CR 2.2
Meneve, I. BE 14.1
Menezes, D. BR 10.1
Menezes, E.B. BR 19.2
Menezes, T.J.B. BR 20.2
Menges, T. US-H 12.1
Mengestu, G. ET 5.1
Menguc, A. TR 6.i
Menhenett, R. GB 18.1.1
Menniti, A.M. IT 3.3
Mensah, R.K. GA 4.4
Menschoy, A.B. BR 30.2
Menye, D. KE 1.7
Menzies*, A.R. AU 3.1
Menzies, D.R. CA 25.2
Menzies, J.A. CA 14.1
Menzies, S.A. NZ 2.3.3
Merayo, A. CR 2.2
Merchat*, L. FR 24.1

Mercier, S. FR-2.1.3
Merckx, R. NL 9.5
Merdith, P. US-D 16.2
Meredith, C.P. US-K 6.1.5
Merin, U. IL 1.3.2
Mermier, M. FR 3.1.4
Merrett, B. AU 63.1
Merriman, P. AU 27.1
Merritt, R.H. US-B 14.1.1
Mertia, R.S. IN 18.4
Meseguer, M. CR 2.2
Meslin, G. FR 25.1
Mesquita, A. Mindenberg M. BR 12.1
Messina, M.A. AR 6.1
Messina, R. IT 25.1
Mészáros, Z. HU 2.5
Meszleny, A. HU 2.2.8
Metcalf, J.G. CA 24.1
Metera, P. PL 32.1.9
Metzidakis, J. GR 8.1.1
Meulemans, M. BE 6.8.7
Meunier, S. BE 12.1
Mével, A. FR 19.1
Mexal, J.G. US-H 19.1
Meyer, F.G. US-C 4.1
Meyer, J. BE 12.1
Meyer, J. DE 15.1.7
Meyer, J.R. US-D 8.1.3
Meyer* Jr., M.M. US-F 12.2.2
Meynet, J. FR 2.1.6
Meynhardt*, J.T. ZA 3.1
Mezei, G. HU 2.1.2
Mguni, C. ZI 3.1.2
Mibuta, T. JP 16.1
Miccolis, V. IT 2.2
Michael, A.M. IN 2.1
Michael, S.H. ET 9.1
Michaelidis, E. GR 6.1.2
Michalak, E. PL 28.1.13
Michalak, Z. PL 28.1.6
Michalczuk, B. PL 28.1.17
Michalczuk, L. PL 28.1.5
Michalek, K. CS 8.1, 13.1
Michalek, Wl. PL 14.1.5
Michalica, K.T. CA 40.1
Michalik*, B. PL 11.1.10
Michalik, H. PL 28.2.10
Michalojc, Z. PL 14.1.3
Michałowski, K. PL 28.2.8
Michalska, E. PL 20.2.11
Michaud, S.G. CA 40.1
Michel, H. FR 34.1
Michel, M. FR 41.2
Michelakis, N. GR 8.1.1
Michelakis, S. GR 8.1.1
Michelési, J.C. FR 1.1.1
Michelmore, R.W. US-K 6.1.4
Michelutti, R. CA 17.1
Michieka, A.O. KE 5.1
Michieka, D. KE 9
Michos, B. GR 18.1.1
Mićić, N. YU 1.2
Micović, M. YU 10.1
Micu, C. RO 9.1
Micu, I. RO 25.1
Micu, V. RO 8.4.2
Miczulski, B. PL 14.1.8
Miczyński, K. PL 11.1.8
Midcap, L.T. US-G 20.1
Middlefell Williams, J.E. GB 15.1.2
Middleton, S.G. AU 25.1
Miele, A. BR 31.1
Mielke*, E.A. US-J 16.1
Mierwinska, J. PL 28.2.11
Migdalska, D. PL 32.3.3
Miginiac, E. FR 30.1.1
Migliori, A. FR 14.1.2
Mihăescu, Gr. RO 8.3
Mihăescu, O. RO 32.1
Mihăilescu, N. RO 8.4.1
Mihăilescu, S. RO 8.5
Mihalache, M. RO 8.1
Mihalca, Gh. RO 8.4.2
Mihályffy, J. HU 2.1.2
Mihira, T. JP 4.3
Mii, M. JP 4.4
Mijačika, M. YU 14.1
Mijanović, O. YU 8.2
Mijas, M. PL 28.2.6
Mijatović, D. YU 1.2
Mijatović, K. YU 8.1
Mijatović, M. YU 13.1
Mijušković, M. YU 7.1
Mika*, A. PL 28.1.5
Mikami, T. JP 3.3
Miki, E. JP 12.1
Mikros, E. GR 25.2.1
Mikulás, J. HU 5.2
Mikuriya, H. JP 34.2
Miladinović, Ž. YU 13.1
Milaire, H. FR 17.1.3
Milanez de Rezende, G. BR 15.1
Milanov, B. BG 5.1
Milas, A. RO 2.1
Milas, D. RO 2.1
Milbocker, D.C. US-C 12.1
Milbourn, G.M. GB 6.1
Milburn, P.H. CA 40.3
Miles, N.W. CA 25.1
Milholland, R.D. US-D 8.1.2
Milia, M. IT 23.1
Milića, C. RO 23.1
Milinković, I. YU 14.2
Milinković, S. YU 8.2
Militiu*, A. RO 8.1
Militiu, I. RO 19.2
Milivojević, J. YU 14.1
Miljković, I. YU 3.1
Milkova, L. BG 13.1
Miller, A. US-F 7.1.2
Miller, C.H. US-D 8.1.1
Miller, D. ZI 5.1
Miller, D. US-F 3.1
Miller, D.G. US-I 18.1
Miller, D.M. CA 18.1
Miller, G.W. US-I 5.1
Miller, H.J. NL 9.8
Miller*, Jr., J.C. US-H 6.1
Miller, L.C. US-D 13.1
Miller, M. US-H 15.1
Miller, P. AU 30.1
Miller, P.A. CA 4.2
Miller, P.C.H. GB 31.2.2
Miller, R.J. US-I 14.1
Miller, R.M. AU 65.2
Miller, S. US-C 8.2
Miller, S.A. NZ 9.1
Miller, S.R. CA 24.1
Miller, W.M. US-D 30.1
Millette, J. CA 35.1
Millies, K.D. DE 12.1.4
Millim, K. RO 8.4.1
Millington, R.J. AU 65.7
Millo, E.P. ES 14.1
Millot, P. FR 2.1.4
Mills, H. US-D 16.1
Mills, R.A. NZ 5.1
Milne, K.S. NZ 9.1, 9.6
Milosavljević, M. YU 14.1
Milošević, D. YU 11.1
Milošević, T. YU 10.1
Milotay, P. HU 5.3.1
Milovanović, N. YU 14.2
Milutinović, M. YU 14.1
Mimata, T. JP 30.1
Minami, K. BR 24.1
Minamibori, K. JP 28.1
Minamide, T. JP 33.2
Minárik, V. CS 1.1
Mine, H. JP 30.1
Minegishi, M. JP 28.1
Minegishi, N. JP 39.1
Minegishi, T. JP 12.1
Miner, J.R. US-J 14.1
Minev, K. YU 4.2
Minks, A.K. NL 9.4.1
Minner, D.D. US-E 18.1
Mino, Y. JP 12.4
Minoiu, N. RO 14.1
Minotti, P.L. US-B 5.1.3
Miranda, P. BR 10.1

Miranda Trejo, S. MX 4.5
Miranović, K. YU 5.1
Mircea, I. RO 8.4.4
Mircetich, S.M. US-K 6.1.1
Mirghis, E. RO 1.1.1
Mirghis, R. RO 1.1.4
Mirihagalla, K. LK 3.7
Mirin, L.J. US-B 5.1.1
Mirkovic, M. YU 9.1
Miron, V. RO 1.1.3
Mirošević, N. YU 3.1
Mirzayev, M.M. SU 18.1
Misaka, T. JP 39.1
Misener, G.C. CA 40.3
Mishra, H.N. IN 16.2
Mišić, P.D. YU 9.1
Misik, S. HU 5.2
Miske, T. DE 14.1
Misra, A.P. IN 24.1.2
Misra, A.K. IN 22.1
Misra, B.K. IN 22.1
Misra, K.K. IN 20.1
Misra, P.C. IN 4.1
Misra, R.S. IN 16.2, 22.1, 24.1.1
Misra, S.P. IN 25.1
Missonnier, C. FR 17.1.4
Missonnier, J. FR 14.1.3
Misurák, E. HU 5.1.2
Mita, T. JP 38.2
Mitchell, C.A. US-F 9.1.4
Mitchell, D.S. AU 65.6
Mitchell, P. AU 32.1
Mitchell, R.E. NZ 2.1.4
Mitchell, W. US-A 1.1
Mitchell, W.W. US-L 2.1
Mitchnick, Z. IL 1.1.1
Mitevski, Z. YU 4.1
Mitić, S. YU 10.1
Mitić-Mužina, N. YU 8.1
Mitidieri, A. AR 19.1
Mito, K. JP 38.1
Mitobe, M. JP 35.2
Mitov, P. BG 8.3
Mitra, S.K. IN 28.1
Mitrache, D. RO 7.1
Mitrakos, C. GR 7.1
Mitrović, M. YU 10.2
Mitrut, T. PL 14.1.4
Mitsou, I. GR 18.1.3
Mitsue, S. JP 20.1
Mitsui, N. JP 38.2
Mitteau, M. FR 17.2
Mitteau, Y. FR 33.1
Mityga, H.G. US-C 5.2
Miura, H. JP 41.5
Miura, Y. JP 19.1
Miwa, S. JP 38.1
Miyagawa, M. JP 29.1
Miyagawa, T. JP 1.1.1
Miyagawa, Y. JP 14.1
Miyagi, M. JP 32.1
Miyagi, S. JP 32.1
Miyahara, M. JP 7.2
Miyahara, T. JP 6.2
Miyaji, R. JP 18.1
Miyajima, D. JP 6.3
Miyajima, H. JP 7.4
Miyajima, Y. JP 26.4
Miyake, S. JP 39.1
Miyakoda, H. JP 41.1
Miyamatsu, K. JP 6.2
Miyamoto, K. JP 44.2
Miyasato, T. JP 32.1
Miyashita, S. JP 18.1
Miyata, A. JP 46.1
Miyata, M. JP 41.3
Miyawaki, K. JP 17.1
Miyazaki, S. JP 34.4
Miyazaki, T. JP 17.2
Miyazawa, Y. JP 26.4
Miyoshi, T. JP 10.1
Miziołck, Z. PL 28.1.18
Mizrachi, Y. IL 4.1
Mizu, O. JP 22.1
Mizubuti, A. BR 17.1
Mizukoshi, Y. JP 2.1
Mizumura, H. JP 35.2
Mizuno, N. JP 2.2, 19.2
Mizuno, S. JP 13.3
Mizushima, S. JP 12.2, 18.2
Mizushima, T. JP 31.1
Mizuta, Y. JP 13.2
Mizutani*, F. JP 5.3
Mizutani, N. JP 33.3
Mizutani, S. JP 15.3
Mladenov, M. BG 6.1
Mladin, Gh. RO 9.1
Mladir, P. RO 9.1
M'laïki, A. TN 1.1
Mlambo, S.S. ZI 3.1.2
M'lika, M. TN 1.1
Mnzava, N.A. TZ 2.1
Mochizuki, M. JP 19.1
Mochizuki, T. JP 16.3, 47.2
Mocquot, B. FR 4.1.4
Moczulska, U. PL 22.1
Modic, D. YU 18.2
Moe, M. NO 10.3
Moe, R. NO 10.1.4
Moeller, C. US-A 14.1
Moes, E. DK 7.1
Moffett, M.L. AU 15.4
Mogal, D.B. IN 14.4
Mogdam, S.B. IN 14.3
Mogos, G. RO 5.1
Mohammed, W/o. K. ET 6.1
Mohan, E. IN 3.1
Mohan, I. IN 19.3
Mohandas, S. IN 1.1.8
Mohanty, B. IN 16.2
Mohanty, C.R. IN 16.2
Mohapatra, P. IN 16.2
Moharana, T. IN 16.2
Mohr, H.C. US-E 15.1
Mohr, W.P. CA 24.1
Moiceanu, D. RO 9.1
Mok, I.G. KR 23.1.6
Mokashi, A.N. IN 11.3
Mokrá, V. CS 24.1
Mokrzecka, E. PL 14.1.3
Molano, M. CO 4.1
Moldovan, I. RO 28.1
Moldovan, V. RO 6.1
Molea, I. RO 8.1
Molhova, E. BG 13.1
Molin, G. FR 17.1.2
Molina, Jr. A.C. PH 1.4
Molina, G.C. PH 1.6
Moline, H.E. US-C 3.1
Molitor, H.D. DE 12.1.8
Molkup, V. CS 20.1
Moll, J.N. ZA 1.1
Møller, B.L. DK 3.4
Möller, H. DE 15.3.5
Møller, I. DK 3.4
Molnár, B. HU 2.1
Molnar*, J.M. CA 1.1
Molot, C. FR 4.1.2
Molot, P-M. FR 3.1.5
Monaco, T.J. US-D 8.1.1
Monastra, F. IT 19.1
Moncaster, M.E. GB 31.2.7
Mondeshka, P. BG 14.1
Monet, R. FR 4.1.1
Monette, P.J. CA 2.1
Mongour, J. FR 41.1
Monin, A. BE 6.1, 6.5, 6.6
Monk, P.R. AU 37.1
Monma, S. JP 16.3
Monnerat, P.H. BR 17.1
Monnier, G. FR 3.1.7
Monro, J.A. NZ 8.2
Monroig, M. US-N 1.2
Monselise*, S.P. IL 2.1
Monsion, M. FR 4.1.6
Monszpart-Sényi, J. HU 2.2.1
Montalv, R. PE 1.1
Montecillo, L.A. PH 1.5
Monteiro*, A.A. PT 5.1
Monteiro, A.A.T. BR 8.1
Monteiro Sobral, C.A. BR 4.1
Monteith, J.L. GB 25.1.2
Montelongo, H. MX 15.1

Montemurro, P. IT 2.2
Monterroso, D. CR 2.2
Montes Martínez, E. MX 5.3
Montfort, M. FR 19.2
Monti, L. IT 18.1, 18.2
Montorsi, F. IT 19.2
Montoya, J.L. ES 20.1
Moody, P.D. GB 19.1.4
Mooi, J. NL 9.4.5
Moon, D.Y. KR 10.1
Moon, D.K. KR 10.3
Moon, J.W. US-H 27.1
Moon, J.Y. KR 23.1.4
Moon, N.P. US-D 20.1
Moór, A. HU 5.3.2
Moorat, A.E. GB 12.1
Moore, C.S. GB 8.1.3
Moore, D. GB 20.1
Moore, D.E. US-B 6.1
Moore, E.L. US-D 30.2
Moore, F. US-K 15.1
Moore, F.D. US-I 2.2
Moore, J. US-E 19.1
Moore, J.F. IE 4.1
Moore*, J.N. US-E 11.1
Moore, K.G. GB 3.1
Moore, N.J. US-L 2.2
Moore, P.H. US-K 11.1
Mor, Y. IL 2.2
Mora, A. CR 2.2
Mora, I. CR 2.2
Mora Blancas, E.E. MX 17.1
Moraes, W.B.C. BR 21.2
Moraes Guimarães, C. BR 26.1
Morales, A. US-N 1.2
Morales, E.A.V. BR 25.3
Morales, G. PE 2.4
Morales, L. CR 2.2
Morali, K. TR 3.1
Moran, P. GB 31.2.4
Morand, J.C. FR 1.1.3
Moras, P. FR 45.1
Moravec, J. CS 20.1, 24.2
More, B.B. IN 14.3
Moreau, B. FR 46.1
Moreau, J.-P. FR 17.1.3
Moreau, M. FR 23.1
Morehart, A.L. US-C 2.1
Moreira, C.S. BR 24.1
Moreira, R.S. BR 20.1
Morel, Ph. FR 15.1.4
Morelet, M. FR 11.1.4
Morelock, T.E. US-E 11.1
Moreno, R.O. AR 18.1
Moretti, G. IT 7.1
Morgan, D. US-H 7.1
Morgan*, J.V. IE 3.1
Morgan, R.P.C. GB 31.1
Morgan, W.C. AU 28.1
Morgans, A.S. AU 28.1
Morgaś, H. PL 28.1.5
Mori, G. JP 33.2
Mori, I. JP 34.1
Mori, M. JP 33.1
Mori, N. JP 27.1
Mori, S. JP 5.2, 36.1, 40.2
Mori, Y. JP 6.3
Morice, J. FR 14.1.1
Moriconi, D. AR 7.1
Moriguchi, T. JP 14.2
Morikawa, S. JP 6.2
Morimoto, J. JP 44.2
Morin, J.P. BR 11.1
Morinaga, K. JP 17.2
Morini, S. IT 17.2
Morioka, K. JP 17.1
Morioka, S. JP 4.3
Morishita, M. JP 33.1
Morishita, T. JP 44.1
Morisot, A. FR 2.1.1
Morita, A. JP 7.1, 27.2
Morita, M. JP 1.1.6
Morita, T. JP 21.1, 36.2
Moriya, O. JP 35.2
Moriya, S. JP 14.2
Morlat, R. FR 1.1.4
Morley, H.V. CA 18.1
Morley-Bunker, M.J.S. NZ 13.5
Morone Fortunato, I. IT 2.2
Morosanu, V. RO 21.1
Morot-Gaudry, J.-F. FR 17.1.5
Morotomi, Y. JP 30.2
Morris, E. GB 31.1
Morris, G.E.L. GB 33.1.7
Morris, J.R. US-E 11.1
Morris, T.J. US-K 3.1.2
Morris-Krsinich, B.A.M. NZ 2.3.4
Morrison, B. AU 30.1
Morrison, C.A. US-F 1.1
Morrison, F.D. US-G 31.1
Morrison, J.C. US-K 6.1.5
Morse, R.D. US-C 10.1
Mortensen, J.A. US-D 32.1
Mortensen, L. NO 10.1.4
Mortensen, P. DK 8.1
Morton, J.G. CA 21.1
Morton, J.S. US-A 11.1
Morton, R.S. CA 42.1
Moruzini, S. BR 29.1
Morvan*, G. FR 3.1.5
Mosch, W.H.M. NL 9.4.4
Moschetti, C.J. AR 11.1
Moschini*, E. IT 17.2
Mosegaard, J. DK 3.1
Moser, B.C. US-F 9.1.4
Moser, E. DE 28.1.2
Moser, L. IT 19.1
Moshra, S.N. IN 16.2
Moss, D.E. AU 39.1
Moss, G.I. AU 65.6
Mossop, D.W. NZ 2.3.4
Mota Barroso, J. PT 4.1
Mota da Costa, D. BR 30.1
Motegi, T. JP 10.1
Motes, D.R. US-E 13.1
Motes, J.E. US-H 3.1
Motoda, K. JP 15.1
Motohashi, I. JP 41.1
Motomura*, Y. JP 24.3
Motosugi, H. JP 22.5
Motozu, Y. JP 14.1
Motta, E. IT 19.2
Motta Macedo, M.C. BR 12.1
Motta Miranda, R. BR 19.2
Mottl, J. CS 6.1
Mouches, M. FR 2.1.5
Mouchtouri, E. GR 19.1
Moulin, E.O. US-B 6.1
Mourikis, P.A. GR 14.1
Moussa, I. EG 4.1
Moussallem Pantoja Pimentel, A.A. BR 2.3
Moustache, A.M. SC 1.1
Moustafa, A. EG 6.1
Moustakas, N.G. GR 20.1
Movileanu, M. RO 21.1
Mowat, W.P. GB 15.1.3
Mower*, R.G. US-B 5.1.1
Moye, H.A. US-D 27.3
Moyer, J.W. US-D 8.1.2
Moyls, A.L. CA 3.1
Mozsár, K. HU 2.2.6
Mpithas, G. GR 23.1
Mreta, R.A.D. TZ 1.2
Mrozowski, M. PL 16.1
Msabaha, M. TZ 7.1
Msika, R.L. ZI 4.1
Muddappa Gowda, P. IN 11.2
Mudge, K.W. US-B 5.1.1
Mudrak, L.Y. US-B 5.1.1
Muehmer*, J.K. CA 21.1
Mueller, S. BR 28.1
Mueller, W.C. US-A 14.1
Muermans, L. BE 4.2
Muftuoglu, T. TR 14.1
Mugniery, D. FR 14.1.3
Muhadja, R. YU 15.1
Muhadjeri, R. YU 15.1
Muhammed Kunju, U. IN 5.5
Muigai, S.G.S. KE 1.8
Muirhead, I.F. AU 15.4
Mujica, F. AR 9.1
Mukai, B. JP 35.2

Mukai, H. JP 23.2.5
Mukhopadhyay, A. IN 1.1.3
Mukobata, H. JP 43.2
Muli, Z.R. KE 4
Muliyar, M.K. IN 3.1
Mulkey, W. US-D 54.1
Mullaly, J.V. AU 26.2
Müller, A.A. BR 2.1
Müller, A.T. AU 53.2
Muller, C. FR 11.1.3
Müller, C.H. BR 2.1
Müller, E. AT 22.1
Muller, F.M. NL 9.7.4
Müller, G. DE 15.1.7
Müller, G.W. BR 20.1
Müller, H. DE 27.1
Muller, J.J. BR 28.1
Muller, L. CR 2.2
Muller, M.W. BR 14.1
Muller, P.J. ZA 1.1
Mullin, R. US-G 7.1
Muller, Th. ZI 3.1.2
Müller, W. CH 1.1
Müller-Haslach, W. DE 33.1
Mullins, C.A. US-E 1.1
Mullins*, M.G. AU 9.3
Mulrooney, R.P. US-C 2.1
Mumo, J.M. KE 1.3
Munakata, T. JP 8.2
Munda, I.M. IC 1.1
Munday, V. US-D 16.1
Mundt, C.A. AR 3.1.1
Munene, D.N. KE 5.1
Munene, S.N. KE 5.1
Munger*, H.M. US-B 5.1.3
Muniz, J.O. de L. BR 8.1
Munk, L. DK 3.5
Muñoz, J. AR 7.1
Munro, D. AU 46.5
Munro Olmos, D. MX 3.2
Munson, S.T. US-G 7.1
Munteanu, N. RO 3.1
Munteanu, T. RO 1.1.6
Munthe, T. NO 10.4.1
Munyoro, G.F. ZI 3.2
Murabi, M.A. LY 2.1
Murai, M. JP 4.2
Murakami, J. JP 12.3
Murakami, K. JP 40.2
Murakami, M. JP 1.1.3
Murakami, T. JP 20.2
Muramatsu, H. JP 12.1, 27.2
Muramatsu, T. JP 39.2
Muramatsu, Y. JP 38.1
Murao, K. JP 42.1
Muraoka, K. JP 10.1
Muraoka, M. JP 34.2
Muraoka, N. JP 14.3
Muraro, R.P. US-D 30.2
Murase, S. JP 14.2
Murashige, T. US-K 12.1.1
Murata, H. JP 27.3
Murata, K. JP 42.1
Murata, R. JP 36.1
Murata*, T. JP 38.4
Muratović, A.S. YU 1.1
Murczek, P. PL 28.2.8
Murdoch, C.L. US-M 2.2
Murfitt, R.F.A. GB 31.1
Murg, S. RO 13.1
Murisier, F. CH 2.3
Muriuki, S.J.N. KE 1.4
Muroi, E. JP 39.1
Murozono, M. JP 7.1
Murphy, A.M. CA 40.3
Murphy, R.F. IE 4.1
Murphy, T. IE 3.1
Murr, D.P. CA 15.1
Murray, R. AR 16.1
Murray, R.A. CA 42.1
Murray, R.A. GB 8.1.2
Murray-Prior, R. AU 7.1
Murtas, A. IT 23.1
Murthy, V.K. IN 6.1
Murti, G.S.R. IN 1.1.8
Murty, P.R.K. IN 22.1
Murvai, M. RO 8.1
Musa, T.M. ZI 3.1.2
Musard, M. FR 21.1
Musgrave, D.R. NZ 13.3
Müssel, H. DE 11.2.8
Mussillon, P. FR 7.1.10
Mustaffa, M.M. IN 1.1.1
Muszyńska, J. PL 28.1.8
Muszyński, A. PL 20.2.10
Mut, M. ES 23.3
Mutagami, S. JP 25.1.3
Mutascu, N. RO 16.1
Muthamia, J. KE 1.8
Muthappa Rai, B.G. IN 11.2
Muto, K. JP 16.1
Mutschler, M.A. US-B 5.1.7
Mutton, L. AU 8.2
Muzik, P. HU 2.2.6
Mwajumwa, E.B. KE 1.2
Mwamba, D. KE 1.2
Mwangi, K. KE 3.1
Mwasya, W.N.P. KE 13
Mwathi, P. KE 1.4
Myczkowski, J. PL 11.1.9
Myers, J.M. US-D 27.9
Mynett*, K. PL 28.1.12
Myrianthousis, T. CY 1.1
Mzira, C.N. ZI 3.1.2

Naber, H. NL 9.8
Nacu, F. RO 1.1.2
Nadal, A. ES 5.1
Nádas, P. HU 4.2
Nadaždin, V. YU 2.1
Nador, M. MA 3.1
Naess-Schmidt, K. DK 3.3
Nagabe, T. JP 16.1
Nagahama, M. JP 18.2
Nagai, H. BR 20.1
Nagai, K. JP 13.1, 14.2, 31.1, 40.1
Nagai, M. JP 12.3
Nagai, N. JP 6.2
Nagai, Y. JP 40.1
Nagaishi, S. JP 17.2
Nagami, K. JP 46.1
Nagamine, Y. JP 32.1
Nagamura*, S. JP 28.1
Nagano, M. JP 27.2
Nagao, A. JP 12.1
Nagao, M.A. US-M 1.1
Nagaoka, K. JP 17.1
Nagaoka, M. JP 23.2.5
Nagara, M. JP 42.1
Nagarajah, S. AU 31.1
Nagasawa, J. JP 22.2
Nagasawa, K. JP 6.1
Nagasawa, M. JP 41.1
Nagata, R. JP 25.1.2
Nagato, T. JP 4.1
Nagatomo, E. JP 25.1.1
Nagawekar, D.D. IN 15.1
Nagda, C.L. IN 19.1
Nagel, D. US-H 10.1
Nagel, G. DE 15.2.2
Nagel, R. AT 5.1
Nagib, M. EG 2.1
Nagl, K. AT 8.7
Nagórska, E. PL 28.5
Nagwade, R.N. IN 14.2
Nagy, B. HU 2.1.5, 2.2.4, 2.5
Nagy, Gy. HU 7.1
Nagy, Gy.Zs. HU 2.5
Nagy, I. HU 3.1.1, 5.1.1
Nagy, J. HU 2.2.12, 5.1.4
Nagy, J.I. HU 2.2.12
Nagy, L.Sz. HU 2.2.10
Nagy, P. HU 2.1.2
Nagy, S. HU 2.2.5
Nagy, S. US-D 30.1
Nagyjános, Zs. HU 2.7
Nagy-Radnai, E. HU 5.2
Nahata, K. JP 43.2
Nahir, D. IL 1.4.4
Naidu, R. IN 3.1
Naik, B. IN 16.2
Naik, L.B. IN 1.1.2
Nair, M.K. IN 3.1

Nair, P.C.S. IN 5.3
Nair, R.R. IN 5.4
Naito, F. JP 23.2.5
Naito, K. JP 28.1
Naito, R. JP 37.2
Naito, Y. JP 31.1
Najasabadi, F.H. MX 7.5
Naji, S.A. LY 1.2
Nakada, A. JP 46.1
Nakada, H. JP 15.2
Nakada, T. JP 36.1
Nakada, Y. JP 35.1
Nakagaki, N. JP 14.1
Nakagami, K. JP 1.1.1
Nakagawa, M. JP 40.2
Nakagawa, O. JP 23.2.3
Nakagawa*, S. JP 33.2
Nakagawa, Y. JP 23.2.5, 37.1
Nakagawara, I. JP 3.3.1
Nakagome, T JP 1.1.1
Nakahara, M. JP 42.2
Nakai, S. JP 4.3
Nakaji, M. JP 21.2
Nakajima, F. JP 12.3
Nakajima, M. JP 1.2
Nakajima, Y. JP 20.4
Nakalemic, A. YU 14.1
Nakama, M. AR 19.1
Nakamura, B. JP 26.4
Nakamura, H. JP 34.2
Nakamura, K. JP 26.4
Nakamura, M. JP 3.1, 9.4, 16.1, 24.2, 26.2, 46.1
Nakamura, R. JP 31.2
Nakamura, S. JP 20.2, 38.1, 46.2
Nakamura, T. JP 7.2, 46.1
Nakamura, Y. JP 15.3, 22.3
Nakanishi, T. JP 13.3
Nakano, M. JP 31.2
Nakano, T. JP 2.3, 23.1, 29.1
Nakao, E. JP 44.2
Nakao, T. JP 17.1, 27.2
Nakashima, T. JP 23.2.3
Nakashima, Y. JP 7.2
Nakasu, B.H. BR 30.1
Nakata, H. JP 5.2
Nakatani, F. JP 16.1
Nakatani, M. JP 11.2
Nakatomi, Y. JP 29.1
Nakaya, H. JP 44.2
Nakayama, M. JP 26.5, 44.2
Nakayama, T. JP 26.3
Nakayama, T. US-D 20.1
Nakazawa, H. JP 26.4
Nakazawa, N. JP 3.2.3
Nakazawa, T. JP 20.2
Nalawadi, U.G. IN 11.3
Nalbant, M. TR 4.1
Nalli, R. IT 19.2
Nambiar, K.K.N. IN 3.1
Namiki, T. JP 22.4
Nampoothiri, K.U.K. IN 3.1
Namuco, L.O. PH 1.2.1
Nanaya, K.A. IN 1.1.1, 1.1.9
Nanba, H. JP 44.1
Nandihalli, B.S. IN 11.10
Nangniot, P. BE 6.8.9
Nankar, J.T. IN 12.1
Naphade, A.S. IN 14.4
Napiórkowska-Kowalik, J. PL 14.1.9
Naragund, V.R. IN 1.1.4
Narase Gowda, N.C. IN 11.9
Narayana Gowda, J.V. IN 11.2
Narayanan, K. IN 1.1.7
Narayanappa, IN 1.1.1
Narayanappa, M. IN 1.1.6
Narayana Reddy, M.A. IN 11.9
Narayana Reddy, P. IN 11.5
Nardin, C. IT 11.1
Narikawa, T. JP 23.2.2
Narimatsu, J. JP 19.1
Narita, H. JP 2.2, 3.2.4
Narkiewicz, B. PL 28.2.13
Narkiewicz-Jodko*, J. PL 28.2.7
Narro, L. PE 2.3
Nartea, R.N. PH 1.5
Narusawa, H. JP 26.4
Narwadkar, P.R. IN 12.1
Nascimento, D. BR 10.1
Nassell, R. US-F 2.1
Nassi, M.O. IT 24.4
Nasu, H. JP 3.3, 31.1
Nasu, K. JP 10.1
Nasuda, K. JP 6.3
Natali, S. IT 17.2
Natarella, N.R. US-E 18.1
Natsumi, K. JP 44.2
Natural, M.P. PH 1.4
Natyal, R.K. IN 10.3
Nauer, E.M. US-K 12.1.1
Naumann, G. DE 6.1
Naumann, W.-D. DE 15.1.4
Nautiyal, M.C. IN 22.1
Navara, J. CS 1.1
Navarro, S. IL 1.3.3
Navarro Lucas, L. ES 14.1
Navatel, J.C. FR 21.1
Nawale, R.N. IN 15.1
Nawata, E. P 22.5
Nayar, G.G. IN 5.1
Nayar*, N.M. IN 4.1
Nazar, C. AR 7.1
Nazar, R.A. BR 15.1
Nazareno, N.R.X. BR 27.2
Nazca, A.J. AR 22.2
Nazrala, M.L. AR 12.1
Nchanjala, R. TZ 3.1
Ndegwa, A.M.M. KE 1.8
Ndegwe, N.A. NG 10.1
Ndoria, E.N. KE 1.7
Ndubizu, T.O.C. NG 5.1
Neagu, I. RO 27.1
Neal, J. US-C 3.1
Neal, J.C. US-B 5.1.1
Neamtu, Gh. RO 2.1
Neamtu, I. RO 9.1
Neculae, L. RO 9.1
Nedev, N. BG 8.2
Nedstam, B. SE 1.4
Nee, C.C. TW 7.1
Neely, D. US-F 12.1.1
Neergaard, M.K. NO 10.1.7
Negash, T. ET 4.1
Negi, S.S. IN 1.1.3
Negm, F.B. US-B 5.1.1
Negrel, J. FR 7.1.10
Negrilă, A. RO 8.3
Negulescu, P. RO 1.1.6
Nehdi, H. TN 1.1
Nehmat-Allah, M.H. LY 4.1
Nei Briançon Busquet, R. BR 19.2
Neild, R.E. US-G 23.1
Neill, B. CA 9.1
Neilsen, G.H. CA 3.1
Neilson, C. US-J 4.1
Neilson, W.T.A. CA 41.1
Nell*, T.A. US-D 27.5
Nellist, M.E. GB 31.2.2
Nelson, D.C. US-G 12.1
Nelson, G.A. CA 8.1
Nelson, J.E. US-I 18.1
Nelson, J.M. US-H 25.1
Nelson, M. US-H 27.1
Nelson, P.V. US-D 8.1.1
Nelson, P.E. US-F 9.1.2
Nelson, S. CA 20.2
Nelson, S. US-D 16.2
Nelson*, S.H. CA 11.1
Nelson, W.R. US-F 12.2.2
Nelson, W.W. US-G 3.1
Nemec, S. US-D 34.1
Némethy, L. HU 4.2
Némethy, Zs. HU 2.2.8
Nemoto, H. JP 35.2
Nenić, P. YU 14.1
Nerd, A. IL 4.1
Nerson*, H. IL 1.1.1
Nes*, A. NO 6.1
Nesm, X. FR 1.1.3
Nestby, R. NO 8.1

Netland, J. NO 10.4.2
Neto, M.J.A. BR 15.1
Netto, A.V. de M. BR 10.1
Neubeller, J. DE 28.1.1
Neuendorf, M. SE 2.1
Neumaier, N. BR 27.1
Neumann, R. US-H 22.1
Neuray, G. BE 6.8.2, 6.10
Neururer, H. AT 8.11
Neuvel, J.J. NL 6.1
Nevins, D.J. US-K 6.1.4
Newman, H.P. AU 65.2
Newman, J.S. US-H 13.1
Newman, R. US-F 23.1, 23.1.3
Newman, S. US-D 50.1
Newmann, P. CA 31.2, 31.3
Newton, F.J. GB 2.1
Newton*, P. GB 22.1
Neyroud, J.A. CH 2.1
Ng, T.J. US-C 5.2
Nguyen*, V. AU 8.2
Nguyen The, C. FR 3.1.8
Ni, W.T. TW 11.1.1
Niazi, D.M. LY 1.2
Niblett, C.L. US-D 27.6
Nicholas, P. AU 38.2
Nichols, J.R. US-C 10.1
Nichols*, M.A. NZ 9.1
Nichols*, R. GB 18.1.3
Nicholson*, J.A.H. GB 35.1.1
Nickerson*, N.L. CA 41.1
Nicklow, C.W. US-A 10.1.1
Nicolaescu, I. RO 26.1
Nicolaou, N. GR 25.1.5
Nicolas, J. FR 3.1.8
Nicolas, J.C. FR 3.1.3
Nicolas, M.G. FR 3.1.8
Nicolaus, L. BE 4.1
Nicolet, J.L. FR 46.1
Nicotra, A. IT 19.1
Niculescu, F. RO 8.4.1
Niehaus, M.H. US-I 2.1
Nielsen, B.J. DK 4.2
Nielsen, G.A. US-I 18.1
Nielsen, G.C. DK 4.1
Nielsen, O.F. DK 1.2
Nielsen, P.C. DK 3.3
Nielsen, P.Chr. DK 3.3
Nielsen, S.L. DK 4.2
Nieman, T. US-E 15.1
Niemann, E-G. DE 15.3.1
Niemczyk*, E. PL 28.1, 28.1.2
Niemeyer, R. DE 15.3.3
Niemi, H. HU 8.1
Niemirowicz-Sczytt, K. PL 32.1.8
Niemirski, W. PL 32.1.11
Nieuwhof, M. NL 9.6, 9.6.1, 9.6.5
Nievas, J.J. AR 17.1
Niezborala, E. PL 27.1
Niezgodzinski, P. PL 33.2
Nigg, H.N. US-D 30.2
Niggli, U. CH 1.1
Nigh, E.L. US-H 30.1
Nightingale, A.E. US-H 6.1
Nii, N. JP 1.2
Niimi, Y. JP 11.4
Niimi*, Y. JP 29.2
Niitsu, A. JP 47.1
Niiya, T. US-K 2.1
Niiyama, T. JP 43.1
Niizawa, T. JP 18.2
Niizuma, T. JP 2.2
Nijensohn, L. AR 15.2.2
Nijskens, J. BE 6.8.5
Nikčević, V. YU 5.1
Niketić-Aleksić, G. YU 14.1
Nikolantonakis, M. GR 10.1
Nikolić, M. YU 10.2
Nikoloff, A. NZ 13.2
Nikolov, N. BG 12.1
Nikolova, N. BG 8.3
Nikosavić, Ž. YU 13.1
Niksić, M. YU 10.1
Nilangakar, R.G. IN 12.2
Nilovankić, M. YU 17.1
Nilsen*, J. NO 9.2
Nilsen, S. NO 7.2
Nilsson*, B. SE 1.4, 6.1
Nilsson, F. SE 1.1.1
Nilsson*, T. SE 1.1.2
Ninkovski, I. YU 9.1
Ninomiya, T. JP 5.2
Nisen, A. BE 6.5, 6.8.1
Nishibe, S. JP 17.2
Nishida, F. JP 21.1
Nishida, K. JP 11.2
Nishide, A. JP 10.1
Nishiguchi, I. JP 23.1
Nishiguchi, Y. JP 6.1
Nishihara, T. JP 18.3
Nishikawa, Y. JP 46.1
Nishimata, D. JP 28.1
Nishimori, H. JP 44.1
Nishimoto, F. JP 21.1
Nishimoto, N. JP 18.2
Nishimoto, R.K. US-M 2.2
Nishimura, J. JP 13.2
Nishimura, N. JP 33.1
Nishimura, S. JP 23.2.2
Nishimura, T. JP 22.1
Nishino, H. JP 22.1
Nishino, T. JP 47.1
Nishio, J. JP 1.1.3
Nishio, K. JP 23.2.3
Nishio, T. JP 23.2.2
Nishio*, T. JP 36.2
Nishioka, M. JP 1.1.6
Nishitani, K. JP 17.1
Nishitani, N. JP 13.1
Nishitani, T. JP 44.1
Nishiyama, T. JP 5.2
Nishiyama, Y. JP 14.2, 17.1
Nishizima, T. JP 47.2
Nissen, G. DK 10.1
Nissinen, O. FI 8.1
Nito*, N. JP 34.4
Nitranský, M. CS 31.1
Nitschke, L. AU 36.5
Nitschke, S. DE 23.1
Nitta, A. JP 42.2
Nitta, H. JP 8.1, 11.2
Nitta, S. JP 1.2
Niveyro, J.R. AR 8.1
Niwata, H. JP 3.3.1
Niwicka, M. PL 28.2.13
Nix, J.S. GB 35.1.1
Nizamoglu, A. TR 13.1
Njenga, D.N. KE 5.1
Njenga, K. KE 2.1
Njoroge, I.N. KE 5.1
Njoroge, J. KE 5.1
Njugunah, S.K. KE 1.1
Nkoloma, L.E. ML 1.1
No, J.H. KR 5.1
Nobeli, C. GR 3.1.2
Nobile, R. AR 7.1
Noble, A.C. US-K 6.1.5
Nóbrega, A.C. BR 18.1
Nobuta, M. JP 9.1
Noda, H. BR 1.2
Noda, H. JP 17.1
Noda, M. JP 7.2
Nøddegaard, E. DK 4.2
Noga, G. DE 6.1
Nogata, T. JP 34.2
Nogawa, T. JP 9.2
Noguchi, K. JP 45.1
Noguchi, S. JP 27.1
Noguchi, Y. JP 2.1, 7.2, 7.3, 42.2
Nokoyama, S. BR 28.1
Noland, D.A. US-F 12.2.2
Nolas, I. GR 18.1.4
Nollen, H.M. NL 9.8
Nolting, J. AR 10.1
Noma, F. JP 25.1.2
Nome, F. AR 7.1
Nomi, Y. JP 44.2
Nomoto, Y. JP 29.1
Nomura, T. JP 28.1
Nomura, Y. JP 7.3, 15.1
Nonaka, M. JP 23.2.3
Nonami, H. JP 20.2

Nonnecke, I.L. CA 15.1
Nonomura, S. JP 37.1
Nonoyama, Y. JP 23.2.5
Nordal, I. NO 7.3
Nordby, A. NO 10.2
Nordin, I. SE 2.1
Nordstrom, P.E. US-G 19.1
Noriega, D. PE 1.3
Norman, S.M. US-K 11.1
Noro, S. JP 3.2.1, 3.3.1
Noro, T. JP 41.1
Norstog, K. US-D 33.1
Northover, J. CA 25.2
Nortje, B.K. ZA 4.1
Norton, C.R. CA 4.2
Norton, G. GB 25.1.3
Norton, J.D. US-D 38.1
Norton*, R.A. US-J 3.1
Nosek, J. CS 22.3
Nothmann*, I. IL 1.1.1
Nothnagel, E.A. US-K 12.1.1
Noto, G. IT 5.2
Notton, B.A. GB 19.1.2
Novak, B. YU 3.1
Novak, J. US-H 6.1
Novák, L. CS 22.2
Novák, V. CS 20.1
Nováková, E. CS 24.3
Novello, V. IT 24.2
Novo, R. AR 7.1
Novotná, M. CS 24.1
Novotný, J. CS 5.1
Nowacka*, H. PL 28.1.2
Nowacka, W. PL 20.2.2
Nowacki, J. PL 28.1.4
Nowaczyk, P. PL 4.1
Nowak, B. PL 28.2.13
Nowak, J. CA 42.2
Nowak, J. PL 28.1.17, 32.3.3
Nowak, T. PL 33.3
Nowakowski, W. PL 32.1.13
Nowakowski, Z. PL 28.1.2
Nowicki, B. PL 32.1.6
Nowitzki, R. DE 19.1
Nowosielska, B. PL 28.2.15
Nowosielski*, O. PL 28.2.6
Noyé, G. DK 5.1
Nozaki, S. JP 6.2
Nozuka, M. JP 7.1
Nsanjama, R. ML 1.1
Nsekela, J. TZ 4.1
Nsowah, G.F. GA 2.2.1
Ntungakwa, C.N. ZI 3.1.2
Nugent, P.E. US-D 12.2
Nukaya, A. JP 38.4
Nuland, D.S. US-G 26.1
Numa, S. JP 8.1
Nunes de Pinho, L. BR 8.1
Nunes e Nunes, L. BR 30.1
Nunes Faria, A.R. BR 12.1
Nunes Fernandes, R. BR 6.1
Nuñez Elisea, R. MX 3.3
Nunomiya, T. JP 45.1
Nur, M. ID 2.1
Nurena S., M. PE 3.2
Nurzyński, J. PL 14.1
Nus, J.L. US-G 31.1
Nuss, J.R. US-B 13.1.1
Nuzum, K.R. NZ 2.2
Nye, W.P. US-I 5.1
Nyéki, J. HU 2.2.2
Nygaard, A. NO 10.1.2
Nyman*, I. FI 10.1.1
Nyomora, A.M.S. TZ 4.1
Nyrop, J.P. US-B 2.1.3
Nys, L. BE 10.1
Nyujtó, F. HU 2.1.1
Nyujtó, S. HU 4.1.1
Nyvall, B. US-G 4.1
Nzima, M.D.S. ZI 2.1

Oag, D.R. AU 65.3
Oancea, M. RO 8.4.1
Oba, Y. JP 7.1
Obama, T. JP 23.2.1
Obara, H. JP 18.1
Obara, N. JP 3.2.1
O'Barr, R. US-D 59.1
Obayashi, H. JP 5.1
Obdržálek, J. CS 24.1
O'Beirne, D. IE 4.1
Oberholtzer, W.R. US-B 14.1.1
Oberländer, H.-E. AT 8.10
Oberly, G.H. US-B 5.1.2
Oberthova, K. CS 19.1
Obiefuna, J.C. NG 9.1
Obisesan, I.O. NG 3.1
Oblinger, J.L. US-D 27.3
Obradović, A. YU 9.1
Obradović, D. YU 14.2
O'Brien, R.G. AU 15.4
O'Callaghan, T.F. IE 2.1
Ochatt, S. AR 3.1.2
Ochiai, M. JP 8.2
Ochoa, L.H. AR 21.1
Ochorowicz, M. PL 32.3.5
Ockendon, D.J. GB 33.1.6
O'Connell, K.P. NZ 5.1
Ocvirk de Simon, M.M. AR 15.2.3
Oda, M. JP 23.2.3
Oda, S. JP 23.1
Oda*, Y. JP 33.2
Odabas, F. TR 21.1
Odagiri, F. JP 29.1
Odagiri, M. JP 24.1
Odagiri, T. JP 15.2
O'Dell, C.R. US-C 10.1
Odenyo, W.O. KE 1.3
O'Dogherty, M.J. GB 31.2.5
O'Donoghue, E.M. NZ 10.1
Ødum, S. DK 3.3
Oduro, K.A. GA 1.1
Oebker*, N.F. US-H 27.1
Oen, R.D. AU 6.1
Oetarakov, K.N. SU 1.1
O'Flaherty, T.T. IE 4.1
Ogano, R. JP 12.2
Ogašanović, D. YU 10.2
Ogasawara, H. JP 20.3
Ogasawara, N. JP 29.2
Ogasawara, S. JP 11.2
Ogata, I. JP 21.1
Ogata*, R. JP 24.3
Ogata, T. BR 26.2
Ogata, T. JP 22.5
Ogawa, J. JP 40.1
Ogawa, J.M. US-K 6.1.1
Ogawa, K. JP 11.2
Ogawa, M. JP 44.2
Ogawa, T. JP 41.1
Ogawa, Y. JP 23.3
Ogazi, P.O. NG 1.1
Ogbuji, R.O. NG 5.1
Oggioni, L. IT 6.1
Ogihara, H. JP 5.2
Ogiso, M. JP 1.1.1
Ogiwara, I. JP 10.1
Ogle, W.L. US-D 13.1
Ogo, T. JP 31.2
Oguchi, T. JP 26.1
Ogunfidodo, A. NG 1.1
Ogunlana, M.O. NG 1.1
Ogura, H. JP 18.3
Ogura, T. JP 22.3
Ogura, Y. JP 20.4
Oh, D.G. KR 23.1.1
Oh, E.Y. KR 23.2
Oh, J.H. KR 1.1
Oh, J.Y. KR 7.1
Oh, S.D. KR 11.1
O'Hair, S.K. US-D 29.1
Ohara, G. JP 23.2.5
Ohara, M. JP 28.1, 30.1
O'Hare, P.J. IE 1.1
Ohba, S. JP 3.3.2
Ohbayashi, N. JP 19.2
Ohe, S. JP 15.2
Ohekar, G.B. IN 13.2
Ohhashi, H. JP 44.2
Ohhashi, K. JP 13.1
Ohhashi, T. JP 19.2
Ohi, M. JP 22.5
Ohira, Y. JP 16.2

Ohishi, K. JP 1.1.1
Ohishi, R. JP 36.1
Ohkawa, K. JP 19.2, 38.4
Ohki, S. JP 6.3
Ohkoshi, K. JP 4.2
Ohkubo, N. JP 27.2
Ohm, Y.H. KR 23.1.1
Ohmori, Y. JP 13.1
Ohnishi, T. JP 13.1
Ohno, H. JP 1.3
Ohno, S. JP 9.3
Ohno, Y. JP 23.2.4
Ohnuma, Y. JP 45.1
Ohshiro, M. JP 32.1
Ohta, H. JP 14.1
Ohta, S. JP 9.4
Ohta, Y. JP 23.2.3
Ohtani, K. JP 13.1
Ohtsuka, C. JP 35.1
Oikawa, T. JP 24.1
Oishi, A. JP 38.4
Oishi, K. JP 27.2
Oishi, R. JP 36.1
Oiyama, I. JP 11.3
Oizumi, H. JP 4.4
Oizumi, T. JP 4.3
Ojima, M. BR 20.1
Okabe, M. JP 19.2
Okada, I. JP 43.2
Okada, J. JP 35.1
Okada, M. JP 38.2
Okada, N. JP 38.2
Okada, S. JP 21.2
Okada, T. JP 23.2.4
Okaichi, T. JP 17.3
Okajima, H. JP 12.4
Okamoto, G. JP 31.2
Okamoto, M. JP 3.2.1
Okamoto, N. JP 25.1
Okamoto, Y. JP 31.1
Okamura, K. JP 46.1
Okano, K. JP 27.1
Okawa, I. JP 3.2.3
Okay, A.N. TR 12.1
Okayasu, T. JP 35.2
Okazaki, A. JP 8.1
Okazaki, H. JP 38.4
Okazoe, A. JP 5.2
Okeda, G. JP 36.1
O'Keefe, R.B. US-G 26.1
O'Keeffe, L.E. US-I 14.1
Okereke, O.U. NG 5.1
Okishima, H. JP 36.1
Okitsu, S. JP 7.3
Okoko, A. KE 1.8
Okon, Y. IL 2.6
Okongo, A.O. KE 1.2
Okpala, E.U. NG 5.1
Okreglicki, A. PL 32.3.2
Oktar, A. TR 15.3
Oku, T. JP 16.2
Okuba*, H. JP 7.4
Okuchi, S. JP 5.2
Okuda, Y. JP 33.1
Okudai, N. JP 27.3
Okumo, T. JP 35.2
Okumura, M. JP 3.4
Okunishi, T. JP 44.2
Okurano, H. JP 18.2
Okuse, I. JP 3.4
Oladiran, J.O. NG 1.1
Olaifa, J.I. NG 3.1
Olalokun, NG 1.2
Olaru, M. RO 8.4.5
Oldenkamp, A. NL 9.8
Oleksyn, G. PL 10.1.3
Olesen, B.D. DK 3.7
Olien*, W.C. US-A 1.1
Olin, P. US-G 1.1
Oliva, R.N. AR 13.1
Oliveira, C. PT 5.1
Oliveira, C.A.S. BR 25.5
Oliveira, J.C. BR 23.1
Oliveira Accioly, L.J. BR 10.1
Oliveira da Conceicao, H.E. BR 1.1
Oliveira de Carvalho, A. BR 19.2
Oliveira Goedert, C. BR 25.3
Oliveira Lima, T.S. BR 2.3
Oliver, A. US-D 52.1
Oliver, C. AU 59.2
Olivier, J.M. FR 4.1.3
Ollenu, L.A. GA 4.6
Olmos, F.S. AR 15.2.2
Olocco, L.E. AR 7.1
Olofsson, B. SE 1.4
Olokoshe, I.O. NG 1.4
O'Loughlin, J.B. AU 46.2
Olsen, B. DK 9.1
Olsen, B.G. NZ 10.1
Olsen, J. US-B 5.1.8
Olsen*, K.L. US-J 10.2
Olsen, O.B. NO 10.6
Olsen, R.A. US-I 18.1
Olson, H. US-G 9.1
Olson, S.M. US-D 37.1
Olszak, M. PL 28.1.2
Olszak, R. PL 28.1.2
Olszewski, D.K. PL 16.1
Olszewski, T. PL 28.1.5
Olthof, T.H.A. CA 25.2
Olufolaji, O.A. NG 1.1
Olver, R.C. ZI 3.1.1
Olympio, N.S. GA 2.1
Olympios*, Ch.M. GR 6.1.3
Omidiji, M.O. NG 1.2
Omiecińska, B. PL 5.1
Omori, S. JP 41.3
Omoso, M. JP 37.1
Omueti, O. NG 1.2
Omunyin, M. KE 1.8
Omura, M. JP 14.2
Omura, S. JP 42.2
Onaha, A. JP 32.1
Ondřej, J. CS 24.1
Ondřejová, V. CS 24.1
Onea, Gh. RO 11.1
Onea, I. RO 9.1
O'Neill, D.H. GB 31.2.1
O'Neill, F. IE 3.2
O'Neill, S. GB 20.1
Onesirosan, P.T. NG 3.1
Onesto, J.P. FR 29.1
Onillon, J.-C. FR 17.1.3
Onillon, J.C. FR 2.1.4
Onkarayya, H. IN 1.1.9
Onke, A.A. ET 10.1
Ono, K. JP 45.1
Ono, M. JP 21.1
Ono, N. JP 29.1
Ono, S. JP 11.4, 27.3
Ono, T. JP 31.1
Ono, Y. JP 28.1, 31.1
Ono, Z. JP 44.1
Onoda, K. JP 16.1
Onogi, S. JP 4.3
Onsando, J.M. KE 1.4
Onuma, K. JP 24.1
Onyango, J. KE 1.7
Oogaki*, C. JP 14.6
Ookuma, M. JP 17.1
Oomasa, Y. JP 5.2
Oomori, H. JP 5.2
Ooraikul, B. CA 7.2
Oosawa, S. JP 17.1
Oosterbaan, G.A. NL 9.2
Oota, T. JP 27.2
Ootani, M. JP 17.1
Ootsuka, T. JP 38.1
Oowada, A. JP 5.2
Ooya, R. JP 17.1
Oparka, K.J. GB 15.1.4
Opatrná, J. CS 24.1
Opatrný, K. CS 22.2
Opeke, L.K. NG 6.1
Operio, Jr. E.C. PH 1.2.1
Opina, O.S. PH 1.4
Opoku Ameyaw, K. GA 4.1
Oppong, F.K. GA 4.1
Oprea, St. RO 17.1, 17.2
Oprean, N. RO 29.1
Oprel, L. NL 1.1.2

Orchard, B. GB 12.1
Ordaz Ordaz, F. MX 3.2
Ördögh, G. HU 2.2.8
Orell, R. AR 22.3
Orengo, E. US-N 1.2
Orffer, C.J. ZA 4.3.2
Orillo, F.T. PH 1.4
Oriolani, E.J.A. AR 15.1
Oriolani, M.J.C. AR 15.1
O'Riordain, F. IE 4.1
Orlikowska*, T. PL 28.1.1
Orlikowski*, L.B. PL 28.1.16
Orlowski, M. PL 31.1
Ormachea, E. PE 4.3
Ormatella, M. MA 5.1
Ormrod*, D.P. CA 15.1
Oron, G. IL 5.1.2
O'Rourke, E.N. US-D 52.1
Orozco Romero, J. MX 3.3
Orozco Santos, M. MX 3.3
Orr, P. US-G 11.1
Ortega G., C.A. MX 16.2
Orth, A.I. BR 28.1
Orth, P.G. US-D 29.1
Ortiz, F.H. US-N 1.2
Ortiz, V. PE 4.2
Ortman, E.E. US-F 9.1.3
Orton, Jr. E.R. US-B 14.1.1
Orzolek, M.D. US-B 13.1.1
Osaer, A. FR 21.1
Osakabe, M. JP 11.3
Osanai, Y. JP 3.2.1
Osara*, K. FI 10.1.1
Osawa, T. JP 33.2
Osborn, R. AU 27.1
Osborne, L.E. GB 31.2.8
Oseni, T.O. NG 1.1
Oshiro, M. JP 43.1
Osiecki, M. PL 28.1.12
Osińska, M. PL 33.1
Osipov, J.V. SU 15.1
Oso, B.A. NG 8.1
Osonubi, O.O. NG 8.1
Osorio, J. CO 7.2
Ostendorf*, H.D. DE 11.1.2
Oster, G.F. US-K 3.1.1
Ostojić, N. YU 8.1, 14.1
Oštrec, Lj. YU 3.2
Ostrowski, W. PL 31.1.1
Ostrzycka, J. PL 28.2.12
Osuch, H. PL 9.1
O'Sullivan, J. CA 3.1
Osumi, K. JP 36.1
Osumi, S. JP 2.2
Oswiecimski, W. PL 32.1.4
Oszkinis, K. PL 20.2.5
Ota, A. JP 39.1
Ota, H. JP 10.1
Ota, J. JP 21.1
Ota, T. JP 46.1
Otake, A. JP 14.2
Otake, S. JP 29.1
Otani, F. JP 26.4
Otani, H. JP 36.1
Otieno, E. KE 1.4
Otoi, N. JP 40.2
Otomo, J. JP 11.1
Otsubo, T. JP 41.3
Otsuka, K. JP 23.2.3
Otteanu, A. RO 9.1
Ottenwaelter, M. FR 4.1.8
Otto, C. FR 2.1.1
Otto, K. DE 12.1.4
Ottosen, C.O. DK 1.2
Ottosson, L. SE 1.1.2
Ottosson, B. SE 1.1.2
Otvös, Z. RO 17.1
Ou, S.K. TW 7.3
Ouchi, Y. JP 35.2
Oudenaarde, T. NL 15.1
Ouellette, J.A. CR 1.1
Ough, C.S. US-K 6.1.5
Overaa, P. NO 10.3
Overman, A.J. US-D 24.1
Overstyns, A. BE 17.1
Øvstedal, D.O. NO 1.1
Owczarek*, M. PL 28.2.5, 28.4
Owens, J. US-H 19.1
Owens, R.A. US-C 3.1
Owuor, J.O. KE 2.1
Owusu, G.K. GA 4.6
Owusu-Manu, E. GA 4.4
Oyamada, M. JP 45.1
Oyamada, T. JP 1.2
Oyolu, C. NG 5.1
Ozahci, E. TR 15.3
Ozaki, T. JP 13.3
Ozarowska, K. PL 28.2.11
Ozawa, K. JP 26.2
Ozawa, T. JP 26.3, 47.2
Ozbun, J.L. US-J 6.1
Ozcagiran, R. TR 15.1
Ozcariz, M.E. AR 23.1
Ozdemir, E. TR 13.1
Ozelkok, Z. TR 14.1
Ozen, H. TR 5.1
Ozen, Y. TR 5.1
Ozeri, Y. IL 1.1.2
Ozgen, S. TR 4.1
Ozkahya, D. TR 15.2
Ozsan, M. TR 1.1
Ozyılmaz, N. TR 15.3

Pacayut, F.M. AR 8.1
Pace, C.A.M. BR 19.2
Pacheco, C.S. CR 2.1
Pacheco, J. PE 4.2
Pacheco, M. PE 4.2
Pacholak, E. PL 20.2.9
Packee, E.C. US-L 1.1
Pacucci, G. IT 2.2
Padi, B. GA 4.4
Padilla, E. AR 22.2
Padmanabham, V. IN 6.1
Padmasiri, I.S. LK 3.2
Pados, P. HU 5.3.2
Padule, D.N. IN 14.1
Pădureanu, T. RO 19.2
Paek, K.Y. KR 3.2
Paet, C. PH 1.7
Páez, V.N. AR 5.1
Page, P.E. AU 15.1
Pagliarini, N. YU 3.2
Paglietta, R. IT 24.1
Pai, S.A. TW 10.1
Paine, M.D. US-H 3.1
Pair, J.C. US-G 30.1
Pais, I. HU 2.2.3
Paiva, M. BR 16.1
Paixão Passos, L. BR 31.1
Pajunen, A. FI 10.1.1
Pak, B.I. KR 19.1
Pak, D.S. KR 9.1
Pak, H.Y. KR 23.1.2
Pak, M.K. KR 19.1
Pak, S.J. KR 19.1
Paket, R. AT 8.9
Pal, A.B. IN 1.1.2
Pal, D.K. IN 1.1.8
Pal, P. IN 28.1
Pal, R.N. IN 26.1
Pal, T.K. IN 5.1
Pala, E. PL 28.1.15
Pala, J. PL 32.1.8
Palacio, I.P. PH 1.3
Palacios, J. AR 22.2
Palacios, S. MX 16.2
Palacios, Y. CO 4.2
Palade, L. RO 23.1
Palanaippan, R. IN 1.1.9
Palczyński, J. PL 28.2.4
Paleg, L.G. AU 42.1
Pálfi, D. HU 2.7
Palladini, L.A. BR 28.1
Paller, E.C. PH 1.6
Palloix, A. FR 3.1.2
Palm, G. DE 17.1
Palmer, J.W. GB 8.1.1
Palmertree, H. US-D 48.1
Paloukis, S. GR 25.2.2
Paludan, N. DK 4.1
Palzkill*, D.A. US-H 27.1
Pamir, M. TR 8.1
Panagiotopoulos, L. GR 22.1.1

Panagiotou, C. GR 18.1.1
Panagopoulos, C.G. GR 6.1.7
Panajototić, J. YU 13.1
Panapoulos, N.J. US-K 3.1.2
Panayiotou, P. GR 22.1.1
Panda, C.S. IN 16.2
Panda, J.M. IN 16.2
Pandey, D. IN 24.1.8
Pandey, I.C. IN 25.1
Pandey, J.C. IN 24.1.5
Pandey, M.P. IN 21.1
Pandey, N.C. IN 22.1
Pandey, R.M. IN 2.1, 2.1.2, 2.1.3
Pandey, R.S. IN 22.1
Pandey, S.C. IN 1.1.2
Pandey, S.K.N. IN 22.1, 25.1
Pandey*, S.N. IN 2.1.1
Pandey, S.N. IN 2.1.3, 2.1.5
Pandey, U.S. IN 22.1
Panelo, M. AR 16.1
Panić, M. YU 14.1
Paningbatan, Jr. E.P. PH 1.5
Panis, A. FR 2.1.4
Panizzi, A.R. BR 27.1
Panizzi, M.C.C. BR 27.1
Pankiewicz, T. PL 17.1
Panneton, B. CA 35.1
Panoutsos, D. GR 6.1.7
Pant, C.C. IN 24.1.2
Pant, N. IN 24.1.7
Pant, T. IN 19.4
Pantastico, E.B. PH 1.2.5
Papadopoulos*, A.P. CA 17.1
Papadopoulou, R. GR 18.1.2
Papafitsoros, G. GR 23.1
Papageorgiou, Chr. GR 5.1
Papandreou, A. GR 13.1
Papanicolaou, E. GR 3.1.2
Papanicolaou, X. GR 24.1
Paparozzi, E.T. US-G 23.1
Papavizas, G.C. US-C 3.1
Papeians, C. BE 6.8.6
Papenhagen, A. DE 7.1
Papke, W. DE 19.1
Papp, J. HU 2.2.2
Pappas, A. GR 14.1.1
Paprić, Dj. YU 17.1
Paraschiv, Gh. RO 4.1
Paraskakis, N. GR 8.1.1
Parchomchuck, P. CA 3.1
Pardal Nogueira, D.J. BR 15.1
Pardee, W.D. US-B 5.1.7
Pardo, A. ES 11.1
Paredes, L.A. CR 2.2
Paredes G., J PE 2.2
Pareek, L.P. IN 19.4
Pareja, M. CR 2.2
Parent, J.-G. CA 38.1
Pares, R.D. AU 9.1.1
Parfitt, D.E. US-K 6.1.3
Paris, H. IL 1.1.1
Parish, L. US-J 10.2
Parisi, J.L. CR 2.1
Park, C.S. KR 18.1.3
Park, D.M. KR 7.1
Park, D.Y. KR 23.1.1
Park, H.B. KR 11.1
Park, H.G. KR 23.3
Park, H.S. KR 13.2
Park*, H.S. KR 13.2
Park, J.B. KR 21.1
Park, J.C. KR 12.2
Park, J.S. KR 8.4
Park, K.H. KR 1.1
Park*, K.W. KR 21.2
Park, P.S. KR 12.3
Park, S.H. KR 12.3
Park, S.K. KR 23.1.2
Park, T.D. KR 13.1
Park, Y.B. KR 10.3
Parker, N. AU 59.2
Parker, R.D. US-A 13.1
Parkerson, R.H. US-L 2.2
Parliman, B. US-C 3.1
Parlitz, M. DE 15.1.7
Parmar, C. IN 10.3
Parmentier, A. BE 11.1
Parmentier, G. BE 6.2
Parmeter, Jr., J.R. US-K 3.1.2
Parnia, C. RO 9.1
Parnia, P. RO 9.1
Parra G., D. MX 16.2
Parraga, M.S. BR 19.2
Parry, R.H. CA 40.2
Parsons, L.R. US-D 30.2
Partis, M.D. GB 18.1.3
Partyka, Z. PL 9.1
Parvin*, P.E. US-M 4.1
Pasc, E. RO 22.1
Paschoalino, J.E. BR 20.2
Pascoe, I. AU 27.1
Pasini, C. IT 22.1
Pasiut, Z. PL 3.1
Paspatis, E.A. GR 14.1.3
Pasqual, M. BR 28.1
Pasqualotto, H.L. AR 13.1
Passalacqua, S.A. AR 14.1
Passos, F.A. BR 20.1
Paster, N. IL 1.3.3
Pasterkamp, H.P. NL 9.9
Pasternak, D. IL 4.1
Pastrana, C.P. AR 17.1
Pasture, E. BE 9.2
Paszkowska, I. PL 28.2.1
Patel, L.R. IN 25.1
Paterno, E.S. PH 1.5, 1.7
Paternotte, E. BE 19.1
Paterson, A.P. CA 16.1
Paterson, D. US-H 12.1
Paterson, D.D. CA 4.2
Paterson, J.W. US-B 14.1.7, 18.1
Pathak, C.P. IN 24.1.7
Pathak, C.S. IN 1.1.2
Pathak, R.A. IN 25.1
Pathak*, R.K. IN 21.1
Pathak, S.P. IN 19.1
Pathirana, A. LK 3.12
Patil, A.A. IN 11.3
Patil, B.A. IN 14.2.1
Patil, B.C. IN 14.2
Patil, J.D. IN 14.11
Patil, J.G. IN 14.9
Patil, J.J. IN 14.2.1
Patil, J.L. IN 15.1
Patil, K.B. IN 14.2
Patil, K.D. IN 3.1
Patil, L.K. IN 14.7
Patil, M.M. IN 15.1
Patil, M.T. IN 14.2
Patil, P.Y. IN 14.1
Patil, S.S.D. IN 14.2.1
Patil, V.K. IN 12.1
Patil, Y.S. IN 14.4
Patkai, Gy. HU 2.2.1
Patnaik, A.K. IN 16.2
Patnaik, K.K. IN 16.2
Patnaik, N.B. IN 16.2
Paton, F.J. GB 35.1.2
Patra, S.N. IN 16.2
Patrick, Z.A. CA 25.1
Patsakos, P.G. GR 14.1.4
Pattanshetti, H.V. IN 11.4
Pattantyus, K. RO 14.1
Patten, K. US-H 12.1
Patterson, B.D. AU 65.4
Patterson, D.E. GB 31.2.5
Patterson*, K.J. NZ 2.1.1
Patthong, P. TH 15.1
Patzer, H. DE 5.2.1
Paul, J.L. US-K 6.1.2
Paul-Bădescu, A. RO 9.1
Paulić, N. YU 3.1
Paulieri Sabino, N. BR 20.1
Paulin, A. FR 38.1
Paulin*, J.P. FR 1.1.3
Paulin, R. AU 56.1
Paulson, A. CA 3.1
Paulson, W.H. US-F 26.1
Paunović, R. YU 14.1
Paunović, S. YU 10.2
Paunović, S.A. YU 9.1, 10.1

Pavan, M.A. BR 27.2
Pavek, J.J. US-I 13.1
Pavel, A. RO 19.1
Paviani, T. BR 25.4
Pavlík, J. CS 24.1
Pavlišta, S. CS 27.1
Pavlović, Lj. YU 9.1
Pavlović, V. YU 9.1
Pavlusek, A. CS 26.1
Pawar, A.B. IN 14.11
Pawar, D.B. IN 14.1
Pawley, R.R. GB 18.1.2
Pawłowska, M. PL 32.1.11
Payan Jacobo, M. MX 3.5
Payne, C.B. ZI 6.1
Payne, C.C. GB 18.1.2
Payne, J. US-D 19.1
Payne, R.N. US-H 3.1
Paynot, M. FR 7.1.10
Peacock*, B.C. AU 15.2
Peacock, F. AU 47.2
Pearce, R.H.T. AU 56.2
Pearson, H. CA 16.1
Pearson, J.A. AU 65.4
Pearson, J.L. US-A 14.1
Pearson, R.C. US-B 2.1.4
Peat, W.E. GB 35.1.3
Peavy, W. US-H 8.1
Pécaut, P. FR 3.1.2
Peck, N.H. US-B 2.1.2
Pecknold, P.C. US-F 9.1.1
Pedersen*, I.S. DK 7.1
Pedersen, P.A. NO 10.1.5
Pedersen, T. DK 10.1
Pediaditakis, G. GR 2.1
Pedryc, A. HU 2.2.6
Peel, C.H. GB 4.1
Peet*, M.M. US-D 8.1.1
Pegazzano, F. IT 9.3
Pegg, G.F. GB 29.1
Pegg, K.G. AU 15.4
Peggie*, I. AU 30.1
Pehar, J. YU 2.1
Peiper, U.M. IL 1.4.9
Peiretti, D. AR 7.1
Peiris, R.R.A. LK 2.1
Peiris, S. LK 3.8
Peiris, T.S.G. LK 2.1
Peixoto Vernette, V. BR 30.2
Pejčinovski, F. YU 4.2
Pejkić, B. YU 14.1
Pejković, M. YU 14.1
Pejović, Lj. YU 7.1
Pekáry, I. HU 5.1.4
Pektekin, T. TR 17.1
Pele, M. RO 8.4.3
Pelekassis, K. GR 6.1.6
Pelerents, C. BE 7.5
Pelet, F. CH 2.1
Pelisse, C. FR 3.1.8
Pelissier, C. FR 3.1.9
Pellet, H. US-G 7.1
Pellet, N. US-A 3.1
Pelletier, G. CA 40.2
Pelletier, J. FR 26.1, 26.1.3
Pelletier, J.R. CA 35.1
Pelletier, Y. CA 40.3
Pelliciari, R. AR 15.2.1
Peloquin, S.J. US-F 23.1.3
Pelzmann*, H. AT 22.1
Pemberton, A.W. GB 13.1.4
Pemberton, B. US-H 12.1
Pemberton, L.A. GB 16.1
Peng, A. US-F 3.1
Peng, C.H. TW 7.1
Penna, R.J. GB 12.1
Penney*, B.G. CA 44.1
Penninck, R. BE 7.4
Pennone, F. IT 19.1.3
Pennycooke, S.R. NZ 2.3.2
Penrose, L.J. AU 7.1
Pénzes, B. HU 2.2.8
Peper, H. DE 32.1
Peperkamp, G. NL 9.7.5
Pepin, H.S. CA 4.1
Peralta, M. AR 3.1.2
Perdok, U.D. NL 9.3
Pereira, A.L. BR 19.2
Pereira, A.R. BR 20.1
Pereira, A.V. BR 1.1, 18.1
Pereira, F.M. BR 23.1
Pereira, J. BR 25.2
Pereira, J.L.M. BR 14.1
Pereira, O. PT 1.1
Pereira, P.C. PT 2.1
Pereira, W. BR 25.5
Pereira da Costa, N. BR 27.1
Pereira das Neves, B. BR 26.1
Pereira de Oliveira, I. BR 26.1
Pereira Rios, G. BR 26.1
Pérennec, P. FR 9.1
Perera, H.A.S. LK 4.1
Perera, K.D.A. LK 3.2
Perera, P.A.C.R. LK 2.1
Peres, J.R.R. BR 25.2
Pereverzev, I.N. SU 6.1
Perez, D. PE 4.2
Pérez, G. MX 15.1
Perez-Aranda, L. BE 4.2
Perez Garrido, J.L. BR 13.2
Perez Gonzalez, S. MX 7.2
Pergola, G. IT 22.1.1
Perianu, A. RO 9.1
Perić, Dj. YU 10.1
Perim, S. BR 25.2
Perjés, A. HU 2.2.4
Perkins, D.Y. US-D 38.1
Perkins, F. US-B 14.1.2
Perley, A.F. CA 40.2
Perłowska, M. PL 28.2.9
Perng, D.C. TW 3.1
Pernkopf, AT 9.1
Perombelon, M.C.M. GB 15.1.1
Péron*, J.Y. FR 19.5
Perrella, C. IT 20.1
Perret, P. CH 1.1
Perrin, P.W. CA 1.1
Perrin, R. FR 7.1.3
Perring, M.A. GB 8.1.3
Perry, D.A. GB 15.1.1
Perry, F.B. US-D 38.1
Perry, K.B. US-D 8.1.1
Perry*, L.P. US-A 3.1
Perry*, R.L. US-F 14.1.1
Perry, V.G. US-D 27.1
Perscheid, M. DE 12.1.3
Persoon, J.G.M. NL 9.7.4
Persson*, A.R. NO 10.1.7
Persson, J. SE 2.1
Pertuit, Jr., A.J. US-D 13.1
Pervan, E. AU 59.2
Pesis, E. IL 1.3.1
Pessala*, R. FI 1.1
Pessala*, T. FI 1.1
Pessanha, G.G. BR 19.2
Pesti, M. HU 8.1
Peterlunger, E. IT 25.1
Peters, J.A. BR 30.3
Peters, R.A. US-A 13.1
Peterson, C. US-F 23.1.3
Peterson, D. US-C 8.2
Peterson, F.J. US-D 55.1
Peterson, J.C. US-F 3.1
Peterson, J.F. CA 37.1
Peterson, J.K. US-D 12.2
Peterson, J.L. US-B 14.1.6
Peterson, L. US-F 23.1.3
Peterson, N.C. US-F 14.1.1
Peterson, R.A. AU 21.1
Peterson, R.M. US-G 19.1
Pethő, F. HU 3.1.1
Petit, E. BE 13.1
Petkov, G. BG 7.1
Petkov, M. BG 2.1
Petkov, Ts. BG 8.3
Petkova, T. BG 2.1
Petlic, W. PL 32.3.2
Petrakiev, A. BG 7.1
Petrakis, M. GR 10.1
Petralia, S. IT 20.1
Petrányi, I. HU 5.1.4
Petre, Gh. RO 33.1
Petre, O. RO 13.1

Petres, J. HU 2.4
Petrescu, C. RO 8.1
Petriča, V. RO 31.1
Petró-Turza, M. HU 2.4
Petrov, A. BG 8.2
Petrov, Ch. BG 8.3
Petrová, E. CS 24.1
Petrovic, A.M. US-B 5.1.1
Petrović, D. YU 9.1
Petrović, G. YU 12.1
Petrović, M. YU 14.1
Petru, O. RO 21.1
Petruccioli, G. IT 8.1
Petrus, D.R. US-D 30.2
Petsas, S. GR 13.1
Petsikou, N. GR 14.1.4
Pevenage, G. BE 4.2
Pevná, E. CS 19.1
Peyrière, J. FR 3.1.14
Pfadenhauer, J. DE 11.1.5
Pfaff, F. DE 23.1
Pfammatter, W. CH 2.2
Pfeufer, B. DE 15.1.3
Phal*, R. IN 2.2
Phan, C.T. CA 31.3
Pharr, D.M. US-D 8.1.1
Phatak, S.C. US-D 22.1
Phelps, K. GB 33.1.7
Philbrick, R.N. US-K 16.1
Philippon, J. FR 38.1
Philippou, G. CY 1.1
Phillips, D.J. US-K 7.3
Phillips, D.R. AU 59.1
Phillips, E.L. US-C 11.1
Phillips, G. US-H 19.1
Phillips, L. GB 32.1
Phillips, M.S. GB 15.1.1
Phillips, T. US-D 50.1
Phillips, V.R. GB 31.2.4
Phillips, W. CR 2.2
Philouze, J. FR 3.1.2
Philp, B. AU 36.1
Phiri, I.M.G. ML 1.1
Piatkowski*, M. PL 5.1
Picard, J. FR 7.1.5
Piccolo, R.J. AR 13.1
Picha, D.H. US-D 52.1
Picha, J. CS 15.1
Pichard, B. FR 23.1
Pickard, J.A. GB 19.1.1
Pidek, A. PL 28.1.6
Pieber*, K. AT 8.3
Piech, J. PL 14.1.7
Piechna, K. PL 32.1.12
Piedice, J.R. BR 21.2
Piekarska, H. PL 20.2.2
Piekło, A. PL 1.1
Pielat, H. PL 28.2.11
Piens, G. BE 22.1
Piepers, R. NL 6.1
Pierce, L.C. US-A 2.1
Pierik*, R.L.M. NL 9.7.1
Pieringer, A.P. US-D 30.2
Pierre, J.N. FR 30.1.2
Pierzga, K. PL 3.1
Pieta, D. PL 14.1.8
Pieterse, B.J. ZA 3.1
Pietizak, B. PL 6.1
Pigati, P. BR 21.2
Piggott, T. AU 62.1
Piironen, P. FI 14.1
Pike, L. US-H 6.1
Pilas, J.M. FR 25.1
Pilgaard*, A. DK 7.1
Pill, W.G. US-C 2.1
Pillai, K.S. IN 5.1
Pillai, N.G. IN 5.1
Pilowsky, M. IL 1.1.3
Pimentel, J.P. BR 19.2
Pimienta Barrios, E. MX 6.5
Pimpli, D. GR 18.1.1
Pinc, M. CS 24.1
Pinchinat, B. PL 28.2.2
Pindel, Z. PL 11.1.1
Piñeiro, M.E. CR 1.1
Pinese, B. AU 21.1
Ping, C.-L. US-L 2.1
Pinguet, A. FR 3.1.4
Pinheiro, P.A. BR 25.3
Pinheiro, R.V.R. BR 17.1
Pinheiro Dantas, A. BR 10.1
Pinheiro Maciel, R.F. BR 8.2
Pinheiro Nunes, G. BR 30.3
Pinho, A.F. de S. BR 14.1
Pinilla, J.E. AR 5.1
Pink, D.A.C. GB 33.1.6
Pinnegar, M.A. GB 32.1
Pinochet, J. CR 2.2
Pinpiban, T.Y. TH 18.1
Pintilie, I. RO 7.1
Pinto, C.M.F. BR 15.1
Pinto, M.E. PT 6.1
Pinto, M.E.R. LK 3.3
Pinto da Cunha, G. BR 12.1
Pinto de Araujo, E. BR 10.1
Pinto P., M. PE 4.1
Pio, R.M. BR 20.1
Pionnat, J.C. FR 2.1.3
Piotrowski, S. PL 21.1
Pipa, R.L. US-K 3.1.1
Pippous, B. GR 18.1.1
Pirc, H. AT 8.1
Pires de Matos, A. BR 12.1
Pireva, M. YU 16.1
Piringer, A.A. US-C 3.1
Pirnat, M. YU 1.1
Piróg*, J. PL 20.2.11
Pîrvulescu, D. RO 8.4.5
Pisani, P.L. IT 9.1
Pisharody, P.N. IN 5.3
Pisklová, A. CS 1.2
Piskor, E. PL 30.1
Piskor, T. PL 30.1
Piskornik, Z. PL 11.1.9
Pisulewska, E. PL 28.4
Pisulewski, T. PL 28.4
Piszczek, P. PL 4.1
Pitblado, R.E. CA 21.1
Pitera*, E. PL 32.1.1
Pitner, J.B. US-D 15.1
Pitrat, M. FR 3.1.2
Pitt, D.G. US-C 5.2
Pitt, J.I. AU 65.4
Pitts, C.W. US-B 13.1.1
Pitts, J.A. US-D 41.1
Pizzolato, T.D. US-C 2.1
Placek, S. PL 28.1.6
Plages, J.N. FR 34.1
Plaisted, R.L. US-B 5.1.7
Plamenac, M. YU 5.1
Planton, G. FR 21.1
Plastira, V.A. GR 14.1.1
Platt, H.W. CA 43.1
Platter, K. IT 11.1
Plattner, H.J. DE 15.3.4
Plavšić, M. YU 9.1
Plazinić, R. YU 10.2
Pleskus, P. AU 59.2
Plich, H. PL 28.1.9
Plich, M. PL 28.1.2
Pliszka*, K. PL 32.1.1
Płocharski*, W. PL 28.1.3
Ploegman, C. NL 9.2
Ploper, D.L. AR 22.3
Ploper, J. AR 22.2
Plowman, L. AU 59.2
Plumier, W. BE 6.1
Plunkett, S. AR 23.1
Pluta, S. PL 28.1.1
Poapst*, P.A. CA 41.1
Poască, C. RO 1.1.2
Pobudkiewicz, A. PL 28.1.12
Pochard, E. FR 3.1.2
Pochon, I. FR 1.1.1
Pociecha, A. PL 32.3.2
Podczaska, A. PL 28.2.13
Podgorska, H. PL 6.1
Podkoński, M. PL 32.1.2.1
Podoleanu, M. RO 1.1.6
Podoler, H. IL 2.5
Podsednik, AT 8.8
Podwyszyńska, M. PL 28.1.17
Poesse, G.J. NL 9.3
Poessel, J.L. FR 3.1.3

Pogroszewska, E. PL 14.1.2
Pohl, R.K. US-I 18.1
Pohronezny, K.L. US-D 29.1
Poinar, G.O. US-K 3.1.1
Pointer, J.L. US-E 3.1
Poitou, N. FR 4.1.3
Poitout, S. FR 3.1.9
Pojnar, E. PL 11.1, 11.1.8
Pokorny, F.A. US-D 16.1
Polách, J. CS 20.1
Połaski, E. PL 32.1.5
Polero, H. AR 3.1.2
Polglase, A.R. AU 34.1
Poli, V. RO 1.1, 7.1
Poling, E.B. US-D 8.1.1
Polito, V.S. US-K 6.1.3
Politycka, B. PL 20.2.7
Poll, J.T.K. NL 6.1
Pollard, J.E. US-A 2.1
Pollatschek, AT 8.9
Pollock, M.R. GB 5.1
Pollock*, R.D. GB 12.1
Polo, A. PE 3.2
Polok, A. PL 23.1
Pölös, E. HU 5.2
Poltronieri, L.S. BR 3.1
Polyák, D. HU 2.2.10
Polyák, L. HU 2.2.1
Pomeroy, N. NZ 9.8
Pommer, C.V. BR 20.1
Pompeu Jr., J. BR 20.1
Ponce, R.T. CR 2.1
Ponce Herrera, V. MX 8.4
Ponchet*, J. FR 2.1.3
Ponchet, M. FR 2.1.3
Ponchia, G. IT 12.1
Poncu, J. RO 1.1.1
Ponder, H.G. US-D 38.1
Poniedziałek, M. PL 11.1.1
Poniedziałek, W. PL 11.1.3
Pontikis, V. GR 13.1
Pool*, R.M. US-B 2.1.2
Poole, R.T. US-D 31.1
Pop, A. RO 12.1
Popa, D. RO 8.4.2
Popa, E. RO 8.4.1
Popa, I. RO 12.1
Popandron, N. RO 1.1.1
Popenoe, J. US-D 33.1
Popescu, RO 8.4.4
Popescu, A. RO 9.1
Popescu, C. RO 8.4.2
Popescu, D. RO 2.1
Popescu, H. RO 2.1
Popescu, I. RO 9.1
Popescu*, M. RO 19.2
Popescu, S. RO 9.1
Popescu, V. RO 8.1, 20.1
Popkowicz, S. PL 28.1.6
Popov, D. BG 8.1
Popov, S. BG 8.3
Popov, S. YU 4.3
Popova, L. BG 6.2
Popova, M. BG 13.1
Popović, M. YU 13.1, 14.1
Popović, V. YU 14.2
Popovici, D. RO 27.1
Popovski, H. YU 4.1
Poppe, J. BE 7.7
Populer, C. BE 6.2
Porcelli, S. IT 20.1
Porlingis*, I.C. GR 25.1.4
Poros, H. PL 19.1
Porpáczy, A. HU 2.1.3
Porpáczy, S. HU 2.1.3
Porras, V. CR 2.2
Porretta, A. IT 14.1
Porreye*, W. BE 19.1
Porś, J. CS 29.1
Porta Puglia, A. IT 19.2
Portas*, C. PT 5.1
Porter, G. US-A 1.1
Porter, I. AU 27.1
Porter, J.R. GB 19.1.2
Porter, L.A. NZ 10.1
Porter, W.C. US-D 54.1
Porterfield, N.H. US-B 13.1.1
Porto, M.J.F. BR 6.1
Poryazov, I. BG 8.1
Posadas, R. MX 16.2
Pośpiech, J. PL 22.1
Pospieszny, H. PL 20.1.2
Pospišil, F. CS 14.1.2, 22.2
Pospišil, J. CS 24.3
Pospišilová, D. CS 1.1
Pospišilová, J. CS 20.1
Possingham*, J.V. AU 65.1, 65.2, 65.3
Post, R.L. US-I 11.2
Posthumus, A.C. NL 9.4.5
Postiglione, L. IT 18.2
Postma, G. NL 9.3
Postma, J. NL 9.5
Potaczek, H. PL 28.2.1
Potec, I. RO 23.3
Potocka, M. PL 21.1
Pott, M. NL 9.7.5
Potter, J.W. CA 25.2
Pottinger, R.P. NZ 5.1
Pouget, J. FR 4.1.8
Poulain, C. FR 19.2
Poulsen, E. DK 1
Poulsen, N. DK 1.1
Poupet, A. FR 2.1.3
Poupet, R. FR 29.1
Povolný, M. CS 22.1
Powell, C.C. US-F 3.1
Powell, C.L.I. NZ 10.1
Powell, J.A. US-K 3.1.1
Powell, L.E. US-B 5.1.2
Powell, W. US-D 16.1
Poysa, V. CA 17.1
Pozo Campodónico, O. MX 12.2
Poszgay, O. HU 2.2.9
Pozzo Ardizzi, M.C. AR 23.1
Pozzoli, R. AR 15.2.1
Prabhakar, B.S. IN 1.1.2
Pradet, A. FR 4.1.4
Prado, E. AR 16.1
Pralavorio, M. FR 2.1.4
Prance, G.T. US-B 1.1
Prange, R.K. CA 42.2
Prasad, A. IN 22.1
Prasad, B. IN 25.1
Prasad, G. IN 22.1
Prasad*, J. IN 21.1
Prasad, L. IN 22.1
Prasad, M. NZ 10.1
Prasad, M.B.N.V. IN 1.1.2
Prasad, R. IN 22.1
Prashar, D.P. US-G 19.1
Prataviera, A.G. AR 5.1
Pratella, G. IT 3.3
Pratt, E.T. CA 40.1
Precheur, R. US-F 7.1.2
Predescu, Gh. RO 35.1
Prędki, S. PL 3.1
Pree, D.J. CA 25.2
Preece, D.A. GB 8.1.3
Preece, R.D. GB 18.1.1
Preller, E. DE 5.2.1
Premaratne, W.H.E. LK 4.1
Prendiville, M.D. IE 4.1
Pressey, R. US-D 16.2
Pressman, E. IL 1.1.1
Preston, D. US-G 13.1
Preston, N.W. GB 19.1.4
Preuter, H. NL 6.1
Prevatt, J.W. US-D 24.1
Preziosi, P. IT 15.1
Přibyl, S. CS 24.2
Pricá, D. RO 22.1
Prica, V. YU 1.2
Price, C. US-E 11.1
Price, D. GB 18.1, 18.1.2
Price, H. US-F 18.1
Price*, H.C. US-F 14.1.1
Price, J.F. US-D 24.1
Price*, R.D. AU 26.2
Priebe, G. GR 7.1
Prijić, Lj. YU 14.1
Přikazský, J. CS 10.1
Prince*, T.A. US-F 3.1

Pringle, G.J. NZ 2.1.1
Pringle, J.S. CA 16.1
Prinsen, J.D. NL 9.4.1
Prior, L.D. AU 4.1
Pritchard*, M.K. CA 14.1
Pritts*, M.P. US-B 5.1.2
Procházková, A. CS 22.4
Proctor*, J.T.A. CA 15.1
Prodanović, T. YU 10.1
Profic-Alwasiak*, H. PL 3.1
Profitou, D. GR 25.1.6
Prohászka, K. HU 5.3.1
Protacio, C.M. PH 1.2.2
Proto, D. IT 14.1
Protopapadakis, E. GR 8.1.1
Proudfoot, K.G. CA 44.1
Provedo, J. ES 11.1
Provenzale, M.G. IT 22.1.1
Provine, J. US-K 2.1
Provvidenti, R. US-B 2.1.4
Prpić, B. YU 3.3
Prudek, M. CS 14.2
Prunier, J-P. FR 3.1.5
Prus-Głowacka, D. PL 32.3.1
Prusky, D. IL 1.3.1
Prussová, H. CS 8.1
Pruszyńska, M. PL 20.1.2
Pruszyński, S. PL 20.1.4
Prychodzien, Z. PL 28.1.8
Przedpełska, M. PL 28.2.11
Przeradzki, D. PL 32.1.4
Przybecki, Z PL 32.1.8
Przybyl, K. PL 10.1.3
Przybyła, A. PL 28.1.1
Przybyłska, M. PL 20.1.2
Przybyszewska, B. PL 18.1
Psallidas, P.G. GR 14.1.1
Psiyllakis, N. GR 8.1
Pua, A.R. PH 1.7
Puchwein, W. AT 10.2
Pudelski*, T. PL 20.2.11
Puente Perez, C. MX 11.2
Pufan, M. RO 24.1
Pühringer, A. AT 12.1
Puiatti, A.E. AR 15.1
Puite, K.J. NL 9.5
Pujari, M.M. IN 8.1
Pujari, P.D. IN 14.10
Pukacka, S. PL 10.1.4
Pukacki, P. PL 10.1.3
Pulcinska, M. PL 28.2.13
Pulfer, D. CH 1.1
Pundir, J.P.S. IN 19.2
Pundrik, K.C. IN 19.1
Purcell, A.H. US-K 3.1.1
Purcifull, D.E. US-D 27.6
Purdy, D.W. US-B 7.1
Purohit, S.P. IN 19.3
Purvis, A.C. US-D 30.1, 52.1
Puscásu, N. RO 9.1
Puskulcu, G. TR 15.3
Pussemier, L. MA 1.1
Putnam, A.R. US-F 14.1.1
Putz, C. FR 6.1.2
Pyke, N.B. NZ 11.1
Pyo, H.K. KR 23.3
Pyon, J.Y. KR 8.2
Pytlewski, Cz. PL 28.1.13
Pyzik, T. PL 32.1.9

Quacquarelli, A. IT 19.2
Quagliotti, L. IT 24.4
Quak, F. NL 9.4, 9.4.4
Quamme*, H.A. CA 3.1
Quarta*, R. IT 19.1
Quast, D.C. BR 20.2
Quast, P. DE 17.1
Quayle, M. CA 4.2
Quebedeaux, B. US-C 5.2
Quebral, F.C. PH 1.4
Queiroz, C. FR 30.1.2
Queiroz, M.O. FR 30.1.2
Quesada, J.R. CR 2.2
Quevedo Willis, S. PE 4.2
Quijandria, M. PE 1.1
Quimio, A.J. PH 1.4, 1.6
Quimio, T.H. PH 1.4
Quin, B.F.C. NZ 5.1
Quinche, J.P. CH 2.1
Quinderé, M-A.W. BR 8.1
Quinlan*, J.D. GB 8.1.1
Quinn, C.E. GB 9.1.2
Quiñones, J.A. US-N 1.2
Quintana, E.G. PH 1.2.2
Quintanilla, J. PE 4.1
Quirk, J.P. AU 42.1
Quiroga de Oriolani, M.E. AR 15.1
Quiros, C. US-K 6.1.4

Ra, S.W. KR 8.1
Raabe, R.D. US-K 3.1.2
Raasch, Z.S. BR 28.1
Raatikainen, T. FI 13.1
Rabasse, J.M. FR 2.1.4
Rabcewicz, J. PL 28.1.6
Rabinovitch, H.D. IL 2.3
Racca, R. AR 7.1
Race, S.R. US-B 14.1.4
Rachwał, L. PL 10.1.3
Raciti, G. IT 1.1
Racka, M. PL 32.1.10
Rácz, E. HU 2.2.1
Rácz, V. HU 2.5
Radaczynska, Z. PL 1.1
Radajewska, B. PL 20.2.9
Radev, R. BG 8.2
Radford, R.W. AU 40.1
Radha, K. IN 3.1
Radhakrishna, S. LK 3.2
Radhakrishna Murthy, P. IN 6.1
Radley, R.W. GB 31.1
Radoi, V. RO 1.1.3
Radoičić, M. YU 14.1
Radovanović, B. YU 14.1
Radu, A. RO 17.2
Radu, Gr. RO 19.1
Radulov, L. BG 8.3
Radwan-Pytlewski, P. PL 28.1.1
Radzikowska A. PL 28.2.11
Rae*, A.N. NZ 9.2
Raese, T. US-J 10.2
Rafael, B.S. LK 3.16
Rafael, G. LK 3.16
Raff, J. AU 30.1
Raftopoulou, E. GR 6.1.5
Ragab, B.A. HU 2.2.11
Ragab, M.T.H. CA 41.1
Ragala, P. CS 1.1
Ragan, P. CA 6.1
Rageau, R. FR 5.1
Ragetli, H.W.J. CA 4.1
Raghava, S.P.S. IN 1.1.3
Raghavendra Rao, N.N. IN 1.1.6
Ragone, M.L. AR 6.1
Rahman, A. NZ 5.1
Rahn, R. FR 14.1.3
Rahović, D. YU 14.1
Rai, K.M. IN 24.1.4
Rai, R.M. IN 24.1.7, 24.1.8
Raičević, D. YU 14.1
Raifer, B. IT 11.1
Raijadhav, S.B. IN 14.1
Raimbault, P. FR 19.5
Raine, J. CA 4.1
Raison, J.K. AU 65.4
Raja, M.E. IN 1.1.9
Rajagopal, R. DK 3.4
Rajashekar, C.B. US-G 31.1
Rajput, C.B.S. IN 27.1
Rajput, M.S. IN 26.1
Rajput, S.G. IN 12.1
Rák, I. HU 2.2.1
Rak, J. PL 28.1.13
Rakićević, M. YU 10.2
Rakich, V. AU 61.1
Rakočević, C. YU 10.1
Ralsgård, K. SE 1.1.1
Ram, A. BR 14.1
Ram*, M. IN 2.1.6
Ram, R.Y. IN 28.1

Ram, S. IN 20.1
Ramachander, P.R. IN 1.1.10
Ramachandra Murthy, N. IN 6.1
Ramakers, P.M.J. NL 9.4.1
Ramallo, C. AR 22.2
Ramanujam, T. IN 5.1
Rama Rao, B.V. IN 6.1
Rama Rao, M. IN 6.1
Rama Rao, R. IN 6.1
Rama Rao, T. IN 6.1
Rama Rao, V. IN 1.1.4
Ramborg, S.O. DK 7.2
Rameau, C. FR 2.1.6
Ramelot, P. BE 6.8.12
Ramina, A. IT 12.1
Ramirez, D.A. PH 1.7
Ramírez, M. PE 3.1
Ramírez, O.D. US-N 1.2
Ramírez Delgado, M. MX 5.6
Ramírez Díaz, J.M. MX 2.1
Ramírez Merás, M. MX 12.3
Ramiro, Z.A. BR 21.2
Ramming, D.R. US-K 7.3
Ramos, C.S. PH 1.7
Ramos, D.P. BR 19.2
Ramos, J.V. BR 14.1
Ramos, R.M. BR 29.1
Ramos Bononi, V.L. BR 21.1
Ramos Cano, J. MX 6.2
Ramos de Siqueira, E. BR 11.1
Ramsay, M. AU 21.1
Ramsdell, D. US-F 14.1.3
Ran, C. BR 11.1
Rana, M.S. IN 4.1
Ranade, D.B. IN 14.2
Rancillac, M. FR 4.1.4
Rand, R.E. US-F 24.1
Randall, J.M. GB 31.2.4
Randell, R. US-F 12.1.2
Randhawa, J.S. IN 17.1
Randhawa, S.S. IN 2.1.6
Randjelović, V. YU 14.1
Randle, P.E. GB 18.1.1
Randle, W.M. US-D 52.1
Randles, J.W. AU 42.1
Rane, D.A. IN 14.2, 14.5
Ranfft, K. DE 11.3
Ranga Reddy, B. IN 6.1
Rangel Jiménez, P. MX 6.6
Rangelov, B. BG 9.1
Rangel Ramírez, J.L. MX 2.2
Rankov, V. BG 8.1
Ranković, M. YU 10.2
Rao, B. IN 11.8
Rao, D.P. IN 16.2
Rao, H. IN 6.1
Rao, M.A. US-B 2.1.5
Rao, M.H.P. IN 1.1.9
Rao, M.M. IN 11.3
Rao, P. IN 6.1
Rao, R. IT 18.1
Rapp, K. NO 9.1
Rappaport*, L. US-K 6.1.4
Rasa, L. RO 27.1
Râscanu, M. RO 1.1.3
Rasco, E.T. PH 1.2.4, 1.7
Rasmussen, A.N. DK 4.2
Rasmussen, E. DK 1.1, 1.2
Rasmussen, G.K. US-D 34.1
Rasmussen, H.P. US-J 6.1
Rasmussen, K. DK 1.2
Rasmussen, P.M. DK 1.1, 1.3
Rast, A.T.B. NL 9.4.4
Rastad, L. DK 3.6
Rastogi, P.P. IN 20.1
Rastoin, F. FR 41.2
Rat, B. FR 1.1.3
Rath, S. IN 16.2
Rati, V. RO 11.1
Ratnapala, G.S. LK 3.7
Ratnayake, P.A.N. LK 2.1
Rattana, V. TH 5.1
Ratti, H.A. AR 5.1
Rattigan*, K. AU 65.7
Rattink, H. NL 1.1.5 9.4.2
Raturi, G.B. IN 1.1.1
Ratynski, M. PL 32.1.8
Rauch, F.D. US-M 2.2
Raulston, J.C. US-D 8.1.1
Raup, A.A.A. BR 30.2
Rauš, D. YU 3.3
Raven, P.W.J. NL 6.1
Ravid, M. IL 1.1.2
Raw, G.R. GB 13.1.2
Rawal, R.D. IN 1.1.6
Rawat, T.S. IN 19.5
Raworth, D.A. CA 4.1
Ray, A.K. IN 3.1
Ray, D. GB 35.1.1
Rayment, A.F. CA 44.1
Raymond, M. CA 38.1
Raymond, P. FR 4.1.4
Razera, L.F. BR 20.1
Read, A.J. NZ 12.2
Read, D.C. CA 43.1
Read, G.S. CA 40.1
Read, J.C. AU 29.1
Read*, P.E. US-G 7.1
Readshaw, J.L. AU 65.5
Reali, G. IT 9.3
Rearte, A. AR 15.3
Reátegui, K. PE 3.1
Reay, P.F. NZ 8.2
Rebandel*, Z. PL 20.2.9
Rebechi, M.D. AU 10.1
Rebeiz, C.A. US-F 12.2.2
Reboucas Lins, A.C. BR 5.1, 25.3
Rechnio, H. PL 27.1
Reck, S.R. BR 29.1
Recupero, S. IT 19.1.3
Rédai, I. HU 2.2.9
Redalen*, G. NO 10.1.6
Reddy, B.M. IN 1.1.1
Reddy, B.M.C. IN 1.1.1
Reddy, K.R. US-D 36.1
Reddy, P.P. IN 1.1.7
Reddy, S. IN 6.1
Reddy, Y.N. IN 1.1.1
Reddy, Y.T.N. IN 1.1.1
Reding, J.M. MX 12.1
Redondo Juárez, E. MX 12.2
Reed, D.W. US-H 6.1
Reed, J.F. US-B 1.1
Reeh, T. AT 8.1
Reeleder, R.D. CA 37.1
Rees, A.R. GB 18.1.1
Reeves, A. US-A 1.1
Reforgiato Recupero, G. IT 1.1
Regel, P.A. AU 49.1
Regenberg, P. DK 6.1
Regulski, F.J. US-D 35.1
Rehalia, Sh.A.S. IN 17.1
Rehbogen, J. DE 10.1
Rehkugler, G.E. US-B 5.1.4
Reich, P. FR 3.1.4
Reichart, R. AR 9.1
Reichmann, E. AT 9.1
Reid, A. AU 59.1
Reid, C.S.W. NZ 8.2
Reid, G.W. US-I 2.2
Reid, M.S. US-K 6.1.2
Reid, W.R. US-G 27.1
Reidy, J. IE 3.1
Reif, H. AT 8.1
Reifschneider, F.J.B. BR 25.5
Reighard, G.L. US-D 14.1
Reijmerink, A. NL 9.11.2
Reijnders, M.A.C. NL 1.1.4
Reijnhoudt, P.J.A.C. NL 19.1
Reimann-Philipp*, R. DE 1.1
Reimher, P. DE 33.1
Reina, A. IT 2.1
Reinert, J. US-H 7.1
Reinert, J.A. US-D 25.1
Reinert, R.A. US-D 8.1.2
Reinganum, C. AU 27.1
Reinink, K. NL 9.6.1
Reinoso, J. PE 4.4
Reinschmidt, J. PL 28.2.11
Reis, E.L. BR 14.1
Reis, G. PT 6.1
Reis, P.R. BR 15.1

Reisch, B.I. US-B 2.1.2
Reisegg, F. NO 7.1
Reissig, W.H. US-B 2.1.3
Reissmann, H.J. CA 25.1
Reist*, A. CH 2.2
Reitsma, T. NL 1.1
Reitzel, J. DK 4.1
Rejesus, B.M. PH 1.3
Rejman*, A. PL 32.1.1, 34.1
Rejman, R. PL 1.1
Rejman, S. PL 28.1.10
Rejnus, M. PL 5.1
Rekowska, E. PL 31.1.2
Relf, P.D. US-C 10.1
Relias, N. GR 25.2, 25.2.1
Remah, M. MA 5.1
Rempel, H.E. DD 1.1
Remphrey, W.R. CA 14.1
Rémy, M. FR 28.1
Remy-Paquay, BE 6.10
Renaud, R. FR 4.1.1
Renault, P. FR 3.1.7
Rengwalska*, M. PL 28.2.1
Rennes, A.M.D. NZ 1.1
Rennie, W.J. GB 9.1.1
Renquist, A.R. US-I 3.1
Repstad, K. NO 7.1
Requena, J.M. AR 10.1
Resh, V.H. US-K 3.1.1
Ress, M. HU 3.1.1
Restad, S.H. US-L 2.1
Restaino, F. IT 20.1
Restuccia, G. IT 5.3, 5.4
Reta, A.J. AR 12.1
Retes Cazares, E. MX 2.3
Rethy, R. BE 7.12
Retta, A.D. ET 10.2
Reuling, G. NL 15.1
Reuter, C. DE 15.1.7
Reuther, G. DE 12.1.10
Reuveni, O. IL 1.2.5
Revillard, E. AR 15.3
Revillard, E.B. AR 15.2.5
Revzin, B. IL 1.1
Rewinkel-Jansen, M.J.H. NL 1.1.3
Rex, B.L. CA 13.1
Reyes, A.A. CA 25.2
Reyes, E.D. PH 1.5
Reyes, I. US-N 1.2
Reyes, M. PE 2.2
Reyes, T.T. PH 1.4
Reyes Castaneda, P. MX 14.1
Reynolds, A.G. CA 3.1
Reynolds, T. GB 16.1
Rezende Fontes, P.C. BR 15.1
Rezende Fontes, R. BR 25.5
Rhoades, H.L. US-D 36.1
Rhoads, F.M. US-D 37.1
Rhodes*, B.B. US-D 11.1
Rhodes, D. US-F 9.1.4
Rhodes, P.J. NZ 13.1
Rhodus*, W.T. US-F 3.1
Riahi, S. TN 1.1
Riba, G. FR 17.1.6
Ribeiro, E.C. BR 19.2
Ribeiro, J.F. BR 6.1
Ribeiro, P.A. BR 28.1
Ribeiro, S.I. BR 4.1
Ribeiro dos Santos, J.H. BR 8.2
Ricard, B. FR 4.1.4, 34.1
Ricci, J.R. AR 22.3
Rice, B. IE 1.1
Rice, P.F. CA 16.1
Richard, M.A. CA 33.1
Richards, D. AU 30.1
Richards, G.D. AU 8.1
Richards, G.N. AU 8.1
Richards, M. NZ 9.6
Richards, R. AU 46.2
Richardson, A. US-I 5.1
Richardson, A.C. NZ 1.1
Richardson, M.J. GB 9.1.2
Richardson, N.L. AU 41.1
Richardson, P.M. US-B 1.1
Richardson, P.N. GB 18.1.2
Richer-Leclerc, C. CA 29.2
Richey, F. US-F 14.1.4
Richitearu, A. RO 9.1
Richmond, A. IL 5.1.6
Richter, G. DE 11.2.3, 15.3.3
Richter, J. CS 32.1
Richter, J. DE 15.4
Richter, K. IC 1.1
Richwine, P. US-H 10.1
Rick, C.M. US-K 6.1.4
Ricketson*, C.L. CA 41.1
Ricón, A. VE 4.1
Ridé, M. FR 1.1.3
Ridley, D.R. CA 19.1
Ridomi, A. IT 7.1
Ridray, G. FR 3.1.14
Riedl, A.T. 8.9
Riekels, J.W. CA 15.1
Ries, S. US-F 14.1.1
Ries, S.K. US-F 12.2.3, 14.1.1
Ries, W. DE 12.1.2
Riesen, W. CH 1.1
Rietstra, I.P. NL 1.1.5
Rieux, R. FR 3.1.9
Riger-Stein, A. IL 1.1.2
Riggs, T.J. GB 33.1.6
Righi, F. IT 6.1
Rij, R.E. US-K 7.3
Rijkenbarg, G.J.H. NL 9.10
Rijkenberg, F.H.J. ZA 2.1
Rijtema, P.E. NL 9.2
Riley, H. NO 6.1
Rimando, T.J. PH 1.2.3
Rimóczi, I. HU 2.2.7
Rinaldelli, E. IT 9.1
Rincón, A. VE 4.1
Rincón Valdés, C.A. MX 9.1
Ringeisen, J.G. AR 3.1.1
Ringoet, A. NL 9.5
Rinker, D.L. CA 25.1
Rinne, K. FI 6.1
Rinnerbauer, O. AT 8.2
Riordan, T.P. US-G 23.1
Rios de Alvarenga, L. BR 15.1
Rioux, J.-A. CA 38.1
Riov, J. IL 2.1
Rippen, H. DE 15.1.3
Riquelme, A.H. AR 15.1
Risch, S.J. US-K 3.1.1
Risk, W.H. NZ 16.1
Risse, L.A. US-D 34.1
Risser, G. FR 3.1.2
Risser, P. US-F 12.1
Ristevski, B. YU 4.1
Ristić, B. YU 9.1, 14.1
Risvik, E. NO 10.5.2
Ritchie, D.F. US-D 8.1.2
Ritchie, J.Y. GB 25.1.2
Ritchot, C. CA 29.1, 34.1
Ritiu, C. RO 8.1
Riva, E. AR 19.1
Rivadeneyra, V. PE 1.1
Rivaldo, O.F. BR 25.1
Rivalta, L. IT 19.1.2
Rivière*, L.M. FR 19.5
Rivoiza, G. IT 23.1
Riyad, M. EG 1.2
Rizzo, R. AR 4.1.4
Rizzon, L.A. BR 31.1
Roán, J.E. AR 14.1
Roa Durán, R. MX 8.4
Robak, J. PL 28.2.3
Robak, M. PL 28.1.14
Robb, A.R. NZ 10.1
Robbins, L.M. US-D 54.1
Robbins, R.C. US-D 27.3
Robbs, Ch.F. BR 19.2
Röber, R. DE 11.2.9
Robert, CR 2.2
Robert, P. FR 6.1.3
Robert, Y. FR 14.1.3
Roberts, C.R. US-E 15.1
Roberts, H.A. GB 33.1.8
Roberts, J. US-A 2.1
Roberts, K. GB 24.1.3
Roberts, R. US-H 9.1
Roberts, R.D. US-B 10.1

Roberts, W.J. US-B 14.1.3
Robertse, P.J. ZA 3.2
Robertsen, B. NO 9.2
Robertson, G.I. NZ 2.3.2
Robertson, L. US-G 28.1
Robinson, A.S. NL 9.5
Robinson*, D.W. IE 4.1
Robinson, H.D. GB 18.1.1
Robinson*, J.B. AU 36.2
Robinson, L.W. GB 3.1
Robinson, R.W. US-B 2.1.2
Robinson, S.P. AU 65.1
Robinson, T.L. US-B 2.1.2
Robison, D.G. US-I 9.1
Robledo, C. FR 35.1
Robles, A. AR 6.1
Robles, P. US-I 11.2
Roby, H. AR 15.2.5
Rocha Pena, M.A. MX 9.2
Rocha Rodriguez, R. MX 7.2
Rochon, D'A. CA 4.1
Rock, G.C. US-D 8.1.3
Rod, J. CS 20.1
Rodeva, V. BG 8.1
Rodic, J. YU 14.1
Roditakis, N. GR 6.1.6, 10.2
Rodrigues, J.A. BR 12.1
Rodrigues, J.E. BR 4.1
Rodriguez, A. CR 2.2
Rodriguez, D.S. AR 2.1
Rodriguez, E. CR 2.2
Rodriguez, W. CR 2.2
Rodriquez, J.P. AR 19.1
Rodriquez, M.D. AR 15.1
Rodriquez, R. AR 10.1
Rodriquez, S.B. US-B 2.1.5
Rodriquez A., J. MX 16.3
Rodriquez Escalante, A. MX 2.3
Rodriquez Olmos, B. MX 9.4
Røed, H. NO 10.4.1
Roeggen, O. NO 10.1.7
Roelofs, W.L. US-B 2.1.3
Roemer, K. DE 15.1.4
Roer, P. NO 10.1.2
Roessing, A.C. BR 27.1
Roest, C.W.J. NL 9.2
Roest, S. NL 9.5
Rogers, I.S. AU 36.1
Rogers, M.N. US-E 18.1
Rogers, O.M. US-A 2.1
Rogers, S. US-E 3.1
Roggemans, J. BE 6.9
Rognes, S.E. NO 7.2
Rogoyski, M.K. US-I 3.2
Rogstad, N.Chr. NO 10.3
Roh, M. US-C 3.1
Rohde, J. DE 12.1.8
Rohlfing, H.R. DE 23.1
Rohrbach, J.F. NZ 11.1
Rohrbach, K. US-M 2.2
Roiseman, Y. IL 1.2.5
Rojas, E. AR 22.3
Rokhade, A.K. IN 11.3
Rokicka-Kieliszeska, B. PL 10.1.4
Rokosza, J. PL 32.1.3
Rolca, J.P. AR 23.1
Rolewska, Z. PL 28.1.18
Rollemberg Fontes, H. BR 11.1
Roller, W.I. US-F 7.1.4
Rollins, Jr., H.A. US-C 14.1
Rolston, L.H. US-D 52.1
Rom, R.C. US-E 11.1
Roman, C. FR 19.2
Román, J. US-N 1.1
Roman, L. RO 1.1.6
Roman, R. RO 9.1, 31.1
Roman, R.F. AR 17.1
Roman, T. RO 1.1.7
Romani, R.J. US-K 6.1.3
Romankow, W. PL 20.2.8
Romanowska, M. PL 28.3
Romankow-Zmudowska, A. PL 20.1.1
Romero, R. PE 4.2
Rommel, M. DE 31.1
Ronan, G. AU 36.1
Roncador, I. IT 21.1
Ronco, B.L. AR 14.1
Rondomanski*, W. PL 28.2.7
Roodzant, M.H. NL 6.1
Roos, E.E. US-I 2.2
Roose, M.L. US-K 12.1.1
Roosjen, M.G. NL 9.8
Roque, A. US-N 1.2
Rosa, E. PT 8.1
Rosa, J.F.L. BR 13.2
Rosales, F. CR 2.2
Rosand, F.P.C. BR 14.1
Rosario, E.L. PH 1.8
Rosario, T.L. PH 1.2.3, 1.7
Rosati*, P. IT 3.1
Rosca, O. RO 12.1
Rose Carneiro, E. BR 1.3
Roselli, G. IT 9.2
Rosen, C. US-G 7.1
Rosen, D. IL 2.5
Rosenau, W.J. US-A 5.1.1
Rosenberger, D.A. US-B 4.1, 2.1.4
Rosenfeld, H.J. NO 10.5.2
Rosenthal, I. IL 1.3.2
Rosi, J.F. BR 28.1
Rosini, M. AR 10.1
Roslycky, E.B. CA 18.1
Roson, J.P. FR 16.1
Rosowski, S. PL 32.3
Ross, D. US-L 2.2
Ross, H.A. GB 15.1.4
Rossall, S. GB 25.1.2
Rossetti, A.G. BR 1.1
Rossetti, V.V. BR 21.2
Rossetto, C.J. BR 20.1
Rossi, L.A. AR 4.1.4
Rossignol, J.P. FR 19.5
Rossini, R. AR 23.1
Rosu, F. RO 23.3
Rosu, L. RO 19.2
Rosu, N. RO 7.1
Rosulescu, R. RO 1.1.3
Roth, L.F. US-H 3.1
Roth, R. DE 15.5
Rothan, Ch. FR 3.1.8
Rothenberger, R.R. US-E 18.1
Rothenburger*, W. DE 11.1.2
Rothschild, G.H.L. AU 65.5
Roubos, I. GR 26.1
Rouchaud, J. BE 12.1
Rouet, M.A. FR 38.1
Roughan, P.G. NZ 2.1.4
Roughley, R.J. AU 9.1.1
Roulund, H. DK 3.3
Rouse, D. US-F 23.1.4
Rouse, R. US-H 15.1
Rouseff, R.L. US-D 30.2
Rouskas, D. GR 15.1
Rousselle, F. FR 9.1
Rousselle*, G.L. CA 39.1
Rousselle, G.L. CA 35.1
Rousselle, P. FR 9.1
Routledge, R.E. CA 40.2
Routley, D.G. US-A 2.1
Roux, J.C. FR 31.1
Roux, L. FR 17.1.5
Rouxel, F. FR 14.1.2
Roversi, A. IT 16.1
Rowe, R.C. US-F 7.1.1
Rowe*, R.N. NZ 13.5
Rowland, P.W. AU 41.1
Rowse, H.R. GB 33.1.2
Roy, A. IN 28.1
Roy, A.J. IN 24.1.5
Roy, N. IN 22.1
Roy*, S.K. IN 2.1.6
Roy, S.K. IN 22.1
Roychoudhury, N. IN 28.1
Royle, D.J. GB 19.1.1
Rozner, M. PL 23.1
Różański, Z. PL 28.3
Rozpara, E. PL 28.1.1
Rozpara, W. PL 28.2.5
Rozsnyay, Zs.D. HU 2.5
Rubino, P. IT 2.3

Rückenbauer, W. AT 25.1
Rudawska, M. PL 10.1.4
Rudelle, M. FR 33.1
Rudi, E. RO 9.1
Rudich*, J. IL 2.3
Rudd-Jones, D. GB 18.1
Rudd-Jones, D.R. GB 18.1.2
Rudnicki*, R. PL 28.1
Rudnicki*, R.M. PL 28.1.11
Rudolphij, J.W. NL 9.10
Rudy, P. PL 29.1
Rueda, S. MX 15.1
Ruesink, W.G. US-F 12.1.2
Ruf, M.E. US-D 42.1
Rüger, H. DE 20.1
Rugerri, A. IT 5.2
Ruggiero, C. BR 23.1
Rugini, E. IT 15.1
Rühling, W. DE 12.1.14
Rukasz, I. PL 14.1.5
Rumayor Rodríguez, A. MX 6.2
Rumbo, E.R. AU 65.5
Rumine*, P. IT 22.1.1
Rumińska*, A. PL 32.1.2.2
Rumpel*, J. PL 28.2.3 Rumpho, M.E. US-F 3.1
Rundqvist, K. SE 2.1
Runeckles, V.C. CA 4.2
Ruschel, A.P. BR 24.2
Rushdy, R. CA 27.1
Rushing, J.W. US-D 12.1
Rusmini, B. IT 6.1
Rusňák, V. CS 13.1
Russ, K. AT 8.11
Russell, R. NZ 12.3
Russo, F. IT 1.1
Rusu, E. RO 17.2
Rusu, N. RO 8.4.1
Ruszkowski, A. PL 28.1.8
Rüther, M. DE 15.1.7
Ruthven, A.D. GB 9.1.3
Rutkowski, S. PL 32.1.11
Rutledge, A.D. US-E 3.1
Ruto, J.K. KE 1.3, 11
Rutzke, M.A. US-B 5.1.2
Ruwende, N. ZI 3.1.2
Ružić, A. YU 8.1
Ružić, Dj. YU 10.2
Ryan, E.W. IE 4.1
Ryan, W.J. AU 12.1
Rybak, H. PL 28.1.8
Rybak, M. PL 28.1.8
Rychnovská, M. CS 24.2
Ryder, E.J. US-K 13.1
Ryder, J.M. NZ 2.1.3
Rygg, T. NO 10.4.3
Rykalin, F.N. SU 8.1
Rylke, J. PL 32.1.11
Rylski, I. IL 1.1.1
Rymal, K.S. US-D 38.1
Ryohei Kato, O. BR 3.1
Rypáček, V. CS 24.2
Ryser, J.P. CH 2.1
Rystedt, J. DK 7.1
Ryu, D.D.Y. US-K 6.1.5
Ryu, O.H. KR 23.1.6
Ryugo*, K. US-K 6.1.3
Ryynänen, V. FI 11.1
Rzedzinska, E. PL 19.1
Rzymowska, R. PL 12.1

Sá, P. BR 8.1
Saayman, D. ZA 4.2
Saba, S. PL 30.1
Sabbahg, N.K. BR 20.2
Sable, M.B. IN 14.7
Sabovljević, R. YU 14.1
Sacalis*, J.N. US-B 14.1.1
Sacamano, C.M. US-H 27.1
Sacco, M. IT 19.1.1
Sacerdote, S. IT 24.1
Sachan, V.K. IN 22.1
Sachs, R.M. US-K 6.1.2
Sackmann, G. DE 11.1.1
Sada, A. JP 44.1
Sada, M. JP 38.1
Sadai, K. JP 11.2
Sadamatsu, M. JP 34.2
Sadanandan, N. IN 5.2
Sadiki, D.A. LY 2.1
Sadowska*, A. PL 32.1.10
Sadowski, S. PL 32.1.1
Safak, A. TR 14.1
Safley, L. US-E 5.1
Safran, H. IL 1.2.3
Safriel, U. IL 5.1.5
Saga, K. JP 3.4
Sagane, S. JP 40.2
Sagawa, M. JP 5.2
Sagawa*, Y. US-M 2.2, 2.3
Saglamer, M. TR 13.1
Saglio, P. FR 4.1.4
Saguy, IL 1.3.2
Sah, S.C. IN 24.1.8
Sahs, W.W. US-G 24.1
Said-Allah, M.H. EG 5.1
Saida, Y. JP 15.2
Saijo, R. JP 23.2.3
Saiki, Y. JP 5.1
Saint-Lebe, L. FR 44.1
Saito, A. JP 3.2.3
Saito, H. JP 34.1
Saito, M. JP 24.1
Saito, S. JP 3.2.1, 20.2
Saito, T. JP 1.3, 19.4, 42.2, 45.2
Sakagami, T. JP 11.3
Sakaguchi, H. JP 44.1
Sakaguchi, T. JP 18.2, 40.1
Sakai, H. JP 7.2
Sakai, K. JP 1.1.4, 11.2, 21.2
Sakai, S. JP 21.2
Sakai, T. JP 26.2
Sakai, Y. JP 3.1, 17.1, 35.2
Sakairi, H. JP 19.2
Sakaki, H. JP 21.2
Sakakibara, I. JP 35.1
Sakamoto, M. JP 32.1
Sakamoto, N. JP 20.2
Sakane, M. BR 21.1
Sakanishi, Y. JP 33.2
Sakashita, T. JP 1.1.1
Sakata, Y. JP 18.3, 23.2.2
Sakaue, H. JP 18.1
Sakaue, O. JP 16.3
Sakiyama*, R. JP 41.5
Sako, I. JP 42.2
Säkö, J. FI 1.1
Sakr, R.A. LY 1.1
Sakshaug, K. SE 1.1.1
Sakuma, S. JP 23.2.5
Sakuma, T. JP 16.2
Sakurada, S. JP 3.2.4
Sakurai, H. JP 29.1
Sakurai, S. JP 29.1
Sakurai, T. JP 36.2, 47.2
Sakurai, Y. JP 1.1.2
Salac, S.S. US-G 23.1
Salacha, M. GR 18.1.4
Salamon, O. PL 28.1.2
Salamon, S.P. HU 2.5
Salamon, Z. PL 28.1.6
Salamończyk, K. PL 28.4
Salas, G.M. AR 22.3
Salas Franco, A. MX 5.3
Salatti, E. BR 24.2
Salazar, J. CR 2.3
Salazar, R. CO 4.1
Salazar, S. CR 2.3
Salazar Cavero, E. BR 30.3
Salazar Valdez, M.A. MX 10.3
Sale*, P.J.M. AU 65.6
Saleck, H.C. BR 19.1
Saleh, A. LY 2.1
Salembier, J-F. BE 6.3
Sales, A. BR 19.2
Salesses, G. FR 4.1.1
Salette, J. FR 1.1.4
Salgado, L.O. BR 16.1
Salibe, A.A. BR 22.1
Salinas, A. AR 16.1
Salinas González, J.G. MX 7.2

Salja, F. YU 6.1
Sallam, S.H. LY 1.2
Sallay-Horváth, Zs. HU 2.4
Salomon, E.A.G. BR 20.2
Salter, P.J. GB 33.1
Salter, P.J. GB 33.1.5
Saltveit, Jr., M.E.
US-K 6.1.4
Saludez, J.D. PH 1.2.2
Salvarredi, E.M. AR 12.1
Salvi, M.J. IN 15.1
Salvi, P.V. IN 15.1
Samancı, H. TR 14.1
Samaipo Seixas, B.L. BR 12.2
Samanta, B.D. IN 16.2
Samaras, F. GR 18.1.2
Samaratunga, H. LK 3.12
Samek, M. HU 3.1.1
Sameshima, K. JP 18.1
Samiullah, R. IN 11.2
Samonte, H.P. PH 1.5
Samotus, B. PL 11.1.8
Sampaio, A.F. BR 21.2
Sampaio, J.B.R. BR 25.2
Sampaio, J.M.M. BR 12.1
Sampaio, M.J.A.M. BR 25.3
Sampaio, S. BR 12.1
Sampaio, V.R. BR 24.1
Sampaio, L. Soares de V.
BR 12.2
Sampson, P.J. AU 46.5
Sams, C. US-E 3.1
Sams, W.D. US-E 2.1
Samsøe-Pedersen, L. DK 4.1
Samson, C. FR 10.1.3
Samson, R. FR 1.1.3
Samuel, J. EG 1.4
Samuels, G.J. NZ 2.3.2
Samuelsen, R.T. NO 9.1
Samuelson, T.J. GB 8.1.1
Samuilă, J. RO 25.1
Samyn, G. BE 7.7
San Antionio, J.P. US-C 3.1
Sanada, T. JP 45.1
Sanchez, F. AR 10.1
Sanchez, F.F. PH 1.3, 1.6
Sanchez, L.A. CO 4.1
Sanchez, N.F. BR 12.1
Sanchez Acosta, M. AR 6.1
Sanderlin, R.S. US-D 59.1
Sanders, C. NL 1.1.5
Sanders, D.C. US-D 8.1.1
Sanderson, J.B. CA 43.1
Sanderson, K.C. US-D 38.1
Sandhu, A.S. IN 17.1
Sandhu, S.S. IN 17.1
Sandler, D. IL 1.1.2
Sandoval, J. CR 2.2
Sandru, V. RO 29.1
Sandved*, G. NO 10.1.4
Sandved*, M. NO 10.1.5
Sandvol, US-I 13.1
Sandwell, I. GB 17.1
Sanford, G. US-A 12.1
Sanford, K.H. CA 41.1
Sanford, L.L. US-C 3.1
Sang, C.K. KR 7.3
Sangalang, J.B. PH 1.2.2
Sanghavi, K.V. IN 14.4
Sangster*, D.M. CA 42.1
Saniewska, A. PL 28.1.16
Saniewski*, M. PL 28.1.17
Sannamarappa, N. IN 3.1
Sano, K. JP 18.2
Sansavini*, S. IT 3.1
Sansdrap, A. BE 6.9
Sansinanea, A. AR 10.1
Sánta, Cs. HU 2.7
Santamour, F.S. US-C 4.1
Santana, M.B.M. BR 14.1
Santiago, M. US-N 1.2
Santokhi, R. SR 1.1
Santoro, R. IT 20.1
Santos, A. PT 8.1
Santos, A. Soares dos A.
BR 25.3
Santos, L. PT 3.1
Santos, M.M. BR 1.1
Santos Cabral, J.R. BR 12.1
Santos Fernandes, H. BR 30.3
Santos Gama, M.I.C. BR 25.3
Santos Neta, J.C.P. BR 13.2
Santos Oliveira, M.A. BR 4.1
San Valentin, G.O. PH 1.5
Saponaro, A. IT 20.1
Sapundjieva, S. BG 11.1
Sapundżić, M. YU 17.1
Saraiva Teixeira, L.M. BR 8.1
Saraiva, I. PT 1.1
Sarakun, N. TH 14.1
Saralp, D. TR 5.1
Sarananda, K.H. LK 3.18
Sáray, T. HU 2.2.1
Saray Meza, C.R. MX 7.2
Sarcinella, L. AR 15.2.1
Sargent*, M.J. GB 3.1
Sarić, A. YU 3.2
Sarifakioglu, C. TR 2.2
Sarig, Y. IL 1.4.2
Sarioglu, H. TR 19.1
Sarode, S.D. IN 14.1
Sarooshi*, R. AU 5.1
Sárosi-Tánczos, E. HU 2.3
Sarrazyn, R. BE 17.1
Sartorato, A. BR 26.1
Sasaki, A. JP 11.2
Sasaki, H. JP 16.1, 19.4
Sasaki, K. JP 19.2, 46.1
Sasaki, N. JP 3.4
Sasaki, S. JP 28.1
Sasaki, T. JP 16.3, 24.1
Sasaki, Y. JP 17.1
Sase, T. JP 4.2
Sasiplarin, T. TH 8.1
Sass, P. HU 2.2.2
Sasser, M. US-C 2.1
Sato, A. JP 46.2
Sato, H. JP 2.2, 7.3, 35.2,
45.1
Sato, I. JP 42.3
Sato, K. JP 12.2
Sato, M. JP 2.3, 8.1, 30.1,
35.2, 41.5
Sato, N. JP 3.3.1
Sato, R. JP 3.1, 8.2
Sato, S. JP 10.1
Sato, S. JP 6.2, 8.1, 16.1
Sato, T. JP 3.1, 3.2.2, 14.2,
16.1, 23.2.2, 30.1, 41.1,
45.1, 47.2
Sato, Y. JP 3.3.1, 14.2,
24.1, 26.5, 45.1
Satoh, K. JP 22.3
Satoh, N. JP 19.2
Saturnino, H.M. BR 15.1
Satyan, S.H. AU 65.4
Satyanarayana, G. IN 6.1
Satyanarayana Rao, D.V.
IN 6.1
Sauer, M.R. AU 65.2
Sauls, J. US-H 15.1
Saunders, J. CR 2.2
Saunier, R. FR 4.1.1
Sauthoff, W. DE 5.4
Savage, H. AU 59.3
Savić, M. YU 1.2
Savić, R. YU 14.2
Savić, S. YU 11.1, 14.1
Savić, V. YU 16.1
Savio, A. FR 32.1
Savitzchi, P. RO 23.1
Savopoulou, M. GR 25.1.6
Savory, B.M. GB 26.1
Savos, M.G. US-A 13.1
Savu, A. RO 1.1.2
Sawa, Y. JP 20.4
Sawabe, H. JP 43.1
Sawachi, N. JP 41.1
Sawada, H. JP 11.3
Sawada, K. JP 12.2
Sawada, Y. JP 8.2
Sawahata, K. JP 14.1
Sawamura, Y. JP 28.1
Sawano, M. JP 13.3

Sawant, D.M. IN 14.6
Sawczuk, E. PL 32.1.13
Saxena, P.K. IN 25.1
Saxena, S.K. IN 2.1.2, 2.1.4
Sayed, A.A.M. IN 3.1
Sayin, A. TR 15.3
Saylor, J.L. US-F 14.1.1
Scaff, L.M. BR 2.2
Scaife, M.A. GB 33.1.2
Scalabrelli, IT 17.2
Scalla, R. FR 7.1.9
Scanlan, R. US-J 14.1
Scannell, R.J. US-B 5.1.1
Scaramuzzi, F. IT 9.1
Scărlătescu, S. RO 34.1
Scerbo, W.F. US-B 14.1.1
Schaad, N. US-I 15.1
Schabort, J.C. ZA 3.3
Schach, H. US-E 15.1
Schache, M. AU 65.2
Schachtschabel, P. DE 15.4
Schadegg, E. DK 4.2
Schaefer, G.H. US-K 3.1.1
Schaefers, G.A. US-B 2.1.3
Schaeffer, J.R. US-I 18.1
Schaer, T. DE 15.1.4
Schales, F.D. US-C 7.1
Schall, R. ET 10.1
Schaller, K. DE 12.1.11
Schapendonk, A.H.C.M. NL 9.1
Schaper, P.M. NL 5.1
Schaplowsky, T. US-G 33.1
Scharpf, H.C. DE 15.6
Schatz, A.S. AR 6.1
Schaufler, E.F. US-B 5.1.1
Schaupmeyer, C.A. CA 6.1
Scheau, V. RO 13.1
Scheenen, T. NL 15.1
Scheepens, P.C. NL 9.1
Schelstraete, A. BE 22.1
Schembecker, F. DE 5.1.1
Schenk, E.-W. DE 15.1.2
Schenk*, M. DE 15.1.5
Schenk, W. DE 12.1.2
Scheuerpflug, G. DE 33.1
Schiavi, M. IT 20.1
Schiffers, B. BE 6.8.4
Schijvens, E.P.H.M. NL 9.10
Schiller, P. AT 8.8
Schilling, A.D. GB 16.1
Schimmelpfeng, H. DE 11.1.1
Schinzel, R.L. BR 27.2
Schiva, T. IT 22.1
Schlamp, H. DE 23.1
Schlegel, D.E. US-K 3.1.2
Schlegel*, G. DE 11.1.1
Schlimme, D.V. US-C 5.2
Schlinger, E.I. US-K 3.1.1
Schlough, D.A. US-F 23.2
Schlüter, U. DE 15.2.4
Schmalscheidt, W. DE 3.1
Schmelcz, A. HU 2.7
Schmid, G. CH 1.1
Schmid, P. DE 11.1.1
Schmidle, A. DE 9.1
Schmidt, G. HU 2.2.4
Schmidt, G.H. DE 15.3.6
Schmidt, H. DE 23.1
Schmidt, H.L. DE 11.1.4
Schmidt, J. HU 7.1
Schmidt, J.C. US-F 12.2.2
Schmidt, J.P. DK 3.2
Schmidt, K. DE 11.2.7
Schmidt, L. US-G 13.1
Schmidt, St. PL 20.2.10
Schmoldt, U. DE 14.1
Schneider, A. IT 24.2
Schneider, C. FR 6.1.4
Schneider, F. NL 9.9
Schneider, H.G. AU 29.1
Schneider, J.E. BR 25.1
Schobinger, U. CH 1.1
Schoch, P-G. FR 3.1.4
Schoenemann, J.A. US-F 23.1.3
Schoeneweiss, D.F. US-F 12.1.1
Schofield, C.P. GB 31.2.4
Schoinas, G. GR 25.2.3
Scholefield, P.B. AU 65.3
Schollenberger, M. PL 32.1.6
Scholten, G. NL 1.1 1.1.5
Scholten, K.F. NL 9.8
Scholtz, E. US-B 6.1
Scholz, E.W. US-G 12.1
Schomaker, C.H. NL 9.4.3
Schönbeck, F. DE 15.1.6
Schönbeck, H. AT 8.11
Schoneveld, J.A. NL 6.1
Schönfeld, P. DE 5.2.1
Schoorl, D. AU 18.1
Schotte, Th.P. NL 12.1
Schouten*, S.P. NL 9.10
Schrage, W.W.M. CA 40.2
Schrale, G. AU 36.3
Schroeder, D.B. US-A 13.1
Schroeder*, R.A. US-E 18.1
Schroeder, W. CA 9.1
Schroeder, W.J. US-D 34.1
Schroth, M.N. US-K 3.1.2
Schrothmayer, J. AT 9.1
Schubert, A. IT 24.1
Schuder, D.L. US-F 9.1.3
Schüepp, H. CH 1.1
Schulz, R.K. US-K 3.1.3
Schumacher, R. CH 1.1
Schumm*, A. DE 11.2.1
Schunke, E. IC 1.1
Schürmer, E. DE 11.2.4
Schüssler, H. SE 1.1.3
Schuster, D.J. US-D 24.1
Schuster, F. IT 11.1
Schuster, V. HU 2.1.2
Schut, B. NL 18.2
Schütz, E. SE 4.1
Schutzki, R.E. US-F 14.1.1
Schvester, D. FR 3.2.2
Schwab, D.P. US-H 3.1
Schwabe, W.F.S. ZA 4.1
Schwabe*, W.W. GB 35.1.2
Schwarz, A. CH 2.2
Schwarz, K.-G. DE 22.1
Schweisguth, B. FR 17.1.1
Schwemmer, E. DE 16.1
Schweppenhauser, M.A. ZI 3.3
Schwertmann, U. DE 11.1.3
Scibisz*, K. PL 32.1.1
Sciortino, A. IT 13.2
Scogin, R. US-K 5.1
Scopes, N.E.A. GB 18.1.2
Scora, R.W. US-K 12.1.1
Scorza, R. US-C 8.2
Scott, D.H. US-F 9.1.1
Scott, D.N. NZ 2.1.3
Scott, H.B. AU 30.2
Scott, J. US-E 20.1
Scott, J.W. US-D 24.1
Scott, N.R. US-B 5.1
Scott, N.S. AU 65.1
Scott, R.M. AU 65.7
Scott, S. US-D 14.1
Scott, S.J. US-E 11.1
Scotter, D.R. NZ 9.5
Scotti, C.A. BR 27.2
Scotti, P.D. NZ 2.2
Scotto la Massesse, C. FR 2.1.5
Scountridakis, M. GR 8.1.1
Scriven, F.M. AU 9.5
Scudamore-Smith, P.D. AU 15.2
Scudder, W.T. US-D 36.1
Scuderi, A. IT 1.1
Scurtu, I. RO 1.1.1
Scurtu, M. RO 1.1.1
Seabrook, J.E.A. CA 40.3
Seager, J.C.R. IE 4.1
Seaton, R.D. GB 9.1.1
Seberry*, J.A. AU 9.2
Sebesta, Z. NO 10.1.2
Sechi, A. IT 4.1
Seck, A. SN 1.1
Seck, O. SN 1.1
Seck, P. SN 1.1
Sedgley*, M. AU 65.1

Sediyama, T. BR 17.1
Sédrati, M. MA 1.5
Seeley*, J.G. US-B 5.1.1
Seeley*, S.D. US-I 5.1
Seelye, J.F. NZ 10.1
Seem, R.C. US-B 2.1.4
Seemüller, E. DE 9.1
Segalin, D.L. BR 28.1
Segawa, K. JP 3.3.1
Segers, J.J.P.M. NL 9.7.5
Seguin, B. FR 3.1.4
Seibold, H.W. DE 15.3.2
Seidel, K. DE 11.2
Seif, A.A. KE 1.4
Seiler, J. CH 1.1
Seita, I. JP 19.2
Seito, M. JP 3.2.4
Seki, E. JP 4.1
Sekiguchi, A. JP 26.1
Sekioka, T.T. US-M 6.1
Sekita, N. JP 3.2.3
Sekiya, H. JP 39.2
Sekiya, K. JP 16.2
Sekiyama, K. JP 4.3
Sękowski, B. PL 20.2.10
Sekse, L. NO 5.1
Šekularac, G. YU 10.1
Sela, I. IL 2.5
Selaru, E. RO 8.1
Self*, B.F. GB 8.1.3
Selhime, A.G. US-D 34.1
Selleck*, G.W. TW 5.1
Selvaraj, Y. IN 1.1.8
Selwa, J. PL 14.1.1
Semal, J. BE 6.8.3
Semb, L. NO 10.4.1
Semel, M. US-B 8.1
Semeniuk, P. US-C 3.1
Seminario, C. PE 1.1
Semjon, A. CS 21.1
Sen, B. TR 13.1
Sen, S.K. IN 28.1
Sen, S.M. TR 21.1
Sena Gomes, A.R. BR 14.1
Senanayake, L. LK 3.7
Senanayake, M. LK 4.1
Senanayake, U. LK 1.1
Seneta, W. PL 32.1.3
Senetiner, A.C. AR 13.1
Senin, V.I. SU 9.1
Seniz, V. TR 6.1
Šenk, L. CS 28.1
Senmache, J. PE 2.2
Senter, S. US-D 16.2
Sento, T. JP 5.3
Seon, S.J. KR 9.1
Seppänen, E. FI 2.1.1
Sequeira, J.C. PT 6.1
Sequeira, L. US-F 23.1.4
Sequeira, M. PT 6.1
Sequeira, O. PT 6.1
Sequi, P. IT 25.1
Serboiu, A. RO 33.1
Serboiu, L. RO 33.1
Serief, A.M. LY 5.1
Serizawa, S. JP 38.2
Serra, G. IT 4.1
Serralheiro, J. PT 2.1
Serrano, J.R. PT 3.1
Šestović, M. YU 14.1
Seszták, L. HU 2.2.10
Seth, J.N. IN 24.1.2
Sethi*, V. IN 2.1.6
Seutin, E. BE 6.3
Severin, C. AR 16.1
Sévert, M. FR 28.1
Sewell, G.W.F. GB 8.1.2
Sexana, R.K. IN 24.1.1
Seyama, N. JP 23.2.3
Sfakiotakis, E. GR 25.1.1
Sfredo, G.J. BR 27.1
Shaaya, E. IL 1.3.3
Shachak, M. IL 5.1.5
Shafer, S.R. US-D 8.1.2
Shah, A. IN 24.1.2
Shaked, A. IL 1.2.3
Shaladan, M.S. LY 1.2
Shalk, J.M. US-D 12.2
Shallenberger, R.S. US-B 2.1.5
Shama Bhat, K. IN 3.1
Shamasundaran, K.S. IN 1.1.10
Shandoufi, F. LY 2.1
Shaner, G.E. US-F 9.1.1
Shang Fa Yang, US-K 6.1.4
Shankar, G. IN 23.1
Shankarnarayan, K.A. IN 18.1, 18.4
Shankland, D.L. US-D 27.2
Shanks, C.H. US-J 9.1
Shannon, E. US-H 19.1
Shannon, J.C. US-B 13.1.1
Shannon, P. CR 2.2
Shanthappa, P.B. IN 11.6
Sharma, B.B. IN 2.1.4
Sharma, C.P. IN 22.1
Sharma, D.K. IN 2.1.1
Sharma, G.K. IN 1.1.3
Sharma, H.C. IN 2.1.2, 19.2
Sharma, I.P. IN 22.1
Sharma, J.N. IN 17.2
Sharma, J.R. IN 10.2
Sharma, K.L. IN 10.3
Sharma, K.P. IN 4.1
Sharma, R.K. IN 8.1
Sharma, R.L. IN 10.3
Sharma, S.C. IN 22.1
Sharma, S.D. IN 10.1
Sharma, S.K. IN 18.2
Sharma, S.R. IN 1.1.6
Sharma, S.S. IN 9.1
Sharma, V.K. IN 2.1.2, 10.3, 22.1
Sharma, V.P. IN 2.1.1
Sharma, Y. IN 10.2
Sharma, Y.K. IN 2.1.3
Sharp, A.K. AU 65.4
Sharp, E.L. US-I 18.1
Sharpe, R.H. US-D 27.4
Sharples, R.O. GB 8.1.3
Shattuck, V. CA 15.1
Shaw, B.D. NZ 8.1
Shaw, G.J. NZ 8.2
Shaw, J. US-A 9.2
Shaw, L.S. US-D 27.9
Shaw, M.W. GB 19.1.1
Shaw, P. GB 24.1.3
Shaw, R.J. US-A 14.1
Shawa*, A.Y. US-J 2.1
Shaw Ashton, P. US-A 7.1
Shea, P.F. AU 26.2
Sheahan, B.T. AU 41.1
Shearman, R.C. US-G 23.1
Sheehan, T.J. US-D 27.5
Sheehy, S.J. IE 3.1
Sheen, T.F. TW 2.1
Shekhawat, J.S. IN 19.2
Shelke, B.D. IN 13.3
Shelp, B. CA 15.1
Shelton, A.M. US-B 2.1.3
Shelton, J.E. US-D 6.1
Shennan, C. US-K 6.1.4
Shepardson, E.S. US-B 5.1.4
Shepherd, H.R. GB 8.1.1
Shepherd, K. BR 12.1
Sherif, A. ET 3.1
Sherman, I.W. US-K 12.1
Sherman*, M. US-D 27.7
Sherman, W.B. US-D 27.4
Sherry, W.J. US-F 12.2.2
Sheu, C.C. TW 11.1.1
Sheu, D.H. TW 11.1.1
Shewale, B.S. IN 14.6
Shiba, H. JP 26.1
Shiba, S. JP 30.1
Shibano, K. JP 23.2.4
Shibata, H. JP 2.1
Shibata, K. JP 21.1, 39.1
Shibata, M. JP 23.2.2
Shibata, T. JP 4.1
Shibata, Y. JP 34.2, 40.2
Shibukawa, S. JP 35.2
Shibuya, M. JP 14.6
Shichijo, T. JP 14.2

Shieh, S.W. TW 11.1.1
Shifriss, Ch. IL 1.1.3
Shiga, Y. JP 12.1
Shigehara, I. JP 26.1
Shigenaga, S. JP 22.5
Shigeoka, H. JP 21.2
Shigeta, D. US-M 4.1
Shigeta, K. JP 46.1
Shigeta, M. JP 31.1
Shigeta, T. JP 19.2
Shii*, C.T. TW 9.1
Shikano, E. JP 38.2
Shikano, S. JP 24.2
Shikhamany, S.D. IN 1.1.1
Shikur, G.M. ET 4.1
Shikura, A. JP 18.2
Shildrick, J.P. GB 4.1
Shillo, R. IL 2.2
Shim, J.S. KR 8.3
Shim, S.C. KR 21.5
Shimabukuro, Y. JP 32.1
Shimamura, K. JP 31.2
Shimamura, Y. JP 20.2
Shimanaka, T. JP 32.1
Shimazaki, Y. JP 35.2
Shimazu, K. JP 44.2
Shimazu, T. JP 26.3
Shimizu, K. JP 47.1
Shimizu, M. JP 5.1
Shimizu, N. JP 35.1
Shimizu, S. JP 26.2, 47.1
Shimizu, T. JP 42.2
Shimizu, Y. JP 1.1.5
Shimmin, J.D. GB 32.1
Shimo, M. JP 18.1
Shimogoori, Y. JP 25.1.3
Shimojyo, S. JP 26.4
Shimokawa, K. JP 25.2
Shimomura, T. JP 23.2.4, 34.1
Shimonaka, M. JP 42.2
Shimoohsako, M. JP 7.1
Shimura*, I. JP 41.4
Shimura, K. JP 23.2.5
Shin, B.W. KR 8.1
Shin, D.G. KR 8.1
Shin, E.P. KR 24.1
Shin, G.Y. KR 18.1.3
Shin, I.Y. KR 1.1
Shin, K.C. KR 23.1.4
Shin, U.D. KR 8.4
Shin, Y.C. KR 8.3
Shin, Y.U. KR 23.1.3
Shinde, G.S. IN 12.1
Shinde, N.N. IN 12.1
Shindo, M. JP 47.1
Shindoo, T. JP 34.2
Shinkaji, N. JP 4.4
Shinoda, K. JP 23.2.5
Shinohara, K. JP 5.1
Shinohara*, Y. JP 14.6
Shinto, H. JP 44.1
Shiobara, K. JP 29.1
Shiomi, T. JP 23.2.4
Shiono, I. JP 35.2
Shioya, T. JP 39.1
Shiozaki, Y. JP 3.4
Shiozawa, J. JP 26.3
Shiozawa, K. JP 12.2, 15.3
Shipway, M.R. GB 21.1
Shirai, T. JP 38.1
Shirai, Y. JP 17.1, 23.2.4
Shiraishi, M. JP 5.3
Shiraishi*, S. JP 7.4
Shiraishi, T. JP 10.1, 30.1
Shirasaki, S. JP 3.2.3
Shirasaki, T. JP 4.3
Shiratori, T. JP 26.4
Shireman, R.B. US-D 27.3
Shirsath, N.S. IN 14.2.1
Shishido, Y. JP 23.2.3
Shoda, K. JP 7.1
Shoemaker, P.B. US-D 6.1, 8.1.2
Shoemaker, W.H. US-F 11.1
Shoji, H. JP 45.1
Shokes, F.M. US-D 37.1
Sholberg, P. CA 3.1
Shomer, I. IL 1.3.2
Shopland, J. US-E 21.1
Short, A.J.W. AU 41.1
Short, K.C. US-D 41.1
Short, T.H. US-F 7.1.4
Shorter, A.J. AU 65.4
Shorter, N.H. AU 58.1
Shreve, L.W. US-H 14.1
Shrimal, M.L. IN 19.1
Shu, Z.H. TW 2.1.2
Shukla, C.P. IN 22.1
Shūkla, J.P. IN 22.1
Shukla, L.M. IN 22.1
Shukla, R.K. IN 22.1
Shukla, V. IN 1.1.2
Shulman, M.D. US-B 14.1.5
Shulman, Y. IL 1.2.4
Shultz, P. US-C 12.1
Shumaker, J.R. US-D 28.1
Shurafa, M.Y. LY 1.2
Shyoka, M. JP 23.1
Siański, J. PL 28.5
Sibal, W.F. PH 1.7
Sibley, J.D. CA 42.1
Sieczka, J.B. US-B 5.1.3, 8.1
Siedler, A.J. US-F 12.2.1
Sieg, R. AT 9.1
Sierber, J. DE 11.2.8
Siewniak, M. PL 32.1.12
Siggeirsson, E.I. IC 1.1
Signorelli, P. IT 5.3, 5.4
Sigrist, J.M.M. BR 20.2
Sigurbjornsson, G. IC 1.1
Sihleanu, D. RO 29.1
Sijaona, M.E.R. TZ 6.1
Sikinyi, S.O. KE 1.6
Sikkhamondhol, B. TH 19.1
Sikoparia, D. YU 17.1
Sikora, D. PL 32.3.2
Sikora, E. PL 28.2.13, 28.2.3
Sikora, J. PL 30.1
Sikorska, B. PL 28.4
Sikurajapathy, M. LK 3.12
Silberbush, M. IL 5.1.2
Silbereisen, R. DE 28.1.3
Silberstein, B.A. IL 1.4.8
Šilhan, J. CS 22.3
Silochi, M. RO 21.1
Šiltar, A. YU 18.3
Silva, F.C.C. BR 17.1
Silva, J. BR 25.3
Silva, K.P.U.D. LK 3.2
Silva, L. BR 15.1
Silva, L. PT 6.1
Silva, L.C. BR 9.1
Silva, M.C. BR 28.1
Silva, R.E.C. BR 14.1
Silva Filho, J. BR 30.3
Silva Freire, M. BR 26.1
Silva Melo, L.A. BR 1.1
Silva Prazeres, E.A. BR 6.1
Silva Ríos, J. MX 2.2
Silva Salazar, J.J. MX 3.3
Silva Vara, S. MX 2.3
Silva Vendruscok, J.L. BR 30.1
Silveira, R. Bergmann de A. BR 21.1
Silveira Filho, A. BR 26.1
Silver, C.N. NL 9.8
Silvey, V. GB 6.1
Silvy, A. FR 44.1
Simão, S. BR 24.1
Simard, R.E. CA 38.1
Simedrea, C. RO 6.1
Šimek, F. CS 22.3
Simeone*, A.M. IT 19.1
Simeoni, A.E. US-I 2.2
Simeria, Gh. RO 16.1
Simitchiev, Ch. BG 8.1
Simmonds, J.A. CA 20.2
Simmons, J.B.E. GB 16.1
Simón, A. ES 11.1
Simon, G. FR 35.1
Simon, I. HU 2.1.3
Simon*, J. US-F 9.1.4
Simon, J.-L. CH 2.3

Simon, P. BE 19.1
Simon, T. HU 5.2
Simon, V. BE 6.10
Simone, G.W. US-D 27.6
Simonis, L. GR 25.2.3
Simonov, D. YU 4.3
Simons*, R.K. US-F 12.2.2
Simović, S. YU 10.1
Simpson, B. US-H 7.1
Simpson, D.W. GB 8.1.2
Simpson, R.F. AU 37.1
Simpson, W.R. US-I 15.1
Sims, A. AU 59.2
Sims*, Jr. E.T. US-D 13.1
Sims, J.R. US-I 18.1
Sims, R.E.H. NZ 9.3
Sims, T.J. US-C 2.1
Simsek, G. TR 14.1
Simser, W.L. CA 16.1
Sinclair, P.T. AU 12.1
Sinden, S.L. US-C 3.1
Sindile, N. RO 8.5
Sing, C. IN 4.1
Singh, A. IN 17.1, 18.5
Singh, B. IN 22.1, 25.1
Singh, B.N. IN 25.1
Singh, B.P. IN 26.1, 27.1
Singh, C. IN 2.1.2, 2.1.3, 2.1.4
Singh, C.P. IN 20.1
Singh, D. IN 25.1
Singh, D.B. IN 23.1
Singh, D.K. IN 20.1
Singh, D.P. IN 1.1.2
Singh, D.R. IN 25.1
Singh, D.S. IN 25.1
Singh, F. IN 1.1.3
Singh, G. IN 17.1, 24.1.6
Singh, G.N. IN 22.1
Singh, H.P. IN 18.4
Singh, I.S. IN 21.1
Singh, J. IN 4.1, 8.1
Singh, J.B. IN 22.1
Singh, J.N. IN 27.1
Singh, J.P. IN 23.1
Singh, K. IN 2.1.1, 25.1
Singh, K.P. IN 27.1
Singh, M. IN 10.2, 22.1
Singh, M.M. IN 19.5, 22.1
Singh, M.P. IN 17.1, 18.2, 20.1, 25.1
Singh, N.P. IN 20.1
Singh, O.P. IN 21.1
Singh, P. IN 19.2
Singh, P. NZ 2.2
Singh, P.V. IN 22.1
Singh, R. IN 1.1.1, 8.1, 9.1, 20.1
Singh*, R. IN 17.1
Singh, R.K. IN 3.1, 8.1
Singh, R.N. IN 24.1.6
Singh, R.P. IN 1.1.10, 19.1, 20.1, 22.1
Singh, R.S. IN 20.1
Singh, R.Y. IN 25.1
Singh, S. IN 17.1, 25.1
Singh, S.B. IN 24.1.8
Singh, S.J. IN 1.1.6
Singh, S.N. IN 17.1, 25.1
Singh, S.P. IN 1.1.7, 27.1
Singh, T. IN 8.1
Singh, U.P. IN 8.1
Singh, Y.P. IN 25.1
Singha, S. US-C 9.1
Singh Verma, H. IN 10.2
Singletary, C.C. US-D 50.1
Singleton, V.L. US-K 6.1.5
Sinha, A.K. IN 2.1.4
Sinha, M.M. IN 24.1.8
Sinha, T.D. NG 2.1
Sink, K.C. US-F 14.1.1
Sinkai, K. JP 1.1.1
Šinžar, B. YU 14.1
Sipa, B. RO 2.1
Sipahioglu, O. TR 6.2
Siqueira, P.R. BR 14.1
Sirianidis, G. GR 19.1
Sirisena, J.A. LK 3.5
Siriwardena, T.D.W. LK 3.3
Sirjacobs, M. MA 1.1
Sirohi, S.C. IN 20.1
Šišaković, V. YU 18.1
Sissay, A.A. ET 10.2
Sitren, H.S. US-D 27.3
Sitterly, W.R. US-D 12.1
Sittolin, I.M. BR 15.1
Sitton, D. IL 4.1
Sivakov, L. YU 4.1
Sivasithamparam, K. AU 59.3
Sive*, A. IL 3.1
Siwecki, R. PL 10.1.3
Siwulski, M. PL 20.2.11
Sjerps, J. FR 35.1
Sjøseth, H. NO 10.3
Sjöstedt, B. SE 1.1.1
Skage, O.R. SE 1.2
Skalli, M. MA 5.2
Skalski, J. PL 32.1.11
Skapski*, H. PL 32.1.2
Skarlou, V. GR 3.1.2
Skelly, J.M. US-B 13.1.1
Skeltis, J. US-F 15.1
Skene, D.S. GB 8.1.1
Skene, K.G.M. AU 65.1
Skibinska, A. Pl 10.1.5
Skierkowska, M. PL 28.2.14
Skierkowski*, J. PL 28.2
Skiredj, A. MA 5.1
Skirvin, R.M. US-F 12.2.2
Skogley, C.R. US-A 14.1
Skogley, E.O. US-I 18.1
Skorupska, A. PL 20.1.1
Skorupska, U. PL 9.1
Skov*, O. DK 3.1
Skowronek, J. PL 28.1.8
Skowronek, W. PL 28.1, 28.1.8
Skowroński, Z. PL 20.2.9
Skroch, W.A. US-D 8.1.1
Skrzypczak, Cz. PL 28.1.16
Skulberg, A. NO 10.5
Skuterud, A. NO 10.4.2
Skwiercz, A. PL 25.1
Slack, C.R. NZ 8.1
Slack, D.A. US-E 11.1
Slack, G. GB 18.1.1
Slack, J. AU 12.1
Slack, S. US-F 23.1.4
Slâmnoiu, T. RO 20.1
Slater, W.G. AU 15.5
Slavík, B. CS 24.2
Slavik, V. CS 25.1
Slavković, S. YU 11.1
Sletten, A. NO 10.4.1
Slinde, E. NO 10.5.2
Sliwiak, M. PL 13.1
Sloane, R.T.J. AU 36.1
Slor, E. IL 1.2.5
Sloth, A. DK 2.1
Słowik, B. PL 28.1.7
Słowik, K. PL 28.1.5
Slusarski, C. PL 28.2.5
Sly, J.M.A. GB 13.1.1
Smagula, J.M. US-A 1.1
Smal, J-P. BE 6.5
Smale, P.E. NZ 11.1
Smardas, C. GR 10.1
Smart, G.C. US-D 27.2
Smart, N.A. GB 13.1.2
Smart, R.E. NZ 5.1
Smeal, P.L. US-C 10.1
Smedegaard-Petersen, V. DK 3.5
Smeets, L. NL 9.6.5
Smekal, V. CS 15.1
Smetana, J. CS 23.1
Smets, B. FR 42.1
Smiech, M. PL 32.1.8
Smierzchalska, K. PL 28.2.12
Smillie, R.M. AU 65.4
Smit, C.J.B. US-D 16.1
Smith, A. GB 24.1.5
Smith, A. US-E 8.1
Smith, A.M. AU 9.1
Smith, B.D. GB 19.1.1

Smith, B.M. GB 33.1.6
Smith, C.B. US-B 13.1.1
Smith, C.E. CA 40.2
Smith, C.W. US-I 18.1
Smith, D. AU 23.1
Smith, D.A. US-D 38.1
Smith, D.L. GB 12.1
Smith, E.M. US-F 3.1
Smith, G.E. US-D 16.1
Smith, G.S. NZ 5.1
Smith, J.W. US-D 1.1
Smith, J.C. NZ 12.3
Smith, J.F. GB 18.1.1
Smith, J.H.F. GB 2.2
Smith, J.R. GB 5.1
Smith, L. US-G 2.1
Smith, L.A. US-D 38.1
Smith, L.G. AU 15.2
Smith, M. AU 2.1
Smith, M.A.L. US-F 12.2.2
Smith, M.W. US-H 3.1
Smith, N.G. AU 2.1
Smith, P.R. AU 27.1
Smith, R.A. GB 8.1.2
Smith*, R.B. CA 25.1
Smith, R.D. GB 16.1
Smith, S.J. GB 19.1.4
Smith, S.M. GB 8.1.3
Smith, T.A. GB 19.1.2
Smitley, D. US-F 14.1.2
Smits, N.J. CA 24.1
Smits, P.H. NL 1.1.5
Smittle, S. US-D 22.1
Smolarek, Cz. PL 28.1.10
Smolarz, K. PL 28.1.7
Smolarz, St. PL 28.1.2
Smole, J. YU 18.2.1
Smoot, J.J. US-D 34.1
Smyrl, T. CA 6.1
Snelgar, W.P. NZ 2.1.1
Snir, I. IL 1.2.2
Snoek, N.J. NL 6.1
Snook, S. AU 59.2
Snyder, J. US-E 15.1
Snyder, Jr., L.C. US-E 18.1
So, I.S. KR 10.3
Soares, M.F. BR 14.1
Soares, N.B. BR 20.1
Soares Campos, I. BR 5.1
Soares da Silva, G. BR 6.1
Soares da Silva, P.H. BR 7.1
Soares de Melo, A. BR 9.1
Soave, J. BR 20.1
Sobajima, Y. JP 22.4
Sobczykiewicz*, D. PL 28.1.2
Sobiczewska, M. PL 28.1.10
Sobiczewski, P. PL 28.1.2
Sobieralski, K. PL 20.2.11
Sobolewska, J. PL 28.1.9
Sobolewski, J. PL 28.2.7
Sobotkowski, Z. PL 32.1.11
Sobral, L.F. BR 11.1
Sobrinho, A.P. de C. BR 12.1
Soczek, U. PL 28.1.17
Soczek*, Z. PL 28.1, 28.1.5
Soderlund, D.M. US-B 2.1.3
Soderlund, R.O. AU 30.2
Soegaard, B. DK 3.3
Soehngen, U. CA 6.1
Soejima, J. JP 14.2
Soemori, H. JP 32.1
Soenarso, I.D. 1.6
Soerodimedjo, F.W. SR 1.1
Sofer, H. IL 1.1.2
Sogo, K. JP 17.1
Sohajda, I. HU 5.1.4
Sohi, B.S. IN 17.1
Sohi, H.S. IN 4.1, 17.1
Soichi, N. JP 17.1
Soing, P. FR 21.1
Sokołowska, A. PL 28.2.2
Sokołowska, G. PL 32.1.11
Sokolowski, G. PL 28.2.11
Solanki, S.S. IN 24.1.2, 25.1
Solecka, M. PL 32.1.3
Solheim, B. NO 9.2
Solomon, E. IL 1.2.3
Solomon, L.T. US-D 45.1
Solomon, M.G. GB 8.1.2
Solórzano, A. PE 3.1
Solomos, T. US-C 5.2
Soltész, M. HU 5.1.2, 5.1.4
Solvang, J. DK 7.1
Solzea, C. RO 15.1
Soma, M. JP 3.2
Soma, S. JP 12.2
Somapala, H. LK 3.13
Somasiri, L.L.W. LK 2.1
Somers, T.C. AU 37.1
Somlyay, G. HU 2.5
Sommer*, N.F. US-K 6.1.3
Somogyi, D. HU 5.1.3
Somogyi, Gy. HU 5.3.3
Son, D.S. KR 16.1.2
Søndergaard, P. NO 1.1
Søndergaard Nielsen, H. DK 7.1
Song, Y.J. KR 3.1
Song, Y.N. KR 5.2
Soni, S.L. IN 19.3
Sonku, Y. JP 7.3
Sonneveld, F. NL 9.11
Sonneveld, J. NL 16.1
Sonoda, R.M. US-D 26.1
Šonský, D. CS 24.1
Sontakke, M.B. IN 12.1
Soong, T.C. TW 11.1.2
Soost*, R.K. US-K 12.1.1
Sørensen, J.N. DK 1.1
Soria, J. CR 1.1
Soriano, J.M. PH 1.2, 1.2.4
Sorrenson, W.J. BR 28.1
Sorte, G. SE 1.2
Sosa, H. AR 7.1
Sosa Quiroga, R.D. AR 17.1
Šoškić, A. YU 1.1
Sossountzov, L. FR 30.1.1
Sotiropoulos, S. GR 18.1.4
Sotomayor, J. PE 2.1
Sotta, B. FR 30.1.1
Sottile, I. IT 13.1
Soudain, P. FR 38.1
Soukup, J. CS 24.1
Soule*, J. US-D 27.4
Souliotis, P. GR 14.1.2
Souq, F. FR 37.1
Sourlekov, P. BG 8.1
Sousa, M.A. de L.B. BR 22.1
Southerland, R. US-E 15.1
Southey, J.F. GB 13.1.1
Souto, G.F. BR 12.1
Souto, S. AR 3.1.1
Souty, M. FR 3.1.8
Souty, N. FR 3.1.7
Souvatzis, G. GR 1.1
Souza, A.F. BR 25.5
Souza, E.A.P. BR 6.1
Souza, E.L.S. BR 25.5
Souza, J.C. BR 15.1
Souza, L.F. BR 12.1
Souza, Z.S. BR 28.1
Souza do Nascimento, A. BR 12.1
Souza L. Veiga, A.F. BR 10.1
Souza Machado, V. CA 15.1
Souza Reis, A.C. BR 10.1
Sowokinos, J. US-G 11.1
Sowokinos, J.R. US-G 7.1
Soyer, J-P. FR 4.1.2
Soylu, A. TR 6.1
Soylu, M. DE 5.1.1
Sozzi, A IT 20.1
Špaček, B. CS 22.3
Spanu, IT 23.1
Sparace, S.A. CA 37.1
Sparacino, A.C. IT 10.1
Sparks, D. US-D 16.1
Sparks, D.L. US-C 2.1
Sparks, T. GB 8.1.1
Sparks, W.C. US-I 13.1
Sparnaaij, L.D. NL 9.6.3
Sparrow, S.D. US-L 1.1
Spasojević, M. YU 10.1
Spassov, S. BG 8.1

Specht, H.B. CA 41.1
Spencer, J.H. US-D 50.1
Spiegel-Roy*, P. IL 1.2.1
Spiers, A.G. NZ 8.3
Spiers, J. AU 65.4
Spiers, T.M. NZ 10.1
Spiertz, J.H.J. NL 9.1
Spikman, G. NL 9.7.1
Spikowska, B. PL 20.2.4
Spina, P. IT 1.1
Spirov, N. BG 16.1
Spirovska, R. YU 4.1
Splittstoesser, D.F. US-B 2.1.5
Splittstoesser, W.E. US-F 12.2.2
Spomer, L.A. US-F 12.2.2
Sponholz, W. DE 12.1.5
Spooner-Hart, R.N. AU 8.1
Spoor, G. GB 31.1
Spoorenberg, P.M. NL 6.1, 9.1
Sporns, P. CA 7.2
Sprenkel, R.K. US-D 37.1
Spriggs, T.W. NZ 8.1
Springer, J. US-B 14.1.6
Springett, B.P. NZ 9.8
Sproul, A.N. AU 59.1
Spryropoulos, G.S. GR 14.1.4
Spurling, M.B. AU 41.1
Sqalli, M. MA 2.1
Sree Ramulu, K. NL 9.5
Sridhar, T.S. IN 1.1.6
Srinivas, M. IN 1.1.3
Srinivasan, K. IN 1.1.7
Srinivasan, V.R. IN 1.1.10
Srinivasa Rao, N.K. IN 1.1.8
Sriram, T.A. IN 2.2
Srivastava, B.K. IN 20.1
Srivastava, H.C. IN 1.1.4
Srivastava*, K.C. IN 26.1
Srivastava, K.K. IN 24.1.5
Srivastava, R.B. IN 25.1
Srivastava, R.P. IN 25.1
Stace-Smith*, R. CA 4.1
Stadelmann, E. US-G 7.1
Städler, E. CH 1.1
Stadler, J.D. ZA 4.1
Stafford, J.V. GB 31.2.5
Stahl, E. IC 1.1
Stahlschmidt, P. DK 3.2
Stahly, E. US-J 10.2
Stähr, E. DE 12.1.9
Stainer, R. IT 11.1
Stajić, N. YU 1.1
Stall, W.M. US-D 27.7
Stallknecht, G.F. US-I 21.1
Stamatov, I. BG 8.2
Stambera, J. CS 14.1.1
Stamborowski, M. PL 28.5
Stamenković, S. YU 10.2
Stamenković, T. YU 8.1
Stamer, J.R. US-B 2.1.5
Stamopoulos, D. GR 25.1.6
Stamova, L. BG 8.1
Stan, G. RO 8.5
Stan, Gh. RO 9.1
Stan, I. RO 5.1
Stan, N. RO 23.1
Stan, S. RO 9.1
Stanciu, N. RO 9.1
Stancu, E. RO 7.1
Stancu, T. RO 8.2
Standardi, A. IT 15.1
Standifer, L.C. US-D 52.1
Staněk, J. CS 7.1
Stanek, R. PL 14.1.5
Staněk, Z. CS 27.1
Stanev, D. BG 4.1
Stanford, J.C. US-B 2.1.2
Stang*, E. US-F 23.1.3
Stanisavljević, M. YU 10.2
Stanković, S. YU 12.1
Stanley, C.D. US-D 24.1
Stanley, G. AU 65.4
Stanley, J. GB 24.1.4
Stanley, Jr., R.L. US-D 37.1
Stanley, W.L. LK 3.13
Stanojević, S. YU 14.1
Stapleton, R.R. GB 11.1
Starbuck, C.S. US-E 18.1
Starck, J.R. PL 32.1.4
Stárek, J. CS 22.3
Stark, R. CA 41.1
St-Arnaud, M. CA 31.1
Staropiętka, M. PL 9.1
Starrantino, A. IT 1.1
Starratt, A.M. CA 18.1
Starzyński, A. PL 32.1.7
Staskawics, B.J. US-K 3.1.2
Stathis, P. GR 6.1.7
Stathopoulos, D. GR 25.2.2
Staub, J. US-F 23.1.3
Stäubli, A. CH 2.1
Staunton, L. IE 4.1
Stautemas, E. BE 22.1
Stavast, F.M. NL 9.7.4
Stavrakas, D. GR 25.1.5
Stavraki, H.G. GR 14.1.2
Stavropoulos, N. GR 20.1
Stavropoulou, E. GR 9.1
Stayner, R.M. GB 31.2.1, 31.2.3
Stayrakakis, E. GR 6.1.2
Stead, D.E. GB 13.1.4
Stecki, Z. PL 10.1.1
Steenhuizen, G. US-K 5.1
Steenkamp, J. ZA 4.1
Steer, B.T. AU 65.6
Stefan, I. RO 13.1
Stefan, M. RO 13.1
Stefan, N. RO 8.3
Stefanek, W. PL 20.2.2
Stefanov, S. BG 8.1
Steffens, G. US-C 3.1
Steffensen, K. DK 9.1
Štefula, O. YU 18.1
Stegešek, C. YU 14.1
Stehr, T. AT 1.1
Stein, L. US-H 13.1
Steinberg, B. DE 12.1.1
Steinberger, J. AT 8.7
Steinbuch, E. NL 9.10
Steinbuch, F. NL 1.1.4
Steineck, O. AT 8.4
Steinegger, D.H. US-G 23.1
Steinhübel, G. CS 16.1.2
Steinitz, B. IL 1.1.2
Steinkraus, K.H. US-B 2.1.5
Steinrücken, A. DE 15.1.7
Steinsholt, K. NO 10.1.3
Stell, G. GB 13.1.3
Stelzig, D. US-C 9.1
Stene, J. NO 8.1
Stenersen, J. NO 10.4.3
Stengel, M. FR 6.1.3
Stengel, P. FR 3.1.7
Stenning*, B.C. GB 31.1
Stenseth, T. NO 10.4.3
Stephan, H. NL 1.1.6
Stephens, C. US-F 14.1.3
Stephens, J.G. AU 47.3
Stephens, J.M. US-D 27.7
Stephens, L.C. US-G 20.1
Stephens, R.J. GB 3.1
Stephenson, J.C. US-D 46.1
Stephenson, R.A. AU 23.1
Sterna, W. PL 20.2.1
Sterrett, S.B. US-C 13.1
Stetter, S. DK 4.1
Steubing, L. IC 1.1
Steur, G.G.L. NL 9.11.2
Steurer, M. AT 21.1
Stevanović, D. YU 13.1
Steven, D. NZ 2.2
Stevenson, A.B. CA 25.2
Stevenson, D.S. CA 3.1
Stevenson, W. US-F 23.1.4
Steward, I. US-D 30.2
Stewart*, K.A. CA 37.1
Steward, K.K. US-D 25.1
Stewart, R.K. CA 37.1
Stewart, T.M. NZ 9.1
Stewart, V.R. US-I 24.1

Stiborek, P. CS 27.1
Stiekema, W.J. NL 9.5
Stiles*, H.D. US-C 15.1
Stilinović, S. YU 8.2
Still, S.M. US-F 3.1
Stimart, D.P. US-C 5.2
Stirling, G.R. AU 15.4
Stirrat, A.M. GB 2.1
Stobbs, L.W. CA 25.2
Stock, C.C. US-J 18.1
Støen, M. NO 10.4.3
Stoenescu, A. RO 1.1.7
Stoeva, R. BG 11.1
Stoewsand, G.S. US-B 2.1.5
Stoffella*, P.J. US-D 26.1
Stoffert*, G. DE 15.1.2
Stoian, E. RO 8.2
Stoian, L. RO 3.1
Stoicesu, M. RO 5.1
Stoiljković, B. YU 11.1
Stoilov, G. BG 8.2
Stoilov, M. BG 13.1
Stojanović, B. YU 12.1
Stojanović, Lj. YU 8.1
Stojanović, M. YU 9.1, 14.1
Stojanovski, D. YU 10.1
Stojanowska, J. PL 32.1.4
Stojicević, D. YU 14.1
Stoker, R. NZ 13.1
Stoker, T. NL 9.7.5
Stoklasa, J. CS 24.3
Stol, W. NL 6.1
Stolberg, A. CR 2.3
Štolić, D.N. YU 16.1
Stolp, J. NL 9.11.2
Stoltz, L. US-E 15.1
Stoltz, R.L. US-I 17.1
Stone, D.A. GB 33.1.2
Stone, G.T. GB 31.1
Stoner, A.K. US-C 3.1
Stoppani, M.J. AR 19.1
Storchschnabel, G. AT 8.3
Storck*, H. DE 15.1.2
Storesund, B. NO 10.5.2
Storey*, J.B. US-H 6.1
Storey, R. AU 65.2
Story, R.N. US-D 52.1
Stösser*, R. DE 28.1.1
Stott, K.G. GB 19.1.3
Stouffer, R. US-D 22.1
Stovold, G.E. AU 9.1.1
Stow, J.R. GB 8.1.3
Stoyanov, S. BG 6.2
Stoylova, E. HU 5.3.4
Straatsma, J.P. NL 9.9
Strachan, G.E. CA 3.1
Straka, Fr. BG 6.1
Strang, J. US-E 15.1
Straub, R.W. US-B 4.1, 2.1.3
Străulea, M. RO 17.1
Straver, N.A. NL 1.1.1
Streclec, V. CS 8.1
Streif, J. DE 28.1.3
Strempfl, F. AT 11.1
Stretch, A. US-C 8.2
Stretch, A.W. US-B 14.1.6, 15.1
Streu, H.T. US-B 14.1.4
Strider, D.L. US-D 8.1.2
St"rkova, Y. BG 1.1
Strobel, G.A. US-I 18.1
Strojna, E. PL 28.1.10
Strojny, Z. PL 28.1.13
Strøm*, J.S. DK 1.2
Strom, S. US-B 14.1.1
Strømme*, E. NO 10.1.4
Stropek, M. PL 1.1
Stroński, A. PL 20.2.7
Štruklec, A. YU 18.2.1
Strullu, D.G. FR 19.2
Struppek, G. DE 3.1
Struve, D.K. US-F 3.1
Stryckers, J.M.T. BE 7.9
Strydom*, D.K. ZA 4.3.1
Strzeszewski, M. PL 32.3.6
Stuchliková, E. CS 2.2
Studman, C.J. NZ 9.4
Stumm, G. DE 2.1
Stushnoff, C. CA 11.1
Stutte, G.W. US-C 5.2
Stuurman, F.J. NL 9.11.2
Stylianides, D. GR 19.1
Stylopoulos, E. GR 16.1
Stynes, B.A. AU 59.1
Subba Reddy, K.V. IN 6.1
Subhas Chander, M. IN 1.1.8
Subijanto, ID 1.1
Subrahmanyam, K.V. IN 1.1.10
Subramaniam, M. IN 5.3
Subramanian, T.R. IN 1.1.9
Subramanya*, R. US-D 23.1
Subramanyam, M.D. IN 1.1.1
Suchara, Z. CS 24.1
Suchorska, K. PL 32.1.2.2
Suciu, I. RO 35.1
Suckling, D.M. NZ 13.4
Sud, G. IN 10.3
Suda, T. JP 8.1
Suekane, T. JP 17.1
Suely Santos, D. BR 30.3
Suenaga, Y. JP 21.1
Suesawa, K. JP 17.1
Suetsugu, N. JP 34.2
Suett, D.L. GB 33.1.3
Sugahara, S. JP 1.1.1
Sugata, S. JP 41.1
Sugawara, K. JP 16.1
Sugio, M. JP 25.1.2
Sugiura*, A. JP 22.5
Sugiyama, K. JP 38.2
Sugiyama, N. JP 41.5
Sugiyama, S. JP 3.3.1
Suh, H.D. KR 23.1.2
Suh, H.S. KR 23.1.3
Suh, J.K. KR 2.1.1
Suh, S.K. KR 9.1
Suh, Y.G. KR 7.2
Suhardi, ID 1.3
Suković, I. YU 14.2
Sukumaran, N.P. IN 4.1
Sulikhere, G.S. IN 11.4
Sulladmath, U.V. IN 11.1, 11.2, 11.3
Sulladmath, V.V. IN 11.6
Sulladmuth, V.V. IN 1.1.1
Sullivan, G.H. US-F 9.1.4
Sullivan, J.G. US-F 12.2.2
Sullivan, W.T. US-D 6.1
Sulunkhe, D.K. IN 14.1
Sulyras, F.N.E. BR 12.1
Sulz Gonsalves, R. BR 25.1
Suman, C.L. IN 26.1
Sumi, H. JP 34.3
Sumi, T. JP 7.1
Sumithraarachchi, B. LK 3.4
Sumitomo, A. JP 40.1
Summer, D.R. US-D 22.1
Summers, C.G. US-K 3.1.1
Summers, W.L. US-G 20.1
Sumption, L.J. US-G 25.1
Sun, M.K. US-I 18.1
Sunagawa, S. JP 32.2
Sundaralingam, S. LK 3.11
Sunderland, K.D. GB 18.1.2
Sundheim, L. NO 10.4.1
Sunding, P. NO 7.3
Sundstrom, F.J. US-D 52.1
Sung, Z.R. US-K 3.1.2
Suplicy Filho, N. BR 21.2
Surányi, D. HU 2.1.1
Suresh, E.R. IN 1.1.5
Surmeli, N. TR 14.1
Suryanarayana, V. IN 6.1
Šušić, S. YU 11.1
Suska, J. PL 28.1.10
Suski, Z. PL 32.1.12
Suski, Z.W. PL 28.1.2
Susnovski, M. IL 1.1.1
Suso, L. ES 11.1
Sussman, M. US-F 23.1.3
Susuri, Lj. YU 16.1
Suszka, B. PL 10.1.2
Suta, A. RO 9.1
Suta, E. RO 1.1.1

Suta, V. RO 9.1
Sutapraja, H. ID 1.4
Sutcu, A.R. TR 14.1
Sutherland, J.P. GB 1.3
Sutherland, R.A. GB 33.1.7
Suto, K. JP 23.2.3
Sutter, E. US-K 6.1.3
Sutton, D.L. US-D 25.1
Sutton, M.W. GB 1.1
Sutton, R.K. US-G 23.1
Sutton, T.B. US-D 8.1.2
Sutton, W.G. GB 9.1.1
Suwa, M. JP 18.2
Suzaki, S. JP 1.1.1
Suzuki, A. BR 28.1
Suzuki, A. JP 8.2
Suzuki, C. JP 3.2.2
Suzuki, H. JP 2.2, 4.1, 38.2, 45.1, 45.2
Suzuki, K. JP 14.2, 38.1, 38.3
Suzuki, M. JP 14.1, 19.4
Suzuki, N. JP 3.2.3, 23.2.5, 33.3
Suzuki, S. JP 1.2, 9.3, 22.3, 24.1, 41.3
Suzuki, T. JP 1.1.1, 38.1, 38.2, 38.4, 41.3
Suzuki*, Y. JP 14.6
Suzuki, Y. JP 45.1, 46.2
Sveatchevici, M. RO 1.1.1
Svedelius*, G. SE 1.4
Švejcar, V. CS 14.1.5
Svejda*, F. CA 20.2
Svendsen, A. DK 9.1
Svensson*, R. SE 1.1.3
Swai, F. TZ 4.1
Swai, I.S. TZ 3.1
Swai, R.E.A. TZ 3.1
Swain, D.J. NZ 10.1
Swain, F.G. AU 8.1
Swain, V.A. AU 57.1
Swait, A.A.J. GB 8.1.2
Swamy, K.R.M. IN 1.1.2
Swanson, B.T. US-G 7.1
Swanton, C.J. CA 21.1
Sward, R. AU 27.1
Swart, P.L. ZA 4.1
Swartz, H.J. US-C 5.2
Swarup, B. IN 22.1
Swasey, J.E. US-C 2.1
Sweatt, M. US-H 6.1
Swiader, J.M. US-F 12.2.2, 13.1
Swiątek, G. PL 32.1.9
Swietlik, J. PL 32.3.3
Swift, F.C. US-B 14.1.4
Syamal, M.M. IN 27.1
Sydnor, T.D. US-F 3.1
Syers, J.K. NZ 9.5
Sykes, C.M. AU 65.2
Sykes, S.R. AU 65.2
Sylvester, E.S. US-K 3.1.1
Sylvestre, P. CA 30.2
Synnes, O.M. NO 10.4.2
Synnevaag, G. NO 2.1
Syrrist, S. NO 10.1.3
Sysiak, J. PL 32.3.3
Sytsema*, W. NL 1.1.3
Syvertsen, J.P. US-D 30.2
Syverud, T.D. US-F 21.1
Szabadkai, A. HU 2.2.9
Szabó, G. HU 2.2.9, 5.2
Szabó, Gy. HU 2.6
Szabó, I. HU 2.2.12
Szabó, M. HU 2.1.2
Szabó, Ö. HU 5.1.4
Szabó, S. HU 5.1.4
Szabó, T. HU 2.1.4
Szabó-Murányi, I. HU 5.2
Szafirowska, A. PL 28.2.2
Szalai, J. HU 2.2.6, 5.1.4
Szalay, F. HU 5.3.4
Szalka, A. HU 2.6
Szaloczy, B. HU 2.7
Szántó, M. HU 2.2.4
Szarka, I. HU 5.3.2
Szász-Kozma, I. HU 2.2.4
Szczepanik, Z. PL 5.1
Szczepańska, K. PL 20.1.5
Szczepański*, K. PL 28.1.10
Szczotka, Z. PL 10.1.4
Szczurek, J. PL 34.1
Szczygieł*, A. PL 3.1
Szczygieł, K. PL 12.1
Szczypinska, W. PL 6.1
Szczypinski, L. PL 6.1
Szegedi, E. HU 5.2
Szemiel, M. PL 32.3.5
Szenci, Gy. HU 2.1.2
Szenes, M. HU 2.3
Szenteleki, K. HU 2.2.9
Szentivanyi, P. HU 2.1.2
Szentkiralyi, F. HU 2.5
Szép, K. HU 2.2.9
Szepes, L. HU 4.2
Szepesy, K. HU 5.3.3
Szeto, S.Y.S. CA 4.1
Szewczuk, A. PL 33.1
Szidarovszky, F. HU 2.2.9
Szijjártó, A. HU 5.3.1
Szilágyi, I. HU 2.3
Szilágyi, K. HU 2.1.3
Sziráki, Gy. HU 2.1.2
Szita, I. HU 5.1.2
Szklanowska, K. PL 14.1.6
Szklarska, J. PL 28.2.11
Szlachetka*, W. PL 32.1.3
Szmal, B. PL 20.2.7
Szmidt, A. PL 10.1.1
Szmidt, B. PL 28.2.6
Szmidt, R.A.K. GB 2.2
Szőke, L. HU 5.2
Szopa, K. PL 32.1.9
Szpunar, J. PL 28.1.7
Sztenjberg, A. IL 2.6
Szücs, E. HU 2.1.2
Szudyga, K. PL 28.5
Szufa, A. PL 28.1.2
Szumańska, G. PL 32.3.3
Szumański, M. PL 32.1.11
Szundy, I. HU 5.3.1
Szwedo, J. PL 14.1.1
Szwejda, J. PL 28.2.7
Szwonek, E. PL 28.2.6
Szymańska, M. PL 14.1.5
Szymanski, J. PL 28.5
Szyndel, M. PL 32.1.6

Tabata, K. JP 18.1
Taber, H.G. US-G 20.1
Tabi, M. CA 36.1
Taborda, R. AR 7.1
Tachibana, H. JP 20.3
Tachibana, S. JP 4.3, 23.3
Tachibana, Y. JP 5.2
Taczanowska, E. PL 28.1.18
Taczanowski, K. PL 28.1.13
Taddesse, G. ET 6.1
Tadesse, A.F. ET 10.2
Taguchi, S. JP 39.1
Taguchi, T. JP 2.2
Tagwira, F. ZI 3.1.2
Tahkur, S. IN 8.1
Tahon, S. BE 6.4
Tai, G.C.C. CA 40.3
Tai, W.C. TW 11.1.1
Taimatsu, T. JP 28.1
Taira, S. JP 45.2
Tait, M.E. GB 27.1
Tajthy, S. HU 5.1.1
Takács, M. HU 2.2.11
Takada, K. JP 16.3
Takada, N. JP 13.1
Takaesu, K. JP 32.1
Takagi, A. JP 9.2
Takagi, H. JP 45.2
Takagi, K. JP 14.2
Takagi, N. JP 5.2, 31.1
Takagi, T. JP 38.4
Takaguchi, M. JP 43.1
Takahara, T. JP 27.3
Takahashi, A. JP 38.2
Takahashi, B. JP 19.4

Takahashi, E. JP 4.4
Takahashi, F. JP 12.1
Takahashi, H. JP 2.3, 18.1, 25.1.2
Takahashi, K. JP 6.1, 35.2, 37.1, 38.4
Takahashi, M. JP 19.2, 26.4
Takahashi, R. JP 16.1
Takahashi, S. JP 10.1, 24.1, 24.2
Takahashi, T. JP 16.1, 26.5, 35.2
Takahashi, U. JP 21.2
Takahashi, Y. JP 16.1, 38.2, 45.1
Takahasi, S. JP 2.2
Takahasi, Y. JP 2.2
Takai, T. JP 2.3
Takakuwa, M. JP 12.2
Takala, M. FI 4.1
Takamura, N. JP 8.2
Takanashi, K. JP 14.2
Takano, K. JP 31.1, 39.1
Takano, S. JP 28.1
Takano, T. JP 1.2, 6.2
Takao, M. JP 7.1
Takao, Y. JP 41.1
Takase, N. JP 1.1.1
Takase, S. JP 1.1.1
Takatsu, A. BR 25.4
Takatsu, I. JP 14.1
Takatsuji, T. JP 27.2
Takaya, S. JP 11.3
Takayama, S. JP 47.1
Takayanagi, K. JP 23.2.2
Takeda*, F. US-C 8.2
Takeda, K. JP 26.4
Takeda, K.Y. US-M 2.2
Takeda, M. JP 35.2
Takeda, T. JP 22.3, 27.1
Takeda*, Y. JP 22.5
Takegawa, M. JP 13.2
Takei, A. JP 1.1.1
Takei, K. JP 47.2
Takeichi, Y. JP 4.2
Takematsu, T. JP 39.2
Takemura, T. JP 3.3.2
Takenaga, S. JP 41.4
Takeno, K. JP 24.3
Takesako, H. JP 41.1
Takeuchi, K. JP 25.1.3
Takeuchi, S. JP 23.2.4
Takeuchi, T. JP 38.1
Takeuchi, Y. JP 39.2, 42.3
Takewaka, A. JP 22.1
Takezaki, C. JP 18.1
Takezawa, H. JP 19.1
Takezawa, I. JP 6.2
Takiguchi, Y. JP 13.2
Takihiro, N. JP 11.1
Takim Mascarenhas, M.H. BR 15.1
Takishita, F. JP 16.2
Takizawa, M. JP 41.1
Taksdal*, G. NO 4.1
Takubo, Y. JP 34.2
Talamo, J.C.D. GB 31.2.1
Talay, A. TR 2.2
Talbert, R.E. US-E 11.1
Taljati, E. YU 15.1
Tamada, A. JP 29.1
Tamada, T. JP 3.3.1
Tamaki, E. JP 32.1
Tamaki, I. JP 17.2
Tamás-Nyiri, J. HU 2.2.12
Tamassy, I. HU 2.2.6
Tamayo, P.J. CO 4.1
Tamponi, G.C. IT 19.1
Tamsett, J. GB 8.1.3
Tamura, F. JP 31.1
Tamura, J. JP 33.2
Tamura, M. JP 44.1
Tamura, O. JP 12.2
Tamura, Y. JP 39.1
Tan*, C.S. CA 17.1
Tana, K. JP 32.1
Tanabe*, K. JP 42.3
Tanabe, K. JP 1.1.6, 6.2, 17.1
Tanabe, S. JP 11.1
Tanada, Y. US-K 3.1.1
Tanahashi, K. JP 9.1
Tanaka*, H. JP 41.2
Tanaka*, M. JP 17.3
Tanaka, A. JP 4.3, 11.1, 27.3, 42.1
Tanaka, D. JP 22.1
Tanaka, F. JP 31.1
Tanaka, H. JP 14.2, 46.1
Tanaka, J.S. US-M 2.2
Tanaka, K. JP 4.3, 23.1, 23.2.4, 23.2.5
Tanaka, M. JP 1.1.3, 12.3, 21.1, 34.1, 46.1
Tanaka, R. JP 37.2
Tanaka, S. JP 7.2
Tanaka, T. JP 6.2, 21.3, 25.2, 29.1, 34.1, 39.1, 45.1
Tanaka, Y. JP 1.2, 3.2.3, 7.1, 14.3
Tănănescu, C. RO 1.1.1
Tănăsescu, M. RO 1.1.1
Tănase, I. RO 8.5
Tanasijević, N. YU 14.1
Tandon, D.K. IN 26.1
Tandon, P.L. IN 1.1.7
Tanev, I. BG 4.1
Tangeraas, H. NO 10.3
Tanguy, J.-L. FR 23.1
Tani, K. JP 15.1
Tanigawa, K. JP 28.1
Taniguchi, T. JP 13.2, 42.2
Taniguchi, Y. JP 11.1
Taniuchi, J. JP 16.2
Tan Jun, F.J. AR 22.1
Tanksley, S.D. US-B 5.1.7
Tanner, H. CH 1.1
Tanno, S. JP 2.2
Tanriverdi, F. TR 9.1
Tantau, H.-J. DE 15.1.7
Tantó, B. HU 5.2
Tanyongana, R. ZI 3.1.2
Tao, C.C. TW 12.1
Tappen, W.B. US-D 37.1
Tapsell, C.R. GB 5.1, 33.1.6
Tarantino, E. IT 2.3
Tarjan, A.C. US-D 27.2
Tarjányi, F. HU 2.2.12
Tarkowian, S. PL 18.1
Tarn, T.R. CA 40.3
Tartier, L. CA 34.1
Tartier, L.M. CA 29.1
Tască, Gh. RO 8.4.1
Tashiro, N. JP 34.3
Tashiro, Y. JP 34.4
Tassara, M.A. AR 10.1
Tassinari, H.H. BR 28.1
Tassiopoulos, E. GR 5.1
Taszek, U. PL 10.1.4
Tatár, S. HU 4.2
Tatara, A. JP 38.2
Tătaru, D. RO 8.4.4
Tatarynowicz, B. PL 20.2.6
Tate, R.L. US-B 14.1.1
Tateishi, S. JP 26.5
Tatlioglu, T. DE 15.1.1
Tatsuda, Y. JP 18.2
Tatsumi, Y. JP 28.1, 38.4
Tattini, M. IT 9.2
Tau, J.L. AR 17.1
Tauber, M.J. US-B 5.1.6
Taveirne, W. BE 15.1
Taven, R.E. US-E 18.1
Tavert, G. FR 4.1.6
Tawade, A.B. IN 15.1
Tayama, H.K. US-F 3.1
Taylor, A.G. US-B 2.1.2
Taylor, B.B. US-H 27.1
Taylor*, B.K. AU 31.1
Taylor, C.E. GB 15.1
Taylor, C.M. GB 2.1
Taylor, D.W. CA 21.1

Taylor, H. GB 15.1.4
Taylor, I.B. GB 25.1.2
Taylor, J.L. US-F 14.1.1
Taylor, J.C. GB 31.1
Taylor, J.D. GB 33.1.4
Taylor, O.A. NG 1.1
Taylor, O.C. US-K 12.1.1
Taylor, R.L. US-L 2.1
Taylor, R.W. US-C 2.1
Taylor, R.L. CA 4.2
Taylor, R.M. US-H 8.1
Tchingov, S. BG 11.1
Teel, M.R. US-C 2.1
Teem, D.H. US-D 38.1
Teferedeg, T. ET 2.1
Tefertiller, K.R. US-D 27.1
Tehrani*, G. CA 25.1
Teixeira, J.B. BR 25.3
Teixeira, L.O.A. BR 1.1
Teixeira Liberal, M. BR 19.1
Teixeira Neto, R.O. BR 20.2
Teixeira Vargas, A.A. BR 18.1
Tekin, H. TR 11.1
Telbisz, M. HU 2.7
Telneset, S. NO 10.3
Tembwe, D.N. ML 1.1
Temiz, K. TR 15.2
Tempel, A. NL 9.4.2
Temperli, A. CH 1.1
Temple-Smith*, M.G. AU 47.1
Templeton, M.D. NZ 2.3.3
Ten Cate, J.A. NL 9.1
Ten Cate, J.A.M. NL 9.11.2
Tendille, C. FR 17.1.5
Tenente, R.C.V. BR 25.3
Teng, T.C. TW 2.1.1
Teng, Y.H. TW 2.1.2
Teng, Y.T. TW 12.1
Tennakoon, A. LK 2.1
Tennier, M. CA 26.2
Teodorescu, A. RO 8.4
Teodorescu, G. RO 5.1, 9.1
Teodorescu, V. RO 1.1.3
Tepfer, D. FR 17.1.4
Tepordei, R. RO 1.1.5
Teppaz-Misson, C. FR 28.1
Terán, A. AR 22.2
Terabayashi, S. JP 22.4
Terabun, M. JP 13.3
Terai, H. JP 13.3
Terai, O. JP 27.2
Terai, Y. JP 47.2
Terakado, K. JP 41.1
Terauds, A. AU 46.4
Terbe, I. HU 2.2.12
Terblanche, J.H. ZA 1.1
Teri, F.J. TZ 2.1
Terpó, A. HU 2.2.7
Terra, M.M. BR 20.1
Terra, P. dos S. BR 14.1
Terra Wetzel, C. BR 25.3
Terranova, G. IT 1.1
Terry, N. US-K 3.1.3
Ter Steege, P. NL 6.1
Tertecel, M. RO 8.1
Terzaghi, B.E. NZ 8.1
Tesař, J. CS 1.1
Tesi*, R. IT 17.2
Tešović, Z. YU 10.2
Testes Graziano, T. BR 23.1
Testolin*, R. IT 25.1
Tétényi, P. HU 1.1
Tetik, D. TR 15.3
Tetileanu, C. RO 31.1
Tetileanu, T. RO 31.1
Tette, J.P. US-B 2.2
Tetzlaff, G. DE 15.5
Te Vrede, H.I. SR 1.1
Tewari, G.C. IN 1.1.7
Tewari, G.N. IN 21.1
Tewari, J.C. IN 24.1.3
Tewari, J.D. IN 24.1.7
Tewari, J.P. IN 20.1, 24.1.1
Tewari, L.D. IN 2.1.4
Tewari, R.S. IN 20.1
Tezuka, T. JP 1.3
Thakur, G.C. IN 10.2
Thakur, S.S. IN 10.2
Thal, J. DE 23.1
Thalassinos, S. GR 17.1
Thankappan, M. IN 5.1
Tharmarajah, S.K. LK 3.6
't Hart, M.J. NL 1.1.6
Thavabalachandran, M. LK 3.17
Thayer, D.W. US-B 12.1
Theerakul, A. TH 13.1
Theiler*, R. CH 1.1
Theilmann, D.A. CA 4.1
Theocharis, I. GR 22.1.1
Thériault, R. CA 35.1
Therios, J. GR 25.1.1
Theunissen, J.A.B.M. NL 9.4.1
Thèvenot, C. FR 38.1
Thiam, A. SN 1.1
Thibault*, B. FR 1.1.1
Thibodeau, P.O. CA 33.1
Thicoïpe, J.P. FR 21.1
Thiesz, N. RO 29.1
Thiesz, R. RO 29.1
Thimma Raju, K.R. IN 11.2
Thirukumaran, C. LK 3.3
Thirukumaran, P. LK 3.3
Thistlethwayte, B. AU 41.1
Thistlewood, H.M.A. CA 25.2
Thoen, C.A. NL 17.1
Thomas, B. GB 18.1.3
Thomas, B.J. GB 18.1.2
Thomas, C.E. US-D 12.2
Thomas, C.J.R. GB 33.1.1
Thomas, C.M.S. GB 8.1.1
Thomas, D.S. GB 19.1.4
Thomas, F. BE 14.1
Thomas, G.C. GB 8.1.1
Thomas, G.G. GB 35.2
Thomas, J. AU 15.4
Thomas, M.B. NZ 13.5
Thomas, R.G. NZ 9.8
Thomas, T.H. GB 33.1.5
Thomas, W.C. US-L 1.1
Thomas, W.B. NZ 13.4
Thomazelli, L.F. BR 28.1
Thomé, V.M. BR 28.1
Thompson, A.R. GB 33.1.3
Thompson, C. CA 42.1
Thompson, C. US-I 4.1
Thompson, D. CA 2.1
Thompson, D. US-H 3.1
Thompson, G.J. ZA 3.1
Thompson, J.M. US-D 19.1
Thompson, L.S. CA 43.1
Thompson, N.P. US-D 27.3
Thompson, P. US-D 50.1
Thompson, R. GB 15.1.4
Thompson, R. US-G 5.1
Thompson, T. US-H 16.1
Thompson, W. AU 30.1
Thomsen, A. DK 4.1
Thomson, G. AU 30.1
Thomson, S.V. US-I 5.1
Thomson, W.W. US-K 12.1.1
Thongprab, C. TH 7.1
Thonke, K.E. DK 5.1
Thornton, I.R. AU 31.1
Thornton, M.K. US-I 1.1
Thorpe, E. GB 32.1
Thorup, S. DK 7.2
Thresh, J.M. GB 8.1.2
Throne, R.F. US-K 5.1
Thuesen, A. DK 1.1
Thwaite, W.G. AU 3.1, 7.1
Thygesen, A.-G. NO 10.1.8
Tibbitts*, T.W. US-F 23.1.3
Tiburcio, A. BR 10.1
Tiemann*, K-H. DE 17.1
Tien Liao, F. BR 6.1
Tiernan, P. IE 3.1
Tiessen, H. CA 15.1
Tigchelaar, E.C. US-F 9.1.4
Tigerstedt, P.M.A. FI 10.1.2
Tijskens, L.M.M. NL 9.10
Tikoo, S.K. IN 1.1.2
Tilakasiri, M.A. LK 2.1
Tilbury, L.G. NZ 10.1
Till, M.R. AU 36.3

Tillman, R.W. NZ 9.5
Tilt, K. US-E 3.1
Tim, S. K-M. US-B 6.1
Timberlake, C.F. GB 19.1.4
Timbrook, S.L. US-K 16.1
Timm, G. DE 12.1.13
Timm*, H.H. US-K 6.1.4
Timmer, L.W. US-D 30.2
Timofte, V. RO 3.1
Tindale, C.R. AU 65.4
Tindall*, H.D. GB 31.1
Ting, I.P. US-K 12.1.1
Ting, J.P. US-K 12.1.1
Ting, S.V. US-D 0.2
Tinga, J.H. US-D 16.1
Tiodorović, J. YU 7.1
Tipton, J. US-H 8.1
Tirilly, Y. FR 23.1
Tiscornia, J.R. BR 30.1
Tisma, M. YU 9.1
Tisne, D. FR 15.1.1
Tisserat, B.H. US-K 11.1
Titman, R.D. CA 37.1
Titulaer, H.H.H. NL 6.1
Titus, J.S. US-F 12.2.2
Tivoli, B. FR 14.1.2
Tiwari, R.P. IN 1.1.6
Tiwari, S. US-D 51.1
Tizio Mayer, C.R. AR 15.1
Tjeertes, P. NL 14.2
Tjia, B.O. US-D 27.5
Tnani, T. TN 1.1
Tobback, P. BE 9.1
Tóbiás, I. HU 5.3.2
Tobiášek, J. CS 22.1
Tobutt, K.R. GB 8.1.2
Tocchini, R.P. BR 20.2
Tochigi, H. JP 39.1
Toda, M. JP 38.1
Todd, S.F. GB 19.1.4
Toderi, G. IT 3.2
Tododrov, Y. BG 8.1
Todorović, C. YU 14.1
Todorović, M. YU 14.1
Todorović, N. YU 14.1
Todorović, R. YU 9.1
Togânel, F. RO 25.1
Tognoni*, F. IT 17.2
Togo, H. JP 18.1
Togura, H. JP 4.3
Tojyo, Y. JP 26.2
Tokarski, M. PL 33.3
Tokeshi, G. JP 32.1
Toki, T. JP 4.1
Tokieda, S. JP 13.1
Tokimoto, T. JP 46.2
Tokita, F. JP 26.5
Tokito, T. JP 18.2
Tokokura, T. BR 6.1
Tokoro, S. JP 4.1
Tokudome, H. JP 18.2
Tokuhashi, S. JP 20.3
Tokui, H. JP 17.1
Tokumasu*, S. JP 5.3
Tokuyama, H. JP 42.2
Tokuyasu, M. JP 34.1
Tolman, J.H. CA 18.1
Toma, R. RO 8.4.4
Tomala, K. PL 32.1.1
Tomana, T. JP 22.5
Tomar, C.S. IN 10.2
Tomar, R.P. IN 25.1
Tomas, D. YU 3.1
Tomašivc, A. YU 3.3
Tombesi, A. IT 15.1
Tomcsányi, HU 2.2.9
Tomcsányi, E. HU 5.1.4
Tomcsányi, P. HU 2.7
Tomczyk, A. PL 32.1.7
Tomer, E. IL 1.2.5
Tomescu, A. RO 1.1.7
Tomescu, I. RO 31.1
Tomescu, M. RO 3.1
Tomić, Z. YU 8.2
Tomin, A. YU 14.1
Tominaga, S. JP 18.3
Tomioka, K. JP 4.1
Tomita, E. JP 44.2
Tomita, H. JP 35.2
Tomita, K. JP 6.1
Tomita, T. JP 6.1
Tomitaka, Y. JP 41.3
Tomizawa, T. JP 3.3.2
Tomlin, A.D. CA 18.1
Tomlinson, J.A. GB 33.1.4
Tomlinson, K. US-K 5.1
Tomoskozi, M. HU 2.2.7
Tompa, B. HU 2.2.10
Tompkins, A.R. NZ 5.1
Tompsett, A.A. GB 5.1
Toms, B. CA 42.1
Tomson, R.S. US-B 6.1
Tone, S. JP 46.1
Tonecki, J. PL 32.1.3
Tongova, E. BG 8.1
Tonietto, J. BR 31.1
Tonini, G. IT 3.3
Tonini, M. IT 9.3
Toniolo, L. IT 12.2
Tonneijck, A.E.G. NL 9.4.5
Tono, T. JP 34.4
Tonosaki, T. JP 3.2.1
Tooby, T.E. GB 13.1.3
Toohill, B.L. AU 55.1
Toop*, E.W. CA 7.1
Topalov, V. BG 4.1
Topchiiska, M. BG 8.2
Toplak, M. YU 3.1
Topoleski, L.D. US-B 5.1.3
Topor, E. RO 18.1
Topp, B.L. AU 25.1
Töpperwein, H. DE 15.1.8
Torello, W.A US-A 5.1.1
Torne, C.A. AR 22.1
Tornéus, C. SE 1.4
Toro, E. US-N 1.2
Török, Sz. HU 2.2.1
Torregrossa, J-P. FR 3.1.9
Torres, A.C. BR 25.5
Torres, J.A. CR 1.1
Torres, K.C. US-D 52.1
Torres, L.M. AR 16.1
Torres, R.E. CO 4.1
Torres Brioso, P.S. BR 19.2
Torres Filho, J. BR 8.1
Torres Pacheco, I. MX 7.4
Torres Soares, R. BR 9.1
Torroba, C. AR 19.1
Toscani, H.A. AR 9.1
Tosi, T. IT 26.1
Tošić, M. YU 14.1
Tosoni, D. AR 1.1
Tosun, F. TR 21.1
Totemeier, C.A. US-B 1.1
Tóth, A. HU 2.2.3, 5.1.4
Tóth, Cs. HU 2.2.4
Tóth, E. HU 2.1.1
Toth, J. FR 3.2.1
Toth, J.T. AU 8.2
Tóth, M.G. HU 2.2.2
Tóth, T. CS 8.1
Tóth-Farkas, E. HU 2.2.3
Toth-Surányi, K. HU 2.2.11
Totsuka, M. JP 41.1
Totth, G. HU 2.2.9
Toul, L. CS 20.1
Touron, E. AR 4.1.4
Toussaint, A. BE 6.8.2
Towill, L.E. US-I 2.2
Townsend, A. US-C 4.1
Townsend, C. US-H 6.1
Townsend, H. US-E 19.1
Townsend, R. GB 24.1.4
Townshend, J.L. CA 25.2
Townsley, R.J. NZ 9.2
Toyama, A. JP 42.2
Toyama, M. JP 42.3
Toyama, Y. JP 8.1
Toyokawa, S. JP 3.3.2
Toyotomi, Y. JP 23.1
Tozani, R. BR 19.2
Trachev, G. BG 8.2
Tracy, T. US-F 23.1.1
Tracz, A. PL 17.1

Tracz, J. PL 28.1.14
Trajkovski, K. SE 1.1.1
Trajkovski, V. SE 1.1.1
Tramel, J.L. US-C 15.1
Tramier, R. FR 2.1.3
Trancik, R.T. US-B 5.1.1
Trandaf, F. RO 1.1.7
Trandafirescu, M. RO 18.1
Tran Thanh Van, K. FR 30.1.1
Traquair, J.A. CA 17.1
Travaglini, D.A. BR 20.2
Treble, C.R. GB 5.1
Treder, W. PL 28.1.5
Treece, R.E. US-F 3.1, 7.1.3
Trefois, R. BE 6.1
Treharne, K.J. GB 19.1
Tremaine, J.H. CA 4.1
Tremblay, R. CA 40.1
Tremolières, A. FR 30.1.2
Treshow, M. US-I 7.1
Treur, A. NL 9.8
Treverrow, N.L. AU 11.1
Tribulato, E. IT 5.1
Trigo, I. ES 11.1
Trigui, A. TN 1.1
Trimble, R.J.M. CA 25.2
Trimeri-Makri, E. GR 9.1
Trimmer, W.A. AU 5.1
Trinklein, D.N. US-E 18.1
Tripathi, G.M. IN 24.1.4
Tripathi, N. IN 22.1
Tripathi, S.P. IN 24.1.1
Tripepi, R. US-I 14.1
Trique, B. FR 23.1
Trivedi, D.P. IN 22.1
Trivedi, R.K. IN 22.1
Trivedi, S.K. IN 19.7
Trninić, V. YU 2.1
Trnka, J. CS 2.2
Trochoulias*, T. AU 1.1
Trocmé, O. CR 2.2
Tromp, J. NL 10.1
Tronićková, E. CS 22.4
Trop, M. IL 4.1
Trottin-Caudal, Y. FR 21.1
Trought, M.C.T. NZ 13.1
Trouslot, M.F. FR 28.1
Trowbridge, P.J. US-B 5.1.1
Trudel, M.J. CA 38.1
Trudelle, M. CA 35.1
Trudgill, D.L. GB 15.1.1
Truel, P. FR 10.1.3
Truol, G. AR 7.1
Trushechkin, V.G. SU 13.1
Tryon, A. US-A 6.1
Trzak, S. PL 28.3
Tsai, J.H. US-D 25.1
Tsai, P.L. TW 9.1
Tsai, Y.S. TW 3.1
Tsalikidis, I. GR 25.1.3
Tsampanakis*, J. GR 12.1
Tsantili, H. GR 6.1.1
Tsao, S.J. TW 7.3
Tsay, T.J. TW 11.1.1
Tsekleev, G. BG 8.1
Tsialis, D. GR 25.2.2
Tsiantos, I. GR 26.1
Tsikalas, P. GR 10.1
Tsipoulidis, C. GR 19.1
Tsiracoglou, V. GR 25.1.1
Tsiropoulos, G. GR 3.1.1
Tsirtsis, E. GR 17.1
Tsitsipis, J. GR 3.1.1
Tsoukalas, M. GR 18.1.4
Tsoutsouras, E. GR 18.1.4
Tsuboi, A. JP 31.1
Tsuboi, I. JP 31.1
Tsuchiya, K. JP 23.2.4
Tsuchiya, S. JP 14.2, 15.3
Tsuchiya, T. JP 47.2
Tsuchizawa, M. JP 39.1
Tsuda, K. JP 1.1.1, 9.1
Tsuda, T. JP 24.3
Tsugawa, H. JP 3.3.2
Tsuji, H. JP 33.1
Tsuji, M. JP 40.2
Tsujii, A. JP 28.1
Tsujita, M.J. CA 15.1
Tsukada, M. JP 26.4
Tsukada, S. JP 23.2.1
Tsukada, T. JP 26.4
Tsukahara, K. JP 26.2
Tsukishima, H. JP 18.1
Tsumura, N. JP 14.3
Tsunekane, K. JP 17.1
Tsuneta, M. BR 27.2
Tsurumi, Y. JP 39.1
Tsuruta, T. JP 47.2
Tsutsui, K. JP 12.4
Tsuyama, K. JP 17.1
Tsuyuki, Y. JP 41.2
Tsuzaki, Y. JP 17.1
Tsvetkov, R. BG 4.1
Tu, C.C. TW 8.1
Tu, C.M. CA 18.1
Tu, J.C. CA 17.1
Tu, T.I. TW 11.1.2
Tuchimochi, T. JP 18.2
Tucholska, H. PL 20.2.10
Tucker, D.P.H. US-D 30.2
Tucker, G. GB 25.1.3
Tucker, G.G. GB 9.1.3
Tucknott, O.G. GB 19.1.4
Tucović, YU 8.2
Tudor, M. RO 1.1.1
Tudor, T. RO 8.1
Tudżakov, T. YU 4.3
Tugwell, B.L. AU 36.2
Tukey*, Jr., H.B. US-J 8.1
Tukey*, L.D. US-B 13.1.1
Tulmann Neto, A. BR 24.2
Tupy, J. CS 22.2
Turac, H. TR 10.1
Turcu, E. RO 32.1
Turi, I. HU 2.2.12
Turica, A. AR 4.1.4
Turk, R. TR 6.1
Turkensteen, L.J. NL 9.4.2
Turkes, N. TR 14.1
Turkes, T. TR 14.1
Turkington, C.R. AU 6.1
Turl, L.A.D. GB 9.1.3
Turnbull, S.A. CA 18.1
Turner, J.W. US-D 38.1
Turner, M.A. NZ 9.5
Turudić, V. YU 10.1
Tusken, G.A. US-G 12.1
Tuza, S. HU 2.7
Tuzcu, O. TR 1.1
Tveite, P. NO 10.1.8
Twig, B.A. US-C 5.2
Tyagi, S.S. IN 19.6
Tyksiński, W. PL 20.2.3
Tylkowska, K. PL 20.2.10
Tylkowski, T. PL 10.1.2
Tylus, K. PL 9.1
Tymoszuk, J. PL 28.1.17
Tymoszuk, S. PL 28.1.5
Tyner, F. US-D 49.1
Tyson, W. US-D 16.2
Tytkowski, A. PL 32.1.9
Tzamos, E. GR 14.1.1
Tzanakakis, M. GR 25.1.6
Tzankov, B. BG 8.3
Tzavella-Klonari, C. GR 25.1.7
Tzolov, Tz. BG 8.3
Tzompanakis, J. GR 2.1
Tzourou, E. GR 18.1.4
Tzoutzoukou, S. GR 6.1.1

Uchida, K. JP 14.1
Uchida, M. JP 27.3, 42.1
Uchida, Y. JP 25.1.2, 37.1
Uchie, K. JP 45.1
Uchihara, S. JP 27.3
Uchiyama, F. JP 35.2
Uchiyama, S. JP 8.2
Uchiyama*, Y. JP 46.1
Uda, A. JP 13.1
Uda, M. JP 23.2.2
Udagawa, H. JP 42.1
Udagawa, U. JP 4.1
Udayagiri, S. IN 1.1.7

Udeogalanya, A.C.Ċ. NG 5.1
Uechi, K. JP 32.1
Ueda, H. JP 37.2
Ueda, N. JP 44.1
Ueda, S. JP 28.1
Ueda, Y. JP 4.4, 33.2
Uehara, H. JP 18.1
Uehara, K. JP 32.1
Uehara, Y. JP 23.2.5
Ueki, K. JP 22.5
Uematsu, H. JP 41.3
Uematsu, J. JP 35.2
Uematsu, S. JP 4.3
Uemoto*, S. JP 7.4
Uemura, K. JP 15.3
Uemura, M. JP 21.2
Uemura, N. JP 22.2
Uemura, R. JP 2.1
Ueno, I. JP 38.3
Ueno, W. JP 45.1
Uesato, K. JP 32.2
Ueshima, Y. JP 44.1
Ugolik, M. PL 20.2.9
Uhlinger, R.D. US-G 23.1
Uitermark, C.G.T. NL 1.1.4
Ujike, T. JP 17.2
Ujiye, T. JP 27.3
Ulenberg, S.A. NL 9.8
Uličević, M. YU 7.1
Ullasa, B.A. IN 1.1.6
Ulubelde, M. TR 15.2
Ulusarac, A. TR 11.1
Uluskan, A. TR 15.3
Um, S.K. KR 12.2
Umada, Y. JP 43.2
Umale, S.B. IN 13.7
Umemaru, M. JP 9.2
Umemiya, Y. JP 14.2
Umiecka, L. PL 28.2.9
Umiel, N. IL 1.1.2
Un, S.F. TR 6.2
Unander, D. US-N 1.2
Uncini, L. IT 20.1
Unk, J. HU 2.7
Uno, T. JP 31.1
Unrath, C.R. US-D 6.1, 8.1.1
Uosukainen, M. FI 9.1
Upadhyay, J. IN 24.1.6
Upadhyaya, K.C. IN 20.1
Upadhyaya, M.K. CA 4.2
Upadhyaya, N.P. IN 24.1.1
Upasena, S.H. LK 3.8
Upchurch, B. US-C 8.2
Uppal*, D.K. IN 17.1
Upreti, G.C. IN 4.1
Urabe, Y. JP 4.3
Uragami, A. JP 12.3
Uragari, Y. JP 23.1
Urakami, Y. JP 40.1
Uraki, M. JP 42.1
Urashima, O. JP 43.2
Uraz, E. TR 15.2
Uraz, S. TR 15.2
Urban, L. AT 8.1
Urbányi, Gy. HU 2.2.1
Urbański, P. PL 20.2.4
Urben Filho, G. BR 25.2
Uresti Cerrillos, R. MX 12.4
Uriu*, K. US-K 6.1.3
Urošević, M. YU 14.1
Urushibara, K. JP 42.1
Usanmaz, D. TR 15.3
Ushiro, T. JP 11.1
Ushiyama, K. JP 19.2
Uslu, I. TR 14.1
Usoroh, N.J. NG 1.1
Usuda, A. JP 26.2
Uthaiah, B.C. IN 11.8
Utkhede, R.S. CA 3.1
Utsunomiya*, N. JP 22.5
Utzinger, J. US-F 3.1
Uygur, N. TR 11.1
Uzêda Luna, J.V. BR 13.1
Uziak, Z. PL 14.1.5
Uzo, J.O. NG 5.1
Uzun, A. TR 12.1
Uzunova, E. BG 8.1

Vaccaro, N.C. AR 6.1
Vachún, Z. CS 14.1.2
Václavík, M. CS 5.1
Vadhaw, O.P. US-D 51.1
Vail, P.V. US-K 7.1
Vaindirlis, N. GR 1.1
Vajagic, D. YU 14.1
Vajna, L. HU 2.5
Vakis, N. CY 1.1
Valadares, J.M.A.A. BR 20.1
Valancogne, Ch. FR 18.1
Valdés Zepeda, R. MX 6.2
Valdez Rodriguez, V.M. MX 5.1
Valen, W.R. US-K 14.1
Valencia, E. PE 2.4
Valencia, N. PE 1.2
Valencia Vilarreal, J.A. MX 1.1
Valentin, Ch. DE 11.1.5
Valentine, E.W. NZ 2.2
Valette, J-Ch. FR 3.2.1
Valette, R. BE 6.1, 6.5
Valk, M. CA 25.1
Vallania, R. IT 24.2
Vallebella, G.E. AR 8.1
Valle Garcia, P. MX 6.3
Vallejo, H. AR 3.1.1
Vallejos, C.E. US-D 27.7
Valls, J.F.M. BR 25.3
Valmayor, H.L. PH 1.2.3
Valsangiácomo, F.J. AR 6.1
Valverde S., C. PE 1.1
Vamvoukas, D. GR 8.1.1
Váňa, J. CS 29.1
Van Aartrijk, J. NL 7.1
Vanachter*, A. BE 9.1
Van Adrichem, B. NL 11.1
Van Assche, C. BE 9.1
Van Beek, G. NL 9.10
Van Bentum, G.P.W. NL 11.1
Van Bragt*, J. NL 9.7.1
Van Buren, J.P. US-B 2.1.5
Van Cleve, K. US-L 1.1
Van Cutsem, J. BE 9.1
Van Dam, J.G.C. NL 9.11.1
Van Damme, J-C. BE 6.3
Vandamme, P. BE 6.8.2
Van de Geijn, S.C. NL 9.5
Van de Knaap, B.J. NL 12.1
Van de Laar, H.J. NL 2.1
Van de Lande, H. SR 1.1
Van den Berg, G.A. NL 1.1.4
Van den Berg, J.J. Azn. NL 17.2
Van den Berg, J.J. Jzn. NL 17.2
Van den Berg, M.W. NL 9.11.2
Van den Berg, M.A. ZA 1.1
Van den Berg, P. NL 11.1
Van den Berg, R.G. NL 9.7.4
Van den Berg, Th.J.M. NL 1.1.1
Van den Berkmortel, L.G. NL 17.1
Van den Born, W. CA 7.1
Van den Brand, H.G.M. NL 6.1
Van den Broek, G.J. NL 1.1.4
Van den Ende*, J. NL 8.1
Van den Ende, B. AU 32.1
Van den Eyden, A. BE 18.1
Van den Plas, G. SN 1.1
Van der Boon, J. NL 4.1
Vandercook, C.F. US-K 11.1
Van der Eerden, L.J.M. NL 9.4.5
Van der Giessen, A.C. NL 9.6.5
Van der Ham, M. NL 6.1
Van der Hoeven, G.A. US-G 31.1
Van der Krogt, Th.M. NL 1.1.4
Van der Kruk, L. NL 17.1
Van der Laan, F.M. NL 9.7.4
Van der Laan, R. NL 17.1

Van der Maesen, L.J.G.
NL 9.7.4
Van der Meer, F.A. NL 9.4.4
Van der Meer, M.A. NL 9.10
Van der Meer, Q.P. NL 9.6.1
Van der Mespel, G.J. NZ 10.1
Van der Pol, P.A. NL 9.7.1
Van der Post, C.J. NL 9.3
Van der Pouw, B.J.A.
NL 9.11.2
Van der Schilden, M. NL 1.1.2
Van der Sommen, F.J. AU 41.1
Van der Valk, G.G.M. NL 9.2
Vanderveke, J. BE 6.8.3
Van der Veken, H. SN 1.1
Van der Veken, L. BE 13.1
Van der Vlugt, J.L.F.
NO 10.1.7
Van der Vossen, H.A.M.
NL 14.2
Vanderwaeren, R. BE 21.1
Van der Zande, M.T. NL 1.1.4
Van der Zwaan, A.G. NL 2.1
Van der Zweep, F. NL 9.9
Van der Zwet, T. US-C 8.2
Van de Sanden*, P.A.C.M.
NL 9.1
Van de Scheer, H.A.T. NL 10.1
Van de Toorn, P. NL 15.1
Van de Ven, A.C.M. NL 17.1
Van de Vooren*, J. NL 1.1.2
Van de Vooren, J.G. NL 9.7.4
Van de Vrie, M. NL 1.1.5,
9.4.1
Van de Weg, R.F. NL 9.11.2
Van de Weg, W.E. NL 9.6.2
Van de Werken, H. US-E 3.1
Van Dijk, H. NL 4.1
Van Dilst, F.J.H. NL 9.7.4
Vandiver, V.V. US-D 25.1
Van Doorn, W.G. NL 9.10
Van Dordrecht, E.M. NL 1.1.4
Van Eck, W.H. NL 9.8
Van Eijk, J.P. NL 9.6.3
Vaněk, J. CS 24.3
Van Elsas, J.D. NL 9.5
Van Emden, H.F. GB 29.1
Van Epenhuijsen, C.W. NZ 10.1
Vang-Petersen, O. DK 1.3
Van Griensven, L.J.L.D.
NL 5.1
Van Hal, J.G. NL 11.1
Van Halteren, P. NL 9.8
Van Heek, L. AU 29.1
Van Himme, M. BE 7.9
Van Holst, A.F. NL 9.11.2
Van Hoof, H.A. NL 9.5
Van Huylebroek, L. BE 7.11
Van Huysteun, P. ZA 4.1
Vani, A. IN 1.1.2
Van Keirsbulck, W. BE 14.1
Van Kester, W.M.N. NL 14.2
Van Koninckxloo, M. BE 3.1
Van Kraayenoord, C.W.S.
NZ 8.3
Van Kruistum, G. NL 6.1
Van Labeke, M. BE 7.7
Van Laere, O. BE 15.2
Van Lanen, H.A.J. NL 9.11.1
Van Leeuwen, P. NL 1.1.3
Van Lierde, D. BE 4.1
Van Linden, F. BE 18.1
Van Loon, L.C. NL 9.7.2
Van Marrewijk, N.P.A. NL 9.9
Vannière, H. FR 15.1.4
Van Nieuwenhuizen, G.H.
NL 9.10
Van Noort, G. ZA 4.3.4
Van Onsem, J.G. BE 14.1
Van Oorschot, J.L.P. NL 9.1
Van Oosten, H.J. NL 10.1
Van Oostrom, C.G.J. NL 9.2
Van Oostrum, A. ZA 1.1
Van Os, P. NL 1.1.4
Vanparys, L. BE 17.1
Van Raamsdonk, L.W.D.
NL 9.6.5
Van Rijssel*, E. NL 1.1.2
Van Rooyen, P.C. ZA 4.1
Van Rooyen, T.J. ZA 4.3.3
Van Rosen, H. SE 1.4
Van Sauers-Muller, A. SR 1.1
Van Schoonhoven, M. NL 1.1.4
Van Slobbe, A.M. NL 9.11.2
Van Steekelenburg, N.A.M.
NL 9.4.2
Van Tuyl, J.M. NL 9.6.3
Van Vaerenbergh, J. BE 15.3
Van Veen, J.A. NL 9.5
Van Velsen, R. NL 9.7.5
Van Vlimmeren, R. NL 14.1
Van Vuurde, J.M. NL 9.4.2
Van Waes, J. BE 13.1
Van Wallenburg, C. NL 9.11.2
Van Wambeke, E. BE 9.1
Van Weel, P.A. NL 1.1.2
Van Weerdenburg, J.W. NL 12.1
Vanwetswinkel, G. BE 19.1
Van Wezer, J. BE 22.1
Van Wiemeersch, L. BE 7.12
Van Wijk, A.J.P. NL 19.1
Van Wijk, A.L.M. NL 9.2
Van Wijk, C.A.Ph. NL 6.1
Van Winden, C.M.M. NL 8.1
Van Wyk, C.J. ZA 4.3.3
Van Wymersch, E. BE 7.7
Van Zijl, J.J.B. ZA 3.1
Van Zinderen Bakker, E.M.
CA 2.1
Vanzulli, C. IT 6.1
Van Zyl, H.J.J. ZA 4.1
Van Zyl, J.L. ZA 4.2
Vârâdie, P. RO 2.1
Vardakis, E. GR 10.1
Vardi, A. IL 1.2.1
Varela, F. US-N 1.2
Varga*, A. NL 9.7.2
Varga, Gy. HU 4.1.1
Varga, J. HU 4.2
Varga, L. HU 2.1.2
Varga, S. HU 5.2
Varga, V. YU 17.1
Vargas, J. US-F 14.1.3
Vargas, M. CR 2.3
Vargas Gil, J.R. AR 17.1
Vargas Oliveira, J. BR 10.1
Vargas Ramos, V.H. BR 15.1
Varma, S.P. IN 16.1
Várnay, Zs. HU 5.2
Varns, J. US-G 11.1
Varoquaux, P. FR 3.1.8
Vasconcellos, E.F.C. BR 24.1
Vasconcellos, H. de O.
BR 19.1
Vasconcellos, J.M.O. BR 29.2
Vasconcelos Costa, A. BR 26.2
Vasconcelos Lopes, J.G.
BR 8.1
Vasek, F.C. US-K 12.1.1
Vashishtha, B.B. IN 18.2,
18.4
Vasić, M. YU 14.2
Vasilakakis, M. GR 25.1.1
Vasilatou, A. GR 7.1
Vasilescu, V. RO 9.1
Vasiliou, Z. GR 11.1
Vasiljević, Z. YU 11.1
Vasoncellos, M.E. da C.
BR 1.1
Vásquez, V. PE 2.3
Vass, E. HU 2.2.11
Vassiliou, G.V. GR 14.1.4
Vassiliou, M. GR 18.1.4
Vasvary, L.M. US-B 14.1.4
Vatthaner, R. US-G 6.1
Vaz, R. BR 26.2
Vaz, R.L. BR 25.5
Vaz Parente, T. BR 25.4
Vázquez, D. AR 6.1
Vazquez, J.T. MX 18.1
Vaz Samapio, C. BR 12.2
Veale, J.A. NZ 9.1
Vedel, H. DK 3.6
Veen, B.W. NL 9.1

Veen, H. NL 9.1
Veenenbos, J.A.J. .NL 9.8
Vega, D. CO 2.1
Vega, E. AR 20.1
Vega Piña, A. MX 3.2
Vegh, I. FR 17.1.2
Véghelyi, K. HU 2.1.2
Veilleux, R.E. US-C 10.1
Velarde Ruesta, O. PE 4.4
Velasquez, R. PE 4.1
Velázquez Monrreal, J. MX 3.4
Velázquez Montoya, J. MX 6.3
Velchev, V. BG 6.2
Veldeman, R. BE 15.3
Veldhuyzen van Zanten, J.E. NL 14.2, 16.1
Velema, J.C.J. NL 16.2
Veleva, D. YU 4.1
Velich, I. HU 5.3.2
Velich, S. HU 4.2
Veličković, D. YU 14.1
Veličoković, S. YU 14.1
Velicu, J. RO 18.1
Velimirovic, V. YU 7.1
Veljovic, P. YU 10.1
Velloso, A.C.X. BR 19.2
Vellupillai, V. LK 3.11
Vendelbo, P. DK 10.1
Veneziano, W. BR 4.1
Venis, M.A. GB 8.1.1
Venkat Rao, M. IN 6.1
Venkat Rao, P. IN 6.1
Venkatarayappa*, T. IN 11.2
Venning, J.A. NZ 2.1.3
Venter*, F. DE 11.1.1
Ventura, J.A. BR 18.1
Venturela, L.R.C. BR 29.1
Venturieri, G.A. BR 1.2
Venzke, J.-F. IC 1.1
Venzke, K. IC 1.1
Verbrugghe, M. FR 3.1.4
Verdonck*, O. BE 14.1
Verdonck, R. SN 1.1
Vereś, A. CS 1.1
Vereijken, P.H. NL 6.1
Verellen, M. AR 1.1
Verghese, A. IN 26.1
Vergnes, A. FR 10.1.3
Vergniaud, P. FR 3.1.10
Verhaegen, A. FR 22.1
Verheyden, C. BE 19.1
Verhoef, R. NL 17.1
Verhoeven, H.A. NL 9.5
Verhoeven, P.A.W. NL 2.1
Verhoyen, M. BE 12.1
Verkleij, F.N. NL 9.1
Verlind, A.L. NL 1.1.6
Verloo, M. BE 7.8
Verma, A. IN 26.1
Verma, B. US-D 20.1
Verma, B.S. IN 19.5
Verma, K.D. IN 10.2
Verma, K.L. IN 10.2
Verma, S.M. IN 4.1
Verma, S.R. IN 22.1
Verma, S.S. IN 19.2
Verma, S.V. IN 25.1
Vermeulen, J. AR 10.1
Vernier, M. FR 11.1.3
Vernon, R.S. CA 4.1
Verók, I. HU 5.1.1
Vértesy, J. HU 2.1.2
Versluijs, J.M.A. NL 1.1.5
Vervloet, J.A.J. NL 9.11.1
Vest, H.G. US-H 6.1
Vestrheim, S. NO 10.1.6
Veürshov, B. DK 3.4
Vez, A. CH 2.1
Vezina, D. CA 29.1
Vezina, L. CA 34.1
Viaene, J. BE 7.10
Viana, J.C.A. BR 18.1
Viana Corrêa, J.R. BR 3.1
Vianna Barbosa dos Reis, N. BR 25.5
Vičić, M. YU 3.1
Vickers, R.A. AU 65.5
Vickery, R.K. US-I 7.1
Vidal, P. ES 6.1
Vidales Fernandez, J.A. MX 3.2
Vidalie, H. FR 19.5
Vidaud, J. FR 21.1
Vidéki, L. HU 5.3.1
Videnov, E. BG 6.1
Vidojković, R. YU 17.1
Vieira, A. BR 19.1
Vieira, C. BR 17.1
Vieira, D.J. BR 9.1
Vieira, J.V. BR 25.5
Vieira Pacova, B.E. BR 18.1
Viémont, J.D. FR 19.4
Vierstra, R. US-F 23.1.3
Vieth, J. CA 31.1, 31.2
Vig, P. HU 2.2.9
Vigier, B. CA 35.1
Vigl, J. IT 11.1
Vigliola, M.J. AR 3.1.1
Vigodski-Haas, H. IL 1.1.2
Vij, V.K. IN 17.2
Vijayakumar, S. IN 11.10
Vijaya Kumar, N. IN 11.2
Viju, C. RO 4.1
Vik*, J. NO 2.1
Vilela, L. BR 25.2
Villa-Isa, M. MX 16.3
Villalba R., E. PE 3.1
Villalon, B. US-H 15.1
Villani, M.G. US-B 2.1.3
Villareal*, R.L. PH 1.1
Villarreal Elizondo, H. MX 9.2
Villemur, P. FR 10.1.1
Villers, A. BE 4.1
Vincent, C. CA 35.1
Vincent, G. CA 31.1
Vince-Prue, D. GB 18.1.3
Vinciquerra, H. AR 22.3
Vindimian, E. IT 21.1
Vines, H.M. US-D 16.1
Vinterhalter, D. YU 9.1
Viola, M. HU 5.1.4, 8.1
Viotošević, I. YU 9.1
Virág, J. HU 2.2.4
Viscardi, D. PL 28.2.8
Viscardi, K. PL 28.2
Vise, C. PE 1.1
Viseur, J. BE 6.8.7
Vishwanath, IN 2.1.4
Viska, J. AU 59.2
Vismara, D. US-E 22.1
Visnyovszky, É. HU 2.5
Visscher, H.R. NL 5.1
Visser, A.F. ZA 3.1
Visser, F.R. NZ 8.2
Visser, L. NL 9.5
Visser*, T. NL 9.6.2
Vit, P. CS 7.1
Vitanova, I. BG 1.1
Vithanange, H.I.M.V. AU 65.2
Viti, R. IT 17.2
Vitti, P. BR 20.2
Vizgarra, O.N. AR 22.3
Vizier, C. FR 3.1.2
Vlachos, M. GR 25.1.5
Vlad, C. RO 5.1
Vlădianu, D. RO 14.1
Vladimirov, B. BG 13.1
Vlassak, K. BE 9.1
Vlček, F. CS 20.1
V"lchev, P. BG 10.1
V"lchev, St. BG 8.1
V"lchev, V. BG 7.1
V"lcheva, P. BG 16.1
Vleeshouwer, J.J. NL 9.11.2
V"lkov, G. BG 6.1
Vodak, M. US-B 14.1.1
Vogel, G. DD 3.1
Vogel, M.L. ZI 4.1
Vogel, R. FR 15.1.2
Vogelezang, J.V.M. NL 1.1.4
Voica, E. RO 19.2
Voican, V. RO 8.1

Volin, R.B. US-D 29.1
Völk, J. DE 11.2.2
Volkman, L.E. US-K 3.1.1
Voll, E. BR 27.1
Volpe, B. IT 16.1
Volpi*, L. IT 22.1
Vömel, A. DE 13.2
Von Alvensleben, R. DE 15.1.2
Vondráček, J. CS 26.1
Von Elsner, B. DE 15.1.7
Von Hentig, W.-U. DE 12.1.8
Vonk, C.R. NL 9.1
Vonk Noordegraaf*, C. NL 1.1.4
Von Lindeman, G. CR 2.2
Von Malek, J. DE 16.1
Von Mollendorff, L.J. ZA 4.1
Von Reichenbach, H. DE 15.4
Vonshak, A. IL 5.1.6
Von Zabeltitz, C. DE 15.1.7
Voorrips, R.E. NL 9.6.1
Vörös, J. HU 2.5
Vorsa, N. US-B 14.1.1
Vos, H. NL 9.9
Vostinar, N. RO 10.1
Voth, V. US-K 6.1.3
Votruba, P. CS 15.1
Votruba, R. CS 24.1
Voyiatzis*, D. GR 25.1.4
Vožda, J. CS 14.2
Voždová, G. CS 14.2
Vrain, T.C. CA 4.1
Vrbanc, L. YU 3.1
Vrecenak, A. US-B 14.1.1
Vredenberg, W.J. NL 9.7.3
Vreeken, A. Jr., NL 13.1
Vreeken, A. NL 13.1
Vreeken, W. NL 13.1
Vreugdenhil, D. NL 17.2
Vreugdenhil, D. NL 9.7.2
Vršek, I. YU 3.1
Vruggink, H. NL 9.4.2
Vrugtman*, F. CA 16.1
Vrugtman, I. CA 16.1
Vučinić, Z. YU 7.1
Vujanović, S. YU 5.1
Vukićevic, E. YU 8.2
Vuković, Lj. YU 9.1
Vukowitz, G. AT 8.11
Vuksanović, P.T. YU 1.2
Vulević, D. YU 15.1
Vullin, G. FR 15.1.4
Vulpes, N. RO 16.1
Vulsteke*, G. BE 17.1
Vurai, G. HU 5.3.1
Vural, H. TR 15.1
Vylku, A.V. SU 5.1
Waalwijk, C. NL 9.5
Wabule, M. KE 1.3
Wacquant, C. FR 21.1
Wada, E. JP 39.1
Wada, F. JP 18.1
Wada, H. JP 40.2
Wada, M. JP 7.1
Wada, O. JP 13.1
Wada, T. JP 7.2
Waddill, V.H. US-D 29.1
Wade, G. US-D 16.1
Wade, M. AU 8.1
Wafi, S.M. LY 2.1
Wagenmakers, P.S. NL 10.1
Wagner, A. PL 14.1.8
Wagner, A. US-H 6.2
Wágner, B. CS 14.1.3
Wagner, D.F. US-D 13.1
Wagner, R. FR 10.1.3
Wagner, St. RO 17.1
Wagner, W. DE 22.1
Wahwal, K.N. IN 14.1
Waiganjo, M.M. KE 1.5
Wainaina, J. KE 1.7
Waines, J.G. US-K 12.1.1, 12.1.2
Wainwright, H. GB 3.1
Waister, P.D. GB 15.1.4
Waite, D. AU 23.1
Wakabayashi, H. JP 15.2
Wakabayashi*, S. JP 39.2
Wakabayashi, Y. JP 17.1
Wakana, A. JP 7.4
Wakefield, R.C. US-A 14.1
Wakhonya, H.M. KE 12
Waki, Y. JP 5.2
Wakisaka, I. JP 15.1
Wakou, M. JP 16.2
Wakugami, H. JP 32.1
Wakulinski, L. PL 32.1.6
Walali, L. MA 5.1
Walberg, A. SE 1.1.2
Walden, R. US-C 12.1
Waldron, L.J. US-K 3.1.3
Waleza, W. PL 32.1.9
Waliczek, A. PL 23.1
Walker, A. GB 33.1.8
Walker, A.R. AU 48.1
Walker, D.W. US-D 52.1
Walker, D.R. US-I 5.1
Walker, G. AU 38.2
Walker, J.R.L. NZ 12.1
Walker, J.T.S. NZ 6.1
Walker, M.R. AU 43.1
Walker, R.R. AU 65.2
Walker, T. US-D 20.1
Walkey, D.G.A. GB 33.1.4
Wall, D.A. CA 13.1
Wall, J.K. GB 29.1
Wallace, D.H. US-B 5.1.3 5.1.7
Wallace, H.R. AU 42.1
Wallerstein, I. IL 1.1.2
Walling, L. US-K 12.1.1
Wallner, S.J. US-I 2.2
Walnik, W. PL 10.1.4
Walser, R.H. US-I 6.1
Walsh, C.S. US-C 5.2
Walsh, J.R. CA 40.2
Walter, B. FR 6.1.2
Walter, R.H. US-B 2.1.5
Waltl, K. AT 8.7
Walton, E.F. NZ 1.1
Wambugu, E.W. KE 1.2
Wambugu, F.M. KE 5.1
Wambugu, L.N. KE 5.1
Wang, C.F. TW 8.1
Wang, C.Y. US-C 3.1
Wang, C.Y. TW 8.1
Wang, D.N. TW 2.1.2
Wang, G.M. BR 27.1
Wang, H.L. TW 2.1.3
Wang, S. CA 25.1
Wang, S. TW 10.1
Wang, S. TW 12.1
Wang, S.C. TW 11.1.2
Wang, S.R. BR 27.1
Wang, S.S. TW 11.1.2
Wang, T.Y. TW 7.1
Wang, T. GB 24.1.1
Wang, U.C. TW 2.1.2
Wang, W.Y. TW 7.3
Wang, Y.K. TW 8.1
Wang, Y.T. US-H 15.1
Wang, Z.N. TW 4.3
Wangchareoan, P. TH 2.1
Warade, S.D. IN 14.3
Warawkomska, Z. PL 14.1.6
Ward, A.T. CA 11.1
Ward, D.M. IC 1.1
Ward, E.W.B. CA 18.1
Ward, G.A. AU 59.1
Wardowski, W.F. US-D 30.1
Wardzyńska, H. PL 32.3.4
Warholic, D.T. US-B 5.1.3
Warid, W.A. LY 1.2
Warman, T.M. GB 8.1.2
Warmenhoven, M.G. NL 1.1.1
Warmund, M.R. US-E 18.1
Warner, J. CA 24.1
Warner, J.S. US-B 1.1
Warriner, S. AU 38.2
Warrington, I.J. NZ 8.1
Wasa, N. JP 44.1
Washida, S. JP 1.1.2
Washington, W. AU 27.1
Wasiak, P. PL 6.2

Wasilewska, I. PL 28.2.1
Wasowski, T. PL 28.1.5
Wassenaar, L.M. NL 9.6.1
Wastie, R.L. GB 15.1.2
Watabe, N. JP 19.2
Watada, A.E. US-C 3.1
Watanabe, A. JP 5.1
Watanabe, E. JP 5.2
Watanabe, H. JP 5.2, 12.1, 13.3, 14.6, 21.1, 28.1, 42.1
Watanabe, I. JP 4.4, 20.1
Watanabe, J. JP 5.3
Watanabe, K. BR 21.2
Watanabe, K. JP 19.4, 29.1, 35.2, 36.1
Watanabe, M. JP 3.2.2, 35.1
Watanabe, N. JP 15.2
Watanabe, O. JP 5.1
Watanabe, S. JP 10.1, 35.2, 45.2
Watanabe, T. JP 6.2, 9.1
Watanabe, Y. JP 14.1, 30.1, 47.1
Watanuma, N. JP 35.2
Watatani, H. JP 14.2
Water, J.K. NL 9.8
Waters, E.W. US-D 24.1
Waters, L. US-G 7.1
Waters, W.E. US-K 3.1.1
Watkin, B.R. NZ 9.3
Watkins, C.B. NZ 2.1.2
Watkins, D.A.M. GB 19.1.1
Watkins, J.B. AU 15.2
Watkins, R. GB 8.1.2
Watson, A.K. CA 37.1
Watson, B.J. AU 17.1
Watson, D.R.W. NZ 2.3.1
Watson, G.D. GB 2.1
Watson, K.A. AU 38.2
Watson, R.N. NZ 5.1
Wattanachat, C. TH 10.1
Watters, B.S. GB 20.1
Watts, J.W. GB 24.1.3
Wauchope, D. AU 29.1
Wauters, J. BE 9.2
Wauthy, J.M. CA 32.1
Wave, H. US-A 1.1
Wawrzynczak, P. PL 28.1.6
Waxman, S. US-A 13.1
Wázbinska, J. PL 18.1
Wearing, C.H. NZ 2.2
Weaver, D.J. US-D 19.1
Weaver, G.M. CA 41.1
Weaver, R.J. US-K 6.1.5
Weaver, S.E. CA 17.1
Webb, M.G. AU 56.1
Webb, R.E. US-C 3.1
Weber, W.E. DE 15.1.1
Webster, A.D. GB 8.1.1
Webster, C.A. US-B 14.1.1
Webster, D.H. CA 41.1
Webster, J.F. US-B 14.1.1
Webster, W.B. US-D 39.1
Wedberg, J. US-F 23.1.2
Weeden, N.F. US-B 2.1.2
Weeks, J.C. US-K 3.1.1
Weerasinghe, G.W.K. LK 3.8
Weglarz, Z. PL 32.1.2.2
Węgorek, W. PL 20.1
Wehner, D.J. US-F 12.2.2
Wehner, T.C. US-D 8.1.1
Wehrmann*, J. DE 15.1.5
Wehunt, E.J. US-D 19.1
Weibull, P. SE 4.1
Weichmann*, J. DE 11.1.1
Weigl, F. AT 15.3
Weigle, J.L. US-G 20.1
Weiler*, T.C. US-B 5.1.1
Weimann, G. DE 15.1.7
Weimarck, G. SE 2.1
Weinbaum, S.A. US-K 6.1.3
Weinberg, Z. IL 1.3.2
Weingartner, D.P US-D 28.1
Weinhold, A.R. US-K 3.1.2
Weinmann, C.J. US-K 3.1.1
Weintraub, M. CA 4.1
Weir, R.G. AU 9.1.2
Weires, Jr., R.W. US-B 2.1.3, 4.1
Weis, G.G. US-F 22.1
Weiser*, C.J. US-J 14.1
Weiser, J. AT 16.1
Weissich, P. US-M 2.4
Welander*, M. SE 1.1.3
Welander*, T. SE 1.1.3
Welch, W.C. US-H 6.2
Welker, W. US-C 8.2
Welkerling de Tacchini, E.M.L. AR 15.2.1
Wellendorf, H. DK 3.3
Weller, S.C. US-F 9.1.4
Welling, B. DK 4.1
Wellington, W.G. CA 4.2
Wells, C.M. NZ 9.4
Wells, D.W. US-D 57.1
Wells, H.D. US-D 22.1
Wells, J.A. US-D 12.2
Wells, J. US-D 19.1
Wells, J.D. US-F 6.1
Wells, L.W. US-D 45.1
Wells, O.S. US-A 2.1
Welsh, J.R. US-I 18.1
Welter, S.C. US-K 3.1.1
Welvaert, W. BE 7.7
Wen, H.C. TW 2.1.3
Wen, Y.C. TW 7.3
Wendlberger, G. AT 8.6
Wendt, Th. DE 6.1.2
Wendtland, W. AT 23.1
Weng, S.W. TW 7.1
Wenner, D. NO 10.1.2
Wenner, L. DE 11.1.5
Wenzel, J. DE 5.2.1
Wenzel, K. DE 12.1.5
Werner, A. PL 10.1.3
Werner, D.J. US-D 8.1.1
Werner, H. DE 6.1
Werner, M. PL 20.2.6
Wernisch, A. AT 9.1
Wertheim, S.J. NL 10.1
Werthein, J. CR 1.1
West, S.J. NZ 9.7
Westerbeek, G.H. NL 9.11.2
Westerhof, J. NL 1.1.4
Westerling, F.J. NL 9.10
Westerveld, G.J.W. NL 9.11.2
Weston, G.C. NZ 2.1
Weststeijn, G. NL 8.1
Wettasinghe, D.T. LK 2.1
Wetzel, M.M.V.S. BR 25.3
Wetzstein, H. US-D 16.1
Whalley, D.N. GB 18.1.1
Whalon, M. US-F 14.1.2
Whan, J. AU 27.1
Wheatley, G.A. GB 33.1.3
Wheaton*, T.A. US-D 30.2
Wheeler, K. AU 65.4
Wheeler, W.B. US-D 27.3
Whenham, R.J. GB 33.1.1
Whett, S. AR 22.2
While, J.A. NZ 13.1
Whiley, A.W. AU 23.1
Whipps, J.M. GB 18.1.2
Whisson, D.L. AU 65.1
Whitaker, B. US-C 3.1
Whitcomb, C.E. US-H 3.1
White, A.G. NZ 2.1.1
White, D.B. US-G 7.1
White, G.A. US-C 3.1
White, G.A. CA 18.1
White, G.C. GB 8.1.1
White, J.W. US-B 13.1.1
White, J.A. NZ 11.2
White, J.G. GB 33.1.4
White, J.M. US-D 36.1
White, P.F. GB 18.1.2
White, R. CA 9.1
White, R.P. CA 43.1
Whitehead, M. AU 36.5
Whitely, K.T. AU 59.1
Whiteside, J.O. US-D 30.2
Whitfield, F.B. AU 65.4

Whiting, J.R. AU 31.1
Whitlow, T.H. US-B 5.1
Whitman, R.J. CA 42.1
Whitney, J.D. US-D 30.1
Whitten, M.J. AU 65.5
Whittington, W.J. GB 25.1.2
Whittle, C.P. AU 65.5
Whitwell, J.D. GB 30.1
Whitwell, T. US-D 13.1
Wicker, G. US-D 30.2
Wickramaratne, M.R.T. LK 2.1
Wickramasinghe, P. LK 4.1
Wicks, T.J. AU 36.2
Wicksell, U. DK 3.9
Wickson, R. AU 3.1
Widders, I.E. US-F 14.1.1
Widmer, A. CH 1.1
Widmer*, R.E. US-G 7.1
Widomyer, F.B. US-H 19.1
Widrlechner, M.P. US-G 20.1
Wielgolaski, F.-E. NO 7.2
Wien*, H.C. US-B 5.1.3
Wienhaus, H. DE 12.1.10
Wieniarska, J. PL 14.1.1
Wierzbicka, B. PL 18.1
Wiesner, L.E. US-I 18.1
Wiest, S.C. US-G 31.1
Wigley, P.J. NZ 2.2
Wignall, W.R. GB 31.2.7
Wigrem, K. US-A 4.1
Wijeratnam, S.W. LK 1.1
Wijeratne, D.B.T. LK 3.2
Wijnands, D.O. NL 9.7.4, 9.7.5
Wijnands, F.G. NL 6.1
Wilcox, D.A. US-B 5.1.3
Wilcox, G.E. US-F 9.1.4
Wilcox, H.J. GB 13.1.4
Wilcox, W.F. US-B 2.1.4
Wilczyński, M. PL 21.1
Wildbolz, Th. CH 1.1
Wilders, N. NL 9.7.4
Wiley, R.C. US-C 5.2
Wilfret, G.J. US-D 24.1
Wilhelm, E. DE 16.1
Wiliams, C.H. US-A 2.1
Wilkaniec, B. PL 20.2.2
Wilkerson, D. US-H 6.2
Wilkins*, H.F. US-G 7.1
Wilkins, J.P.G. GB 13.1.2
Wilkinson, A.G. NZ 8.3
Wilkinson, A.T.S. CA 4.1
Wilkinson, H.T. US-F 12.2.3
Wilkinson, I. AU 30.1
Wilkońska, A. PL 28.1.12
Wilkowica, M. PL 14.1.5
Willaert, G. BE 7.8
Willer, C.C. US-F 4.1
Williams, A. US-D 16.2
Williams, A.A. GB 19.1.4
Williams, A.G. GB 31.2.4
Williams, A.S. US-E 15.1, 16.1, 17.1
Williams, B.D. AU 41.1
Williams, B.R. US-D 57.1
Williams, C.M.J. AU 36.5
Williams, D.B. US-E 3.1
Williams, D.J. US-F 12.2.2
Williams*, F. US-I 6.1
Williams, J.M. US-C 10.1
Williams, L.E. US-F 3.1
Williams, M. AU 46.4
Williams*, M.W. US-J 10.2
Williams, P. US-F 23.1.4
Williams, P.H. NZ 13.1
Williams, P.J. AU 37.1
Williams, R.A.D. IC 1.1
Williams, R.N. US-F 7.1.3
Williams, R.R. GB 19.1.3
Williamson, B. GB 15.1.1
Willink, E. AR 22.3
Wills, R.B.H. AU 9.5
Willumsen, J. DK 1.2
Willumsen*, K.J. NO 10.1.7
Wilmers, F. DE 15.5
Wilson, C. US-C 8.2
Wilson, D.B. CA 8.1
Wilson, F. GB 1.2
Wilson, G.J. NZ 3.1
Wilson, H. US-F 3.1
Wilson, H.P. US-C 13.1
Wilson, P.W. US-D 52.1
Wilson*, S.J. AU 46.2
Wilson, T.M.A. GB 24.1.4
Wilson, W.C. US-D 30.1
Wiltbank*, W.J. US-D 27.4
Wilton, B. GB 25.1.1
Wilusz, K. PL 13.1
Wimalajeewa, S. AU 27.1
Winandy, A.L.P. BR 29.1
Winarno, M. ID 3.1
Windham, S.L. US-D 47.1
Windrich, W. NL 9.4.3
Winks, C.W. AU 23.1
Winoto-Suatmadji, R. AU 27.1
Winston, E.C. AU 17.1
Winston, R. US-E 4.2
Winter*, F. DE 28.1.3
Winter, M.J. FR 31.1
Wiseman, J.S. GB 15.1.4
Wishart*, R.L. AU 38.1
Wiśniewska, H. PL 28.2.6
Wisniewska-Grzeszkiewicz, H. PL 28.1.13
Wisser, R. PE 1.2
Wit, A.K.H. NL 1.1.5
Witaszek, W. PL 19.1
Witcombe, R. AU 33.1
Witham, F.H. US-B 13.1.1
Withers, A.C. GB 18.1.3
Withers, B.J. GB 6.1
Withey, R.K. AU 10.1
Witkowska, D. PL 20.1.5
Witkowski, W. PL 20.1.5
Witomska, M. PL 32.1.3
Witt, H. DD 4.1
Witt, H.H. DE 3.1
Witt, M. US-E 15.1
Witte, W.T. US-E 3.1
Witters, R.E. US-J 14.1
Włodek, J. PL 18.1.10
Wnuk, A. PL 11.1.4
Wöbse, H.-H. DE 15.2.4
Woción, St. PL 14.1.1
Wodner, M. IL 1.2.4
Woess, F. AT 8.3
Woets, J. NL 10.1
Wohanka, W. DE 12.1.12
Wojcik, J. PL 28.1.8
Wojdyła, A. PL 28.1.16
Wojłkiewicz, A. PL 33.1
Wojniakiewicz, A. PL 28.1.10
Wojniakiewicz, E. PL 28.2.12
Wojniakiewicz, T. PL 28.1.6
Wojtas, B. PL 15.1
Wojtaszek*, T. PL 11.1.1
Wojtatowica, J. PL 32.1.12
Wojtkowiak, D. PL 20.2.7
Wojtowicz, M. PL 28.1.15
Wold, A. NO 10.3
Wolde-Giorgis, A.F. ET 10.2
Wolf, E.A. US-D 23.1
Wolf, M. CS 24.1
Wolf, T.K. US-C 14.1
Wolfe, D.W. US-B 5.1.3
Wolfe, F. CA 7.2
Wolff, W.Y. DE 15.2.1
Wolffhardt, D. AT 8.7
Wolgast, L.J. US-B 14.1.1
Wollin, A. AU 30.1
Wolnick, D.J. US-B 13.1.1
Wolski, P. PL 32.1.11
Wolstenholme, B.N. ZA 2.1
Wolters, A. NL 15.1
Wolting, H.G. NL 9.4.5
Woltz, S.S. US-D 24.1
Wong, J. AU 46.5
Wong, W.C. AU 15.4
Wonkyi-Appiah, J.B. GA 3.4
Woo, J.K. KR 23.1.1
Wood, B.J. US-K 11.1
Wood, C.C. GB 33.1
Wood, D.A. GB 18.1.2
Wood, D.E.S. NZ 9.1

Wood, D.L. US-K 3.1.1
Wood, F.A. US-D 27.1
Wood, F.H. NZ 5.1
Woodford, J.A.T. GB 15.1.1
Wooding, F. US-L 1.1
Woodson, W.R. US-D 52.1
Woodson, W.R. US-F 9.1.4
Woolhouse, A.R. GB 4.1
Woolhouse, H.W. GB 24.1, 24.1.5
Woolley*, D.J. NZ 9.1
Worf, G. US-F 23.1.4
Workman, M. US-I 2.2
Workman, R.B. US-D 28.1
Worku, Z. KE 1.3
Worley, R. US-D 22.1
Worlock, P.A. AR 20.1
Worrall*, R.J. AU 5.1
Worthington, J. US-H 13.1
Wösten, J.H.M. NL 9.11.1
Wouters, O.D. AR 12.1
Wouts, W.M. NZ 2.2
Woychik, J.H. US-B 12.1
Woyke*, H. PL 28.2.2
Wredin*, A. SE 1.3, 1.1.2
Wricke, G. DE 15.1.1
Wright*, C.I. US-L 2.2
Wright, C.J. GB 25.1.1
Wright, D. US-B 7.1
Wright, D.L. US-D 37.1
Wright, D.N. AU 47.1
Wright, H.M. GB 24.1.2
Wright, M.E. US-D 52.1
Wright, N.S. CA 4.1
Wright, R. US-B 8.1
Wright, R.D. US-C 10.1
Wright, R.M. AU 14.1
Wright, S.J. US-L 2.2
Wright, S.E. GB 35.1.2
Wright, T.W.J. GB 35.1.2
Wrigley, M.P. NZ 9.2
Wróblewska, A. PL 14.1.6
Wróblewska, C. PL 15.1
Wrolstad, R.E. US-J 14.1
Wu, L.L. US-K 6.1.2
Wu, M.T. TW 7.3
Wucherpfennig, K. DE 12.1.4
Wuest, P.J. US-B 13.1.1
Wulster, G.J. US-B 14.1.1
Wünsch, A. DE 11.1.3
Wurr*, D.C.E. GB 33.1.5
Würschmidt, E. AR 22.3
Wutscher, K. US-D 34.1
Wyatt, I.J. GB 18.1.2
Wyatt, J. US-E 2.1
Wyles, J.W. US-E 14.1
Wyman, J. US-F 23.1.2
Wypych, M. PL 13.1
Wysocki, C. PL 32.1.13

Xiloyannis, C. IT 17.2

Yabe, K. JP 1.1.2
Yadav, D.S. IN 25.1
Yadav, E.D. IN 14.1
Yadav, G.R. IN 25.1
Yadav*, I.S. IN 26.1
Yadav, J.L. IN 20.1
Yadav, L.P. IN 28.1
Yadav, S.N. IN 19.1
Yadava, S.S. IN 22.1
Yadukumar, N. IN 3.1
Yae, B.W. KR 23.1.3
Yagi, Y. JP 34.1
Yaginuma, K. JP 14.2
Yahata, S. JP 4.3
Yahel, H. IL 1.1.2
Yai, H. JP 9.1
Yair, A. IL 5.1.5
Yajima, H. JP 35.2
Yakut, Y. TR 12.1
Yakuwa, T. JP 12.4
Yalcın, O. TR 3.1
Yalcın, T. TR 8.1
Yamabe, M. JP 15.1
Yamabe, T. JP 23.1
Yamada, H. JP 22.5
Yamada, K. JP 7.2, 33.1, 37.1
Yamada, M. JP 3.2.2, 3.2.3, 11.3, 13.1
Yamada, S. JP 15.2
Yamada, T. JP 3.3.1, 47.2
Yamada, Y. JP 19.1, 38.3
Yamaga, H. JP 8.2
Yamagata, H. JP 22.5
Yamagishi, H. JP 23.2.2
Yamaguchi, A. JP 14.2
Yamaguchi, H. JP 6.1
Yamaguchi, K. JP 45.1
Yamaguchi, M. JP 14.2
Yamaguchi, S. JP 7.3, 17.1, 45.1
Yamaguchi, T. JP 6.1, 23.2.3
Yamakawa, K. JP 23.2.1
Yamakawa, S. JP 7.3
Yamakawa, T. JP 15.2
Yamaki, Y. JP 41.5
Yamamoto, F. JP 4.1
Yamamoto, H. JP 14.3
Yamamoto, K. JP 6.2, 19.2, 37.1
Yamamoto, M. JP 14.1, 29.1
Yamamoto, S. JP 21.2, 25.2, 44.2
Yamamoto*, T. JP 3.1, 17.1, 25.1.3, 27.1, 43.2, 45.2
Yamamoto, Y. JP 1.3, 4.2, 46.1
Yamamura, B. JP 44.2
Yamamura, H. JP 37.2
Yamamuro, K. JP 14.1
Yaman, C. TR 7.1
Yamanaka, A. JP 39.1
Yamanaka, T. JP 5.2
Yamane, M. JP 42.3
Yamane, Y. JP 11.3
Yamanishi, H. JP 26.2
Yamao, M. JP 40.2
Yamasaki, K. JP 17.2
Yamasaki, N. JP 20.3
Yamasaki, S. JP 44.2
Yamashita, F. JP 1.1.4
Yamashita, H. JP 38.1
Yamashita, K. JP 25.2
Yamashita, S. JP 7.2, 34.3, 44.2
Yamato, H. JP 40.2
Yamauchi, M. JP 17.1
Yamauchi, S. JP 8.1
Yamaya, H. JP 3.2.1
Yamaya, T. JP 2.3
Yamazaki, H. JP 6.1
Yamazaki, K. JP 19.2, 39.1
Yamazaki, T. JP 11.2, 14.2
Yamdagni, R. IN 9.1
Yamvrias, Ch. N. GR 14.1.2
Yanagibashi, Y. JP 14.1
Yanagida, M. JP 3.3.2
Yanagisawa, M. JP 44.1
Yanagita, A. JP 41.1
Yanai, T. JP 20.2
Yanase, H. JP 14.2
Yandell, B. US-F 23.1.3
Yañez, C.E. AR 17.1
Yang, K.T. TW 12.1
Yang, S.P. TW 12.1
Yang, S.R. TW 8.1
Yang, S.Y. KR 22.1
Yang, T.S. TW 12.1
Yang, W.M. KR 22.1
Yang, W.Z. TW 7.3
Yang, Y.S. TW 7.1
Yangen, J.A. US-J 17.1
Yankulova, M. BG 6.1
Yano, K. JP 17.1
Yano, M. JP 23.2.3
Yano, R. JP 47.2
Yano, S. JP 27.1, 44.1
Yano, T. JP 5.2
Yaron, A. IL 4.1
Yasuda, H. JP 9.2
Yasuda, M. JP 14.2
Yasui, H. JP 23.2.5
Yasui, K. JP 31.2

Yasui, S. JP 28.1
Yasuoka, K. JP 20.1
Yasutomi, N. JP 32.1
Yatabe, K. JP 39.1
Yatou, O. JP 14.4
Yayla, F. TR 22.1
Yazawa, S. JP 22.4
Yazawa, T. JP 26.4
Yazgan, A. TR 23.1
Yeam, D.Y. KR 23.3
Yee, J. CA 25.2
Yeh, C.Y. TW 7.3
Yeh, Y.Y. TW 9.1
Yelboga, K. TR 3.1
Yelenosky, G. US-D 34.1
Yen, C.R. TW 1.2
Yen, Y.F. TW 8.1
Yermanos, D.M. US-K 12.1.1
Yiem, M.S. KR 23.1.4
Yılmaz, B. TR 2.2
Yılmaz, K. TR 8.1
Yilmaz, M. TR 1.1
Yim, Y.J. KR 6.1
Yli-Rekola, Aman. M. FI 12.1
Yoda, S. JP 31.1
Yoder, K.S. US-C 14.1
Yokoi, K. JP 28.1
Yokoi, M. JP 4.4
Yokomizo, K. JP 27.2
Yokomizo, Y. BR 20.2
Yokota, K. JP 16.4
Yokota, M. JP 32.1
Yokoyama, H. US-K 11.1
Yokoyama, T. JP 24.1
Yokoyama, Y. JP 29.1
Yoncheva, M. BG 1.1
Yoneda, K. JP 19.4
Yonemori, K. JP 23.3
Yonemori, S. JP 32.2
Yonemura, K. JP 1.1.1
Yoneyama, S. JP 14.1
Yoneyama, T. JP 41.1
Yoneyasu, A. JP 41.3
Yoo, K.C. KR 5.2
Yoon, C.J. KR 2.1.3
Yoon, H.M. KR 23.1.1
Yoon, J.Y. KR 23.1.1
Yoon, K.I. KR 3.1
Yoon, T.M. KR 7.1
Yoon, Y.K. *KR 24.1
Yordanov, M. BG 8.1
Yorinori, J.T. BR 27.1
York, A.C. US-F 9.1.3
Yoshida, A. JP 45.1
Yoshida, F. JP 41.2
Yoshida, M. JP 7.1, 14.2, 16.4
Yoshida, S. JP 4.2, 30.2
Yoshida, T. JP 7.3, 27.2
Yoshida, Y. JP 16.2, 22.5
Yoshihara, T. JP 7.3
Yoshiike, T. JP 16.1
Yoshikawa, H. JP 12.3
Yoshikawa, K. JP 38.2
Yoshikawa, M. JP 22.1
Yoshimatsu, K. JP 46.1
Yoshimoto, H. JP 44.1
Yoshimura, F. JP 33.3
Yoshimura, Y. JP 25.1.3
Yoshinaga, K. JP 11.3
Yoshinaga, T. JP 20.1
Yoshino, A. JP 4.2
Yoshino, S. JP 37.2
Yoshino, T. JP 10.1
Yoshioka, H. JP 16.2, 23.2.1
Yoshioka, K. JP 8.1
Yoshitake, S. JP 7.1
Yoshizawa, E. JP 26.2
Yoshizawa, K. JP 36.1
Yossen, V. AR 7.1
Yossif, M.Y. LY 2.1
Yost, G.E. US-J 10.2
Young, B.W. ZA 3.1
Young*, C. AU 46.2
Young, D. CA 40.3
Young, D.A. US-B 5.1.8
Young, E. US-D 8.1.1
Young, F. US-H 16.1
Young, G. AU 32.1
Young, H. NZ 2.1.4
Young, J.M. NZ 2.3.1
Young, P. GB 24.1.1
Young, R. US-C 8.1
Young, R.A. US-B 5.2
Young, R.H. US-D 34.1
Young, W.A. US-D 53.1
Yovchev, I. BG 8.2
Ystaas*, J. NO 5.1
Yu, I.C. KR 2.1.1
Yu, K.B. KR 1.1
Yu, K.S. US-I 3.2
Yu, R.C. TW 2.1.4
Yu, S.Y. KR 18.1.3
Yu, S.J. KR 9.1
Yu, Y.S. KR 7.3
Yuasa, T. JP 11.2
Yuce, B. TR 15.3
Yucel, A. TR 14.1
Yuda*, E. JP 33.2
Yuhashi, K. JP 4.1
Yui, S. JP 23.2.2
Yukawa, Y. JP 44.2
Yukinari, M. JP 40.2
Yukita, K. JP 3.2.3
Yuksel, I. TR 19.1
Yule, W.N. CA 37.1
Yumoto, T. JP 42.2
Yunculer, G. TR 13.1
Yunculer, O. TR 13.1
Yurdakul, S. TR 11.1
Yureturk, M. TR 14.1
Yusa, Y. JP 24.1
Yutani, U. JP 10.1
Yuza, T. JP 3.1

Zabel, A. YU 8.1
Zablotowicz, R.M. US-D 30.2
Zabtocka, L. PL 28.1.6
Zacharakis, K. GR 6.1.1
Zachariah, G.L. US-D 27.9
Zachariasse, L.C. NL 3.1
Zachos, D. GR 25.1.7
Zackel, E. HU 2.2.1
Zafiriou, C.Z. GR 14.1.4
Zafośnik, T. YU 18.1
Zagórska, M. PL 22.1
Zagórski, Z. PL 21.1
Zagaja*, S.W. PL 28.1
Zaglul, J.A. CR 2.1
Zahara, M. US-K 6.1.4
Zaharia, C. RO 24.1
Zaharia, D. RO 17.1
Zaharia, I. RO 24.1
Zahn, F.-G. DE 17.1
Zaina, E.M. AR 15.2.1
Zaizen, T. JP 30.1
Zając, J. PL 13.1
Zając, L. PL 13.1
Zając, R. PL 28.1.2
Zajmi, A. YU 16.1
Zaki, M. EG 1.2
Zakrzewska, M. PL 32.1.13
Zala, G. HU 5.3.1
Zalewska*, M. PL 4.1
Zamora, A.B. PH 1.7
Zamarrón Loredo, L. MX 6.4
Zamboni, M. IT 16.1
Zambrano, D.H. CO 8.1
Zambrano, R. PE 1.1
Zambrowicz, E. RO 8.2
Zamir*, N. IL 1.1.2, 1.4.7
Zamorski, C. PL 32.1.6
Zamudio, N. AR 22.3
Zamudio Guzmán, V. MX 8.4
Zandstra, B.H. US-F 14.1.1
Zanin, A.C.W. BR 22.1
Zankov, Z. BG 8.3
Zapletal, M. AR 19.1
Zaprzalek*, P. PL 28.1.6
Záruba, J. CS 1.2
Zatykó, E. HU 5.3.2
Zatykó, F. HU 2.2.12
Zatykó, I. HU 2.1.4

Zatykó, J. HU 2.1.3
Zatykó, K. HU 2.1.5
Zatykó, L. HU 5.3.2
Zauberman, G. IL 1.3.1
Zavadil, J. CS 25.1
Zawadzka*, B. PL 28.1.2
Zbinden, W. CH 1.2
Zbroszczyk, J. PL 28.1.3
Zdravković, M. YU 13.1
Zdyb, H. PL 13.1
Zea, J. ES 10.1
Zečević, J. YU 6.1
Zee, J. CA 38.1
Zegbe Dominguez, J. MX 6.2
Zehetner, J. AT 9.1
Zehnder, G. US-C 13.1
Zehnder, H.J. CH 1.1
Zeid, E.A. LY 4.1
Zelechowski, W. PL 32.1.3
Zelenović, J. YU 9.1
Zelleke, A. ET 9.1
Żembery, A. CS 1.1
Zembo, J.C. AR 16.1
Zenbayashi, R. JP 35.2
Zens, A. DE 15.1.8
Zepp, A. PL 3.1
Zepúlveda Torres, J.L. MX 3.4
Zerecero, O.A. MX 16.2
Zeroni, M. IL 5.1.1
Zervas, G. GR 3.1.1
Zetelaki-Horváth, K. HU 2.4
Zettler, F.W. US-D 27.6
Zevallos, A.C. BR 14.1
Zgórkiewicz, A. PL 20.1.2
Zglobiu, V. RO 18.1
Ziarkiewica, T. PL 14.1.9
Zielińska, D. PL 20.2.7
Zieliński, J. PL 10.1.6
Ziemnicka, J. PL 20.1.3
Zieslin*, N. IL 2.2
Zika, V. CS 7.1
Zilai, J HU 5.2
Zimandl, F. CS 26.1
Zimmer*, K. DE 15.1.8
Zimmerman, A. ET 9.1
Zimmermann, J. CA 22.2
Zimmermann, M.J. de O. BR 26.1
Zimmerman, R.H. US-C 3.1
Zimny, H. PL 32.1.13
Zink*, F.W. US-K 6.1.4
Zinkernagel*, V. DE 11.1.3
Zinno, M. JP 5.2
Ziogas, B.N. GR 6.1.7
Ziombra, M. PL 20.2.11
Zioziou, E. GR 25.1.5
Zirojević, D. YU 12.1
Zislowsky, W. AT 8.11
Zitnak, A. CA 15.1
Zitsanza, E. ZI 3.1.2
Živković, M. YU 14.1
Živković, S.Z. YU 14.2
Zizzo, G. IT 22.1.2
Zjawiony, W. PL 23.1
Zmarlicki, C. PL 28.1.8
Zmuda, E. PL 14.1.1
Zobel, R.W. US-B 5.1.5
Zocca, A. IT 3.1
Zocchi, R. IT 9.3
Zoecklein, B.K. US-C 10.1
Zong, T.M. TW 1.2
Zorilla, R.A. PH 1.6
Zorzić, M. YU 17.1
Zraly, B. PL 28.1.9
Zuang, H. FR 21.1
Zubrzycki, A.D. AR 2.1
Zubrzycki, H.M. AR 2.1
Zuccardi, R. AR 22.2
Zuchowska, J. PL 32.1.1
Zuckermann, B.M. US-A 11.1
Zuloaga Albarrán, A. MX 8.1
Zuluaga, E.M. AR 15.2.3
Zumelzú, G. AR 7.1
Żupnik, M. CS 12.1
Zurawicz, E. PL 28.1.1
Zurawicz, Z. PL 28.1.10
Zürn, F. DE 12.1.3
Zutić, I YU 3.1
Žváček, O. CS 22.3
Zwaan, J. NL 16.2
Zwaan, M. NL 16.2
Zwierz, J. PL 28.1.10
Zwierz, W. PL 28.1.6
Zwinkels, J.H.M. NL 11.1
Zwolschen, J.W. NL 9.11.2
Zygadlewica, W. PL 28.2.14